P9-AEV-301

Physical Metallurgy Principles

UNIVERSITY SERIES IN BASIC ENGINEERING

REED-HILL, ROBERT E. – Physical Metallurgy Principles, Second Edition

ROSENTHAL, DANIEL – Introduction to Properties of Materials, Second Edition

HAGERTY, WILLIAM W., AND HAROLD J. PLASS, JR. – Engineering Mechanics

CHOU, PEI CHI, AND NICHOLAS PAGANO – Elasticity: Tensor, Dyadic, and Engineering Approaches

Physical Metallurgy Principles

Second Edition

University Series in Basic Engineering

Robert E. Reed-Hill

D. VAN NOSTRAND COMPANY
NEW YORK CINCINNATI TORONTO LONDON MELBOURNE

D. Van Nostrand Company Regional Offices:
New York Cincinnati Chicago

D. Van Nostrand Company International Offices:
London Toronto Melbourne

Library of Congress Catalog Card Number: 79-181094
ISBN: 0-442-06864-6

Manufactured in the United States of America

Published by D. Van Nostrand Company
450 West 33rd Street, New York, N.Y. 10001

Published simultaneously in Canada by Van Nostrand Reinhold Ltd.

15 14 13 12 11 10 9 8 7 6 5 4 3 2 1

Preface

The basic plan and philosophy of the original edition are continued in this volume. The major changes in the new edition are largely the result of constructive suggestions and advice by Professor Richard W. Heckel, of Drexel University, Dean Walter S. Owen, of Northwestern University, and Professor Marvin Metzger, of the University of Illinois. One result of these suggestions is the inclusion of a chapter on nucleation and growth kinetics. The outline of this chapter was also inspired by a set of class notes kindly loaned to the author by Professor Heckel. The considerable assistance of Dr. John Kronsbein in revising and expanding Chapter 3, Elementary Theory of Metals is also gratefully acknowledged.

As a consequence of requests for the inclusion of topics either missing or too lightly covered in the first edition, the new book has been increased in size by about ten percent. In a broad sense, the additional material fits into two classifications. First are the topics that have recently become significant in the field of metallurgy. The second group consists of well-established subjects not covered in the first edition, but which, from use of the text, were found to be needed for a more unified presentation. Among the former subject areas are electron microscopy, fracture mechanics, superconductivity, superplasticity, dynamic recovery, dynamic strain aging, electrotransport, thermal migration, and emissary dislocations. In the latter category belong the new chapter on nucleation and growth kinetics and such topics as magnetism, the zone theory of alloy phases, the five degrees of freedom of a grain boundary, the phase rule, true stress and true strain, coring and homogenization of castings, work hardening, and diffusion in nonisomorphic systems.

The number of problems is substantially increased over that in the original book, in conformity with the current trend in engineering to place more emphasis on problem solving. Problems have been written with the aim of both illustrating points covered in the text and exposing the student to material and concepts not covered directly in the book.

The helpful assistance of Dr. John Hren, Dr. Robert T. DeHoff, Dr. Derek Dove, Dr. Ellis Verink, and Dr. F. N. Rhines, all of the University of Florida, who either reviewed sections of the book or gave suggestions, is acknowledged with thanks.

Robert E. Reed-Hill

Preface

Contents

Preface v

1 THE STRUCTURE OF METALS 1

1.1 The Structure of Metals, 1 1.2 Metallographic Specimen Preparation, 2 1.3 The Crystal Structure of Metals, 5 1.4 Unit Cells, 5 1.5 The Body-Centered Cubic Structure, 6 1.6 Coordination Number of the Body-Centered Cubic Lattice, 7 1.7 The Face-Centered Cubic Lattice, 7 1.8 Comparison of the Face-Centered Cubic and Close-Packed Hexagonal Structures, 9 1.9 Coordination Number of the Systems of Closest Packing, 11 1.10 The Unit Cell of the Close-Packed Hexagonal Lattice, 11 1.11 Anisotropy, 12 1.12 Textures or Preferred Orientations, 13 1.13 Miller Indices, 15 1.14 Crystal Structures of the Metallic Elements, 21 1.15 The Stereographic Projection, 22 1.16 Directions That Lie in a Plane, 24 1.17 Planes of a Zone, 24 1.18 The Wulff Net, 25 1.19 Standard Projections, 31 1.20 The Standard Stereographic Triangle for Cubic Crystals, 33 Problems, 34

2 DIFFRACTION METHODS 38

2.1 The Bragg Law, 39 2.2 Laue Techniques, 43 2.3 The Rotating-Crystal Method, 45 2.4 The Debye-Scherrer or Powder Method, 46 2.5 The X-Ray Spectrometer, 50 2.6 The Transmission Electron Microscope, 51 Problems, 58

3 ELEMENTARY THEORY OF METALS 60

3.1 The Internal Energy of a Crystal, 60 3.2 Ionic Crystals, 61 3.3 The Born Theory of Ionic Crystals, 62 3.4 Van der Waals Crystals, 67 3.5 Dipoles, 67 3.6 Inert Gases, 68 3.7 Induced Dipoles, 69 3.8 The Lattice Energy of an Inert-Gas Solid, 71 3.9 The Debye Frequency, 71 3.10 The Zero Point Energy, 73 3.11 Dipole-Quadrupole and Quadrupole-Quadrupole Terms, 75

3.12 Molecular Crystals, 75 3.13 Refinements to the Born Theory of Ionic Crystals, 76 3.14 Covalent and Metallic Bonding, 77 3.15 The Uncertainty Principle, 81 3.16 The Dual Nature of Matter, 83 3.17 Free-Electron Theory, 88 3.18 The Density of States, 98 3.19 The Effect of Temperature on the Distribution of Electrons in the Energy Levels, 99 3.20 Electronic Specific Heat, 100 3.21 The Zone Theory, 101 3.22 The Dependence of the Energy on the Wave Number, 107 3.23 The Density of States Curve, 111 3.24 Application of the Zone Theory to Alloy Phases, 112 3.25 Conductors and Insulators, 116 3.26 Semiconductors, 120 3.27 Impurity Semiconductors, 121 3.28 Magnetism, 124 3.29 Ferromagnetism, 130 3.30 Antiferromagnetism and Ferrimagnetism, 133 3.31 The Electrical Resistivity as a Function of Temperature, 134 3.32 Superconductivity, 135 Problems, 141

4 DISLOCATIONS AND SLIP PHENOMENA 144

4.1 The Discrepancy Between the Theoretical and Observed Yield Stresses of Crystals, 144 4.2 Dislocations, 146 4.3 The Burgers Vector, 156 4.4 The Frank-Read Source, 158 4.5 Nucleation of Dislocations, 159 4.6 Bend Gliding, 163 4.7 Rotational Slip, 166 4.8 Climb of Edge Dislocations, 169 4.9 Slip Planes and Slip Directions, 171 4.10 Slip Systems, 173 4.11 Critical Resolved Shear Stress, 173 4.12 Slip on Equivalent Slip Systems, 177 4.13 The Dislocation Density, 177 4.14 Slip Systems in Different Crystal Forms, 178 Problems, 184

5 DISLOCATIONS AND GRAIN BOUNDARIES 190

5.1 Cross-Slip, 190 5.2 Slip Bands, 192 5.3 Double Cross-Slip, 193 5.4 Crystal Structure Rotation During Tensile and Compressive Deformation, 195 5.5 The Notation for the Slip Systems in the Deformation of F.C.C. Crystals, 197 5.6 Vector Notation for Dislocations, 200 5.7 Dislocations in the Face-Centered Cubic Lattice, 201 5.8 Intrinsic and Extrinsic Stacking Faults in Face-Centered Cubic Metals, 206 5.9 Extended Dislocations in Hexagonal Metals, 207 5.10 Extended Dislocations and Cross-Slip, 207 5.11 Dislocation Intersections, 209 5.12 Grain Boundaries, 213 5.13 Dislocation Model of a Small-Angle Grain Boundary, 214 5.14 The Five Degrees of Freedom of a Grain Boundary, 216 5.15 Grain-Boundary Energy, 219 5.16 Surface Tension of the Grain Boundary, 219 5.17 Boundaries Between Crystals of Different Phases, 222

5.18 The Grain Size, 226 5.19 The Effect of Grain Boundaries on Mechanical Properties, 228 Problems, 230

6 VACANCIES 238

6.1 Thermal Behavior of Metals, 238 6.2 Internal Energy, 240 6.3 Entropy, 240 6.4 Spontaneous Reactions, 241 6.5 Gibbs Free Energy, 242 6.6 Statistical Mechanical Definition of Entropy, 244 6.7 Vacancies, 250 6.8 Vacancy Motion, 257 6.9 Interstitial Atoms and Divacancies, 260 Problems, 263

7 ANNEALING 267

7.1 Stored Energy of Cold Work, 267 7.2 The Relationship of Free Energy to Strain Energy, 268 7.3 The Release of Stored Energy, 269 7.4 Recovery, 271 7.5 Recovery in Single Crystals, 272 7.6 Polygonization, 274 7.7 Dislocation Movements in Polygonization, 277 7.8 Recovery Processes at High and Low Temperatures, 282 7.9 Dynamic Recovery, 282 7.10 Recrystallization, 284 7.11 The Effect of Time and Temperature on Recrystallization, 286 7.12 Recrystallization Temperature, 288 7.13 The Effect of Strain on Recrystallization, 288 7.14 The Rate of Nucleation and the Rate of Nucleus Growth, 290 7.15 Formation of Nuclei, 291 7.16 Driving Force for Recrystallization, 294 7.17 The Recrystallized Grain Size, 294 7.18 Other Variables in Recrystallization, 297 7.19 Purity of the Metal, 297 7.20 Initial Grain Size, 298 7.21 Grain Growth, 298 7.22 Geometrical Coalescence, 302 7.23 Three-Dimensional Changes in Grain Geometry, 303 7.24 The Grain Growth Law, 304 7.25 Impurity Atoms in Solid Solution, 310 7.26 Impurities in the Form of Inclusions, 311 7.27 The Free-Surface Effects, 315 7.28 The Limiting Grain Size, 316 7.29 Preferred Orientation, 318 7.30 Secondary Recrystallization, 318 7.31 Strain-Induced Boundary Migration, 319 Problems, 321

8 SOLID SOLUTIONS 326

8.1 Intermediate Phases, 327 8.2 Interstitial Solid Solutions, 327 8.3 Solubility of Carbon in Body-Centered Cubic Iron, 328 8.4 Substitutional Solid Solutions and the Hume-Rothery Rules, 332 8.5 Interaction of Dislocations and Solute Atoms, 333 8.6 The Stress Field of a Screw Dislocation, 333 8.7 The Stress Field of an Edge Dislocation, 334 8.8 Dislocation Atmospheres, 336 8.9 The Orowan Equation, 338 8.10 The Drag of Atmospheres on Moving

Dislocations, 339 8.11 The Sharp Yield Point and Lüders Bands, 341 8.12 The Theory of the Sharp Yield Point, 344 8.13 Strain Aging 346 8.14 Dynamic Strain-Aging, 347 8.15 The Role of the Drag-Stress in Dynamic Strain-Aging, 351 Problems, 353

9 PRECIPITATION HARDENING 358

9.1 The Significance of the Solvus Curve, 359 9.2 The Solution Treatment, 360 9.3 The Aging Treatment, 361 9.4 Nucleation of Precipitates, 365 9.5 Heterogeneous Versus Homogeneous Nucleation, 368 9.6 Theories of Hardening, 370 9.7 Additional Factors in Precipitation Hardening, 374 Problems, 375

10 DIFFUSION IN SUBSTITUTIONAL SOLID SOLUTIONS 378

10.1 Ideal Solutions, 378 10.2 Nonideal Solutions, 379 10.3 Diffusion in an Ideal Solution, 381 10.4 The Kirkendall Effect, 386 10.5 Porosity, 390 10.6 Darken's Equations, 392 10.7 Fick's Second Law, 397 10.8 The Matano Method, 400 10.9 Determination of the Intrinsic Diffusivities, 404 10.10 Self-Diffusion in Pure Metals, 406 10.11 Temperature Dependence of the Diffusion Coefficient, 409 10.12 Chemical Diffusion at Low-Solute Concentration, 412 10.13 The Study of Chemical Diffusion Using Radioactive Tracers, 414 10.14 Diffusion Along Grain Boundaries and Free Surfaces, 418 10.15 Fick's First Law in Terms of a Mobility and an Effective Force, 422 10.16 Electrotransport and Thermomigration, 425 Problems, 428

11 INTERSTITIAL DIFFUSION 431

11.1 Measurement of Interstitial Diffusivities, 432 11.2 The Snoek Effect, 433 11.3 Experimental Determination of the Relaxation Time, 441 11.4 Experimental Data, 450 11.5 Anelastic Measurements at Constant Strain, 450 Problems, 451

12 PHASES 454

12.1 Basic Definitions, 454 12.2 The Physical Nature of Phase Mixtures, 456 12.3 Thermodynamics of Solutions, 457 12.4 Equilibrium Between Two Phases, 460 12.5 The Number of Phases in an Alloy System, 461 12.6 Two-Component Systems Containing Two Phases, 470 12.7 Graphical Determinations of Partial-Molal Free Energies, 471 12.8 Two-Component Systems with Three Phases in

Equilibrium, 474 12.9 The Phase Rule, 475 12.10 Ternary Systems, 478 Problems, 478

13 NUCLEATION AND GROWTH KINETICS 480

13.1 Nucleation of a Liquid from the Vapor, 480 13.2 The Becker-Döring Theory, 488 13.3 Freezing, 490 13.4 Solid State Reactions, 493 13.5 Heterogeneous Nucleation, 496 13.6 Growth Kinetics, 501 13.7 Diffusion Controlled Growth, 504 13.8 Interference of Growing Precipitate Particles, 510 13.9 Interface Controlled Growth, 511 13.10 Transformations That Occur on Heating, 514 13.11 Dissolution of a Precipitate, 515 Problems, 518

14 BINARY PHASE DIAGRAMS 522

4.1 Isomorphous Alloy Systems, 522 14.2 The Lever Rule, 524 14.3 Equilibrium Heating or Cooling of an Isomorphous Alloy, 527 14.4 The Isomorphous Alloy System from the Point of View of Free Energy, 530 14.5 Maxima and Minima, 532 14.6 Superlattices, 534 14.7 Miscibility Gaps, 539 14.8 Eutectic Systems, 541 14.9 The Equilibrium Microstructures of Eutectic Systems, 542 14.10 The Peritectic Transformation, 549 14.11 Monotectics, 552 14.12 Other Three-Phase Reactions, 552 14.13 Intermediate Phases, 553 14.14 The Copper-Zinc Phase Diagram, 556 14.15 Diffusion in Non-Isomorphic Alloy Systems, 559 Problems, 565

15 SOLIDIFICATION OF METALS 568

15.1 The Liquid Phase, 568 15.2 Nucleation, 572 15.3 Crystal Growth from the Liquid Phase, 575 15.4 Stable Interface Freezing, 578 15.5 Dendritic Growth, 579 15.6 Dendritic Freezing in Alloys, 583 15.7 Freezing of Ingots, 588 15.8 The Grain Size of Castings, 593 15.9 Segregation, 594 15.10 Homogenization, 596 15.11 Inverse Segregation, 602 15.12 Porosity, 603 Problems, 608

16 DEFORMATION TWINNING AND MARTENSITE REACTIONS 611

16.1 Deformation Twinning, 611 16.2 Formal Crystallographic Theory of Twinning, 614 16.3 Identification of Deformation Twins, 622 16.4 Nucleation of Twins, 624 16.5 Twin Boundaries, 625 16.6 Twin Growth, 627 16.7 Accommodation of the Twinning Shear, 629 16.8 Martensite, 635 16.9 The Bain Distortion, 638 16.10 The Martensite Transformation in an Indium-Thallium Alloy,

640 16.11 Reversibility of the Martensite Transformation, 642 16.12 Athermal Transformation, 642 16.13 Wechsler, Lieberman, and Read Theory, 643 16.14 Irrational Nature of the Habit Plane, 650 16.15 Multiplicity of Habit Planes, 651 16.16 The Iron-Nickel Martensitic Transformation, 651 16.17 Isothermal Formation of Martensite, 653 16.18 Stabilization, 653 16.19 Nucleation of Martensite Plates, 654 16.20 Growth of Martensite Plates, 655 16.21 The Effect of Stress, 655 16.22 The Effect of Plastic Deformation, 656 Problems, 657

17 THE IRON-CARBON SYSTEM 661

17.1 The Transformation of Austenite to Pearlite, 663 17.2 Pearlite, 665 17.3 The Effect of Temperature on the Pearlite Transformation, 670 17.4 Time-Temperature-Transformation Curves, 679 17.5 The Bainite Reaction, 681 17.6 The Complete *T-T-T* Diagram of an Eutectoid Steel, 687 17.7 Slowly Cooled Hypoeutectoid Steels, 690 17.8 Slowly Cooled Hypereutectoid Steels, 693 17.9 Isothermal Transformation Diagrams for Noneutectoid Steels, 695 Problems, 698

18 THE HARDENING OF STEEL 701

18.1 Continuous Cooling Transformations, 701 18.2 Hardenability, 704 18.3 The Variables that Determine the Hardenability of a Steel, 710 18.4 Austenitic Grain Size, 711 18.5 The Effect of Austenitic Grain Size on Hardenability, 712 18.6 The Influence of Carbon Content on Hardenability, 713 18.7 The Influence of Alloying Elements on Hardenability, 714 18.8 The Significance of Hardenability, 719 18.9 The Martensite Transformation in Steel, 720 18.10 The Hardness of Iron-Carbon Martensite, 725 18.11 Dimensional Changes Associated with the Formation of Martensite, 729 18.12 Quench Cracks, 730 18.13 Tempering, 733 18.14 Spheroidized Cementite, 739 18.15 The Effect of Tempering on Physical Properties, 742 18.16 The Interrelation Between Time and Temperature in Tempering, 744 18.17 Secondary Hardening and the Fifth Stage of Tempering, 745 Problems, 747

19 FRACTURE 749

19.1 Failure by Easy Glide, 749 19.2 Rupture by Necking (Multiple Glide), 751 19.3 The Effect of Twinning, 752 19.4 Cleavage, 753 19.5 Fracture in Glass, 755 19.6 The Griffith Theory,

759 19.7 Griffith Cracks in Glass, 760 19.8 Crack Velocities, 761 19.9 The Griffith Equation, 762 19.10 The Nucleation of Cleavage Cracks, 763 19.11 Crack Nucleation in Iron, 765 19.12 Crack Nucleation in Zinc, 768 19.13 Propagation of Cleavage Cracks, 769 19.14 The Effect of Grain Boundaries 774 19.15 The Effect of the State of Stress, 777 19.16 The Impact Test, 780 19.17 The Significance of the Impact Test, 782 19.18 Ductile Fractures, 786 19.19 Intercrystalline Brittle Fracture, 794, 19.20 Temper Brittleness, 795 19.21 Blue Brittleness, 796 19.22 Fracture Mechanics, 797 19.23 The Stress Intensity Factor, 800 19.24 Fracture Toughness Measurements, 801 19.25 Fatigue Failures, 809 19.26 The Macroscopic Character of Fatigue Failure, 810 19.27 The Rotating-Beam Fatigue Test, 811 19.28 The Microscopic Aspects of Fatigue Failure, 814 19.29 Fatigue Crack Growth, 818 19.30 The Effect of Nonmetallic Inclusions, 821 19.31 The Effect of Steel Microstructure on Fatigue, 822 19.32 Low-Cycle Fatigue, 823 19.33 Certain Practical Aspects of Fatigue, 824 Problems, 824

20 CREEP **827**

20.1 Work Hardening, 827 20.2 Considère Criterion, 829 20.3 The Relation Between Dislocation Density and the Stress, 830 20.4 Taylor's Relation, 832 20.5 The Nature of the Flow-Stress Components, 834 20.6 The Effect of Alloying on the Flow-Stress Components, 835 20.7 The Relative Roles of the Flow-Stress Components in Face-Centered Cubic and Body-Centered Cubic Metals, 837 20.8 Superplasticity, 840 20.9 The Nature of the Time-Dependent Strain, 841 20.10 Creep Mechanisms, 847 20.11 Creep When More than One Mechanism is Operating, 859 20.12 Grain-Boundary Shear, 865 20.13 Intercrystalline Fracture, 867 20.14 The Creep Curve, 871 20.15 Practical Applications of Creep Data, 877 20.16 Creep-Resistant Alloys, 880 20.17 Alloy Systems, 882 Problems, 884

Appendices 888
List of Important Symbols 901
List of Greek Letter Symbols 903
Index 904

1 The Structure of Metals

The most important aspect of any engineering material is its structure, because its properties are closely related to this feature. To be successful, a materials engineer must have a good understanding of this relationship between structure and properties. By way of illustration, wood is a very easy material in which to see the close interaction between structure and properties. A typical structural wood, such as southern yellow pine, is essentially an array of long hollow cells or fibers. These fibers, which are formed largely from cellulose, are aligned with the grain of the wood and are cemented together by another weaker organic material called lignin. The structure of wood is thus roughly analogous to that of a bundle of drinking straws. It can be split easily along its grain; that is, parallel to the cells. Wood is also much stronger in compression (or tension) parallel to its grain than it is in compression (or tension) perpendicular to the grain. It makes excellent columns and beams, but it is not really suitable for tension members required to carry large loads, because the low resistance of wood to shear parallel to its grain makes it difficult to attach end fastenings that will not pull out. As a result, wooden bridges and other large wooden structures are often constructed containing steel tie rods to support the tensile loads.

1.1 The Structure of Metals. The structure in metals is of similar importance to that in wood, although often in a more subtle manner. Metals are crystalline when in the solid form. While very large single crystals can be prepared, the normal metallic object consists of an aggregate of many very small crystals. Metals are therefore *polycrystalline*. The crystals in these materials are normally referred to as its grains. Because of their very small sizes, an optical microscope, operating at magnifications between about 100 and 1000 times, is usually required in order to examine the structural features associated with the grains in a metal. Structures requiring this range of magnification for their examination fall into the class known as *microstructures*. Occasionally, metallic objects, such as castings, may have very large crystals that are discernible to the naked eye or are easily resolved under a low-power microscope. Structure in this category is called *macrostructure*. Finally, there is the basic structure inside the grains themselves: that is, the atomic arrangements inside the crystals. This form of structure is logically called the *crystal structure*.

Of the various forms of structure, microstructure (that visible under the optical microscope) has been historically of the greatest use and interest to the metallurgist. Because the metallurgical microscope is normally operated at magnifications where its depth of field is extremely shallow, the metallic surface to be observed must be very flat. At the same time, it must reveal accurately the nature of the structure inside the metal. One is therefore presented with the problem of preparing a very smooth flat and undistorted surface, which is by no means an easy task. The procedures required to obtain the desired goal fall under the general heading of metallographic specimen preparation.

1.2 Metallographic Specimen Preparation. To a large extent, metallographic specimen preparation is an art. The techniques tend to vary somewhat from one laboratory to another. Variations in the procedures may also be required depending on the nature of the metal to be examined, because metals vary widely in hardness and texture. Nevertheless, the basic operations tend to be similar: for the purpose of illustrating the nature of metallographic specimen preparation, let us consider a technique suitable for iron and steels. What follows is only a simple outline, and for further details a suitable reference [1] should be consulted.

Let us suppose that a small specimen has been cut out of a steel object and that a suitable flat surface has been prepared on one side of this specimen by sawing and grinding. The normal procedure is to mount the specimen in a small plastic disc (about 1 in. in diameter and ½ in. thick) with the surface of the specimen to be polished exposed on one side of the disc, as shown in Fig. 1.1. One technique to form this disc consists of first placing the specimen inside a

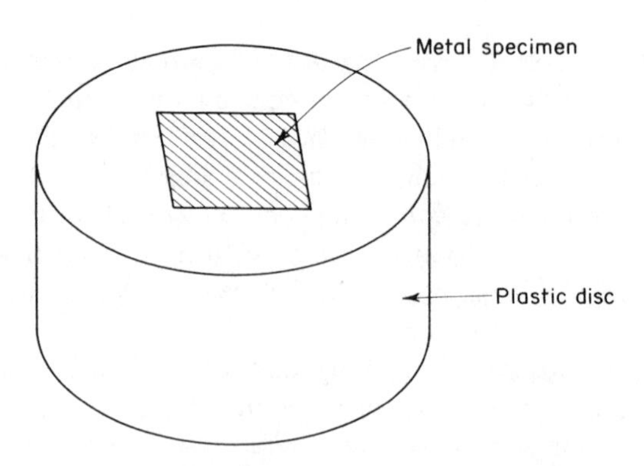

Fig. 1.1 A metallographic specimen.

1. *Polishing the Micro Section*, Parts 1 and 2. *The AB Metal Digest,* **11**, Nos. 2 and 3, (1965). Buehler, Ltd., Evanston, Illinois.

simple ring mold, and then pouring liquid epoxy resin into the mold to fill it. The epoxy hardens in a matter of hours to yield a strong, convenient handle for holding the specimen during the subsequent steps in surface preparation. These steps involve four basic types of operations: (1) fine grinding, (2) rough polishing, (3) final polishing, and (4) etching. In the first three stages, the basic objective is to reduce the thickness of the deformed layer lying below the specimen surface. All cutting and grinding operations badly deform the metal next to the surface. The true structure of the metal will become visible only when one has completely removed the deformed layer. Since each stage in the preparation of the specimen itself tends to deform the surface, it is necessary to use a succession of finer and finer abrasives. Each abrasive acts to remove the deformed layer left from the coarser stage preceding it, while it in turn leaves a distorted layer of reduced depth.

Fine grinding. In this stage, the specimen surface is ground using silicon carbide powders bonded onto specially prepared papers. The specimen may be hand-rubbed against the abrasive paper, which is laid over a flat surface, such as a piece of plate glass. Alternatively, the abrasive paper may be mounted on the surface of a flat, horizontally rotating wheel and the metallographic specimen held against it. In either case, the surface is usually lubricated with water, which provides a flushing action to carry away the particles cut from the surface. Three grades of abrasive are used: 320 grit, 400 grit, and 600 grit. The corresponding particle sizes of the silicon carbide are about 33, 23, and 17 microns respectively, where one micron is 10^{-4} cm. In each of these fine grinding stages the specimen is moved over the surface so that the scratches are formed in only one direction. In proceeding from one paper to the next, the specimen is rotated through an angle of about 45°, so that the new scratches are placed on the surface at an angle to those formed during the preceding stage. Grinding is then continued until well past the point at which the scratches from the preceding stage have disappeared.

Rough polishing. This is probably the critical stage. At the present time the abrasive extensively used for the rough polishing operation is powdered diamond dust, with a particle size of about 6 microns. The diamond powder is carried in a paste that is oil-soluble. A very small quantity of this paste is needed, and this is placed on the nylon cloth-covered surface of a rotating polishing wheel. The lubricant used during the polishing operation is a specially prepared oil. The specimen is pressed against the cloth of the rotating wheel with considerable pressure. In the polishing stages, the specimen is not held in a fixed position on the polishing wheel, but is moved around the wheel in the direction opposite to the rotation of the wheel itself. This insures a more uniform polishing action. The diamond particles have a strong cutting action and can be very effective in removing the deep layer of deformation remaining from the fine grinding operation. The fact that 6 micron diamond powder is able to remove the effects resulting from the 17 micron silicon carbide abrasive in the last stage of fine grinding speaks for itself.

Final polishing. Here the fine scratches and very thin distorted layer remaining from the rough polishing stage are removed. The polishing compound usually used is alumina (Al_2O_3) powder (gamma form), with a particle size of 0.05 microns. This is placed on a cloth-covered wheel, and distilled water is used as a lubricant. In contrast to the napless nylon cloth used in rough polishing, the cloth used in this step normally contains a nap. If this and the preceding stages have been carefully carried out, a scratch-free surface, with almost no detectable layer of distorted metal, will be obtained.

Etching. The granular structure in a metallographic specimen usually cannot be seen under the microscope after the final polishing operating has been completed. The grain boundaries in a metal have a thickness of the order of a few atom diameters at best, and the resolving power of a microscope is much too low to reveal their presence. Only if a metal contains several crystals of different colors that touch each other will boundaries be visible. In a pure metal this is, of course, impossible. In order to make the boundaries visible, metallographic specimens are usually etched. In most cases this involves immersing the polished surface in a weak acidic or alkaline etching solution. The most commonly used solution for steels is called *nital* and consists of a 2 percent solution of nitric acid in alcohol. In some cases the etching solution is applied to the surface by rubbing it gently with a cotton swab wetted with the agent. In either case, the resultant effect is to dissolve metal away from the specimen surface. With a good etching solution, the removal of metal from the surface will not occur uniformly. Sometimes the etching agent will attack the grain boundaries more rapidly than the surface of the grains. Other etching solutions will dissolve the various grains according to the orientation of the surface that they present to the etching solution. The nature of this effect is shown schematically in Fig. 1.2. Here the boundaries are made visible because they appear as shallow steps on the surface of the specimen. These nearly vertical surfaces do not reflect light into the objective lens of a microscope in the same fashion as the smooth horizontal crystal surfaces that lie between them, and, as a result, the locations of the crystal boundaries should be observed under a microscope.

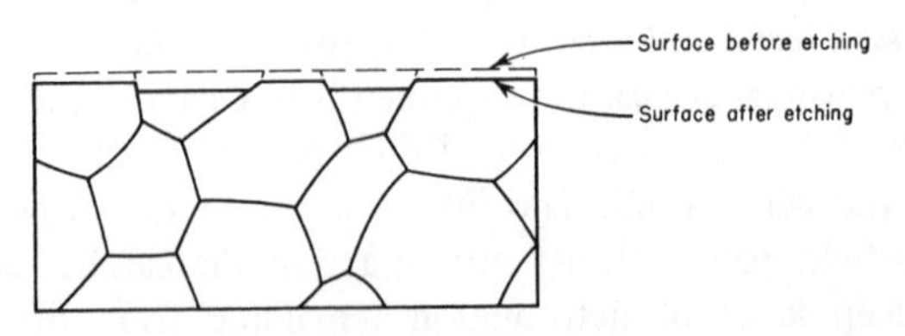

Fig. 1.2 The etching solution reveals the crystal boundaries.

Electro-polishing and electro-etching. There are a number of metals, such as stainless steel, titanium, and zirconium, from which it is very difficult to remove the distorted surface layers. Mechanical polishing is not very successful with these materials. As a result, they are often polished by an electro-polishing technique in the final polishing stage. In this case, the specimen is made the anode in a suitable electolytic bath, while the cathode is an insoluble material. If the proper current density is used, it is possible to dissolve away the specimen surface in a manner that produces a fine surface polish. In electro-polishing, the bath and current are deliberately controlled so as to produce a surface that is smooth and without relief. On the other hand, it is also possible to alter the composition of the bath and the current density conditions so as to produce the surface relief required in etching. The corresponding procedure is called electro-etching.

1.3 The Crystal Structure of Metals. A *crystal* is defined as an orderly array of atoms in space. There are many different types of crystal structures, some of which are quite complicated. Fortunately, most metals crystallize in one of three relatively simple structures: the face-centered cubic, the body-centered cubic, and the close-packed hexagonal.

1.4 Unit Cells. The *unit cell* of a crystal structure is the smallest group of atoms possessing the symmetry of the crystal which, when repeated in all directions, will develop the crystal lattice. Figure 1.3A shows the unit cell of the body-centered cubic lattice. It is evident that its name is derived from the shape of the unit cell. Eight unit cells are combined in Fig. 1.3B in order to show how the unit cell fits into the complete lattice. Note that atom *a* of Fig. 1.3B does not belong uniquely to one unit cell, but is a part of all eight unit cells that surround it. Therefore, it can be said that only one-eighth of this corner atom

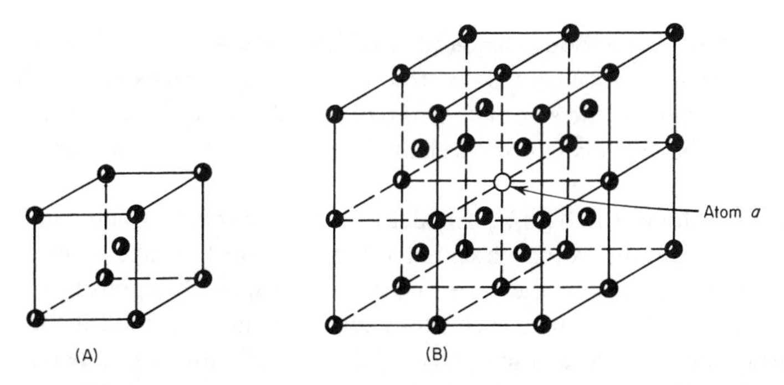

Fig. 1.3 (A) Body-centered cubic unit cell. (B) Eight unit cells of the body-centered cubic lattice.

belongs to any one unit cell. This fact may be used to compute the number of atoms per unit cell in a body-centered cubic crystal. Even a small crystal will contain billions of unit cells, and the cells in the interior of the crystal must greatly exceed in number those lying on the surface. Therefore, surface cells may be neglected in our computations. In the interior of a crystal, each corner atom of a unit cell is equivalent to atom *a* of Fig. 1.3B and contributes one-eighth of an atom to a unit cell. In addition, each cell also possesses an atom located at its center that is not shared with other unit cells. The body-centered cubic lattice thus has two atoms per unit cell; one contributed by the corner atoms, and one located at the center of the cell.

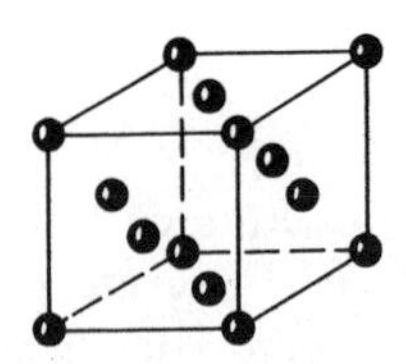

Fig. 1.4 Face-centered cubic unit cell.

The unit cell of the face-centered cubic lattice is shown in Fig. 1.4. In this case, the unit cell has an atom in the center of each face.

The number of atoms per unit cell in the face-centered cubic lattice can be computed in the same manner as in the body-centered cubic lattice. The eight corner atoms again contribute one atom to the cell. There are also six face-centered atoms to be considered, each a part of two unit cells. These contribute six times one-half an atom, or three atoms. The face-centered cubic lattice has a total of four atoms per unit cell, or twice as many as the body-centered cubic lattice.

1.5 The Body-Centered Cubic Structure. It is frequently convenient to consider metal crystals as structures formed by stacking together hard spheres. This leads to the so-called *hard-ball model* of a crystalline lattice, where the radius of the spheres is taken as half the distance between the centers of the most closely spaced atoms.

Figure 1.5 shows the hard-ball model of the body-centered cubic unit cell. A study of the figure shows that the atom at the center of the cube is colinear with each corner atom; that is, the atoms connecting diagonally opposite corners of the cube form straight lines, each atom touching the next in sequence. These linear arrays do not end at the corners of the unit cell, but continue on through the crystal much like a row of beads strung on a wire (see Fig. 1.3B). These four cube diagonals constitute the close-packed directions of the body-centered cubic crystal, directions that run continuously through the lattice on which the atoms are as closely spaced as possible.

Further consideration of Figs. 1.5 and 1.3B reveals that all atoms in the body-centered cubic lattice are equivalent. Thus, the atom at the center of the cube of Fig. 1.5 has no special significance over those occupying corner positions. Each of the latter could have been chosen as the center of a unit cell, making all corner atoms of Fig. 1.3B centers of cells, and all centers of cells corners.

Fig. 1.5 Hard-ball model of the body-centered cubic unit cell.

1.6 Coordination Number of the Body-Centered Cubic Lattice. The coordination number of a crystal structure equals the number of nearest neighbors that an atom possesses in the lattice. In the body-centered cubic unit cell, the center atom has eight neighbors touching it (see Fig. 1.5). We have already seen that all atoms in this lattice are equivalent. Therefore, every atom of the body-centered cubic structure not lying at the exterior surface possesses eight nearest neighbors, and the coordination number of the lattice is eight.

1.7 The Face-Centered Cubic Lattice. The hard-ball model assumes special significance in the face-centered cubic crystal, for in this structure the atoms or spheres are packed together as closely as possible. Figure 1.6A shows a complete face-centered cubic cell, and Fig. 1.6B shows the same unit cell with a corner atom removed to reveal a close-packed plane (octahedral plane) in which the atoms are spaced as tightly as possible. A larger area from one of these close packed planes is shown in Fig. 1.7. Three close-packed directions lie in the octahedral plane (the directions *aa, bb* and *cc*). Along these directions the spheres touch and are colinear.

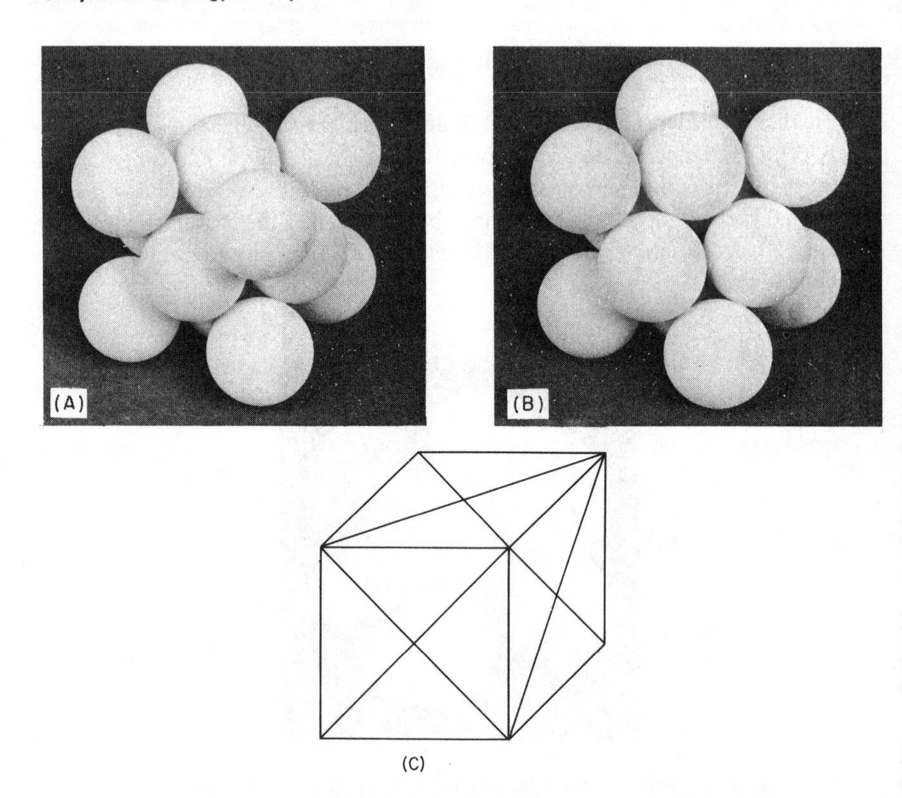

Fig. 1.6 (A) Face-centered cubic unit cell (hard-ball model). (B) Same cell with a corner atom removed to show an octahedral plane. (C) The six-face diagonal directions.

Returning to Fig. 1.6A, we see that the close-packed directions of Fig. 1.7 correspond to diagonals that cross the faces of the cube. There are six of these close-packed directions in the face-centered cubic lattice, as shown in Fig. 1.6C. Face diagonals lying on the reverse faces of the cube are not counted in this total because each is parallel to a direction lying on a visible face, and, in considering crystallographically significant directions, parallel directions are the same. It should also be pointed out that the face-centered cubic structure has four close-packed or octahedral planes. This can be verified as follows. If an atom is removed from each of the corners of a unit cell in a manner similar to that of Fig. 1.6B, an octahedral plane will be revealed in each instance. There are eight of these planes, but since diagonally opposite planes are parallel, they are crystallographically equal. This reduces the number of different octahedral planes to four. The face-centered cubic lattice, however, is unique in that it contains as many as four planes of closest packing, each containing three

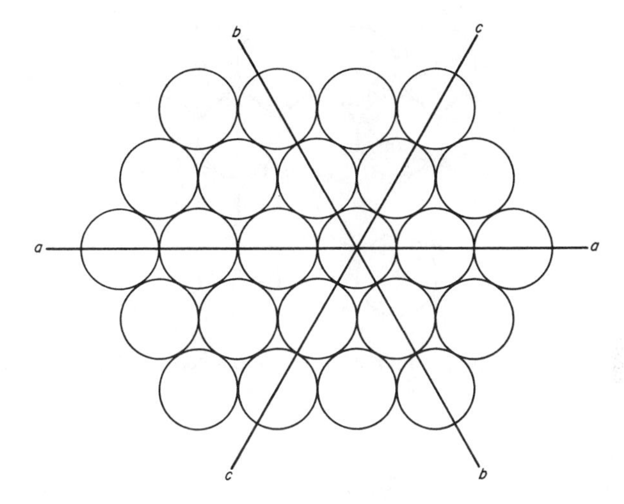

Fig. 1.7 Atomic arrangement in the octahedral plane of a face-centered cubic metal. Notice that the atoms have the closest possible packing. This same configuration of atoms is also observed in the basal plane of close-packed hexagonal crystals. The close-packed directions are *aa, bb,* and *cc.*

close-packed directions. No other lattice possesses such a large number of close-packed planes and closed-packed directions. This fact is important, since it gives face-centered cubic metals physical properties different from those of other metals, among which is the ability to undergo severe plastic deformation.

1.8 Comparison of the Face-Centered Cubic and Close-Packed Hexagonal Structures. The face-centered cubic lattice can be constructed by first arranging atoms into a number of close-packed planes, similar to that shown in Fig. 1.7, and then by stacking these planes over each other in the proper sequence. There are a number of ways in which planes of closest packing can be stacked. One sequence gives the close-packed hexagonal lattice, another the face-centered cubic lattice. That there is more than one way of stacking close-packed planes is because any one plane can be.set down on a previous one in two different ways. For example, consider the close-packed plane of atoms in Fig. 1.8. The center of each atom in the figure is indicated by the symbol A. Now, if a single atom is placed on top of the configuration of Fig. 1.8, it will be attracted by interatomic forces into one of the natural pockets that occur between any three contiguous atoms. Suppose that it falls into the pocket marked B_1 at the upper left of the figure; then a second atom cannot be dropped into either C_1 or C_2 because the atom at B_1 overlaps the pocket at these two points. However, the second atom can fall into B_2 or B_3 and start the formation of a second close-packed plane consisting of atoms occupying all B

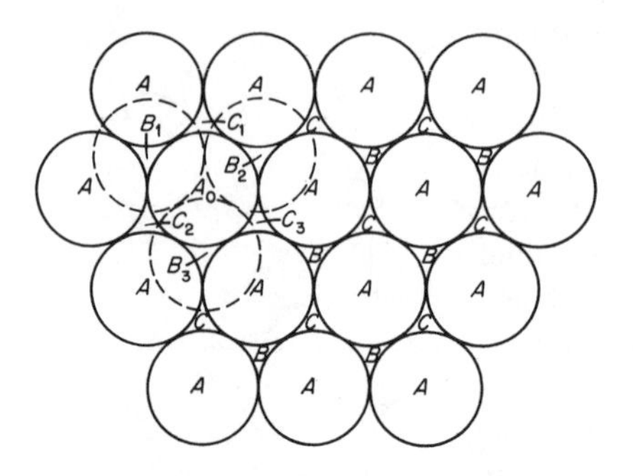

Fig. 1.8 Stacking sequences in close-packed crystal structures.

positions. Alternatively, the second plane could have been set down in such a way as to fill only *C* positions. Thus if the first close-packed plane occupies *A* positions, the second plane may occupy *B* or *C* positions. Let us assume that the second plane has the *B* configuration. Then the pockets of the second plane fall half over the centers of the atoms in the first plane (*A* positions) and half over the *C* pockets in the first plane. The third plane may now be set down over the second plane into either *A* or *C* positions. If set down into *A* positions, the atoms in the third layer fall directly over atoms in the first layer. This is not the face-centered cubic order, but that of the close-packed hexagonal structure. The face-centered cubic stacking order is: *A* for the first plane, *B* for the second plane, and *C* for the third plane, which may be written as *ABC*. The fourth plane in the face-centered cubic lattice, however, does fall on the *A* position, the fifth on *B*, and the sixth on *C*, so that the stacking order for face-centered cubic crystals is *ABCABCABC* etc. In the close-packed hexagonal structure, the atoms in every other plane fall directly over one another, corresponding to the stacking order *ABABAB* . . . .

There is no basic difference in the packing obtained by the stacking of spheres in the face-centered cubic or the close-packed hexagonal arrangement, since both give an ideal close-packed structure. There is, however, a marked difference between the physical properties of hexagonal close-packed metals (such as cadmium, zinc, and magnesium) and the face-centered cubic metals, which is related directly to the difference in their crystalline structure. The most striking difference is in the number of close-packed planes. In the face-centered cubic lattice there are four planes of closest packing, the *octahedral planes*; but in the close-packed hexagonal lattice only one plane, the *basal plane*, is equivalent to the octahedral plane. The single close-packed plane of the hexagonal lattice engenders, among other things, plastic deformation properties that are much more directional than those found in cubic crystals.

1.9 Coordination Number of the Systems of Closest Packing. The *coordination number* of an atom in a crystal has been defined as the number of nearest neighbors that it possesses. This number is 12 for both face-centered cubic and close-packed hexagonal crystals, as may be verified with the aid of Fig. 1.8. Thus, consider atom A_0 lying in the plane of atoms shown as circles drawn with continuous lines. Six other atoms lying in the same close-packed plane as A_0 are in nearest neighbor positions. Atom A_0 also touches three atoms in the plane directly above. These three atoms could occupy B positions, as is indicated by the dashed lines around pockets B_1, B_2, and B_3, or they could occupy positions C_1, C_2, and C_3. In either case, the number of nearest neighbors in the plane just above A_0 is limited to three. In the same manner, it may be shown that A_0 has three nearest neighbors in the next plane below the close-packed plane containing A_0. The number of nearest neighbors of atom A_0 are thus twelve in number: six in its own plane, three in the plane above it, and three in the plane below it. Since the argument is valid no matter whether the atoms in the close-packed planes just above or below atom A_0 are in B or C positions, it holds for both face-centered cubic and close-packed hexagonal stacking sequences. We conclude, therefore, that the coordination number in these lattices is 12.

1.10 The Unit Cell of the Close-Packed Hexagonal Lattice. The configuration of atoms most frequently used to represent the close-packed hexagonal structure is shown in Fig. 1.9. This group of atoms contains more than the minimum number of atoms needed to form an elementary building block for the lattice; therefore it is not a true unit cell. However, because the arrangement of Fig. 1.9 brings out important crystallographic features, including

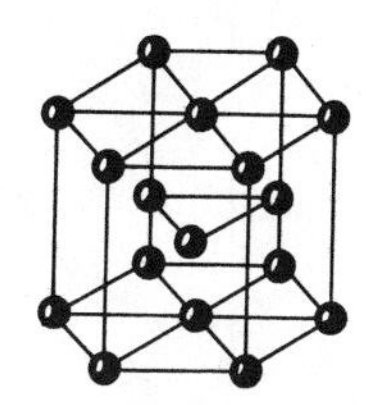

Fig. 1.9 The close-packed hexagonal unit cell.

the sixfold symmetry of the lattice, it is commonly used as the unit cell of the close-packed hexagonal structure. A comparison of Fig. 1.9 with Fig. 1.7 shows that the atoms in the planes at the top, bottom, and center of the unit cell belong to a plane of closest packing, the basal plane of the crystal. The figure also shows that the atoms in these basal planes have the proper stacking sequence for the hexagonal close-packed lattice (*ABA* . . .); atoms at the top of the cell are directly over those at the bottom, while atoms in the central plane have a different set of positions.

The basal plane of a hexagonal metal, like the octahedral plane of a face-centered cubic metal, has three close-packed directions. These directions correspond to the lines *aa, bb, cc* of Fig. 1.7.

1.11 Anisotropy. When the properties of a substance are independent of direction, the material is said to be isotropic. Thus, in an ideal isotropic material, one should expect to find that it has the same strength in all directions. Or, if its electrical resistivity were to be measured, the same value of this property would be obtained irrespective of how a resistivity specimen was cut from a quantity of the material. The physical properties of crystals normally depend strongly on the direction along which they are measured. This means that, basically, crystals are not isotropic, but anisotropic. In this regard, consider a body-centered cubic crystal of iron. The three most important directions in this crystal are the directions labeled *a, b,* and *c* in Fig. 1.10. That these directions are not equivalent can be recognized from the fact that the spacings of the atoms along the three directions are different, being equal respectively, in terms of the lattice parameter *a* (the length of one edge of the unit cell), to a, $\sqrt{2}a$, and $\sqrt{3/2}\,a$. The physical properties of iron, measured along these three directions, also tend to be different. As an example, consider the *B-H* curve for the

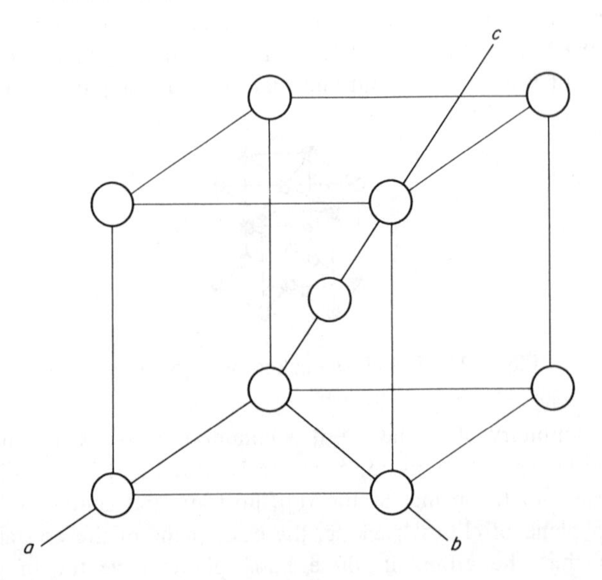

Fig. 1.10 The most important directions in a body-centered cubic crystal.

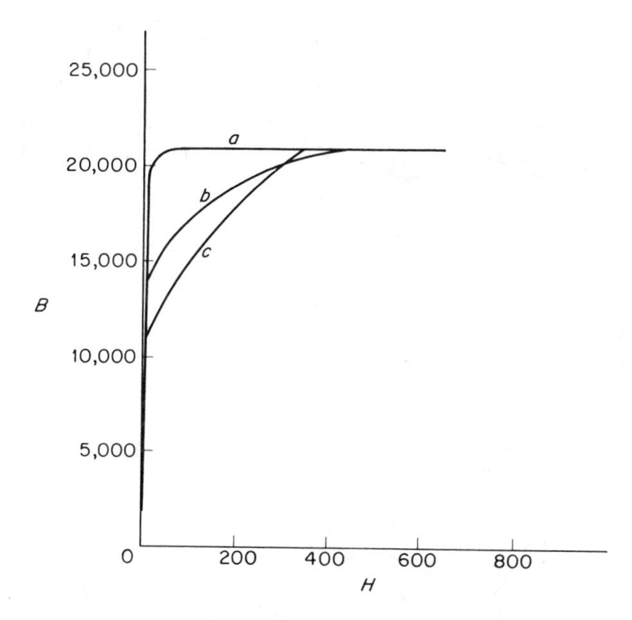

Fig. 1.11 An iron crystal is much easier to magnetize along an *a* direction of Fig. 1.10 than along a *b* or *c* direction. (After Barrett, C. S., *Structure of Metals,* p. 453. New York: McGraw-Hill Book Co., 1943. Used by permission.)

magnetization of iron crystals. As may be seen in Fig. 1.11, the magnetic induction B rises most rapidly with the magnetic field intensity H along the direction a, at an intermediate rate along b, and least rapidly along c. Interpreted in another way, we may say that a is the direction of easiest magnetization, while c is correspondingly the most difficult.

Ideally, a polycrystalline specimen might be expected to be isotropic if its crystals were randomly oriented, for then, from a macroscopic point of view, the anisotropy of the crystals should be averaged out. However, a truly random arrangement of the crystals is seldom achieved, because manufacturing processes tend to align the grains in a metal so that their orientations are not uniformly distributed. The result is what is known as a texture or a preferred orientation. Because most polycrystalline metals have a preferred orientation, they tend to be anisotropic, the degree of this anisotropy depending on the degree of crystal alignment.

1.12 Textures or Preferred Orientations. Wires are formed by pulling rods through successively smaller and smaller dies. In the case of iron, this kind of deformation tends to align a b direction of each crystal parallel to the wire axis. About this direction the crystals are normally considered to be

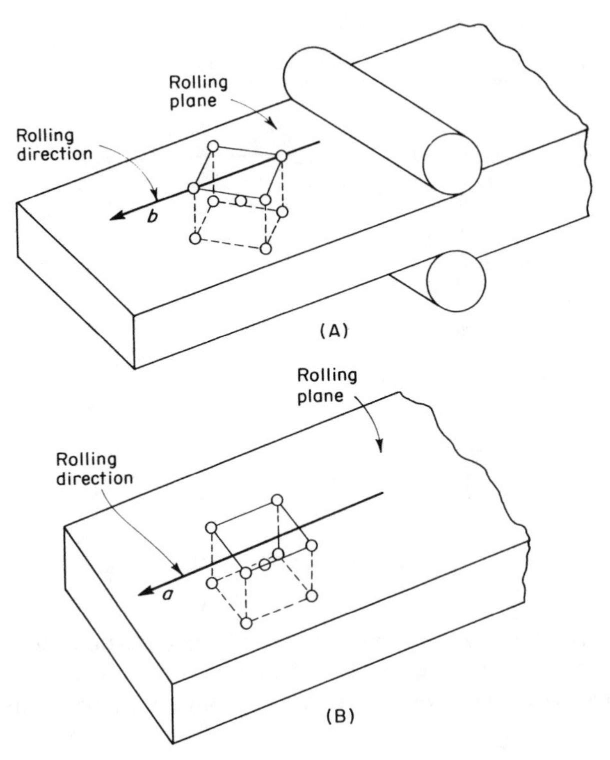

Fig. 1.12 Two basic crystalline orientations that can be obtained in rolled plates of body-centered cubic metals.

randomly arranged. This type of preferred arrangement of the crystals in an iron or steel wire is quite persistent. Even if the metal is given a heat treatment[2] that completely reforms the crystal structure, the crystals tend to keep a *b* direction parallel to the wire axis. Because the deformation used in forming sheets and plates is basically two-dimensional in character, the preferred orientation found in them is more restrictive than that observed in wires. As indicated in Fig. 1.12A, not only does one tend to find a *b* direction parallel to the rolling direction or length of the plate, but there is also a strong tendency for a cube plane, or face of the unit cell, to be aligned parallel to the rolling plane or surface of the sheet or plate.

There are a number of reasons why an understanding of crystal properties is important to the metallurgist. One of these is that the basic anisotropy of crystalline materials is reflected in the polycrystalline objects of commerce. It should be noted also that this is not always undesirable. Preferred orientations

2. Recrystallization following cold work is discussed in Chapter 7.

can often result in materials with superior properties. An interesting example of this sort is found in the alloy of iron with 4 percent of silicon, used for making transformer coils. In this case, by a complicated combination of rolling procedures and heat treatments, it is possible to obtain a very strong preferred orientation in which an *a* direction of the crystals is aligned parallel to the rolling direction, while a cube plane, or face of the unit cell, remains parallel to the rolling plane. This average orientation is shown schematically in Fig. 1.12B. The significant feature of this texture is that it places the direction of easy magnetization parallel to the length of the sheet. In manufacturing transformers, it is then a rather simple matter to align the plates in the core so that this direction is parallel to the direction along which the magnetic flux runs. When this occurs, the resultant hysteresis loss can be made very small.

1.13 Miller Indices. As one becomes more and more involved in the study of crystals, the need for symbols to describe the orientation in space of important crystallographic directions and planes becomes evident. Thus, while the directions of closest packing in the body-centered cubic lattice may be described as the diagonals that traverse the unit cell, and the corresponding directions in the face-centered cubic lattice as the diagonals that cross the faces of a cube, it is much easier to define these directions in terms of several simple integers. The Miller system of designating indices for crystallographic planes and directions is universally accepted for this purpose. In the discussion that follows, the Miller indices for cubic and hexagonal crystals will be considered. The indices for other crystal structures are not difficult to develop, but space considerations preclude their discussion.

Direction Indices in the Cubic Lattice. Let us take a cartesian coordinate system with axes parallel to the edge of the unit cell of a cubic crystal. (See Fig. 1.13.) In this coordinate system, the unit of measurement along all three

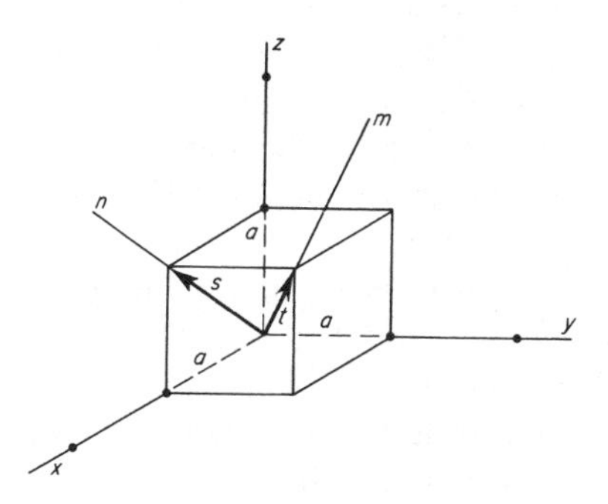

Fig. 1.13 The [111] and [101] directions in a cubic crystal; directions *m* and *n*, respectively.

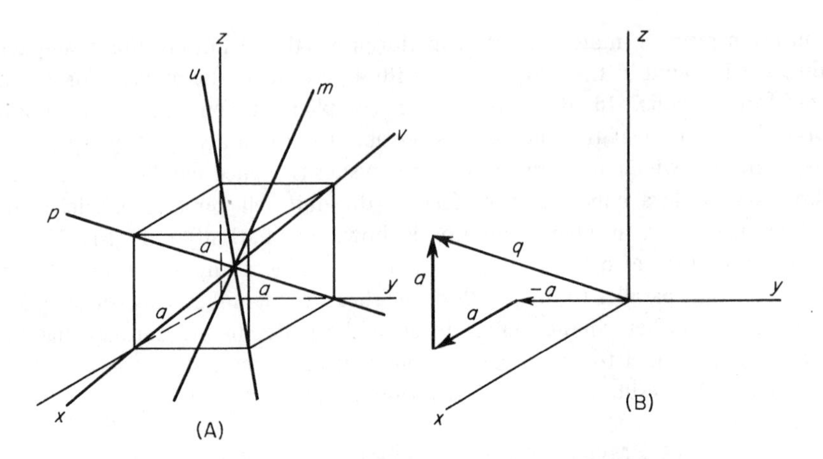

Fig. 1.14 (A) The four cube diagonals of a cubic lattice, *m, n, u,* and *v*. (B) The components of the vector *q* that parallels the cube diagonal *p* are *a*, −*a*, and *a*. Therefore, the indices of *q* are $[1\bar{1}1]$.

axes is the length of the edge of a unit cell, designated by the symbol a in the figure. The Miller indices of directions are introduced with the aid of several simple examples. Thus the cube diagonal m of Fig. 1.13 has the same direction as a vector t with a length that equals the diagonal distance across the cell. The component of the vector t on each of the three coordinate axes is equal to a. Since the unit of measurement along each axis equals a, the vector has components 1, 1, and 1 on the x, y, and z axes respectively. The Miller indices of the direction m are now written [111]. In the same manner, the direction n, which crosses a face of the unit cell diagonally, has the same direction as a vector s the length of which is the face diagonal of the unit cell. The x, y, and z components of this vector are 1, 0, and 1 respectively; the corresponding Miller indices are [101]. The indices of the x axis are [100], the y axis [010], and the z axis [pp1].

A general rule for finding the Miller indices of a crystallographic direction can now be stated. Draw a vector from the origin parallel to the direction whose indices are desired. Make the magnitude of the vector such that its components on the three coordinate axes have lengths that are simple integers. These integers must be the smallest numbers that will give the desired direction. Thus, the integers 1, 1, and 1, and 2, 2, and 2 represent the same direction in space, but, by convention, the Miller indices are [111] and not [222].

Let us apply the above rule to the determination of the Miller indices of a second cube diagonal; that indicated by the symbol p in Fig. 1.14. The vector q (which starts at the origin in Fig. 1.14B) is parallel to the direction p. The components of q are 1, −1, and 1, and, by the above definition, the

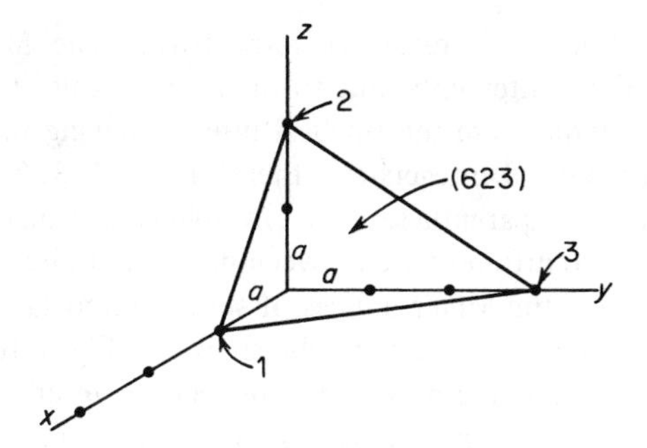

Fig. 1.15 The intercepts of the (623) plane with the coordinate axes.

corresponding Miller indices of p are $[1\bar{1}1]$ where the negative sign of the y index is indicated by a bar over the corresponding integer. The indices of the diagonal m in Fig. 1.14A have already been shown to be [111], and it may also be shown that the indices of the diagonals u and v are $[11\bar{1}]$ and $[\bar{1}11]$. The four cube diagonals thus have indices [111], $[\bar{1}11]$, $[1\bar{1}1]$, and $[11\bar{1}]$.

When a specific crystallographic direction is referred to, the Miller indices are enclosed in square brackets as above. However, it is sometimes desirable to refer to all of the directions of the same form. In this case, the indices of one of these directions are enclosed in carets ⟨111⟩, and the symbol is read to signify all four directions ([111], [111], [111], and [111]), which are considered as a class. Thus, one might say that the close-packed directions in the body-centered cubic lattice are ⟨111⟩ directions, whereas a specific crystal might be stressed in tension along its [111] direction and simultaneously compressed along its $[1\bar{1}1]$ direction.

Cubic Indices for Planes. Crystallographic planes are also identified by sets of integers. These are obtained from the intercepts that the planes make with the coordinate axes. Thus, in Fig. 1.15 the indicated plane intercepts the x, y, and z

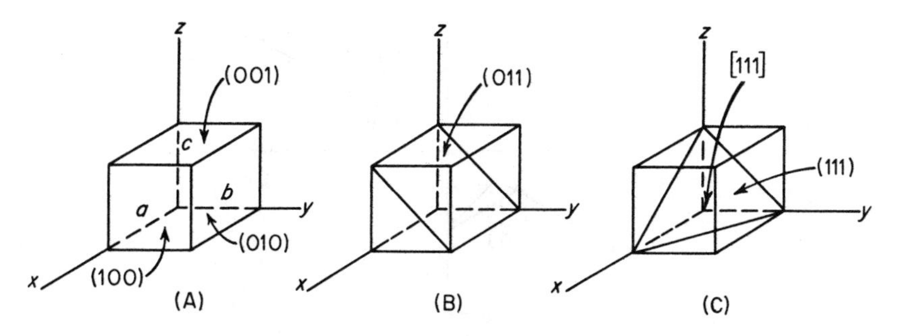

Fig. 1.16 (A) Cube planes of a cubic crystal: *a* (100); *b* (010); *c* (001). (B) The (011) plane. (C) The (111) plane.

axes at 1, 3, and 2 unit-cell distances respectively. The Miller indices are proportional not to these intercepts, but to their reciprocals $1/1$, $1/3$, $1/2$ and, by definition, the Miller indices are the smallest integers having the same ratios as these reciprocals. The desired integers are, therefore, 6, 2, 3. The Miller indices of a plane are enclosed in parentheses, for example (623), instead of brackets, thus making it possible to differentiate between planes and directions.

Let us now determine the Miller indices of several important planes of cubic crystals. The plane of the face *a* of the cube shown in Fig. 1.16A is parallel to both the *y* and *z* axes and, therefore, may be said to intercept these axes at infinity. The *x* intercept, however, equals 1, and the reciprocals of the three intercepts are $1/1$, $1/\infty$, $1/\infty$. The corresponding Miller indices are (100). The indices of the face *b* are (010), while those of *c* are (001). The indicated plane of Fig. 1.16B has indices (011), and that of Fig. 1.16C (111). The latter plane is an octahedral plane, as may be seen by referring to Fig. 1.7. Other octahedral planes have the indices $(\bar{1}11)$, $(1\bar{1}1)$, and $(11\bar{1})$, where the bar over a digit represents a negative intercept. By way of illustration, the $(\bar{1}11)$ plane is shown in Fig. 1.17, where it may be seen that the *x* intercept is negative, whereas the *y* and *z* intercepts are positive. This figure also shows that the $(1\bar{1}\bar{1})$ plane is parallel to the $(\bar{1}11)$ plane and is, therefore, the same crystallographic plane. Similarly, the indices $(\bar{1}1\bar{1})$ and $(\bar{1}\bar{1}1)$ represent the same planes as $(1\bar{1}1)$ and $(11\bar{1})$.

The set of planes of a given form, such as the four octahedral planes (111), (111), $(\bar{1}11)$, and $(11\bar{1})$, are represented as a group with the aid of braces enclosing one of the indices, that is, {111}. Thus, if one wishes to refer to a specific plane in a crystal of known orientation, parentheses are used, but if the class of planes is to be referred to, braces are used.

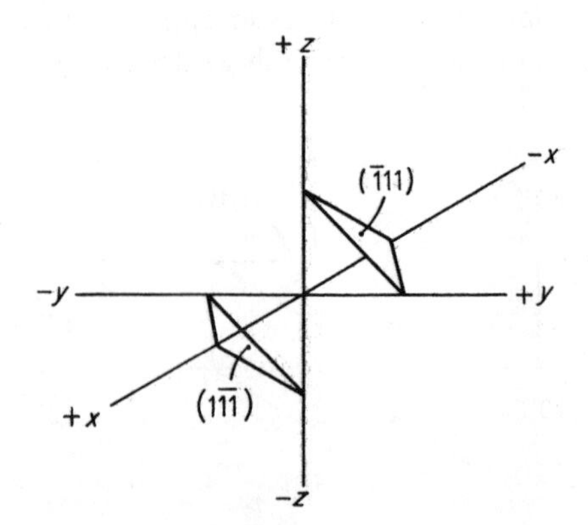

Fig. 1.17 The $(\bar{1}11)$ and $(1\bar{1}\bar{1})$ planes are parallel to each other and therefore represent the same crystallographic plane.

An important feature of the Miller indices of cubic crystals is that the integers of the indices of a plane and of the direction normal to the plane are identical. Thus, face a of the cube in Fig. 1.16A has indices (100), and the x axis, perpendicular to this plane, has indices [100]. In the same manner, the octahedral plane of Fig. 1.16C and its normal, the cube diagonal, have indices (111) and [111] respectively. Noncubic crystals do not, in general, possess this equivalence between the indices of planes and normals to the planes.

Miller Indices for Hexagonal Crystals. Planes and directions in hexagonal metals are defined almost universally in terms of Miller indices containing four digits instead of three. The use of a four-digit system gives planes of the same form similar indices. Thus, in a four-digit system, the planes $(11\bar{2}0)$ and $(1\bar{2}10)$ are equivalent planes. The three-digit system, on the other hand, gives equivalent planes indices that are not similar. Thus, the above two planes would have indices (110) and $(1\bar{2}0)$ in the three-digit system.

Four-digit hexagonal indices are based on a coordinate system containing four axes. Three axes correspond to close-packed directions and lie in the basal plane of the crystal, making 120° angles with each other. The fourth axis is normal to the basal plane and is called the c axis, whereas the three axes that lie in the basal plane are designated the a_1, a_2, and a_3 axes. Figure 1.18 shows the hexagonal unit cell superimposed upon the four-axis coordinate system. It is customary to take the unit of measurement along the a_1, a_2, and a_3 axes as the distance between atoms in a close-packed direction. The magnitude of this unit is indicated by the symbol a. The unit of measurement for the c axis is the height of the unit cell that is designated as c.

Let us now determine the Miller indices of several important close-packed hexagonal lattice planes. The uppermost surface of the unit cell in Fig. 1.19

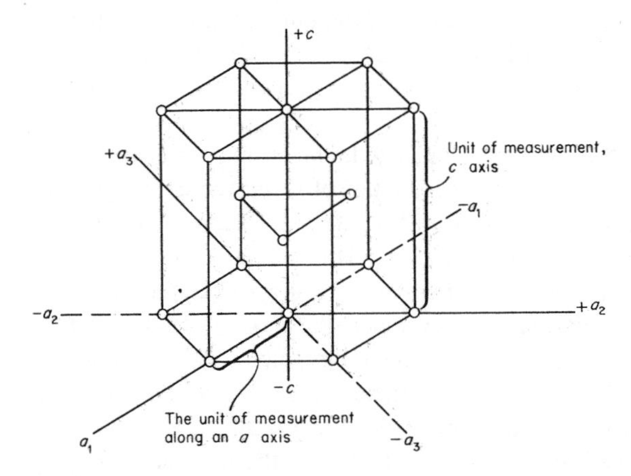

Fig. 1.18 The four coordinate axes of a hexagonal crystal.

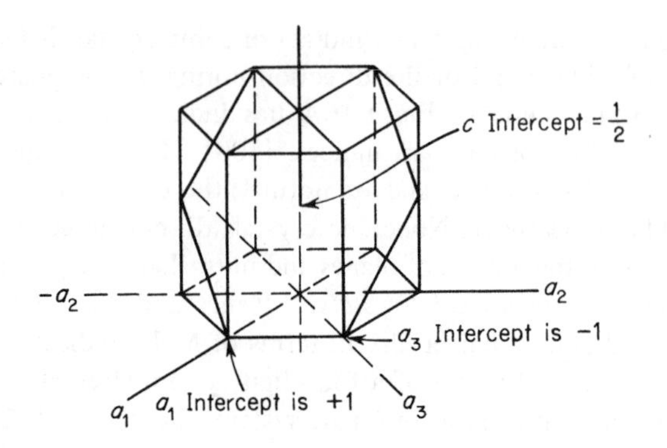

Fig. 1.19 The (10$\bar{1}$2) plane of a hexagonal metal.

corresponds to the basal plane of the crystal. Since it is parallel to the axes a_1, a_2, and a_3, it must intercept them at infinity. Its c axis intercept, however, is equal to 1. The reciprocals of these intercepts are $1/\infty$, $1/\infty$, $1/\infty$, and $1/1$. The Miller indices of the basal plane are, therefore, (0001). The six vertical surfaces of the unit cell are known as *prism planes* of Type I. Consider now the prism plane that forms the front face of the cell, which has intercepts as follows: a_1 at 1, a_2 at ∞, a_3 at -1, and c at ∞. Its Miller indices are, therefore, (10$\bar{1}$0). Another important type of plane in the hexagonal lattice is shown in Fig. 1.19. The intercepts are a_1 at 1, a_2 at ∞, a_3 at -1, and c at ½, and the Miller indices are accordingly (10$\bar{1}$2).

Miller indices of directions are also expressed in terms of four digits. In writing direction indices, the third digit must always equal the negative sum of the first two digits. Thus, if the first two digits are 3 and 1, the third must be −4, that is [31$\bar{4}$0].

Let us investigate directions lying only in the basal plane, since this will simplify the presentation. If a direction lies in the basal plane, then it has no component along the c axis, and the fourth digit of the Miller indices will be zero.

As our first example, let us find the Miller indices of the a_1 axis. This axis has the same direction as the vector sum of three vectors (Fig. 1.20), one of length +2 along the a_1 axis, another of length -1 parallel to the a_2 axis, and the third of length -1 parallel to the a_3 axis. The indices of this direction are, accordingly [2$\bar{1}\bar{1}$0]. This unwieldy method of obtaining the direction indices is necessary in order that the relationship mentioned above be maintained between the first two digits and the third. The corresponding indices of the a_2 and a_3 axes are [$\bar{1}$2$\bar{1}$0] and [$\bar{1}\bar{1}$20]. These three directions are known as the digonal axes of Type I. Another important set of directions lying in the basal plane are the digonal axes of Type II; a set of axes perpendicular to the digonal axes of Type I. Fig. 1.21 shows one of the axes of Type II and indicates how its direction indices are

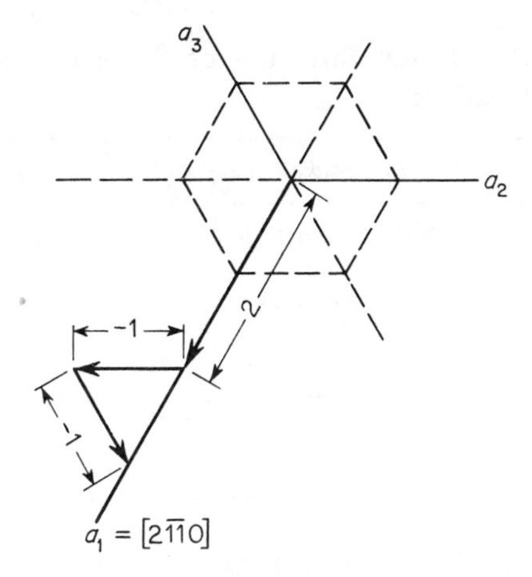

Fig. 1.20. Determination of indices of a digonal axis of Type I – [$2\bar{1}\bar{1}0$].

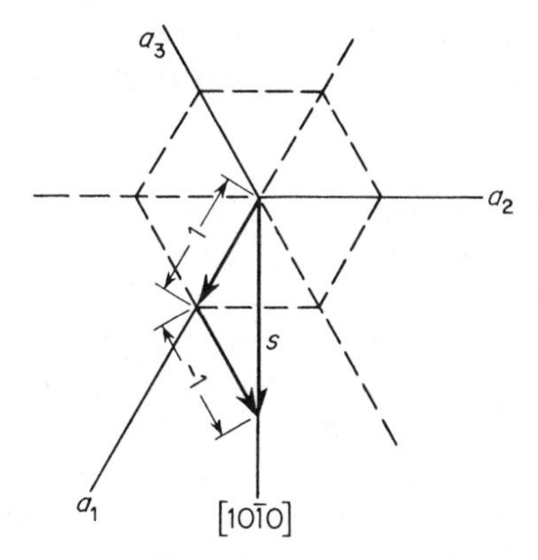

Fig. 1.21 Determination of indices of a digonal axis of Type II – [$10\bar{1}0$].

determined. The vector s in the figure determines the desired direction and equals the vector sum of a unit vector lying on a_1, and another parallel to a_3, but measured in a negative sense. The indices of the digonal axis of Type II are thus [$10\bar{1}0$]. In this case, the second digit is zero because the projection of the vector s on the a_2 axis is zero.

1.14 Crystal Structures of the Metallic Elements. Some of the most important metals are classified according to their crystal structures in Table 1.1.

A number of metals are polymorphic, that is, they crystallize in more than

Table 1.1 Crystal Structure of Some of the More Important Metallic Elements.

Face-centered cubic	*Close-packed hexagonal*	*Body-centered cubic*
Iron (910°C to 1400°C)	Magnesium	Iron (below 910°C and 1400°C to 1539°C)
Copper	Zinc	Tungsten
Silver	Titanium	Vanadium
Gold	Zirconium	Molybdenum
Aluminum	Beryllium	Chromium
Nickel	Cadmium	Alkali Metals (Li, Na, K, Rb, Cs)
Lead		
Platinum		

one structure. The most important of these is iron, which crystallizes as either body-centered cubic or face-centered cubic, with each structure stable in separate temperature ranges. Thus, at all temperatures below 910°C and above 1400°C to the melting point, the preferred crystal structure is body-centered cubic, whereas between 910°C and 1400°C the metal is stable in the face-centered cubic structure.

1.15 The Stereographic Projection. The stereographic projection is a useful metallurgical tool, for it permits the mapping in two dimensions of crystallographic planes and directions in a convenient and straightforward manner. The real value of the method is attained when it is possible to visualize crystallographic features directly in terms of their stereographic projections. The purpose of this section is to concentrate on the geometrical correspondence between crystallographic planes and directions and their stereographic projections. In each case, a sketch of a certain crystallographic feature, in terms of its location in the unit cell, is compared with its corresponding stereographic projection.

Several simple examples will be considered, but before this is done, attention will be called to several pertinent facts. The stereographic projection is a two-dimensional drawing of three-dimensional data. The geometry of all crystallographic planes and directions are accordingly reduced by one dimension. Planes are plotted as great circle lines, and directions are plotted as points. Also, the normal to a plane completely describes the orientation of a plane.

As our first example, consider several of the more important planes of a cubic lattice: specifically the (100), (110), and (111) planes. All three planes are treated in the three parts of Fig. 1.22. Notice that the stereographic projection of each plane can be represented either by a great circle or by a point showing the direction in space that is normal to the plane.

Many crystallographic problems can be solved by considering the stereo-

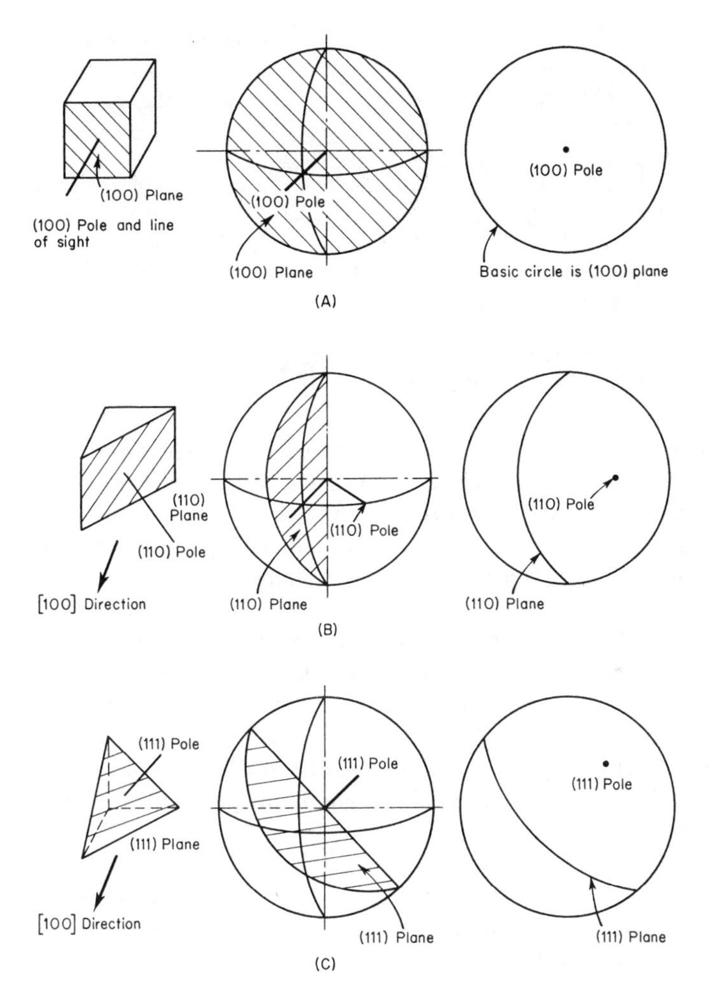

Fig. 1.22 Stereographic projections of several important planes of a cubic crystal. (A) Cubic system, the (100) plane, line of sight along the [100] direction. (B) Cubic system, the (110) plane, line of sight along the [100] direction. (C) Cubic system, the (111) plane, line of sight along the [100] direction. direction.

graphic projections of planes and directions in a single hemisphere, that is, normally the one in front of the plane of the paper. The three examples given in Fig. 1.22 have all been plotted in this manner. If the need arises, the stereographic projections in the rear hemisphere can also be plotted in the same diagram. However, it is necessary that the projections in the two hemispheres be distinguishable from each other. This may be accomplished if the stereographic

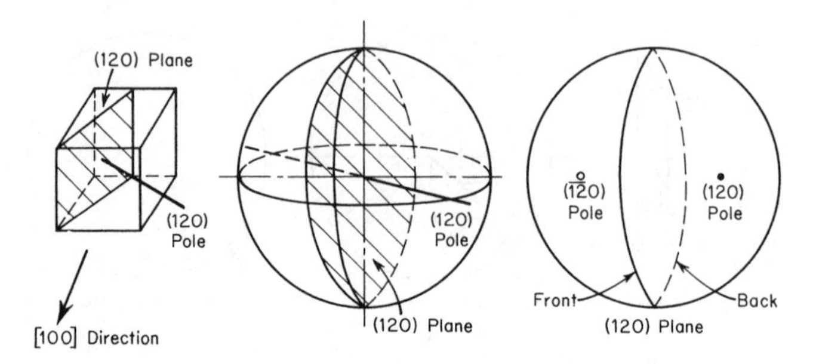

Fig. 1.23 Cubic system, the (120) plane, showing the stereographic projections from both hemispheres, line of sight the [100] direction.

projections of planes and directions in the forward hemisphere are drawn as solid lines and dots respectively, while those in the rear hemisphere are plotted as dotted lines and circled dots respectively. As an illustration, consider Fig. 1.23, in which the projections in both hemispheres of a single plane are shown. The (120) plane of a cubic lattice is used in this example.

1.16 Directions That Lie in a Plane. Frequently one desires to show the positions of certain important crystallographic directions that lie in a particular plane of a crystal. Thus, in a body-centered cubic crystal one of the more important planes is {110} and in each of these planes one finds two close-packed ⟨111⟩ directions. The two that lie in the (101) plane are shown in Fig. 1.24, where they appear as dots lying on the great circle representing the (101) plane.

1.17 Planes of a Zone. Those planes that mutually intersect along a common direction form the planes of a zone, and the line of intersection is called the *zone axis*. In this regard, consider the [111] direction as a zone axis.

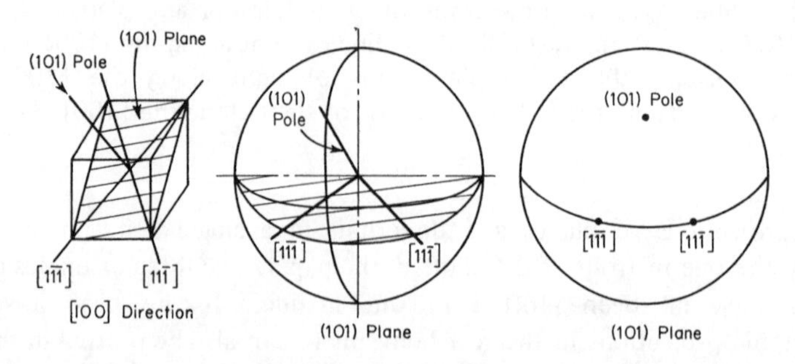

Fig. 1.24 Cubic system, the (101) plane and the two ⟨111⟩ directions that lie in this plane, line of sight [100].

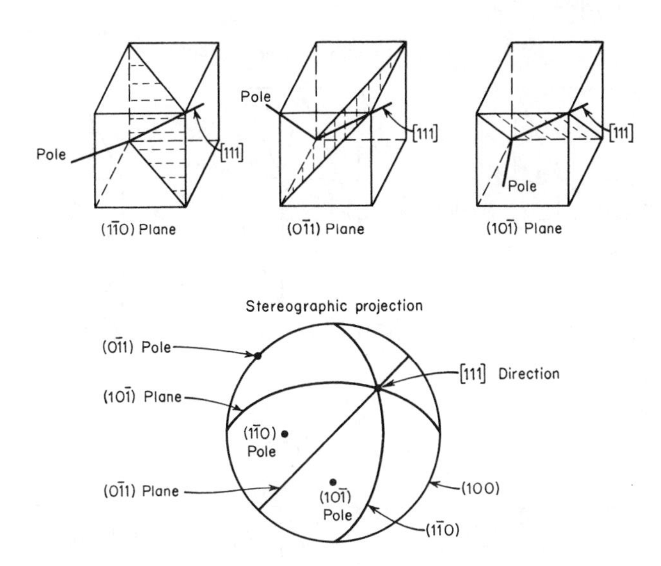

Fig. 1.25 Cubic system, zone of planes the zone axis of which is the [111] direction. The three {110} planes that belong to this zone are illustrated in the above figures.

Figure 1.25 shows that there are three {110} planes that pass through the [111] direction. There are also three {112} planes and six {123} planes, as well as a number of higher indice planes that have the same zone axis. The pertinent {112} and {123} planes are shown in Fig. 1.26, whereas the stereographic projection of these and the above-mentioned {110} planes are shown in Fig. 1.27. Notice that in this latter figure only the poles of the planes are plotted, and it is significant that all of the poles fall on the great circle representing the stereographic projection of the (111) plane.

1.18 The Wulff Net. The *Wulff net* is a stereographic projection of latitude and longitude lines in which the north-south axis is parallel to the plane of the paper. The latitude and longitude lines of the Wulff net serve the same function as the corresponding lines on a geographical map or projection; that is, they make possible graphical measurements. However, in the stereographic projection we are primarily interested in measuring angles, whereas in the geographical sense it is distance that is usually more important. A typical Wulff, or meridional, net drawn to 2° intervals is shown in Fig. 1.28.

Attention is now called to several facts about the Wulff net. First, all meridians (longitude lines), including the basic circle, are great circles. Second, the equator is a great circle. All other latitude lines are small circles. Third, angular distances between points representing directions in space can be

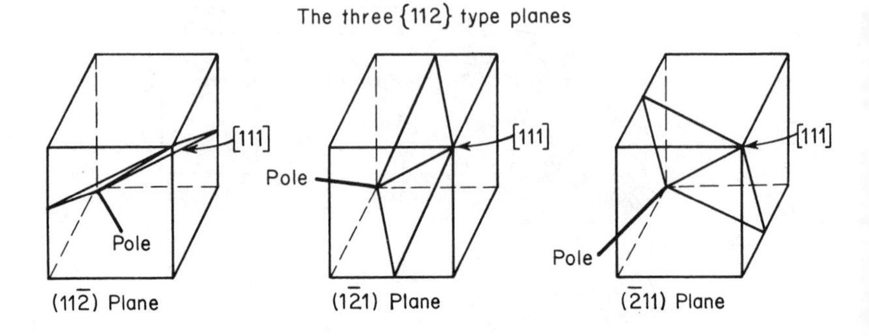

The six $\{123\}$ type planes

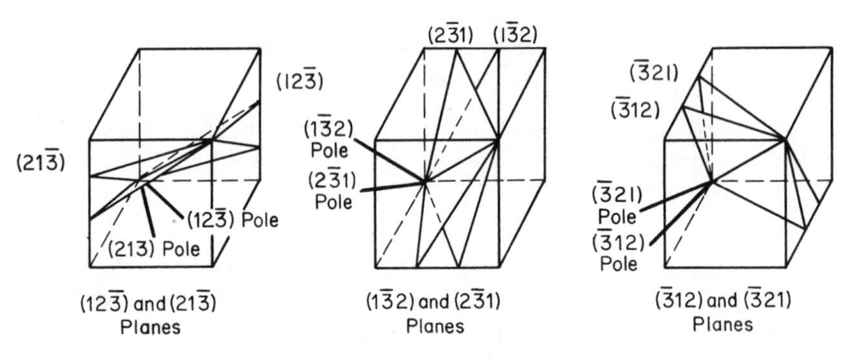

Fig. 1.26 The $\{112\}$ and $\{123\}$ planes that have $[111]$ as their zone axis.

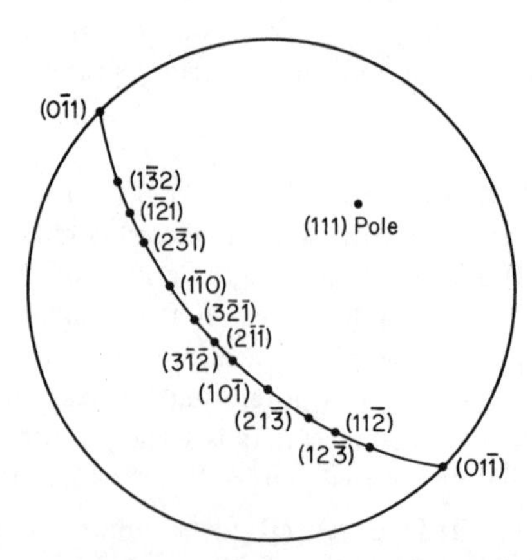

Fig. 1.27 Stereographic projection of the zone containing the 12 planes shown in Figs. 1.25 and 1.26. Only the poles of the planes are plotted. Notice that all of the planar poles lie in the (111) plane.

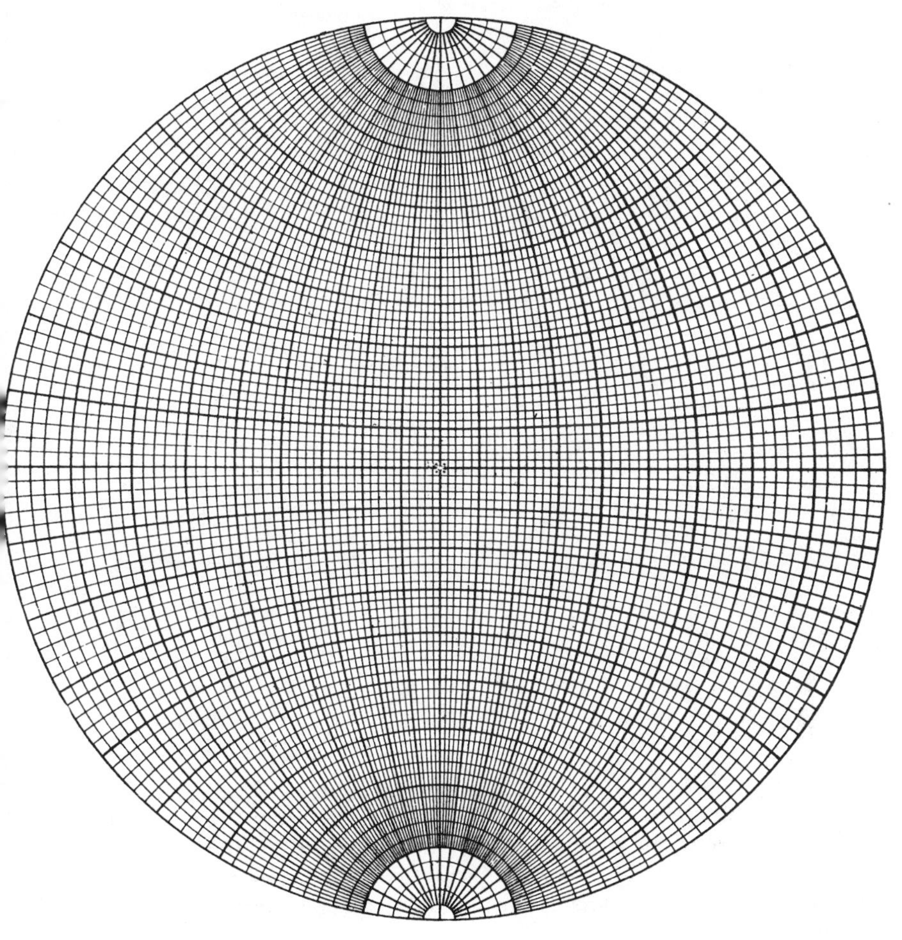

Fig. 1.28 Wulff, or meridional, stereographic net drawn with 2° intervals.

measured on the Wulff net only if the points are made to coincide with a great circle of the net.

In the handling of many crystallographic problems, it is frequently necessary to rotate a stereogrpahic projection corresponding to a given crystal orientation into a different orientation. This is done for a number of reasons. One of the most important is to bring experimentally measured data into a standard projection where the basic circle is a simple close-packed plane such as (100) or (111). Deformation markings, or other experimentally observed crystallographic phenomena, usually can be more readily interpreted when studied in terms of standard projections.

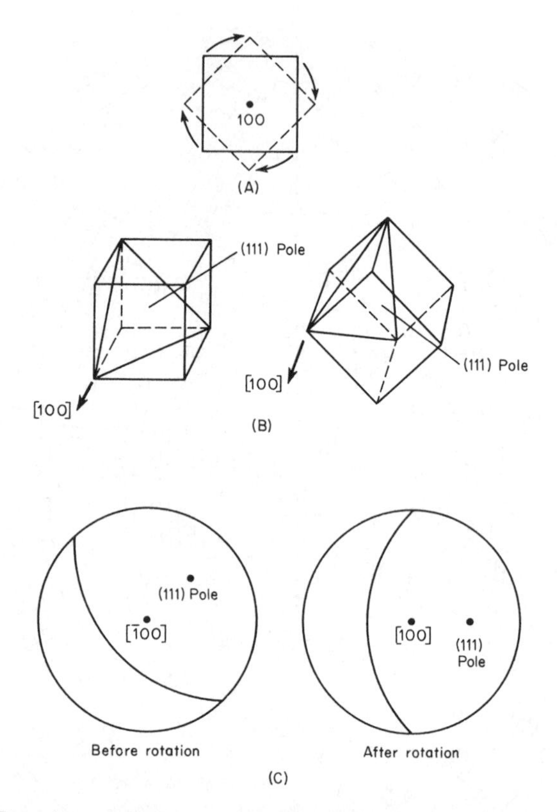

Fig. 1.29 Rotation about the center of the Wulff net. (A) The effect of the disired rotation on the cubic unit cell. Line of sight [100]. (B) Perspective view of the (111) plane before and after the rotation. (C) Stereographic projection of the (111) plane and its pole before and after rotation. Rotation clockwise 45° about the [100] direction.

In solving problems with the aid of the Wulff net, it is customary to cover it with a piece of tracing paper. A common pin is then driven through the paper and into the exact center of the net. The paper, thus mounted, serves as a work sheet on which crystallographic data are plotted. The following two types of rotation of the plotted data are possible.

Rotation about an Axis in the Line of Sight. This rotation is easily performed by merely rotating the tracing paper, relative to the net, about the pin. As an example, let us rotate a cubic lattice 45° clockwise around the [100] direction as an axis. This rotation has the effect of placing the pole of the (111) plane, as plotted in Fig. 1.22C, on the equator of the Wulff net. Figure 1.29A shows the effect of the desired rotation on the orientation of the cubic unit cell

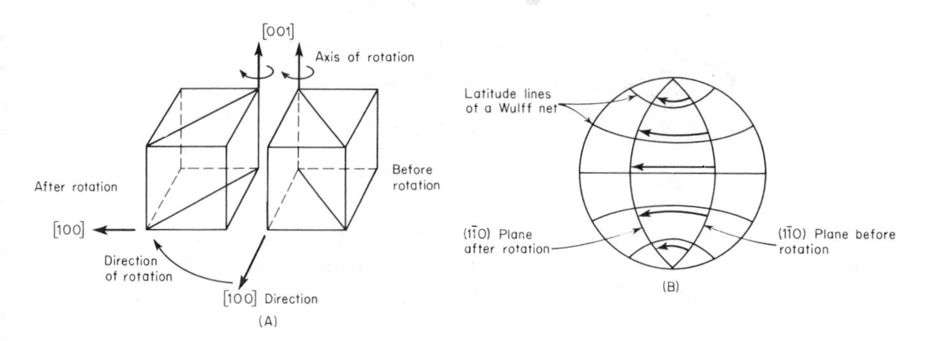

Fig. 1.30 Rotation about the north-south axis of the Wulff net. (A) Perspective views of the unit cell before and after the rotation showing the orientation of the (1$\bar{1}$0) plane. (B) Stereographic projection showing the above rotation. For the sake of clarity of presentation, only the (1$\bar{1}$0) plane is shown. The rotation of the pole is not shown. Also, the meridians of the Wulff net are omitted.

when the cell is viewed along the [100] direction. Note that because the basic circle represents the (100) plane, a simple rotation of the paper by 45° about the pin produces the desired rotation in the stereographic projection.

Rotation about the North-South Axis of the Wulff Net. This rotation is not as simple to perform as that given above, which can be accomplished by merely rotating the work sheet about the pin. Rotations of this second type are accomplished by a graphical method. The data are first plotted stereographically and then rotated along latitude lines and replotted in such a manner that each point undergoes the same change in longitude. The method will be quite evident if one considers the drawings of Fig. 1.30. In the present example, it is assumed that the forward face (100) of the unit cell is rotated to the left about the [001] direction as an axis. Consider now the effect of this rotation on the spacial orientation and stereographic projection of the (1$\bar{1}$0) plane. In Fig. 1.30A, the right- and left-hand drawings represent the cubic unit cell before and after the rotation respectively. The effect of the rotation on the stereographic projection of the (1$\bar{1}$0) plane is shown in Fig. 1.30B. Each of the curved arrows shown in this drawing represents a change in longitude of 90°. In these drawings, the pole of the (1$\bar{1}$0) plane is not shown in order to simplify the presentation. The rotation that the (110) pole undergoes is shown, however, in Fig. 1.31.

By using simple examples, the two basic rotations that can be made when the Wulff net is used have been pointed out above. All possible rotations of a crystal in three dimensions can be duplicated by using these rotations on a stereographic projection.

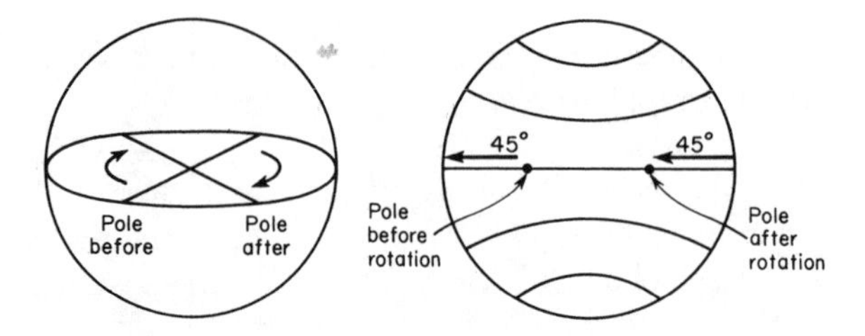

Fig. 1.31 The rotation of the pole of the ($1\bar{1}0$) plane is given above. The diagram on the left shows the rotation in a perspective figure, whereas that on the right shows the motion of the pole along a latitude line of a stereographic projection which, in this case, is the equator.

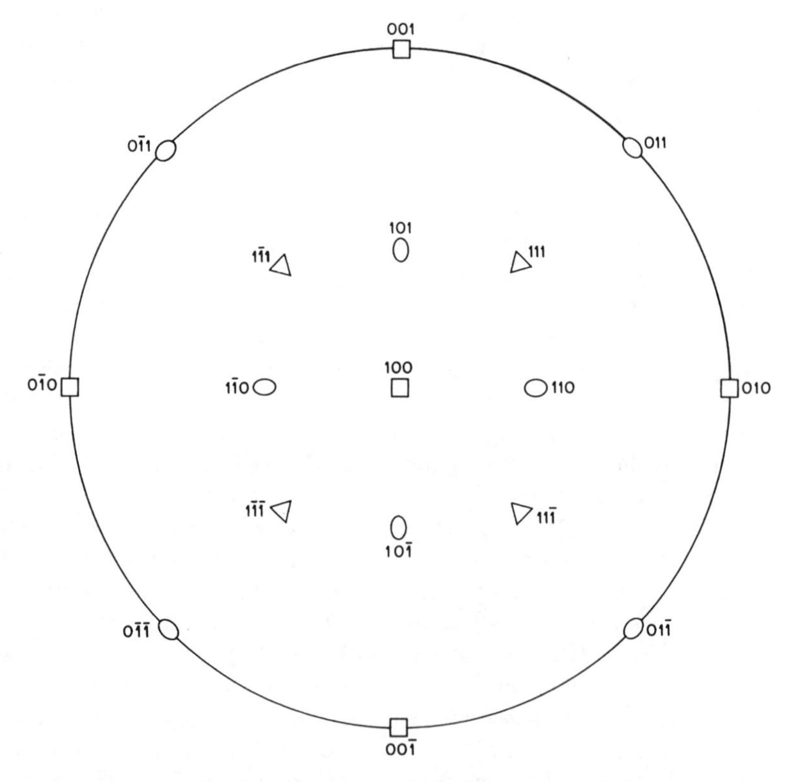

Fig. 1.32 A 100 standard stereographic projection of a cubic crystal.

1.19 Standard Projections. A stereographic projection, in which a prominent crystallographic direction or pole of an important plane lies at the center of the projection, is known as a standard stereographic projection. Such a projection for a cubic crystal is shown in Fig. 1.32, where the (100) pole is assumed to be normal to the plane of the paper. This figure is properly called a standard 100 projection of a cubic crystal. In this diagram, note that the poles of all the {100}, {110}, and {111} planes have been plotted at their proper orientations. Each of these basic crystallographic directions is represented by a characteristic symbol. For the {100} poles, this is a square, signifying that these poles correspond to four-fold symmetry axes. If the crystal is rotated 90° about any one of these directions, it will be returned to an orientation exactly equivalent to its original orientation. In a 360° rotation about a {100} pole, the crystal reproduces its original orientation four times. In the same fashion, a ⟨111⟩ direction corresponds to a three-fold symmetry axis, and these directions are

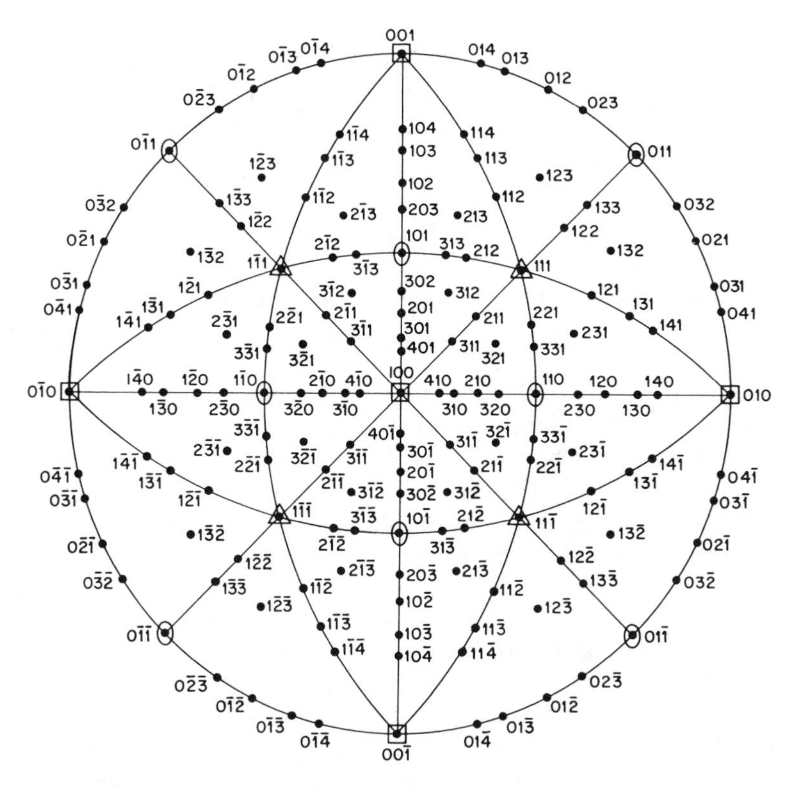

Fig. 1.33 A 100 standard stereographic projection of a cubic crystal showing additional poles.

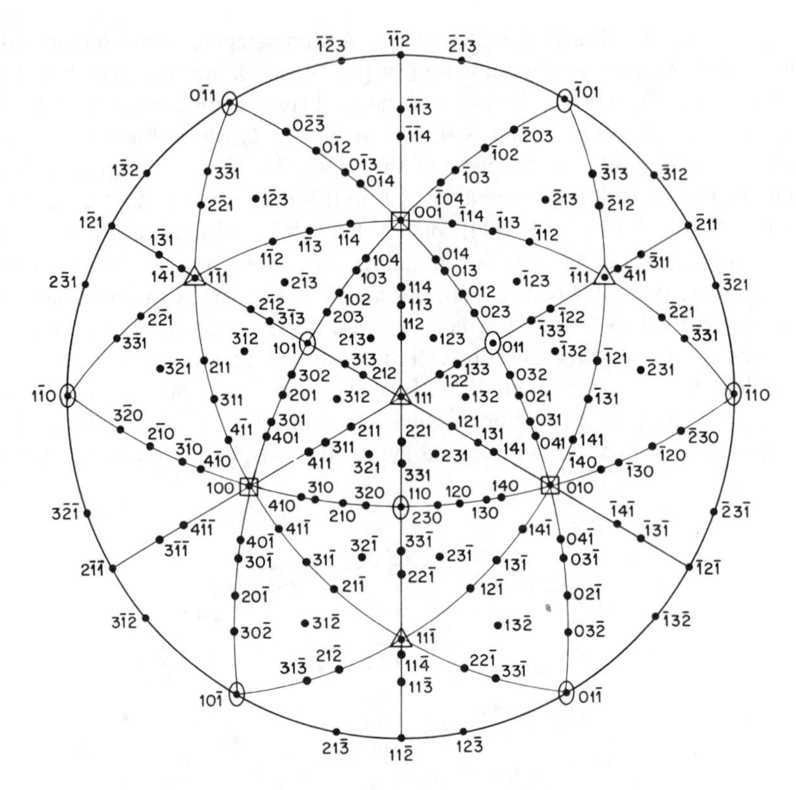

Fig. 1.34 A 111 standard projection of a cubic crystal.

indicated in the stereographic projection by triangles. Finally, the two-fold symmetry of the ⟨110⟩ directions is indicated by the use of small ellipses to designate their positions in the stereographic projection.

A more complete 100 standard projection of a cubic crystal is shown in Fig. 1.33. This includes the poles of other planes of somewhat higher Miller indices. Fig. 1.33 can be considered to be either a projection showing the directions in a cubic crystal or the poles of its planes. This is because in a cubic crystal, a plane is always normal to the direction with the same Miller indices. In a hexagonal close-packed crystal, however, the projection showing the poles of the planes is not the same as one showing crystallographic directions.

A 111 standard projection is shown in Fig. 1.34 that contains the same poles as the 100 projection in Fig. 1.33. The three-fold symmetry of the crystal structure about the pole of a {111} plane is clearly evident in this figure. At the same time, attention is called to the fact that the 100 projection of Fig. 1.33 also plainly reveals the four-fold symmetry about a {100} pole.

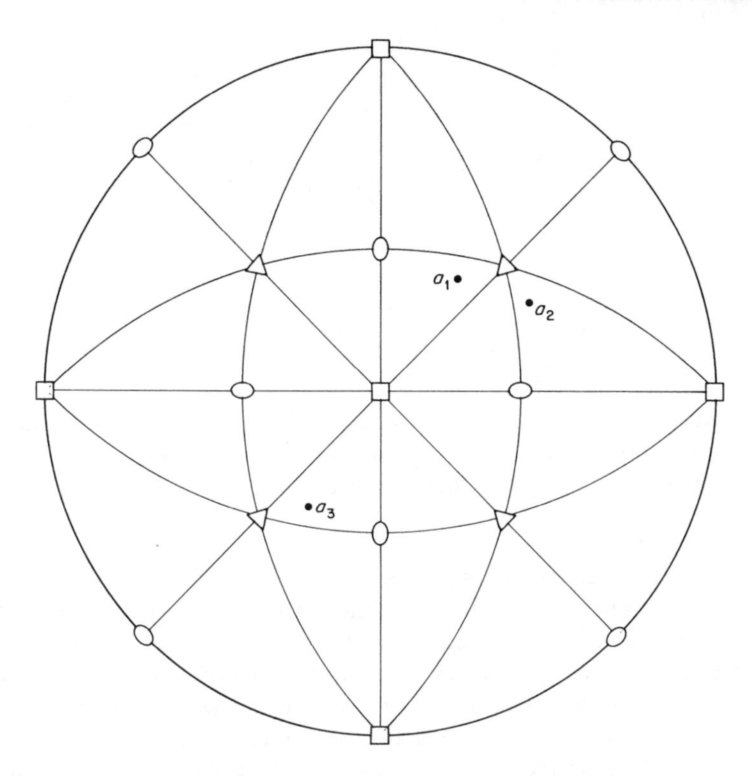

Fig. 1.35 The crystallographic directions a_1, a_2, and a_3 shown in this standard projection are equivalent because they lie in similar positions inside their respective standard stereographic triangles.

1.20 The Standard Stereographic Triangle for Cubic Crystals. The great circles corresponding to the {100} and {110} planes of a cubic crystal are also shown in the standard projections of Figs. 1.33 and 1.34. These great circles pass through all of the poles shown on the diagram except those of the {123} planes. At the same time, they divide the standard projection into 24 spherical triangles. These spherical triangles all lie in the forward hemisphere of the projection. There are, of course, 24 similar triangles in the rear hemisphere. An examination of the triangles outlined in Figs. 1.33 and 1.34 shows an interesting fact: in every case, the three corners of the triangles are formed by a ⟨111⟩ direction, a ⟨110⟩ direction, and a ⟨100⟩ direction. This is a highly significant observation, since it means that each triangle corresponds to a region of the crystal that is equivalent. In effect, this signifies that the three lattice directions, marked a_1, a_2, and a_3, and shown in Fig. 1.35, are crystallographically equivalent, because they are located at the same relative positions inside three stereographic triangles. To illustrate this point, let us assume that it is

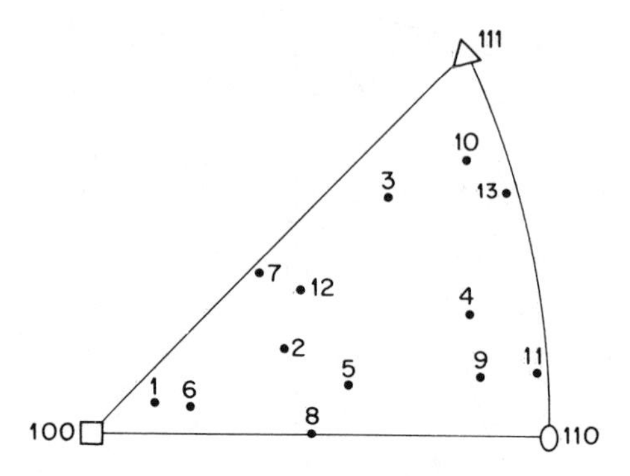

Fig. 1.36 When it is necessary to compare the orientations of a number of crystals, this often can be done conveniently by plotting the crystal axes in a single stereographic triangle, as indicated in this figure.

possible to cut three tensile specimens, with axes parallel to a_1, a_2, and a_3, out of a very large single crystal. If now tensile tests were to be performed on these three smaller crystals, one would expect to get identical stress-strain curves for the three specimens. A similar result should be obtained if some other physical property, such as the electrical resistivity, were to be measured along these three directions. The plotting of crystallographic data is often simplified because of the equivalence of the stereographic triangles. For example, if one should have a large number of long, cylindrical crystals and wishes to plot the orientations of the individual crystal axes, this can be done conveniently in a single stereographic triangle, as shown in Fig. 1.36.

Problems

1. Make a drawing of the cubic unit cell in which are shown the close-packed directions of the body-centered cubic crystal. Label all of these directions with their proper direction indices.
2. In the same manner as in Problem 1, draw a unit cell on which are superimposed the close-packed directions of the face-centered cubic crystal structure. Identify each direction with its proper direction indices.
3. Fig. 1.25 shows three planes of the form {110}. How many planes of this type are there in all? In sketches similar to those of Fig. 1.25 show the remaining {110} planes and label each with its Miller indices.
4. There are 12 planes of the form {112}. Three of these are shown inside a cubic unit cell in Fig. 1.26. Draw similar sketches showing the other nine. Label each with its proper Miller indices.

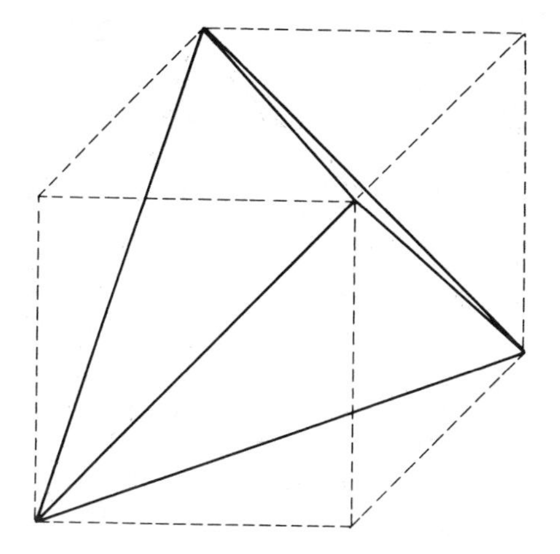

5. This diagram shows the Thompson Tetrahedron, a geometrical figure formed by the four cubic {111} planes. Identify each plane by its proper Miller indices. Would this figure be of more significance in regard to showing important crystallographic features in the body-centered cubic crystal structure or the face-centered cubic crystal structure? Explain, considering both the surfaces and the edges of the tetrahedron.
6. The hexagonal close-packed planes of the type $\{10\bar{1}0\}$ are called prism planes of Type I. How many of them are there and what are their indices? What is the significance of these planes relative to the hexagonal unit cell shown in Fig. 1.19?
7. Draw a hexagonal unit cell and show the orientation in this cell of the plane $\{11\bar{2}0\}$. This is a prism plane of Type II. How many prism planes of Type II are there and what are their indices? Show that a six-sided prism can also be formed using prism planes of Type II.
8. Note that the $\langle 11\bar{2}0\rangle$ direction is normal to the $(11\bar{2}0)$ plane. However, $\langle 10\bar{1}2\rangle$ is not perpendicular to $(10\bar{1}2)$. Show that this is true geometrically.
9. In a hexagonal unit cell, draw in the $(10\bar{1}2)$ and the $(10\bar{1}1)$ planes. Also draw in the $[10\bar{1}\bar{2}]$ and $[10\bar{1}\bar{1}]$ directions. How are these planes and directions related?
10. The $\{11\bar{2}1\}$ and $\{11\bar{2}2\}$ planes are important with regard to plastic deformation mechanisms in certain hexagonal close-packed metals, such as titanium and zirconium. Show the orientation in the unit cell of one plane of each type.

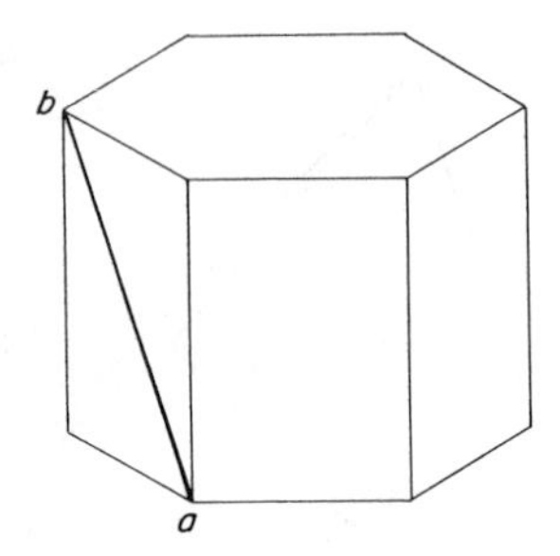

11. Identify the direction indices of the line $a-b$ shown in this unit cell. This line lies in three important planes. Identify them.

Stereographic Projection

The following problems involve the plotting of stereographic projections and require the use of the Wulff net, shown in Fig. 1.28, and a sheet of tracing paper. In each case, first place the tracing paper over the Wulff net and then mark the center of the net on the tracing paper with a dot. Next, trace the outline of the basic circle and place a small vertical mark at the top of the circle to serve as an index.

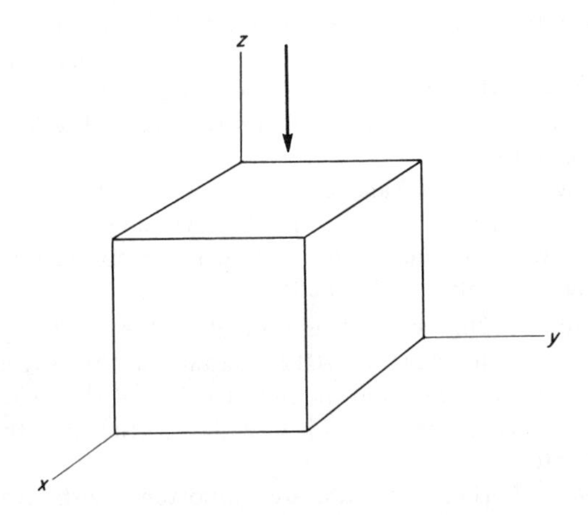

12. Place the tracing paper so that the basic circle is aligned with the perimeter of the Wulff net and the index mark falls over the north pole of the net. Now assume that a cubic crystal is viewed from above along the z axis, as indicated in the figure. Plot the position of the poles of all the {100} planes and label them properly on the diagram.
13. The basic circle in Problem 12 corresponds to what {100} plane? Draw in the great circles corresponding to the other two {100} planes and label them with their Miller indices.
14. Plot the poles of all six {110} planes and label them. Now draw on the

projection the great circles corresponding to these planes. At this point you will have plotted a figure similar to that of Fig. 1.32, which is a standard 100 projection of the cubic crystal. What standard projection have you plotted?

15. How can you rotate the crystal shown in the projection obtained at the end of Problem 14 so that it will be in a standard 100 orientation? Show the paths that each {100} and {110} pole will follow during this rotation.
16. Plot a standard (0001) stereographic projection of a hexagonal close-packed crystal showing the poles of all prism planes of Type I, $\{10\bar{1}0\}$, and Type II, $\{11\bar{2}0\}$, as well as the basal plane pole (0001).
17. In the drawing of Problem 16, connect the basal plane pole to the prism plane poles with great circles (straight lines). This will divide the stereographic projection into a set of spherical triangles. Examine these triangles and note if there is any similarity to those shown in Fig. 1.31 for a cubic crystal.
18 Reconstruct the standard projection of Problem 16, but add onto this diagram the poles of all the $\{10\bar{1}1\}$, $\{10\bar{1}2\}$, $\{10\bar{1}3\}$, $\{10\bar{1}4\}$, $\{11\bar{2}1\}$, $\{11\bar{2}2\}$, $\{11\bar{2}3\}$, and $\{11\bar{2}4\}$, using the data for magnesium given in Appendix B. This will produce a pole figure for a hexagonal metal that shows most of the important planes of this crystal structure.

2 Diffraction Methods

Because crystals are symmetrical arrays of atoms containing rows and planes of high atomic density, they are able to act as three-dimensional diffraction gratings. If light rays are to be efficiently diffracted by a grating, then the spacing of the grating (distance between ruled lines on a grating) must be approximately equal to the wavelength of the light waves. In the case of visible light, gratings with line separations between 10,000 and 20,000 Å are used to diffract wavelengths in the range from 4000 to 8000 Å. In crystals, however, the separation between equally spaced parallel rows of atoms or atomic planes is much smaller and of the order of a few Å units. Fortunately, low-voltage X-rays have wavelengths of the proper magnitudes to be diffracted by crystals; that is, X-rays produced by tubes operated in the range between 20,000 and 50,000 volts, as contrasted to those used in medical applications, where voltages exceed 100,000 volts.

When X-rays of a given frequency strike an atom, they interact with its electrons, causing them to vibrate with the frequency of the X-ray beam. Since the electrons become vibrating electrical charges, they reradiate the X-rays with no change in frequency. These reflected rays come off the atoms in any direction. In other words, the electrons of an atom "scatter" the X-ray beam in all directions.

When atoms spaced at regular intervals are irradiated by a beam of X-rays, the scattered radiation undergoes interference. In certain directions constructive interference occurs; in other directions destructive interference occurs. For instance, if a single atomic plane is struck by a parallel beam of X-rays, the beam undergoes constructive interference when the angle of incidence equals the angle of reflection. Thus, in Fig. 2.1, the rays marked a_1 to a_3 represent a parallel beam of X-rays. A wave front of this beam, where all rays are in phase, is shown by the line *AA*. The line *BB* is drawn perpendicular to rays reflected by the atoms in a direction such that the angle of incidence equals the angle of reflection. Since *BB* lies at the same distance from the wave front *AA* when measured along any ray, all points on *BB* must be in phase. It is, therefore, a wave front, and the direction of the reflected rays is a direction of constructive interference.

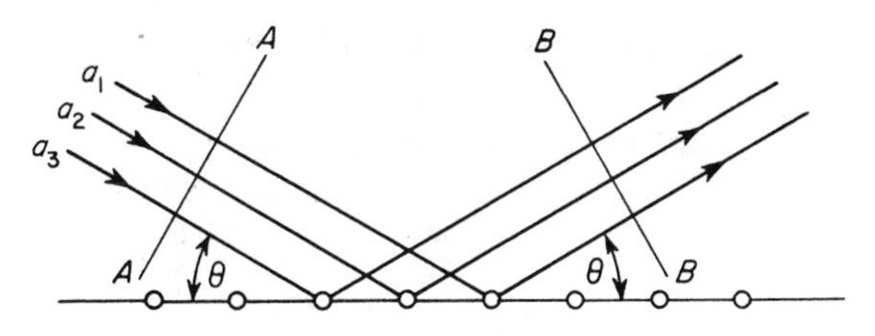

Fig. 2.1 An X-ray beam is reflected with constructive interference when the angle of incidence equals the angle of reflection.

2.1 The Bragg Law. The above discussion does not depend on the frequency of the radiation. However, when the X-rays are reflected, not from an array of atoms arranged in a solitary plane, but from atoms on a number of equally spaced parallel planes, such as exist in crystals, then constructive interference can occur only under highly restricted conditions. The law that governs the latter case is known as Bragg's law. Let us now derive an expression for this important relationship. For this purpose, let us consider each plane of atoms in a crystal as a semitransparent mirror; that is, that each plane reflects a part of the X-ray beam and also permits part of it to pass through. When X-rays strike a crystal, the beam is reflected not only from the surface layer of atoms, but from atoms underneath the surface to a considerable depth. Figure 2.2 shows an X-ray beam that is being reflected simultaneously from two parallel lattice planes. In an actual case, the beam would be reflected not from just two lattice planes, but from a large number of parallel planes. The lattice spacing, or distance between planes, is represented by the symbol d in Fig. 2.2. The line oA_i is drawn perpendicular to the incident rays and is therefore a wave front. Points o and m, which lie on this wave front, must be in phase. The line oA_r is drawn perpendicular to the reflected rays a_1 and a_2, and the condition for oA_r to be a wave front is that the reflected rays must be in phase at points o and n.

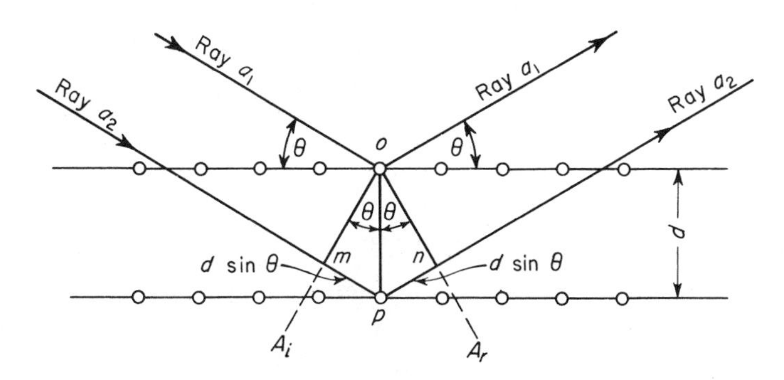

Fig. 2.2 The Bragg law.

This condition can only be satisfied if the distance *mpn* equals a multiple of a complete wavelength, that is, it must equal λ or 2λ or 3λ or $n\lambda$ where λ is the wavelength of X-rays and n an arbitrary integer.

An examination of Fig. 2.2 shows that both the distances *mp* and *pn* equal $d \sin \theta$. The distance *mpn* is, accordingly, $2d \sin \theta$. If this quantity is equated to $n\lambda$, we have Bragg's law:

$$n\lambda = 2d \sin \theta$$

where $n = 1, 2, 3, \ldots$
λ = wavelength in Å
d = interplanar distance in Å
θ = angle of incidence or reflection of X-ray beam

When the above relationship is satisfied, the reflected rays a_1 and a_2 will be in phase and constructive interference will result. Furthermore, the angles at which constructive interference occur when a narrow beam of X-rays strikes an undistorted crystal are very sharply defined, because the reflections originate on many thousands of parallel lattice planes. Under this condition, even a very small deviation from the angle θ satisfying the above relationship causes destructive interference of the reflected rays. As a consequence, the reflected beam leaves the crystal as a narrow pencil of rays capable of producing sharp images of the source on a photographic plate.

Let us now consider a simple example of an application of the Bragg equation. Let us assume that the {110} planes of a body-centered cubic crystal have a separation of 1.181 Å. If these planes are irradiated with X-rays from a tube with a copper target, the strongest line of which, the $K_{\alpha 1}$, has a wavelength 1.540 Å, first-order ($n = 1$) reflection will occur at an angle of

$$\theta = \sin^{-1}\left(\frac{n\lambda}{2d}\right) = \sin^{-1} \frac{(1)\ 1.540}{2\ (1.181)} = 40.8°$$

A second-order reflection from these {110} planes is not possible with radiation of this wavelength because the argument of the arc sin $(n\lambda/2d)$ is

$$\frac{2(1.540)}{2(1.181)} = 1.302$$

a number greater than unity, and therefore the solution is impossible. On the other hand, a tungsten target in an X-ray gives a $K_{\alpha 1}$ line with a wavelength of 0.2090 Å. Eleven orders of reflections are now possible. The angle θ, corresponding to several of these reflections, is shown in Table 2.1, and Fig. 2.3 shows a schematic representation of the same reflections.

Table 2.1

Order of reflection	*θ, angle of incidence or reflection*
1	5° 5′
2	10° 20′
5	26° 40′
11	80°

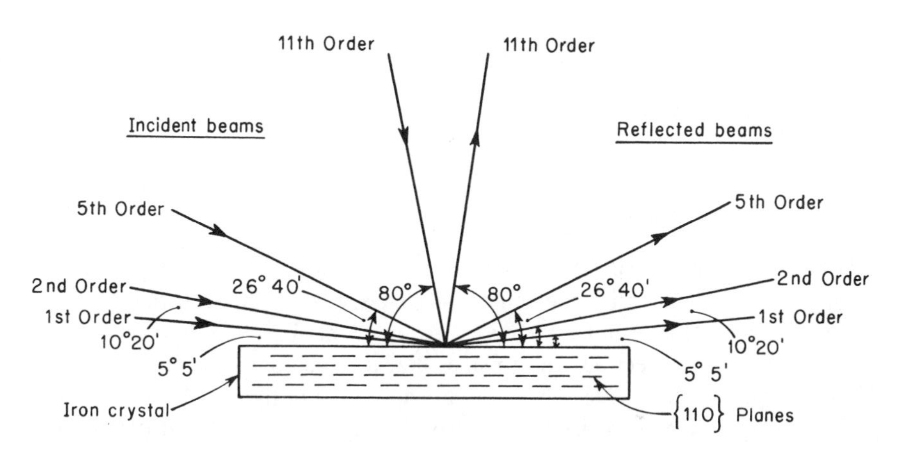

Fig. 2.3 Certain angles at which Bragg reflections occur using a crystal with an interplanar spacing of 1.181 Å and X-rays of wavelength 0.2090 Å (WK_{α_1}).

In considering the above example, it is important to notice that, although there are eleven angles at which a beam of wavelength 0.2090 Å will be reflected with constructive interference from the {110} planes, only a very slight shift in the angle θ away from any of these eleven values causes destructive interference and cancellation of the reflected beam. Whether a beam of X-rays is reflected from a set of crystallographic planes is thus a sensitive function of the angle of inclination of the X-ray beam to the plane, and a constructive reflection should not be expected to occur every time a monochromatic beam impinges on a crystal.

Suppose that a crystal is maintained in a fixed orientation with respect to a beam of X-rays and that this beam is not monochromatic but contains all wavelengths longer than a given minimum value λ_0. This type of X-ray beam is called a *white* X-ray beam, since it is analogous to white light, which contains all the wavelengths in the visible spectrum. Although the angle of the beam is fixed with respect to any given set of planes in the crystal, and the angle θ of Bragg's law is therefore a constant, reflections from all planes can now occur as a result of the fact that the X-ray beam is continuous. The point in question can be illustrated with the aid of a simple cubic lattice.

Let the X-ray beam have a minimum wavelength of 0.5 Å, and let it make a 60° angle with the surface of the crystal, which, in turn, is assumed to be parallel to a set of {100} planes. In addition, let the {100} planes have a spacing of 1 Å. Substituting these values in the Bragg equation gives

$$n\lambda = 2d \sin \theta$$

or

$$n\lambda = 2(1) \sin 60^\circ = 1.732$$

Thus, the rays reflected from {100} planes will contain the wavelengths

1.732 Å	for first-order reflection
0.866 Å	for second-order reflection
0.546 Å	for third-order reflection

All other wavelengths will suffer destructive interference.

In the above examples, the reflecting planes were assumed parallel to the crystal surface. This is not a necessary requirement for reflection; it is quite possible to obtain reflections from planes that make all angles with the surface. Thus, in Fig. 2.4, the incident beam is shown normal to the surface and a (001) plane, while making an angle θ with two {210} planes – (012) and $(0\bar{1}2)$. The reflections from these two planes are shown schematically in Fig. 2.4. It may be concluded that when a beam of white X-rays strikes a crystal, many reflected beams will emerge from the crystal, each reflected beam corresponding to a reflection from a different crystallographic plane. Furthermore, in contrast to the incident beam that is continuous in wavelength, each reflected beam will contain only discrete wavelengths as prescribed by the Bragg equation.

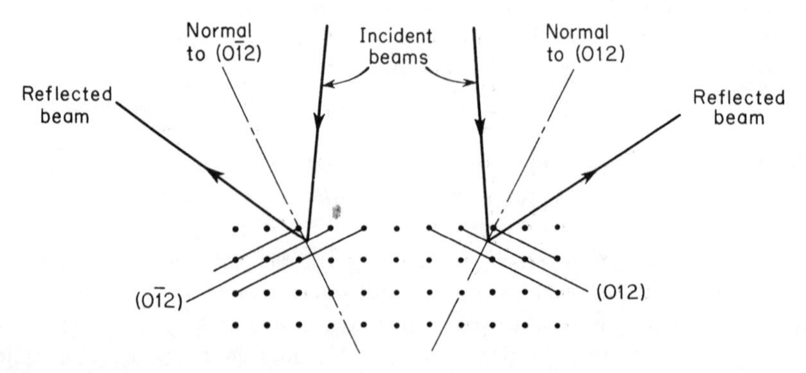

Fig. 2.4 X-ray reflections from planes not parallel to the surface of the specimen.

2.2 Laue Techniques. The Laue X-ray diffraction methods make use of a crystal with an orientation that is fixed with respect to a beam of continuous X-rays, as described in the preceding section. There are two basic Laue techniques: in one, the beams reflected back in directions close to that of the incident X-ray beam are studied; in the other, the reflected beams that pass through the crystal are studied. Clearly the latter method cannot be applied to crystals of appreciable thickness (1 mm or more) because of the loss in intensity of the X-rays by their absorption in the metal. The first method is known as the *back-reflection Laue technique*; the latter as the *transmission Laue technique.*

The back-reflection Laue method is especially valuable for determining the orientation of the lattice inside crystals when the crystals are large and therefore opaque to X-rays. Many physical and mechanical properties vary with direction inside crystals. The study of these anisotropic properties of crystals requires a knowledge of the lattice orientation in the crystals.

Figure 2.5 shows the arrangement of a typical Laue back-reflection camera. X-rays from the target of an X-ray tube are collimated into a narrow beam by a tube several inches long with an internal diameter of about 1 mm. The narrow beam of X-rays impinges on the crystal at the right of the figure, where it is diffracted into a number of reflected beams that strike the cassette containing a photographic film. The front of the cassette is covered with a thin sheet of material, for example, black paper, opacque to visible light, but transparent to the reflected X-ray beams. In this way, the positions of the reflected beams are recorded on the photographic film as an array of small dark spots.

Figure 2.6A shows the back-reflection X-ray pattern of a magnesium crystal oriented so that the incident X-ray beam was perpendicular to the basal plane of the crystal. Each spot corresponds to a reflection from a single crystallographic plane, and the sixfold symmetry of the lattice, when viewed in a direction perpendicular to the basal plane, is quite apparent. If the crystal is rotated in a

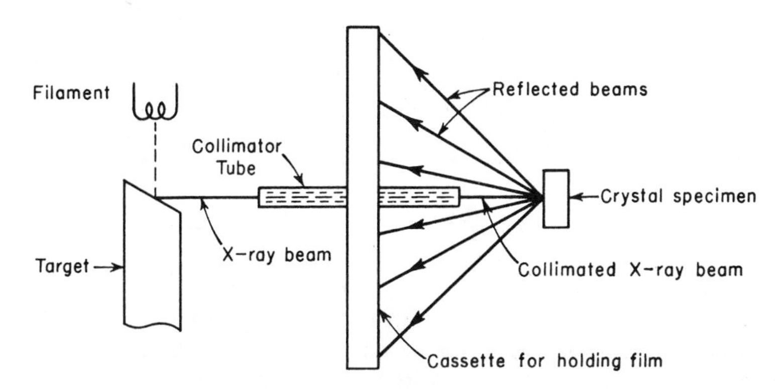

Fig. 2.5 Laue back-reflection camera.

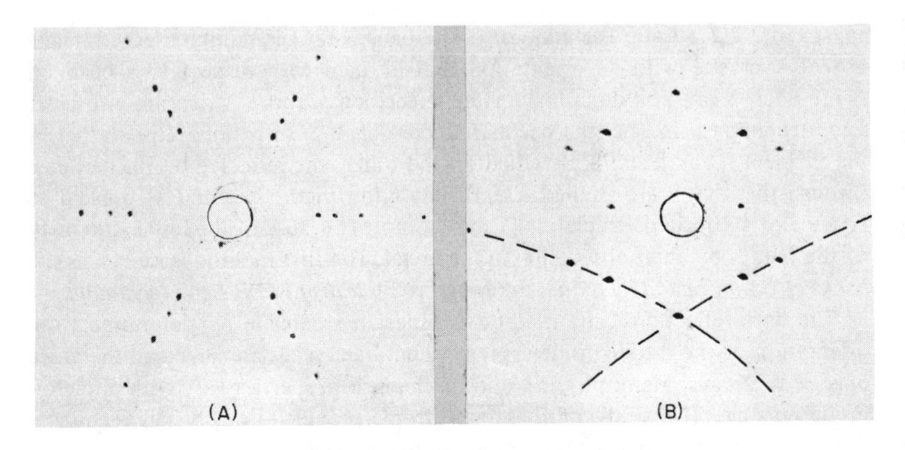

Fig. 2.6 Laue back-reflection photographs. (A) Photograph with X-ray beam perpendicular to the basal plane (0001). (B) Photograph with X-ray beam perpendicular to a prism plane $(11\bar{2}0)$. Dashed lines on the photograph are drawn to show that the back-reflection spots lie on hyperbolas.

direction away from the one that gives the pattern of Fig. 2.6A, the pattern of spots changes (Fig. 2.6B); nevertheless it still defines the orientation of the lattice in space. Therefore, the orientation of the crystal can be determined in terms of a Laue photograph.

Transmission Laue patterns can be obtained with an experimental arrangement similar to that for back-reflection patterns, but the film is placed on the opposite side of the specimen from the X-ray tube. Specimens may have the shape of small rods or plates, but must be small in their dimension parallel to the

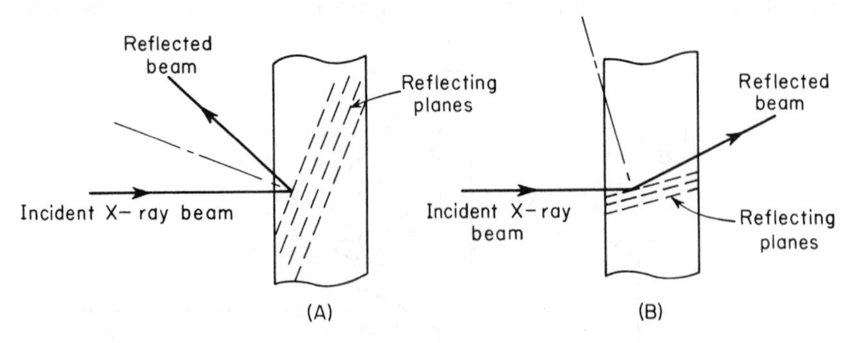

Fig. 2.7 (A) Laue back-reflection photographs record the reflections from planes nearly perpendicular to the incident X-ray beam. (B) Laue transmission photographs record the reflections from planes nearly parallel to the incident X-ray beam.

X-ray beam. Whereas the back-reflection technique reflects the X-ray beam from planes nearly perpendicular to the beam itself, the transmission technique records the reflections from planes nearly parallel to the beam, as can be seen in Fig. 2.7.

Laue transmission photographs, like back-reflection photographs, consist of arrays of spots. However, the arrangements of the spots differ in the two methods: transmission patterns usually have spots arranged on ellipses, back-reflection on hyperbolas. (See Fig. 2.6B.)

The Laue transmission technique, like the back-reflection technique, is also used to find the orientation of crystal lattices. Both methods can be used to study a phenomenon called *asterism*. A crystal that has been bent, or otherwise distorted, will have curved lattice planes that act in the manner of curved mirrors and form distorted or elongated spots instead of small circular images of the X-ray beam. A typical Laue pattern of a distorted crystal is shown in Fig. 2.8. In many cases, analysis of the asterism, or distortion, of the spots in Laue photographs leads to valuable information concerning the mechanisms of plastic deformation.

In the above examples (Laue methods), a crystal is maintained in a fixed orientation with respect to the X-ray beam. Reflections are obtained because the beam is continuous, that is, the wavelength is the variable. Several important X-ray diffraction techniques that use X-rays of a single frequency or wavelength will now be considered. In these methods, since λ is no longer a variable, it is necessary to vary the angle θ in order to obtain reflections.

2.3 The Rotating-Crystal Method. In the rotating-crystal method, crystallographic planes are brought into reflecting positions by rotating

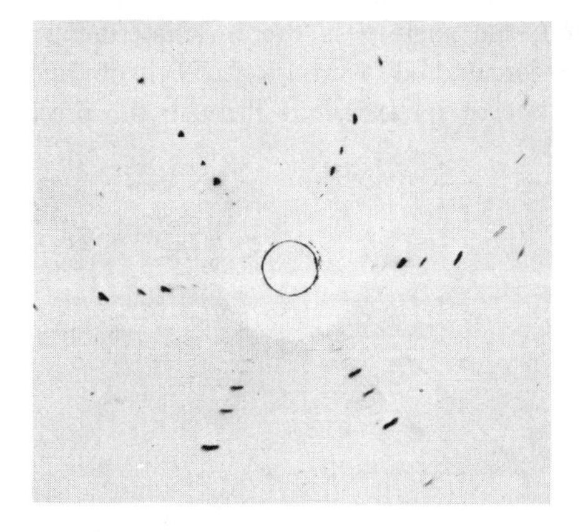

Fig. 2.8 Asterism in a Laue back-reflection photograph. The reflections from distorted- or curved-crystal planes form elongated spots.

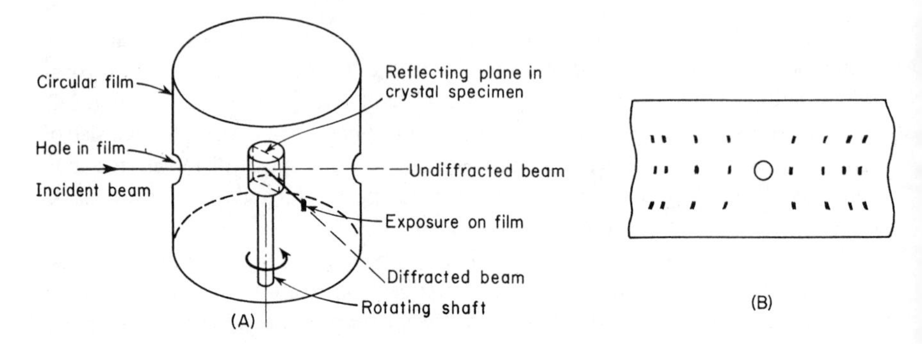

Fig. 2.9 (A) Schematic diagram of a rotating single-crystal camera. (B) Schematic representation of the diffraction pattern obtained from rotating crystal camera. Reflected beams make spots lying in horizontal rows.

a crystal about one of its axis while simultaneously radiating it with a beam of monochromatic X-rays. The reflections are usually recorded on a photographic film that is so curved as to surround the specimen. (See Fig. 2.9 for a schematic view of the method.)

2.4 The Debye-Scherrer or Powder Method. In this method care is taken to see that the specimen contains not one crystal, but more than several hundred randomly oriented crystals. The specimen may be either a small polycrystalline metal wire, or a finely ground powder of the metal contained in a plastic, cellulose, or glass tube. In either case, the crystalline aggregate consists of a cylinder about 0.5 mm in diameter with crystals approximately 0.1 mm in diameter or smaller. In the Debye-Scherrer method, as in the rotating single-crystal method, the angle θ is the variable; the wavelength λ remains constant. In the powder method, a variation of θ is obtained, not by rotating a single crystal about one of its axes, but through the presence of many small

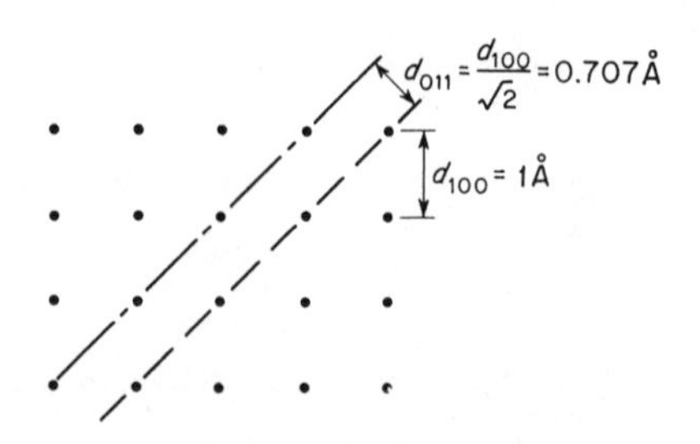

Fig. 2.10 Simple cubic lattice. Relative interplanar spacing for {100} and {110} planes.

crystals randomly oriented in space in the specimen. The principles involved in the Debye-Scherrer method can be explained with the aid of an example.

For the sake of simplicity, let us assume a crystalline structure with the simple cubic lattice shown in Fig. 2.10, and that the spacing between {100} planes equals 1 Å. It can easily be shown that the interplanar spacing for planes of the {110} type equals that of {100} planes divided by the square root of two, and is, therefore, 0.707 Å. (See Fig. 2.10.) The {110} spacing is, therefore, smaller than the {100} spacing. In fact, all other planes in the simple cubic lattice have a smaller spacing than that of the cube, or {100}, planes, as is shown by the following equation for the spacing of crystallographic planes in a cubic lattice, where h, k, and l are the three Miller indices of a plane in the crystal, d_{hkl} the interplanar spacing of the plane, and a the length of the unit-cell edge

$$d_{hkl} = \frac{a}{\sqrt{h^2 + k^2 + l^2}}$$

In the simple cubic structure, the distance between cubic planes, d_{100}, equals a. Therefore the above expression is written

$$d_{hkl} = \frac{d_{100}}{\sqrt{h^2 + k^2 + l^2}}$$

Now, according to the Bragg equation

$$n\lambda = 2d \sin \theta$$

and for first-order reflections, where n equals one, we have

$$\theta = \sin^{-1}\left(\frac{\lambda}{2d}\right)$$

This equation tells us that planes with the largest spacing will reflect at the smallest angle θ. If now it is arbitrarily assumed that the wavelength of the X-ray beam is 0.4 Å, first-order reflections will occur from {100} planes (with the assumed 1 Å spacing) when

$$\theta = \sin^{-1} \frac{0.4}{2(1)} = \sin^{-1} \frac{1}{5} = 11° \, 30'$$

On the other hand, {100} planes with a spacing 0.707 Å reflect when

$$\theta = \sin^{-1} \frac{0.4}{2(0.707)} = 16° \, 28'$$

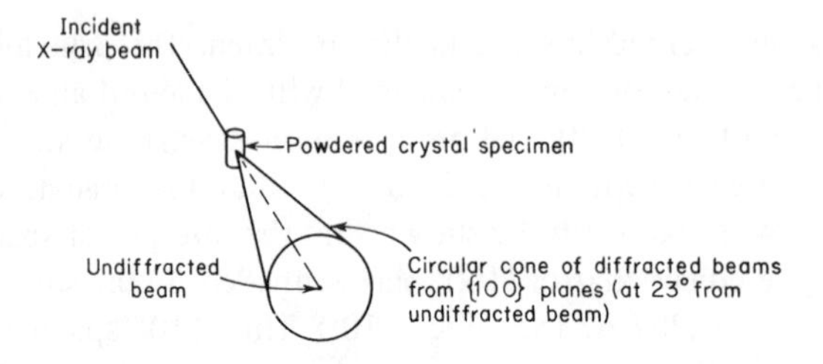

Fig. 2.11 First-order reflections from {100} planes of a hypothetical simple cubic lattice. Powdered crystal specimen.

All other planes with larger indices (that is, {111}, {234}, etc.) reflect at still larger angles.

Figure 2.11 shows how the Debye-Scherrer reflections are found. A parallel beam of monochromatic X-rays coming from the left of the figure is shown striking the crystalline aggregate. Since the specimen contains hundreds of randomly oriented crystals in the region illuminated by the incident X-ray beam, many of these will have {100} planes at the correct Bragg angle of 11°30′. Each of these crystals will therefore reflect a part of the incident radiation in a direction that makes an angle twice 11°30′ with the original beam. However, because the crystals are randomly oriented in space, all the reflections will not lie in the same direction, but, instead, along the surface of a cone that makes an angle of 23° with the original direction of the X-ray beam. In the same manner, it can be shown that first-order reflections from {110} planes form a cone the

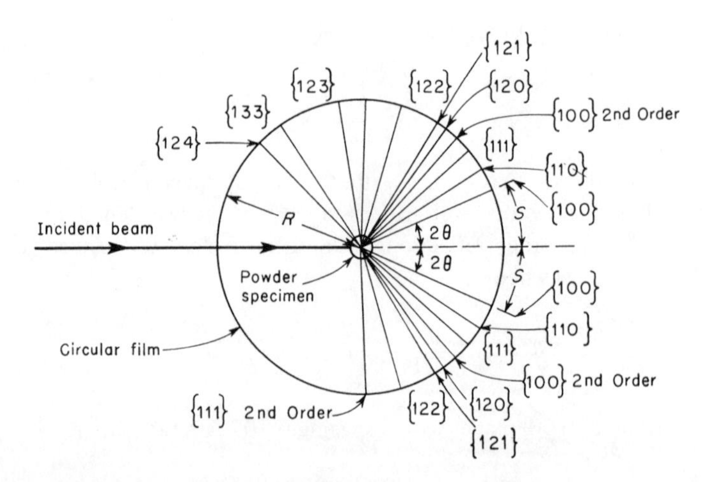

Fig. 2.12 Schematic representation of the Debye or powder camera. Specimen assumed to be simple cubic. Not all reflections are shown.

surface of which makes an angle of twice 16°28′, or 32°56′ with the original direction of the beam, and that the planes of still higher indices form cones of reflected rays making larger and larger angles with the original direction of the beam.

The most commonly used powder cameras employ a long strip film that is curved into the form of a circular cylinder surrounding the specimen, as shown in Fig. 2.12. A schematic view of a Debye-Scherrer film after exposure and development is shown in Fig. 2.13.

On a Debye-Scherrer film, the distance $2S$ between the two circular segments of the {100} cone is related to the angular opening of the cone and therefore to the Bragg angle θ between the reflecting plane and the incident beam. Thus, the angle in radians between the surface of the cone and the X-ray beam equals S/R, where R is the radius of the circle formed by the film. However, this same angle also equals 2θ, and, therefore,

$$2\theta = \frac{S}{R}$$

or

$$\theta = \frac{S}{2R}$$

This last relationship is important because it is possible to measure the Bragg angle θ with it. In the above example, the spacing between parallel lattice planes was assumed to be known. This assumption was made in order to explain the principles of the Debye-Scherrer method. In many cases, however, one may not know the interplanar spacings of a crystal, and measurements of the Bragg angles can then be used to determine these quantities. The powder method is, accordingly, a powerful tool for determining the crystal structure of a metal. In complicated crystals, other methods may have to be used in conjunction with the powder method in order to complete an identification. In any case, the Debye-Scherrer method is probably the most important of all methods used in

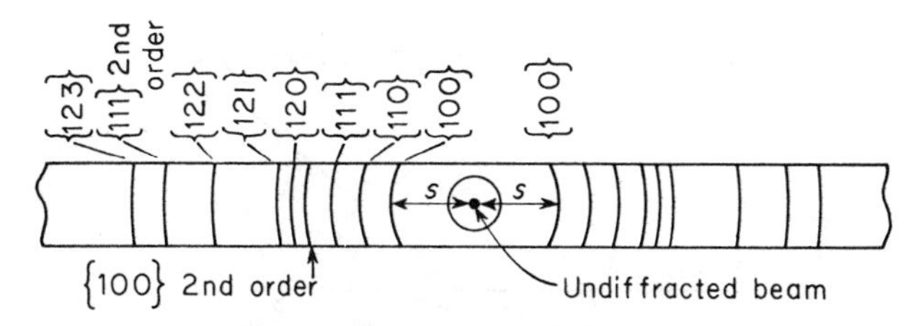

Fig. 2.13 Powder camera photograph. Diffraction lines correspond to the reflections shown in Fig. 2.12.

the determination of crystal structures. Another very important application of the powder method is based upon the fact that each crystalline material has its own characteristic set of interplanar spacings. Thus, while copper, silver, and gold all have the same crystal structure (face-centered cubic), the unit cells of these three metals are different in size, and, as a result, the interplanar spacings and Bragg angles are different in each case. Since each crystalline material has its own characteristic Bragg angles, it is possible to identify unknown crystalline phases in metals with the aid of their Bragg reflections. For this purpose, a card file system (X-ray Diffraction Data Index) has been published listing for approximately a thousand elements and crystalline compounds, not only the Bragg angle of each important Debye-Scherrer diffraction line, but also its relative strength or intensity. The identification of an unknown crystalline phase in a metal can be made by matching powder pattern Bragg angles and reflected intensities of the unknown substance with the proper card of the index. The method is analogous to a fingerprint identification system and constitutes an important method of qualitative chemical analysis.

2.5 The X-Ray Spectrometer. The X-ray spectrometer is a device that measures the intensity of the X-ray reflections from a crystal with an electronic device, such as a Geiger countertube or ionization chamber, instead of a photographic film. Figure 2.14 shows the elementary parts of a spectrometer – a crystalline specimen, a parallel beam of X-rays, and a Geiger countertube. The apparatus is so arranged that both the crystal and the intensity measuring device (Geiger countertube) rotate. The countertube, however, always moves at twice the speed of the specimen, which keeps the intensity recording device at the proper angle during the rotation of the crystal so that it can pick up each Bragg reflection as it occurs. In modern instruments of this type, the intensity measuring device is connected through a suitable amplification system to a chart recorder, where the intensity of the reflection is recorded on a chart by a pen. In this manner one obtains an accurate plot of intensity against Bragg angles. A typical X-ray spectrometer plot is shown in Fig. 2.15.

The X-ray spectrometer is most commonly used with a powder specimen in

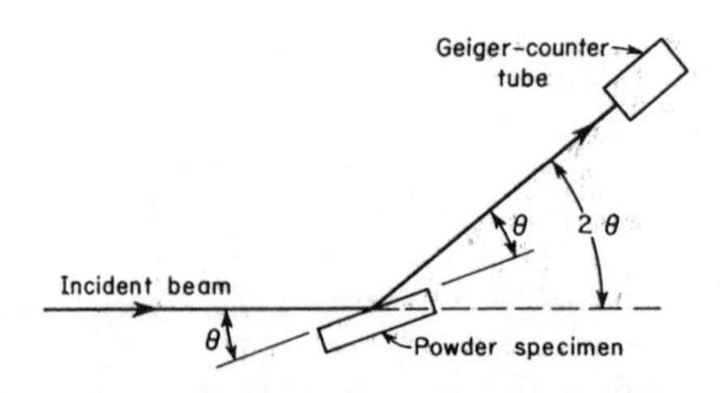

Fig. 2.14 X-ray spectrometer.

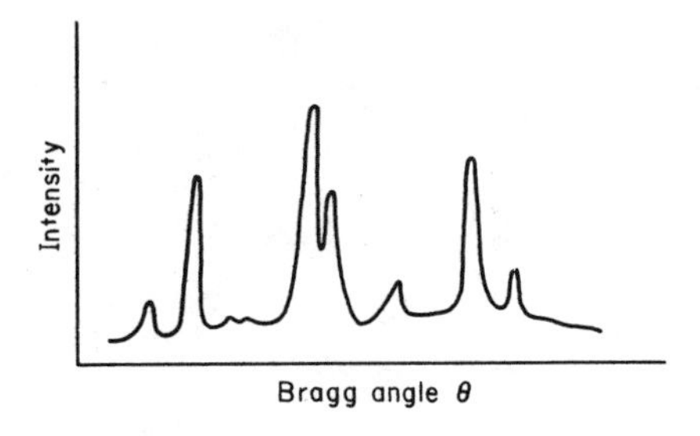

Fig. 2.15 The X-ray spectrometer records on a chart the reflected intensity as a function of Bragg angle. Each intensity peak corresponds to a crystallographic plane in a reflecting position.

the form of a rectangular plate with dimensions about 1 in. long and ½ in. wide. The specimen may be a sample of a polycrystalline metal, and it should be noted that, in contrast to the Debye-Scherrer method where the specimen is a fine wire (approximately 0.5 mm diameter), the spectrometer sample has a finite size, which makes the specimen much easier to prepare and is therefore advantageous. Because the X-ray spectrometer is capable of measuring the intensities of Bragg reflections with great accuracy, both qualitative and quantitative chemical analyses can be made by this method.

2.6 The Transmission Electron Microscope. Within the past few years a very powerful technique has become available to metallurgists. It involves the use of the electrom microscope to study the internal structure of thin crystalline films or foils. These foils, which can be removed from bulk samples, are normally only several thousand Å thick. The thickness is dictated by the voltage at which the microscope is operated. The normal instrument is rated at about 100,000 volts, and electrons accelerated by this voltage can give an acceptable image if the foil is not made thicker than the indicated value. Thinner foils, on the other hand, tend to be less useful in revealing the nature of the structure in the metal. Some instruments have been developed that operate at much higher voltages (of the order of a million volts), and in them the foils can be proportionally thicker. However, equipment costs are also much greater, and relatively few of these instruments are available.

In the transmission electron microscope, the detail in the image is formed by the diffraction of electrons from the crystallographic planes of the object being investigated. The electron microscope is, in many respects, analogous to an optical microscope. The source is an electron gun instead of a light filament. The lenses are magnetic, being composed normally of a current-carrying coil surrounded by a soft iron case. The lenses are energized by direct current. An excellent, easy to read description of the electron microscope is given in the

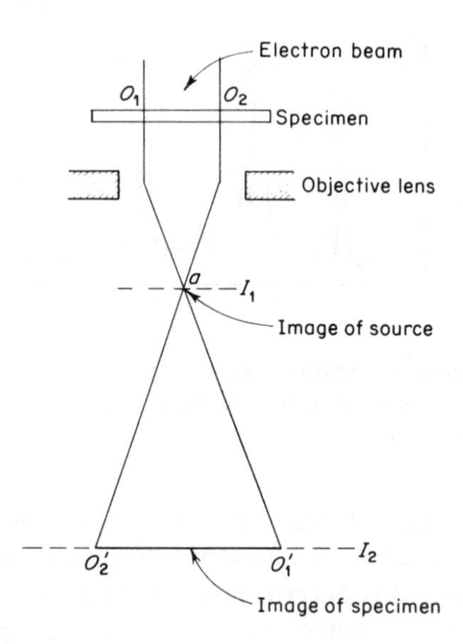

Fig. 2.16 Schematic drawing of a transmission electron microscope.

book by Smallman and Ashbee[1]. For our present purposes we shall concentrate on that part of the microscope containing the specimen and the objective lens. This region is indicated schematically in Fig. 2.16. In this diagram, the electron beam is shown entering the specimen from above. This beam originated in the electron gun and has passed through a set of condensing lenses before it reaches the specimen. On emerging from the specimen, the beam passes through the rear element of the instrument's objective lens. Shortly beyond this lens element, the rays converge to form a spot at point a in plane I_1. This spot is equivalent to an image of the source. Somewhat beyond this point, the image of the specimen is formed at plane I_2. Similar double image effects are observable in simple optical instruments where it is possible to form images of the light source at one position and images of a lantern slide or other object at other positions.

Because the image formation in the transmission electron microscope depends on the diffraction of electrons, it is necessary to consider some elementary facts about this type of diffraction. As demonstrated in Chapter 3, electrons not only have many of the attributes of particles, but they also possess wavelike properties. It will also be shown that the wavelength of an electron is related to its velocity v by the relation

$$\lambda = \frac{h}{mv}$$

1. Smallman, R. E., and Ashbee, K. H. G. *Modern Metallography.* Pergamon Press, Oxford, 1966.

where λ is the wavelength of the electron, m is its mass, and h is Planck's constant equal to 6.63×10^{-27} erg sec. This expression shows that the wavelength of an electron varies inversely as its velocity. The higher the velocity, the shorter its wavelength.

Now let us assume that an electron is accelerated by a potential of 100,000 volts. It can easily be shown that this will give the electron a velocity of about 2×10^{10} cm/sec and, by the above equation, a wavelength of about 4×10^{-10} cm, or about 4×10^{-2} Å. This is about two orders of magnitude smaller than the average wavelength used in X-ray diffraction studies of metallic crystals. This causes a corresponding difference in the nature of the diffraction, as can be deduced by considering Bragg's law. Suppose that we are concerned with first order diffraction, where $n = 1$. Then, by Bragg's law, we have

$$\lambda = 2d \sin \theta$$

If d, the spacing of the parallel planes from which the electrons are reflected, is assumed to be about 2 Å, we have

$$\theta \approx \sin \theta = 0.01$$

The angle of incidence or reflection of a diffracted beam is thus only of the order of 10^{-2} radians, or about 30′. This means that when a beam of electrons is passed through a thin layer of crystalline material, only those planes nearly parallel to the beam can be expected to contribute to the resulting diffraction pattern.

Let us now consider how the image is formed in the electron microscope as the result of diffraction. In this regard, consider Fig. 2.17. Here it is assumed that some of the electrons, in passing through the specimen, are diffracted by one of the sets of planes in the specimen. In general, only part of the electrons will be diffracted, and the remainder will pass directly through the specimen without being diffracted. These latter electrons will form a spot at position a and an image of the specimen ($O_2' - O_1'$) at the plane I_2, as indicated in Fig. 2.16. On the other hand, the diffracted electrons will enter the objective lens at a slightly different angle and will converge to form a spot at point b. These rays that pass through point b will also form an image of the specimen at I_2 that is superimposed over that from the direct beam. In the above, it has been assumed that the crystal is so oriented that the electrons are reflected primarily from a single crystallographic plane. This should cause the formation of a single pronounced spot at point b as a result of the diffraction. It is also possible to have simultaneous reflections from a number of planes. In this case, instead of a single spot appearing in I_1 at point b a typical array of spots or a diffraction pattern will form on plane I. A characteristic diffraction pattern will be described presently.

The electron microscope is so constructed that either the image of the

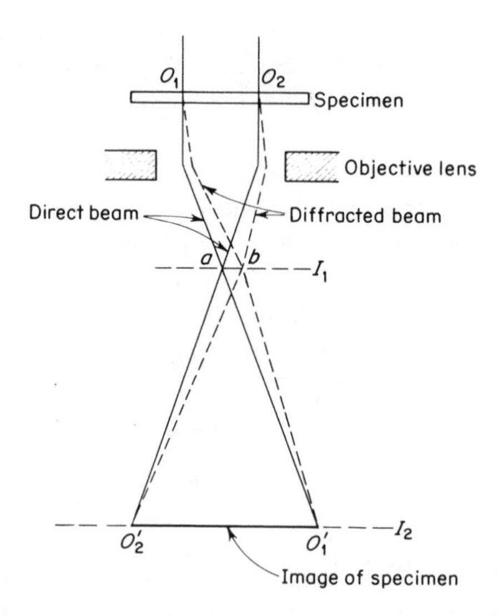

Fig. 2.17 Images can be formed in the transmission electron microscope corresponding to the direct beam or to a diffracted beam. (Images from more than one diffracted beam are also possible).

diffraction pattern (formed at I_1) or the image of the detail in the specimen (formed at I_2) can be viewed on the fluorescent screen of the instrument. Alternatively, both of these images can be photographed on a plate or film. This is possible because a projection lens system (not shown) is located in the microscope below that part of the instrument shown in Fig. 2.16. This lens system can be adjusted to project either the image of the diffraction pattern at plane I_1, or that of the specimen at plane I_2, onto the fluorescent screen or photographic emulsion.

In operating the instrument as a microscope, one has the choice of using either the image formed by the direct beam or the image formed by diffraction from a particular set of planes. The elimination of the beam causing either of these two types of images is made possible by the insertion of an aperture diaphragm at plane I_1 that allows only one of the corresponding beams to pass through it, as shown in Fig. 2.18. In this diagram, the diffracted beam is shown intercepted by the diaphragm, while the direct beam is allowed to pass through the aperture. When the specimen is viewed in this manner, a bright-field image is formed. Imperfections in the crystal will normally appear as dark areas in this image. These imperfections could be small inclusions of different transparency from the matrix crystal, and therefore visible in the image as a result of a loss in intensity of the beam where it passes through the more opaque particles. Of more general interest, however, is the case where the imperfections are faults in

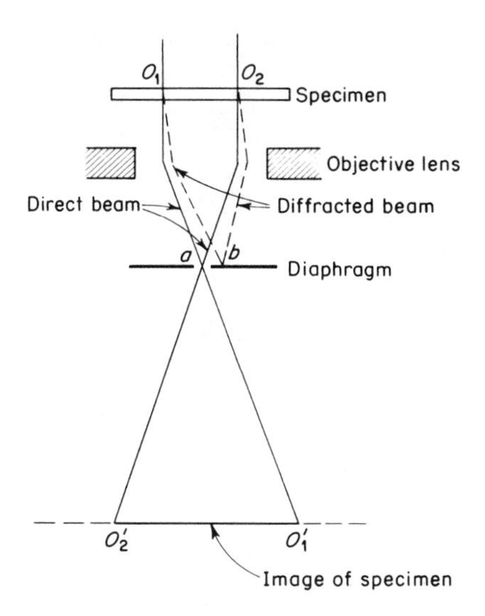

Fig. 2.18 The use of a diaphragm to select the desired image.

the crystal lattice itself. A very important defect of this type, that will be of considerable concern to us in subsequent chapters, is a dislocation. Without delving too deeply into the nature of dislocations at this time, it is necessary to point out that dislocations involve distortions in the arrangement of the crystal planes. Such local distortions will have effects on the diffraction of electrons, because the angle of incidence between the electron beam and the lattice planes around the dislocation are altered. In some cases this may cause an increase in the number of diffracted electrons, and in others a decrease. Since the direct beam can be considered to be the difference between the incident beam and the diffracted beam, a local change in the diffracting conditions in the specimen will be reflected by a corresponding alteration in the intensity recorded in the specimen image. Dislocations are thus visible in the image because they affect the diffraction of electrons. In a bright-field image, dislocations normally appear as dark lines. A typical bright-field photograph is shown in Fig. 4.9.

The alternate method of using the electron microscope is to place the aperture so that a diffracted ray is allowed to pass, while the direct beam is cut off. The image of the specimen formed in this case is of the dark-field type. Here, dislocations appear as white lines lying on a dark background. Bright-field illumination is normally used, because dark-field images are usually more subject to distortion. This is because (as may be seen in Fig. 2.17) the diffracted beam, after leaving the specimen, does not travel down the microscope axis.

An important feature of the transmission electron microscope is the stage that holds the specimen. As indicated above, diffraction plays a very important

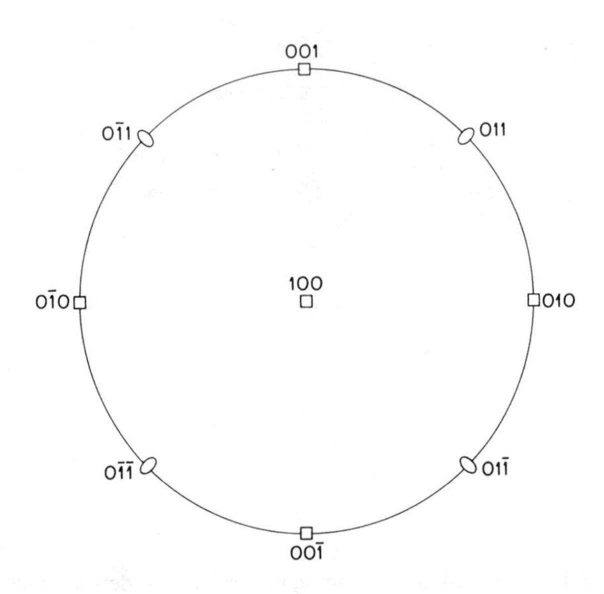

Fig. 2.19 This stereographic projection of a cubic crystal shows the principle planes whose zone axis is [100].

role in making the defects in the crystal structure visible in the image. In order that the specimen may be capable of being aligned so that a suitable crystallographic plane can be brought into a reflecting condition, it is usually necessary for the specimen to be tilted with respect to the electron beam. The stage of an electron microscope used for metallurgical investigations is normally constructed so that the specimen may be rotated or tilted.

With regard to the diffraction patterns observable in the microscope, an interesting diffraction pattern is obtained when the specimen is tilted in its stage so that an important zone axis is placed parallel to the microscope axis. When this is done, a pattern is obtained whose spots correspond to the planes of the zone whose axis parallels the electron beam. By way of illustration, let us assume that the specimen is so oriented that a $<100>$ direction is parallel to the instrument axis. Figure 2.19 shows a stereographic projection in which the zone axis is located at the center of the projection. The poles of the planes belonging to this zone should therefore lie on the basic circle of the stereographic projection. In the figure, only the planes of low indices are shown. The diffraction pattern corresponding to this zone is shown in Fig. 2.20. The Miller indices of the planes responsible for each spot are indicated alongside the corresponding spots.

The most significant feature of the diffraction pattern in Fig. 2.20 is that all spots correspond to planes parallel to the electron beam. Furthermore, as may

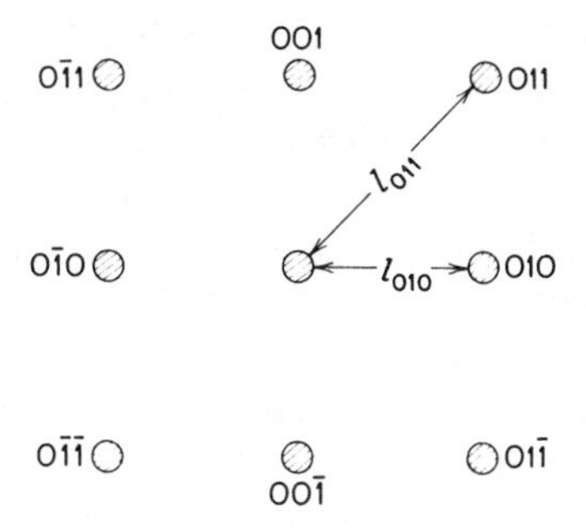

Fig. 2.20 The diffraction pattern corresponding to a beam directed along [100] in a cubic crystal.

be seen in Fig. 2.20, spots are indexed at both 100 and $\bar{1}00$. This implies that the electrons are reflected from both sides of the same planes. Obviously, the simple Bragg picture shown in Fig. 2.2, where the angle of incidence equals the angle of reflection, does not apply in this case. The reasons for this are not easy to understand. However, several factors are undoubtedly involved. Of some importance is the small value of the Bragg angle θ, which is about 10^{-2} radians. Another is the fact that the transmission specimen is very thin, so that the electron beam, in traversing the specimen, sees a lattice that is nearly two-dimensional. This tends to relax the diffraction conditions. Finally, the electron microscope, with the specimen located inside a system of lenses, is not a simple diffracting device. For our present purposes, however, it is more important to note the nature of the diffraction pattern than the causes for it.

Attention is now called to the spacing of the spots in the diffraction pattern in Fig. 2.20. In this figure, the distance from the spot corresponding to the direct beam to that of a reflection from a {100} plane is indicated as l_{010}, while the corresponding distance to a {110} reflection is l_{011}. As can be deduced from the figure, $l_{001} = \sqrt{2}\, l_{010}$. Attention is now called to Fig. 2.10, where it is shown that the spacing between the two respective planes varies inversely as $\sqrt{2}$. This indicates that the spacing of the spots in the diffraction pattern is inversely proportional to the interplanar spacing. This result, unlike the relationship of the incident beam to the planes from which it is reflected, is in good agreement with the Bragg law. In the present case, where the angle θ is small, $\sin\theta \approx \theta$ and Bragg's law reduces to

$$n\lambda = 2d\theta$$

or

$$\theta = \frac{n\lambda}{2d}$$

Since the angle θ is small, tan θ is also approximately equal to θ and we should expect the diffracted spots to be deviated through distances that are inversely proportional to the interplanar spacing d.

It is clear from the above that with the electron microscope it is possible both to investigate the internal defect structure of a crystalline specimen using the instrument as a microscope, and to determine a considerable degree of information about the crystallographic features of the specimen using it as a diffraction instrument. With regard to the latter application, the diffraction patterns can yield information about the nature of the crystal structure and about the orientation of the crystals in a specimen. Furthermore, the electron microscope has a diaphragm in its optical path that controls the size of the area that is able to contribute to the diffraction pattern. As a result, it is possible to obtain information about an area of the specimen that has a radius as small as 0.5μ. The diffraction patterns are therefore called *selected area diffraction patterns.*

Problems

1. The interplanar spacing of the {110} planes in an iron crystal is 2.024 Å. At what Bragg angle, θ, will silver, K_{α_1}, radiation (0.558 Å) suffer first order reflection from these planes?
2. How many orders of reflection are possible when Cu K_{α_1}, radiation is diffracted from the {110} planes of an iron crystal?
3. At what angles will the reflections of Problem 2 occur?
4. Given that the interplanar spacing of the {110} planes in iron is 2.024 Å, compute the lattice parameter a for this crystal sttucture.
5. Suppose that a silver single crystal has been prepared so that one of its surfaces is parallel to a {111} plane. At what angle to the axis of the collimator tube should this surface be inclined if a second order reflection is desired using copper K_{α_1} radiation, given that the lattice parameter a for Ag is 4.078 Å?
6. Make a sketch of a face-centered cubic unit cell. Show that there exists a (010) plane that lies midway between the outside (010) faces of this cell. In terms of the lattice parameter, what is therefore the true interplanar spacing between the {100} planes in this crystal structure? Now consider the relation

$$d = \frac{a}{\sqrt{h^2 + k^2 + l^2}}$$

   What values of h, k, and l will give the (010) interplanar spacing that you have deduced?
7. Turn to Appendix C. This table lists the indices or *hkl* values for the most important reflecting planes in the simple cubic, body-centered cubic, and face-centered cubic structures. Note that indices of the form {220} correspond to a second order reflection from {110} planes. On the basis of your answer to Problem 6, why is no reflection listed for {100} in the face-centered cubic structure?

8. Why is no reflection listed in Appendix C for the indices {110}? What may be deduced in general about the missing reflections in the table of Appendix C?
9. Assume that two face-centered cubic crystals, copper and aluminum respectively, are identically oriented with a {111} face normal to the X-ray beam in a Laue back-reflection camera. Would there be any difference in the arrangement or positions of the spots in the respective photographs?
10. Assume that the distance between the specimen and the film in a Laue back-reflection camera is 5 cm, and that a copper crystal is photographed with the (111) plane normal to the X-ray beam. For simplicity, also assume that the film is circular in shape, with a diameter of 20 cm. On this film, would it be possible to see reflections from any of the {100} planes?
11. In the diffraction pattern of Problem 10, would it be possible to see any reflections from {110} planes? If so, make a sketch to scale showing the pattern of the spots.
12. In a powder camera photograph of a face-centered cubic nickel specimen the separation between the two lines lying closest to the position of the undiffracted beam is 7.80 cm. What planes of the crystal structure are responsible for these lines? If the camera radius is 5 cm and the radiation is Cu K_{α_1}, what is the lattice parameter of nickel?
13. With the aid of Appendix C determine the data in order to make a sketch like Fig. 2.13 that will show the first four pairs of lines in a powder pattern photograph of an iron specimen. Assume a camera radius of 5 cm, Cu K_{α_1} radiation (1.537 Å), and a_{Fe} = 2.861 Å.
14. Compute the velocity of an electron after it has been accelerated by a potential of 100,000 volts.
15. Determine the wavelength of the electron corresponding to Problem 14.
16. Determine the wavelength corresponding to the electrons in a million-volt electron microscope.
17. Assuming that the wavelength of the electrons in an electron microscope is 3.9×10^{-2} Å, compute the Bragg angle for the reflection of these electrons from the {100} planes of an iron crystal.
18. Examine Fig. 1.34 showing a {111} standard projection of a cubic crystal. What can you deduce from this pole figure about the indices of the planes whose spots should appear nearest the center of the selected area diffraction pattern of a crystal with its (111) plane normal to the electron beam?
19. Make a sketch to scale similar to Fig. 2.21 showing the diffraction pattern of Problem 18.

3 Elementary Theory of Metals

CRYSTAL BINDING

Crystalline solids are empirically grouped into four classifications: (*a*) ionic, (*b*) van der Waals, (*c*) covalent, and (*d*) metallic. This is not a rigid classification, for many solids are of an intermediate character and not capable of being placed in a specific class. Nevertheless, this grouping is very convenient and greatly used in practice to indicate the general nature of the various solids.

3.1 The Internal Energy of a Crystal. The internal energy of a crystal is considered to be composed of two parts. First, there is the lattice energy U that is defined as the potential energy due to the electrostatic attractions and repulsions that atoms exert on one another. Second, there is the thermal energy of the crystal, associated with the vibrations of the atoms about their equilibrium lattice positions. It consists of the sum of all the individual vibrational energies (kinetic and potential) of the atoms. In order to most conveniently study the nature of the binding forces that hold crystals together (the lattice energy U), it is desirable to eliminate from our considerations, as far as possible, complicating considerations of the thermal energy. This can be done most conveniently by assuming that all cohesive calculations refer to zero degrees absolute. Quantum theory tells us that at this temperature the atoms will be in their lowest vibrational energy states and that the zero-point energy associated with these states is small. For the present, it will be assumed that the zero-point energy can be neglected and that all calculations refer to 0° K.

In setting up quantitative relationships to express the cohesion of solids, it is customary to work with cohesive energies rather than cohesive forces. The energy concept is preferred because it is more conveniently compared with experimental data. Thus, the heat of sublimation and/or the heat of formation of a compound are both related to cohesive energies. In fact, the heat of sublimation at 0° K, which is the energy required to dissociate a mole of a substance into free atoms at absolute zero, is a particularly convenient measure of the cohesive energy of a simple solid such as a metal.

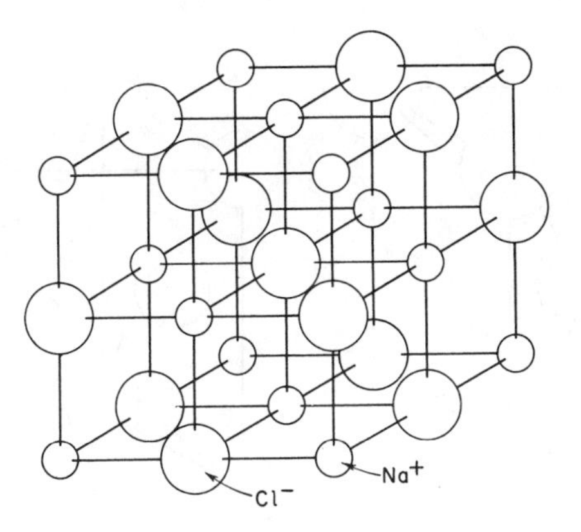

Fig. 3.1 The sodium chloride lattice.

3.2 Ionic Crystals. The sodium chloride crystal serves as a good example of an ionic solid. Figure 3.1 shows the lattice structure of this salt, which is simple cubic with alternating lattice positions occupied by positive and negative ions. This lattice can also be pictured as two interpenetrating face-centered cubic structures made up of positive and negative ions respectively. Figure 3.2 shows another form of ionic lattice, that of cesium chloride. Here each ion of a given sign is surrounded by eight neighbors of the opposite sign. In the sodium chloride lattice, the corresponding coordination number is 6. An example in which this number is 4 is shown in Fig. 3.3. In general, all three of the above structures are characteristic of two-atom ionic crystals in which the positive and negative ions both carry charges of the same size. This is equivalent to saying that the atoms that make up the crystal have the same valence. Ionic crystals can also be formed from atoms that have different valences, for example, CaF_2 and TiO_2. These, of course, form still different types of crystal lattices[1] that will not be discussed.

Ideally, ionic crystals are formed by combining a highly electropositive metallic element with a highly electronegative element such as one of the halogens, oxygen, or sulfur. Certain of these solids have very interesting physical properties that are of considerable importance to the metallurgist. In particular, the study of plastic deformation mechanisms in ionic crystals, for example, LiF, AgCl, and MgO, has greatly added to our understanding of similar processes in metals.

1. Seitz, F., *Modern Theories of Solids.* McGraw-Hill Book Co., Inc., New York, 1940. p. 49.

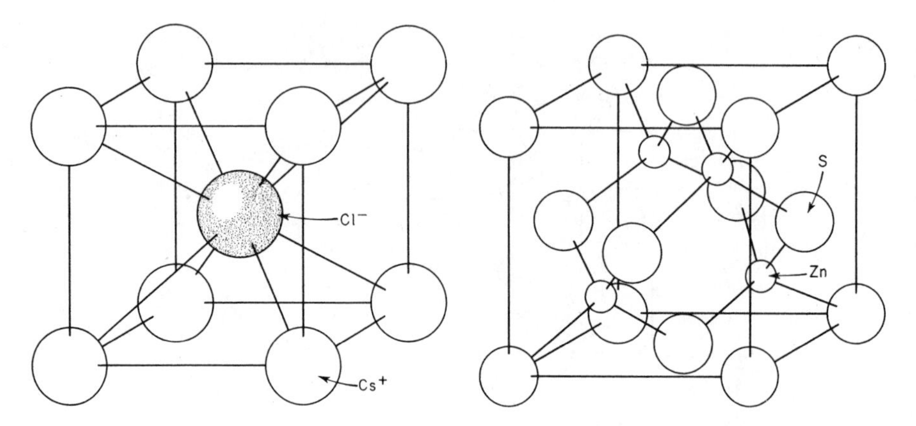

Fig. 3.2 The cesium chloride lattice.

Fig. 3.3 The zincblende lattice, ZnS.

3.3 The Born Theory of Ionic Crystals. The classical theory developed by Born[2] and Madelung[3] gives a simple and rather understandable picture of the nature of the cohesive forces in ionic crystals. It is first assumed that the ions are electrical charges with spherical symmetry and that they interact with each other according to simple central-force laws. In ionic crystals, these interactions take two basic forms, one long range and the other short range. The first is the well-known electrostatic, or coulomb, force that varies inversely with the square of the distance between a pair of ions, or

$$f = \frac{e_1 e_2}{r_{12}^2}$$

where e_1 and e_2 are the charges on the ions, and r_{12} is the center-to-center distance between the ions. The corresponding coulomb potential energy for a pair of ions is

$$\phi = \frac{e_1 e_2}{r_{12}}$$

The other type of interaction is a short-range repulsion that occurs when ions are brought so close together that their outer electron shells begin to overlap. When this happens, very large forces are brought into play that force the ions away from each other. In a typical ionic crystal, such as NaCl, both the positive and negative ions have filled electron shells characteristic of inert gases. Sodium, in

2. M. Born's work is summarized in the *Handbuch der Physik*, **XXIV/2**.
3. Madelung, E., *Zeits. für Physik*, **11** 898 (1910).

giving up an electron, becomes a positive ion with the electron configuration of neon ($1s^2$, $2s^2$, $2p^6$), while chlorine, in gaining an electron, assumes that of argon ($1s^2$, $2s^2$, $2p^6$, $3s^2$, $3p^6$). On a time-average basis, an atom with an inert-gas arrangement of electrons may be considered as a positively charged nucleus surrounded by a spherical volume of negative charge (corresponding to the electrons). Inside the outer limits of this negatively charged region all the available electronic energy states are filled. It is not possible to introduce into this volume another electron without drastically changing the energy of the atom. When two closed-shell ions are brought together so that their electron shells begin to overlap, the energy of the system (the two ions taken together) rises very rapidly, or, as it might otherwise be stated, the atoms begin to repel each other with large forces.

According to the Born theory, the total potential energy of a single ion in an ionic crystal of the NaCl type, due to the presence of all the other ions, may be expressed in the form

$$\phi = \phi_M + \phi_R$$

In this expression, ϕ is the total potential energy of the ion, ϕ_M is its energy due to coulomb interactions with all the other ions in the crystal, and ϕ_R is the repulsive energy. This expression may also be written

$$\phi = -\frac{z^2 e^2 A}{r} + \frac{Be^2}{r^n}$$

Here e is the electron charge, r is the distance between centers of an adjacent pair of negative and positive ions (Fig. 3.4), n is a large exponent, usually of the order of 9, and A and B are constants. If we now think in terms of the potential energy of a crystal containing one mole of NaCl, rather than in terms of a single ion, the above equation becomes

$$U = -\frac{Nz^2 e^2 A}{r} + \frac{NBe^2}{r^n}$$

where N is Avogadro's number and U is the total lattice potential energy. The first term on the right-hand side of this equation represents the electrostatic energy due to simple coulomb forces between ions, whereas the second term is that due to the repulsive interactions that arise when ions closely approach each other. It is a basic assumption of the Born theory that the repulsive energy can be expressed as a simple inverse power of the interionic distance. While quantum theory tells us that a repulsive term of the type Be^2/r^n is not rigorously correct,

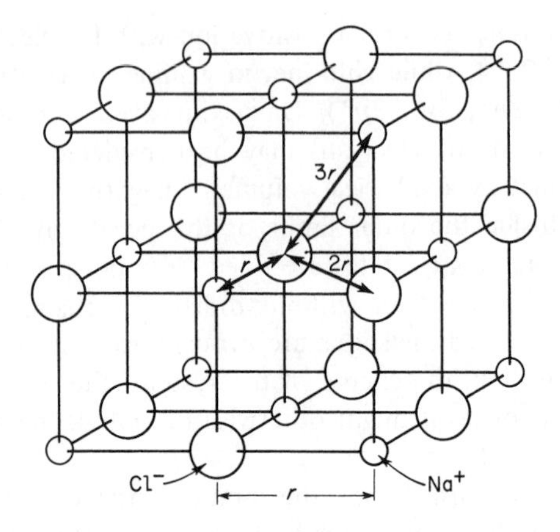

Fig. 3.4 Interionic distances in the sodium chloride lattice.

it is still a good approximation[4] for small variations of r from the equilibrium separation between atoms r_0.

We shall presently consider in more detail the individual terms of the Born equation, but before doing this let us look at the variation of the lattice energy with respect to the interionic distance r. This can be done conveniently by plotting each of the two terms on the right of the equation separately. The cohesive energy U is then obtained as a function of r by summing the curves of the individual terms. This is done in Fig. 3.5 for an assumed value of 9 for the exponent n. Note that the repulsive term, because of the large value of the exponent, determines the shape of the total energy curve at small distances, whereas the coulomb energy, with its smaller dependence on r, is the controlling factor at large values of r. The important factor in this addition is that the cohesive energy shows a minimum, U_0, at the interionic distance r_0, where r_0 is the equilibrium separation between ions at 0° K. If the separation between ions is either increased or decreased from r_0, the total energy of the crystal rises. Corresponding to this increase in energy is the development of restoring forces acting to return the ions to their equilibrium separation r_0.

Let us now consider the coulomb energy term of the Born equation, which for a single ion is

$$\phi_M = -\frac{z^2 e^2 A}{r}$$

4. Seitz, F., *The Modern Theory of Solids*. McGraw-Hill Book Co., Inc., New York, 1940.

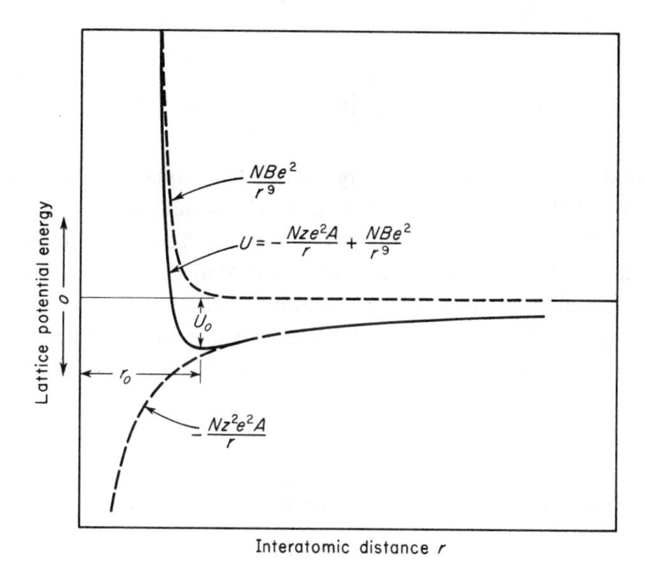

Fig. 3.5 Variation of the lattice energy of an ionic crystal with the spacing between ions.

or

$$\phi_M = -\frac{e^2 A}{r}$$

assuming that we are specifically interested in a sodium chloride crystal where there is a unit charge on each ion and $z^2 = 1$. Because the coulomb energy varies inversely as the first power of the distance between charged ions, coulomb interactions act over large distances, and it is not sufficient to consider only the coulomb energy between a given ion and its immediate neighbors. That this is true may be seen in the following. Around each negative chlorine ion there are 6 positive sodium atoms at a distance of r. This may be confirmed by studying Fig. 3.4. There is an attractive energy between each of the six sodium ions and the chlorine ions equal to $-e^2/r$, or in total $-6e^2/r$. The next closest ions to a given chlorine ion are 12 other negatively charged chlorine ions at a distance $\sqrt{2}r$. The interaction energy between these ions and the given ion is accordingly $12e^2/\sqrt{2}r$. Following this there are 8 sodium ions at a distance of $\sqrt{3}r$, 6 chlorine ions at $\sqrt{4}r$, 24 sodium atoms at $\sqrt{5}r$, etc. It is evident, therefore, that the coulomb energy of a single ion equals a series of terms of the form.

$$\phi_M = -\frac{6e^2}{\sqrt{1}r} + \frac{12e^2}{\sqrt{2}r} - \frac{8e^2}{\sqrt{3}r} + \frac{6e^2}{\sqrt{4}r} - \frac{24e^2}{\sqrt{5}r} \cdots$$

or

$$\phi_M = -\frac{Ae^2}{r} = -\frac{e^2}{r}[6 - 8.45 + 4.62 - 3.00 + 10.7\ldots]$$

The constant A of the coulomb energy, of course, equals the sum of the terms inside the brackets in the above expression. This series, as expressed above, does not converge because the terms do not decrease in size as the distance between ions is made larger. There are other mathematical methods of summing the ionic interactions[5] and it is quite possible to evaluate the constant A, which is called the *Madelung number*. For sodium chloride, the Madelung number is 1.7476 and the coulomb, or Madelung energy, for one ion in the crystal is, accordingly,

$$\phi_M = -\frac{1.7476e^2}{r}$$

For a single ion, the repulsive energy term is

$$\phi_R = \frac{Be^2}{r^n}$$

In this term the two quantities, B and n, must be evaluated. This can be accomplished with the aid of two experimentally determined quantities: r_0, the equilibrium interionic separation at 0° K; and K_0, the compressibility of the solid at 0° K. At r_0 the net force on an ion due to the other ions is zero, so that the first derivative of the total potential energy with respect to the distance, which equals the force on the ion, is also zero or

$$\left(\frac{d\phi}{dr}\right)_{r=r_0} = \frac{d}{dr}\left(-\frac{Ae^2}{r} + \frac{Be^2}{r^n}\right) = 0$$

Since the quantity A is already known, the above expression produces an equation relating n and B. A second equation is obtained from the fact that the compressibility is a function of the second derivative of the cohesive energy $(d^2\phi/dr^2)_{r=r_0}$ at $r = r_0$. In making these computations, the equilibrium separation of ions r_0 can be experimentally obtained from X-ray diffraction measurements of the lattice constant extrapolated to 0° K. In the NaCl crystal, this quantity equals 2.81 Å. The compressibility is defined by the expression

$$K_0 = \frac{1}{V}\left(\frac{\partial V}{\partial p}\right)_T$$

5. Slater, J. C., *Introduction to Chemical Physics*. McGraw-Hill Book Co., Inc., New York, 1939.

where K_0 is the compressibility, V is the volume of the crystal, and $(\partial V/\partial p)_T$ is the rate of change of the volume of the crystal with respect to pressure at constant temperature. The compressibility is thus the relative rate of change of volume with pressure at constant temperature, and is a quantity capable of experimental evaluation and extrapolation to 0° K.

When the calculations outlined above are made,[6,7] it is found that the Born exponent for the sodium chloride lattice is 8.0. In terms of this exponent, the computed cohesive energy is 180.4 kcal per mole. The latter is actually the energy of formation of a mole of solid NaCl from a mole of Na^+ ions, in the vapor form, and a mole of gaseous Cl^- ions. It is not possible experimentally to measure this quantity directly, but it can be evaluated from measured values of the heat of formation of NaCl from metallic sodium and gaseous Cl_2 in combination with measured values of the energy to sublime sodium, the energy to ionize sodium, the energy to dissociate molecular chlorine to atomic chlorine, and the energy to ionize chlorine. When all of these values are considered, the experimental value of U for the NaCl lattice turns out to be 182 Kcal per mole.

The rather good correspondence between the measured value of the cohesive energy for NaCl and the value computed with the Born equation shows that the latter gives a good first approximation of the cohesive energy for a typical ionic solid.

3.4 Van der Waals Crystals. In the final analysis, the cohesion of an ionic crystal is the result of its being composed of ions; atoms carrying electrical charges. In forming the crystal, the atoms arrange themselves in such a way that the attractive energies between ions with unlike charges is greater than all of the repulsive energies between ions with charges of the same signs. We shall now consider another type of bonding that makes possible the formation of crystals from atoms or even molecules that are electrically neutral and possess electron configurations characteristic of inert gases. The forces that hold this type of solid together are usually quite small and of short range. They are called van der Waals forces and arise from nonsymmetrical charge distributions. The most important component of these forces can be ascribed to the interactions of electrical dipoles.

3.5 Dipoles. An electrical dipole is formed by a pair of oppositely charged particles ($+e_1$ and $-e_1$) separated by a small distance. Let us call this distance a. Because the charges are not concentric, they produce an electrostatic field that is capable of exerting a force on other electrical charges. Thus, in Fig. 3.6, let l_1 and l_2 be the respective distances from the two charges

6. Slater, J. C., *Physics Review*, 23 488 (1924).
7. Seitz, F., *Modern Theory of Solids*. McGraw-Hill Book Co., Inc., New York, 1940, p. 80.

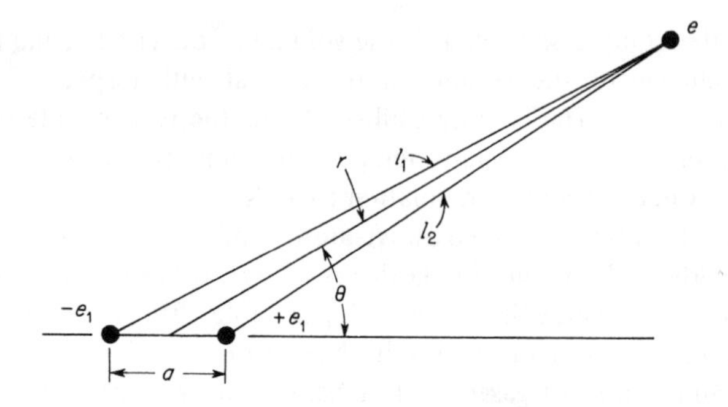

Fig. 3.6 An electrical dipole.

of the dipole to another charge e lying at a distance r from the center of the dipole. This latter will experience a force

$$F = -\frac{e_1 e}{l_1^2} + \frac{e_1 e}{l_2^2}$$

It can be shown that, if the distance r is large relative to the dipole separation a, the above expression is equivalent to

$$F = \frac{2e_1 a \cos\theta}{r^3} \cdot e$$

In this latter, r is the distance from the dipole to the position of the single charge e, and θ is the angle between the axis of the dipole and the direction r. It is customary to call $e_1 a$ the product of one of the dipole charges and the distance between the dipole charges the dipole moment. Let us designate it as μ. The electrical field intensity (force on a unit charge) due to a dipole can now be written as

$$\mathbf{E} = \frac{\mu \cos\theta}{r^3}$$

which shows that the field of a dipole varies as the inverse cube of the distance, whereas the field of a single charge varies as the inverse square of the distance.

3.6 Inert Gases. Let us turn to a consideration of the inert-gas atoms such as neon or argon. The solids formed by these elements serve as the prototype for the van der Waals crystals, just as crystals of the alkali halides (NaCl, etc.) are the prototype for ionic solids. It is interesting that they crystallize (at low temperatures) in the face-centered cubic system. In these atoms, as in all others, there is a positively charged nucleus surrounded by electrons traveling in orbits. Because of their closed shell structures, we can consider that over a period of time the negative charges of the electrons are distributed about the nucleus with complete spherical symmetry. The center of gravity of the negative charge on a time-average basis therefore coincides with the center of the positive charge on the nucleus, which means that the inert gas atoms have no average dipole moment. They do, however, have an instantaneous dipole moment because their electrons, in moving around the nuclei, do not have centers of gravity that instantaneously coincide with the nuclei.

3.7 Induced Dipoles. When an atom is placed in an external electrical field, its electrons are, in general, displaced from their normal positions relative to the nucleus. This charge redistribution may be considered equivalent to the formation of a dipole inside the atom. Within limits, the size of the induced dipole is proportional to the applied field, so that we write

$$\mu_I = \alpha \mathbf{E}$$

where μ_I is the induced dipole moment, $\mathbf{E}$ is the electrical field intensity, and α is a constant known as the *polarizability*.

When two inert-gas atoms are brought close together, the instantaneous dipole in one atom (due to its electron movements) is able to induce a dipole in the other. This mutual interaction between the atoms results in a net attractive force between the atoms. Figure 3.7 represents two inert-gas atoms of the same kind (perhaps argon atoms) separated by the distance r. Now let it be assumed that the atom on the left possesses an instantaneous dipole moment μ due to the movement of the electrons around the nucleus. This moment will produce a field $\mathbf{E}$ at the position of the second atom, which, in turn, induces a dipole moment in the latter equal to

$$\mu_I = \alpha \mathbf{E}$$

where μ_I is the induced dipole moment, α is the polarizability, and $\mathbf{E}$ is the field at the right-hand atom due to the dipole moment in the left-hand atom.

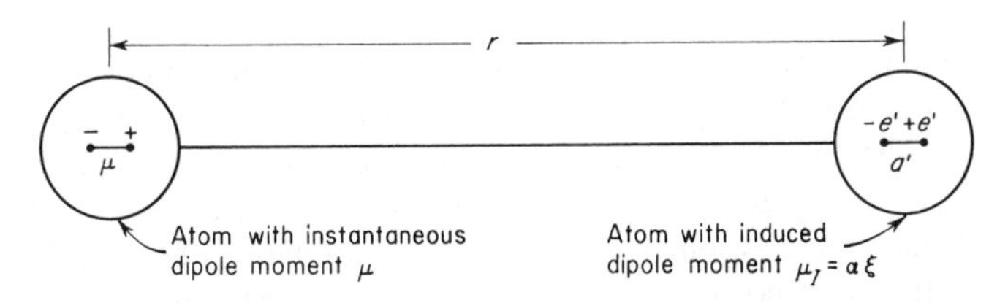

Fig. 3.7 Dipole-dipole interaction in a pair of inert-gas atoms.

First, consider the force exerted by the left dipole on the right dipole, which may be evaluated as follows. Let us assume, as indicated in Fig. 3.7, that the induced dipole in the right-hand atom is equivalent to the pair of charges $-e'$ and $+e'$ separated by the distance a'. According to this, the induced dipole moment is $e'a'$. Now let $\mathbf{E}$ be the field intensity due to the instantaneous dipole on the left atom at the negative charge $(-e')$ of the induced dipole. The corresponding field at the position of the positive charge $(+e')$ of the induced dipole is $\mathbf{E} + d\mathbf{E}/dr \cdot a'$. The total force on the induced dipole due to the field of the other dipole is:

$$f = -e'\mathbf{E} + e'\left(\mathbf{E} + \frac{d\mathbf{E}}{dr}a'\right) = e'a'\frac{d\mathbf{E}}{dr} = \mu_I \frac{d\mathbf{E}}{dr}$$

However,

$$\mu_I = \alpha \mathbf{E}$$

and therefore

$$f = \alpha \mathbf{E} \frac{d\mathbf{E}}{dr}$$

but in general the field of a dipole is proportional to the inverse cube of the distance, or

$$\mathrm{E} \simeq \frac{\mu}{r^3}$$

This leads us to the result

$$f \simeq \alpha \frac{\mu}{r^3} \frac{d}{dr} \frac{\mu}{r^3} \simeq \alpha \frac{\mu^2}{r^7}$$

The energy of a pair of inert-gas atoms due to dipole interactions can now be evaluated as follows:

$$\phi \simeq \int_{\infty}^{r} \alpha \frac{\mu^2}{r^7} dr \simeq \alpha \frac{\mu^2}{r^6}$$

It can be seen, therefore, that the van der Waals energy between a pair of inert-gas atoms due to dipole interactions varies as the square of the dipole moment and the inverse sixth power of their distance of separation; that the force varies as the square of the dipole moment is significant. Over a period of time the average dipole moment for an inert-gas atom must be zero. The square of this quantity does not equal zero, and it is on this basis that inert-gas atoms can interact.

3.8 The Lattice Energy of an Inert-Gas Solid. When the atoms of a rare-gas solid have their equilibrium separation, the van der Waals attraction is countered by a repulsive force. The latter is of the same nature as that which occurs in ionic crystals and is due to the interaction that occurs when closed shells of electrons start to overlap. The cohesive energy of an inert-gas solid may therefore be expressed in the form

$$U = -\frac{A}{r^6} + \frac{B}{r^n}$$

where A, B, and n are constants. It has been shown[8] that if n equals 12 the above equation correlates well with the observed properties of rare-gas solids. The first term on the right-hand side represents the total energy for one mole of the crystal caused by the dipole-dipole interactions between all the atoms of the solid. It may be obtained by first computing the energy of a single atom caused by its interactions with its neighbors. This quantity is then summed over all the atoms of the crystal. The computations are lengthy and will not be discussed. The second term in the above equation is the molar repulsive energy.

As might be surmised from the second-order nature of the van der Waals interaction, the cohesive energies of the inert-gas solids are quite small, being of the order of $\frac{1}{100}$ of those of the ionic crystals. The rare gases have also very low melting and boiling points, which is to be expected because of their small cohesive energies. Table 3.1 gives these properties for the inert-gas elements, with the exception of helium.

Table 3.1 Experimental Cohesive Energies,[9] Melting Points, and Boiling Points of Inert Gas Elements.

Element	*Cohesive Energy* Kcal/mol	*Melting Point* °C	*Boiling Point* °C
Ne	0.450	–248.6	–246.0
A	1.850	–189.4	–185.8
Kr	2.590	–157	–152
Xe	3.830	–112	–108

3.9 The Debye Frequency. The zero-point energy of a crystal is its thermal energy when the atoms are vibrating in their lowest energy states. When theoretical and experimental cohesive energies are compared at an

8. Lennard-Jones, J. E., *Physica*, 4 941 (1937).
9. Dobbs, E. R., and Jones, G. O., *Reports on Prog. in Phys.*, 20 516 (1957).

assumed temperature of 0° K, this energy, which has previously been neglected, should be included along with other terms. In a crystalline solid there are three vibrational degrees of freedom per atom. To each of these degrees a mode of vibration can be assigned so that there are three modes per atom. A crystal of N atoms is considered as equivalent to $3N$ oscillators of various frequencies ν.

The basic reasoning that brought Debye to these conclusions is as follows. First, he assumed that the forces of interaction between a neighboring pair of atoms were roughly equivalent to a linear spring. Pushing the atoms together would have the effect of compressing the spring, and in so doing, a restoring force would be developed that would act to return the atoms to their rest positions. Pulling the atoms apart would produce an equivalent opposite result. On this basis, Debye concluded that the entire lattice might be considered to be a three-dimensional array of masses interconnected by springs. In effect, assuming a simple cubic crystal, each atom would be held in space by a set of three pairs of springs, as indicated in Fig. 3.8*A*. He next considered how such an array might be able to vibrate. To simplify the presentation, a one-dimensional crystal will be considered, as indicated in Fig. 3.8*B*, and following Debye, the existence of longitudinal lattice vibrations will be ignored because they are of less significance. The vibration modes of such an array are analogous to the standing waves that can be set up in a string. A set of these latter are shown in Fig. 3.17. In a simple string, the number of possible harmonics is theoretically infinite, and there is no lower limit for the wavelengths that might be obtained. According to Debye, this is not true when a series of masses connected by springs are caused to vibrate. Here, as shown in Fig. 3.9, the minimum

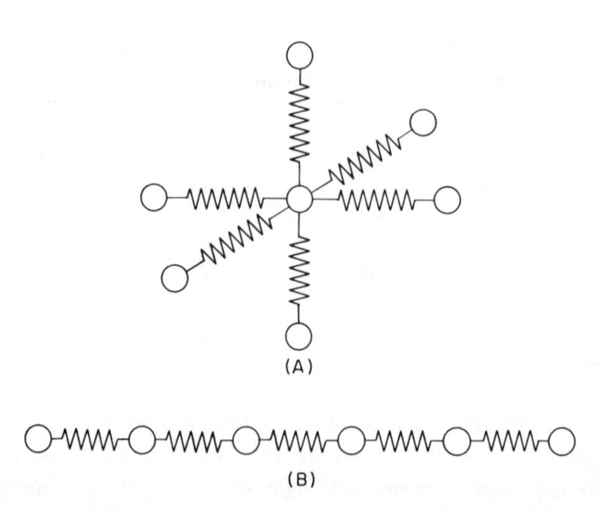

Fig. 3.8 (A) Debye model of a simple cubic crystal pictures an atom as a mass joined to its neighbors by springs. (B) A one-dimensional crystal model.

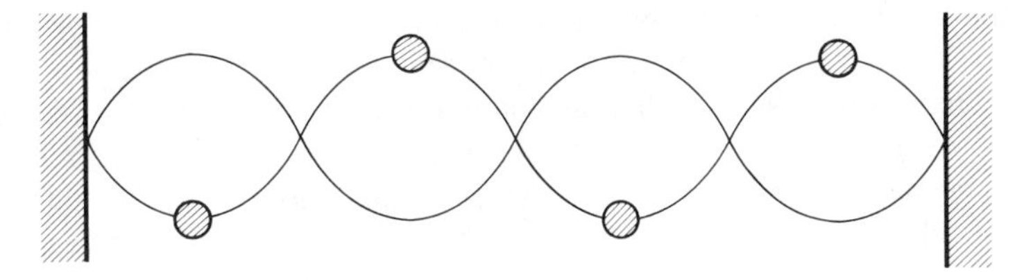

Fig. 3.9 The highest frequency vibration mode for an array of four masses.

wavelength or the mode of maximum frequency is obtained when neighboring atoms vibrate against each other. As may be seen in the drawing, the minimum wavelength corresponds to twice the spacing between the atoms, or $\lambda_{min} = 2a$, where a is the interatomic spacing. The (maximum) vibrational frequency associated with this wavelength is

$$\nu_m = \frac{v}{\lambda}$$

where v is the velocity of the shortest sound waves. This latter is normally of the order of 5×10^5 cm/sec. At the same time, the interatomic spacing in metals is about 2.5 Å, so that

$$\nu_m = \frac{5 \times 10^5}{2(2.5 \times 10^{-8})} = 10^{13} \text{ vib/sec}$$

The value, $\nu_m = 10^{13}$ vib/sec, is often used in simple calculations to represent the vibration frequency of an atom in a crystal. Since these calculations ordinarily are accurate to only about an order of magnitude (factor of 10), the use of the maximum vibration frequency for the average vibration frequency does not cause serious problems.

3.10 The Zero Point Energy. In standing waves such as those indicated in Fig. 3.17, the order of the harmonic corresponds to the number of half wavelengths in the standing wave pattern. Observe that in Fig. 3.9, where there are four atoms, there are four half wavelengths, and this system of four atoms will be able to vibrate in only four modes. In the case of a linear array of N_x atoms, there will be N_x half wavelengths when the array vibrates at its maximum frequency. Since this latter frequency corresponds to the N_xth harmonic, the system will have N_x transverse modes of vibration in the vertical plane, which is the assumed plane of vibration in the drawing.

In a three-dimensional crystal of N atoms, each atom inside the crystal can undergo transverse vibrations in three independent directions, as can be deduced by examining Fig. 3.8*A*; and by reasoning similar to that above, it is possible to show that there are $3N$ independent transverse modes of vibrations.

In a linear crystal, such as that implied in Fig. 3.8*B*, the density of vibrational modes is the same in any frequency interval $d\nu$. However, in a three-dimensional array or crystal the vibrational modes are three-dimensional, and the multiplicity of the standing wave patterns increases with increasing frequency. As a result, in the three-dimensional case the number of modes possessing frequencies in the range ν to $\nu + d\nu$ is given by

$$f(\nu)d\nu = \frac{9N}{{\nu_m}^3}\nu^2\, d\nu$$

where $f(\nu)$ is a density function, N is the number of atoms in the crystal, ν is the vibrational frequency of an oscillator, and ν_m is the maximum vibrational frequency.

Figure 3.10 is a schematic plot of the Debye density function $f(\nu)$ as a function of ν. The area under this curve from $\nu = 0$ to $\nu = \nu_m$ equals $3N$, the total number of oscillators. According to the quantum theory, the zero-point energy of a simple oscillator is $h\nu/2$. The total vibrational energy of the crystal at absolute zero is, accordingly,

$$E_z = \int_0^{\nu_m} f(\nu)\,\frac{h\nu}{2}\,d\nu = \frac{9}{8}Nh\nu_m$$

This quantity should normally be added to experimentally determined cohesive energies (Table 3.1) in comparing them with computed static lattice energies. The correction due to the zero-point energy is about 31 percent, or about 0.14

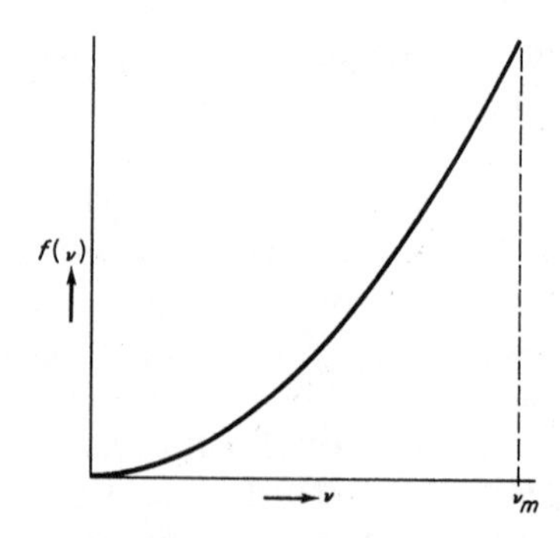

Fig. 3.10 Frequency spectrum of a crystal according to Debye. The maximum lattice frequency is ν_m.

Kcal per mol in the case of neon,[10] so that the lattice energy U_0 should be about 0.59 Kcal per mol rather than 0.45 Kcal per mol, as shown in Table 3.1. The importance of this correction decreases as the atomic number of the rare-gas element rises, so that for Xe it amounts to about 3 percent.

3.11 Dipole-Quadrupole and Quadrupole-Quadrupole Terms. The van der Waals attractive energy is caused by a synchronization of the motion of the electrons on the various atoms of a solid. As a first approximation it might be considered that this interaction is equivalent to the development of synchronized dipoles on the atoms. The summation of the dipole interactions throughout the crystal then leads to an attractive energy that varies as the inverse sixth power of the distance. Actually, the complex charge distributions that exist in real atoms cannot be accurately represented by picturing them as simple dipoles. Modern quantum mechanical treatments generally use an expression for the van der Waals attractive energy (expressed in terms of a single ion) of the type

$$\phi_{(r)} = -\left(\frac{c_1}{r^6} + \frac{c_2}{r^8} + \frac{c_3}{r^{10}}\right)$$

where c_1, c_2, and c_3 are constants. The first term of this expression is the dipole-dipole interaction already considered. The second term in the inverse eighth power of the distance is called the *dipole-quadrupole term* because the interaction between a dipole on one atom and a quadrupole on another will lead to an energy that varies as the inverse eighth power of the distance. A quadrupole is a double dipole consisting of four charges. The last term, varying as the inverse tenth power of the distance, is called the *quadrupole-quadrupole term*. It is, in general, small and amounts to less than 1.3 percent of the total van der Waals attractive energy for all of the inert-gas solids.[11] The dipole-quadrupole term, on the other hand, equals approximately 16 percent of the total attractive energy, indicating that, whereas the dipole-dipole term makes up most of the van der Waals attractive energy, the second term is also significant.

3.12 Molecular Crystals. Many molecules form crystals which are held together by van der Waals forces. Among these are N_2, H_2, and CH_4; typical covalent molecules in which the atoms share valence electrons to effectively obtain closed shells for each atom in the molecule. The forces of attraction between such molecules are very small and of the order of those in the inert-gas crystals.

10. *Ibid.*
11. *Ibid.*

The molecules mentioned in the above paragraph are nonpolar molecules; they do not have permanent dipole moments. Thus, the attractive force between two hydrogen molecules comes in large measure from the synchronism of electron movements in the two molecules, or from instantaneous dipole-dipole interactions. In addition to these, there are also polar molecules, such as water (H_2O), which possess permanent dipoles. The interaction between a pair of permanent dipoles is, in general, much stronger than that between induced dipoles. This leads to much stronger binding (van der Waals binding) in their respective crystals with correspondingly higher melting and boiling points.

3.13 Refinements to the Born Theory of Ionic Crystals. In the inert-gas and molecular solids considered in the preceding section, van der Waals forces are the primary source of the cohesive energy. These forces exist in other solids, but when the binding due to other causes is strong they may contribute only a small fraction of the total binding energy. This is generally true in ionic crystals, although some types, like the silver halides, may have van der Waals contributions of more than 10 percent. The alkali halides, as may be seen in Table 3.2, have van der Waals energies that amount to only a small percentage of the total energy. This table is of particular interest because it lists the contribution of five terms to the total lattice energy. The first column gives the Madelung energy, or the first term in the simple Born equation. The second is the repulsive energy, caused by the overlapping of closed ion shells. The third and fourth columns are van der Waals terms: dipole-dipole and dipole-quadrupole. The fifth column lists zero-point energies; the energy of vibration of the atoms in their lowest energy levels. Finally, the last column is that

Table 3.2 Contributions to the Cohesive Energies of Certain of the Alkali Halides.*

Alkali Halide Crystal	*Madelung*	*Repulsive*	*Dipole-Dipole*	*Dipole-Quadru-pole*	*Zero Point*	*Total*
LiF	285.5	−44.1	3.9	0.6	−3.9	242.0
LiCl	223.5	−26.8	5.8	0.1	−2.4	200.2
LiBr	207.8	−22.5	5.9	0.1	−1.6	189.7
LiI	188.8	−18.3	6.8	0.1	−1.2	176.2
NaCl	204.3	−23.5	5.2	0.1	−1.7	184.4
KCl	183.2	−21.5	7.1	0.1	−1.4	167.5
RbCl	175.8	−19.9	7.9	0.1	−1.2	162.7
CsCl	162.5	−17.7	11.7	0.1	−1.0	155.6

*All values given above are expressed in Kilo calories per mole. From *The Modern Theory of Solids*, by Seitz, F. Copyright 1940, McGraw-Hill Book Company, Inc., New York, p. 88. Used by permission.

corresponding to the sum of all five terms, which should equal the internal energy of the crystals at zero degrees absolute and at zero pressure.

3.14 Covalent and Metallic Bonding. In both the ionic and the inert-gas crystals that have been considered, the crystals are formed of atoms or ions with closed-shell configurations of electrons. In these solids, the electrons are considered to be tightly bound to their respective atoms inside the crystal. Because of this fact, ionic and rare-gas solids are easier to interpret in terms of the laws of classical physics. It is only when greater accuracy is needed in computing the physical properties of these crystals that one needs to turn to the more modern quantum mechanical interpretation. Quantum, or wave mechanics, is, however, an essential when it comes to studying the bonding in covalent or metallic crystals. In each of the latter, the bonding is associated with the valence electrons that are not considered to be permanently bound to specific atoms in the solids. In other words, in both of these solids, the valence electrons are shared between atoms. In valence crystals, the sharing of electrons effects a closed-shell configuration for each atom of the solid. The prototype of this form of crystal is the diamond form of carbon. Each carbon atom brings four valence electrons into the crystal. The coordination number of the diamond structure is also four, as shown in Fig. 3.11. If a given carbon atom shares one of its four valence electrons with each of its four neighbors, and the neighbors reciprocate in turn, the carbon atom will, as a result of this sharing of eight electrons, effectively achieve the electron configuration of neon ($1s^2, 2s^2, 2p^6$). In this type of crystal, it is often convenient to think of the pairs of electrons which are shared between nearest neighbors as constituting a chemical bond between a pair of atoms. On the other hand, according to the band theory of solids, the electrons are not fixed to specific bonds, but can interchange between bonds. Thus, the valence electrons in a valence crystal can also be thought of as belonging to the crystal as a whole.

The binding associated with these covalent, or *homopolar*, linkages as they are called, is very strong so that the cohesive energy of a solid such as diamond is very large and, in agreement with this fact, valance solids are usually very hard and have high melting points.

The covalent bond is also responsible for the cohesion of many well-known molecules, for example, the hydrogen molecule. An idea of how the binding energy develops can be obtained by an elementary consideration of the hydrogen molecule. The lowest atomic energy state is associated with the $1s$ shell. Two electrons can be accommodated in this state, but only if they have opposite spins. Thus, an unexcited helium atom will have both of its electrons in the $1s$ state, but only if the spin vectors of the electrons are opposed. The fact that two electrons can occupy the same quantum state only if the spins are oppositely directed is known as the *Pauli exclusion principle*. Now suppose that

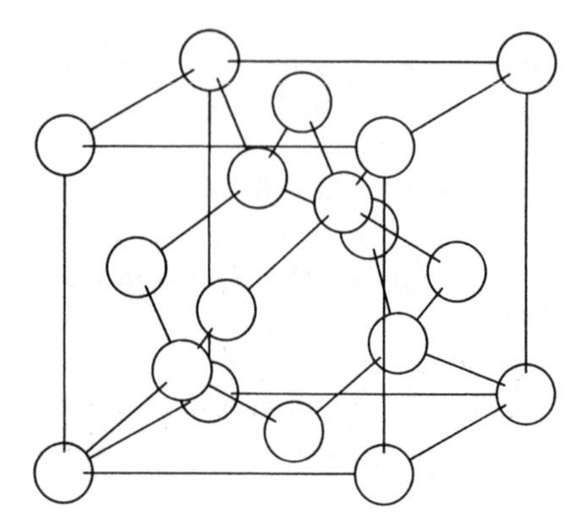

Fig. 3.11 The diamond structure. Each carbon atom is surrounded by four nearest neighbors. *Note:* this structure is the same as that of zincblende (ZnS), Fig. 3.3, except that this lattice contains one instead of two kinds of atoms.

two hydrogen atoms are made to approach each other. Then there are two cases to be considered: when the spins of the electrons on the two atoms are parallel, and when the spins are opposed. First consider the latter case. As the atoms come closer and closer together, the electron on either atom begins to find itself in the field of the charge on the nucleus of the other atom. Since the spins of the electrons are opposed, each nucleus is capable of containing both electrons in the 1*s* ground state. Under these conditions, there is a strong probability that the electrons will spend more time in the neighborhood of one nucleus than the other, and the hydrogen molecule becomes a pair of charged ions – one positive and the other negative. This structure is unstable, especially when the hydrogen atoms are far apart, as may be estimated from the energy required to form a positive and a negative pair of hydrogen ions (–295.6 Kcal per mol).[12] At the normal distance of separation of the atoms in a hydrogen molecule, the ionic structure exists for limited periods of time and contributes about 5 percent of the total binding energy.[12]

A much more important type of electron interchange occurs when both electrons simultaneously exchange nuclei. The resulting shifting of the electrons back and forth between the nuclei, which occurs at a very rapid rate, is commonly known as a *resonance effect*, and about 80 percent[12] of the binding energy of the hydrogen molecule is attributed to it. With the aid of quantum mechanics, the total binding energy of a hydrogen molecule has been computed and, as a result, an insight into the nature of the binding energy associated with this exchange of electrons has been obtained. Quantum mechanics shows that, on the average, the electrons spend more of their time in the region

12. Pauling, L., *The Nature of the Chemical Bond.* Cornell University Press, Ithaca, N.Y., 1940, p. 22.

between the two protons than they do on the far sides of the protons. From a very elementary point of view, we may consider that the binding of the hydrogen molecule results from the attraction of the positively charged hydrogen nuclei to the negative charge which exists between them.

Attention is called to the interrelation between space and time implied in this discussion concerning the time-average position of the electrons. These ideas are normally expressed in terms of a phase space. This is a coordinate system that includes both the positions of the particles and their momenta (that is, velocities). Thus for a single particle free to move along a single direction (the x-axis) there will be two dimensions in phase space; its position along the axis, and its momentum. For n particles capable of moving in a single direction, there will be $2n$ linear dimensions in the phase space associated with these particles. These are $x_1, x_2, \ldots x_n$, the positions of the particles; and $p_1, p_2, \ldots p_n$, the momenta of the particles. For particles capable of moving in three dimensions, there will be $6n$ degrees of freedom and therefore a corresponding number of dimensions in phase space.

An important theorem in statistical mechanics that bears on the subject of phase space states that the position and momentum average in phase space coincides with the same average over an infinite time. As an example of the application of this theorem, consider the following. The energy of a set of particles is a function of their positions in space and their momenta (velocities). The average energy can be computed by averaging a finite number of position and velocity measurements. The greater the number of these sets of readings that are taken, the closer to the true average will be the result. On the other hand, the positions and velocities are also a function of time. Therefore their energy can also be considered as a function of time, and the average could be taken from a set of readings made over a very long period of time. This average should agree with that made inside a very short time interval.

In the above, it was assumed that the electrons of the two hydrogen atoms had oppositely directed spins. Now consider the case where two hydrogen atoms with parallel spins are made to approach each other. Here, as the electron on either of the atoms comes within the range of effectiveness of the field on the nucleus of the other atom, it is found that the energy level it would normally occupy is already filled. The situation is similar to that when an electron is brought within the limits of the closed shell of an inert-gas configuration. The normal electronic orbits become badly distorted, or else the second electron moves into a higher energy state, such as $2s^1$. In either case, bringing together two hydrogen atoms with electrons that have parallel spins increases the energy of the system. A stable molecule cannot be formed in this fashion. This is shown in Fig. 3.12, where the uppermost curve represents the hydrogen molecule with parallel spins and both electrons in $1s$ orbits. The lower curve is for opposed electron spins, and it can be observed that this curve has a pronounced

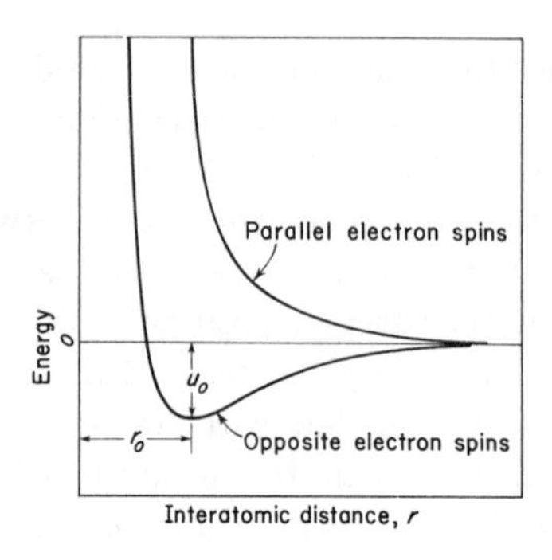

Fig. 3.12 Interaction energy of two hydrogen atoms.

minimum, indicating that in this case a stable molecule can be formed. Notice that the curves of Fig. 3.12 represent the total energy of the hydrogen molecule at corresponding values of r. In addition to the ionic and the exchange, or resonance energy, there are other more complicated electrostatic interactions between the two electrons and the two protons. These contribute the remaining 15 percent of the binding energy of the hydrogen molecule.

Valence electrons are also shared between atoms in a typical metallic crystal. Here, however, they are best considered as free electrons and not as electron pairs forming bonds between neighboring atoms. This difference between a metal and a covalent solid is probably a matter of degree, for it is known that the valence electrons in a covalent solid can move from one bond to another and thus are able to move throughout the crystal. It will be shown presently that the zone or band theory of solids is able to explain adequately the difference between a valence crystal and a metallic one. However, for the present let us assume that in a metal the valence electrons are able to move at will through the lattice, while in a valence crystal the electrons form directed bonds between neighboring atoms. One of the results of this difference is that metals tend to crystallize in close-packed lattices (face-centered cubic and close-packed hexagonal structures) in which the directionality of the bonds between atoms is of secondary importance, while valence crystals form complicated structures so that the bonds between neighboring atoms will give each atom the effect of having a closed-shell configuration of electrons. Thus, carbon, with four valence electrons per atom, crystallizes in the diamond lattice with four nearest neighbors, so that each carbon atom has a total of eight shared electrons. In the same manner, arsenic, antimony, and bismuth, which have five valence electrons per atom, need to share electrons with three neighbors in order to attain the eight electrons needed for a closed-shell configuration. The latter substances therefore crystallize with three nearest neighbors. In general, covalent crystals follow what is known as the $(8 - N)$ rule, where N is the number of valence electrons, and the factor $(8 - N)$ gives the number of nearest neighbors in the structure.

On the assumption that inside a metal the valence electrons are free, we arrive at the following elementary idea of a metal. A metal consists of an ordered array of positively charged ions between which the valence electrons move in all directions with high velocities. Over a period of time, this movement of electrons is equivalent to a more or less uniform distribution of negative electricity which might be thought of as an electron gas that holds the assembly together. In the absence of this gas, the positively charged nuclei would repel each other and the assembly would disintegrate. On the other hand, the electron gas itself could not exist without the presence of the array of positively charged nuclei. This is because the electrons would also repel each other. The cooperative interaction between the electron gas and the positively charged nuclei forms a structure that is stable. The binding forces that hold a metallic crystal together can be assumed to come from the attraction of the positively charged ions for the cloud of negative charge that lies between them. It is also interesting to note that, as the distance between nuclei is made smaller (as the volume of the metal is decreased), the velocities of the free electrons increase with a corresponding rise in their kinetic energy. This leads to a repulsive energy term which becomes large when a metal is compressed. The simplest metals to understand are such alkali metals as sodium and potassium. In them there is only a single valence electron and the positive ions are well separated in the solid so that there is little overlapping of ion shells, and the repulsion caused by the overlapping of shells is therefore small. In these metals, the repulsive energy is chiefly due to the electronic kinetic energy term. In other metallic elements, the theories of the cohesive energies are much more involved. They will not be discussed here.

ELECTRON THEORY OF METALS

3.15 The Uncertainty Principle. One of the cornerstones of modern atomic theory and, in particular, the electron theory of metals, is the *Heisenberg uncertainty principle.* The older Bohr theory of the atom was based on simple Newtonian mechanics. In principle, it assumed that an electron traveled about the nucleus in orbits similar to those of a planet about the sun. In each Bohr orbit, the electron was assumed to have a known velocity so that, theoretically, it should be possible to measure the position of an electron in its orbit at two different times and thereby determine the distance traveled during the given time interval. It is now recognized that accurate measurements of this nature on a particle the size of an electron are impossible both in theory and in practice. The uncertainty principle, or the condition which defines these limitations will now be briefly discussed.

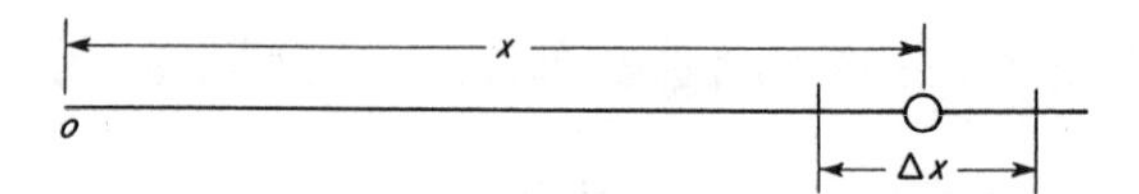

Fig. 3.13 The uncertainty in position of an electron.

Suppose that an electron is moving along a straight line, as in Fig. 3.13. Let the distance of the electron from the center of coordinates be x. When we experimentally determine this quantity there is always an experimental error $\mp\Delta x/2$ so that the uncertainty in the position of the electron is Δx. Now let $p = mv$, the momentum of the electron. Any experimental measurement of the momentum of the electron will also involve an uncertainty Δp. Heisenberg's principle states that the product of these two uncertainties cannot be smaller than Planck's constant h, or

$$\Delta p \cdot \Delta x \geqslant h$$

The uncertainty principle has little significance in measuring the position or momentum of large objects. This is because of the small size of Planck's constant relative to the measured quantities. However, for particles as small as an electron or an atom, it is extremely important, as will now be shown.

Suppose that the electron has a zero momentum signifying that it is at rest. Then we have

$$\Delta x = \frac{h}{\Delta p} = \frac{h}{o} = \infty$$

which tells us that the uncertainty in the position of the electron is infinite, or that we would be entirely unable to locate it. On the other hand, if we should ever be able to determine the exact location of an electron, we would be entirely ignorant of its velocity. Consequently, even though the position of the electron might be known at a given moment, we would not be able to predict its position at some future instant in time.

The above is strong evidence that the Bohr theory is not based on strong theoretical foundations, for the latter implies that the position and momentum of an electron are always capable of accurate measurement at any instant.

The uncertainty principle is consistent with the assumption that the vibrational energy of a solid is not zero at absolute zero. If the zero-point energy vanished, then each atom of the solid would be at rest in its equilibrium lattice position. Its momentum would be zero and its position would be fixed and determinable. This reasoning is clearly not in agreement with the uncertainty principle.

3.16 The Dual Nature of Matter. Not only are we faced with the problem that we are not able to accurately determine both the position and velocity of an electron, but we must realize that there is also considerable doubt as to just what an electron is. Both matter and light have a dual nature: corpuscular and wavelike.

Let us consider an electron. It is a well-known fact that a single electron carries an electrical charge that is finite and capable of measurement ($e = 1.60 \times 10^{-19}$ coulombs), which is a corpuscular property, as is the fact that an electron has a finite rest mass ($m_o = 9.11 \times 10^{-28}$ gm) and an effective size of approximately 10^{-13} cm. On the other hand, when an electron beam consisting of a large number of electrons is made to strike crystalline material, it can be diffracted according to the same laws as electromagnetic X-rays, provided the wavelength of an electron is expressed by the de Broglie relationship,

$$\lambda = \frac{h}{mv} = \frac{h}{p}$$

where λ is the wavelength of the electron, m its mass, v its velocity, and p its momentum. This ability of electrons to be diffracted shows that they also possess wavelike characteristics.

An important principle concerning the duality of matter is the *principle of complementarity* as enunciated by Bohr. This states that there is a particle and wave duality to matter that are complementary to each other. However, this complementarity is so arranged in nature that when one wants to make numerical calculations about events that occur, one must either use only particle properties or only wave properties for the calculations. In other words, reliable results cannot be obtained by mixing wave and particle concepts in the same set of computations.

If the electrons are actually waves, we are faced with the problem of deciding what kind of waves they are. This we do not know. Our basic problem is that the uncertainty principle does not allow us to use the concept of the electron as a particle, as is implied in the Bohr theory. On the other hand, we are unable to define an electron completely in terms of waves. Fortunately there is a mathematical approach to this matter which obviates the need to define an electron exactly. The approach is through the use of wave mechanics, the results of which are usually expressed in terms of the energy of an electron. Energy is, of course, an important property. As we know, the energy of radiation from an atom (perhaps a hydrogen atom) is related by Planck's law to the difference in the energies of the electron in two different states in the atom, or

$$h\nu = E_{II} - E_I$$

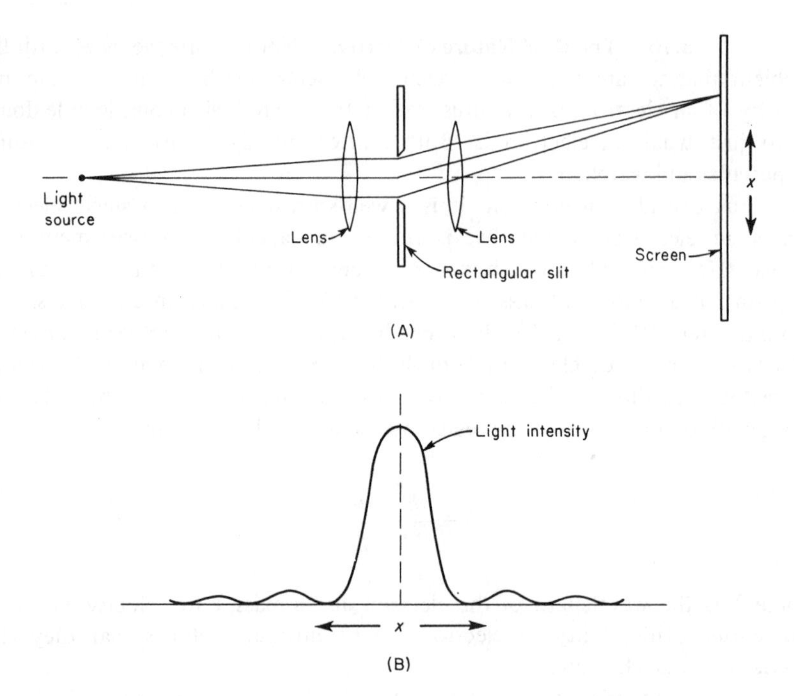

Fig. 3.14 (A) Experimental arrangement for Fraunhofer diffraction using a single slit. (B) The Fraunhofer diffraction pattern for a single slit.

where h is Planck's constant, ν is the frequency of the radiation, E_{II} and E_I are the energies of the electron in two different atomic-energy levels. While wave mechanics can render accurate estimates of the possible energies of an electron, it cannot give similar information about the positions of the electrons. What it does give is an idea of the regions in space where an electron will most probably be found. Wave mechanics is, therefore, closely related to the concept of probability.

In order to see how the probability concept enters into the problem of electron dynamics, let us first consider a simple optical experiment. The diffraction of light through a slit, under the experimental conditions shown in Fig. 3.14A produces a diffraction pattern like that of Fig. 3.14B where the abscissae correspond to distances along the pattern, and the height of the pattern represents the intensity of the radiation falling on the photographic plate or some other device for recording the pattern. By definition, the intensity of light is the energy per second which falls on a square centimeter, so that the ordinate in Fig. 3.14B represents the rate at which energy falls at various points along the pattern. Now let us consider this same experiment from a corpuscular point of view. In this case, the height of the pattern must be proportional to the number of light quanta falling per second on a unit area of the screen at any given point. However, there is still another way that this diffraction pattern can be viewed. Suppose that only a single quantum passes through the slit, then the ordinate of the diffraction pattern is considered to represent the chance, or probability, that the quantum will fall at a given point on the screen. According to this point of view, the area under the entire curve must be unity, that is, the quantum must land somewhere along the screen.

It can be seen from the above diffraction experiment that the ordinate can be considered to measure either the light intensity at a given position, or the probability that a single quantum will strike it. There is, accordingly, a direct correspondence between the light intensity and the probability as defined above. In wave mechanics one treats the wave properties of electrons from this probability point of view.

Before considering the subject of wave mechanics, it should be mentioned that the diffraction of electrons through a slit differs from the diffraction of light photons through a slit. When electrons are made to pass through a single slit, only a single maximum is observed in the diffraction pattern. To obtain fringes, electrons must be made to pass through at least a double slit. In general, only light quanta or photons are able to produce fringes when passed through a single slit. On the other hand, both *leptons* (particles with a small mass, such as electrons and positrons) and *baryons* (particles with a large mass, such as protons and neutrons) require a double slit in order to form a diffraction pattern with fringes.

Now let us consider a very simple wave that conforms to the following relation:

$$y = y_0 \sin\left(2\pi\nu t - \frac{2\pi x}{\lambda}\right)$$

This might represent a wave in an infinitely long string excited by the vibration of an electric tuning fork attached to one end. As indicated in Fig. 3.15, λ is the wavelength of the wave, y_0 is its amplitude, and the wave travels in the positive x direction. At the same time, ν is the frequency, or number of vibrations per second, x is the distance in the direction of travel, and t is the time. This equation can also be written in the simpler form

$$y = y_0 \sin(\omega t - kx)$$

where $\omega = 2\pi\nu$ is the angular frequency, and $k = 2\pi/\lambda$ is called the wave number.

It is a well-known law of physics that the intensity of a wave is proportional to the square of the wave amplitude, or

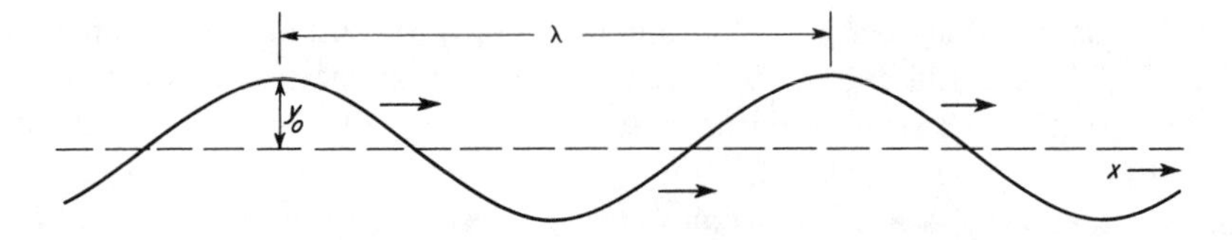

Fig. 3.15 Wave traveling in an infinitely long string.

$$I \propto {y_0}^2$$

where I is the intensity, and y_0 is the amplitude. However, we have seen above that the intensity is directly related to the probability that a given quantum (or electron) will fall at a given point. By analogy, therefore, let us define the function ψ such that

$$P = \psi^2$$

Here P is the probability and ψ is a function which, when multiplied by itself, yields the probability P. Actually, more accurately because the function ψ is often a complex number, we should write the above expression as follows:

$$P = \psi\psi^*$$

where ψ and ψ^* are complex conjugates, and their product, the probability, is a real number.

The quantity ψ is evidently equivalent to the amplitude of a wave, so that we may write (by analogy with our expression for a wave given above):

$$\psi = \psi_0 \sin(\omega t - kx)$$

Now let us find the second derivative of this function with respect to the distance

$$\frac{d\psi}{dx} = +k\psi_0 \cos(\omega t - kx)$$

and

$$\frac{d^2\psi}{dx^2} = -k^2 \psi_0 \sin(\omega t - kx) = -k^2 \psi$$

or

$$\frac{d^2\psi}{dx^2} + k^2 \psi = 0$$

This differential equation is the basic equation for wave motion in one dimension. One of its solutions is, of course,

$$\psi = \psi_0 \sin(\omega t - kx)$$

By definition, the quantity k in the above differential equation is the wave number and equals $2\pi/\lambda$. Now we can make the assumption that we can substitute the de Broglie wavelength of the electron, $\lambda = h/mv$, into the wave number k so that

$$k = \frac{2\pi}{\lambda} = \frac{2\pi}{h} mv$$

It is common practice in wave mechanics to substitute the symbol $\hbar$ for $h/2\pi$. The equation for the energy $h\nu$ may then be expressed in terms of the angular frequency ω. This is shown in the following equation

$$\text{Energy} = h\nu = \frac{h}{2\pi}(2\pi\nu) = \hbar\omega$$

If the substitution of $\hbar$ for $h/2\pi$ is made, then we have

$$k^2 = \frac{m^2 v^2}{\hbar^2} = \frac{2}{\hbar^2} m(\tfrac{1}{2} mv^2)$$

where $\frac{1}{2}mv^2$ is clearly the kinetic energy of an electron. The above expression for k^2 may now be substituted into the wave equation with the following result:

$$\frac{d^2\psi}{dx^2} + \frac{2}{\hbar^2} m(\tfrac{1}{2} mv^2)\psi = 0$$

This is the well-known Schrödinger equation in one dimension. In three dimensions (x, y, and z), it has the form

$$\frac{\partial^2\psi}{\partial x^2} + \frac{\partial^2\psi}{\partial y^2} + \frac{\partial^2\psi}{\partial z^2} + \frac{2}{\hbar^2} m(\tfrac{1}{2} mv^2)\psi = 0$$

Because the kinetic energy of an electron is equal to its total energy less its potential energy, the Schrödinger equation is usually written:

$$\frac{\partial^2\psi}{\partial x^2} + \frac{\partial^2\psi}{\partial y^2} + \frac{\partial^2\psi}{\partial z^2} + \frac{2}{\hbar^2} m(W - V)\psi = 0$$

where $W - V = \frac{1}{2}mv^2$ and W is the total energy of the electron and V is its potential energy.

3.17 Free-Electron Theory. The Sommerfeld free-electron theory of a metal will now be briefly discussed. Its basic assumptions are that the valence electrons in a metal are free and that they are confined to move within the boundaries of a metal object, that is, they move back and forth inside the metal, but do not leave it. In effect, the electrons in the metal constitute a gas and are commonly referred to as the electron gas. As such they do not obey the statistical laws (Boltzmann statistics) of an ordinary gas in a container where $N = N_o e^{-q/kT}$ and N is the number of gas molecules with an energy equal to q, N_o is the total number of gas molecules in the container, k is Boltzmann's constant (1.38×10^{-16} erg per °K), and T is the absolute temperature. Rather, the electrons conform to the Fermi statistics that will be discussed shortly.

This idea conforms to the concept that the electrons have a lower potential energy inside the metal than outside of it, or that it takes work to remove an electron from a piece of metal. In the Sommerfeld theory, it is also assumed that the potential of an electron inside a metal is constant. The statement is an oversimplification because an electron actually moves in a very complicated potential field because of the presence of both the other electrons and the periodic arrangement of positively charged ions within the metal. In spite of this simplification, the elementary theory yields certain important results.

In developing it, let us first consider what is known as the particle-in-a-box problem. Here we assume that an electron is confined within a box and that its potential energy is zero (constant) within the box and infinite outside of the box. These conditions on the potential assure us that the probability of finding the electron outside of the box will be zero, which, in turn, tells us that any solution ψ[13] of the equation

$$\frac{\hbar^2}{2m}\left(\frac{\partial^2 \psi}{\partial x^2} + \frac{\partial^2 \psi}{\partial y^2} + \frac{\partial^2 \psi}{\partial z^2}\right) + (W - V)\psi = 0$$

must be zero at all points outside the box and, in particular, at the boundaries of the box. For simplicity, let us assume that the electron is able to move in only one direction, so that we are effectively working with a one-dimensional box. The basic problem is now illustrated in Fig. 3.16 where the length of the box has been chosen equal to L. By the condition that ψ must be zero at the boundaries, we have

13. In general, the ψ function must have the following properties: it should be (1) single valued, (2) continuous, (3) smoothly varying except at infinities of the potential function V, (4) finite everywhere, (5) not zero everywhere, and (6) $\psi\psi^*$ should be integrable (i.e., $\int \psi\psi^*$ is finite).

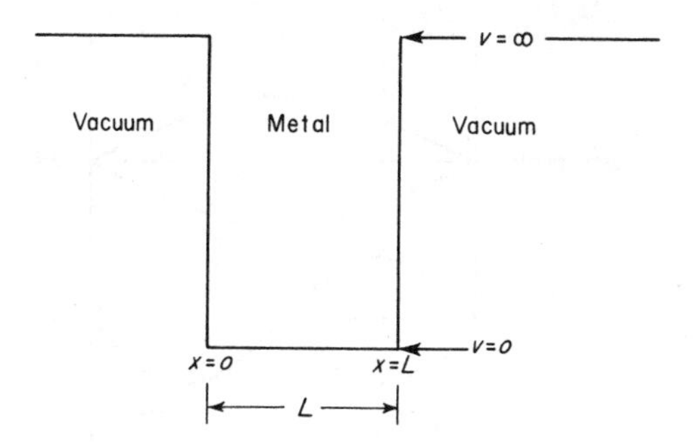

Fig. 3.16 The potential well inside a one-dimensional box as assumed in the free-electron theory.

$$\psi = 0 \text{ at } x = 0$$

$$\psi = 0 \text{ at } x = L$$

These are the same boundary conditions that apply to the amplitude of vibration of waves traveling in a string of length L the ends of which are fixed. Because the same basic wave equation controls the motion of electrons in one dimension and the vibration of a string, we can use the known solutions of the string with fixed ends in order to define the solutions of the electron in the box. It is well known that a string of this type is capable of vibrating at certain natural frequencies. These frequencies, of course, are determined by the possible standing wave patterns which can develop in a string of length L. Thus, in Fig. 3.17, the fundamental, or lowest, frequency of the vibrating string has a wavelength equal to $2L$. The second mode of vibration corresponds to a wavelength L, the third to $\frac{2}{3}L$, etc., and, in general, we can write

$$\lambda = \frac{2}{n_x}L \qquad n_x = 1, 2, 3, 4, \ldots$$

where n_x assumes integer values from one to infinity. This same condition holds for the standing waves of the ψ function of the electron in the one-dimensional box and, by definition, $k = 2\pi/\lambda$, so that

$$k = \frac{n_x \pi}{L}$$

and

$$k^2 = \frac{n_x^2 \pi^2}{L^2}$$

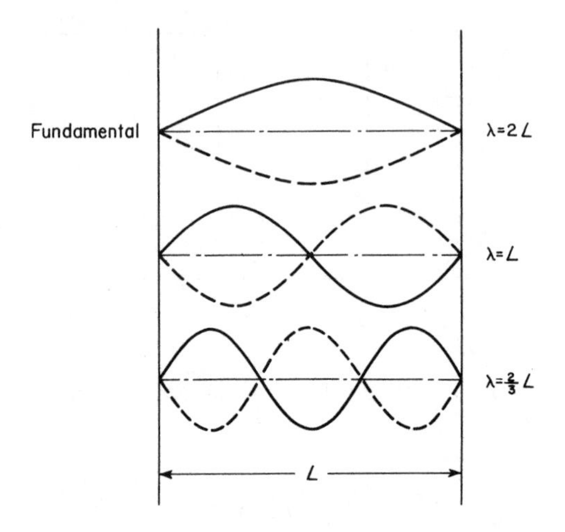

Fig. 3.17 Standing waves in a string with fixed ends.

However, we have already seen that, in general,

$$k^2 = \frac{2}{\hbar^2}m(W - V)$$

but inside the box where the standing waves exist, we have assumed V to be zero, so that

$$k^2 = \frac{2}{\hbar^2}mW = \frac{2}{\hbar^2}mE$$

where W is the total energy of the electron and E is the kinetic energy of the electron. Equating the two values given above for k^2 and solving for E yields

$$E = \frac{\hbar^2 \pi^2 n_x{}^2}{2mL^2} \qquad n_x = 1, 2, 3, \ldots$$

This result is extremely important because it shows that the energy of an electron in a metal is quantized; it can have only certain fixed values corresponding to integer values of the quantum number n_x.

In a three-dimensional box, standing waves of the type described above can be set up along each of the three primary axes, resulting in a three-dimensional standing wave. Now if it is assumed that the box is a cube with each side of the length L, then it can be shown that the energy of the resulting three-dimensional standing wave pattern is also quantized and given by the expression

$$E = \frac{\hbar^2 \pi^2}{2mL^2}(n_x{}^2 + n_y{}^2 + n_z{}^2) \qquad \begin{aligned} n_x &= 1, 2, 3, \ldots \\ n_y &= 1, 2, 3, \ldots \\ n_z &= 1, 2, 3, \ldots \end{aligned}$$

The lowest energy state that an electron can occupy in a three-dimensional box occurs when n_x n_y, and n_z all have the value one. Its value is:

$$E = \frac{\hbar^2 \pi^2}{2mL^2}(1^2 + 1^2 + 1^2) = \frac{3\hbar^2 \pi^2}{2mL^2}$$

Following the lowest energy state, there are three states of equal energy. In general, it is common practice to call different states with equal energies *degenerate states*. The present three degenerate states occur when two quantum numbers equal one and the other is two: (112), (121), and (211). The energy of these states is:

$$E = \frac{\hbar^2 \pi^2}{2mL^2}(1^2 + 1^2 + 2^2) = \frac{6\hbar^2 \pi^2}{2mL^2}$$

All of the electronic energy levels in a three-dimensional box with an energy less than $45\hbar^2\pi^2/2mL^2$ are given in Table 3.3. The first column is proportional to the energy of the various states. For simplicity, values of $(n_x{}^2 + n_y{}^2 + n_z{}^2)$ are given in this column. These values are proportional to the energy because

$$(n_x{}^2 + n_y{}^2 + n_z{}^2) = \frac{2mL^2}{\hbar^2 \pi^2} E$$

The second column indicates the quantum number combinations which give the values in column one. Notice that, as the energy rises, the frequency with which more than one combination of quantum numbers is able to yield the same energy increases. The third column lists the degeneracy, or the number of states with the same value of the energy. Only when all the quantum numbers of a given state have the same value [that is, (222)] is it possible to have nondegenerate states: only one state for a given value of the energy. The last column represents the sum of all the energy states up to a given value of the energy; actually a given value of $(n_x{}^2 + n_y{}^2 + n_z{}^2)$. Thus, when $(n_x{}^2 + n_y{}^2 + n_z{}^2)$ is 9 there are seven states with an energy equal to or less than $9\hbar^2\pi^2/2mL^2$: one with an energy equal to $3\hbar^2\pi^2/2mL^2$, three with $6\hbar^2\pi^2/2mL^2$, and three with $9\hbar^2\pi^2/2mL^2$.

In an average-sized piece of metal, the number of electrons is very large. For example, consider a monovalent metal, such as silver, where there is one free electron per atom. Here 107.9 gm (1 mole) of silver will contain Avogadro's number 6.02×10^{23} of free electrons. In the Sommerfeld theory the Pauli exclusion principle is applied, which says that no two electrons can be in the same state, including the spin state. That is, no two electrons can have identical positions, momenta, and spin states. Thus, only two electrons with oppositely directed spins can occupy any given quantum state.

Table 3.3 Low-Lying Energy Levels for an Electron in a Box.

$(n_x^2 + n_y^2 + n_z^2)$ or $\frac{2mL^2}{\pi^2\hbar^2}E$	*Quantum Number Combinations Giving the State*	*Degeneracy or Number of States with Same Energy*	*Total Number of States*
3	(111)	1	1
6	{112}	3	4
9	{122}	3	7
11	{113}	3	10
12	(222)	1	11
14	{123}	6	17
17	{223}	3	20
18	{114}	3	23
19	{133}	3	26
21	{124}	6	32
22	{233}	3	35
24	{224}	3	38
26	{134}	6	44
27	{115} and (333)	4	48
29	{234}	6	54
30	{125}	6	60
32	{334}	3	63
33	{144} and {225}	6	69
35	{135}	6	75
36	{244}	3	78
38	{235} and {116}	9	87
41	{344} and {126}	9	96
42	{145}	6	102
43	{335}	3	105
44	{226}	3	108
45	{245}	6	114

At absolute zero the electrons will seek the lowest energy states in the metal. It can be expected that at this temperature (starting from the state of lowest energy), all the states will be filled with a pair of electrons in each state up to some state which contains the electron with the highest energy.

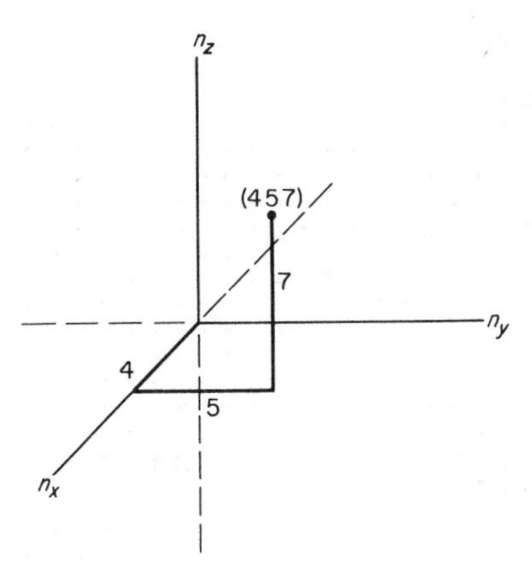

Fig. 3.18 Quantum number space. Since the quantum numbers are positive integers, only the upper forward right-hand octant is used to designate the quantum state of an electron.

As an example, suppose that we have a small piece of metal containing only 120 free electrons. At absolute zero the 120 electrons occupy the 60 lowest energy levels in the metal. All the levels up to and including those with an energy of $30\hbar^2\pi^2/2mL^2$ are filled. This may be confirmed by examining Table 3.3. All energy states with higher energies than $30\hbar^2\pi^2/2mL^2$ will be empty.

In an actual piece of metal, the magnitude of the energy of the electron lying in the highest energy state is very large even at 0° K. This is a result of the very large number of electrons involved and of the large number of quantum states required to hold all of them. We shall now attempt to determine the magnitude of this energy. In order to do this we must first derive a relationship for the number of energy states with an energy equal to or less than that of a given state.

Let us form a three-dimensional cartesian coordinate system, using as coordinates for the three axes values of n_x, n_y, and n_z. Because each energy state of an electron in a box corresponds to a particular set of quantum numbers, a state can be represented by a single point in quantum-number space. Thus, in Fig. 3.18, the point representing the state with the quantum numbers $n_x = 4$, $n_y = 5$, and $n_z = 7$ is specifically designated.

When all of the electronic energy states are considered together, they form a three-dimensional network of points that resembles a simple cubic-space lattice. Because the values of n_x, n_y, and n_z are always positive whole numbers, the network has to lie in the same octant of the coordinate system as that occupied by the given point in Fig. 3.18 (the upper right-hand forward octant.)

To simplify the explanation, the number of coordinates of the metal will be reduced to two. In this case, we need two quantum numbers, n_x and n_y, and the

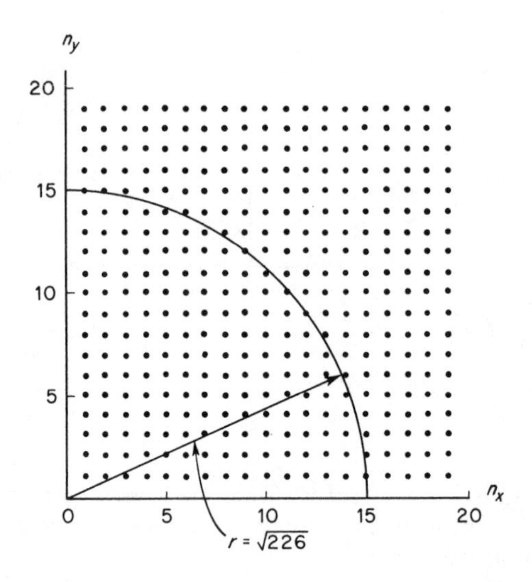

Fig. 3.19 A two-dimensional quantum number space.

three-dimensional network of points representing the energy states is reduced to a two-dimensional square grid of the type shown in Fig. 3.19. Our basic problem is to find how many energy states possess an energy equal to or less than some maximum value E_m. In two dimensions, the energy of any state is given by

$$E = \frac{\hbar^2 \pi^2}{2mL^2}(n_x^2 + n_y^2)$$

This relationship tells us that the energy of a state varies directly as the sum of the squares of the two quantum numbers. A state will have a lower energy than a given state only if its quantum numbers yield a smaller value of $(n_x^2 + n_y^2)$ than the given state. The following simple example will aid in understanding the significance of this last statement.

Suppose that E_m corresponds to a state whose quantum numbers are $n_x = 15$ and $n_y = 1$. Then we have

$$(n_x^2 + n_y^2)_{\text{max}} = (15^2 + 1^2) = 226$$

In Fig. 3.19 a quadrant of a circle has been drawn with a radius

$$r = \sqrt{(n_x^2 + n_y^2)_{\text{max}}} = \sqrt{226}$$

All of the points that lie inside this arc represent quantum states with energies below E_m. Thus, consider the state with quantum numbers $n_x = 11$ and $n_y = 10$. Here $n_x^2 + n_y^2 = 221$ which is smaller than 226. The number of points (states) inside the given area when counted proves to be 163. This same number can be estimated by merely computing the area enclosed by the quadrant of a circle (OAB). This is

$$\text{Area} = \frac{\pi}{4} r_{\text{max}}^2 = \frac{\pi}{4}(226) = 177 \text{ states}$$

This latter value, while differing from the counted value of 163 states, is a fair approximation. The agreement between counted and computed values will be much closer when the total number of states is larger, and for very large numbers of states (corresponding to very large values of E_m) the number of states can be predicted with a high degree of accuracy by this type of computation. Thus, for large values of E_m

$$\eta = \frac{\pi}{4}(n_x^2 + n_y^2)_{\text{max}} = \frac{\pi}{4}\frac{2mL^2}{\hbar^2 \pi^2} E_m$$

where η is the number of states with an energy equal to or less than E_m.

Let us return to a consideration of a metal with three dimensions. Here all of the states with an energy less than E_m lie inside an octant of a sphere, as defined in Fig. 3.20. For large values of E_m by analogy with the two-dimensional case,

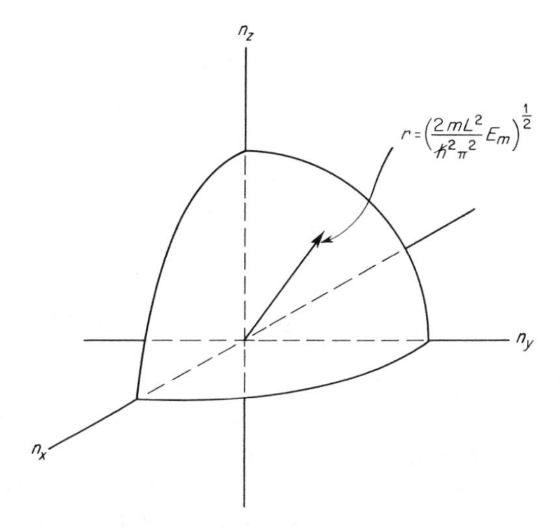

Fig. 3.20 At absolute zero, the occupied states may be assumed to lie within this octant of a sphere in quantum number space.

we may take the volume of this octant as a measure of the number of quantum states with an energy equal to or less than E_m. The radius of the sphere is:

$$r_{max} = (n_x^2 + n_y^2 + n_z^2)^{1/2} = \left(\frac{2mL^2}{\hbar^2 \pi^2}\right)^{1/2}$$

and therefore we have

$$\eta = \text{volume of octant} = \frac{1}{8}\frac{4\pi}{3} r_{max}^3 = \frac{\pi}{6}\left(\frac{2mL^2}{\hbar^2 \pi^2} E_m\right)^{3/2}$$

where η is the number of states with an energy equal to or less than E_m.

If N is the number of electrons in a piece of metal, then at absolute zero there should be $\frac{N}{2}$ filled quantum states, or

$$\frac{N}{2} = \frac{\pi}{6}\left(\frac{2mL^2}{\pi^2 \hbar^2} E_F\right)^{3/2}$$

In this expression, E_F, known as the *Fermi energy,* is the maximum electronic energy at absolute zero. Solving the above relation for the Fermi energy gives us:

$$E_F = \frac{\hbar^2 \pi^2}{2mL^2}\left(\frac{3N}{\pi}\right)^{2/3}$$

or

$$E_F = \frac{\hbar^2 \pi^2}{2m}\left(\frac{3N}{\pi L^3}\right)^{2/3} = \frac{\hbar^2 \pi^2}{2m}\left(\frac{3}{\pi}\frac{N}{V}\right)^{2/3}$$

where $V = L^3$ is the volume of the assumed metal box. This last result is of interest because it shows that the Fermi energy depends only on the number of electrons per unit volume in the metal.

With the aid of the above equation, it is possible to estimate the Fermi energy of a metal. As an example, consider silver which has a gram molecular volume of 10.28 cm^3. Since silver is monovalent, a gram mole should contain 6.02×10^{23} free electrons. In addition, $\hbar$ is 1.06×10^{-27} erg sec, and m, the mass of the electron, is 9.11×10^{-28} gm. Substituting these values in the equation gives

$$E_F = \frac{(1.06 \times 10^{-27})^2 \pi^2}{2(9.11 \times 10^{-28})} \left(\frac{3}{\pi} \frac{6.02 \times 10^{23}}{10.28} \right)^{2/3}$$

$$E_F = 8.9 \times 10^{-12} \text{ ergs} = 5.5 \text{ electron volts}$$

This energy is perhaps better expressed in calories per mole of electrons (at the Fermi level) because then we are able to compare it with atomic thermal energies. Thus,

$$E_F = \frac{8.9 \times 10^{-12} \times 6.02 \times 10^{23}}{4.19 \times 10^7} = 130{,}000 \text{ cal/mole}$$

This energy is extremely high. It may be shown that the average electron in a metal at absolute zero possesses an energy which is $\frac{3}{5}$ of the Fermi value, so that in silver

$$E_{av} = \tfrac{3}{5}(130{,}000) = 78{,}000 \text{ cal/mole}$$

As mentioned earlier we can think of the electrons in a metal as a gas. The thermal energy (kinetic) of a mole of a monatomic gas, for example, helium, according to classical theory, is of the order of $\frac{3}{2}RT$ cal per mole. Since R is approximately 2 cal-per-mole degree, at room temperature ($T = 300°$ K) the thermal energy of such a gas should be about 900 cal per mole. It can be seen then that the atoms of an ordinary gas possess at room temperature only about $\frac{1}{80}$ of the mean kinetic energy of the electrons in a metal at absolute zero.

The Sommerfeld theory states that the electrons in a metal move with extremely high velocities and that they occupy definite energy levels with two electrons of opposite spins in each state. At absolute zero the electrons fill all of the lowest lying states up to a level known as the *Fermi level.* These statements are illustrated in Fig. 3.21. In this figure, V_s represents the difference in potential between a point inside the metal and a point outside the metal in a vacuum. In our previous derivation, this potential difference was assumed to be infinite, but in an actual metal it is finite. The simplifying assumption that we made, however, does not affect our final results. Assuming that the potential in the metal is zero, then the space around the metal lies at a potential that is higher by an amount V_s. The volume of the metal can therefore be considered to be in a potential well. The electronic energy states can be thought of as lying at various distances above the bottom of the well. In entering these states, the electrons fill each state before the next one is started. The highest energy state at absolute zero, the Fermi level, is designated in the figure as E_F. The average energy lies at $\frac{3}{5}E_F$. Finally, the symbol ϕ represents the energy required to

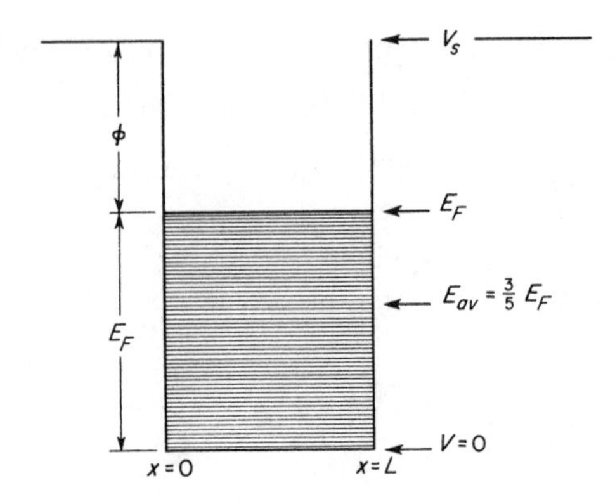

Fig. 3.21 Energy relationships according to the Sommerfeld theory.

remove an electron from the Fermi level and take it out of the metal. This is called the *work function* and is equal to the energy which is normally measured when an electron is removed from the surface of a metal (when extrapolated to 0°K). This term "work function" was introduced by Einstein in his paper on the photoelectric effect. Notice that at 0 °K an electron in the Fermi level possesses the highest energy of all the electrons in the metal and is the easiest to remove. The experimentally determined value of the work function for the metal silver is 4.46 electron volts. The Fermi energy of silver is about 5.5 electron volts. Thus, the difference in potential V_s between a point inside and a point outside the metal is about 10 electron volts.

3.18 The Density of States. It is frequently convenient, in thinking of the energy states in a metal, to use the concept of the density of states. For any given value of the energy E, the number of states with an energy equal to or less than this value is

$$\eta = \frac{\pi}{6}\left(\frac{2mL^2}{\hbar^2 \pi^2} E\right)^{3/2}$$

Because the number of states in a metal is very large, we may consider that both η and E are continuous functions in spite of the fact that they vary by discrete steps. With this assumption, we shall take the derivative of η with respect to E, which gives us a new function $N(E)$ where

$$N(E) = \frac{d\eta}{dE} = \frac{\pi}{4}\left(\frac{2mL^2}{\hbar^2 \pi^2}\right)^{3/2} E^{1/2}$$

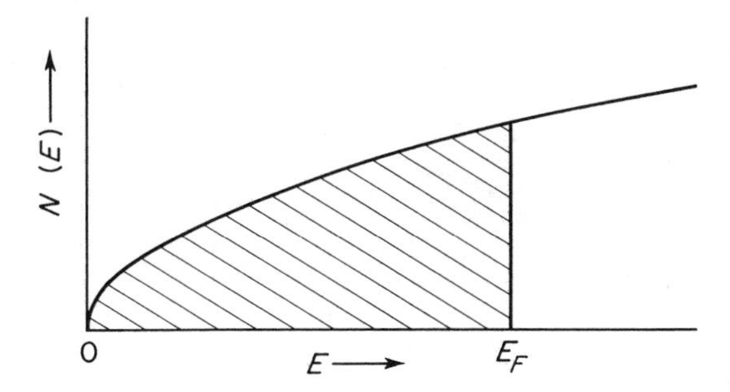

Fig. 3.22 The density of states as a function of the energy at 0° K according to the free-electron theory.

Here $N(E)$ is the density of energy states at a given energy level; the number of energy states with an energy between E and $E + dE$ is $N(E)\,dE$. Figure 3.22 shows $N(E)$ plotted as a function of the energy E. The resulting curve is a parabola. The area under this parabola, starting at the origin and running to a given value of the energy E, represents the number of states with an energy equal to or less than E. The cross-hatched area in Fig. 3.22, accordingly, represents the number of states with an energy equal to or less than the Fermi energy. By definition of the Fermi energy, this area measures the number of states which are occupied by the electrons at absolute zero. Thus, Fig. 3.22 may be considered as showing the nature of the distribution of the electrons in the various energy levels at 0° K according to the Sommerfeld theory.

3.19 The Effect of Temperature on the Distribution of Electrons in the Energy Levels. Figure 3.22 can be taken to represent the distribution of electrons in the various energy levels at 0 °K. The question is how is this distribution affected by an increase in temperature. In general, a rise in temperature signifies that the atoms of a solid vibrate with larger amplitudes. The energy associated with these enhanced vibrations can be passed back and forth between the free electrons and vibrating atoms. The magnitude of the energy that can be given to an average electron by a vibrating atom is approximately kT(RT per mole of electrons), where k is Boltzmann's constant 1.38×10^{-16} ergs/°K. At room temperature, kT is about 4×10^{-14} ergs or $\frac{1}{40}$ eV. However, we have seen that in a typical metal (silver) E_F is about 9×10^{-12} ergs or about 5 eV. Thus, thermal vibrations are able to change the energy of electrons by values of the order of $\frac{1}{200}$ of the energy of the Fermi level. This is a significant fact when considered in conjunction with the Pauli exclusion principle. The latter can be interpreted as prohibiting the movement of an electron from one state into another that is already filled. Thus, thermal energy cannot take an electron in one of the lowest states and raise it into

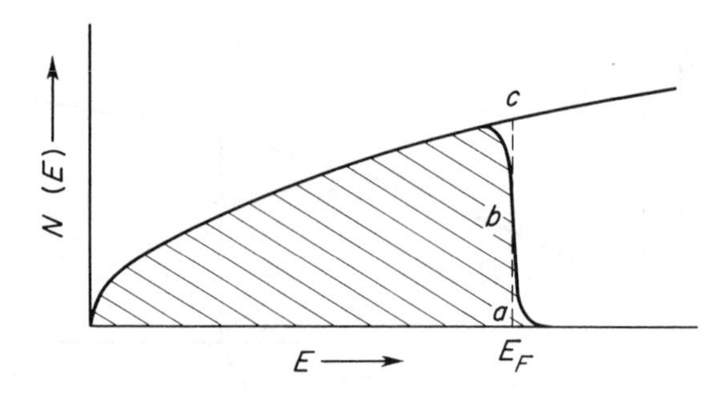

Fig. 3.23 Thermal energy, at any normal temperature, only slightly changes the distribution of the electrons in the energy states.

another level which is already occupied. Because of the small size of kT relative to E_F, this means that only the electrons lying in energy levels close to the Fermi level can be affected by the thermal energy of the atoms. The distribution in the energy states at any ordinary temperature (T) may be assumed, according to the Fermi-Dirac statistics, to be as shown in Fig. 3.23. Here it can be observed that some states above E_F are occupied and a corresponding number below E_F are unoccupied. This, of course, is a statistical distribution, with the electrons moving in and out of the states lying near E_F. The change in the distribution of the electrons in the energy levels is only appreciable over a range of several values of kT above and below E_F.

The Fermi energy is more properly defined in terms of Fig. 3.23. We have previously defined it as the maximum energy of an electron at absolute zero. More properly it should be the energy of the level, at any temperature, which has a 50 percent chance of being occupied by an electron. Thus, in Fig. 3.23 the Fermi energy corresponds to the energy where the actual density of states (line *ab*) has a value equal to one-half of the maximum density of states (line *ac*).

3.20 Electronic Specific Heat. Because only a relatively few electrons are affected by thermal energy, the electrons make only a small contribution to the specific heat of a metal. Consequently, as a metal is heated, the majority of the energy expended in raising the temperature of the metal goes to increasing the vibrational energy of the atoms. In fact, the specific heat of metals may be computed with reasonable accuracy on the assumption that the specific heat depends only on atomic vibrations. The fact that the free-electron theory is able to explain why the electronic contribution to the specific heat is so small is one of its greatest successes. The problem could not be solved by the Boltzmann statistics, which implied the equipartition of energy, or that the energy was approximately uniformly distributed to the amount of $\frac{1}{2}kT$ per degree of freedom per particle.

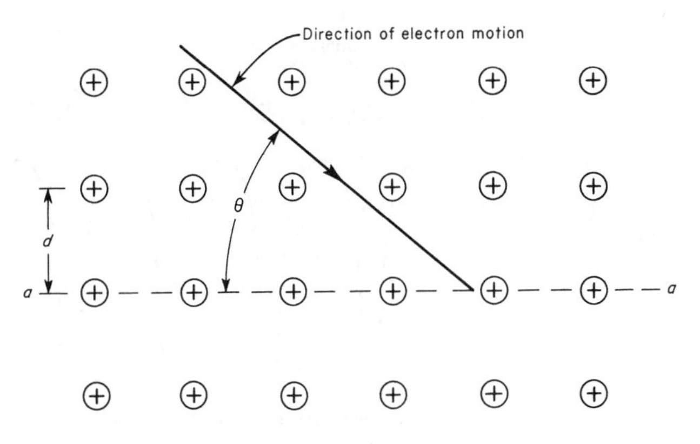

Fig. 3.24 Electrons cannot move through the lattice when Bragg's law is satisfied; $n\lambda = 2d \sin \theta$.

3.21 The Zone Theory. In the simple free-electron theory, the electrons are assumed to move in a region of constant potential. Actually, however, inside any real metal there is a periodic arrangement of positively charged ions through which the electrons must move (Fig. 3.24). Consequently, the potential along a line such as *aa*, which passes through the centers of the ions, must vary in the fashion of Fig. 3.25. A wave mechanical treatment of the motion of electrons in a periodic potential was first carried out by Bloch.[14] Since this analysis is too complicated for us to consider, we shall concern ourselves only with certain important conclusions of this analysis. One of the most significant of these is that the movement of electrons in a crystal cannot occur under conditions which satisfy the Bragg diffraction law,

$$n\lambda = 2d \sin \theta$$

where n is an integer, λ is the wavelength of the electrons, d is the spacing of a given lattice plane, and θ is the angle of incidence of the electrons relative to the given lattice plane. Thus, if in Fig. 3.24 *aa* represents a lattice plane and θ is the angle between the direction of motion of the electrons and this plane, an electron is incapable of moving through the crystal in the indicated direction when $\lambda = 2d \sin \theta / n$.

The fact that electrons cannot move inside a crystal when they satisfy the Bragg law is not surprising. The Bragg relation merely tells us the conditions under which the planes of a crystal reflect electrons. Thus, electrons moving in specific directions and with certain energies are capable of being reflected from their original direction of motion. In general, this may be interpreted as

14. Bloch, F., *Zeits. fur Physik*, **52** 555 (1928); **59** 208 (1930).

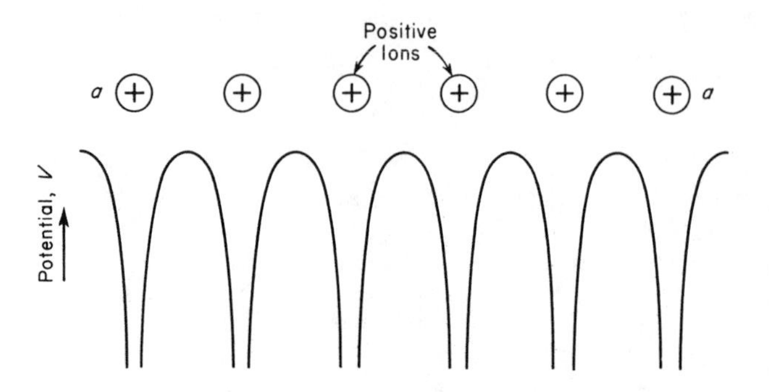

Fig. 3.25 Periodic potential inside a crystal.

signifying that the electrons are not capable of possessing those energies and directions of movement that produce internal reflections. The Bragg law, therefore, has an important effect on the energy states that an electron can possess in a metal. It is our present purpose to develop these relations, but first the Bragg law itself must be considered in greater detail.

The wavelength of an electron is related to its momentum by the equation

$$\lambda = \frac{h}{mv} = \frac{h}{p} = \frac{2\pi\hbar}{p}$$

or, on substituting the momentum for the wavelength in Bragg's law, we have

$$p = \frac{n\hbar\pi}{d \sin \theta}$$

This statement of Bragg's law tells us that reflections of electrons will occur at certain combinations of a direction of motion and magnitude of the momentum of the electron, that is, at specific values of the momentum vector of an electron. It is customary in wave mechanics to use the wave number k instead of the momentum p in describing the results of the present type of work; the two vectors differing only by a constant $1/\hbar$, or

$$k = \frac{p}{\hbar}$$

When expressed in terms of the wave number k, the Bragg relation becomes

$$k = \frac{n\pi}{d \sin \theta}$$

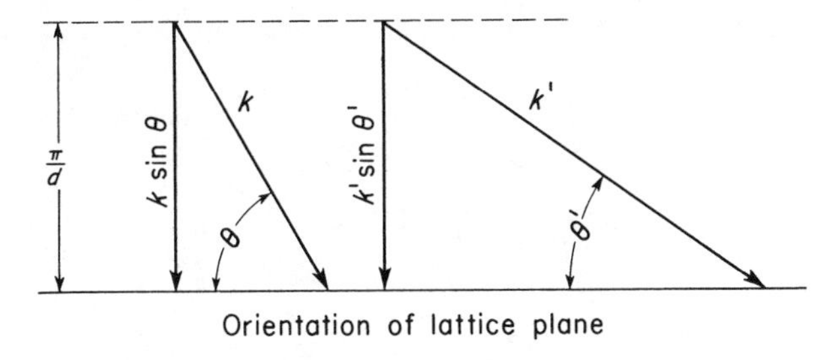

Fig. 3.26 The first Bragg reflections occur when $k \sin \theta = \pi/d$, or when the component of k normal to the reflection plane is a constant (π/d). In the above figure, two different values of k which satisfy the Bragg condition are shown.

We emphasize that k is a vector whose direction is the same as that in which the electron moves, and whose magnitude is directly proportional to the momentum of the electron. Rearranging the above equation gives us

$$k \sin \theta = \frac{n\pi}{d}$$

The quantity $k \sin \theta$, however, is the component of the wave-number vector normal to the Bragg reflecting plane, as shown in Fig. 3.26 for $n = 1$ (first-order reflection). The above expression tells us that, whenever the component of k resolved normal to a lattice plane equals $n\pi/d$, Bragg's condition will be satisfied. Since the critical values of this normal component of the wave number k vary inversely with the interplanar spacing d, it is evident that the first planes to have an influence on the motion of the electrons will be those with the largest values of d: planes with the lowest Miller indices.

Thus, consider as an example a simple cubic lattice of two dimensions, as shown in Fig. 3.27. The Bragg condition is first satisfied by the {100} planes of this lattice. Electrons occupying very low-lying energy states, with correspondingly very small values of k, will, in general, be able to move in any direction through the crystal without reflection. However, as we consider electrons in higher and higher energy states, a point is eventually reached where the Bragg condition with respect to {100} planes influences the motion of the electrons. After this, the next set of planes for which the Bragg condition will be satisfied is {110}, the second most widely spaced planes in the simple cubic structure. The reflections from {100} planes will now be considered in more detail. Figure 3.28A represents a coordinate system which will be called k space. Any point, such as m in this figure, corresponds to a specific wave-number vector. Figure 3.28B shows that the vector in question may be obtained by drawing an arrow from the origin of coordinates to the point in question, and that if we call this

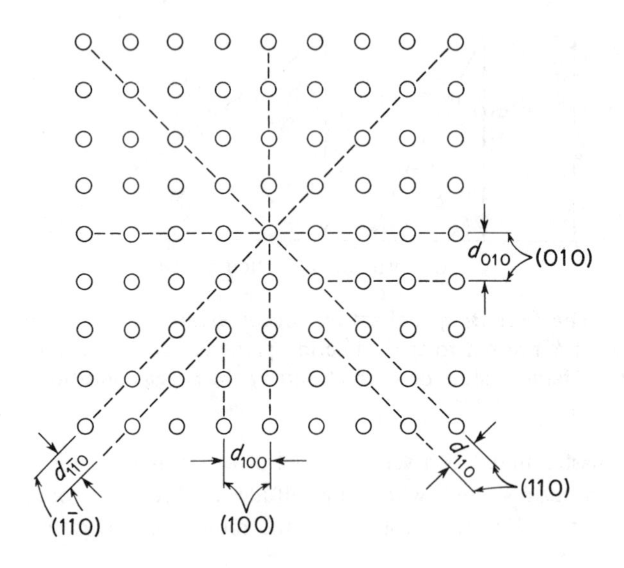

Fig. 3.27 A simple cubic lattice.

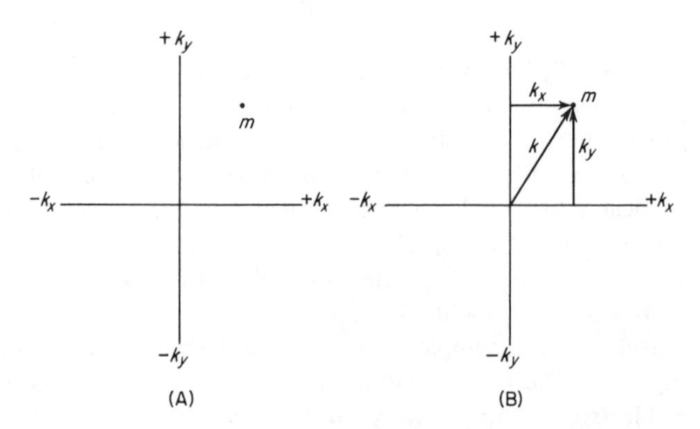

Fig. 3.28 Two-dimensional wave-number space. Point m in (A) represents the vector k in (B).

vector k then its x component will be k_x and its y component will be k_y. In a two-dimensional lattice such as Fig. 3.27, {100} planes – (010) horizontal and (100) vertical. The Bragg condition is satisfied for these two types of planes by all the k vectors corresponding to the line $ABCD$ in Fig. 3.29. In this respect, consider the arbitrary vector k' that runs from the origin of the k space coordinate system to a point on line AB. This vector makes an angle θ' with the horizontal and thus with the (010) planes in Fig. 3.27 as well. According to the drawing, the vertical component of this vector, which is normal to the (010)

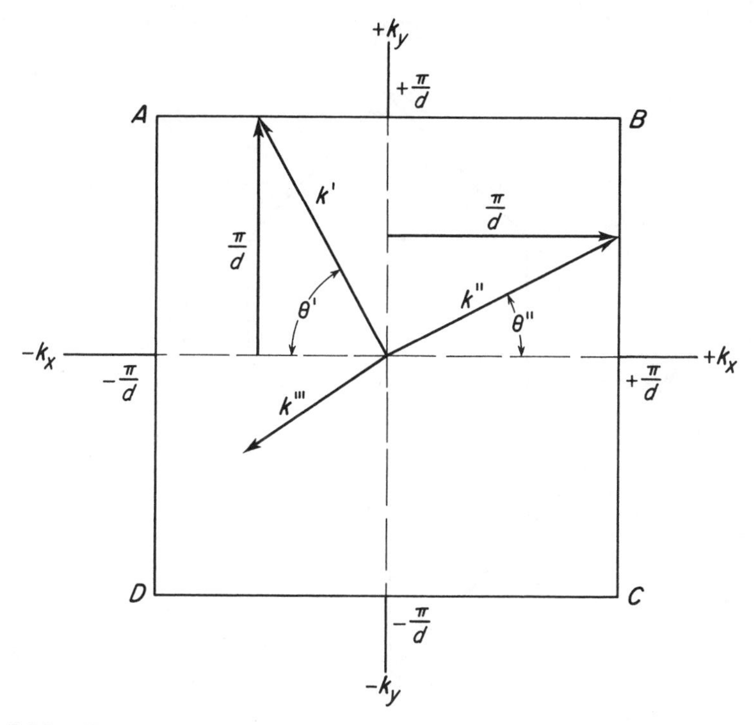

Fig. 3.29 The first Brillouin zone of a two-dimensional simple cubic lattice.

planes, is π/d. The vector k', thus satisfies the Bragg condition with respect to (010) planes and the same may be said for all values of k which fall on line AB and on line CD as well. Also, it can be seen that the vector k'', which ends on line BC, satisfies the Bragg condition for (100) planes. Lines AB and DC, accordingly, represent k vectors for reflections from horizontal planes, and lines BC and AD those for vertical planes. All vectors such as k' and k'' that touch $ABCD$ represent wave numbers which satisfy the Bragg condition. On the other hand, any vector such as k''' which does not extend to the boundary $ABCD$ represents a motion of an electron which does not suffer reflection. The area in wave-number space enclosed by the line $ABCD$ is known as a *Brillouin zone.* In the case of the two-dimensional simple cubic lattice presently being considered, $ABCD$ is the first zone. The second is formed by reflections from $\{110\}$ planes, and is shown in conjunction with the first zone in Fig. 3.30. The outer limits of the second zone are given by the lines connecting points E, F, G, and H. The zone itself is represented by the four crosshatched triangular areas. An interesting property of Brillouin zones is that the different zones for a given crystal all have the same area. Thus the areas of zone one and zone two in

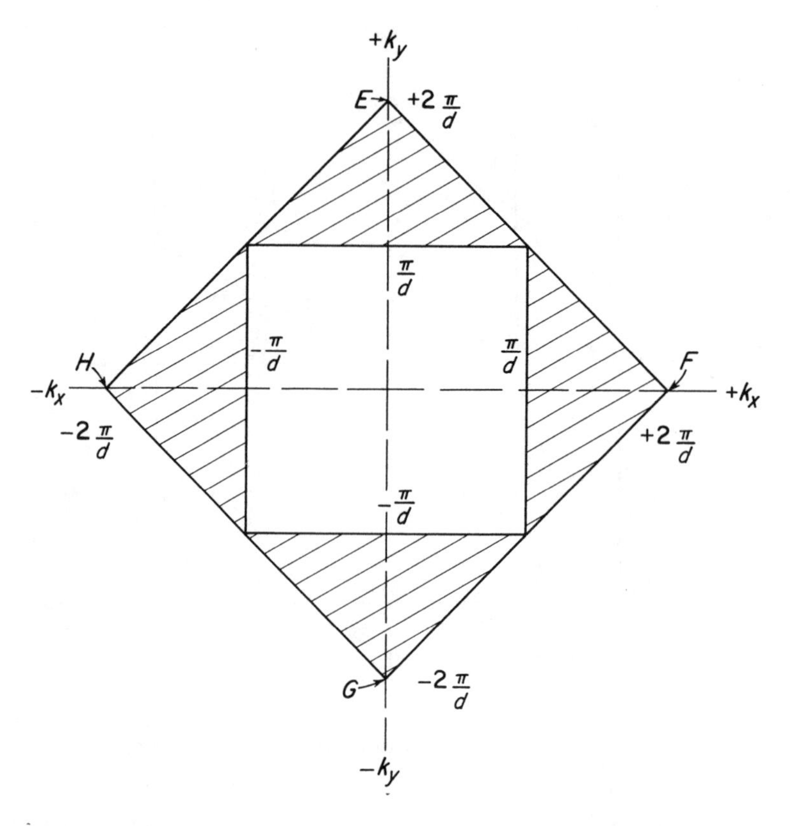

Fig. 3.30 The first and second Brillouin zones of a simple cubic lattice. Second zone is crosshatched.

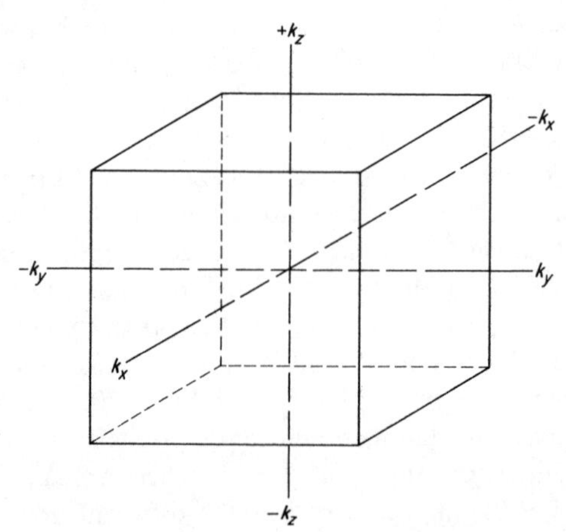

Fig. 3.31 First Brillouin zone of a three-dimensional simple cubic lattice.

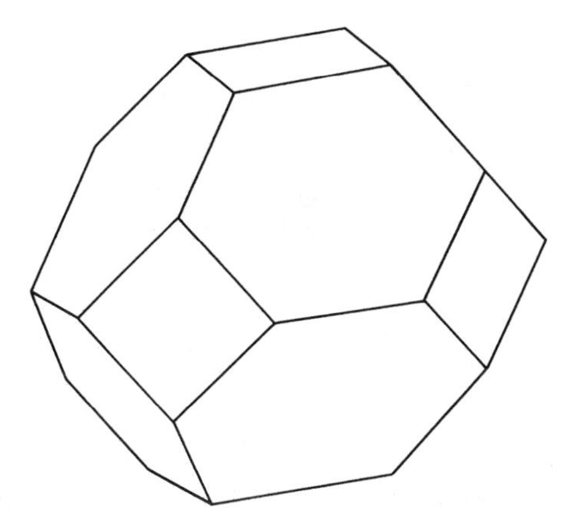

Fig. 3.32 The first Brillouin zone of the face-centered lattice. Surfaces are parallel to either {100} or {111}.

Fig. 3.30 are equal. The second zone has the same significance as the first. Any k vector which falls on the boundary of this zone corresponds to a reflection of electrons and an impossible type of motion of the electrons.

In three dimensions, the first Brillouin zone of a simple cubic lattice is a cube (Fig. 3.31). The faces of this cube are parallel to cube planes {100} and represent critical values of the wave-number vector in three dimensions. The Brillouin zones of the regular lattice forms, in which metals normally crystallize, can be readily deduced in the same manner as that for the simple cubic lattice. In the face-centered cubic system, for example, it turns out that in certain directions reflections occur first from {200} planes, and in other directions from {111} planes. The first Brillouin zone of this lattice, which appears in Fig. 3.32 is therefore formed from surfaces parallel to both of these types of planes.

3.22 The Dependence of the Energy on the Wave Number. The energy of an electron, according to the Sommerfeld theory, is related to the wave number k of the electron by the equation

$$E = \frac{\hbar^2 k^2}{2m}$$

where E is the energy of the electron, m is its mass, k is its wave number, and $\hbar$ is Planck's constant.

Now let us confine our attention to electrons traveling in only one direction, say parallel to the x axis and therefore normal to (100) planes. The wave number

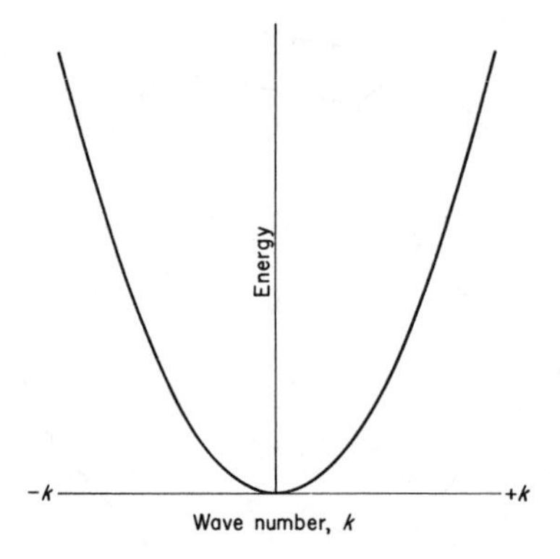

Fig. 3.33 The energy of an electron (moving in a single direction) plotted as a function of its wave number according to the free-electron theory.

of such electrons is represented by the symbol k_x and we have

$$E = \frac{\hbar^2 k_x{}^2}{2m}$$

This parabolic relation between the energy and the wave number is shown diagrammatically in Fig. 3.33 and is based on the Sommerfeld assumption that the electrons move in a constant potential. When we consider the movement of electrons in a periodic potential, Fig. 3.33 still represents a good approximation of the energy-wave-number relationship over most of the energy spectrum. However, as we have just seen, at certain critical values of the wave number the normal motion of the electron is disrupted, which has the effect of introducing discontinuities into the energy-wave-number curve. Thus, for electrons traveling normal to (100) planes, we have

$$k_x = \pm \frac{n\pi}{d \sin \theta}$$

but here θ is 90° and therefore $\sin \theta = 1$, so that $k_x = n\pi/d$, where n is an integer and d is the interplanar spacing of the (100) planes. The discontinuities in the energy-wave number relationship therefore occur at intervals of π/d along the k_x axis, as is shown in Fig. 3.34. Notice that as k approaches a multiple of π/d, the energy deviates from the simple parabola of Fig. 3.33. In particular, consider a specific discontinuity such as that at π/d. Here, as k_x approaches the critical value from either direction, the energy begins to change more and more slowly, so that the curve representing the energy state tends to turn and run parallel to the k_x or wave-number axis. At the specific value π/d, the energy has two values

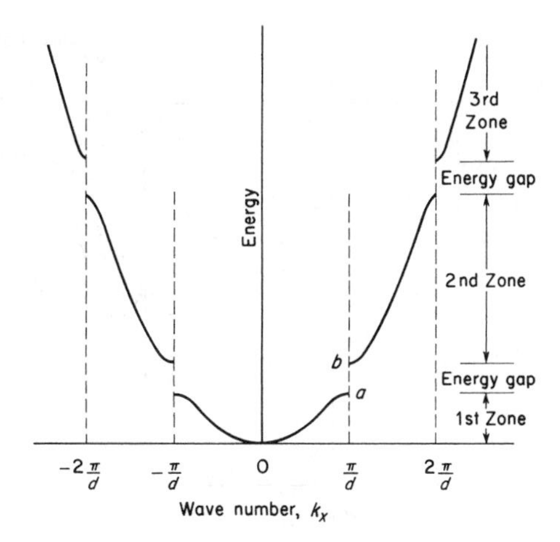

Fig. 3.34 Variation of the energy with wave number for electrons moving in a single direction according to the zone theory.

represented by points *a* and *b* respectively in Fig. 3.34. Electrons traveling normal to (100) planes cannot have an energy that falls in the interval between *a* and *b*.

Let us now refer to another diagram showing the first Brillouin zone of a two-dimensional simple cubic lattice – Fig. 3.35A. In the latter diagram, two small dots marked *a* and *b* have been placed along the k_x axis. These points correspond to similar points in Fig. 3.34 and are intended to show that when the boundary of a Brillouin zone is crossed, a discontinuity in the energy levels occurs. Such discontinuities occur no matter where the boundary of a Brillouin zone is crossed. However, the position of the energy gap, as seen in the plot of energy as a function of wave number, varies with the direction of motion of the electron in the crystal, and it therefore varies with the point where the zone boundary is crossed. This is because the absolute magnitude of the critical wave number changes as we move along the zone boundary. Thus, in Fig. 3.35A a critical wave-number vector such as k_1, which is parallel to one of the principal axes of the crystal, ⟨100⟩, corresponds to the smallest value of the wave number at which the electrons will be reflected in the simple two-dimensional cubic lattice under consideration. All other critical wave numbers for the first zone are larger than k_1 as, for example, k_2 and k_3, which correspond to electrons moving in other directions than normal to a {100} plane. In fact, it is easily seen that k_3 represents the largest critical wave number for a first Brillouin-zone reflection. The critical wave number, therefore, increases from a minimum value at the center of one of the sides of the two-dimensional zone to a maximum value at the corners of the zone. Corresponding to this change in the

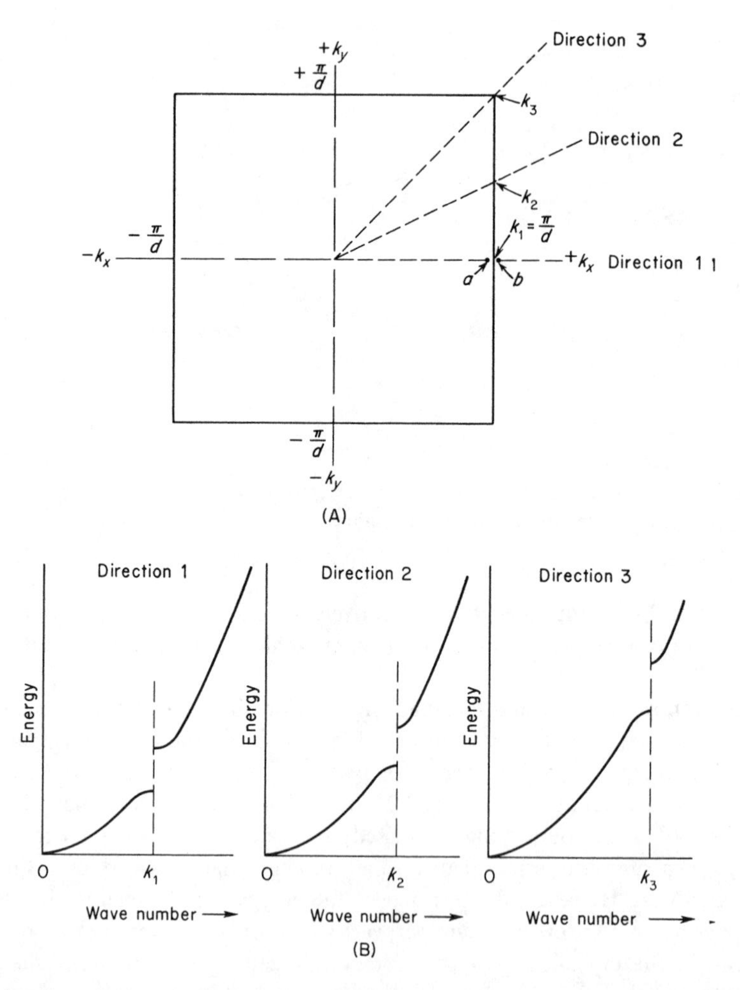

Fig. 3.35 The energy gap varies with the direction of motion of the electrons. (Notice that only the positive half of the energy-wave number parabola is shown.)

value of the wave number along the boundary of a zone is a shift in the position of the energy gap, as is shown in Fig. 3.35B where the energy as a function of wave number is plotted for the three indicated positions where the zone boundary is crossed. Notice, however, that only half of the energy-wave-number curve is given in each instance. These curves clearly show how the position of the energy gap changes as the direction of motion of the electron is varied. The energy gaps occur at the lowest values of the energy when the electrons move in a direction normal to {100} planes, and at the highest values when the electrons travel at 45° to these planes.

As previously mentioned, the Brillouin zone of a simple cubic lattice in three dimensions is a cube. Here, too, it can be demonstrated that the energy gaps will occur at the smallest values of k when the electrons move in a ⟨100⟩ direction. The maximum values of k for crossing the first Brillouin zone in this case, however, occur at the corners of the cube which, in three dimensions, correspond to electrons moving in ⟨111⟩ directions

3.23 The Density of States Curve. Because the boundaries of a Brillouin zone represent a gap in the allowed energy levels for all directions of motion of the electrons, it can be concluded that the number of energy levels inside a given zone is finite or bounded. Actually, it can be shown[15] that the number of energy states in either a body-centered or a face-centered cubic metal is exactly equal to the number of atoms in the metal. By the Pauli principle, each zone of these cubic metals should, therefore, be capable of containing twice as many electrons as there are atoms in the metal. In monovalent metals, for example, gold, silver, and copper, there are, accordingly, just enough valence electrons to fill half of the first zone.

As a result of the fact that there are a definite number of energy levels in a given zone and of the fact that there are energy gaps between zones, the simple parabolic relationship between the density of states in a metal and the energy of the states in a metal cannot hold when the existence of the zones is considered. The parabolic law which is predicted by the Sommerfeld theory is shown in Fig. 3.36A. The corresponding curve, according to the zone theory, is shown in Fig. 3.36B. The latter curve refers specifically only to the first zone of a simple two-dimensional cubic lattice, but it may be taken as representative of the nature of the shape of the density of states curve as a function of energy for other more complex crystals. The area under this curve represents the total number of energy states in the entire first zone. This curve has several points of interest. First, for small values of the energy, the zone-theory curve is identical with the parabolic relationship of the free-electron theory. Second, as the energy of the states is increased, a point is eventually reached where the electrons traveling in ⟨100⟩ directions begin to suffer Bragg reflections from {100} planes. At this time, the density of the states increases and the curve climbs above the simple parabolic relationship. For further increases in the energy, however, the density of the states decreases and eventually falls to zero. This last effect is due to the fact that, as the energy is made higher and higher, the number of states remaining in the zone becomes fewer and fewer.

15. Raynor, G. V., *An Introduction to the Electron Theory of Metals*. The Institute of Metals, London, 1947.

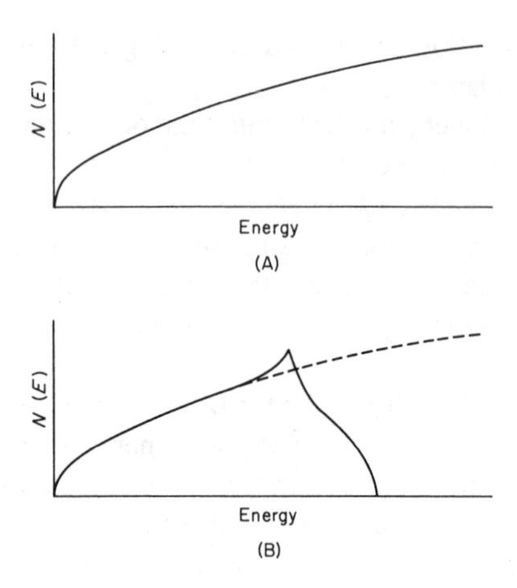

Fig. 3.36 The density of states as a function of the energy (A) according to the free electron theory, (B) the zone theory.

3.24 Application of the Zone Theory to Alloy Phases. Brass is an alloy of copper and zinc. Most commercial brasses contain less than 50 atomic percent zinc, and if they contain less than about 45 percent zinc, the alloy is face-centered cubic. However, at about 50 atomic percent zinc, a new crystal structure is formed. This is body-centered cubic and is designated the beta phase. Since copper crystallizes in the face-centered cubic structure and zinc in the close-packed hexagonal, the beta structure is different from that of either pure metal. Other important intermediate crystal structures are also observed in the copper-zinc alloy system. These are the gamma (γ) phase, whose composition range is centered near 62 atomic percent zinc, and the epsilon (ϵ) phase, which is stable for compositions near 75 atomic percent zinc. The gamma crystal structure is cubic, but it is very complex and contains 52 atoms per unit cell. The epsilon structure is close-packed hexagonal.

Similar intermediate crystal structures or phases are found when copper is alloyed with a variety of other metals. They are also observed in many other alloy systems where one of the components is a face-centered cubic metal. This is particularly true of the alloys of gold and silver. Table 3.4 lists most of the alloy systems in which the brass-type crystal structures are observed.

Now consider the beta-brass structure. In 1926, Hume-Rothery[16] pointed out that the compositions of the beta phase in many alloys correspond to a ratio of valence electrons to atoms of 1.5. Thus in the copper-zinc system, the body-centered cubic phase is observed when one copper atom containing one valence electron is combined with one zinc atom containing two valence

16. Hume-Rothery, W., J. Inst. Metals, **35** 309 (1926).

electrons; a ratio of 3 valence electrons per pair of atoms in the alloy. Similarly, the body-centered cubic structure is obtained near the stoichiometric formula Cu_3Al in the aluminum-copper system where the respective valences are one and three. Here six valence electrons correspond to four atoms. Finally, in the copper-tin system the beta phase corresponds to Cu_5Sn, or to a ratio of 6 atoms to 9 primary valence electrons, taking the valence of tin as four.

Table 3.4 Some Electron Compounds.*

Beta Brass Type (Electron to Atoms ≈ 3/2)	*Gamma Brass Type (Electron to Atoms ≈ 21/13)*	*Epsilon Brass Type (Electron to Atoms ≈ 7/4)*
Ag–Al	Ag–Cd	Ag–Al
Ag–Cd	Ag–Hg	Ag–Cd
Ag–In	Ag–In	Ag–Zn
Ag–Mg	Ag–Zn	Au–Al
Ag–Zn	Au–Cd	Au–Cd
Al–Co	Au–In	Au–Sn
Al–Fe	Au–Zn	Au–Zn
Al–Ni	Cu–Al	Cu–Cd
Au–Cd	Cu–Cd	Cu–Si
Au–Mg	Cu–Ga	Cu–Sn
Au–Zn	Cu–Hg	
Cu–Al	Cu–In	
Cu–Be	Cu–Si	
Cu–Ga	Cu–Sn	
Cu–In	Ni–Be	
Cu–Si	Ni–Cd	
Cu–Sn	Ni–Zn	
Ni–In	Pb–Na	
Pd–In	Pd–Zn	
	Pt–Be	
	Pt–Zn	

* For a more complete list, see Massalski, T. B., ASM Seminar, *Theory of Alloy Phases*, p. 70. Am. Soc. for Metals, Cleveland, Ohio (1956).

In a similar manner, it may be shown that the gamma brass (complex cubic) type of structure corresponds to a ratio of about 21 electrons to 13 atoms, and the epsilon brass (close-packed hexagonal) structure is obtained when the ratio of electrons to atoms is nearly 7/4. Because of this close correspondence of the structures to the ratio of valence electrons to atoms, these intermediate crystal phases are commonly called *electron compounds.*

A rational understanding was first given to the electron compounds by Jones[17]. In brief, his analysis of the transformation from the alpha (f.c.c.) to the beta (b.c.c.) structures in brass (Cu-Zn system) is based on the fact that the

17. Jones, H., Proc. Phys. Soc., **49** 250 (1937).

density of energy states as a function of the energy level is not the same in the face-centered cubic and body-centered cubic crystal structure, and is as follows. First, consider the face-centered cubic crystal density of states curve shown schematically in Fig. 3.37A. In pure copper with a single valence electron, only half of the total number of energy levels (two per atom) will be occupied.

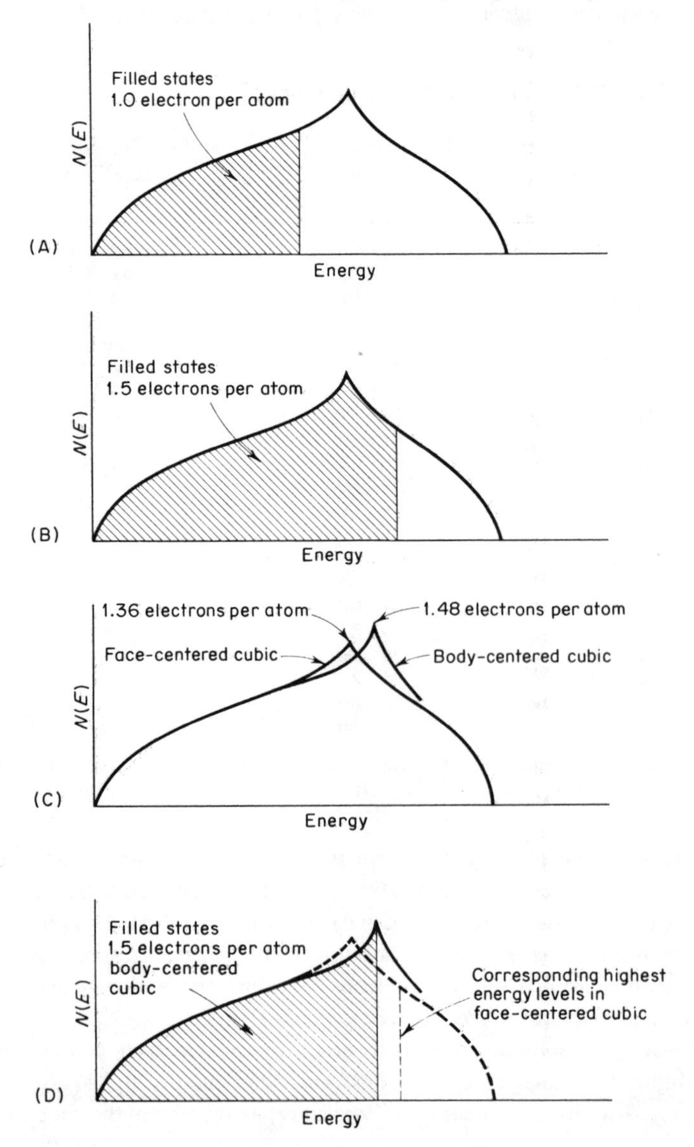

Fig. 3.37 The role of the zones in the shift in crystal structure from face-centered cubic to body-centered cubic in a metal such as brass.

This is indicated by the cross-hatched area in this diagram. Alloying copper with bivalent zinc raises the number of electrons per atom and causes a larger fraction of the available energy levels in the first zone to be filled. When the electron to atom ratio is around 1.5, the filled zones will occupy a number of states roughly like that shown in Fig. 3.37B. At an electron to atom ratio of 1.38, the most energetic electrons begin to be reflected from {100} planes. The peak in the curve, therefore, corresponds approximately to this ratio.[18] For ratios above 1.38, the energy of the electrons tends to rise rapidly because the density of available states falls rapidly above the peak. In other words, the smaller the density of the states, the more rapidly they will be filled up with additional electrons, forcing further electrons rapidly to increase their energies.

Now consider the density of states curve for the body-centered cubic structure shown schematically in Fig. 3.37C, where it is superimposed over the corresponding curve for the face-centered cubic structure. That these curves are not identical is because the first Brillouin zone for the body-centered cubic crystal is not the same as that for the face-centered cubic crystal. In the body-centered cubic structure, the first Bragg reflections of the electrons occur from the six {110} planes, and the first Brillouin zone is therefore a 12-sided polyhedron. This difference in the shapes of the Brillouin zones shifts the peak on the density of states curve to a higher energy in the body-centered cubic case. This means that the peak on the density of states curve appears at a higher energy in the case of the body-centered cubic crystal than it does for the face-centered cubic crystal. This peak corresponds approximately[19] to a ratio of electrons to atoms of 1.48.

The significant feature of Fig. 3.37C is that when the energy levels in the face-centered cubic zone are filled to the extent shown in Fig. 3.37B, a corresponding filling of the states in the body-centered cubic zone (shown in Fig. 3.37D) results in a lowering of the energy level of the electrons in the highest energy levels. This means that the body-centered cubic structure should become energetically more favorable at an electron to atom ratio of about 1.5.

Actually, the zone theory shows that the electron to atom ratio, where the body-centered cubic structure becomes the more stable structure, is not exactly 1.5. The fact that empirically this phase was observed at a composition near CuZn led the original workers in this field to assume that the ratio of 3/2 was the significant feature of this phase.

The other two phases, gamma and epsilon, can also be explained by the zone theory, and for further reading in this area, a review paper by Massalski[20] is recommended.

18. Raynor, G. V., *An Introduction to the Electron Theory of Metals*, p. 55. The Institute of Metals, London (1947).

19. Raynor, G. V., *An Introduction to the Electron Theory of Metals*, p. 55. The Institute of Metals, London (1947).

20. Massalski, T. B., ASM Seminar, *Theory of Alloy Phases*, p. 63. Am. Soc. for Metals, Cleveland, Ohio (1956).

3.25 Conductors and Insulators. We are now in a position where we can apply the principles of the zone theory to explain the differences between several well-known types of solids. Let us first consider the case where the energy gap between the first and second zones is large enough so that there is no overlapping of the energy levels in the first two zones.

To illustrate our point, the example of the two-dimensional simple cubic lattice will be considered again. In addition it will be assumed that the energy-wave-number curves corresponding to the lowest and highest position of the energy gaps are as shown in Fig. 3.38A. Here, because of the large size of the gap in any specific direction, the energy level at the bottom of the second zone lies well above the uppermost energy level of the first zone. The energy levels in the second zone are thus all considerably higher than those of the first zone and an electron in the first zone will not usually be able to move into the second

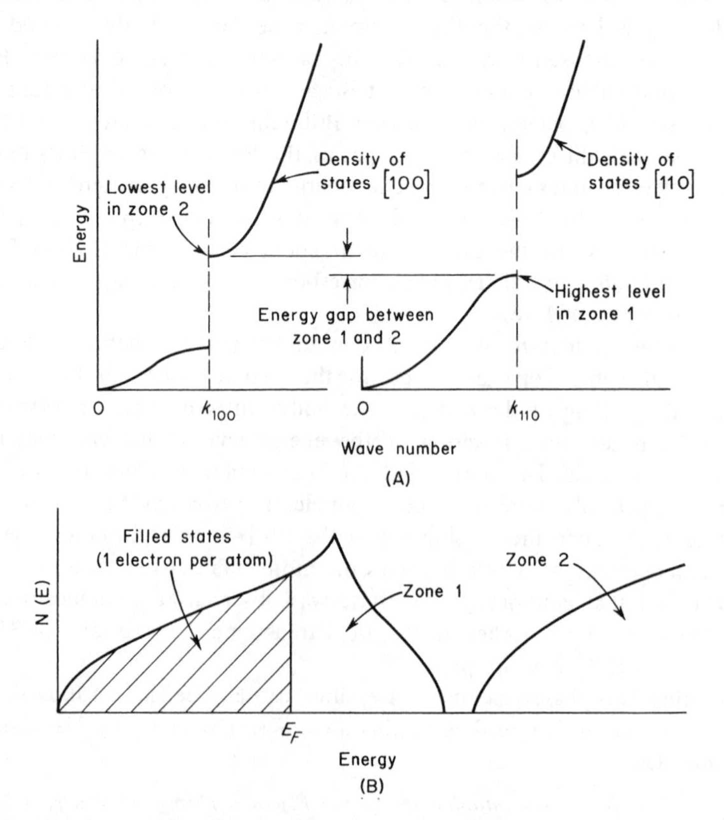

Fig. 3.38 When the energy gap for electrons moving in a single direction is large (as in [A]), a gap in the energy levels between the zones may result (as in [B]).

zone unless it receives a relatively large excitation energy. In general, it may be assumed that the needed energy is much higher, since it is of the order of several electron volts, than would be generally supplied by thermal vibrations at any normal temperature. This condition is due to the fact that kT is approximately $\frac{1}{40}$ electron volts (room temperature). An alternative method of picturing the energy level relationship with respect to the two zones is shown in Fig. 3.38B. Here the density of states as a function of the energy is plotted for the first two zones, and the separation between the two zones is clearly shown.

Now assume that we are concerned with a monovalent metal and that each of the indicated zones is capable of holding, at most, two electrons per atom. Since the atoms of the metal contain only a single valence electron per atom, there are only enough electrons to fill half of the first zone. At absolute zero it can be assumed that the electrons will occupy the bottom half of the energy levels of the first zone. This is implied in Fig. 3.38B, where the lower half of the area is crosshatched, representing the filled energy states.

When a zone is only partly filled, as shown in Fig. 3.38, the solid is a normal metal or electrical conductor. In order that an electrical current can move through a solid, electrons must be capable of moving into higher energy levels. This is certainly the case in the present example, where half of the energy levels are vacant and any small applied electrical potential can cause electrons to be accelerated into higher energy levels.

We can visualize the partly filled zone of a metallic conductor in the manner of Fig. 3.39A, where the circular area in the center of the Brillouin zone represents all of the filled energy states. The significance of this area is as follows. Remembering that we are dealing with a two-dimensional example, the

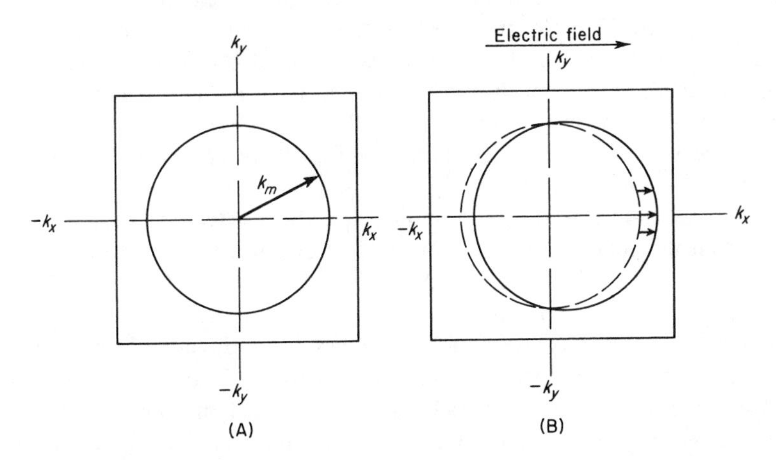

Fig. 3.39 Effect of an electric field on the distribution of energy states in the first Brillouin zone (two-dimensional simple cubic lattice).

circle shows that the electrons possess all possible values of k up to k_m which corresponds to the radius of the circle and to the Fermi level. Because the direction of a k vector (the momentum vector) is the same as the direction of motion of an electron, the symmetry of the distribution of the k values implies that the electrons move in all directions in such a manner that there is no net flow of charge.

Now let us assume that an electric field is applied to our two-dimensional metal. The effect of this field will be to accelerate some of the electrons and move them into higher energy levels in the direction of the field. Thus, if the field is assumed to be directed in the positive x direction, electrons with wave vectors lying at the right-hand side of Fig. 3.39A will move up into energy levels with larger components of k in the x direction, as is shown in Fig. 3.39B. The states that these electrons vacate will, in turn, be filled by electrons in those states with k values just behind the first group. This process of promotion of electrons into energy states with larger values of k_x can be considered to extend throughout the entire distribution of electrons in k space. As a result, the entire distribution is shifted, in the manner shown in Fig. 3.39B. What effectively happens is that the circle moves over to the right. In this new arrangement, for each electron with an increased value of k in the direction of the field (on the right of the circle) there is another with a decreased value (on the left of the circle). This unsymmetrical distribution of electron velocities results in a net flow of charge in the direction of the field and constitutes an electrical current.

Now consider a solid which contains just enough electrons to fill the first band completely. In the present example, this would be two electrons per atom. Here, if we still assume a wide separation between the energy levels of the first band and those of the second, the solid constitutes an insulator. Electrons cannot move into higher energy levels in the first band because all of the states are filled. In addition, no normal electrical potential is capable of raising them across the energy gap from the first to the second zone. If it is not possible to change the energy states of the electrons, then it will not be possible to change the average velocity of the electrons, therefore there can be no net electrical current.

The above simple explanation shows that the electron theory, which was founded on the concept of free electrons in a metal, is also able to explain why certain solids are insulators. These materials are, in general, those with electrons that completely fill a Brillouin zone and, at the same time, have wide energy gaps between the filled zones and the zones lying directly above them. An excellent example is furnished by the diamond form of carbon, which has four valence electrons per atom in each zone. This is just the number of valence electrons needed to completely fill the first zone in the diamond structure. Since there is a considerable energy gap between the first and second zones, diamond is an insulator.

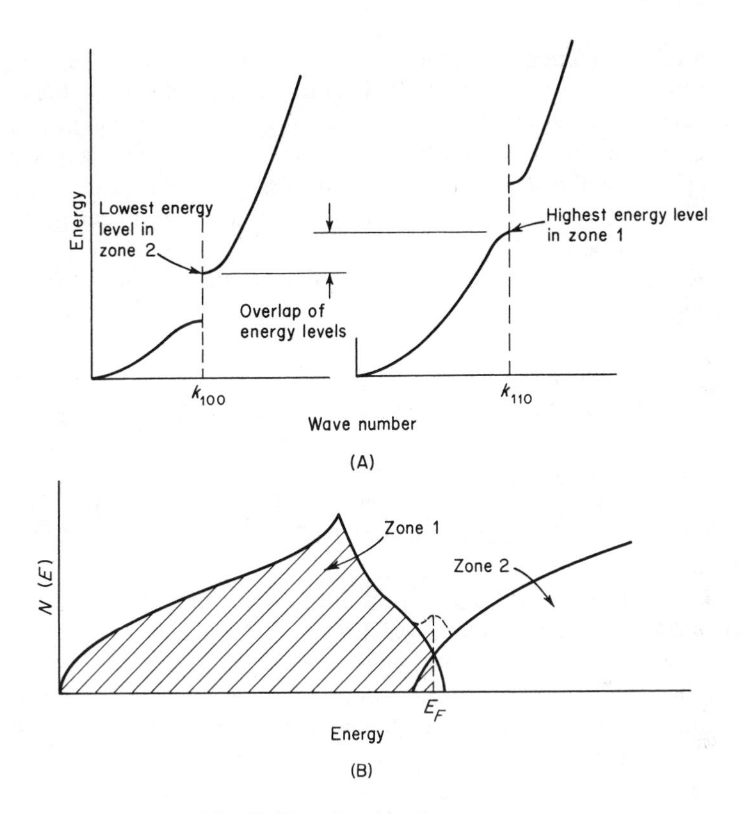

Fig. 3.40 Overlapping zones.

When a solid contains enough electrons to fill a Brillouin zone completely it may still possess the properties of a metallic conductor provided there is an overlapping of zones. This overlapping may be explained with the aid of Fig. 3.40A which again shows the energy-wave-number curves for different directions of motion in a two-dimensional simple cubic crystal. In the present case, the energy gap for electrons, moving in any given direction, is assumed to be small and the states at the top of the first zone have a higher energy than the states at the bottom of the second zone. Figure 3.40B shows the corresponding zones plotted on a density of states diagram. In this figure, the dotted line lying above the region where the zones overlap represents the sum of the individual density curves.

If it is assumed that there are just enough electrons to fill the first zone, then because of the overlap in the energy levels, some of the electrons will overflow into the second zone, with the result that both zones will be filled to the same level of energy. This is shown in Fig. 3.40 where the crosshatched area indicates the filled states.

Electrical conduction is possible in a solid of the type indicated in Fig. 3.40 because there are empty energy states in both zones directly above the electrons in the Fermi level (highest filled states). Another fact that should be mentioned is that electrons can easily shift from one zone to the other in this type of solid. An electron occupying a high level in the first zone can shift to a state of equivalent energy in the second zone without a large change in energy if only its direction of motion is changed, that is, from [100] to [110]. Such changes in the direction of motion can easily be caused by the scattering effect of the lattice vibrations.

3.26 Semiconductors. In the preceding section it was shown that insulators possess filled zones which are separated from empty zones by relatively large energy gaps. At the same time it was also pointed out that if the zones overlap (energy gap zero) the solid has the properties normally associated with a metal even though it has enough electrons to fill the first zone. Semiconducting solids represent an intermediate case, one between these two elementary types.

Semiconductors have two basic forms: intrinsic and impurity.

An *intrinsic semiconductor*, which we shall treat first, has the zone arrangement of a typical insulator, only the energy gap is smaller in magnitude than in a typical insulator. Thus, with reference to Fig. 3.38, an intrinsic semiconductor will have a filled first zone with a narrow gap between the first and second zones. At absolute zero this kind of solid has the properties of a typical insulator. However, as the temperature is raised, thermal energy makes it possible for electrons to jump over the narrow energy gap and enter the empty zone. When this happens, the solid becomes a weak conductor of electricity. The ability of an intrinsic semiconductor to carry electrical currents depends on two factors. Thus a vacant energy state, or *hole*, is left in the lower band for every electron that moves into the upper band, and the ability of an intrinsic semiconductor to conduct electricity depends on both the electrons that are raised into the upper band and the holes left behind in the lower band. The fact that the electrons when elevated into an otherwise empty Brillouin zone can conduct current is self-evident.

The contribution of the holes to the current can be explained as follows. In general, when there is an empty state in an otherwise filled zone, this state can be filled almost at will by electrons of almost equal energy in adjacent states. However, when an electron moves into the empty state, the state that it leaves is made into the empty state. When an electrical field is applied to this form of solid, the successive movements of electrons into empty states is influenced by the applied field. In general, the field will tend to cause the electrons to move into vacant states in such a direction as to contribute an element of current in the direction of the applied field. However, for every movement that an electron makes into a hole in the direction of the applied field, there is a corresponding reverse movement of the empty state or hole. It is, therefore, quite possible to consider a hole in the sense of a carrier of positive charge, for it moves in the opposite direction to that of a negatively charged electron. The contribution of the holes in the lower band to the conductivity of a semiconductor is frequently described in terms of the movement of these fictitious positively charged particles.

In an *intrinsic semiconductor*, the conductivity increases with increasing temperature because the number of charge carriers increases with increasing temperature; the number of electrons that move over into the upper band grows with rising temperature. The number of electrons that are thermally excited into the upper band, or *conduction band* as it is called, varies[21] approximately with a Boltzmann factor of the form $e^{-\Delta E/kT}$, where ΔE is the size of the energy gap and k is Boltzmann's constant. This sort of functional relationship will appear in later sections of this book in connection with a number of other phenomena that occur in metals. At this time it is sufficient to say that it represents a very rapid rise in the conductivity with temperature. Still, it must be remembered that, in general, the conductivities of semiconductors are very small when compared to the conductivities of metals. Only if kT is of the order of ΔE would we expect a semiconductor to have a conductivity comparable to a normal metal, for if we assume that ΔE is about ½ electron volt (a reasonable value for a semiconductor), a temperature of the order of 6000°K is required in order to make kT equal to this value of ΔE. This temperature, however, is sufficient to both melt and vaporize all known elements.

An interesting feature of intrinsic semiconductors is that their conductivity increases with increasing temperature. This is opposite to the variation of the conductivity in most metals, where increasing the temperature normally decreases the conductivity. In the case of metallic conductors, a perfect lattice should offer no resistance to the movement of electrons at absolute zero. At an elevated temperature, however, the thermal vibrations of the lattice are constantly shifting the electrons from one state into another in those states near the Fermi level. This has the effect of tending to destroy the orderly motion of electrons imparted by an electrical field and lead to a resistance to the flow of electrical current. Thermal vibrations, accordingly, increase the resistance of a metal, whereas they decrease it in a semiconductor.

3.27 Impurity Semiconductors. In speaking of semiconductors, it is the usual practice to call the lower band the valence band and the upper band the conduction band. Impurity semiconductors are characterized by having discrete energy levels that lie in the energy gap between the valence and

21. Slater, J. C., *Quantum Theory of Matter*, McGraw-Hill Book Company, Inc., New York, 1951, p. 292.

conduction bands. These energy levels are associated with the presence of impurity atoms in the lattice. The most important of the impurity semiconductors are silicon and germanium. Both elements belong to the same group of the periodic table as carbon, and crystallize in the covalent diamond lattice. As a result, they have a zone structure similar to that of a diamond, but the energy gaps between the conduction band and the valence band are smaller than the diamond gap. The respective gap sizes are about 0.74 electron volts for germanium,[22,23] 1.10 electron volts for silicon,[24] and 7 electron volts for diamond.[25] Diamond is classed as an insulator because of its large energy gap, while the moderate size of the energy gaps in silicon and germanium make it possible to class these elements as intrinsic semiconductors. At room temperature, however, their intrinsic conductivity is extremely low and only becomes appreciable when they are heated to temperatures several hundred degrees above room temperature. Accordingly, the conductivity normally associated with these elements at room temperature can be considered to be due primarily to the presence of impurity atoms in these elements.

In all, there are four elements that crystallize in the diamond structure: carbon (diamond), silicon, germanium, and tin (gray). Each of these elements possesses four valence electrons. There are two ways that their crystal structure can be interpreted. First, from the covalent point of view, we can assume that each atom is bound to its four nearest neighbors by covalent linkages, each linkage corresponding to a pair of shared electrons. Alternately, from the zone theory, we know that the linkages possess just enough electrons to fill the first zone and that in diamond, silicon, and germanium there is an energy gap separating the valence zone from the conduction zone. It is interesting to note, however, that it is believed that these zones overlap in gray tin; an effect which gives this structure the properties of a metallic conductor.

Now let us consider specifically the elements silicon and germanium. Suppose that an atom with a valence of three (B, Al, Ga, In) is substituted in the crystal of one of these elements for one of the silicon or germanium atoms. The impurity atom is thus assumed to occupy one of the lattice positions which would normally be held perhaps by a silicon (or germanium) atom. Since the impurity atom contributes only three valence electrons to the crystal, there will be one less electron in the solid than the number required to give each atom four pairs of shared electrons. As a result, one covalent bond of the solid must have only a single valence electron. From the zone concept, this has the effect of creating an unoccupied energy level which, in general, lies slightly above the highest energy state in the valence band. The position of this energy level relative

22. Morin, F. J., and Maita, J. P., *Phys. Rev.*, **94** 1525 (1954).
23. Johnson, E. R., and Christian, S. M., *Phys. Rev.*, **95** 560 (1954).
24. Morin, F. J., and Maita, J. P., *Phys. Rev.*, **96** 28 (1954).
25. Kimball, G. E., *Jour. Chem. Phys.*, **3** 560 (1935).

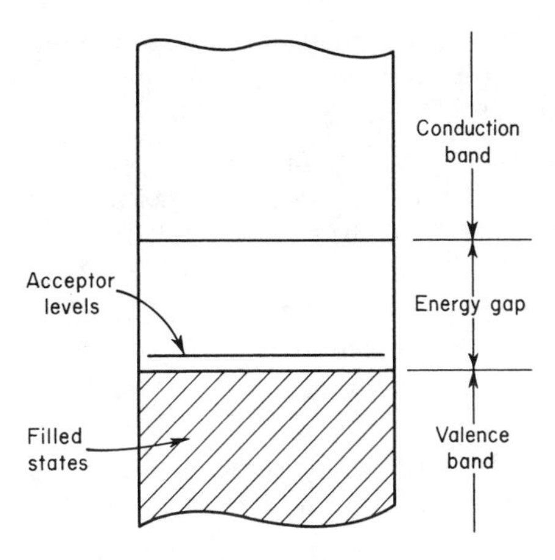

Fig. 3.41 Acceptor levels in an impurity semiconductor.

to the valence band is shown schematically in Fig. 3.41. The magnitude of the energy gap between the top of the valence band and this unfilled level, which is generally known as an *acceptor level*, is given in Table 3.5 for a number of trivalent elements. The Table shows that the magnitude of the gaps is about 0.01 electron volts, which is of the same order as the value of kT at room temperature – about 1⁄40 electron volts. Since the latter gives an estimate of the energy which thermal vibrations may impart to an electron, it is quite evident that thermal energy may excite electrons from the valence band into the acceptor levels at room temperature. When this happens, a vacant energy level or hole is left behind in the valence band. Since holes in the valence band are able to

Table 3.5 Energy Gaps in Impurity Semiconductors.*

Impurity	*In Germanium*	*In Silicon*
Energy gap between acceptor level and valence band		
B	0.0104	0.045
Al	0.0102	0.057
Ga	0.0108	0.066
In	0.0112	0.16
Energy gap between donor level and conductor band		
P	0.0120	0.044
As	0.0127	0.049
Sb	0.0096	0.039

* After Hobstetter, J. N., Data of Burton, J. A., *Physica,* 20 845 (1954).

conduct electrical currents, the excitation of electrons into the acceptor levels gives the solids the ability to conduct current. A semiconductor that exhibits conductivity caused by the movements of holes in the valence band is called a *p-type semiconductor*, on the assumption that the holes are considered as positive charges. An alternative way that impurity atoms can contribute semiconduction properties should be pointed out here.

Suppose an atom with five valence electrons (P, As, Sb, Bi) is substituted into the lattice of either a germanium or silicon crystal. In either case, each impurity atom will contribute one more electron than is needed to form the four covalent bonds per atom in the solid. From the zone or band theory, this extra electron will lie in an energy level which is well above the levels associated with the valence band and which, in general, falls just below the conduction band. The position of these energy levels is shown in Fig. 3.42 where they are designated as donor levels. The energy difference between the bottom of the conduction band and the donor levels is also given in Table 3.4 for some of the elements with a valence of five. These gaps are also comparable to the value of kT at room temperature, so we can assume that the electrons which would normally occupy these levels are excited in large measure at room temperature into the levels of the conduction band. In the conduction band, the excited electrons give the material the ability to carry an electrical current. Semiconductors, with current carrying properties that are due primarily to the presence of electrons excited from donor levels into the conduction band, are known as *n-type*, or *negative, semiconductors.*

3.28 Magnetism. Magnetic phenomena in metals and alloys tend to be very complicated and rightfully constitute an important area of study.

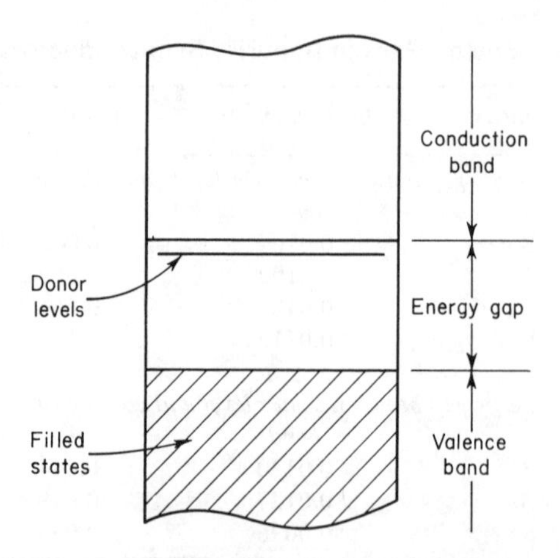

Fig. 3.42 Donor levels in an impurity semiconductor.

Space considerations do not permit us to do more than indicate the nature of the basic phenomena.

The most characteristic property of any magnetic material is its *magnetic moment*. The magnetic moment is a measureable quantity and is designated by the symbol μ. When one is considering magnetic effects arising from atoms and electrons that are in bulk, as in a crystal, it becomes convenient to think in terms of a magnetic moment per unit volume. This quantity is given the symbol M. In most materials, M is zero when there is no magnetic field. However, if these substances are placed in a magnetic field, it is found that

$$M = \chi H$$

where the proportionality factor χ is known as the magnetic susceptibility per unit volume, and H is the magnetic field intensity. The established literature in the field of magnetism tends to use the Gaussian unit gauss for both M and H.[26]

The susceptibility χ can be either positive or negative. If it is negative, a material is classed as *diamagnetic*. Diamagnetic susceptibilities are normally small. If it is positive and small, the material is *paramagnetic*. A few materials, such as iron, have very large susceptibilities and are classed as *ferromagnetic*. Furthermore, ferromagnetic materials are spontaneously magnetized and are able to retain a permanent magnetic moment. When the field H is removed from these materials, M does not necessarily go to zero. This is not true in the case of diamagnetic and paramagnetic materials. In these, when the field is removed, the magnetization disappears.

The magnetic susceptibility of a material is a convenient way of characterizing its magnetic behavior. A diamagnetic material with a negative susceptibility tends to reduce the magnetic flux B associated with a magnetic field intensity H, while a positive susceptibility tends to raise it. Most of the important metallic elements are classified in Table 3.6 according to their susceptibilities.

There are three important ways that magnetic moments are developed on an atomic scale. An electron orbiting about a nucleus has a magnetic dipole moment. The spin of an electron is associated with a magnetic dipole moment. Finally, the nucleus itself also possesses angular momentum and produces a nuclear dipole moment.

Nuclear magnetic moments are much smaller than those resulting from the electron spin or the orbital motion of electrons. The angular momentum in all three cases is of the same order. However, the angular momentum varies as $\mathcal{M}\omega$ where $\mathcal{M}$ is a generalized mass (i.e., moment of inertia in a rotating system) of the particle, and ω is its angular velocity. Because the mass of the neutron is of the order of 10^3 greater than that of the electron, the angular velocity associated

26. Kittel, C., *Introduction to Solid State Physics*, John Wiley and Sons, New York, 1966, p. 429.

with the motion of charges on the nucleus has to be much smaller. As a result, we shall not consider the nuclear magnetic moment further.

Diamagnetism is produced by an effect analogous to that which occurs in a simple air-cored transformer when the current is changed in the primary coil. The change this induces in the magnetic field intensity H produces a current in the secondary coil that opposes the change in flux (Lenz's Law). When moving electrons are subjected to a magnetic field, their angular momenta may be changed in such a manner as to produce a field component that opposes the

Table 3.6 Mass Susceptibilities of Some Metallic Elements, $\chi \times 10^6$ cgs units.*

Paramagnetic		*Diamagnetic*		*Ferromagnetic*
Element	*Susceptibility*	*Element*	*Susceptibility*	*Element*
Cesium	0.22	Copper	−0.09	Iron
Tungsten	0.32	Silicon	−0.11	Cobalt
Potassium	0.53	Lead	−0.13	Nickel
Magnesium	0.54	Gold	−0.14	
Aluminum	0.61	Mercury	−0.17	
Thorium	0.66	(solid)		
Sodium	0.70	Zinc	−0.18	
Platinum	1.04	Silver	−0.18	
Zirconium	1.30	Cadmium	−0.31	
Uranium	1.7	Indium	−0.56	
Molybdenum	1.8	Antimony	−1.10	
Niobium	2.20	Bismuth	−1.30	
Technetium	2.9			
Titanium	3.0			
Chromium	3.1			
Vanadium	5.0			
Palladium	5.3			
Manganese	11.50			

* Data abstracted from *Handbook of Physicochemical Properties of the Elements, edited by Samsonov, G. V. Plenum Publishing Corp., New York, 1968, pp. 338–353.*

applied field. This effect produces a force that acts to move a diamagnetic material away from a region where the magnetic field is strong and into a region where the field is weaker. A thin rod of a diamagnetic material such as bismuth, when hung in a magnetic field, tends to set itself across the lines of force. It was for this reason that Faraday called materials of this type diamagnetic (Greek *dia* means through).

There are two basic sources of diamagnetism in metals. First, an applied field can induce an observable change in the dipole moment of an electron that orbits a nucleus by causing the electron orbit to precess about the direction of the field. This is equivalent to a change in its angular momentum. Second, in a metal with free electrons, the applied magnetic field can cause the electrons in the electron gas to alter their motion in such a manner as to produce a diamagnetic effect.

Paramagnetism is associated with magnetic moments of a different kind. It involves an interaction between the applied magnetic field and the magnetic moments that an electron possesses as a result of its spin or from its orbital motion about a nucleus. In this case, the elementary magnets tend to align themselves with the magnetic field. Thus, consider an atom with an electron traveling in an orbit. If the applied field causes the electron to precess, the result is diamagnetic; while if the atomic orbit rotates so that the component of its angular momentum along the field direction increases, the result is paramagnetic. Both effects may occur simultaneously. In the absence of a magnetic field, the net paramagnetic dipole moment is zero because the elementary magnets will have a random orientation. Application of the field tends to align the elementary magnets in the direction of the field, thereby increasing the flux.

It is quite possible for an atom in a solid to have no net paramagnetic dipole moment component. Thus, consider a noble gas element such as argon. In this atom with its closed shells, the electrons in their orbits produce magnetic moments that cancel each other. At the same time, the paramagnetic component due to the spin of the electrons is also zero. This is because in the filled shells there are equal numbers of electrons with the two possible types of spin vectors. Such an atom is therefore diamagnetic, as may be seen by examining Table 3.6.

The next atom after argon in the periodic table is potassium, which has a single valence electron outside the closed shell characteristic of argon. In the solid form, the valence electrons form an electron gas that occupies the space between the potassium ions. In this case, the susceptibility is the net result of the paramagnetism of the conduction or valence electrons, the diamagnetism of these same electrons, the diamagnetism of the ions, and several other effects.[27] It is obvious that even in this very simple metal, the complexity of the magnetic phenomena is large.

As can be deduced from Table 3.6, because potassium is paramagnetic, the major contribution to the magnetism of potassium is the paramagnetism associated with the conduction electrons. The magnitude of this paramagnetic effect is, however, much smaller than one would predict on the basis of the classical free electron theory. Pauli[28] has shown that the Fermi-Direc

27. Kittel, C., *Introduction to Solid State Physics*, John Wiley and Sons, New York, 1966, p. 447.
28. Pauli, W., *Phyzik*, 41 81 (1927).

distribution of the electrons into energy states gives an answer in much better agreement with the observed value. Since this represents another simple but significant application of the band theory, it will be considered briefly.

The magnetic dipole moment of a single electron is

$$\mu_B = \frac{e\hbar}{2mc} = 0.93 \times 10^{-20} \text{ ergs/gauss}$$

where $\hbar$ is $h/2\pi$, e and m are the charge and mass of the electron respectively, and c is the velocity of light. The quantity $e\hbar/2mc$ serves as a unit of magnetic dipole strength and is called the *Bohr magneton*. In a magnetic field, when the spin vector is parallel to the magnetic field, the energy of the electron is decreased by an amount $\mu_B H$, where H is the magnetic field intensity. When the spin vector of an electron is antiparallel to the magnetic field, the energy is decreased by an equivalent amount. In an electron gas containing N electrons at 0°K, the electrons will occupy the $N/2$ energy states of lowest energy. Half of these electrons will have negative spins, and half will have positive spins. This fact is represented in Fig. 3.43A where the cross-hatched area above the

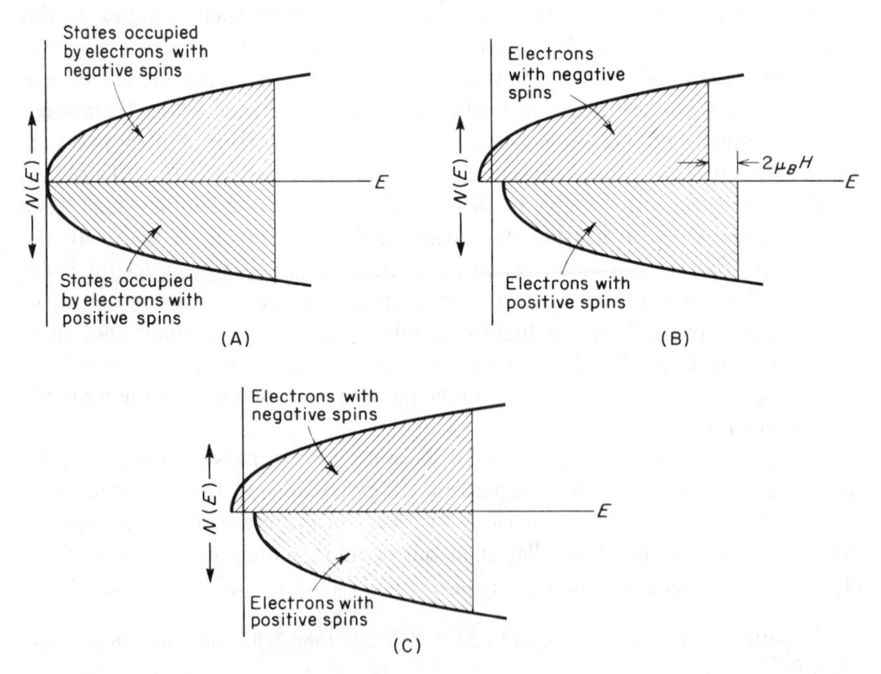

Fig. 3.43 The effect of a magnetic field on the distribution of the electrons into states with positive and negative spins.

horizontal axis represents the states occupied by electrons with negative spins, and the corresponding area below the axis represents the electrons with positive spins. In the presence of a magnetic field, those electrons with negative spins will align themselves parallel to the magnetic field, while those with positive spins will align themselves in the antiparallel direction. The energy of each electron with a negative spin will thus be decreased by $\mu_B H$ and the energy of each positive electron will be increased by $\mu_B H$. This is illustrated in Fig. 3.43B by shifting the two areas, representing the energy states of the electrons with negative and positive spins relative to each other, by an amount $2\mu_B H$. The situation represented by Fig. 3.43B is unstable, because there are some electrons with positive spins occupying energy levels that lie above the highest occupied energy level in the states associated with the electrons of negative spins. This situation is rectified by some of the electrons with positive spins changing their spin quantum numbers and moving into the energy levels available in the upper band in Fig. 3.43B. The resultant distribution of the electrons in the energy states is shown in Fig. 3.43C. Note that it corresponds to an excess of electrons with a negative spin; that is, electrons with spin vectors aligned in the direction of the magnetic field. For every electron that changes its spin vector there is a change in the magnetic dipole moment of $2\mu_B = \{-(-\mu_B) + \mu_B\}$ and the total change in magnetic dipole moment is equal to $2\mu_B n_e$, where n_e is the number of electrons that alter their spins. It can be shown by a more comprehensive treatment[29] of this problem that in the case of an electron gas, the band theory yields a paramagnetic susceptibility given by the equation

$$\chi = \frac{3}{2} N \mu_B{}^2 / 2kT_F$$

where N is the number of electrons per unit volume, μ_B is the magnetic moment of an electron, k is Boltzmann's constant, and T_F is the Fermi temperature. This result is in good agreement with experimental measurements. It also predicts a susceptibility that is temperature independent because T_F is a constant.

A corresponding equation can be derived using classical Boltzmann statistics for the electron gas. This gives the paramagnetic susceptibility as

$$\chi = N \mu_B{}^2 / kT$$

The difference in these equations lies primarily in the fact that Pauli's result involves T_F, the Fermi temperature, while the classical free electron model involves the temperature at which the susceptibility is measured. The Fermi temperature in many metals is of the order of 10^5 °K, while most experiments

29. Kittel, C., *Introduction to Solid State Physics*, John Wiley and Sons, New York, 1966.

are conducted at temperatures of the order of 10^3 °K. Thus the classical free electron theory answer is about 100 times larger than the Pauli one. Also, the classical model predicts a susceptibility that is temperature dependent, which is not in agreement with experimental observations.

3.29 Ferromagnetism. Ferromagnetism is a very remarkable phenomenon in which, as a result of the spontaneous alignment of dipole moments, very large magnetic effects are produced. These are of great technological significance. The elementary magnets that are responsible for ferromagnetism could be either the electrons in their orbits or, alternatively, the spinning electrons. This problem has been resolved experimentally by measuring the ratio of the magnetic dipole moment to the angular momentum of the elementary magnets in ferromagnetic materials. This ratio is called the *gyromagnetic ratio* and is given the symbol ρ. Theory predicts that for a spinning electron, $\rho = m/e$, and for an orbiting electron, $\rho = 2m/e$. In ferromagnetic materials, the most reliable measurements give a value of the gyromagnetic ratio that is close to m/e, so it may be concluded that ferromagnetism results primarily from the dipole moments of spinning electrons.

At a given temperature, the magnetization that can be obtained by increasing the field strength has a maximum value. This is known as the *saturation magnetization*. The saturation magnetization is also a function of temperature, as shown in Fig. 3.44, and reaches a maximum value at absolute zero. With increasing temperature, the saturation magnetization decreases; and the decrease becomes progressively more rapid as the temperature approaches a temperature

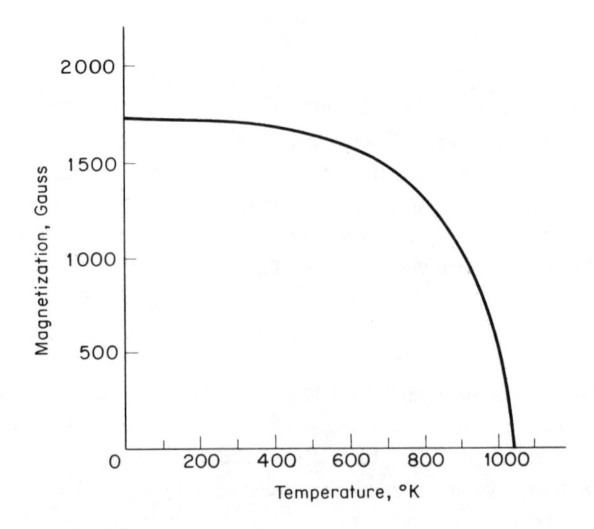

Fig. 3.44 The saturation magnetization of iron as a function of temperature.

T_C, known as the Curie temperature. Above the Curie temperature the metal is no longer ferromagnetic, but paramagnetic.

Table 3.7 lists some pertinent magnetic data for the three ferromagnetic elements: iron, cobalt, and nickel. It includes the saturation magnetization (at 0°K and room temperature), the magnetic moment per atom expressed in terms of the Bohr magneton (at 0°K), and the Curie temperature. Note that when iron is completely magnetized, each atom has a magnetic dipole moment equivalent to that due to the spins of 2.2 electrons. These ferromagnetic elements are transition metals and possess a set of $3d$ energy states whose energy levels overlap those of the $4s$ states. As a result, the electrons tend to enter the $4s$ levels before they fill up the $3d$ states. This results in an atomic structure that is much more complex than that of an ordinary atom. Ferromagnetism is a direct consequence of this complex structure, and the theories dealing with ferromagnetism are also complex and beyond our present consideration.

Table 3.7 Data Concerning the Ferromagnetic Elements Fe, Co, and Ni.*

Element	*Saturation Magnetization M, in gauss*		n_e(0°K)	*Curie Temperature* (°K)
	0°K	*Room Temp.*		
Iron	1740	1707	2.22	1043
Cobalt	1446	1400	1.72	1400
Nickel	510	485	0.61	631

* After Kittel, C., *Introduction to Solid State Physics*, John Wiley and Sons, New York, 1966.

The large difference in the magnetic behavior between the ferromagnetic metals and those that behave paramagnetically is not hard to understand. In a metal that shows paramagnetism, an applied field, as shown in Fig. 3.43, reverses the dipole moments of only a relatively small number of conduction electrons. This amounts to an effective dipole moment per atom that is a very small fraction of a Bohr magneton. On the other hand, in a metal such as iron, when its magnetization is saturated, the net dipole moment can be as large as 2.2 Bohr magnetons. The accepted view of ferromagnetism is that each elementary magnet in the material is aligned parallel to all of the other neighboring elementary magnets. This means that all of the atomic dipole moments are completely ordered. A strong cooperative interaction between large numbers of elementary magnets is implied. Suppose that one small magnet in such a structure were to reverse its orientation. This action would be opposed by the driving force that causes ordering. Thermal vibrations do, however, tend to destroy this ordered arrangement of the elementary magnets. That the saturation magnetization decreases with increasing temperature is a direct result of this fact. At any given temperature, the degree of order represents a balance between the tendency for the ions to form an ordered magnetic structure, and the thermal vibrations that tend to destroy the order. Above the Curie temperature, the thermal vibrations are too strong to allow the ordering of the elementary magnets associated with ferromagnetism to exist.

According to the above, at very low temperatures the elementary magnets in a ferromagnetic substance are completely aligned on an atomic scale. However, the fact exists that it is still possible to demagnetize a metal such as iron. This means that on a macroscopic scale, the magnetic moment per unit volume can be greatly reduced below the saturation value. An explanation of this apparent anomaly was first proposed by Weiss,[30] who suggested that a zero magnetic moment should be possible on a macroscopic scale if a ferromagnetic material actually consisted of a number of separate regions, called *domains*, in which the magnetization was saturated, but also in which the direction of the magnetization was not the same. The alignment of the direction of magnetization in these zones would then be such that their fields canceled each other outside the specimen. Experimental techniques are available for showing the domains in a ferromagnetic material, and it is possible to rationalize the domain structure in complete accord with Weiss' hypothesis. A schematic illustration of the domains as they might appear in a single crystal is shown in Fig. 3.45.

When a magnetic field is applied to a ferromagnetic material, the domain structure becomes altered in response to the field. This change in the zone

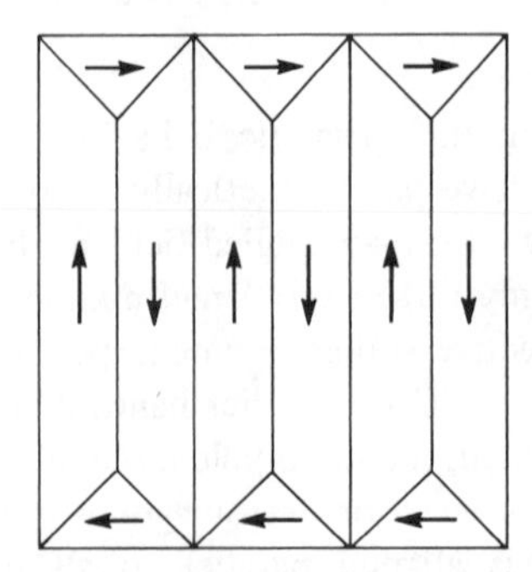

Fig. 3.45 A schematic representation of the magnetic domains in a single crystal of a ferromagnetic metal. The directions of the magnetization in the zones are indicated by the arrows.

30. Weiss, P., *L-Hypothese du Champ Moleculaire et la Propriete ferromagnetique*, *J. de Physique*, 4th Series, 6 661–90 (1907).

structure results in the development of a net magnetic dipole moment. Under a weak field, the domains whose directions of magnetization are favorably aligned with respect to the field tend to grow at the expense of the neighboring domains whose directions of magnetization are less favorably oriented. As the field becomes stronger, it can also produce a rotation of the direction of magnetization inside the domains toward the direction of the field. Both effects will increase the magnetic moment of the specimen.

3.30 Antiferromagnetism and Ferrimagnetism. In a ferromagnetic material the elementary magnetic dipoles inside a domain are all oriented in a direction parallel to each other, as shown schematically in Fig. 3.46A. Another possibility is shown in Fig. 3.46B. Here the elementary magnets are antiparallel. When this happens, *antiferromagnetism* is said to exist. The net magnetic dipole moment is zero in an antiferromagnetic material. A typical antiferromagnetic material is FeO. In this substance the oxygen atoms, which do not have a magnetic moment, act to bind the iron atoms together in an antiferromagnetic arrangement.

A third possible ordered arrangement of the magnetic dipoles is shown in Fig. 3.46C. In this case, the magnetic dipoles are antiparallel, but a difference exists in the magnitudes of the oppositely directed dipole moments. This situation can arise as a result of several causes. In one case, the size of the magnetic moment of one type of dipole may be greater than that of the other. The possibility also exists that in a crystal structure the number of ions with a magnetic moment pointed in one direction may be greater than the number of ions with a magnetic moment pointed in the other direction. The general type of structure being described is said to be *ferrimagnetic*. As can be deduced from the above, a ferrimagnetic substance possesses a net magnetic dipole moment. A typical ferrimagnetic material is Fe_3O_4. This is magnetite, or the magnetic form of iron oxide. In this material the iron atoms ionize to form both Fe^{++} as well as Fe^{+++} ions. The orderly arrangement of these two types of iron atoms in the

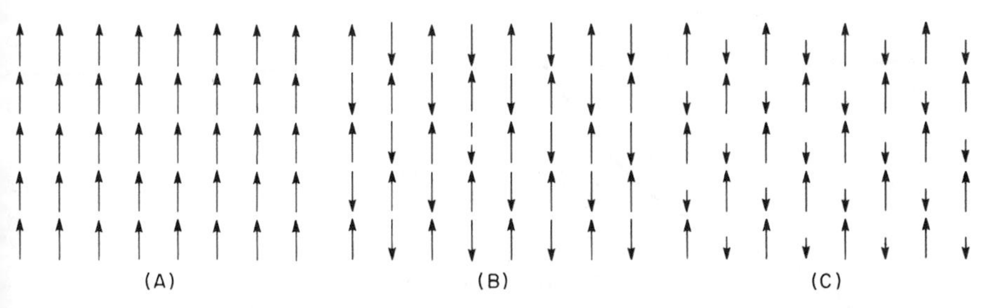

Fig. 3.46 Several possible ordered arrangements of the magnetic dipole moments. (A) ferromagnetic, (B) antiferromagnetic, (C) ferrimagnetic.

Fe_3O_4 crystal structure is responsible for the ferrimagnetic character of the material.

In recent years, ferrites have become important electronic substances because they are ferromagnetic, but at the same time they have a high resistivity. This means that when a ferrimagnetic material is used in the core of a transformer, the eddy current loss is small. As a consequence, ferrimagnetic materials are used as cores in radio frequency transformers where the eddy current loss is a serious problem. Ferrites are also used as magnets in computers.

3.31 The Electrical Resistivity as a Function of Temperature. According to the Sommerfeld free electron theory, a simple metal such as sodium or copper should have a zero resistance. This implies that electrical resistivity results from phenomena that disturb the normal motions of the electrons. These disturbances can come from two basic sources. These are (1) collisions of electrons with *phonons* (quantized lattice vibrational energy), and (2) collisions of electrons with solute atoms and mechanical imperfections in a crystal lattice. Ordinarily we may write

$$\rho = \rho_T + \rho_I$$

where ρ is the resistivity, ρ_T is the thermal (phonon) component of the

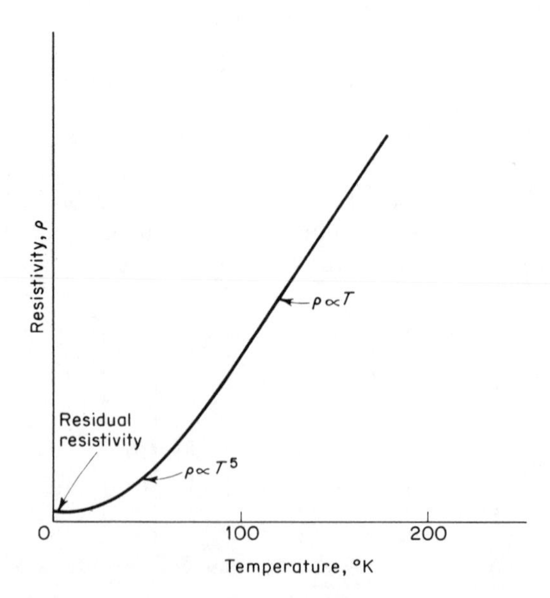

Fig. 3.47 Variation of the resistivity with temperature of a simple metal below room temperature.

resistivity, and ρ_I is that part of the resistivity due to imperfections in the structure. Assuming that the imperfections are impurity atoms and the concentration of these latter is small, and ignoring superconductivity effects, then ρ_I is independent of the temperature. This is known as *Matthiessen's rule.* On the other hand, ρ_T normally approaches zero as the temperature approaches zero. The dependence of ρ_T on the temperature is nearly linear above about 100°K, but below this temperature ρ varies as T^5. Figure 3.47 shows schematically the variation of the resistivity ρ with temperature below room temperature. At very low temperatures, such as that of liquid He (4.2°K), the resistivity of such a metal will be largely determined by ρ_I, whereas at room temperature it will be primarily a measure of ρ_T. Since ρ_I depends on the purity of a specimen and ρ_T is independent of it, the ratio of the resistivity at room temperature to that at 4°K often serves as a good practical measure of the purity and perfection of a metal. In a very pure metal that is free of distortion, this *resistivity ratio* can be as high as 10^4 to 10^5. In impure metals, it may be below 10.

3.32 Superconductivity. In 1911, Kamerlingh-Onnes[31] made the startling discovery that, as far as he could determine experimentally, the resistivity of solid mercury dropped to zero when it was cooled below about 4.2°K. Many experiments since that time have verified this fact. The phenomenon in question is known as *superconductivity*. It has been observed in a large number of elements, as may be seen in Table 3.8. The notable exceptions are the monovalent metals and the ferromagnetic metals. Over 1000 alloys and intermetallic compounds have also been observed to become superconducting at very low temperatures.

The interest in superconductivity is large[32] because there are important technological advantages to be gained from a material in which the current can flow without energy loss due to resistivity. As an example, the transmission of electrical power through superconducting cables is an intriguing possibility. Another use of superconductors that is actually taking place is in the solonoids of high field-strength magnets. Strong magnetic fields are needed for devices such as masers, lasers, and infra-red detectors. Permanent magnets are often too bulky and heavy for some of these applications. On the other hand, iron-cored magnets are also limited by the saturation magnetization they can support. For large, very high field-strength magnets one is therefore forced to employ air-cored magnets through which high currents are passed. The power loss in such magnets can be considerable. The advantage of superconductors in these magnets is obvious. The above two examples are not the only applications that

31. Kamerlingh-Onnes, H., *Commun. Phys. Lab.*, Univ. of Leiden, No. 122*b* (1911).
32. Dew-Hughes, D., *Mater. Sci. Eng.*, 1 2 (1966).

Table 3.8 Critical Temperatures, T_c, and Critical Field, H_0, for some Superconducting Elements.*

Element	T_c(°K)	H_o (Oersteds)
Iridium	0.14	19
Titanium	0.39	100
Cadmium	0.54	29
Zirconium	0.55	47
Zinc	0.86	53
Molybdenum	0.92	98
Aluminum	1.18	99
Thorium	1.37	162
Indium	3.40	293
Tin	3.72	309
Mercury	4.15	412
Vanadium	5.30	1020
Lead	7.19	803
Technetium	8.22	–

* Data abstracted from *Handbook of Chemistry and Physics*, 50th Edition, pp. E-93–106, The Chemical Rubber Co., Cleveland, Ohio, 1964.

have been conceived for superconductors. This subject has been covered in several books.[33,34]

Because of these commercial applications, the research on superconductors has been intensive. However, there are several basic problems inherent in superconductors. The first is the fact that there is a fixed temperature T_c above which a metal loses its superconducting properties and acts like a normal resistive metal, as may be seen in Fig. 3.48. The highest value of this transition temperature that has been achieved is about 20°K, and in most materials, T_c lies well below this value. For economic reasons, it is desirable to find superconducting materials with higher critical temperatures. A large effort is being made in this direction, but the outlook is not considered to be favorable.

Second, the magnetic phenomena associated with superconductivity must be considered. There are serious problems associated with the industrial use of superconductors that bear directly on this fact. When a simple superconductor such as mercury, which is called a *Type I superconductor*, is placed in a magnetic field, currents are induced in the material producing a magnetic field that completely cancels the applied field inside the specimen. A Type I superconductor is therefore a perfect diamagnet, and the flux density B inside the specimen is zero. This is known as the Meissner effect, and the state of perfect diamagnetism inside the specimen is called the *Meissner state*. Although

33. Newhouse, V. L., *Applied Superconductivity*, John Wiley and Sons, New York, 1964.
34. Bremer, J. W., *Superconducting Devices*, McGraw-Hill Book Co., New York, 1962.

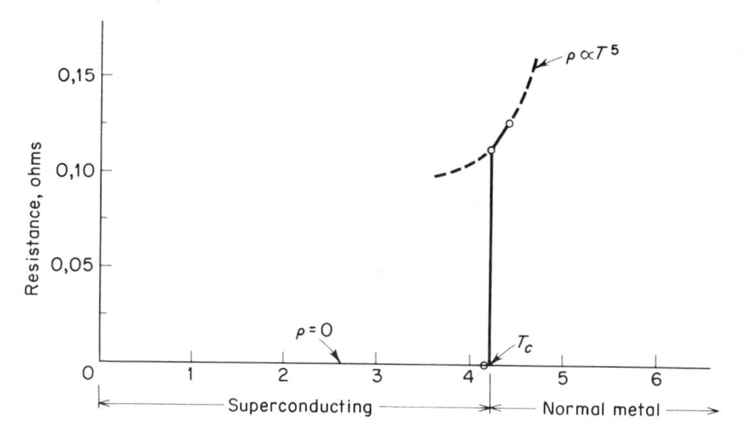

Fig. 3.48 The resistance of a specimen of solid mercury as a function of temperature. (After the data of Kammerling-Onnes, H., *Commun. Phys. Lab.*, Univ. of Leiden, No. 122b 1911).

superconductivity was discovered in 1911, it was not until 1957 that an accepted theory[35] for the phenomenon was developed. This theory is well beyond our scope, but an elementary interpretation is that below T_c a small fraction of the conduction electrons, with energies at or close to the Fermi level, enter an ordered state involving pairs of correlated electrons. Associated with this correlation is a small energy gap of the system which is of the order of kT_c. At the critical temperature T_c, the thermal energy becomes strong enough to completely break up this ordering of the electron pairs.

If the magnetic field applied to a metal in the superconducting state is made too large, it will destroy the superconducting state. This breakdown of the superconducting state is shown schematically in Fig. 3.49A, where the flux density B is plotted as a function of H, the magnetic field intensity. Note that B is zero until the field reaches H_c, the critical value, but above this point the metal loses its superconducting properties and the flux penetrates the material so that the superconductive current is pinched off. It should be mentioned that Fig. 3.49A applies strictly to the interior of a long thin specimen of a Type I superconductor. Actually there is a thin transition layer at the specimen surface called the *penetration depth*, inside which the flux drops from its value at the surface to the zero value in the specimen interior. The current flows in this penetration depth.[36]

An alternative method of representing the concepts in Fig. 3.49A is shown in Fig. 3.49B. Here the magnetization is plotted as a function of H, using the parameter $-4\pi M$ to represent the magnitude of the magnetization. In general,

35. Bardeen, J., Cooper, N. L., and Schrieffer, J. R., *Phys. Rev.*, **108** 1175 (1957).
36. Dew-Hughes, D., *Mater. Sci. Eng.*, 1 2 (1966).

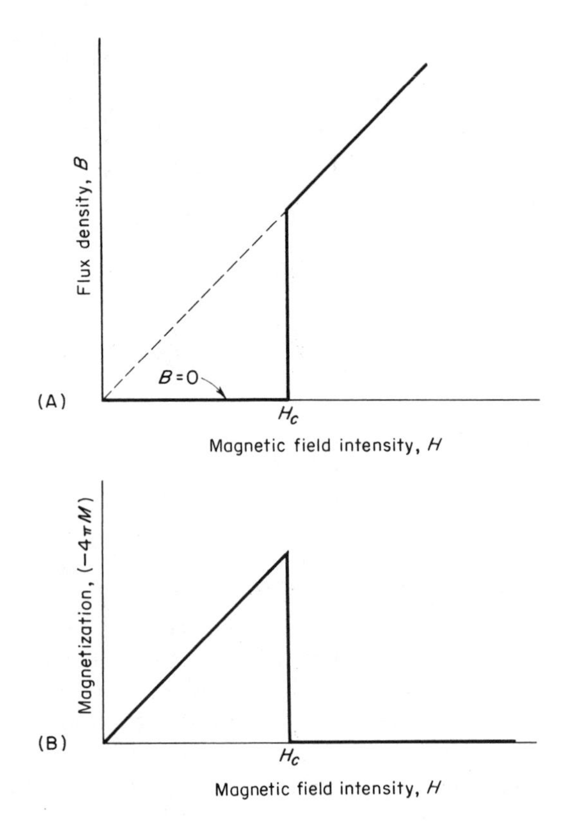

Fig. 3.49 (A) The variation of the flux density with the magnetic field strength H in a Type I superconductor. (B) The variation of the magnetization with the field strength in a Type I superconductor.

the flux density is given by the relation

$$B = H + 4\pi M$$

Therefore, with B equal to zero in the superconducting state, H must equal $-4\pi M$. Note that at H_c the diamagnetic magnetization associated with the superconducting state vanishes and in the normal state the magnitude of the magnetization is too small to be shown with the scale used in the drawing.

The critical value of H is a function of temperature. This is shown in Fig. 3.50. At the critical temperature T_c, H_c is zero but rises with decreasing temperature. Thus, for the specimen to support a reasonable value of H, the temperature must be lowered below T_c. However, a temperature difference of a

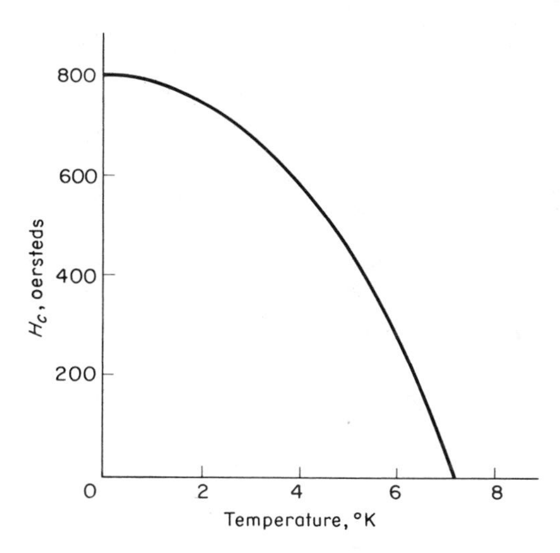

Fig. 3.50 The critical magnetic field H_c as a function of temperature for lead. (After data of Bremer, J. W., *Elec. Mfg.,* Feb. 1958, p. 78. Copyright © Industrial Research Inc.)

few degrees is important at temperatures near absolute zero. It is significant that the magnetic field, produced by a current flowing in a superconductor, itself can act as an externally applied field and destroy the superconducting state. This means that a Type I superconductor has a limited ability to pass an electrical current. For this reason, interest is now centered on a different type of superconductor called a *Type II, or hard superconductor*. These are able to support much larger magnetic fields and to carry much heavier currents. In general, this type of superconductor is either an alloy or an intermetallic compound.

The magnetization curve of a typical Type II superconductor is shown in Fig. 3.51. In this form of material the behavior is essentially that of a Type I superconductor up to a field strength H_{c_1}. This means that the flux is expelled completely, except for the penetration depth, from the inside of the specimen as long as H is below H_{c_1}. When the field is increased above H_c, the flux begins to penetrate the specimen. However, this penetration is only partial and can be thought of as occurring along a number of cylindrical or tubular paths, as indicated in Fig. 3.52. The material in between these tubes maintains its superconducting properties until the field reaches H_{c_2}, at which point the superconductivity is lost for all practical purposes.

Because Type II superconductors are able to withstand much higher magnetic fields, they are employed almost exclusively in the manufacture of magnets with

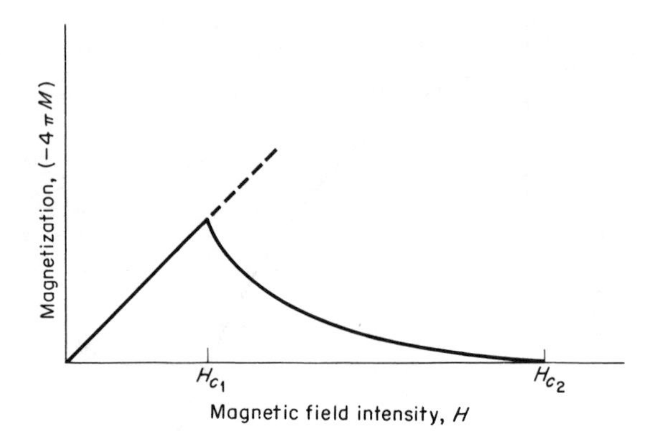

Fig. 3.51 Schematic representation of the magnetization curve of a Type II superconductor.

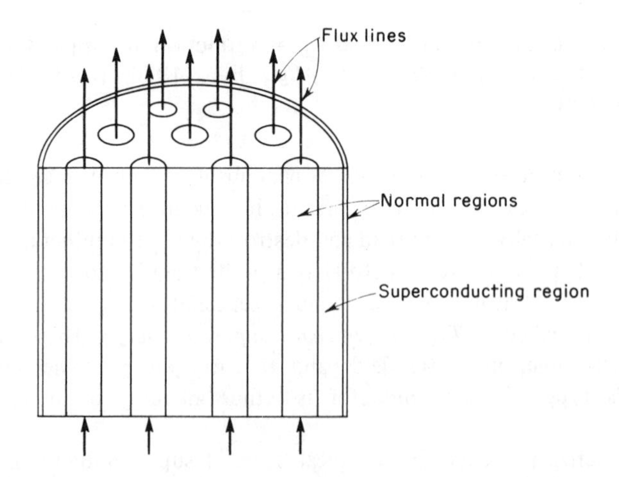

Fig. 3.52 A simplified schematic representation of the flux penetration in a Type II superconductor. Note that this type of material represents a mixed superconducting and normal state. (Drawing after Seeger, A., *Met. Trans.*, 1 2987, 1970.

superconducting coils. There are several good review papers concerning this type of superconductor to which the reader is referred.[37,38,39] In brief, the mixed state (normal and superconducting) that exists between H_{c_1} and H_{c_2} requires some type of pinning action to maintain the "flux lines" or flux line bundles in

37. Dew-Hughes, D., *Mater. Sci. Eng.*, **1** 2 (1966).
38. Seeger, A., *Met. Trans.*, **1** 2987 (1970).
39. Livingston, J. D., and Schadler, H. W., *Prog. Mat. Sci.*, **12** 183 (1964).

place. If this pinning action is missing, the structure shown in Fig. 3.52 will disintegrate. There are a number of ways that pinning occurs. One is believed to be a result of variations in the chemical composition of these alloy superconductors. Another is related to a nonuniform distribution of the effects of plastic deformation of the specimen.

Problems

1. The general equation for the force between electrical charges is

$$f = k\frac{e_1 e_2}{r_{12}^{\,2}}$$

where k is a constant with units force x distance^2 ÷ charge^2. In the electrostatic system, $k = 1$ dyne cm^2/(statcoulombs)^2.
(*a*) Compute the force between a positive and a negative charged ion pair separated by a distance equal to the average spacing of atoms in a crystal, or about 3 Å.
(*b*) Compute the potential energy ϕ of these charges. Express your answer in ergs, in electron volts, and in calories per mole.

2. Compute the coulomb interaction energy for the KCl lattice (this has the same structure as the NaCl lattice). Express the answer in calories per mole and compare with the value given in Table 3.2. Assume $r = 3.15$ Å.

3. Compute the repulsive energy in calories per mole for the KCl lattice assuming that the Born exponent is 8.75. B can be determined by taking the derivative of the Born equation with respect to r and assuming that at r_0 the forces on the ions are zero. (Coulomb force equals repulsive force).

4. Prove that the force of a simple dipole on an electric charge e is given by

$$f = \frac{2e_1 a \cos\theta e}{r^3}$$

Note: with regard to Fig. 3.6, it can be assumed that if $r >> a$, then $l_1 \approx r + \frac{a}{2}\cos\theta$, and $l_2 \approx r - \frac{a}{2}\cos\theta$.

5. Compute the force exerted between an ion with a single charge and a simple dipole. Assume $r = 3 \times 10^{-8}$ cm and $a = 10^{-10}$ cm. Let $\theta = 0°$. Compare the answer to this problem with that of Problem 1.

6. The Debye temperature is assumed to represent the temperature where the lattice vibrational energy, as given by kT, equals the energy associated with the highest vibrational mode, or when $kT_D = h\nu_m$, where k is Boltzmann's constant, T_D is the Debye temperature, h is the Planck's constant, and ν_m is the maximum vibrational frequency of the crystal. The Debye temperature for copper, as determined by specific heat measurements, is 315°K. Compute a value for ν_m. How does this value compare with $\nu_m = 10^{13}$ as estimated in the text?

7. Assuming that the maximum vibration frequency is 10^{13}, compute a value in calories per mole for the zero point energy of a crystal. Compare the answer to this problem with the zero point energies listed in Table 3.2.
8. The Debye temperature (see Problem 6) for neon is 63°K. Compute the zero point energy for solid neon and compare with the value given in Section 3.10.
9. If an electron is accelerated from rest by a potential of 50,000 volts, what will be its
   (*a*) kinetic energy in ergs and in calories per mole;
   (*b*) velocity and momentum;
   (*c*) DeBroglie wavelength in Å; and
   (*d*) wave number k?
10. When the sum of the squares of the quantum numbers $(n_x^2 + n_y^2 + n_z^2)$ is 50, what will be the degeneracy in the problem of an electron in a box? See Table 3.3.
11. The second Brillouin zone in a single cubic lattice corresponds to reflections from what crystallographic planes? Make a sketch of the second Brillouin zone of this structure.
12. (*a*) What planes determine the first Brillouin zone of the body-centered cubic crystal structure? Hint: see Appendix C.
   (*b*) What planes should form the second zone?
   (*c*) Is there a correspondence between the body-centered cubic zones and the simple cubic zones?
13. Compute the energy of an electron traveling parallel to the x axis in a simple cubic crystal if it would have an energy equivalent to that of the center of the first energy gap. Assume d is 2.5 Å.
14. Given an atom diameter of 3 Å,
   (*a*) Compute the lattice parameter $a_{\text{f.c.c.}}$ if this atom forms a face-centered cubic crystal.
   (*b*) Compute the corresponding lattice parameter if the atom forms a body-centered cubic crystal.
   (*c*) Compute the energy corresponding to the center of the first energy gap if an electron moves along [100] in the f.c.c. lattice.
   (*d*) Compute the corresponding energy if an electron moves along [110] in the b.c.c. lattice.
   (*e*) Compare your results with Fig. 3.38.
15. Assume a simple metal with free electrons to which is applied a magnetic field whose strength is 1000 oersteds. Compute the energy change in eV associated with the reversal of the spin of one electron under the action of this field.
16. Compute the paramagnetic susceptibility of potassium due to its conduction electrons. The Fermi temperature may be determined by letting kT_F equal the Fermi energy, (see Section 3.19). Take the atomic volume of potassium as 45 cm^3/mole and the Fermi energy as 2.1 eV.
17. The force between two magnetic poles m and m' is given by the equation $F = mm'/r^2$, so that the potential or energy associated with these two poles,

separated by a distance r, is mm'/r. The units of a magnetic pole can thus be expressed as $\text{ergs}^{1/2} \times \text{cm}^{1/2}$. At the same time, the magnetic field intensity H, due to a pole of strength m, is m/r^2.

(*a*) Express the units of H in terms of ergs and centimeters.

(*b*) Prove that μH has the units of ergs.

(*c*) Show that H and M have identical units.

18. The saturation magnetization of iron at 0°K is 1740 gauss. Show that this value corresponds to a dipole moment of 2.2 Bohr magnetons per atom.

19. (*a*) Compute the magnitude of the susceptibility of a metal when it is in the superconductivity state (Meissner state).

(*b*) How many times larger is this than the susceptibility normally observed with diamagnetic materials?

4 Dislocations and Slip Phenomena

4.1 The Discrepancy Between the Theoretical and Observed Yield Stresses of Crystals. The stress-strain curve of a typical magnesium single crystal, oriented with the basal plane inclined at 45° to the stress axis and strained in tension, is shown in Fig. 4.1. At the low-tensile stress of 100 psi, the crystal yields plastically and then easily stretches out to a narrow ribbon which may be four or five times longer than the original crystal. If one examines the surface of the deformed crystal, markings can be seen which run more or less continuously around the specimen in the form of ellipses. (See Fig. 4.2.) These markings, if viewed at very high magnifications, are recognized as the visible manifestations of a series of fine steps that have formed on the surface. The nature of these steps is shown schematically in Fig. 4.3. Evidently, as a result of the applied force, the crystal has been sheared on a number of parallel planes. Crystallographic analyses of the markings, furthermore, show that these are basal (0002) planes and, therefore, the closest packed plane of the crystal. When this type of deformation occurs, the crystal is said to have undergone "slip," the visible markings on the surface are called *slip lines*, or *slip traces*, and the crystallographic plane on which the shear has occurred is called the *slip plane*.

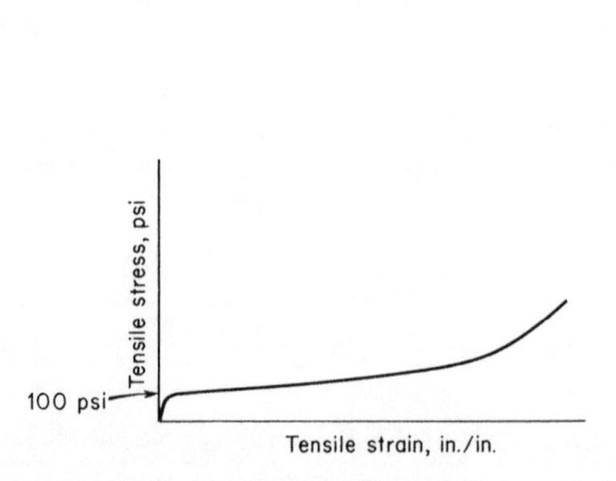

Fig. 4.1 Tensile stress-strain curve for a magnesium single crystal.

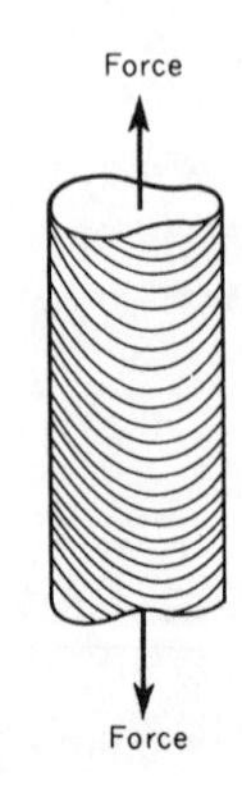

Fig. 4.2 Slip lines on magnesium crystal.

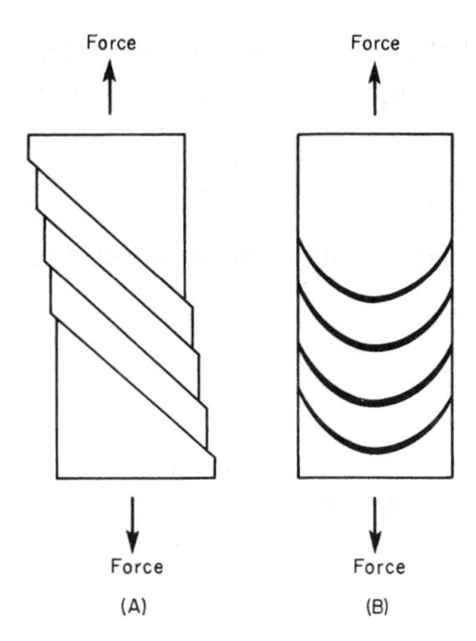

Fig. 4.3 (A) Magnified schematic view of slip lines (side view). (B) Magnified schematic view of slip lines (front view).

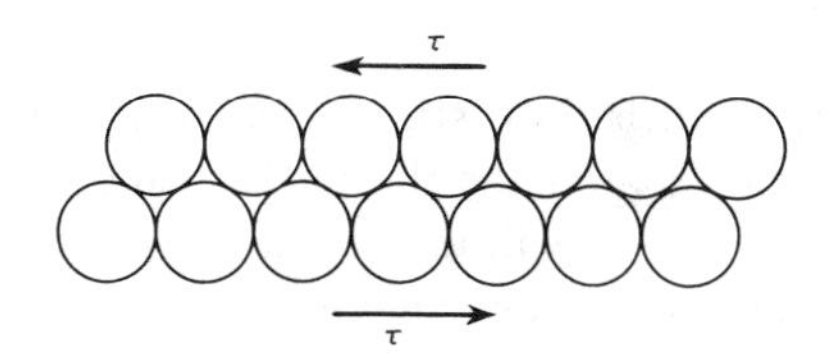

Fig. 4.4 Initial position of the atoms on a slip plane.

The shear stress at which plastic flow begins in a single crystal is amazingly small when compared to the theoretical shear strength of a perfect crystal (computed in terms of cohesive forces between atoms). An estimate of this strength can be obtained in the following manner. Figure 4.4 shows two adjacent planes of a hypothetical crystal. A shearing stress, acting as indicated by the vectors marked τ tends to move the atoms of the upper plane to the left. Each atom of the upper plane rises to a maximum position (Fig. 4.5) as it slides over its neighbor in the plane below. This maximum position represents a saddle point, for continued motion to the left will now be promoted by the forces which pull the atom into the next well. A shear of one atomic distance requires that the atoms of the upper plane in Fig. 4.4 be brought to a position equivalent to that in Fig. 4.5, after which they move on their own accord into the next equilibrium position. To reach the saddle point, a horizontal movement of each atom is required equal to an atomic radius. This movement is shown in Fig. 4.6.

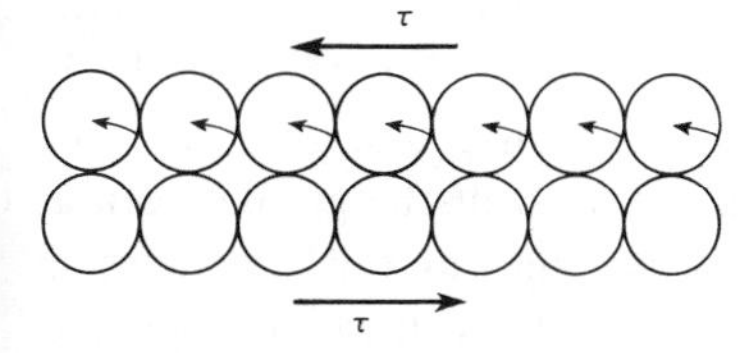

Fig. 4.5 The saddle point for the shear of one plane of atoms over another.

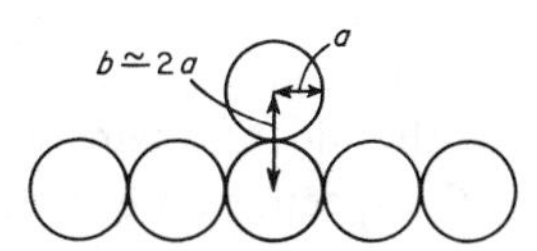

Fig. 4.6 The shear at the saddle point is approximately a/b or 1/2.

Since the separation of the two planes is of the order of two atomic radii, the shear strain at the saddle point is approximately equal to one half. That is,

$$\gamma \simeq \frac{a}{2a} \simeq \frac{1}{2}$$

where γ = shear strain. In a perfectly elastic crystal, the ratio of shear stress to shear strain is equal to the shear modulus:

$$\frac{\tau}{\gamma} = \mu$$

where γ is shear strain, τ is shear stress, and μ is shear modulus. Substituting the value ½ for γ the shear strain, and the value 2,500,000 psi for μ, which is of the order of magnitude of the shear modulus for magnesium, we obtain for τ the stress at the saddle point,

$$\tau = \frac{2.5 \times 10^6}{2} \simeq 10^6 \text{ psi}$$

The ratio of the theoretical stress to start the shear of the crystal to that observed in a real crystal is, therefore, approximately

$$\frac{10^6}{10^2} = 10{,}000$$

In other words, the crystal deforms plastically at stresses $\frac{1}{10{,}000}$ of its theoretical strength. Similarly with other metals, real crystals deform at small fractions of their theoretical strengths ($\frac{1}{1000}$ to $\frac{1}{10{,}000}$).

4.2 Dislocations. The explanation for the discrepancy between the computed and real yield stresses lies in the fact that real crystals are not perfect but contain defects. Experimental proof for the existence of these defects can be obtained with the aid of the electron microscope. Thus, suppose that a crystal has been deformed so as to form visible slip lines, as indicated in Fig. 4.3. Let us now assume that it is possible to cut from the deformed crystal a transmission electron microscope foil containing a portion of a slip plane. In preparing such a foil, great care is normally required so that the preparation of the specimen does not introduce additional deformation into the metal. A convenient machine for this purpose is a spark machining unit that can cut slices with the aid of a wire slicer. In this case, electrical discharges between a wire and the specimen are used to cut the metal by removing small particles of metal from the surface of the specimen as the wire slices down through the specimen. This leaves a thin, highly distorted layer near the cut surface that can be later polished away by chemical or electrochemical means. A typical section obtained by this type of spark machining may be about 0.2 mm thick. This is

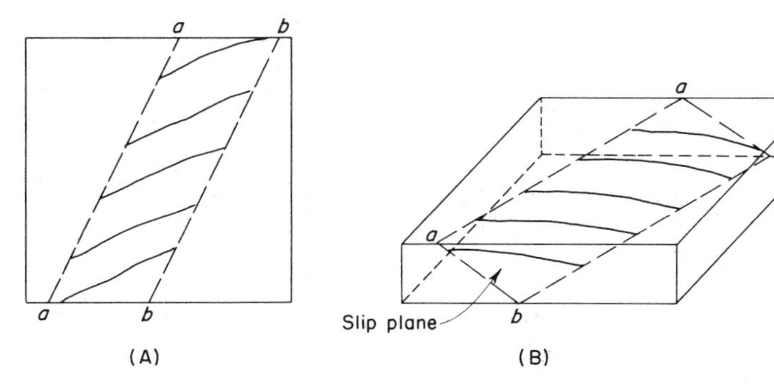

Fig. 4.7 (A) Schematic representation of an electron microscope photograph showing a section of a slip plane. (B) A three-dimensional view of the same slip plane section.

much too thick to be transparent to the electrons in an electron microscope. Therefore, it is necessary to thin the specimen to the desired thickness of several thousand Å using a chemical or electrochemical polishing technique. The details of the procedure can be found in standard texts dealing with electron microscopy.[1]

If the transmission foil has been prepared properly and contains a section of a slip plane, when it is examined in the microscope one may obtain a photograph of the type shown schematically in Fig. 4.7A. In this photograph a set of dark lines may be seen that start and end at the two dashed lines *a-a* and *b-b*. These latter have been drawn on the figure to indicate the positions where the slip plane intersects the foil surfaces. It should be noted that the drawing in Fig. 4.7A is a two-dimensional projection of a three-dimensional specimen. So that the geometrical relations involved in this figure may be better understood, a three-dimensional sketch of the specimen is shown in Fig. 4.7B. This diagram demonstrates that the dark lines in the photograph run across the slip plane from the top to the bottom surfaces of the foil. The fact that these lines are visible in an electron microscope photograph implies that they represent defects in the crystal structure.

We may conclude from the above that in a crystal which has undergone slip, lattice defects tend to accumulate along the slip planes. These defects are called *dislocations*. The presence of dislocations can also be made evident in another fashion. The points where they intersect a specimen surface can often be made visible by etching the surface with a suitable etching solution. The fact that the dislocations are defects in the crystal structure tends to make the places where they intersect the surface preferred positions for the attack of the etching

1. Thomas, G., *Transmission Electron Microscopy of Metals*, John Wiley and Sons, Inc., New York, 1962.

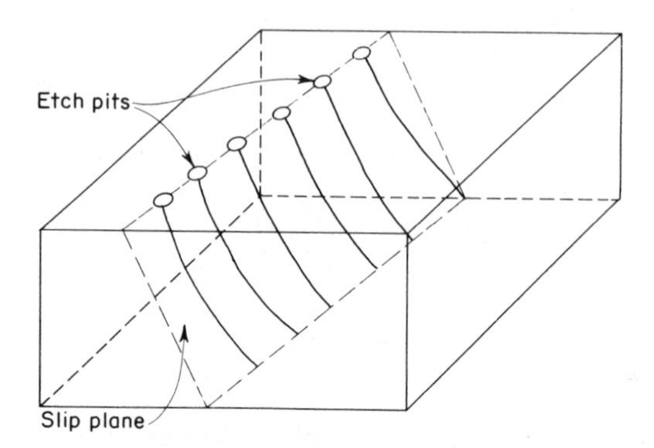

Fig. 4.8 Dislocations can also be revealed by etch pits.

solution. As a result, etch pits may form. This is indicated schematically in Fig. 4.8. Photographs showing etch pits associated with dislocations may be seen in Figs. 4.25 and 5.23.

An actual electron microscope photograph is shown in Fig. 4.9. This picture shows a portion of a foil of an aluminum specimen with a grain containing a slip plane with dislocations. The specimen was polycrystalline. The dark region at the upper right-hand corner represents a second grain. Its orientation was such that it did not diffract as strongly as the larger grain. As a consequence, this crystal appears black. Note that the dislocations apparently have originated at the grain boundary between the two crystals.

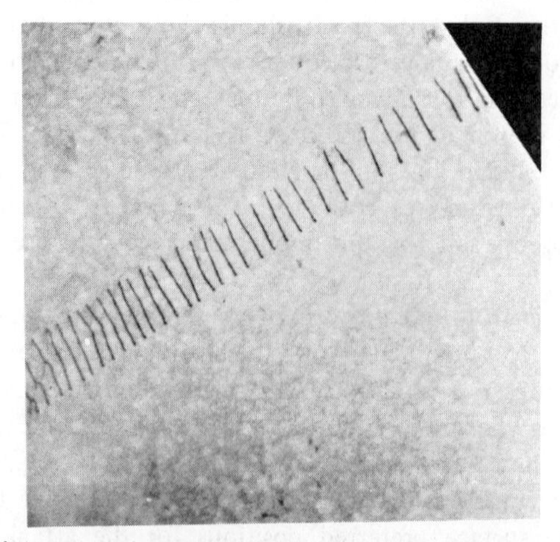

Fig. 4.9 An electron micrograph of a foil removed from an aluminum specimen. Note the dislocations lying along a slip plane, in agreement with Fig. 4.7. (Photograph courtesy of E. J. Jenkins and J. Hren).

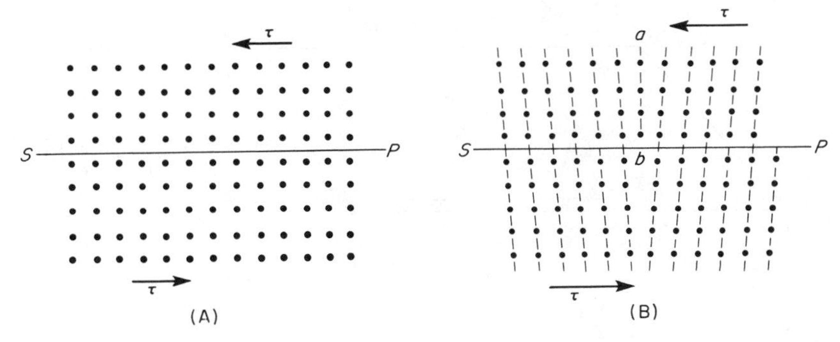

Fig. 4.10 An edge dislocation. (A) A perfect crystal. (B) When the crystal is sheared one atomic distance over part of the distance *S–P*, an edge dislocation is formed.

The fact that dislocations are visible in a transmission electron microscope, and that they may also be revealed by etching a specimen surface, agrees with the assumption that they represent disturbances in the crystal structure. The best evidence now indicates that they are boundaries on the slip planes where a shearing operation has ended. Let us look into the nature of these small shears. Figure 4.10A represents a simple cubic crystal that is assumed to be subjected to shearing stresses, τ, on its upper and lower surfaces, as indicated in the diagram. The line *SP* represents a possible slip plane in the crystal. Suppose that as a result of the applied shear stress, the right-hand half of the crystal is displaced along *SP* so that the part above the slip plane is moved to the left with respect to the part below the slip plane. The amount of this shear is assumed to equal one interatomic spacing in a direction parallel to the slip plane. The result of such a shear is shown in Fig. 4.10B. As may be seen in the figure, this will leave an extra vertical half-plane *cd* below the slip plane at the right and outside the crystal. It will also form an extra vertical half-plane *ab* above the slip plane and in the center of the crystal. All other vertical planes are realigned so that they run continuously through the crystal.

Now let us consider the extra half-plane *ab* that lies inside the crystal. An examination of Fig. 4.10B clearly shows that the crystal is badly distorted where this half-plane terminates at the slip plane. It can also be deduced that this distortion decreases in intensity as one moves away from the edge of this half-plane. This is because at large distances from this lower edge of the extra plane, the atoms tend to be arranged as they would be in a perfect crystal. The distortion in the crystal is thus centered around the edge of the extra plane. This boundary of the additional plane is called an *edge dislocation*. It represents one of the two basic orientations that a dislocation may take. The other is called a *screw dislocation* and will be described shortly.

Figure 4.11 represents a three-dimensional sketch of the edge dislocation of Fig. 4.10. The figure clearly shows that the dislocation has the dimensions of a

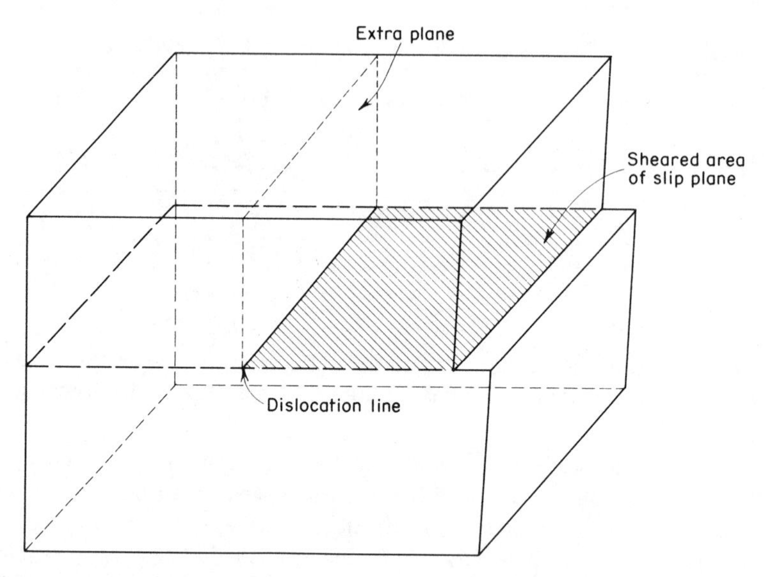

Fig. 4.11 This three-dimensional view of a crystal containing an edge dislocation shows that the dislocation forms the boundary on the slip plane between a region that has been sheared and a region that has not been sheared.

line, in agreement with our discussion of Fig. 4.10. Another important fact shown in Fig. 4.11 is that the dislocation line marks the boundary between the sheared and unsheared parts of the slip plane. This is a basic characteristic of a dislocation. In fact, a dislocation may be defined as a line that forms a boundary on a slip plane between a region that has slipped and one that has not.

Figure 4.12 illustrates how the above dislocation moves through the crystal under an applied shear stress which is indicated by the vectors τ. As a result of the applied stress, atom c may move to the position marked c' in Fig. 4.12B. If it does, the dislocation moves one atomic distance to the left. The plane x, at the top of the figure, now runs continuously from top to bottom of the crystal, while plane y ends abruptly at the slip plane. Continued application of the stress will cause the dislocation to move by repeated steps along the slip plane of the crystal. The final result is that the crystal is sheared across the slip plane by one atomic distance, as is shown in Fig. 4.12C.

Each step in the motion of the dislocation, as can be seen in Figs. 4.12A and B, requires only a slight rearrangement of the atoms in the neighborhood of the extra plane. As a result, a very small force will move a dislocation. Theoretical calculations show that this force is of the correct order of magnitude to account for the low-yield stresses of crystals.

The existence of dislocations was postulated at least a quarter of a century before experimental techniques were available to make them visible. In 1934,

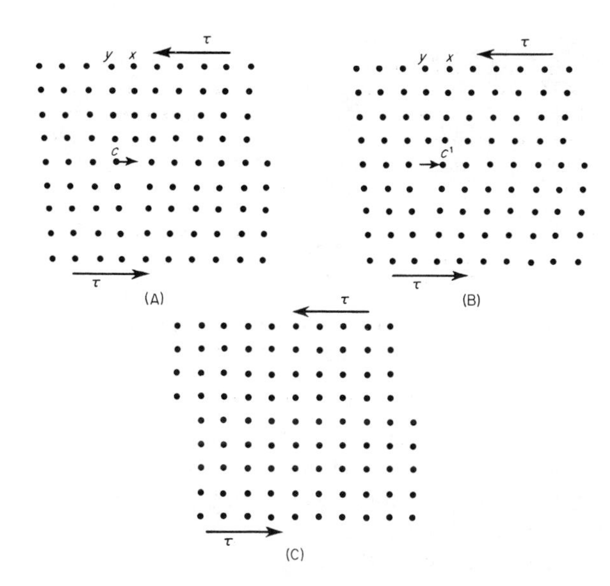

Fig. 4.12 Three stages in the movement of an edge dislocation through a crystal.

Orowan,[2] Polyani,[3] and Taylor[4] presented papers which are said to have laid the foundation for the modern theory of slip due to dislocations.[5] This early work in the field of dislocations in metal crystals had as its basis an effort to explain the large discrepancy between the theoretical and observed shear strengths of metal crystals. It was felt that the observed low yield strengths of real crystals could best be explained on the basis that the crystal contained defects in the form of dislocations.

The movement of a single dislocation completely through a crystal produces a step on the surface the depth of which is one atomic distance. Since an atomic distance in metal crystals is of the order of a few Angstrom units, such a step will certainly not be visible to the naked eye. Many hundreds or thousands of dislocations must move across a slip plane in order to produce a visible slip line. A mechanism will be given presently to show how it is possible to produce this number of dislocations on a single slip plane inside a crystal. First, however, it is necessary to define a screw dislocation, which is shown schematically in Fig. 4.13A where each small cube can be considered to represent an atom. Figure 4.13B represents the same crystal with the position of the dislocation line

2. Orowan, E., *Z. Phys.*, 89 634 (1934).
3. Polanyi, *Z. Phys.*, 89 660 (1934).
4. Taylor, G. I., *Proc. Roy. Soc.*, **A145** 362 (1934).
5. Nabarro, F. R. N., *Theory of Crystal Dislocations*, p. 5, Oxford University Press, London, 1967.

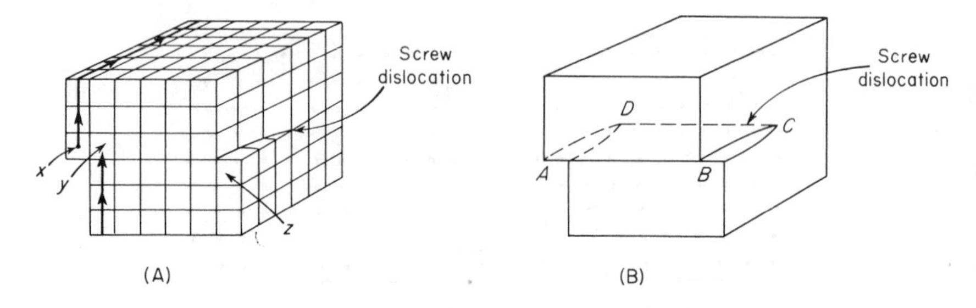

Fig. 4.13 Two representations of a screw dislocation. Notice that the planes in this dislocation spiral around the dislocation like a left-hand screw.

marked by the line *DC*. The plane *ABCD* is the slip plane. The upper front portion of the crystal has been sheared by one atomic distance to the left relative to the lower front portion. The designation "screw" for this lattice defect is derived from the fact that the lattice planes of the crystal spiral the dislocation line *DC*. This statement can be proved by starting at point x in Fig. 4.13A then proceeding upward and around the crystal in the direction of the arrows. One circuit of the crystal ends at point y; continued circuits will finally end at point z.

Figure 4.13B plainly shows that a dislocation in a screw orientation also represents the boundary between a slipped and an unslipped area. Here the dislocation, centered along line *DC*, separates the slipped area *ABCD* from the remainder of the slip plane in back of the dislocation.

The edge dislocation shown in Fig. 4.10B has an incomplete plane which lies above the slip plane. It is also possible to have the incomplete plane below the slip plane. The two cases are differentiated by calling the former a *positive edge dislocation*, and the latter a *negative edge dislocation*. It should be noted that this differentiation between the two dislocations is purely arbitrary, for a simple rotation of the crystal of 180° will convert a positive dislocation into a negative one, and vice versa. Symbols representing these two forms are ⊥ and ⊤ respectively, where the horizontal line represents the slip plane and the vertical line the incomplete plane. There are also two forms of screw dislocations. The screw dislocation shown in Fig. 4.13 has lattice planes that spiral the line *DC* like a left-hand screw. An equally probable screw dislocation is one in which the lattice spirals in a right-hand fashion around the dislocation line. Both forms of the edge and the screw dislocations respectively are shown in Fig. 4.14. The figure also illustrates how the four types move under the same applied shear stress (indicated by the vectors τ). As previously mentioned, Fig. 4.14 shows that a positive edge dislocation moves to the left when the upper half of the lattice is sheared to the left. On the other hand, a negative edge dislocation moves to the right, but produces the identical shear of the crystal. The figure also demonstrates that the right-hand screw moves forward and the left-hand screw moves to the rear, again producing the same shear of the lattice.

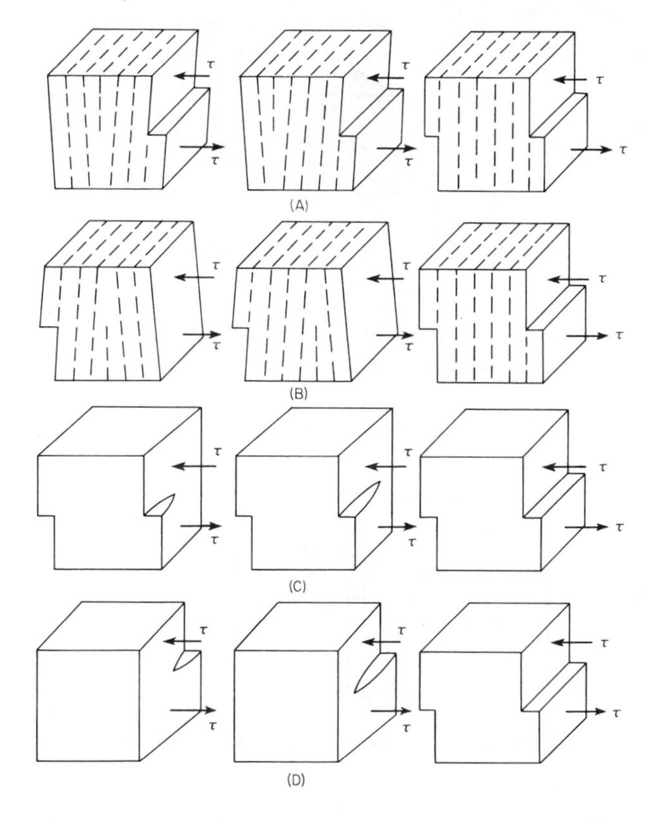

Fig. 4.14 The ways that the four basic orientations of a dislocation move under the same applied stress: (A) Positive edge, (B) Negative edge, (C) Left-hand screw, and (D) Right-hand screw.

In the above examples, the dislocation lines have been assumed to run as straight lines through the crystal. It is a consequence of the basic nature of dislocations that they cannot end inside a crystal. Thus, the extra plane of an edge dislocation may extend only part way through the crystal, as is shown in Fig. 4.15, and its rear edge *b* then forms a second edge dislocation. The two dislocation segments *a* and *b* thus form a continuous path through the crystal from front to top surfaces.

It is also possible for all four edges of an incomplete plane to lie inside a crystal, forming a four-sided closed edge dislocation at the boundaries of the plane. Furthermore, a dislocation that is an edge in one orientation can change to a screw in another orientation, as is illustrated in Fig. 4.16. Figure 4.17 shows the same dislocation viewed from above. Open circles represent atoms lying in the plane just above the slip plane, while dots represent the atoms just below the slip plane. Notice that the lattice is sheared by an atomic distance in the region

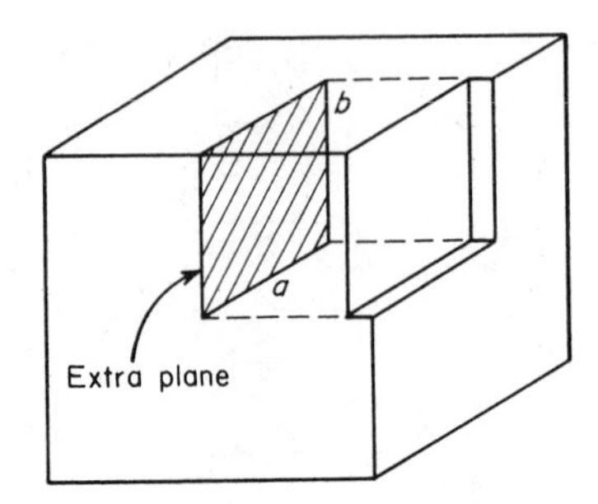

Fig. 4.15 Dislocations can vary in direction. This crosshatched extra plane forms a dislocation with edge components *a* and *b*.

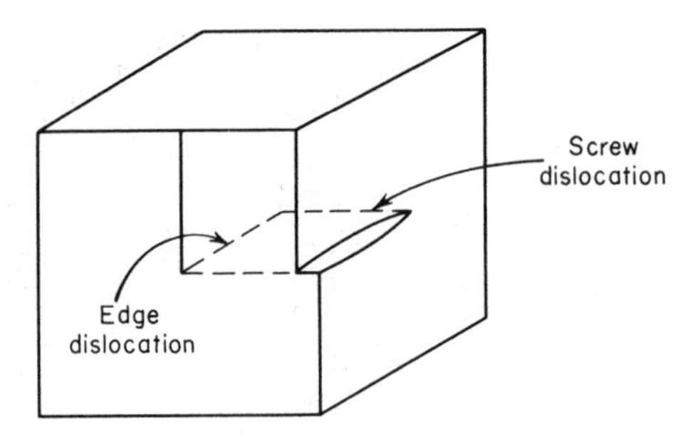

Fig. 4.16 A two-component dislocation composed of an edge and a screw component.

at the lower right-hand quarter of the figure bounded by the two dislocation segments. Finally, a dislocation does not need to be either pure screw or pure edge, but may have orientations intermediate to both. This fact signifies that dislocation lines do not have to be straight, but can be curved. An example is shown in Fig. 4.18. The drawing, like Fig. 4.17, shows a change in orientation from edge to screw, but here the change is not abrupt.

Consider the closed rectangular dislocation of Fig. 4.19, consisting of the four elementary types of dislocations shown in Fig. 4.14. Sides *a* and *c* are positive and negative edge dislocations respectively, while *b* and *d* are right- and left-hand screws respectively. Figure 4.14 shows that, under the indicated shear-stress sense, dislocations *a* and *c* move to the left and right respectively, while *b* and *d* move forward and to the rear respectively. The dislocation loop thus opens, or becomes larger, under the given stress. (It would close, however, if the sense of the stress were reversed.) From what has been stated above, it is evident that the dislocation loop *abcd* does not need to be rectangular in order to open under the given shearing stress. A closed curve, such as a circle, would also expand in a similar manner and shear the crystal in the same way.

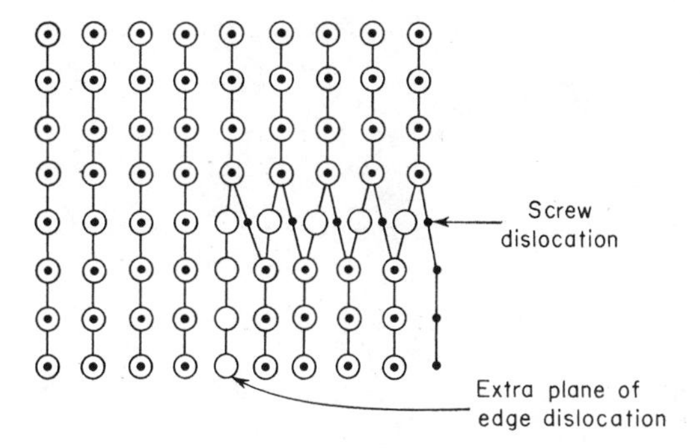

Fig. 4.17 Atomic configuration corresponding to the dislocation of Fig. 4.16 viewed from above. Open-circle atoms above slip plane, dot atoms below slip plane.

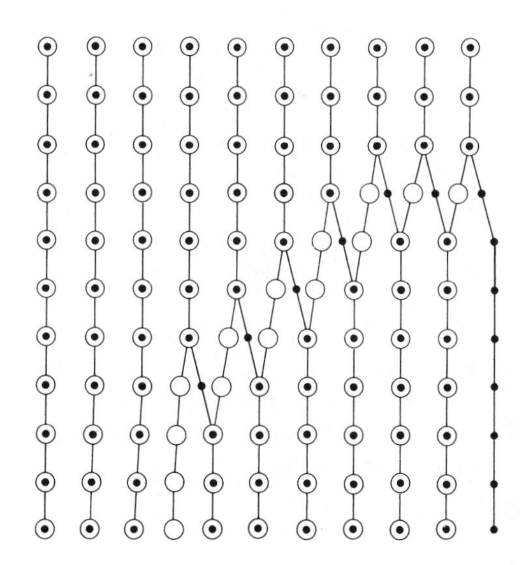

Fig. 4.18 A dislocation that changes its orientation from a screw to an edge as viewed from above looking down on its slip plane.

It has been mentioned that a dislocation cannot end inside a crystal. The reason for this is easy to understand. This is because a dislocation represents the boundary between a slipped area and an unslipped area. If the slipped area on the slip plane does not touch the specimen surface, as in Fig. 4.20, then its boundary is continuous and the dislocation has to be a closed loop. Only when the slipped area extends to the specimen surface, as in Fig. 4.15, is it possible for a single dislocation to have an end point.

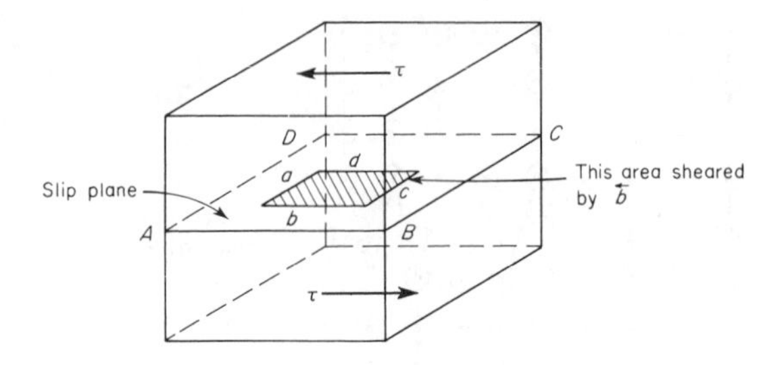

Fig. 4.19 A closed dislocation loop consisting of (*A*) Positive edge, (*B*) Right-hand screw, (*C*) Negative edge, and (*D*) Left-hand screw.

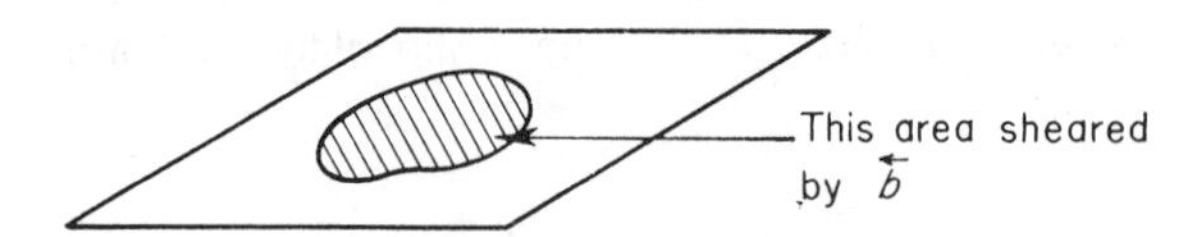

Fig. 4.20 A curved dislocation loop lying in a slip plane.

4.3 The Burgers Vector. The area inside the rectangle *abcd* of Fig. 4.19, or inside the closed loop of a more general curved dislocation loop, such as that in Fig. 4.20, is sheared by one atomic distance, that is, inside this region the lattice lying above the slip plane (*ABCD*) has slipped one atomic distance to the left relative to the lattice below the slip plane. The direction of this shear is indicated by the vector $\overleftarrow{b}$, the length of which is one atomic distance. Outside of the dislocation loop shown in Fig. 4.19, the crystal is not sheared. The dislocation is, therefore, a discontinuity at which the lattice shifts from the unsheared to the sheared state. Although the dislocation varies in orientation in the slip plane *ABCD*, the variation in shear across the dislocation is everywhere the same, and the slip vector $\overleftarrow{b}$ is therefore a characteristic property of the dislocation. By definition, this vector is called the *Burgers vector of the dislocation*.

The Burgers vector of a dislocation is an important property of a dislocation because, if the Burgers vector and the orientation of the dislocation line are known, the dislocation is completely described. Figure 4.21 shows a method of determining the Burgers vector applied to a positive edge dislocation.[6] It is first necessary to choose arbitrarily a positive direction for the dislocation. In the present case, let us assume it to be the direction out of the paper. In Fig. 4.21A a counterclockwise circuit of atom-to-atom steps in a perfect crystal closes, but

6. There are several conventions for defining the Burgers vector. This gives what is known as the local Burgers vector. For a more detailed discussion of Burgers vectors see J. P. Hirth and J. Lothe, *Theory of Dislocations*, pp. 19–22, McGraw-Hill Book Co., New York, 1968.

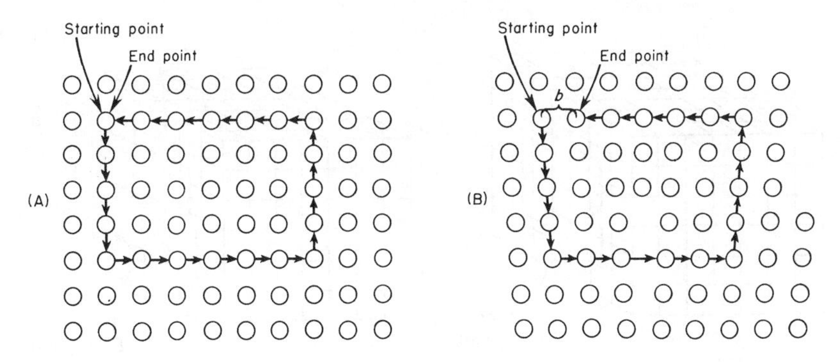

Fig. 4.21 The Burgers circuit for an edge dislocation: (A) Perfect crystal and (B) Crystal with dislocation.

when the same step-by-step circuit is made around a dislocation in an imperfect crystal (Fig. 4.21B) the end point of the circuit fails to coincide with the starting point. The vector *b* connecting the end point with the starting point is the Burgers vector of the dislocation. The above procedure can be used to find the Burgers vectors of any dislocation if the following rules are observed:

1. The circuit is traversed in the same manner as a rotating right-hand screw advancing in the positive direction of the dislocation.
2. The circuit must close in a perfect crystal and must go completely around the dislocation in the real crystal.
3. The vector that closes the circuit in the imperfect crystal (by connecting the end point to the starting point) is the Burgers vector.

Figure 4.22 shows a Burgers circuit around a left-hand screw dislocation. In Fig. 4.22A, the circuit is indicated for the perfect crystal. Figure 4.22B shows the same circuit transferred to a crystal containing a screw dislocation.

It is now possible to summarize certain characteristics of both edge and screw dislocations.

1. *Edge dislocations:*
   (*a*) An edge dislocation lies perpendicular to its Burgers vector.
   (*b*) An edge dislocation moves (in its slip plane) in the direction of the Burgers vector (slip direction). Under a shear-stress sense ⇄ a positive dislocation ⊥ moves to the right, a negative one ⊤ to the left.

2. *Screw dislocations:*
   (*a*) A screw dislocation lies parallel to its Burgers vector.
   (*b*) A screw dislocation moves (in the slip plane) in a direction perpendicular to the Burgers vector (slip direction).

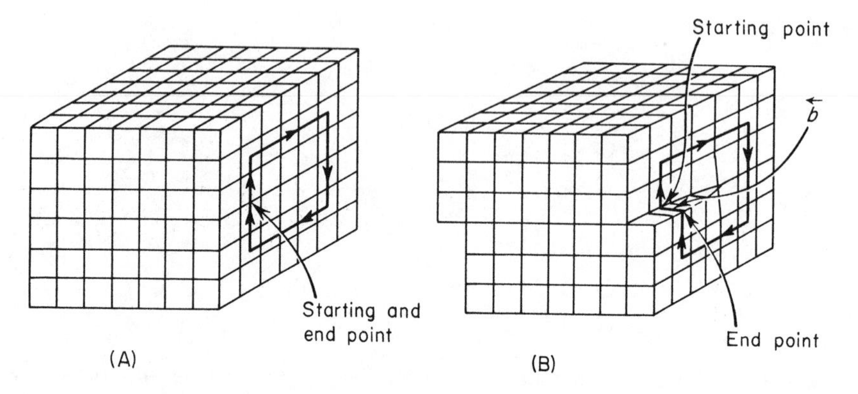

Fig. 4.22 The Burgers circuit for a dislocation in a screw orientation. (A) Perfect crystal and (B) Crystal with dislocation.

A useful relationship to remember is that the slip plane is the plane containing both the Burgers vector and the dislocation. The slip plane of an edge dislocation is thus uniquely defined because the Burgers vector and the dislocation are perpendicular. On the other hand, the slip plane of a screw dislocation can be any plane containing the dislocation because the Burgers vector and dislocation have the same direction. Edge dislocations are thus confined to move in a unique plane, but screw dislocations can glide in any direction as long as they move parallel to their original orientation.

4.4 The Frank-Read Source. Suppose that a positive edge dislocation, lying in plane *ABCD* of Fig. 4.23 is connected to two other edge dislocations running vertically to the upper surface of the crystal. The two vertical edge dislocations cannot move under the applied shear stress. (They move, however, under a shear stress applied to the front and rear of the crystal.) Since the dislocation segment *xy* is a positive edge, the stress τ will tend to move

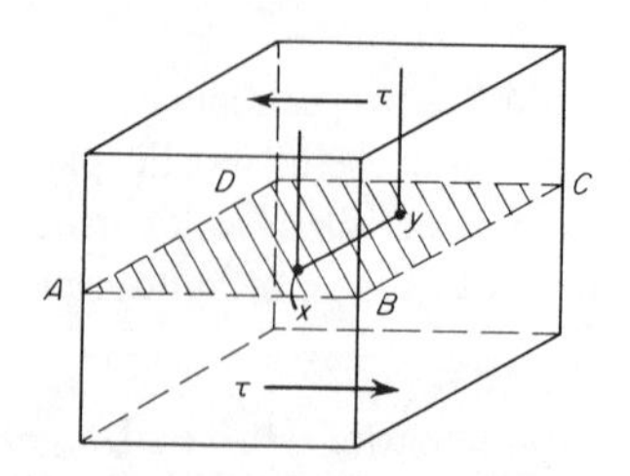

Fig. 4.23 Frank-Read source. The dislocation segment *xy* may move in plane *ABCD* under the applied stress. Its ends, *x* and *y*, however, are fixed.

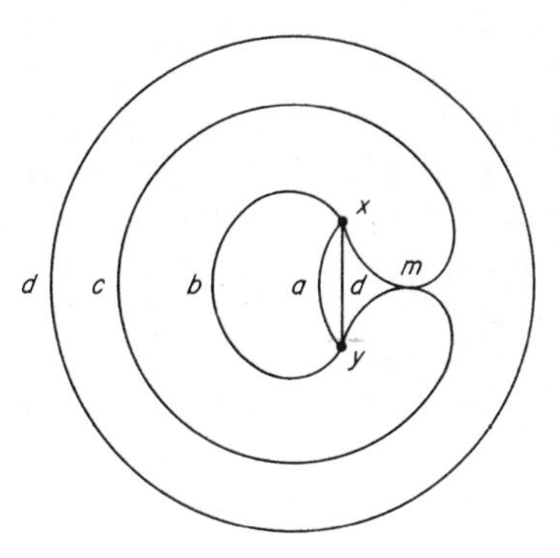

Fig. 4.24 Various stages in the generation of a dislocation loop at a Frank-Read source.

it to the left, causing the line to form an arc with ends at the fixed endpoints *x* and *y*. This arc is indicated in Fig. 4.24 by the symbol *a*. Further application of the stress causes the curved dislocation to expand to the successive positions *b* and *c*. At *c* the loop intersects itself at point *m*, but since one intersecting segment is a left-hand screw and the other a right-hand screw, the segments cancel each other at the point of intersection. (A cancellation always occurs when opposite forms of dislocations lying in the same plane intersect. This fact can be easily demonstrated for the case of positive and negative edge dislocations since their intersections form a complete lattice plane from two incomplete planes.) The cancellation of the dislocation segments at the point of contact *m* breaks the dislocation into two segments marked *d*, one of which is circular and expands to the surface of the crystal, producing a shear of one atomic distance. The other component remains as a regenerated positive edge, lying between points *x* and *y* where it is in a position to repeat the cycle. In this manner many dislocation loops can be generated on the same slip plane, and a shear can be produced that is large enough to account for the large size of observed slip lines. A dislocation generator of this type is called a *Frank-Read source.*

4.5 Nucleation of Dislocations Experimental evidence shows that Frank-Read sources actually exist in crystals.[7] How important these dislocation generators are in the plastic deformation of metals is not known, but recent evidence shows that dislocations can also be formed without the aid of Frank-Read or similar sources.[8] If dislocations are not formed by dislocation generators, then they must be created by a nucleation process. As with all

7. Dash, W. C., *Dislocations and Mechanical Properties of Crystals*, John Wiley and Sons, Inc., 1957, p. 57.
8. Gilman, J. J., *Jour. Appl. Phys.*, **30** 1584 (1959).

phenomena of this nature, dislocations can be created in two ways: homogeneously or heterogeneously. In the case of dislocations, *homogeneous nucleation* means that they are formed in a perfect lattice by the action of a simple stress, no agency other than stress being required. *Heterogeneous nucleation*, on the other hand, signifies that dislocations are formed with the help of defects present in the crystal, perhaps impurity particles. The defects make the formation of dislocations easier by lowering the applied stress required to form dislocations. It is universally agreed that homogeneous nucleation of dislocations require extremely high stresses, stresses that theoretically are of the order of $\frac{1}{10}$ to $\frac{1}{20}$ of the shear modulus of a crystal.[9] Since the shear modulus of a metal is usually about 10^6 to 10^7 psi, the stress to form dislocations should be of the order of 10^5 psi. However, the actual shear stress at which metal crystals start to deform by slip is usually around 100 psi. This evidence certainly favors the opinion that if dislocations are not formed by Frank-Read sources then they must be nucleated heterogeneously.

Metal crystals are not particularly suited to an investigation of nucleation phenomena. When prepared by solidification or other means, they usually possess a relatively high density of dislocations in the form of a more-or-less random network which extends throughout the specimen. Our present interest does not lie in these networks, but in the creation of new, independent dislocation loops. A high concentration of grown-in dislocations, however, complicates the observation of nucleation phenomena. Also, in metals, slip occurs readily so that once the yield point is reached many dislocations usually form at the same time. These experimental difficulties are almost eliminated if dislocation nucleation is studied in crystals of lithium fluoride[10] which is an ionic salt that crystallizes in the simple cubic (rock salt) lattice. This material may be prepared in single-crystal form with a high degree of perfection so that it has a low density of grown-in dislocations ($\simeq 5 \times 10^4/\text{cm}^2$). In addition, crystals of LiF are rigid enough at room temperature to be handled without distortion, and they are only slightly plastic at this temperature. Thus, with a small stress ($\simeq$700 to 1000 psi) applied for a short period of time, dislocations can be formed in them in controlled small numbers.

One of the best and simplest ways of observing dislocations in crystals is through the use of an etching reagent which forms an etch pit on the surface of a crystal at each point where a dislocation intersects the surface. This method is not without its difficulties for there is often no way of knowing whether the etch reveals all dislocations, or whether some of the pits are due to other defects. In lithium fluoride, the etch-pit method seems to be highly reliable.[11] Several

9. Kelly, A., Tyson, W. R., and Cottrell, A. H., *Can. Jour. Phys.*, **45** No. 2, Part 3, p. 883 (1967).
10. Gilman J. J., *Jour. Appl. Phys.*, **30** 1584 (1959).
11. *Ibid.*

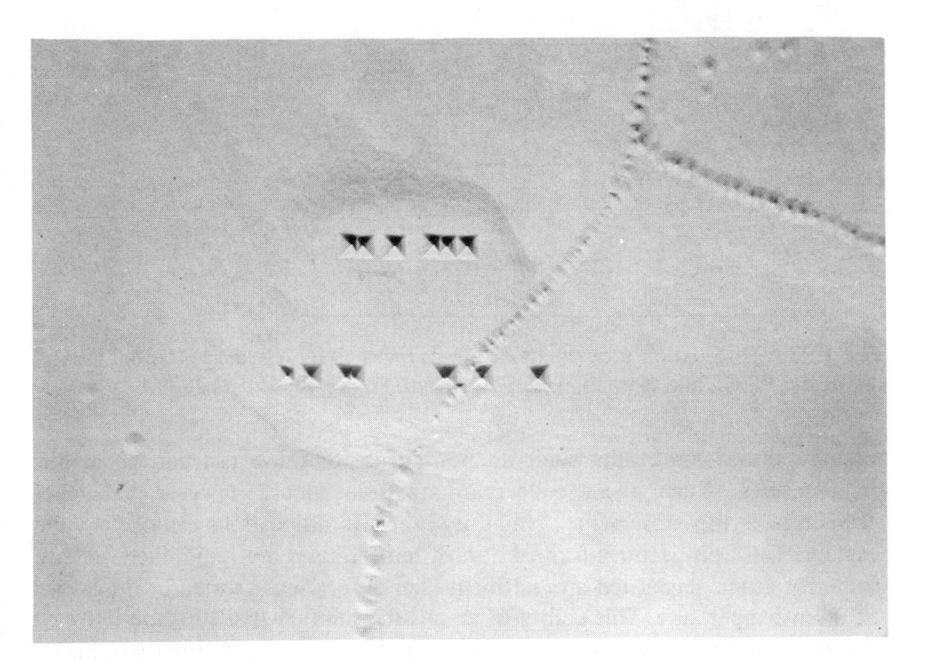

Fig. 4.25 The large square etched pits in horizontal rows correspond to dislocations formed in LiF at room temperature, while the smaller, closely spaced pits lying in curved rows were grown into the crystal when it was manufactured. (Gilman, J. J., and Johnson, W. G., *Dislocations and Mechanical Properties of Crystals*, p. 116, John Wiley and Sons, Inc., New York, 1957.)

etching solutions have been developed [12] for use in LiF, one of which is capable of distinguishing between grown-in dislocations and newly formed dislocations. The action of this solution can be seen in Fig. 4.25. The large square pits which run in two horizontal rows are associated with newly formed dislocations. The horizontal rows define the intercept of the slip plane of these dislocations with the surface. In addition to the large pits, there are two intersecting curved rows of closely spaced smaller pits. The latter outline what is known as a *low-angle grain boundary*. (See Chapter 5.) The boundaries actually consist of a number of closely spaced dislocations. It should be noted that the given etching solution forms large pits at new dislocations and small pits at network dislocations. The reason for this ability of the etch to distinguish between the two types of dislocations is not understood, but it might be related [11] to the fact that impurity atoms tend to collect around dislocations. This segregation of impurity atoms cannot normally occur in a reasonable length of time at low temperatures

12. Gilman, J. J., and Johnston, W. G., *Dislocations and Mechanical Properties of Crystals*. John Wiley and Sons, Inc., New York, 1957.

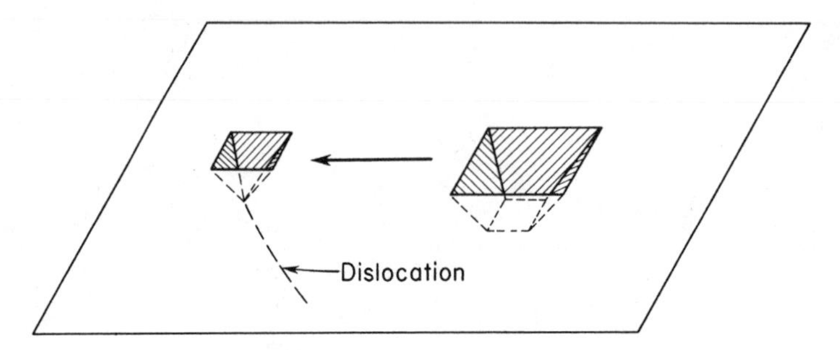

Fig. 4.26 Dislocation movement in LiF as revealed by repeated etching. (After Johnston, W. G., and Gilman, J. J., *Jour. Appl. Phys.*, **30**, 129 [1959].)

because the atoms of the solid do not diffuse or move fast enough at low temperatures to let them collect around dislocations. However, at higher temperatures the movement of impurity atoms to dislocations can occur quite rapidly. Thus, dislocations formed at high temperatures are more likely to have impurity atoms segregated around them than dislocations created in the lattice at room temperature. This ability of an etching solution to distinguish between the grown-in and newly formed dislocations is one of the distinct advantages offered by the use of LiF in studying nucleation phenomena.

Another interesting facet to the use of etch pits in observing dislocations in LiF is that with the proper technique one can follow the movement of a dislocation under the action of an applied stress. This usually requires several repetitions of the etching process. Thus, the surface of a specimen is first etched to reveal the positions of the dislocations at a given time. The pits that form are usually observed on {001} surfaces: LiF crystals are easily split along {001} planes. On this type of surface, the pits form as four-sided pyramids with a sharp point at their lower extremity. If now a stress is applied to the specimen, the dislocations will move away from their pits. A second etch will both reveal the new positions of the dislocations and enlarge the old pits representing the original positions. The two sets of pits have a distinct difference in appearance, however. The pits actually connected with the dislocation always have pointed extremities, while those from which the dislocations have moved have flat bottoms. See Fig. 4.26.

The work on lithium fluoride has shown that, in this particular material, the grown-in dislocations are usually firmly anchored in place and do not take part in the plastic deformation processes. The immobility of network dislocations can be credited to the presence of atmospheres of impurity atoms that segregate around each dislocation. This subject is considered in more detail in Chapter 8.

It has also been demonstrated [13] experimentally that very large stresses can

13. Gilman, J. J., *Jour. Appl. Phys.*, **30** 1584 (1959).

be applied to LiF crystals without homogeneously nucleating dislocations. For example, when a small, carefully cleaned sphere (made of glass) is pressed on a dislocation-free region of the surface of a LiF crystal, it is possible to attain shear stresses estimated to be as large as 110,000 psi, and still not create dislocations. This stress, it is to be noted, is more than 100 times larger than the stress normally required to cause yielding in this material. It is further believed[13] that even this high stress, which was limited by experimental difficulties caused by breaking the glass ball, does not define the stress required to homogeneously nucleate dislocations in LiF. At any rate, it is clear that nucleation of dislocations by an unaided stress is very difficult. Because the yield stress, or the stress at which dislocations normally start to move, is much lower than that required to homogeneously nucleate dislocations, it is clear that the majority of dislocations must be nucleated heterogeneously. Gilman concludes[14] that in LiF small foreign heterogeneities cause most of the dislocation nucleation. The most important of these are probably small impurity particles. Experimental evidence[14] for the formation of dislocations at inclusions in LiF crystals has actually been attained.

4.6 Bend Gliding. It has been known for a good many years that crystals can be plastically bent and that this form of plastic deformation occurs as a result of slip. The bending of crystals can be explained in terms of Frank-Read or other sources.

Let equal couples (of magnitude M) be applied to the ends of the crystal shown in Fig. 4.27. The effect of these couples is to produce a uniform bending moment (M) throughout the length of the crystal. Until the yield point of the

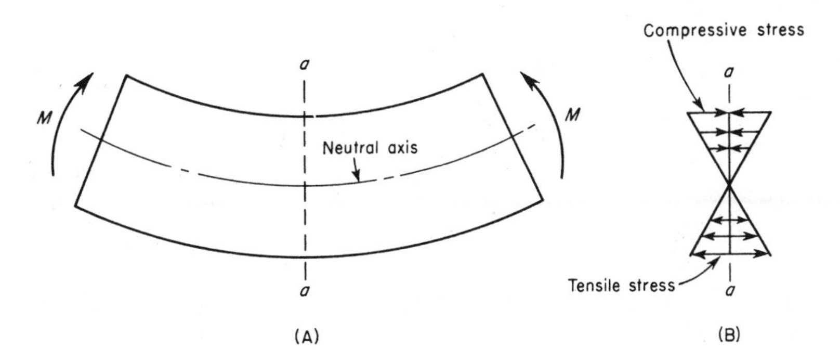

Fig. 4.27 (A) The elastic deformation of a crystal subject to two equal moments (M) applied at its ends. (B) The normal stress distribution on a cross section such as *aa*.

14. *Ibid.*

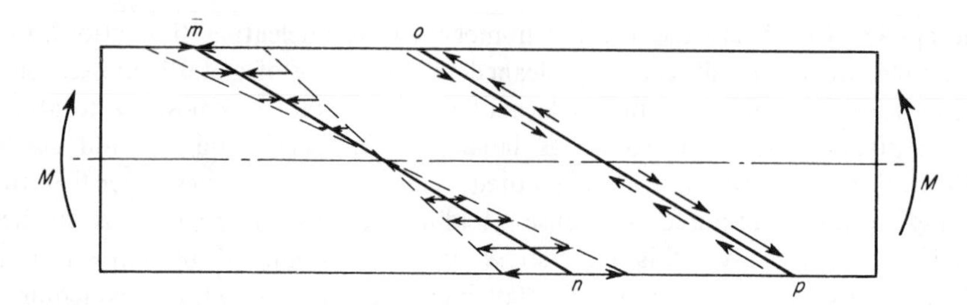

Fig. 4.28 The stress distribution on slip planes corresponding to the elastic deformation shown in Fig. 4.27.

crystal is exceeded, the deformation will be elastic. The stress distribution across any cross section, such as *aa* is given by the equation

$$\sigma_x = \frac{My}{I}$$

where y is the vertical distance measured from the neutral axis of the crystal (dotted horizontal center line), M is the bending moment, and I the moment of inertia of the cross section of the crystal ($\pi r^4/4$ for crystals of circular cross section). Figure 4.27B shows the stress distribution which varies uniformly from a maximum compressive stress at the upper surface, to zero at the neutral axis, and then to a maximum tensile stress at the lower surface. Figure 4.28 represents the same crystal in which the line *mn* is assumed to represent the trace of a slip plane. The curvature of the crystal is not shown in order to simplify the figure. For convenience, it is further assumed that the slip plane is perpendicular to the plane of the paper and that the line *mn* is also the slip direction. The horizontal vectors associated with line *mn* represent the same stress distribution as that of Fig. 4.27B. Figure 4.28 also shows, along line *op*, the shear-stress component (parallel to the slip plane) of the stress distribution. Notice that the sense of the shear stress changes its sign as it crosses the neutral axis. Furthermore, the shear stress is zero at the neutral axis and a maximum at the extreme ends of the slip plane. Because of the shear-stress distribution, the first dislocation loops will form at Frank-Read or other sources close to either the upper or lower surface. The manner in which these dislocation loops move, however, depends on whether the dislocation lies above or below the neutral axis of the specimen. In both cases, the positive edge components of all dislocation loops move toward the surface, while the negative edge components move toward the specimen's neutral axis. (See Fig. 4.29.) The negative edge dislocations move toward a region of decreasing shear stress and must eventually stop. The positive edge dislocations, on the other hand, are in a region of high stress, as are the right- and left-hand screw components of each dislocation loop, that move under the applied stress

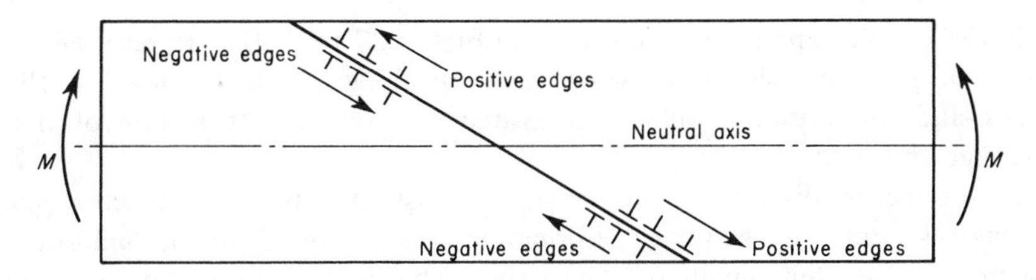

Fig. 4.29 The effect of the stress distribution on the movement of dislocations. Positive-edge components move toward the surface; negative edges toward the neutral axis.

in a direction either into or out of the paper. All three of these last components (positive edge and right- and left-hand screws) can be assumed to move to the surface and leave the crystal. For example, under the applied bending-stress distribution, closed dislocation loops originating at Frank-Read sources eventually become negative edge dislocations that move toward the neutral axis of the crystal. As the crystal is bent further and further, the negative edge dislocations will be driven further and further along the slip planes toward the center of the crystal. Eventually an orderly sequence of dislocations will be found on each active slip plane, the dislocations having a more or less uniform separation, the minimum spacing of which is dependent on the fact that dislocations of the same type and same sign mutually repel if they lie on the same slip plane. Figure 4.30 illustrates the general nature of the dislocation distribution. The narrow section surrounding the neutral axis is presented free of dislocations, in agreement with the fact that this region, under moderate bending stresses, will not be stressed above the elastic limit, and the deformation will be elastic and not plastic.

Now each negative edge dislocation of a sequence lying on a given slip plane represents an extra plane that ends at the slip plane. Each of these extra planes lies on the left of its slip plane. In order to accommodate these extra planes (all on one side of a given slip plane), the slip planes must assume a curvature that is convex downward and to the left, and the crystal as a whole a convex curvature downward.

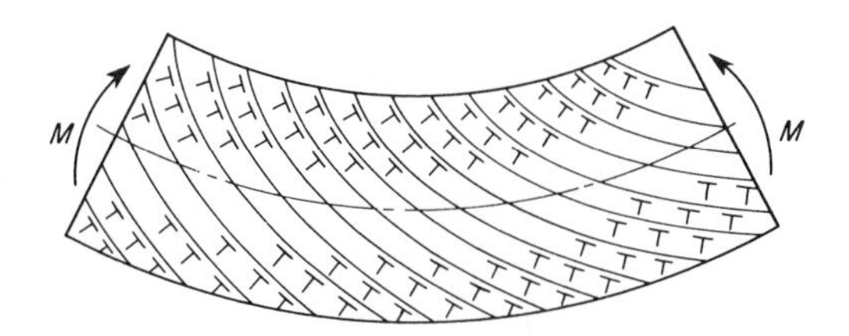

Fig. 4.30 Distribution of the excess edge dislocations in a plastically bent crystal.

If the couples applied to the crystal in Figs. 4.27 to 4.30 were reversed, an excess of positive dislocations would develop along the slip planes. The slip planes and the crystal would then assume a curvature the reverse of that described above.

In the above discussion, the crystal was assumed to be of macroscopic dimensions and the bending deformation was assumed to be uniformly distributed over the length of the crystal. The phenomenon that has been described is not, however, restricted to large crystals. Bending of crystal planes through accumulation of an excess number of edge dislocations of the same sign may occur in quite small crystals, or even in extremely small areas of crystals. A number of related phenomena that have been observed in metal crystals can be explained in terms of localized lattice rotations. In each case, there is an accumulation of edge dislocations of the same sign upon slip planes of the crystal. Among these are kinks, bend planes, and deformation bands. A detailed description of the latter is beyond the scope of the present text, but the fact that they are frequently observed emphasizes the fact that plastic bending of metal crystals is an important deformation mechanism.

4.7 Rotational Slip. Thus far we have seen that dislocations are capable of producing simple shear (as in Fig. 4.14) and bending (as in Fig. 4.30).

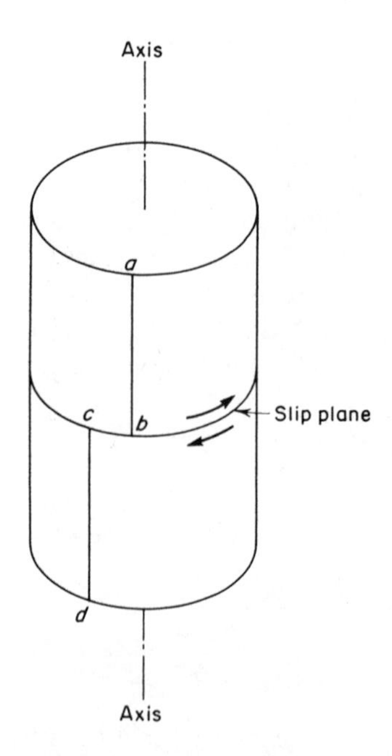

Fig. 4.31 A single crystal can be rotated about an axis normal to a slip plane that contains several slip directions.

A third type of deformation that can be developed by dislocations is shown in Fig. 4.31. In this sketch, the crystal is assumed to be a cylinder with an active slip plane perpendicular to its axis. As may be seen in the drawing, the top half of the crystal has been rotated clockwise relative to the bottom half. This is indicated by the horizontal displacement of the line *abcd* between the points *b* and *c*. This type of deformation therefore corresponds to a rotation of a crystal on its slip plane about an axis normal to the slip plane. Torsional deformation such as this can be explained in terms of screw dislocations lying on the slip plane. However, unlike the case of bending, more than one set of dislocations is required. This, in turn, signifies that the slip plane must contain more than one possible slip direction. The basal plane in a hexagonal close-packed metal and the {111} planes in face-centered cubic metals, with their three slip directions, are almost ideal for producing this type of deformation.

The need for more than one set of screw dislocations in order to explain rotational slip can be seen with the aid of Figs. 4.32 and 4.33. These diagrams

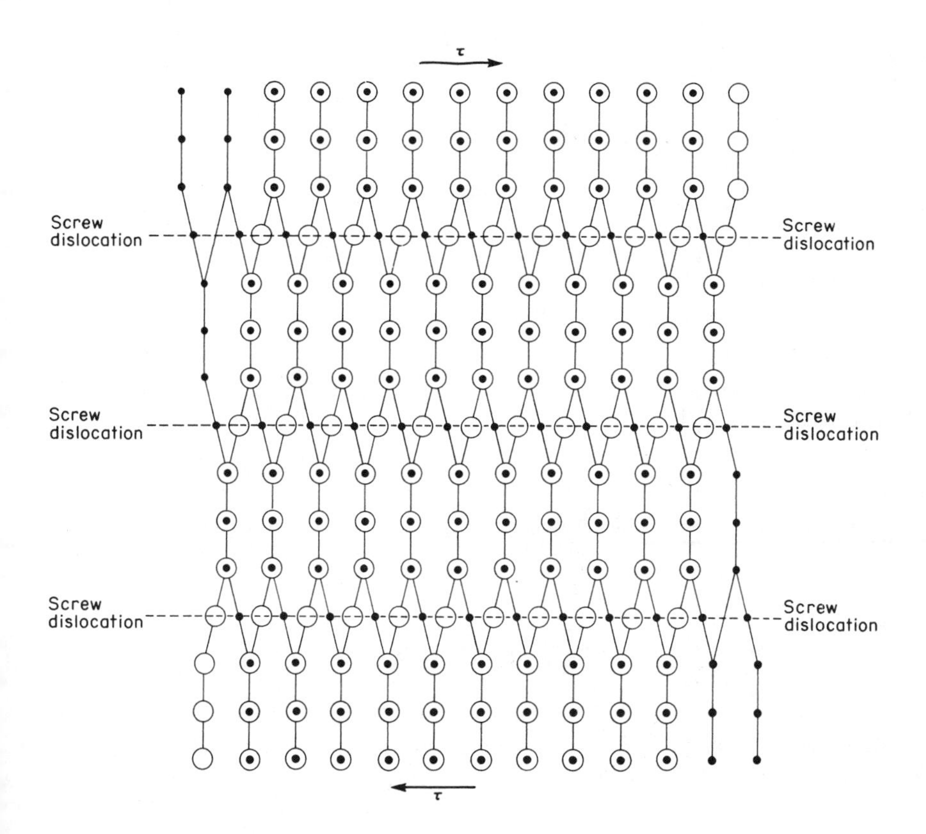

Fig. 4.32 An array of parallel screw dislocations.

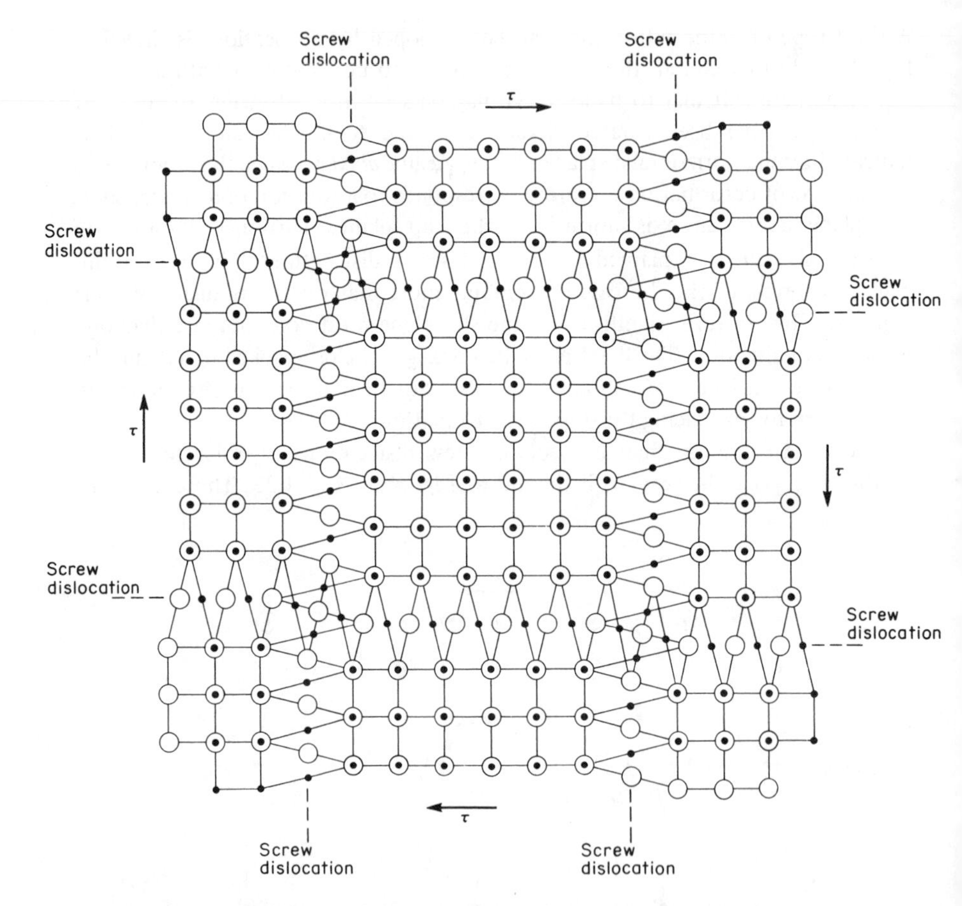

Fig. 4.33 A double array of screw dislocations. This array does not have a long-range strain field.

are drawn in a manner similar to Figs. 4.17 and 4.18 and therefore correspond to a view looking down from above on the slip plane in a simple cubic lattice. As before, open circles represent atoms just above the slip plane, while dots correspond to atoms just below it. Figure 4.32 corresponds to a single array of (horizontal) parallel screw dislocations. As may be seen in the illustration, such a dislocation arrangement shears the material above the slip plane relative to that below it only in the horizontal direction. For a true rotation one needs a similar component of shear at 90° to this direction. A simple double array of screw dislocations is shown in Fig. 4.33. This gives the desired rotational deformation. Further, it should be noted that the two arrays of screw dislocations at 90° to each other have strain fields that tend to compensate each other, either above or below the slip plane. As a result, the strain field of the double array tends to be very small at any reasonable distance from the slip plane; and the array, therefore, corresponds to one of low strain energy. This is not true, however, of a single array, as shown in Fig. 4.32. Here the strain fields of the individual parallel dislocations are additive and the array is not one of low energy.

While rotational slip has not been extensively studied, it does represent a basic way in which a crystal can be deformed. The amount of slip deformation one can obtain by this mechanism can be very large. In fact, it is possible to twist a 1 cm diameter zinc crystal about its basal plane pole through as many as 10 or more revolutions per inch. This deformation, of course, occurs on many slip planes distributed over the gage section and not on one plane, as shown in Fig. 4.31.

4.8 Climb of Edge Dislocations. The slip plane of a dislocation is defined as the plane that contains both the dislocation and its Burgers vector. Since the Burgers vector is parallel to a screw dislocation, any plane containing the dislocation is a possible slip plane. (See Fig. 4.34A). On the other hand, the Burgers vector of an edge dislocation is perpendicular to the dislocation, and there is only one possible slip plane. (See Fig. 4.34B.) A screw dislocation may move by slip or glide in any direction perpendicular to itself, but an edge dislocation can only glide in its single slip plane. There is, however, another method, fundamentally different from slip, by which an edge dislocation can move. This process is called *climb* and involves motion in a direction perpendicular to the slip plane.

Figure 4.35A represents a view of an edge dislocation with the extra plane perpendicular to the plane of the paper and designated by filled circles. In this diagram, a vacancy or vacant lattice site, has moved up to a position just to the right of atom a, one of the atoms forming the edge or boundary of the extra

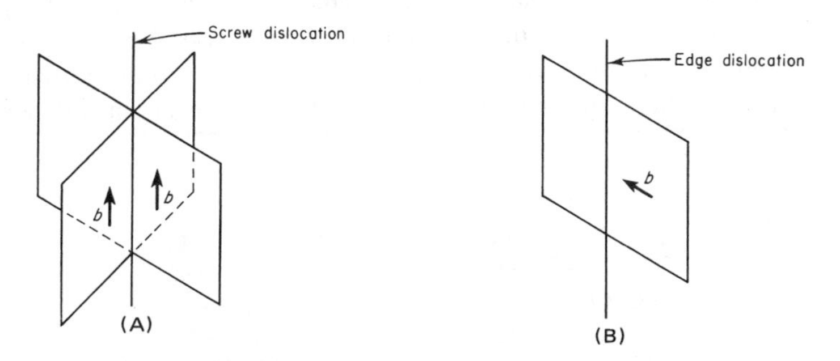

Fig. 4.34 (A) Any plane containing the dislocation is a slip plane for a screw dislocation. (B) There is only one slip plane for an edge dislocation. It contains both the Burgers vector and the dislocation.

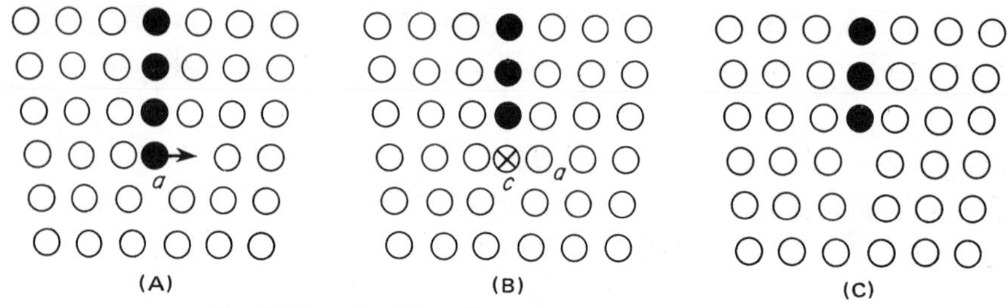

Fig. 4.35 Positive climb of an edge dislocation.

plane. If atom *a* jumps into the vacancy, the edge of the dislocation loses one atom, as is shown in Fig. 4.35B, where atom *c*, designated with a crossed circle, represents the next atom of the edge (lying just below the plane of the paper). If atom *b* and all others that formed the original edge of the extra planes move off through interaction with vacancies, the edge dislocation will climb one atomic distance in a direction perpendicular to the slip plane. This situation is shown in Fig. 4.35C. Climb, as illustrated in the above example, is designated as positive climb and results in a decrease in size of the extra plane. Negative climb corresponds to the opposite of the above in that the extra plane grows in size instead of shrinking. A mechanism for negative climb is illustrated in Fig. 4.36A and Fig. 4.36B.

In this case, let us suppose that atom *a* of Fig. 4.36A moves to the left and joins the extra plane, leaving a vacancy to its right, as is shown in Fig. 4.36B. This vacancy then moves off into the crystal. Notice that this is again an atom by atom procedure and not a cooperative movement of the entire row of atoms lying behind atom *a*. Thus, atom *c* (crossed circle), shown in Fig. 4.36B, represents the atom originally behind atom *a*. A cooperative movement of all atoms in the row behind *a* corresponds to slip and not to climb.

Because we are removing material from inside the crystal as the extra plane itself grows smaller, the effect of positive climb on the crystal is to cause it to shrink in a direction parallel to the slip plane (perpendicular to the extra plane). Positive climb is therefore associated with a compressive strain and will be

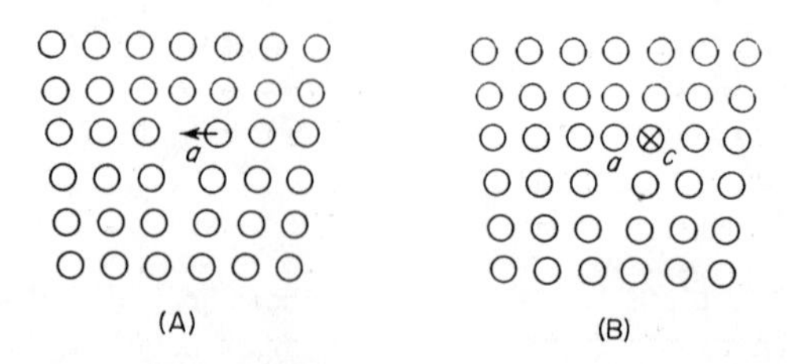

Fig. 4.36 Negative climb of an edge dislocation.

promoted by a compressive stress component perpendicular to the extra plane. Similarly, a tensile stress applied perpendicular to the extra plane of an edge dislocation promotes the growth of the plane and thus negative climb. A fundamental difference therefore exists between the nature of the stress that produces slip and that which produces climb. Slip occurs as the result of shear stress; climb as the result of a normal stress (tensile or compressive).

Both positive and negative climb require that vacancies move through the lattice, toward the dislocation in the first case and away from it in the second case. If the concentration of vacancies and their rate of jump is very low, then it is not expected that edge dislocations will climb. As we shall see, vacancies in most metals are practically immobile at low temperatures (one jump in eleven days in copper at room temperature), but at high temperatures they move with great rapidity, and their equilibrium number increases by many powers of ten. Climb, therefore, is a phenomenon that becomes increasingly important as the temperature rises. Slip, on the other hand, is only slightly influenced by temperature.

4.9 Slip Planes and Slip Directions. It is an experimental fact that in metal crystals slip, or glide, occurs preferentially on planes of high-atomic density. It is a general rule that the separation between parallel lattice planes varies directly as the degree of packing in the planes. Therefore, crystals are sheared most easily on planes of wide separation. This statement does not mean that slip cannot occur in a given crystal on planes other than the most closely packed plane. Rather, it means that dislocations move more easily along planes of wide spacing where the lattice distortion due to the movement of the dislocation is small.

Not only does slip tend to take place on preferred crystallographic planes, but the direction of shear associated with slip is also crystallographic. The slip direction of a crystal (shear direction) has been found to be almost exclusively a close-packed direction, a lattice direction with atoms arranged in a straight line, one touching the next. This tendency for slip to occur along close-packed directions is much stronger than the tendency for slip to occur on the most closely packed plane. For practical purposes, it can usually be assumed that slip occurs in a close-packed direction.

The fact that the experimentally determined slip direction coincides with the close-packed directions of a crystal can be explained in terms of dislocations. When a dislocation moves through a crystal, the crystal is sheared by an amount equal to the Burgers vector of the dislocation. After the dislocation has passed, the crystal must be unchanged in the geometry of the atoms; that is, the symmetry of the crystal must be retained. The smallest shear that can fulfill this condition equals the distance between atoms in a close-packed direction.

In order to explain this point more clearly, let us consider a hard-ball model

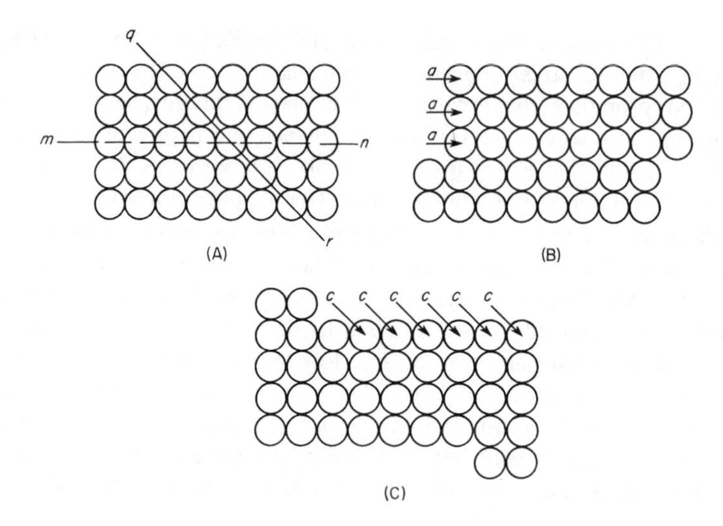

Fig. 4.37 Two ways in which a simple cubic lattice can be sheared while still maintaining the lattice symmetry: (A) Crystal before shearing, (B) Shear in a close-packed direction, and (C) Shear in a nonclose-packed direction.

of a simple cubic crystal structure. Line *mn* of Fig. 4.37A is a close-packed direction. In Fig. 4.37B the upper half of the lattice has been sheared to the right by a, the distance between atoms in the direction *mn*. The shear, of course, has not changed the crystal structure. Consider now an arbitrarily chosen nonclose-packed direction such as *qr* in Fig. 4.37A. Figure 4.37C shows that a shear of c (the distance between atom centers in this direction) also preserves the lattice. However, c is larger than a ($c = 1.414a$). Furthermore, c and a equal the respective sizes of the Burgers vectors of the dislocation capable of producing the two shears. The dislocation corresponding to the shear in the close-packed direction thus has the smallest Burgers vector. However, the lattice distortion and strain energy associated with a dislocation are functions of the size of the Burgers vector, and it has been shown by Frank that the strain energy varies directly as the square of the Burgers vector. In the present case, the strain energy of a dislocation of Burgers vector c is twice that of a dislocation of Burgers vector a (that is, $c^2 = (1.414)^2 a^2$). Thus, a dislocation with a Burgers vector equal to the spacing of atoms in a close-packed direction would be unique. It possesses the smallest strain energy of all dislocations the movement of which through the crystal does not disturb the crystal structure. The fact that it possesses the least strain energy should make this form of dislocation much more probable than forms of higher strain energy. It should also account for the experimentally observed fact that the slip direction in crystals is almost always a close-packed direction.

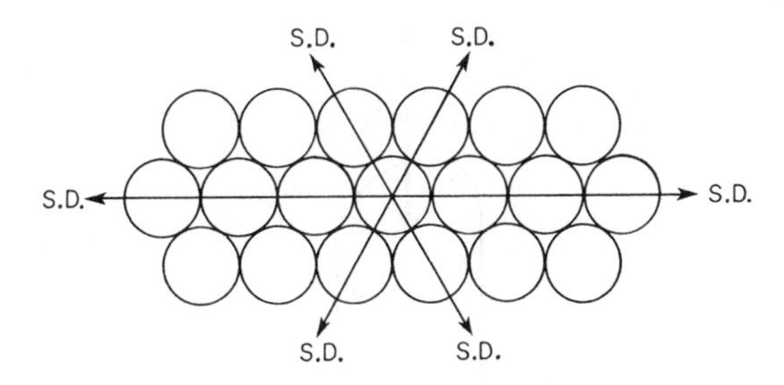

Fig. 4.38 The three slip directions in a plane of closest packing. Notice that this type of plane occurs in both the hexagonal close-packed and the face-centered cubic lattices.

4.10 Slip Systems. The combination of a slip plane and one of its close-packed directions defines a possible *slip mode* or *slip system.* If the plane of the paper in Fig. 4.38 is considered to define a slip plane, then there will be three slip systems associated with the indicated close-packed plane, one mode corresponding to each of the three slip directions. All of the modes of a given slip plane are crystallographically equivalent. Further, all slip systems in planes of the same form [(111), $(1\bar{1}1)$, $(\bar{1}11)$, and $(11\bar{1})$] are also equivalent. However, the ease with which slip can be produced on slip systems belonging to planes of different forms [(111) and (110)] will, in general, be greatly different.

4.11 Critical Resolved Shear Stress. It is a well-known fact that polycrystalline metal specimens possess a yield stress that must be exceeded in order to produce plastic deformation. It is also true that metal single crystals need to be stressed above a similar yield point before plastic deformation by slip becomes macroscopically measurable. Since slip is caused by shear stresses, the yield stress for crystals is best expressed in terms of a shear stress resolved on the slip plane and in the slip direction. This stress is called the *critical resolved shear stress.* It is the stress that will cause sufficiently large numbers of dislocations to move so that a measurable strain can be observed. Most crystal specimens are not tested directly in shear, but in tension. There are good reasons for this. The most important is that it is almost impossible to test a crystal in direct shear without introducing bending moments where the specimen is gripped. The effect of these bending moments is to produce shear stress components on slip planes other than that on which it is desired to study slip. If slip on these planes is not measurably more difficult than on the plane to be tested, one obtains a condition where slip occurs on several slip planes over those parts of the specimen near the grips. The effect of this deformation may be to cause bending of the specimen near the grips, and one is thus left with a deformation that is far from homogeneous. There are also problems associated with the use of single

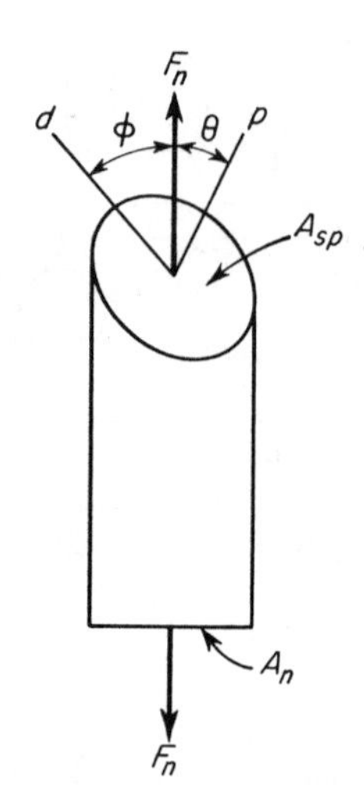

Fig. 4.39 A figure for the determination of the critical resolved shear-stress equation.

crystal tensile specimens, but these are less serious and, by proper design of the grips, they may be largely eliminated.

An equation will now be derived which relates the applied tensile stress to the shear stress resolved on the slip plane and in the slip direction.

Let the inclined plane at the top of the cylindrical crystal in Fig. 4.39 correspond to the slip plane of the crystal. The normal to the slip plane and the slip direction are indicated by the lines p and d respectively. The angle between the slip plane normal and the stress axis is represented by θ, and that between the slip direction and the stress axis by ϕ. The axial tensile force in pounds applied to the crystal is designated by f_n.

The cross-section area of the specimen perpendicular to the applied tensile force is to the area of the slip plane as the cosine of the angle between the two planes. This angle is the same as the angle between the normals to the two planes in question and is the angle θ in the figure. Thus,

$$\frac{A_n}{A_{sp}} = \cos\theta$$

or

$$A_{sp} = \frac{A_n}{\cos\theta}$$

where A_n is the cross-section area perpendicular to the specimen axis, and A_{sp}

the area of the slip plane. The stress on the slip plane equals the applied force divided by the area of the slip plane:

$$\sigma_A = \frac{f_n}{A_{sp}} = \frac{f_n}{A_n} \cos \theta$$

where σ_A is the stress on the slip plane in the direction of the original force f_n. This is not, however, the shear stress which acts in the slip direction, but the total stress acting on the slip plane. The component of this stress parallel to the slip direction is the desired shear stress and may be obtained by multiplying σ_A by $\cos \phi$, where ϕ is the angle between σ_A and τ the resolved shear stress. As a result of the above, we can now write

$$\tau = \sigma_A \cos \phi = \frac{f_n}{A_n} \cos \theta \cos \phi$$

where τ is the shear stress resolved on the slip plane and in the slip direction. Finally, since f_n/A_n is the applied tensile force divided by the area normal to this force, this term may be replaced by σ the normal tensile stress:

$$\tau = \sigma \cos \theta \cos \phi$$

Several important conclusions may be drawn from the above equation. If the tensile axis is perpendicular to the slip plane, the angle ϕ is 90° and the shear stress is zero. Similarly, if the stress axis lies in the slip plane, the angle θ is 90° and the shear stress is again zero. Thus, it is not possible to produce slip on a given plane when the plane is either parallel or perpendicular to the axis of tensile stress. The maximum shear stress that can be developed equals 0.5 σ and occurs when both θ and ϕ equal 45°. For all other combinations of these two angles, the resolved shear stress is smaller than one-half the tensile stress.

It has been experimentally verified that the critical resolved shear stress for a given crystallographic plane is independent of the orientation of the crystal for some metals. Thus, if a number of crystals, differing only in the orientation of the slip plane to the axis of tensile stress, are pulled in tension, and the shear stress at which they yield is computed with the above equation, it will be found that the yield stress is a constant. Figure 4.40 shows the critical resolved shear-stress data of Burke and Hibbard for magnesium single crystals of 99.99 percent purity. The ordinate of this curve is the tensile stress at which yielding was observed, while the abscissae give corresponding values of the function $\cos \theta$ $\cos \phi$. A smooth curve is plotted through the data corresponding to a constant yield stress (shear stress) of 44 gm per mm^2 (62 psi). The experimental points

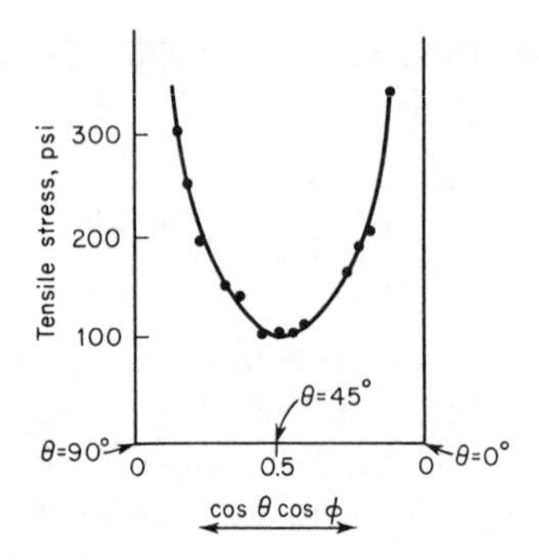

Fig. 4.40 The tensile yield point for magnesium single crystals of different orientations. Abscissae are values of the function cos θ cos ϕ. Smooth curve is for an assumed constant critical resolved shear stress of 63 psi. (Burke, E. C., and Hibbard, W. R., Jr., *Trans. AIME,* **194**, 295 [1952].)

fall on this curve with remarkable accuracy. Recent work[15,16,17] on b.c.c. metals has indicated that, in these metals, the critical resolved shear stress may be a function of orientation as well as of the type of stress. In other words, the yield stress for this type of crystal can be different depending upon whether the applied stress is tensile or compressive.

In some metals the critical resolved shear stress for slip on a given type of plane is remarkably constant for crystals of the same composition and previous

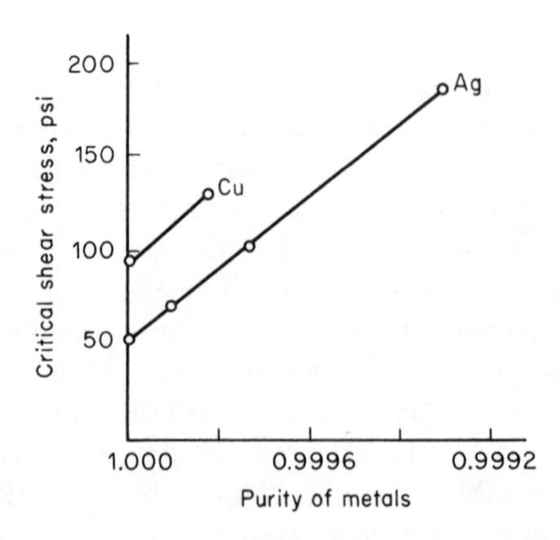

Fig. 4.41 Variation of the critical resolved shear stress with purity of the metal. (After Rosi, F. D., *Trans. AIME,* **200**, 1009 [1954].)

15. Stein, D. F., *Canadian J. Phys.*, **45**, No. 2, Part 3, 1063 (1967).
16. Sherwood, P. J., Guiu, F., Kim, H. C., and Pratt, P. L., *Ibid*, p. 1075.
17. Hull, D., Byron, J. F., and Noble, F. W., *Ibid*, p. 1091.

treatment. However, the critical resolved shear stress is sensitive to changes in composition and handling. In general, the purer the metal the lower the yield stress, as may be seen quite clearly from the curves of Fig. 4.41 for silver and copper single crystals. The silver data in particular show that changing the composition for a purity of 99.999 to 99.93 percent raises the critical resolved shear stress by a factor or more than three.

The critical resolved shear stress is a function of temperature. In the case of face-centered cubic crystals, this temperature dependence may be small. Metal crystals belonging to other crystal forms (body-centered cubic, hexagonal, and rhombohedral) show a larger temperature effect. The yield stress in these crystals increases as the temperature is lowered, with the rate of increase generally becoming greater as the temperature drops. Figure 4.42 shows this effect for a number of different non face-centered cubic metals.

4.12 Slip on Equivalent Slip Systems. It has been empirically determined that when a crystal possesses several crystallographically equivalent slip systems, slip will start first on the system having the highest resolved shear stress. It has also been found that if several equivalent systems are equally stressed, slip will usually commence simultaneously on all of these systems.

4.13 The Dislocation Density. Even in a deformed single crystal, only a small fraction of the dislocations formed during deformation come to the surface and are lost. This means that with continued straining, the number of dislocations in a metal increases. As will be shown later, this increase in the

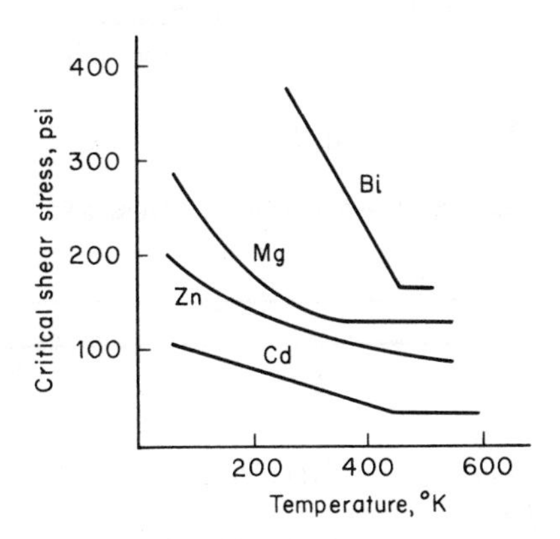

Fig. 4.42 Effect of temperature on the critical shear stress. *Note*: The data on which these curves are based predate that of Table 4.2. The higher critical stresses in this case correspond to crystals of lower purity. (Schmid, E., and Boas, W., *Kristallplastizität*, Julius Springer, Berlin, 1935.)

number of retained dislocations results in a strengthening of the metal. In other words, the increase in hardness or strength of a metal with deformation is closely associated with an increase in the concentration of the dislocations. The parameter commonly used to express this quantity is ρ, the dislocation density, which is defined as the total length of all the dislocation lines in a unit volume. Its dimensions are thus cm/cm^3 or cm^{-2}.

4.14 Slip Systems in Different Crystal Forms.

Face-Centered Cubic Metals. The close-packed directions are the ⟨110⟩ directions in the face-centered cubic structure. These ae directions that run diagonally across the faces of the unit cell. Figure 4.38 shows a segment of a plane of closest packing. There are four of these planes in the face-centered cubic lattice, called *octahedral planes*, with indices (111), $(1\bar{1}1)$, $(11\bar{1})$, and $(\bar{1}11)$. Each octahedral plane contains three close-packed directions, as can be seen in Fig. 4.38, and, therefore, the total number of octahedral slip systems is 4 x 3 = 12. The number of octahedral slip systems can also be computed in a different manner. There are 6⟨110⟩ directions and, since each close-packed direction lies in two octahedral planes, the number of slip systems is therefore 12.

The only important slip systems in the face-centered cubic structure are those associated with slip on the octahedral plane. There are several reasons for this. First, slip can occur much more easily on a plane of closest packing than on planes of lower atomic density; that is, the critical resolved shear stress for octahedral slip is lower than for other forms. Second, there are twelve different ways that octahedral slip can occur, and the twelve slip systems are well distributed in space. It is, therefore, almost impossible to strain a face-centered cubic crystal and not have at least one {111} plane in a favorable position for slip.

Table 4.1 lists the critical resolved shear stress, measured at room tempera-

Table 4.1 Critical Resolved Shear Stresses for Face-Centered Cubic Metals.

Metal	Purity	Slip System	Critical Resolved Shear Stress Psi
Cu*	99.999	{111} ⟨110⟩	92
Ag†	99.999	{111} ⟨110⟩	54
Au‡	99.99	{111} ⟨110⟩	132
Al§	99.996	{111} ⟨110⟩	148

* Rosi, F. D., *Trans. AIME*, 200, 1009 (1954).
† daC. Andrade, E. N., and Henderson, C., *Trans. Roy. Soc.* (London), 244, 177 (1951).
‡ Sachs, G., and Weerts, J., *Zeitschrift für Physik*, 62, 473 (1930).
§ Rosi, F. D., and Mathewson, C. W., *Trans. AIME*, 188, 1159 (1950).

ture, for several important face-centered cubic metals. Table 4.1 clearly shows that the critical resolved shear stresses of face-centered cubic metals in the nearly pure state are very small.

In general, plastically deformed face-centered cubic crystals slip on more than one octahedral plane because of the large number of equivalent slip systems. In fact, even in a simple tensile test it is very difficult to produce strains of over a few percent without inducing glide simultaneously on several planes. However, when slip occurs at the same time on several intersecting slip planes, the stress required to produce additional deformation rises rapidly. In other words, the crystal strain-hardens. Figure 4.43 shows typical tensile stress-strain curves for a pair of face-centered cubic crystals. Curve *a* corresponds to a crystal whose original orientation lies close to ⟨100⟩. In this crystal several slip systems have nearly equal resolved shear stresses. As a consequence, plastic deformation occurs by slip on several slip planes and the curve has a steep slope from the beginning of the deformation. On the other hand, curve *b* corresponds to a crystal whose orientation falls in the center of the stereographic triangle and is representative of crystals in which one slip plane is more highly stressed than all the others at the start of deformation. The region marked 1 of this curve corresponds to slip on this plane only; the other slip planes are inactive. The

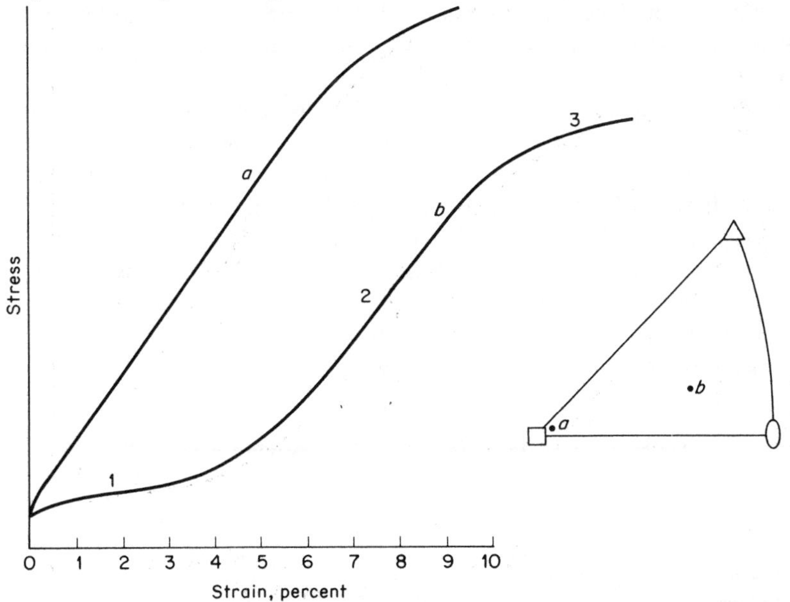

Fig. 4.43 Typical face-centered cubic single crystal stress-strain curves. Curve *a* corresponds to deformation by multiple glide from start of deformation; curve *b* corresponds to multiple glide after a period of single slip (easy glide). Crystal orientations are shown in the stereographic triangle.

small slope of the curve in stage 1 shows that the strain-hardening is minor when slip occurs on a single crystallographic plane. Stage 2 of curve *b*, which appears after strains of several percent, has a much steeper slope, and the crystal hardens rapidly with increasing strain. In this region, slip on a single plane ceases, and multiple glide on intersecting slip planes begins. Here the dislocation density grows rapidly with increasing strain. Finally, stage 3 represents a region where the rate of strain-hardening progressively decreases. In this region, the rate of increase of the dislocation density becomes smaller with increasing strain.

Region 1 of curve *b*, where slip occurs on a single plane, is called the region of *easy glide.* The extent of the region of easy glide depends on several factors, among which are size of specimen and purity of the metal. When the cross-section diameter of a crystal specimen is large, or the metal very pure, the region of easy glide tends to disappear. In any case, the region of easy glide, or single slip, rarely exceeds strains of several percent in face-centered cubic crystals, and, for all practical purposes, it can be assumed that these metals deform by multiple glide on a number of octahedral systems. This deformation is especially true in the case of polycrystalline face-centered cubic metals.

The plastic properties of pure face-centered cubic metals are as follows. Low critical resolved shear stresses for slip on octahedral planes signifies that plastic deformation of these metals starts at low stress levels. Multiple slip on intersecting slip planes, however, causes rapid strain-hardening as deformation proceeds.

Hexagonal Metals. Since the basal plane of the close-packed hexagonal crystal and the octahedral {111} plane of the face-centered cubic lattice have identical arrangements of atoms, it would be expected that slip on the basal plane of hexagonal metals would occur as easily as slip on the octahedral planes of face-centered cubic metals. In the case of the three hexagonal metals, zinc, cadmium, and magnesium, this is actually the case. Table 4.2 lists the critical resolved shear stress for basal slip of these metals as measured at room temperature. The hexagonal Miller indices of the basal plane are (0001), and the close-packed, or slip, directions are $\langle 11\bar{2}0 \rangle$.

Table 4.2 Critical Resolved Shear Stress for Basal Slip.

Metal	*Purity*	*Slip Plane*	*Slip Direction*	*Critical Resolved Shear Stress* Psi
Zinc*	99.999	(0001)	$\langle 11\bar{2}0 \rangle$	26
Cadmium†	99.996	(0001)	$\langle 11\bar{2}0 \rangle$	82
Magnesium‡	99.95	(0001)	$\langle 11\bar{2}0 \rangle$	63

* Jillson, D. C., *Trans. AIME*, 188, 1129 (1950).
† Boas, W., and Schmid, E., *Zeits. für Physik*, 54, 16 (1929).
‡ Burke, E. C., and Hibbard, W. R., Jr., *Trans. AIME*, 194, 295 (1952).

Table 4.2 definitely confirms that plastic deformation by basal slip in these three hexagonal metals begins at stresses of the same order of magnitude as those required to start slip in face-centered cubic metals.

Two other hexagonal metals of interest are titanium and beryllium, in which the room-temperature critical resolved shear stress for basal slip is very high (approximately 16,000 psi in the case of titanium,[18] and 5700 psi for beryllium[19]). Furthermore, it has been established that in titanium the critical resolved shear stress for slip on $\{10\bar{1}0\}$ prism planes in $\langle 11\bar{2}0\rangle$ close-packed directions is about 7100 psi.[18] The latter plane is therefore the preferred slip plane in titanium. The basal slip of zirconium, still another hexagonal metal, has thus far not been observed. It appears that slip occurs primarily on $\{10\bar{1}0\}\langle 11\bar{2}0\rangle$ slip systems. The critical resolved shear stress for this form of slip is about 900 psi.[20] The question now arises as to how the differences in the slip behavior of magnesium, zinc, and cadmium on the one hand, and beryllium, titanium, and zirconium on the other can be explained. A complete solution to this problem is not at hand, but the following is undoubtedly related to this effect.

Figure 1.18 of Chapter 1 shows the unit cell of the hexagonal lattice. In this figure, distance a equals the distance between atoms in the basal plane, while c is the vertical distance between atoms in every other basal plane. The ratio c/a is, therefore, a dimensionless measure of the spacing between basal planes. If the atoms of hexagonal metals were truly spherical in shape, the ratio c/a would be the same in all cases (1.632). Table 4.3 shows, however, that this value is not the same, but that it varies from 1.886 in the case of cadmium to 1.586 in the case of beryllium. Only magnesium has an atom that approaches a true spherical shape, $c/a = 1.624$. Cadmium and zinc have a basal-plane separation greater than that of packed spheres, while beryllium, titanium, and zirconium have a smaller one. It is significant that the hexagonal metals with small separations between basal planes are those with very high critical resolved shear stresses for basal slip.

Table 4.3 The c/a Ratio for Hexagonal Metals.

Metal	c/a
Cd	1.886
Zn	1.856
Mg	1.624
Zr	1.590
Ti	1.588
Be	1.586

18. Anderson, E. A., Jillson, D. C., and Dunbar, S. R., *Trans. AIME*, **197** 1191 (1953).
19. Tuer, G. R., and Kaufmann, A. R., *The Metal Beryllium*, ASM Publication, Novelty, Ohio (1955) p. 372.
20. Rapperport, E. J., and Hartley, C. S., *Trans. Metallurgical Society, AIME,* **218** 869 (1960).

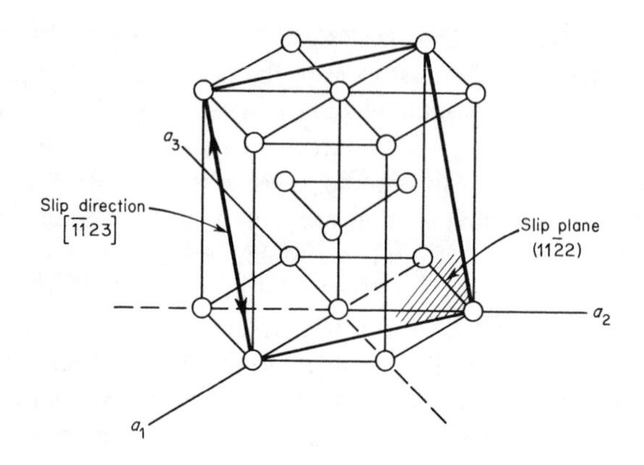

Fig. 4.44 $\{11\bar{2}2\}\langle 11\bar{2}\bar{3}\rangle$ slip in hexagonal metals.

The hexagonal metals zinc and cadmium have been observed to slip on a unique slip system when deformed in such a manner as to make the resolved shear stress on the basal plane very small. This deformation can be accomplished, for example, by placing the tensile stress axis nearly parallel to the basal plane. The observed slip plane for this type of deformation is $\{11\bar{2}2\}$, while the slip direction is $\langle 11\bar{2}\bar{3}\rangle$. Figure 4.44 shows the position of one of these slip planes and one of these slip directions in a hexagonal unit cell. The significant characteristic of this mode of deformation is that the $\langle 11\bar{2}\bar{3}\rangle$ direction is not the closest packed direction in the hexagonal crystal structure. Prior to the observation of $\{11\bar{2}2\}$ $\langle 11\bar{2}\bar{3}\rangle$ slip, the slip direction in all metals had been almost universally observed as the direction of closest packing. This second-order pyramidal glide was first observed by Bell and Cahn[21] on macroscopic zinc crystals, but it has also been verified by Price[22, 23] in both zinc and cadmium, who used the transmission electron microscope. The work of Price has not only confirmed the existence of this kind of slip, but has actually shown the dislocations responsible for it. For photographs of these dislocations, one is referred to the original publications.[22, 23]

Easy Glide in Hexagonal Metals. The metals zinc, cadmium, and magnesium are unique in that they possess both a low critical resolved shear stress and a single primary slip plane, the basal plane. Provided that this slip plane is suitably oriented with respect to the stress axis, strains of very large magnitude can be developed by basal slip. Simultaneous slip on intersecting primary slip planes is not a problem in these metals with a single slip plane, and the rate of strain-hardening is therefore much smaller than in face-centered cubic crystals.

21. Bell, R. L., and Cahn, R. W., *Proc. Roy. Soc.*, **A239** 494 (1957).
22. Price, P. B., *Phil. Mag.*, **5** 873 (1960).
23. Price, P. B., *Jour. Appl. Phys.*, **32** 1750 (1961).

The region of easy glide in a tensile stress-strain curve, instead of extending only several percent, may exceed values of 100 percent. In fact, it is quite possible to stretch a magnesium crystal ribbonlike four to six times its original length.

The very large plasticity of single crystals of these three hexagonal metals does not carry over to their polycrystalline form. Polycrystalline magnesium, zinc, or cadmium have low ductilities. In the previous paragraph it was pointed out that the large ductility of the single crystals is due to the fact that slip occurs on a single crystallographic plane. However, in polycrystalline material, plastic deformation is much more complicated than it is in single crystals. Each crystal must undergo a deformation that allows it to conform to the changes in the shape of its neighbors. Crystals with only a single slip plane do not have enough plastic degrees of freedom for extensive deformation under the conditions that occur in polycrystalline metals.

Body-Centered Cubic Crystals. The body-centered cubic crystal is characterized by four close-packed directions, the ⟨111⟩ directions, and by the lack of a truly close-packed plane such as the octahedral plane of the face-centered cubic lattice, or by the basal plane of the hexagonal lattice. Figure 4.45 shows a bard-ball model of the body-centered cubic (110) plane, the most closely packed plane in this lattice. Two close-packed directions lie in this plane, the $[\bar{1}11]$ and the $[1\bar{1}1]$. A comparison of Fig. 4.45 and Fig. 4.38 (face-centered cubic octahedral plane) confirms the fact that the {110} body-centered cubic planes are not planes of closest packing.

The slip phenomena observed on body-centered cubic crystals corresponds closely to that expected in crystals with close-packed directions and no truly close-packed plane. The slip direction in the body-centered cubic crystals is the close-packed direction, ⟨111⟩; the slip plane, however, is not well defined. Body-centered cubic slip lines are wavy and irregular, often making the identification of a slip plane extremely difficult. The {110}, {112}, and {123} planes have all been identified as slip planes in body-centered cubic crystals, but

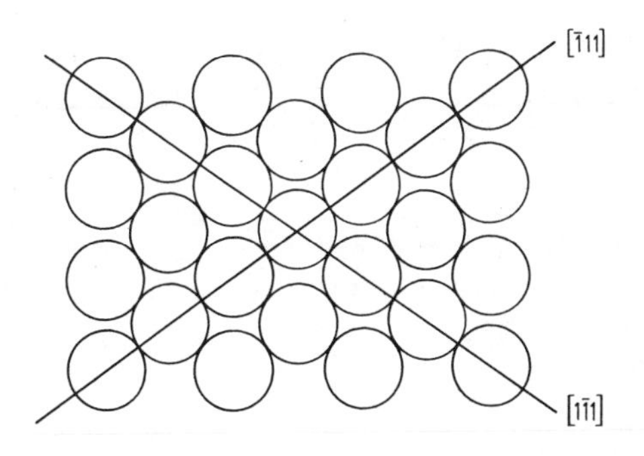

Fig. 4.45 The (110) plane of the body-centered cubic lattice.

recent work on iron single crystals indicates that any plane that contains a close-packed ⟨111⟩ direction can act as a slip plane. In further agreement with the lack of a close-packed plane is the high critical resolved shear stress for slip in body-centered cubic metals. In iron it is approximately 4000 psi at room temperature.

Problems

1. The shear modulus for iron is about 10×10^6 psi. Compute the theoretical shear stress for slip in this material. How many times larger is this than its observed shear stress for slip of 4000 psi?

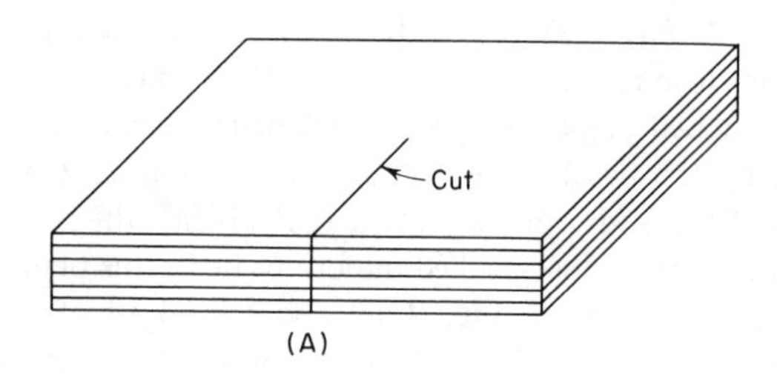

(A)

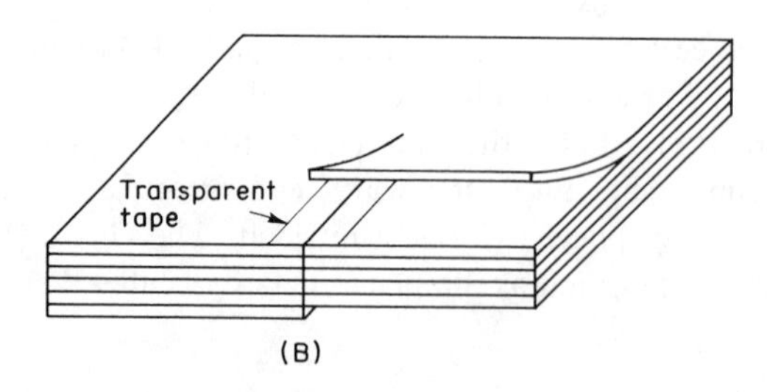

(B)

2. Make a model of a screw dislocation using a stack of filing cards, or small sheets of paper, and a roll of transparent tape. First, cut halfway through the stack of cards, as indicated in Fig. *A*. Then tape (along the cut) the left-hand front half of the top card to the right-hand front half of the card just below it. This is indicated in Fig. *B*. Now repeat this process through the entire stack. The result should be a spiral plane that goes entirely through the pack. What type of screw dislocation is this?

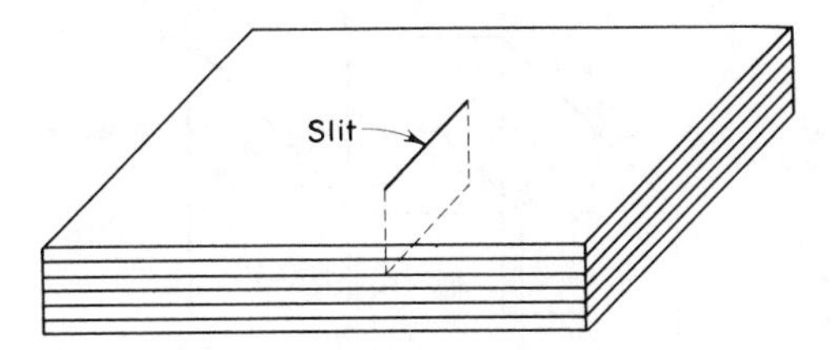

3. Make a model of the two opposite kinds of screw dislocations, in the manner of the preceding exercise, by cutting a slit through the stack of cards with a sharp knife, as indicated in the accompanying figure. If extra cards are placed over both the top and bottom of this stack, what does the model then represent?

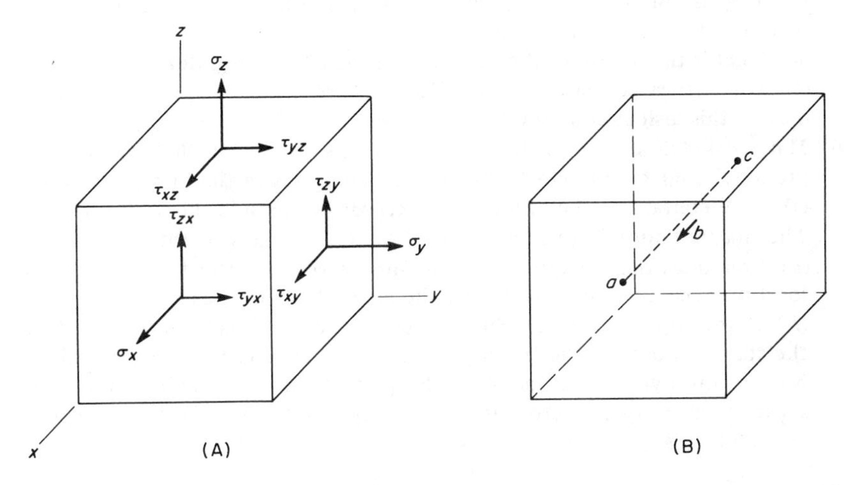

4. In Fig. A, a crystal in the form of a cube is considered to be aligned with the cartesian coordinate axes. With this arrangement, σ_x, σ_y, and σ_z are the three normal stresses, and τ_{yx}, τ_{xy}, τ_{zx}, τ_{xz}, τ_{yz}, and τ_{zy} are the six possible shear stresses. Assuming that (B) represents a screw dislocation passing directly through the crystal from its front face to its rear face, which of the above stress components would be capable of moving the dislocation?

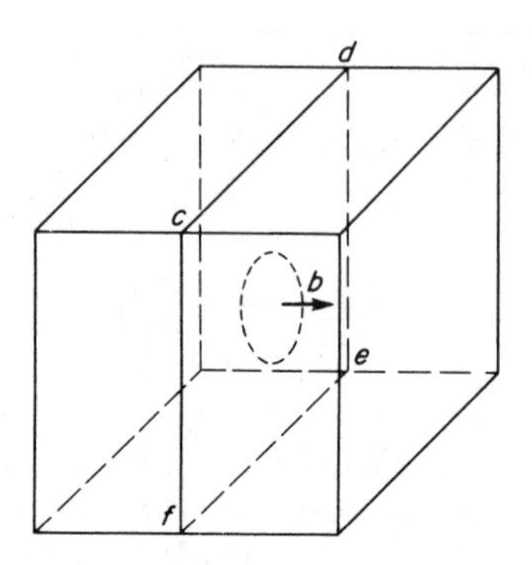

5. The dashed line represents what is known as a prismatic dislocation. It lies in the plane *cdef*, but its Burgers vector is normal to this plane. Assume that this loop is able to move by slip and that its motion is not restricted to any specific crystallographic plane.
   (*a*) What is the nature of the surface along which it can glide?
   (*b*) What stress components (see Fig. *A* of Problem 4) would be capable of making this dislocation move?
6. Make a sketch similar to that shown in Fig. 4.21, but in this case make an atom-by-atom counterclockwise circuit that closes in the imperfect crystal (the one containing the dislocation). Repeat this path in the perfect crystal. The closure failure in this case is called the true Burgers vector.
   (*a*) How does the magnitude of this Burgers vector differ from that of the local Burgers vector determined by the method of Fig. 4.21?
   (*b*) If the direction of the Burgers vector is defined as the direction from the starting point to the finish point in the circuit with the closure failure, how do the two definitions of the Burgers vector differ with regard to the sense of the Burgers vector? Is this a really significant difference? Explain.

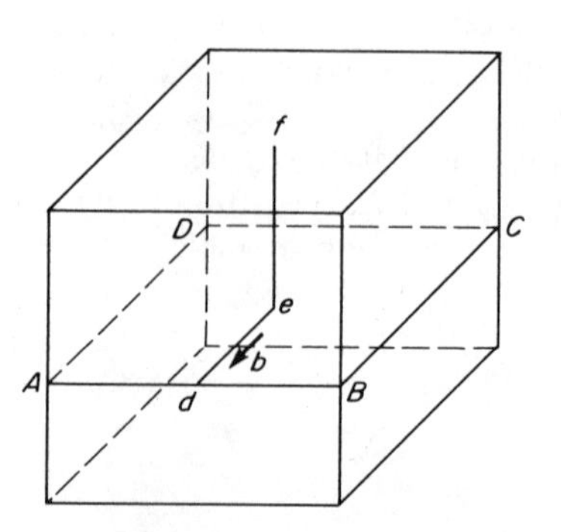

7. Lines *de* and *ef* represent a dislocation whose Burgers vector is parallel to line *de*.

(*a*) Describe the stress system that will allow this dislocation to move in the plane *ABCD* but will not cause the segment *ef* to move.
(*b*) Is this dislocation equivalent to a Frank-Read source?

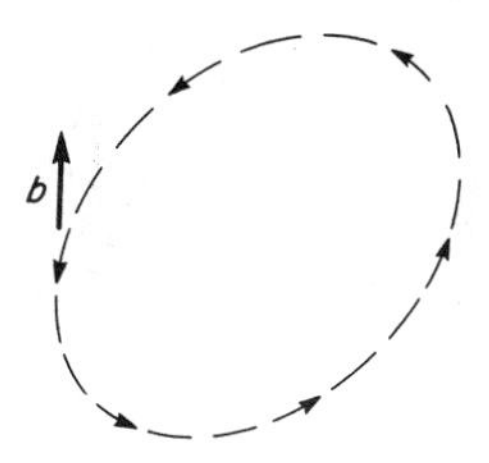

8. The dislocation loop in the accompanying diagram corresponds to that in the figure for Problem 5. The indicated sense of the Burgers vector was obtained on the assumption that the positive direction of the dislocation is that indicated by the arrows on the loop in the drawing.
(*a*) Does this loop correspond to the boundary of a circular extra plane, or to part of a missing plane? Explain. *Note:* use the method for determining the Burgers vector given in Section 4.3.
(*b*) If a tensile stress is applied to the crystal of Problem 5 at an elevated temperature, will the loop expand or contract?

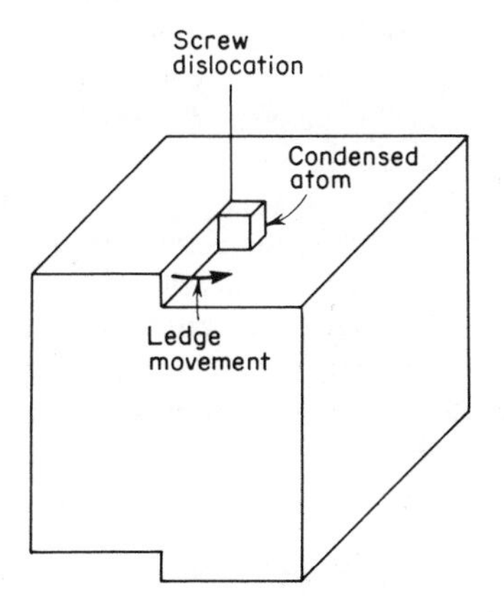

9. The ledge that exists at the surface of a crystal when a screw dislocation comes to the surface is believed to be very important with respect to the growth of crystals from the vapor. The vapor atoms are believed to attach themselves to the crystal along this ledge. The result is that the ledge tends to spiral around the surface as the crystal grows. In light of the above

consideration, discuss the dislocation configuration of Problem 7 as a possible Frank-Read source that operates by climb and not slip. Indicate the stress that would promote the operation of this mechanism so as to increase the depth of this crystal from front to back. Assume the positive sense of this dislocation is in the direction *d* to *e* to *f*.

10. Make a 001 standard stereographic projection of a cubic crystal showing the four {111} slip planes as great circles. Indicate with small dots the ⟨110⟩ slip directions that lie in these planes. Next, plot in the poles of the {111} planes. Identify all these crystallographic features with their proper indices. Finally, locate on this diagram the (321) pole. The position of this pole is indicated in Fig. 1.33. For the purpose of plotting it accurately, note that the angle between (111) and (321) is 22°, and between (110) and (321) the angle is 19°.
11. The angle between two directions in a cubic crystal is defined by the relationship

$$\cos \phi = \frac{h_1 \cdot h_2 + k_1 \cdot k_2 + l_1 \cdot l_2}{\sqrt{h_1^2 + k_1^2 + l_1^2} \cdot \sqrt{h_2^2 + k_2^2 + l_2^2}}$$

where h_1, k_1, and l_1 are the direction indices of one of the directions, and h_2, k_2, and l_2 are the direction indices of the other direction.
(*a*) Now with the aid of this equation, compute the angle between [321] and $[11\bar{1}]$. You can check your answer by graphically measuring the angle on the stereographic projection of Problem 10 and by the table in Appendix A.
(*b*) In the same manner, compute the angle between [321] and [101].
12. If the axis of a face-centered cubic crystal lies along [321] and it is loaded in tension to a stress of 100 gm/mm^2, what will be the resolved shear stress on the $(11\bar{1})$ slip plane and parallel to the [101] slip direction? *Note:* use the data from the preceding problem.
13. What is the resolved shear stress for the crystal of Problem 12 for slip on the system $(111)[10\bar{1}]$? Determine your answer by the method of Problem 11.
14. With the aid of the stereographic projection of Problem 10, determine the resolved shear stresses on all 12 of the {111}⟨110⟩ slip systems of the face-centered cubic crystal structure, assuming the stress axis and loading conditions of Problem 12. Is the resolved shear stress on the system $(11\bar{1})[101]$ unique? Explain.
15. With the aid of the stereographic projection of Problem 10, determine the resolved shear stress on all 12 of the {111}⟨110⟩ slip systems when the stress axis lies along [100] and the applied stress is 200 gm/mm^2. How many systems are subject to a maximum resolved shear stress?
16. Repeat Problem 15, assuming first a stress axis orientation parallel to [111], and then to [110].

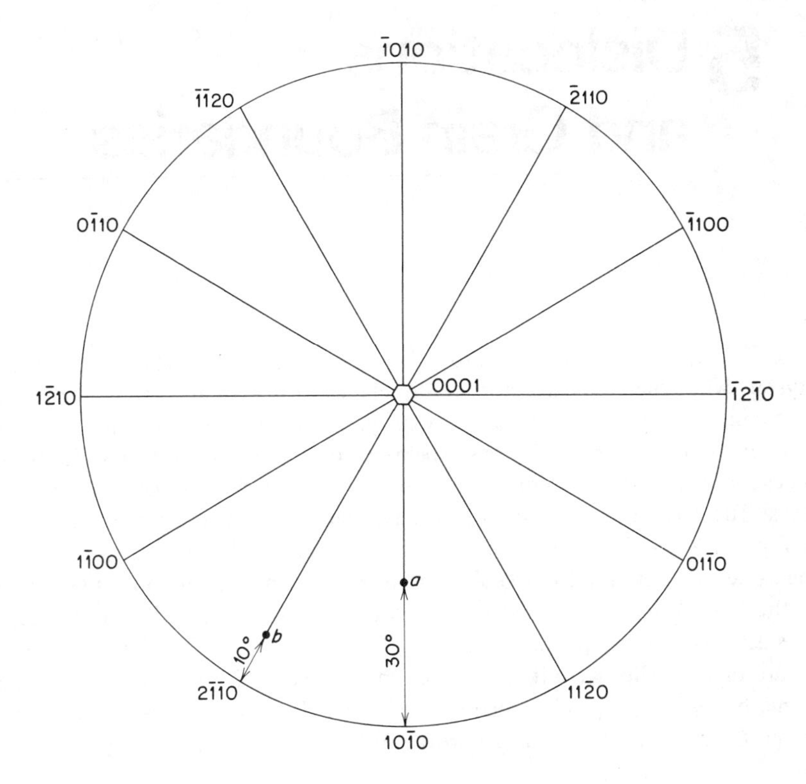

17. This figure represents the (0001) standard projection of a hexagonal close-packed metal. Assume that for this metal the critical resolved shear stress for basal slip is 100 gm/mm^2, and for prism slip it is 250 gm/mm^2.
(*a*) Consider the stress axis position marked *a* in the figure. If basal slip were to occur, along what slip direction or directions would it be expected to occur?
(*b*) Compute the Schmid factor ($\cos\theta\ \cos\phi$) for basal slip at orientation *a*.
(*c*) If prism slip were to occur for a stress axis orientation *a*, on what slip system or systems should the slip occur?
(*d*) Compute the Schmid factor ($\cos\theta\ \cos\phi$) for prism slip when the stress axis lies at point *a*.

18. (*a*) If the stress axis in the figure of Problem 17 falls at point *b*, will the crystal start to deform by prism slip or basal slip?
(*b*) What tensile stress would be needed to cause the crystal to yield?

19. Much of the crystal deformation literature records stress in gm/mm^2. The conversion factor to psi is approximately 1.4. Prove this relation.

5 Dislocations and Grain Boundaries

5.1 Cross-Slip. Cross-slip is a phenomenon that can occur in crystals when there are two or more slip planes with a common slip direction. As an example, take the hexagonal metal magnesium in which, at low temperatures, slip can occur either on the basal plane or on $\{10\bar{1}0\}$ prism planes. These two types of planes have common slip directions – the close-packed $\langle 11\bar{2}0 \rangle$ directions. The relative orientations of the basal plane and one prism plane are shown in Fig. 5.1A and 5.1B, where each diagram is supposed to represent a crystal in the same basic orientation (basal plane parallel to the top and bottom surfaces). In the first sketch, the crystal is sheared on the basal plane, while in the second it is sheared on the prism plane. The third illustration, Fig. 5.1C, shows the nature of cross-slip. Here it is observed that the actual slip surface is not a single plane, but is made up of segments, part of which lie in the basal plane and part in the prism plane. The resulting profile of the slip surface has the appearance of

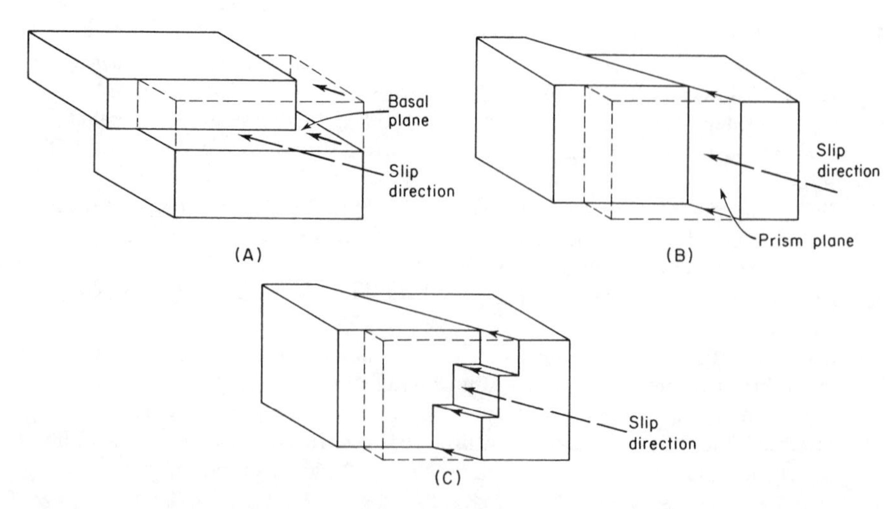

Fig. 5.1 Schematic representation of cross-slip in a hexagonal metal: (A) Slip on basal plane, (B) Slip on prism plane, and (C) Cross-slip on basal and prism planes.

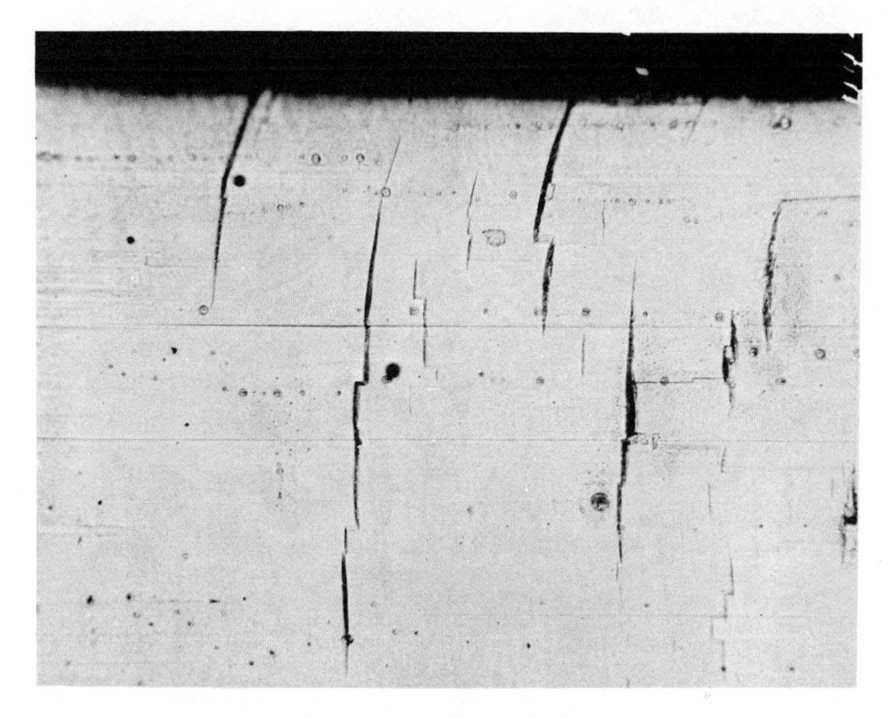

Fig. 5.2 Cross-slip in magnesium. The vertical slip plane traces correspond to the $\{10\bar{1}0\}$ prism plane, whereas the horizontal slip plane traces correspond to the basal plane (0002). 290 x. (Reed-Hill, R. E., and Robertson, W. D., *Trans. AIME,* **209** 496 [1957].)

a staircase. A simple analogy for cross-slip is furnished by a drawer in a piece of furniture. The sliding of the sides and bottom of the drawer relative to the frame of the piece is basically similar to the shearing motion that results from cross-slip.

A photograph of a magnesium crystal showing cross-slip is given in Fig. 5.2, where the plane of the photograph is equivalent to the forward faces of the schematic crystals shown in Fig. 5.1. Notice that while the slip in this specimen has occurred primarily along a prism plane, cross-slip segments on the basal plane are clearly evident.

During cross-slip the dislocations producing the deformation must, of necessity, shift from one slip plane to the other. In the example given above, the dislocations move from prism plane to basal plane and back to prism plane. The actual shift of the dislocation from one plane to another can only occur for a dislocation in the screw orientation. Edge dislocations, as was pointed out earlier, have their Burgers vector normal to their dislocation lines. Since the active slip plane must contain both the Burgers vector and the dislocation line,

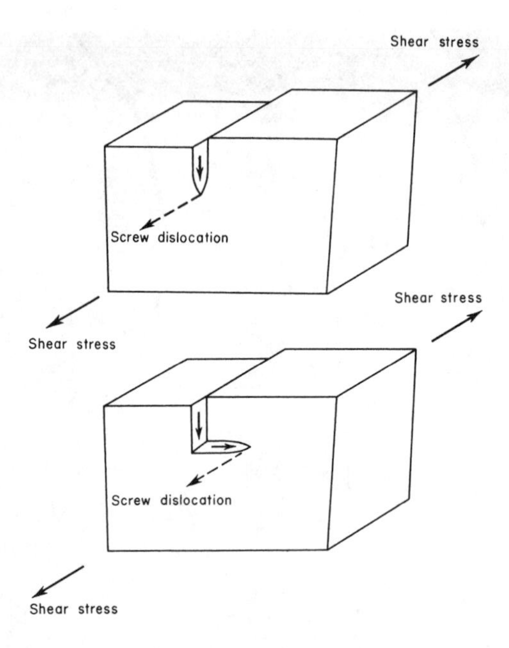

Fig. 5.3 Motion of a screw dislocation during cross-slip. In the upper figure the dislocation is moving in a vertical plane, while in the lower figure it has shifted its slip plane so that it moves horizontally. (From *Dislocations in Crystals,* by Read, W. T., Jr. Copyright 1953. McGraw-Hill Book Company, Inc., New York. Used by permission.)

edge dislocations are confined to move in a single slip plane. Screw dislocations, with their Burgers vectors parallel to the dislocation lines, are capable of moving in any plane that passes through the dislocation line. The manner in which a screw dislocation can produce a step in a slip plane is shown schematically in Fig. 5.3.

5.2 Slip Bands. A slip band is a group of closely spaced slip lines that appears, at low magnification, to be a single large slip line. In many metals slip bands tend to be wavy and irregular in appearance which is evidence that the dislocations that produce the bands in these metals are not so confined as to move in a single plane. The shifting of the dislocations from one slip plane to another is usually the result of the cross-slip of screw dislocations.

The resolution of the individual slip lines in a slip band is normally a task requiring the use of an electron microscope. However, when a slip band is observed on a surface that is nearly parallel to the active slip plane, it is sometimes possible to partially resolve the components of a slip band with a light microscope.

Fig. 5.4 Slips bands in LiF. Bands formed at −196°C and 0.36 per cent strain. (Johnston, W. G., and Gilman, J. J., *Jour. Appl. Phys.*, **30** 129 [1959].)

5.3 Double Cross-Slip. Another very important aspect of the work done on LiF crystals[1] is that it has shown that moving dislocations can multiply. The mechanism that seems most adequately to explain this dislocation multiplication is *double cross-slip*. As mentioned on p. 163, dislocations in LiF appear to nucleate at impurity particles. The slip process that develops from these nucleated dislocations first forms narrow slip planes that grow into slip bands of finite width with continued straining. These latter, in turn, can expand so as effectively to cover the complete crystal. Several rather narrow bands, as well as some that have grown to a moderate width, may be seen in the photograph of Fig. 5.4. Note that the presence of the slip bands is revealed by the rows of etch pits.

The development of the double cross-slip mechanism, which was originally proposed by Koehler[2] and later by Orowan,[3] is shown in Fig. 5.5. In sketch (A), a dislocation loop is assumed to be expanding on the primary slip plane. In part (B), a segment of the dislocation loop that is in the screw orientation is shown cross-slipping onto the cross-slip plane. Finally, in (C), the dislocation moves back into a plane parallel to its original slip plane. Attention is now called to the

1. Johnston, W. G., and Gilman, J. J., *Jour. Appl. Phys.*, **31** 632 (1960).
2. Koehler, J. S., *Phys. Rev.*, **86** 52 (1952).
3. Orowan, E., *Dislocations in Metals*, p. 103, American Institute of Mining, Metallurgical and Petroleum Engineers, New York, 1954.

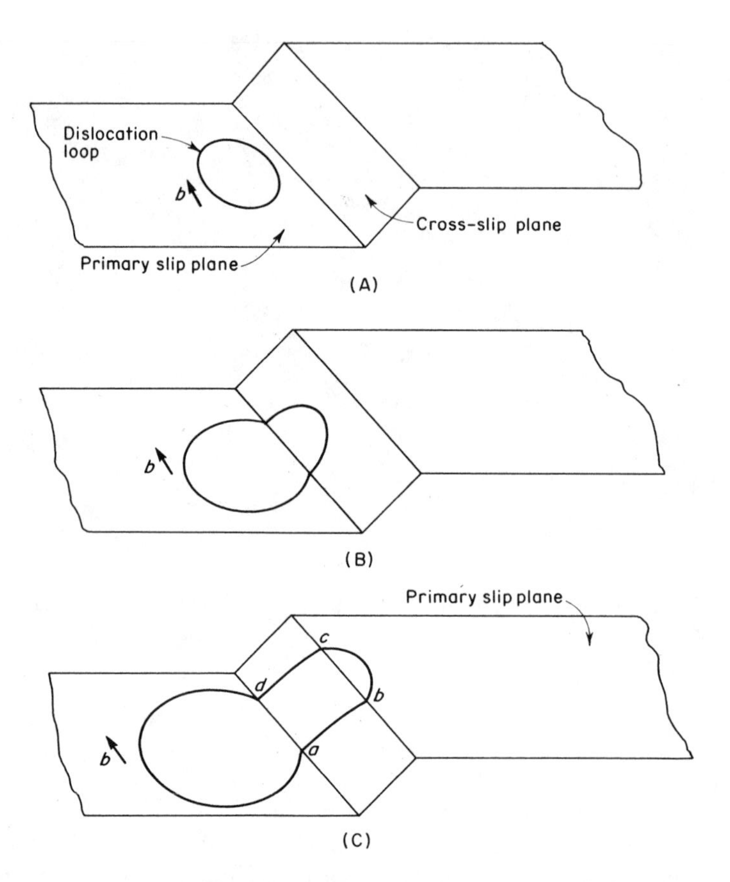

Fig. 5.5 Double cross-slip.

similarity of the dislocation configuration, starting at point b and ending at point c, to the Frank-Read source configuration shown in Fig. 4.23. This is, in fact, a classical dislocation generator, and dislocations can be created with it on the new slip plane. It is also true that this dislocation configuration can also act as a Frank-Read source in the slip plane of the original loop. Proof of this last statement is left as an exercise (Problem 5.4).

An important feature of the double cross-slip mechanism is that Frank-Read sources associated with this mechanism involve freshly created dislocations; that is, dislocations which, in general, will not have had time enough to become pinned by impurity atoms. The operation of a Frank-Read source with this type of dislocation is much more probable than that involving grown-in dislocations.

5.4 Crystal Structure Rotation During Tensile and Compressive Deformation. When a single crystal is deformed in either tension or compression, the crystal lattice usually suffers a rotation, as indicated schematically in Fig. 5.6. In tension, this tends to align the slip plane and the active slip direction parallel to the tensile stress axis. It is often possible to identify the active slip plane and slip direction as a result of this rotation. Thus, suppose that, before it is deformed, a f.c.c. crystal has the orientation marked a_1 shown in the standard projection of Fig. 5.7. If it is now strained by a small amount and its orientation is redetermined by a Laue back-reflection photograph, the stress axis should plot (in the standard stereographic projection) at a new position such as a_2. A second similar deformation should produce a third stress axis orientation at a_3. These three stress axis positions will normally fall along a great circle and, if this great circle is projected ahead, it should pass through the active slip direction. In the present case, as may be seen in Fig. 5.7, the slip direction is $[10\bar{1}]$. As shown in Fig. 5.8, this slip direction lies in both the (111) and the $(1\bar{1}1)$ planes. Since the (111) pole makes an angle of nearly 45° with the stress axis, while the $(1\bar{1}1)$ pole makes a corresponding angle closer to 90°, one can deduce that the resolved shear stress should be larger on the former. This means that the active slip plane should be (111). Normally it should also be possible to verify this fact by measuring the orientation of the slip line traces on the specimen surface.

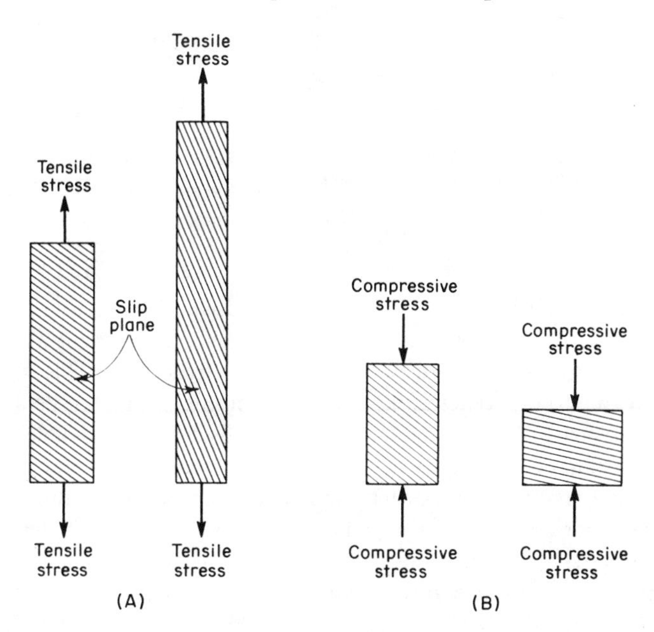

Fig. 5.6 Rotation of the crystal lattice in tension and compression.

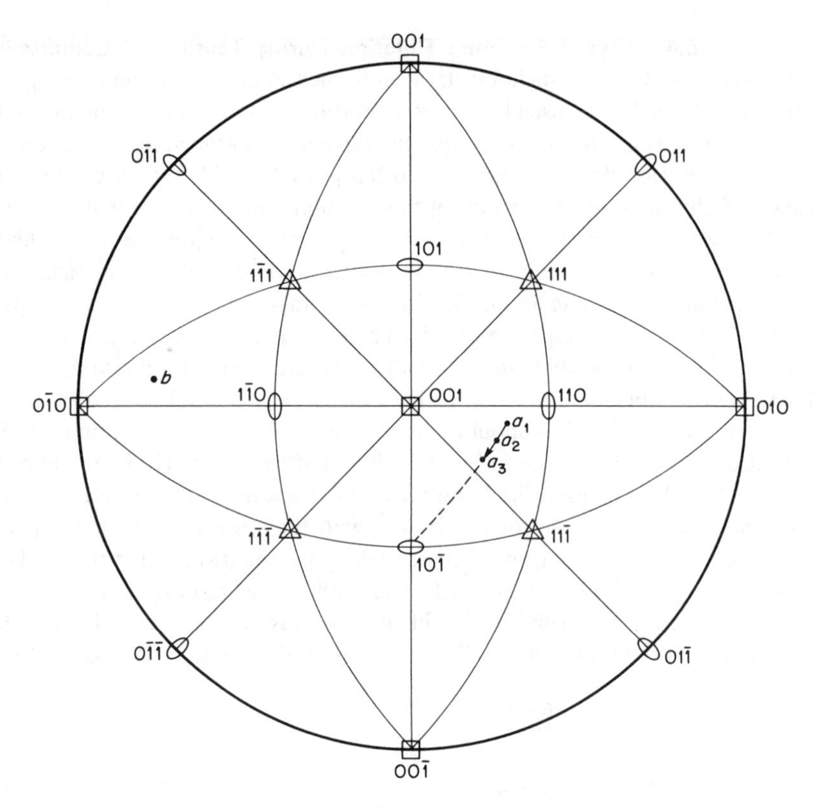

Fig. 5.7 In tension the lattice rotation is equivalent to a rotation of the stress axis (a) toward the slip direction. This stereographic projection shows this rotation in a face-centered cubic crystal.

Note that in the example of Fig. 5.7 the active slip direction is the closest ⟨110⟩ direction to the stress axis that can be reached by crossing a boundary of the stereographic triangle that contains the stress axis. This fact can be considered to define a general rule that is applicable to the tensile deformation of f.c.c. crystals. As an added example, consider the possible stress axis orientation, marked b, lying in the stereographic triangle at the left of the figure and near the $(0\bar{1}0)$ pole. Following the above rule, the slip direction should be $[0\bar{1}1]$. The corresponding slip plane pole should be $(1\bar{1}\bar{1})$. Once the slip direction has been obtained, the pole of the slip plane can easily be determined in the case of f.c.c. crystals. It should lie on the other side of the stress axis from the slip direction and all three directions should lie roughly on a common great circle. Finally, the indicated combinations of slip directions and slip planes yield the active slip systems because in each case they represent the slip system that

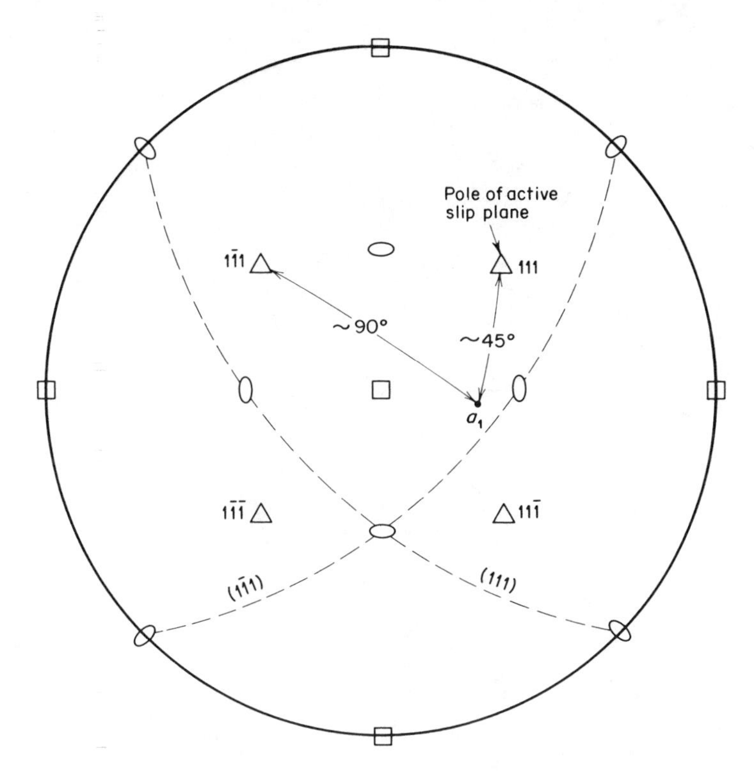

Fig. 5.8 The original stress axis orientation in Fig. 5.7 lies about 45° from the pole of the (111) plane and about 90° from the pole of $(1\bar{1}1)$. These are the two slip planes that contain the active slip direction.

has the highest resolved shear stress acting on it in a given stereographic triangle. (Problems 11–14 in Chapter 4 actually demonstrate this fact.)

Now let us consider the case of a crystal deformed in compression. As shown in Fig. 5.6B, the slip plane rotates so that it tends to become perpendicular to the stress axis. In this case, if the orientation of the stress axis is plotted in a standard stereographic projection, the stress axis will be found to follow a path that passes through the position of the slip plane normal. This is illustrated in Fig. 5.9, using once again as an example a single crystal of a face-centered cubic metal.

5.5 The Notation for the Slip Systems in the Deformation of F.C.C. Crystals. The preceding section has shown that, for an orientation in the center of a stereographic triangle, one slip system is favored because the resolved shear stress is highest on this system. This slip system is called the *primary slip system.* In the example of Fig. 5.7, the primary slip system for the

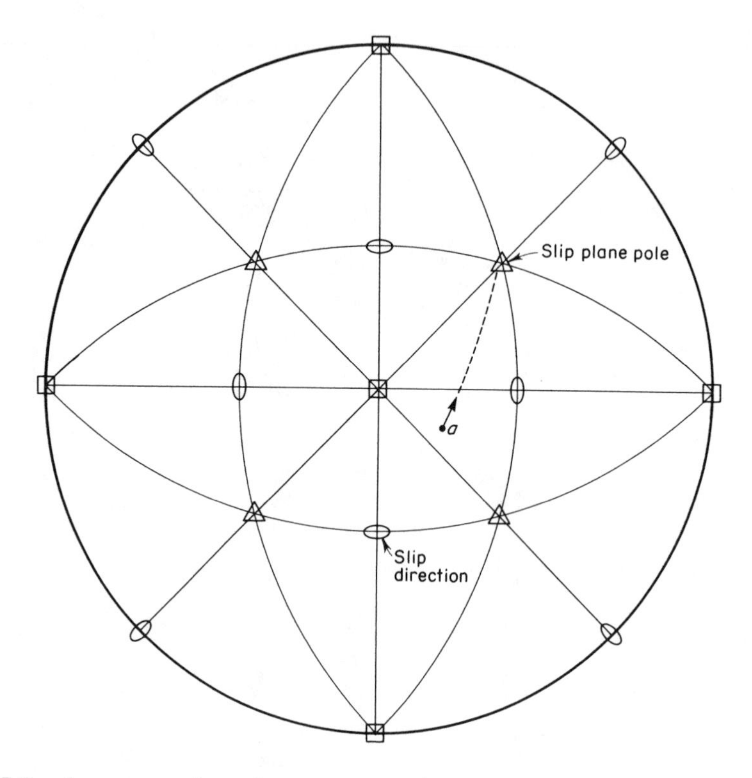

Fig. 5.9 In compression, the stress axis (a) rotates toward the pole of the active slip plane.

original stress axis orientation is $(111)[10\bar{1}]$. If the dislocations of this system were to cross-slip, they would have to move onto the other slip plane that contains the $[10\bar{1}]$ slip direction. As may be seen in Fig. 5.8, this is the $(1\bar{1}1)$ plane. The cross-slip system for the stress axis orientation a_1 is, accordingly, $(1\bar{1}1)[10\bar{1}]$.

Another important slip system is called the *conjugate slip system.* This is the system that becomes the preferred slip system once the rotation of the crystal structure, relative to its tensile stress axis, results in moving the orientation of the stress axis out of its original stereographic triangle into the one adjoining it. This movement of the stress axis on the stereographic projection is shown in Fig. 5.10, which is an enlarged view of the standard projection showing only the two stereographic triangles of present interest.

As indicated above, the stress axis position on the stereographic projection tends to move toward the active slip direction through the positions a_1, a_2, and a_3. Continued motion of this type carries it to a position such as a_4, where it lies

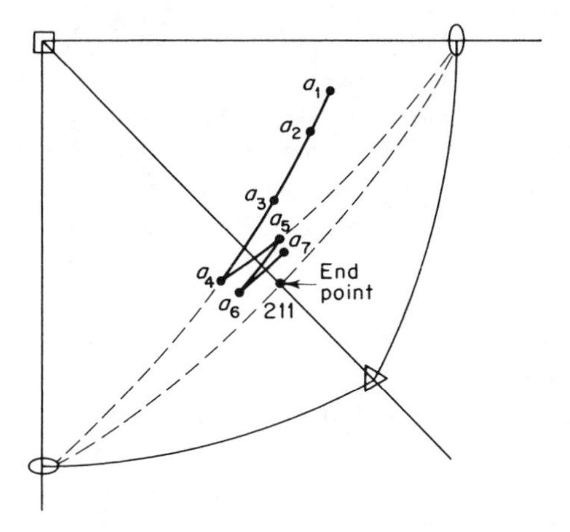

Fig. 5.10 When the stress axis leaves its original stereographic triangle, a second or conjugate slip system becomes more highly stressed than the primary slip system. After this occurs, deformation occurs alternately on both systems.

in another stereographic triangle where the resolved shear stress is greater on the $(1\bar{1}\bar{1})$ [110] slip system. As a result, slip will be instituted on this slip system, called the conjugate slip system, and the stress axis will now move along a path toward the [110] slip direction. Theoretically, while this shift in the slip system should occur once the stress axis crosses the boundary separating the two stereographic triangles, there is always some degree of overshoot before the conjugate slip system takes over. The degree of this overshoot depends on a number of factors that will not be discussed here. The diagram shows clearly that slip on the second (or conjugate) system will bring the stress axis back into the original stereographic triangle where the primary slip system is again favored. It is therefore to be expected that the primary slip system will eventually predominate at a point such as a_5 and the stress axis will again move toward the slip direction of the primary system. This oscillation of the stress axis will be repeated a number of times, with the result (shown in the figure) that the axis eventually reaches the [211] direction, a direction that lies on the same great circle as the conjugate and primary slip directions and midway between them. This is a stable end orientation for the crystal, and once it has obtained this orientation, further deformation will not change the orientation of the crystal relative to the tensile stress axis.

In the above discussion, the primary, conjugate, and cross-slip slip systems have been identified for a given starting orientation of the axis of a crystal. These have involved three of the possible slip planes in the face-centered cubic structure. The fourth plane, which is $(11\bar{1})$ in the example in Fig. 5.7, is called the *critical plane.*

5.6 Vector Notation for Dislocations. Up to this point we have been interested in dislocations only in a general sense and, for purposes of illustration, only the most elementary type of structure – a simple cubic lattice – has been considered. In an actual crystal, because of the more complex spatial arrangement of atoms, the dislocations are complicated and often difficult to visualize. It is sometimes convenient to neglect all considerations of the geometrical appearance of a complicated dislocation and define the dislocation by its Burgers vector. The vector notation for Burgers vectors is especially convenient here. It has already been pointed out that, in any crystal form, the distance between atoms in a close-packed direction corresponds to the smallest shear distance that will preserve the crystal structure during a slip movement. Dislocations with Burgers vectors equal to this shear are energetically the most favored in a given crystal structure. With regard to the vector notation, the direction of a Burgers vector can be represented by the Miller indices of its direction, and the length of the vector can be expressed by a suitable numerical factor placed in front of the Miller indices. Several explanatory examples need to be considered.

Figure 5.11 represents the unit cell for cubic structures. In a simple cubic lattice, the distance between atoms in a close-packed direction equals the length of one edge of a unit cell. Now consider the symbol [100]. This quantity may be taken to represent a vector, such as is shown in Fig. 5.11, with a unit component in the x direction, and is, accordingly, the distance between atoms in

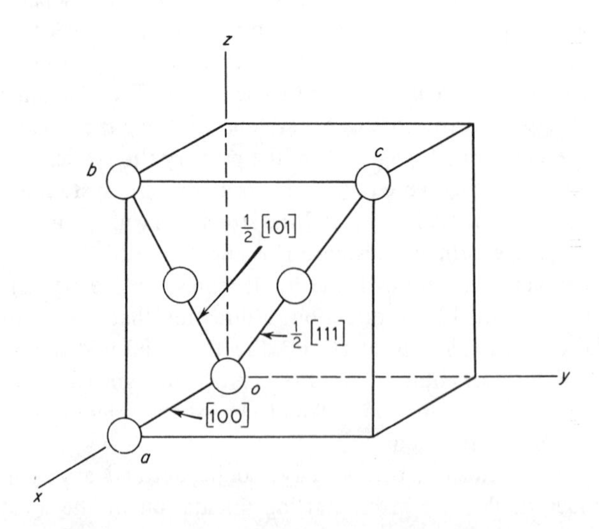

Fig. 5.11 The spacing between atoms in the close-packed directions of the different cubic systems: face-centered cubic, body-centered cubic, and simple cubic.

the x direction. A dislocation with a Burgers vector parallel to the x direction in a simple cubic lattice is therefore represented by [100]. Now consider a face-centered cubic lattice. Here the close-packed direction is a face diagonal, and the distance between atoms in this direction is equal to one-half the length of the face diagonal. In Fig. 5.11, the indicated face diagonal ob has indices [101]. This symbol also corresponds to a vector with unit components in the x and z directions and a zero component in the y direction. As a result, a dislocation in a face-centered cubic lattice having a Burgers vector lying in the [101] direction should be written ½[101]. In the body-centered cubic lattice, the close-packed direction is a cube diagonal, or a direction of the form ⟨111⟩. The distance between atoms in these directions is one-half the length of the diagonal, so that a dislocation having a Burgers vector parallel to [111], for example, is written ½[111].

5.7 Dislocations in the Face-Centered Cubic Lattice. The primary slip plane in the face-centered cubic lattice is the octahedral plane {111}. Figure 5.12 shows a plane of this type looking down on the extra plane of an edge dislocation. The darkened circles represent the close-packed (111) plane at which the extra plane of the edge dislocation ends, while the white circles are the atoms in the next following close-packed (111) plane. Notice that in the latter plane a zigzag row of atoms is missing. This corresponds to the missing plane of the edge dislocation. Now consider the plane of atoms (lying directly over the

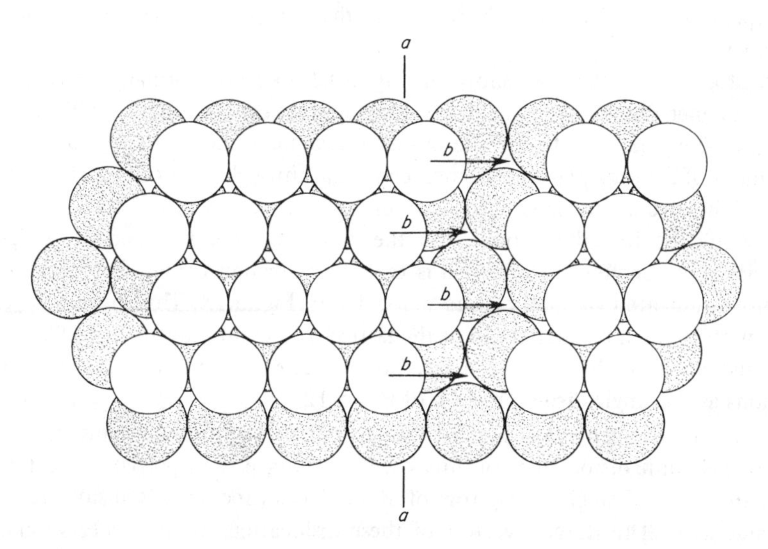

Fig. 5.12 A total dislocation (edge orientation) in a face-centered cubic lattice as viewed when looking down on the slip plane.

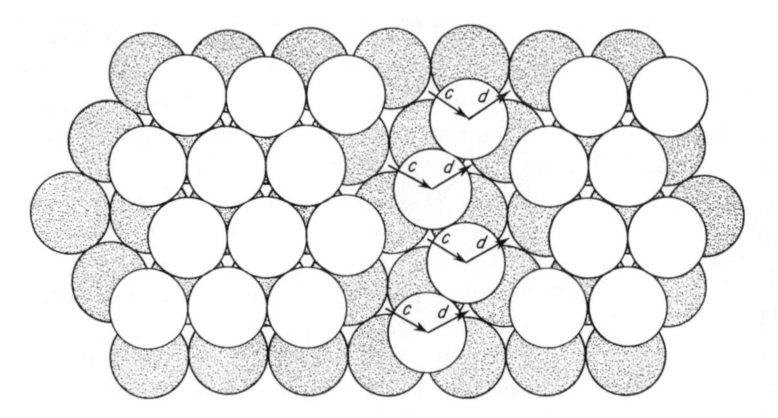

Fig. 5.13 Partial dislocation in a face-centered cubic lattice.

zigzag row of white atoms) immediately to the left of the missing plane of atoms. The movement of the atoms of this plane through the horizontal distance *b* displaces the dislocation one unit to the left. The vector *b* thus represents the Burgers vector of the dislocation which is designated as $\frac{1}{2}[\bar{1}10]$, according to the method outlined in the preceding section. Similar movements in succession of the planes of atoms that find themselves to the left of the dislocation will cause the dislocation to move across the entire crystal. As is to be expected, this dislocation movement shears the upper half of the crystal (above the plane of the paper) one unit *b* to the right relative to the bottom half (below the plane of the paper).

A dislocation of the type shown in Fig. 5.12 does not normally move in the simple manner discussed in the above paragraph. As can be deduced with the aid of a ping-pong-ball model of the atom arrangement shown in Fig. 5.12, the movement of a zigzag plane of atoms, such as *aa*, through the horizontal distance *b* would involve a very large lattice strain, because each white atom at the slip plane would be forced to climb over the dark atom below it and to its right. What actually is believed to happen is that the indicated plane of atoms makes the move indicated by the vectors marked *c* in Fig. 5.13. This movement can occur with a much smaller strain of the lattice. A second movement of the same type, indicated by the vectors marked *d*, brings the atoms to the same final positions as the single displacement *b* of Fig. 5.12.

The atom arrangement of Fig. 5.13 is particularly significant because it shows how a single-unit dislocation can break down into a pair of partial dislocations. Thus, the isolated single zigzag row of atoms has an incomplete dislocation on each side of it. The Burgers vectors of these dislocations are the vectors *c* and *d* shown in the figure. The vector notation for these Burgers vectors can be deduced with the aid of Fig. 5.14. This figure shows the (111) surface of a

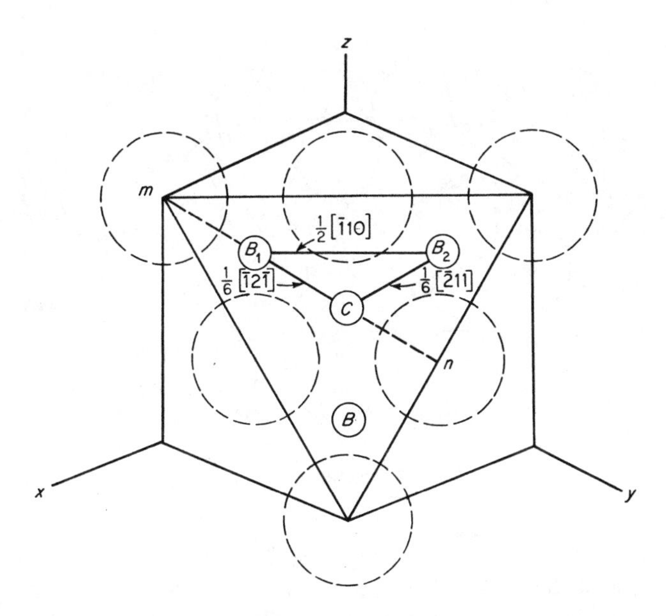

Fig. 5.14 The orientation relationship between the Burgers vectors of a total dislocation and its partial dislocations.

face-centered cubic crystal in relation to the unit cell of the structure. The positions of the atoms in this surface are indicated by circles drawn with dashed lines. The places which would be occupied by the atoms in the next plane above the given one are designated by small circles with the letter *B* inside them. In the center of Fig. 5.14 is another small circle marked *C*. The Burgers vector of the total dislocation equals the distance $B_1 B_2$, while the Burgers vectors of the two partial dislocations *c* and *d* of Fig. 5.13 are the same as the distances B_1C and CB_2. The Burgers vector of the total dislocation is $\frac{1}{2}[\bar{1}10]$, which follows from the previous discussion. The line B_1C lies in the $[\bar{1}2\bar{1}]$ direction. The symbol $[\bar{1}2\bar{1}]$ represents a vector with unit negative components in the *x* and *z* directions and a component of 2 in the *y* direction. This is a vector the length of which is twice the distance *mn* in Fig. 5.14. Since B_1C is just one-third of line *mn*, the Burgers vector for this dislocation is $\frac{1}{6}[\bar{1}2\bar{1}]$. In the same manner, it can be shown that the vector CB_2 can be represented by $\frac{1}{6}[\bar{2}11]$. The total face-centered cubic dislocation $\frac{1}{2}[\bar{1}10]$ is thus able to dissociate into two partial dislocations according to the relation

$$\tfrac{1}{2}[\bar{1}10] = \tfrac{1}{6}[\bar{1}2\bar{1}] + \tfrac{1}{6}[\bar{2}11]$$

When a total dislocation breaks down into a pair of partials, the strain energy of the lattice is decreased. This results because the energy of a dislocation is

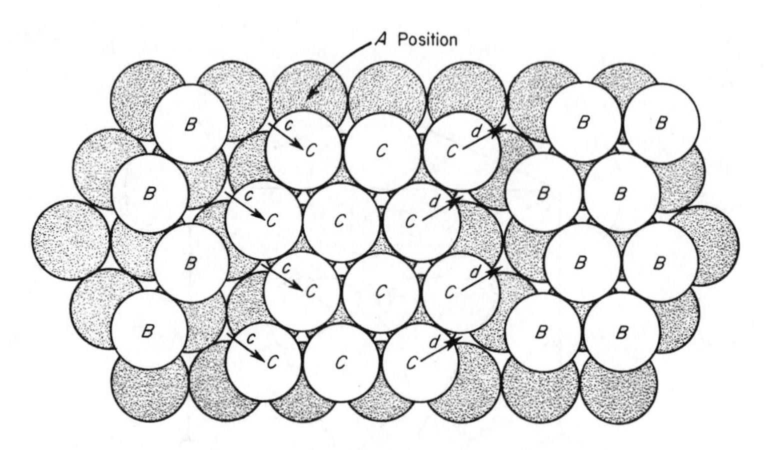

Fig. 5.15 An extended dislocation.

proportional to the square of its Burgers vector and because the square of the Burgers vector of the total dislocation is more than twice as large as the square of the Burgers vector of a partial dislocation. Because the partial dislocations of Fig. 5.13 represent approximately equal lattice strains, a repulsive force exists between them which forces the partials apart. Such a separation will add additional planes of atoms to the single zigzag plane of Fig. 5.13, as shown in Fig. 5.15. A total dislocation that has dissociated into a pair of separated partials like those in the latter figure is known as an *extended dislocation.* An important fact to notice about Fig. 5.15 is that the white atoms that lie between the two partial dislocations have stacking positions which differ from those of the atoms on the far side of either of the partial dislocations. Thus, if we assume that the dark-colored atoms occupy *A* positions in a stacking sequence and the white atoms at either end of the figure, *B* positions, then the white atoms between the two partial dislocations lie on *C* positions. In this region, the *ABCABCABC* . . . stacking sequence of the face-centered cubic lattice suffers a discontinuity and becomes *ABCA*↓↑*CABCA* . . . . The arrows indicate the discontinuity. Discontinuities in the stacking order of the {111}, or close-packed planes, are called *stacking faults.* In the present example, the stacking fault occurs on the slip plane (between the dark and white atoms) and is bounded at its ends by Shockley partial dislocations. In a face-centered cubic lattice[4] stacking faults may arise in a number of other ways. In all cases, if a stacking fault terminates inside a crystal, its boundaries will form a partial dislocation. The partial dislocations of stacking faults, in general, may be either of the Shockley type, with the Burgers vector of the dislocation lying in the plane of the fault, or of

4. Read, W. T., Jr., *Dislocations in Crystals.* McGraw-Hill Book Co., Inc., New York, 1953.

the Frank type, with the Burgers vector normal to the stacking fault. The partial dislocations presently being considered, however, are only those associated with slip and are of the Shockley form.

Since the atoms on either side of a stacking fault are not at the positions they would normally occupy in a perfect lattice, a stacking fault possesses a surface energy which, in general, is small compared with that of an ordinary grain boundary, but nevertheless finite. This stacking-fault energy plays an important part in determining the size of an extended dislocation. The larger the separation between the partial dislocations, the smaller is the repulsive force between them. On the other hand, the total surface energy associated with the stacking fault increases with the distance between partial dislocations. The separation between the two partials thus represents an equilibrium between the repulsive energy of the dislocations and the surface energy of the fault. Seeger and Schoeck.[5,6] have shown that the separation of the pair of partial dislocations in an extended dislocation depends on a dimensionless parameter $\gamma_I c/Gb^2$, where γ_I is the specific surface energy of the stacking fault, G is the shear modulus in the slip plane, c is the separation between adjoining slip planes, and b is the magnitude of the Burgers vector. In certain face-centered cubic metals typified by aluminum, this parameter is larger than 10^{-2} and the separation between dislocations is only of the order of a single atomic distance. These metals are said to have high stacking-fault energies. When the parameter is less than 10^{-2}, a metal is said to have a low stacking-fault energy. A typical example of a metal with a low stacking-fault energy is copper. The computed[6] separation of the partial dislocations in copper are of the order of twelve interatomic spacings if the extended dislocation is in the edge orientation, and about five interatomic spacings if it is in the screw orientation.

A word should be said about the movement of an extended dislocation through a crystal. The actual movement may be quite complicated for several reasons. First, if the moving dislocation meets obstacles, such as other dislocations, or even second-phase particles, the width of the stacking fault should vary. Second, thermal vibrations may also cause the width of the stacking fault to vary locally along the dislocation, the variation being a function of time. On the assumption that these and other complicating effects can be neglected, an extended dislocation can be pictured as a pair of partial dislocations, separated by a finite distance, which move in consort through the crystal. The first partial dislocation, as it moves, changes the stacking order, while the second restores the order to its proper sequence. After both partials have passed a given point in the lattice, the crystal will have been sheared (at the slip plane) by an amount equal to the Burgers vector b of the total dislocation.

5. Seeger, A., and Schoeck, G., *Acta Met.* **1** 519 (1953).
6. Seeger, A., *Dislocations and Mechanical Properties of Crystals.* John Wiley and Sons, Inc., New York, 1957.

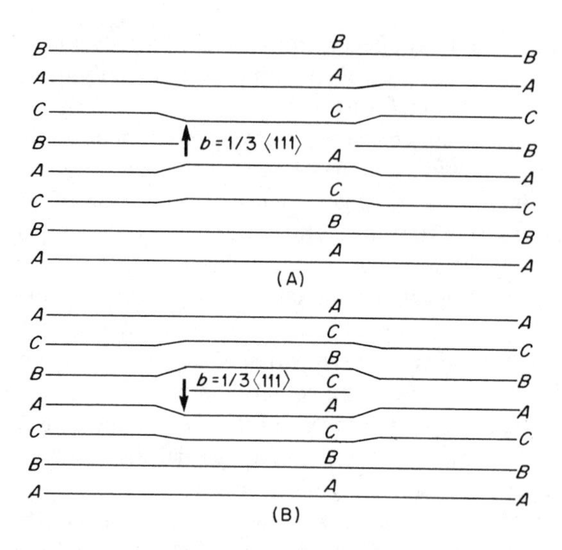

Fig. 5.16A An intrinsic stacking-fault can also be formed in a face-centered cubic crystal by removing part of a close-packed plane.

Fig. 5.16B The addition of a portion of an extra close-packed plane to a face-centered cubic crystal produces an extrinsic stacking fault.

5.8 Intrinsic and Extrinsic Stacking Faults in Face-Centered Cubic Metals. The movement of a Shockley partial across the slip plane of a f.c.c. metal has been shown to produce a stacking sequence *ABCACABCA* . . . . In this case the normal stacking sequence can be observed to exist right up to the plane of the fault. A fault of this type, following Frank, is called an *intrinsic stacking fault.* An intrinsic stacking fault may also be developed in a f.c.c. crystal by removing part of a close-packed plane, as shown in Fig. 5.16A. Such a method of developing an intrinsic stacking fault is physically quite possible and can occur by the condensation of vacancies on an octahedral plane. While the fault produced in this manner is the same as that resulting from the slip of a Shockley partial, the partial dislocation surrounding the fault is not the same. In this case it is normal to the {111} slip plane and therefore of the Frank type. Its Burgers vector is equal to one-third of a total dislocation and therefore may be written $\frac{1}{3}\langle 111\rangle$.

The addition of a portion of an octahedral plane produces a different type of stacking sequence which is *ABCACBCABC* . . . . In this fault (Fig. 5.16B) a plane has been inserted that is not correctly stacked with respect to the planes on either side of the fault. This type of fault is called an *extrinsic* or double *stacking fault.* The Burgers vector for the extrinsic fault, shown in Fig. 5.16B is also $\frac{1}{3}\langle 111\rangle$. An extrinsic stacking fault could be formed by the precipitation of interstitial atoms on an octahedral plane. This is believed to be much less probable than the condensation of vacancies to produce an intrinsic fault. It is also possible to form extrinsic stacking faults by the slip of Shockley partials. However, to do this one normally has to assume that this type of slip occurs on two neighboring planes.[7]

5.9 Extended Dislocations in Hexagonal Metals. Because the basal plane of a hexagonal metal has exactly the same close-packed arrangement of atoms as the octahedral $\{111\}$ plane of face-centered cubic metals, extended dislocations also occur in these metals. In the hexagonal system, the dissociation of a total dislocation into a pair of partials on the basal plane is expressed in the following fashion:

$$\tfrac{1}{3}[\bar{1}2\bar{1}0] = \tfrac{1}{3}[01\bar{1}0] + \tfrac{1}{3}[\bar{1}100]$$

The above equation represents exactly the same addition of Burgers vectors as was given in the previous section for the face-centered cubic system, namely,

$$\tfrac{1}{2}[\bar{1}10] = \tfrac{1}{6}[\bar{1}2\bar{1}] + \tfrac{1}{6}[\bar{2}11]$$

The only difference between the two equations lies in the present use of the four-digit Miller indices for hexagonal metals. The stacking fault associated with extended dislocations in the hexagonal metals is similar to that of the face-centered metals. The movement of the first partial dislocation through the crystal changes the stacking sequence *ABABABABAB* . . . to *ABACBCBCBC* . . . on the assumption that the fourth plane slips relative to the third. Notice that the sequence *CBCBCB* . . . is a perfectly good hexagonal sequence with alternate planes lying above one another. The stacking fault occurs between the third and fourth planes: between $A{\downarrow\uparrow}C$ . . . . As in the face-centered cubic example, the movement of the second partial dislocation restores the crystal to the proper stacking sequence *ABABAB* . . . .

5.10 Extended Dislocations and Cross-Slip. While a total $\tfrac{1}{2}\langle 110\rangle$ dislocation in a f.c.c. metal can readily cross-slip between a pair of octahedral planes, this is not true of an extended dislocation. The reason for this is shown in the diagrams of Fig. 5.17. In part A of this illustration, a total $\tfrac{1}{2}\langle\bar{1}10\rangle$ dislocation is shown cross-slipping from the primary slip plane (111) to the cross-slip plane $(11\bar{1})$. Note that the Burgers vector of this dislocation lies along a direction common to both slip planes. In part B of this diagram, a corresponding extended dislocation is shown moving downward on the (111) primary slip plane. Assume, as shown in Fig. 5.17C, that the leading partial

7. Hirth, J. P., and Lothe, J., *Theory of Dislocations*, p. 291, McGraw-Hill Book Company, New York, 1968.

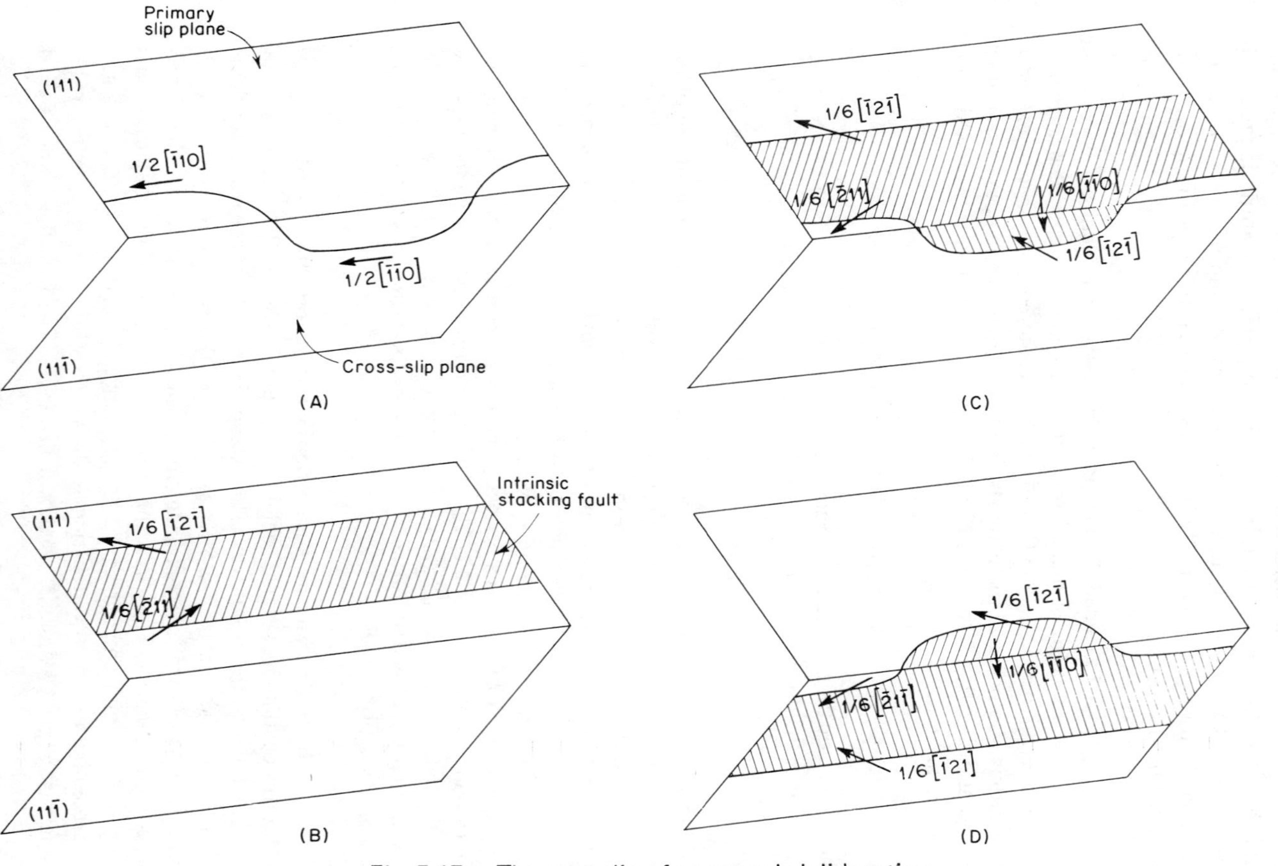

Fig. 5.17 The cross-slip of an extended dislocation.

dislocation of this extended dislocation moves off on the cross-slip plane. Since the Burgers vector of this partial dislocation is $\frac{1}{6}[\bar{2}11]$ and this vector has a direction that does not lie in the $(11\bar{1})$ cross-slip plane, a dislocation reaction has to occur that results in the creation of two new dislocations. This reaction is

$$\tfrac{1}{2}[\bar{2}11] \rightarrow \tfrac{1}{6}[\bar{1}21] + \tfrac{1}{6}[\bar{1}\bar{1}0]$$

One of these dislocations is a Shockley partial that is free to move on the cross-slip plane and has a Burgers vector $\frac{1}{6}[\bar{1}21]$. The other is immobile (sessile) and remains behind along the line of intersection between the primary and cross-slip planes. This latter dislocation is called a *stair-rod* dislocation, and its Burgers vector is $\frac{1}{6}[\bar{1}\bar{1}0]$. Stair-rod dislocations, of which the present example represents only one of a number of different types, are always found when a stacking fault runs from one slip plane over into another, as shown in Fig. 5.17C. Note that the $\frac{1}{6}[\bar{1}\bar{1}0]$ of the stair-rod dislocation in Fig. 5.17C is perpendicular to the line of intersection of the two slip planes and that it also does not lie in either slip plane.

In the cross-slip of an extended dislocation, the stair-rod dislocation is removed when the trailing partial dislocation arrives at the line of intersection of the two slip planes and then moves off on the cross-slip plane, as shown by the following equation

$$\tfrac{1}{6}[\bar{1}2\bar{1}] + \tfrac{1}{6}[\bar{1}\bar{1}0] = \tfrac{1}{6}[\bar{2}1\bar{1}]$$

This reaction is illustrated in Fig. 5.17C.

Since additional strain energy is required to create the stair-rod dislocation associated with the cross-slip of an extended dislocation, it should be much easier for a total dislocation than for an extended dislocation to cross-slip.

5.11 Dislocation Intersections. The dislocations in a metal constitute a three-dimensional network of linear faults. On any given slip plane, there will be a certain number of dislocations that lie in this plane and are capable of producing slip along it. At the same time, there will be many other dislocations that intersect it at various angles. Consequently, when a dislocation moves it must pass through those dislocations that intersect its slip plane. The cutting of dislocations by other dislocations is an important subject because, in general, it requires work to make the intersections. The relative ease or difficulty with which slip occurs is thus determined in part by the intersection of dislocations.

For a simple example of the result of a dislocation intersection, see Fig. 5.18. In the drawing it is assumed that a dislocation has moved across the slip plane *ABCD* thereby shearing the top half of the rectangular crystal relative to the bottom half by the length of its Burgers vector b. A second (vertical) dislocation,

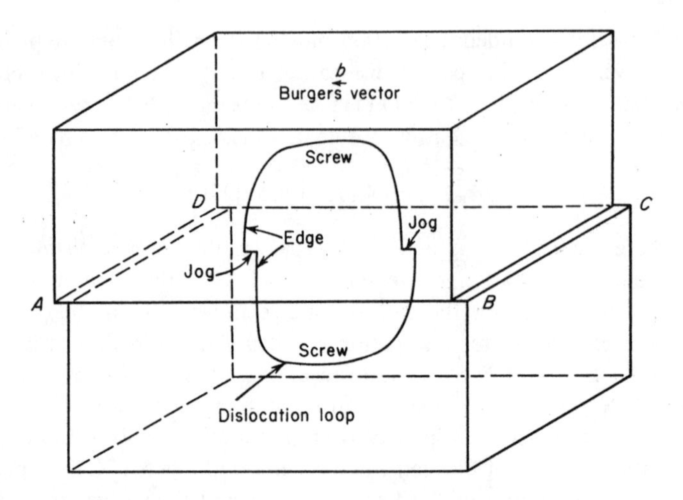

Fig. 5.18 In the figure a dislocation is assumed to have moved across the horizontal plane *ABCD* and, in cutting through the vertical-dislocation loop, it forms a pair of jogs in the latter.

having a loop that intersects the slip plane at two points, is shown in Fig. 5.18. It is assumed, for the sake of convenience, that this loop is in the edge orientation where it intersects the slip plane. The indicated displacement of the crystal, shown in Fig. 5.18, also shears the top half of the vertical-dislocation loop relative to its bottom half by the amount of the Burgers vector b. This shearing action cannot break the loop into two separated half-loops because, according to rule, a dislocation cannot end inside a crystal. The only alternative is that the displacement lengthens the vertical-dislocation loop by an amount equal to the two horizontal steps shown in Fig. 5.18. This result is characteristic of the

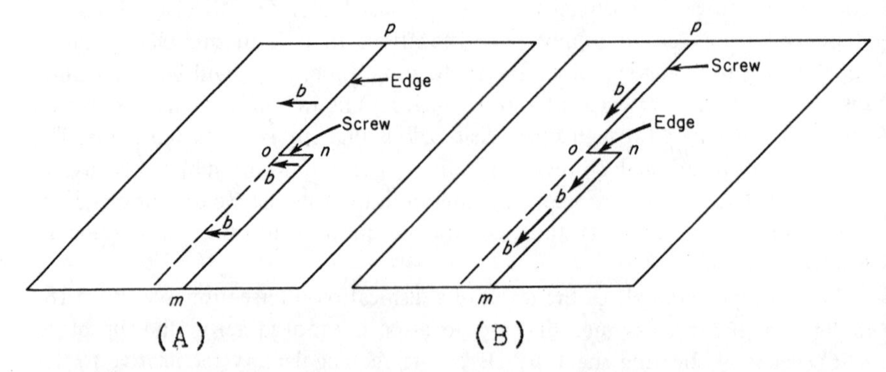

Fig. 5.19 Dislocations with kinks that lie in the slip plane of the dislocations.

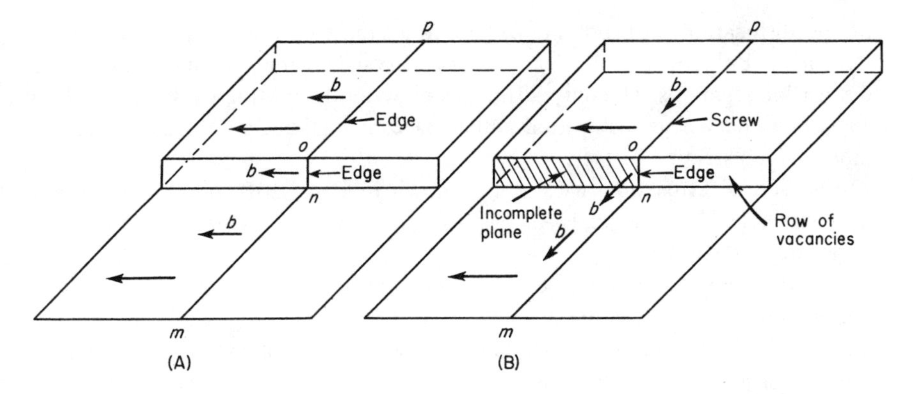

Fig. 5.20 Dislocations with jogs normal to their slip planes.

intersection of dislocations, for whenever a dislocation cuts another dislocation, both dislocations acquire steps of a size equal to the other's Burgers vector.

Let us now consider some of the simpler types of steps, formed by dislocation intersections. The two basic cases are, first, where the step lies in the slip plane of a dislocation and, second, where the step is normal to the slip plane of a dislocation. The first type is called a *kink*, while the second is called a *jog*. The first is treated in Fig. 5.19. Here (A) represents a dislocation in the edge orientation, and (B) a dislocation in the screw orientation, both of which have received kinks (*on*) as a result of intersections with other dislocations. The kink in the edge dislocation has a screw orientation (Burgers vector parallel to line *on*), while the step in the screw dislocation has an edge orientation (Burgers vector normal to line *on*). Both of these steps can easily be eliminated, for example, by moving line *mn* over to the position of the dashed line. This movement in both cases can occur by simple slip. Since the elimination of a step lowers the energy of the crystal by the amount of the strain energy associated with a step, it can be assumed that steps of this type may tend to disappear.

An edge and a screw dislocation, with steps normal to the primary slip plane, are shown in Figs. 5.20A and 5.20B. This type of discontinuity is called a jog. It should be noted that the jog of Fig. 5.20B is also capable of elimination if the dislocation is able to move in a plane normal to the indicated slip plane, that is, in a vertical plane. This follows from the fact that if this stepped dislocation is viewed as lying in a vertical plane, then we have the same step arrangement as that shown in Fig. 5.19B. However, let us assume that both dislocations of Fig. 5.20 are not capable of gliding in a vertical plane. What then is the effect of the steps on the motion of the dislocations in their horizontal-slip surfaces? In this respect, it is evident that the edge dislocation with a jog, shown in Fig. 5.20A, is free to move on the stepped surface shown in the figure, for all three segments of the dislocation, *mn, no,* and *op*, are in a simple edge orientation with their respective Burgers vectors lying in the crystal planes which contain the dislocation segments. The only difference between the motion of this dislocation and an ordinary edge dislocation is that instead of gliding along a single plane it moves over a stepped surface.

The screw dislocation with a jog, shown in Fig. 5.20B, represents quite a different case. Here the jog is an edge dislocation with an incomplete plane lying in the stepped surface.

For the sake of argument, let us assume that its extra plane lies to the left of line *no* and thus corresponds to the cross-hatched area in Fig. 5.20B. Now there are two basic ways that an incomplete plane that is only one atomic spacing high can be formed. In the first case, it can be assumed that the crosshatched area corresponds to a row of interstitial atoms which ends at *no.* Alternatively, the crosshatched step may be considered a part of a normal continuous lattice plane, and then the stepped surface to the right of *no* would have to be a row of vacancies. The latter case is the one indicated in the drawing in Fig. 5.20B. Which of these two alternatives eventuates depends on the relative orientation of the Burgers vectors of the two dislocations, the intersection of which caused the jog. In this respect, it should be mentioned that these jogs are formed when one screw dislocation intersects another screw dislocation. The other steps illustrated in Fig. 5.19 and 5.20A are formed by various intersections involving edge dislocations with other edges, or edge dislocations with screws. Read[8] discusses these different possibilities in detail and shows in particular how the intersection of two screw dislocations can lead, at least in principle, to the production of either a row of vacancies or a row of interstitial atoms. Relative to these rows of vacancies, or interstitial atoms, it should be pointed out that after one screw dislocation has intersected another, and if it moves on away from the point of intersection, the row of point defects (vacancies or interstitial atoms) stretches in a line from the moving dislocation back to the stationary one. This, of course, is on the assumption that thermal energy does not cause the point defects to diffuse off into the lattice. In terms of Fig. 5.20B, this signifies that if the indicated dislocation is moving to the left, the row of vacancies to its right extends back to another screw dislocation, whose intersection by the moving dislocation caused the row of vacancies to be formed.

Whereas the edge dislocation with a jog normal to its slip plane is capable of moving by simple glide along the stepped surface of Fig. 5.20A, this is not the case of the stepped screw dislocation of Fig. 5.20B. Here the jog (line *no*), which is in an edge orientation, is not capable of gliding along the vertical surface shown in the figure because its Burgers vector is not in the surface of the step but is normal to it. The only way that the jog can move across the surface of the

8. Read, W. T., Jr., *Dislocations in Crystals.* McGraw-Hill Book Company, Inc., New York, 1953.

step is for it to move by dislocation climb. Thus, in Fig. 5.20B, if the jog is to move to the left with the rest of the dislocation, additional vacancies will have to be added to the row of vacancies (to the right of line *no*). Alternatively, if the extra plane had been formed by a row of interstitial atoms to the left of the jog, the movement of the jog (in this case to the right) requires the creation of additional interstitial atoms.

5.12 Grain Boundaries. In the previous chapters we have been concerned primarily with a study of the properties of very pure metals in the form of single crystals. Single crystal studies are important because they lead to quicker understanding of many basic phenomena. However, single crystals are essentially a laboratory tool and are rarely found in commercial metal objects which, as a rule, consist of many thousands of small microscopic metallic crystals. Figure 5.21 shows the crystalline structure of a typical *polycrystalline* (many crystal) specimen magnified 500 times. The average diameter of the

Fig. 5.21 A polycrystalline zirconium specimen photographed with polarized light. In this photograph, individual crystals can be distinguished by a difference in shading, as well as by the thin dark lines representing grain boundaries. 350 x. (Photomicrograph by E. R. Buchanan.)

crystals is very small, approximately 0.05 mm, and each crystal is seen to be separated from its neighbors by dark lines, the grain boundaries. The grain boundaries appear to have an appreciable width, but this is only because the surface has been etched in an acid solution in order to reveal their presence. A highly polished polycrystalline specimen of a pure metal which has not been etched will appear entirely "white" under the microscope; that is, it will not show grain boundaries. The width of a grain boundary is, therefore, very small.

The grain boundaries play an important part in determining the properties of a metal. For example, at low temperatures the grain boundaries are, as a rule, quite strong and do not weaken metals. In fact, heavily strained pure metals, and most alloys, fail at low temperatures by cracks that pass through the crystals and not the boundaries. Fractures of this type are called *transcrystalline.* However, at high temperatures and slow strain rates, the grain boundaries lose their strength more rapidly than do the crystals, with the result that fractures no longer traverse the crystals but run along the grain boundaries. Fractures of the later type are called *intercrystalline fractures.* A number of other important grain-boundary aspects will be covered in the following paragraphs.

5.13 Dislocation Model of a Small-Angle Grain Boundary. In 1940, both Bragg and Burgers introduced the idea that boundaries between crystals of the same structure might be considered as arrays of dislocations. If two grains differ only slightly in their relative orientation, it is possible to draw a dislocation model of the boundary without difficulty. Figure 5.22A shows an

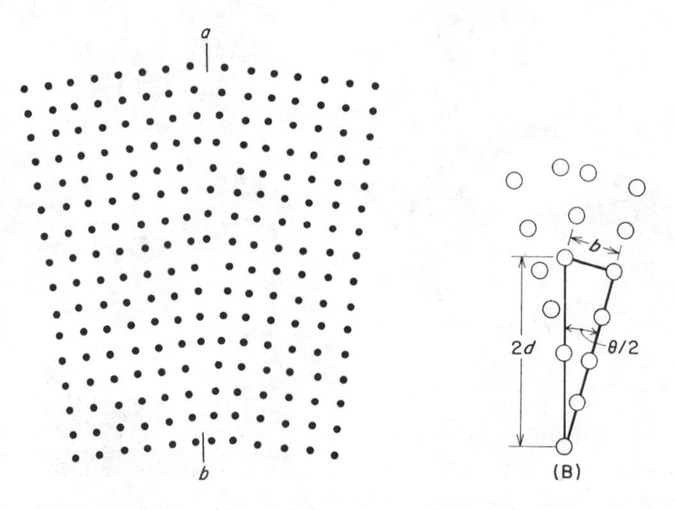

Fig. 5.22 (A) Dislocation model of a small-angle grain boundary. (B) The geometrical relationship between θ, the angle of tilt, and d, the spacing between the dislocations.

example of such an elementary boundary in a simple cubic lattice in which the crystal on the right is rotated with respect to that on the left about the [100] direction (the direction perpendicular to the plane of the paper). The vertical line between points *a* and *b* corresponds to the boundary. The lattice on both sides of the boundary is seen to be inclined downward toward the boundary, with the result that certain nearly vertical lattice planes of both crystals terminate at the boundary as positive-edge dislocations. The greater the angular rotation of one crystal relative to the other, the greater is the inclination of the planes that terminate as dislocations at the boundary, and the closer is the spacing of the dislocations in the vertical boundary. The dislocation spacing in the boundary thus determines the angular inclination between the lattices. It is, in fact, easy to derive a simple expression for this relationship. Thus, with reference to Fig. 5.22B, it can be seen that

$$\sin \theta/2 = b/2d$$

where b is the Burgers vector of a dislocation in the boundary, and d is the spacing between the dislocations. If the angle of rotation of the crystal structure across the boundary is assumed to be small, then $\sin \theta/2$ may be replaced by $\theta/2$, and the equation relating the angle of tilt across a boundary composed of simple edge dislocations becomes

$$\theta = b/d$$

Experimental evidence for small-angle boundaries has been obtained through the use of suitable etching reagents that locally attack the surface of metals at positions where dislocations cut the surface. After etching with one of these reagents, small-angle boundaries, consisting of rows of edge dislocations, appear as arrays of well-defined etch pits. According to the discussion in the preceding paragraph, the separation of the etch pits (dislocations) should determine the angular rotation of the lattice on one side of the boundary relative to that on the other. The fact that X-ray diffraction determinations of the same lattice rotation are in good agreement with the value predicted by dislocation spacings serves as an excellent check on the validity of the dislocation model for small angle boundaries. Figure 5.23 shows an actual photograph of low-angle boundaries in a magnesium crystal. Low-angle boundaries may also be observed by transmission electron microscopy; an example is shown in Fig. 5.24.

The grain boundary represented in Fig. 5.22 is a very special boundary in which the two lattices are considered to be rotated only slightly with respect to each other about a common lattice direction (a cube edge [100]). This, of course, does not represent the general case of a grain boundary. More complicated grain boundaries with large angles of misfit between grains must

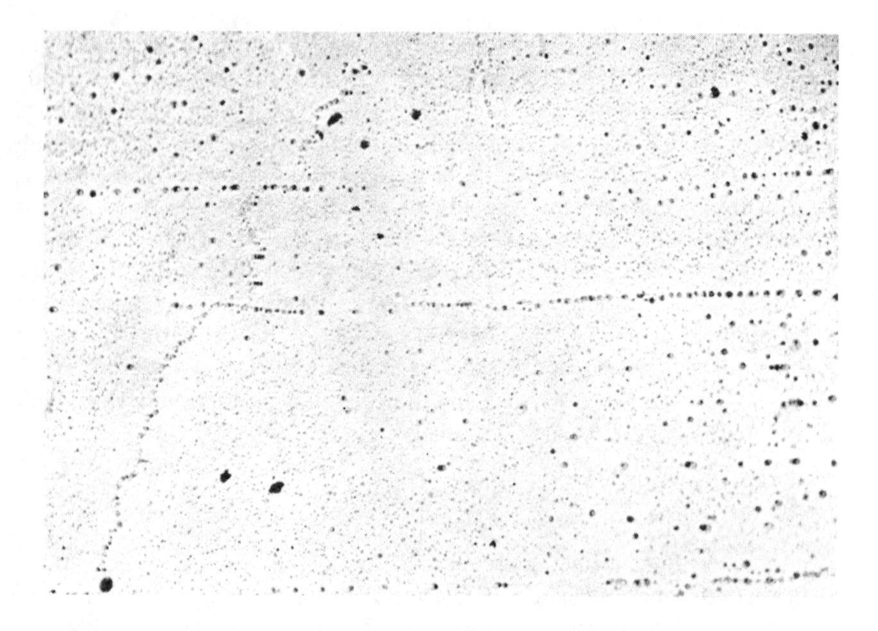

Fig. 5.23 Low-angle boundaries in a magnesium specimen. The rows of etch pits correspond to positions where dislocations intersect the surface.

involve very complicated arrays of dislocations, and no simple picture of their structure has yet been worked out.

The principal difficulty in postulating a suitable dislocation model of a large angle boundary lies in the fact that as the angle of mismatch between grains becomes larger, the dislocations must come closer together and, in so doing, tend to lose their identity.

5.14 The Five Degrees of Freedom of a Grain Boundary. The boundary shown in Fig. 5.22 is special in more than one sense. Not only is the angle of misorientation across the boundary small, but the boundary has only a single degree of freedom. Actually, there are five degrees of freedom of a grain boundary, illustrated in Fig. 5.25. The boundary of Fig. 5.22 is shown again in part *A* of this illustration. Note that it is symmetrically positioned between the two crystals that are tilted with respect to each other about a horizontal axis that runs out of the plane of the paper. This is called a simple symmetrical *tilt boundary*. In part *B*, a symmetrical tilt boundary with a vertical tilt axis is shown. Part *C* shows a basically different way of orienting two crystals relative to each other. In this case they are rotated about an axis normal to the boundary, instead of about an axis lying in the boundary, as in the tilt boundaries in *A* and *B*. This is called a *twist boundary*. Such a boundary will normally contain at

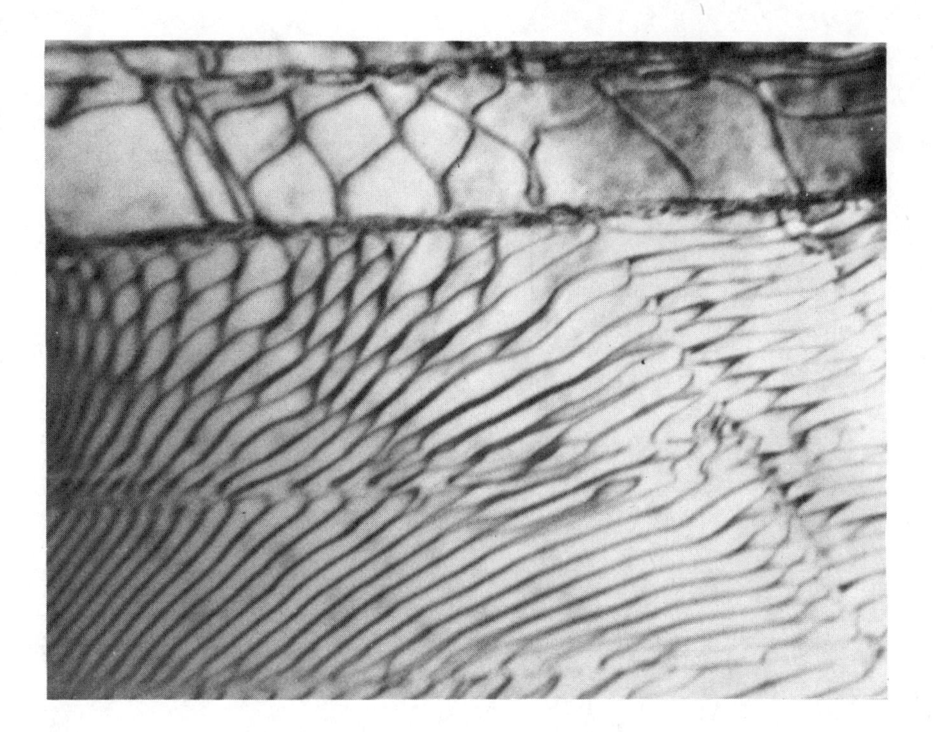

Fig. 5.24 A low-angle boundary in a copper–13.2 atomic percent aluminum specimen deformed 0.7 percent in tension as observed in the transmission electron microscope. This boundary has both a tilt and a twist character. Magnification: 32,000 times. (Photograph courtesy of J. Kastenbach and E. J. Jenkins.)

least two different arrays of screw dislocations, in agreement with our earlier discussion on rotational slip.

The examples just shown represent the three basically different ways that two crystals can be oriented with respect to each other. In each case, the boundary was assumed to be symmetrically positioned with respect to the two crystals. In addition, the boundary itself has two degrees of freedom, as indicated in Figs. 5.25D and 5.25E. These show that the boundary does not have to be in a symmetry position between the two crystals, but can be rotated about either of two axes at 90° to each other. Thus, there are three ways that we can tilt or twist one crystal relative to another, and two ways that we can align the boundary between the crystals. An average grain boundary in a polycrystalline metal will normally involve, in varying degree, all five of these degrees of freedom. It is obvious that the general grain boundary is complex.

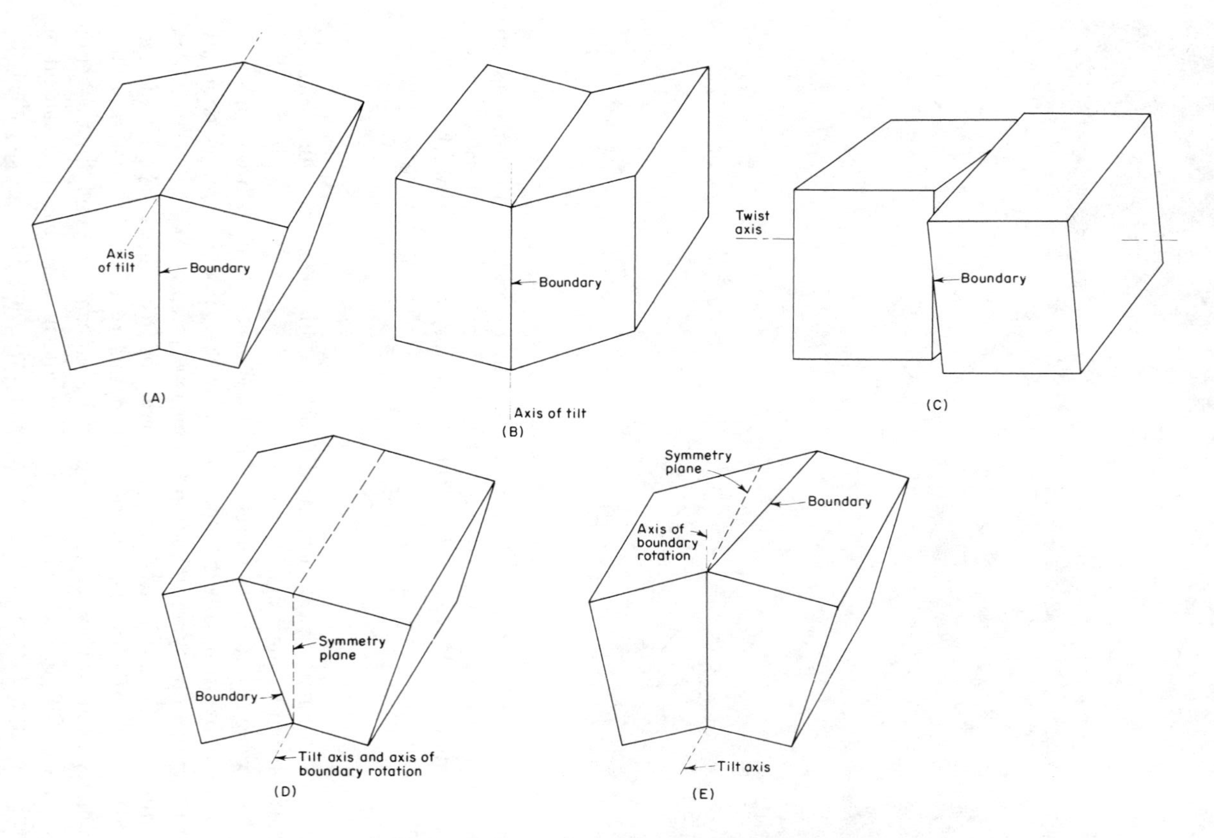

Fig. 5.25 The five degrees of freedom of a grain boundary.

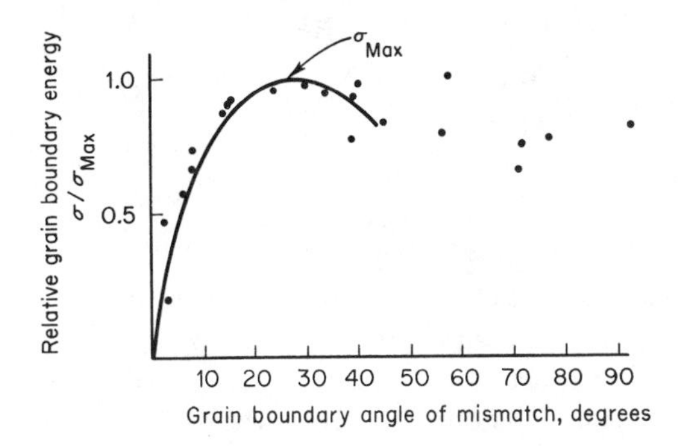

Fig. 5.26 Relative grain-boundary energy as a function of the angle of mismatch between the crystals bordering the boundary. Solid-line theoretical curve; dots experimental data of Dunn for silicon iron. (From *Dislocations in Crystals*, by Read, W. T., Jr. Copyright 1953. McGraw-Hill Book Co., Inc., New York. Used by permission.)

5.15 Grain-Boundary Energy. A dislocation in a crystal represents a linear defect in the crystal structure, with atoms adjacent to dislocations pulled or pushed out of their normal lattice positions. This displacement of large numbers of atoms from positions they would normally occupy in a perfect crystal requires an increase in the energy of the crystal. Strain energy is consequently associated with every dislocation. If there is an energy associated with each dislocation, there must also be an energy associated with an array of dislocations forming a boundary between two crystals. Because of the inherent two-dimensional nature of a grain boundary, this energy is best expressed as an energy per unit area (ergs/cm^2). Read and Shockley have used elasticity theory to derive a theoretical expression for the grain-boundary energy of a low-angle boundary as a function of the angle of mismatch between adjacent crystals. The fact that their equation is in excellent agreement with experimental data is shown in Fig. 5.26. The solid line in the figure represents the Shockley-Read equation, while the data points are experimental values.

The equation derived on assumptions valid only for mismatch angles less than several degrees is in agreement with experimental data for angles as high as 20°. The fact that theoretical and experimental data agree so well over this large range may be fortuitous, but it should not overshadow the fact that the dislocation model for small-angle boundaries has been successfully used to predict the energy dependence of the boundary on the angle of mismatch.

5.16 Surface Tension of the Grain Boundary. The surface energy of a grain boundary has the units of ergs/cm^2. That is,

$$\gamma_G = \frac{\text{ergs}}{\text{cm}^2}$$

which is equivalent to

$$\gamma_G = \frac{\text{dynes} \cdot \text{cm}}{\text{cm}^2} = \frac{\text{dynes}}{\text{cm}}$$

but the units dynes/cm are those of a surface tension. It is often reasonable to assume that solid grain boundaries possess a surface tension equivalent to that of liquid surfaces. The surface tension of crystal boundaries is an important metallurgical phenomenon. The experimental data of Fig. 5.26 show that the grain-boundary surface tension is an increasing function of the angle of mismatch between grains to an angle of approximately 20°, and then it is essentially constant for all larger angles.

Absolute or numerical values of the surface tension of metal surfaces are difficult to determine. However, measurements have been made on copper, silver, and gold which show that their surface tensions of free surfaces (external surfaces) are of the order of 1200 to 1800 dynes/cm. External metal surfaces, therefore, possess relatively large surface tensions; they are approximately 20 times larger than that of liquid water. It has been determined further that the surface tensions of large-angle grain boundaries equal approximately one-third of the free surface values, and are therefore of the order of 300 to 500 dynes/cm.

While absolute values of the surface tension of grain boundaries are difficult to measure, relative values of boundary-surface energy may be estimated with the aid of a simple relationship. Let the three lines of Fig. 5.27 represent grain boundaries that lie perpendicular to the plane of the paper and meet in a line the projection of which is o. The three vectors γ_a, γ_b, and γ_c originating at point o represent, by their directions and magnitudes, the surface tensions of the three boundaries. If it is assumed that these three force vectors are in static equilibrium, then the following relationship must be true:

$$\frac{\gamma_a}{\sin a} = \frac{\gamma_b}{\sin b} = \frac{\gamma_c}{\sin c}$$

where a, b, and c are the dihedral angles between boundaries.

Since crystal boundaries are regions of misfit or disorder between crystals, it is to be expected that atom movements across and along boundaries should occur quite easily. The boundary is caused to move by the simple process whereby atoms leave one crystal and join another crystal on the other side of a boundary. Crystal boundaries in solid metals can move and should not be thought of as fixed in space. The speed with which crystal boundaries move depends on a number of factors. The first of these is the temperature, as the energy that makes it possible for an atom to move from one equilibrium grain-boundary position to another comes from thermal vibrations. We have, therefore, an analogous situation to that of atom movements inside crystals

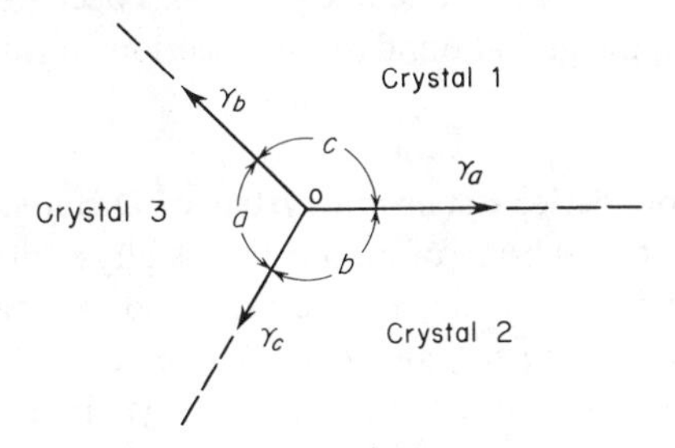

Fig. 5.27 The grain-boundary surface tensions at a junction of three crystals.

where the atoms diffuse into lattice vacancies as a result of thermally activated jumps. The rate of movement must increase rapidly with rising temperature. However, other things being equal, it is to be expected that the same number of atoms would cross a boundary in one direction as in the other, and that the boundary would be relatively fixed in space. Actually, grain-boundary movements occur only if the energy of the metal as a whole is lowered by a greater movement of atoms in one direction than in the other. One way that the energy of a specimen can be lowered by the motion of a grain boundary occurs when it moves into a deformed crystal, leaving behind a strain-free crystal. Another driving force for movement lies in the energy of the boundaries themselves. A metal can approach a more stable state by reducing its grain-boundary area. There are two ways in which this may be achieved. First, boundaries may move so as to straighten out badly curved regions; and, second, they may move in such a way that some crystals are caused to disappear, while others grow in size. The latter phenomenon, which results in a decrease in the total number of grains, is called *grain growth*.

One of the consequences of grain-boundary movement is that if a metal is heated at a sufficiently high temperature for a long enough time, the equilibrium relationship between the surface tensions and the dihedral angles can actually be observed. (See Equation, p. 220 .) In pure metals with randomly oriented crystals, low-angle boundaries occur infrequently, and it may be assumed that the grain-boundary energy is constant for all boundaries. (See Fig. 5.26.) If the surface energies (surface tension) are equal in each of three boundaries that meet at a common line, and if the boundaries have attained an equilirbium configuration, the three dihedral angles must be equal. An examination of the junctions where three boundaries meet in the photograph, Fig. 5.21 shows a large number of 120° dihedral angles. This fact is surprising when it is considered that many of the grain boundaries are not normal to the plane of the photograph. In fact, if it were possible to cut the surface so that all the boundaries were perpendicular to the surface, a good agreement between the experimental and predicted angles would be observed. It may be concluded that in a well-annealed pure metal, that is, one which has been heated for a long time at a high temperature, the grain-boundary intersections form angles very close to 120°.

5.17 Boundaries Between Crystals of Different Phases. A *phase* is defined as a homogeneous body of matter that is physically distinct. The three states of matter (liquid, solid, and gas) all correspond to separate phases. Thus, a pure metal, for example, copper, can exist in either the solid, liquid, or gaseous phase, each stable in a different temperature range. In this respect, it would appear that there is no difference between the concepts of phase and state. However, a number of metals are allotropic (polymorphic); that is, they are able to exist in different crystal structures, each stable in a different temperature range. Each crystal structure in these metals corresponds to a separate phase, and allotropic metals, therefore, possess more than three possible phases. When pure metals are combined to form alloys, additional crystal structures can result in certain composition and temperature ranges. Each of these crystal structures in itself constitutes a separate phase. Finally, it should be pointed out that a solid solution (a crystal containing two or more types of atoms in the same lattice) also satisfies the definition of a phase. A good example of a solid solution occurs when copper and nickel are melted together and then frozen slowly into the solid state. The crystals that form contain both nickel and copper atoms in the same ratio as the original liquid mixture; both types of atoms occupy the lattice sites of the crystal indiscriminately.

For the present, the concept of phases has been introduced in order to explain additional surface-tension phenomena. The study of the phases of alloy systems is a topic that will be considered in a later chapter.

The grain boundaries of alloys containing crystals of only a single phase should behave in a manner analogous to those of a pure metal. The crystals of a well-annealed solid-solution alloy, like the copper-nickel alloy mentioned above, appear under the microscope like those of a pure metal. At all points where three grains meet there is a mean dihedral angle of 120°.

In alloys of two phases, two types of boundaries are possible: boundaries separating crystals of the same phase, and boundaries separating crystals of the

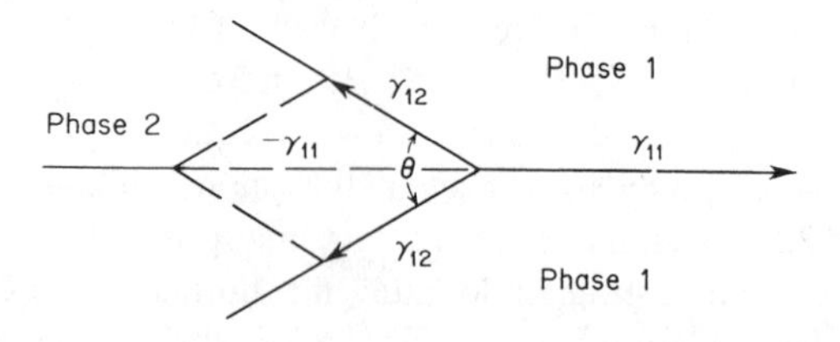

Fig. 5.28 The grain-boundary surface tensions at a junction between two crystals of the same phase with a crystal of a different phase.

two phases. Three grains of a single phase can still intersect on a line, but there is the additional possibility of two grains of one phase meeting a grain of the other phase at a common intersection. A junction of this type is shown in Fig. 5.28. If the surface tensions in the boundaries are in static equilibrium, then

$$\gamma_{11} = 2\gamma_{12} \cos \frac{\theta}{2}$$

where γ_{11} is the surface tension in the single-phase boundary, γ_{12} the surface tension in the two-phase boundary, and θ the dihedral angle between the two boundaries that separate phase 2 from phase 1.

Let us solve the above expression for the ratio of the surface tension of the two-phase boundary to that of the single-phase boundary. This yields

$$\frac{\gamma_{12}}{\gamma_{11}} = \frac{1}{2 \cos \frac{\theta}{2}}$$

The ratio γ_{12}/γ_{11} is plotted as a function of the dihedral angle θ in Fig. 5.29. Notice that as the surface tension of the boundary between two phases approaches the value 0.5, the dihedral angle falls rapidly to zero. The significance of small dihedral angles is readily apparent in Fig. 5.30, where the shape of the intersection is shown for angles of 10° and 1°. It is apparent that as the angle approaches zero, the second phase moves to form a thin film between the crystals of the first phase. Further, if the surface tension of the two-phase boundary (γ_{12}) falls to a value less than ½ of the single-phase surface tension γ_{11}, static equilibrium of the three forces is not possible and point o moves to the left, which is equivalent to saying that when the dihedral angle becomes zero, the second phase penetrates the single-phase boundaries and isolates the crystals of the first phase. This effect may occur even though the second phase is present in almost negligible quantities. A good example occurs in the case of bismuth in copper. The surface tension of a bismuth-copper interface is so low that the dihedral angle is zero and a minute quantity of bismuth is capable of

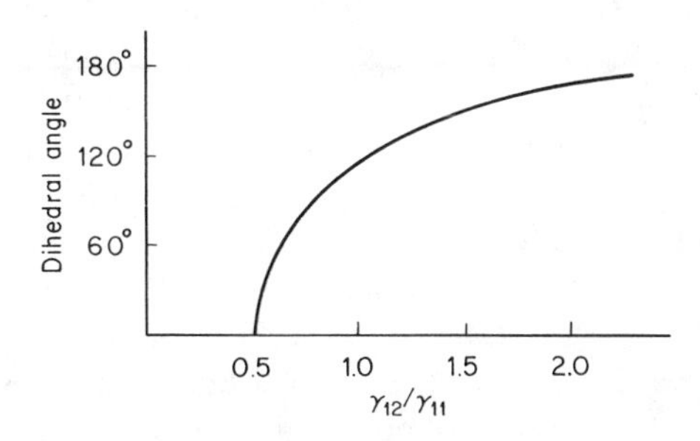

Fig. 5.29 The dependence of the two-phase dihedral angle on the ratio of the two-phase surface tension to the single phase surface tension.

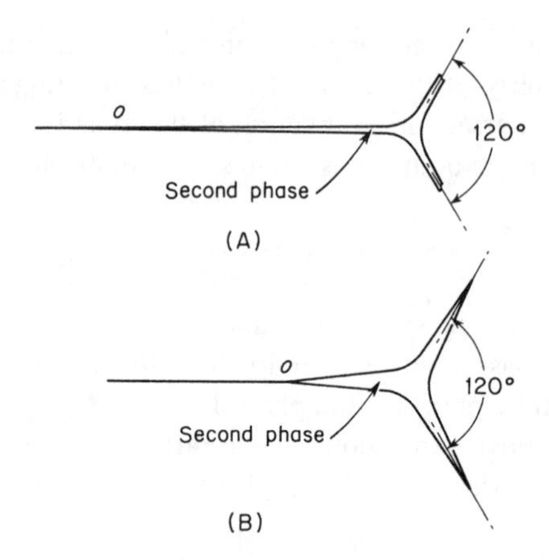

Fig. 5.30 When the dihedral angle is small, the second phase (even if present in small quantities) tends to separate crystals of the first phase. (A) dihedral angle 1°, (B) dihedral angle 10°.

forming a thin film between the copper crystals. Whereas copper is ordinarily a metal of high ductility and capable of extensive plastic deformation, bismuth is not. In fact, bismuth is a very brittle metal, and when it forms a continuous film around copper crystals the copper loses its ductility, even though the total amount of the bismuth impurity is less than 0.05 per cent. This loss in ductility is observed at all temperatures at which copper is worked.

In a number of important metallurgical cases, second-phase impurities remain in the liquid state until a temperature is reached well below the freezing point of the major phase. The amount of harm that these impurities (in small percentages) can do to the plastic properties of metals is a function of the surface tension between liquid and solid. If this interfacial energy is high, the liquid tends to form discrete globules with little effect on the hot working properties of metals. On the other hand, low interfacial energies lead to liquid grain-boundary films. These, of course, are very harmful to the plastic properties of metals.

Let us take the case of iron containing small quantities of sulphur as an impurity. Iron sulfide is liquid at temperatures well below the freezing point of iron. This range includes the temperature range normally used for hot rolling iron and steel products. Unfortunately, the surface energy of the iron sulfide to iron boundary is very close to one-half that of the boundary between iron crystals, and the liquid sulfide forms a grain-boundary film that almost completely separates the iron crystals. Since a liquid possesses no real strength, iron or steel in this condition is brittle and cannot be hot-worked without disintegrating. In such a case as this, when a metal becomes brittle at high temperatures, the metal is said to be "hot short," that is, its properties are deficient at hot-working temperatures.

It has already been explained that a second phase in small quantities forms thin intergranular films when the dihedral angle is zero. For dihedral angles greater than zero, but less than 60°, the equilibrium configuration is one where small quantities of the second phase run as a continuous network along the grain edges of the first phase; that is, along the lines where three grains of the first phase would meet in the absence of the second phase. According to Fig. 5.29, a dihedral angle of 60° corresponds to a ratio of interphase to single-phase surface tensions of 0.582. Many liquid-solid interfaces possess surface tensions less than 0.600 of the solid-boundary surface tension. Therefore, when liquids and solids coexist, there is a strong possibility that the liquid will form a continuous network throughout the metal. Since diffusion in liquids is more rapid than in solids, this network forms a convenient path for rapid diffusion of elements (both good and bad) into the center of the metal.

When the dihedral angle between the interphase boundaries is greater than 60°, the second phase no longer forms a continuous network unless it is present as the major phase. The second phase, if present in small quantities, now forms discrete globular particles, usually along the boundaries of the first-phase crystals. See Fig. 5.31.

Below approximately 1000° C, iron sulfide is no longer a liquid, but a solid. Solid iron sulfide forms an equilibrium angle with solid iron greater than 60° and, therefore, tends to crystallize as discrete particles of FeS instead of a continuous grain-boundary film. Small discrete particles of a second phase, even though they may be brittle in themselves, are far less damaging to the properties of a metal than are continuous brittle grain-boundary films. Iron containing sufficient FeS to embrittle it at hot-working temperatures will retain much of its ductility at room temperature.

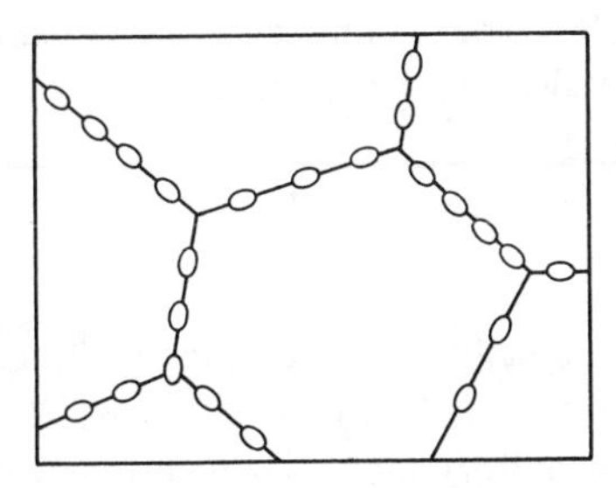

Fig. 5.31 When the dihedral angle is large (above 60°), the second phase (when present in small quantities) tends to form small discrete particles, usually in the boundaries of the first phase.

5.18 The Grain Size. The size of the average crystal is a very important structural parameter in a polycrystalline aggregate of a pure metal or single-phase metal. Unfortunately, this is a difficult variable to define precisely. In many cases, the three-dimensional shape of a grain is very complex, and even in those cases where the grains appear to be of nearly equal size in a two-dimensional micrograph, the grains can vary through a very wide range of sizes. Hull[9] has actually demonstrated this by allowing mercury to penetrate along the grain boundaries of a high zinc brass (beta brass) which embrittled the boundaries and allowed the individual grains to be separated from each other. Fig. 5.32 shows a set of his photographs corresponding to three of some 15 size classifications into which he grouped the grains. The upper picture shows the largest grains and the lowest the smallest. The middle photograph represents an average size. Most metallographic structures are observed on planar sections, and linear measurements made on such surfaces are not normally capable of being accurately related to the grain diameter, which is a property of a complex, three-dimensional quantity.

In spite of the problems indicated above, some sort of measurement to define the size of the structural unit in a polycrystalline material is badly needed, and the concept of an average grain size is widely used in the literature.

Perhaps the most useful parameter for indicating the relative size of the grains in a microstructure is $\bar{l}$ determined by the linear intercept method. This quantity is called the mean grain intercept and is the average distance between grain boundaries along a line laid down on a photomicrograph. In making this measurement, one can lay a straight edge of perhaps 10 cm in length down on the photograph and then count the number of boundaries that the edge of the instrument crosses. This measurement should then be repeated several times, placing the straight edge down on the photograph in a random fashion. The total number of intersections, when divided by the total line length over which the linear measurement was made, and multiplied by the magnification in the photograph, yields the quantity $\bar{N}_l$, the average number of grain boundaries intercepted per centimeter. The reciprocal of this quantity is $\bar{l}$, which is often used as a parameter for indicating the approximate grain size. Thus we have

$$\bar{l} = \frac{1}{\bar{N}_l}$$

Even though there is actually no geometrical relationship between $\bar{l}$ and the actual average grain diameter d, this quantity is widely used to represent the grain diameter. In this regard, it might be mentioned that such a relationship would exist if all grains had the same shape and size. On the other hand,

9. Hull, F., Westinghouse Electric Corporation, Research and Development Center, Pittsburgh, Pa. These experimental results were demonstrated at the Quantitative Microscopy Symposium, Gainesville, Fla., Feb., 1961.

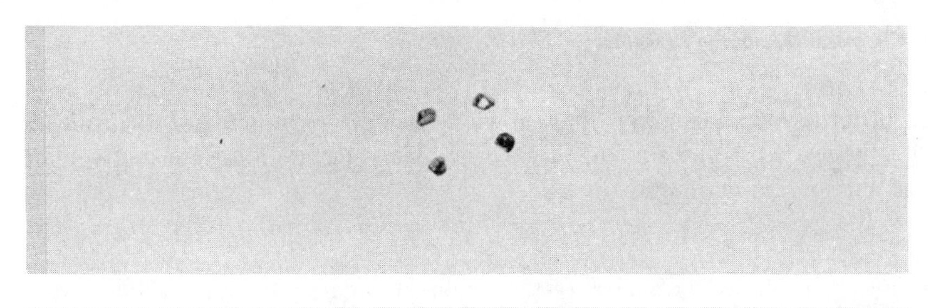

Fig. 5.32 Grains removed from a beta brass specimen that appeared to be of a nearly uniform grain size on a metallographic section. Three of some 15 size classifications are shown in these photographs. They represent the smallest, the largest, and a size from the middle of the range. (The original photographs are from F. Hull, Westinghouse Electric Company Research Laboratories, Pittsburgh. Copies were furnished courtesy of K. R. Craig.)

quantitative metallography[10] has shown[11] that the reciprocal of $\bar{l}$, that is $\bar{N}_l$, is directly related to the amount of grain boundary surface area in a unit volume The relationship in question is

$$S_v = 2\bar{N}_l$$

where S_v is the surface area of the grain boundaries per unit volume. It is therefore well to recognize that when one determines $\bar{l}$ what is actually determined is the reciprocal of the grain boundary surface area in a unit volume.

5.19 The Effect of Grain Boundaries on Mechanical Properties. Except possibly at very elevated temperatures, polycrystalline metals almost always show a strong effect of grain size on hardness and strength. The smaller the grain size the greater the hardness or flow-stress, where the flow-stress is the stress in a tension test corresponding to some fixed value of strain. Figs. 5.33 and 5.34 are offered in support of this statement. In the first of these two

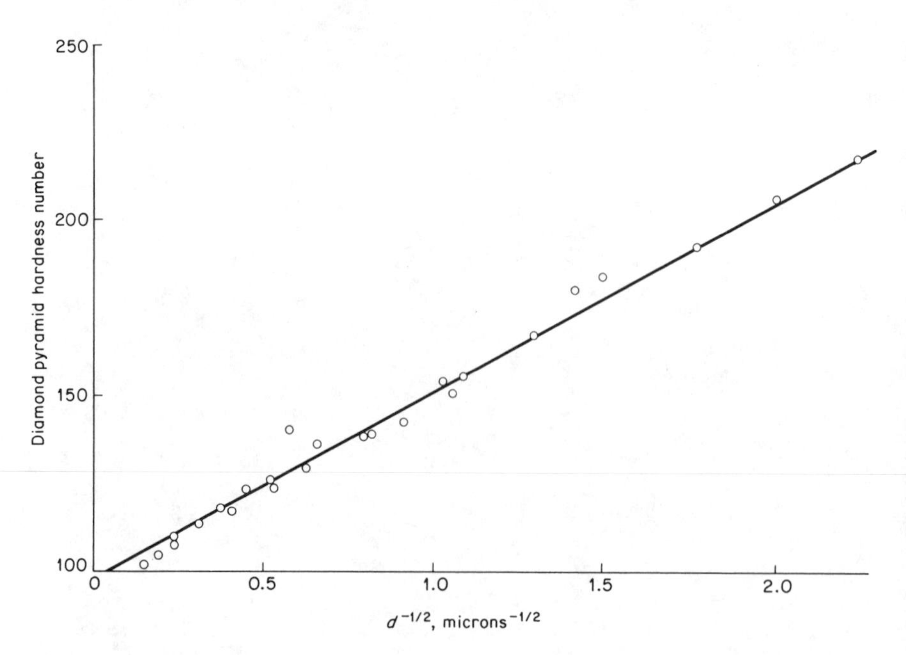

Fig. 5.33 The hardness of titanium as a function of the reciprocal of the square root of the grain size. (From the data of H. Hu and R. S. Cline, *TMS-AIME*, **242** 1013 [1968]. This data has been previously presented in this form by R. W. Armstrong and P. C. Jindal, *TMS-AIME*, **242** 2513 [1968].)

10. DeHoff, R. T., and Rhines, F. N., *Quantitative Microscopy*, McGraw-Hill Book Company, New York, 1968.
11. Fullman, R. L., *Trans. AIME*, 197 447–53 (1953).

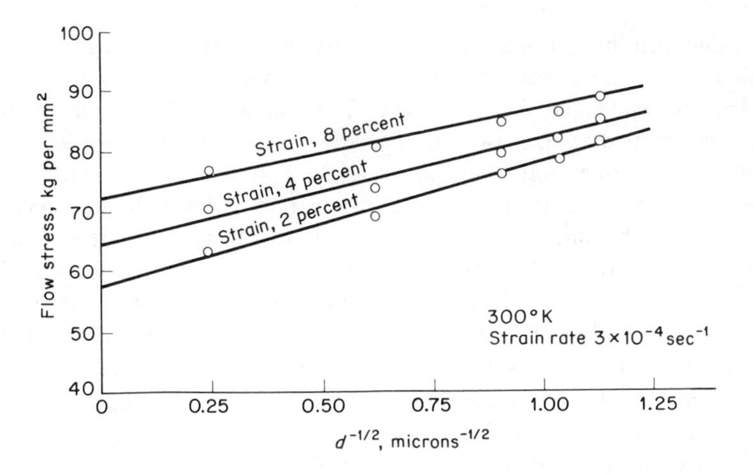

Fig. 5.34 The flow-stress of titanium as a function of the reciprocal of the square root of the grain size. (After Jones, R. L. and Conrad, H., *TMS-AIME*, **245** 779 [1969].)

illustrations, the hardness of titanium is plotted as a function of the reciprocal of the square root of the grain size. The hardness measurements corresponding to this data were obtained using the 138° diamond-pyramid indentor.[12] According to the illustration, it is possible to write an empirical relationship of the form

$$H = H_o + k_H d^{-1/2}$$

where H is the hardness, d is the average grain diameter, k_H is the slope of the straight line drawn through the data, and H_o is the intercept of the line with the ordinate axis and corresponds to the hardness expected at a hypothetical infinite grain size. It is not necessarily the hardness of a single crystal.

In the other diagram, Fig. 5.34, tensile test data is plotted for titanium specimens tested at room temperature. Here the flow-stress corresponding to three different strains (2 percent, 4 percent, and 8 percent) are plotted against $d^{-1/2}$, and straight lines are drawn through the data corresponding to linear relationships of the form

$$\sigma = \sigma_o + kd^{-1/2}$$

where σ is the flow-stress and σ_o and k are constants equivalent to H_o and k_H in the previous equation.

A linear relationship between the flow-stress and square root of the dislocation density was originally proposed by both Hall[13] and Petch[14] and, as a consequence, it is now generally known as the Hall-Petch equation. Such a

12. The diamond pyramid hardness test is discussed further in Sec. 18.10.
13. Hall, E. O., *Pro. Phys. Soc. London*, **B64** 747 (1951).
14. Petch, N. J., *J. Iron and Steel Inst.*, **174** 25 (1953).

relationship can be rationalized by dislocation theory assuming that grain boundaries act as obstacles to slip dislocations, causing dislocations to pile up on their slip planes behind the boundaries. The number of dislocations in these pile-ups is assumed to increase with increasing grain size and magnitude of the applied stress. Furthermore, these pile-ups should produce a stress concentration in the grain next to that containing a pile-up that varies with the number of dislocations in the pile-up and the magnitude of the applied stress. Thus, in coarse-grained materials, the stress multiplication in the next grain should be much greater than that in fine-grained materials. This means that in the fine-grained materials a much larger applied stress is needed to cause slip to pass through the boundary than is the case with coarse-grained materials.

Although the Hall-Petch relation has been widely accepted, it has by no means been completely verified. While in many cases grain size data can be plotted to yield an apparent linear relationship between σ and $d^{-1/2}$, as in Fig. 5.34. Baldwin[15] has shown that the general scatter normally observed with this type of data can, in many cases, give equally good linear plots when σ is plotted against d^{-1} or $d^{-1/3}$.

Overriding the question of whether a variation of σ or H with $d^{-1/2}$ has true physical significance is the fact that plots such as those in Figs. 5.33 and 5.34 clearly demonstrate the dependence of the hardness and flow-stress on the grain size. Note, for example, that the tensile flow-stress of titanium at 2 percent strain increases from about 65,000 psi at a grain size of 17 microns to about 82,000 at a grain size of 0.8 microns (10^{-4} cm).

Problems

1. Consider the $(\bar{1}\bar{1}1)$ plane of a face-centered cubic crystal. What are the Burgers vectors of the primary (total) slip dislocations that can move on this plane? Indicate the other possible octahedral planes on which the respective dislocations can move.
2. Cross-slip involving the basal plane and the $\{10\bar{1}0\}$ or prism planes in magnesium has been discussed in the text. In this metal, cross-slip between (0001) and $\{10\bar{1}1\}$ has also been observed.
   (*a*) Show, with a sketch of the unit cell, why this is possible.
   (*b*) Is cross-slip involving $(10\bar{1}1)$ and $(10\bar{1}0)$ geometrically feasible? Explain.
3. It has been pointed out in the text that {110}, {112}, and {123} have all been identified as slip planes in body-centered cubic crystals. On how many possible slip planes does this imply that a single ½[111] dislocation might be able to move?
4. Prove that the dislocation configuration in Fig. 5.5 can act as a Frank-Read source in its original slip plane.

15. Baldwin, W. M., Jr., *Acta Met.*, **6** 141 (1958).

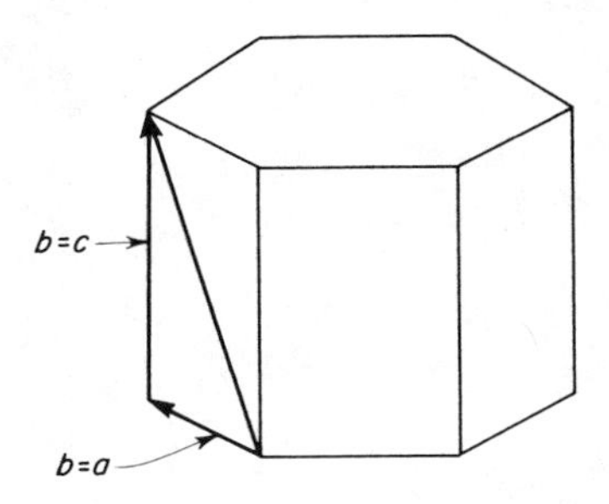

5. The line $(a-b)$ of Problem 11, Chapter 1 is shown again in this sketch. The length of this line and its direction define a possible slip Burgers vector in zinc and cadmium. It can be formed by adding a basal slip dislocation a to a c dislocation [0001]. Deduce the Burgers vector of this dislocation by making the proper vector addition.
6. Between what planes can the Burgers vector of Problem 4 cross-slip?

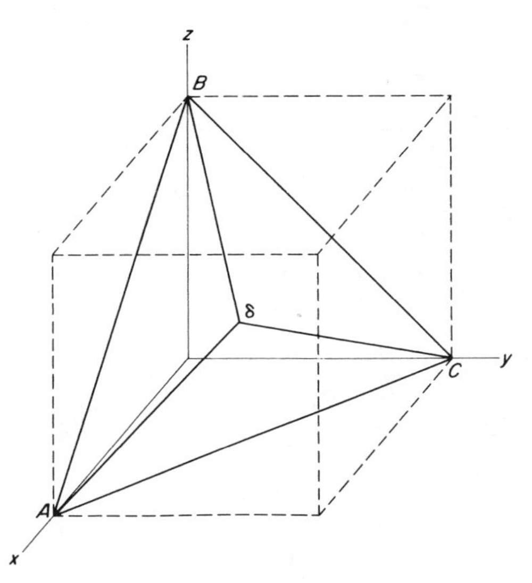

7. Assume that the triangle in the drawing lies on the (111) plane of a face-centered cubic crystal, and that its edges are equal in magnitude to the Burgers vectors of the three total dislocations that can glide in this plane. Then, if δ lies at the centroid of this triangle, lines $A\delta$, $C\delta$, and $B\delta$ accordingly correspond to the three possible partial dislocations of this plane.
(*a*) Identify each line (AB, $A\delta$, etc.) with its proper Burgers vector expressed in the vector notation.
(*b*) Demonstrate by vector addition that

$$A\delta + \delta B = AB$$

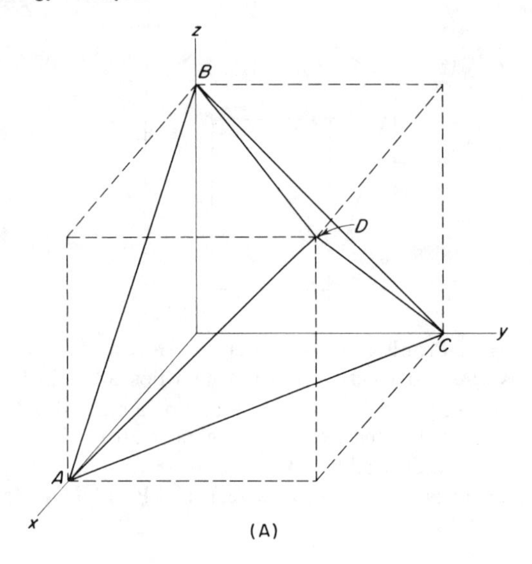

(A)

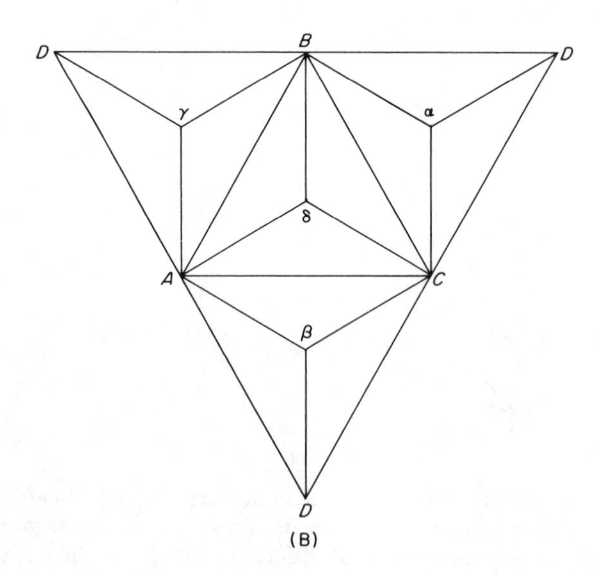

(B)

8. Figure A represents the Thompson tetrahedron as seen in three dimensions. In Fig. B., the sides of the tetrahedron have been hypothetically folded out so that all four surfaces can be easily viewed. The symbols α, β, γ, and δ represent the centroid of each face respectively. As in Problem 6, lines such as BD represent total dislocations, that is, $\frac{1}{2}[110]$, and lines such as $B\gamma$ represent partial dislocations, that is, $\frac{1}{6}[21\bar{1}]$.
   (*a*) Write the Miller indices symbols for CD and DC.
   (*b*) Do the same for BD and DB.
   (*c*) Show that $BD + DC = BC$.

9. Show graphically what the vector $BD + CD$ represents. Give its Burgers vector in the Miller indices form.
10. Frank's rule considers that the energy per unit length of a dislocation is proportional to its Burgers vector squared. This signifies that if

$$b = a[hkl]$$

then

$$\text{Energy/cm} \approx a^2(h^2 + k^2 + l^2)$$

where a is a simple numerical factor.
(*a*) Compare the energy per unit length of a dislocation with $b = BD + CD$ with that of a dislocation with $b = BD + DC$. Which reaction is the more probable?
11. Show that the dissociation of an f.c.c. total dislocation, such as AC, into its two partial dislocations, $A\delta$ plus δC, is energetically feasible.
12. Assume that the dislocation configuration shown in this figure corresponds to dislocations lying on the basal plane of a h.c.p. metal and that, for the dislocation marked A, b is ⅓ $[2\bar{1}\bar{1}0]$, and for the dislocation marked B, b is ⅓ $[\bar{1}2\bar{1}0]$. In this case, the dislocation C can be assumed to be the sum of A plus B. Add the Burgers vectors of A and B to obtain that of C. *Note: C* is assumed to be a unit or total basal slip dislocation.

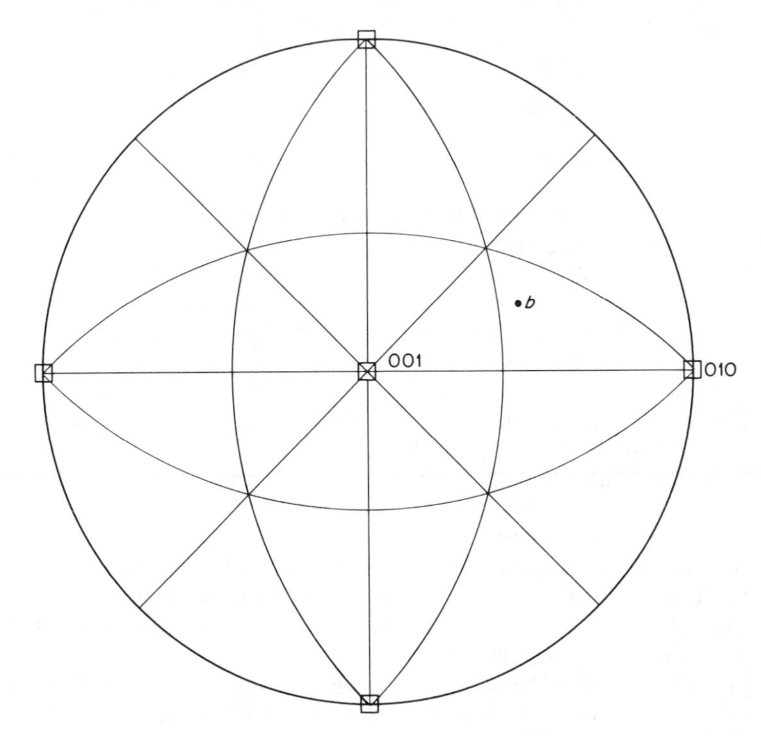

13. Assume that the diagram corresponds to a 001 standard projection of a face-centered cubic crystal (see Fig. 1.31), and that point b represents the

original axis of a single crystal. Indicate on a copy of this diagram the path that the crystal axis will follow in tension. Give the Miller indices of the final (average) end orientation of the crystal.

14. Alternatively, consider that the crystal of Problem 13 is stressed in compression. Diagram the path that the axis will follow and identify the indices of the final end orientation. Note that in this case, as in the tensile case, the axis alternates its direction of rotation as it passes out of its original stereographic triangle into the next one and then returns to the first triangle, and so forth. Also the end orientation is symmetrically oriented along a great circle midway between the two active slip plane poles.
15. For the stress axis orientation of Problem 13 identify the indices of the slip plane and slip direction for (*a*) the primary slip system, (*b*) the cross-slip system, (*c*) the conjugate system, and (*d*) the critical slip plane.

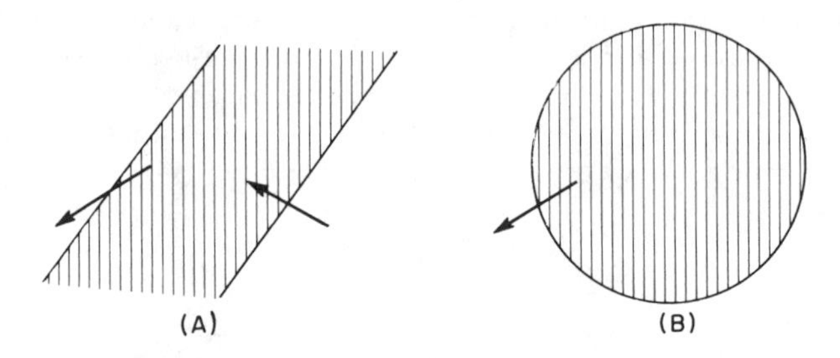

16. Figure A represents an extended dislocation in a f.c.c. metal, while Fig. B corresponds to a closed dislocation loop whose Shockley partial dislocation is the same as that of the left-hand dislocation in Fig. A.
    (*a*) Are the stacking faults the same in both cases? Explain.
    (*b*) If both dislocation arrangements move to the left without changing their basic size or shape, what will be the effect of the movement of each arrangement with regard to the shearing of the crystal across the slip plane?
17. What stacking fault will result in a hexagonal close-packed crystal if a portion of an extra basal plane is inserted into the structure? Give your answer in terms of the resulting stacking sequence of the close-packed plane. (*Hint*: see Fig. 5.17 for the f.c.c. case.)
18. In analogy with Fig. 5.16 for the f.c.c. case, consider that:
    (*a*) A vacancy disc forms on a basal plane of a h.c.p. metal. What stacking sequence would result?
    (*b*) Why would the resulting stacking fault be of high energy?
    (*c*) Explain how a simple shear (equal to that of a Shockley partial) along the basal plane could eliminate this high energy stacking fault and replace it with one of lower energy.
    (*d*) What would be the resulting stacking arrangement of the basal planes?
    (*e*) Is the result in (*d*) unique, or are there two basic possibilities? Explain.

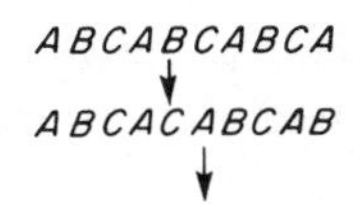

19. As demonstrated in the text, a Shockley slip partial dislocation in a f.c.c. metal changes the stacking sequence in the manner shown in the figure. Note that the shear of the atoms in the *B* layer, indicated by the upper small vertical arrow, also results in the displacement of all the planes to the right of this plane. Now consider that plane *A* undergoes a similar displacement so that it becomes a *B* plane. Demonstrate that this will result in a stacking sequence equivalent to an extrinsic stacking fault (see Fig. 5.16B).

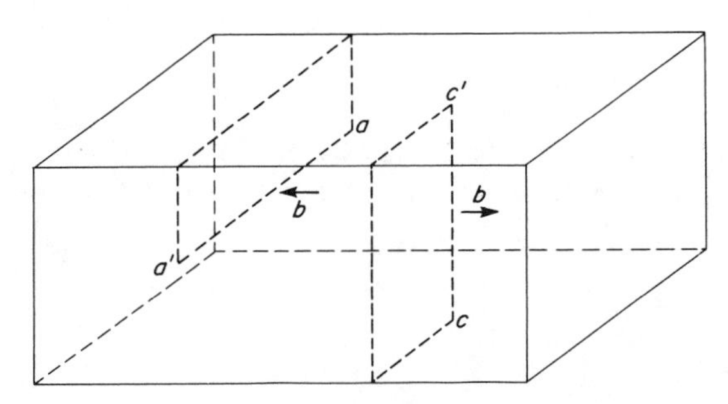

20. $a - a'$ and $c - c'$ represent two edge dislocations whose extra half-planes are outlined in the diagram. Assume that dislocation $a - a'$ moves to the right in its slip plane and cuts through dislocation $c - c'$. With the aid of a sketch, show what would happen to each dislocation and its associated half-plane. Will the intersection result in the formation of kinks, or jogs, or both?

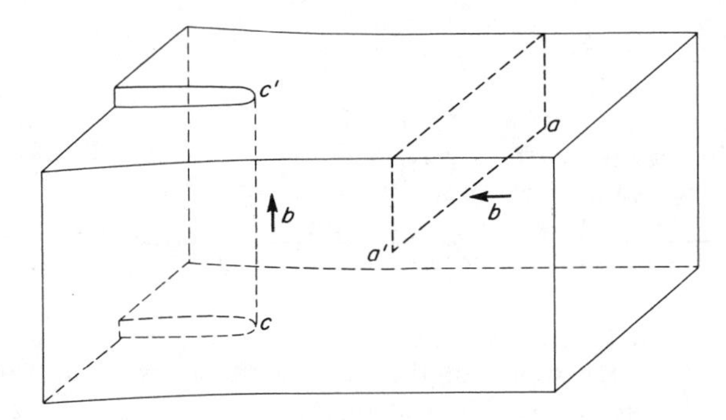

21. In this drawing, $a - a'$ is an edge dislocation, as in Problem 20, but $c - c'$ is now a screw dislocation. Assume that $c - c'$ moves on its slip plane (plane parallel to the plane of the drawing) and cuts through $a - a'$. Sketch the effect of the intersection on each dislocation and indicate whether a jog or a kink results in each dislocation.

22. Now assume that, in Problem 21, $a - a'$ moves on its slip plane and cuts through $c - c'$. Describe the results of this intersection.

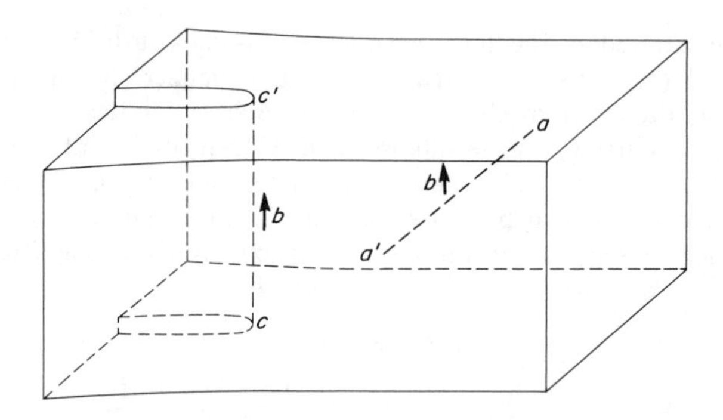

23. In this case, assume that the Burgers vector of the edge dislocation $a - a'$ lies parallel to the vertical direction. Discuss the aspects of this intersection.

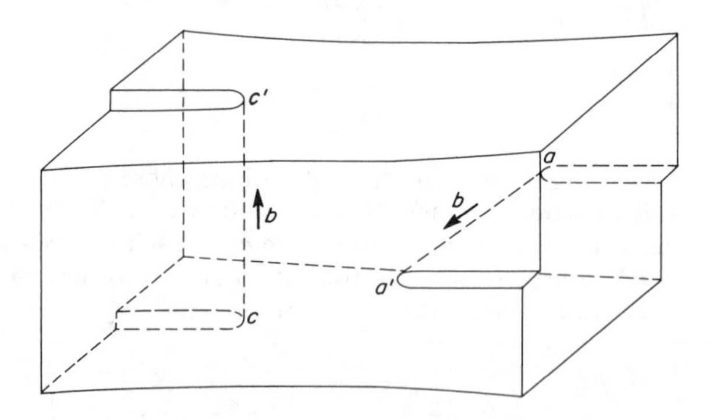

24. Analyze the geometrical aspects of the intersection of the two screw dislocations shown in this figure. After the intersection, would the steps on the two dislocations be connected by a row of interstitials or a row of vacancies? *Note:* the Burgers vectors drawn on the figure correspond to the right-hand, start to finish criterion for defining the Burgers vector given on p. 156.
25. In Fig. 5.23 let it be assumed that the nearly horizontal small angle boundary near the center of the figure corresponds to an array of edge $\langle 11\bar{2}0 \rangle$ basal slip dislocations, all of the same sign. If $b = 3.2$ Å and the magnification of the photograph is 200 times, what is the angle of tilt across the boundary?
26. According to quantitative metallography, $\overline{N_l}$, the average number of grain boundary intercepts per unit length along a line laid over a microstructure is

related to S_v, the surface area per unit volume, by the relation

$$S_v = 2\bar{N}_l$$

Estimate $\bar{N}_l$ for Fig. 5.21. Note that the magnification is 350x. Next, compute S_v. Finally, assuming that the grain boundary surface energy for zirconium is about 1000 ergs/cm^2, determine the energy per unit volume associated with the grain boundaries in this specimen. Express the answer in calories per cm^3.

27. The average grain diameter, as measured using the mean grain intercept $\bar{l}$, is about 1 micron in a very fine-grained material. What would be the approximate surface energy per unit volume for the zirconium of Problem 26 if its grain size were reduced to 1 micron (10^{-4} cm)?

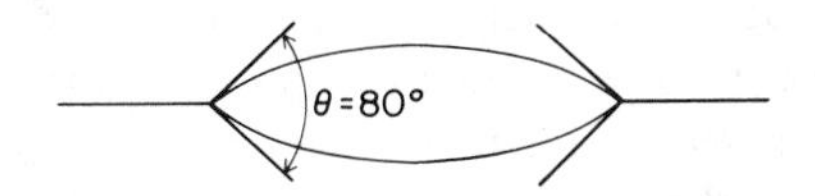

28. The drawing represents a second-phase particle (nonmetallic inclusion) lying along the grain boundary between a pair of iron crystals. The grain boundary surface energy between iron crystals is about 780 ergs/cm^2. Compute the surface energy between the iron and the inclusion.

29. (*a*) If, in the preceding problem, θ had been 1°, what would be the surface energy of the interface between the iron and the nonmetallic inclusion?
(*b*) What value of surface energy between the iron and the nonmetallic inclusion would be required to prevent the formation of a continuous network of the inclusion that would run along the edges where three grains meet?

30. The Table accompanying this problem gives the published data[16] corresponding to the effect of grain size on the flow-stress measured at 4 percent strain in very high purity titanium. Make a plot of σ against $d^{-1/2}$ and determine the Hall-Petch parameters k and σ_o. Express k in kg/mm$^{3/2}$.

Grain Size in Microns, μ	*Stress, σ in kg/mm²*
1.1	32.7
2.0	28.4
3.3	26.0
28.0	19.7

31. Plot the data in Problem 30 showing σ as a function of d^{-1}, as well as of $d^{-1/3}$. Is there some justification for Baldwin's comments on the Hall-Petch relationship (see p. 230)?

16. Jones, R. L., and Conrad, H., *TMS-AIME*, **245** 779 (1969).

6 Vacancies

6.1. Thermal Behavior of Metals. Many important metallurgical phenomena depend strongly on the temperature at which they occur. An important example of practical importance occurs in the softening of hardened metals by heating. A brass specimen, hardened by hammering, can be softened to its original hardness in a few minutes if exposed to a temperature of 1000° F. At 600°F it may take several hours to achieve the same loss of hardness, while at room temperature it might easily take several thousand years. For all practical purposes it can therefore be said that brass will not soften, or anneal, at room temperature.

In recent years, large advances have been made toward theoretical explanations of such highly temperature-dependent phenomena. Each of the three sciences of heat – *thermodynamics, statistical mechanics*, and *kinetic theory* – has contributed to this knowledge and each approaches the subject of heat differently.

Thermodynamics is based upon laws that are postulated from experimental evidence. Since the experiments that lead to the laws of thermodynamics were performed on bodies of matter containing very large numbers of atoms, thermodynamics is not directly concerned with what happens on an atomic scale; it is more concerned with the average properties of large numbers of atoms, and mathematical relationships are developed between such thermodynamical functions as temperature, pressure, volume, entropy, internal energy, and enthalpy, without consideration of atomic mechanisms. This neglect of atomic mechanisms in thermodynamics has both advantages and disadvantages. It makes computations easier and more accurate, but, unfortunately, it tells us nothing about what makes things happen as they do. As a simple example, consider the *equation of state* for an ideal gas,

$$PV = nRT$$

where P is the pressure in psi, V is the volume in cubic inches, n is the number of moles of gas, R the universal gas constant, and T the absolute temperature. In pure thermodynamics, this equation is derived from experiments (Boyle's Law,

Law of Gay-Lussac, and Avogadro's Law). No explanation as to the reasons for the existence of this relationship is given.

Kinetic theory, in contrast to thermodynamics, attempts to derive relationships such as this equation of state, starting with atomic and molecular processes. In college physics textbooks classical derivations of the above equation which use a simple kinetic approach can be found. In these derivations, the gas atoms are assumed to behave as elastic spheres, to move at high speeds with random velocities, and to be separated by distances that are large compared with the size of the atoms. Using these assumptions, it can be shown that the pressure exerted by the gas equals the time rate of change of momentum due to the collision of gas atoms with the walls of the container. Thus, the pressure is merely the average force exerted on the walls by the collision of the gas atoms with the walls. It can also be shown that the average kinetic energy of the atoms is directly proportional to the absolute temperature. Kinetic theory thus gives us an insight into the significance of two important thermodynamical functions: temperature and pressure. Thermodynamics does not give us this same ability to understand gas phenomena.

The third science of heat, *statistical mechanics,* applies the realm of statistics to heat problems. Kinetic theory is concerned with the explanation of heat phenomena in terms of the mechanics of individual atoms. The mechanics of the disordered atomic motions that are ascribed to heat phenomena are attacked from probability considerations. This approach is feasible because most practical problems involve quantities of matter containing large numbers of atoms, or molecules. It is thus possible to think in terms of the behavior of the group as a whole, as does a life insurance actuarial who predicts the vital statistics of a large population.

With the use of statistical mechanics, it is possible to derive the first and second laws of thermodynamics. While these two laws are the foundation of thermodynamics, the latter science assumes that they are derived from experiments and gives no explanation as to their mechanical significance. However, statistical mechanics not only explains the basic laws of thermodynamics, but explains the entire field of thermodynamics. In one of the following sections, a physical interpretation will be given for the thermodynamical function called entropy. When approached from the point of view of statistical mechanics, this quantity is given a real significance that is not apparent in thermodynamics.

It should now be mentioned that thermodynamics and statistical mechanics are only applicable to problems involving equilibrium, and cannot predict the speed of a chemical or metallurgical reaction. This latter is the special province of the kinetic theory.

As a simple example of a system in equilibrium, consider a liquid metal and its equilibrium vapor in which the average number of metal atoms leaving the liquid to join the vapor equals the corresponding number traveling in the opposite direction. The concentration of atoms in the vapor, and therefore the vapor pressure, is a constant with respect to time. Under conditions such as these, thermodynamics and statistical mechanics are able to produce much useful information; for example, how equilibrium-vapor pressure changes with a change in temperature. Suppose, however, that liquid metal is placed inside the bell jar of a vacuum system so that the vapor is swept away as fast as it forms. In this case, there can be no equilibrium because atoms will leave the liquid at a much faster rate than they return to it. Because the liquid-vapor system is no longer in equilibrium, thermodynamics and statistical mechanics can no longer be used. Questions relating to how fast the metal atoms evaporate belong in the realm of the kinetic theory. Kinetic theory is thus most useful when the rates at which atomic changes take place are being studied.

6.2 Internal Energy. The preceding paragraphs have discussed the inter-relationships of the three branches of the science of heat. The principal objective of this discussion was to show that physical meanings can be given to thermodynamical functions. Let us now consider a solid crystalline material. An important thermodynamical function that will be needed in the following sections is *internal energy*, that shall be denoted by the symbol E. This quantity represents the total kinetic and potential energy of all the atoms in a material body, or system. In the case of crystals, a large part of this energy is associated with the vibration of the atoms in the lattice. Each atom can be assumed to vibrate about its rest position with three degrees of freedom (x,y, and z directions). According to the Debye theory, the atoms do not vibrate independent of each other, but rather as the result of random elastic waves traveling back and forth through the crystal, and, because they have three degrees of vibrational freedom, the lattice waves can be considered to be three independent sets of waves traveling along the x,y, and z axes respectively. When the temperature of the crystal is raised, the amplitudes of the elastic waves increase, with a corresponding increase in internal energy. The intensity of the lattice vibrations is therefore a function of temperature.

6.3 Entropy. In thermodynamics, the entropy S may be defined by the following equation:

$$\Delta S = S_B - S_A = \left[\int_A^B \frac{dQ}{T}\right]_{\text{rev}}$$

where S_A = entropy in state A

S_B = entropy in state B

T = absolute temperature

dQ = element of heat added to system

and where the integration is assumed to be taken over a reversible path between the two equilibrium states A and B.

The entropy is a function of state; that is, it depends only on the state of the system. This signifies that the entropy difference $(S_B - S_A)$ is independent of the way that the system is carried from state A to state B. If the system moves from A to B as a result of an irreversible reaction, the entropy change is still $(S_B - S_A)$. However, the entropy change equals the right-hand side of the above equation only if the path is reversible. Thus, to measure the entropy change of a system (in going between states A and B), one needs to integrate the quantity dQ/T over a reversible path. The integral of this same quantity over an irreversible path does not equal the entropy change. In fact, it is shown in all standard thermodynamical textbooks that

$$\Delta S = S_B - S_A > \int_A^B \frac{dQ}{T}\bigg]_{\text{irrev}}$$

for an irreversible path between A and B.

Differentiating the above two equations yields the following:

$$dS = \frac{dQ}{T} \quad \text{(reversible change)}$$

$$dS > \frac{dQ}{T} \quad \text{(irreversible change)}$$

for an infinitesimal change in state.

6.4 Spontaneous Reactions. Consider the transformation of water from liquid to solid. The equilibrium temperature for this reaction at atmospheric pressure is 0°C. At the equilibrium temperature, liquid water and ice can be maintained in the same isolated container for indefinite periods of time as long as heat is neither added nor taken away from the system. Ice and water under these conditions furnish a good example of a system at equilibrium. Now, if heat is slowly applied to the water, some of the ice will melt and become liquid; or if heat is abstracted from the container, more ice will form at the expense of the liquid. In either case, a reversible exchange of heat (heat of fusion) between the liquid-solid system and its surroundings is necessary in order to change the ratio of liquid to solid. This transformation of liquid to solid at 0°C is an example of an equilibrium reaction.

Consider now the case where liquid water is supercooled below the equilibrium freezing point (0°C). Even though ice may be thermally isolated from its surroundings, freezing can begin spontaneously without loss of heat to the surroundings. The heat of fusion, released by the portion of the water that freezes, will, in this case, raise the temperature of the system back toward the equilibrium freezing temperature. The difference between freezing at the equilibrium freezing point and at temperatures below it is an important one. In one case, freezing occurs spontaneously; in the other, it is spontaneous only if the heat of fusion is removed from the system. It should also be pointed out that the reverse transformation, in which ice melts at a temperature above the equilibrium freezing point, also occurs spontaneously. In this case, if the system is isolated (so that heat cannot be added to the system during the melting of the ice), the heat of fusion will cause the temperature to fall back toward the equilibrium temperature (0°C).

Reactions that occur spontaneously are always irreversible. Liquid water will transform to ice at −10°C, but the reverse reaction is, of course, impossible. Spontaneous reactions occur frequently in metallurgy, sometimes with drastic results, often with quite beneficial results. In either case, it is important to know the conditions that bring about spontaneous reactions, and to have a yardstick for measuring the driving force of this type of reaction. The yardstick that is most valuable for this purpose is called *Gibbs free energy*.

6.5 Gibbs Free Energy. Gibbs free energy is defined by the following equation:

$$F = E + PV - TS$$

where F = Gibbs free energy
P = pressure
V = volume
T = absolute temperature
S = entropy

Most metallurgical processes of interest (that is, freezing) take place at contant temperature and constant pressure (atmospheric). Furthermore, since we are primarily interested in solid or liquid metals, the volume change in metallurgical reactions is usually very small. For example, the quantity PV in the above equation can be assumed to be constant. Because changes of free energy are more important to us than the absolute value of the free energy, the above definition is frequently simplified to read

$$F = E - TS$$

In the following presentation, this simplified equation for free energy is used.

Let us consider again water in equilibrium with its solid form (ice). Let F_2 be the free energy of a mole of solid water (ice), and F_1 that of a mole of liquid water. When a mole of water changes to ice, the free-energy change is

$$\Delta F = F_2 - F_1 = (E_2 - TS_2) - (E_1 - TS_1)$$

where E_2 and E_1 are the internal energies of the solid and liquid water respectively, S_2 and S_1 the respective entropies, and T the temperature (that remains constant during the reaction). The above equation may also be written

$$\Delta F = \Delta E - T\Delta S$$

Water freezing to ice at the equilibrium freezing point is a reversible reaction, and under these conditions we have seen that the entropy change is given by

$$\Delta S = \int_A^B \frac{dQ}{T}$$

which, in this case, reduces to the following equation

$$\Delta S = \frac{\Delta Q}{T}$$

where ΔQ is the latent heat of freezing for water. Also, by the first law of thermodynamics, we have

$$\Delta E = \Delta W + \Delta Q$$

where ΔE = change in internal energy
ΔW = work done on system
ΔQ = heat added to system

In the freezing of water, however, the only external work done is that against the pressure of the atmosphere due to the expansion that occurs when water changes from liquid to solid. This pressure can be neglected in the present case because of its small size, and setting ΔW equal to zero, we have

$$\Delta E = \Delta Q$$

Substituting ΔS and ΔE in the free-energy equation gives us

$$\Delta F = \Delta Q - T\frac{\Delta Q}{T} = \Delta Q - \Delta Q = 0$$

The free energy in this reversible reaction (the freezing of water at 0° C) is zero. It may also be shown, with the aid of thermodynamics, that the Gibbs free-energy change is zero for any reversible reaction that takes place at constant temperature and pressure. If liquid water is now cooled to a temperature well below 0° C and allowed to freeze isothermally, the transformation will be made under irreversible conditions. But, in an irreversible reaction,

$$\Delta S > \frac{\Delta Q}{T}$$

or

$$T\Delta S > \Delta Q$$

The free-energy equation tells us that for this reaction

$$\Delta F = \Delta E - T\Delta S$$

where ΔE again equals ΔQ as in the previous example. Therefore,

$$\Delta F = \Delta Q - T\Delta S$$

If $T\Delta S$ is greater than ΔQ, however, ΔF must be negative. This fact is important. The free-energy change for this spontaneous reaction is negative, which means that the system reacts so as to lower its free energy. This result is true not only in the above simple system, but for all spontaneous reactions. A spontaneous reaction occurs when a system can lower its free energy.

While the free energy tells us whether or not a spontaneous reaction is possible, it cannot predict the speed of the reaction. An excellent example of this fact can be seen in the case of the two phases of carbon – diamond and graphite. Graphite is the phase with the lower free energy, and diamond should, therefore, transform spontaneously into graphite. The rate is so slow, however, that there is no need to consider it. Rate problems such as these are treated in the kinetic theory of matter and are not in the realm of thermodynamics.

6.6 Statistical Mechanical Definition of Entropy. The significance of the entropy will now be considered. Let us take a two-chamber box, each chamber filled with a different monatomic ideal gas. Gas A is in chamber I, and gas B in chamber II. Let the partition between the chambers be removed and the gases will mix by diffusion. Such mixing occurs at constant temperature and constant pressure, if the gases have the same original temperature and pressure.

No work is done and no heat is transferred to or from the gases, therefore, the internal energy of the gaseous system does not change. This fact is in agreement with the law of conservation of energy (first law of thermodynamics), which is expressed in the following equation:

$$dE = dQ + dW$$

$$dE = 0$$

where dQ = heat absorbed by the system (gases)
dE = change in internal energy of the system (gases)
dW = work done on gases by the surroundings

A fundamental change in the system occurs as a result of the diffusion. That this is true can be judged from the fact that considerable effort is required to separate this mixture of gases into its components again. Like the freezing of water at temperatures below 0° C, the mixing of gases is a spontaneous, or irreversible reaction, and, as in all irreversible reactions, the free energy must decrease. The free-energy equation states that

$$dF = dE - TdS$$

However, it was shown above that dE is zero and, therefore

$$dF = -TdS$$

A decrease in free energy can only mean that dS must be positive. In other words, the entropy of the system has increased by the diffusing of the gases. The entropy increase involved in this reaction is known as *an entropy of mixing*. It is only one of a number of forms of entropy. All forms, however, have one thing in common. When the entropy of a system increases, the system becomes more disordered.

In the above example, a disorder in the spatial distribution of two kinds of gas atoms has been considered, but entropy can also be associated with disorder in the motion of atoms, with respect to the directions and to the magnitudes of the velocities of the atoms.

As a hypothetical example, let us consider that it is possible to introduce gas molecules into a chamber in such a manner that all of them begin to travel in the same direction and with the same speed back and forth between two opposite walls of the box. Collisions of the molecules with each other and with the walls quickly bring about a random distribution in the directions and magnitudes of the molecular velocities. An increase in entropy necessarily accompanies this irreversible change from uniformly ordered motion to random motion. The question now arises: why do two unmixed gases seek a random distribution? The answer lies in the fact that, on removal of the partition, each gas atom is free to move through both compartments and has equal probability of being found in either. Once the barrier has been removed, the probability is extremely small that all of the A atoms will be found in compartment A and, at the same time, all of the B atoms in compartment B. Furthermore, the probability of the atoms maintaining such segregation is even more remote. On the other hand, the chance of finding a random distribution throughout the box is almost a certainty.

Since a shift from a state of low probability (two unmixed gases in contact) to a state of high probability (random mixture) accompanies an entropy increase, it would appear that there is a close relationship between entropy and probability. This relationship does exist and was first expressed mathematically by Boltzmann, who introduced the following equation:

$$S = k \log_e P$$

where S is the entropy of a system in a given state, P is the probability of the state, and k is Boltzmann's constant (1.38×10^{-16} ergs/deg (K°)).

The change in entropy (mixing entropy) resulting from the mixing of gas A and gas B may be expressed in terms of the Boltzmann equation:

$$\Delta S = S_2 - S_1 = k \log_e P_2 - k \log_e P_1$$

$$\Delta S = k \log_e \frac{P_2}{P_1}$$

where S_1 = entropy of unmixed gases
S_2 = entropy of mixed gases
P_1 = probability of unmixed state
P_2 = probability of mixed state

The probability of finding the atoms in the unmixed, or segregated, state is computed as follows:

Let V_A = the volume originally occupied by the atoms of gas A
V_B = the volume originally occupied by the atoms of gas B
V = the total volume of the box

If one A atom is introduced into the undivided box, the probability of finding it in V_A is V_A/V. If a second A atom is now added to the box, the chance of finding both at the same time in V_A is $(V_A/V) \times (V_A/V)$. This problem is similar to that of producing two heads on the toss of a pair of coins, where the chance of a head on either coin is 1/2, but for the pair it is $1/2 \times 1/2 = 1/4$. A third A atom reduces the probability of finding all A atoms in V_A to $(V_A/V)^3$, and if n_A is the total number of atoms of gas A, the probability of finding all n_A in V_A is $(V_A/V)^{n_A}$. Now if an atom of B is added to the box, the chance of finding it in V_B is (V_B/V), and the chance of finding it in V_B, and at the same time finding all the A atoms in V_A, is $(V_A/V)^{n_A} \times (V_B/V)$. Finally, the probability of finding all atoms of gas A in V_A, while all atoms of gas B are in V_B is

$$P_1 = \left(\frac{V_A}{V}\right)^{n_A} \cdot \left(\frac{V_B}{V}\right)^{n_B}$$

where n_A = number of A atoms
n_B = number of B atoms

It is now necessary to consider the probability of the homogeneous mixture. Thought must first be given to the meaning of the term *homogeneous mixture.* A very accurate experimental analysis might possibly be able to detect a variation in composition of the mixture of the order of one part in 10^{10}, but be unable to detect a smaller variation. Therefore, an experimentally homogeneous mixture is not one with a perfectly constant ratio of A to B atoms, but one with a ratio that does not vary sufficiently from the mean value to be detectable. Such a mixture is extremely probable when the number of atoms is large, as in most real systems, where the number usually exceeds 10^{20} (approximately 10^{-3} moles). The importance of large numbers in statistics can easily be seen in a somewhat analogous case: that of flipping coins and counting the number that come "heads up." If 10 coins are tossed together, it may be shown that the chance of getting 5 heads is 0.246, while that of obtaining any of the numbers in the group of 4, 5, or 6 heads is 0.666. Thus, even with 10 coins, it is apparent that a number close to the mean distribution is quite probable. If the number of tossed coins is increased to 100, the probability of finding somewhere between 40 and 60 heads is 0.95. Increasing the number of coins by one factor of 10 clearly increases the probability of finding a distribution near the mean. Finally, let us increase the number of coins to approximately that found in a real system of atoms (10^{20}). In this case, the mean distribution is 5×10^{19} heads. The probability of finding a number of heads within 3.5 parts in 10^{10} of this number is 0.999999999997. In other words, statistics tells us that the chance of finding between

50,000,000,035,000,000,000

and

49,999,999,965,000,000,000

heads is within 3 parts in 10^{12} of unity. It also tells us that the chance of finding a distribution containing a number of heads that lies outside the above range is just about nil.

Consider again the box containing the mixture of A and B gas atoms. The chance of finding a head on the flip of a coin is exactly analogous to the chance of finding an A atom in one half of the box. It is also analogous to the chance of finding a B atom in the same half. It can be concluded, therefore, that both atom forms, when present in very large numbers, will seek a mean distribution in which the atoms are uniformly distributed in the box. Deviations from this mean value will be very small and the probability of an experimentally homogeneous mixture extremely high. It can be assumed, for the purposes of calculation, that this probability is equal to one.

Returning to the mixing-entropy equation, namely:

$$\Delta S = k \log_e \frac{P_2}{P_1}$$

where ΔS = mixing entropy
k = Boltzmann's constant
P_1 = probability of unmixed state
P_2 = probability of mixed state

As a result of the above considerations, P_2 can be taken as unity. Thus,

$$\Delta S = k \log_e \frac{1}{P_1} = -k \log_e P_1$$

but it has already been shown that

$$P_1 = \left(\frac{V_A}{V}\right)^{n_A} \cdot \left(\frac{V_B}{V}\right)^{n_B}$$

where V_A = volume originally occupied by gas A
V_B = volume originally occupied by gas B
n_A = number of A atoms
n_B = number of B atoms
V = total volume $(V_A + V_B)$

and therefore,

$$\Delta S = -k \log_e \left(\frac{V_A}{V}\right)^{n_A} \cdot \left(\frac{V_B}{V}\right)^{n_B}$$

$$= -k \log_e \left(\frac{V_A}{V}\right)^{n_A} - k \log_e \left(\frac{V_B}{V}\right)^{n_B}$$

$$= -kn_A \log_e \left(\frac{V_A}{V}\right) - kn_B \log_e \left(\frac{V_B}{V}\right)$$

But since we have assumed ideal gases at the same temperature and pressure, the volumes occupied by the gases must be proportional to the numbers of atoms in the gases. Thus, we have

$$\frac{n_A}{n} = \frac{V_A}{V}$$

and

$$\frac{n_B}{n} = \frac{V_B}{V}$$

where n is the total number of atoms of both kinds. However, the ratios n_A/n and n_B/n are the mean chemical concentrations of atoms A and B in the box, and thus we may set

$$\frac{n_A}{n} = \frac{V_A}{V} = C$$

and

$$\frac{n_B}{n} = \frac{V_B}{V} = (1 - C)$$

where C is the concentration of A, and $(1 - C)$ is the concentration of B. The entropy of mixing can now be expressed in terms of concentrations as follows:

$$\Delta S = -kn \left(\frac{n_A}{n}\right) \log_e C - kn \left(\frac{n_B}{n}\right) \log_e (1 - C)$$

$$= -knC \log_e C - kn(1 - C) \log_e (1 - C)$$

If it is now assumed that we have one mole of gas, then the number of atoms n

becomes equal to N where N is Avogadro's number. Also, k (Boltzmann's constant) is the gas constant for one atom; that is

$$k = \frac{R}{N}$$

where R is the gas constant (2 cal per mol) and N is Avogadro's number. As a result, we have

$$kn = kN = R$$

We therefore write the mixing-entropy equation in its final form

$$\Delta S = -R[C \log_e C + (1 - C) \log_e (1 - C)]$$

6.7 Vacancies. Metal crystals are never perfect. It is now well understood that they may contain many defects. These are classified according to a few important types. One of the most important is called a *vacancy*. The existence of vacancies in crystals was originally postulated to explain solid-state diffusion in crystals. Because of the mobility of gas atoms and molecules, gaseous diffusion is easy to comprehend, but the movement of atoms inside crystals is more difficult to grasp. Still, it is a well-known fact that two metals, for example, nickel and copper, will diffuse into each other if placed in intimate contact and heated to a high temperature. It is also known that atoms in a pure metal diffuse. Several mechanisms have been proposed to explain these diffusion phenomena, but the most generally acceptable has been the vacancy mechanism.

Diffusion in crystals is explained, in terms of vacancies, by assuming that the vacancies move through the lattice, thereby producing random shifts of the atoms from one lattice position to another. The basic principle of vacancy diffusion is illustrated in Fig. 6.1, where three successive steps in the movement of a vacancy are shown. In each case, it can be seen that the vacancy moves as a result of an atom jumping into a hole from a lattice position bordering the hole.

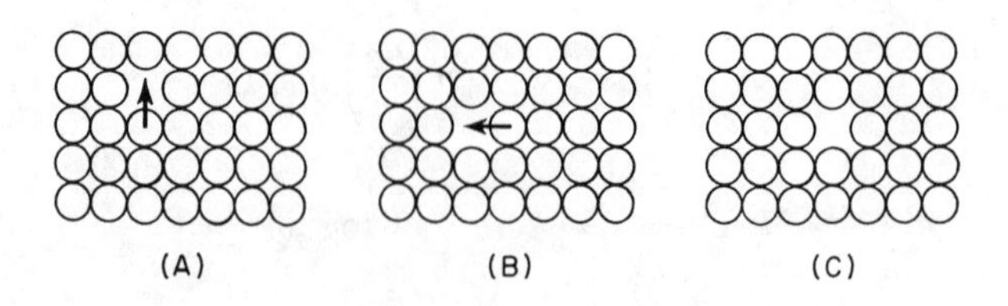

Fig. 6.1 Three steps in the motion of a vacancy through a crystal.

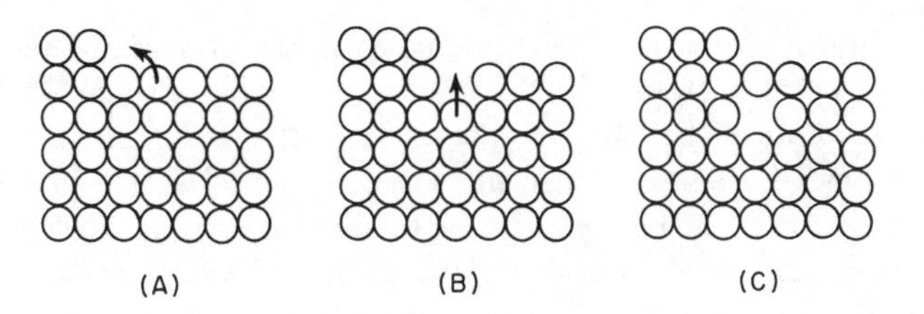

Fig. 6.2 The creation of a vacancy.

In order to make the jump, the atom must overcome the net attractive force of its neighbors on the side opposite the hole. Work is therefore required to make the jump into the hole, or, as it may also be stated, an energy barrier must be overcome. Energy sufficient to overcome the barrier is furnished by the thermal or heat vibrations of the crystal lattice. The higher the temperature, the more intense the thermal vibrations, and the more frequently are the energy barriers overcome. Vacancy motion at high temperatures is very rapid and, as a consequence, the rate of diffusion increases rapidly with increasing temperature. An equation will now be derived that gives the equilibrium concentration of vacancies in a crystal as a function of temperature.

Let us assume that in a crystal containing n_o atoms there are n_v vacant lattice sites. The total number of lattice sites is, accordingly, $n_o + n_v$, or the sum of the occupied and unoccupied positions. Suppose that vacancies are created by movements of atoms from positions inside the crystal to positions on the surface of the crystal, in the manner shown in Fig. 6.2. When a vacancy has been formed in this manner, *a Schottky defect* is said to have been formed. Let the work required to form a Schottky defect be represented by the symbol w. A crystal containing n_v vacancies will therefore have an internal energy greater than that of a crystal without vacancies by an amount $n_v w$.

The free energy of a crystal containing vacancies will be different from that of a crystal free of vacancies. This free-energy increment may be written as follows:

$$F_v = E_v - TS_v$$

where F_v = the free energy due to vacancies
E_v = the internal energy increase due to the vacancies
S_v = the entropy due to the vacancies
but, according to the above

$$E_v = n_v w$$

and thus

$$F_v = n_v w - TS_v$$

Now the entropy of the crystal is increased in the presence of vacancies for two reasons. First, the atoms adjacent to each hole are less restrained than those completely surrounded by other atoms and can, therefore, vibrate in a more irregular or random fashion than the atoms removed from the hole. Each vacancy contributes a small amount to the total entropy of the crystal.

Let us designate the vibrational entropy associated with one vacancy by the symbol s. The total increase in entropy arising from this source is $n_v s$, where n_v is the total number of vacancies. While a consideration of this vibrational entropy is important in a thorough theoretical treatment of vacancies, it will be omitted in our present calculations because its effect on the number of vacancies present in the crystal is of secondary importance.

The other entropy form arising in the presence of vacancies is an *entropy of mixing*. The entropy of mixing has already been derived for the mixing of two ideal gases, and is expressed by the equation

$$S_m = \Delta S = -nk[C \log_e C + (1 - C) \log_e (1 - C)]$$

where
S_m = mixing entropy
n = total number of atoms $(n_A + n_B)$
k = Boltzmann's constant
C = concentration of atom $A = n_A/n$
$(1 - C)$ = concentration of atom $B = n_B/n$

The above equation applies directly to the present problem if we consider mixing of lattice points, instead of atoms, in which there are two types of lattice points: one occupied by atoms, the other unoccupied. If there are n_o occupied sites and n_v unoccupied, the unmixed state will correspond to one in which a lattice of $n_o + n_v$ sites has all positions on one side filled and all on the other empty. This is equivalent to a box problem, shown schematically in Fig. 6.3A, in which there are two compartments, one filled with occupied lattice positions

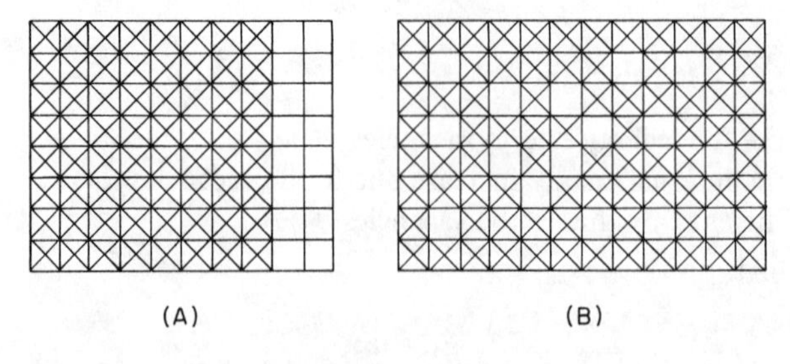

Fig. 6.3 The box analogy of a crystal. (A) Vacancies and atoms in the segregated state. Atoms to the left, vacancies to the right. (B) The mixed state.

and the other filled with empty positions. The corresponding mixed state in the box is shown in Fig. 6.3B.

The present problem consists of mixing n_o objects of one kind with n_v of another kind, with a total number $(n_o + n_v)$ to be mixed. Therefore, we may make the following substitutions in the mixing-entropy equation:

$$n = n_o + n_v$$

$$C = C_v = \frac{n_v}{n_o + n_v}$$

$$(1 - C) = C_o = \frac{n_o}{n_o + n_v}$$

where C_v = the concentration of vacancies
C_o = the concentration of occupied lattice positions

If the above quantities are set into the mixing-entropy equation, we have

$$S_m = -(n_o + n_v)k\left[\frac{n_v}{n_o + n_v}\log_e\frac{n_v}{n_o + n_v} + \frac{n_o}{n_o + n_v}\log_e\frac{n_o}{n_o + n_v}\right]$$

which becomes after simplification

$$S_m = k[(n_o + n_v)\log_e(n_o + n_v) - n_v \log_e n_v - n_o \log_e n_o]$$

The free-energy equation for vacancies may now be written

$$\begin{aligned} F_v &= n_v w - S_m T \\ &= n_v w - kT[(n_o + n_v)\log_e(n_o + n_v) - n_v \log_e n_v - n_o \log_e n_o] \end{aligned}$$

This free energy must be a minimum if the crystal is in equilibrium; that is, the number of vacancies (n_v) in the crystal will seek the value that makes F_v a minimum at any given temperature. As a result, the derivative of F_v with respect to n_v must equal zero, the temperature being held constant. Thus,

$$\frac{dF_v}{dn_v} = w - kT\left[(n_o + n_v)\frac{1}{(n_o + n_v)} + \log_e(n_o + n_v) - n_v\frac{1}{n_v} - \log_e n_v - 0\right]$$

$$0 = w + kT\left[\log_e\frac{n_v}{n_o + n_v}\right]$$

which, when expressed in exponential form, becomes

$$\frac{n_v}{n_o + n_v} = e^{-w/kT}$$

In general, it has been found that the number of vacancies in a metal crystal is very small when compared with the number of atoms (the number of occupied sites n_o). Therefore the above equation can be written in terms of n_v/n_o, the ratio of vacancies to atoms in the crystal.

$$\frac{n_v}{n_o} = e^{-w/kT}$$

where w = work to form one vacancy
k = Boltzmann's constant
T = the temperature, °K
n_v = number of vacancies
n_o = number of atoms

If both the numerator and the denominator of the exponent of the above equation are now multiplied by N, Avogadro's number (6.03×10^{23}), the equation will be unaltered with respect to the functional relationship between the concentration of vacancies and the temperature. However, the quantities in the exponent will then correspond to standard thermodynamical notation. Therefore, let

$$Q_f = Nw$$
$$R = kN$$

$$\frac{n_v}{n_o} = e^{-Nw/NkT} = e^{-Q_f/RT}$$

where Q_f = the heat of activation; that is, the work required to form one mole of vacancies, in calories per mole
N = Avogadro's number
k = Boltzmann's constant
R = gas constant = 2 cal per mole – °K

and therefore

$$\frac{n_v}{n_o} = e^{-Q_f/RT}$$

The experimental value for the activation energy for the formation of vacancies in copper is approximately 20,000 cal per mole of vacancies. This value may be substituted into the above equation in an attempt to determine the effect of temperature on the number of vacancies. Remembering that $R = 2$ cal per mole – K°

$$\frac{n_v}{n_o} = e^{-Q_f/RT} = e^{-20,000/2T} = e^{-10,000/T}$$

At absolute zero the equilibrium number of vacancies should be zero, for in this case

$$\frac{n_v}{n_o} = e^{-10,000/0} = e^{-\infty} = 0$$

At 300° K, approximately room temperature

$$\frac{n_v}{n_o} = e^{-10,000/300} = e^{-33} = 4.45 \times 10^{-15}$$

However, at 1350° K, six degrees below melting point,

$$\frac{n_v}{n_o} = e^{-10,000/1350} = e^{-7.40} = 6.1 \times 10^{-4} \simeq 10^{-3}$$

Therefore, just below the melting point there is approximately one vacancy for every 1000 atoms. While at first glance this appears to be a very small number, the mean distance between vacancies is only about 10 atoms. On the other hand, the room-temperature equilibrium concentration of vacancies (4.45×10^{-15}) corresponds to a mean separation between vacancies of the order of 100,000 atoms. These figures show clearly the strong effect of temperature on the number of vacancies.

Two questions present themselves: first, why should there be an equilibrium number of vacancies at a given temperature, and, second, why does the equilibrium number change with temperature? Figure 6.4 answers the first of these questions in terms of curves of the functions F_v, $n_v w$, and $-TS_m$ as functions of the number of vacancies n_v. In this figure, the temperature is assumed close to the melting point. Whereas the work to form vacancies ($n_v w$) increases linearly with the number of vacancies, at low concentrations the entropy component ($-TS$) increases very rapidly with n_v, but less and less rapidly as n_v grows larger. At the value marked n_1 in the figure, the two quantities $n_v w$ and $-TS$ become equal, and, at this point, the free energy F_v,

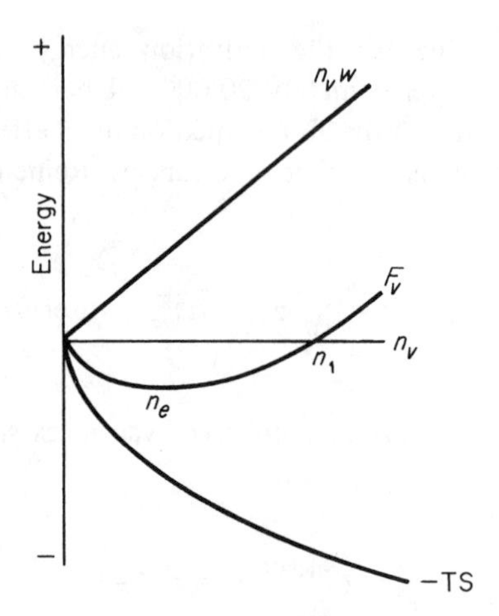

Fig. 6.4 Free energy as a function of the number of vacancies, n_v, in a crystal at a high temperature.

the sum of $(n_v w - TS)$, equals zero. For all values of n_v greater than n_1, the free energy is positive, and for all values of n_v smaller than n_1, the free energy is negative. Furthermore, in the region of negative free energy, a minimum occurs corresponding to the concentration marked n_e in the figure. This is the equilibrium concentration of vacancies.

Figure 6.5 shows a second set of curves similar to those in Fig. 6.4, except that the temperature is assumed to be lower $(T_B < T_A)$. The lowering of the temperature does not change the $n_v w$ curve, but makes all ordinates of the $-TS$

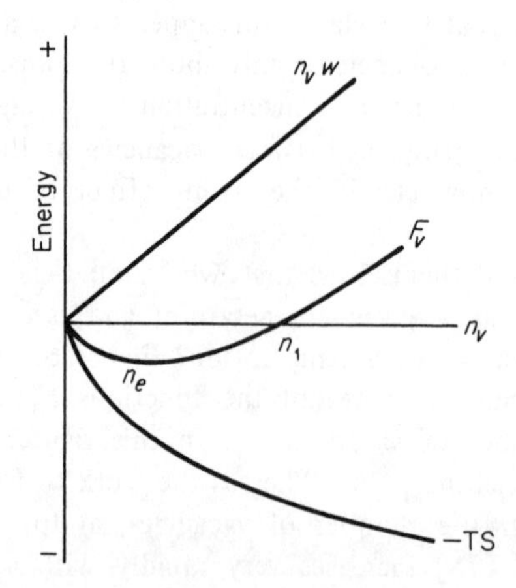

Fig. 6.5 The free-energy curve for a lower temperature than that of Fig. 6.4. Notice that n_e, the equilibrium number of vacancies, has decreased.

curve smaller in the ratio T_B/T_A. As a result, both n_1 and n_e of the free-energy curve (F_v) are shifted to the left. Thus, with decreasing temperature, the equilibrium number of vacancies becomes smaller because the entropy component ($-TS$) decreases.

6.8 Vacancy Motion. We have determined that the equilibrium ratio of vacancies to atoms at a given temperature T is

$$\frac{n_v}{n_o} = e^{-Q_f/RT}$$

where n_v = number of vacancies
n_o = number of atoms
Q_f = work to form one mole of vacancies
R = gas constant (2 calories per mole per °K)

Nothing, however, has been said about the time required to attain an equilibrium number of vacancies. This may be very long at low temperatures or very rapid at high temperatures. Since the movement of vacancies is caused by successive jumps of atoms into vacancies, a study of the fundamental law governing the jumps is important. It was mentioned earlier that an energy barrier shown schematically in Fig. 6.6, must be overcome when a jump is made, and that the required energy is supplied by the thermal, or heat, vibrations of the crystal.

If q_o is the height of the energy barrier that an atom, such as atom a in Fig. 6.6, must overcome in order to jump into a vacancy, then the jump can only occur if atom a possesses vibrational energy greater than q_o. Conversely, if the

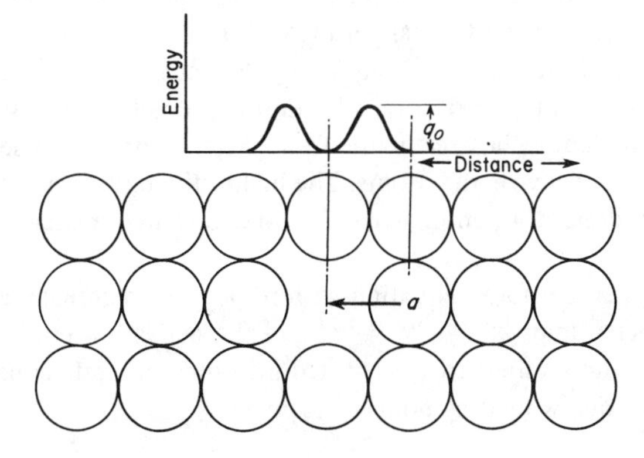

Fig. 6.6 The energy barrier that an atom must overcome in order to jump into a vacancy.

vibrational energy is lower than q_o, the jump cannot occur. The chance that a given atom possesses an energy greater than q_o has been found to be proportional to the function $e^{-q_o/kT}$, or

$$p = \text{const } e^{-q_o/kT}$$

where p is the probability that an atom possesses an energy equal to or greater than a given energy q_o, k is Boltzmann's constant, and T is the absolute temperature. The above equation was originally derived for the energy distribution of atoms in a perfect gas (Maxwell-Boltzmann Distribution). However, this same function has also been found to predict quite accurately the vibrational energy distribution of the atoms in a crystalline solid.

Since the above function is the probability that a given atom has an energy greater than that required for a jump, the probability of jumping must be proportional to this function. Therefore, we can write the following equation:

$$r_v = Ae^{-q_o/kT}$$

where r_v is the number of atom jumps per second into a vacancy, A is a constant, q_o is the activation energy per atom (height of the energy barrier), and k and T have the usual significance. If the numerator and denominator of the exponent in the above equation are both multiplied by N, Avogadro's number 6.02×10^{23}, we have

$$r_v = Ae^{-Q_m/RT}$$

where Q_m is the activation energy for the movement of vacancies in calories per mole, and R the gas constant (2 cal per mole °K).

The constant A in the above equation depends on a number of factors. Among these is the number of atoms bordering the hole. The larger the number of atoms able to jump, the greater the frequency of jumping. A second factor is the vibration frequency of the atoms. The higher the rate of vibration, the more times per second an atom approaches the hole, and the greater is its chance of making a jump.

Let us consider the above equation with respect to an actual metal. The value of A for copper is approximately 10^{15}, and the activation energy Q_m, 29,000 cal per mole. If these values are substituted into the jump-rate equation, we have at 1350° K (just below melting point of copper)

$$r_v \simeq 3 \times 10^{10} \text{ jumps/sec}$$

at 300° K (room temperature)

$$r_v \simeq 10^{-6} \text{ jumps/sec.}$$

The tremendous difference in the rate with which vacancies move near the melting point, compared with their rate at room temperature, is quite apparent. In one second at 1350° K the vacancy moves approximately 30 billion times, while at room temperature a time interval of about 10^6 sec, or 11 days, occurs between jumps.

While the jump rate of vacancies is of some importance, we are really more interested in how many jumps the average atom makes per second when the crystal contains an equilibrium number of vacancies. This quantity equals the fractional ratio of vacancies to atoms n_v/n_o times the number of jumps per second into one vacancy, or

$$r_a = \frac{n_v}{n_o} A e^{-Q_m/RT}$$

where r_a is the number of jumps per second made by an atom, n_v is the number of vacancies, n_o the number of atoms. However, we have seen that

$$\frac{n_v}{n_o} = e^{-Q_f/RT}$$

where Q_f is the activation energy for the formation of vacancies. Therefore,

$$r_a = A e^{-Q_m/RT} \times e^{-Q_f/RT} = A e^{-(Q_m + Q_f)/RT}$$

The rate at which an atom jumps, or moves from place to place in a crystal, thus depends on two energies: Q_f, the work to form a mole of vacancies, and Q_m, the energy barrier that must be overcome in order to move a mole of atoms into vacancies. Since the two energies are additive, the atomic jump rate is extremely sensitive to temperature. This last statement can be made in a different form. It has previously been shown that in copper the ratio of vacancies to atoms is about one to a thousand at 1350° K, while at 300° K, the ratio is one to approximately 5×10^{15}; a decrease by a factor of about 10^{12}. During the same temperature interval, the number of jumps per second into a vacancy decreases by a factor of approximately 10^{16}. Between the melting point and room temperature, the average rate of atomic movement thus decreases by a factor of about 10^{28}. From a more simple point of view, at 1350° K the vacancies in copper are approximately 10 atoms apart, and atoms jump into these vacancies at the rate of about 30 billion jumps per second; whereas at 300° K the vacancies are 100,000 atoms apart, and atoms jump into them at the rate of one jump in 11 days.

The above discussion leads to one firm conclusion: those physical properties of copper that can be changed through diffusion of copper atoms can be considered unchangeable at room temperature.

Similar conclusions can be drawn for other metals, but the degree of change will depend upon the metal in question. For example, vacancies in lead at room temperature may be shown, by calculations similar to the above, to jump at a rate of approximately 22 jumps per second, with a mean spacing between vacancies of about 100 atoms. Appreciable atomic diffusion therefore occurs in lead at room temperature.

6.9 Interstitial Atoms and Divacancies. Next to a vacancy, the most important point defects in a metal crystal are an *interstitial atom* and a *divacancy*. Let us briefly consider the first of these.

An interstitial atom is one that occupies a place in a crystal that normally would be unoccupied. Such places are the holes or interstitial positions that occur between the atoms lying on the normal lattice sites. Such a hole occurs even in the close-packed face-centered cubic lattice at the center of the unit cell, as indicated in Fig. 6.7. An equivalent hole exists between any pair of atoms lying at the corners of the unit cell. The hole between the two corner atoms along the right-hand front vertical edge of the cell is indicated in Fig. 6.7.

It is necessary to differentiate between two basically different types of atoms that can occupy these interstitial sites. The first of these is composed of atoms

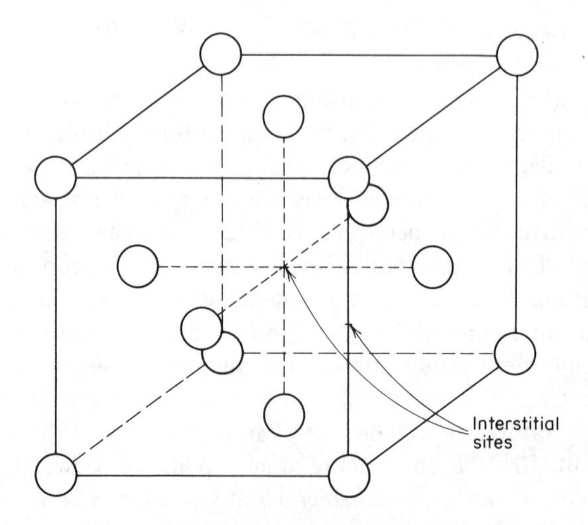

Fig. 6.7 Interstitial sites in a face-centered cubic crystal.

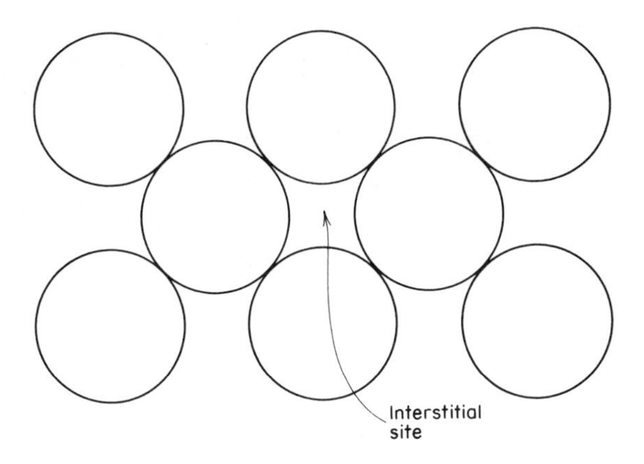

Fig. 6.8 The size of the interstitial site is much smaller than the size of the solvent atoms.

whose sizes are small, such as carbon, nitrogen, hydrogen, and oxygen. Atoms of this type can sometimes occupy a small fraction of the interstitial sites in a metal. When this occurs, it is said to be an interstitial solid solution. (This subject is considered further in Chapter 8.) The other type of interstitial atom is our present concern: this consists of an atom that would normally occupy a regular lattice site. In the case of a pure metal like copper, such an interstitial atom would be a copper atom. The hard-ball model of the {100} plane of a f.c.c. crystal in Fig. 6.8 clearly shows that the interstitial sites are too small to hold such a large atom without badly distorting the lattice. As a consequence, while the activation energy to form vacancies in copper is a little less than 1 eV, that for the formation of interstitital copper atoms is probably about 4 eV. This difference in activation energy means that interstitial atoms produced by thermal vibrations should be very rare in a metal such as copper. This can easily be shown by a computation using the same form of equation as that developed earlier in this chapter to compute the equilibrium ratio of vacancies to atoms (see Problem 12).

While interstitial defects of the type under consideration do not normally exist in sufficient concentrations in a metal to be significant, they can be readily produced in metals as a result of radiation damage. Collisions between fast neutrons and a metal can result in knocking atoms out of their normal lattice sites. The removal of an atom from its lattice position of course produces both an interstitial atom and a vacancy. It has been estimated [1] that each fast neutron can result in the creation of about 100–200 interstitials and vacancies.

While it is very difficult to create an interstitial defect of the type presently

1. Cottrell, A. H., *Vacancies and Other Point Defects in Metals and Alloys*, p. 1. The Institute of Metals, London, 1958.

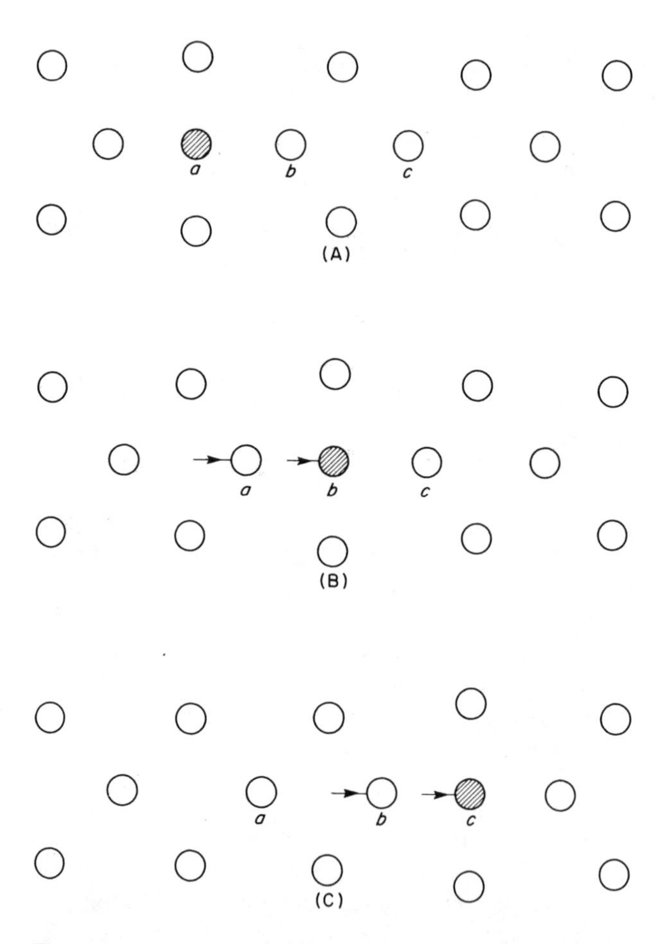

Fig. 6.9 The configuration associated with an interstitial atom can move readily through the crystal.

being considered, they are very mobile once they are produced. The energy barrier for their motion is small and, in the case of copper, it has been estimated [2] to be of the order of 0.1 eV. The reason for this high mobility is shown in Fig. 6.9. In this figure it is assumed that one is observing the {100} plane of a f.c.c. crystal. In Fig. 6.9A the interstitial atom is shown as a cross-hatched circle and is designated by the symbol *a*. Suppose that this atom were to move to the right. This motion would push atom *b* into an interstitial position, as shown in Fig. 6.9B. In this process atom *a* would return to a normal lattice site. Further motion of *b* in the same direction would make an interstitial atom out of *c*. By this type of motion, the configuration associated with an

2. Huntington, H. B., *Phys. Rev.*, **91**, 1092 (1953).

interstitial atom can readily move through the lattice. Any given interstitial atom only moves a small distance, and because the distortion of the lattice is severe around the interstitial atom, this type of motion requires little energy. The effect of the lattice distortion around the defect in making the motion easier is exactly equivalent to that observed in the motion of dislocations.

If a pair of vacancies combine to make a single point defect, a *divacancy* is said to be created. It is difficult to estimate or measure the binding energy of this type of defect. One calculation,[3] however, places it at about 0.3–0.4 eV in the case of copper. In a metal where the vacancies and divacancies are in equilibrium, one can approximately compute[4] the ratio of divacancies to vacancies using the equation

$$\frac{n_{dv}}{n_v} = 1.2ze^{-\frac{q_b}{kT}}$$

where n_{dv} is the concentration of divacancies, z is the coordination number, and q_b is the binding energy of a divacancy.

Problems

1. The heat of fusion of lead is 6.2 cal/gm and its melting temperature is 327°C. Determine the entropy change associated with the freezing of one gram of lead.
2. In the equilibrium freezing of water
   (*a*) Does the internal energy of the system (water) increase or decrease? Explain.
   (*b*) Does the entropy increase or decrease? Explain.

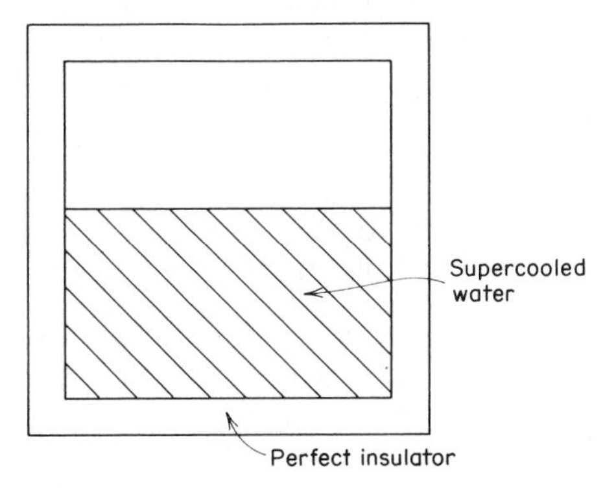

3. Let us assume that it is possible to supercool water to a temperature T_s so that when the water freezes the release of the heat of fusion is just able to

3. Seeger, A., and Bross, H., *Z. Physik,* **145** 161 (1956).
4. Cottrell, A. H., *Vacancies and Other Point Defects in Metals and Alloys*, p. 1. The Institute of Metals, London, 1958.

bring the ice back to 0°C. Further, let the freezing process occur in a perfect insulator.

(*a*) In this freezing process, how large will be the change in the internal energy?

(*b*) Will the entropy increase or decrease? Discuss your answer in terms of the answer to part (*b*) of the preceding problem.

4. Now assume that the ice obtained at the end of Problem 3 is returned reversibly to its original state at the start of Problem 3 by melting it at 0°C and then, by very slow cooling, returning it to T_S, the original supercooling temperature.

   (*a*) What is the entropy change associated with the melting at 0°C? Is it positive or negative?

   (*b*) Indicate the procedure for computing the entropy change associated with reversibly cooling the water from 0°C to T_S. Will this entropy change be positive or negative?

   (*c*) Will the sum of the above two entropies have to be positive or negative? Explain.

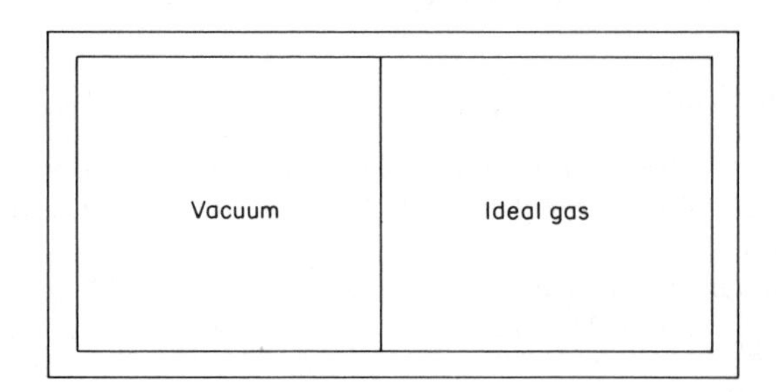

5. This figure represents a two-sided insulated chamber with a removable partition separating the two halves. The left-hand side of the chamber is assumed to be evacuated, while the right-hand side contains one mole of an ideal gas at standard atmospheric conditions. If the partition is removed suddenly

   (*a*) Will the temperature of the gas change? Explain.

   (*b*) Will the entropy change? Explain.

   (*c*) If the entropy changes, indicate how one could evaluate the magnitude of this change.

6. The number of possible combinations of n coins that can come up with a number of heads equal to r is given by the algebraic relation

$$n^{c}r = \frac{n!}{r!\,(n-r)!}$$

If this number is divided by the total number of permutations, which is equal to 2^n, the result is the probability of obtaining r heads in a toss of n coins. Compute the probability of obtaining either 4, 5, or 6 heads in a toss of 10 coins.

7. Compute the probability of obtaining a number of heads between 8 and 12 on a toss of 20 coins.

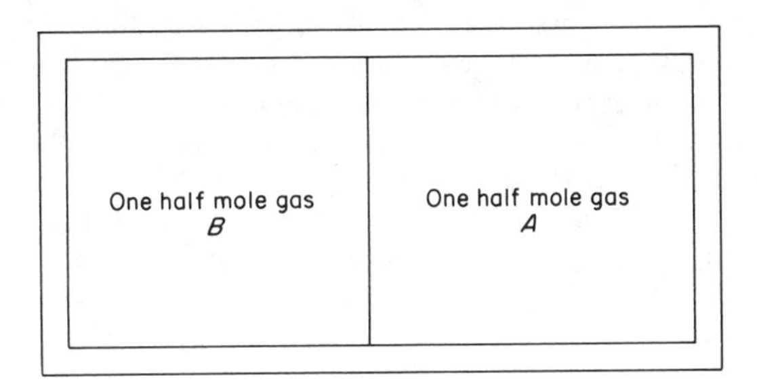

8. (*a*) If the partition in this box is removed, what will be the magnitude of the entropy change associated with the resulting process of mixing, assuming that both gases are at atmospheric pressure?
   (*b*) What will be the corresponding change in free energy if the temperature is 27°C?
   (*c*) Is this change in free energy a significant number (that is, compare it to heats of freezing, etc.)?
9. How large would the change in free energy be if, in Problem 8, the quantity of gas *A* had been 0.9 moles, and of gas *B* 0.1 moles?
10. (*a*) Determine the magnitude of the internal energy associated with the vacancies in one mole of copper at room temperature.
    (*b*) Make the same computation for copper held at 1350°K long enough for the equilibrium number of vacancies to have been achieved.
    (*c*) What is the magnitude of the corresponding entropy of mixing at 1350°K?
    (*d*) If the metal could be instantly heated from 0°K to 1350°K, what decrease in its free energy due to the creation of vacancies might be expected if it is held at this temperature?
11. From the work of Simmons and Balluffi[5] it is deduced that the activation energies for the formation of vacancies in aluminum and silver are 0.76eV and 1.09eV, respectively. The corresponding melting points are 660°C and 961°C, respectively. Compute the ratio of vacancies to atoms in these metals at their melting points. Compare your answers to that computed for copper in the text. Do you think there is a significant relationship between vacancy concentration and the breakdown of a crystal to form a liquid?
12. Huntington[6] estimates that the activation energy to form an interstitial atom in copper is about 4eV. What is the equilibrium concentration of these defects at 300°K and at 1350°K? How do these numbers correspond to those relating to vacancies?
13. The activation energy for the movement of an interstitial atom in copper has also been estimated by Huntington to be of the order of 0.1eV. Assume

5. Simmons, R. O., and Balluffi, R. W., *Phys. Rev.*, **117** 52,160 (1960).
6. Huntington, H. B., *Phys. Rev.*, **91** 1092 (1953).

that the pre-exponential factor A in the jump rate equation for interstitials is the same as for vacancies, that is 10^{15} sec^{-1}, and estimate the jump frequency of an interstitial atom at room temperature.

14. Now compute the jump frequency of an interstitial atom at 40°K. Is this jump rate consistent with experimental observations that indicate interstitials are mobile even at this low a temperature?
15. Compute the ratio of the concentration of vacancies to divacancies in copper at 300°K and 1350°K, assuming $q_B = 0.25\text{eV}$. If copper is quenched from 1350°K to 300°K, is it reasonable to expect that an appreciable fraction of the point defects will be divacancies?

7 Annealing

7.1 Stored Energy of Cold Work. When a metal is plastically deformed at temperatures that are low relative to its melting point, it is said to be *cold worked.* The temperature defining the upper limit of the cold working range cannot be expressed exactly, for it varies with composition as well as the rate and the amount of deformation. A rough rule-of-thumb is to assume that plastic deformation corresponds to cold working if it is carried out at temperatures lower than one-half of the melting point measured on an absolute scale.

Most of the energy expended in cold work appears in the form of heat, but a finite fraction is stored in the metal as strain energy associated with various lattice defects created by the deformation. The amount of energy retained depends on the deformation process and a number of other variables, for example, composition of the metal as well as the rate and temperature of deformation. A number of investigators have indicated that the fraction of the energy which remains in the metal varies from a low percentage to somewhat over 10 percent. Figure 7.1 shows the relationship between the stored energy and the amount of deformation in a specific metal (polycrystalline 99.999 percent pure copper) for a specific type of deformation (tensile strain). The data, from the work of Gordon,[1] show that the stored energy increases with increasing deformation, but at a decreasing rate, so that the fraction of the total energy stored decreases with increasing deformation. The latter effect is shown by a second curve plotted in Fig. 7.1.

The maximum value of the stored energy in Fig. 7.1 is only 6 cal per mole, which represents the strain energy left in a very pure metal after a moderate deformation (30 percent extension) at room temperature. The amount of stored energy can be greatly increased by increasing the severity of the deformation, lowering the deformation temperature, and by changing the pure metal to an alloy. Thus, metal chips formed by drilling an alloy (82.6 percent Au – 17.4 percent Ag) at the temperature of liquid nitrogen are reported[2] to have a stored-energy content of 200 cal per mole.

1. Gordon, P., *Trans. AIME* 203 1043 (1955).
2. Greenfield, P., and Bever, M. B., *Acta Met.*, 4 433 (1956).

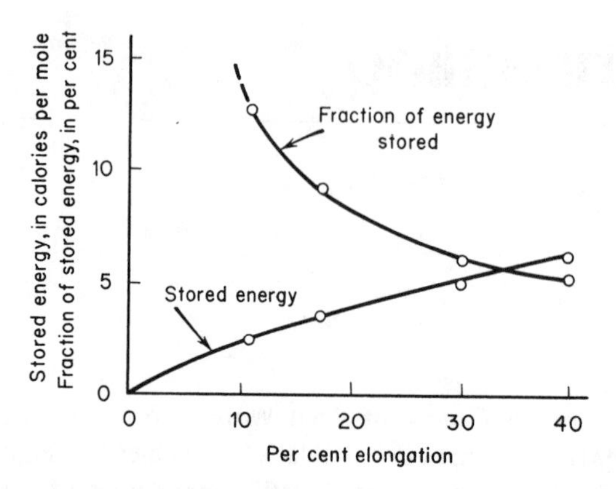

Fig. 7.1 Stored energy of cold work and fraction of the total work of deformation remaining as stored energy for high purity copper plotted as functions of tensile elongation. (From data of Gordon, P., *Trans. AIME*, **203** 1043 [1955].)

Let us consider the nature of the stored energy of plastic deformation. Cold working is known to increase greatly the number of dislocations in a metal. A soft annealed metal can have dislocation densities of the order of 10^6 to 10^8 cm^{-2}, and heavily cold worked metals can have approximately 10^{12}. Accordingly, cold working is able to increase the number of dislocations in a metal by a factor as large as 10,000 to 1,000,000. Since each dislocation represents a crystal defect with an associated lattice strain, increasing the dislocation density increases the strain energy of the metal.

The creation of point defects during plastic deformation is also recognized as a source of retained energy in cold worked metals. One mechanism for the creation of point defects in a crystal has already been described. In Chapter 5 it was mentioned that a screw dislocation that cuts another screw dislocation may be capable of generating a close-packed row of either vacancies or interstitials as it glides, with the type of point defect produced depending on the relative Burgers vectors of the intersecting dislocations. Since the strain energy associated with a vacancy is much smaller than that associated with an interstitial atom, it can be assumed that vacancies will be formed in greater numbers than interstitial atoms during plastic deformation.

7.2 The Relationship of Free Energy to Strain Energy. The free energy of a deformed metal is greater than that of an annealed metal by an amount approximately equal to the stored strain energy. While plastic deformation certainly increases the entropy of a metal, the effect is small compared to the increase in internal energy (the retained strain energy). The term $-TS$ in the free-energy equation may, therefore, be neglected and the free-energy increase equated directly to the stored energy. Therefore

$$F = E - TS$$

becomes

$$F = E$$

where F is the free energy associated with the cold work, E is the internal, or stored strain energy, S is the entropy increase due to the cold work, and T is the absolute temperature.

Since the free energy of cold worked metals is greater than that of annealed metals, they may soften spontaneously. A metal does not usually return to the annealed condition by a single simple reaction because of the complexity of the cold worked state. A number of different reactions occur, the total effect of which is the regaining of a condition equivalent to that possessed by the metal before it was cold worked. Many of these reactions involve some form of atom, or vacancy, movement and are, therefore extremely temperature sensitive. The reaction rates of these reactions may usually be expressed as simple exponential laws similar to that previously written for vacancy movement. Heating a deformed metal, therefore, greatly speeds up its return to the softened state.

7.3 The Release of Stored Energy. Valuable information about the nature of the reactions that occur as a cold worked metal returns to its original state may be obtained through a study of the release of its stored energy. There are several basically different methods of accomplishing this. Two of the more important will now be briefly indicated. In the first, the *anisothermal anneal* method, the cold worked metal is heated continuously from a lower to a higher temperature and the energy release is determined as a function of temperature. One form of the anisothermal anneal measures the difference in the power required to heat two similar specimens at the same rate. One specimen of the two is cold worked before the heating cycle, while the other serves as a standard and is not deformed. During the heating cycle, the cold worked specimen undergoes reactions that release heat and lower the power required to heat it in comparison with that required to heat the standard specimen. Measurements of the difference in power give direct evidence of the rate at which heat is released in the cold worked specimen. Figure 7.2 shows a typical anisothermal anneal curve[3] for a commercially pure copper (99.97 percent copper). It is noteworthy that some heat is released at temperatures only

3. Clarebrough, H. M., Hargreaves, M. E., and West, G. W., *Proc. Roy. Soc.* (London), A232 252 (1955).

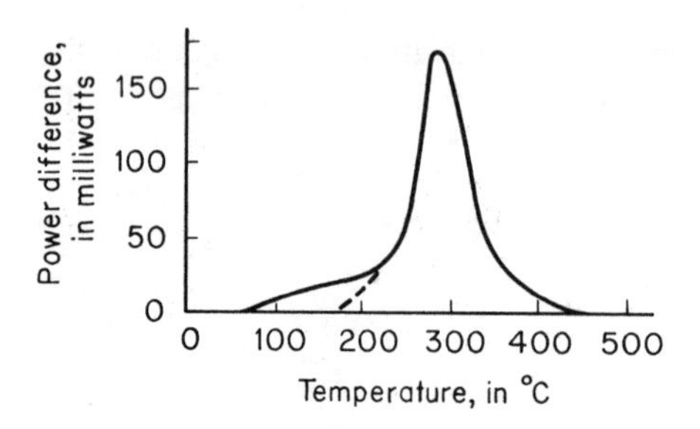

Fig. 7.2 Anisothermal anneal curve. Electrolytic copper. (From Clarebrough, H. M., Hargreaves, M. E., and West, G. W., *Proc. Roy. Soc.*, London, **232A**, 252 [1955].)

slightly above room temperature. The significance of this, and of the pronounced maximum that appears in the curve, will be discussed presently.

The other method of studying energy release involves *isothermal annealing*. Here the freed energy is measured while the specimen is maintained at a constant temperature. Figure 7.3 is representative of the type of curves obtained in an isothermal anneal. The data[4] for this particular curve were obtained with the aid of a microcalorimeter with a sensitivity capable of measuring a heat flow as low as 0.003 cal per hr.

Both the anisothermal anneal and the isothermal anneal curves of Figs. 7.2 and 7.3 show maxima corresponding to large energy releases. Metallographic specimens prepared from samples annealed by either method show that an interesting phenomenon occurs in the region of maximum energy release. These large energy releases appear simultaneously with the growth of an entirely new set of essentially strain-free crystals which grow at the expense of the original badly deformed crystals. The process by which this occurs is called *recrystalliza-*

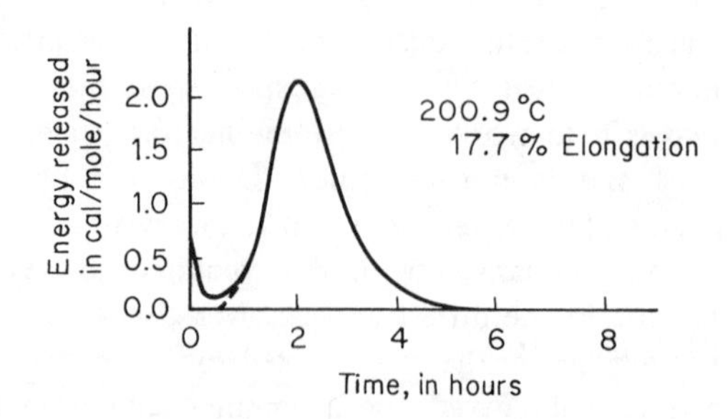

Fig. 7.3 Isothermal anneal curve. High purity copper. (From Gordon, P., *Trans. AIME*, **203** 1043 [1955].)

4. Gordon, P., *Trans. AIME*, 203 1043 (1955).

tion and may be understood as a realignment of the atoms into crystals with a lower free energy.

While the major energy release of the curves of Fig. 7.2 and 7.3 correspond to recrystallization, both curves show that energy is released before recrystallization. In this regard, dashed lines have been drawn schematically on both curves in order to delineate the recrystallization portion of the energy release. The area under each solid curve that lies to the left and above the dashed lines represents an energy release not associated with recrystallization. In the anisothermal anneal curve, this freeing of strain energy starts at temperatures well below those at which recrystallization starts. Similarly, in the isothermal anneal curve it begins at the start of the annealing cycle and is nearly completed before recrystallization starts. The part of the annealing cycle that occurs before recrystallization is called *recovery*. It should be recognized, however, that the reactions that occur during the recovery stages are able to continue during the progress of recrystallization; not in those regions which have already recrystallized, but in those that have not yet been converted into new crystals. Recovery reactions will be described in the next section. First, however, it is necessary to define the third stage of annealing – *grain growth*. Grain growth occurs when annealing is continued after recrystallization has been completed. In grain growth, certain of the recrystallized grains continue to grow in size, but only at the expense of other crystals which must, accordingly, disappear.

The three stages of annealing – recovery, recrystallization, and grain growth – have now been defined. Some of the important aspects of each will now be considered.

7.4 Recovery. When a metal is cold worked, changes occur in almost all of its physical and mechanical properties. Working increases strength, hardness, and electrical resistance, and it decreases ductility. Furthermore, when plastically deformed metal is studied using X-ray diffraction techniques, the X-ray reflections become characteristic of the cold worked state. Laue patterns of deformed single crystals show pronounced asterism corresponding to lattice curvatures. Similarly, Debye-Scherrer photographs of deformed polycrystalline metal exhibit diffraction lines that are not sharp, but broadened, in agreement with the complicated nature of the residual stresses and deformations that remain in a polycrystalline metal following cold working.

In the recovery stage of annealing, the physical and mechanical properties that suffered changes as a result of cold working tend to recover their original values. For many years, the fact that hardness and other properties could be altered without an apparent change in the microstructure, as signified by recrystallization, was regarded as a mystery. That the various physical and mechanical properties do not recover their values at the same rate in indicative of the complicated nature of the recovery process. Figure 7.4 shows another

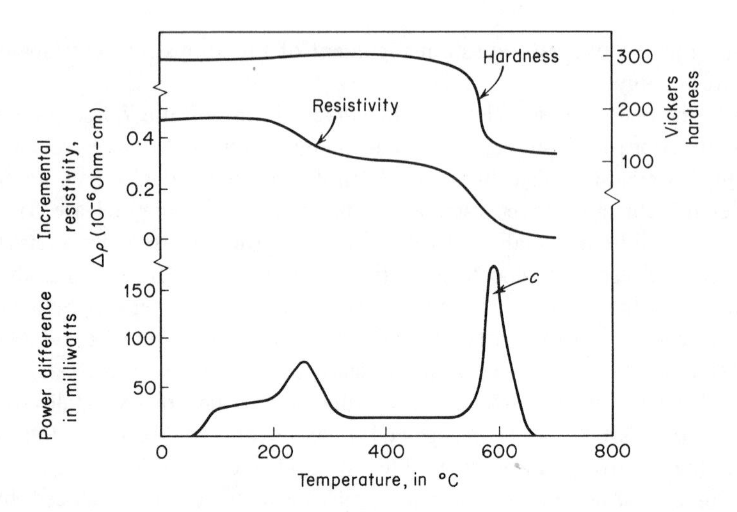

Fig. 7.4 Anisothermal-anneal curve for cold worked nickel. At the top of the figure curves are also drawn to show the effect of annealing temperature on the hardness and incremental resistivity of the metal. (From the work of Clarebrough, H. M., Hargreaves, M. E., and West, G. W., *Proc. Roy. Soc.*, London, **232A** 252 [1955].)

anisothermal anneal curve corresponding to the energy released on heating cold worked polycrystalline nickel. The peak at point *c* defines the region of recrystallization. The fraction of the energy released during recovery in this metal is much larger than that in the example of Fig. 7.2. Plotted on the same diagram are curves indicating the change in electrical resistivity and hardness as a function of the annealing temperature. Notice that the resistivity is almost completely recovered before the state of recrystallization. On the other hand, the major change in the hardness occurs simultaneously with recrystallization of the matrix.

7.5 Recovery in Single Crystals. The complexity of the cold worked state is directly related to the complexity of the deformation that produces it. Thus, the lattice distortions are simpler in a single crystal deformed by easy glide than in a single crystal deformed by multiple glide (simultaneous slip on several systems), and lattice distortions may be still more severe in a polycrystalline metal.

If a single crystal is deformed by easy glide (slip on a single plane) in a manner that does not bend the lattice, it is quite possible to completely recover its hardness without recrystallization of the specimen. In fact, it is generally impossible to recrystallize a crystal deformed only by easy glide, even if it is heated to temperatures as high as the melting point. Figure 7.5 shows

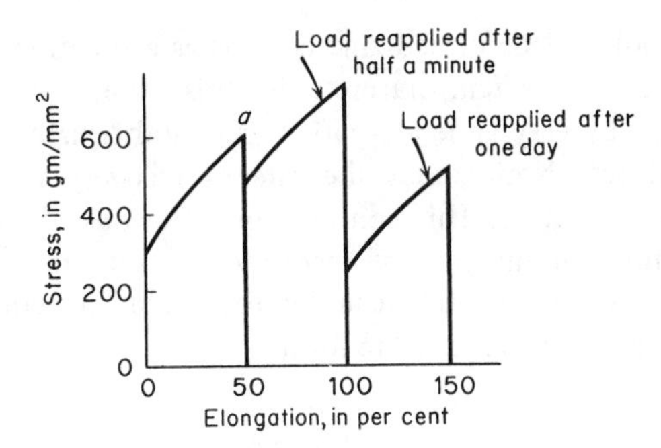

Fig. 7.5 Recovery of the yield strength of a zinc single crystal at room temperature. (After Schmid, E., and Boas, W., *Kristallplastizität,* Julius Springer, Berlin, 1935.)

schematically a stress-strain curve for a zinc single crystal strained in tension at room temperature where it deforms by basal slip.

Let us suppose that the crystal was originally loaded to point *a* and then the load was removed. If the load is reapplied after only a short rest period (about half a minute), it will not yield plastically until the stress almost reaches the value attained just before the load was removed on the first cycle. There is, however, a definite decrease in the stress at which the crystal begins to flow the second time. This *flow stress* would only equal that attained at the end of the previous loading cycle if it could be unloaded and reloaded without loss of time. Recovery of the yield point thus begins very rapidly. The yield point can be completely recovered in a zinc crystal at room temperature in a period of a day. This is shown by the third loading cycle in Fig. 7.5. These stress-strain diagrams indicate a well-known fact: the rate at which a property recovers isothermally is a decreasing function of the time. Figure 7.6 illustrates this effect graphically by

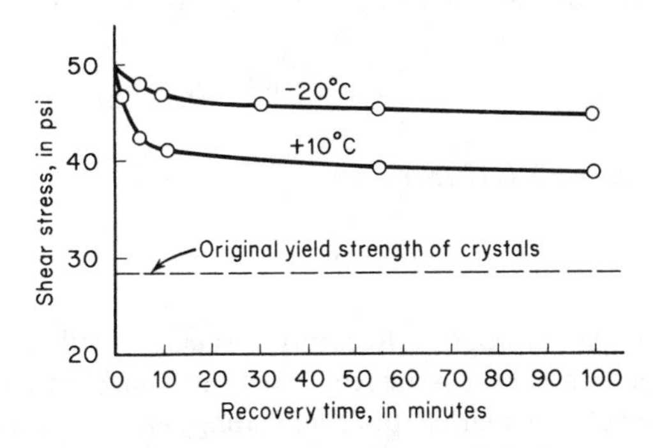

Fig. 7.6 Recovery of the yield strength of zinc single crystals at two different temperatures. (From the data of Drouard, R., Washburn, J., and Parker, E. R., *Trans. AIME,* **197** 1226 [1953].)

plotting the yield point of deformed zinc crystals as a function of the time for several different recovery temperatures. In this case,[5] the crystals were plastically deformed by easy glide at – 50° C and isothermally annealed at the indicated temperatures. Notice that the rate of recovery is much faster at + 10°C than it is at – 20°C. This confirms our previous observations of the effect in temperature on the rate of recovery. In fact, the recovery in zinc crystals deformed by simple glide can be expressed in terms of a simple activation, or Arrhenius-type law, of the form

$$\frac{1}{\tau} = Ae^{-Q/RT}$$

where τ is the time required to recover a given fraction of the total yield point recovery: Q the activation energy; R the universal gas constant; T the absolute temperature; A a constant.

Let us now assume that the reaction has proceeded to the same extent at two different temperatures. Then we have

$$Ae^{-Q/RT_1}\tau_1 = Ae^{-Q/RT_2}\tau_2$$

which is equivalent to

$$\frac{\tau_1}{\tau_2} = \frac{e^{-Q/RT_2}}{e^{-Q/RT_1}} = e^{-\frac{Q}{R}\left(\frac{1}{T_2} - \frac{1}{T_1}\right)}$$

According to Drouard, Washburn, and Parker, the activation energy Q for recovery of the yield point in zinc is 20,000 cal per mole. Thus, if a strained zinc crystal recovers one-fourth of its original yield point in 5min at 0°C (273°K), we would expect that the same amount of recovery at 27°C (300°K) would take

$$\tau_1 = 5e^{-\frac{20{,}000}{2}\left(\frac{1}{273} - \frac{1}{300}\right)} = 0.275 \text{ min}$$

On the other hand, at –50°C (223°K)

$$\tau_1 = 25{,}000 \text{ min or } 17 \text{ days}$$

7.6 Polygonization. Recovery associated with a simple form of plastic deformation has been considered in the preceding section. Recovery in this case is probably a matter of annihilating excess dislocations. Such annihilation can occur by the coming together of dislocation segments of

5. Drouard, R., Washburn, J., and Parker, E. R., *Trans. AIME*, 197 1226 (1953).

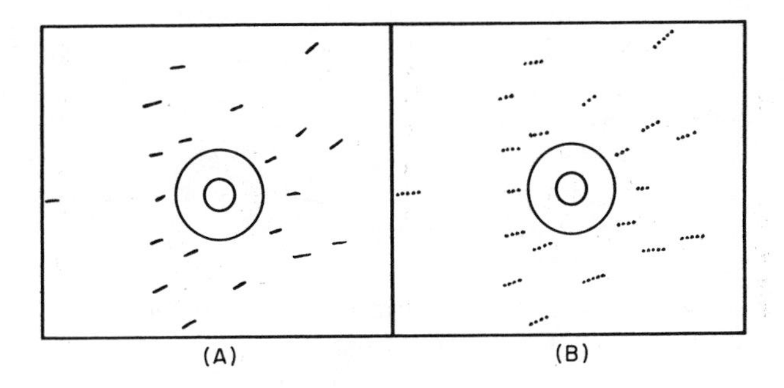

Fig. 7.7 Schematic Laue patterns showing how polygonization breaks up asterated X-ray reflections into a series of discrete spots. The diagram on the left corresponds to reflections from a bent single crystal; that on the right corresponds to the same crystal after an anneal that has polygonized the crystal.

opposite sign (that is, negative edges with positive edges and left-hand screws with right-hand screws). In this process it is probable that both slip and climb mechanisms are involved.

Another recovery process is called *polygonization.* In its simplest form it is associated with crystals that have been plastically bent. Because the X-ray beam is reflected from curved planes, bent crystals give Laue photographs with elongated, or asterated, spots. (See Chapter 2.) Many workers[6] have shown that the Laue spots of bent crystals assume a fine structure after a recovery anneal (an anneal which does not recrystallize the specimen). This is shown schematically in Fig. 7.7 where the left-hand figure represents the Laue pattern of a bent crystal before annealing, and the right-hand figure the pattern after annealing. Each of the elongated, or asterated spots of the deformed crystal is replaced by a set of tiny sharp reflections in the annealed crystal.

In a Laue photograph, each spot corresponds to the reflection from a specific lattice plane. When a single crystal is exposed to the X-ray beam of a Laue camera, a finite number of spots is obtained with a pattern on the film characteristic of the crystal and its orientation. If a Laue beam straddles the boundary between two crystals, a double pattern of spots is observed, each characteristic of the orientation of its respective crystal. Furthermore, if the two crystals have nearly identical orientations, as is the case when the boundary is a low angle boundary, the two patterns will almost coincide, and the photograph will show a set of closely spaced double spots. Finally, if the X-ray beam falls on a number of very small crystalline areas, each separated from its neighbors by low-angle boundaries, a pattern such as that in Fig. 7.7B can be expected. Evidently, when a bent crystal is annealed, the curved crystal breaks up into a

6. Cahn, R. W., *Prog. in Metal Phys.,* 2 151 (1950).

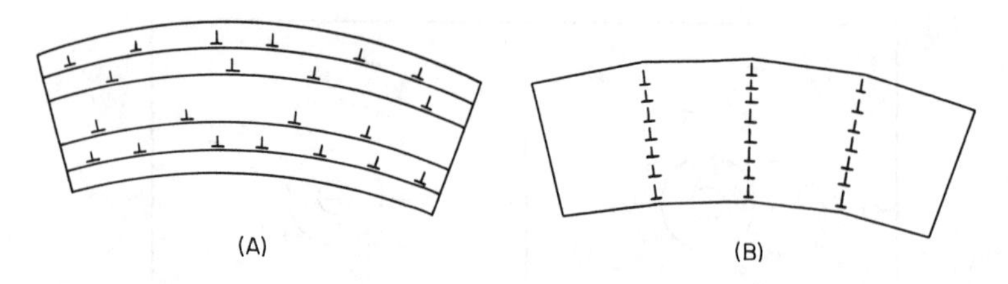

Fig. 7.8 Realignment of edge dislocations during polygonization (A) The excess dislocations that remain on active slip planes after a crystal is bent. (B) The rearrangement of the dislocations after polygonization.

number of closely related small perfect crystal segments. This process has been given the name *polygonization.*[7]

The polygonization phenomenon can also be explained with drawings such as those of Fig. 7.8. The left-hand figure represents a portion of a plastically bent crystal. For simplicity, the active slip plane has been assumed parallel to the top and the bottom surface of the crystal. A plastically bent crystal must contain an excess of positive edge dislocations that lie along active slip planes in the manner suggested by the figure. The dislocation configuration of Fig. 7.8A is one of high strain energy. A different arrangement of the same dislocations possessing a lower strain energy is shown in Fig. 7.8B. Here the excess edge dislocations are found in arrays that run in a direction normal to the slip planes. Configurations of this

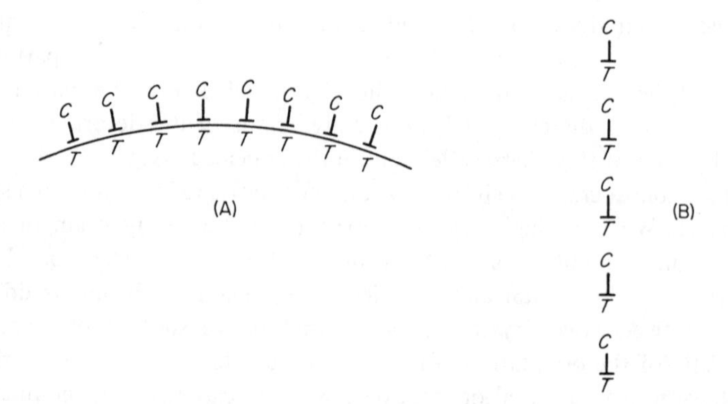

Fig. 7.9 Simple explanation showing why a vertical array of edge dislocations corresponds to a lower state of strain energy than does an array of the same dislocations, all on a single slip plane. The letters T and C in (A) and (B) correspond to the tensile and compressive strains associated with each dislocation.

7. Orowan, E., *Communication to the Congres de la Société Française de la Metallurgie d'Octobre*, 1947.

nature constitute low-angle grain boundaries. (See Chapter 5.) When edge dislocations of the same sign accumulate on the same slip plane, their strain fields are additive, as indicated in Fig. 7.9A where the local nature of the strain field of each dislocation is suggested by the appropriate letter – *C* for compression and *T* for tension. Obviously, the regions just above and below the slip planes in Fig. 7.8A are areas of intense tensile and compressive strain respectively. However, if the same dislocations are arranged in a vertical sequence (perpendicular to the slip plane), as shown in Fig. 7.8B, the strain fields of adjacent dislocations partly cancel each other, for the tensile strain in the region below the extra plane of one dislocation overlaps the compressive strain field of the next lower dislocation. This arrangement is shown schematically in Fig. 7.9B.

In addition to lowering the strain energy, the regrouping of edge dislocations into low-angle boundaries has a second important effect. This is the removal of general lattice curvature. As a result of polygonization, crystal segments lying between a pair of low-angle boundaries approach the state of strain-free crystals with flat uncurved planes. However, each crystallite possesses an orientation slightly different from its neighbors because of the low-angle boundaries that separate them from each other. When an X-ray beam strikes the surface of a polygonized crystal it falls on a number of small, relatively perfect crystals of slightly different orientation. The result is a Laue pattern of the type shown in Fig. 7.7B.

It is customary to call low-angle boundaries, such as develop in polygonization, *subboundaries*, and the crystals that they separate *subgrains*. The size, shape, and arrangement of the subgrains constitute the substructure of a metal. The difference between the concepts of grains and subgrains is an important one: subgrains lie inside grains.

7.7 Dislocation Movements in Polygonization. An edge dislocation is capable of moving either by slip on its slip plane or by climb in a direction perpendicular to its slip plane. Both are required in polygonization, as shown schematically in Fig. 7.10, where the indicated vertical movement of each dislocation represents climb, and the horizontal movement slip. The driving

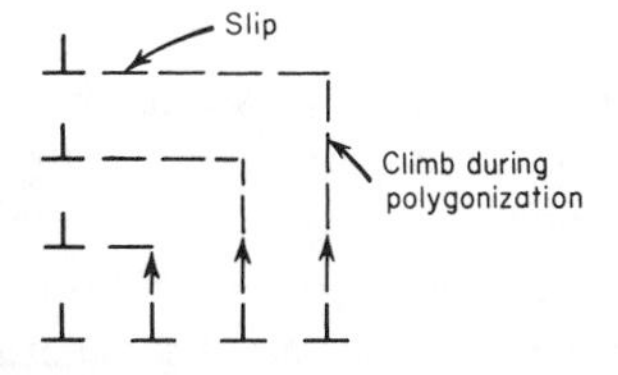

Fig. 7.10 Both climb and slip are involved in the rearrangement of edge dislocations.

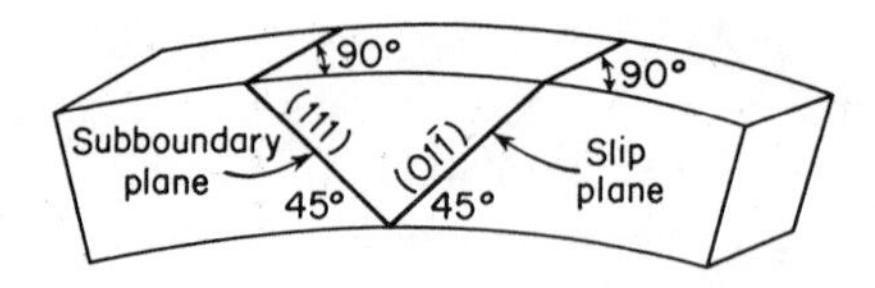

Fig. 7.11 Orientation of the iron-silicon crystal shown in the photographs of Fig. 7.12.

force for these movements comes from the strain energy of the dislocations, which decreases as a result of polygonization. From an equivalent viewpoint, we may say that the strain field of dislocations grouped on slip planes produces an effective force that makes them move into subboundaries. This force exists at all temperatures, but at low temperatures edge dislocations cannot climb. However, since dislocation climb depends on the movement of vacancies (an activated process), the rate of polygonization increases rapidly with temperature. Increasing temperature also aids the polygonization process in another manner, for the movement of dislocations by slip also becomes easier at high temperatures. This fact is observable in the fall of the critical resolved shear stress for slip with rising temperature.

The photographs of Fig. 7.12 are of special interest because they show the polygonization process in an actual metal (a silicon-iron alloy with 3.25 percent Si). The four photographs show the surface of crystals plastically bent to a fixed radius of curvature, and then annealed. Each specimen was given a 1-hr anneal at a different temperature in order to bring out the various stages of polygonization. The higher the temperature the more complete is the polygonization process. The plane of the photograph is perpendicular to the axis of bending and corresponds to the front face of the crystal, shown diagrammatically in Fig. 7.11. The surface in question is perpendicular to the slip plane $(01\bar{1})$ of this body-centered cubic metal, and also to the plane along which the subboundaries form (111). The orientations of both planes are shown in the figure and it is clear that they make 45° angles with the horizontal: the slip plane having a positive slope and the subboundary a negative slope. Figure 7.12A shows the effect of a 1-hr anneal at 700°C. Each black dot of the illustration is a pit developed by etching the specimen in a suitable etching reagent, and shows the intersection of a dislocation with the specimen surface. Notice that the dislocations are, for the most part, associated with the slip planes, although several well-defined subboundaries can be seen near the top of the photograph. Many small groups containing three or four dislocations aligned perpendicular to the slip planes can also be seen in all areas of the picture. Figure 7.12B shows a more advanced stage in the polygonization process corresponding to an anneal at a higher temperature (775°C). In this photograph, all the dislocations lie in the subboundaries, or polygon walls.

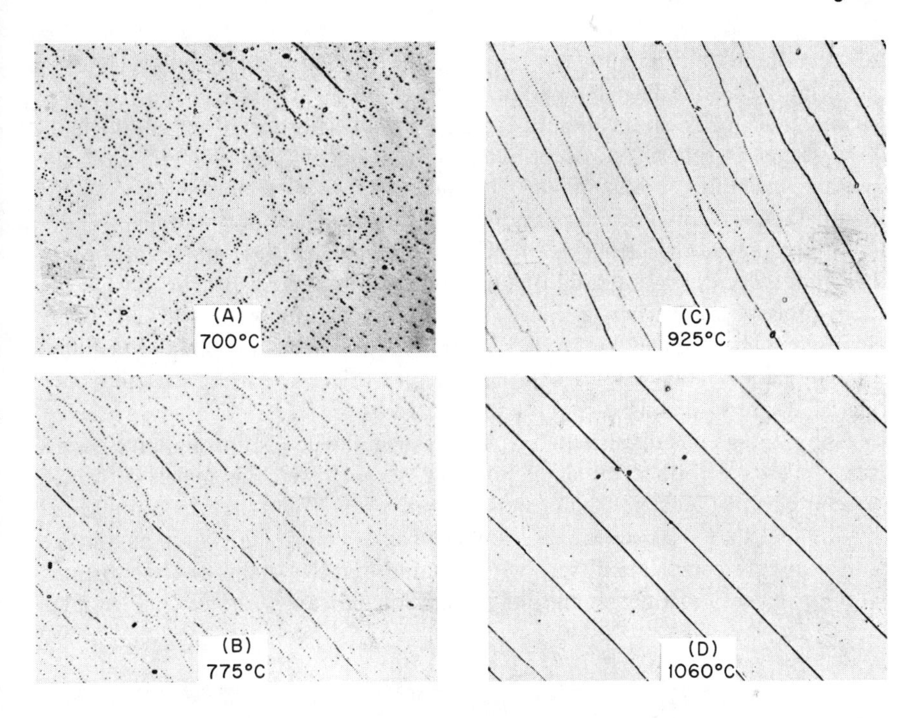

Fig. 7.12 Hibbard and Dunn's photographs showing the polygonization process in bent and annealed iron-silicon single crystals. All specimens annealed for one hour at the indicated temperatures. Note that polygonization is more complete the higher the temperature of the anneal. (From Hibbard, W. R., Jr., and Dunn, C. G., *Acta Met.,* **4** 306 [1956] and ASM Seminar, *Creep and Recovery,* 1957, p. 52.)

When all of the dislocations have dissociated themselves from the slip planes and have aligned themselves in low-angle boundaries, the polygonization process is not complete. The next stage is a coalescence of the low-angle boundaries where two or more subboundaries combine to form a single boundary. The angle of rotation of the subgrains across the boundary must, of course, grow in this process. Well-defined subboundaries formed by coalescence can be seen in Fig. 7.12C, where the surface corresponds to a crystal annealed for an hour at 925°C. Notice that, although this photograph and that of Fig. 7.12B were taken at the same magnification, the number of subboundaries is now much less. On the other hand, the density of dislocations in the boundaries is much higher, so that it is not generally possible to see individual dislocation etch pits.

The coalescence of subboundaries results from the fact that the strain energy associated with a combined boundary is less than that associated with two separated boundaries. The movement of subboundaries that occurs in coalescence is not difficult to understand, for the boundaries are arrays of edge dislocations which, in turn, are fully capable of movement by either climb or glide at high-annealing temperatures. In Fig. 7.12C, several junctions of subboundary pairs can be seen in the upper central portion of the photograph. Coalescence is believed to occur by the movement of these *y*-junctions. In the present example, movement of the junctions toward the bottom of the photograph will combine the pairs of branches into single polygon walls.

As the polygon walls become more widely separated, the rate of coalescence becomes a decreasing function of time and temperature so that the polygonization process approaches a more or less stable state with widely spaced, approximately parallel (in a single crystal deformed by simple bending) subboundaries. This state is shown in Fig. 7.12D and corresponds to a 1-hr anneal of a bent silicon-iron crystal at 1060°C.

The photographs discussed above have shown the polygonization process in a single crystal deformed by simple bending. In polycrystalline metals deformed by complex methods, polygonization may still occur. The process is complicated by the fact that slip occurs on a number of intersecting slip planes, and lattice curvatures are complex and vary with position in the crystal. The effect of such a complex deformation on the polygonization process is shown in Fig. 7.13,

Fig. 7.13. Complex polygonized structure in a silicon-iron single crystal that was deformed 8 percent by cold rolling before it was annealed 1 hr at 1100°C. (Hibbard, W. R., Jr., and Dunn, C. G., ASM Seminar, *Creep and Recovery*, 1957, p. 52.)

which reveals a substructure resulting from deforming a single crystal in a small rolling mill after being annealed at a very high temperature (1100°C). The boundaries in this photograph are subboundaries and are not grain boundaries. No recrystallization is involved in this highly polygonized single crystal. Substructures such as this can be expected in cold worked polycrystalline metals that have been annealed at temperatures high enough to cause polygonization, but not high enough, or long enough, to cause recrystallization.

A typical substructure, as revealed by the transmission electron microscope, is shown in Fig. 7.14. This iron plus 0.8 percent copper specimen was originally cold-rolled 60 percent and then annealed at elevated temperatures. Rolling tends to form subboundaries that are nearly parallel to the rolling plane. This specimen was obtained by sectioning the rolled plate in a direction normal to the rolling plane and, as a result, many of the subboundaries are nearly normal to the plane of the photograph and are therefore sharply defined. A substructure viewed normal to the rolling plane is not so easily resolved because the subboundaries tend to lie parallel to the transmission electron microscope foil. The range of misorientation in these subgrains is of the order of ± 10° and the subboundaries

Fig. 7.14 An electron micrograph of a polygonized iron specimen. For details, see text. (Leslie, W. C., Michalak, J. T., and Aul, F. W., AIME Conf, Series Vol., *Iron and Its Dilute Solid Solutions,* p. 161. Interscience Publishers, New York, 1963.) Photograph courtesy of J. T. Michalak and the United States Steel Corporation.

are therefore no longer simple low angle boundaries. An important feature of this photograph is that the dislocations resulting from this large strain (60 percent) have almost completely entered the cell walls. As a result, most subgrains have interiors that are basically free of dislocations. Where groups of dislocations are visible, it is probable that one is looking through a part of a cell wall.

7.8 Recovery Processes at High and Low Temperatures. Polygonization is too complicated a process to be expressed in terms of a simple rate equation, such as that used to describe the recovery process after easy glide. Because polygonization involves dislocation climb, relatively high temperatures are required for rapid polygonization. In deformed polycrystalline metals, high-temperature recovery is considered to be essentially a matter of polygonization and annihilation of dislocations.

At lower temperatures other processes are of greater importance. At these temperatures, current theories picture the recovery process as primarily a matter of reducing the number of point defects to their equilibrium value. The most important point defect is a vacancy which may have a finite mobility even at relatively low temperatures.

7.9 Dynamic Recovery. As may be seen in Figs. 7.13 and 7.14, the basic effect of high temperature recovery is the movement of the dislocations resulting from plastic deformation into subgrain or cell boundaries. In many cases, this process can actually start during plastic deformation. When this happens, the metal is said to undergo *dynamic recovery*. The tendency for dislocations to form a cell structure is quite strong in many pure metals and exists even to very low temperatures, as can be seen in Fig. 7.15. This illustration shows two transmission electron microscope foils taken from nickel specimens deformed at the temperature of liquid nitrogen (77°K). Note that for small strains (9 percent) the dislocation arrangement tends to be roughly uniform. However, when the strain becomes large (26 percent) a definite trend toward a cell structure becomes clearly evident. The size of the cells in this structure is small.

At more elevated temperatures, the effects of dynamic recovery naturally become stronger because the mobility of the dislocations increases with increasing temperature. As a result, the cells tend to form at smaller strains, the

---

Fig. 7.15 These transmission electron micrographs illustrate dynamic recovery in nickel deformed at 77°K. Note that even at this low temperature there is a definite tendency to form a cell structure that increases at high strains. (Longo, W. P., *Work Softening in Polycrystalline Metals*, University of Florida, Thesis, 1970.)

(A) 9 percent strain

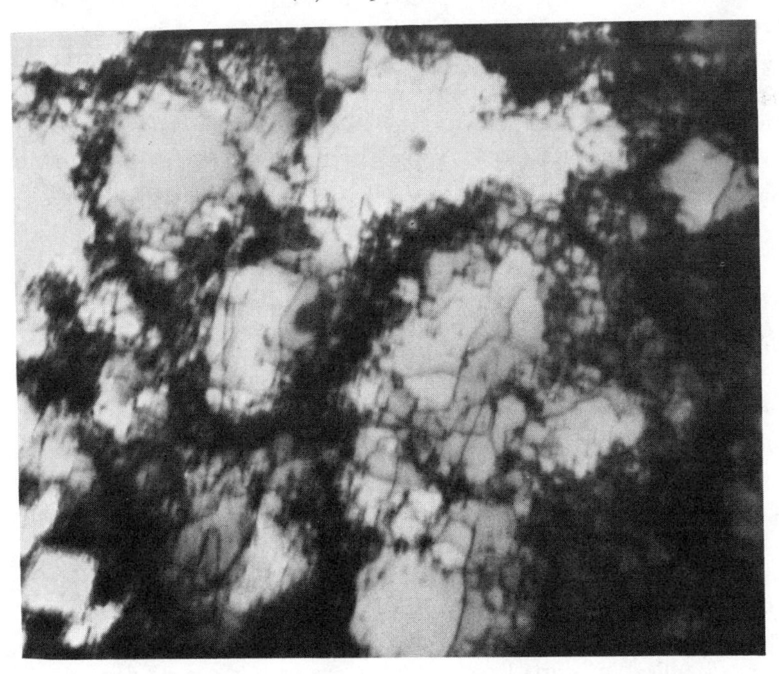

(B) 26 percent strain

cell walls become thinner and much more sharply defined, and the cell size becomes larger. Dynamic recovery is therefore often a strong factor in the deformation of metals under hot working conditions.

Dynamic recovery has a strong effect on the shape of the stress-strain curve. This is because the movement of dislocations from their slip planes into walls lowers the average strain energy associated with the dislocations. The effect of this is to make it less difficult to nucleate the additional dislocations that are needed to further strain the material. Dynamic recovery thus tends to lower the effective rate of work-hardening.

The role of dynamic recovery is not the same in all metals. Dynamic recovery occurs most strongly in metals of high stacking-fault energies and is not readily observed in metals of very low stacking-fault energy. These latter are normally alloys such as brass (copper plus zinc). The dislocation structure resulting from cold work in these latter materials often shows the dislocations still aligned along their slip planes. An example is shown in Fig. 7.16. In this case the alloy is one of nickel and aluminum.

The correspondence between the ability of a metal to undergo dynamic recovery and the magnitude of its stacking-fault energy strongly suggests that the primary mechanism involved in dynamic recovery is thermally activated cross-slip. This mechanism is considered in Chapter 20. At present, it is important to note the basic underlying difference between dynamic recovery and static recovery, such as occurs in annealing after cold work. In static recovery the movement of the dislocations into the cell walls occurs as a result of the interaction stresses between the dislocations themselves. In dynamic recovery, the applied stress causing the deformation is added to the stresses acting between the dislocations. As a result, dynamic recovery effects may be observed at very low temperatures, and at these temperatures the applied stresses can be very large.

7.10 Recrystallization. Recovery and recrystallization are two basically different phenomena. In an isothermal anneal, the rate at which a recovery process occurs always decreases with time; that is, it starts rapidly and proceeds at a slower and slower rate as the driving force for the reaction is expended. On the other hand, the kinetics of recrystallization are quite different, for it occurs by a nucleation and growth process. In agreement with other processes of this type, recrystallization during an isothermal anneal begins very slowly, and builds up to a maximum reaction rate, after which it finishes slowly. This difference between the isothermal behavior of recovery and recrystallization is clearly evident in Fig. 7.3, where recovery processes start at the beginning of the annealing cycle and account for the initial energy release, while recrystallization starts later and accounts for the second (larger) energy release.

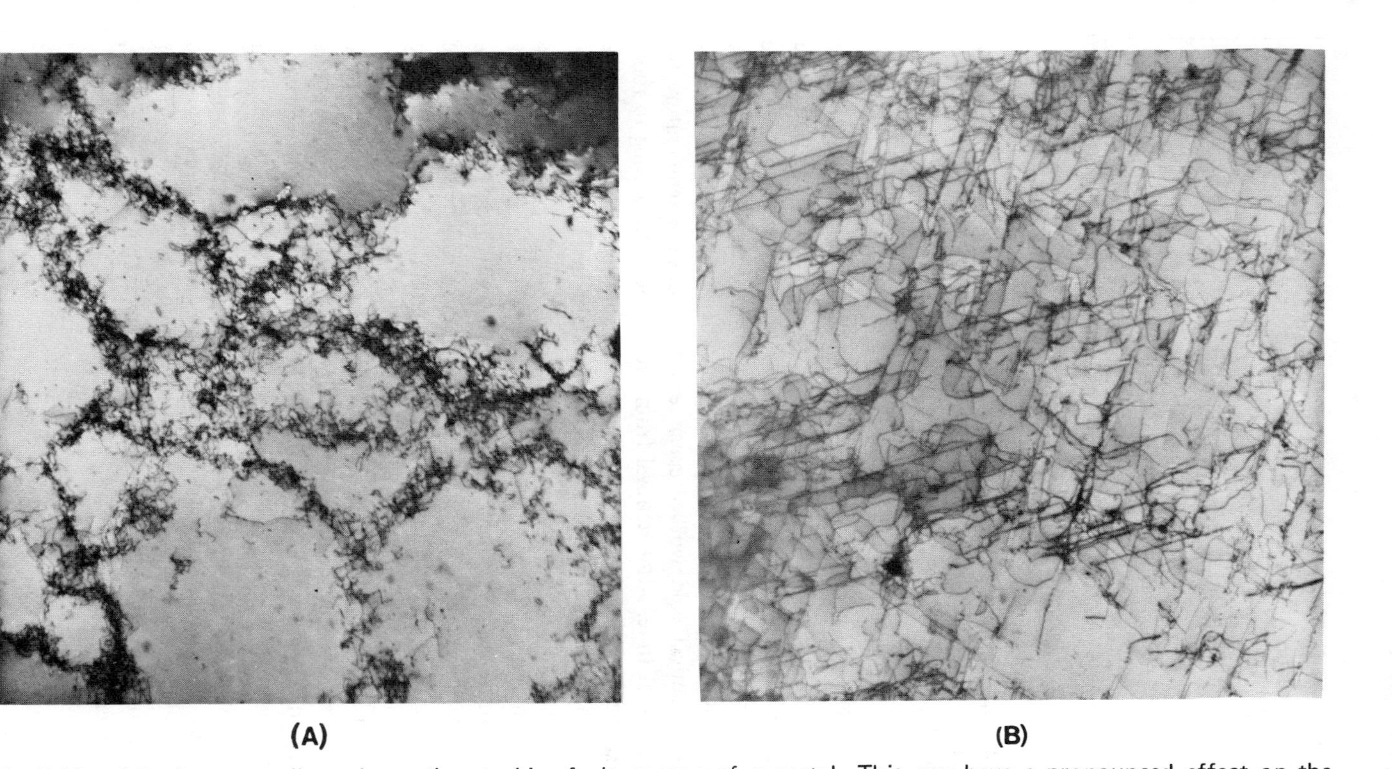

(A) (B)

Fig. 7.16 Alloying normally reduces the stacking-fault energy of a metal. This can have a pronounced effect on the dislocation structure, as can be seen in these two electron micrographs. (A) Pure nickel strained 3.1 percent at 293°K. Magnification: 25,000x. (B) Nickel – 5.5 wt. percent aluminum alloy strained 2.7 percent at 293°K. Magnification: 37,500x. (Photographs courtesy of J. O. Stiegler, Oak Ridge National Laboratories, Oak Ridge, Tenn.)

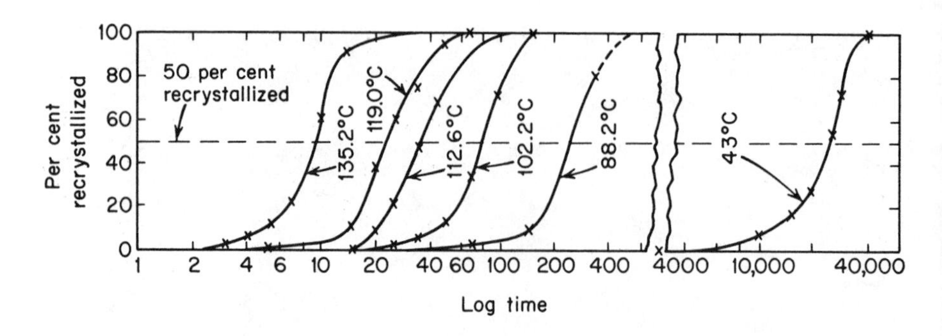

Fig. 7.17 Isothermal transformation (recrystallization) curves for pure copper (99.999 percent Cu) cold-rolled 98 percent. (From Decker, B. F., and Harker, D., *Trans. AIME,* **188** 887 [1950].)

7.11 The Effect of Time and Temperature on Recrystallization. One way to study the recrystallization process is to plot isothermal recrystallization curves of the type shown in Fig. 7.17. Each curve represents the data for a given temperature and shows the amount of recrystallization as a function of time. Data for each curve of this type are obtained by holding a number of identical cold worked specimens at a constant temperature tor different lengths of time. After removal from the furnace and cooling to room temperature, each specimen is examined metallographically to determine the extent of recrystallization. This quantity (in percent) is then plotted against the logarithm of the time. The effect of increasing the annealing temperature is shown clearly in Fig. 7.17. The higher the temperature, the shorter the time needed to finish the recrystallization. The S-shaped curves of Fig. 7.17 are typical of nucleation and growth processes. A similar curve will be shown tor the nucleation and growth reaction involved in the precipitation of a second phase from a supersaturated solid solution. (See Fig. 9.4 in Chapter 9.)

A horizontal line drawn through the curves of Fig. 7.17 corresponds to a constant fraction of recrystallization. Let us arbitrarily draw such a line corresponding to 50 percent recrystallization. The intersection of this line with each of the isothermal recrystallization curves gives the time at a given temperature required to recrystallize half of the structure. Let us designate this time interval by the symbol τ. Figure 7.18 shows τ plotted as a function of the reciprocal of the absolute temperature. The curve of Fig. 7.18 is a straight line, which shows that the recrystallization data for this metal (pure copper) can be expressed as an empirical equation of the form.

$$\frac{1}{T} = K \log_{10} \frac{1}{\tau} + C$$

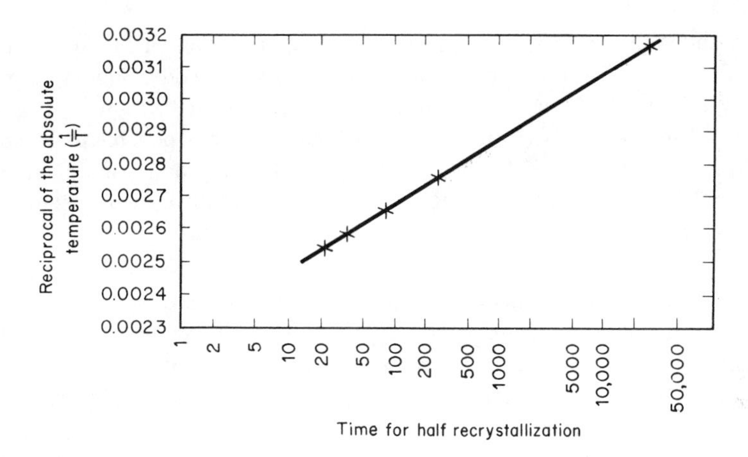

Fig. 7.18 Reciprocal of absolute temperature (°K) vs. time for half recrystallization of pure copper. (From Decker, B. F., and Harker, D., *Trans. AIME,* **188** 887 [1950].)

where K (the slope of the curve) and C (the intercept of the curve with the ordinate axis) are both constants. This last equation can also be expressed in the form

$$\frac{1}{\tau} = Ae^{-Q_r/RT}$$

where $1/\tau$ is the rate at which 50 percent of the structure is recrystallized, R is the gas constant (2 cal/mole – °K), and Q_r is called the activation energy for recrystallization.

It is necessary to point out the difference between the activation energy Q_r for recrystallization and the previously discussed activation energy for the motion of vacancies. The latter quantity can be directly related to a simple physical property: the height of the energy barrier that an atom must cross to jump into a vacancy. In the present case, the physical significance of Q_r is not completely understood. There is good reason to believe that more than one process is involved in recrystallization, so that Q_r cannot be related to a single simple process. It is best, therefore, to consider the recrystallization activation energy as an empirical constant.

The quantity τ in the above equation is not restricted to representing the time to recrystallize half of the matrix. It may represent the time to complete any fraction of the recrystallization process, such as the time to start the formation of new grains (several percent), or the time to complete the process (100 percent).

The above rate equation is exactly equivalent to the empirical equation that holds for the recovery of zinc single crystals after easy glide. Similar empirical rate equations have been found to describe the recrystallization process of a number of metals besides pure copper. Although it is not possible to generalize this equation and state that it accurately applies to recrystallization phenomena in all metals, we can consider that it is roughly descriptive of the relationship between time and temperature in recrystallization.

7.12 Recrystallization Temperature. A frequently used metallurgical term is *recrystallization temperature*. This is the temperature at which a particular metal with a particular amount of cold deformation will completely recrystallize in a finite period of time, usually 1 hr. Of course, in light of the above rate equation, the recrystallization temperature has no meaning unless the time allowed for recrystallization is also specified. However, because of the large activation energies encountered, recrystallization actually appears to occur at some definite minimum temperature. Suppose for a given metal Q_r = 50,000 cal per mole, and that recrystallization is completed in 1 hr. at 600°K (323°C). Then it may be shown, with the aid of the rate equation, that if the annealing is carried out at a 10°C lower temperature (313°C), complete recrystallization will require slightly over 2 hr. A specimen of this particular metal will only be partly recrystallized at the end of an anneal of 1 hr at 10°C below its recrystallization temperature (323°C). On the other hand, an hour's anneal is more than enough to recrystallize the metal at any temperature above 323°C. In fact, a 10°C rise in temperature to 333°C shortens the recrystallization time to half an hour, and a 20°C rise to approximately 15 min. To the practical man, this sensitivity of the recrystallization process to small changes in temperature makes it appear as though the metal has a fixed temperature, below which it will not recrystallize, and for this reason, there is a tendency to regard the recrystallization temperature as a property of the metal and to neglect the time factor in recrystallization.

7.13 The Effect of Strain on Recrystallization. Two rate-of-crystallization curves similar to that of Fig. 7.18 are plotted in Fig. 7.19. These differ only in that they represent the recrystallization of zirconium instead of copper, and the rate at which complete recrystallization is observed instead of half-recrystallization. The data are plotted in the usual sense (log $1/\tau$ vs · $1/T$), but for convenient reading, the abscissae and ordinates are expressed in degrees centigrade and hours respectively. The two curves represent data from specimens cold worked by different amounts. In both cases, the zirconium metal was cold worked by swaging. *Swaging* is a means of mechanical deformation used on

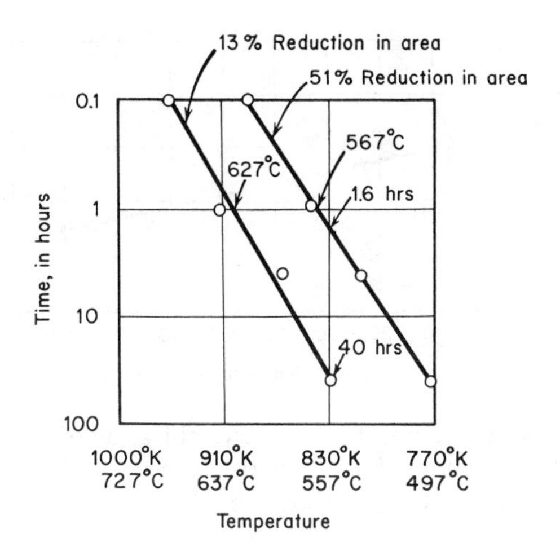

Fig. 7.19 Temperature-time relationships for recrystallization of zirconium (iodide) corresponding to two different amounts of prior cold work. (Treco, R. M., Proc., 1956, *AIME* Regional Conference on Reactive Metals, p. 136.)

cylindrical rods in which the diameter of the rod is reduced uniformly by a mechanical hammer equipped with rotating dies. The amount of this cold work is measured in terms of the percentage reduction in area of the cylindrical cross-section. The left-hand curve of Fig. 7.19 corresponds to specimens with a cross-section area reduced 13 percent, while the right-hand curve represents specimens that suffered a larger reduction in area (51 percent). The two curves show clearly that recrystallization is promoted by increasing amounts of cold work. When annealed at the same temperature, the metal with the larger amount of cold work recrystallizes much faster than that with the lesser reduction. As an example, at 553°C the times for completion of recrystallization are 1.6 and 40 hr for the larger and the smaller reductions in area respectively. Similarly, the temperature at which the metal recrystallizes completely within an hour is lower for a greater amount of cold work – 567°C as compared to 627°C.

A close examination of the two curves of Fig. 7.19 shows that these straight lines do not have the same slope. This means that the temperature dependence of recrystallization varies with the amount of cold work, or that the activation energy for recrystallization is a function of the amount of deformation. This fact is further emphasized by the data of Fig. 7.20, which show the variation of the activation energy for this same metal (zirconium) as a function of the percent reduction in area over the range of deformation from less than 10 percent to greater than 90 percent. The fact that the activation energy Q_r varies with the

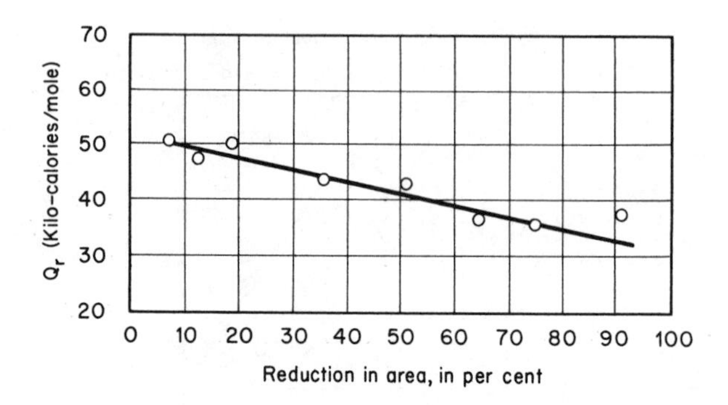

Fig. 7.20 Activation energy for the recrystallization of zirconium (iodide) as a function of the amount of cold work. (Treco, R. M., Proc., 1956, *AIME* Regional Conference on Reactive Metals, p. 136.)

amount of cold work lends additional confirmation to the previous statement about the complex nature of Q_r.

7.14 The Rate of Nucleation and the Rate of Nucleus Growth. The preceding sections have shown that the recrystallization reaction in many metals can be described by a simple activation equation having the form

$$\text{rate} = Ae^{-Q_r/RT}$$

Unfortunately, this empirical equation reveals very little about the atomic mechanisms that occur during recrystallization. This is because of the dual character of a nucleation and growth reaction. The rate at which a metal recrystallizes depends on the rate at which nuclei form, and also on the rate at which they grow. These two rates also determine the final grain size of a recrystallized metal. If nuclei form rapidly and grow slowly, many crystals will form before their mutual impingement completes the recrystallization process. In this case, the final grain size is small. On the other hand, it will be large if the rate of nucleation is small compared to the rate of growth. Since the kinetics of recrystallization can often be described in terms of these two rates, a number of investigators have measured these quantities under isothermal conditions in the hope of learning more about the mechanism of recrystallization. This requires the introduction of two parameters: N, the rate of nucleation and G, the rate of growth.

It is customary to define the nucleation frequency, N, as the number of nuclei that form per second in a cubic centimeter of unrecrystallized matrix. This parameter is referred to the unrecrystallized matrix because the recrystallized portion is inactive with regard to further nucleation. The linear rate of growth, G, is defined as the time rate of change of the diameter of a recrystallized grain. In practice, G is measured by annealing for different lengths of time a number of identical specimens at a chosen isothermal temperature. The diameter of the largest grain in each specimen is measured after the specimens are cooled to room temperature and prepared metallographically. The variation of this diameter with isothermal annealing time gives the rate of growth, G. The rate of nucleation can be determined from the same metallographic specimens by counting the number of grains per unit area on the surface of each. These surface-density measurements can then be used to give the number of recrystallized grains per unit volume. Of course, each determination must be corrected for the volume of the matrix that has recrystallized.

Several equations [8] have been derived starting with the parameters N and G which express the amount of recrystallization as a function of time. (See Fig. 7.17.) Because the theories on which these equations are based diverge to some extent, and because space is limited, they will not be discussed further. The concepts of nucleation rate and growth rate are useful, however, in explaining the effects of several other variables on the recrystallization process.

7.15 Formation of Nuclei. In recrystallization, an entirely new set of grains is formed. New crystals are nucleated at points of high-lattice-strain energy such as slip-line intersections, deformation twin intersections, and in areas close to grain boundaires. In each case, it appears that nucleation occurs at points of strong lattice curvature. In this regard it is interesting to note that bent, or twisted single crystals recrystallize more readily [9] than do similar crystals that have been bent, or twisted, and then unbent, or untwisted.

Because nuclei form in regions of severe localized deformation, the sites where they appear are apparently predetermined. Nuclei of this type are called *preformed nuclei.*[10] A number of models have been proposed to show how it is possible to form a small, strain-free volume that can grow out and consume the deformed matrix around it. These models are in general agreement on two points. First, a region of a crystal can become a nucleus and grow only if its size exceeds some minimum value. For example, Detert and Zieb [11] have computed that in a deformed metal with a dislocation density of 10^{12} cm^{-2}, a nucleus has to have a diameter greater than about 150 Å for it to be able to expand. (This

8. Avramic, M. (now M. A. Melvin), *Jour. Chem. Phys.,* 7 1103 (1939); *ibid.,* 8 212 (1940); *ibid.,* 9 177 (1941). Also Johnson, W. A., and Mehl, R. F., *Trans. AIME,* **135** 416 (1939).

9. Schmid, E., and Boas, W., *Crystal Plasticity*. F. A. Hughes and Co., London, 1950.

10. Cahn, R. W., *Recrystallization, Grain Growth and Texture*, ASM Seminar Series, pp. 99–128, American Society for Metals, Metals Park, Ohio, 1966.

11. Detert, K., and Zieb, J., *Trans. AIME*, **233** 51 (1965).

general concept of the critical size of a nucleus will be considered in section 9.5 of Chapter 9, and in greater detail in Chapter 13.)

The other condition for the formation of a nucleus is that it become surrounded, at least in part, by the equivalent of a high-angle grain boundary. This condition is required because the mobility of an arbitrary low-angle grain boundary is normally very low. Beyond these two points, the various models of the nucleation process vary to a considerable degree. It is possible that most of these mechanisms may operate and that the preferred one in a given situation will depend largely on the nature of the deformed specimen being recrystallized. In this regard, a single crystal lacks the sites along grain boundaries and along lines where three grains meet that are available for nucleation in a polycrystalline metal. Both grain boundaries and these triple lines are regions where high-angle boundaries already exist, so that one of the criterions for the formations of a nucleus is effectively satisfied. A typical mechanism applicable to polycrystals is that of Bailey and Hirsch,[12] who propose that if a difference in dislocation density exists across a grain boundary in a cold worked metal, then during annealing a portion of the more perfect grain might migrate into the less perfect grain under the driving force associated with the strain energy difference across the boundary. This would be accomplished by the forward movement of the grain boundary so as to form a bulge, as indicated in Fig. 7.21. This boundary movement should effectively sweep up the dislocations in its path, thereby creating a small, relatively strain-free volume of crystal. If this bulge exceeds the critical nucleus size, both primary conditions for the formation of a nucleus would be satisfied.

It is not possible to consider in detail all of the proposed nucleation models. We will discuss briefly two mechanisms that are probably more applicable to single crystals than to polycrystals. The first of these is that due to Cahn[13] and

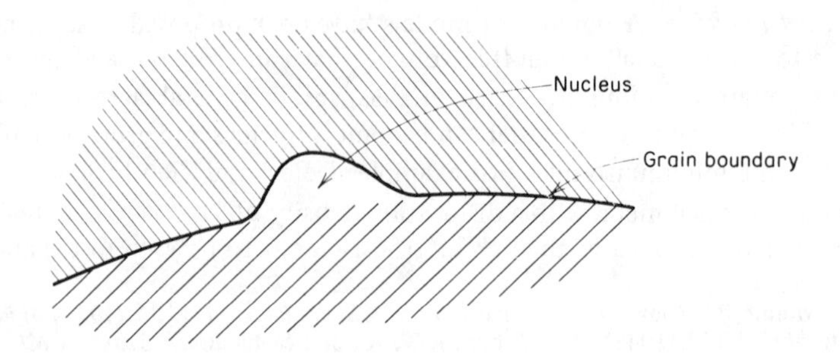

Fig. 7.21 The bulge mechanism for the formation of a nucleus at a grain boundary, after Bailey and Hirsch.

12. Bailey, J. E., and Hirsch, P. B., *Proc. Roy. Soc.*, **267A** 11 (1962).
13. Cahn, R. W., *Proc. Phys. Soc. London*, **63A** 323 (1950).

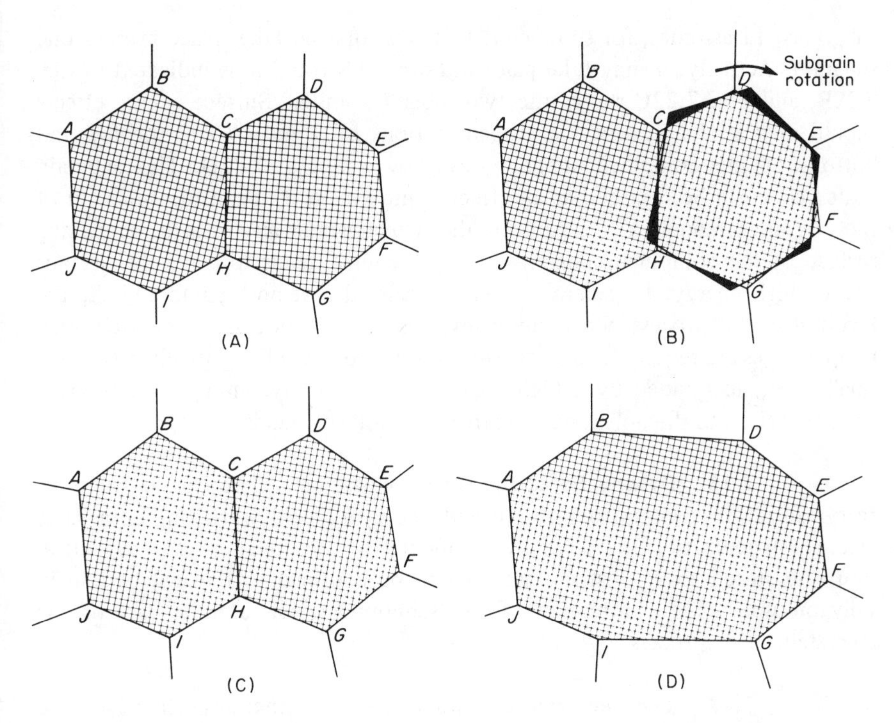

Fig. 7.22 Subgrain coalescence by a rotation of one subgrain. (From Li, J. C. M., *J. Appl. Phys.,* **33** 2958 (1962).

to Beck.[14] This mechanism, which predates the transmission electron microscopy studies of recrystallization, simply proposed that, as a result of polygonization, it might be possible to produce a subgrain capable of growing out into the surrounding polygonized matrix.

The other mechanism that should be applicable to single crystals involves the concept of subgrain coalescence, or the combination of subgrains to form a strain-free region large enough in size to grow. As postulated in this theory,[15] the elimination of a subgrain boundary must result in a relative rotation of the two subgrains that are combined. This process is shown schematically in Fig. 7.22 taken from a paper by Li.[16] In part A of this illustration, two subgrains about to coalesce are shown. Note that they do not have identical orientations. In order to remove the subboundary there must be a relative rotation of the two

14. Beck, P. A., *J. Appl. Phys.,* **20** 633 (1949).
15. Hu, Hsun, *Recovery and Recrystallization of Metals*, AIME Conference Series, pp. 311–362, Interscience Publishers, New York, 1963.
16. Li, J. C. M., *J. Appl. Phys.,* **33** 2958 (1962).

subgrains. Li assumes, for simplicity, that this rotation takes place in only one subgrain. Actually, it may take place in both. This rotation is indicated in Fig. 7.22B, and Fig. 7.22C shows the two subgrains united. Surface energy effects should now straighten out the cusped sections *BCD* and *GHI* at the top and bottom of the combined subgrain, as shown in Fig. 7.22D. This subgrain coalescence can be regarded as an effective movement of the dislocations out of the original boundary *CH* separating the two subgrains and into the remaining surface of the combined subgrains. This, of course, tends to make this surface one of high energy. In general it can be assumed that both climb and slip are involved in this process. Since climb involves vacancy motion, moderately high temperatures are required. An extension of this process will eventually result in a small grain surrounded by a high-angle boundary of large enough size to grow progressively into the polygonized matrix that still surrounds it.

7.16 Driving Force for Recrystallization. The driving force for recrystallization comes from the stored energy of cold work. In those cases where polygonization is essentially completed before the start of recrystallization, the stored energy can be assumed to be confined to the dislocations in polygon walls. The elimination of the subboundaries is a basic part of the recrystallization process.

7.17 The Recrystallized Grain Size. Another important factor in recrystallization studies is the recrystallized grain size. This is the crystal size immediately at the end of recrystallization, that is, before grain growth proper has had a chance to occur. Figure 7.23 shows that the recrystallized grain size depends upon the amount of deformation given to the specimens before annealing. The significant part of this curve is that the grain size grows rapidly with decreasing deformation. Too little deformation, however, will make recrystallization impossible in any reasonable length of time. This leads to the concept of the critical amount of cold work, which may be defined as the minimum amount of cold deformation that allows the specimen to recrystallize (within a reasonable time period). In the above figure it corresponds to about 3 percent elongation of the polycrystalline brass in a tensile test. The critical deformation, like the recrystallization temperature, is not a property of a metal, since its value varies with the type of deformation (tension, torsion, compression, rolling, etc.). In single crystals of hexagonal metals, when deformation occurs by easy glide the critical deformation may exceed several hundred percent. Twisting the same crystal to a few percent strain, however, may make it possible to recrystallize ths specimen.

The concept of a critical deformation is important because the very large grain size associated with it is usually undesirable in metals that are to be further

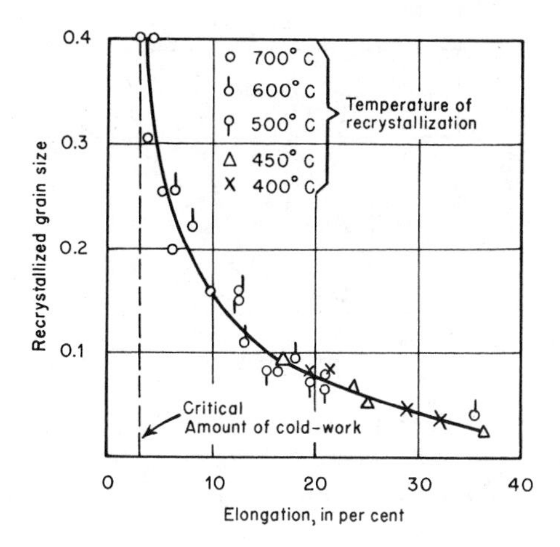

Fig. 7.23 Effect of prior cold-work on the recrystallized grain size of alpha-brass. Notice that the grain size at the end of recrystallization does not depend on the temperature of recrystallization. (Smart, J. S., and Smith, A. A., *Trans. AIME,* **152** 103 [1943].)

deformed. This is particularly true for sheet metal that is to be cold-formed into complicated shapes. If the grain size of a metal is very small (less than about 0.05 mm in diameter), plastic deformation occurs without appreciable roughing of the surface (assuming that deformation does not occur by movement of Lüders bands). On the other hand, if the diameter of the average grain is large, cold working produces a roughened, objectionable surface. Such a phenomenon is frequently identified by the term *orange-peel effect* because of the similarity of the roughened surface to that of the peel of a common orange. The anisotropic nature of plastic strain inside crystals is directly responsible for the orange-peel effect, and the larger the crystals the more evident will be the nonhomogeneous nature of the deformation.

In metal that is cold-rolled (sheets) or drawn cold through dies (wires, rods, and pipe), it is relatively easy to avoid a critical amount of cold work because the metal is more or less uniformly deformed. On the other hand, if only a portion of a metallic object is deformed cold, a region containing a critical amount of cold work must exist between the worked and unworked areas. Annealing in this case can easily lead to a localized, very coarse-grain growth.

The ratio of the rate of nucleation to the rate of growth, N to G, is frequently used in the interpretation of recrystallization data. If it is assumed that both N and G are constant or are average values for an isothermal recrystallization process, then the recrystallized grain size can be deduced from this ratio. If the

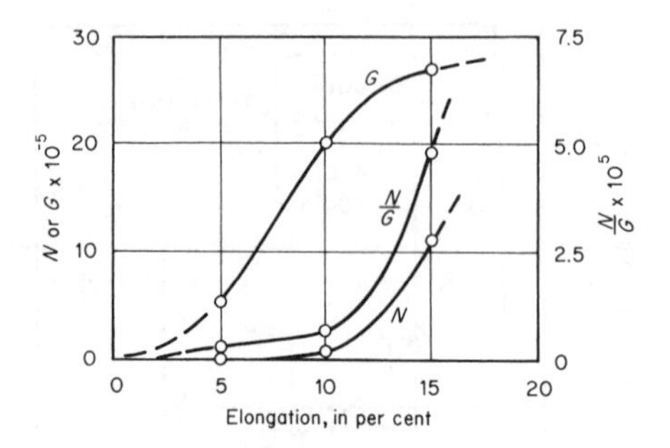

Fig. 7.24 Variation of the rate of nucleation (*N*), the rate of growth (*G*), and their ratio (*N/G*) as a function of deformation before annealing. (Data for aluminum annealed at 350°C.) (From Anderson, W. A., and Mehl, R. F., *Trans. AIME,* **161** 140 [1945].)

ratio is high, many nuclei will form before the recrystallization process is completed, and a fine-grain size will result. On the other hand, a low ratio corresponds to a slow rate of nucleation relative to the rate of growth, and to a coarse crystal size in a recrystallized specimen. In Fig. 7.24, the rate of nucleation, N, the rate of growth, G, and the ratio of N to G are all plotted as a function of strain for a specific metal (aluminum). These curves show that, as the deformation before annealing is reduced to smaller and smaller values, the rate of nucleation falls much faster than the rate of growth. As a consequence, the ratio N to G decreases in magnitude with decreasing strain and, for the data of Fig. 7.24, is effectively zero at several percent elongation.

Thus it can be concluded that a critical amount of cold work corresponds to an amount just capable of forming the nuclei needed for recrystallization. This agrees well with the fact that nuclei form at points of high-strain energy in the lattice. The number of these points certainly should increase with the severity of the deformation and at low strains the number should almost vanish.

Another very important factor concerning the recrystallized grain size (at the end of recrystallization and before the beginning of grain growth) is apparent in Fig. 7.23: the recrystallized grain size in the case of the brass specimens used for the data of Fig. 7.23 is independent of the temperature of recrystallization. Notice that these data correspond to five different annealing temperatures and that all temperatures give values that fall on a single curve. This relationship also holds, within limits, for many other metals [17] and to a first approximation we can assume that the grain size of a metal at the end of recrystallization is independent of the recrystallization temperature.

17. Sachs, G., and Van Horn, K. R., *Practical Metallurgy.* American Society for Metals, Cleveland, Ohio, 1940.

7.18 Other Variables in Recrystallization. It has been shown that the rate of recrystallization is dependent upon two variables: (1) temperature of annealing and (2) amount of deformation.

Similarly, it has been shown that, for many metals, the recrystallized grain size is independent of the annealing temperature, but sensitive to the amount of strain. The recrystallization process is also dependent upon several other variables. Two of the more important of these are: (1) purity, or composition of the metal; and (2) the initial grain size (before deformation). These factors will be considered briefly.

7.19 Purity of the Metal. It is a well-known fact that extremely pure metals have very rapid rates of recrystallization. This is apparent in the sharp dependence of the recrystallization temperature on the presence of solute. As little as 0.01 percent of a foreign atom in solid solution can raise the recrystallization temperature by as much as several hundred degrees. Conversely, a spectroscopically pure metal recrystallizes in a fixed interval at much lower temperatures than a metal of commercial purity.

The effect of solute atoms on the rate of recrystallization is most apparent at very small concentrations. This is shown clearly in Fig. 7.25 for aluminum of various degrees of purity. It has also been observed that the increase in the recrystallization temperature caused by the presence of foreign atoms depends markedly upon the nature of the solute atoms. Table 7.1 shows the increase in the recrystallization temperature corresponding to the addition of the same amount (0.01 atomic percent) of several different elements in pure copper.

That very small numbers of solute atoms have such a pronounced effect on

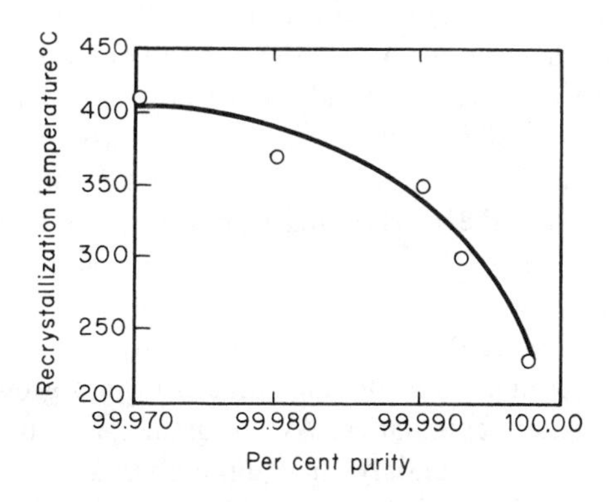

Fig. 7.25 Effect of impurities on the recrystallization temperature (30 minutes of annealing) of aluminum cold-rolled 80 percent. (From Perryman, E. C. W., ASM Seminar, *Creep and Recovery,* 1957, p. 111.)

Table 7.1 Increase in the Recrystallization Temperature of Pure Copper by the Addition of 0.01 Atomic Percent of the Indicated Element.*

Added Element	*Increase in Recrystallization Temperature* °C
Ni	0
Co	15
Fe	15
Ag	80
Sn	180
Te	240

* Data of Smart, J. S., and Smith, A. A., *Trans. AIME*, 147 48 (1942); 166 144 (1946).

recrystallization rates is believed to indicate that the solute atoms interact with grain boundaries,[18] The proposed interaction is similar to that between dislocations and solute atoms. When a foreign atom migrates to a grain boundary, both its elastic field, as well as that of the boundary, are lowered. In recrystallization, grain-boundary motion occurs as the nuclei form and grow. The presence of foreign atoms in atmospheres associated with these boundaries strongly retards their motion, and therefore lowers the recrystallization rates.

7.20 Initial Grain Size. When a polycrystalline metal is cold worked, the grain boundaries act to interrupt the slip processes that occur in the crystals. As a consequence, the lattice adjacent to the grain boundaries is, on the average, much more distorted than in the center of the grains. Decreasing the grain size increases the grain-boundary area and, as a consequence, the volume and uniformity of distorted metal (that adjacent to the boundaries). This effect increases the number of possible sites of nucleation and, therefore, the smaller the grains of the metal before cold work, the greater will be the rate of nucleation and the smaller the recrystallized grain size for a given degree of deformation.

7.21 Grain Growth. It is now generally recognized that in a completely recrystallized metal the driving force for grain growth lies in the surface energy of the grain boundaries. As the grains grow in size and their numbers decrease, the grain boundary area diminishes and the total surface energy is lowered accordingly. The growth of cells in a foam of soap also occurs as a result of a decrease in surface energy: the surface energy of the soap film. Because a number of complicating factors that influence the growth of metal

18. Lücke, K., and Detert, K., *Acta Met.*, 5 628 (1957).

crystals do not apply in the case of soap films, the growth of soap bubbles may be taken as a rather ideal case of cellular growth. For this reason, the growth of soap cells will be considered before the more complicated case of metallic grain growth.

First, consider a single spherical soap bubble. The gas enclosed by the soap film is always at a greater pressure than that on the outside of the bubble because of the surface tensions in the soap film. In an elementary physics course it is shown that this pressure difference between the inside and outside of the soap bubble can be expressed by the simple equation

$$\Delta p = \frac{4\gamma}{R} = \frac{8\gamma}{D}$$

where γ is the surface tension of one surface of the film (soap films have two surfaces), R is the radius of the soap bubble and D is its diameter. This equation shows clearly that the smaller the bubble, the greater the excess pressure inside the bubble.

Because of the pressure difference that exists across a curved soap film, gaseous diffusion occurs; the net flow occurring through the film from the high- to the low-pressure side. In other words, the atoms diffuse from the inside to the outside of the bubble, resulting in a decrease in size of the bubble and a movement of its walls inward toward their center of curvature.

The above discussion can now be carried over to the more general example, a soap froth. In the froth, the cells contain curved walls, and the curvature varies from cell to cell and within the film surrounding any one cell, depending on the relative size and shape of the neighboring cells. (These factors will be discussed in more detail presently.) In all cases, however, a pressure difference exists across each curved wall, with the greater pressure on the concave side. Gas diffusion resulting from this pressure difference, in turn, causes the walls to move but always in a direction toward their center of curvature.

Let us find out how the movement of cell walls causes the cells in the soap froth to grow in size. For the sake of simplicity, consider a network of two-dimensional cells, those with walls perpendicular to the plane of view. A froth of this type can be formed between two closely spaced parallel plates of glass. This simplification greatly reduces geometrical complexity, and still permits the more important principles of cellular growth to be observed.

Figure 7.26 presents a sequence of pictures from the work of C. S. Smith, showing the growth of the cells in a two-dimensional soap froth formed in a small flat glass cell. The figure at the lower right-hand corner of each photograph represents the number of minutes from the time that agitation of the cell to form the froth was ended. It also represents the time during which cell growth has taken place. In several of the photographs, small, three-sided cells can be

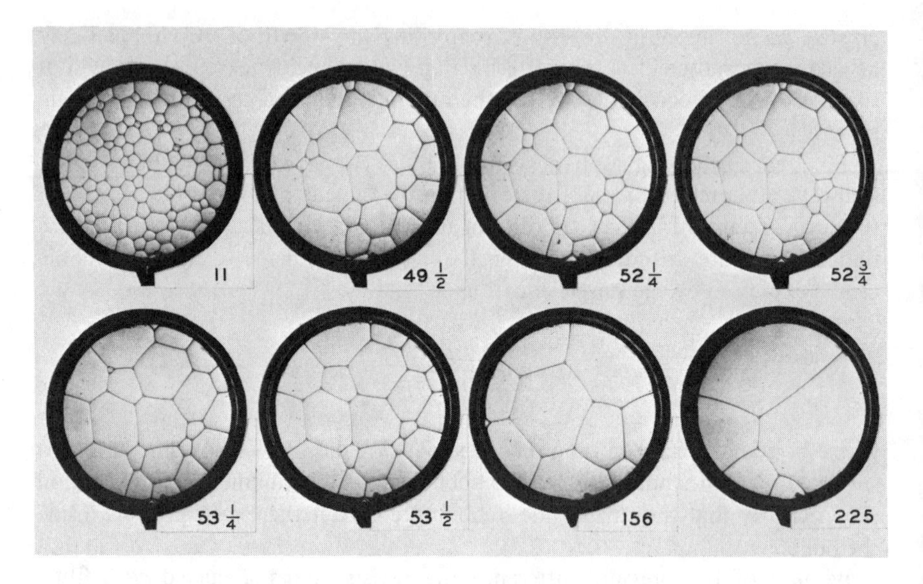

Fig. 7.26 Growth of soap cells in a flat container. (From Smith, C. S., ASM Seminar, *Metal Interfaces,* 1952, p. 65.)

observed. In the third photograph from the left in the upper row one appears at approximately 9 o'clock, and another at about 10 o'clock in the first photograph from the left in the lower row. An enlarged sketch of one of these cells is shown in Fig. 7.27. Notice that in order to maintain the equilibrium angle of 120° required when three surfaces with equal surface tensions meet at a common junction, the cell walls of the three-sided cell have been forced to assume a rather pronounced curvature. Since this curvature is concave toward the center of the cell, the wall can be expected to migrate, thus decreasing the volume of the cell and causing it to disappear completely. That this actually

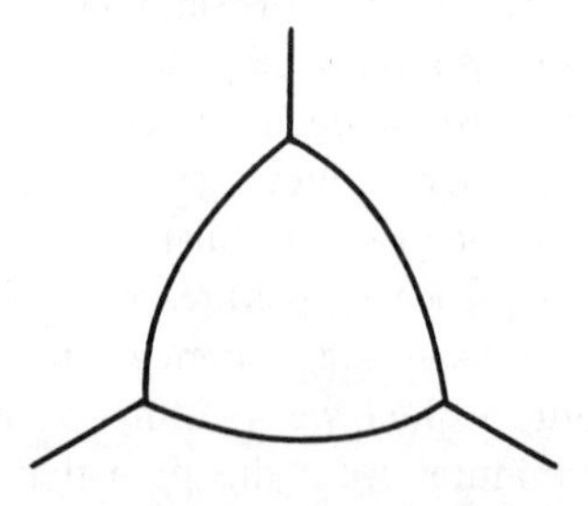

Fig. 7.27 Sketch of a three-sided soap cell. Notice the pronounced curvature of the boundaries, which is concave toward the center of the cell.

occurs can be seen by studying the photographs just to the right of the two photographs indicated above. In each case the triangular grains are no longer visible.

Further study of the photographs of Fig. 7.26 reveals that cells with less than six sides have walls that are primarily concave toward their centers, while those with more than six sides have walls convex toward their centers, this effect being more pronounced the larger the number of sides above six. This confirms the fact that the only two-dimensional geometrical figure formed by straight lines that can have an average internal angle of 120° is the hexagon. In a two-dimensional structure, such as that shown in the photographs of Fig. 7.26, all cells with less than six sides are basically unstable and tend to shrink in size, while those with more than six sides tend to grow in size. Another interesting fact is that there is a definite correspondence between the size of the cells and the number of sides that they contain. The smaller cells usually have the fewest number of sides. It is no wonder that three-sided cells disappear so rapidly, for both their small size and minimum number of sides require that their walls have very large curvatures, with accompanying high-pressure differentials, diffusion rates, and rates of wall migration. The photographs show that, in general, four- and five-sided cells do not disappear as a unit, but first change to three-sided cells which rapidly disappear.

Another important aspect of cellular growth can be seen in the photographs of Fig. 7.26. Over a period of time the number of sides that any given grain possesses continually changes. The number of sides may increase or decrease, as can be seen by considering the mechanism[19] illustrated in Fig. 7.28. Because of the curvature of the boundaries that separate cells B and D from A and C respectively, the boundaries migrate, thus eliminating the boundary between

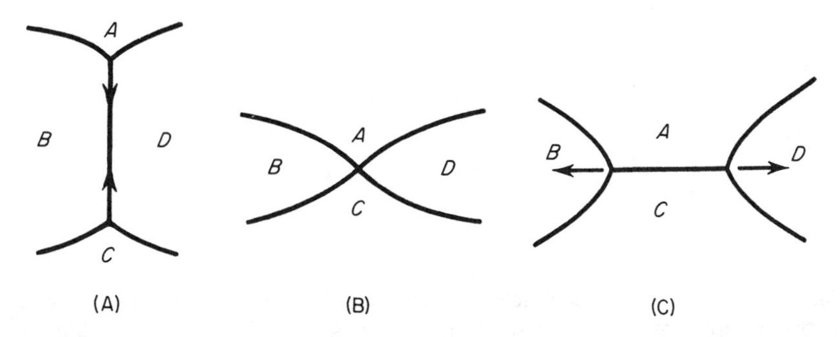

Fig. 7.28 A mechanism that changes the number of sides of a grain during grain growth.

19. Burke, J. E., and Turnbull, D., *Recrystallization and Grain Growth, Prog. in Metal Phys.*, 3 220 (1952).

cells B and D, and then create a new boundary between A and C. These steps are indicated in Figs. 7.28B and 7.28C respectively. As a consequence of this process, cells B and D each lose a side, while cells A and C each gain a side. Another method by which the number of sides of a cell can be changed has already been described. Each time a three-sided cell disappears, each of its neighboring cells loses one side.

7.22 Geometrical Coalescence. Nielsen[20] has proposed that the geometrical coalescence of grains is an important phenomenon in recovery, recrystallization, and grain growth. Subgrain coalescence has already been discussed in relation to the formation of nuclei during recrystallization. The mechanism that Nielsen favors does not require that the subgrains or grains rotate through an appreciable angle relative to each other, as in the Hu and Li mechanism. Geometrical coalescence can be simply described as an encounter of two grains whose relative orientations are such that the boundary formed between the two grains is one of much lower surface energy, γ_G, than that of

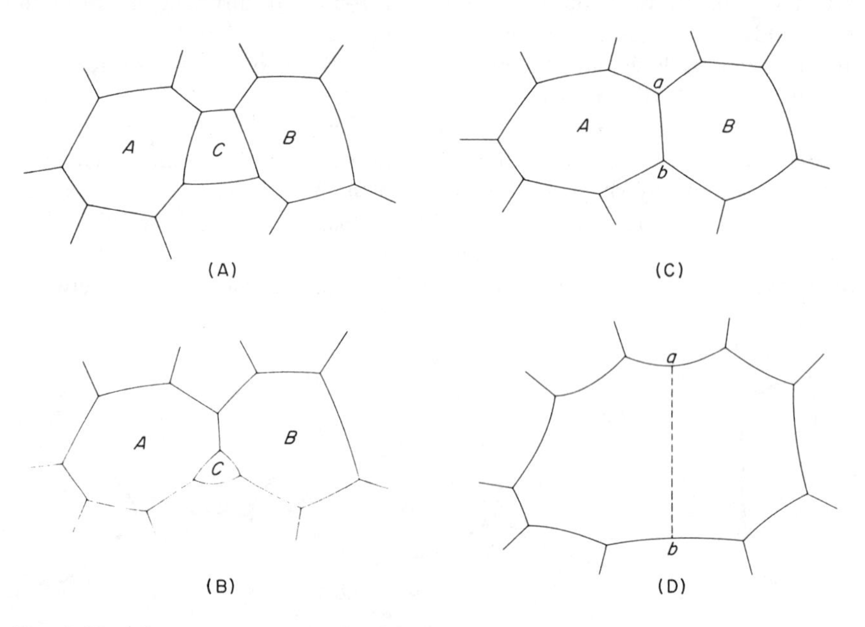

Fig. 7.29 Geometrical coalescence. Two grains, A and B, encounter as a result of the disappearance of grain C. If grains A and B have nearly identical orientation, then boundary ab becomes the equivalent of a subboundary and grains A and B may be considered the equivalent of a single grain.

20. Nielsen, J. P., *Recrystallization, Grain Growth and Texture*, ASM Seminar Series, pp. 141–64, American Society for Metals, Metals Park, Ohio, 1966.

the average boundary. In a polycrystalline metal, such a boundary would be equivalent to a subgrain boundary. The effect of producing this type of boundary on the microstructure is indicated schematically in Fig. 7.29. First, let us imagine that grains *A* and *B* encounter each other during a process of grain growth in a metal where it is assumed that the boundaries have the two-dimensional character of the soap froth in a flat cell. Before the encounter, the two grains are separated, as shown in Fig. 7.29A. If the boundary that is produced when they meet is a typical high-angle boundary, the grain boundary surface energy γ_G will be effectively the same as that of the other boundaries, and a boundary configuration such as that in Fig. 7.29C is expected. On the other hand, if a very low-energy boundary is formed, then, because of the low value of γ_G in the boundary *ab*, the effective surface tension forces along *ab* will be very small and the grain boundary configuration (shown in Fig. 7.29D) should result. From this illustration it can be seen that a geometrical coalescence should result in the sudden development of a much larger grain. Note that in this two-dimensional example this large grain has nine sides. Grains formed by geometrical coalescence should thus have the possibility of continued rapid growth.

Geometrical coalescence, if it occurs to any extent, should have a strong effect on grain-growth kinetics. In a metal containing a more or less random set of crystal orientations, geometrical coalescence should probably occur infrequently. On the other hand, geometrical coalescence may be an important phenomenon in a highly textured metal: one that has a strong preferred orientation. In this case the chances of two grains of nearly identical orientation encountering each other during grain growth is certainly much greater. It is also much greater in the case of subgrain growth during recrystallization.

7.23 Three-Dimensional Changes in Grain Geometry. In a metal the grains are not two-dimensional in character, but three-dimensional. The five basic mechanisms by which the geometrical properties of the three-dimensional grains can change have been outlined by Rhines.[21] A brief summary of these mechanisms due to DeHoff is shown in Fig. 7.30. The analog of the three-sided two-dimensional grain of Fig. 7.27 is shown in Fig. 7.30A. It is a four-sided or tetrahedral grain. Its disappearance results in the loss of four grain boundaries. Figure 7.30B is the corresponding three-dimensional analog of the mechanism illustrated in Fig. 7.28. Note that the grain boundary *BD* in Fig. 7.30A becomes a three-grain juncture or triple line in the three-dimensional example. If the upper and lower grains should come together, this line is removed and replaced by a horizontal grain boundary lying between the upper and lower grains. The result is a gain of one grain boundary. The third case, in Fig. 7.30C, is simply the inverse of that which has just been discussed.

21. Rhines, F. N., *Met. Trans.*, 1 1105–20 (1970).

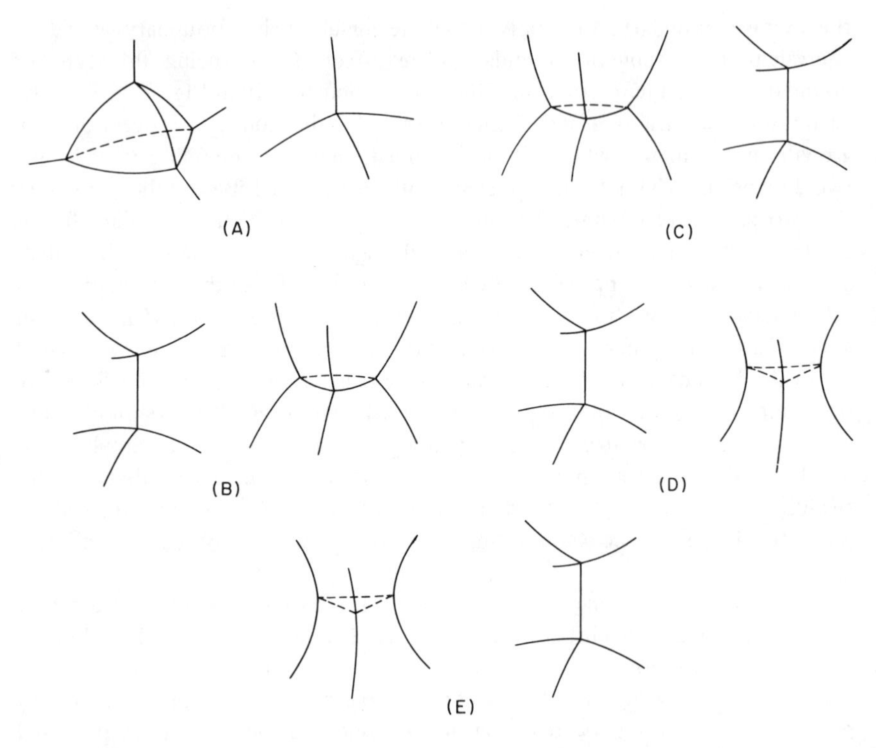

Fig. 7.30 The five basic three-dimensional geometrical processes in grain growth. (After Rhines, F. N., and DeHoff, R. T.)

An interesting mechanism is shown in Fig. 7.30D. This is the three-dimensional example of the geometrical coalescence of grains discussed in the preceding section. In this case, the upper and lower grains are again assumed to be able to approach each other as in the case of Fig. 7.30B, but in this case the boundary formed between the two grains is a low-energy boundary. The result is an effective coalescence of the upper and lower grains. Finally, Fig. 7.30E shows the inverse of geometrical coalescence. Here one grain necks and separates into two grains.

7.24 The Grain Growth Law. Another interesting aspect of the grain growth pictures in Fig. 7.24 is that, while the number of cells keeps decreasing as time goes on, the cellular arrays are geometrically similar at all times. This would be even more evident if a larger sample of soap froth were available for our study. At any rate, it is apparent that at any given instant the cells vary in size about a mean and that this mean size grows with time. The mean cell diameter serves as a convenient measure of the cell size of an aggregate.

Therefore, when one refers to the cell size of a froth, it is the diameter of the average cell that is meant. This statement also holds true for metals where the commonly used term "grain size" refers, in general, to the mean diameter of an aggregate of grains. It follows that grain growth, or cellular growth, refers to the growth of the average diameter of the aggregate. Attention is called, however, to the difficulties involved in accurately defining this concept of an average grain size in a metal, as pointed out in Chapter 5.

Let us now derive an expression for a soap froth that relates the size of the average cell to the time. In this derivation we shall not limit ourselves to the two-dimensional case, but shall consider that we are working in three dimensions. We shall first assume that the rate of growth of the cells is proportional to the curvature of the cell walls of the average cell at a given instant of time. This is in agreement with the fact that boundaries move as a result of gaseous diffusion caused by a pressure difference, from one side of a soap film to the other, that is proportional to the curvature of the boundary. If the symbol D represents the mean diameter of the average-sized soap cell, and c the curvature of the cell walls, we have

$$\frac{dD}{dt} = K'c$$

where t is the time, and K' a constant of proportionality. Let us also assume that the curvature of the average-sized cell is inversely proportional to its diameter, and rewrite the above equation in the following form:

$$\frac{dD}{dt} = \frac{K}{D}$$

where K is another constant of proportionality. Integration of this equation leads to the following result:

$$D^2 = Kt + c$$

Assuming that D_0 is the size of the average cell at the start of the observation ($t = 0$), an evaluation of the constant of integration gives

$$D^2 - D_0^2 = Kt$$

Although several broad assumptions have been made in deriving the above equation, experimental measurements of the growth of cells in a soap froth have shown that this expression fits the observed data quite closely.[22] It can

22. Fullman, R. L., ASM Seminar *Metal Interfaces*, p. 179 (1952).

therefore be concluded that the above equation is essentially correct for the growth of soap cells under the action of surface-tension forces.

If it is assumed that the grain size is very small at the beginning of cell growth, then it is possible to neglect D_0^2 in relation to D^2, with the result that the equation relating the cell size to the time can be expressed in the somewhat simpler form

$$D^2 = Kt$$

or

$$D = kt^{1/2}$$

where $k = \sqrt{K}$. According to this relationship, the mean diameter of the cells in a soap froth grows as the square root of the time. Figure 7.31 shows a sketch of this relationship – clearly, as time progresses, the rate of cellular growth diminishes.

Let us consider the case of grain growth in metals. Here, as in soap froth, the driving force for the reaction lies in the surface energy of the grain boundaries. Grain boundary movements in metals are, in many respects, perfectly analogous to those of the cell walls of soap froth. In both cases, the boundaries move toward their centers of curvature, and the rate of this movement varies with the amount of curvature. On the other hand, while it is known that the growth of soap cells can be explained in terms of a simple diffusion of gas molecules through the walls of the soap film, less is known about the mechanism by which atoms on one side of a grain boundary cross the boundary and join the crystal on the other side.

It has been proposed[23] and generally accepted that the boundary atoms in

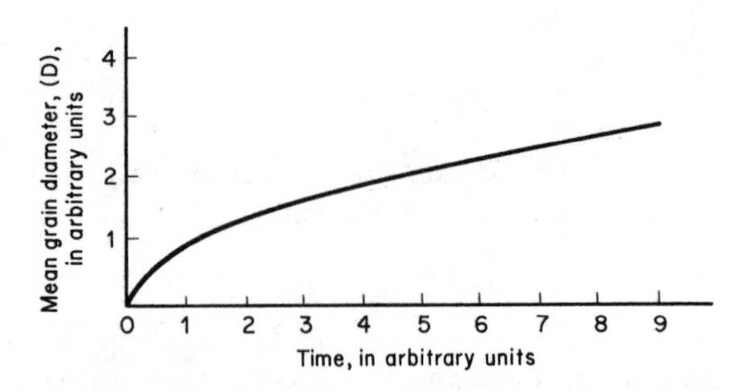

Fig. 7.31 Graphical representation of the ideal grain growth law, $D = kt^{1/2}$.

23. Harker, D., and Parker, E. A., *Trans. ASM*, **34** 156 (1945).

the crystal on the concave side of the boundary are more tightly bound than the boundary atoms in the crystal on the convex side, because they are more nearly surrounded by neighboring atoms of the same crystal. This tighter binding of the atoms on the concave side of the boundary should make the rate at which atoms jump across the boundary from the convex to the concave crystal greater than that in the opposite direction. The greater the curvature of the boundary then, the greater should be this effect, and the faster the movement of the crystal boundary. However, because detailed knowledge of the structure of metallic grain boundaries is lacking, the exact nature of the transfer mechanism by which atoms cross a boundary is still not known, and it is not possible to explain quantitatively the apparently irrational results obtained when studying the growth of crystals in a metal.

While a quantitative theory capable of explaining grain growth in pure metals and alloys is lacking, much is known about the causes for the apparent abnormal nature of metallic crystal growth. A number of these will now be considered, but first let us reconsider some of the factors that have been brought out in the study of soap froths. If metallic grain growth is assumed to occur as a result of surface-energy considerations and the diffusion of atoms across a grain boundary, then it is to be expected that at any given temperature a grain-growth law of the same form as that found in a soap froth might be observed, namely,

$$D^2 - D_0^2 = Kt$$

Also, if the diffusion of atoms across a grain boundary is considered to be an activated process, then it can be shown that the constant K in the above equation can be replaced by the expression

$$K = K_0 e^{-Q/RT}$$

where Q is an empirical heat of activation for the process, T is the temperature in degrees Kelvin, and R is the international gas constant. The grain-growth law can therefore be written as a function of both temperature and time in the following manner:

$$D^2 - D_0^2 = K_0 t e^{-Q/RT}$$

Most early experimental studies of metallic growth have failed to confirm the above relationship. However, some work[24] has given results that conform quite closely to the above equation. Figure 7.32 shows part of these results for brass

24. Feltham, P., and Copley, G. J., *Acta Met.*, **6** 539 (1958).

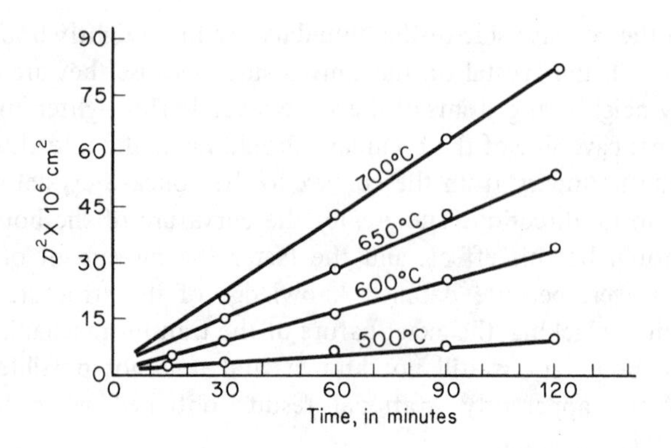

Fig. 7.32 Grain-growth isotherms for alpha-brass (10 percent Zn–90 percent Cu). Notice that the grain diameter squared (D^2) varies directly as the time. (From Feltham, P., and Copley, G. J., *Acta Met.*, **6** 539 [1958].)

(10 percent zinc and 90 percent copper). Note that in each case, straight lines are obtained when D^2 is plotted against the time. Let us now rearrange the grain-growth equation as follows:

$$\frac{D^2 - D_0{}^2}{t} = K_0 e^{-Q/RT}$$

Taking logarithms of both sides of this equation gives

$$\log \frac{D^2 - D_0{}^2}{t} = -\frac{Q}{2.3\,RT} + \log K_0$$

This relationship tells us that the quantity $\log (D^2 - D_0{}^2)/t$ should vary directly as the reciprocal of the absolute temperature $(1/T)$ and that the slope of this linear relationship is $Q/2.3R$. Now, the quantity $(D^2 - D_0{}^2)/t$ equals the slope of a grain-growth isotherm, such as those shown in Fig. 7.32. Figure 7.33 shows the logarithm of the slopes of the four lines in Fig. 7.32 ($\log [D^2 - D_0{}^2]/t$) plotted as functions of the reciprocal of the absolute temperature $(1/T)$. The data in question give an excellent straight line from which it can be deduced that the activation energy Q for grain growth in alpha-brass containing 10 percent Zn is 17.6 Kcal per gm atom.

Most reported experimental work does not conform so well to the grain-growth equation. In order to compare these results with the grain-growth equation, let us assume that $D_0{}^2$ may be neglected in comparison to D^2, so that

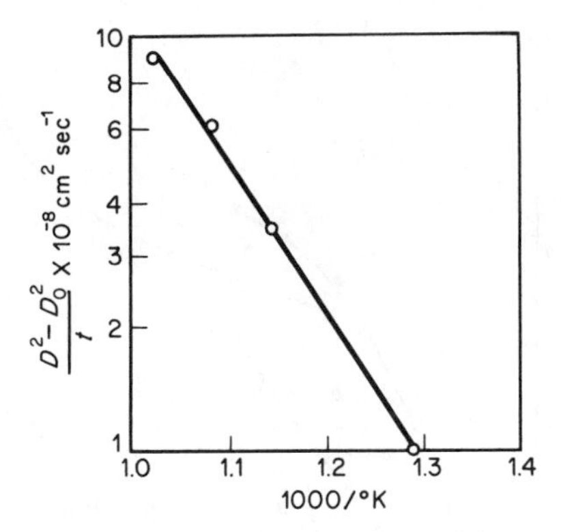

Fig. 7.33 The logarithms of the slope of the isotherms of Fig. 7.35 vary directly as the reciprocal of the absolute temperature. (From Feltham, P., and Copley, G. J., *Acta Met.*, **6** 539 [1958].)

the ideal grain-growth law can be stated in the simpler form

$$D = kt^{1/2}$$

where k equals $\sqrt{K}$ and should be a function of temperature of the form

$$k = \sqrt{K} = \sqrt{K_0 e^{-Q/RT}} = k_0 e^{-Q/2RT}$$

where $k_0 = \sqrt{K_0}$.

Many of the experimental isothermal grain-growth data correspond to empirical equations of the form

$$D = kt^n$$

where the exponent n is, in most cases, smaller than the value ½ predicted by the grain-growth equation. Furthermore, the exponent n is not usually constant for a given metal or alloy, if the isothermal reaction temperature is changed. Figure 7.34 shows the grain-growth exponent plotted as a function of the temperature for the grain-growth data obtained prior to 1951. Notice that, as a general rule, the exponent increases with increasing temperature and approaches the value ½. Experimental grain-growth data also, as a rule, fail to conform to a simple activation law, for the temperature dependence of grain growth does not usually

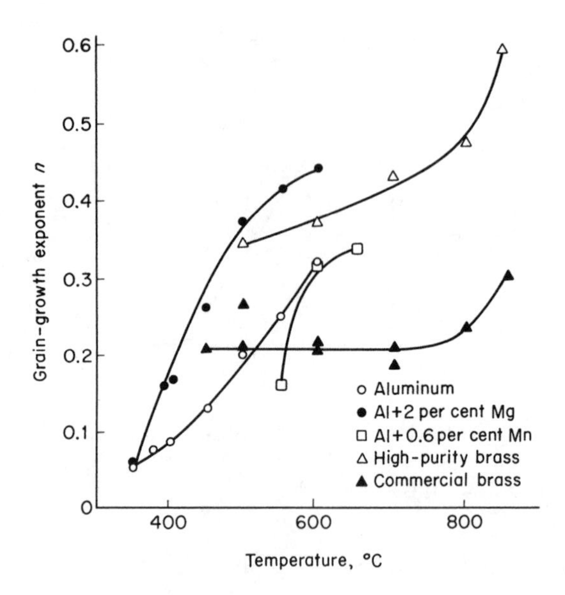

Fig. 7.34 Grain-growth exponent as a function of temperature for several metals. (From Fullman, R. L., ASM Seminar *Metal Interfaces*, 1952, p. 179.)

give a constant value for Q in the expression

$$k = k_0 e^{-Q/2RT}$$

7.25 Impurity Atoms in Solid Solution. Of considerable importance in grain growth is that foreign atoms in a lattice can interact with grain boundaries. This interaction is analogous to the interaction between impurity atoms and dislocations, previously mentioned as a factor in recrystallization (that is, nucleus growth). If the size of a foreign atom and that of the parent crystal are different, then there will be an elastic stress field introduced into the lattice by each foreign atom. However, since grain boundaries are regions of lattice misfit, the strain energy of the boundary, as well as that of the lattice surrounding a foreign atom, can be reduced by the migration of the foreign atom to the neighborhood of the grain boundary. In this manner, we can conceive of grain-boundary atmospheres just as we can dislocation atmospheres. That these atmospheres can effectively hinder the motion of grain boundaries has been verified experimentally. Figure 7.35 shows the grain-growth exponent as a function of impurity content for copper containing small percentages of aluminum in solid solution. Notice that as the metal approaches 100-percent purity, the grain-growth exponent increases toward the theoretical value ½. Also, note that at the higher temperature (600°C) the rate of approach is faster, a fact

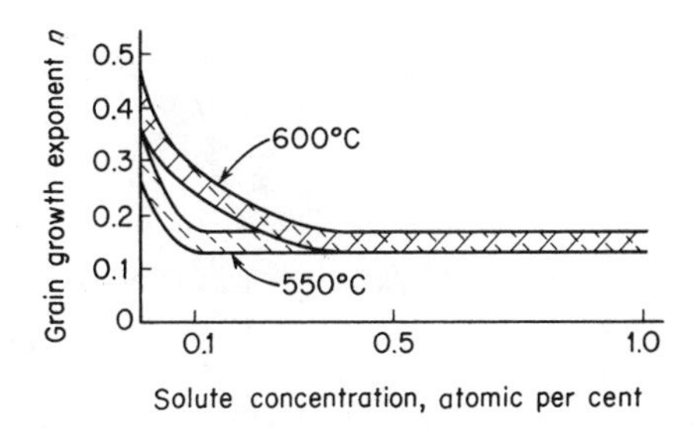

Fig. 7.35 Variation of the grain-growth exponent of copper with the concentration of aluminum in solid solution. (From Weinig, S., and Machlin, E. S., *Trans. AIME,* **209** 843 [1957].)

that is in agreement with the general trend of the curves in Fig. 7.34, where raising the annealing temperature has the effect of increasing the magnitude of the grain-growth exponent. This temperature dependence of the grain-growth exponent can be explained by assuming that the grain-boundary solute atmospheres are broken up by thermal vibrations at high temperatures.

The effect of solutes in retarding grain growth varies with the element concerned. While solid solutions of zinc in copper may behave normally (provided other impurities are eliminated) with respect to the grain-growth law ($n \simeq ½$), even as small a quantity as 0.01 percent of oxygen will effectively retard grain growth in copper.[25] This difference is probably due to the different magnitudes of strain that various elements produce in the lattice: those elements which distort the lattice structure the most have the largest effect on the rate of grain growth.

7.26 Impurities in the Form of Inclusions. Solute atoms not in solid solution are also capable of interacting with grain boundaries. It has been known for more than forty years[26] that impurity atoms in the form of second-phase inclusions or particles can inhibit grain growth in metals. The inclusions referred to are frequently found in commercial alloys and so-called commercially pure metals. For the most part, they may consist of very small oxide, sulfide, or silicate particles that were incorporated in the metal during its manufacture. In the present discussion, however, they may be considered any second-phase particle that is finely divided and distributed throughout the metal.

Let us consider a simplified theory of the interaction between inclusions and grain boundaries, due to Zener.[27] Figure 7.36A is a schematic sketch of an

25. Wood, D. L., *Trans. AIME,* **209** 406 (1957).
26. Jefferies, Zay, and Archer, R. S., *The Science of Metals*, p. 95. McGraw-Hill Book Co., Inc., New York, 1924.
27. As quoted by Smith, C. S., *Trans. AIME,* **175** 15 (1948). (See Ref. 24.)

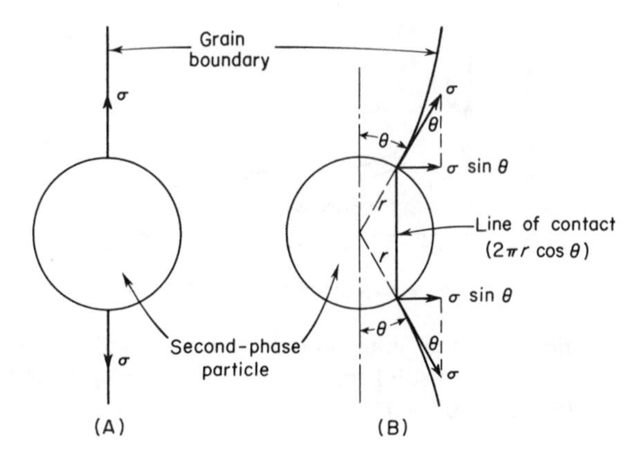

Fig. 7.36 Interaction between a grain boundary and a second-phase inclusion.

inclusion located in a grain boundary represented as a vertical straight line. For convenience, the particle has been made spherical. As shown in the left-hand sketch, the inclusion and boundary are in a position of mechanical equilibrium. If the boundary is moved to the right, as indicated in Fig. 7.36B, the grain boundary assumes a curved shape in which the boundary (because of its surface tension) strives to maintain itself normal to the surface of the particle. The vectors marked σ in the figure indicate the direction and magnitude of the surface-tension stress at the circular line of contact (in three dimensions) between the grain boundary and the surface of the inclusion. The total length of the line of contact is $2\pi r \cos\theta$, where r is the radius of the spherical particle, and θ is the angle between the equilibrium position of the boundary and the vector σ. The product of the horizontal component of this vector $\sigma \sin\theta$ and the length of the line of contact (between particle and boundary) gives the pull of the boundary on the particle:

$$f = 2\pi r\sigma \cos\theta \sin\theta$$

This force, by Newton's Second Law, is also the drag of the particle on the grain boundary; it is a maximum when θ is 45°. Substituting this value gives for the maximum force

$$f = \pi r\sigma$$

Notice that the drag of a single particle varies directly as the radius of the particle. Since the volume of each particle varies as the cube of its radius, the effect of second-phase inclusions in hindering grain boundary motion will be

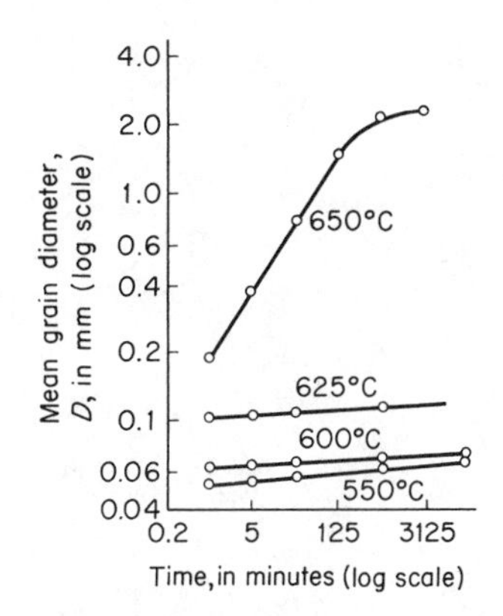

Fig. 7.37 The effect of second-phase inclusions on grain growth in a manganese-aluminum alloy (1.1 percent Mn). Grain growth is severely inhibited at the temperatures below 650°C because of the presence of second-phase precipitate particles ($MnAl_6$). (From Beck, P. A., Holzworth, M. L., and Sperry, P., *Trans. AIME,* **180** 163 [1949].)

greater the smaller and more abundant are the particles (assuming particles of the same shape).

In many cases, second-phase particles tend to dissolve at high temperatures. It is also generally true that second-phase particles tend to coalesce at high temperatures and form fewer large particles. Both of these effects – the decrease in amount of the second phase and the tendency to form larger particles – remove the retarding effect of the inclusions on grain growth in metals. An excellent example is shown in Fig. 7.37 where grain-growth data are plotted on log-log coordinates. A plot of this type for data that conform to an equation of the form

$$D = k(t)^n$$

gives a straight line, the slope of which is the grain-growth exponent n. In this alloy of aluminum containing 1.1 percent manganese, a second phase ($MnAl_6$) exists up to a temperature of 625°C, but dissolves at temperatures above 650°C, so that only a single solid solution remains. The effect of the inclusions on the grain growth is evident. At 625°C and below, grain growth is almost completely stopped (n is of the order of 0.02); on the other hand, at 650°C, grain growth occurs very rapidly, with a grain-growth exponent (0.42) close to the theoretical 0.5. The curved upper part of the curve, corresponding to the 650°C data, is the result of a geometrical effect that will be explained in the next section.

Holes or pores in a metal can have the same effect on grain-boundary motion as second-phase inclusions. This is clearly shown in the photographs of Fig. 7.38

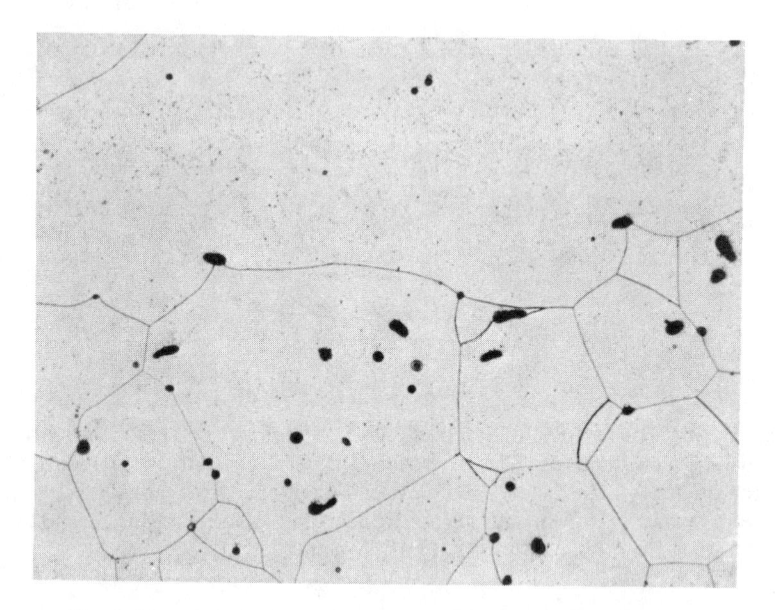

Fig. 7.38 Interaction between pores and grain boundaries.

where the cross-section of a small metal magnet made of Remalloy (12 percent Co, 17 percent Mo, and 71 percent Fe) can be seen. This specimen was made by the powder metallurgy technique involving the heating of a compressed metal compact to a temperature below the melting point of the metals contained in the compact, but sufficiently high to weld and diffuse the particles together. The sample presented in Fig. 7.38 has a structure consisting of a single homogeneous solid solution. A number of pores are clearly visible in the photograph, and at a number of them it can be seen that crystal boundaries have been held up by the pores. As an example, the nearly horizontal boundary of the very large grain that occupies the upper part of the picture passes through three pores. The motion of this boundary was toward the bottom. Observe how the boundary is curved back toward the pore in each case.

The effect of impurities on grain growth can be summarized as follows. Solute atoms in solid solution can form grain-boundary atmospheres, the presence of which retards the normal surface-tension induced boundary motion. In order for the boundary to move, it must carry its atmosphere along with it. On the other hand, solute atoms in the form of second-phase impurities can also interact with the boundaries. In this case, the boundary must pull itself through the inclusions that lie in its path. In either case, an increase in temperature lowers the retarding effect of the solute atoms and grain growth occurs under conditions more closely resembling the growth of soap cells.

7.27 The Free-Surface Effects. Specimen geometry may play a part in controlling the rate of grain growth. Grain boundaries near any free surface of a metal specimen tend to lie perpendicular to the specimen surface, which has the effect of reducing the net curvature of the boundaries next to the surface. This means that the curvature becomes cylindrical rather than spherical and, in general, cylindrical surfaces move at a slower rate than spherical surfaces with the same radius of curvature. The reason for this difference may be understood in terms of the soap-bubble analogy where it is easily shown that the pressure difference across a cylindrical surface is $2\gamma/R$ and across a spherical surface $4\gamma/R$, where γ is the surface tension of a soap film and R the radius of the curvature.

Mullins[28] has pointed out the importance of another phenomenon associated with grain boundaries that meet a free surface. It probably is of greater importance in its effect on grain growth than the reduction of the curvature of boundaries at a surface. The phenomenon in question is *thermal grooving*. At the high temperatures normally associated with annealing, grooves may form on the surfaces where grain boundaries intersect the specimen surface. These channels are a direct result of surface-tension factors. (See Fig. 7.39). Point *a* in this figure represents the line where three surfaces meet: the grain boundary and the free surfaces to the right and left of point *a* respectively. In order to balance the vertical components of the surface tensions in the three surfaces that meet at this line, a groove must form with a dihedral angle θ that satisfies the relation

$$\gamma_b = 2\gamma_{fs.} \cos \frac{\theta}{2}$$

where γ_b is the surface tension of the grain boundary, and $\gamma_{f.s.}$ that of the two free surfaces. Diffusion of atoms over the free surfaces is believed[29] to be the most important factor involved in the transport of atoms out of the grooved regions during the formation of the grooves.

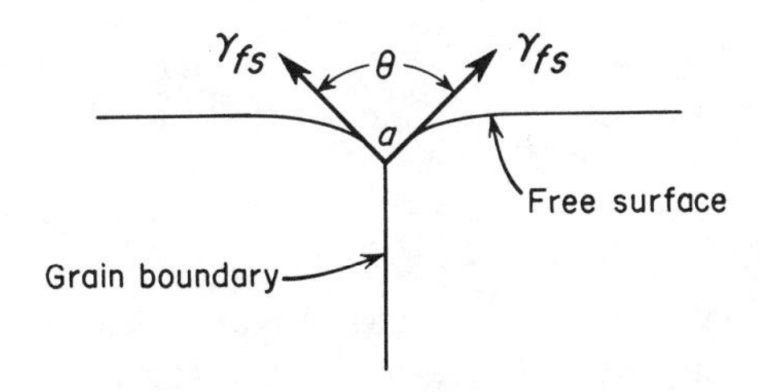

Fig. 7.39 A thermal groove.

28. Mullins, W. W., *Acta Met.*, 6 414 (1958).
29. Mullins, W. W., *Jour. Appl. Phys.*, 28 333 (1957).

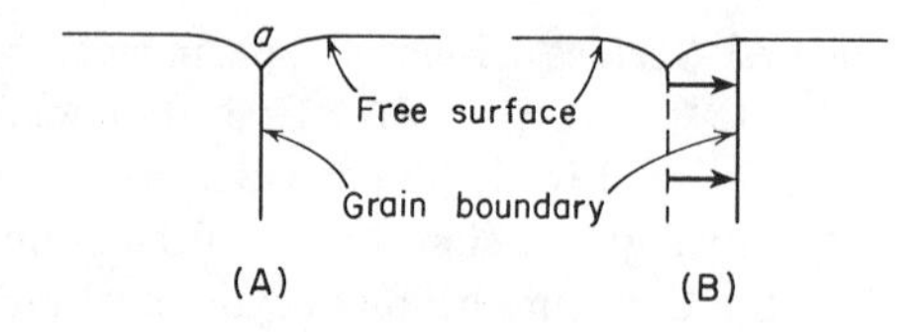

Fig. 7.40 Moving a grain boundary away from its groove increases its surface, if the boundary is nearly normal to the free surface.

Grain-boundary grooves are important in grain growth because they tend to anchor the ends of the grain boundaries (where they meet the surface), especially if the boundaries are nearly normal to the surface. This anchoring effect can be explained in a very qualitative fashion with the aid of Fig. 7.40. The left-hand figure represents a boundary attached to its groove, while the right-hand figure shows the same boundary moved to the right and freed from its groove. Freeing this boundary from its groove increases the total surface area and, therefore, the total surface energy. To free the boundary from its groove requires work and, as a result, the groove restrains the movement of the boundary.

When the average grain size of a metal specimen is very small compared to the dimensions of the specimen, thermal grooving, or the lack of curvature in the surface grains, has little effect on the overall rate of growth. However, when the grain size approaches the dimensions of the thickness of the specimen, it can be expected that grain-growth rates will be decreased. In this respect, it has been estimated[30] from experimental data that when the grain size of metal sheets becomes larger than one-tenth the thickness, the growth rate decreases. It is just this effect which explains the divergence of the curve, for the 650°C data in Fig. 7.37 from a straight line at times greater than 625 min. The size of the specimens used in these experiments was small. When the average grain size became larger than approximately 1.8 mm, grain growth was retarded because of the free-surface effect.

7.28 The Limiting Grain Size. In the preceding section it was pointed out that the specimen dimensions can influence the rate of grain growth when the average crystal size approaches the thickness of the specimen. In many cases, this situation can have the effect of putting a practical upper limit on the grain size; that is, the growth may be slowed down to the point where it appears that no further growth is possible. In extreme cases, this free-surface effect can completely stop grain growth. Consider the case of a wire in which the grains have become so large that their boundaries cross the crystal in the manner shown in Fig. 7.41. Boundaries of this nature have no curvature and cannot migrate

30. Beck, P. A., *Phil. Mag. Supplement,* 3 245 (1954).

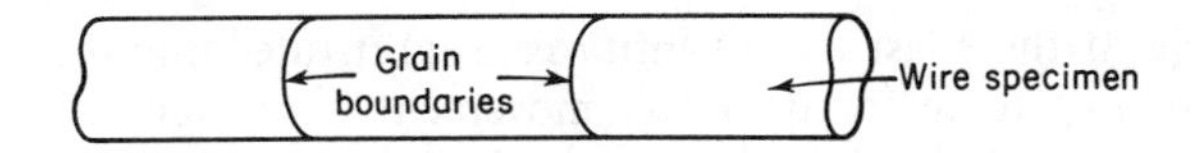

Fig. 7.41 One example of a stable grain boundary configuration.

under the action of surface-tension forces. Further grain growth is then not possible.

Second-phase inclusions are also known to put an upper limit on the grain size of a metal. Here it can be considered that the boundary has trapped so many inclusions that the surface-tension force, which is small due to its lack of sufficient curvature, cannot overcome the restraining force of the inclusions.

Metal grain boundaries differ from soap films for they possess only a single surface, whereas the latter have two surfaces. With this in mind, the driving force per unit area for grain-boundary movement (Δp in the analogous soap bubble case) can be written in the following manner:

$$\eta = 2\frac{\gamma}{R}$$

where η is the force per unit area, γ the surface tension of the grain boundary, and R the net radius of curvature of the boundary. At the point where the grain boundary is no longer able to pull itself away from its inclusions, the restraining force of the inclusions must equal the above force. This restraining force is equal to that of a single particle times the number of particles per unit area, or

$$\eta = 2\frac{\gamma}{R} = n_s \pi r \gamma$$

where n_s is the number of inclusions per unit area, and r the radius of the particles. Let us assume that the inclusions are uniformly distributed throughout the metal, then the approximate number of these inclusions which can be expected to be holding back a surface of area A are those whose centers lie inside a volume with an area A and whose thickness equals twice the radius of the particles. This volume $2Ar$ will hold $2n_v Ar$ particles, where n_v is the number of particles per unit volume, and it is concluded that the number of particles per unit area equals $2n_v r$. The number of particles per unit volume, n_v, may be expressed:

$$n_v = \frac{\zeta}{4/3\,\pi r^3}$$

where ζ is the volume fraction of the second phase, and $4/3\,\pi r^3$, the volume of a single particle. If these last two quantities are substituted into the equation that relates the driving force for boundary movement to the retarding force of the inclusions, Zener's relationship[31] is obtained:

$$\frac{r}{R} = \frac{3}{4}\zeta$$

or

$$R = \frac{4}{3}\frac{r}{\zeta}$$

where r is the radius of the inclusions, R the radius of curvature of the average grain, and ζ the volume fraction of inclusions. This relationship assumes spherical particles and a uniform distribution of particles, neither of which can be realized in an actual metal. Nevertheless, it gives us a first approximation of the effect of inclusions on grain growth. Since we can assume that the radius of curvature is directly proportional to the average grain size, this equation shows that the ultimate grain size to be expected in the presence of inclusions is directly dependent on the size of the inclusions.

7.29 Preferred Orientation. Other factors in addition to grain-boundary atmospheres, second-phase inclusions, and the free-surface effects are known to affect the measured rate of grain growth. Among these is a preferred orientation of the crystal structure. By a preferred orientation one signifies a nearly identical orientation in all the crystals of a given sample of metal. When this situation occurs, it has been generally observed that grain-growth rates are reduced.

7.30 Secondary Recrystallization. In a previous section it was pointed out that it is frequently possible to obtain a limiting grain size in a metal. When this grain-growth inhibition occurs as a result of the presence of inclusions, the size effects, or from the development of a strong preferred orientation, a secondary recrystallization is often possible. This second recrystallization behaves in much the same manner as the primary one and is usually induced by raising the annealing temperature above that at which the original grain growth occurred. After a nucleation period, some of the crystals start to grow at the expense of their neighbors. The causes for the nucleation are not entirely clear but grain coalescence could well produce this effect. It is relatively easy to understand the growth that occurs once it starts, for the enlarged crystals quickly become grains with a large number of sides as they consume their smaller

31. Quoted by Smith, C. S., *Trans. AIME,* 175 15 (1948). (See Ref. 24.)

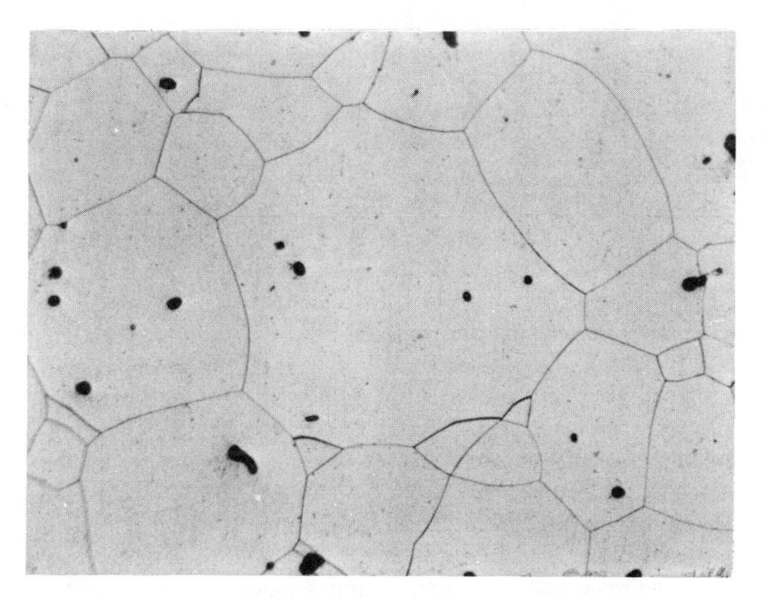

Fig. 7.42 A specimen undergoing secondary recrystallization. Notice that the central grain has thirteen sides.

neighbors. As mentioned earlier, grains with many sides possess boundaries that are concave away from their centers and, consequently, the grains become larger in size. Secondary recrystallization is really a case of exaggerated grain growth occurring as a result of surface-energy considerations, rather than as a result of the strain energy of cold work that is responsible for primary recrystallization.

Figure 7.42 is a good illustration of secondary recrystallization. In the center of the photograph, one sees a large grain with 13 sides, all of which are concave away from the center of the crystal. The specimen shown in the figure is the same as that of Fig. 7.38. The latter also shows a large grain growing at the expense of its neighbors (that occupying the top of the figure).

Very large grains, or even single crystals, can sometimes be grown by secondary recrystallization because the number of grains that finally results depends only on the number of secondary nuclei. Growth-inhibiting factors that control the grain growth after primary recrystallization, such as inclusions and free surfaces, do not control the growth of crystals in secondary recrystallization. In the latter case, as in primary recrystallization, the nuclei grow until the matrix is completely recrystallized.

7.31 Strain-Induced Boundary Migration. While it is generally conceded that normal grain growth occurs as a result of the surface energy stored in the grain boundaries, it is also possible for crystals to grow as a result of strain energy induced in the lattice by cold work.[32,33] Strain-induced boundary migration should not be expected in a metal after complete recrystallization unless it has been distorted either by handling after recrystallization, or by residual strains caused by uneven rates of recrystallization in different parts of the specimen.

Strain-induced boundary movements differ from recrystallization in that no new crystals are formed. Rather, the boundaries between pairs of grains move so as to increase the size of one grain of a pair while causing the other to disappear. As the movement occurs, the boundary leaves behind a crystalline region which is lower in its strain energy. In contrast to surface tension-induced grain-boundary migration, this form of boundary movement occurs in such a manner that the boundary usually moves away from its center of curvature. This movement is shown schematically in Fig. 7.43, where the moving grain boundary is shown with an irregular curved shape. The irregular form of a boundary that grows as a consequence of strain energy can be explained [34] by assuming that the rate of motion is a function of the strain magnitude in the metal. The boundary should move faster into those regions where the distortion has been greatest. One of the interesting aspects of strain-induced boundary migrations is that instead of the boundary of the moving grain lowering its surface energy through the movement, it may actually increase it by increasing its area.

Strain-induced migration of boundaries only occurs after relatively small or moderate amounts of cold work. Too great a degree of deformation will bring about normal recrystallization. On the other hand, this form of boundary

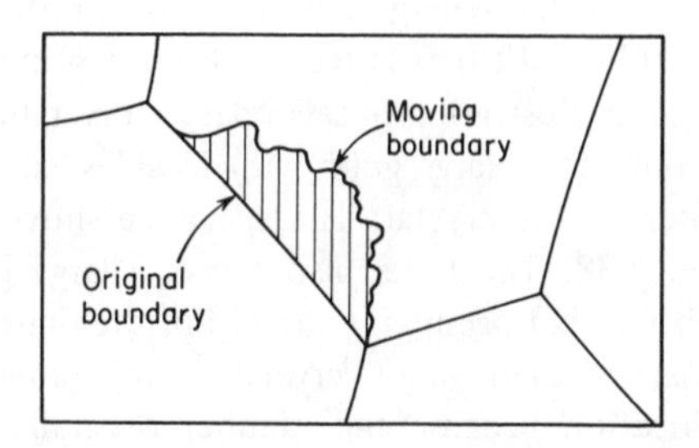

Fig. 7.43 Schematic representation of strain-induced boundary migration. In this case, the boundary moves away from its center of curvature, which is in the opposite direction to the movement in surface-tension induced boundary migration. (After Beck, P. A., and Sperry, P. R., *Jour. Appl. Phys.*, **21** 150 [1950].)

32. Crussard, C., *Etude du Recuit de L'Aluminum, Revue de Metallurgie,* **41** 240 (1944).
33. Beck, P. A., and Sperry, P. R., *Jour. Appl. Phys.*, **21** 150 (1950).
34. Beck, P. A., ASM Seminar (1951), *Metal Interfaces*, p. 208.

movement can be induced in sheet specimens where the normal grain growth has been inhibited by the size effects previously discussed.

Problems

1. The activation energy for the formation of vacancies in gold is about 1 eV. Smith and Bever[35] have measured the energy stored in gold (cold worked at 78°K) and find that it is about 180 calories per mole at a strain of about 180 percent. This energy they attribute to the accumulation in the metal of dislocations, point defects, and stacking faults.
   (*a*) Assume that 10 percent of the stored energy is in the form of vacancies. What would be the ratio of vacancies to atoms?
   (*b*) What is the equilibrium concentration of vacancies at 78°K?
   (*c*) How does the answer to part (*a*) of this problem compare to the equilibrium concentration of vacancies at the melting point of gold (1063°K)?
2. According to isotropic elasticity theory, the energy per unit length of a screw dislocation is $\mu b^2/4\pi$, where μ is the shear modulus and b is the Burgers vector. The corresponding energy per unit length of an edge dislocation is $\mu b\ /4\pi\,(1 - \nu)$, where ν is Poisson's ratio, which is normally about 0.3. An interesting computation is to determine very roughly what dislocation density might be consistent with the stored energy measurement of Smith and Bever. To do this, assume that all dislocations are screws and that there is no interaction energy between the dislocations. The atomic volume of gold is 10.2 cm^3 per mole, b is 2.9 Å, and μ is approximately 4×10^6 psi.
   (*a*) Compute the energy per unit length of the screw dislocation.
   (*b*) On the arbitrary assumption that perhaps three-fourths of the 180 cal/mole observed by Smith and Bever is associated with the dislocations, compute the required dislocation density.
   (*c*) Is the computed dislocation density reasonable considering the amount of strain and the temperature of deformation?
3. At a strain of 180 percent at room temperature, Smith and Bever report that the stored energy in gold is about 25 cal per mole. At this temperature they believe that nearly all of the stored energy is due to dislocations. Smith and Bever also report that the flow-stress of gold at 180 percent strain is about 30,000 psi at room temperature, and 75,000 psi at 78°K. An approximate assumption that can be made in this case is that the flow-stress σ varies directly as the square root of the dislocation density.

   Assuming that the stored energy due to dislocations varies directly as the dislocation density, what estimate can be made with regard to the contribution of the dislocations to the stored energy at 78°K?
4. The data of Drouard, Washburn, and Parker for the recovery of zinc single crystals corresponding to a recovery of 25 percent of the yield stress is given in the accompanying table.

35. Smith, J. H., and Bever, M. B., *TMS-AIME,* **242** 880 (1968).

(*a*) Plot the logarithm of time against the reciprocal of the absolute temperature.

Time in Seconds	*Temperature* °C
1.5	+10
5.0	0
28.0	−10
110.0	−20

(*b*) From the slope of the straight line drawn through the data of this curve, determine the activation energy Q for the recovery of zinc single crystals.
(*c*) Determine the value of the pre-exponential coefficient A in the equation

$$\frac{1}{\tau} = Ae^{-Q/RT}$$

5. (*a*) With the aid of the equation whose constants were evaluated in Problem 4, determine the time needed to obtain 25 percent recovery of the yield stress at − 5°C.
   (*b*) How long would it take to obtain 25 percent recovery at − 30°C?
6. The radius of curvature in a bent crystal is related to the excess edge dislocation density by the relationship

$$r = \frac{1}{\rho b}$$

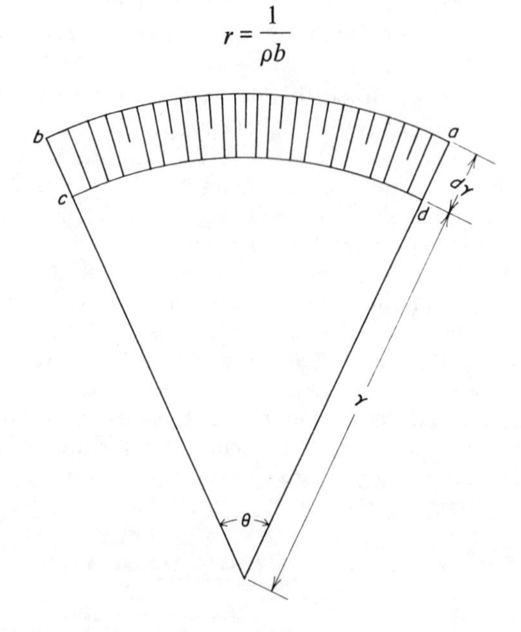

where r is the radius of curvature, ρ is the dislocation density, and b is the Burgers vector. Derive this expression using the steps outlined below. *Note:* in this case consider ρ as the number of dislocations per cm^2 and the area of the segment of a bent crystal in the figure to be $cd \cdot dr$.

(*a*) Make a Burgers circuit going from a to b to c to d to a. What is the net Burgers vector and how is it related to the two arc lengths ab and cd?

(*b*) Next, express the angle θ in terms of both of the arcs ab and cd and their respective radii of curvature. Equate these two angles.

(*c*) Simplify the resulting expression and show that it can be reduced to the desired relationship.

7. A rough estimate gives about 60 dislocations per cm^2 in the photograph of Fig. 7.14A. The magnification of this photograph is 750x. Assume $b = 2.5$ Å.

(*a*) Compute the dislocation density.

(*b*) Assuming these are all edge dislocations of the same sign, what is the radius of curvature to which the crystal has been bent?

8. (*a*) What is the radius of curvature corresponding to a dislocation density of 10^{12} where all dislocations are edges and of the same sign?

(*b*) The subgrain size in a heavily cold worked metal is often of the order of 1 micron (10^{-6} meters). How does the subgrain diameter correspond to the above radius of curvature?

9. According to isotropic elasticity theory, the energy per unit length of an edge dislocation is $\mu b^2/4\pi(1 - \nu)$, while the surface energy of a small-angle tilt-boundary of tilt-angle θ is given by[36]

$$\gamma = \frac{\mu b}{4\pi(1 - \nu)} \theta(A - \log_e \theta)$$

where A is a constant equal in magnitude to about 0.5.

(*a*) Estimate the fractional loss of energy when a set of n edge dislocations polygonize to form a subgrain boundary in which the dislocations are separated from each other by $1/n$. Assume the angle of tilt is 10^{-3} radians and that there is no interaction energy between the edge dislocations before they polygonize.

(*b*) Does the answer to part (*a*) of this question signify anything about the relative driving forces for recovery and recrystallization?

10. Assume that two small-angle tilt-boundaries coalesce to form a single tilt-boundary. Let θ for each of the two original boundaries be 10^{-4} radians.

(*a*) What will be the fractional loss of surface energy resulting from this coalescence?

(*b*) Make the same computation assuming $\theta = 10^{-3}$ radians.

11. Plot the logarithm of time against the reciprocal of the absolute temperature for complete recrystallization (99 percent) using the data in Fig. 7.16. Determine the activation energy for recrystallization.

36. Cottrell, A. H., *Dislocations and Plastic Flow in Crystals*, Oxford Press, London, 1953.

12. (*a*) Determine the recrystallization temperature (if the annealing time is 5 hours) for the copper of Harker and Parker in Fig. 7.16.
    (*b*) How long would it take to completely recrystallize this high purity copper at room temperature?
13. Using the data shown in Figs. 7.19 and 7.20, determine the value of A in the rate of recrystallization equation

$$\frac{1}{T} = Ae^{-Q/RT}$$

    for zirconium swaged 51 percent.
14. According to Fig. 7.19, zirconium deformed 51 percent will crystallize in one hour at 567°C. Compute the temperature at which recrystallization occurs in 30 minutes.
15. The surface energy for the grain boundaries in copper is about 650 ergs/cm^2.
    (*a*) Compute the effective excess internal pressure in a hypothetical spherical copper grain of diameter 1 mm.
    (*b*) Make the same computation for a spherical grain of diameter 1 micron.
16. If the grain size of a metal varies according to the ideal grain-growth law, how should the driving force for grain growth vary with time? *Note:* assume the grain size to be measured by the linear intercept method and that $D_o = 0$.
17. The isothermal anneal curve in Fig. 7.3 was obtained using a calorimeter with a sensitivity capable of reading a heat flow as low as 0.003 cal per hr.
    (*a*) On the assumption that the grain-boundary energy of brass is of the order of 500 ergs per cm^2, determine if this instrument, at 700°C, is sensitive enough to measure the rate of heat release in the brass specimens of Feltham and Copley (see Fig. 7.32). Assume that their grain size diameter data were obtained by the linear intercept method, that the grain size at the end of recrystallization was 0.02 mm, and that the specimen volume is 1 cm^3.
    (*b*) Discuss the significance of your answer.
18. Suppose that you were given the task of heat-treating some alpha brass specimens of the same composition as those used by Feltham and Copley, and which had been worked to the same extent as theirs. Could you use their data, given in Fig. 7.32, to determine the time and temperature of anneal required to obtain a grain size of 0.06 mm?
19. According to Fig. 7.23, the grain size of alpha brass at the end of recrystallization is independent of the annealing temperature. It is also true that raising the annealing temperature shortens the time for recrystallization. Economic considerations imply that annealing at very high temperatures for very short periods of time might be the best thing to do, since more parts could be annealed per day with a given investment in furnaces. What other factors should be considered with regard to selecting the annealing time and temperature?

20. The surface energy of a free surface of gold is about 1500 ergs/cm^2, while the grain-boundary energy is about 365 ergs/cm^2. What groove angle should form where a grain boundary meets the surface in a well-annealed gold specimen?
21. Suppose that you are given a metal with a stable precipitate whose volume fraction is 2×10^{-2} and whose average diameter is about 0.5 microns. You are asked by a customer to furnish this metal in one-inch thick plates with an average grain diameter of 1 mm. Do you think that it might be possible to satisfy this request?

8 Solid Solutions

When homogeneous mixtures of two or more kinds of atoms occur in the solid state, they are known as *solid solutions.* These crystalline solutions are quite common and equivalent to liquid and gaseous forms, for the proportions of the components can be varied within fixed limits, and the mixtures do not separate naturally. In addition, the term *solvent* refers to the more abundant atomic form, and *solute* to the less abundant.

Solid solutions occur in either of two distinct types. The first is known as a *substitutional solid solution.* In this case, a direct substitution of one type of atom for another occurs so that solute atoms enter the crystal to take positions normally occupied by solvent atoms. Figure 8.1A shows schematically an example containing two kinds of atoms (Cu and Ni). The other type of solid solution is shown in Fig. 8.1B. Here the solute atom (carbon) does not displace a solvent atom, but, rather, enters one of the holes, or interstices, between the solvent (iron) atoms.

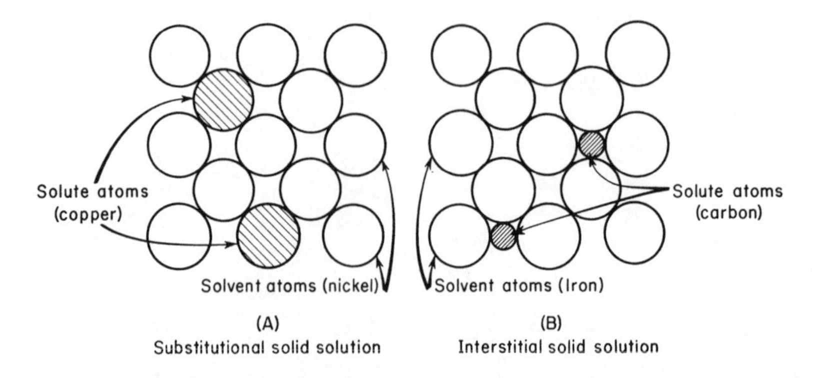

Fig. 8.1 The two basic forms of solid solutions. *Note:* In the interstitial example on the right, carbon is shown in solid solution in the face-centered cubic form of iron. Fig. 8.2 considers carbon dissolved interstitially in the body-centered cubic form of iron.

8.1 Intermediate Phases. In many alloy systems, crystal structures or phases are found that are different from those of the elementary components (pure metals). If these structures occur over a range of compositions, they are, in all respects, solid solutions. However, when the new crystal structures occur with simple whole-number fixed ratios of the component atoms, they are intermetallic compounds.

The difference between intermediate solid solutions and compounds can be more easily understood by actual examples. When copper and zinc are alloyed to form brass, a number of new structures are formed in different composition ranges. Most of these occur in compositions which have no commercial value whatsoever, but that which occurs at a ratio of approximately one zinc atom to one of copper is found in some useful forms of brass. The crystal structure of this new phase is body-centered cubic, whereas that of copper is face-centered cubic, and zinc is close-packed hexagonal. Because this body-centered cubic structure can exist over a range of compositions (it is the only stable phase at room temperature between 47 and 50 weight percent of zinc), it is not a compound, but a solid solution. On the other hand, when carbon is added to iron, a definite intermetallic compound is observed. This compound has a fixed composition (6.67 weight percent of carbon) and a complex crystal structure (orthorhombic, with 12 iron atoms and 4 carbon atoms per unit cell) which is quite different from that of either iron (body-centered cubic) or carbon (graphite).

8.2 Interstitial Solid Solutions. An examination of Fig. 8.1B shows that solute atoms in interstitial alloys must be small in size. The conditions which determine the solubilities in both interstitial and substitutional alloy systems have been studied in great detail by Hume-Rothery and others. According to their results, extensive interstitial solid solutions occur only if the solute atom has an apparent diameter smaller than 0.59 that of the solvent. The four most important interstitial solute atoms are carbon, nitrogen, oxygen, and hydrogen, all of which are small in size.

Atomic size is not the only factor that determines whether or not an interstitial solid solution will form. Small interstitial solute atoms dissolve much more readily in transition metals than in other metals. In fact, we find that carbon is so insoluble in most nontransition metals that graphite-clay crucibles are frequently used for melting them. Some of the commercially important transition metals are

Iron	Vanadium	Tungsten
Titanium	Chromium	Thorium
Zirconium	Manganese	Uranium
Nickel	Molybdenum	

The ability of transition elements to dissolve interstitial atoms is believed to be due to their unusual electronic structure. All transition elements possess an incomplete electronic shell inside of the outer, or valence, electron shell. The nontransitional metals, on the other hand, have filled shells below the valency shell.

The extent to which interstitial atoms can dissolve in the transition metals depends on the metal in question, but it is usually small. On the other hand, interstitial atoms can diffuse easily through the lattice of the solvent and their effects on the properties of the solvent are larger than might otherwise be expected. Diffusion, in this case, occurs not by a vacancy mechanism, but by the solute atoms jumping from one interstitial position to another.

8.3 Solubility of Carbon in Body-Centered Cubic Iron. Let us consider a specific interstitial solid solution, carbon in body-centered cubic iron, and confine our attention to temperatures below 723° C. This restriction is applied so that we do not have to concern ourselves with considerations of the face-centered cubic form of iron, which is stable at temperatures as low as 723° C in the presence of carbon.

The solubility of carbon in body-centered cubic iron is low. In fact, the number of carbon atoms in an iron crystal is roughly equivalent to the number of vacancies found in crystals. It is therefore possible to deduce an equation for the equilibrium number of carbon atoms in an iron crystal by analogy with our earlier vacancy computations. This will be done in the following section.

The positions marked with an *x* in Fig. 8.2 are those that carbon atoms take when they enter the iron lattice. They lie either midway between two corner atoms or in the centers of the faces of the cube, where they are midway between two atoms located at the centers of unit cells. In the diamond lattice, the carbon

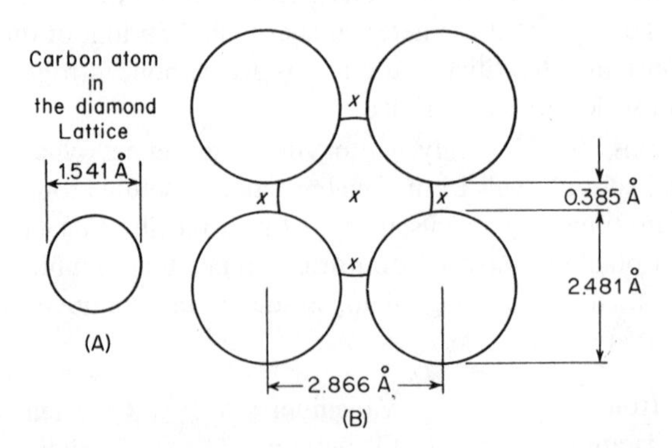

Fig. 8.2 The interstitial positions in the body-centered cubic iron unit cell that may be occupied by carbon atoms.

atom has an apparent diameter of 1.541 Å. However, Fig. 8.2B shows that the lattice constant of iron (length of one edge of the unit cell) is 2.866 Å. The diameter of the iron atom is, accordingly, 2.481 Å, and another simple calculation shows that the width of the hole occupied by carbon atoms is only 0.385 Å. It is quite apparent that the carbon atom does not fit well into the body-centered cubic crystal of iron. The lattice in the vicinity of each solute atom is badly strained, and work must be done in order to introduce the interstitial atom into the crystal.

Let us designate this work by the symbol w_c. If n_c is the number of carbon atoms in an iron crystal, the internal energy of the crystal (in the presence of carbon atoms) will be increased by the amount $n_c w_c$.

Each interstitial atom increases the entropy of the crystal in the same manner as does each vacancy. This form of entropy increment is known as an *intrinsic entropy*. This intrinsic entropy arises from the fact that the introduction of an interstitial atom affects the normal modes of lattice vibrations. The solute atom distorts the orderly array of iron atoms and this, in turn, makes the thermal vibrations of the crystal more random, or irregular. The total entropy contribution from this source is $n_c s_c$ where n_c is the number of carbon atoms.

In the derivation of the equation for the equilibrium number of vacancies in a crystal, the intrinsic entropy of the vacancies $n_v s_v$ was neglected. However, if it had been incorporated in the expression for the free energy associated with the presence of vacancies, the solution of the problem would have been no more difficult. In that case, the final result would have had the form

$$\frac{n_v}{n_o} = e^{-(w_v - Ts_v)/kT}$$

and not

$$\frac{n_v}{n_o} = e^{-w_v/kT}$$

where n_v = number of vacancies
n_o = number of atoms
w_v = work to form a vacancy
T = absolute temperature °K
k = Boltzmann's constant

The above, more exact, equation, including the intrinsic entropy, can also be expressed as follows:

$$\frac{n_v}{n_o} = e^{s_v/k} e^{-w_v/kT}$$

However, since the exponent s_v/k is a constant, this expression reduces to

$$\frac{n_v}{n_o} = Be^{-w_v/kT} = Be^{-Q_f/RT}$$

where B is a constant, Q_f is the work to introduce a mole of vacancies into the lattice, R is the universal gas constant, T the temperature in degrees Kelvin.

Now, with reference to carbon atoms in an iron crystal, it should be noted that the mixing of carbon atoms and iron atoms to form a solid solution involves an entropy of mixing analogous to that resulting from the mixing of vacancies and atoms.

$$S_m = k\{(n_{Fe} + n_c)\log_e (n_{Fe} + n_c) - n_c \log_e n_c - n_{Fe} \log_e n_{Fe}\}$$

where S_m = mixing entropy
n_{Fe} = number of iron atoms
n_c = number of carbon atoms

The analogy between interstitial carbon atoms in an iron crystal and vacancies in crystals is complete. The internal energy is increased in the presence of vacancies, or interstitial atoms, by amounts $n_v w_v$ and $n_c w_c$ respectively. The total intrinsic entropy of vacancies is $n_v s_v$; that of solute atoms $n_c s_c$, and, in both cases, there are equivalent mixing entropies. Because of this one-to-one correspondence between vacancies and interstitial atoms, the relationship between the temperature and the equilibrium number of carbon atoms in the iron lattice can be written directly as

$$C_c = \frac{n_c}{n_{Fe}} = e^{s_c/k}e^{-w_c/kT} = Be^{-w_c/kT} = Be^{-Q_c/RT}$$

where B is a constant and Q_c is the work to introduce a mole of carbon atoms into the iron lattice, R is 2 cal per mole, and T the temperature in degrees Kelvin.

Now let us consider the physical significance of w_c, which has been defined as the work required to introduce a carbon atom into an interstitial position. This element of work depends upon the source of the carbon atoms, which in alloys of pure iron and carbon can be either graphite crystals or iron-carbide crystals. In either case, the crystals that supply the carbon atoms are assumed to be in intimate contact with the iron. The most important of these two sources of carbon is iron-carbide, for commercial steels are almost universally aggregates of iron and iron-carbide. Graphite rarely appears in steels, even though it represents a more stable phase than iron-carbide. (Iron-carbide is a metastable phase and there is a free-energy decrease when it decomposes to form graphite and iron. However, the rate of this decomposition is extremely slow in the temperature range of normal steel usage, and, for all practical purposes, Fe_3C may be considered a stable phase.)

In metallurgical terminology, iron-carbide is called *cementite*, and the interstitial solid solution of carbon in body-centered cubic iron is known as *ferrite*. These terms will be used in the following sections of the book.

According to Wert,[1] the experimentally determined activation energy Q_c for the transfer of a mole of carbon atoms from cementite to ferrite is 9700 cal per mole. On the other hand, Darken and Gurry[2] give the corresponding value for graphite to ferrite as 14,800 cal per mole. The work to take a carbon atom from graphite and place it interstitially in iron is therefore greater than that to take it from iron-carbide and place it in iron. This, of course, has an effect on the equilibrium number of carbon atoms in the iron lattice. The solubility of carbon in ferrite is lower when the iron is in equilibrium with graphite than where it is in equilibrium with cementite, because a greater energy is required to remove an atom from graphite and place it in iron. The solubility equation

$$C_c = Be^{-Q_c/RT}$$

can be evaluated in terms of the experimental results of Wert and others, which specifically apply to the case of ferrite and cementite in equilibrium (that is, where the interstitial atoms are supplied by iron-carbide). If this is done, we obtain

$$C_c = 0.119\ e^{-9700/RT}$$

where C_c is the equilibrium concentration of carbon atoms (n_c/n_{Fe}). This equation is easily converted to read the carbon concentration in weight percent, thus

$$C'_c = 2.55\ e^{-9700/RT}$$

Figure 8.3 shows a plot of the equilibrium carbon concentration between room temperature and 723° C. The curve clearly shows that the solubility of carbon in ferrite is very small. The equilibrium value at room temperature is only 2.3×10^{-7} weight percent, or one part in 2.3×10^{-9}. This is equivalent to approximately one carbon atom in every 10^8 iron atoms, or a mean separation between solute atoms of about 100 solvent atoms. Figure 8.3 also shows the temperature dependence of the equilibrium number of carbon atoms. At 723° C

1. Wert, C. A., *Trans. AIME,* 188 1242 (1950).

2. Darken, L. S., and Gurry, R. W., *Physical Chemistry of Metals.* McGraw-Hill Book Co., Inc., New York, 1953.

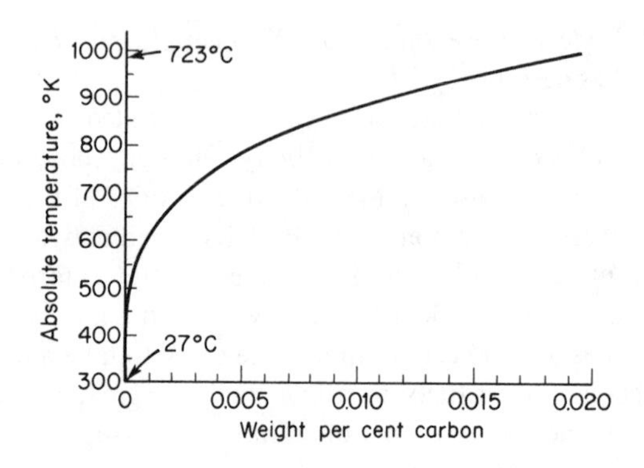

Fig. 8.3 Solubility of carbon in alpha-iron (body-centered cubic iron).

the carbon concentration reaches its maximum value, 0.02 percent. In this case, there is about one carbon atom for every 1000 iron atoms, or a separation equivalent to approximately 10 solvent atoms between carbon atoms.

8.4 Substitutional Solid Solutions and the Hume-Rothery Rules. In Figure 8.1A, the copper and nickel atoms are drawn with the same diameters. Actually, the atoms in a crystal of pure copper have an apparent diameter (2.551 Å) about 2 percent larger than those in a crystal of pure nickel (2.487 Å). This difference is small and only a slight distortion of the lattice occurs when a copper atom enters a nickel crystal, or vice versa, and it is not surprising that these two elements are able to crystallize simultaneously into a face-centered cubic lattice in all proportions. Nickel and copper form an excellent example of an alloy series of complete solubility.

Silver, like copper and nickel, crystallizes in the face-centered cubic structure. It is also chemically similar to copper. However, the solubility of copper in silver, or silver in copper, equals only a fraction of 1 percent at room temperature. There is, thus, a fundamental difference between the copper-nickel systems and the copper-silver systems. This dissimilarity is due primarily to a greater difference in the relative sizes of the atoms in the copper-silver alloys. The apparent diameter of silver is 2.884 Å, or about 13 percent larger than that of copper. It is thus very close to the limit noted first by Hume-Rothery, who pointed out that an extensive solid solubility of one metal in another only occurs if the diameters of the metals differ by less than 15 percent. This criterion for solubility is known as the *size factor* and is directly related to the strains produced in the lattice of the solvent by the solute atoms.

The size factor is only a necessary condition for a high degree of solubility. It is not a sufficient condition, since other requirements must be satisfied. One of the most important of these is the relative positions of the elements in the electromotive series. Two elements, which lie far apart in this series, do not, as a rule, alloy in the normal sense, but combine according to the rules of chemical valence. In this case, the more electropositive element yields its valence electrons to the more electronegative element, with the result that a crystal with ionic bonding is formed. A typical example of this type of crystal is found in NaCl. On the other hand, when metals lie close to each other in the electromotive series, they tend to act as if they were chemically the same, which leads to metallic bonding instead of ionic.

Two other factors are of importance, especially when one considers a completely soluble system. Even if the size factor and electromotive series positions are favorable, such a system is only possible when both components (pure metals) have the same valence, and crystallize in the same lattice form.

8.5 Interaction of Dislocations and Solute Atoms. A dislocation is a crystal defect the presence of which in a crystal means that large numbers of atoms have been displaced from their normal lattice positions. This displacement of the atoms around the center of a dislocation results in a complex two-dimensional strain pattern with the dislocation line at its center.

8.6 The Stress Field of a Screw Dislocation. The elastic strain of a screw dislocation is shown in Fig. 8.4. In this form of dislocation, the lattice spirals around the center of the dislocation, with the result that a state of shear strain is set up in the lattice. This strain is symmetrical about the center of the dislocation and its magnitude varies inversely as the distance from the dislocation center. Vector diagrams are drawn on the front and top of the cylindrical crystal to indicate the shear sense.

That the strain decreases as one moves away from the dislocation line is deduced as follows. Consider the circular Burgers circuit shown in Fig. 8.4. Such a path results in an advance (parallel to the dislocation line) equal to the Burgers

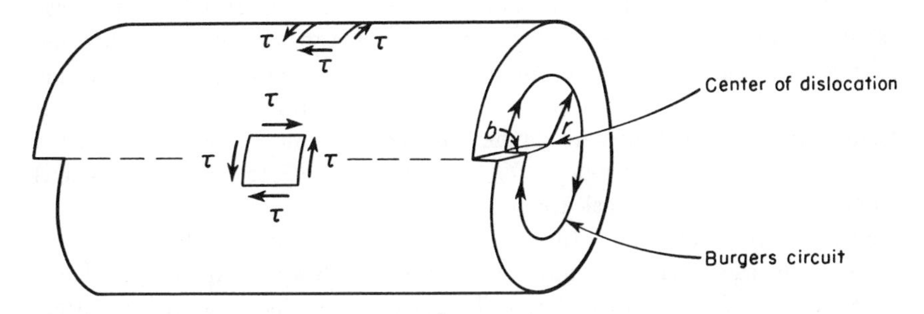

Fig. 8.4 Shear strain associated with a screw dislocation.

vector b. The strain in the lattice, however, is the advance divided by the distance around the dislocation. Thus,

$$\gamma = \frac{\mu b}{2\pi r}$$

where r is the radius of the Burgers circuit. This strain is accompanied by a corresponding state of stress in the crystal. A great deal of useful information about the nature of the stress fields produced by dislocations has been obtained by assuming the crystals to be homogeneous isotropic bodies. If this is done, the elastic stress field surrounding a screw dislocation is written:

$$\tau = \mu\gamma = \frac{\mu b}{2\pi r}$$

where μ is the shear modulus of the material of the crystal. This equation gives a reasonable approximation of the stress at distances greater than several atomic distances from the center of the dislocation. As the center of the dislocation is approached, however, representation of the crystal as a homogeneous and isotropic medium becomes less and less realistic. At positions close to the dislocation, atoms are displaced long distances from their normal lattice positions. Under these conditions, the stress is no longer directly proportional to strain, and it becomes necessary to think in terms of forces between individual atoms. The analysis of the stresses close to the center of the dislocation is extremely difficult, and no completely satisfactory theory has yet been developed. In this respect, it should be mentioned that the infinite stress predicted by the above equation at zero radius has no meaning. While the exact stress at the center of the dislocation is not known, it cannot be infinite.

8.7 The Stress Field of an Edge Dislocation. Thelattice strain around a dislocation in the edge orientation is more complicated than that of a dislocation in a screw orientation. The nature of the lattice distortion in this case is shown in Fig. 8.5, where small, free-body diagrams are superimposed upon the drawing of an edge dislocation. The diagrams indicate the nature of the lattice strain at four positions 90° apart about the dislocation. At these particular locations the strain patterns are simple because of symmetry considerations. On the other hand, similar free-body diagrams at intermediate positions (angular) exhibit both shear and normal stresses.

Figure 8.5B shows the stress patterns, corresponding to the strains of Fig. 8.5A. In each free-body diagram, the appropriate stress equation is placed alongside its respective stress vector. These equations, like that for the stress field around a screw dislocation, are based on the assumption of a homogeneous

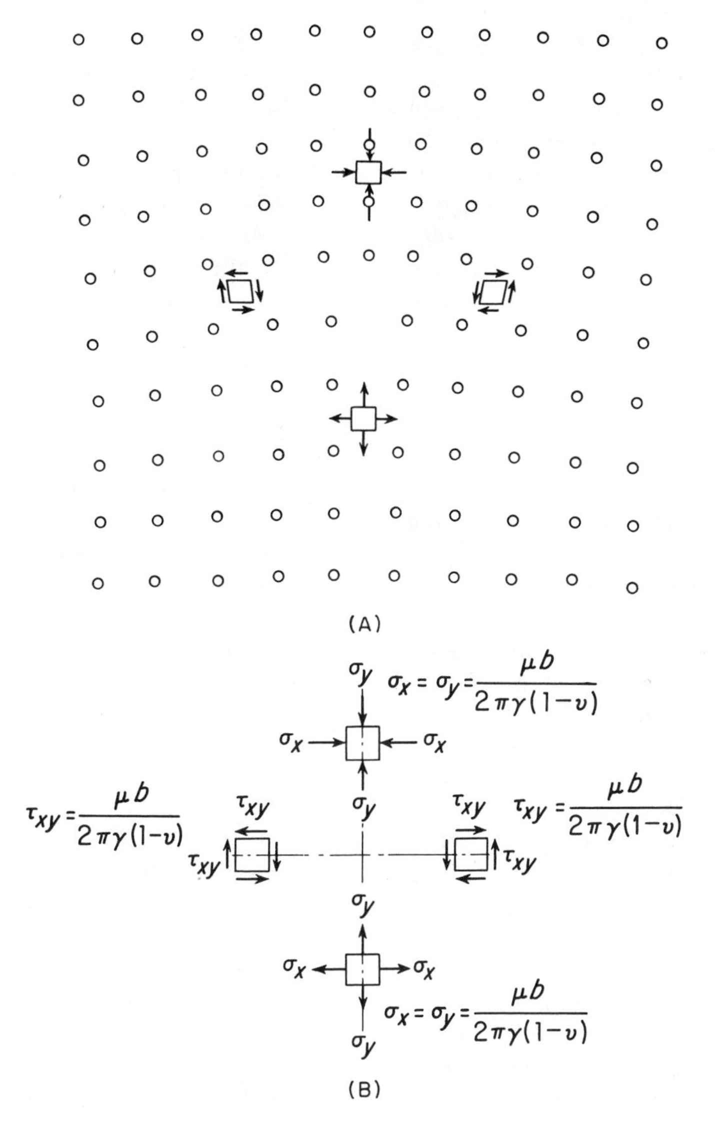

Fig. 8.5 Stress and strain associated with an edge dislocation. (In the above equations, G is the shear modulus in psi, b the Burgers vector of the dislocation, μ is Poisson's ratio, and r the distance from the center of the dislocation.)

and isotropic material, and it is also necessary to point out that they are not valid at distances close to the center of the dislocation. The uppermost diagram shows that the lattice above the edge is in a state of biaxial compressive stress, while that below the edge is subjected to a biaxial tensile stress. The left- and right-hand figure show that along the slip plane the lattice is in a state of simple shear. It is important to notice that in all cases the magnitude of the stress depends only on r and varies as $1/r$, where r is the distance from the center of the dislocation. This is true not only for the four angular positions shown in Fig. 8.5, but at any intermediate angle. It can be concluded that the nature of the stress pattern around an edge dislocation varies as a function of angular position about the center of the dislocation, but that the magnitude of the stress at any given angle depends only on the distance from the center of the dislocation.

8.8 Dislocation Atmospheres. When a crystal contains both dislocations and solute atoms, interactions may occur. Of particular interest is the interaction between substitutional solutes and dislocations in the edge orientation. If the diameter of a solute atom is either larger or smaller than that of the solvent atom, the lattice of the latter is strained. A large solute atom expands the surrounding lattice, while a small one contracts it. These distortions may be largely relieved if the solute atom finds itself in the proper place close to the center of a dislocation. Thus, the free energy of the crystal will be lowered when a small solute atom is substituted for a larger solvent atom in the compressed region of a dislocation in, or close to, the extra plane of the dislocation. (See Fig. 8.5). In fact, it can be shown by computations that the stress field of the dislocation attracts small solvent atoms to this area. Similarly, large solvent atoms are drawn to lattice positions below the edge; the expanded region of the dislocation.

Substitutional atoms do not react strongly with dislocations in the screw orientation where the strain field is nearly pure shear. The lattice distortion associated with substitutional atoms can be assumed to be spherical in shape. Figure 8.6 shows that a state of pure shear is equivalent to two equal normal strains (principal strains) – one tensile and one compressive. It is clear that lattice strains of this type will not react strongly with the spherical strain associated with substitutional solute atoms.

Now consider the interaction of interstitial atoms and dislocations. Interstitial atoms can react with the edge orientations of dislocations since they normally cause an expansion of the solvent lattice. As a result, interstitial atoms are drawn toward the expanded regions of edge dislocations. Also, because of the nonspherical lattice distortions that interstitial atoms produce in body-centered cubic metals, they are capable of reacting with the screw components of dislocations in this type of lattice. An example has already been cited when it was shown that the carbon atom enters a restricted space between a pair of iron atoms and shoves them apart. (See Fig. 8.2). Since, in the body-centered cubic lattice, the atom pairs referred to always lie along the edges of the unit cells or at

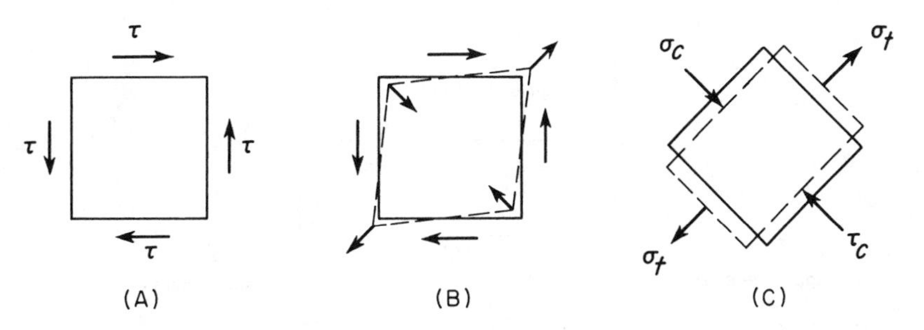

Fig. 8.6 (A) Stress vectors on a unit cube corresponding to a state of pure shear. (B) Deformation of a unit cube corresponding to shear stresses of Fig. 8.6A. (C) A state of pure shear is equivalent to two equal normal stresses, one compressive and one tensile, aligned at 45° to the shear stresses.

the centers of two adjacent unit cells, the distortions (expansions) are in ⟨100⟩ directions. A nonspherical lattice distortion such as this will react with the shear strain field of a screw dislocation as follows. In those regions close to a screw dislocation where a ⟨100⟩ direction lies close to the principal tensile strain of the dislocation's strain field (Fig. 8.6), the holes separating iron-atom pairs lying along this direction will be enlarged. Carbon atoms will naturally seek these enlarged interstitial positions.

Solute atoms are drawn toward dislocations as a result of the interactions of their strain fields. However, the rate at which the solute atoms move under these attractive forces is controlled by the rate at which they can diffuse through the lattice. At high temperatures, diffusion rates are rapid, and solute atoms concentrate quickly around dislocations. If the solute atoms have a mutual attraction, the precipitation of a second crystalline phase may start at dislocations. In this case, dislocations act as sinks for solute atoms, and the flow can be assumed to be in one direction – toward the dislocations. Movements of this nature can be expected to continue until the solute concentration in the crystal is depleted (lowered to the point where it is in equilibrium with the newly formed phase). On the other hand, if the solute atoms do not combine to form a new phase, an equilibrium state should develop, with equal numbers of solute atoms entering and leaving a finite volume containing a dislocation. Under these conditions, a steady-state concentration of solute atoms builds up around the dislocation which is higher than that of the surrounding lattice. This excess of solute atoms associated with a dislocation is known as its *atmosphere.* The number of atoms in the atmosphere depends on the temperature. Increasing the temperature tends to tear solute atoms away from dislocations and to increase the entropy of the crystal. Increasing temperatures thus lower the solute concentrations around dislocations, and at a sufficiently high temperature, the

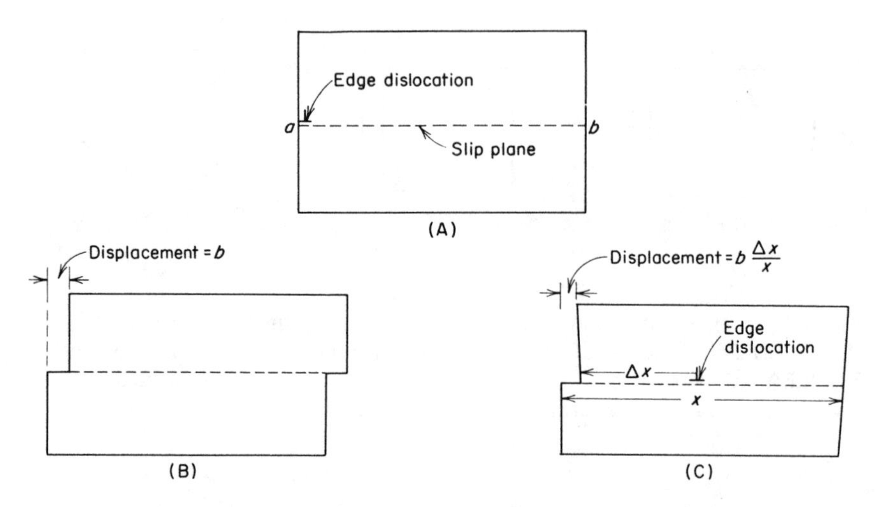

Fig. 8.7 The displacement of the two halves of a crystal is in proportion to the distance that the dislocation moves on its slip plane.

concentrations can be lowered to the point where it can be considered that dislocation atmospheres no longer exist.

8.9 The Orowan Equation. A relationship between the velocity of the dislocations in a test specimen and the applied strain rate will now be derived. This expression is known as the Orowan equation.

As shown in Figs. 8.7A and 8.7B, when an edge dislocation moves completely across its slip plane, the upper half of the crystal is sheared relative to the lower half by an amount equal to one Burgers vector. It can also be deduced and rigorously proved[3] (Fig. 8.7C) that if the dislocation moves only through a distance Δx, then the top surface of the crystal will be sheared by an amount equal to $b(\Delta x/x)$ where x is the total distance across the slip plane. In other words, the displacement of the upper surface will be in proportion to the fraction of the slip plane surface that the dislocation has crossed, or to $b(\Delta A/A)$, where A is the area of the slip plane and ΔA is the fraction of it passed over by the dislocation. Since the shear strain γ given to the crystal equals the displacement $b(\Delta A/A)$ divided by the height z of the crystal, we have

$$\Delta\gamma = \frac{b\Delta A}{Az} = \frac{b\Delta A}{V}$$

3. Cottrell, A. H., *Dislocations and Plastic Flow in Crystals*, p. 45, Oxford Press, London, 1953.

since Az is the volume of the crystal. For the case where n edge dislocations of length l move through an average distance $\Delta\bar{x}$, this relation becomes

$$\Delta\gamma = \frac{bnl\Delta\bar{x}}{V} = \rho b\Delta\bar{x}$$

where ρ, the dislocation density, is equal to nl/V. If, in a time interval Δt, the dislocations move through the average distance $\Delta\bar{x}$, we have

$$\frac{\Delta\gamma}{\Delta t} = \dot{\gamma} = \rho b\bar{v}$$

where $\dot{\gamma}$ is the shear strain rate and $\bar{v}$ is the average dislocation velocity.

This expression, derived for the specific case of parallel edge dislocations, is a general relationship, and it is customary to consider that ρ represents the density of all the mobile dislocations in a metal whose average velocity is assumed to be $\bar{v}$. Furthermore, if $\dot{\epsilon}$ is the tensile strain rate in a polycrystalline metal, a reasonable assumption is that

$$\dot{\epsilon} = \frac{1}{2}\dot{\gamma} = \frac{1}{2}\rho b\bar{v}$$

where the factor 1/2 is an approximate Schmid orientation factor.

8.10 The Drag of Atmospheres on Moving Dislocations. Consider an edge dislocation in a body-centered cubic metal such as iron. Directly above the dislocation line the compressive stress that exists there tends to lower the carbon or nitrogen atom concentration to a value below that existing in the crystal as a whole. At the same time, the tensile stress below the dislocation attracts these interstitial solute atoms. The dislocation atmosphere around an edge dislocation therefore consists of an excess concentration of interstitial atoms below the edge, and a deficiency of these atoms above the edge.

When such a dislocation moves, its atmosphere tends to move with it. The movement of a dislocation away from its atmosphere creates an effective stress on the solute atoms that draws them back toward their equilibrium distribution. This motion can only occur by thermally activated jumps of the atoms from one interstitial position to another. As a result, the atmosphere tends to lag behind the dislocation. At the same time, the distribution of the atoms in the atmosphere also changes. This is because the structure of the atmosphere is now influenced by several additional factors. The most important of these is probably that the movement of the dislocation through the crystal tends to bring additional solute atoms into the atmosphere. At the same time, a corresponding number of solute atoms must leave the atmosphere on the side away from the direction of motion. In this process, it can be considered that the movement of the dislocation through the crystal tends to realign those solute atoms lying just above its slip plane into positions below its slip plane. The atmosphere associated with a moving dislocation is thus a dynamic concept, but its existence can have a strong influence on the motion of a dislocation.

Now let us consider how the atmosphere affects the motion of a dislocation. The interaction stress between the solute atoms in the atmosphere and the dislocation makes it more difficult to move the dislocation, and this stress has to be overcome in order to advance the dislocation. The drag-stress due to a dislocation atmosphere is therefore one of the important components of the flow-stress of a metal. This stress component is a function of the dislocation velocity. The qualitative nature of its dependence on the velocity can be easily visualized. At both very high and very low velocities the drag-stress has to be very small. At extremely high dislocation velocities the dislocation passes by the solute atoms at such a fast rate that there is insufficient time for the atoms to rearrange themselves. Under these conditions, the solute atoms can be considered as a set of fixed obstacles through which the dislocation moves. An atmosphere of solute atoms should not exist under these conditions. On the other hand, when the dislocation is at rest there exists no net stress between the dislocation and its atmosphere. If the dislocation is now given a small velocity, the center of its atmosphere will move to a position behind the dislocation. The distance of separation between the dislocation and the effective center of its atmosphere increases with increasing velocity. Computations[4] indicate that this causes an increase in the drag-stress that is proportional to the dislocation velocity. However, eventually a maximum drag-stress should be reached, since at very high velocities the atmosphere itself becomes less and less well-defined. Figure 8.8A shows the nature of this dependence of the drag-stress on the dislocation velocity.

It is appropriate now to consider the effect of temperature on the drag-stress dislocation velocity relationship. The drag-stress is, in effect, a manifestation of a coupling between a moving dislocation and a set of interstitial atoms that also must move in order to both form and maintain the atmosphere. Accordingly, it can be assumed that the maximum drag-stress corresponds to a direct relationship between the dislocation velocity and the diffusion rate of the solute atoms. Increasing the temperature increases the rate at which the solute atoms move and, as a result, the dislocation velocity corresponding to a maximum drag-stress also has to increase. This is shown in Fig. 8.8B. Note that at the higher temperature the critical velocity v_c is higher. However, the maximum

4. Cottrell, A. H., and Jaswon, M. A., *Proc. Roy. Soc.*, A199 104 (1949).

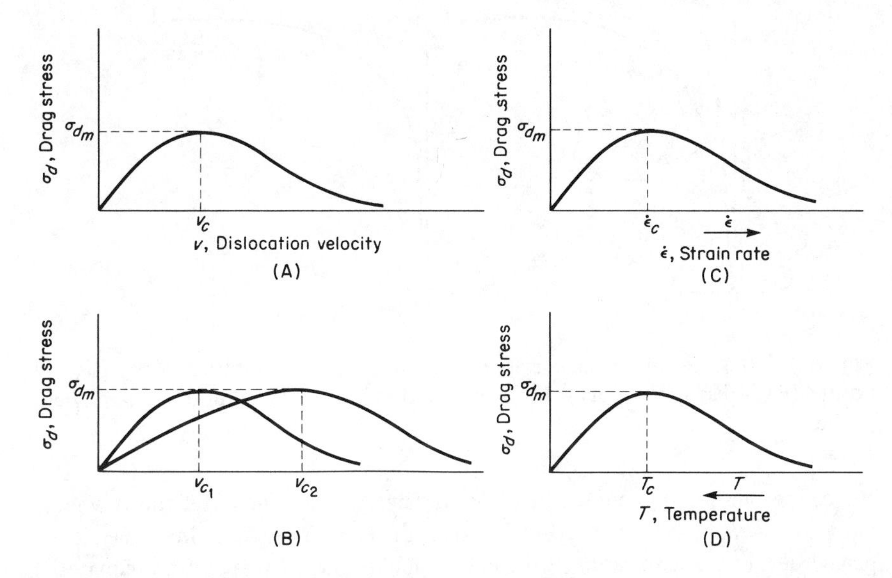

Fig. 8.8 Variation of the drag-stress with, (A) the dislocation velocity, (B) the dislocation velocity at two different temperatures, (C) the strain rate, and (D) the temperature at constant strain rate.

drag-stress σ_{d_m} remains the same. This is in agreement with the theoretical treatment of Cottrell and Jaswon.[5]

By the Orowan equation we have $\dot{\gamma} = \rho b \bar{v}$. Therefore, at a constant dislocation density, the average dislocation velocity should be directly proportional to the strain rate. Therefore, it is reasonable to assume that when deformation occurs at a nearly constant value of ρ, the relationship between the strain rate and the drag-stress component of the flow-stress should be similar to that shown in Fig. 8.8C. Finally, because of the interrelationship between strain rate and temperature, it may be deduced that a similar relationship exists between the temperature and the drag-stress when the strain rate is held constant. This is illustrated in Fig. 8.8D. Note that in this case the temperature decreases as the abscissa coordinate increases. This is because a very low temperature at a constant strain rate is equivalent to a high strain rate at a constant temperature. In either case, the solute becomes immobile with respect to the moving dislocation.

8.11 The Sharp Yield Point and Lüders Bands. When the stress-strain curves of metals deformed in tension are plotted, two basic types of

5. Cottrell, A. H., and Jaswon, M. A., *Proc. Roy. Soc.*, **A199** 104 (1949).

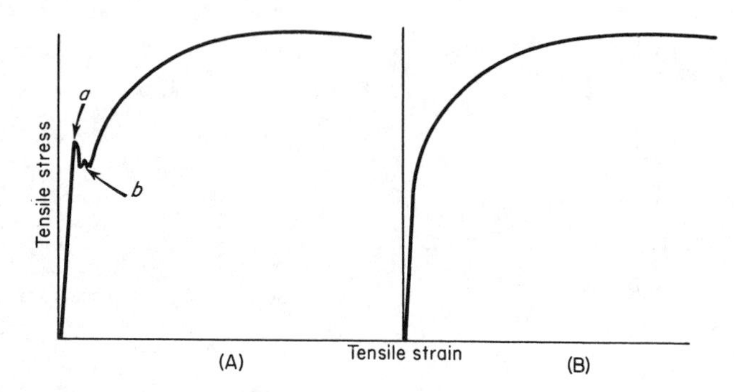

Fig. 8.9 (A) Tensile stress-strain curve for a metal exhibiting a sharp yield point. (B) Stress-strain curve for a metal that does not exhibit a sharp yield point.

curves are observed, as shown in Fig. 8.9. The curve on the left exhibits what is known as a *sharp yield point.* In this curve, the stress rises with almost negligible plastic deformation to point *a,* the upper yield point. At this point, the material begins to yield, with a simultaneous drop in the flow-stress required for continued deformation. This new yield stress, point *b*, is called the lower yield point and corresponds to an appreciable plastic strain at an almost constant stress. Eventually the metal starts to harden with an increase in the stress necessary for additional deformation. After this occurs, there is little difference between the appearance of the stress-strain curves for metals with a yield point and those without it.

The sharp yield point is an especially important effect because it occurs in iron and in low-carbon steels. Its existence is of considerable concern to manufacturers who stamp or draw thin sheets of these materials in forming such objects as automobile bodies. The significance of the yield point is this: once plastic deformation starts in a given area, the metal at this point is effectively softened and suffers a relatively large plastic deformation. This deformation then spreads into the material adjoining the region which has yielded because of the stress concentration at the boundary between the deformed and undeformed areas. In general, deformation starts at positions of stress concentration as discrete bands of deformed material, called *Lüders bands.* In the usual tensile test specimen (Fig. 8.10), the fillets are stress raisers and points at which Lüders bands form.[6] The edges of these bands make approximately 50° angles with the stress axis and are designated *Lüders lines.* Lüders bands should not be confused with slip lines. It is quite possible to have hundreds of crystals co-operating to form a Lüders band, each slipping in a complicated fashion on its own slip planes. Once a Lüders band has formed at one fillet of a tensile test specimen, it

6. Liss, R. B., *Acta Met.*, 5 342 (1957).

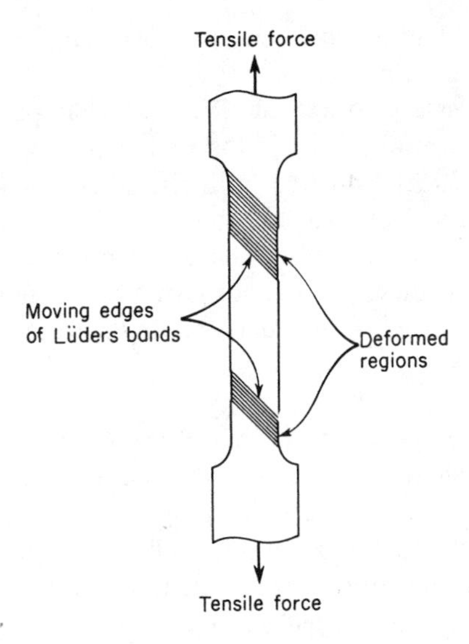

Fig. 8.10 Lüders bands in a tensile-test specimen.

can then move through the gage length of the specimen. The bands can form simultaneously at both ends of a tensile test specimen, or even, under certain conditions, at a number of positions throughout the specimen gage length. In any case, the deformation starts at localized areas and spreads into undeformed areas. This occurs at an almost constant stress and explains the horizontal part of the stress-strain curve at the lower yield point. In fact, the lower yield stress can be viewed as the stress required to propagate the Lüders bands. The velocity with which a Lüders front moves increases with an increase in the applied stress. Most testing machines tend to deform a specimen at a fixed rate. If two Lüders bands move through a specimen's gage section, the front of each Lüders band will move at approximately half the rate of the front of a single band. This means that the lower yield stress required to move two fronts will be less than that required to move a single front and, in general, an increase or decrease in the number of moving Lüders fronts must be accompanied by a corresponding fluctuation in the lower yield stress. This accounts for the fluctuating lower yield stress shown schematically in Fig. 8.9. Only after the Lüders deformation has covered the entire gage section does the stress-strain curve start to rise again.

In the previously mentioned example of low-carbon-steel sheet used in forming automobile bodies, the effect of using metals containing a yield point is to develop roughened surfaces. These surfaces result from the uneven spread of Lüders bands which leave striations on the surface commonly called *stretcher strains*. On the other hand, when material without a yield point is used, work hardening instead of softening sets in as soon as plastic deformation starts.

This tends to spread deformation uniformly over large areas and to produce smooth surfaces after deformation.

It may not be necessary to exceed all of the yield point strain in order to remove most of the harmful effects of the Lüders phenomenon. Annealed steel sheet is often given a slight reduction in thickness by rolling which amounts to about a 1 percent strain. This is called a *temper roll*, and it produces a very large number of Lüders band nuclei[7] in the sheet. When the metal is later deformed into a finished product, these small bands grow; but, because of their small size and close proximity to each other, the resulting surface roughening is greatly reduced.

8.12 The Theory of the Sharp Yield Point. It has been proposed by Cottrell[8] that the sharp yield point which occurs in certain metals is a result of the interaction between dislocations and solute atoms. According to this theory, the atmosphere of solute atoms that collect around dislocations serves to pin down, or anchor the dislocations. Additional stress, over that normally required for movement, is needed in order to free a dislocation from its atmosphere. This results in an increase in the stress required to set dislocations in motion, and corresponds to the upper yield-point stress. The lower yield point in the original Cottrell theory then represents the stress to move dislocations that have been freed from their atmospheres.

All evidence points to the correctness of Cottrell's assumption that the increase in yield stress associated with the upper yield point is caused by the interaction of dislocations with solute atoms. Whether or not the yield point is associated with a simple breaking away of dislocations from their atmospheres is still in doubt. The original Cottrell theory does not satisfactorily explain certain experimental data.

A theory of the yield point originally proposed by Johnston and Gilman[9,10] to explain yielding in LiF crystals is worth noting. In effect, this theory postulates that in a metal with a very low initial dislocation density, the first increments of plastic strain cause an extremely large relative increase in the dislocation density. To be explicit, suppose that the dislocation density in a metal were to increase directly with the strain at a rate of about 10^8 cm^{-2} per each one percent strain. If the initial dislocation density was only 10^4 cm^{-2}, the dislocation density would increase by a factor of about 10,000 times in the first one percent of strain. However, in the next one percent of strain, the density

7. Butler, J. F., *Flat Rolled Products III*, p. 65. AIME Metallurgical Conf. Series, Vol. 16, Interscience Publishers, New York, 1962.

8. Cottrell, A. H., *Dislocations and Plastic Flow in Crystals*, p. 140. Oxford University Press, London, 1953.

9. Johnston, W. G., *J. Appl. Phys.*, 33 2716 (1962).

10. Johnston, W. G. and Gilman. J. J., *J. Appl. Phys.*, 30 129 (1959).

would only increase by a factor of two; and for each succeeding one percent increment, the relative increase would be progressively smaller.

The Johnston-Gilman analysis shows how this rapid dislocation multiplication at the beginning of a tensile test can produce a yield drop. The average testing machine tries to deform a tensile specimen at a constant rate. When the dislocation density is very low, the resultant strain is largely elastic, and the stress rises rapidly. Associated with this resultant large stress is a very high dislocation velocity. At the same time, it should be remembered that the dislocation density is increasing rapidly. A high dislocation velocity and an increasing number of mobile dislocations eventually produce a point of instability where the specimen deforms plastically at the same rate as the machine. Beyond this point the load has to fall and, in so doing, lowers the dislocation velocity, with the result that the rate of decrease in the load, while rapid at first, becomes slower and slower. This continued decrease in flow-stress with strain beyond the yield point is normally masked by the work hardening that occurs in the metal. While the Johnston-Gilman theory does not account for the Lüders deformation, it does give a very interesting analysis of the phenomenon of yielding.

In the case of iron and steel, the room-temperature yield point has been shown to be due to either carbon or nitrogen in interstitial solid solution. An important question is how much carbon or nitrogen is needed to form atmospheres around the dislocations. This can be roughly estimated in the following fashion. The number of carbon atoms in an atmosphere is not known with certainty, but can be assumed to be of the order of 1 carbon atom per atomic distance along the dislocation. The iron atom in the body-centered cubic structure has a diameter of about 2.5 Å. In a dislocation 1 cm long, there are 10^8 Å units, or 4×10^7 atomic distances. It can, accordingly, be assumed that there will be 4×10^7 carbon atoms per cm of dislocation.

The density of dislocations in a soft annealed crystal is usually about 10^8 cm^{-2}. The corresponding number in a highly cold worked metal is 10^{12}, or 10,000 times as many as in unstrained material. In the soft state, the total length of all dislocations in a cubic centimeter is 10^8 cm, and with 4×10^7 carbon atoms per cm, the total number of carbon atoms is 4×10^{15}. Let us compare this with the total number of iron atoms in the same crystal. The length of the body-centered cubic unit cell is 2.86 Å. Thus there are 4.3×10^{22} unit cells in a cubic centimeter. Since the body-centered cubic structure has 2 atoms per unit cell, the number of iron atoms is 8.6×10^{22}, or about 10^{23}. The concentration of carbon atoms required to form an atmosphere of 1 carbon atom per atomic distance along a dislocation is thus:

$$\frac{n_c}{n_c + n_{Fe}} \simeq \frac{n_c}{n_{Fe}} = \frac{4 \times 10^{15}}{10^{23}} = 4 \times 10^{-8}$$

or four carbon atoms in every hundred million iron atoms. In a heavily cold worked crystal, the same ratio would be

$$\frac{n_c}{n_{Fe}} = 4 \times 10^{-4}$$

or about 0.04 percent. The significance of the above figures is plain: very little carbon (or nitrogen) is required in order to have sufficient interstitial solute available to form dislocation atmospheres.

8.13 Strain Aging. Figure 8.11A shows a tensile stress-strain curve for a metal with a sharp yield point in which loading was stopped at point *c* and then the load was removed. During the unloading stage, the stress-strain curve follows a linear path parallel to the original elastic portion of the curve (line *ab*). On reloading within a short time (hours), the specimen behaves elastically to approximately point *c*, and then deforms plastically, and no yield point is observed. On the other hand, if the specimen is not retested for a period of some months and during this period is allowed to age at room temperature, the yield point reappears, as is shown in Fig. 8.11B. The aging period has raised the stress at which the specimen yields and, as a result, the specimen is strengthened and hardened. This type of phenomenon, where a metal hardens as a result of aging after plastic deformation, is called *strain aging*. The yield point that reappears during strain aging is also associated with the formation of solute atom atmospheres around dislocations. Those dislocation sources which were active in the deformation process just before the specimen was unloaded are pinned down as a result of the aging process. Because solute atoms must diffuse through the lattice in order to accumulate around dislocations, the reappearance of the yield point is a function of time. It also depends on the temperature,

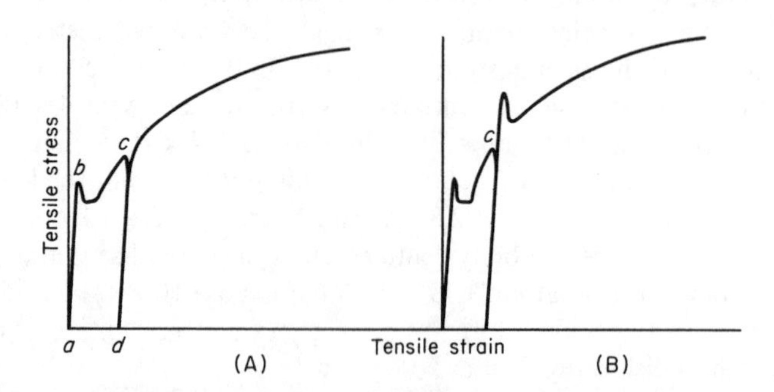

Fig. 8.11 Strain aging. (A) Load removed from specimen at point *c* and specimen reloaded within a short period of time (hours). (B) Load removed at point *c* and specimen reloaded after a long period of time (months).

inasmuch as diffusion is a temperature-dependent function. The higher the temperature the faster the rate at which the yield point will reappear. The yield point is not normally observed in iron and steels tested at elevated temperatures (above approximately 400°C). This fact can be explained by the tendency for dislocation atmospheres to be dispersed as a result of the more intense thermal vibrations found at elevated temperatures.

The yield point and strain aging phenomena are most closely associated with iron and low-carbon steel. However, they are also observed in many metals including other body-centered cubic, face-centered cubic, and hexagonal close-packed metals. In many cases the phenomena are not as pronounced as they are in steel.

8.14 Dynamic Strain Aging. As indicated above, the higher the temperature the faster the yield point reappears. At a sufficiently high temperature, the interaction between the impurity atoms and the dislocations should occur during deformation. When aging occurs during deformation, the phenomenon has been callled *dynamic strain aging*. There are many physical manifestations of dynamic strain aging, some of which will now be described.

First, it is important to note that dynamic strain aging tends to occur over a wide temperature range and that the temperature interval in which it is observed depends on the strain rate. Increasing the strain rate raises both the upper and lower temperature limits associated with the dynamic strain aging phenomena. Thus, in steel deformed at a normal cross-head speed of 0.02 in. per min, dynamic strain aging effects are observed between approximately 100°C and 350°C. However, at a strain rate 10^6 faster, these same limits occur at about 450°C and 700°C.

An interesting aspect of dynamic strain aging is that when it occurs the yield stress (or the critical resolved shear stress) of a metal tends to become independent of temperature. This is shown in Fig. 8.12, where the 0.2 percent yield stress of titanium is plotted as a function of temperature. Note that from slightly above 600°K to about 800°K the yield stress is almost constant. At the same time, the flow-stress becomes almost independent of strain rate. In many metals the flow-stress can be related to the strain rate by a simple power law

$$\sigma = A(\dot{\epsilon})^n$$

where A is a constant and the exponent n is called the strain rate sensitivity. This equation can also be written in the form

$$\frac{\sigma_2}{\sigma_1} = \left(\frac{\dot{\epsilon}_2}{\dot{\epsilon}_1}\right)^n$$

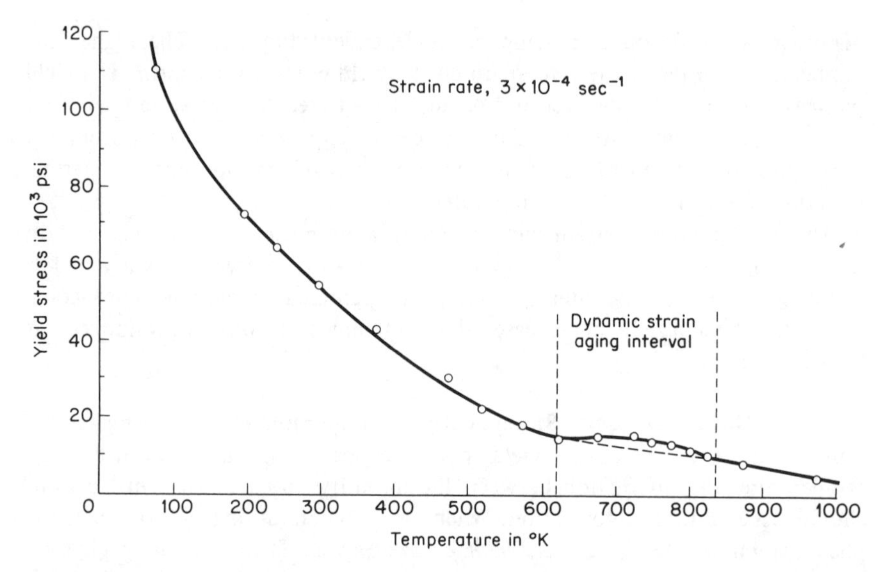

Fig. 8.12 The variation of the yield stress (0.2 percent strain) with the temperature for commercial purity titanium. Note that in the dynamic strain aging interval the flow-stress becomes approximately constant. Data courtesy of A. T. Santhanam.

and taking the logarithms of both sides we have

$$n = \frac{\log_e \dfrac{\sigma_2}{\sigma_1}}{\log_e \dfrac{\dot{\epsilon}_2}{\dot{\epsilon}_1}}$$

In most modern testing machines very rapid changes in the applied strain rate may be made during a tensile test. If this is done and the corresponding values of strain rate and flow-stress (measured just before and after the rate change) are substituted into this equation, one is able to determine a value of n. When measurements of this type, corresponding to a change in strain rate between two rates differing by a simple ratio, are made on a metal that is not subject to dynamic strain aging, the value of n tends to rise linearly with the absolute temperature. However, when dynamic strain aging is observed, the value of n becomes very low in the temperature range of strain aging, because, in this interval, the flow-stress becomes nearly independent of the strain rate. This is illustrated for the case of an aluminum alloy in Fig. 8.13.

Inside the dynamic strain aging temperature range the plastic flow often tends to become unstable. This is manifested by irregularities in the stress-strain

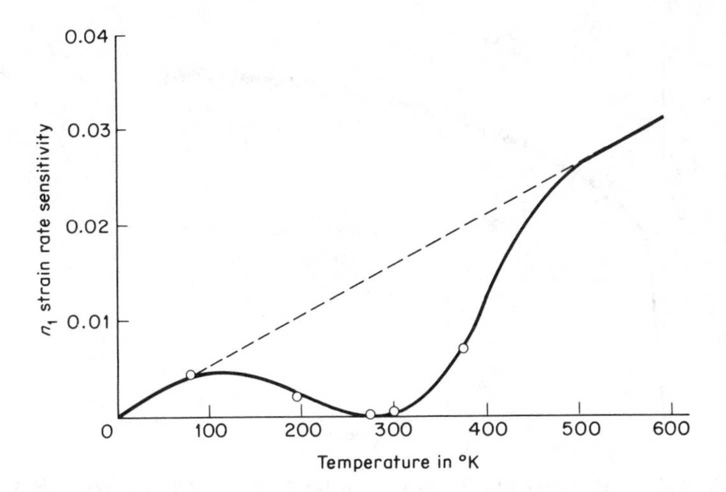

Fig. 8.13 Aluminum alloys are subject to strong dynamic strain aging effects near room temperature. Note that the strain rate sensitivity of 6061S-T aluminum shows a minimum strain rate sensitivity just below room temperature. After Lubahn, J. D., *Trans. AIME,* **185** 702 (1949).

diagram. These discontinuities may be of several types. In some cases the load tends to rise abruptly and then fall. In others the plastic flow is best described as jerky. However, actual sharp load drops are often observed, as indicated schematically in Fig. 8.14. These serrations, as they appeared in aluminum alloys, were first studied in detail by Portevin and LeChatelier,[11] and it is now common to call the phenomenon associated with these serrations the *Portevin-LeChatelier effect.*

One of the most significant aspects of dynamic strain aging, observed primarily in metals containing interstiitial solutes, is that the work hardening rate can become abnormally high during dynamic strain aging and, at the same time, it can become strain-rate and temperature dependent. This effect is illustrated for commercially pure titanium in Fig. 8.15. The three stress-strain curves shown in the figure correspond to specimens strained at the same temperature but at three different strain rates. Note that the specimen deformed at the intermediate rate was subject to a much higher degree of work hardening than either of the specimens deformed at a rate 20 times slower or 10 times faster. This illustration suggests that, at the temperature in question, there is a maximum work hardening rate corresponding to a specific strain rate. A similar maximum work hardening rate will be observed if the temperature is raised or lowered, provided that the strain rate is adjusted accordingly. Thus, if the temperature is raised, the strain rate at which maximum work hardening is observed also rises.

11. Portevin, A., and LeChatelier, F., *Comp. Rend. Acad. Sci., Paris,* **176** 507 (1923).

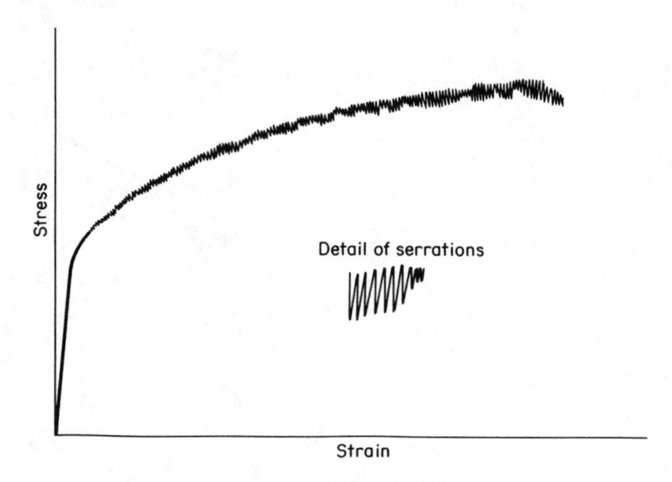

Fig. 8.14 Discontinuous plastic flow is a common aspect of dynamic strain aging. The above diagram indicates one form of serrations that may be observed.

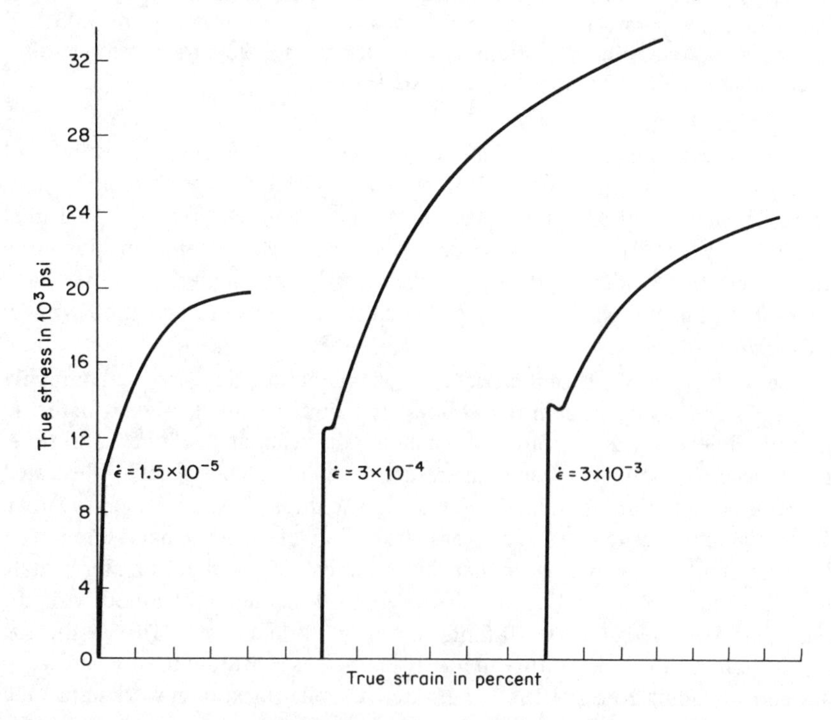

Fig. 8.15 In the temperature range of dynamic strain aging, the work-hardening rate may become strain-rate dependent. This figure shows the stress-strain curves for three titanium specimens deformed at 760°K at different rates.

Finally, one of the other well-known manifestations of dynamic strain aging is the phenomenon called *blue brittleness* when it occurs in steel. In approximately the center of the temperature range where the other phenomena of dynamic strain aging are observed, it has been found that the elongation, as measured in a tensile test, becomes very small or passes through a minimum on a curve of elongation plotted against the temperature. This subject is considered in more detail in Section 19.21.

The various dynamic strain aging phenomena do not appear to the same degree in all metals. However, they are commonly observed and it is probably safe to state that, in general, dynamic strain aging is the rule rather than the exception in metals.

8.15 The Role of the Drag-Stress in Dynamic Strain Aging. All of the phenomena of dynamic strain aging are believed to be associated with interactions between moving dislocations and solute atoms. It is probable that, in most cases, dynamic strain aging effects can be related to the drag that dislocation atmospheres place on moving dislocations. Several authors[12,13,14] have suggested that the drag-stress might be directly added to a normal strain-rate dependent flow-stress. This latter type of flow-stress component is generally assumed to increase continuously with increasing strain rate in the manner shown in Fig. 8.16A. The corresponding strain-rate dependence of the drag-stress is reproduced in Fig. 8.16B. How the sum of these two curves should appear depends on the relative magnitude of the two components, and two possibilities are illustrated in Figs. 8.16C and 8.16D. In the first of these drawings it is assumed that the drag-stress is relatively small compared to the normal strain-rate dependent component. In Fig. 8.16D the drag-stress is assumed to be of larger significance. In either case, it is apparent that the drag-stress could have a strong effect on the flow-stress strain-rate relationship. Thus, consider the case represented by Fig. 8.16C where the resultant curve has an extended region in which the flow-stress is independent of strain rate. Such a rate-independent region is in good agreement with yield stress measurements in the dynamic strain aging temperature interval. Furthermore, the temperature independence of the flow-stress in the same interval can be rationalized in a similar manner by adding the drag-stress to the rate (that is, temperature) dependent component when the variation of the latter with temperature is considered. A curve similar to that obtained experimentally and shown in Fig. 8.12 should result.

12. Nabarro, F. R. N., *Theory of Crystal Dislocations,* p. 431, Oxford Press, London, 1967.
13. Hirth, J. P., and Lothe, J., *Theory of Dislocations*, p. 621, McGraw-Hill Book Company, New York, 1968.
14. Hahn, G. T., Reid, C. N., and Gilbert, A., *Acta Met.*, **10** 747 (1962).

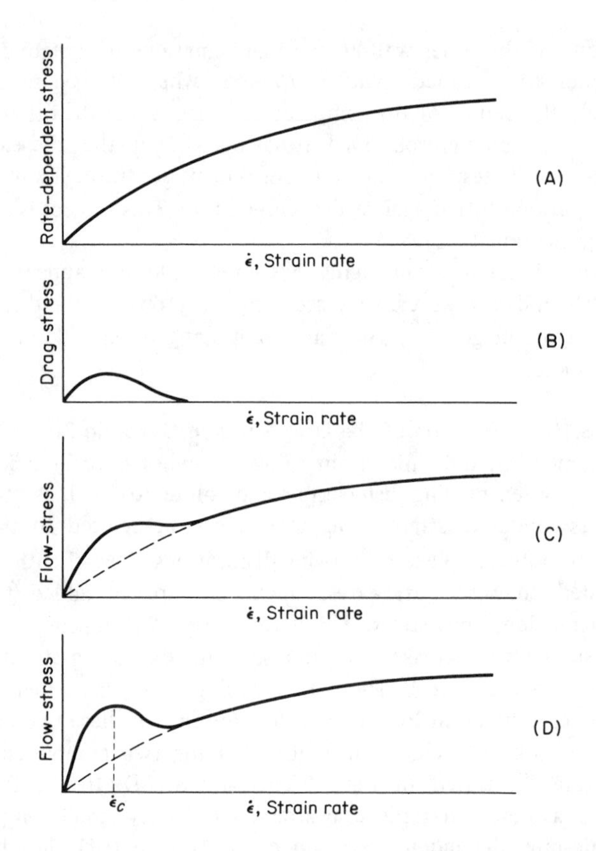

Fig. 8.16 The addition of the components of the flow stress that depend on strain rate, (A) the normal strain-dependent flow-stress, (B) the drag-stress, (C) the case where the drag-stress is relatively small, and (D) the case where the drag-stress is large.

If the drag-stress is larger, as suggested in Fig. 8.16D, it is possible to rationalize the serrations that appear in many metals when they are subject to dynamic strain aging. Note that, as shown in the diagram, the flow-stress drops with increasing strain rate. This represents a state of instability. A metal, deforming at a strain rate $\dot{\epsilon}_c$, can lower its flow-stress by increasing its strain rate. This increase in strain rate corresponds to an effective softening of the metal and is conducive to the initiation of a Lüders band. That is, as in all similar cases, the softening will occur in a localized region. When this happens in a hard testing machine, the load drops. In this process Nabarro[15] suggests that the dislocations, by increasing their velocities, effectively free themselves from their

15. Nabarro, F. R. N., *Theory of Crystal Dislocations*, p. 397, Oxford Press, London, 1967.

atmospheres. However, these dislocations will eventually become stopped by obstacles (such as grain boundaries), and in so doing will gather fresh atmospheres. For further deformation the load must once again rise, making possible a repetition of the cycle.

Problems

1. In spite of the fact that the f.c.c. lattice is more closely packed than the b.c.c. lattice, carbon is many times more soluble in face-centered cubic iron than it is in body-centered cubic iron. Rationalize this fact by computing the size of a hole in a f.c.c. iron crystal using the atom diameter given in Fig. 8.2. Assume the carbon atoms lie between iron atoms along ⟨100⟩ directions.
2. The solubility of carbon in face-centered cubic iron (γ-phase) is 2.06 weight percent at 1147°C (see Fig. 16.1).

   Assuming that the equation given in Section 8.3 is valid to this high a temperature, determine the carbon concentration in body-centered cubic iron at 1147°C. (*Note:* as can be seen in Fig. 17.1, b.c.c. iron is not stable at 1147°C, so this is a hypothetical question.)
3. A portion of Fig. 17.1 is reproduced in this drawing. Point A represents the maximum solubility of carbon in f.c.c. cubic iron. It is 2.06 wt. percent at 1147°C. At point B the solubility is 0.80 wt. percent when the temperature

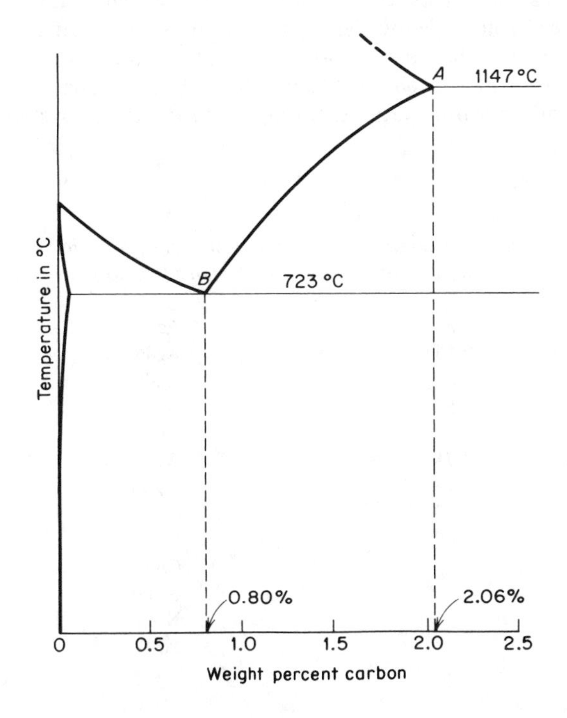

is 723°C. Assuming that the line AB is equivalent to the curve drawn in Fig. 8.3 for carbon in b.c.c. iron and that the equations given in Section 8.3 can be used with concentrations of this large size, determine an activation energy for the transferral of a mole of carbon atoms from cementite to face-centered cubic iron (gamma iron). Assume that you can use concentrations in weight percent in your calculations.

4. When the concentrations are as high as those in Problem 3, a ratio of the two concentrations in weight percent is not equivalent to the same ratio expressed in atomic percent.
   (*a*) Compute the ratios of the two concentrations expressed in atomic percent.
   (*b*) Using this ratio, recompute the activation energy for the transferral of carbon atoms to gamma iron.
   (*c*) What is the *approximate* average separation between carbon atoms in a solution containing 2.06 wt. percent? Express your answer in terms of the diameter of an iron atom.
   (*d*) The answer to part (*c*) should suggest another probable error involved in the use of the equations in Section 8.3 when the concentrations are of the order of several percent. What is this factor?
5. Using the activation energy deduced in Problem 4, make a hypothetical estimate of the equilibrium concentration of carbon atoms in gamma iron at room temperature (300°K).
6. Table A gives the atom diameters of the most important face-centered cubic metals. Table B lists the binary systems which form isomorphous or completely soluble alloy systems, as well as those binary systems where only a limited solubility is observed. Determine the size factor for each of the systems listed in Table B and rationalize the results with Hume-Rothery's observations. Discuss in particular the system Ag-Al.

(A)

Element	*Atomic Diameter in Å*
Aluminum	2.86
Copper	2.55
Gold	2.89
Lead	3.50
Nickel	2.49
Palladium	2.75
Platinum	2.78
Silver	2.89
Thorium	3.60

(B)

Completely Soluble Systems (Isomorphic)	*Systems of Limited Solubility*
Ag-Au	Ag-Al
Ag-Pd	Ag-Cu
Ag-Pt	Ag-Ni
Au-Cu	Ag-Pb
Au-Ni	Au-Th
Au-Pd	Cu-Pb
Au-Pt	Ni-Pb
Cu-Ni	Pb-Pd
Cu-Pd	Pb-Pt
Cu-Pt	
Ni-Pd	
Ni-Pt	
Pd-Pt	

7. Consider Fig. 8.5 and, in particular, the nature of the stress that an edge dislocation develops on its own slip plane.
   (*a*) If another positive edge dislocation was placed on this slip plane and to the right of the given dislocation, in what direction would this stress tend to move the second dislocation?
   (*b*) At the same time, would the second dislocation have any effect on the first? Explain.
   (*c*) A pair of parallel edge dislocations react to each other's stress field in the same manner as they react to an externally applied stress. Thus, each can move under the action of the other's stress. With this in mind, determine how far apart two parallel edge dislocations, lying on the same slip plane in a copper crystal, would have to be so that they would not move under each other's stress field. Take the critical resolved shear-stress of copper as 90 psi and the shear modulus as 4×10^{11} dynes/cm^2. Express your answer in terms of the number of atom diameters (Burgers vectors). Let $\nu = 0.3$.
8. (*a*) If two edge dislocations lie directly above each other, as shown in this sketch, what will be the shear stress acting to cause each dislocation to move on its slip plane?

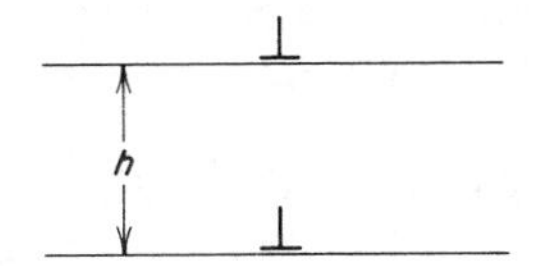

   (*b*) If the dislocations are able to move, in which direction will they move? Describe the mechanism that could make this possible.
9. Two parallel screw dislocations of opposite sign exert an attractive stress on each other that acts along the plane containing the two dislocations. Each dislocation feels a stress equal to $\mu b/2\pi r$ where r is the distance between the two dislocations. Two screws of opposite sign repel each other in the same manner.

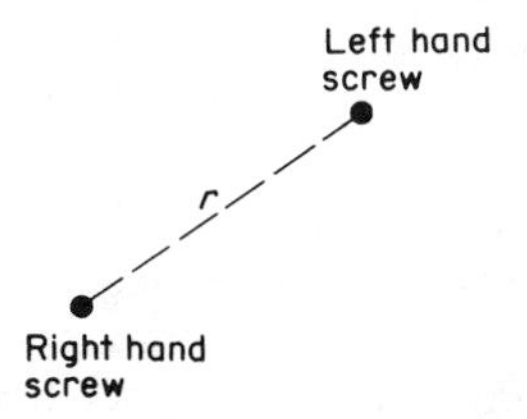

   Rationalize the above statement by considering the basic orientation relationship between a screw dislocation, the shear stress that makes it move, and the direction in which it moves in relation to the shear-stress field around another screw dislocation, as shown in Fig. 8.4.

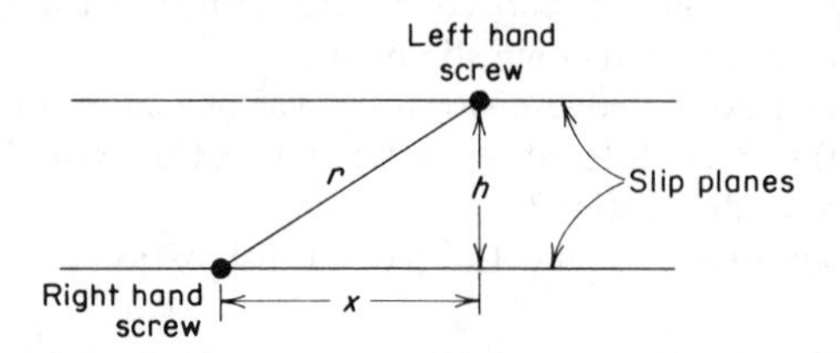

10. Consider two screw dislocations lying on parallel slip planes separated by the distance h. Assume that the dislocations can only move along these planes as in the case of basal slip in a h.c.p. metal like zinc. Let x be the projected horizontal separation between the two dislocations.
(*a*) Resolve the attractive stress between the two dislocations into its component parallel to the respective slip planes. This is the shear stress that can cause these dislocations to move.
(*b*) Let x vary from $-\infty$ to $+\infty$ and sketch the shape of the shear-stress distance curve.
(*c*) At what distance x is the interaction stress between the dislocations a maximum?
(*d*) How large is this maximum interaction stress?
(*e*) Might this type of stress play a part in work hardening? Explain.

11. The strain rates normally employed in tensile tests vary from about 1 $\sec^{-1}$ to about 10^{-6} $\sec^{-1}$. While it is quite possible to measure the total dislocation density in a metal, the mobile dislocation density (that is, the density of dislocations actually participating in the strain) is normally not known. A good estimate is that it is about 10^8 cm^{-2}. Using this value, determine the probable range of the average dislocation velocity in these tensile tests. Let $b = 4$ Å.

12. In an explosively loaded specimen, the stress wave lasts for about 10^{-6} secs. If a specimen deformed in this manner to a strain of about 10 percent has a mobile dislocation density of 10^{10} cm^{-2}, what average distance will these dislocations have moved? Let $b = 4$ Å.
What would be the corresponding average dislocation velocity?

13. How large a strain would be required before an iron specimen containing 10 parts per million in weight percent would be unable to show a yield-point return? Assume $b = 2.5$ Å and that the dislocation density varies with the strain according to $\rho = 4 \times 10^{11}\, \epsilon$.

14. Like other manifestations of dynamic strain-aging, whether or not serrations are observed on a stress-strain curve depends on the strain rate and the temperature. At a given strain rate there is a specific temperature at which they appear, and a correspondingly higher temperature above which they disappear. The table gives data obtained by Keh, Nakada, and Leslie[16]

16. Keh, A. S., Nakada, Y., and Leslie, W. C., *Dislocation Dynamics*, Ed. Rosenfield, A. R., Hahn, G. T., Bement, A. L., Jr., and Jaffee, R. I., p. 381, McGraw-Hill Book Company, New York, 1968.

corresponding to the lower temperature limit for the appearance of serrations in a 0.03 percent carbon steel.

(*a*) Plot the logarithm of the strain rate against $1/T$ and obtain an activation energy associated with the formation of serrations.

(*b*) The activation energy for diffusion of carbon and nitrogen in iron are about 20,000 and 18,000 cal/mole respectively. Compare these values with the result of your computation.

Temperature °*C*	*Strain Rate*, $\dot{\epsilon}$, *Sec*$^{-1}$
60	1.0×10^{-4}
75	3.3×10^{-4}
85	7.0×10^{-4}
100	2.3×10^{-3}
115	6.5×10^{-3}

9 Precipitation Hardening

The equation for the equilibrium concentration of carbon in body-centered cubic iron as a function of temperature, $C = Be^{-Q/RT}$, is not typical only of the iron-carbon system. Similar relationships are observed for nitrogen, or hydrogen in interstitial solution in iron, or for all three elements (carbon, nitrogen, and hydrogen) in interstitial solution in other metals, providing that the maximum solubility is under approximately 1 atomic percent. This 1-percent limitation is usually unimportant in interstitial solid solutions as they are almost always very dilute.

The above law may also be used to predict the maximum solubility in substitutional solid solutions. Here again it is quite accurate[1] as long as the maximum solubility at high temperature does not exceed 1 percent. In those cases where the solubility exceeds 1 percent, the equation still gives a curve that is a useful approximation of experimentally observed solubility relations. Since there are many binary alloys (two-component) with limited solubilities of the component metals in each other, curves like that in Fig. 9.1 are quite common. This figure, incidentally, is the same as Fig. 8.3 in Chapter 8 and shows the solubility of carbon in alpha-iron.

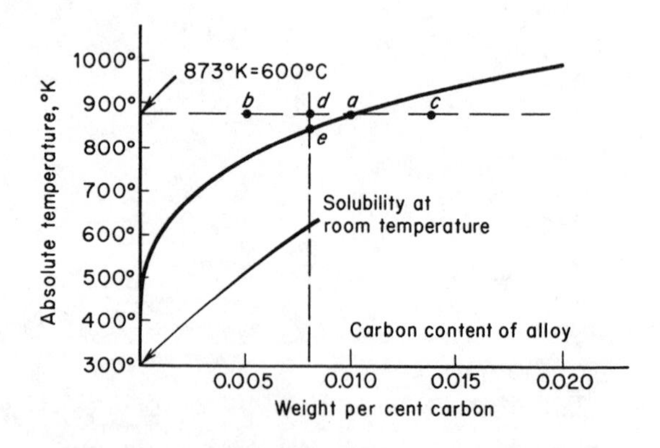

Fig. 9.1 Solubility of carbon in alpha-iron.

1. Freedman, J. F., and Nowick, A. S., *Acta Met.*, **6** (1958), p. 176.

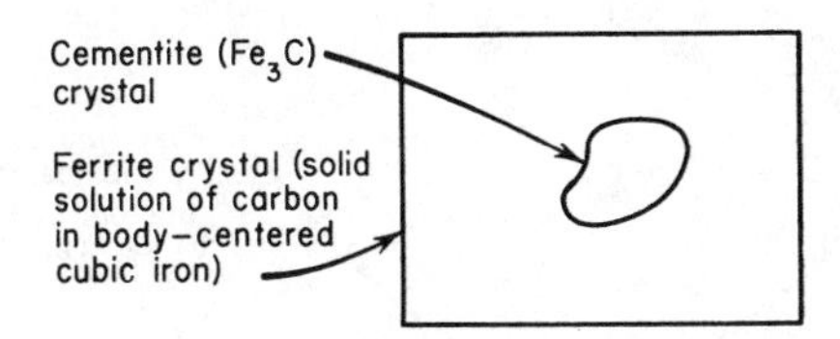

Fig. 9.2 Crystal of cementite in contact with a ferrite crystal.

Solubility relationships of the type shown in Fig. 9.1 have great practical significance, for they make possible the precipitation, or age-hardening, of metals – an extremely important method of hardening of metals. This type of hardening is used most often in commercial strengthening of nonferrous alloys, especially aluminum and magnesium alloys. For the sake of convenience, the age-hardening process in iron alloyed with small quantities of carbon will be taken up here. The basic principles developed, however, are applicable to other more general alloy systems.

9.1 The Significance of the Solvus Curve. Curves of the type shown in Fig. 9.1 are called *solvus lines.* The significance of this line will now be investigated. For this purpose, consider a specific temperature: 600°C. Assume that at this temperature a crystal of ferrite is in contact with a crystal of cementite, as shown in Fig. 9.2. In a system of this type, it is possible for carbon atoms to leave the solid solutions (ferrite) and enter the cementite. When this happens, however, three iron atoms of the solution must also join the cementite lattice in order to maintain the strict stoichiometric ratio characteristic of iron carbide: three atoms of iron to one of carbon. Similarly, when a carbon atom leaves Fe_3C to enter the solution, three iron atoms must leave the compound. In the very dilute solutions presently considered (approximately 0.01 percent carbon), the effect on the concentration of the solid solution of the addition, or removal of solvent atoms (iron) simultaneously with the carbon is negligible. It can, therefore, be assumed that concentration changes are due only to transfers of carbon atoms between the two phases. It must be kept in mind, however, that as carbon enters iron carbide, the latter phase grows in volume, and that the composition of this phase does not change. On the other hand, when carbon enters the solid solution (ferrite), the composition of the latter changes.

Figure 9.3 shows a hypothetical plot of the free energy of the ferrite plus cementite system as a function of the carbon concentration in the ferrite. Point *a* on this curve represents the composition at which the free energy of the ferrite-cementite system is a minimum. It is also the same composition (0.010 percent) that one obtains from the solvus curve at 600°C. The ferrite in a ferrite-cementite aggregate will thus seek the composition of the solvus curve at any given temperature. If for any reason the solid solution should have the

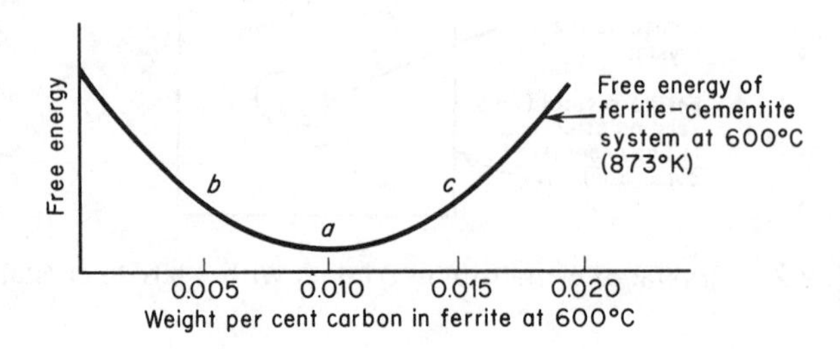

Fig. 9.3 Hypothetical free-energy curves at 600°C for the ferrite-cementite system of Fig. 9.2 as a function of carbon concentration in the ferrite.

composition *b* in Figs. 9.1 and 9.3, carbon will leave the cementite to increase the concentration of the solid solution. The free energy of the system is, of course, lowered by this spontaneous reaction. On the other hand, if the ferrite is supersaturated with carbon and has a composition such as that of point *c*, more cementite will form. This spontaneous reaction also lowers the carbon concentration of the ferrite (and the free energy of the system).

9.2 The Solution Treatment. Consider a specific dilute iron-carbon alloy, one with a total carbon content of 0.008 percent. If this alloy is in equilibrium at room temperature, nearly all of the carbon will be in the form of cementite because the solubility of carbon in ferrite at 24°C is only 2.3×10^{-7} percent (see Fig. 9.1). Suppose that this same alloy is heated to 600°C, indicated by point *d* in Fig. 9.1. At this temperature the equilibrium concentration of carbon in the solid solution is 0.010 percent, which is more than the total amount of carbon in the metal. The cementite stable at room temperature is no longer stable at 600°C and dissolves by yielding its carbon atoms to the solid solution. Because the equilibrium concentration is greater than the total carbon content of the alloy, the cementite must disappear completely if the alloy is maintained for a sufficiently long time at the elevated temperature. The alloy which originally contained two phases (cementite and ferrite) is thus converted to a single phase (ferrite). However, the solid solution attained by maintaining the specimen at 600°C is not a saturated solution because it contains less than the equilibrium concentration. On the other hand, it cannot lower its free energy and assume the equilibrium concentration for a mixture of ferrite and cementite because there is no extra carbon available for this purpose.

Let us study the effect of rapid cooling on the above alloy after it has been transformed into a homogeneous solid solution at 600°C. Very rapid cooling of heated metal specimens can be accomplished by immersing them in a liquid cooling medium, for example, water. This operation is generally known as a *quench.* In the present case, a very rapid quench will prevent appreciable diffusion of carbon atoms, so it can be assumed that the solid solution that existed at 600°C is brought down to room temperature essentially unchanged. The alloy which was slightly unsaturated at the higher temperature will now be extremely supersaturated. Its 0.008 percent carbon in solution is roughly 4000 times greater than the equilibrium value (2.3×10^{-7}). The alloy is, accordingly, in a very unstable condition. Precipitation of this excess carbon through the formation of cementite will markedly lower the free energy and can be expected to occur spontaneously. The conditions under which this may occur will be discussed in the next section, but at this point it should be mentioned that the first stage in a precipitation-hardening heat treatment has just been outlined. A suitable alloy is heated to a temperature at which a second phase (usually present in small quantities) dissolves in the more abundant phase. The metal is left at this temperature until a homogeneous solid solution is attained, and then it is quenched to a lower temperature to create a supersaturated condition. This heat-treating cycle is known as the *solution treatment*, while the second stage, which is to be discussed, is called the *aging treatment*.

9.3 The Aging Treatment. The precipitation of cementite from supersaturated ferrite occurs by a nucleation and growth process. First, it is necessary to form the beginnings of the cementite crystals; a process called *nucleation*. Following nucleation, cementite particles grow in size as a result of the diffusion of carbon from the surrounding ferrite toward the particles. This is called *growth*. No precipitation can occur until nucleation begins, but once it has started the solid solution can lose its carbon in two ways, either to the growing particles already formed or in the formation of additional nuclei. In other words, nucleation may continue simultaneously with the growth of particles previously formed. The progress of precipitation at a given temperature is shown in Fig. 9.4, where the amount of precipitate, in percentage of the maximum, is plotted as a function of time. Logarithmic units are used for the time scale because spontaneous reactions of this nature usually start rapidly and finish slowly. In general, precipitation does not begin immediately, but requires a finite time t_0 before it is detectable. This time interval is called the *incubation period* and represents the time necessary to form stable visible nuclei. The curve also shows that the precipitation process finishes very slowly, an effect that is to be expected in light of the continued loss of solute from the solution.

The speed at which precipitation occurs varies with temperature. This is shown qualitatively in Fig. 9.5. At very low temperatures, long times are required to complete the precipitation because the diffusion rate is very slow. Here the rate of the reaction is controlled by the rate at which atoms can migrate. The rate of precipitation is also very slow at temperatures just below the solvus line (point *e*, Fig. 9.1). In this case, the solution is only slightly oversaturated and the free-energy decrease resulting from precipitation is very

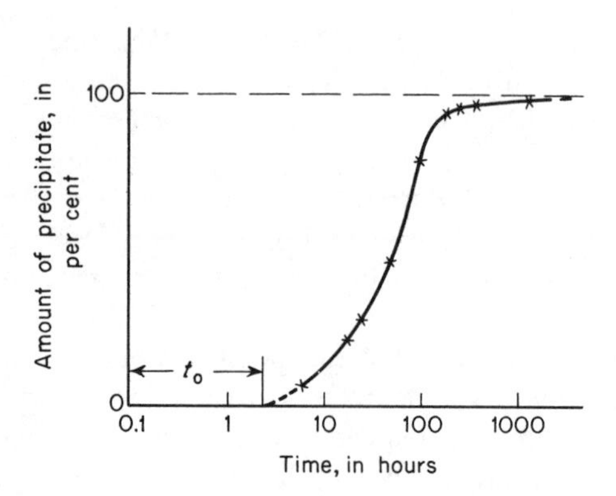

Fig. 9.4 Amount of precipitate as a function of time in an iron-carbon alloy (0.018 percent C) allowed to precipitate from a supersaturated solution at 76°C. (Data of Wert, C., ASM Seminar, *Thermodynamics in Physical Metallurgy,* 1950.)

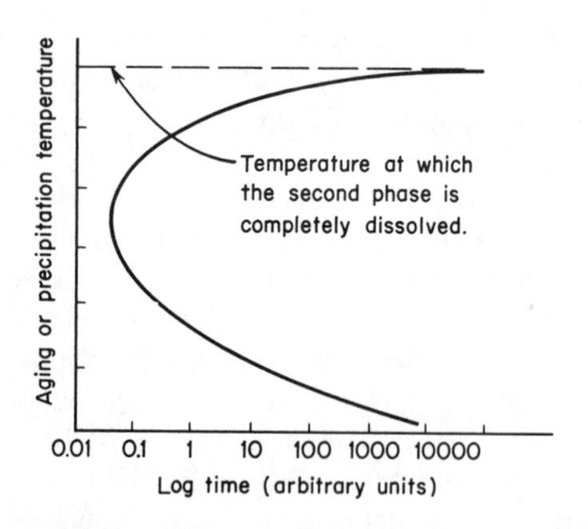

Fig. 9.5 Time for 100 percent of the precipitate to form in a supersaturated alloy.

small. Nucleation is, accordingly, slow, and precipitation is controlled by the rate at which nuclei can form. The high diffusion rates that exist at these temepratures can do little if nuclei do not form. At intermediate temperatures, between the two above-mentioned extremes, the precipitation rate increases to a maximum, so that the time to complete the precipitation is very short. In this range, the combination of moderate diffusion and nucleation rates makes precipitation rapid.

The most important effect of the precipitation of the second phase (cementite) is that the matrix (ferrite) is hardened. Figure 9.6 shows a typical

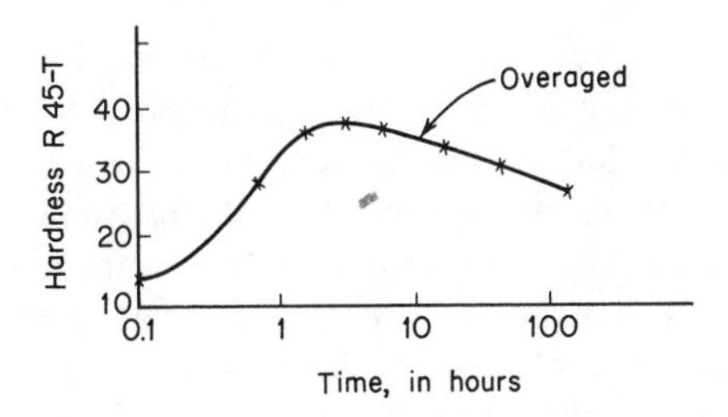

Fig. 9.6 Change in hardness during the aging treatment. Alloy is iron plus 0.015 percent C, and aging temperature 90°C. (Data from Wert, C., ASM Seminar, *Thermodynamics in Physical Metallurgy*, 1950.)

hardening curve for a dilute iron-carbon alloy. To obtain curves of this nature, a number of specimens are first given a solution heat treatment in order to convert their structures into supersaturated solid solutions. Immediately following the quench of the solution heat treatment, the specimens are placed in a suitable furnace maintained at an intermediate constant temperature (a temperature above room temperature, but below the solvus temperature). They are then removed from the furnace at regular intervals, cooled to room temperature, and tested for hardness. The data obtained in this manner are then plotted to show the effect of time on hardness. The significant feature of this curve is the maximum it possesses. Holding, or aging, the specimens for too long a period at a given temperature causes them to lose their hardness. This effect is known as *overaging*.

The shape of the aging curve is primarily a function of two variables: the temperature at which aging occurs and the composition of the metal. Let us consider the first of these variables. Figure 9.7 shows three curves, each

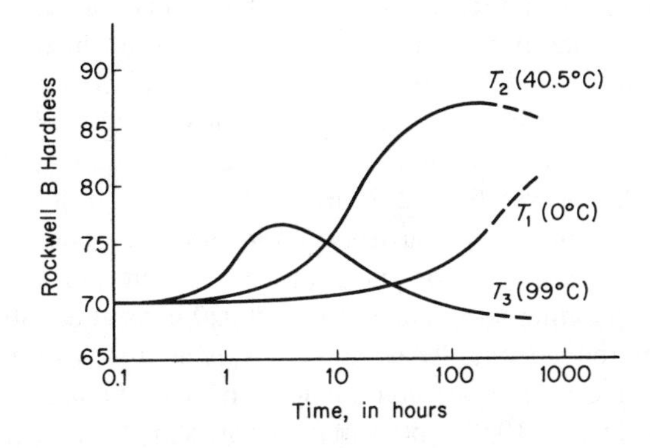

Fig. 9.7 Effect of temperature on the aging curves during precipitation hardening. Curves are for a 0.06 percent C steel. (After Davenport, E. S., and Bain, E. C., *Trans. ASM,* **23** 1047 [1935].)

corresponding to aging at different temperatures. The curve marked T_1 represents aging at too low a temperature. In this case, atomic motion is so slow that no appreciable precipitation occurs and hardening occurs slowly. A further lowering of the aging temperature below T_1 will effectively stop all precipitation and prevent hardening. Use is made of this fact in the aircraft industry when aluminum-alloy rivets, that normally harden at room temperature, are kept in deep-freeze refrigerators until they are driven. In this way the rivets, which have been previously solution-treated in the supersaturated condition, are prevented from aging until the rivets are incorporated in the article being fabricated.

Temperature T_2 corresponds to an optimum temperature, a temperature at which maximum hardening occurs within a reasonable length of time. This temperature lies above T_1, but below T_3. At T_3, hardening occurs quickly due to rapid diffusion. However, softening effects also are accelerated, resulting in a lower maximum hardness. The mechanism explaining these effects will be discussed in Theories of Hardening in Section 9.6.

The effect of changing the composition, the other variable which influences the aging curve, will now be considered. For low-solute concentrations, the degree of supersaturation is small at the end of the solution treatment, and the free energy of the system is, at best, only slightly higher than that of the equilibrium concentration. Under these conditions it is difficult to nucleate the second phase, and hardening occurs slowly at constant temperatures. Furthermore, the maximum hardness that can be obtained will be small because the total amount of precipitate is not large and, in general, the smaller the amount of precipitate, the smaller the maximum hardness. On the other hand, increasing the total solute concentration makes possible a greater maximum hardness at a given aging temperature. The more solute that is available, the more the precipitate and the greater the hardness. In addition, at higher solute concentrations the maximum hardness will be attained in a shorter time, for both the rate of nucleation and the rate of growth are increased. Nucleation rates rise because of a greater difference in free energy between the supersaturated and equilibrium states, while growth rates are increased because of the greater amount of solute available for the formation of the precipitate. These effects are, however, limited to the extent that it is possible to dissolve the solute in the solvent during the solution treatment. Thus, in Fig. 9.1, the maximum solubility of carbon in iron (0.020 percent) occurs at 723°C. Compositions containing carbon in excess of 0.020 percent can still only dissolve this amount in the ferrite, the remainder remains in the form of cementite. Low-carbon-steel compositions containing more than the maximum amount of carbon soluble in iron (0.020 percent) are still able to undergo precipitation hardening as long as the solution-treating temperature is not raised above 723°C. Figure 9.7 shows a number of aging curves for a steel of 0.06 percent carbon.

9.4 Nucleation of Precipitates. How nuclei form and start to grow during precipitation is a subject of great interest. It is, however, very complicated and difficult to resolve in specific commercially important alloys. In many cases the precipitate phase does not originate in its final structure, but may form a number of intermediate crystal structures before the final stable precipitate is developed. Thus, for example, an alloy of aluminum containing 4 percent of copper, which forms the basis of some important aluminum alloys, may pass through three intermediate stages of precipitation before the final θ phase ($CuAl_2$) is obtained. The initial stages of precipitation are the most difficult to analyze because of the extremely small size of the precipitate particles and their relatively uniform distribution. Figure 9.8 shows the starting

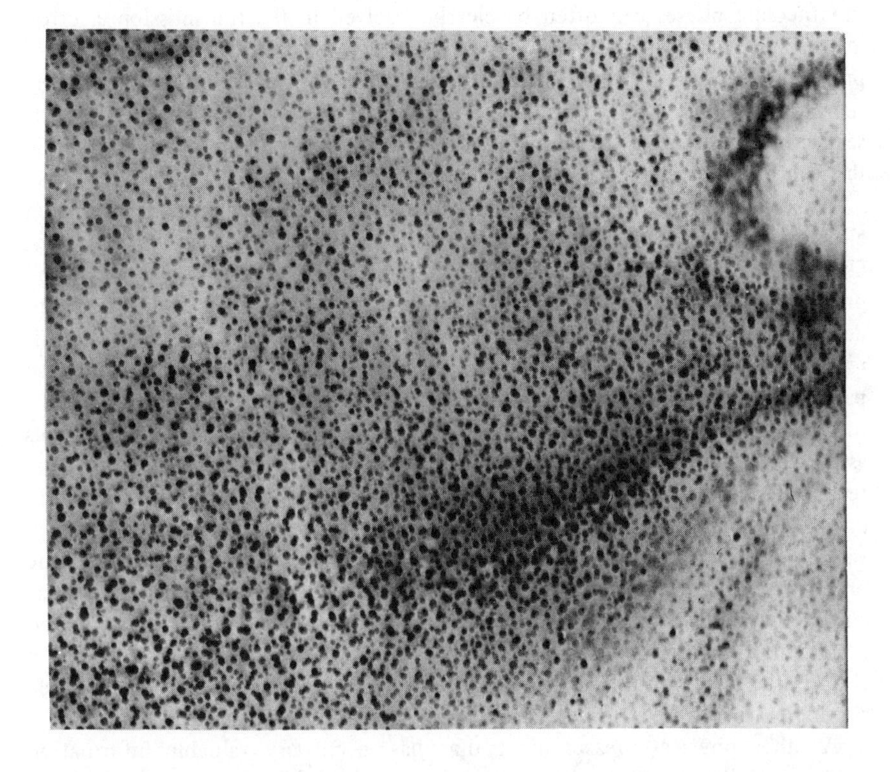

Fig. 9.8 Guinier Preston zones in an aluminum 16 percent silver alloy. (From Nicholson, R. B. and Nutting, J., *Acta Met.,* 9 332 [1961]. Photograph courtesy of R. B. Nicholson.)

precipitate structure in an aluminum-silver alloy containing 16 percent silver. Note that it consists of a set of very small, difficult-to-resolve particles that appear as dark spots on this transmission electron microscope photograph. These particles, which are known as *Guinier-Preston zones (G.P. zones)*, have a definite composition and structure[2] that is not the same as that of the final stable precipitate. Evidently these particles are easier to nucleate than the final precipitate and, as a result, form first. Eventually they disappear as later, more stable (usually intermediate) phases appear. These also, as in the case of the aluminum-copper alloy mentioned above, may again disappear, to be replaced by other still more stable phases. In this sequence, whether or not a later phase is formed directly from an earlier phase, or is independently nucleated, has not been determined in many cases.

Usually the precipitate particles become visible only in an optical microscope when the metal is well advanced into the overaged stage. However, some of the intermediate phases can often be clearly resolved in the transmission electron microscope. Thus, for example, Fig. 9.9 shows very clearly the plate-like γ' precipitate particles in an Al-Ag-Zn alloy. The γ' precipitate in this alloy represents the second of three stages of precipitation. The dark lines that may be seen crossing some of these precipitate particles represent dislocations lying in the interface between the particle and the matrix.

As evidenced by the photographs in Figs. 9.8 and 9.9, the transmission electron microscope is a very useful tool for the study of precipitation phenomena. In addition to making possible the direct observation of very small precipitate structures, it is sometimes possible to use selected area diffraction to deduce the nature of the crystal structure in the precipitate particles. This requires that the particles be large enough to be able to give a diffraction pattern. Very small particles cannot be studied in this manner. For their analysis, the best techniques still rely heavily on X-ray diffraction. Since the problems associated with precipitation are complex, it is often necessary to invoke other techniques in addition to those mentioned above. These involve changes in physical properties such as hardness and electrical resistivity, induced by the various phenomena of precipitation. With regard to electrical resistivity, the orderly motion of electrons through a crystal, which constitutes an electric current, is badly disturbed when an extremely small and uniformly distributed precipitate, such as the G.P. zones, forms. Thus, when precipitation starts there is an initial large rise in resistivity, which then decreases as the average particle size increases during the progress of the precipitation phenomena.

Another physical measurement that has given very valuable information about precipitation mechanisms is based on internal friction measurements. By *internal friction* one means the ability of a metal to absorb vibrational energy.

2. Christian, J. W., *The Theory of Transformations in Metals and Alloys*, p. 609, Oxford Press, London, 1965.

Fig. 9.9 An intermediate precipitate, γ', in an aluminum-silver-zinc alloy. Magnification 35,000 times. (Photograph courtesy of S. R. Bates.)

One of the ways this can be accomplished is related to the presence of solute in solid solution in a metal, provided that the solute strains the lattice nonisotropically. An excellent example of such an alloy is furnished by dilute iron-carbon alloys. In this system, the interstitial positions between iron atoms (along ⟨100⟩ directions) are either expanded or contracted as a result of an applied stress. In a vibrating system, the stress field alternates during each cycle of the vibration and, as a result, a given interstitial position is first enlarged and then contracted during each cycle. Adjacent interstitial sites, lying along different ⟨100⟩ directions, can be expected to be strained in the same fashion, but to be out of phase with the first set. The net result is that the carbon atoms tend to follow the strain pattern by jumping from contracted sites into enlarged ones. An energy loss accompanies this stress-induced atomic motion, which reaches a maximum for a given carbon concentration when the vibration frequency of the metal is close to the normal atomic jump rate of the carbon atoms. At room temperature in iron, this maximum occurs at a vibration frequency of about 1 cycle per sec. If iron is induced to vibrate at this critical frequency, the energy losses will be proportional to the carbon in solid solution. Measurements made in this fashion give only the concentration of carbon in the ferrite and are, accordingly, independent of carbon contained in cementite particles. It is interesting to note that Wert's data, used in plotting Fig. 9.1, were obtained through internal friction measurements. In those alloy systems where this method is applicable, it forms an excellent means of studying precipitation effects.

9.5 Heterogeneous Versus Homogeneous Nucleation. A precipitate particle can be nucleated in two basic ways. It can form at internal lattice defects such as: dislocations, dislocation nodes (intersections of two or more dislocations), impurity particles, or discontinuities in grain boundaries. This process is known as *heterogeneous nucleation,* wherein the formation of a second-phase particle is made easier by lattice defects. *Homogeneous nucleation*, on the other hand, is the spontaneous formation of nuclei through composition fluctuations of the solute. Here solute atoms cluster in the lattice of the matrix and start a second-phase particle growing in an otherwise perfect crystal.

Homogeneous nucleation occurs only with considerable difficulty. For example, the freezing of a liquid to form a solid is also a nucleation and growth process. Nucleation in this case is predominantly heterogeneous, starting at mold walls, or at impurity particles in the liquid itself. Thus, very pure water does not freeze homogeneously until supercooled to temperatures approaching –40°C. The principal difficulty in the forming of nuclei homogeneously is that a surface is created when a second-phase particle is nucleated.

Consider Fig. 9.2, which shows a second-phase particle embedded in a matrix of the more abundant phase. In a dilute iron-carbon alloy, the crystal particle would be cementite (Fe_3C) and the matrix a ferrite crystal.

A free-energy change of the system accompanies the formation of a particle, like that in Fig. 9.2, which may be expressed by the following equation:

$$\Delta F = -\Delta F_v + \Delta F_s + \Delta F_m$$

where ΔF_v is the free energy associated with the formation of the volume of cementite. The second term, ΔF_s, is the energy of the surface created between the cementite and the ferrite, which is always positive. The last term, ΔF_m, represents the strain energy arising from the formation of the particle. This includes both the strain energy in the matrix (ferrite) and in the particle and is caused by the fact that the volume of the cementite particle does not necessarily equal the volume of ferrite that it displaces. The strain-energy term varies directly as the volume of the precipitate particle. This is also true of the volume free-energy change, ΔF_v. The sign of the former is positive, while that of the latter is negative, so that the strain-energy term can be considered merely to

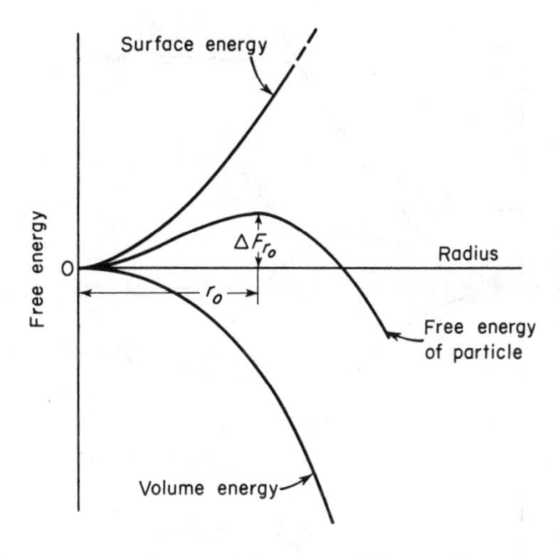

Fig. 9.10 Free energy of a precipitate particle as a function of its radius.

modify the value of the volume free-energy change. In the present qualitative discussion it can, therefore, be assumed that ΔF_m is zero, without altering the basic conclusions that are derived. Since the surface free energy depends on the area of the particle, and ΔF_v on the volume of the particle, we may write the following expression, assuming a spherical particle,

$$\Delta F = -A_1 r^3 + A_2 r^2$$

where A_1 and A_2 are constants and r is the particle radius. Figure 9.10 is a plot of the above equation.

At small radii, the surface free energy ($A_2 r^2$) is larger than the volume free energy ($A_1 r^3$), and the total free energy is positive. The situation changes, however, as the radius grows in size, so that with large radii the free energy becomes negative. The radius r_0 is known as the *critical radius*. Below this value a particle lowers its free energy by decreasing its size, so that particles with radii smaller than r_0 tend to dissolve and go back in solution. On the other hand, particles with radii larger than r_0 undergo a decrease in free energy with increasing radius. For this reason they are stable and should continue to grow.

Homogeneous nucleation requires that thermal fluctuations produce particles large enough to exceed r_0, otherwise the second phase cannot nucleate. It is customary to call a nucleus which is subcritical ($r < r_0$) in size an *embryo*. Let us now consider how the size of a stable nucleus varies with temperature. Referring to the above free-energy equation, it can be assumed, as a first approximation, that the surface-energy term does not change with temperature. On the other

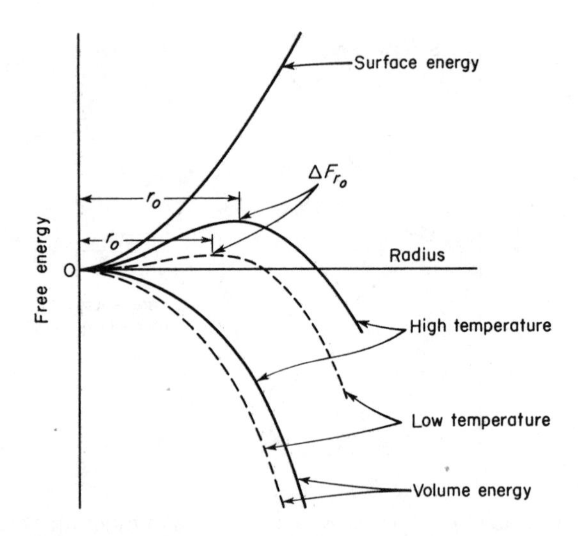

Fig. 9.11 Effect of temperature of precipitation on the free energy of a precipitate particle as a function of its radius.

hand, the volume free energy varies with temperature, becoming larger at low temperatures because the degree of supersaturation increases (solubility decreases) as the temperature falls. This effect is shown qualitatively in Fig. 9.11, where it can be seen that the critical radius decreases with falling temperature. At temperatures just below the solvus line, r_0 is very large (approaching infinity in size). The rate of homogeneous nucleation is, therefore, very small at this temperature. As the temperature is lowered, the critical radius rapidly decreases in size and so does the energy necessary to form a critical embryo. This latter is measured by the height (ΔF_{r_0}) of the free-energy-curve maxima in Fig. 9.11. Homogeneous nucleation is thus made easier by a drop in temperature, which explains why ice must be supercooled to such a low temperature if heterogeneous nucleation is precluded. On the other hand, in precipitation reactions, nucleation also depends on the ability of the solute to diffuse through the lattice of the solvent. This becomes the controlling factor at very low temperatures. When diffusion effectively stops, so does precipitation.

9.6 Theories of Hardening. The crystallographic nature of the precipitate particles that form during the various stages of precipitation are now much better understood than they were a few years ago. However, the exact nature of the hardening process is still not completely resolved. It appears that there are at least several hardening mechanisms, and that what is predominant in one alloy is not necessarily important in another. In general, however, it may be said that an increase in hardness is synonymous with an increased difficulty of

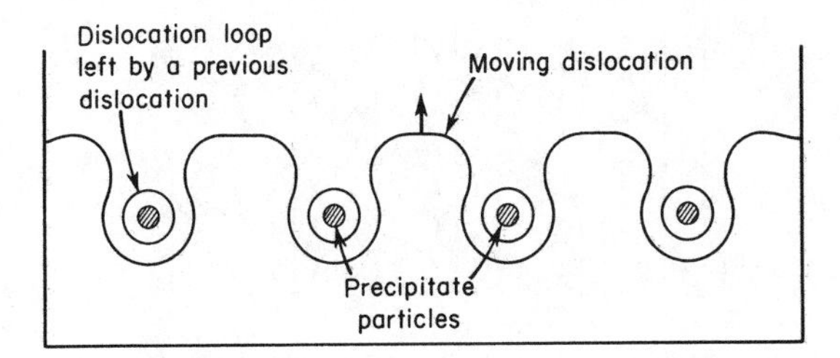

Fig. 9.12 Orowan's mechanism for the movement of dislocations through a crystal containing precipitate particles.

moving dislocations. Either a dislocation must cut through the precipitate particles in its path, or it must move between them. In either case, it can be shown that a stress increase is needed to move the dislocations through a lattice containing precipitate particles. The mechanism of Fig. 9.12 has been proposed by Orowan[3] and, in this case, the dislocation is assumed to bend in the form of expanding loops around the precipitate particles. When adjacent loops intersect on the far side of the particles, they cancel in the same manner as at a Frank-Read source. This cancellation permits the dislocation to continue, but leaves a dislocation ring surrounding the particle, the stress field of which increases the resistance to the motion of the next dislocation.

It is believed by many that an important factor in the interaction between precipitate particles and dislocations is the presence of stress fields surrounding precipitate particles. This is especially true when the precipitate particle is coherent with the matrix. Coherency is understood with the aid of Fig. 9.13. The upper diagram represents a supersaturated solid solution of atoms A and B. The A species is assumed to be the solvent, and the B the solute.

For the purpose of simplification, let us assume that the solubilities of A in B and of B in A, are both small enough at the temperature of precipitation so that they can be assumed to equal zero. Let it also be assumed that the precipitate is not a compound, such as Fe_3C, but the β phase; the crystal structure of the B atoms. In the present case, the B atoms will be attracted to each other, and the first stage in the formation of a precipitate particle will be the formation of a cluster of B atoms. The lattice planes in this cluster will, in general, be continuous with the planes of the matrix, and the cluster is said to be a *coherent particle.* If, as shown in Fig. 9.13, the diameter of the solute atom differs from that of the solvent, then the matrix and the nucleus will both be strained by the presence of the latter. The strain associated with a nucleus will enlarge as its size increases, but its size cannot increase indefinitely. It is possible that the particle may break away from the lattice of the matirx, and when this occurs, a surface,

3. Orowan, E., *Dislocations in Metals*, p. 69. AIME Publication, 1954.

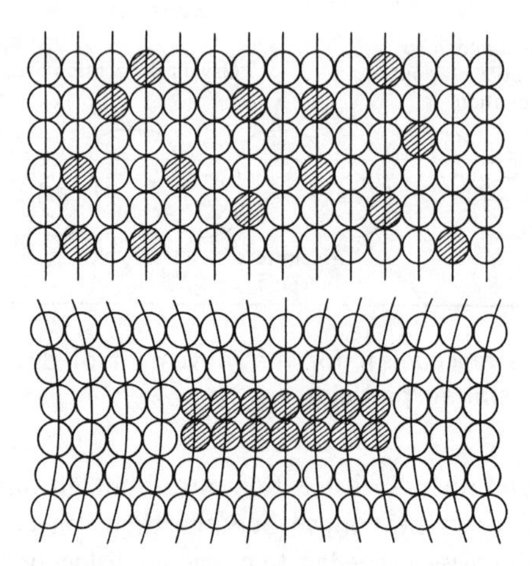

Fig. 9.13 Coherency. The upper figure represents a supersaturated solid solution of *B* atoms (dark circles) in a matrix of *A* atoms (light circles). The lower figure shows a coherent precipitate particle formed by clustering of the *B* atoms.

or grain boundary, forms between the two phases. Such a loss of coherency would greatly reduce the state of strain associated with the precipitate particles. There exists an alternate possibility, however, that now appears to be more probable. This is that the strains associated with coherency are lost or reduced when a new phase that is noncoherent or less coherent forms. This point of view is consistent with the fact that multiple precipitate structures are observed in many alloys.

Overaging is softening resulting from prolonged aging (see Fig. 9.6). In some age-hardening alloys, it appears concurrently with the loss of coherency by the precipitate. In any case, it may be stated that it is connected with the continued growth of precipitate particles. Growth will continue as long as the metal is maintained at a fixed temperature. This does not mean that all particles continue to grow, because this is an impossibility once the solute has attained an equilibrium concentration. It merely means that certain particles (the larger ones) continue to grow, while others (the smaller ones), disappear. As aging progresses, the size of the average particle increases, but the number of particles decreases. Maximum hardening is associated with an optimum small-particle size and a corresponding large number of particles; while overaging is associated with few relatively large particles.

The growth of precipitate particles is directly related to the surface tension at the interface between the matrix and the particles. Because of the boundary-surface energy, the free energy per atom in a large precipitate particle is lower than that in a small particle. This free-energy difference is the driving force that

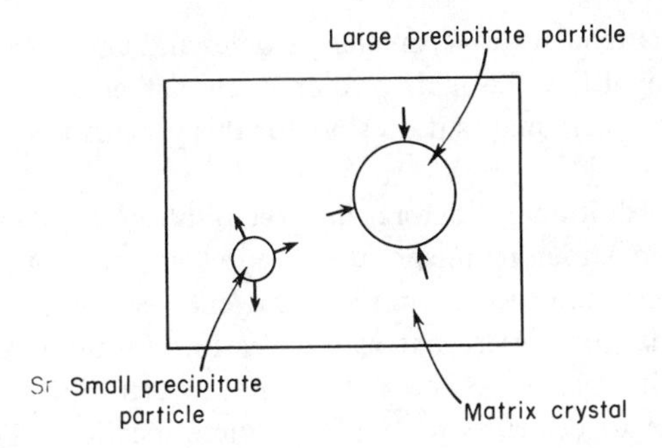

Fig. 9.14 Growth of precipitate particles. Small arrows at surfaces of particles indicate the *net* direction of the flow of solute atoms.

causes the dissolution of small particles and the growth of large ones. A simple derivation of this relationship will now be given.

First, assume that the precipitate has a spherical shape, like that shown in Fig. 9.14. The free energy of a particle can then be expressed by the equation of the previous section

$$\Delta F = -A_1 r^3 + A_2 r^2$$

This is the total free energy of a given amount of precipitate. The free energy per unit volume of the precipitate is the above quantity divided by the volume of the particle, or

$$\Delta F' = \frac{\Delta F}{\frac{4}{3}\pi r^3} = -A'_1 + \frac{A'_2}{r}$$

where $\Delta F'$ is the free energy per unit volume,

$$A'_1 = A_1/\tfrac{4}{3}\pi \text{ and } A'_2 = A_2/\tfrac{4}{3}\pi$$

The free energy per atom of the precipitate is proportional to the free energy per unit volume, so that

$$F_a \approx -A'_1 + \frac{A'_2}{r}$$

where F_a is the free energy per atom. This quantity varies inversely as the radius of the particle. The larger the radius, the more negative the free energy of the second-phase atom and, therefore, the more stable it is in the precipitate. Conversely, the smaller the radius, the less stable it is. Under conditions such as these, solute atoms tend to leave smaller particles and enter the matrix, while at the same time they leave the matrix to enter the larger particles. Diffusion of solute through the matrix makes it possible for this process to continue.

9.7 Additional Factors in Precipitation Hardening. In many alloys, precipitation-hardening phenomena are made even more complicated by the fact that nucleation occurs both homogeneously and heterogeneously. Preferred locations for heterogeneous nucleation in these alloys are grain boundaries and slip planes. Since heterogeneous signifies easier nucleation, precipitation tends to occur more rapidly at these locations. This introduces a time lag between the aging responses in areas undergoing heterogeneous and homogeneous nucleations, and overaging frequently occurs at the grain boundaries long before precipitation in the matrix has had a chance to develop fully. Another effect of rapid precipitation at grain boundaries is that precipitate particles may grow large in size and, as a result, deplete the solute from the areas adjacent to the boundaries. A band of metal, free of precipitation, is then left on each side of the boundary (see Fig. 9.15A). This effect can be greatly magnified if the alloy is heated to the solution-treating temperature and slowly cooled. On slow cooling, nucleation starts at temperatures just below the solvus line at points of easy nucleation, as for example, grain boundaries. At the same time, homogeneous nucleation is prevented because of the negligible rate for this form of nucleation at temperatures close to the solvus line. On continued slow cooling, the grain-boundary precipitate continues to grow through diffusion of solute from the matrix to the precipitate. At the same time, the solute concentration in the matrix is continually reduced and the solution never

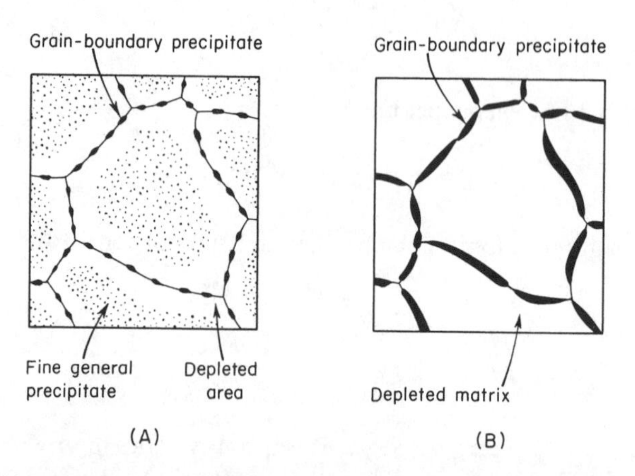

Fig. 9.15 Heterogeneous nucleation at grain boundaries. (A) Moderate rate of cooling may result in both heterogeneous nucleation at grain boundaries and homogeneous nucleation in the centers of the grains. (B) Very slow cooling may result in the precipitate occurring only at grain boundaries.

Fig. 9.16 Schematic representation of a Widmanstätten structure. Short, dark lines represent plate-shaped precipitate particles that are aligned on specific crystallographic planes of the crystals of the matrix.

becomes greatly supersaturated. In this manner, nearly all of the solute finds its way to the second phase in the boundaries and general precipitation in the matrix does not occur. Figure 9.15B is representative of a slowly cooled alloy with grain-boundary precipitation.

Heterogeneous nucleation on slip planes is frequently induced by quenching stresses that develop when the metal is rapidly cooled from the solution-treating temperature. Relief of these stresses through plastic deformation by slip leaves large numbers of dislocation segments along slip planes which act as loci for heterogeneous nucleation.

Precipitation phenomena are sometimes further complicated by the occurrence of recrystallization during the formation of the precipitate. When this occurs, it is the matrix that recrystallizes; matrix atoms regroup to form new crystals.

Finally, it should be pointed out that precipitate particles are not always spherical in shape. Frequently, the precipitate has a platelike, or even a needlelike, form. In many cases plate- or needle-shaped precipitate particles grow in such a manner that they are aligned along specific crystallographic planes or directions of the matrix crystals. Interesting geometrical patterns may result from this type of precipitate growth. It is customary to call such formations *Widmanstätten structures.* Figure 9.16 shows schematically a typical Widmanstätten structure.

Problems

1. An alloy of iron containing 0.01 weight percent carbon is heated to 700°C and allowed to come to equilibrium. It is then quenched in ice water.
   (*a*) Determine the amount of carbon in solid solution after the quench.

(*b*) Would it be possible to store this specimen indefinitely in the ice water and keep the carbon in solid solution?

2. An iron carbon alloy containing 0.018 weight percent carbon is heated to 723°C, where it is allowed to come to equilibrium. Following this, it is cooled to 500°C and then quenched to room temperature and aged at 100°C for 10 hours.

   Will the precipitate be uniformly distributed? If not, describe quantitatively the nature of the distribution.

3. This diagram shows the aluminum end of the copper-aluminum phase diagram. With regard to precipitation hardening, this diagram is analogous

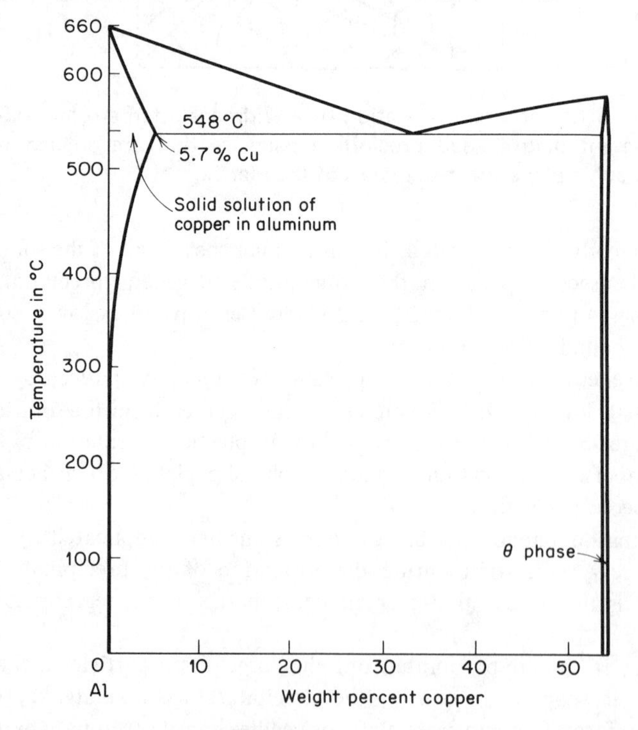

to the iron-carbon diagram shown in Fig. 9.1. There is a solid solution containing copper dissolved in aluminum similar to that of carbon dissolved in iron, and a stable second phase that can precipitate. This is the θ phase, or $CuAl_2$, that is analogous to cementite. While θ does not form directly from the solid solution because there are three intermediate precipitate phases, this is unimportant from an elementary point of view. Finally, an alloy containing 4.5 percent of copper will harden at room temperature in a matter of several days.

(*a*) Indicate the complete procedure that would be required to harden this 4.5 percent copper alloy if aging is to take place at room temperature and maximum hardness is desired.

(*b*) Do you think that the increase in strength given to the metal by the treatment outlined in (*a*) above would be affected if the metal was subjected to a temperature of 400°C for an extended period of time? Explain.

4. Assume that at a given temperature, the volume free energy change associated with the formation of a spherical precipitate particle is 10 calories per cm^3 and that the surface energy of the boundary separating the matrix from the precipitate particle is 500 ergs per cm^2.
   (*a*) Sketch a curve showing the change in the free energy associated with the formation of a particle as a function of the particle radius. Assume that the strain-energy term can be neglected and note that the constants A_1 and A_2, given in the equation in Section 9.5, contain shape factors which, for a spherical particle, are $4/3\pi$ and 4π respectively.
   (*b*) On this curve indicate the magnitudes of r_0, ΔF_{r_0}, and the radius at which ΔF becomes zero.
5. Assume that the temperature is lowered so that the decrease in the volume free energy per unit volume increases to 100 cal per cm^3 while the surface energy remains unchanged.
   (*a*) Recompute the value of the critical radius and the magnitude of the energy barrier for nucleation.
   (*b*) What is the relative change in these quantities compared to the answers in Problem 4?
6. (*a*) Suppose that the alloy of Problem 4 had a total volume fraction of precipitate equal to 2 percent and that the average particle had a radius equal to 4 times the critical radius of Problem 4. How many particles would there be in one cubic centimeter of the alloy?
   (*b*) What would be the total change in free energy due to the precipitate? Express your answer in calories.
7. Repeat Problem 6 for the critical radius of Problem 5.
8. Consider the illustrations shown in Fig. 9.13. Indicate a procedure that might convert the structure shown in Fig. 9.13A into that shown in the accompanying figure.

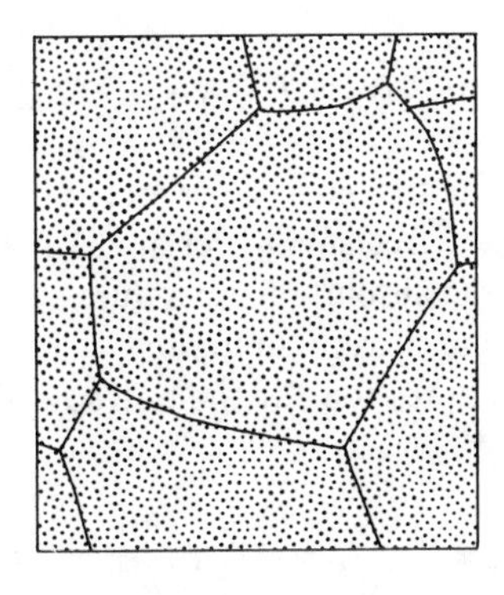

10 Diffusion in Substitutional Solid Solutions

The field of diffusion studies in metals is of great practical, as well as theoretical importance. By *diffusion* one means the movements of atoms within a solution. In general, our interests lie in those atomic movements that occur in solid solutions. This chapter will be devoted in particular to the study of diffusion in substitutional solid solutions and the following chapter will be concerned with atomic movements in interstitial solid solutions.

Before taking up the subject of diffusion proper, the differences between ideal and nonideal solutions will be considered. This is necessary in order that certain complicated phenomena that are observed in metallic diffusion may be better understood.

10.1 Ideal Solutions. Assume that a solution is formed between atoms of two kinds, A and B, and that any given atom of either A or B exhibits no preference as to whether its neighbors are of the same kind or of the opposite kind. In this case, there will be no tendency for A atoms or for B atoms to cluster together or for opposite types of atoms to attract each other. Such a solution is said to be an *ideal solution*, and its free energy per mole is expressed as the sum of the free energy of N_A moles of A atoms, plus N_B moles of B atoms, less a decrease in free energy due to the entropy associated with mixing A and B atoms, where N_A and N_B are the mole, or atom, fractions of A and B respectively.

$$F = F^\circ{}_A N_A + F^\circ{}_B N_B - T\Delta S_M$$

In this expression: $F^\circ{}_A$ = the free energy per mole of pure A
$F^\circ{}_B$ = the free energy per mole of pure B
T = the absolute temperature
ΔS_M = the entropy of mixing

If the previously derived expression for the entropy of mixing is substituted in the above equation,

$$F = F^\circ{}_A N_A + F^\circ{}_B N_B + RT(N_A \ln N_A + N_B \ln N_B)$$

Rearranging the above equation gives us:

$$F = N_A(F^\circ_A + RT \ln N_A) + N_B(F^\circ_B + RT \ln N_B)$$

It is customary, in chemical thermodynamics, to express an extensive property of a solution, such as its free energy, in the following form:

$$F = N_A\bar{F}_A + N_B\bar{F}_B$$

where F is the free energy per mole of solution, N_A and N_B are the mole fractions of the components A and B in the solution, and $\bar{F}_A$ and $\bar{F}_B$ are called the *partial molal free energies* of components A and B respectively. In the ideal solution discussed in the preceding paragraph, the partial molal free energies are equal to

$$\bar{F}_A = F^\circ_A + RT \ln N_A$$

$$\bar{F}_B = F^\circ_B + RT \ln N_B$$

These relationships can also be written:

$$\Delta\bar{F}_A = \bar{F}_A - F^\circ_A = RT \ln N_A$$

$$\Delta\bar{F}_B = \bar{F}_B - F^\circ_B = RT \ln N_B$$

The quantities $\Delta\bar{F}_A$ and $\Delta\bar{F}_B$ represent the increase in free energy when one mole of A, or one mole of B, is dissolved at constant temperature in a very large quantity of the solution. These free-energy changes are functions of the composition, and it is necessary to define them in terms of the addition of a quantity of pure A or pure B to a very large volume of the solution in order that the composition of the solution remains unchanged.

10.2 Nonideal Solutions. In general, most liquid and solid solutions are not ideal and, in solid solutions, it is not to be expected that two atomic forms, chosen at random, will show no preference either for their own kind, or for their opposites. If either of these events occurs, then a larger or smaller free-energy change ($\Delta\bar{F}_A$, $\Delta\bar{F}_B$) will be observed than that expected in an ideal solution. For a nonideal solution, the change in molal free energy is not given by the expression:

$$\Delta\bar{F}_A = RT \ln N_A$$

which holds for an ideal solution. However, in order to retain the form of this simple relationship, we define a quantity "a", known as the activity, in such a fashion that for a nonideal solution

$$\Delta \overline{F}_A = RT \ln a_A$$

$$\Delta \overline{F}_B = RT \ln a_B$$

where a_A is the activity of component A, and a_B the activity of component B. The activities of the components of a solution are useful indicators of the extent to which a solution departs from an ideal solution. They are functions of the composition of the solution, and Fig. 10.1 shows typical activity curves for two types of alloy systems. In both Figs. 10.1A and 10.1B the straight lines marked N_A and N_B correspond to the atom fractions of the components A and B respectively. Figure 10.1A shows an alloy system in which activities a_A and a_B are greater than the corresponding atom fractions N_A and N_B at an arbitrary composition, as indicated by the dotted vertical line marked x. The significance of this positive deviation will now be considered. Since both activities are greater than the corresponding mole fractions, we need consider only one component, which we shall arbitrarily choose as B. According to the figure, N_B is 0.70 while a_B is 0.80. Substituting these values and solving for $\Delta \overline{F}$ leads to the following results for an ideal solution:

$$\Delta \overline{F} = RT \ln 0.70 = -0.356\, RT$$

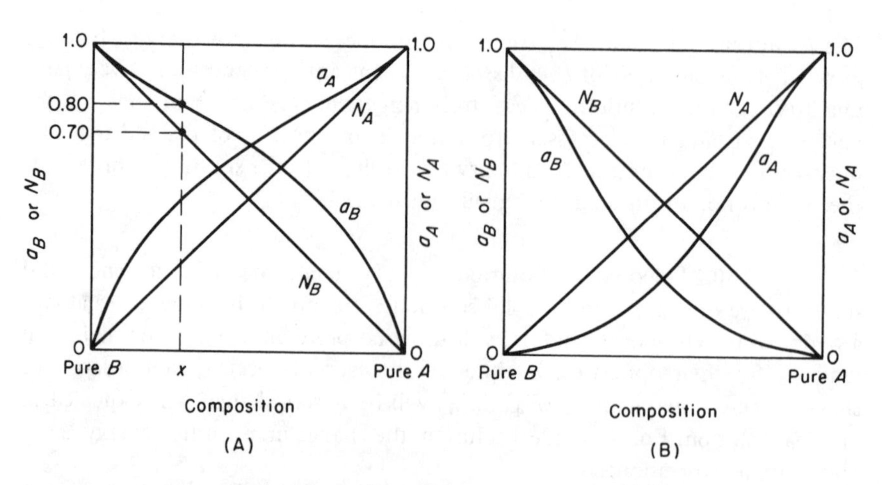

Fig. 10.1 Variation of the activities with concentration: (A) positive deviation, and (B) negative deviation.

for the non-ideal solution:

$$\Delta\overline{F} = RT \ln 0.80 = -0.223\, RT$$

A solution which behaves in the manner shown in Fig. 10.1A has a smaller decrease in free energy as a result of the formation of the solution than if an ideal solution had formed. In situations such as this, the attraction between atoms of the same kind is greater than the attraction between dissimilar atoms. The extreme example would be one in which the two components were completely insoluble in each other, in which case the activities would be equal to unity for all ratios of A and B. The other set of curves in Fig. 10.1B shows the opposite effect: unlike atoms are more strongly attracted to each other than are those of the same kind. Here we find negative deviations of the activity curves from the mole fraction lines.

It is apparent from the above that when the activities of the components of a solution are compared to their respective atom fractions, they indicate roughly the nature of the interactions of the atoms in the solution. For this reason, it is frequently convenient to use quantities known as *activity coefficients*, which are the ratios of the activities to their respective atom fractions. The activity coefficients of a binary solution of A and B atoms are thus defined as

$$\gamma_A = \frac{a_A}{N_A} \text{ and } \gamma_B = \frac{a_B}{N_B}$$

10.3 Diffusion in an Ideal Solution. Consider a solid solution composed of two forms of atoms A and B respectively. Let us arbitrarily designate the A component as the solute and the B component as the solvent, and consider that the solution is ideal. This, of course, implies that there is no interaction between solute and solvent atoms or that the two forms act inside the crystal as though they were a single chemical species.

Experimental work, that will be discussed, has shown that the atoms in face-centered cubic, body-centered cubic, and hexagonal metals move about in the crystal lattice as a result of vacancy motion. Let it now be assumed that the jumps are entirely random; that is, the probability of jumping is the same for all of the atoms surrounding a given vacancy. This statement implies that the jump rate does not depend on the concentration.

Figure 10.2 represents a single crystal bar composed of a solid solution of A and B atoms in which the composition of the solute varies continuously along the length of the bar, but is uniform over the cross-section. For the sake of simplifying the argument, the crystal structure of the bar is assumed to be simple cubic with a ⟨100⟩ direction along the axis of the bar. It is further assumed that the concentration is greatest at the right end of the bar and least at the left end,

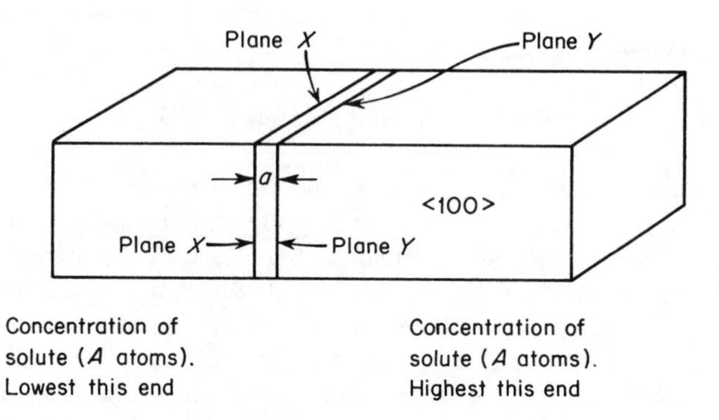

Fig. 10.2 Hypothetical single crystal with a concentration gradient.

and that the macroscopic concentration gradient dn_A/dx applies on an atomic scale so that the difference in composition between two adjacent transverse atomic planes is:

$$(a)\,\frac{dn_A}{dx}$$

where a is the interatomic, or lattice spacing (see Fig. 10.3). Let the mean time of stay of an atom in a lattice site be τ. The average frequency with which the

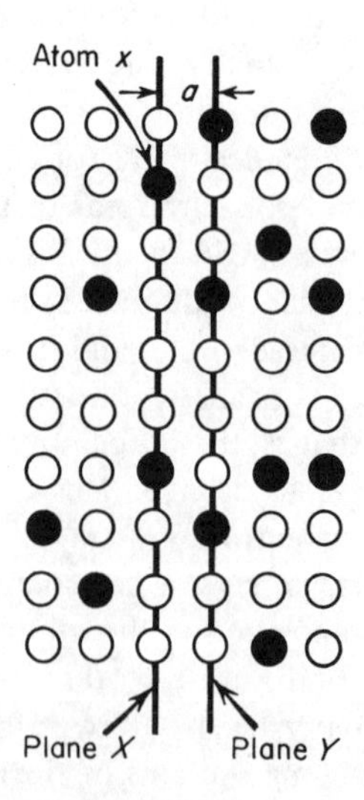

Fig. 10.3 Atomistic view of a section of the hypothetical crystal of Fig. 10.2.

atoms jump is therefore $1/\tau$. In the simple cubic lattice pictured in Fig. 10.3, any given atom, such as that indicated by the symbol x, can jump in six different directions: right or left, up or down, or into or out of the plane of the paper. The exchange of A atoms between two adjacent transverse atomic planes, such as those designated X and Y in Fig. 10.3, will now be considered. Of the six possible jumps which an A atom can make in either of these planes, only one will carry it over to the other indicated plane, so that the average frequency with which an A atom jumps from X to Y is $1/6\tau$. The number of these atoms that will jump per second from plane X to plane Y equals the total number of the atoms in plane X times the average frequency with which an atom jumps from plane X to plane Y. The number of solute atoms in plane X equals the number of solute atoms per unit volume (the concentration n_A) times the volume of the atoms in plane X, (Aa), so that the flux of solute atoms from plane X to plane Y is

$$J_{X \to Y} = \frac{1}{6\tau}(n_A aA)$$

where $J_{X \to Y}$ = flux of solute atoms from plane X to plane Y
τ = mean time of stay of a solute atom at a lattice site
n_A = number of A atoms per unit volume
A = cross-section area of specimen
a = lattice constant of crystal

The concentration of A atoms in plane Y may be written:

$$(n_A)_Y = n_A + (a)\frac{dn_A}{dx}$$

where n_A is the concentration at plane X, and a is the lattice constant, or distance between planes X and Y. The rate at which A atoms move from plane Y to X is thus

$$J_{Y \to X} = \left[n_A + (a)\frac{dn_A}{dx}\right]\frac{aA}{6\tau}$$

where $J_{Y \to X}$ represents the flux of A atoms from plane Y to plane X. Because the flux of solute atoms from right to left is not the same as that from left to right, there is a net flux (designated by the symbol J) which can be expressed mathematically as follows:

$$J = J_{X \to Y} - J_{Y \to X} = \frac{aA}{6\tau}(n_A) - \left[n_A + (a)\frac{dn_A}{dx}\right]\frac{aA}{6\tau}$$

or

$$J = -\frac{a^2 A}{6\tau} \cdot \frac{dn_A}{dx}$$

Notice that, in the above equation, the flux (J) of A atoms is negative when the concentration gradient is positive (concentration of A atoms increases from left to right in Fig. 10.3). This result is general for diffusion in an ideal solution; the diffusion flux is down the concentration gradient. Notice that if one considers the flow of B atoms instead of A atoms, the net flux will be from left to right, in agreement with a decreasing concentration of the B component as one moves from left to right. Again, the flux (in this case of B atoms) is down the concentration gradient.

Let us now make the substitution

$$D = \frac{a^2}{6\tau}$$

in the equation for the net flux, which gives:

$$J = -DA \frac{dn_A}{dx}$$

This equation is identical with that first proposed by Adolf Fick[1] in 1855 on theoretical grounds for diffusion in solutions. In this equation, called *Fick's first law*, J is the flux, or quantity per second, of diffusing matter passing normally through an area A under the action of a concentration gradient dn_A/dx. The factor D is known as the *diffusivity*, or the *diffusion coefficient*.

As originally conceived, the diffusivity D in Fick's first law was assumed to be constant for measurements made at a fixed temperature. However, only in the case of a solution composed of two gases has it been verified experimentally that

Table 10.1 Variation of the Diffusivity with Composition in a Binary Gas Solution of Oxygen and Hydrogen. (Data of Deutsch.)

Atom Fraction $\frac{n_1}{n_1 + n_2}$	*Diffusivity* D
0.25	0.767
0.50	0.778
0.75	0.803

1. Fick, Adolf, "Über Diffusion," *Poggendorff's Annalen,* 94 59 (1855).

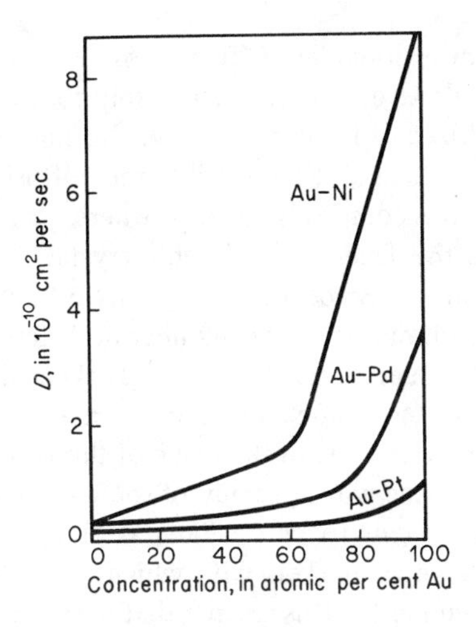

Fig. 10.4 Variation of the diffusion coefficient (D) with composition in Au-Ni, Au-Pd, and Au-Pt alloys. (From Matano, C., *Proc. Phys. Math. Soc. Japan,* **15** 405 [1933]; *Japan Jour. Phys.,* **8** 109 [1933].)

the diffusivity approaches a constant value at a fixed temperature. An example is shown in Table 10.1 for binary mixtures of the gases oxygen and hydrogen, which have been selected as an example because of their rather large molecular mass differences (16 to 1) and a correspondingly large difference in their arithmetic mean speed (4 to 1). The table indicates that the diffusivity (D) changes by less than 5 percent when the atom fraction $n_1/(n_1 + n_2)$ changes from 0.25 to 0.75. When the molecular masses are more nearly equal, the variation in the diffusivity with composition becomes even smaller and harder to detect in gaseous solutions.

In contrast to gaseous solutions, the diffusivity in both liquid and solid solutions is seldom constant. A typical example of this variation of the diffusion coefficient with composition can be seen in Fig. 10.4, taken from the work of Matano.[2] The figure shows the variation of the diffusivity in three types of solid solutions formed by alloying gold with nickel, palladium, and platinum respectively. The variation of diffusivity with composition is especially marked in the case of gold-nickel solid solutions, changing by a factor of about 10, while the composition varies from 20 to 80 atomic percent. The variation in the Au-Pt solutions is much smaller, but still considerably greater than the variation of D observed in gases. It must be concluded, therefore, that in metallic solid solutions, the diffusivity is usually not constant.

2. Matano, C., *Proc. Phys. Math. Soc., Japan,* **15** 405 (1933); *Japan J. Phys.,* **8** 109 (1933).

10.4 The Kirkendall Effect. An experiment will now be discussed which shows that, in a binary solid solution, each of the two atomic forms can move with a different velocity. In the original experiment, as performed by Smigelskas and Kirkendall,[3] the diffusion of copper and zinc atoms was studied in the composition range where zinc dissolves in copper and the alloy still retains the face-centered cubic crystal structure characteristic of copper (alpha-brass range). Since their original work, many other investigators have found similar results using a large number of different binary alloys. Figure 10.5 is a schematic representation of a Kirkendall diffusion couple: a three-dimensional view of a block of metal formed by welding together two metals of different compositions. In the plane of the weld, shown in the center of Fig. 10.5, a number of fine wires (usually of some refractory metal that will not dissolve in the alloy system to be studied) are incorporated in the diffusion couple. These wires serve as markers with which to study the diffusion process. For the sake of the argument, let us assume that the metals separated by the plane of the weld are originally pure metal A and pure metal B; that on the right side of the weld is pure A, while that on the left is pure B. In order that a total amount of diffusion, which is large enough to be experimentally measurable, can be obtained in a specimen of this type, it is necessary that it be heated to a temperature close to the melting point of the metals comprising the bar, and maintained there for a relatively long time, usually of the order of days, for diffusion in solids is much slower than in gases or liquids. Upon cooling the specimen to room temperature, it is placed in a lathe and thin layers parallel to the weld interface are removed from the bar. Each layer is then analyzed chemically and the results plotted to give a curve showing the composition of the bar as a function of distance along the bar. Such a curve is shown schematically in Fig. 10.6, from which it is easily deduced that there has been a

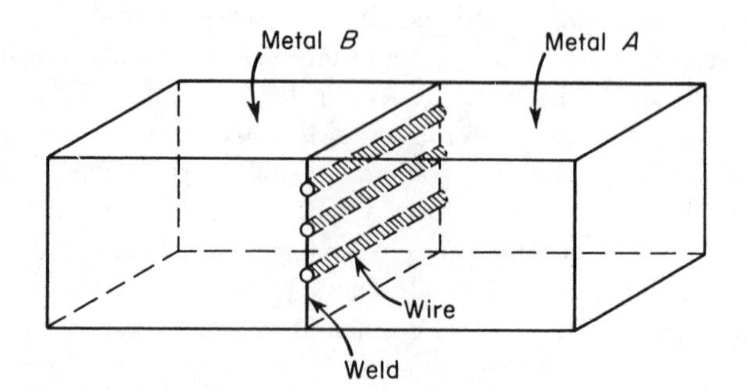

Fig. 10.5 Kirkendall diffusion couple.

3. Smigelskas, A. D., and Kirkendall, E. O., *Trans. AIME,* **171** 130 (1947).

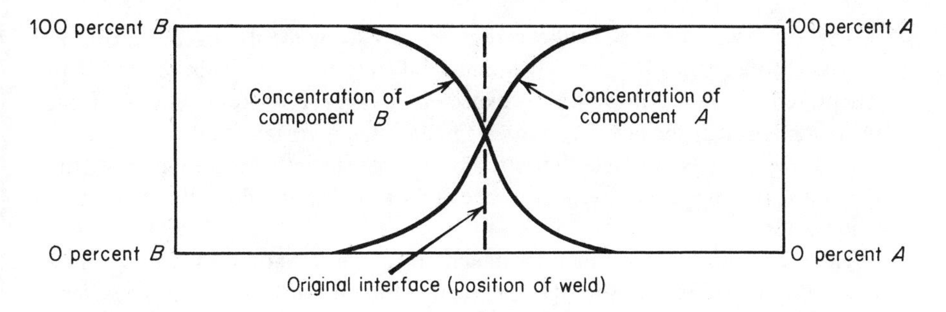

Fig. 10.6 Curves showing concentration as a function of distance along a diffusion couple. Curves of this type are usually called *penetration curves.*

flow of B atoms from the left side of the bar toward the right, and a corresponding flow of A atoms in the other direction.

Curves such as those shown in Fig. 10.6 were obtained for diffusion couples many years before the Smigelskas and Kirkendall experiment was performed. The original feature to their work was the incorporation of marker wires between the members of the diffusion couple. The interesting result which they obtained was that the wires moved during the diffusion process. The nature of this movement is shown in Fig. 10.7, where the upper figure represents the diffusion couple before the isothermal treatment (anneal), and the lower, the same bar after diffusion has occurred. The latter figure shows that the wires have moved to the right through the distance x. This distance, while small, is measurable, and for those cases where the markers have been placed at the weld between two different metals, the distance has been found to vary as the square root of the time during which the specimen was maintained at the diffusion temperature.

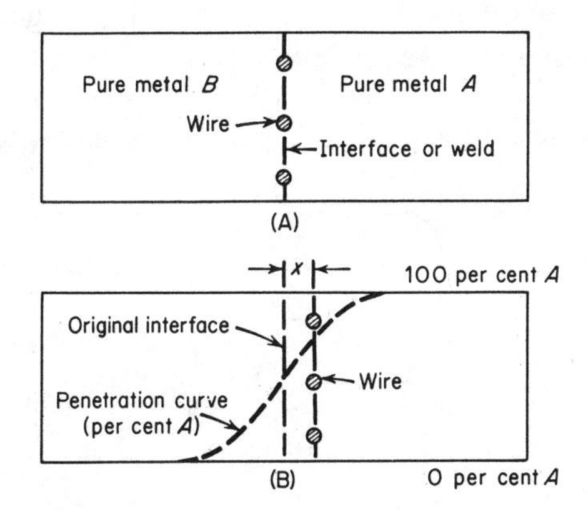

Fig. 10.7 Marker movement in a Kirkendall diffusion couple.

The only way to explain the movement of the wires during the diffusion process (in Fig. 10.7) is for the A atoms to diffuse faster than the B atoms. In this present example, it would be expected that more A atoms than B atoms must pass through the cross-section (defined by the wires) per unit time, causing a net flow of mass through the wires from right to left. Measurement of the position of the wires with respect to one end of the bar will show the movement of the wires.

The Kirkendall effect can be taken as a confirmation of the vacancy mechanism of diffusion. Through the years a number of mechanisms have been proposed to explain the movement of atoms in a crystal lattice. These can be grouped roughly into two classifications: those involving the motion of a single atom at a time, and those involving the cooperative movement of two or more atoms. As examples of the former, we have diffusion by the vacancy mechanism and diffusion of interstitial atoms (such as carbon in the iron lattice where the carbon atoms jump from one interstitial position to an adjacent one). While interstitial diffusion is recognized as the proper mechanism to explain the movement of small interstitial atoms through a crystal lattice, it is generally conceded that a mechanism of diffusion which involves placing large atoms (which normally enter into substitutional solutions) into interstitial positions is not feasible on energy considerations. The distortion of the lattice caused by placing one of these atoms in an interstitial position is very large, requiring a very large activation energy. Of these two possibilities based on the motion of individual atoms, the vacancy mechanism is much to be preferred in explaining diffusion in substitutional solid solutions.

The simplest conceivable cooperative movement of atoms is a direct interchange, as is illustrated schematically in Fig. 10.8. Here two adjacent atoms jump past each other and exchange positions. However, this involves the outward displacement of the atoms surrounding the jumping pair during the period of the transfer. Theoretical computations[4] of the energy required to make a direct interchange indicate that it is much larger than that required for the jump of an atom into a vacancy in metallic copper, and it is commonly

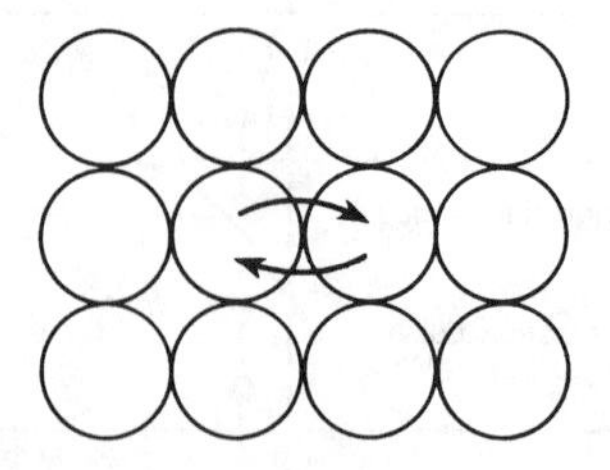

Fig. 10.8 Direct interchange diffusion mechanism.

4. Huntington, H. B., and Seitz, F., *Phys. Rev.*, **61** 315, 325 (1942).

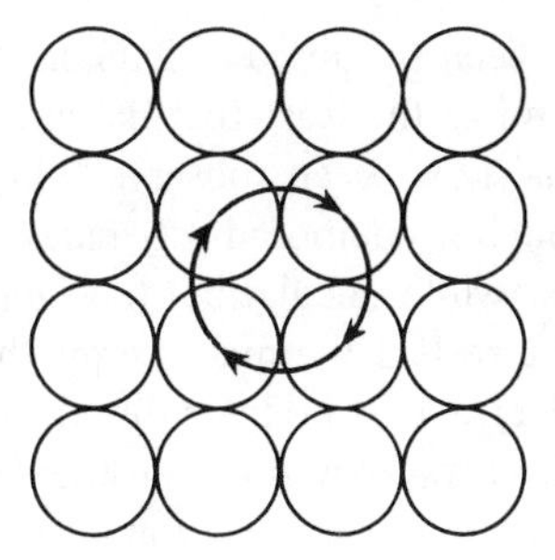

Fig. 10.9 Zener ring mechanism for diffusion.

believed that this conclusion can be applied to other metals. For this reason the direct interchange is usually ruled out as an important mechanism in the diffusion of metals.

Another possible mechanism that explains diffusion in substitutional solid solutions is the Zener[5] ring mechanism. In this case, it is assumed that thermal vibrations are sufficient to cause a number of atoms, which form a natural ring in a crystal, to jump simultaneously and in synchronism in such a manner that each atom in the ring advances one position around the ring. This mechanism is illustrated in Fig. 10.9, where a ring of four atoms is shown and the arrows indicate the motion of the atoms during a jump. Zener has suggested, on the basis of theoretical computations, that a ring of four atoms might be the preferred diffusion mechanism in body-centered cubic metals because their structure is more open than that of close-packed metals (face-centered cubic and close-packed hexagonal). The more open structure would require less lattice distortion during the jump. It has also been proposed as a more probable mechanism than the direct interchange because the lattice distortion that occurs during the jump is smaller, requiring less energy for the movement. However, the principal objection to the acceptance of the ring mechanism, even in body-centered cubic metals, is that it has been shown conclusively by diffusion experiments involving couples composed of a number of different body-centered cubic metals, that a Kirkendall effect occurs. It is possible to explain different rates for the diffusion of two atoms A and B by a vacancy mechanism; the rate at which the two atoms jump into the vacancies need only be different. In a ring mechanism, or in a direct exchange, the rate at which A atoms move from left to right must equal the rate at which B atoms move from right to left. Thus, in those alloy systems where a Kirkendall effect has been observed, one must rule out interchange mechanisms.

In summation, the vacancy mechanism is generally conceded to be the correct mechanism for diffusion in face-centered cubic metals. First, because of the

5. Zener, C., *Acta Cryst.*, 3 346–354 (1950).

various methods that have been proposed to explain the movements of atoms in this type of crystal, it requires the least thermal energy to activate it. Second, the Kirkendall effect has now been observed in a great many diffusion experiments involving couples composed of face-centered cubic metals. In body-centered cubic metals, while calculations have indicated that diffusion by a ring mechanism might require less thermal energy than vacancy diffusion, the discovery of a Kirkendall effect in body-centered cubic metals indicates that these metals also diffuse by a vacancy mechanism. Finally, while the picture is not quite as clear in the case of hexagonal metals, results of diffusion experiments are in accord with a vacancy mechanism. In this regard, it should be mentioned that, due to the asymmetry of the hexagonal lattice, the rate of diffusion is not the same in all directions through the lattice. Thus, diffusion in the basal plane occurs at a different rate from diffusion in a direction perpendicular to the basal plane. If a vacancy mechanism is assumed for hexagonal metals, this implies that the jump rate of an atom into vacancies lying in the same basal plane as itself will differ from the rate at which it jumps into vacancies lying in the basal planes directly above or directly below it.

10.5 Porosity. The Kirkendall experiment demonstrates that the rate at which the two types of atoms of a binary solution diffuse is not the same. Experimental measurements have shown that the element with the lower melting point diffuses faster. Thus, in alpha-brass (a mixture of copper and zinc atoms) zinc atoms move at the faster rate. On the other hand, diffusion couples formed of copper and nickel show that copper moves faster than nickel, in agreement with the fact that copper melts at a lower temperature than nickel.

In the case of a copper-nickel couple, shown schematically in Fig. 10.10, there is a greater flow of copper atoms toward the nickel side than nickel atoms

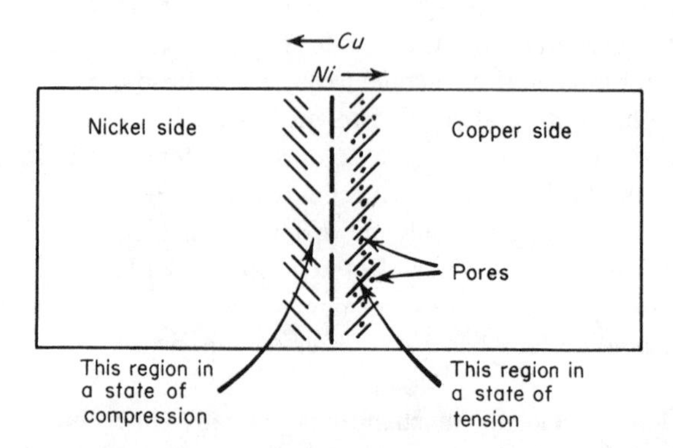

Fig. 10.10 In a copper-nickel diffusion couple the copper diffuses faster into the nickel than the nickel does into the copper.

toward the copper side. The right-hand side of the specimen suffers a loss of mass because it loses more atoms than it gains, while the left-hand side gains in mass. As a result of this mass transfer, the right- and left-hand portions of the specimen shrink and expand respectively. In cubic metals, these volume changes would be isotropic [6] except for the fact that in a diffusion couple of any considerable size the diffusion zone containing the regions that undergo the volume changes may be only a small part of the total specimen. Shrinkage or expansion in a direction perpendicular to the weld interface occurs without an appreciable restraint from the rest of the specimen, but dimensional changes parallel to the weld interface are resisted by the bulk of the metal that lies outside of the diffusion zone. The net effect of this restraining action is twofold: the dimensional changes are essentially one-dimensional (along the axis of the bar lying perpendicular to the weld interface) and a state of stress is set up in the diffusion zone. The region to the right of the weld which suffers a loss of mass is placed under a two-dimensional tensile stress, while that on the side that gains mass is placed in a state of two-dimensional compressive stress. These stress fields may bring about plastic flow, with the resulting structural changes normally associated with plastic deformation[7] and high temperatures; formation of substructures, recrystallization, and grain growth.

In addition to the effects mentioned in the previous paragraph, another phenomenon is encountered when two diffusing atom forms move with greatly different rates during diffusion, and it is directly connected with the vacancy motion associated with diffusion in metals. It has frequently been observed that voids, or pores, form in that region of the diffusion zone from which there is a flow of mass. Since every time an atom makes a jump a vacancy moves in the opposite direction, an unequal flow in the two types of atoms must result in an equivalent flow of vacancies in the reverse direction to the net flow of atoms. Thus, in the copper-nickel couple shown in Fig. 10.10, more copper atoms leave the area just to the right of the weld interface than nickel atoms arrive to take their place, and a net flow of vacancies occurs from the nickel-rich side of the bar toward the copper-rich side. This movement reduces the equilibrium number of vacancies on the nickel-rich side of the diffusion range, while increasing it on the copper side. The degree of supersaturation and undersaturation is, however, believed to be small. On the side of a diffusion couple that loses mass, it has been estimated by several workers[8,9] that the vacancy supersaturation is of the order of 1 percent, or less, of the equilibrium number of vacancies. Since the concentration of vacancies is not greatly affected by the conditions of flow, and the number is, at best, a small fraction of the number of atoms, vacancies must

6. Balluffi, R. W., and Seigle, L. I., *Acta Met.*, **3** 170 (1955).
7. Doo, V. Y., and Balluffi, R. W., *Acta Met.*, **6** 428 (1958).
8. Barnes, R. S., and Mazey, D. J., *Acta Met.*, **6** 1 (1958).
9. Balluffi, R. W., *Acta Met.*, **2** 194 (1954).

be created on the side of a couple that gains mass, and absorbed on the side that loses mass. Various sources and sinks have been proposed for the maintenance of this flux of vacancies. Grain boundaries and exterior surfaces are probable positions for both the creation and elimination of vacancies, but it is now generally agreed that the experimental facts agree best on the concept that the most important mechanism for the creation of vacancies is *dislocation climb*, and that dislocation climb and void formation account for most of the absorbed vacancies, positive climb being associated with the removal of vacancies and negative climb with the creation of vacancies.

The formation of voids, as a result of the vacancy current that accompanies the unequal mass flow in a diffusion couple, is influenced by several factors. It is generally believed that voids are heterogeneously nucleated,[10,8] that is, they form on impurity particles. The tensile stress that exists in the region of the specimen where the voids form is also recognized as a contributing factor in the development of voids. If this tensile stress is counteracted by a hydrostatic compressive stress placed on the sample during the diffusion anneal, the voids can be prevented from forming.[11]

10.6 Darken's Equations. The Kirkendall effect shows that in a diffusion couple composed of two metals, the atoms of the components move at different rates, and the flux of atoms through a cross-section defined by markers is not the same for both atom forms. It is, thus, more logical to think in terms of two diffusivities D_A and D_B corresponding to the movement of the A and B atoms respectively. These quantities may be defined by the following relationships.

$$J_A = -D_A \frac{A\,\partial n_A}{\partial x} \quad \text{and} \quad J_B = -D_B \frac{A\,\partial n_B}{\partial x}$$

where J_A and J_B = the fluxes (number of atoms per second passing through a given cross-section of area A) of the A and B atoms respectively
D_A and D_B = the diffusivities of the A and B atoms
n_A and n_B = the numbers of A and B atoms per unit volume respectively

The diffusivities D_A and D_B are known as the intrinsic diffusivities and are functions of composition and therefore of position along a diffusion couple.

Darken's equations[12] which make it possible to determine the intrinsic diffusivities experimentally will now be derived. Several important assumptions are made in this derivation. First, it is assumed that all of the volume expansion and contraction during diffusion, due to unequal mass flow, occurs only in the direction perpendicular to the weld interface. (The cross-section area A does not

10. Balluffi, R. W., and Seigle, L. I., *Acta Met.*, 5 449 (1957).
11. Barnes, R. S., and Mazey, D. J., *Acta Met.*, 6 1 (1958).
12. Darken, L. S., *Trans. AIME,* 175 184 (1948).

change during the diffusion.) As pointed out earlier, this condition is effectively realized when the diffusion couple is large in size compared to the diffusion zone. It is also assumed that the total number of atoms per unit volume is a constant or that

$$n_A + n_B = \text{constant}$$

This is equivalent to a requirement that the volume per atom be independent of concentration, which is an assumption that has not been realized experimentally, but deviations from this relationship can generally be compensated for by computation. The following derivation is also based on the condition that porosity does not occur in the specimen during the diffusion process.

Let us first consider the velocity of the Kirkendall markers through space. In Fig. 10.11, the cross-section at the distance x from the left end of the bar represents the position of the markers at the time t_0, while the cross-section at the distance x' represents their position after a time interval dt. At the same time the origin of coordinates (left end of the bar) is assumed to be far enough away from the weld that its composition is not affected by the diffusion. The velocity v of the Kirkendall markers may be written

$$v = \frac{x - x'}{dt}$$

The marker velocity is also equal in magnitude, but opposite in direction to the volume of matter flowing past the markers per second divided by the cross-section area A of the bar at the markers.

$$v = -\frac{\text{volume}}{\text{seconds}} \times \frac{1}{\text{area}}$$

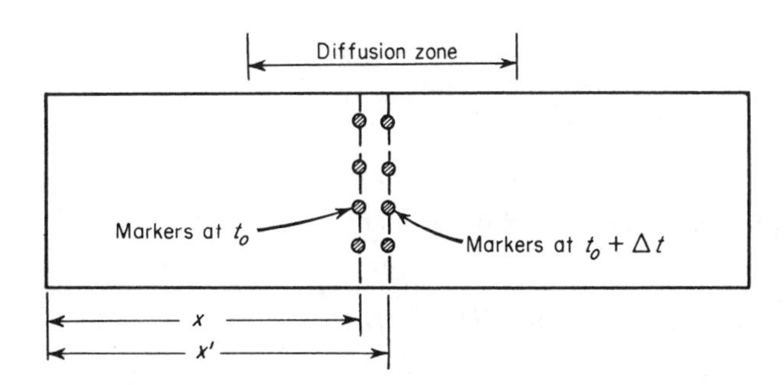

Fig. 10.11 Marker movement is measured from a point outside the zone of diffusion (one end of bar).

The volume of matter that flows through the markers per second is equal to the net flux of atoms (net number of atoms per second) passing the wires times the volume per atom, or

$$\frac{\text{Volume}}{\text{Seconds}} = \frac{J_{\text{net}}}{n_A + n_B}$$

where $1/(n_A + n_B)$ is the volume per atom by the definition of n_A and n_B. The net flux equals the sum of the fluxes of the A and B atoms, or

$$J_{\text{net}} = J_A + J_B = -D_A \frac{A\partial n_A}{\partial x} - D_B \frac{A\partial n_B}{\partial x}$$

Substituting the above expressions into the equation for the marker velocity gives us the following equation:

$$v = -\frac{\text{volume}}{\text{seconds}} \times \frac{1}{\text{area}} = + \frac{\left(D_A \dfrac{\partial n_A}{\partial x} + D_B \dfrac{\partial n_B}{\partial x}\right) A}{(n_A + n_B)A}$$

or

$$v = + \frac{\left(D_A \dfrac{\partial n_A}{\partial x} + D_B \dfrac{\partial n_B}{\partial x}\right)}{n_A + n_B}$$

Remembering that $n_A + n_B$ is a constant, and that by definition

$$N_A = \frac{n_A}{n_A + n_B}, \qquad N_B = \frac{n_B}{n_A + n_B}$$

$$N_B = 1 - N_A \quad \text{and} \quad \frac{\partial N_B}{\partial x} = -\frac{\partial N_A}{\partial x}$$

where N_A and N_B are the atom fractions of the A and B atoms, we may write the velocity of the markers in the following manner:

$$v = (D_A - D_B) \frac{\partial N_A}{\partial x}$$

In principle it is possible to insert an imaginary set of markers on any cross-section along a diffusion couple. The above equation may therefore be used to compute the velocity with which any lattice plane, normal to the diffusion flux, moves. However, since D_A and D_B are normally functions of composition, v is a function of the composition of the specimen at the cross-section where v is measured. This means that the value of v is normally a function of position along the diffusion couple.

The above equation is one of our desired relationships, for it relates the two intrinsic diffusivities to the marker velocity and the concentration gradient $(\partial N_A/\partial x)$, two quantities that can be experimentally determined. However, another equation is also needed in order that we may solve it and the above equation simultaneously for the two unknowns D_A and D_B. This needed equation can be obtained by considering the rate at which the number of one of the atomic forms (A or B) changes inside a small-volume element.

Figure 10.12 is another figure representing the diffusion couple of Fig. 10.11. In this case, however, the cross-section (designated mm) at a distance x from the left end of the bar represents a cross-section fixed in space; that is, fixed with respect to the end of the bar, which is assumed outside of the diffusion zone. The cross-section at x (mm), accordingly, is not an instantaneous position of the markers as was the case in Fig. 10.11. A second cross-section of the bar (nn) is also shown, the distance of which from the end of the bar is $x + dx$. The two cross-sections define a volume element $A(dx)$.

Let us consider the rate at which the number of A atoms changes inside this volume. This quantity equals the difference in the number of A atoms moving into and out of the volume per second, or to the difference in the flux of A atoms across the surface at x and at $x + dx$. Because a cross-section fixed in the lattice moves relative to our cross-sections fixed in space (at x and $x + dx$), the flux of A atoms through one of these space-fixed boundaries is due to two effects. First, because the metal is moving with a velocity v, a number $(n_A vA)$ of A atoms is carried each second through the cross-section at x, where n_A is the number of A atoms per unit volume, and vA is the volume of metal flowing past the cross-section at x in a second. To this flux must be added the usual diffusion

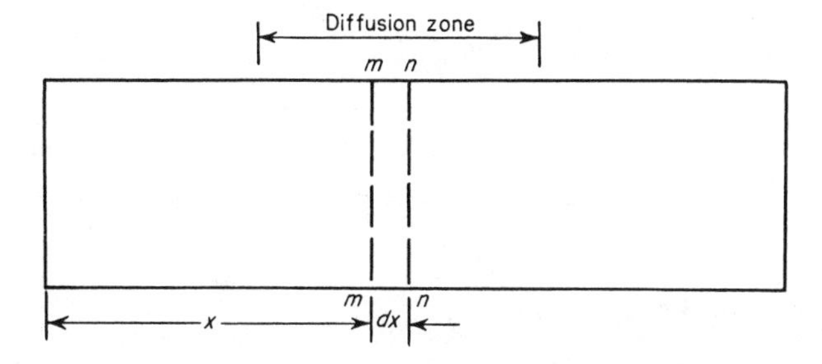

Fig. 10.12 The cross-sections mm and nn are assumed to be fixed in space; that is, they do not move with respect to the left end of the bar.

flux, so that the total number of atoms per second crossing the boundary at x is:

$$(J_A)_x = -AD_A \frac{\partial n_A}{\partial x} + n_A v A$$

The flux of A atoms through the cross-section at $x + dx$ is

$$(J_A)_{x+dx} = (J_A)_x + \frac{\partial (J_A)_x}{\partial x} \cdot dx$$

The rate at which the number of atoms inside the volume Adx changes is therefore

$$(J_A)_x - (J_A)_{x+dx} = \frac{\partial}{\partial x}\left[D_A \frac{\partial n_A}{\partial x} - n_A v\right] \cdot Adx$$

or the rate at which the number of A atoms per unit volume changes is

$$\frac{(J_A)_x - (J_A)_{x+dx}}{Adx} = \frac{\partial n_A}{\partial t} = \frac{\partial}{\partial x}\left[D_A \frac{\partial n_A}{\partial x} - n_A v\right]$$

This equation may also be written:

$$\frac{\partial N_A}{\partial t} = \frac{\partial}{\partial x}\left[D_A \frac{\partial N_A}{\partial x} - N_A v\right]$$

because $n_A + n_B$ is assumed constant and the division of each term of the previous equation by this quantity makes it possible to convert concentration units from numbers of A atoms per unit volume to atom fractions. The expression

$$v = (D_A - D_B) \frac{\partial N_A}{\partial x}$$

is now substituted into the above equation, and with the aid of the relationship

$$N_A = 1 - N_B$$

one obtains the final relationship

$$\frac{\partial N_A}{\partial t} = \frac{\partial}{\partial x}[N_B D_A + N_A D_B] \frac{\partial N_A}{\partial x}$$

This equation is in the form of Fick's second law, which is generally written

$$\frac{\partial N_A}{\partial t} = \frac{\partial}{\partial x} \tilde{D} \frac{\partial N_A}{\partial x}$$

where the quantity $\tilde{D}$ is seen to be equal to $[N_B D_A + N_A D_B]$. In fact, it is just this relationship that we have been seeking:

$$\tilde{D} = N_B D_A + N_A D_B$$

for the quantity $\tilde{D}$ may be evaluated experimentally, as can the atom fractions N_A and N_B. This is the second of Darken's equations for the determination of the intrinsic diffusivities D_A and D_B.

10.7 Fick's Second Law. Fick's second law

$$\frac{\partial N_A}{\partial t} = \frac{\partial}{\partial x} \tilde{D} \frac{\partial N_A}{\partial x}$$

is the basic equation for the experimental study of isothermal diffusion. Solutions of this second-order partial differential equation have been derived corresponding to the boundary conditions found in many types of diffusion samples. Most metallurgical specimens, such as the diffusion couple shown in Fig. 10.6, involve only a (net) one-dimensional flow of atoms and the assumption that the specimen is long enough in the direction of diffusion (so that the diffusion process does not change the composition at the ends of the specimen). There are two standard methods of measuring the diffusion coefficient when using this type of specimen. In one case, the diffusivity is assumed constant, and in the other, it is taken as a function of composition. The former method, known as the *Grube method,*[13] is strictly applicable only to those cases in which the diffusivity varies very slightly with composition. However, it may be applied to diffusion in alloy systems where the diffusivity varies moderately with composition if the two halves of a couple are made of metals differing slightly in composition. Thus, the couple in Fig. 10.5, instead of consisting of a bar of pure *A* welded to another of pure *B*, might have consisted of an alloy of perhaps 60 percent of *A* and 40 percent of *B* welded to one of 55 percent of *A* and 45 percent of *B*. Our preceding analysis would hold just as well for this couple as for the couple composed of pure metals. Over a small range of composition such as this, the diffusivity is essentially constant and the measurement effectively gives an average value of the diffusivity for the interval.

13. Grube, G., and Jedele, A., *Zeit, Elektrochem.* **38** 799 (1932).

If the diffusivity $\tilde{D}$ is assumed to be a constant, then Fick's second law can be written:

$$\frac{\partial N_A}{\partial x} = \frac{\partial}{\partial x}\tilde{D}\frac{\partial N_A}{\partial x} = \tilde{D}\frac{\partial^2 N_A}{\partial x^2}$$

The solution of this equation for the case of a diffusion couple consisting originally of two alloys of the elements A and B, one having the composition N_{A_1} (atom fraction) and the other the composition N_{A_2} at the start of the diffusion process, is:

$$N_A = N_{A_1} + \frac{(N_{A_2} - N_{A_1})}{2}\left[1 + \text{erf}\frac{x}{2\sqrt{\tilde{D}t}}\right] \quad \text{for } -\infty < x < \infty$$

where N_A is the composition or atom fraction at a distance x (in cm) from the weld interface, t is the time in seconds, and $\tilde{D}$ is the diffusivity. The symbol erf $x/2\sqrt{\tilde{D}t}$ represents the error function, or probability integral with the argument $y = x/2\sqrt{\tilde{D}t}$. This function is defined by the equation

$$\text{erf}(y) = \frac{2}{\sqrt{\pi}}\int_0^y e^{-y^2}\,dy$$

This function is tabulated in many mathematical tables[14] in much the same manner as trigonometric and other frequently used functions. Table 10.2 gives a few values of the error function for some corresponding values of the argument y. In applying this table, the error function becomes negative when y (or x in $x/2\sqrt{\tilde{D}t}$) is negative.

Figure 10.13 shows the theoretical penetration curve (distance versus composition curve) obtained when the solution of Fick's equation is plotted as a

Table 10.2 Error Function Values.

y	erf (y)
0	0
0.2	0.2227
0.4	0.42839
0.477	0.50006
0.6	0.60386
0.8	0.74210
1.0	0.84270
1.4	0.95229
2.0	0.99532
3.0	0.99998

14. Pierce, B. O., *A Short Table of Integrals*. Ginn and Company, Boston, 1929.

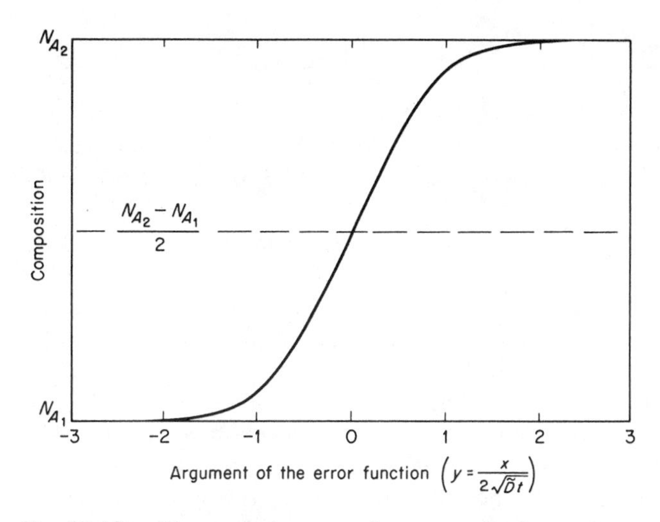

Fig. 10.13 Theoretical penetration curve, Grube method.

function of the variable $x/2\sqrt{\tilde{D}t}$, using the data of Table 10.2. Note that this curve is obtained under the assumption that $\tilde{D}$ is constant, or that $\tilde{D}$ varies only slightly within the composition interval N_{A_1} to N_{A_2} (original compositions of the two halves of the diffusion couple).

The curve in Fig. 10.13 shows that, under the stated conditions, the concentration is a single-valued function of the variable $x/2\sqrt{\tilde{D}t}$. Thus, if a diffusion couple has been maintained at some fixed temperature for a given period of time (t) so that diffusion can occur, then a single determination of the composition at an arbitrary distance (x) from the weld permits the determination of the diffusivity $\tilde{D}$. Thus, suppose that a diffusion couple is formed by the welding together of two alloys of the elements A and B having the compositions N_{A_1}, 40 percent of element A (the alloy to the right of the weld), and N_{A_2}, 50 percent of element A (left of the weld). Let the couple be heated quickly to some temperature T_1 and held there 40 hr (144,000 sec), and let it be assumed that after cooling to room temperature, chemical analysis shows that at a distance 0.20 cm to the right of the weld the composition N_A is 42.5 percent A. Then in the equation

$$N_A = N_{A_1} + \frac{N_{A_2} - N_{A_1}}{2}\left[1 - \operatorname{erf}\frac{x}{2\sqrt{\tilde{D}t}}\right]$$

we have on substituting the assumed data

$$0.425 = 0.400 + \frac{(0.500 - 0.400)}{2}\left[1 - \operatorname{erf}\frac{0.20}{2\sqrt{\tilde{D}(144{,}000)}}\right]$$

or

$$\text{erf}\,\frac{0.1}{\sqrt{144{,}000\,\tilde{D}}} = 0.500$$

Table 10.2 shows, however, that when the error function equals 0.500 (i.e., 0.50006), the value of the argument $y = (x/2\sqrt{\tilde{D}t})$ is 0.477, and so

$$\tilde{D} = 3.04 \times 10^{-7}\ \text{cm}^2/\text{sec}$$

In general, the Grube method is based on the evaluation of the error function in terms of the composition of the specimen at an arbitrary point and then, with the aid of the error-function table, the argument y is found. This, in turn, permits the value of the diffusivity to be determined.

In the numerical computation just considered, a composition 42.5 percent A was assumed at a distance of 0.20 cm from the weld when the diffusion time was 40 hr. With the assumption of a constant diffusivity ($\tilde{D}$) and the same diffusion temperature, let us compute the length of time needed to obtain the same composition, 42.5 percent A, at twice the distance from the weld interface. Since the composition N_A remains the same, this problem requires that the argument of the probability integral has the same value as in the last example. That is,

$$\frac{x_1}{2\sqrt{\tilde{D}t_1}} = \frac{x_2}{2\sqrt{\tilde{D}t_2}}$$

or

$$\frac{x_1{}^2}{t_1} = \frac{x_2{}^2}{t_2}$$

where $x_1 = 0.2$ cm
$x_2 = 0.4$ cm
$t_1 = 40$ hr
t_2 = time to reach a composition of 42.5 percent at 0.40 cm from weld interface

Substituting these values into the above equation and solving for t_2 gives a value for the latter of 160 hr. The equation $x_1{}^2/t_1 = x_2{}^2/t_2$ is often used to indicate the relationship between distance and time during isothermal diffusion even when $\tilde{D}$ is not a constant. In this latter case, it still serves as a rough but very convenient approximation.

10.8 The Matano Method. The second well-known method of analyzing experimental data from metallurgical diffusion samples was devised by

Matano,[15] who first proposed it in 1933. It is based on a solution of Fick's second law, originally proposed by Boltzmann[16] in 1894. In this method, the diffusivity is assumed to be a function of concentration, which requires the solution of Fick's second law in the form:

$$\frac{\partial N_A}{\partial t} = \frac{\partial}{\partial x} \tilde{D}(N_A) \frac{\partial N_A}{\partial x}$$

This mathematical operation is much more difficult than when $\tilde{D}$ is constant and, as a consequence, the Matano-Boltzmann method of determining the diffusivity $\tilde{D}$ uses graphical integration. The first step in this procedure, after the diffusion anneal and chemical analysis of the specimen, is to plot a curve of concentration versus distance along the bar measured from a suitable point of reference, say from one end of the couple. For the purpose of simplifying the following discussion, it will be assumed that the number of atoms per unit volume $(n_A + n_B)$ is constant. The second step is to determine that cross-section of the bar through which there have been equal total fluxes of the two atomic forms (A and B). This cross-section is known as the Matano interface and lies at the position where areas M and N in Fig. 10.14 are equal. The position of the Matano interface is determined by graphical integration, but, in general, it has also been experimentally determined that, in the absence of porosity, the Matano interface lies at the position of the original weld. (This is not the position of the weld after diffusion has occurred since, as we have seen, markers placed at the weld move during diffusion.) Once the Matano interface has been located, it serves as the origin of the x coordinate. In agreement with the normal sign convention, distances to the right of the interface are considered positive, while those to the left of it are negative. With the coordinate system thus defined, the Boltzmann solution of Fick's equation is

$$\tilde{D} = -\frac{1}{2t} \frac{\partial x}{\partial N_A} \int_{N_{A_1}}^{N_A} x \, dN_A$$

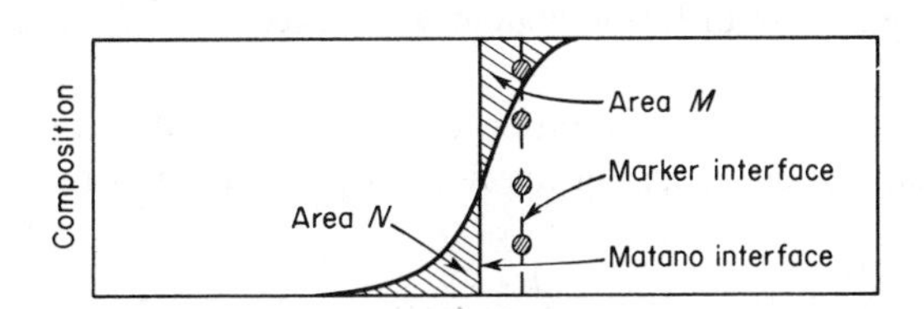

Fig. 10.14 The Matano interface lies at the position where area M equals area N.

15. Matano, C., *Japan Jour. Phys.*, 8 109 (1933).
16. Boltzmann, L., *Ann. Phys., Leipzig,* 53 959 (1894).

Table 10.3 Assumed Diffusion Data to Illustrate the Matano Method.

Composition Atomic Percent Metal A	*Distance from the Matano Interface* cm
100.00	0.508
93.75	0.314
87.50	0.193
81.25	0.103
75.00	0.051
68.75	0.018
62.50	−0.007
56.25	−0.027
50.00	−0.039
43.75	−0.052
37.50	−0.062
31.25	−0.072
25.00	−0.087
18.75	−0.107
12.50	−0.135
6.25	−0.182
0.00	−0.292

where t is the time of diffusion, N_A is the concentration in atomic units at a distance x measured from the Matano interface, and N_{A_1} is the concentration of one side of the diffusion couple at a point well removed from the interface where the composition is constant and not affected by the diffusion process.

In a manner similar to that used to explain the Grube method, we shall take arbitrarily assumed diffusion data and solve them by the Matano method. Table 10.3 represents this concentration-distance data corresponding to no actual alloy system, but representative, in a broad general way, of diffusion data obtained in actual experiments. The diffusion couple is assumed to be formed from pure A and pure B.

Figure 10.15 shows the penetration curve obtained when the data of Table 10.3 are plotted. Let us reconsider the Boltzmann solution of Fick's second law:

$$\tilde{D} = -\frac{1}{2t}\frac{\partial x}{\partial N_A}\int_{N_{A_1}}^{N_A} x\,dN_A$$

Suppose that we desire to know the diffusivity at a particular concentration, which we shall arbitrarily take as 0.375. This concentration corresponds to the

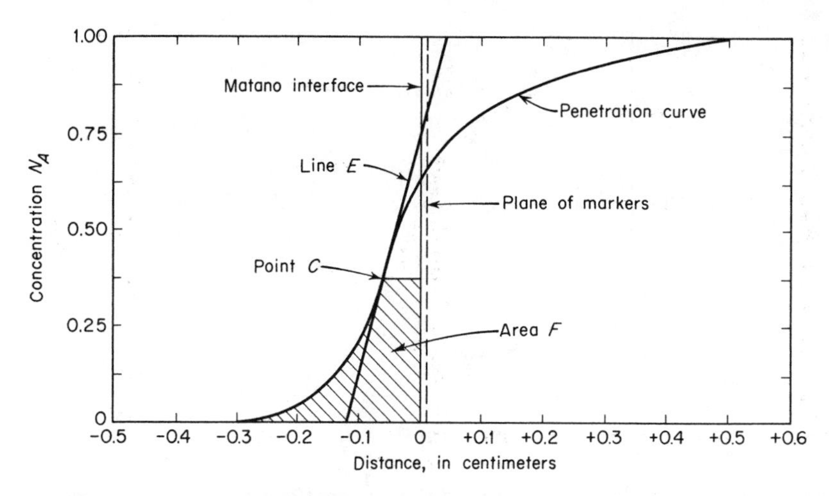

Fig. 10.15 Plot of hypothetical diffusion data (Matano method).

point marked C in Fig. 10.15. In order to compute the diffusivity at this point, we must first evaluate two quantities with the aid of Fig. 10.15. The first of these is the derivative $\partial x/\partial N_A$, the reciprocal of the slope of the penetration curve at point C. The tangent to the curve at this point is shown in the figure as line E and its slope is 6.10 cm^{-1}. The other quantity is the integral, the integration limits of which are $N_{A_1} = 0$ and $N_A = 0.375$. The indicated integral corresponds to the cross-hatched area (F) of Fig. 10.15. Evaluation of this area by a graphical method (Simpson's Rule) yields the value 0.0466 cm. The diffusivity at the composition 0.375 is, accordingly,

$$\tilde{D}\,(0.375) = \frac{1}{2t}\left(\frac{1}{\text{slope at } 0.375}\right) \times (\text{area from } N_A = 0 \text{ to } N_A = 0.375)$$

If the diffusion time is now assumed to be 50 hr (18,000 sec), the complete evaluation of the diffusivity is:

$$\tilde{D}(0.375) = \frac{1}{2(18{,}000)} \times \frac{1}{6.10} \times 0.0466 = 2.1 \times 10^{-8}\ \frac{\text{cm}^2}{\text{sec}}$$

Computations similar to the above may be made to determine the diffusivity $\tilde{D}(N_A)$ at any concentration not too close to the terminal compositions ($N_A = 0$ and $N_A = 1$). In any case, the slope at the desired concentration and the area (between the penetration curve, the vertical line representing the position of the Matano interface and within the composition limits zero to N_A) must be

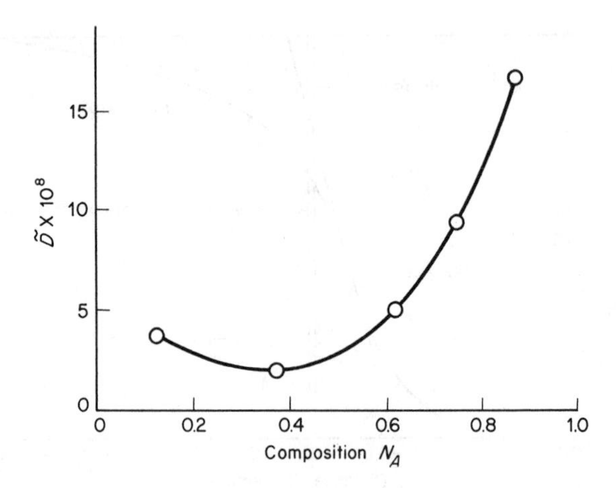

Fig. 10.16 Variation of the interdiffusion coefficient $\tilde{D}$ with composition. (Data of Table 10.2.)

determined. Since the desired area approaches zero as the composition approaches one of the terminal compositions, the accuracy of the determination falls off as N_A becomes very close to either zero or one.

Fig. 10.16 shows the variation of $\tilde{D}(N_A)$ with the concentration N_A for the data of Fig. 10.15, as determined by computation at several compositions. Notice that large values of $\tilde{D}$ are obtained as one approaches the concentration $N_A = 1$. There is also a minimum in this curve in the middle of the concentration range. A diffusivity-concentration curve of this form has been reported[17] for the diffusion of zirconium and uranium. However, the curves of Fig. 10.4 are more typical of the diffusion data reported to date.

10.9 Determination of the Intrinsic Diffusivities. The determination of the intrinsic diffusivities will now be illustrated with the use of the assumed data of Table 10.3. First we must derive an expression for the marker velocity v in terms of the marker displacement and the time of diffusion t. Experimentally, it has been determined that the markers move in such a manner that the ratio of their displacement squared to the time of diffusion is a constant. Thus

$$\frac{x^2}{t} = k$$

17. Adda, Y., and Philbert, J., *La Diffusion dans Les Metaux*. Eindhoven, Holland Bibliothèque Technique Philips, 1957.

where k is a constant. The marker velocity is, accordingly,

$$v = \frac{\partial x}{\partial t} = \frac{k}{2x}$$

but k equals x^2/t, so that

$$v = \frac{x}{2t}$$

In Fig. 10.15, an arbitrarily assumed position of the marker interface is shown at a distance $x = 0.01$ cm from the Matano interface. The diffusion time t taken for the data is 50 hr, or 18,000 sec. These numbers correspond to a marker velocity

$$v = \frac{0.01}{2(18{,}000)} = 2.78 \times 10^{-7} \text{ cm/sec}$$

At the position of the markers we also have $N_A = 0.65$ and $N_B = 0.35$ and

$$\tilde{D}_M = 5.5 \times 10^{-8}; \quad \frac{\partial N_A}{\partial x} = 2.44 \text{ cm}^{-1}$$

The value of $\tilde{D}_M$ is obtained from Fig. 10.16, while dN_A/dx is the slope of the penetration curve in Fig. 10.15 at the position of the markers, and N_A and N_B are the atom fractions of A and B respectively at the position of the markers. The above values can now be substituted into the Darken equations:

$$\tilde{D} = N_B D_A + N_A D_B$$

$$v = (D_A - D_B) \frac{\partial N_A}{\partial x}$$

yielding

$$5.5 \times 10^{-8} = 0.35\, D_A + 0.65\, D_B$$

$$2.78 \times 10^{-7} = (D_A - D_B)\, 2.44$$

The solution of this pair of simultaneous equations has as a result

$$D_A = 13 \times 10^{-8} \text{ cm}^2/\text{sec}$$

$$D_B = 1.5 \times 10^{-8} \text{ cm}^2/\text{sec}$$

These values tell us that the flux of A atoms through the marker interface from right to left is approximately nine times that of the flux of B atoms moving from left to right.

The preceding section has demonstrated that it is possible to determine experimentally the intrinsic diffusivities of a binary diffusion system (D_A and D_B). These quantities are valuable because they measure the speed with which the individual atomic forms move during diffusion. However, there has been, to date, very little success in the development of a theory capable of predicting the numerical values of the intrinsic diffusivities starting from a consideration of atomic processes. While it is generally agreed that diffusion in metallic substitutional solid solutions is the result of the movement of vacancies, the factors that control jump rates into vacancies of the two different atomic forms are complex and not completely understood. Thus, in our previous derivation of Fick's first law, several simplifying assumptions were made which do not hold for real metallic substitutional solid solutions. First, it was assumed that the solution was ideal, but, as we have seen, most metallic solutions are not ideal, and in a nonideal solution the diffusion rates are influenced by a tendency for like atoms either to group together, or to avoid each other. Second, it was assumed that the rate of jumping was independent of composition, that is, whether the jump was made by an A or a B atom. The assumption that the rate of jumping is independent of the composition is certainly not true, as may be judged by the fact that measured diffusivities vary widely with composition.

As discussed above, theoretical interpretation of substitutional diffusion is difficult because of the number of variables that must be considered. For this reason, the major effort in diffusion studies has been directed toward the investigation of diffusion in relatively simple systems that are more amenable to interpretation. In general, these consist of the study of diffusion in very dilute substitutional solid solutions or the study of diffusion using radioactive tracers.

10.10 Self-Diffusion in Pure Metals. In self-diffusion studies, one investigates the diffusion of a solute consisting of a radioactive isotope in a solvent which is a nonradioactive isotope of the same metal. In such a system, both atomic forms are identical except for the small mass difference between the isotopes. The principal effect of this mass difference is to cause the solute isotope to vibrate about its rest point in the lattice with a frequency slightly different from that of the solvent isotope, giving the two isotopes a slightly different jump rate. This difference is easily calculated because the vibration frequency is proportional to the reciprocal of the square root of the mass, and since the jump rate into vacancies is proportional to the vibration frequency, we have

$$\frac{1}{\tau^*} = \sqrt{\frac{m}{m^*}}\left(\frac{1}{\tau}\right)$$

where $1/\tau$ and $1/\tau^*$ are the jump-rate frequencies of normal and radioactive isotopes respectively (τ and τ^* are the mean times of stay of the respective atoms at lattice positions, and m and m^* are the masses of the two isotopes [m^* radioactive]).

Except for the mass difference, solute and solvent are chemically identical and the solid solution is truly ideal. Considerations of the effect of the departure of a solution from ideality may thus be neglected. Furthermore, the mass correction is usually small so that, to a good approximation, we may assume that the intrinsic diffusivity of the radioactive isotope is the same as that of the nonradioactive isotope. When the intrinsic diffusivities are equal, the interdiffusion coefficient equals the intrinsic diffusivities, as may be seen by considering the Darken equation:

$$\tilde{D} = N_B D_A + N_A D_B = (N_A + N_B) D = D$$

where $\tilde{D}$ is the interdiffusion coefficient, $D = D_A = D_B$, the intrinsic diffusivity of either the radioactive or nonradioactive isotopes, and $(N_A + N_B)$ is unity by the definition of atom fractions. Because the intrinsic coefficients do not depend on composition, it is also true that the interdiffusion coefficient does not depend on composition. Therefore experimental determinations of self-diffusivities may be made by using the simpler Grube method.

Because self-diffusion in pure metals occurs in an ideal solution and with a diffusivity that is independent of concentration, experimentally determined self-diffusion coefficients of pure metals are usually of high accuracy. Furthermore, because the diffusion process occurs in a relatively simple system, the measured diffusivities are capable of theoretical interpretation. The assumption made in our derivation of Fick's first law are those actually observed in self-diffusion experiments, and the following relationship is correct for self-diffusion in a simple cubic system:

$$D = \frac{a^2}{6\tau}$$

where D is the diffusivity, a is the lattice constant, and τ the mean time of stay of an atom in a lattice position. While only polonium is believed to crystallize in a simple cubic lattice, similar relationships can be derived for other metallic lattices.[18] As examples, we have for face-centered cubic metals

$$D_{\text{F.C.C.}} = \frac{a^2}{12\tau}$$

18. LeClaire, A. D., *Phil. Mag.*, **42** 673–688 (1951).

and body-centered cubic metals

$$D_{\mathrm{B.C.C.}} = \frac{a^2}{8\tau}$$

and, in general, for any lattice,

$$D = \frac{\alpha a^2}{\tau}$$

where α is a dimensionless constant depending on the structure.

In the chapter on vacancies, it was shown that

$$r_a = Ae^{-(Q_m + Q_f)/RT}$$

where r_a, the number of jumps made per second by an atom in a pure metal crystal, is identical to our quantity $1/\tau$; Q_f is the enthalpy change or work to form a mole of vacancies; Q_m is the enthalpy change or energy barrier that must be overcome to move a mole of atoms into vacancies; R is the universal gas constant (2 cal per mole); T the absolute temperature in degrees Kelvin.

In the above expression, the coefficient A may be replaced by $Z\nu$ and the equation rewritten in the form

$$r = Z\nu e^{-Q_m/RT} e^{-Q_f/RT}$$

where Z is the lattice coordination number and ν the lattice vibration frequency. This relationship may be interpreted as follows. The jump rate of atoms into vacancies (r_a) varies directly as (1) the number of atoms (Z) that are next to a vacancy; (2) the frequency (ν) or number of times per second that an atom moves toward a vacancy; (3) the probability ($e^{-Q_m/RT}$) that an atom will have sufficient energy to make a jump; (4) the concentration of vacancies in the lattice ($e^{-Q_f/RT}$). The above equation neglects entropy changes associated with the formation and movement of vacancies and should be more correctly written:

$$\frac{1}{\tau} = r_a = Z\nu e^{-(\Delta F_m + \Delta F_f)/RT}$$

where ΔF_m and ΔF_f are the free-energy changes associated with the movement and formation of vacancies respectively. These quantities may be expressed as

$$\Delta F_m = \Delta Q_m - T\Delta S_m$$

$$\Delta F_f = \Delta Q_f - T\Delta S_f$$

where ΔS_m is the entropy change per mole resulting from the strain of the lattice during the jumps, and ΔS_f the increase in entropy of the lattice due to the introduction of a mole of vacancies. Thus, the self-diffusion coefficient is:

$$D = \alpha a^2 Z \nu e^{-(\Delta F_m + \Delta F_f)/RT}$$

In the body-centered cubic lattice, α is $1/8$, while Z is 8, and, similarly, in the face-centered cubic lattice, α is $1/12$ and Z is 12, so that for both forms of cubic crystals

$$D = a^2 \nu e^{-(\Delta F_m + \Delta F_f)/RT}$$

or

$$D = a^2 \nu e^{(\Delta S_m + \Delta S_f)/R} \times e^{-(Q_m + Q_f)/RT}$$

This expression will be discussed further when we consider the temperature dependence of the diffusion coefficient. For the present, it suffices to point out that considerable success has been obtained in the theoretical interpretation of the various factors in this equation. A detailed discussion, such as is necessary to treat these quantities adequately, is beyond our scope and for this reason one is referred to advanced papers covering this subject.[19]

10.11 Temperature Dependence of the Diffusion Coefficient. It has already been seen that the diffusion coefficient is a function of composition. It is also a function of temperature. The nature of this temperature dependence is shown clearly in the equation for the self-diffusion coefficient stated in the previous section,

$$D = a^2 \nu e^{(\Delta S_m + \Delta S_f)/R} e^{-(Q_m + Q_f)/RT}$$

Suppose that we set

$$D_0 = a^2 \nu e^{(\Delta S_m + \Delta S_f)/R}$$

and

$$Q = \Delta H_m + \Delta H_f$$

where D_0 and Q are constants since all the quantities from which they are

19. LeClaire, A. D., *Prog. in Metal Phys.*, **1** 306 (1949); **4** 305 (1953); *Phil. Mag.*, **3** 921 (1958).

formed are effectively constant. The quantity Q is the activation energy of diffusion, and D_0 is called the *frequency factor*. The self-diffusion coefficient may now be written in the simplified form

$$D = D_0 e^{-Q/RT}$$

In this form, the equation can be applied directly to the study of experimental data.

Taking the common logarithm $\{2.3 \log_{10}(x) = ln\ (x)\}$ of both sides of the above expression gives us

$$\log D = -\frac{Q}{2.3RT} + \log D_0$$

This is an equation in the form

$$y = mx + b$$

where the dependent variable is log D, the independent variable $1/T$, the ordinate intercept log D_0, and the slope $-Q/(2.3\ R)$.

In light of the above, it is evident that if logarithms of experimental values of a self-diffusion coefficient yield a straight line when plotted against the reciprocal of the absolute temperature, then the data conform to the equation

$$D = D_0 e^{-Q/RT}$$

The slope of the experimentally determined straight line determines the activation energy Q since $m = -Q/2.3R$ or $Q = -2.3Rm$. At the same time, the intercept of the line with the ordinate designated by b yields the frequency factor D_0, since $b = \log D_0$ or $D_0 = 10^b$.

Table 10.4 Assumed Data to Show the Temperature Dependence of Self-Diffusion.

Temperature °K	*Self-Diffusion Coefficient* D	$\frac{1}{T}$	log D
700	1.9×10^{-11}	1.43×10^{-3}	−10.72
800	5.0×10^{-10}	1.25×10^{-3}	−9.30
900	6.58×10^{-9}	1.11×10^{-3}	−8.12
1000	5.00×10^{-8}	1.00×10^{-3}	−7.30
1100	2.68×10^{-7}	0.91×10^{-3}	−6.57

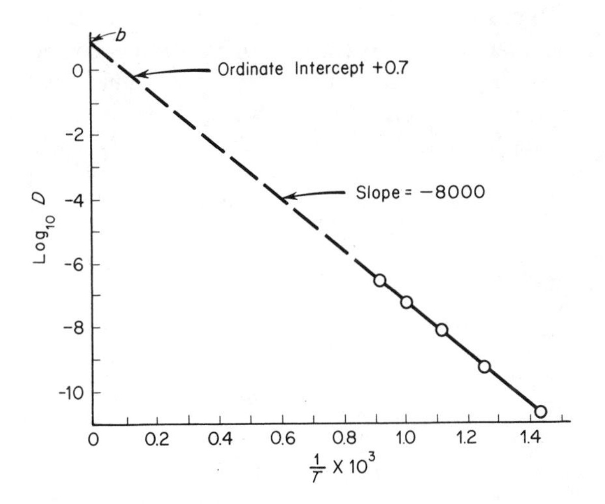

Fig. 10.17 Experimental diffusion data plotted to obtain the activation energy Q and frequency factor D_0.

The above method of determining experimental activation energies and frequency factors can be illustrated with the use of some representative data given in Table 10.4.

The data of Table 10.4 are plotted in Fig. 10.17. The slope of the resulting straight line is –8000, or

$$m = -\frac{Q}{2.3R} = -8000$$

Solving for Q and remembering that R is 2 cal per mole – °K, gives

$$Q = 2.3(2)8000 = 36{,}000 \text{ cal/mole}$$

The ordinate intercept of the experimental curve has the value 0.7. The value of D_0 is, accordingly,

$$D_0 = 10^b = 10^{0.7} = 5 \frac{\text{cm}^2}{\text{sec}}$$

The experimentally determined equation for the self-diffusion coefficient is, accordingly,

$$D = 5e^{-36{,}000/RT} \frac{\text{cm}^2}{\text{sec}}$$

The above discussion has been concerned only with the temperature dependence of self-diffusion coefficients. However, experimentally determined values of chemical interdiffusion coefficients $\tilde{D}$, and of their component intrinsic diffusivities D_A and D_B, also show the same form of temperature dependence. Thus, speaking in general, all diffusion coefficients tend to follow an empirical activation law, so that we have for self-diffusion

$$D^* = D_0{}^* e^{-Q/RT}$$

and for chemical diffusion

$$\tilde{D} = \tilde{D}_0 e^{-Q/RT}$$

and

$$D_A = D_{A_0} e^{-Q_A/RT}$$

$$D_B = D_{B_0} e^{-Q_B/RT}$$

where $\tilde{D} = N_B D_A + N_A D_B$.

Whereas the activation energy in the case of self-diffusion has a significance capable of explanation in terms of atomic processes, the meanings of the activation energies Q, Q_A, and Q_B for chemical diffusion when the solute concentration is high are vague and not clearly understood. Therefore, except when the solute concentration is very low, these activation energies should only be considered as empirical constants.

10.12 Chemical Diffusion at Low-Solute Concentration. The chemical interdiffusion coefficient $\tilde{D}$ also assumes a simple form when the solute concentration becomes very small, as can be seen by referring again to Darken's equation:

$$\tilde{D} = N_B D_A + N_A D_B$$

Suppose that component B is taken as the solute. Let us assume that at all points in the diffusion couple the concentration of B is very low. Then

$$N_A \simeq 1 \quad N_B \simeq 0$$

and

$$\tilde{D} \simeq D_B$$

Thus, at very low solute concentrations, the chemical interdiffusion coefficient approaches the intrinsic diffusivity of the solute.

If the solute concentration is much smaller than the solubility limit, solute atoms can be considered to be uniformly and widely dispersed throughout the lattice of the solvent. Considerations relative to the interaction between individual solute atoms can be neglected and each solute atom assumed to have an equivalent set of surroundings. All neighbors of each solute atom will be solvent atoms. Under these conditions, it is possible to give a theoretical interpretation to the frequency factor D_{B_0} and the activation energy for diffusion Q_B. In fact, for cubic crystals we can write[20] an expression for D_B that is entirely equivalent to that which we have previously derived for the self-diffusion coefficient

$$D_B = a^2 \nu e^{(\Delta S_{Bm} + \Delta S_{Bf})/R} e^{-(Q_{Bm} + Q_{Bf})/RT}$$

In this expression, ν is the vibration frequency of the solute atom in the solvent lattice, ΔS_{Bm} the entropy change per mole associated with the jumping of solute atoms into vacancies, Q_{Bm} the energy barrier per mole for the jumping of solute atoms, ΔS_{Bf} the entropy increase of the lattice for the formation of a mole of vacancies adjacent to solute atoms, and Q_{Bf} the work to form a mole of vacancies in positions next to solute atoms. Notice that, while the above expression has the same form as the self-diffusion equation, most of the quantities on the right-hand side of the equation have somewhat different meanings.

The frequency factor D_{B_0} and activation energy Q_B for chemical diffusion at low solute concentrations are

$$D_{B_0} = a^2 \nu e^{(\Delta S_{Bm} + \Delta S_{Bf})/R}$$

$$Q_B = Q_{Bm} + Q_{Bf}$$

Table 10.5 contains experimental data for the diffusion of several different solutes (at low concentrations) in nickel. The chemical diffusion data shown in this table were obtained with the use of diffusion couples consisting of a plate of pure nickel welded to another composed of an alloy of nickel containing the indicated element in an amount of the order of 1 atomic percent. For the systems studied, these couples conform to the condition that the solute concentration be low.

Column one of Table 10.5 indicates the diffusing solute atom. The symbol Ni in this column indicates that the values on the lowest line correspond to self-diffusion in pure nickel. Columns two and three give values of the frequency factor D_{B_0} and activation energy Q_B respectively, while column four lists computed values of the diffusivity D_B for the temperature 1470°K.

20. Swalin, R. A., *Acta Met.*, 5 443 (1957).

Table 10.5 Solute Diffusion in Dilute Nickel-Base Substitutional Solutions.*

Solute	*Frequency Factor,* D_{B_0}	*Activation Energy,* Q_B	*Diffusivity* $D_B = D_{B_0} e^{-Q_B/RT}$ *at* 1470°K
Mn	7.50	67,000	8.78×10^{-10}
Al	1.87	64,000	5.63×10^{-10}
Ti	0.86	61,000	6.82×10^{-10}
W	11.10	76,800	0.25×10^{-10}
Ni	1.27	66,800	1.68×10^{-10}

* Values for Mn, Al, Ti, and W from the data of Swalin and Martin, *Trans. AIME,* **206,** 567 (1956). Self-diffusion values for nickel from R. E. Hoffman (see Ref. 4 in above).

Table 10.5 shows that the diffusivities of the elements Mn, Al, Ti, and W in dilute solid solution in nickel differ, but not in large measure, from the nickel self-diffusion coefficient.

10.13 The Study of Chemical Diffusion Using Radioactive Tracers. Consider the diffusion couple shown in Fig. 10.18, consisting of two halves with the same chemical composition, but with a fraction of the *B* atoms in the right-hand member radioactive. If such a couple is heated and the atoms allowed to diffuse, there should be no detectable change in chemical composition throughout the length of the bar. However, there will be a redistribution of the radioactive atoms. This change in concentration of tracer atoms along the axis of the bar can be determined in the following manner. After the diffusion anneal, the specimen is placed in a lathe and thin layers of equal thickness removed parallel to the weld interface. Measurement of the radioactive radiation intensity from each set of lathe turnings indicates the concentration of the radioactive *B* atoms in the corresponding layer. A plot of these intensity values as a function of position along the bar is equivalent to a normal penetration curve. In a specimen of this type, one measures the diffusion of *B* atoms in a homogeneous alloy of atoms *A* and *B*. Since the specimen is chemically homogeneous, the composition is everywhere the same and there can be no variation of the diffusivity with composition, which means that the

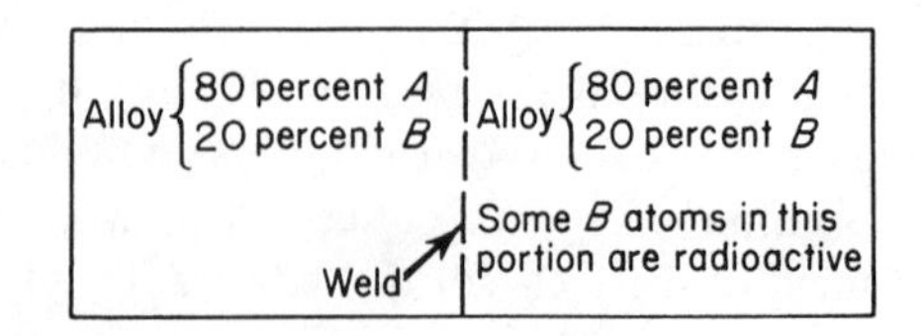

Fig. 10.18 Schematic representation of a diffusion couple using the tracer technique.

penetration curve can be analyzed for the diffusivity by the Grube method. The diffusion coefficient found in this manner is like a self-diffusion coefficient, but indicates the rate at which B atoms diffuse in an alloy of A and B atoms rather than in a matrix of pure B atoms.

In some binary systems it is possible to find radioactive isotopes of both of the component elements (A and B) which are suitable for use as tracers. When this is possible, measurements of the tracer diffusion coefficients for both elements can be made over the complete range of solubility. It is customary to give these quantities the symbols $D_A{}^*$ and $D_B{}^*$. (The tracer diffusivities are denoted by asterisks to differentiate them from the intrinsic diffusion coefficients D_A and D_B.) The coefficients $D_A{}^*$ and $D_B{}^*$, like the intrinsic diffusivities D_A and D_B, are functions of composition, which means that the rate of diffusion of either the A or the B atoms in a homogeneous crystal containing both atomic forms is not the same as it is in a pure metal of either component. We now have two different types of diffusion coefficients describing the diffusion process of the two atomic forms – tracer and intrinsic – in a substitutional solid solution. It is not unreasonable to wonder if they are related. A relationship does exist and was first derived by Darken[21] and has been fully verified by experiment.[22] According to Darken, the relationship between the intrinsic diffusivities and the self-diffusion coefficients are:

$$D_A = D_A{}^* \left(1 + N_A \frac{\partial \ln \gamma_A}{\partial N_A}\right)$$

$$D_B = D_B{}^* \left(1 + N_B \frac{\partial \ln \gamma_B}{\partial N_B}\right)$$

where D_A and D_B are the intrinsic diffusion coefficients, $D_A{}^*$ and $D_B{}^*$ are the tracer-diffusion coefficients, γ_A and γ_B are the activity coefficients of the two components, and N_A and N_B are the atom fractions of the two components. With the aid of the Gibbs-Duhem equation (a well-known physical chemistry relationship),

$$N_A \partial \ln \gamma_A = -N_B \partial \ln \gamma_B$$

and from the fact that $\partial N_A = -\partial N_B$ we have

$$1 + \frac{N_A \partial \ln \gamma_A}{\partial N_A} = 1 + \frac{(-N_B \partial \ln \gamma_B)}{-\partial N_B} = 1 + \frac{N_B \partial \ln \gamma_B}{\partial N_B}$$

21. Darken, L. S., *Trans. AIME,* **175** 184 (1948).
22. LeClaire, A. D., *La Diffusion dans Les Metaux.* Eindhoven, Holland, Bibliothèque Technique Philips, 1957.

The two factors which give the intrinsic diffusivities, when multiplied by the respective tracer-diffusion coefficients, are actually equal and it is customary to call this quantity the *thermodynamic factor*.

Let us consider the significance of the thermodynamic factor. In an ideal solution, the activity (a_A) equals the concentration of the solution N_A, and the activity coefficient γ_A (which is the ratio of these two quantities), is one. Since the logarithm of unity is zero, the thermodynamic factor becomes

$$1 + \frac{N_A \partial \ln \gamma_A}{\partial N_A} = (1 + 0) = 1$$

and the intrinsic-diffusion coefficient equals the tracer-diffusion coefficient. The tracer-diffusion coefficient can thus be considered a measure of the rate at which atoms would diffuse in an ideal solution, and the thermodynamic factor can be considered a correction that takes into account the departure of the crystal from ideality.

It is possible to express the chemical diffusivity in terms of the tracer-diffusion coefficients, for by Darken's equation

$$\tilde{D} = N_B D_A + N_A D_B$$

When one expresses the intrinsic diffusivities D_A and D_B in terms of the

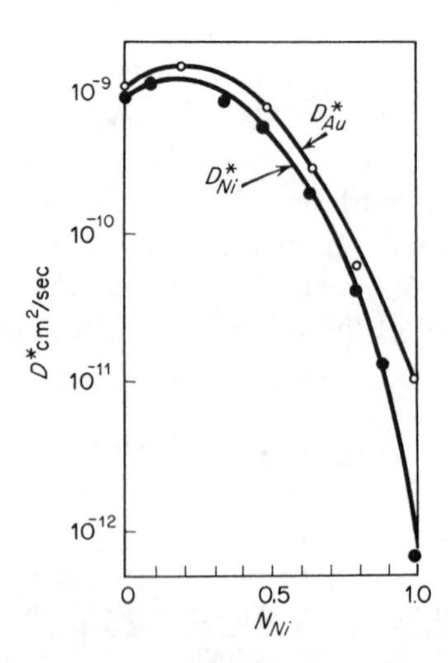

Fig. 10.19 Self-diffusion coefficients of Au and Ni in gold-nickel alloys at 900°C. (From Reynolds, J. E., Averbach, B. L., and Cohen, Morris, *Acta Met.*, **5** 29 [1957].)

self-diffusion coefficients, we have

$$\tilde{D} = N_B D_A{}^* \left(1 + N_A \frac{\partial \ln \gamma_A}{\partial N_A}\right) + N_A D_B{}^* \left(1 + N_B \frac{\partial \ln \gamma_B}{\partial N_A}\right)$$

or

$$\tilde{D} = (N_B D_A{}^* + N_A D_B{}^*) \left(1 + N_A \frac{\partial \ln \gamma_A}{\partial N_A}\right)$$

since the two forms of the thermodynamic factor are equal.

Figures 10.19 through 10.21 contain experimental data for gold-nickel diffusion at 900°C. At this temperature, gold and nickel dissolve completely in each other and form a completely soluble series of alloys. The significance of this experimental information is that it gives experimental confirmation of the Darken relationships.

In Fig. 10.19, the tracer-diffusion coefficients are plotted as a function of composition, and the very large variation of $D_A{}^*$ and $D_B{}^*$ is apparent. Notice that the tracer-diffusion rate of nickel atoms in pure gold is about 1000 times larger than nickel atoms in pure nickel. Figure 10.20 shows the thermodynamic factor

$$\left(1 + N_{\text{Ni}} \frac{\partial \ln \gamma_{\text{Ni}}}{\partial N_{\text{Ni}}}\right)$$

as a function of composition for gold-nickel alloys at 900°C. The variation of the interdiffusion coefficient as a function of composition is given in Fig. 10.21. Two curves are shown, that marked $\tilde{D}$ (calculated) is derived from the self-diffusion coefficients (Fig. 10.19) and the thermodynamic factor (Fig. 10.20), the other marked $\tilde{D}$ (observed) is obtained from direct chemical-diffusion measurements using Matano analysis. Quite good agreement is found

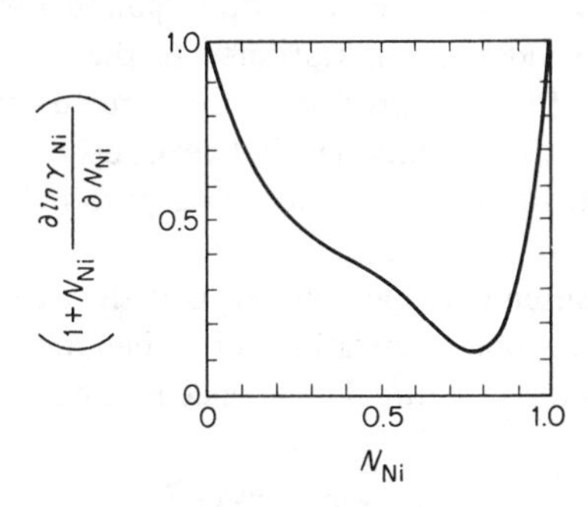

Fig. 10.20 Thermodynamic factor for interdiffusion at 900°C. (From Reynolds, J. E., Averbach, B. L., and Cohen, Morris, *Acta Met.,* **5** 29 [1957].)

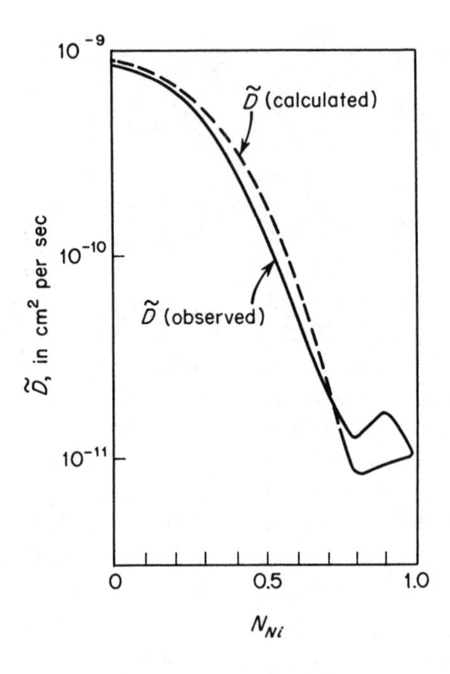

Fig. 10.21 Calculated and observed interdiffusion coefficients in gold-nickel alloys at 900°C. (From Reynolds, J. E., Averbach, B. L., and Cohen, Morris, *Acta Met.*, **5** 29 [1957].)

between the calculated and observed curves. The slight divergence between the two curves at high nickel concentrations can be explained on the basis of experimental errors.

10.14 Diffusion Along Grain Boundaries and Free Surfaces. Atom movements in solids are not restricted to the interiors of crystals. It is a well-known fact that diffusion processes also occur on the surfaces of metallic specimens and along the boundaries between crystals. In order to reduce complexity, these two forms of diffusion have purposely been neglected in our previous discussions.

Experimental measurements have shown that the surface and grain-boundary forms of diffusion also obey activation, or Arhennius-type laws, so that it is possible to write their temperature dependence in the form

$$D_s = D_{s_0} e^{-Q_s/RT}$$

$$D_b = D_{b_0} e^{-Q_b/RT}$$

where D_s and D_b are the surface and grain-boundary diffusivities, D_{s_0} and D_{b_0} are constants (frequency factors), and Q_s and Q_b are the experimental activation energies for surface and grain-boundary diffusion.

Sufficient evidence has been accumulated[23] to conclude that diffusion is more rapid along grain boundaries than in the interiors of crystals, and that free-surface diffusion rates are larger than either of the other two. These observations are understandable because of the progressively more open structure found at grain boundaries and on exterior surfaces. It is quite reasonable that atom movements should occur most easily on free metallic surfaces, with more difficulty in boundary regions, and least easily in the interior of crystals.

Because of the very rapid movements of atoms on free surfaces, surface diffusion plays an important role in a large number of metallurgical phenomena. However, grain-boundary diffusion is of more immediate concern because, in the average metallic specimen, the grain-boundary area is many times larger than the surface area. Furthermore, grain boundaries form a network that passes through the entire specimen. It is this latter property that often causes large errors to appear in the measurement of lattice, or volume diffusion, coefficients. When the diffusivity of a metal is measured with polycrystalline samples, the results are usually representative of the combined effect of volume and grain-boundary diffusion. What is obtained, therefore, is an apparent diffusivity, D_{ap}, which may not correspond to either the volume- or the grain-boundary-diffusion coefficient. However, under certain conditions, the grain-boundary component may be small, so that the apparent diffusivity equals the volume diffusivity. On the other hand, if the conditions are right, the grain-boundary component may be so large that the apparent diffusivity diverges considerably from the lattice diffusivity. Let us investigate these conditions.

Diffusion in a polycrystalline specimen cannot be described as a simple summation of diffusion through the crystals and along the boundaries. Diffusion in the boundaries tends to progress more rapidly than diffusion through the crystals, but this effect is counteracted because as the concentration of solute atoms builds up in the boundaries, a steady loss of atoms occurs from the boundaries into the metal on either side of the boundary. The nature of this process can be visualized with the aid of Fig. 10.22, which represents a diffusion couple composed of two pure metals A and B. Both halves of the couple are assumed to be polycrystalline, but, for the sake of convenience, grain boundaries are indicated only on the right side of the couple. A number of short arrows are in the figure to represent the nature of the movement of A atoms into the B matrix. That group of parallel arrows perpendicular to the weld interface represents the volume component of diffusion. Other arrows parallel to the grain

23. Turnbull, D., ASM Seminar (1951), *Atom Movements*, p. 129.

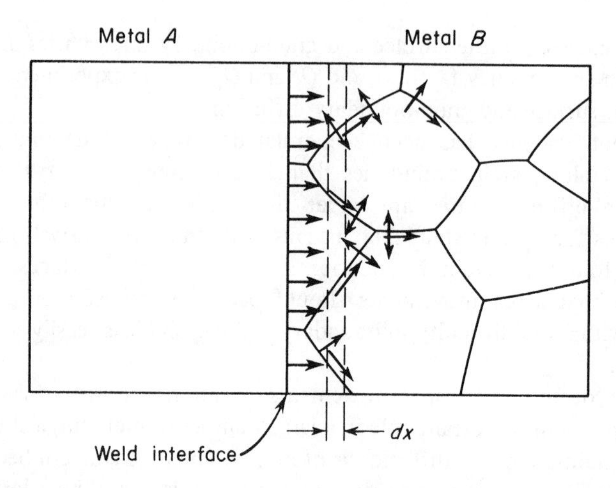

Fig. 10.22 The combined effect of grain boundary and volume diffusions.

boundaries indicate the movement of atoms along boundaries, and a third set of arrows perpendicular to the boundaries represent the diffusion from the boundaries into the crystals.

In the usual diffusion experiment, one removes thin layers of metal parallel to the weld interface (like that indicated in Fig. 10.22 by two vertical dashed lines separated by the distance dx). These layers are chemically analyzed to obtain the penetration curve. The concentration of A atoms in the layer dx depends on how many A atoms reach this layer by direct-volume diffusion, and how many reach the layer by the short-circuiting grain-boundary paths. The problem is quite complex, but for a given ratio of the grain-boundary and lattice diffusivities (D_b/D_l), the relative number of A atoms that reach the layer dx by traveling along grain boundaries and by direct lattice diffusion is a function of the grain size. The smaller the grain size, the greater the total grain-boundary area available for boundary diffusion and, therefore, the greater the importance of boundaries in the diffusion process.

The importance of grain-boundary-diffusion phenomena in diffusion measurements is also a function of temperature. In Fig. 10.23, two sets of self-diffusion data are plotted for silver specimens:[24] that at the upper right for grain-boundary measurements and that at the lower left for lattice measurements (single-crystal specimens). Both groups of points plot as straight lines on the log $D - 1/T_{\text{Abs}}$ coordinate system. For grain-boundary self-diffusion the equation of the line is:

$$D_b = 0.025e^{-20{,}200/RT}$$

24. *Ibid.*

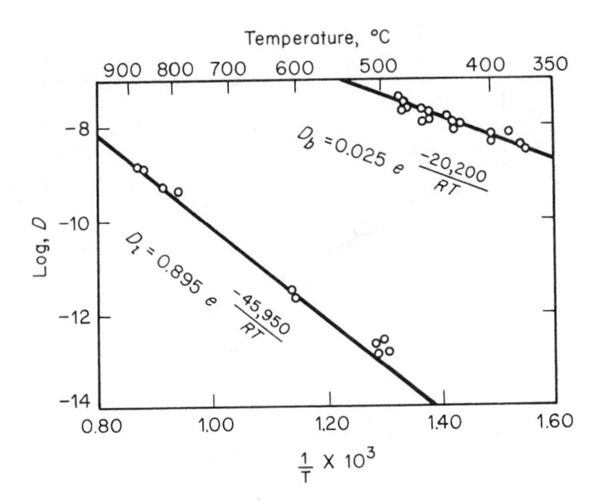

Fig. 10.23 Lattice and grain-boundary diffusion in silver. (From data of Hoffman, R. E., and Turnbull, D. J., *Jour. Appl. Phys.*, **22** 634 [1951].)

For volume or lattice self-diffusion,

$$D_l = 0.895e^{-45,950/RT}$$

Notice that in this example the activation energy for grain-boundary diffusion is only about half that for volume diffusion. This fact is significant for two reasons: first, it emphasizes the fact that diffusion is easier along grain boundaries and, second, it shows that grain-boundary and volume diffusions have a different temperature dependence. Volume diffusion, when compared with boundary diffusion, is more sensitive to temperature change. Thus, as the temperature is raised, the rate of diffusion through the lattice increases more rapidly than the rate of diffusion along the boundaries. Conversely, as the temperature is lowered, the rate of diffusion along the boundaries decreases less rapidly. The net effect is that at very high temperatures diffusion through the lattice tends to overpower the grain-boundary component, but at low temperatures diffusion at the boundaries becomes more and more important in determining the total, or apparent, diffusivity.

The above facts are illustrated in Fig. 10.24, where the curves of Fig. 10.23 are redrawn as dashed lines, and another curve (solid line) is also shown which corresponds to self-diffusion measurements made on fine-grain polycrystalline silver specimens (grain size 35 microns before diffusion anneal). This latter curve consists of two segments, one part with the equation

$$D_{ap} = 2.3 \times 10^{-5} e^{-26,400/RT}$$

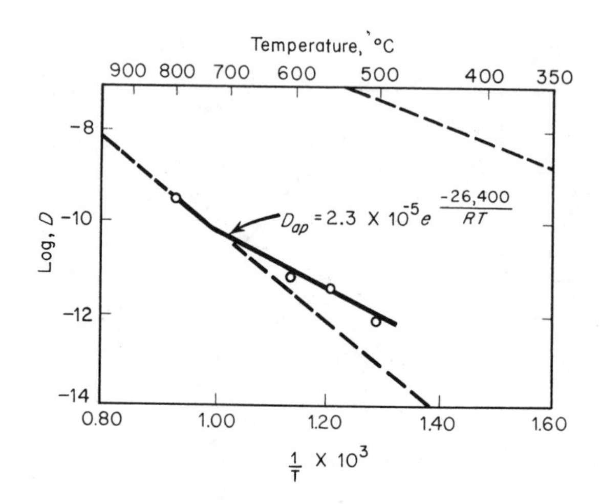

Fig. 10.24 Diffusion in polycrystalline silver. (From data of Hoffman, R. E., and Turnbull, D. J., *Jour. Appl. Phys.,* **22** 634 [1951].)

for temperatures below about 700°C, and the other coinciding with the single-crystal, or volume-diffusion, data for temperatures above about 700°C. It can be concluded that at temperatures above this value the volume-diffusion component is overpowering in silver specimens with a 35-micron grain size. Below this temperature, the grain-boundary component is a factor in determining the measured diffusivity.

The above results have general implications: (*a*) diffusivities determined with polycrystalline specimens are more liable to be representative of lattice diffusion if they are measured at high temperatures and (*b*) the reliability of the data can be increased by controlling the grain size of the specimens. The larger the grain size, the smaller the grain-boundary contribution to the diffusivity. Thus, for accurate measurements of lattice diffusivities using polycrystalline specimens, high temperatures and large-grained specimens should be used.

10.15 Fick's First Law in Terms of a Mobility and an Effective Force. Fick's first law is often expressed by using a different set of variables. This form of the equation will now be developed by considering the diffusion of one component in an ideal binary solution. As shown earlier, Fick's first law in this case may be written

$$J_A = -AD_A{}^* \frac{\partial n_A}{\partial x}$$

where J_A is the number of A atoms passing an interface of area A per second,

n_A is the number of A atoms per unit volume, and $D_A{}^*$ is the diffusivity of A in an ideal solution. Note that this diffusion coefficient is equivalent to that measured with radioactive tracers in a solution of constant chemical composition (see Section 10.13). This expression may be reexpressed in the form

$$J'_A = -D_A{}^* (n_A + n_B) \frac{\partial N_A}{\partial x}$$

where J'_A equals J_A/A and represents the flux of A atoms per cm^2, $(n_A + n_B) \times N_A$ equals n_A by the definition of the mole fraction, and $(n_A + n_B)$ is assumed to be a constant.

In Section 10.1, $\bar{F}_A$, the partial molal free energy of the A component in an ideal solution, was defined by the expression

$$\bar{F}_A = F^\circ{}_A + RT \ln N_A$$

where $F^\circ{}_A$ is the free energy of a mole of pure A at the temperature T and N_A is the mole fraction of A. Taking the derivative of $\bar{F}_A$ with respect to x, the distance along a diffusion couple, and noting that by definition $\partial F^\circ{}_A/\partial x$ is zero, we have

$$\frac{\partial \bar{F}_A}{\partial x} = \frac{RT}{N_A} \frac{\partial N_A}{\partial x}$$

Solving this expression for $\partial N_A/\partial x$ and substituting the result into Fick's equation and noting that $(n_A + n_B)N_A$ equals n_A (the number of atoms of A per unit volume), gives us

$$J'_A = -\frac{D_A{}^*}{RT} n_A \frac{\partial \bar{F}_A}{\partial x}$$

Now let us make the substitution $B = D_A{}^*/RT$ and we obtain the result

$$J'_A = -n_A B \frac{\partial \bar{F}_A}{\partial x}$$

where J'_A is the flux per cm^2, B is a parameter called the mobility, n_A is the concentration of A atoms in number of atoms per cm^3, and $\partial \bar{F}_A/\partial x$ is the partial derivative of the partial molal free energy of the A component in the solution with respect to the distance x. Both $\partial \bar{F}_A/\partial x$ and B have a physical significance worth noting. The partial molal free energy $\bar{F}_A$ has the dimensions of an energy, and its derivative with respect to the distance x can be considered as an effective "force" causing diffusion to occur in this direction. B has the dimensions of velocity divided by force. In terms of its dimensions, the above equation may be written

$$\frac{\text{Number of atoms}}{\text{cm}^2 \times \text{sec}} = \frac{\text{Number of atoms}}{\text{cm}^3} \times \frac{\text{cm}}{\text{sec} \times \text{force}} \times \text{force}$$

While the above relationship was derived for the special case of diffusion in an ideal solution, the result is of general use and may be applied to the case of diffusion in a nonideal solution where the partial molal free energy is given (see Section 10.2) by

$$\bar{F}_A = F^{\circ}{}_A + RT \ln a_A$$

where a_A is the activity of the A component in the solution. Let us now consider the derivative of this partial molal free energy with respect to x, the distance along a diffusion couple. Since by the definition of the activity coefficient γ_A (see Section 10.2), $a_A = \gamma_A N_A$, we have

$$\frac{\partial \bar{F}_A}{\partial x} = RT \frac{\partial}{\partial x} (\ln \gamma_A N_A) = RT \frac{\partial}{\partial N_A} (\ln \gamma_A N_A) \frac{\partial N_A}{\partial x}$$

or

$$\frac{\partial \bar{F}_A}{\partial x} = RT \left(\frac{1}{N_A} + \frac{\partial \ln \gamma_A}{\partial N_A} \right) \frac{\partial N_A}{\partial x} = \frac{RT}{N_A} \left(1 + N_A \frac{\partial \ln \gamma_A}{\partial N_A} \right) \frac{\partial N_A}{\partial x}$$

Substituting this relationship back into the equation

$$J'_A = -\frac{D_A{}^*}{RT} n_A \frac{\partial \bar{F}_A}{\partial x}$$

and simplifying, using the relationship $N_A = n_A/(n_A + n_B)$, we obtain

$$J'_A = -\left(1 + N_A \frac{\partial \ln \gamma_A}{\partial N_A} \right) D_A{}^* \frac{\partial n_A}{\partial x}$$

where $(1 + N_A \, \partial \ln \gamma_A / \partial N_A) D_A{}^*$ is the intrinsic diffusion coefficient D_A as defined in section 10.13. We have therefore derived the Darken relationship $D_A = (1 + N_A \, \partial \ln \gamma_A / \partial N_A) D_A{}^*$ given in this section.

10.16 Electrotransport and Thermomigration. Up to this point we have considered only the case of diffusion in solid metals where the driving force for the redistribution of the atomic species has come from compositional gradients. Temperature gradients and electric potential gradients can also produce diffusion phenomena. The subject area associated with the study of atomic diffusion under an electric field is known as *electrotransport*. That dealing with atomic rearrangements due to a temperature gradient is called *thermomigration*.

The study of both electrotransport and thermal diffusion is simpler when only a single component of a binary solid solution can be assumed to be mobile or to diffuse. The iron-carbon system at high temperatures, where the iron is in the face-centered cubic or γ phase and the carbon is in solid solution, corresponds to this condition. Here the iron atoms can be considered to be much less mobile than the carbon atoms, and it is reasonable to assume that the iron atoms remain fixed in space, while the carbon atoms diffuse through the iron lattice. Figure 10.25 shows the effect of an electric field on the distribution of

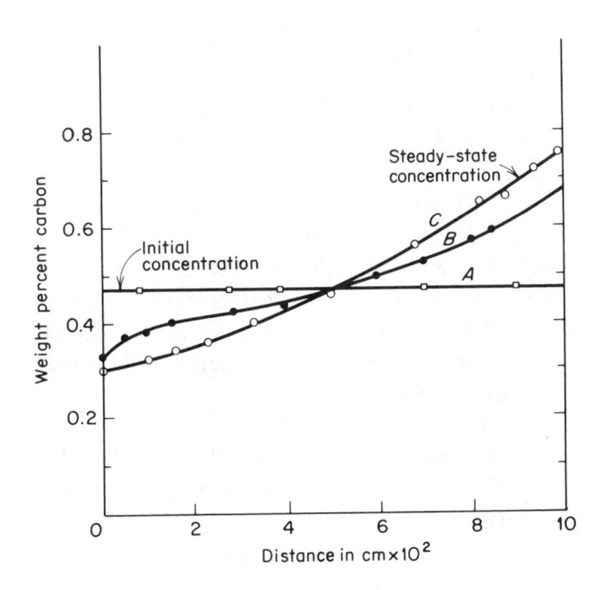

Fig. 10.25 The carbon distribution after electrotransport in a steel specimen containing 0.47 wt. percent carbon. Curve *A* shows the original carbon concentration distance curve. Curve *B* shows the corresponding curve after a one-hour anneal under an electric field that produced a current density of 2170 amp per sq. cm, and curve *C* shows the carbon distribution after 8 hours, which represents the steady-state distribution. Temperature 927°C. (After Okabe, T., and Guy, A. G., *Met Trans.* 1 2705 [1970].)

the carbon atoms in a 0.48 wt. percent thin carbon steel plate. Note that before the electric field was applied across the plate, the concentration of the carbon was a constant from one side of the specimen to the other. This is demonstrated by curve *A*. After an anneal of one hour under the electric field, the concentration profile changed to that shown by curve *B*. Finally, curve *C* represents the steady-state distribution attained in about 8 hours. Increasing the annealing time above 8 hours had no measurable effect on this latter distribution.

Let us now consider the nature of the diffusion flux due to an electric field intensity applied to a plate like that corresponding to Fig. 10.25. The effective force that this field applies to a carbon ion may be represented by the product of Avogadro's number *N*, the electric field intensity *E*, and the effective charge on the carbon ion Z^*e. Z^* is the effective valence (not necessarily the same as the chemical valence) of the ion, and *e* is the charge on an electron. The product Z^*eE represents the effective force on one ion, and to convert this to a molar quantity it is necessary to multiply by Avogadro's number *N*. We may therefore write

$$J'_{A_E} = B\, n_A N Z^* eE$$

where *B* is the mobility under the electric field, n_A is the number of carbon atoms per cm^3, Z^*e is the effective charge on a carbon atom (ion), and *E* is the electric field intensity.

It is normally assumed that the effect of an electric field is to bias the jump direction of an atom in the direction of the electric field. However, the small size of this bias should change neither the jump mechanism nor the mean jump frequency at a given temperature.[25] In effect, this means that the electrotransport mobility *B* should be proportional to the diffusivity *D* and therefore to the mobility under a concentration gradient. In practice, *B* is normally assumed to be the same for both diffusion mechanisms.

As may be seen in Fig. 10.25, when the steady state is attained, a composition gradient is also developed across the specimen. This gradient will produce a flux of carbon atoms in the direction opposite to that due to the potential gradient equal to

$$J'_{A_C} = -B\, n_A \frac{\partial \overline{F}}{\partial x}$$

If the steady state exists, these two fluxes are equal and opposite, so that

$$J'_A = J'_{A_C} + J'_{A_E} = -B\, n_A \frac{\partial \overline{F}}{\partial x} + B\, n_A N Z^* eE = 0$$

25. Shewmon, P. G., *Diffusion in Solids*, p. 189. McGraw-Hill Book Company, New York, 1963.

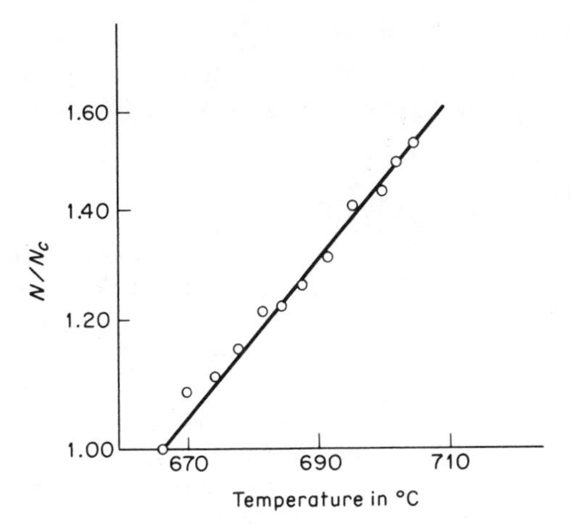

Fig. 10.26 Steady state variation of the carbon concentration with temperature in alpha iron. N/N_c represents the ratio of the concentration at a higher temperature to that existing at the colder end of the specimen. (After Shewmon, P., *Acta. Met.,* **8** 605 [1960].)

If a steady state condition does not exist, the net flux J'_A is not zero, and we have

$$J'_A = -B\, n_A \frac{\partial \bar{F}}{\partial x} + B\, n_A N Z^* eE$$

A similar redistribution of the carbon atoms can also be observed in a steel specimen subjected to a temperature gradient. A typical example is shown in Fig. 10.26. In this case the steel was heated to just below its transformation temperature so that the crystal structure was body-centered cubic and not face-centered cubic. The condition that the carbon atoms diffuse much more rapidly than the iron atoms, however, was still satisfied. The net flux can be represented by an equation of the form

$$J = -B\, n_A \frac{\partial \bar{F}}{\partial x} + B\, n_A \frac{Q^*}{T} \frac{\partial T}{\partial x}$$

where $(Q^*/T)(\partial T/\partial x)$ is the effective force causing the thermal diffusion, B is the mobility, and the parameter Q^* is called the heat of transport. Note that as in the case of the electrotransport relationship, the mobility B is taken to be the same in both the concentration-dependent term and the temperature-dependent term.

The present effect, where one component of a solid solution specimen subjected to a temperature gradient develops a concentration gradient, is also known as thermal diffusion. In general, this type of phenomenon is known as the Ludwig-Soret effect and is observed in gaseous and liquid solutions as well as in solid solutions.

Problems

1. Assume that two metals, A and B, form an ideal solid solution and that at 1000°C the free energy of a mole of pure A is 10,000 cal, while that of a mole of pure B is 14,000 cal.
(*a*) What would be the free energy of one mole of a solution in which the mole fraction of B was 0.25?
(*b*) What is the free energy decrease associated with the formation of this solid solution from its pure components?
2. (*a*) Consider the two alloys whose activity-composition curves are shown in Fig. 10.1. Assume that it is possible to form an alloy of each type with a composition $N_A = N_B = 0.5$. Let $F^\circ_A = 10{,}000$ cal per mole and $F^\circ_B = 14{,}000$ cal per mole. With the aid of Fig. 10.1, estimate the free energy of one mole of each solution at 500°C.
(*b*) What is the physical significance of the fact that one of these free energies is smaller than the other?
3. (*a*) Make a sketch showing the nature of the variation of γ_A (the activity coefficient of the A component) as a function of the concentration of A in the alloy system represented by Fig. 10.1A. This can be done by considering the relative magnitudes of N_A and a_A at various compositions.
(*b*) Discuss the significance of this curve.
4. The atomic volume of silver is 10.28 cm^3 per gm. atom.
(*a*) What is the number of silver atoms in a cubic cm?
(*b*) What is the volume per atom?
(*c*) The handbook value for the effective diameter of the silver atom is 2.89 Å. How does the volume of the atom, based on this diameter, compare with that found in part (*b*) of this problem?
5. Metals A and B, whose atomic volumes are both 10.3 cm^3 per gm atom, form ideal solutions.
(*a*) What is the concentration gradient $\partial n_A / \partial x$ in a diffusion couple formed from these metals in a region where N_A changes from 0.30 to 0.65 in a distance of 0.1 mm?
(*b*) If the diffusion coefficient is 10^{-10} cm^2/sec and the cross-section area of the specimen is 0.5 cm^2, what is the flux of A atoms through a cross-section within the region whose concentration gradient equals that found in part (*a*) of this problem?
(*c*) What is the corresponding flux of B atoms?
6. (*a*) In the preceding problem it was assumed that $D = 10^{-10}$ cm^2/sec. What would be the mean time of stay of an atom at a simple cubic crystal lattice site when the diffusion coefficient is 10^{-10}? Assume $a = 3$ Å.
(*b*) How many times per second would each atom jump?

7. With the aid of the data and equations given in Sections 6.7 and 6.8, compute the number of times per second that the average copper atom should make a jump into a vacancy at 1000°C.
8. It is determined by experiment that the markers placed at the interface of a diffusion couple formed by welding a thin plate of metal A to a similar plate of metal B move with a velocity of 3×10^{-10} cm/sec toward the A component when the concentration is $N_A = 0.35$ and the concentration gradient is 2×10^2 atomic percent per cm. The chemical diffusion coefficient $\tilde{D}$ under these conditions is found to be 1.03×10^{-10} cm^2/sec. Determine the values of the intrinsic diffusivities of the two components.
9. Assume that in the preceding problem the atomic volume of each component is 12 cm^3 per gm atom.
   (*a*) What will be the flux through the cross-section containing the markers in atoms per sec per cm^2 of each component?
   (*b*) What will be the net vacancy flux through this cross-section?
   (*c*) If all of the excess vacancies that pass through the markers in one hour were to combine and form a single spherical pore, what would be the pore radius in mm?
10. (*a*) Compute the flux of A atoms per cm^2 across a reference plane fixed in space; that is, with respect to one end of the diffusion couple of Problem 8. Assume this plane coincided with the plane of the markers at the time the data given in Problem 8 was obtained.
    (*b*) Compute the corresponding flux of B atoms.
11. A thin plate of a binary alloy of composition $N_A = 0.250$ and $N_B = 0.750$ is welded to a similar plate of composition $N_A = 0.300$ and $N_B = 0.700$ so as to form a diffusion couple. After a diffusion anneal of 100 hours at 1000°C, the composition at a point 0.100 mm from the original weld interface on the side whose original composition was $N_A = 0.250$ was observed to be $N_A = 0.260$. Compute the diffusivity by the Grube method. (*Note*: assume $x = 0.100$).
12. With regard to the data in Table 10.3 and Fig. 10.15, determine the interdiffusion coefficients for $N_A = 0.875$ using the following procedure:
    (*a*) By laying a tangent on the penetration curve at $N_A = 0.875$, determine $\partial x/\partial N_B$.
    (*b*) Graphically integrate the data in Table 10.3 from $N_B = 0$ to $N_B = 0.125$ (that is, from $N_A = 1.000$ to $N_A = 0.875$) using Simpson's rule.
    (*c*) Evaluate the Matano expression for $\tilde{D}$ assuming $t = 100$ hours. Check your answer with Fig. 10.16.
13. In Section 10.10 it is stated that in general for any lattice, $D = \alpha a^2/\tau$. For f.c.c. metals, α is given as 1/12. Prove this relationship using a method similar to that outlined in Section 10.3 where the value $\alpha = 1/6$ was obtained for simple cubic metals. The following outline is provided as an aid to making this derivation:
    (*a*) Assume that planes x and y in Fig. 10.3 correspond to two adjacent {111} planes. Determine the separation between these planes in terms of the f.c.c. lattice parameter a (the length of one edge of the unit cell).

(*b*) With the aid of a drawing such as Fig. 1.7, determine the maximum number of directions along which a given atom may jump. How many of these directions would carry the atom from plane x to plane y? What, therefore, is the chance that a given jump will carry the atom from x to y or from y to x?

(*c*) Using the results from parts (*a*) and (*b*), compute the value of α.

14. Show that α for substitutional diffusion in b.c.c. metals is 1/8.

15. The activation energy for the self-diffusion of gold is 1.81 eV, and the lattice parameter is 4.07 Å. Assume $\nu = 10^{13}$ vib/sec and that $\Delta S_m + \Delta S_f = 0$.
(*a*) Compute a value for the self-diffusion coefficient for gold at 900°C.
(*b*) How does this compare with the value given in Fig. 10.19?

16. With the aid of the experimental data given in Table 10.5, determine a value for the self-diffusion coefficient of nickel at 900°C and compare it with the value shown in Fig. 10.19.

17. With the aid of Figs. 10.19 and 10.20, estimate the interdiffusion coefficient for diffusion in a nickel-gold alloy when $N_{Ni} = 0.70$.

18. (*a*) With the aid of the data in Figs. 10.19 and 10.20, determine the mobility of a nickel atom at 900°C in a nickel-gold solid solution of composition $N_A = 0.5$. Express your answer in units cm/dyne-sec.
(*b*) Compute the effective force in dynes on a nickel atom in this solid solution at 900°C.
(*c*) Determine the corresponding flux of nickel atoms per cm^2 when the concentration gradient $\partial N_A/\partial x = 2$ cm^{-1}, assuming $n_{Ni} = 7 \times 10^{22}$.

19. The voltage drop across Okabe's specimen in Fig. 10.25 was .025 volts. Over the same distance and under steady-state conditions the carbon concentration changed from 0.3 to 0.75 wt. percent.
(*a*) Determine the value of the average concentration gradient and the average electric field intensity.
(*b*) On the assumption that Z^*, the effective valence of carbon, is 4, determine an average value for the thermodynamic factor $(1 + N_c\, \partial \ln \gamma_c/\partial N_c)$ involved in this experiment.

20. (*a*) Show that under steady-state conditions of thermomigration one can write the following approximate relationship for an ideal solid solution

$$Q^* = RT^2 \frac{\Delta \ln N_A}{\Delta T} = \frac{RT^2 \ln \dfrac{N_2}{N_1}}{T_2 - T_1}$$

(*b*) Using this relationship and the data shown in Fig. 10.26, determine the average value of the heat of transport of carbon in alpha iron for the temperature interval between 670°C to 710°C. Assume T is the average temperature (that is, 690°C). Express your answer in calories per mole.

11 Interstitial Diffusion

In Chapter 10, the diffusion of atoms in substitutional solid solutions was considered, while in this chapter we shall be concerned with diffusion in interstitial solid solutions. In the former case, atoms move as a result of jumps into vacancies; in the present case diffusion occurs by solute atoms jumping from one interstitial site into a neighboring one. Interstitial diffusion is basically simpler, since the presence of vacancies is not required for the solute atoms to move. The following expression for the diffusivity of the solute atoms in a dilute substitutional solid solution was presented in Chapter 10.

$$D = \alpha a^2 Z \nu e^{-(\Delta F_m + \Delta F_f)/RT}$$

where a is the lattice parameter of the crystal, α is a geometrical factor that depends on the crystal, Z is the coordination number, ν the vibration frequency of a solute atom in a substitutional site, ΔF_f the free-energy change per mole associated with the formation of vacancies, and ΔF_m the free energy per mole required for solute atoms to jump over their energy barriers into vacancies. A similar expression can be written for interstitial diffusivity:

$$D = \alpha a^2 p \nu e^{-\Delta F_m/RT}$$

In this case, p is the number of nearest interstitial sites, α and a have the same meanings as above, ν is the vibration frequency of a solute atom in an interstitial site, and ΔF_m is the free energy per mole needed for solute atoms to jump between interstitial sites. This expression, in contrast to the previous one, contains only one free-energy term; the direct result of the fact that interstitial diffusion is not dependent upon the presence of vacancies. Because a free-energy change is capable of being expressed in the form

$$\Delta F = Q - T\Delta S$$

the expression for the interstitial diffusivity can be written:

$$D = \alpha a^2 p \nu e^{+\Delta S_m/R} e^{-Q_m/RT}$$

where ΔS_m is the entropy change of the lattice (per mole solute atoms), and Q_m is the work (per mole of solute atoms) associated with bringing solute atoms to the saddle point during a jump between interstitial positions.

11.1 Measurement of Interstitial Diffusivities. Interstitial diffusion is often studied, especially when it occurs at high temperatures, with the same experimental techniques (Matano, Grube, etc.) used for the study of diffusion in substitutional solid solutions. On the other hand, a great deal of success has been achieved in the investigation of interstitial diffusion, especially in body-centered cubic metals, with an entirely different technique. This technique has the advantage that it can be used at very low temperatures where normal methods of studying diffusion are inoperative because of very slow diffusion rates. The method will be discussed in the next section and considerable time will be spent upon it, not only because it is an important tool for the study of diffusion, but also because the general field of internal friction in metals is of importance in the study of metallurgical phenomena. While space does not permit a discussion of the use of internal-friction methods to other than the study of diffusion, an appreciation of the advantage of this type of technique can be gained by investigating its application to diffusion studies.

Before taking up the subject of internal-friction measurements, it should be pointed out that, as in the case of substitutional alloys, experimental measurements of interstitial-diffusion coefficients conform to an equation of the type

$$D = D_0 e^{-Q/RT}$$

where D is the diffusivity or diffusion coefficient, D_0 is a constant known as the frequency factor, and Q is the experimental activation energy for diffusion. Comparison of this expresssion with the theoretical one given above shows that

$$Q = Q_m \quad \text{and} \quad D_0 = \alpha a^2 p \nu e^{\Delta S_m/R}$$

Excellent agreement has been found between the quantities Q_m and ΔS_m (which may be computed theoretically for basic atomic considerations) and the experimentally determined quantities Q and D_0. This very good agreement is due to two factors. First, because interstitial-diffusion coefficients may be measured over a much larger temperature range, they are, in general, more accurate than corresponding substitutional values. Second, the interstitial diffusion process does not depend upon the presence of vacancies and is easier to interpret theoretically. It is well to note, however, that we are speaking of dilute interstitial solid solutions. When the concentration of the solute becomes appreciable, so that large numbers of interstitial sites are occupied, solute atoms

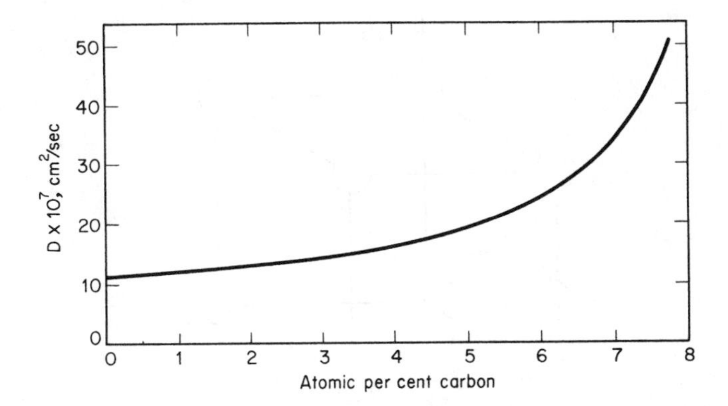

Fig. 11.1 The diffusion coefficient is also a function of composition in interstitial systems. Diffusion of carbon in face-centered cubic iron at 1127°C. (From Wells, C., Batz, W., and Mehl, R. F., *Trans. AIME,* **188** 553 [1950].)

interact, or at least interfere, with each other's jumps. As was found to be the case in substitutional solid solutions, interstitial diffusivities are, in general, functions of composition. For example, see Fig. 11.1.

11.2 The Snoek Effect. The study of interstitial diffusion by internal-friction methods usually makes use of an effect first explained by Snoek.[1] In a body-centered cubic metal like iron, interstitial atoms, such as carbon or nitrogen, take positions either at the centers of the cube edges or at the centers of the cube faces (see Fig. 11.2). Both positions are crystallographically equivalent, as may be deduced from Fig. 11.2. An interstitial atom at either x or w would lie between two iron atoms aligned in a ⟨100⟩ direction (the iron atoms on either side of the w position lie at the center of the unit cell shown in the figure and the center of the next unit cell in front of this cell [not shown in the figure]). It has been previously mentioned (Chapter 8, Fig. 8.2B) that the space available for the solute atom between two iron atoms is smaller than the diameter of the solute atom. The occupancy of one of these positions (such as that designated by x in Fig. 11.2) by a solute atom pushes apart the two solvent atoms a and b. An atom at x or w thus increases the length of the crystal in the [100] direction. Similarly, an atom at y or z increases the length of the crystal in the [010] or [001] directions.

For the sake of convenience, let us define the axis of an interstitial site as that direction along which the solvent atoms (at either side of the interstitial site) are spread when it is occupied by an interstitial solute atom.

When a body-centered cubic crystal containing interstitial atoms is in an

1. Snoek, J., *Jour. Physica,* **6** 591 (1939).

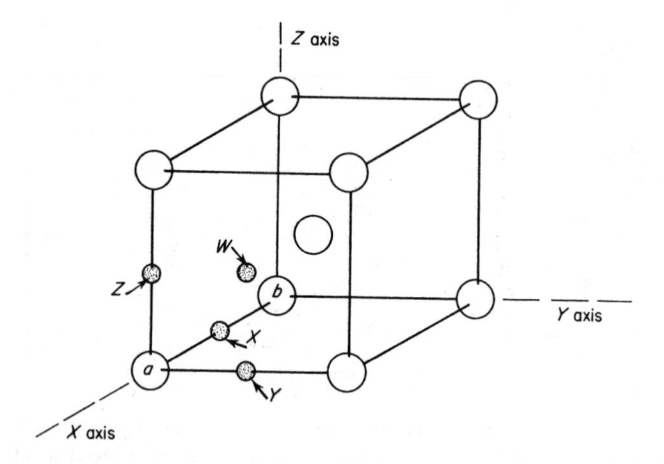

Fig. 11.2 The nature of the sites that interstitial carbon atoms occupy in the body-centered cubic iron lattice.

unstressed state, a statistically equal number of solute atoms will be found in the three types of sites, the axes of which are parallel to the [100], [010], and [001] directions respectively. Now, if an external force is applied to the crystal so as to produce a state of tensile stress parallel to the [100] axis, it will have the effect of straining the lattice in such a manner that those sites with axes parallel to [100] will have their openings enlarged, while those with axes normal to the stress ([010] and [001]) will have their openings diminished. The effect of a stress, therefore, is to give the solute atoms a somewhat greater preference for interstitial sites with axes that are parallel to the stress. After the application of the stress, the number of solute atoms tends to increase in these preferred sites, so that an equal division of solute atoms among the three types of sites ceases to exist.

When the applied stress is small, so that the elastic strain is small (order of 10^{-5} or smaller), the number of excess solute atoms per unit volume that eventually find themselves in interstitial positions, with axes parallel to the tensile-stress axis, is directly proportional to the stress. Thus,

$$\Delta n_p = K s_n$$

where Δn_p is the additional number of solute atoms in preferred positions, K is a constant of proportionality, s_n is the tensile stress. Each of the additional solute atoms in one of the preferred positions adds a small increment to the length of the specimen in the direction of the tensile stress. The total strain of the metal thus consists of two parts: the normal elastic strain, ϵ_{el}, and the anelastic strain, ϵ_{an}, which is caused by the movement of solute atoms into sites with axes parallel to the stress axis

$$\epsilon = \epsilon_{el} + \epsilon_{an}$$

When a stress is suddenly applied, the elastic component of the stress can be considered to develop instantly. The anelastic strain, however, is time dependent and does not appear instantly. The sudden application of a stress to a crystal places the solute atoms in a nonequilibrium distribution, for equilibrium now corresponds to an excess of solute atoms, Δn_p, in sites with axes parallel to the stress. The attainment of the equilibrium distribution occurs as a result of the normal thermal movements of solute atoms. The net effect of the stress is to cause a slightly greater number of jumps to go into the preferred positions than come out of them. However, when equilibrium is attained, the number of jumps per second into and out of the preferred positions will be the same. Obviously, the number of excess atoms in preferred positions and the anelastic strain must both be a maximum at equilibrium.

The rate at which the number of additional atoms in preferred interstitial sites grows depends directly on the number of the excess sites that are still unoccupied at any instant. The rate is greatest, therefore, at the instant that the stress is applied, because at this time the number of excess atoms in preferred sites is zero. As time goes on, however, and the number of excess atoms approaches its maximum value, the rate becomes progressively smaller and smaller. As in all physical problems where the rate of change depends on the number present, an exponential law can be expected to govern the time dependence of the number of additional interstitial atoms. In the present case, this law is:

$$\Delta n_p = \Delta n_{p(\max)}[1 - e^{-t/\tau_\sigma}]$$

where Δn_p is the number of excess solute atoms at any instant, $\Delta n_{p(\max)}$ the maximum number which may be attained under a given tensile stress, t is the time, and τ_σ is a constant known as *relaxation time* at constant stress.

Since the anelastic strain is directly proportional to the number of excess atoms in preferred sites, it can also be expressed by an exponential relation

$$\epsilon_{an} = \epsilon_{an(\max)}[1 - e^{-t/\tau_\sigma}]$$

where ϵ_{an} and $\epsilon_{an(\max)}$ are the instantaneous and maximum (equilibrium values) of the anelastic strain respectively. The relationship between the elastic and anelastic strains is shown in Fig. 11.3.

The effect of removing the stress after the anelastic strain has reached its maximum value is also shown in Fig. 11.3. If the stress is suddenly removed, the

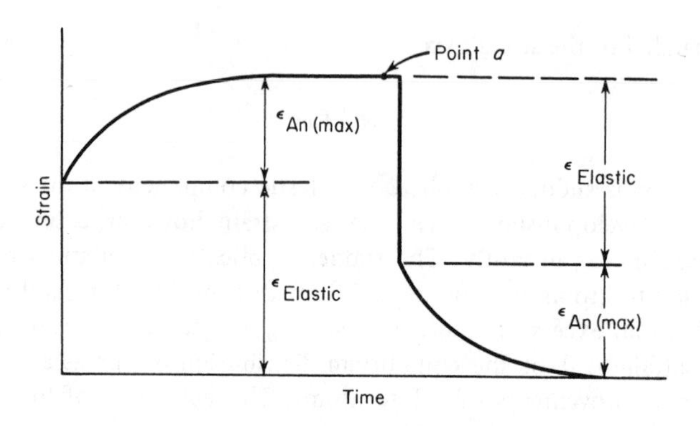

Fig. 11.3 Relationship between elastic and anelastic strains. (After A. S. Nowick.)

elastic strain is recovered instantly, while the anelastic component is time dependent. For the condition of stress removal, the anelastic strain follows a law of the form:

$$\epsilon_{an} = \epsilon_{an(\max)} e^{-t/\tau_\sigma}$$

where ϵ_{an} is the anelastic strain at any instant, $\epsilon_{an(\max)}$ is the ane'astic strain at the instant of the removal of the stress, and t and τ_σ have the same meanings as in the previous equations.

The significance of τ_σ can be seen if the time t is set equal to τ_σ and then substituted into the above equation

$$\epsilon_{an} = \epsilon_{an(\max)} e^{-\tau_\sigma/\tau_\sigma} = \frac{1}{e}\,\epsilon_{an(\max)}$$

The relaxation time τ_σ is thus the time that it takes the anelastic strain to fall to $1/e$ of its original value. If τ_σ is large, the strain relaxes very slowly, and if τ_σ is small, the strain relaxes quickly. The rate at which the strain relaxes is consequently an inverse function of the relaxation time. It is also an inverse function of the mean time of stay of an atom in an interstitial position τ, for small values of τ correspond to large jump rates, $1/\tau$, and rapid strain relaxation. These two essentially different time concepts (relaxation time and mean time of stay of an atom in an interstitial position) are directly related and, in the case of the body-centered cubic lattice, it can be shown that

$$\tau = \tfrac{3}{2}\tau_\sigma$$

This relationship will now be derived following Nowick.[2] For this purpose, let us rewrite the above expression for the anelastic strain

$$\epsilon_{an} = \epsilon_{an\,(\max)} - \epsilon_{an\,(\max)} e^{-t/\tau_\sigma}$$

The derivative of this latter is

$$\frac{d\epsilon_{an}}{dt} = \frac{\epsilon_{an\,(\max)}}{\tau_\sigma} e^{-t/\tau_\sigma} = \frac{\epsilon_{an\,(\max)} - \epsilon_{an}}{\tau_\sigma}$$

As stated earlier, this equation shows that the time rate of change of the anelastic strain equals the difference between the maximum attainable anelastic strain (under a given applied stress) and the instantaneous value of the anelastic strain. A similar relationship holds for Δn_p, the number of excess carbon atoms per unit volume in preferred sites, so that

$$\frac{d(\Delta n_p)}{dt} = \frac{\Delta n_{p(\max)} - \Delta n_p}{\tau_\sigma}$$

where if the stress is assumed to be applied along one of the three ⟨100⟩ axes of the iron crystal, say the z axis, $\Delta n_p = n_z - n/3$, and $\Delta n_{p(\max)} = n_{z(\max)} - n/3$. This is based on the assumption that under zero stress, the carbon atoms will be uniformly distributed in the three possible ⟨100⟩ sites. Therefore, we may also write

$$\frac{dn_z}{dt} = \dot{n}_z = \frac{\left(n_{z(\max)} - \frac{n}{3}\right) - \left(n_z - \frac{n}{3}\right)}{\tau_\sigma}$$

An expression can also be written for $\dot{n}_z$ in terms of the difference in the rates at which carbon atoms enter and leave z sites. This equation is

$$\dot{n}_z = n_x \left(\frac{1}{\tau_{xz}}\right) + n_y \left(\frac{1}{\tau_{yz}}\right) - n_z \left(\frac{1}{\tau_{zx}}\right) - n_z \left(\frac{1}{\tau_{zy}}\right)$$

where n_x, n_y, and n_z are the numbers of carbon atoms per unit volume in x, y, and z sites, respectively, and $1/\tau_{xz}$, $1/\tau_{yz}$, $1/\tau_{zx}$, and $1/\tau_{zy}$ are the jump frequencies of the carbon atoms between the types of sites indicated by the subscripts. Thus, $1/\tau_{xz}$ is the jump rate of a carbon atom from an x to a z site, and $1/\tau_{zx}$ is the jump frequency in the opposite direction.

2. Nowick, A. S., *Prog. in Metal Phys.*, **4** 1 (1953).

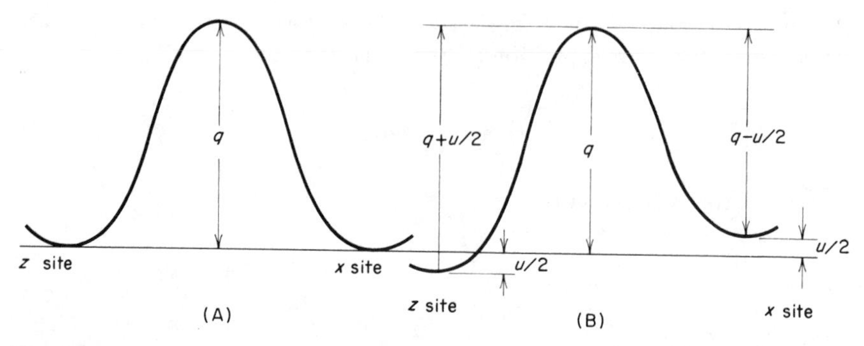

Fig. 11.4 The effect of an applied stress along a z-axis direction in a b.c.c. metal on the energy barrier for the jumping of carbon atoms between x and z sites. (A) The energy barrier when the stress is zero. (B) The barrier when the applied stress is finite.

Under an applied constant stress, the jump frequencies will be different from those obtained when there is no stress. This is because a stress applied along the z axis, as assumed above, lowers the energy barrier for a jump into a z site from an x or y site, while it raises the energy barrier for a jump of the reversed kind. This is indicated schematically in Fig. 11.4 for the case of atoms jumping between x and z sites. Due to the symmetry of the lattice, an identical curve would be obtained for interchanges between y and z sites. Note that as a result of the applied stress, the energy level of the x site is higher by an amount u than the energy level of the z site. In effect, this makes the energy barrier for a jump from an x to a z site $(q - u/2)$, and for a jump from a z to an x site $(q + u/2)$. The jump frequency in the absence of a stress is

$$\frac{1}{\tau} = \frac{1}{\tau_0} e^{-q/kT}$$

and by the symmetry of the lattice we have

$$\frac{1}{\tau_{xz}} = \frac{1}{\tau_{yz}} = \frac{1}{2\tau} = \frac{1}{2\tau_0} e^{-q/kT}$$

where the factor ½ is introduced because an atom in an x site, for example, will normally make half of its jumps into z sites and the other half into y sites. On the other hand, in the presence of the stress along the z axis

$$\frac{1}{\tau_{xz}} = \frac{1}{2\tau_0} e^{-(q - u/2)/kT}$$

and

$$\frac{1}{\tau_{zx}} = \frac{1}{2\tau_0} e^{-(q+u/2)/kT}$$

and by the symmetry of the lattice

$$\frac{1}{\tau_{yz}} = \frac{1}{2\tau_0} e^{-(q-u/2)/kT}$$

and

$$\frac{1}{\tau_{zy}} = \frac{1}{2\tau_0} e^{-(q+u/2)/kT}$$

Substituting these relationships into the equation for $\dot{n}_z$ gives us

$$\dot{n}_z = (n_x + n_y)\frac{e^{-(q-u/2)/kT}}{2\tau_0} - \frac{2n_z}{2\tau_0} e^{-(q+u/2)/kT}$$

or

$$\dot{n}_z = \frac{e^{-q/kT}}{2\tau_0}\left[(n_x + n_y)\, e^{u/2kT} - 2n_z\, e^{-u/2kT}\right]$$

Because u, the difference in the energy levels due to the stress, is normally very small and therefore $u/2 \ll kT$, we may make the substitutions

$$e^{u/2kT} = 1 + \frac{u}{2kT} \quad \text{and} \quad e^{-u/2kT} = 1 - \frac{u}{2kT}$$

so that

$$\dot{n}_z = \frac{e^{-q/kT}}{2\tau_0}\left[(n_x + n_y)\left(1 + \frac{u}{2kT}\right) - 2n_z\left(1 - \frac{u}{2kT}\right)\right]$$

Since $n_x + n_y + n_z = n$, where n is the total number of carbon atoms per cm^3 and $e^{-q/kT}/\tau_0 = 1/\tau$, the jump frequency in the absence of a stress, this equation can also be written

$$\dot{n}_z = \frac{1}{2\tau}\left[(n - 3n_z) + (n + n_z)\frac{u}{2kT}\right]$$

or

$$\dot{n}_z = \frac{3}{2\tau}\left[\left(\frac{n + n_z}{3}\right)\frac{u}{2kT} - \left(n_z - \frac{n}{3}\right)\right]$$

and since $n_z u$ is a small quantity and n_z is not greatly different from $n/3$, we may make the approximation in the first term inside the brackets that $n_z = n/3$, so that

$$\dot{n}_z = \frac{3}{2\tau}\left[\frac{2}{9}\frac{nu}{kT} - \left(n_z - \frac{n}{3}\right)\right]$$

Now let us assume that the stress is applied for a very long time. In this case, $n_z \to n_{z(\max)}$ and $\dot{n}_z \to 0$. Therefore in the limit when $t \to \infty$, we find that

$$\frac{2}{9}\frac{nu}{kT} = \left(n_{z(\max)} - \frac{n}{3}\right)$$

This equation states that the energy level difference u is directly proportional to the final excess number of carbon atoms in z sites. In addition we have

$$\dot{n}_z \frac{3}{2\tau}\left[\left(n_{z(\max)} - \frac{n_z}{3}\right) - (n_{z(\max)} - n_z)\right]$$

which by our earlier equation for $\dot{n}_z$ shows that

$$\frac{1}{\tau_\sigma} = \frac{3}{2\tau}$$

or

$$\tau = \frac{3\tau_\sigma}{2}$$

The above equation is significant in that an experimentally determined value (relaxation time τ_σ) is capable of directly yielding the value of a theoretically important quantity (the mean time of stay of a solute atom in an interstitial position τ). In addition, once the value of τ is determined, it is possible to compute directly the diffusion coefficient for interstitial diffusion from the relation

$$D = \frac{\alpha a^2}{\tau}$$

This equation is identical in form to that used in Chapter 10 in the discussion of substitutional diffusion. In the present case, the quantity a is the lattice constant of the solvent, τ is now the mean time of stay of a solute atom in an interstitial site, and α is a constant which is determined by the geometry of the lattice and the nature of the diffusion process (in this case, jumping of solute atoms between interstitial positions). In interstitial diffusion, the constant α equals $1/24$ in the body-centered cubic lattice, and $1/12$ in the face-centered cubic lattice. Since interstitial diffusion has been most thoroughly investigated in body-centered cubic lattices, the present discussion will be confined to this lattice. Here α is $1/24$ and $\tau = 3/2\ \tau_\sigma$. Therefore,

$$D = \frac{a^2}{24\tau} = \frac{a^2}{36\tau_\sigma}$$

Substitution of the experimentally determined relaxation time τ_σ in the above equation yields the diffusivity directly.

11.3 Experimental Determination of the Relaxation Time. If the relaxation time is very long (of the order of minutes to hours), it can be determined by the elastic after-effect method. In this case, a stress is applied to a suitable specimen and maintained until the anelastic component of the stress has effectively reached its equilibrium value. This corresponds to a point such as a in Fig. 11.3. After this condition has been reached, the stress is quickly removed and the strain measured as a function of time. The data obtained correspond to the curve at the lower right-hand side of Fig. 11.3 from which the relaxation time can be obtained either by determining the time required for the anelastic part of the strain to fall $1/e$ of its value, or by more elaborate methods[3] of handling the data.

The elastic after-effect method is not particularly suitable for use when the relaxation time is of the order of seconds instead of minutes. In this case, a convenient and frequently used method for determining relaxation time makes use of a torsion pendulum of the type[4] shown schematically in Fig. 11.5. The specimen, in the form of a wire, is clamped into two pin vises; the upper one rigidly fixed in the apparatus, the lower one connected to the inertia bar and free to rotate with it. Small iron blocks are located at each end of the inertia bar in order to make possible a smooth release of the pendulum when it is set into oscillation. In placing the pendulum in motion, it is given an initial twist about its axis so that the iron blocks are brought into contact with two small electromagnets that hold it in the twisted position until the current is broken in

3. Nowick, A. S., *Phys. Rev.*, 88 9 (1952).
4. Ké, T. S., *Phys. Rev.*, 71 553 (1947).

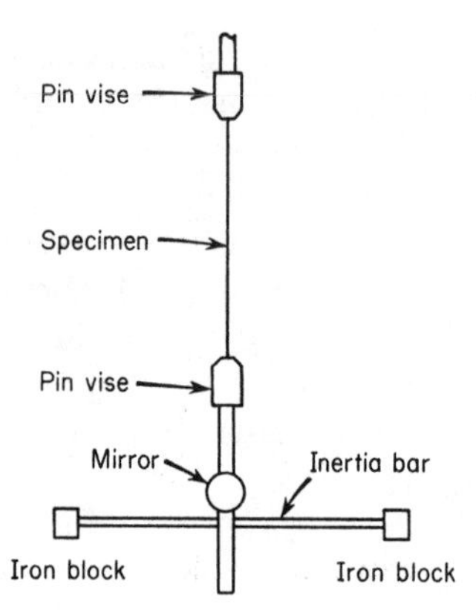

Fig. 11.5 Torsion pendulum. (After Kê, T. S., *Phys. Rev.*, **71** 553 [1947].)

the magnet circuit. When this is done, the pendulum is free to oscillate. A mirror placed on the bar connecting the lower pin vise to the inertia bar makes it possible to follow the oscillations of the pendulum when a light beam is reflected from its surface onto a translucent scale. Figure 11.6 shows a typical trace of the pendulum amplitude as a function of time. Note that the amplitude decreases with increasing time because of vibrational energy losses inside the wire (neglecting air-friction effects on the torsion bar). Such an energy loss is said to be due to internal friction in the metal. There are many sources of internal friction in metals. We are interested in that due to the presence of interstitial solute atoms in body-centered cubic metals.

Three possible cases will be considered. First, let us assume that the period of the torsion pendulum is very small compared with the relaxation time of the metal. In this case, the length of time during a cycle that the wire is subject to a given type of stress is very much shorter than the mean time of stay of a solute

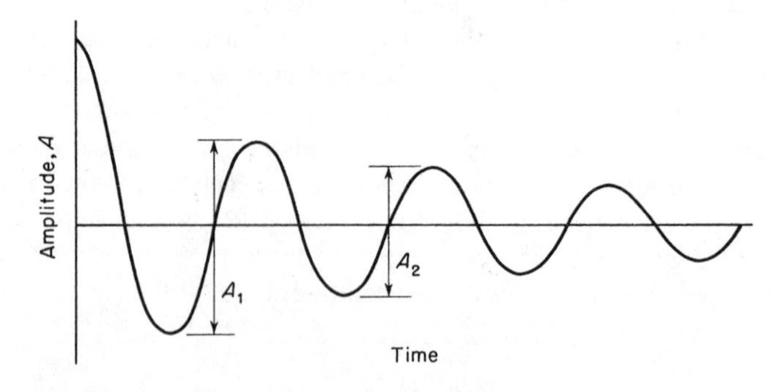

Fig. 11.6 Damped vibration of a torsion pendulum.

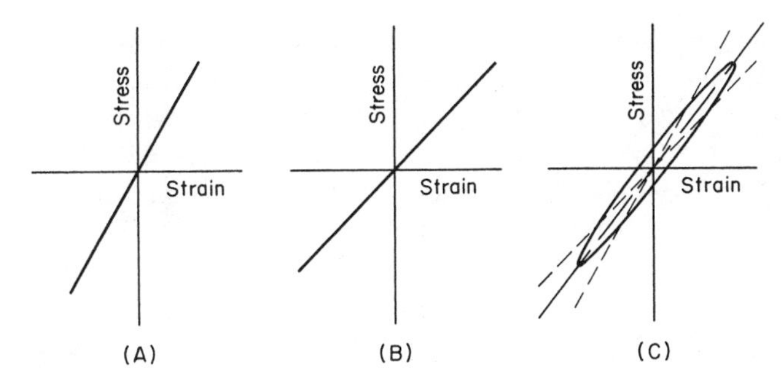

Fig. 11.7 Stress-strain curves for a torsion pendulum: (A) Pendulum period much shorter than the relaxation time, (B) Pendulum period much longer than the relaxation time, and (C) Pendulum period approximately equal to the relaxation time.

atom in an interstitial position. Stated slightly differently, the stress alternates so rapidly that it is impossible for the solute atoms to follow the stress changes. The anelastic component of the strain can be taken as zero and the pendulum considered to vibrate in a completely elastic manner. A stress-strain diagram for this case (Fig. 11.7A) is a straight line with a slope equal to the modulus of elasticity. In anelastic work this slope is normally designated as M_U and is called the *unrelaxed modulus*. The line is drawn so that it passes through the origin of coordinates, in agreement with the fact that a pendulum oscillates to either side of its rest point. The curve of Fig. 11.7A can also be considered to represent the stress-strain diagram for a complete oscillation of the stress.

The second possibility corresponds to the other extreme where the period of the pendulum is very much larger than the relaxation time. In this case, the solute atoms have no difficulty in following the stress alternations and it can be assumed that a state of equilibrium is constantly maintained. At all times, the anelastic strain will have its maximum, or equilibrium, value corresponding to the instantaneous value of the stress. Both the anelastic and elastic components of the strain vary directly as the stress and, therefore, the total strain varies linearly with the stress. However, in contrast to the case where the period of the pendulum is very short, the strain at each value of the stress will now be of greater magnitude due to the finite value of the anelastic strain. Figure 11.7B shows that the stress-strain curve for this second case has a smaller slope than the previous one, signifying that the modulus of elasticity (ratio of stress to strain) is smaller. The modulus of elasticity, measured under these conditions, is known as the *relaxed modulus* and is given the symbol M_R.

The third case is the intermediate one where the period of the torsion pendulum approximately equals the relaxation time. Here the stress cycles are slow enough that the anelastic strain assumes finite values, but the stress variation is not slow enough for a state of equilibrium to be effectively reached. Under these conditions, the anelastic strain does not vary linearly with the stress, and the total strain contains a nonlinear component (anelastic part). The stress-strain curve for a complete cycle of stress now assumes the form of an ellipse (Fig. 11.7C) with the major axis lying between the lines corresponding to the stress-strain curves for very short and very long periods. The area inside this (*hysteresis*) loop has the dimensions of work and represents the energy lost in the specimen per unit volume during a complete cycle. In the two other examples (Figs. 11.7A and 11.7B), the area inside the stress-strain loop is zero. It appears, therefore, that the energy loss per cycle is a function of the period of oscillation, having a zero value when the period is either very long or very short, and a finite value when the period has intermediate values.

The energy loss per cycle in a torsion pendulum can be determined directly from a plot of the pendulum amplitude as a function of time. In Fig. 11.6, A_1 and A_2 represent two adjacent vibration amplitudes and the assumption is made that the difference in the magnitudes of these two amplitudes is small, a condition usually met in work of the type being discussed.

Now in an oscillating system, the energy of vibration is proportional to the square of the vibration amplitude, so that the vibrational energy, when the pendulum has amplitude A_1, is proportional to A_1^2, and when it has the amplitude A_2, the energy corresponds to A_2^2, and the fractional loss of energy per cycle is

$$\frac{\Delta E}{E} = \frac{A_2^2 - A_1^2}{A_1^2}$$

where E is the energy of the pendulum, ΔE is the loss of energy during a cycle, and A_1 and A_2 are the amplitudes at the beginning and end of the cycle.

Factoring the expression on the right-hand side of the above equation leads to

$$\frac{\Delta E}{E} = \frac{(A_1 - A_2)(A_1 + A_2)}{A_1^2}$$

but since it has been assumed that the difference between A_1 and A_2 is small, we can write $A_1 - A_2 = \Delta A$ and $A_1 + A_2 = 2A_1$. Therefore,

$$\frac{\Delta E}{E} = \frac{2\Delta A}{A}$$

This equation states that the fractional energy loss per cycle is twice the fractional amplitude loss per cycle. This latter quantity is easily determined experimentally.

The measure of the internal friction given above, $\Delta E/E$, is usually known as the *specific damping capacity*. This quantity is often used by engineers to express the energy-absorbing properties of materials of construction.

A more commonly used measure of the internal friction in problems of the type currently being discussed is the *logarithmic decrement*, which is the natural logarithm of the ratio of successive amplitudes of vibration. Thus,

$$\delta = \ln \frac{A_1}{A_2}$$

where δ is the logarithmic decrement, and A_1 and A_2 are two successive vibration amplitudes.

Provided that the damping is small, we may write[5]

$$\delta = \frac{1}{2}\frac{\Delta E}{E} = \frac{\Delta A}{A}$$

Still another method of expressing the internal friction in a metal undergoing cyclic strain uses the phase angle α by which the strain lags behind the stress. The tangent of this angle can also be taken as an index of the internal-energy loss. Again, if the damping is small,[5] it can be shown that

$$\tan \alpha = \frac{1}{\pi} \ln \frac{A_1}{A_2} = \frac{\delta}{\pi}$$

This quantity, $\tan \alpha$, is often written as Q^{-1} and called the *internal friction.* This is in analogy with the damping, or energy loss, in an electrical system.

For the purpose of the present discussion, let us use the specific damping capacity as the measure of internal friction.

The energy loss per cycle $\Delta E/E$ is a smoothly varying function of the period, or frequency, of the pendulum capable of being stated mathematically. When expressed in terms of the angular frequency of the pendulum ω

$$\omega = 2\pi\nu = \frac{2\pi}{\tau_p}$$

where ν = pendulum frequency in cycles per second and τ_p = pendulum period in seconds, this expression assumes the following simple and convenient form

$$\frac{\Delta E}{E} = 2\left(\frac{\Delta E}{E}\right)_{\max}\left[\frac{\omega\tau_R}{1 + \omega^2\tau_R{}^2}\right]$$

5. Zener, Clarence M., *Elasticity and Anelasticity of Metals.* The University of Chicago Press, Chicago, Ill., 1948.

where $\frac{\Delta E}{E}$ = fraction of energy lost per cycle, $\left(\frac{\Delta E}{E}\right)_{max}$ = maximum fractional energy loss, ω = angular frequency of pendulum, τ_R = relaxation time for interstitial diffusion.

The relaxation time τ_R, as measured using a torsion pendulum, is not exactly the same as τ_σ, the relaxation time as measured in an elastic after-effect experiment, or other experiments carried out at constant stress. The two quantities are, however, simply related to each other by the expression[6]

$$\tau_R = \tau_\sigma \left(\frac{M_R}{M_U}\right)^{1/2}$$

where M_R is the relaxed modulus and M_U is the unrelaxed modulus. In the case of the interstitial diffusion of carbon in alpha iron, M_R and M_U are very nearly equal, so we may assume $\tau_R = \tau_\sigma$ and also that the mean time of stay of a carbon atom in an interstitial site is $\tau = 3/2\ \tau_R$.

The above equation is symmetrical in both ω and τ_R, so that the variation of the fractional energy loss $\Delta E/E$ is the same no matter whether we hold τ_R constant and vary ω or hold ω constant and vary τ_R. In either case, the given function yields a curve symmetrical with respect to the point of maximum-energy loss, if $\Delta E/E$ is plotted as a function of the log ω, or of the log τ_R. Figure 11.8 shows the shape of this curve and, as indicated in the figure, the

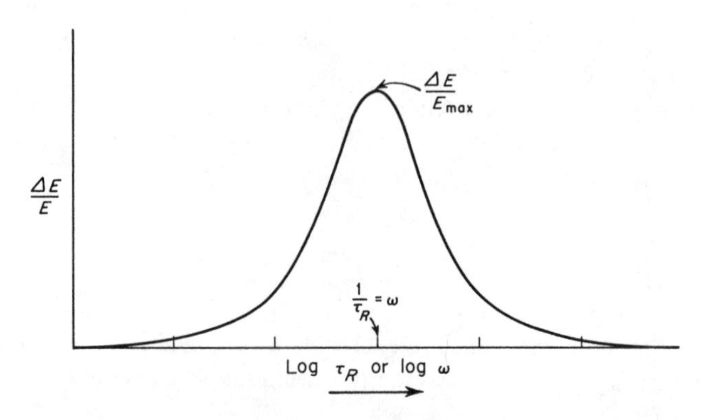

Fig. 11.8 Theoretical relationship between the fractional energy loss per cycle and either log τ_R or log ω. At $\left(\frac{\Delta E}{E}\right)_{max}$ $\frac{1}{\tau_R}$ equals ω.

6. *Ibid.*, p. 44.

maximum energy loss occurs when the following relationship holds:

$$\omega = \frac{1}{\tau_R}$$

The above equation states that the energy loss is a maximum when the angular frequency of the pendulum equals the reciprocal of the relaxation time. This fact is important because it gives a direct relationship between the experimentally measurable frequency of the pendulum and the relaxation time.

The relaxation time is only a function of temperature, while the frequency of the pendulum is a function of its geometry. In theory, there are two basic methods of determining the point of maximum energy loss: the frequency of the pendulum may be varied while the temperature is maintained constant (thereby keeping the relaxation-time constant) or the frequency of the pendulum may be kept constant while the temperature is varied. The latter method is more often used as it is usually more convenient. In this case we have

$$D = D_0 e^{-Q/RT} = \frac{a^2}{36\tau_R}$$

Solving this equation for relaxation time yields:

$$\tau_R = \frac{a^2}{36 D_0 e^{-Q/RT}} = \tau_{R_0} e^{+Q/RT}$$

where τ_{R_0} is a constant equal to $a^2/36D_0$. From this relationship it is clear that the $\log_{10} \tau_R$ varies as $1/T$, where T is the absolute temperature. A plot of the fractional energy loss as a function of $1/T$ should give a curve of the type shown in Fig. 11.8. A set of five of these curves for iron, containing carbon in solid solution, is shown in Fig. 11.9. Each curve was obtained by adjusting the torsion pendulum to operate at a different frequency. The maximum energy loss of each curve occurs at a different temperature. The frequencies corresponding to the curves of Fig. 11.9 are listed in the first column of Table 11.1; the angular frequencies in column 2; the temperatures of the energy-loss maxima in column 3; the reciprocal of the temperature in column 4; and the logarithms of the relaxation times in column 5, as computed with the aid of the equation

$$\tau_R = \frac{1}{\omega}$$

τ_{R_0} and Q may be obtained from these data by plotting the $\log_{10} \tau_R$ as a

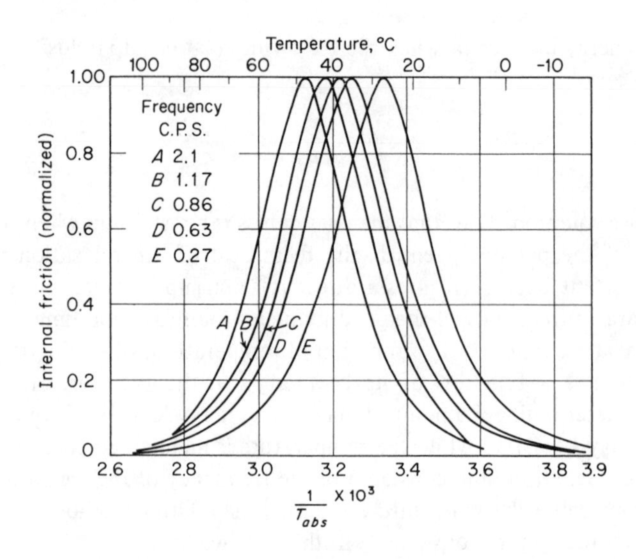

Fig. 11.9 Internal friction as a function of temperature for Fe with C in solid solution at five different pendulum frequencies. (From Wert, C., and Zener, C., *Phys. Rev.,* **76** 1169 [1949].)

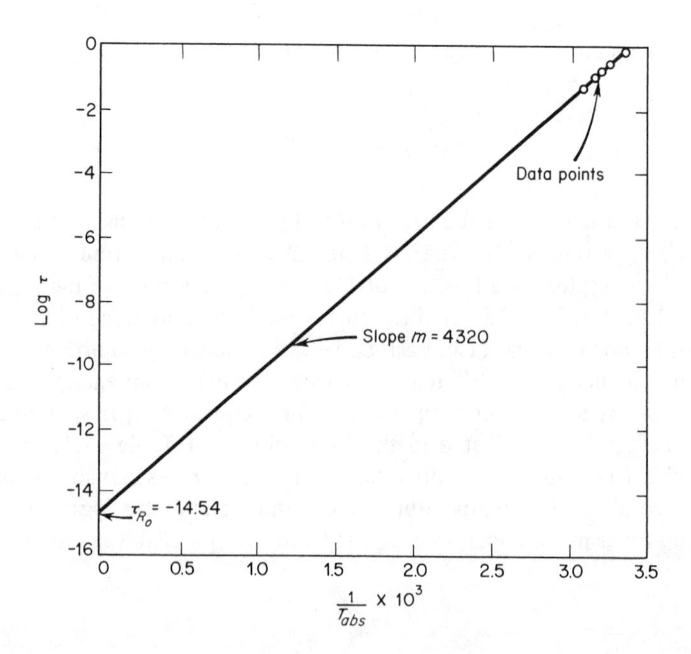

Fig. 11.10 Variation of log τ with $1/T_{abs}$ for data in Fig. 11.9. (Data of Wert, C., and Zener, C., *Phys. Rev.,* **76** 1169 [1949].)

Table 11.1 Data Corresponding to the Five Curves shown in Figure 11.8.*

Frequency cps	*Angular Frequency*	*Absolute Temperature* T	$\frac{1}{T}$	$\text{Log}_{10}\, \tau_R$
2.1	13.2	320	3.125×10^{-3}	−1.120
1.17	7.35	314	3.178×10^{-3}	−0.866
0.86	5.39	311	3.22×10^{-3}	−0.731
0.63	3.95	309	3.25×10^{-3}	−0.595
0.27	1.69	300	3.35×10^{-3}	−0.227

* From the work of Wert, C., and Zener, C., *Phys. Rev.*, 76 1169 (1949).

function of $1/T$. This plotting is done in Fig. 11.10, where the slope of the line is

$$m = \frac{Q}{2.3R} = 4320$$

Note that the factor 2.3 enters the above equation because we are using logarithms to the base 10. The activation energy Q is accordingly

$$Q = 4.6(4320) = 19{,}800 \text{ cal/mol}$$

The value of the constant τ_{R_0} is obtained from the ordinate intercept in Fig. 11.10, which has a value of

$$\log_{10} \tau_{R_0} = -14.54$$

so that

$$\tau_{R_0} = 2.92 \times 10^{-15} \text{ sec}$$

The mean time of stay of a carbon atom in an interstatial position is, accordingly,

$$\tau = {}^3\!/_2\, \tau_R = {}^3\!/_2\, \tau_{R_0} e^{Q/RT} = ({}^3\!/_2) 2.92 \times 10^{-15} e^{19{,}800/RT}$$

and the diffusivity of carbon in body-centered cubic iron (alpha-iron), taking the lattice constant a for iron as 2.86 Å, is:

$$D = \frac{a^2}{36\tau_R} = 0.008\, e^{-19{,}800/RT}$$

Table 11.2 Diffusivity Equations for Interstitial Diffusion in Certain Body-Centered Cubic Metals.

Solvent Metal	Diffusing Element		
	C	N	O
Iron*†	$0.02e^{-20,100/RT}$	$0.0014e^{-17,700/RT}$	
Vanadium‡	$0.0047e^{-27,300/RT}$		$0.019e^{-29,300/RT}$
Tantalum§		$0.004e^{-37,000/RT}$	$0.006e^{-25,900/RT}$
Columbium¶	$0.0046e^{-33,300/RT}$	$0.0072e^{-34,800/RT}$	

* Wert, C., *Phys. Rev.*, **79** 601 (1950).
† Wert, C., and Zener, C., *Phys. Rev.*, **76** 1169 (1949).
‡ Powers, R. W., and Doyle, Margaret V., *Acta Met.*, **6** 643 (1958).
§ Powers, R. W., and Doyle, Margaret V., *Acta Met.*, **4** 233 (1956).
¶ Powers, R. W., and Doyle, Margaret V., *Trans. AIME*, **209** 1285 (1957).

11.4 Experimental Data. A compilation of certain experimentally determined interstitial-diffusivity equations is given in Table 11.2. The value given in the table for the diffusion of carbon in iron is not the same as that given above, for it is a later and somewhat more accurate value. In general, it is noted that in the elements listed, the activation energies of interstitial diffusion are somewhat smaller than those of substitutional diffusion.

11.5 Anelastic Measurements at Constant Strain. In addition to measurements made at constant stress, as exemplified by the technique of the elastic after-effect (zero stress) and vibrational studies, of which the torsion pendulum is only one of a number of techniques, there is a third basic type of measurement. This involves the measurement of the relaxation of the stress under the condition of a constant strain. In a typical experiment of this type, a specimen may be loaded in a universal (constant deformation rate) testing machine to some predetermined stress, whereupon the machine is stopped. Because the specimen is subjected to a stress, it continues to deform anelastically. For the purpose of this discussion it will be assumed that the testing machine and its grips are very *hard* elastically. Under this assumption, the ends of the specimen will be fixed in space so that the total strain in the specimen is constant, or

$$\epsilon = \epsilon_{an} + \epsilon_{el} = 0$$

Thus as the specimen increases its length anelastically there must be a corresponding decrease in its elastic strain. This will, of course, relax the stress in the specimen in a manner like that shown schematically in Fig. 11.11. This curve is similar in its general form to the elastic after-effect curve and follows a similar law. It is, accordingly, possible to measure a relaxation time at constant strain,

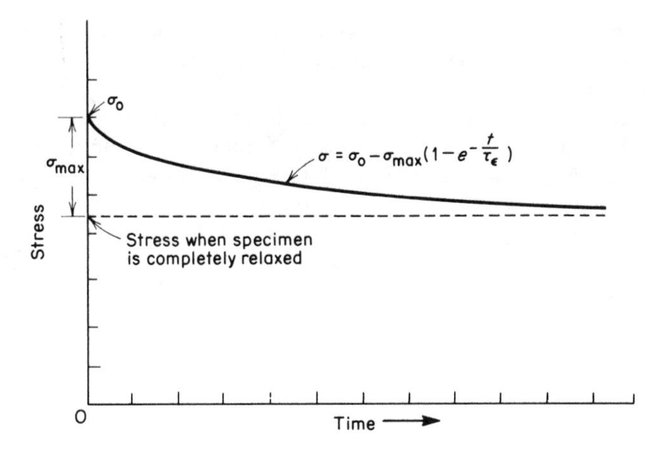

Fig. 11.11 The relaxation of the stress at constant strain.

designated τ_ϵ. It is not the same as τ_σ, the relaxation time at constant stress, but it is related to it by the equation[7]

$$\frac{\tau_\sigma}{\tau_\epsilon} = \frac{M_U}{M_R}$$

where M_U is the unrelaxed modulus and M_R is the relaxed modulus. Both of these relaxation times are related to that obtained in a torsion pendulum as follows[8]

$$\tau_R = (\tau_\sigma \tau_\epsilon)^{1/2}$$

This shows that τ_R, obtained in a torsion pendulum, is the geometric mean of the other two relaxation times.

Problems

1. Prove that the quantity α in the diffusion equation ($D = \alpha a^2/\tau$) equals 1/24 for interstitial diffusion in a body-centered cubic crystal. *Hint*: consider diffusion normal to (010) and determine the fraction of the carbon atoms lying in this plane that can make a jump normal to this plane.
2. (*a*) What experimental relaxation time τ_σ would be expected in an anelastic experiment at room temperature where the anelastic strain was due to the jumping of carbon atoms? Assume $D = 0.008\, e^{19{,}800/RT}$ and the lattice constant for iron is 2.86 Å.

7. *Ibid.*
8. *Ibid.*

(*b*) Below what temperature would it be feasible to make anelastic measurements involving the diffusion of carbon atoms using the elastic after-effect technique?

3. This figure shows a set of five elastic aftereffect curves obtained from the same zirconium specimen. In this case the internal friction is associated with the thermally activated motion of deformation twin boundaries in response to an applied stress. The twins were put into the metal by a small prestrain at 77°K. The curves correspond closely to the simple exponential decay law for strains less than 10×10^{-6}. Lines are drawn across the diagram at strains

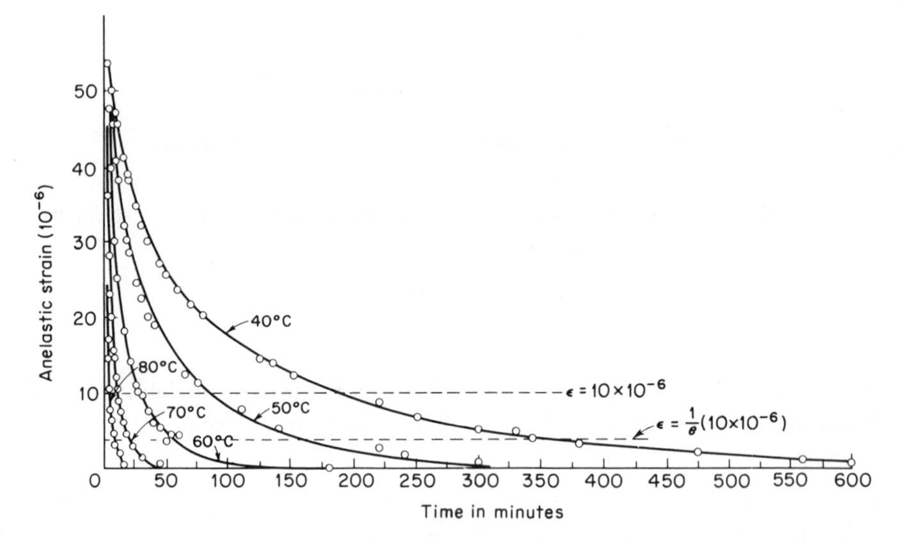

of 10×10^{-6} and $1/e(10 \times 10^{-6})$. At each of the five temperatures determine the relaxation time associated with the twin boundary movement and determine an activation energy for the process.

4. Given a torsion pendulum with a period of 3.14 seconds and an anelastic effect in the wire whose fractional maximum energy loss per cycle is 10^{-2}, plot a curve of the fractional energy loss per cycle as a function of the logarithm (base 10) of the relaxation time. Use semi-log paper of at least four cycles for this problem.

5. In a typical anelastic measurement involving the diffusion of carbon atoms in iron, the internal friction ($\tan \alpha$) has a maximum value of about 10^{-3}. Anelasticity theory[9] states that

$$(\tan \alpha)_{max} = \frac{M_U - M_R}{(M_U M_R)^{1/2}}$$

9. *Ibid.*

where M_U and M_R are the unrelaxed and relaxed moduli, respectively. On this basis, estimate the magnitude of the error made in assuming that $\tau_\sigma = \tau_R$ where τ_σ is the relaxation time in an elastic after-effect experiment and τ_R is the relaxation time as determined by a torsion pendulum.

6. The wire in a torsion pendulum is made of vanadium containing oxygen dissolved in solid solution. The period of the pendulum is 4 sec. At what temperature will the energy loss per second be a maximum? The lattice parameter of vanadium is 3.04 Å.
7. (*a*) What would be the relaxation time at room temperature of the metal in Problem 6?

   (*b*) Would it be feasible to make relaxation measurements with this material around room temperature using the elastic after-effect technique?
8. The zirconium specimen whose data are given in the figure associated with Problem 3 had an unrelaxed modulus of about 14×10^6 psi and a relaxed modulus of about 10×10^6 psi. Determine the three values for the relaxation time that one should obtain at 70°C if measurements are carried out at constant stress, constant strain, and with a torsion pendulum.

12 Phases

12.1 Basic Definitions. The concept of phases is very important in the field of metallurgy. A phase is defined as a *macroscopically homogeneous body of matter*. This is the precise thermodynamic meaning of the word. However this term is often used more loosely in speaking of a solid or other solution which can have a composition that varies with position and still be designated as a phase. For the present this fact will be ignored and the basic definition will be considered to hold. Let us consider a simple system: a single metallic element, for example copper, a so-called one-component system. Solid copper conforms to the definition of a phase, and the same is true when it is in the liquid and gaseous forms. However, solid, liquid, and gas have quite different characteristics, and at the freezing point (or at the boiling point) where liquid and solid (or liquid and gas) can coexist respectively, two homogeneous types of matter are present rather than one. It can be concluded that each of the three forms of copper – solid, liquid, and gas – constitutes a separate and distinct phase.

Certain metals, for example, iron and tin, are polymorphic (allotropic) and crystallize in several structures, each stable in a different temperature range. Here each crystal structure defines a separate phase, so that polymorphic metals can exist in more than one solid phase. As an example, consider the phases of iron, as given in Table 12.1. Notice that there are three separate solid phases for

Table 12.1 Phases of Pure Iron

Stable Temperature Range °C	*Form of Matter*	*Phase*	*Identification Symbol of Phase*
Above 2740	gaseous	gas	gas
1539 to 2740	liquid	liquid	liquid
1400 to 1539	solid	body-centered cubic	(delta)
910 to 1400	solid	face-centered cubic	(gamma)
Below 91C	solid	body-centered cubic	(alpha)

iron, each denoted by one of the Greek symbols: alpha (α), gamma (γ), or delta (δ). Actually, there are only two different solid iron phases since the alpha and delta phases are identical; both are body-centered cubic.

At one time, iron was believed to possess a third solid phase, the beta (β) phase, because heating and cooling curves indicated a phase transformation in the temperature range from 500° to 791°C. It has since been demonstrated that in this temperature range iron changes from the ferromagnetic to the paramagnetic form on heating. Since the crystal structure remains body-centered cubic throughout the region of the magnetic transformation, the β phase is no longer generally recognized as a separate phase. In other words, the magnetic transformation is now considered to occur inside the alpha phase.

Let us now consider alloys instead of pure metals. *Binary alloys*, two-component systems, are mixtures of two metallic elements, while *ternary alloys* are three-component systems: mixtures of three metallic elements. At this time, two terms that have been used a number of times should be clearly defined: the words *system* and *component. System*, as used in the sense usually employed in thermodynamics, or physical chemistry, is an isolated body of matter. The *components* of a system are the metallic elements that make up the system. Pure copper or pure nickel are by themselves one-component systems, while alloys formed by mixing these two elements are two-component systems. Metallic elements are not the only types of components that can be used to form metallurgical systems; it is possible to have systems with components that are pure chemical compounds. The latter type is perhaps of more interest to the physical chemist, but it is also of importance in the field of metallurgy and ceramics. A typical two-component system, with components that are chemical compounds, is formed by mixing common salt, NaCl, and water, H_2O. Another important compound that is usually considered as a component occurs in steels. Steels are normally considered to be two-component systems consisting of iron (an element) and iron carbide (Fe_3C), a compound.

When alloys are formed by mixing selected component metals, the gaseous state, generally speaking, is of little practical interest. In any case, there is only a single phase in the gaseous state, for all gases mix to form uniform solutions. In the liquid state, it sometimes happens, as in alloys formed by adding lead to iron, that the liquid components are not miscible, and then it is possible to have several liquid phases. However, in most alloys of commercial interest, the liquid components dissolve in each other to form a single liquid solution. In the discussion of phases and phase diagrams that follows, our attention will be concentrated on alloys in which the liquid components are miscible in all proportions.

The nature of the solid phases that occur in alloys will now be considered. Certain metals, for example, the lead-iron combination mentioned above, do not dissolve appreciably in each other in either the liquid or the solid state. Since this is the case, there are two solid phases, each of which is extremely close to being an absolutely pure metal (a component). The solid phases in most alloy systems, however, are usually solid solutions of either of two basic types. First, there are the *terminal solid solutions* (phases based on the crystal structures of the components). Thus, in binary alloys of copper and silver, there is a terminal solid solution in which copper acts as the solvent, with silver the solute, and another in which silver is the solvent and copper the solute. Second, in some binary alloys, crystal structures, different from those of either component, may form at certain ratios of the two components. These are called *intermediate phases*. Many of them are solid solutions in every sense of the word, for they do not have a fixed composition (ratio of the components) and appear over a range of compositions. A well-known intermediate solid solution is the so-called β phase in brass (copper-zinc) which is stable at room temperature in the range 47 percent Zn and 53 percent Cu to 50 percent Zn and 50 percent Cu, all measured in weight percentages. This intermediate phase is not the only one that appears in copper-zinc alloys. There are three other intermediate solid solutions, making in all six copper-zinc solid phases: the four intermediate phases and the two terminal phases.

In some alloys intermediate crystal structures are formed that are best identified as compounds. One intermetallic compound has already been discussed, namely, iron-carbide, Fe_3C.

12.2 The Physical Nature of Phase Mixtures. A brief explanation of the physical significance of "a system of several phases" is in order. To illustrate, consider a simple mixture of two phases. In general, these phases will not be separated into two distinct and separate regions, such as oil floating on water. Rather, the usual metallurgical two-phase system is comparable to an emulsion of oil droplets in a matrix of water. In speaking of such systems, it is common to refer to the phase (water) that surrounds the other phase as the *continuous phase*, and the phase (oil) that is surrounded, the *discontinuous phase*. It should be noticed, however, that the structure of a two-phase system may be so interconnected that both phases are continuous. An example was cited in Chapter 5, where it was mentioned that, under the proper conditions, a second phase, present in a limited amount, might be able to run through the structure as a continuous network along grain edges. The factors controlling this type of structure are described in Chapter 5.

In the solid state, a metallurgical system of several phases is a mixture of several different types of crystals. If the crystal sizes are small, surface energy effects become important and should rightfully be included in thermodynamic or energy calculations. In the following presentation, for the sake of simplicity, the surface energy will be neglected. It will also be assumed that all systems are removed from electric- and magnetic-field gradients so that their effects on our systems can also be ignored.

12.3 Thermodynamics of Solutions. It can be concluded that the phases in alloy systems are usually solutions – either liquid, solid, or gaseous. It is true that solid phases do sometimes form, with composition ranges so narrow that they are considered as compounds, but it is also possible to think of them as solutions of very limited solubility. In the discussion immediately following, the latter viewpoint is taken and all alloy phases are spoken of as solutions.

In general, the free energy of a solution is a thermodynamic property of the solution, or a variable that depends on the thermodynamic state of the solution (that is, system). In a one-component system (a pure substance) in a given phase, the thermodynamic state is determined uniquely if any two of its thermodynamic properties (variables) are known. Among the variables that are classed as properties are the temperature (T), the pressure (P), the volume (V), the enthalpy (H), the entropy (S), and the free energy (F). Thus, in a one-component system of a given mass and phase, if the two variables temperature and pressure are specified, the volume of the system will have a definite fixed value. At the same time, its free energy, enthalpy, and other properties will also have values that are fixed and determinable.

Solutions have additional degrees of freedom compared with pure substances, and it is necessary to specify values for more than just two properties to define the state of a solution. In general, temperature and pressure are employed as two of the required variables, while the composition of the solution furnishes the remainder. The number of independent composition variables is, of course, one less than the number of components. This can be seen if the composition is expressed in atom or mole fractions. In a three-component system, for example, the mole fractions are given by

$$N_A = \frac{n_A}{n_A + n_B + n_C} \quad N_B = \frac{n_B}{n_A + n_B + n_C} \quad N_C = \frac{n_C}{n_A + n_B + n_C}$$

where N_A, N_B, and N_C are the mole fractions, and n_A, n_B, and n_C are the actual number of moles of the A, B, and C components respectively. By definition of the mole fraction, we have the condition

$$N_A + N_B + N_C = 1$$

From this expression the value of any one of the mole fractions can be computed once the other two are known. There are only two independent mole fractions in a ternary system.

Most metallurgical processes occur at constant temperature and pressure, and, under these conditions, the state of a solution can be considered to be a function of its composition. Similarly, any of the state functions (properties), such as free energy (F), can be considered a function of only the composition variables. In the case of a three-component system, the total free energy (F) of a solution is written:

$$F = F(n_A, n_B, n_C) \quad \text{(temperature and pressure constant)}$$

where n_A, n_B, and n_C are the number of moles respectively of the A, B, and C components.

By partial differentiation, the differential of the free energy of a single solution of three components at constant pressure and temperature is:

$$dF = \frac{\partial F}{\partial n_A} \times dn_A + \frac{\partial F}{\partial n_B} \times dn_B + \frac{\partial F}{\partial n_C} \times dn_C$$

where the partial derivatives, such as $\partial F/\partial n_A$, represent the change in the free energy when only one of the components is varied by an infinitesimal amount. Thus, for a very small variation of component A, while the amounts of the components B and C in the solution are maintained constant, we have

$$\frac{dF}{dn_A} = \frac{\partial F}{\partial n_A}$$

In the present case, the partial derivatives are the partial molal free energies of the solution and are designated by the symbols $\bar{F}_A$, $\bar{F}_B$, and $\bar{F}_C$, so that the above equation can also be written:

$$dF = \bar{F}_A\, dn_A + \bar{F}_B\, dn_B + \bar{F}_C\, dn_C$$

The total free energy of a solution composed of n_A moles of component A, n_B moles of component B, and n_C moles of component C can be obtained by integrating the above equation.[1] This integration can be made quite easily in the following manner.

Let us start with zero quantity of the solution and form it by simultaneously adding infinitesimal quantities of the three components dn_A, dn_B, and dn_C. Each time that we add the infinitesimals, however, let us make the amounts of the components in the infinitesimals have the same ratio as the final numbers of moles of the components n_A, n_B, and n_C, so that

$$\frac{dn_A}{n_A} = \frac{dn_B}{n_B} = \frac{dn_C}{n_C}$$

1. Darken, L. S., and Gurry, R. W., *Physical Chemistry of Metals*. McGraw-Hill Book Co., New York, 1953.

If the solution is formed in this manner, its composition at any instant will be the same as its final composition. In other words, the composition will be constant at all times, and since the partial-molal free energies are functions of only the composition of the solution (at constant temperature and pressure), they will also be constant during the formation of the solution. Integration of

$$dF = \bar{F}_A\, dn_A + \bar{F}_B\, dn_B + \bar{F}_C\, dn_C$$

starting from zero quantity of solution, with the condition that $\bar{F}_A$, $\bar{F}_B$, and $\bar{F}_C$ are constant, gives us:

$$F = n_A \bar{F}_A + n_B \bar{F}_B + n_C \bar{F}_C$$

where n_A, n_B, and n_C are the number of moles of the three components in the solution.

Let us differentiate the above equation completely to obtain

$$dF = n_A\, d\bar{F}_A + \bar{F}_A\, dn_A + n_B\, d\bar{F}_B + \bar{F}_B\, dn_B + n_C\, d\bar{F}_C + \bar{F}_C\, dn_C$$

But we have already seen that the derivative of the free energy is

$$dF = \bar{F}_A\, dn_A + \bar{F}_B\, dn_B + \bar{F}_C\, dn_C$$

and the only way that both of these expressions for the derivative of the total free energy can be true is for

$$n_A\, d\bar{F}_A + n_B\, d\bar{F}_B + n_C\, d\bar{F}_C = 0$$

This equation gives a relationship between the number of moles of each component in a three-component solution and the derivatives of the partial-molal free energies. Similar relationships can be deduced for a two-component solution and for a solution of more than three components. Thus, for two components,

$$n_A\, d\bar{F}_A + n_B d\bar{F}_B = 0$$

and for four components

$$n_A\, d\bar{F}_A + n_B\, d\bar{F}_B + n_C d\bar{F}_C + n_D\, d\bar{F}_D = 0$$

The significance of these relationships in explaining the phenomena of polyphase systems in equilibrium will be shown in Section 12.6.

12.4 Equilibrium Between Two Phases. A binary (two-component) system with two phases in equilibrium will now be considered. The total free energy for the first phase, which we shall designate as the alpha (α) phase, is

$$F^{\alpha} = n_A{}^{\alpha}\bar{F}_A{}^{\alpha} + n_B{}^{\alpha}\bar{F}_B{}^{\alpha}$$

while that of the beta (β) phase is

$$F^{\beta} = n_A{}^{\beta}\bar{F}_A{}^{\beta} + n_B{}^{\beta}\bar{F}_B{}^{\beta}$$

Let a small quantity (dn_A) of component A be transferred from the alpha phase to the beta phase. As a result of this transfer, the free energy of the alpha phase will be decreased, while that of the beta phase will be increased. The total free-energy change of the system is the sum of these two changes and can be represented by

$$dF = dF^{\alpha} + dF^{\beta} = \bar{F}_A{}^{\alpha}(-dn_A) + \bar{F}_A{}^{\beta}(dn_A)$$

or

$$dF = (\bar{F}_A{}^{\beta} - \bar{F}_A{}^{\alpha})dn_A$$

But we have assumed that the two given phases are at equilibrium. This signifies that the state of the two-phase system is at a minimum with respect to its free energy (total free energy of the two solutions). The variation in the free energy for any infinitesimal change inside the system, such as the shift of a small amount of component A from one phase to the other, must, accordingly, be zero. Therefore,

$$dF = (\bar{F}_A{}^{\beta} - \bar{F}_A{}^{\alpha})dn_A = 0$$

Since dn_A is not zero, we are able to deduce the following important conclusion:

$$\bar{F}_A{}^{\alpha} = \bar{F}_A{}^{\beta}$$

and in the same manner it can be shown that

$$\bar{F}_B{}^{\alpha} = \bar{F}_B{}^{\beta}$$

The above quite general results are not restricted to systems of only two components or systems containing only two phases. In fact, it may be shown that, in the general case where there are M components with μ phases in equilibrium, the partial molal free energy of any given component is the same in all phases, or

$$\begin{array}{ccccccc}
\bar{F}_A^{\alpha} & = \bar{F}_A^{\beta} & = \bar{F}_A^{\gamma} & = & \cdots\cdots\cdots & = & \bar{F}_A^{\mu} \\
\bar{F}_B^{\alpha} & = \bar{F}_B^{\beta} & = \bar{F}_B^{\gamma} & = & \cdots\cdots\cdots & = & \bar{F}_B^{\mu} \\
\bar{F}_C^{\alpha} & = \bar{F}_C^{\beta} & = F_C^{\gamma} & = & \cdots\cdots\cdots & = & \bar{F}_B^{\mu} \\
\cdot\;\cdot\;\cdot & \cdot\;\cdot\;\cdot & \cdot\;\cdot\;\cdot & \cdot & \cdots\cdots\cdots & \cdot & \cdot\;\cdot \\
\bar{F}_M^{\alpha} & = \bar{F}_M^{\beta} & = \bar{F}_M^{\gamma} & = & \cdots\cdots\cdots & = & \bar{F}_M^{\mu}
\end{array}$$

where the superscripts designate the phases (solutions) and the subscripts the components. Further, since the partial molal free energy of any component is the same in all phases, we need no longer use the phase superscripts when specifying partial molal free energies of the components.

12.5 The Number of Phases in an Alloy System.

One-Component Systems. For a complete understanding of alloy systems, it is necessary to know the conditions determining the number of phases in a system at equilibrium. In a one-component system the conditions are well known. A glance at Table 12.1 shows that in the single-component system (pure iron) under the conditions of constant pressure, two phases can coexist only at those temperatures having a phase change: the boiling point, the melting point, and the temperatures at which solid-state phase changes occur. At all other temperatures only a single phase is stable.

Let us now consider the causes for phase changes in a one-component system. In particular, let us consider a solid-state phase change; the interesting allotropic transformation of white tin to gray tin. The former, or beta (β) phase has a body-centered tetragonal crystal structure that can be thought of as a body-centered cubic structure with one elongated axis. This is the ordinary, or commercial, form of tin and possesses a true metallic luster. The other, or alpha (α) phase, that is gray in color is the equilibrium phase at temperatures below 13.2°C. Fortunately, from both a practical and a scientific point of view, the transformation from white to gray tin is very slow at all temperatures below 13.2°C. With regard to the practical aspect, it is fortunate that tin does not turn instantly to gray tin once the temperature falls below the equilibrium temperature, because objects made of tin are usually ruined when the change occurs. Gray tin has a diamond cubic-crystal structure, a basically brittle structure, and this fact, coupled with the large volume expansion (about 27 percent) that accompanies the transformation, can cause the metal to disintegrate into a powder. On the other hand, the fact that the change occurs very slowly makes it possible to study the properties of a single element in two crystalline forms over a very wide range of temperatures. In this respect, the specific heat at constant

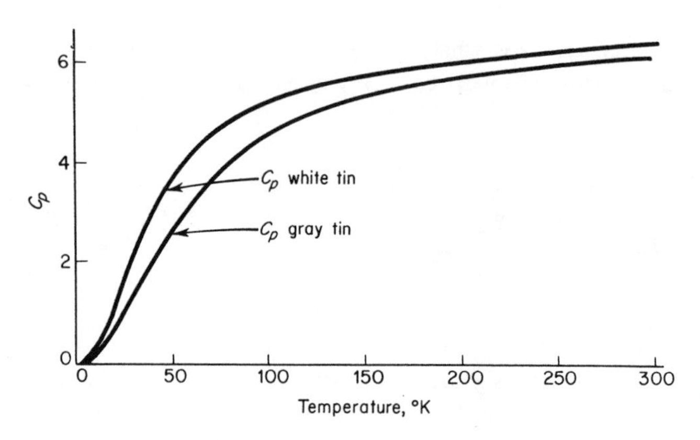

Fig. 12.1 C_p as a function of temperature for both forms of solid tin.

pressure of both the gray and the white forms of tin have been measured from ambient temperatures down to almost absolute zero. A plot of these data is shown in Fig. 12.1. The importance of this information lies in the fact that with it it is possible to compute the free energy of both solid phases as a function of temperature. Let us see how this is done.

By definition, the Gibbs free energy of a pure substance is

$$F = H - TS$$

where F = the free energy in calories per mole
T = the temperature in degrees Kelvin
H = the enthalpy (calories per mole)
S = the entropy in calories per mole per degree Kelvin

To evaluate the free energy of the substance at some temperature T, one needs to know both the enthalpy H and the entropy S. Both of these quantities can be found in terms of the specific heat at constant pressure C_p. Thus, in a reversible process at constant pressure, the heat exchanged between the system and its surrounding equals the enthalpy change of the system, or

$$q = dH$$

where q represents a small transfer of heat into or out of the system and dH is the accompanying enthalpy change of the system. But for one mole of a substance by the definition of specific heat

$$q = \int C_p dT$$

where C_p is the specific heat at a constant pressure (the number of calories required to raise one mole of the substance 1°K). Thus,

$$dH = C_p\,dT$$

and taking the enthalpy of the system at absolute zero as H_0, the enthalpy at any temperature T is, accordingly,

$$H = H_0 + \int_0^T C_p\,dT$$

Similarly, for a reversible process by the thermodynamic definition of entropy

$$dS = \frac{dq}{T} = \frac{C_p\,dT}{T}$$

Integration of this equation from 0°K to the temperature in question leads to

$$S = S_0 + \int_0^T \frac{C_p\,dT}{T} = \int_0^T \frac{C_p\,dT}{T}$$

The term S_0, which is the entropy at absolute zero, may be equated to zero, for at this temperature, assuming perfect crystallinity (a pure substance with no entropy of mixing) and lattice vibrations in their ground levels, there is no disorder of any kind. (This is actually a statement of the third law of thermodynamics which says that the entropy of a pure crystalline substance is zero at the absolute zero of temperature.)

With the aid of the equations stated in the above paragraph, we can now express the free energy of both the alpha and beta phases of tin as functions of the temperature and the specific heat at constant pressure.

$$F^{\alpha} = H_0^{\alpha} + \int_0^T C_p^{\alpha}\,dT - T\int_0^T \frac{C_p^{\alpha}\,dT}{T}$$

$$F^{\beta} = H_0^{\beta} + \int_0^T C_p^{\beta}\,dT - T\int_0^T \frac{C_p^{\beta}\,dT}{T}$$

These equations can be evaluated by means of graphical calculus, using the curves of Fig. 12.1. The results are shown in Fig. 12.2 where the free energies of the two phases are equal at the temperature 286.2°K. At temperatures below this value, gray tin has the lowest free energy, while for temperatures above it, white tin has the lowest free energy. Since the phase with the lowest free energy

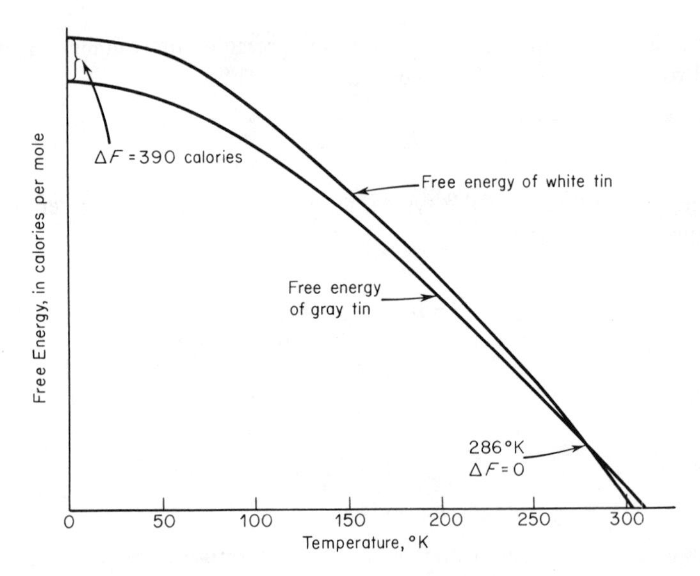

Fig. 12.2 Temperature-free energy curves for two solid forms of tin.

is always the most stable one, we see that gray tin is the preferred phase below 286.2°K, while white tin is the more stable at temperatures higher than this.

An interesting feature to notice in Fig. 12.2 is that the free energy of both phases decreases with increasing temperature. This is a general property of free-energy curves, which demonstrates the importance of the *TS* term in the free-energy equation. Figure 12.3 shows another method of interpreting the information given in Fig. 12.2. The curve marked ΔF in this figure represents the difference in free energy between the two solid phases of tin as a function of temperature. This curve is also the difference of the other two curves designated in the figure as ΔH and $T\Delta S$, where

$$\Delta H = H_0{}^{\beta} + \int_0^T C_p{}^{\beta}\, dT - H_0{}^{\alpha} - \int_0^T C_p{}^{\alpha}\, dT$$

and

$$T\Delta S = T\int_0^T \frac{C_p{}^{\beta}\, dT}{T} - T\int_0^T \frac{C_p{}^{\alpha}\, dT}{T}$$

At the temperature of transformation (286.2°K), ΔH and $T\Delta S$ are equal, so that ΔF is zero.

The above allotropic transformation can also be interpreted in the following manner. At low temperatures, the phase with the smaller enthalpy is the stable

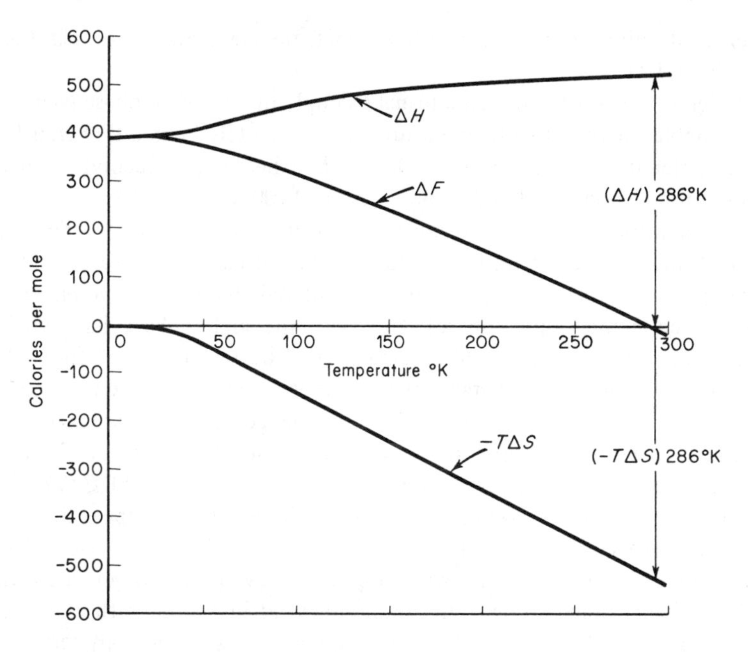

Fig. 12.3 Curves showing the relationship between ΔF, ΔH, and $-T\Delta S$ for the two solid phases of tin, where $\Delta F = \Delta H - T\Delta S$.

phase where, by definition, the enthalpy is

$$H = E + PV$$

where E is the internal energy, P the pressure, and V the volume. In solids, the PV term is usually very small compared with the internal energy. Thus, the stable phase at lower temperatures is that with the lower internal energy or with the tighter binding of the atoms. On the other hand, at high temperatures the $T\Delta S$ becomes more important, and the phase with the greater entropy becomes the preferred phase. Generally, this is a phase with a looser form of binding or with greater freedom of atom movement. In tin, it appears that, although the diamond cubic lattice (alpha phase) has the lower density, the binding forces between atoms in this structure are of such a nature that the atoms are more restrained in their vibratory motion than they are in the tetragonal, or beta, lattice.

In many cases, the structure of closest packing represents the more closely bound phase, and a more open structure the one with the greatest entropy of vibration. In this classification, where the high-temperature stable phase is body-centered cubic and the low-temperature phase a close-packed structure, such as face-centered cubic or close-packed hexagonal, the elements are Li, Na, Ca, Sr, Ti, Zr, Hf, and Tl.

The element iron presents an unusual example of allotropic-phase changes, in that the stable phase at low temperatures is body-centered cubic that transforms to face-centered cubic at 910°C. On further heating, a second solid-state tranformation to body-centered cubic occurs at 1400°C.

An explanation of these allotropic reactions that occur in iron has been given by Zener[2] and can be interpreted in the following broad general terms. The two competing solid phases of iron are face-centered cubic (gamma phase) and body-centered cubic (alpha phase). However, the alpha phase has two basic modifications: ferromagnetic and paramagnetic. The ferromagnetic form of the body-centered cubic phase is stable at low temperatures, while the paramagnetic form is stable at high temperatures. The change from ferromagnetism to paramagnetism takes place by what is known as a *second-order transformation*, occurring over a range of temperatures extending from about 500°C (773°K) to several hundred degrees above this temperature. It would thus appear that at low temperatures the competing phases in iron are actually ferromagnetic alpha and gamma, where the phase of the lowest internal energy and entropy is the former. The alpha phase is, accordingly, the stable phase at low temperatures, but gives way to the gamma phase with increasing temperatures as the entropy term of the difference in free energy between the phases becomes dominant. This transformation takes place at 910°C.

At about the same time that the gamma phase becomes the more stable form, the alpha phase loses its ferromagnetism, therefore for further increases in temperature the competing phases are gamma and paramagnetic alpha. In the paramagnetic form, the alpha phase has a greater internal energy and entropy than it does in the ferromagnetic form. This effect is large enough to reverse the relative positions of the alpha phase and the gamma phase. The gamma phase now becomes the one with the lowest internal energy and entropy and gives way to the alpha phase when the temperature exceeds 1400°C.

Two-Component Systems. As soon as one moves from a consideration of pure substances to systems of more than one component, he finds that he is involved in a study of solutions. The simplest type of multicomponent system is a binary one, and the least complex structure in such a system is a single solution. Some of the aspects of single-phase binary systems will now be considered.

In the discussion of diffusion phenomena it was shown that when a composition gradient exists in a solid solution, diffusion usually tends to occur in such a way that a uniform composition results. Let us briefly review the thermodynamic reasons for this effect.

2. Zener, C., *Trans. AIME,* **203** 619 (1955).

In a solution of n_A moles of component A, and n_B moles of component B, assuming constant pressure and temperature, the total free energy F' is

$$F' = n_A \bar{F}_A + n_B \bar{F}_B$$

or the free energy per mole is

$$F = N_A \bar{F}_A + N_B \bar{F}_B$$

where F is the free energy per mole, $\bar{F}_A$ and $\bar{F}_B$ the partial-molal free energies of the components, and

$$N_A = \frac{n_A}{n_A + n_B}$$

and

$$N_B = \frac{n_B}{n_A + n_B}$$

In the diffusion chapter it was shown that

$$\bar{F}_A = F^\circ_A + RT \ln a_A$$

$$\bar{F}_B = F^\circ_B + RT \ln a_B$$

where F°_A and F°_B are the free energies of the pure components A and B, and a_A and a_B the corresponding activities.

The free energy per mole of the solution can now be written:

$$F = N_A F^\circ_A + N_B F^\circ_B + RT(N_A \ln a_A + N_B \ln a_B)$$

In order to simplify the following presentation, let us assume that we are dealing with an ideal solution. In this case, the activities of the components are equal to the corresponding mole fractions or $a_A = N_A$ and $a_B = N_B$. The free energy of the solution is, accordingly,

$$F = N_A F^\circ_A + N_B F^\circ_B + RT(N_A \ln N_A + N_B \ln N_B)$$

The first two terms $(N_A F^\circ_A + N_B F^\circ_B)$ on the right-hand side of the above equation may be considered to represent the free energy of one total mole of the two components if the components are not mixed, or do not mix. In nonmiscible iron-lead considered previously, these two terms would be all that are required to specify the free energy of one total mole of iron and lead. The

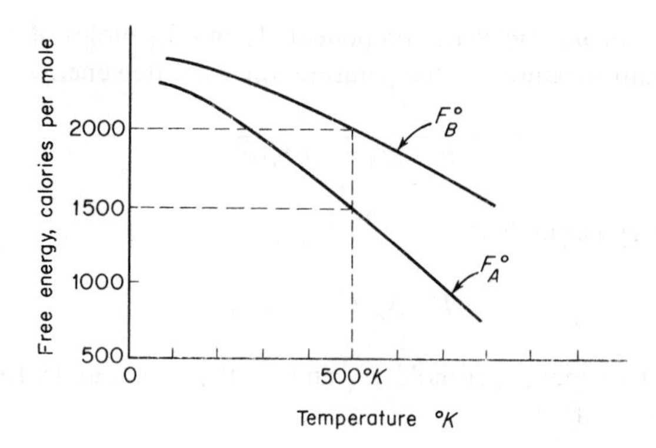

Fig. 12.4 Free energies of the elements are different functions of the absolute temperature.

other term, $RT(N_A \ln N_A + N_B \ln N_B)$, is the contribution of the entropy of mixing to the free energy of the solution. Notice that this term is directly proportional to the temperature and becomes increasingly important as the temperature is increased.

To illustrate the dependence of the free energy of an ideal solution on its composition, let us consider the following assumed data. Free-energy curves for two hypothetical pure elements A and B, which are assumed to be completely soluble in each other in the solid state, are shown in Fig. 12.4. Notice that in both curves the free energy is shown to decrease with increasing temperature, in agreement with the previously considered free-energy curves of the two solid phases of pure tin (Fig. 12.2). At 500° K, the curves of Fig. 12.4 show that the free energy of the A component (F°_A) is 1500 cal per mole, while that of the B

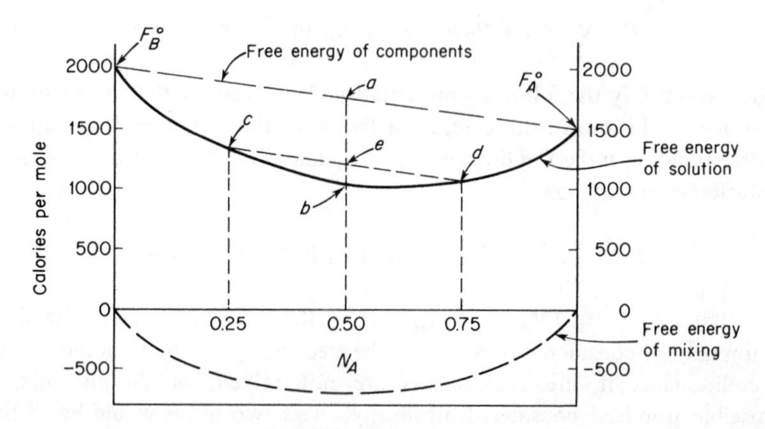

Fig. 12.5 Hypothetical free energy of an ideal solution.

Table 12.2 Data for Computing the Entropy of Mixing Contribution to the Free Energy of an Ideal Solution

Atom Fraction, N_A	$(N_A \ln N_A + N_B \ln N_B)$	$RT(N_A \ln N_A + N_B \ln N_B)$ *for Temperature* 500° K
0.00	0.000	000 cal/mole
0.10	– 0.325	– 325
0.20	– 0.500	– 500
0.30	– 0.611	– 611
0.40	– 0,673	– 673
0.50	– 0.690	– 690
0.60	– 0.673	– 673
0.70	– 0.611	– 611
0.80	– 0.500	– 500
0.90	–0.325	– 325
1.00	0.000	000

component ($F°_B$) is 2000 cal per mole. These two values are plotted in Fig. 12.5 at the right- and left-hand sides of the figure respectively. The dashed line connecting these points represents the terms $N_A F°_A + N_B F°_B = 1500N_A + 2000N_B$, or the free energy of the solution less the entropy-of-mixing term. The latter term can be evaluated with the aid of the data given in Table 12.2, where values of $(N_A \ln N_A + N_B \ln N_B)$ are given as a function of the atom fraction N_A (column 2), while values of the product $RT(N_A \ln N_A + N_B \ln N_B)$ are given in column 3 for the temperature 500° K (assuming R equals 2 cal/mole – °K). The data of column 3 are plotted in Fig. 12.5 as the dashed curve at the bottom of the figure. The free energy of the solution proper is shown as the solid curve of Fig. 12.5 and is the sum of the two dashed curves.

At this point let us assume that we have a diffusion couple, one part of which consists of half a mole of pure component A, and the other part of half a mole of pure component B. The average composition of this pair of metals, when expressed in mole fraction units, is $N_A = N_B = 0.5$, while its free energy at 500°K corresponds to point a in Fig. 12.5 and is 1750 cal per mole. The same quantity of A and B, when mixed in the form of a homogeneous solid solution, has the free energy of point b in Fig. 12.5, or 1060 cal per mole. Clearly, mixing one-half mole of component A with one-half mole of component B reduces the free energy of the pair of metals by 690 calories, the value of the contribution of the entropy of mixing to the free energy of the solution. This same 690 cal per mole is therefore the driving force that is capable of causing the components of the couple to diffuse.

Now suppose that, instead of two pure components, the above couple had consisted of two homogeneous solid solutions: one solid solution with a composition $N_A = 0.25$ and the other with the composition $N_A = 0.75$. At

500°K the free energy of the two portions of this couple are shown in Fig. 12.5 as points c and d respectively. If the average composition of the whole couple is again 0.5, the average free energy of the couple will lie at point e. A single homogeneous solid solution of the same average composition will have the free energy of point b. Here again we see that the homogeneous solid solution has the lower free energy and represents the stable state.

With arguments similar to the above, it is possible to show that any macroscopic nonuniformity of composition in a solution phase represents a state of higher free energy than a homogeneous solution. If the temperature is sufficiently high that the atoms are able to move at an appreciable rate, diffusion will occur that has, as its end result, a homogeneous solid solution. While the arguments used above were based on an average total composition $N_A = 0.5$, any other total or average composition would have given the same final results.

12.6 Two-Component Systems Containing Two Phases. Let us now take up the case of two-component systems that contain not a single phase, but two phases. In either of the two phases, an equation of the form

$$n_A \, d\bar{F}_A + n_B \, d\bar{F}_B = 0$$

must be satisfied. Now, if each term is divided by the quantity $n_A + n_B$, we can rewrite this relationship in the form

$$N_A \, d\bar{F}_A + N_B \, d\bar{F}_B = 0$$

where $n_A/(n_A + n_B) = N_A$ and $n_B/(n_A + n_B) = N_B$. If the phases are labeled alpha (α) and beta (β), then in the respective phases we have

$$\text{alpha phase:} \quad N_A^{\alpha} \, d\bar{F}_A + N_B^{\alpha} \, d\bar{F}_B = 0$$

$$\text{beta phase:} \quad N_A^{\beta} \, d\bar{F}_A + N_B^{\beta} \, d\bar{F}_B = 0$$

where the phase superscripts of the partial-molal free energies are omitted because, assuming equilibrium, the partial-molal free energy of either component is the same in both phases.

This pair of equations restricts the values of the mole fractions of the components in the solutions. These equations will be referred to as *restrictive equations.* In order to understand how restrictive equations work, let us consider an example.

Imagine that an alloy of copper and silver is formed by melting together an equal amount of each component. After the mixture has been formed, let it be frozen into the solid state and then reheated and held at 779°C long enough to attain equilibrium. If, at the end of this heating period, the alloy is cooled very rapidly to room temperature, we can assume that the phases which were stable at the elevated temperatures will be brought down to room temperature unchanged. Metallographic examination of such an alloy will reveal a structure consisting of two solid-solution phases, and a corresponding chemical analysis of the phases will show that the phases have the following compositions:

Alpha (α) *Phase*	*Beta* (β) *Phase*
$N_{Ag} = 0.86$	$N_{Ag} = 0.05$
$N_{Cu} = 0.14$	$N_{Cu} = 0.95$

Substituting these values in the restrictive equations gives us:

$$0.86 d\bar{F}_{Ag} + 0.14 d\bar{F}_{Cu} = 0$$

$$0.05 d\bar{F}_{Ag} + 0.95 d\bar{F}_{Cu} = 0$$

If the first equation is divided by the second, there results

$$\frac{0.86 d\bar{F}_{Ag}}{0.05 d\bar{F}_{Ag}} = \frac{-0.14 d\bar{F}_{Cu}}{-0.95 d\bar{F}_{Cu}}$$

or

$$\frac{0.86}{0.05} = \frac{0.14}{0.95}$$

This impossible result suggests that the only way that the above equations can be true is for both $d\bar{F}_{Ag}$ and $d\bar{F}_{Cu}$ to be zero, which means that at a constant temperature (779°C) and a constant pressure (one atmosphere) there can be no change in the partial-molal free energies when the two phases are in equilibrium. Further, since the partial-molal free energies are functions of only the composition of the phases, this, in turn, implies that the compositions of the two phases must be constant. No matter what the relative amounts of the two phases happen to be, as long as there is some of both phases present the composition in atomic percent of the alpha phase will be 86 percent Ag and 14 percent Cu, while that of the beta phase will be 5 percent Ag and 95 percent Cu.

12.7 Graphical Determinations of Partial-Molal Free Energies. A graphical method for determining the partial-molal free energies of a single binary solution is shown in Fig. 12.6. This figure shows the same molal-free energy curve previously given in Fig. 12.5. Suppose that one desires to determine the partial-molal free energies of the two components of the solution when the

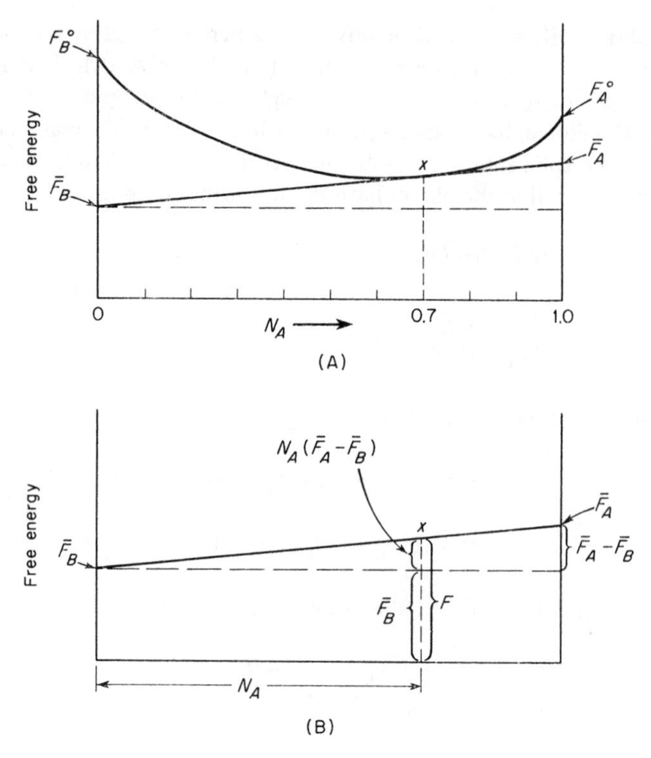

Fig. 12.6 Graphical determination of the partial-molal free energies.

composition has some arbitrary value, perhaps $N_A = 0.7$. A vertical line through this composition intersects the free-energy curve at point x, thereby determining the total free energy of the solution. Now if a tangent is drawn to the free-energy curve at point x, the ordinate intercepts of this tangent with the sides of the diagram (that is, at compositions $N_A = 0$ and $N_A = 1$) give the partial-molal free energies. The intercept on the left is $\bar{F}_B$ and that on the right $\bar{F}_A$. That these relationships are true is easily shown by the geometry of the figure, which is reproduced for clarity in Fig. 12.6B. Thus, the free energy of the solution is

$$F = \bar{F}_B + N_A(\bar{F}_A - \bar{F}_B)$$

or

$$F = N_A \bar{F}_A + (1 - N_A)\bar{F}_B$$

but

$$(1 - N_A) = N_B$$

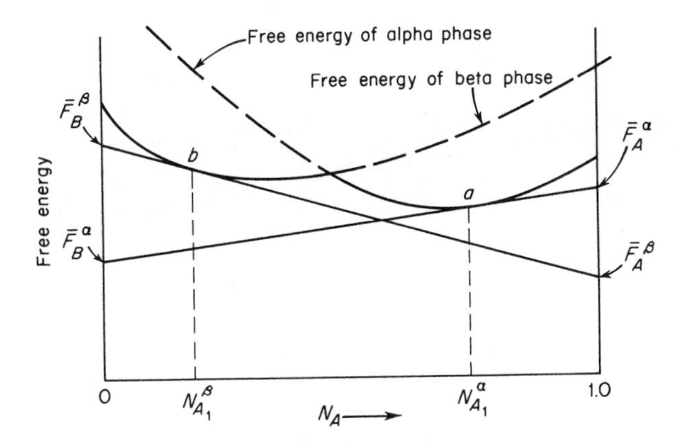

Fig. 12.7 Two phases having compositions of $N_{A_1}{}^\alpha$ and $N_{A_1}{}^\beta$ respectively cannot be in equilibrium.

and therefore

$$F = N_A \bar{F}_A + N_B \bar{F}_B$$

which is the basic equation for the free energy of a binary solution.

The fact that at constant pressure and temperature the compositions of the phases are fixed in a binary two-phase mixture at equilibrium can also be shown with the aid of free-energy diagrams of the phases plus the condition that the partial-molal free energies of each component be the same in both phases at equilibrium (that is, $\bar{F}_A{}^\alpha = \bar{F}_A{}^\beta$ and $\bar{F}_B{}^\alpha = \bar{F}_B{}^\beta$). Figure 12.7 shows hypothetical free-energy curves for two phases of a binary system at a temperature where the free-energy curves intersect. This intersection of the curves is a condition for the existence of two phases at equilibrium. Certainly, if the free-energy-versus-composition curve of one phase lies entirely below that of another phase, the lower one will always be the more stable and there can only be a single equilibrium phase. It is also important to notice that the free-energy curves of the type shown in Fig. 12.7 are functions of temperature. It is entirely possible for a pair of these curves to intersect at one temperature, but at some higher or lower temperature to have no points of intersection. In Chapter 13 more will be said on this subject where phase diagrams are discussed.

Now assume that we have an alloy containing both phases in which the composition of the alpha phase is $N_{A_1}{}^\alpha$ and that of the beta phase is $N_{A_1}{}^\beta$. Points a and b on the respective free-energy curves give the free energies of the alpha and beta phases. The partial-molal free energies of the two phases can be obtained by drawing tangents to points a and b. Their interception with the sides of the figure determine the desired quantities. As may be seen by examining Fig.

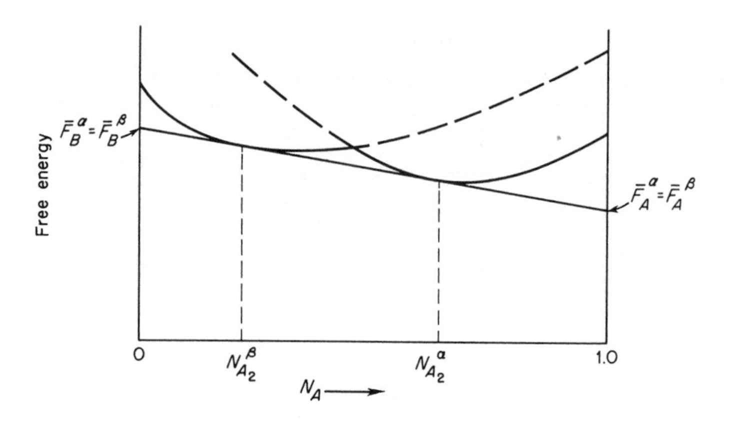

Fig. 12.8 Two phases in equilibrium.

12.7, the arbitrarily chosen compositions of the two phases result in partial-molal free energies of the components which are not the same in the two phases: $\bar{F}_A{}^\alpha \neq \bar{F}_A{}^\beta$ and $\bar{F}_B{}^\alpha \neq \bar{F}_B{}^\beta$. Actually the only way that two phases can have identical partial-molal free energies is for both phases to have the same tangent. In other words, as shown in Fig. 12.8, the composition of the two phases must have values corresponding to the intersection of a common tangent with the two free-energy curves. Since only one common tangent can be drawn to the curves of Fig. 12.8, the compositions of the alpha and beta phases must be $N_{A_2}{}^\alpha$ and $N_{A_2}{}^\beta$.

12.8 Two-Component Systems with Three Phases in Equilibrium. In two-component systems, three phases in equilibrium occur only under very restricted conditions. The reason for this is not hard to see, for the partial-molal free energies of the components must be the same in each of the three phases. In a graphical analysis like that used in the two-phase example, this means that we must be able to draw a single straight line that is tangent simultaneously with the free-energy curves of all three phases. Now it is necessary to notice that each free-energy curve varies with temperature in a manner different from that of the other phases and, in general, there will only be one temperature where it is possible to draw a single straight-line tangent to all three curves. It can be seen, therefore, that three phases in a binary system can only be in equilibrium at one temperature. Furthermore, the compositions of the phases are fixed by the points where the common tangent touches the free-energy curves. We must conclude that in a binary system when three phases are in equilibrium at constant pressure, the temperature and the composition of each phase is fixed.

At this point, it is well to recall the phase relations in single-component systems where it was observed that, under the condition of constant pressure, two phases could be in equilibrium at only those fixed temperatures at which phase changes occur. An analogous situation exists in two-component systems, for the temperatures at which three phases can be in equilibrium are also temperatures at which phase changes take place. Associated with each of these three-phase reactions is a definite composition of the entire alloy. At this composition, a single phase can be converted into another of two phases. In certain alloy systems, the single phase is the stable phase at temperatures above the transformation temperature, while two phases are stable below it. In other systems the reverse is true.

As an example of a three-phase reaction, we may take the lead-tin alloy containing 61.9 percent Sn and 38.1 percent Pb (expressed in weight percent). This alloy forms a simple liquid solution at temperatures above 183°C, and at temperatures below 183°C it is stable in the form of a two-phase mixture. Each of the latter is a terminal solid solution, the compositions of which are alpha phase 19.2 percent Sn and beta phase 97.5 percent Sn. A reaction of this type, when a single liquid phase is transformed into two solid phases, is known as an *eutectic reaction*. The temperature at which the reaction takes place is the eutectic temperature, and the composition which undergoes this reaction is the eutectic composition (in the present example 61.9 percent Sn and 38.1 percent Pb).

The eutectic reaction is only one of several well-known three-phase transformations, as can be seen by examining Table 12.3.

In the four transformation equations in Table 12.3, the reactions proceed to the right when heat is removed from the system, and to the left when heat is added to the system. These reactions will be treated separately in Chapter 14.

12.9 The Phase Rule. We will now summarize most of the pertinent points of the preceding sections. For this purpose, consider Fig. 12.9, a phase diagram of a one-component system where the variables are the pressure and the temperature. A point such as that marked a, lying within the limits of the gaseous phase, represents a gas whose state is determined by the temperature T_a and the pressure P_a. Changing either or both variables so as to bring the gas to

Table 12.3 Types of Three-Phase Transformations that can Occur in Binary Systems.

Type of Transformation	*Nature of the Phase Transformation* Phase A	Phase B	Phase C
Eutectic	Liquid	⇌ Solid Solution	+ Solid Solution
Eutectoid	Solid Solution	⇌ Solid Solution	+ Solid Solution
Peritectic	Liquid Solution	+ Solid Solution	⇌ Solid Solution
Monotectic	Liquid Solution	⇌ Liquid Solution	+ Solid Solution

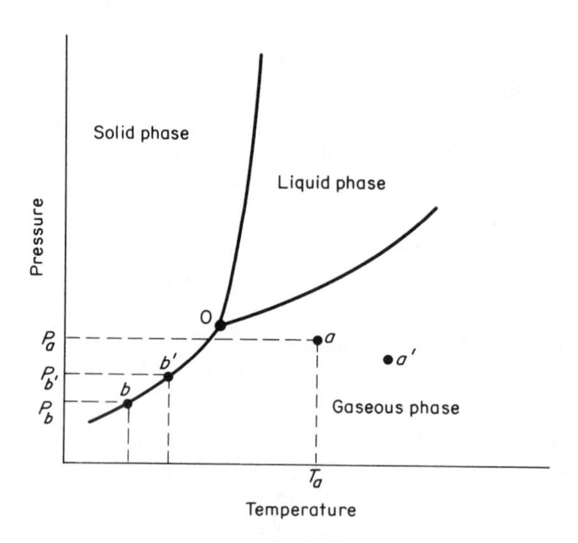

Fig. 12.9 A single-component phase diagram.

some other arbitrary state (a') is entirely possible. As long as one stays within the limits of the region indicated as belonging to the gaseous phase, it is evident that there are two degrees of freedom. That is, both the temperature and the pressure may be arbitrarily varied. This statement also applies to the other two single phase regions, liquid and solid. In summary, when a single-component system exists as a single phase, there are two degrees of freedom. This conclusion is stated in tabular form along the first horizontal line in Table 12.3.

Now consider the case where there are two phases in a single-component system. This can only occur if the state of the system falls along one of the three lines in Fig. 12.9 separating the single-phase fields. For example, point b represents a solid-vapor (gas) mixture in equilibrium at temperature T_b and pressure P_b. If now the temperature is changed to T_b', the pressure has to be changed to exactly P_b'. If this is not done, the two-phase system becomes a single-phase system; either all solid or all gas. It can be concluded that there is only one degree of freedom. This result is tabulated on the second line in Table 12.3.

The last possibility for the single-component system is illustrated by the third line in Table 12.3 and corresponds to the case where three phases are in equilibrium in a single component system. This can only occur at the triple point designated by the symbol 0 in Fig. 12.9. This three-phase equilibrium only occurs at one specific combination of temperature and pressure and there are, accordingly, no degrees of freedom.

Table 12.3 next lists the various possibilities for a two-component system. Note that for a single phase in a two-component system there are three degrees of freedom. These are normally considered to be the temperature, the pressure, and the composition of the phase. In the case of two phases in equilibrium in a binary system there are two degrees of freedom. This conforms to the conclusions of Section 12.6, where it was demonstrated that at a chosen temperature and pressure where two phases could be maintained in equilibrium, the compositions of the phases were automatically determined. Of course, it is possible to vary the temperature and pressure, but such variations bring about determinable changes in the composition of the phases.

Next, consider a binary system with three phases in equilibrium. Section 12.8 has shown that here, if the pressure is fixed, three-phase equilibrium can only occur at one temperature. Varying the pressure naturally changes this temperature. However, this result implies that there is only one degree of freedom.

Finally, the case of four phases in a two-component system is equivalent to a triple point in a single-component system. Four-phase equilibrium can only occur at a single combination of temperature and pressure, and the compositions of all phases are fixed. There are, accordingly, no degrees of freedom.

This type of analysis could be extended to systems with larger numbers of components, but it is simpler to invoke the use of the Gibbs phase rule, named after J. Willard Gibbs.[3] This relationship can be deduced from the data in Table 12.4. Note that if along any one of the horizontal lines in the table one adds the number of phases P (given in column 2) to the number of degrees of freedom F (given in column 3), the result is always the number of components, C, plus a factor of 2. Thus we have

$$P + F = C + 2$$

The phase rule is often of very great value in determining the factors involved in phase equilibria.

Table 12.4 The Relative Number of Phases and Degrees of Freedom in One- and Two-Component Systems.

Number of Components C	*Number of Phases* P	*Degrees of Freedom* F
1	1	2 (T, P)
1	2	1 (T or P)
1	3	0
2	1	3 (T, P, N_A or N_B)
2	2	2 (T, P)
2	3	1 (T or P)
2	4	0

3. *The Collected Works of J. Willard Gibbs*, Vol. 1, p. 88. Longmans, Green and Co., Inc., New York, 1931.

12.10 Ternary Systems. The phase structures of binary alloys are our primary concern here, but the phases in ternary alloys will also be considered. In general, ternary alloys are much more difficult to understand than binary alloys and, as a result, one is usually forced into a study of binary systems, in spite of the fact that many practical problems involve three or more components.

Analogous to the relations already considered for single and two-component systems, at certain fixed temperatures four-phase transformations may occur in which the compositions of the phases are fixed throughout the reactions. Furthermore, in every case there are certain definite alloy compositions at which the entire alloy undergoes these phase transformations. In these reactions, either a single phase is changed into three other phases, or two phases are changed into two different phases. A typical example of the former type of transformation is a ternary eutectic where a single liquid solution changes into three solid solutions.

The above paragraph points out the similarity between four-phase systems in ternary alloys and three-phase systems in binary alloys. In the same manner, three-phase ternary systems are analogous to two-phase binary systems. Thus, at constant temperature and pressure, a three-phase system in a ternary alloy will have fixed compositions of the phases. This fact does not mean that the composition of the alloy as a whole has one value, but that the phases, which may be in any proportion, have fixed compositions.

Three-component alloys may also have structures containing two or even a single phase. It is important to notice that in a ternary alloy containing two phases, the compositions of the phases are not fixed. This statement is true of binary systems, but not of ternary.

Problems

1. (*a*) Figure 17.4 shows the microstructure of a high carbon steel specimen. What are the phases in this system?

   (*b*) What are the phases in the silver-copper alloy shown in Fig. 14.15?
2. Is a fog a single phase? Explain.
3. The elements titanium, zirconium, and hafnium can form a solid solution in all proportions of the elements. In a solid solution containing 100 gm. of each of these elements, what is the mole fraction corresponding to each? The atomic weights of titanium, zirconium, and hafnium are 47.9, 91.2, and 178.6 gm/mole respectively.
4. A titanium-zirconium-hafnium alloy contains 15 atomic percent titanium and 45 atomic percent hafnium. How many grams of zirconium will there be in 500 gms of this metal?
5. Magnesium melts at 650°C with a heat of fusion of 88 cal/gm, and its atomic weight is 24.3 gm/mole.

(*a*) Compute the entropy change associated with the freezing of one mole of magnesium.

(*b*) What is the corresponding change in internal energy?

6. (*a*) With reference to Fig. 12.5 and the data given in this illustration, compute the free energy change associated with the formation of one mole of solid solution starting with 0.25 moles of pure component A and 0.75 moles of pure component B.

   (*b*) What is the internal energy change associated with this formation of a solid solution?
7. At the end of Section 12.5 the case of a diffusion couple consisting of two homogeneous solid solutions is discussed. Using the data given in Fig. 12.5, compute the magnitude of the free energy change associated with converting the original pair of solid solutions of compositions $N_A = 0.25$ and $N_A = 0.75$ respectively into a single homogeneous solid solution of composition $N_A = 0.5$.
8. Graphically determine the partial-molal free energies $\bar{F}_A$ and $\bar{F}_B$ corresponding to a solid solution of composition $N_A = 0.25$. Then, with the aid of the equation $F = N_A\bar{F}_A + N_B\bar{F}_B$ determine the free energy of one mole of the solution. Finally, check your answer against the value of F determined directly from the curve in Fig. 12.5.
9. With the aid of the phase rule, construct a table like Table 12.3 for the case of ternary or three-component systems. As in Table 12.4, indicate the nature of the degrees of freedom.

13 Nucleation and Growth Kinetics

Structural changes in metallic systems usually take place by nucleation and growth whether a change in state, a phase change within one of the three states, or a simple structural rearrangement within a single phase is involved. A number of examples have already been discussed and others will be covered in later chapters. Thus in Section 4.5, the subject of the nucleation of dislocations was considered. In the broadest sense, the nucleation of dislocation loops and their growth during straining represents a structural change. Nucleation and growth during recrystallization is discussed in Section 7.13; in precipitation hardening in Sections 9.4 and 9.5; in solidification in Sections 15.2 and 15.3; in the formation of deformation twins in Sections 16.4 and 16.6; in martensite transformations in Sections 16.19 and 16.20; and in the eutectoid transformation of steel to form pearlite in Sections 17.2 and 17.3.

Because metallurgical transformations depend so strongly on the processes of nucleation and growth, this chapter will consider in greater detail the conditions controlling the formation of nuclei, their rate of formation, and the rate of their growth.

13.1 Nucleation of a Liquid from the Vapor. The simplest case that can be considered is that involving the formation of liquid droplets in a vapor phase. There are a number of reasons why this is so. In an ordinary nucleation process, a small region of a new phase is created within another phase. Associated with such an event is the formation of a boundary separating the two phases. The nature of this boundary is often difficult to define when the newly formed particle is very small and contains very few atoms. Fortunately, the conditions controlling the formation of nuclei are not normally sensitive to the properties of the particles when they are extremely small. As indicated earlier in the precipitation hardening section, there is always a critical particle size above which the particles become stable, and below which they are unstable. Below this critical size the particles are normally called embryos, and above it, nuclei. When an embryo has grown to approximately the critical size it is usually of sufficient size that one can treat the particle as macroscopic. Thus in the region of primary interest, it is often possible to ignore the complications associated with very small groups of atoms and consider the nucleation problem from a simple macroscopic point of view. It is worth noting that the formation of a solid crystalline nucleus from a vapor or a liquid is more complicated than the formation of a liquid droplet from its vapor. This has to do with the shape of the particle as it grows. In a crystal the surface energy may be a function of the orientation of the surface, and a crystal nucleus of minimum surface energy may have a complicated shape. Furthermore, the growth of a crystal nucleus is also a more complicated process than the growth of a liquid droplet. In a solid, the atoms have to fit into the fixed pattern of the crystal lattice. The problem of attachment of atoms to the crystal surface has to consider this fact, and when the crystal grows under a relatively low driving force, the growth may occur by the addition of atoms to steps on the crystal surface.

Another simplifying factor in considering the nucleation of a liquid from a vapor that also applies to the nucleation of crystals from either the vapor or liquid phases is that the nucleation process does not involve strain energy. In solid-solid reactions, the newly formed phase often does not fit well into the matrix surrounding it, with the result that both the nuclei and the matrix usually become strained.

Finally, with regard to vapor-liquid transformations, it may be mentioned that the vapor phase itself is capable of being modeled in a relatively simple fashion using the concept of the ideal gas and the kinetic theory of gaseous collisions.

The nucleation of liquid droplets in a vapor will not be treated in detail. However, the basic principles and primary results will be covered. For a more extensive treatment, one is referred to a standard text on transformations.[1] At this point, attention is called to the fact that, in Section 9.5, a simplified equation was written for the variation of the free energy change, ΔF_r, with increase in radius r of a particle during precipitation hardening. Since this equation did not consider the role of strain energy in the formation of the particle, it is applicable to the case of precipitation of a liquid from a vapor. We may therefore write, by analogy,

$$\Delta F_n = 4/3\pi r^3 \frac{\Delta f^{vl}}{v_l} + 4\pi r^2 \gamma$$

where Δf^{vl} is the chemical free energy change per atom associated with the transfer of atoms from the vapor to the liquid phases, v_l is the volume of an atom in the liquid phase, r is the radius of the droplet, and γ is the specific surface free energy. This expression can also be written in terms of n, the

1. Christian, J. W., *The Theory of Transformations in Metals and Alloys*, Pergamon Press, London, 1965.

number of atoms in the particle, as follows

$$\Delta F_n = \Delta f^{vl} n + \eta \gamma n^{2/3}$$

where $n = 4\pi r^3/v_l$ and η is a shape factor equal to $(4\pi)^{1/3}(3/v_l)^{2/3}$. A diagram showing the variation of all the terms in this equation, with n the number of atoms, is shown in Fig. 13.1. Note that the free energy of the particle ΔF_n passes through a positive maximum and then decreases and eventually becomes negative. As might be expected, this is the same behavior shown for the variation of ΔF_r, with r in Fig. 9.8.

In Fig. 13.1 it is assumed that the vapor is supersaturated. This is equivalent to assuming that it lies at a position on a temperature-pressure diagram like point a in Fig. 13.2. At this temperature and pressure, liquid is the stable phase and

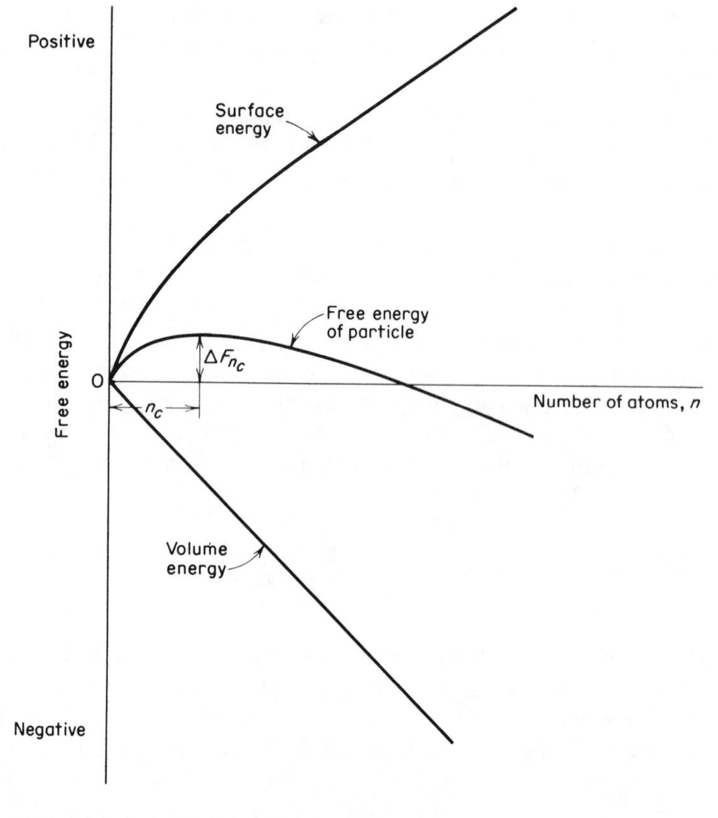

Fig. 13.1 Variation of the free energy per atom in a precipitate particle with the number of atoms in the particle.

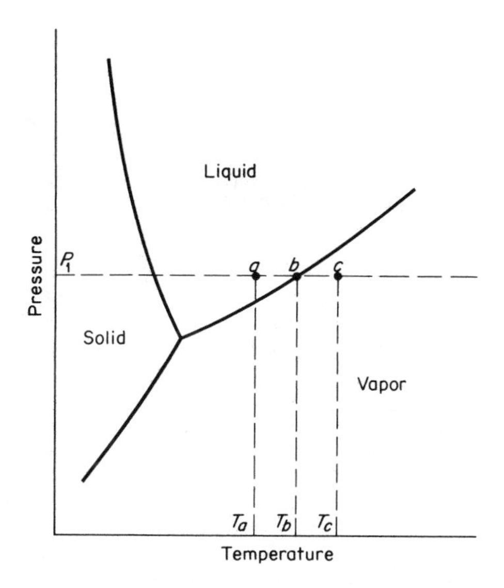

Fig. 13.2 A one-component (temperature-pressure) phase diagram.

the chemical free energy change per atom, Δf^{vl}, is negative. On the other hand, if the vapor lies at point c, the vapor phase is the stable phase and Δf^{vl} is positive, and the free energy of the particle is always positive and increases monotonically with the number of atoms, as shown in Fig. 13.3. It is therefore evident that, in this case, there is no tendency for liquid droplets to continue to grow and to form large stable particles. This does not mean, however, that vapor atoms cannot come together and combine to form a distribution of very small embryos. In the vapor, the random motion of the atoms causes local fluctuations in concentration or density. Certain of these fluctuations may bring a sufficient number of atoms close enough together so that, in a small volume, the atoms may tend to arrange themselves in a structure more characteristic of the liquid phase than of the vapor phase. Such a fluctuation is called a *heterophase fluctuation*. These are basically the sources from which embryos may be considered to derive. In a stable phase, such as at point c in Fig. 13.2, the probability of observing an embryo of a given size containing n atoms may be shown to be proportional to $e^{-\Delta F_n/kT}$, where ΔF_n is the free energy increase associated with the formation of the particle. As a consequence, the numbers of particles containing n atoms is given by

$$N_n = Ce^{-\Delta F_n/kT}$$

where N_n is the number of these particles, ΔF_n is the free energy required to form a particle, C is a slowly varying function of n, and k and T have their usual

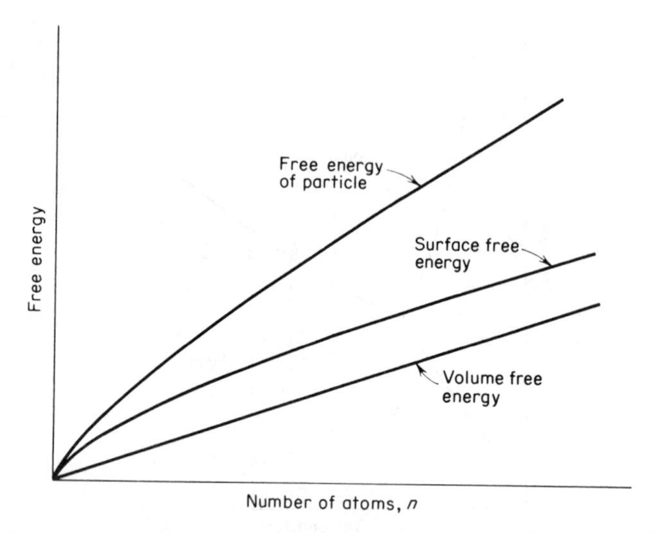

Fig. 13.3 Variation of the total surface and volume free energies with n for a particle at a temperature and pressure corresponding to point c in Fig. 13.2.

significance. To a reasonable approximation, C may be assumed equal to N, the total number of atoms under consideration, provided that the number of particles is small compared to N. We may therefore write

$$N_n = Ne^{-\Delta F_n/kT}$$

Attention is now called to the similarity of this equation to that giving the equilibrium number of vacancies in a crystal, which may be found on page 254. The above equation could be derived in an analogous fashion to that used in obtaining the vacancy equation.[2] To do this one needs to consider the entropy of mixing associated with the division of atoms between the vapor phase and a range of particles of various sizes.

Let us now consider the variation of N_n with n, as predicted by the above equation, when the vapor phase is stable. For convenience, we shall consider that the vapor is very close to being in equilibrium with the liquid phase, so that it lies nearly at point b in Fig. 13.2. In this case, the change in chemical free energy per atom, on going from the vapor to the liquid, is approximately zero, so that we may ignore the term containing Δf^{vl} in the equation for the free energy of the particle, which becomes

$$\Delta F_n = \eta\gamma n^{2/3}$$

2. *Ibid.*

and

$$N_n = Ne^{-\eta\gamma n^{2/3}/kT}$$

This latter expression can be easily evaluated using characteristic approximate values for the various parameters. Those that are given roughly correspond to those for tin. Thus let us assume that γ, the surface free energy, is 500 ergs/cm^2, the boiling temperature is 2550°K, and the diameter of the atom in the liquid phase is 3 Å. In this case η, the geometrical factor $\{(4\pi)^{1/3}(3v_1)^{2/3}\}$, is about 4.3×10^{-15} cm^2, and $\Delta F_n = 2.15 \times 10^{-12} n^{2/3}$ ergs. The corresponding variation of ΔF_n with n is shown as the upper graph in Fig. 13.4. The lower diagram

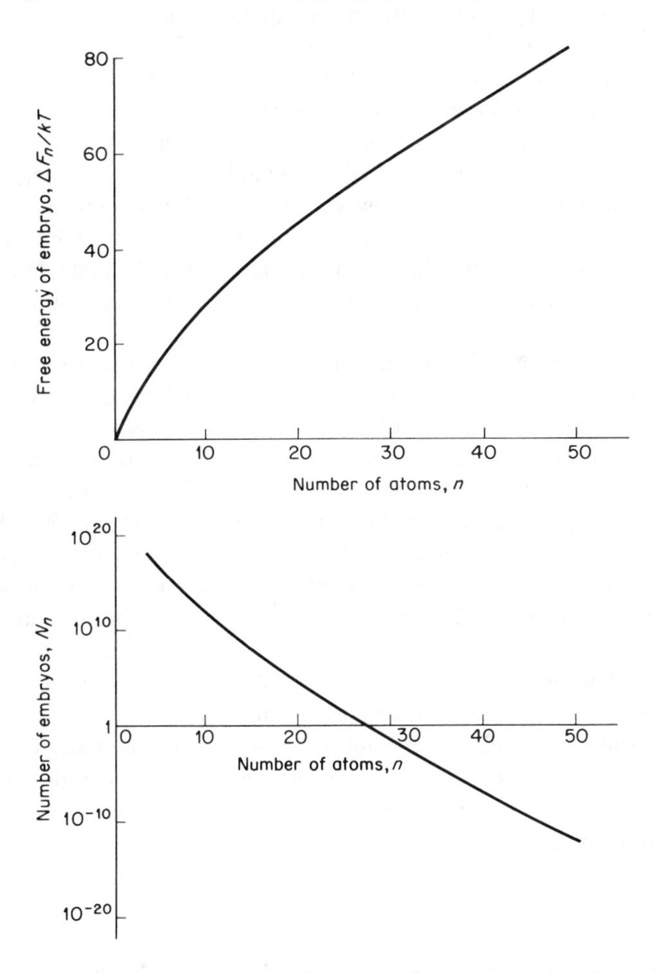

Fig. 13.4 The free energy of an embryo and the number of embryos per mole as a function of n, the number of atoms in an embryo, for a hypothetical metal vapor just above the boiling point of the metal.

corresponds to the variation of N_n with n, assuming that the total number of atoms in the assembly is that of one mole, or about 10^{24}. Note that in this latter diagram semi-logarithmic coordinates have been used. The normal procedure in demonstrating a curve of this type is to use simple linear coordinates. However, as may be readily seen in Fig. 13.4, N_n varies by over 20 orders of magnitude in the interval shown in the diagram. It is very difficult to represent this fact using a linear system of coordinates.

While Fig. 13.4 is basically qualitative, it does show several very interesting features. First, the number of embryos is very large when the size of the embryo is small. Second, as the size of the embryo increases or the value of n becomes larger, the number of embryos falls very rapidly. For the example given in Fig. 13.4, while there should be about 10^{10} embryos containing about 12 atoms in the assembly, there should be only one embryo containing about 25 atoms. For any embryo larger than this there is only a fractional chance of it existing. The significant point is that in a stable assembly, embryos varying through a wide range of sizes can exist. However, this situation is dynamic. Individual embryos are constantly growing and shrinking in size as they either add or lose atoms. Nevertheless, the distribution implied in Fig. 13.4 is stable. Furthermore, it should be mentioned that there is no tendency for the larger embryos to grow to form nuclei.

Now let us consider the supersaturated vapor corresponding to point a in Fig. 13.2. Here the system is metastable. It wants to transform to the liquid state, but the liquid has to be nucleated. The basic problem is how to represent the distribution of particles according to their sizes, as was done for the stable state in Fig. 13.4. At this point we shall introduce the symbol Z_n to represent the number of particles containing n atoms in a metastable assembly. It is equivalent to N_n in the stable assembly. We know that the larger embryos have to grow to form nuclei. This implies, in effect, that there is a constant net rate of particle growth. In the stable state, since the number of embryos of any one size is fixed, as many of these embryos leave a size class as enter it. In the metastable state, more embryos of a given size class will increase their size by gaining an atom than will decrease their size by losing an atom in a given interval of time. The net result is a steady progression of embryos that increase their sizes to eventually become nuclei. While this implies a basic difference in the variation of N_n with n for the two cases, the earlier theories assumed that in both cases the same relation holds, or that

$$Z_n = Ne^{-\Delta F_n/kT}$$

where for the supersaturated vapor the variation of ΔF_n with n is that given in Fig. 13.1. The fact that above n_c this figure shows that ΔF_n decreases and eventually becomes negative implies that given sufficient time, the number of

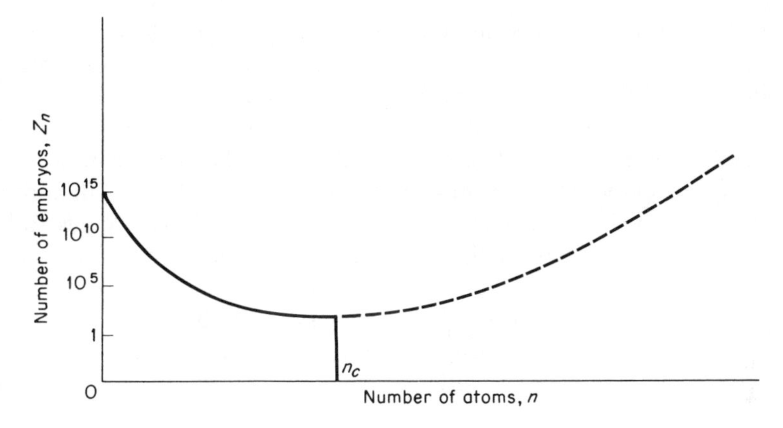

Fig. 13.5 The number of embryos as a function of the number of atoms in an embryo for a system at point *a* in Fig. 13.2 according to the Volmer-Weber theory.

particles of very large size should approach infinity. This, of course, implies that the entire assembly would be effectively in the liquid state. However, in most nucleation problems of interest, one normally starts with a phase that has just been brought into the supersaturated condition. In the present case this would signify a vapor that was originally at point T_c in Fig. 13.2 whose temperature had been lowered to T_a. For this case, it was originally proposed by Volmer and Weber[3] that the distribution of embryos might correspond to the solid curve shown in Fig. 13.5. This curve is equivalent to the assumption that the distribution function is determined by the variation of ΔF_n (shown in Fig. 13.1) up to the critical embryo size at n_c. Above this value of n it is assumed that the particles grow rapidly and are effectively removed from the problem. Thus, a nucleus can be considered to be formed whenever an embryo of the critical size captures an additional atom. The rate of nucleation would, therefore, be proportional to the number of critical-size embryos, Z_{n_c}, and the rate of condensation of vapor atoms on these embryos. The former quantity is given by $Ne^{-\Delta F_{n_c}/kT}$, while the latter is proportional to the surface area of the critical sized embryo and the probability per unit area per unit time of a vapor atom condensing on the surface. Accordingly we have

$$I = q_o O_c N e^{-\Delta F_{n_c}/kT}$$

where I is the number of stable nuclei created per second, N is the total number of atoms in the assembly, O_c is the area of the critical embryo, q_0 is the

3. Volmer, M., and Weber, A., *Z. Phys. Chem.*, **119** 227 (1926).

probability per unit time per unit area of capturing a vapor atom, and ΔF_{nc} is the free energy change associated with forming an embryo of critical size. It should be noted that q_0 can be evaluated in terms of the kinetic theory collision factor $p(2\pi mkT)^{1/2}$, where p is the pressure and m is the atomic mass.

A basic precept of the Volmer-Weber theory is that after a nucleus has formed it no longer has to be considered with respect to the formation of the nuclei that follow it. Of course the atoms that have joined to make this nucleus are no longer available for the formation of other nuclei. In effect, the theory assumes that additional atoms are continually added to the system to compensate for those removed from the system. A theory of this type is therefore designated a quasi-steady state theory. This kind of theory conforms best to the experimental conditions when the number of nuclei is small compared to the total number of atoms.

13.2 The Becker-Döring Theory. The Volmer-Weber theory, which served as the basis of the preceding discussion, assumes that once a nucleus of the critical size obtains an additional atom, it always grows into a stable nucleus. This assumption is not strictly true. The addition of an atom, or even several atoms, to a critical nucleus will certainly tend to make it become more stable. However, this increase in stability has to be small. This is because, at n_c, the free energy reaches a maximum and, at the same time, $d\Delta F_n/dn$ passes through zero. Therefore an embryo that has grown slightly beyond the critical size always has a nearly equal chance of shrinking back and becoming smaller. In formulating their nucleation theory, Becker and Döring recognized this fact. They also postulated that in the quasi-steady state, the number of embryos of a given size should remain effectively constant, although, individually, embryos might either grow or shrink in size. The distribution function Z_n, giving the number of embryos corresponding to a given number of atoms, was assumed to be equal to the number of embryos predicted by the Volmer-Weber theory for very small-sized embryos, and for very large ones, Z_n was assumed to approach zero. Furthermore, unlike the Volmer-Weber theory, Z_n was not assumed to go to zero just above n_c. The difference in the assumptions with regard to Z_n in the two theories is illustrated in simple schematic fashion in Fig. 13.6. Finally, with regard to Z_n, the Becker-Döring theory[4] does not require an exact specification for the distribution function near the critical nucleus size.

The Becker-Döring theory postulates that the nucleation rate I must equal the difference in the two rates at which embryos of a given size containing n atoms grow into embryos containing $n + 1$ atoms, and the inverse rate corresponding to the reversion of embryos containing $n + 1$ atoms into those containing n atoms. This follows from the assumption that Z_n is independent of time. We may

4. Becker, R., and Döring, W., *Ann. Phys.*, **24** 719 (1935).

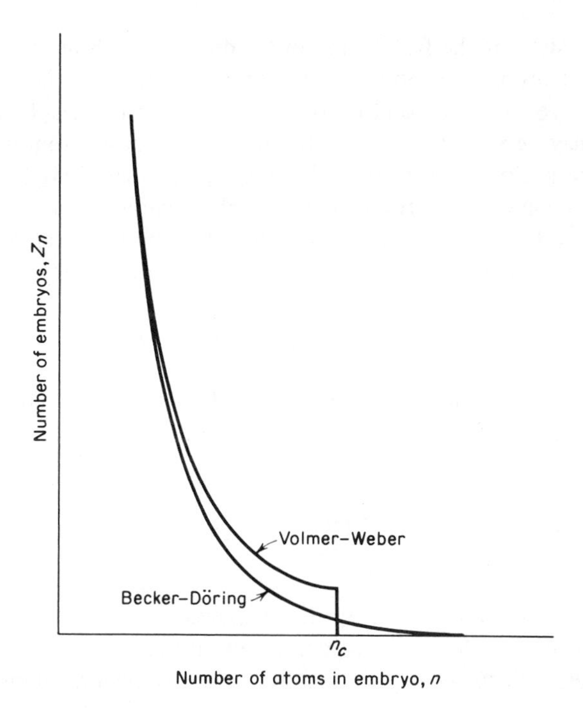

Fig. 13.6 A qualitative comparison of the Becker-Döring and Volmer-Weber distribution functions.

express this relation in the form

$$I = i_{n \to n+1} - i_{n+1 \to n}$$

where I is the number of nuclei formed per sec per cm^3, $i_{n \to n+1}$ is the rate of conversion of embryos of size n to size $n + 1$, and $i_{n+1 \to n}$ is the corresponding opposite rate. For the case of a liquid nucleating in a vapor, this equation may also be written

$$I = q_o O_n Z_n - q_{n+1} O_{n+1} Z_{n+1}$$

where Z_n is the number of embryos containing n atoms, q_0 is the probability per cm^2 per sec of an atom jumping from the vapor to the liquid, O_n is the surface area of a droplet containing n atoms, and the corresponding symbols with a $n + 1$ subscript relate to an embryo containing $n + 1$ atoms that is assumed to lose an atom. It should be noted that whereas q_0 can be assumed to be independent of n, since the chance of a vapor atom striking a droplet should be nearly independent of the free energy of the droplet, this is not true of an atom that leaves the droplet to return to the vapor. In this case, the flux of atoms from the embryo should depend on the free energy of the droplet and thus on its size. As may be seen in Fig. 13.1, the free energy of an atom in an embryo becomes more positive with increasing size of the particle. This means that the rate of evaporation of atoms from a unit area of the surface of an embryo should increase as it becomes larger. An equation similar to the above can be written for all values of the number of atoms in the embryo, as for example

$$I = q_o O_{n+1} Z_{n+1} - q_{n+2} O_{n+2} Z_{n+2}$$

It is thus apparent that there is a large set of these related equations. With the aid of the proper assumptions,[5] these relationships may be solved to yield the following expression for the nucleation current

$$I = \frac{q_o O_c N}{n_c} \left(\frac{\Delta F_{nc}}{3\pi kT} \right)^{1/2} e^{-\Delta F_{nc}/kT}$$

where the subscript c refers to quantities measured at the critical embryo size. It should be noted that the above equation only differs from the Volmer-Weber equation, given in the preceding section, by the pre-exponential term

$$\frac{1}{n_c} \left(\frac{\Delta F_{nc}}{3\pi kT} \right)^{1/2}$$

It will be shown in the next section that for the case of freezing, ΔF_{nc} varies approximately as the square of the supercooling ΔT, or the temperature difference between the temperature of transformation and the equilibrium or freezing temperature. The factor T also appears in the denominator of the above expression. In spite of these facts, the variation of this pre-exponential factor with temperature, in the range of temperatures involved in nucleation problems, is much smaller than the variation due to the exponential term. In addition, the accuracy of experimental determinations of nucleation rate data is not very high. Therefore, the basic Volmer-Weber relationship may still be considered to adequately represent the nucleation current. The principal advantage of the Becker-Döring theory is it treats the nucleation problem on a basis that is physically more satisfying.

13.3 Freezing. Let us now consider the freezing of a pure metal. While a rigorous treatment would require that one should consider the problem of the shape of the embryo and nucleus as well as the problems associated with

5. Christian, J. W., *Op. Cit.*

the attachment of atoms to the surface of a solid, these factors apparently do not strongly influence the rate of nucleation. One can therefore assume, to a reasonable approximation, that the solid embryos take a simple spherical shape and that atoms attach themselves to this particle at all points on its surface. In this case, the nucleation rate should depend on the number of critical nuclei, N_{n_c}, and the rate at which an atom can attach itself to the embryo. The first quantity, in analogy with the vapor-liquid reaction, should be given by $Ne^{-\Delta F_{nc}/kT}$, where N is the total number of atoms in the assembly and ΔF_{nc} is the free energy change associated with the formation of a critical embryo. The rate of jumping of atoms across the boundary between the liquid and the solid may be expressed as the product of an atomic vibration frequency ν and the probability that an atom will make a successful jump. In making this jump there will normally be an energy barrier that the atom must overcome. Let us designate the magnitude of the free energy associated with this barrier as Δf_a. The nucleation rate or nucleation current then becomes

$$I = \nu e^{-\Delta f_a/kT}\ (Ne^{-\Delta F_{nc}/kT})$$

In analogy with the equation for the free energy associated with the formation of a liquid embryo in a vapor, we may set down the following expression for ΔF_n

$$\Delta F_n = n\Delta f^{ls} + \eta\gamma_{ls}n^{2/3}$$

where n is the number of atoms in an embryo, Δf^{ls} is the free energy difference between an atom in the liquid and solid phases, η is a shape factor, and γ_{ls} is the surface free energy of the liquid-solid boundary. It is now possible to evaluate Δf^{ls} approximately in terms of the heat of fusion. Thus at the freezing point

$$\Delta f^{ls} = \Delta h^{ls} - T_0\Delta s^{ls} = 0$$

where Δh^{ls} and Δs^{ls} are the heat of fusion and entropy of fusion per atom, and T_0 is the freezing point temperature. Solving this relation for Δs^{ls} gives $\Delta s^{ls} = \Delta h^{ls}/T_0$. Now both Δs^{ls} and Δh^{ls} are not strongly varying functions of the temperature near the freezing point and may be considered to be constant. Therefore, near the freezing point we have

$$\Delta f^{ls} = \Delta h^{ls} - \frac{T\Delta h^{ls}}{T_0} = \frac{\Delta h^{ls}\Delta T}{T_0}$$

In this last expression, ΔT is the temperature increment measured with respect to the freezing point and therefore represents the degree of supercooling.

The free energy associated with the formation of a nucleus may now be written

$$\Delta F_{nc} = \frac{4\eta^3 \gamma_{ls}{}^3 T_0{}^2}{27(\Delta h^{ls})^2 \Delta T^2} = \frac{A}{\Delta T^2}$$

where $A = \frac{4\eta^3 \gamma_{ls}{}^3 T_0{}^2}{27(\Delta h^{ls})^2}$

This indicates that the free energy needed to form a critical embryo varies inversely as the square of the degree of supercooling. Thus not only does the size of the critical nucleus decrease with the supercooling, but so does the free energy required to form a critical nucleus. Let us now investigate the effect of some of these factors on the rate of nucleation. The nucleation current, when expressed in terms of this simplified expression for ΔF_{nc}, becomes

$$I = \nu N e^{-(\Delta f_a + A/\Delta T^2)/kT}$$

The most important part of this relation is the exponent, because it controls the nucleation rate. As written above, the exponent has two terms which do not depend upon the temperature in the same way. The first term, $\Delta f_a/kT$, represents the effect of the boundary on the jumping of atoms toward the nucleus. To a reasonable approximation, Δf_a may be assumed to be temperature independent and, therefore, this term should vary inversely with the absolute temperature. The nature of this variation is shown schematically in Fig. 13.7. In this diagram, the values of the exponential terms have been plotted using a logarithmic scale. Note that $\Delta f_a/kT$ increases continuously with decreasing temperature. On the other hand, the behavior of the second term, $A/\Delta T^2 kT$, is quite different. At the melting temperature, where ΔT is zero, this quantity is infinite. This implies that at the melting temperature, the nucleation rate must be zero. With decreasing temperature this term falls rapidly and eventually becomes much smaller than the first term. Since the nucleation rate is controlled by the sum of these two terms, it is a maximum where their sum is a minimum. This point is indicated on the diagram. Above this temperature, the rate of nucleation is effectively controlled by the energy barrier associated with forming the critical embryo; below it the nucleation rate is controlled by the jump rate of the atoms across the boundary from the liquid to the solid. Fig. 13.7 is instructive in another fashion: it shows a very rapid rise of the exponent as one approaches the melting point. This is in excellent agreement with the fact that homogeneous nucleation is very difficult to achieve near the melting point.

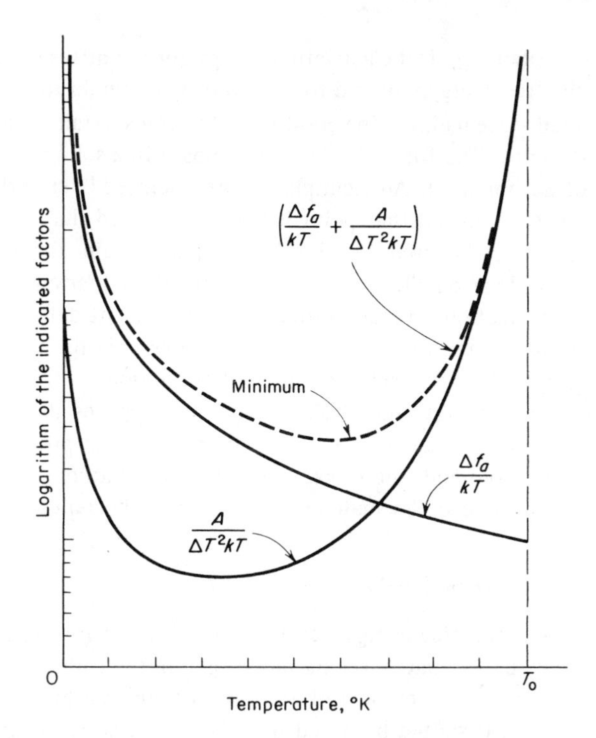

Fig. 13.7 Variation with temperature of the various terms involved in the exponent of the nucleation current equation.

Conversely, it is also in agreement with the fact that in a very pure metal it is often possible to achieve a very high degree of supercooling.

13.4 Solid State Reactions. Reactions that occur between liquid and solid or solid and solid are said to occur in condensed systems. In the preceding section we have considered the simplest possible type of condensed system transformation: the freezing of a pure metal. In a more general case, such as in the precipitation of a new solid phase from a single-phase solid solution, one may have to consider that the reaction may be largely controlled by diffusion. This situation would occur when solute, present in a relatively small concentration, has to diffuse through the matrix in order for the embryo to grow. In 1940 Becker[6] proposed that when this occurs, the nucleation current might be written

$$I = Ae^{-Q_d/kT}\, e^{-\Delta F_{nc}/kT}$$

6. Becker, R., *Proc. Phys. Soc.*, 52 71 (1940).

where A is a constant, Q_d is the activation energy for the diffusion of the solute, and ΔF_{nc} is the free energy required to form the critical embryo.

In condensed systems involving solid state reactions, strain energy is usually an important factor. The formation of a new phase in a solid normally involves some form of deformation. An example will be discussed later in the martensite transformation sections. At that point it will be observed that most martensite transformations are believed to involve a plane-strain deformation. This corresponds to a shear parallel to the habit plane of the martensite plate and an expansion or contraction normal to this plane. Equivalent deformations can be expected in most solid-solid transformations. Strains in both the matrix and the newly formed particles will result from these deformations.

Like the surface energy, the strain energy also opposes the formation of a nucleus. It is normally assumed that the strain energy is proportional to the volume of the embryo and, therefore, to the number of atoms in the embryo. If this is true, the free energy associated with the embryo becomes

$$\Delta F_n = n(\Delta f^{\alpha\beta} + \Delta f_s) + \eta \gamma n^{2/3}$$

where Δf_s is the strain free energy per atom, $\Delta f^{\alpha\beta}$ is the free energy difference between an atom in the matrix (alpha phase) and in the embryo (beta phase), η is the shape factor, n is the number of atoms in the embryo, and γ is the specific surface free energy. It should be noted that $\Delta f^{\alpha\beta}$ is negative, whereas both Δf_s and γ are positive. If Δf_s is larger in absolute magnitude than $\Delta f^{\alpha\beta}$, ΔF_n must increase in magnitude with increasing number of atoms in the embryo. Under these conditions it will be impossible to form a nucleus. In general, it is to be expected that Δf_s will be largest for coherent embryos. When the interface between the embryo and the matrix is coherent, there exists a perfect match between planes and directions across the interface separating the two structures. These crystallographic features may, of course, suffer a change in direction as they cross the interface. This type of boundary is analogous to the coherent twin boundary shown in Fig. 16.16. A coherent boundary can be formed by matching a {111} plane of a face-centered cubic crystal with a (0002) plane of a hexagonal close-packed crystal. Such a boundary is theoretically possible when a f.c.c. metal transforms to a h.c.p. structure. Because the total strain energy associated with the embryo tends to increase with the size of the embryo, the latter may eventually lose its coherency. When this takes place, the interface separating the embryo from the matrix becomes incoherent. This latter boundary is equivalent to a high-angle grain boundary. When a particle is noncoherent, Δf_s may be considered to be reduced, but it does not necessarily become zero. There is thus some tendency for the changes that occur in the two energy terms to compensate when a particle loses its coherency. In effect, when the strain-energy term decreases, the surface-energy term increases.

It is most important that it be recognized that the above discussion is far from complete. The subject of nucleation in the solid state is very complex. In this regard it should be mentioned that there also exists a third type of interface between the matrix and the second phase. This is a semicoherent boundary that is basically a coherent interface containing a grid of dislocations in the boundary. In a coherent boundary, the mismatch between the two crystal structures is small enough to be accommodated by elastic strains. In the semicoherent interface the mismatch is accommodated by dislocations. An extremely simple model for this type of boundary is furnished by the small-angle grain boundary shown in Fig. 5.13. Another model is that of the incoherent twin boundary of Fig. 16.17. Because of space limitations, the complexity of the total problem can only be suggested. In a given transformation, a coherent boundary should be preferred if the mismatch is extremely small, whereas with increasing degree of mismatch the semicoherent boundary might possess the smaller total surface energy. This statement has to be tempered by the fact that the size of the particle also has a bearing on the question. With growing particle size, the total nonchemical energy of the particle (strain plus surface energies) may be reduced if a coherent boundary is replaced by a semicoherent boundary.

At this point we shall ignore further considerations of the semicoherent boundary and consider only coherent and incoherent boundaries, primarily because these two types of boundaries have been treated in somewhat more detail and are perhaps easier to define. In the case of the coherent boundary, it has been shown[7] that the strain energy of the embryo is not very dependent on the shape of the particle. In other words, the strain energy associated with the formation of a spherical embryo is not very different from that associated with a platelike or even needle-shaped embryo. This is not the case for an incoherent embryo. Here the shape may be very important, as has been demonstrated by Nabarro[8] who attacked the problem from the point of view of isotropic elasticity. His computations also involved another basic assumption: that the matrix was much stiffer elastically than the particles, so that the strain energy of the particles could be neglected in comparison to that of the matrix. In effect, Nabarro considered the strain energy associated with expanding a cavity in the matrix. This expansion would be analogous to that resulting from the pumping of an incompressible fluid into a hole. The shapes of these cavities were assumed to correspond to various ellipsoids of revolution. His results are shown schematically in Fig. 13.8, plotted as a function of the ratio r_1/r_2 of the semi-axes of the ellipsoid of revolution, where the three semi-axes were taken to be r_1, r_2, and r_2. If the ratio r_1/r_2 is very small, the ellipsoid approximates a disc. If r_1 and r_2 are equal, the ellipsoid is a sphere; whereas if r_1 is much larger

7. Christian, J. W., *Op. Cit.*
8. Nabarro, F. R. N., *Proc. Phys. Soc.*, **52** 90 (1940); and *Proc. Roy. Soc.*, **A175** 519 (1940).

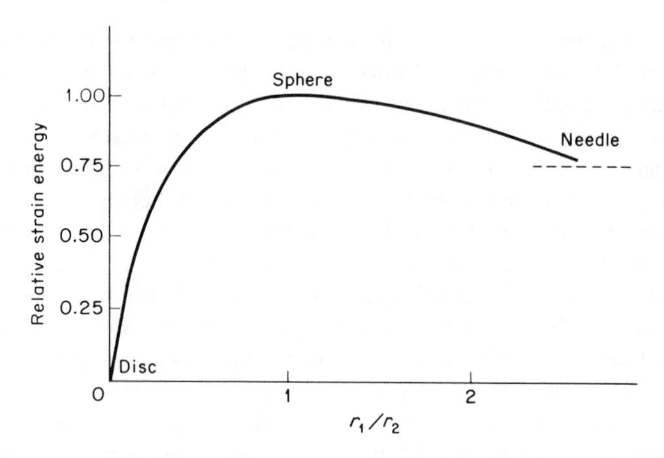

Fig. 13.8 The strain energy of an incoherent nucleus is a function if its shape. (After Nabarro, F. R. N., *Proc. Phys. Soc.,* **52** 90 [1940].)

than r_2, the shape of the ellipsoid approaches a needle. As may be seen in the diagram, the strain energy is a maximum for a sphere and is least for a disc. Furthermore, as the thickness of the disc approaches zero, the strain energy also approaches zero.

In evaluating the results of Nabarro for incoherent embryos, one has to recognize that the total surface energy of the embryo should be least for a sphere. Thus the effect of shape on the two energy factors is opposing. Because the strain energy is a function of the volume of the particle, and the surface energy is dependent on the surface area, one should expect that the tendency to form discs should become more important as the particle grows in size. It should be mentioned that many precipitates certainly assume a plate-like morphology, and this is in good agreement with Nabarro's predictions. However, the tendency for a plate-like habit to develop can also be the result of other factors, such as the nature of the mechanism that controls the growth of the particle. Thus for example, a plate may form if it is easier for atoms to attach themselves to the edge of the particle than to its flat surfaces.

13.5 Heterogeneous Nucleation. In nature most nucleation occurs heterogeneously, as indicated previously at several points in this book. In the preceding sections we have shown that homogeneous nucleation is a rather difficult process. At this time a simple problem will be considered which illustrates one reason why heterogeneous nucleation may be a much easier process to invoke. In freezing, the container or mold walls often offer preferred sites for nucleation. This problem can be placed on a quantitative basis if several simplifying assumptions are made. First, let us assume that solid embryos form

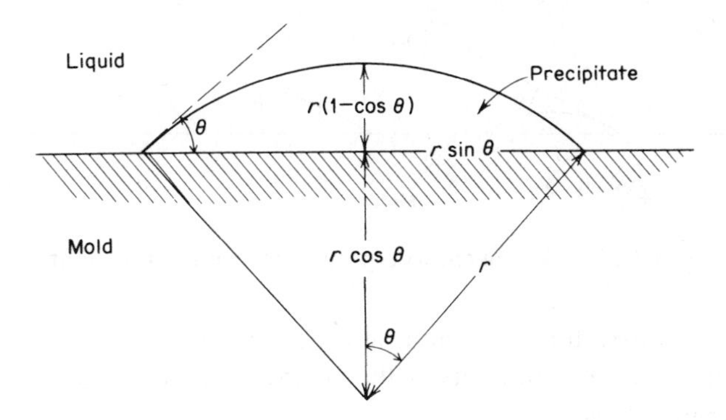

Fig. 13.9 A hypothetical spherical cap embryo.

on the mold walls as spherical caps, as suggested in Fig. 13.9. Second, let us assume that at the positions where the surfaces of the caps make contact with the mold walls, there exists a state of quasi-equilibrium between the surface forces, as indicated in Fig. 13.10. This balance between the surface forces is assumed to take place in the direction parallel to the mold surface. In the direction normal to the mold surface, the surface tensions are not balanced, implying a net traction acting on the surface. With sufficient time to allow diffusion to occur in the mold wall, a complete balance between both sets of components might be obtained. The equation relating the surface forces parallel to the mold walls is

$$\gamma^{lm} = \gamma^{sm} + \gamma^{ls} \cos \theta$$

where γ^{lm}, γ^{sm}, and γ^{ls} are the surface tensions between liquid and mold wall, solid and mold wall, and liquid and solid respectively, and θ is the angle of contact of the embryo surface with the mold wall. Note that θ is a function of only the three surface tensions. This signifies that, no matter how large the size of the particle, the angle of contact should be the same. In effect, this says that as the embryo grows, its shape remains invariant as a spherical cap. This point is illustrated in Fig. 13.11.

We can now write an equation giving the free energy of an embryo. In this

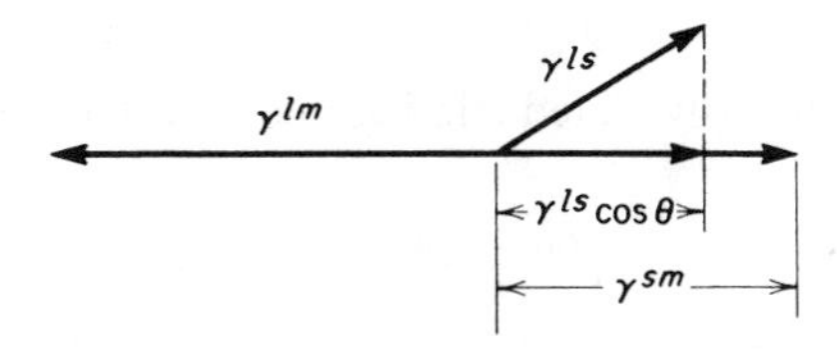

Fig. 13.10 The surface forces associated with the spherical cap embryo.

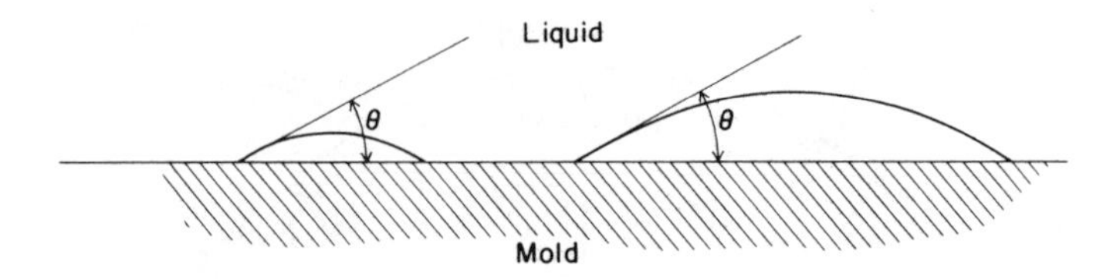

Fig. 13.11 As the embryo grows, its shape is invariant.

case we shall express the equation in terms of the radius of the surface of the spherical cap rather than in terms of the number of atoms in the embryo. The equation in question is

$$\Delta F^{\text{het}} = V_c \Delta f_v + A_{ls}\gamma^{ls} + A_{sm}(\gamma^{sm} - \gamma^{lm})$$

where ΔF^{het} is the free energy associated with the heterogeneously nucleated embryo, V_c is the volume of the cap-shaped embryo, A_{ls} is the area of the cap that faces the liquid, A_{sm} is the area of the interface between the embryo and the mold wall, Δf_v is the free energy per unit volume associated with the freezing process, and γ^{ls}, γ^{sm}, and γ^{lm} are the surface energies as defined above. Note that A_{sm} is multiplied by the difference between γ_{sm} and γ^{lm}. This is because the surface formed between the embryo and the mold replaces an equivalent area of liquid and mold interface.

Now consider Fig. 13.9 which corresponds to a cross-section taken through the center of the embryo. Note that the height of the cap is equal to $r(1 - \cos\theta)$ and that the radius of the circular area, corresponding to the interface between the cap and the mold wall, is $r(\sin\theta)$. We may therefore set down the following relationships

$$A_{ls} = 2\pi r^2(1 - \cos\theta)$$

$$A_{sm} = \pi r^2 \sin^2\theta$$

$$V_c = 1/3\pi r^3(2 - 3\cos\theta + \cos^3\theta)$$

From the equation previously stated relating the surface tensions, we have

$$\gamma^{lm} - \gamma^{sm} = \gamma^{ls}\cos\theta$$

If all of the above relationships are substituted into the equation for the free energy of the embryo and the resulting expression is simplified, one obtains

$$\Delta F^{\text{het}} = \frac{4}{3}\pi r^3 \frac{(2 - 3\cos\theta + \cos^3\theta)}{4} \Delta f_v + 4\pi r^2 \gamma^{ls} \frac{(2 - 3\cos\theta + \cos^3\theta)}{4}$$

or

$$\Delta F^{\text{het}} = (V_{\text{sph}}\Delta f_v + A_{\text{sph}}\,\gamma^{ls}) \frac{(2 - 3\cos\theta + \cos^3\theta)}{4}$$

where V_{sph} and A_{sph} represent the volume and area of a sphere, respectively. In this last equation the quantity inside the first set of parentheses on the right side is the same as the free energy associated with the formation of a spherical embryo by homogeneous nucleation. Therefore, for the present example, we may equate the free energy of the heterogeneously nucleated embryo to the free energy of an equivalent homogeneously nucleated one. This relation is as follows

$$\Delta F^{\text{het}} = \Delta F^{\text{hom}} \frac{(2 - 3\cos\theta + \cos^3\theta)}{4}$$

where ΔF^{hom} is the free energy of a spherical embryo of radius equal to the radius of the cap of the heterogeneously nucleated particle. This expression states that the free energy to form a nucleus heterogeneously at the mold wall varies directly as the homogeneous free energy modified by a factor that is a function only of the angle of contact θ between the mold wall and the embryo. Since this angle is determined by the relative surface energies of the three surfaces in question, it is obvious that the free energy of heterogeneous nucleation is directly dependent on these surface energies. Also, since the above equation holds for all values of r, it must hold when r is equal to r_c, the critical radius. This, in turn, signifies that

$$\Delta F_c^{\text{het}} = \Delta F_c^{\text{hom}} \frac{(2 - 3\cos\theta + \cos^3\theta)}{4}$$

The factor $(2 - 3\cos\theta + \cos^3\theta)/4$ that converts the homogeneous nucleation free energy to the free energy of heterogeneous nucleation is plotted as a function of θ in Fig. 13.12. It is significant that this factor is very small to even rather large values of the contact angle. Thus, when θ is 10 degrees, the multiplying factor is of the order of about 10^{-4}. When θ is 30 degrees, it is only about 0.02, and at 90 degrees, or at the limit of applicability of the above equation, it is still only equal to one-half. The significance of this large decrease in the free energy to form a critical nucleus cannot be overestimated. The effect has to be very large.

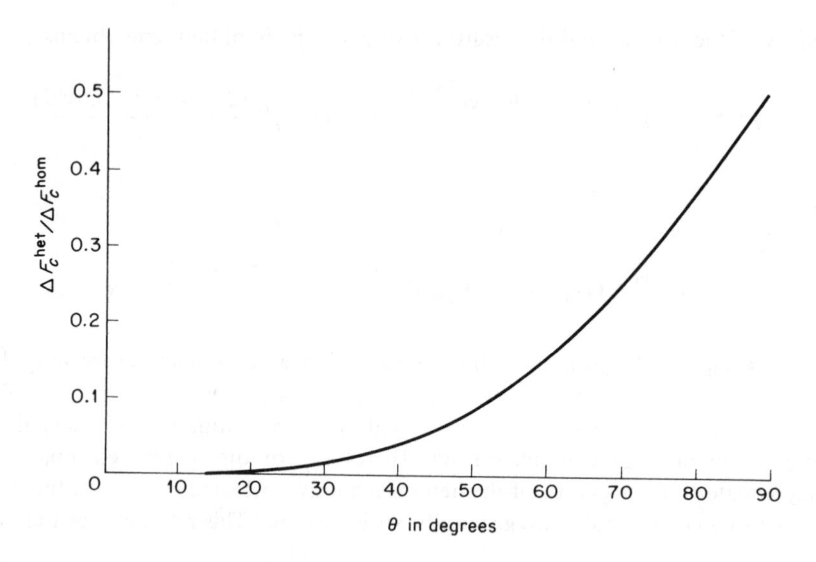

Fig. 13.12 The variation of the ratio of the free energy of heterogeneous nucleation to that for homogeneous nucleation as a function of θ, the angle of contact between the embryo and the mold wall.

In this regard, let us consider the nucleation rate equation for the case of heterogeneous nucleation. The number of critical embryos should be given by the equation

$$N_c^{het} = N^m e^{-\Delta F_c^{het}/kT}$$

where N^m is the number of atoms in the liquid phase facing the mold wall. This is a reasonable assumption, because only the atoms in contact with the mold can form an embryo at the surface. Note that N^m is a much smaller number than N, the number of atoms in the assembly, that appears in the corresponding equation for the embryos undergoing homogeneous nucleation. On the other hand, the exponent in the above equation can easily be enough smaller to more than compensate for this difference in the pre-exponential term. By analogy to our previous expressions for the nucleation rate or nucleation current, we may now write for the case of heterogeneous nucleation

$$I = \nu N^m (e^{-\Delta f_a/kT})(e^{-\Delta F_c^{het}/kT})$$

where ν is a frequency and Δf_a is the free energy associated with the jump of an atom across the interface between the liquid and solid embryo.

13.6 Growth Kinetics. Growth kinetics become important once an embryo has exceeded the critical size and become a stable nucleus. In some cases this may occur at an extremely early stage in the development of the particle. As was the case with nucleation kinetics, there are also many facets to the study of growth kinetics, and only a couple of very simple examples will be considered. The primary purpose of these is to show the nature of the approach used in studying growth. First, it should be noted that reactions involving a large heat of transformation, such as occurs in freezing, present a special problem. Here the growth rate may be largely determined by the rate at which it is possible to remove the heat of fusion. The simplest growth theories, accordingly, ignore this type of problem and correspond to transformations in which the heat of transformation is very small, so that they may be assumed to be isothermal. Certain solid state reactions probably come closest to satisfying this condition. Thus, for example, consider the phase changes in iron.[9] The heat of fusion of iron is 3670 cal per mole, while the heat of transformation from the delta to the gamma phase is 165 cal per mole, and that for the transformation from gamma to alpha is 215 cal per mole.

Let us assume that we are concerned with a transformation in a pure substance that occurs within the solid state and that a particle has grown sufficiently so that it has become a stable nucleus. It is further assumed that the particle is spherical in shape, that no volume change is involved as atoms leave the alpha phase and join the beta phase in the particle, and that surface energy or capillarity effects can be neglected. This means that we shall ignore strain energy. Finally, it is assumed that growth occurs continuously or without the need for steps on the surface where the atoms can successfully attach themselves. The role of growth involving the movement of steps has received considerable attention in recent years. The subject is too lengthy to consider here, so that the reader is referred to other sources.[10,11] In general, it is felt that step-wise growth is probably most important when the driving force for growth is very small.

Under the conditions stated above, it is reasonable to assume that the curve relating the free energy of an atom as it moves across the boundary from the alpha phase matrix onto the beta phase precipitate has the form given in Fig. 13.13. Note that in crossing the boundary, the atom has to pass over an energy barrier equal to Δf_a, and when it joins the beta phase it has a lower energy than it had in the alpha phase. The corresponding difference in free energy is

9. Darken, L. S., and Gurry, R. W., *Physical Chemistry of Metals*, p. 397. McGraw-Hill Book Company, New York, 1953.

10. Christian, J. W., *Op. Cit.*

11. Fine, M. E., *Introduction to Phase Transformations in Condensed Systems*, The Macmillan Company, New York, 1965.

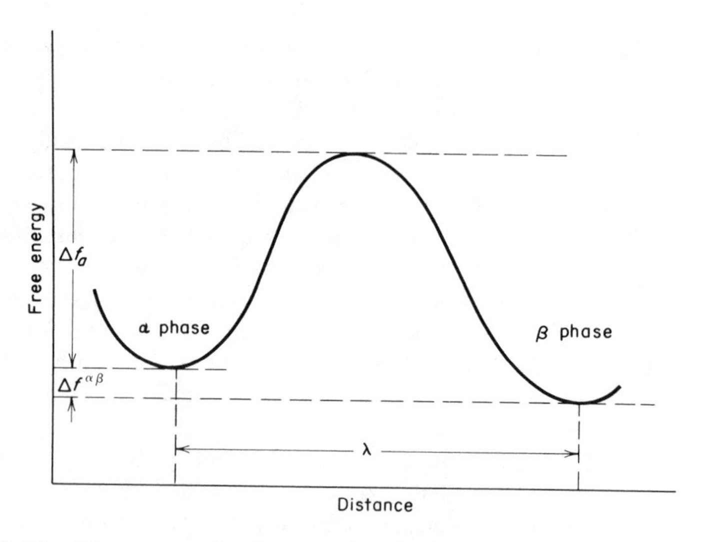

Fig. 13.13 The energy barrier associated with the growth of a precipitate during a solid state phase change in a polymorphic pure metal.

indicated on the diagram by $\Delta f^{\alpha\beta}$. This difference in energy per atom is assumed to be equal to the chemical free energy difference between an atom in the α and β phases.

We may now write an expression for the net rate of atom transfer from the matrix to the particle as equal to the difference in the rate of atom movement to and away from the particle. This is

$$I = S\nu e^{-\Delta f_a/kT} - S\nu e^{-(\Delta f_a + \Delta f^{\beta\alpha})/kT}$$

where S is the number of atoms facing the surface, ν is an atomic vibration frequency, and I is the net number of atoms per second leaving the matrix to join the beta phase. This expression reduces to

$$I = S\nu e^{-\Delta f_a/kT}\left(1 - e^{-\Delta f^{\beta\alpha}/kT}\right)$$

Let us now assume that when the atoms jump they move through an average distance λ. The velocity of the boundary will therefore be given by

$$v = \frac{\lambda I}{S}$$

where I/S represents the average number of jumps per second per atom facing the boundary, and λ is the distance corresponding to each of these jumps. The velocity may now be expressed in terms of the equation for I as

$$v = \lambda \nu e^{-\Delta f_a/kT}\left(1 - e^{-\Delta f^{\beta\alpha}/kT}\right)$$

Now consider $\Delta f^{\beta\alpha}$. This quantity represents the decrease in free energy when an atom joins the nucleus, ignoring surface and strain energy effects. In analogy with Δf^{ls}, this free energy may be expected to vary directly as the degree of supercooling. For a very small supercooling, $\Delta f^{\beta\alpha}$ will be small, and for a sufficiently small supercooling, we may assume $\Delta f^{\beta\alpha} \ll kT$. If this is true, then the exponential term $e^{-\Delta f^{\beta\alpha}}$ may be assumed to be approximately equal to the first two terms in its series expansion, or

$$e^{-\Delta f^{\beta\alpha}/kT} \approx 1 - \Delta f^{\beta\alpha}/kT$$

Under these conditions, the growth velocity becomes

$$v = \lambda \nu \left(\frac{\Delta f^{\beta\alpha}}{kT}\right) e^{-\Delta f_a/kT}$$

Since $\Delta f^{\beta\alpha}$, as indicated above, should vary approximately as ΔT (the supercooling), for small supercooling the growth velocity should be roughly proportional to the degree of supercooling. On the other hand, if the degree of supercooling is large, $\Delta f^{\beta\alpha}$ may become greater than kT. This is also encouraged by the fact that large supercooling normally requires that the temperature be lowered to very small values. When this happens, the exponential term becomes very small and the quantity $(1 - e^{-\Delta f^{\beta\alpha}/kT})$ may be equated to unity. The velocity equation may then be written

$$v \approx \lambda \nu e^{-\Delta f_a/kT}$$

This last expression shows that as T becomes very small, the growth velocity again approaches zero. Since it is also zero at the transformation temperature, the velocity must be a maximum at some intermediate temperature. This has actually been experimentally verified for the transformation of beta tin to alpha tin on cooling, as shown in Fig. 13.14. Note that just below the transformation temperature at 13°C the growth rate is very small but increases with increasing supercooling and then eventually decreases again as the temperature becomes very low. This general trend of the dependence of the growth rate on temperature is similar to that which was shown to be characteristic of nucleation.

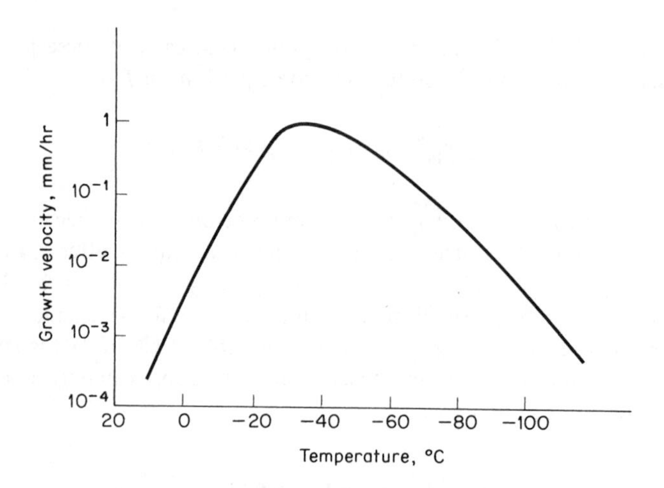

Fig. 13.14 The growth rate in the transformation of beta tin to alpha tin (on cooling). The equilibrium transformation temperature is 13°C. (Data of Becker, J. H., *J. Appl. Phys.*, **29** 1110 [1958].)

13,7 Diffusion Controlled Growth. In the preceding section growth was considered under conditions where a new phase grows out of another phase by a simple transfer of atoms of a single component. We shall now treat the case where the transformation not only involves the formation of a new phase, but where this phase also possesses a composition different from that of the old phase. A relatively simple example of this type of reaction is the precipitation of iron-carbide particles from a supersaturated solid solution of carbon in alpha iron. This reaction has been studied in detail by Wert[12] using torsion pendulum measurements. Some of his results have been described earlier, in Chapter 9. Since the carbide particles contain 6.7 percent carbon, whereas the matrix from which they form always has a small fraction of a percent of carbon, it is obvious that carbon atoms have to diffuse over relatively long distances in order for the carbide particles to grow.

As a result of Wert's observations on the precipitation of carbides in dilute solutions of alpha iron, Zener[13] proposed a simple theory for this type of growth. For our present purposes, however, let us first consider Zener's theory in terms of a precipitate that is not spherical, but which grows as a plate in the direction normal to its surface. This assumption of one-dimensional growth greatly simplifies the problem. Now consider Fig. 13.15A. In this diagram the cross-hatched region represents the precipitate after it has grown to a thickness equal to x. Fig. 13.15B shows a schematic plot of the concentration of the

12. Wert, C., *J. Appl. Phys.*, **20** 943 (1949).
13. Zener, C., *J. Appl. Phys.*, **20** 950 (1949).

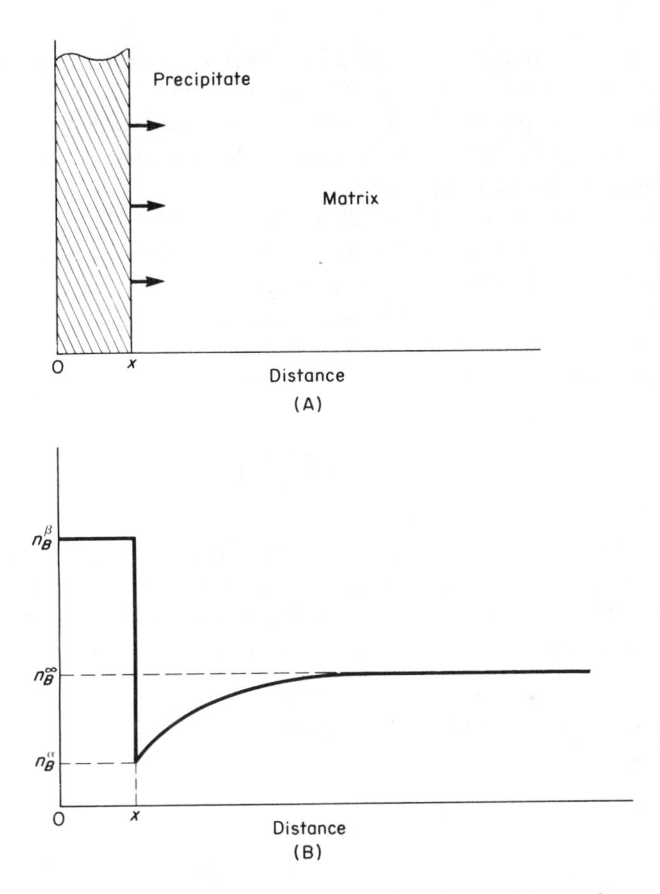

Fig. 13.15 A planar precipitate growing under conditions where growth is controlled by diffusion. The lower drawing shows schematically how the composition varies with distance.

solute B as a function of distance in which the concentration is expressed in atoms per cm^3. Note that, as in the case of iron-carbide precipitating in iron, the concentration of the solute in the precipitate is higher than that in the matrix. This concentration is shown as $n_B{}^\beta$. The assumption is also made that the time for the atoms to cross the boundary between the matrix and the precipitate is small compared to the time for the solute to diffuse up to the precipitate. Under this condition, the concentration of B in that part of the matrix immediately facing the precipitate may be assumed to equal the equilibrium concentration. This is the concentration of B in the alpha phase that would be in equilibrium with the beta phase and is designated $n_B{}^\alpha$ in the diagram. Note that, in effect, one considers that "local" equilibrium exists at the boundary. The figure also shows that as one moves away from the interface, the concentration of B rises in the matrix to the value marked $n_B{}^\infty$. This latter concentration is assumed to be that which the matrix possessed before the precipitation started. The drop in concentration near the interface is the result of the short range jumping of atoms from the matrix into the precipitate.

Now let us assume that in a small time increment dt, the boundary of the precipitate moves forward into the matrix through a distance dx. This has the effect of converting a volume of material, equal to Adx, from a concentration $n_B{}^\alpha$ to a concentration $n_B{}^\beta$. In order for this to occur, $(n_B{}^\beta - n_B{}^\alpha) Adx$ atoms of B would have to diffuse up to the interface and cross over it. By Fick's first law, this number of atoms should also equal

$$-Jdt = AD\left(\frac{dn_B{}^\alpha}{dx}\right)dt$$

where J is the flux or number of B atoms crossing the area per second, D is the diffusion coefficient which is assumed to be independent of concentration, and $dn_B{}^\alpha/dx$ is the concentration gradient of the B component in the matrix at the interface. Equating these two quantities we have

$$(n_B{}^\beta - n_B{}^\alpha)\, Adx = AD\left(\frac{dn_B{}^\alpha}{dx}\right)dt$$

or solving for the interface velocity v

$$v = \frac{dx}{dt} = \frac{D}{(n_B{}^\beta - n_B{}^\alpha)}\,\frac{dn_B{}^\alpha}{dx}$$

Zener has given an approximate solution to this equation based on the drawing of Fig. 13.16. In this diagram the curved concentration distance curve, to the right of the boundary in the matrix in Fig. 13.15B, is assumed to be a straight line. The slope of this straight line is determined by the fact that the two cross-hatched regions shown in the diagram should have equal areas. This is because the square section to the left of the boundary represents the B atoms that have joined the precipitate, while the triangular shaped area to the right of the boundary corresponds to the atoms that have left the matrix to enter the precipitate. Equating these areas gives us

$$\frac{1}{2}\,\Delta n_B{}^\alpha \Delta x = (n_B{}^\beta - n_B{}^\alpha)\, x$$

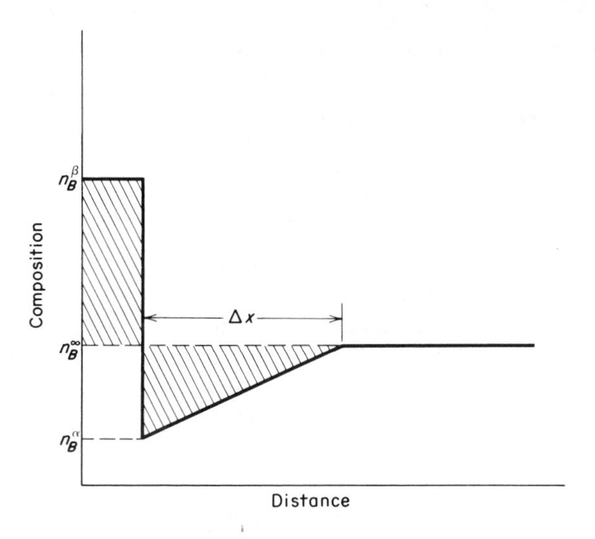

Fig. 13.16 Zener's approximation for the composition-distance curve.

or

$$\Delta x = \frac{2(n_B{}^\beta - n_B{}^\alpha)\, x}{\Delta n_B{}^\alpha}$$

and the slope of the straight line concentration gradient is accordingly

$$\frac{\Delta n_B{}^\alpha}{\Delta x} = \frac{(\Delta n_B{}^\alpha)^2}{2(n_B{}^\beta - n_B{}^\alpha)x} = \frac{(n_B{}^\infty - n_B{}^\alpha)^2}{2(n_B{}^\beta - n_B{}^\alpha)x}$$

Substituting this approximate slope into the velocity equation yields

$$v = \frac{dx}{dt} = \frac{D(n_B{}^\beta - n_B{}^\alpha)^2}{(n_B{}^\beta - n_B{}^\alpha)\cdot(n_B{}^\beta - n_B{}^\infty)x}$$

Integration of this differential equation leads to the following relation for the position of the boundary as a function of time

$$x = \alpha_1{}^* \sqrt{Dt}$$

where

$$\alpha_1{}^* = \frac{(n_B{}^\infty - n_B{}^\alpha)}{\sqrt{(n_B{}^\beta - n_B{}^\alpha)}\ \sqrt{(n_B{}^\beta - n_B{}^\infty)}}$$

The subscript 1 of the parameter $\alpha_1{}^*$ indicates that the approximate solution is for one-dimensional growth. Differentiation of the above equation gives the growth velocity in the simplified form

$$v = \frac{dx}{dt} = \frac{\alpha_1{}^* \sqrt{D/t}}{2}$$

This equation indicates that for one-dimensional growth, the interface position varies as $\sqrt{Dt}$ and its velocity as $\sqrt{D/t}$.

These results have a very general application and, as Zener has shown by dimensional analysis, whenever the growth is controlled by a simple diffusion process of the type indicated above, the interface position varies with the square root of the time, and growth velocity varies inversely as the square root of the time. Thus, in the case of three-dimensional or spherical growth, we may therefore write

$$x = \alpha_3{}^* \sqrt{Dt}$$

However, in this case Zener indicates that the parameter $\alpha_3{}^*$ depends to some degree upon the original concentration of the solute in the alpha phase.

To obtain a better appreciation of the roles of diffusion and the composition gradients in diffusion-controlled growth, let us consider the one-dimensional growth equation of Zener relative to the iron-carbon diagram. This diagram is redrawn in Fig. 13.17 with concentrations expressed in atomic percent. For the purposes of this discussion, we will assume that we can use concentrations in atomic percent to express the various values of n_B in the Zener equation where they will actually correspond to numbers of atoms per cm^3. This will not affect our general conclusions. Suppose an alloy of iron containing 0.08 atomic percent of iron is heated to 723°K and is then quenched to 400°C. The metal will be supersaturated and will contain 0.08 atomic percent carbon, or $n_B{}^\infty = 0.0008$. At the same time, by the phase diagram, the equilibrium concentration $n_B{}^\alpha$ is only 0.0001. Since cementite contains a ratio of one carbon atom to three iron atoms, $n_B{}^\beta$ is 0.25. Substituting these values in the expression for $\alpha_1{}^*$ gives us

$$\alpha_1{}^* = \frac{n_B{}^\infty - n_B{}^\alpha}{\sqrt{n_B{}^\beta - n_B{}^\alpha}\,\sqrt{n_B{}^\beta - n_B{}^\infty}}$$

$$= \frac{0.0008 - 0.0001}{\sqrt{0.25 - 0.0001}\,\sqrt{0.25 - 0.0008}} \approx 2.8 \times 10^{-3}$$

Any further lowering of the temperature to which the metal is quenched will not markedly change the value of $\alpha_1{}^*$. This is because the equilibrium concentration is already small at 400°C, and any further decrease in $n_B{}^\alpha$ can have only a small effect on the numerator in the above equation. On the other

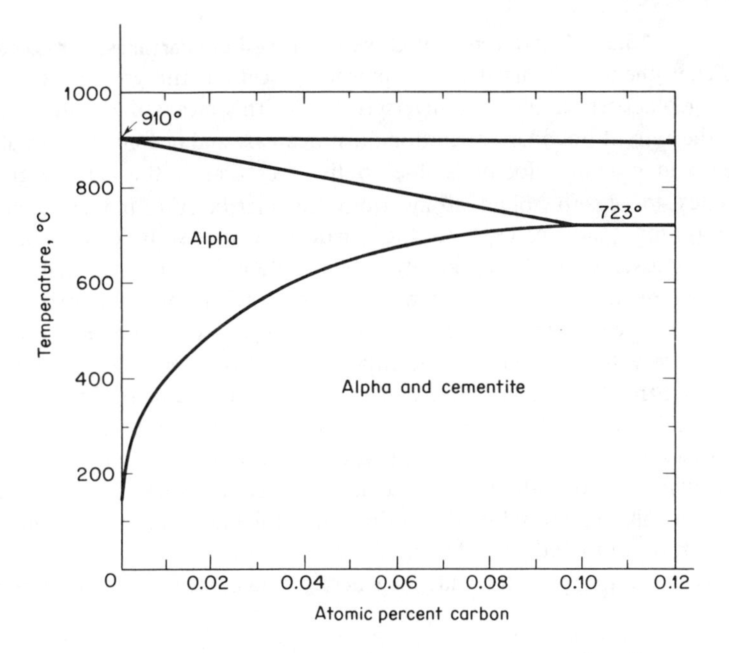

Fig. 13.17 The iron-carbon diagram at the iron-rich end.

hand, v also depends on $\sqrt{D}$ where, as shown in Section 11.4, the diffusion coefficient for carbon in iron may be represented by

$$D = D_0 e^{\frac{-19{,}800}{RT}}$$

Thus we may write

$$v = \alpha_1 * \left(\frac{D_0}{t}\right)^{1/2} e^{\frac{-9{,}900}{RT}}$$

Below 400° C the term $e^{-9{,}900/RT}$ becomes increasingly more important and, as T becomes smaller, the growth velocity can be considered to decrease primarily because of the decreasing diffusion rate of the carbon atoms.

At elevated temperatures, the roles of diffusion and concentration gradient are effectively reversed. Near 700°C for example, a given change in temperature has a stronger effect on the concentration gradient than it does on $D^{1/2}$.

The result of these effects is a growth rate that maximizes at an intermediate temperature. It is small at high temperatures because the concentration gradient tends to diminish, and it is small at low temperatures because of the lowering of the rate of diffusion.

13.8 Interference of Growing Precipitate Particles. According to the Zener theory discussed in the preceding section, the growth velocity for simple geometries should vary inversely as $\sqrt{t}$. This means that with increasing time, the rate of boundary migration must decrease and become very small. This decrease in growth velocity is due to the fact that as the particle grows, it continues to absorb solute atoms from the matrix surrounding it, and the concentration gradient next to the particle decreases. In this theory, each particle is assumed to lie in a matrix of infinite extent. In an actual specimen there will be many particles drawing solute atoms from the matrix, and the distances between them will be finite. In the beginning, as the particles start to grow, there will be no effective competition between particles for the solute atoms, and the Zener assumption is in agreement with the facts. This is indicated schematically for the case of two parallel plate-shaped precipitates in Fig. 13.18A. However, with continued growth, the regions from which the particles are drawing solute atoms will tend to overlap. When this occurs, the maximum value of the solute concentration in the matrix has to fall below n_B^{∞}. This, in turn, must influence the effective value of the concentration gradient at the surface of the particles and, in general, should act to further decrease the

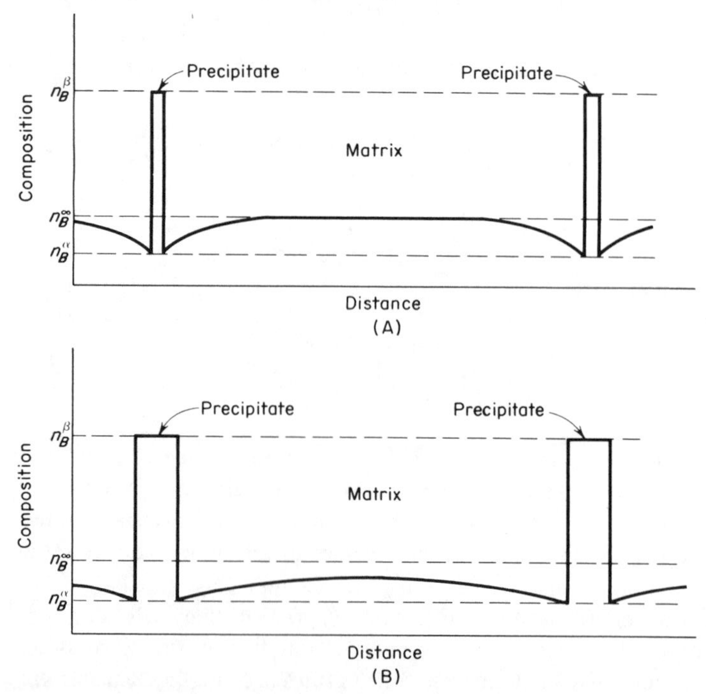

Fig. 13.18 Depletion of the matrix can occur as the precipitate particles grow, as illustrated by this schematic diagram showing planar precipitate particles.

rate of growth. The resultant change in the concentration profile is shown schematically in Fig. 13.18B.

13.9 Interface Controlled Growth. There is still another basic possibility under which growth can occur. In this case the precipitate, as in the preceding sections, is assumed to differ in composition from the matrix. However, here the growth rate is controlled by the mechanism that allows the solute atoms to cross over from the matrix to the precipitate. Thus, suppose that the time required for an atom to jump across the interface is very long compared to that required for it to diffuse to the interface. In this case, as shown schematically in Fig. 13.19, the solute concentration in the matrix will remain effectively constant throughout the matrix. However, with continued growth of

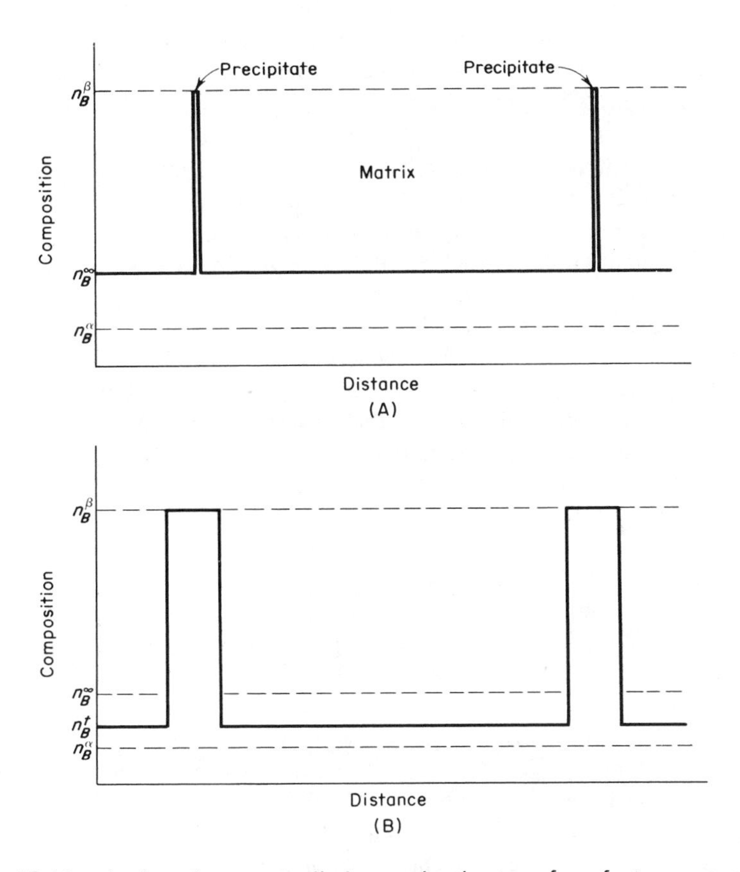

Fig. 13.19 In interface controlled growth, the transfer of atoms across the interface may be so slow that diffusion effectively removes the concentration gradient in the matrix.

the precipitate, the level of the concentration in the matrix has to fall. This means that the driving force for growth also has to fall, since it is directly related to the degree of supersaturation. Attention is called to the difference between this case and that discussed earlier involving the growth of a precipitate where there was no change in composition between the precipitate and the matrix.

Returning to Fig. 13.19, attention is called to the fact that one-dimensional growth is again assumed, and two plate-like precipitates are indicated. Part *A* of this diagram is assumed to correspond to an early period in the growth of the precipitates, while part *B* corresponds to a later time. Note that as a result of the growth of the precipitates, the concentration level in the matrix drops from $n_B{}^\infty$ to $n_B{}^t$. As before, the equilibrium concentration of solute in the matrix is taken as $n_B{}^\alpha$ and the concentration in the precipitate or beta phase as $n_B{}^\beta$.

We have already considered an example of growth controlled by an interface reaction: that involved in a phase transformation of a pure metal in Section 8.6. In some respects the present growth phenomenon is analogous to that already discussed. Thus, with reference to Fig. 13.13, a *B* atom, on leaving the matrix, has to pass over an energy barrier (Δf_a) in order to join the precipitate. After this has occurred, its energy level will have been lowered by the amount $\Delta f_B{}^{\alpha\beta}$. However, there is one important difference: a composition difference exists across the boundary. Thus the number of atoms capable of jumping, that face the surface, is different on the two sides of the boundary. In other words, there now exist two factors, S_1 and S_2, instead of a single factor *S*. However, at equilibrium the jump rate across the boundary has to be equal in the two directions. At the same time, the free energy of a jumping atom is also the same on both sides of the boundary so that $\Delta f^{\alpha\beta} = 0$. Therefore, a reasonable assumption for a reaction occurring near equilibrium is that, effectively, $S_1 = S_2$. With this and the previous assumptions that the strain and surface energies may be neglected, we may write a velocity equation of the form

$$v = \frac{\gamma \nu \Delta f_B{}^{\alpha\beta}}{kT} e^{-\Delta f_a/kT}$$

where λ is a factor proportional to the jump distance of an atom, and ν has the same significance as in the equation in Section 13.6; $\Delta f_B{}^{\alpha\beta}$ is the free energy difference between a *B* atom in the alpha and beta phases; and Δf_a is the energy barrier at the interface that a *B* atom has to pass over in order to join the precipitate. The energy difference, $\Delta f_B{}^{\alpha\beta}$, which is assumed to be much smaller than kT, can now be evaluated in terms of the difference in the partial molal free energy of a *B* atom in the alpha and beta phases. This follows from the definition of the partial molal free energy (*see* p. 379). Thus we have

$$\Delta f_B{}^{\alpha\beta} = \frac{1}{N} \{(\bar{F}_B{}^\alpha)_t - (\bar{F}_B{}^\beta)\}$$

where N is Avogadros number, $(\bar{F}_B{}^\alpha)_t$ is the partial molal free energy of B in the alpha phase, and $(\bar{F}_B{}^\beta)$ is the partial molal free energy of B in the beta phase. Since the partial molal free energies are expressed per mole, their difference is divided by N to yield a free energy difference per atom. The subscript t on the term $(\bar{F}_B{}^\alpha)_t$ signifies that this quantity changes with time. Substitution of the above relation into the velocity equation gives us

$$v = \frac{\gamma \nu e^{-\Delta f_a/kT}}{NkT} \{(\bar{F}_B{}^\alpha)_t - (\bar{F}_B{}^\beta)\}$$

which may also be written

$$v = \frac{\gamma \nu e^{-\Delta f_a/kT}}{RT} \{(\bar{F}_B{}^\alpha)_t - (\bar{F}_B{}^\alpha)_e\}$$

where $(\bar{F}_B{}^\alpha)_e$ is the partial molal free energy of B in the alpha phase when the latter is at equilibrium with the beta phase. The above substitution is possible since, at equilibrium, $(\bar{F}_B{}^\alpha) = (\bar{F}_B{}^\beta)$. The expression can be further simplified if the alpha phase can be assumed to be an ideal solution since, by p. 378, we have

$$\bar{F}_B{}^\alpha = F_B{}^0 + RT \ln N_B{}^\alpha$$

which, on substitution into the velocity equation, yields

$$v = \gamma \nu e^{-\Delta f_a/kT} \{\ln (N_B{}^\alpha)_t - \ln (N_B{}^\alpha)_e\}$$

where $(N_B{}^\alpha)_t$ and $(N_B{}^\alpha)_e$ are the mole fractions of the B component in the alpha phase at time t and at equilibrium, respectively. Furthermore, as the concentration of B in the alpha phase approaches the equilibrium concentration, the above expression may be approximated by

$$v = \gamma \nu e^{-\Delta \tilde{f}_a/kT} \{(N_B{}^\alpha)_t - (N_B{}^\alpha)_e\}$$

It is interesting to compare this relation for the interface-controlled growth of a precipitate with that derived earlier for the growth rate when it is controlled by diffusion. Thus, in the case of one-dimensional, diffusion-controlled growth, it was found that

$$v = \alpha_1{}^* \sqrt{D/t}$$

where $\alpha_1{}^*$ is a function of the compositions of the phases but, within limits, may be assumed to be effectively constant. In comparing these two rate equations, the basic difference between the composition-dependent terms $\{(N_B{}^{\alpha})_t - (N_B{}^{\alpha})_e\}$ and $\alpha_1{}^*$ should be emphasized. Until the solute becomes depleted, $\alpha_1{}^*$ can be considered a constant, whereas the expression $\{(N_B{}^{\alpha})_t - (N_B{}^{\alpha})_e\}$ is not a constant, since $(N_B{}^{\alpha})_t$ decreases with increasing growth of the precipitate. This difference is due to the fact that in diffusion-controlled growth, local equilibrium is attained at the interface. This is not true in interface-controlled growth.

Next, it should be noted that in a precipitation reaction, interface growth rates must generally be much slower than diffusion-controlled growth rates. In a broad sense, the precipitation process involves two mechanisms working in series. An atom has to diffuse up to the interface in order to jump across it. Only when the mean time for an atom to jump across the interface is very long will the interface reaction be the controlling one. Finally, it is evident that both types of reactions have rates that decrease with time. In the diffusion-controlled velocity equation, this is revealed by the fact that $v \propto t^{-1/2}$. In the interface-controlled velocity equation, the time dependence appears in the term $(N_B{}^{\alpha})_t$ which decreases as the solute leaves the matrix.

13.10 Transformations That Occur on Heating. Only phase transformations that occur as a result of cooling have been considered to this point. The corresponding reactions that take place on heating are also significant. There are important differences between the kinetics of the reactions that occur on cooling and those that occur on heating. Thus, consider the melting of a pure metal. Superheating is almost impossible to achieve whereas, under the proper conditions, supercooling can be readily observed. The causes for this difference in behavior are probably related to the fact that it is very easy to heterogeneously nucleate a liquid droplet at an exterior surface or along a grain boundary. In Fig. 13.14 it was shown that the growth rate in the cooling phase transformation from white (β) to grey (α) tin followed a typical nucleation and growth temperature dependence. That is, the growth rate is small just below the transformation temperature, but with decreasing temperature it rises and passes through a maximum and then decreases again. The growth rate of the reversed phase transformation does not show this trend but rises steeply and continuously with increasing temperature, as shown in Fig. 13.20. At very low temperatures, the reaction rate, whether it is nucleation or growth, becomes small because diffusion rates always decrease rapidly with decreasing temperatures. On heating, the diffusion rate is always a steadily increasing function of the temperature. Therefore it is generally true that both the rate of nucleation or the rate of growth will rise continuously with temperatures above a transformation temperature. In other words, the effect of temperature on the diffusion rate always tends to accelerate the reaction as the temperature is raised above the transformation temperature.

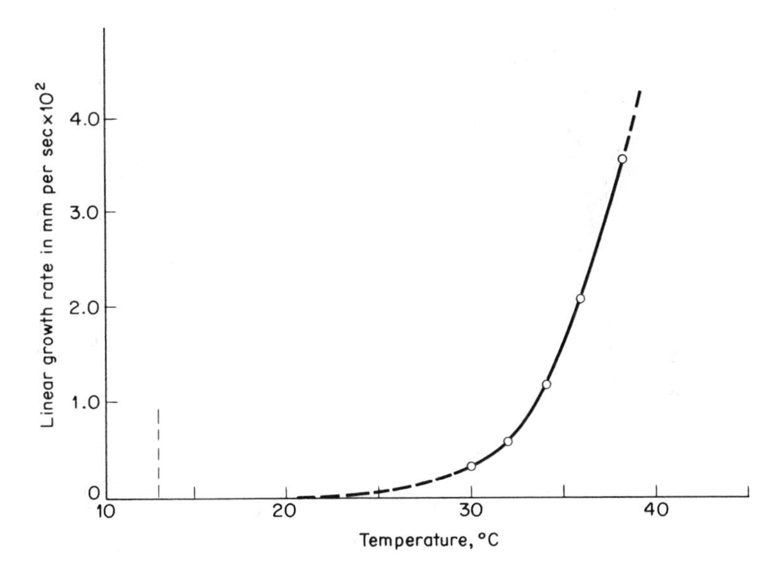

Fig. 13.20 The growth rate in the transformation from alpha tin to beta tin (on heating). After Burgers, W. G., and Groen, L. J., *Disc. Faraday Soc.,* **23** 183 [1957].)

13.11 Dissolution of a Precipitate. Another form of reaction associated with heating is the dissolution of a precipitate. In principle this is the reverse of the precipitation process. The kinetics are usually somewhat different, and this will be discussed briefly below. For the present, let us consider the nature of the process. Fig. 13.21 shows the aluminum-rich end of the aluminum-copper phase diagram. Suppose that one has an alloy, containing about 4 percent copper, that is nearly in equilibrium at 200°C. The structure should consist of small θ particles in a matrix of almost pure aluminum. Such a structure is shown in Fig. 13.22. Since these particles are readily visible under an optical microscope, the metal is actually in the overaged condition. To develop this structure it was necessary to age the alloy for 200 hours at 200°C. Aging at room temperature will not produce a structure of this type within any reasonable time interval, since the reaction rates at room temperature are too slow to produce the equilibrium structure. As can be deduced from Fig. 13.21, heating this specimen to 540°C will increase the solubility of copper in the aluminum to where it can be completely dissolved. Fig. 13.23 shows some measurements made by Batz, Tanzilli, and Heckel[14] that reveal some very interesting data about the solution process of this alloy at 540°C. While the

14. Batz, D. L., Ranzilli, R. A., and Heckel, R. W., *Met. Trans.,* 1 1651 (1970).

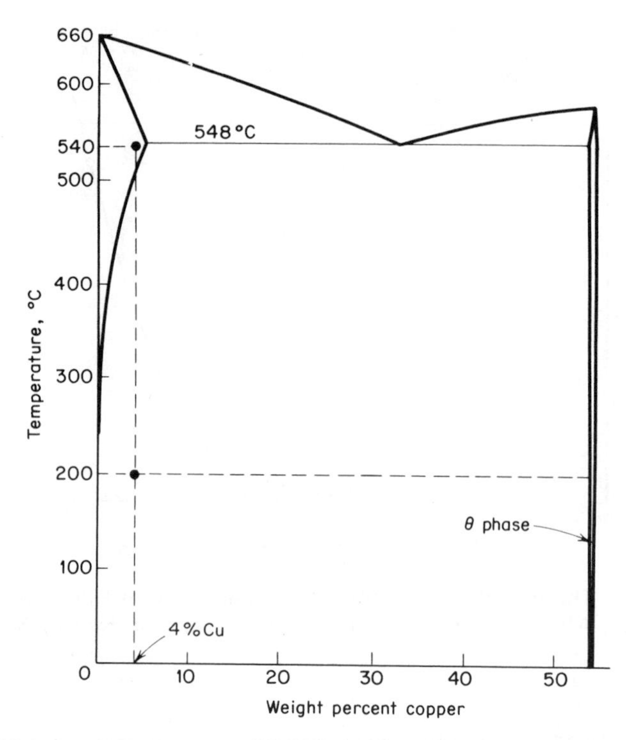

Fig. 13.21 The aluminum end of the aluminum-copper phase diagram.

original average particle size was larger than that shown in Fig. 13.21, this has no real bearing on the results. The abscissa of Fig. 13.22 gives the size of the particles in microns, while the ordinate gives the number per unit volume in a given size class. Each solid line represents the size distribution of the particles. Note that with increasing time these curves are displaced to the left with little change in shape. This implies that the size distribution curve is essentially unaltered in shape as the particles dissolve.

Aaron and Kotler[15] have pointed out a number of practical reasons why the kinetics of dissolution should be better understood. In those cases where the presence of the second-phase particles results in an alloy of superior properties, it is desirable to know how it might be possible to develop a precipitate that will dissolve very slowly if the metal is heated. On the other hand, when the presence of a precipitate is harmful to the properties of an alloy, it is advantageous to know how to decrease the life of the precipitate so that the metal may be more easily homogenized.

The dissolution of a precipitate differs in one fundamental respect from the

15. Aaron, H. B., and Kotler, G. R., *Met. Trans.*, 2 393 (1971).

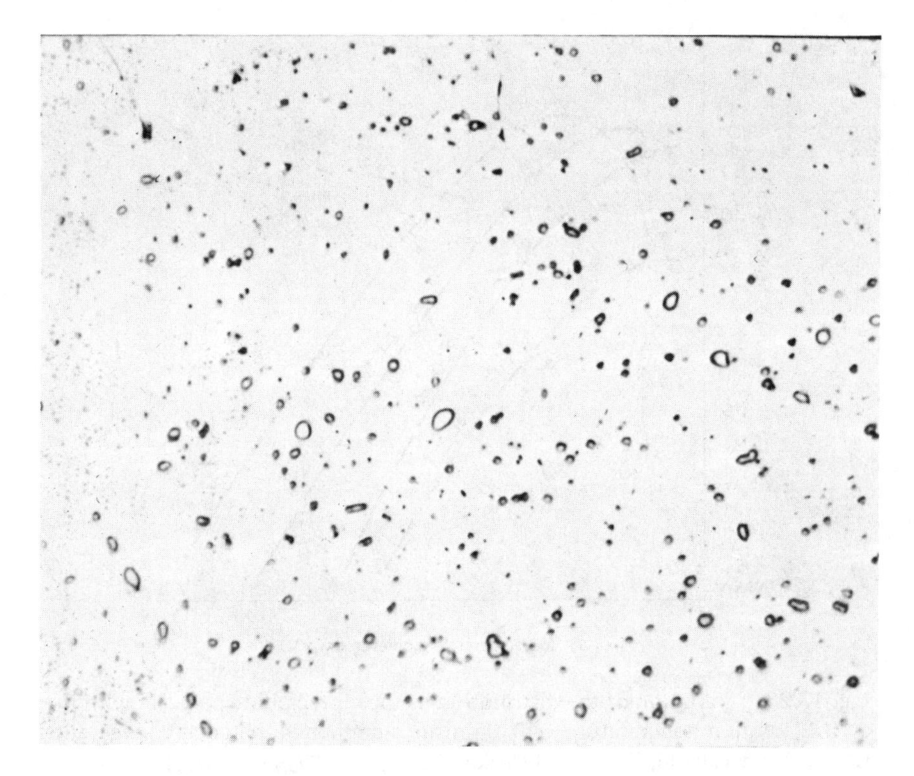

Fig. 13.22 Photomicrograph showing θ or $CuAl_2$ particles in an aluminum-4 percent copper alloy. Magnification 1500 times. (Batz, D. L., Tanzilli, R. A., and Heckel, R. W., *Met. Trans.*, **1** 1651 [1970].) Photograph courtesy of R. W. Heckel.

growth of a precipitate. The process does not require nucleation since the particles that are to be dissolved are already present. The problem, therefore, is essentially one of reversed growth, although the kinetics are usually different. If one limits himself to a one-dimensional analysis, it is possible to show that the kinetics of growth and dissolution are both governed by a $t^{1/2}$ law. Thus Aaron,[16] using an analysis similar to that of Zener in Section 13.6, has shown that the thickness of a precipitate should vary as

$$x(t) = x_o - k\sqrt{Dt}$$

where $x(t)$ is the thickness of the planar precipitate at any time t, x_o is its original thickness, k is a constant, and D is the diffusion coefficient of the solute.

16. Aaron, H. B., *Materials Sci. J.*, **2** 192 (1968).

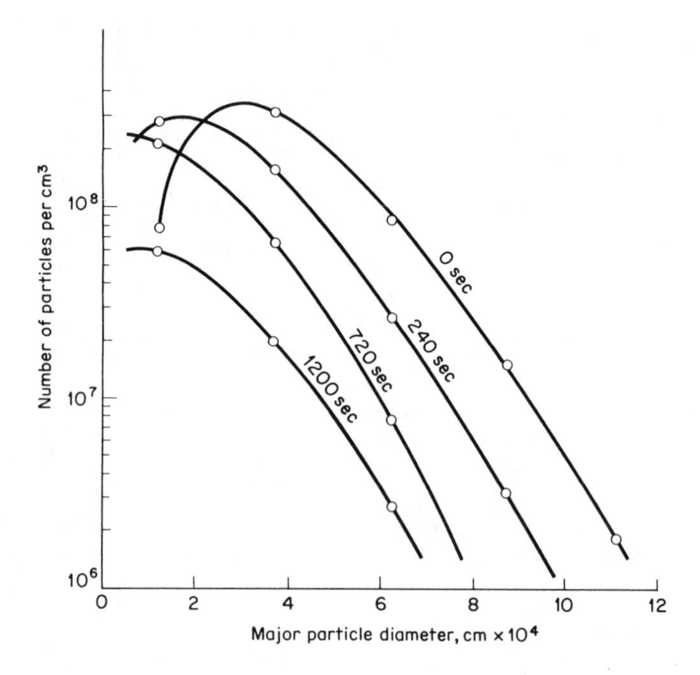

Fig. 13.23 Variation of the distribution of the θ precipitate particles with time at the solution temperature 540°C in an aluminum-4 percent copper alloy. (Baty, D. L., Tanzilli, R. A., and Heckel, R. W., *Met. Trans.,* 1 1651 [1970].)

In the case of a spherical precipitate, the growth and dissolution kinetics are not generally so simply related. This subject is beyond the scope of this book, and for further information, one is referred to the review paper by Aaron and Kotler.[17]

Problems

1. The boiling point of copper is 2600°C and the surface energy of liquid copper is approximately 1100 ergs/cm^2. Assuming that the atomic diameter of copper in the liquid phase equals its diameter in the solid phase (2.55 Å), compute the equilibrium number of embryos containing 10 atoms in one mole of copper vapor just above its boiling point.
2. (*a*) In order to evaluate the significance of the answer to Problem 1, compute the number of these embryos in a volume equal to that of an electron microscope foil where the magnification is 20,000 times and the foil thickness is 2000 Å. Assume that the foil is viewed on a screen 10 cm wide by 10 cm long.

17. Aaron, H. B., and Kotler, G. R., *Met. Trans.*, 2 393 (1971).

(*b*) In a transmission optical microscope where the magnification is 200 times, it may be possible to focus over a depth of 2 microns. What volume would be in the field of view? Assume a field of view of 10 cm by 10 cm. How many embryos might exist in such a volume?

3. Under the conditions of Problem 1, how many more embryos will there be containing 10 atoms than containing 20 atoms?
4. In the text on p. 492, the free energy associated with the formation of a critical nucleus in freezing is given as $\Delta F_{nc} = A/\Delta T^2$, where ΔT is the supercooling. The value of A is stated to be $4\eta^3 \gamma_{ls}{}^3 T_0{}^2/27(\Delta h^{ls})^2$. Prove this relationship.
5. Consider the following data for the freezing of tin: melting point 232°C, heat of fusion 14.5 cal/gm, atomic weight 118.7 gm/mole, and surface energy of the liquid-solid interface 59 ergs/cm^2.
   (*a*) Compute Δh^{ls}, the heat of fusion per atom, in ergs.
   (*b*) How many electron volts is $\Delta h^{ls}{}_T$
6. (*a*) Assuming that $\Delta f^{ls} = \Delta h^{ls} \Delta T/T_0$ and using the data either given or derived in Problem 5, evaluate the coefficients of n and $n^{2/3}$ in the equation

$$\Delta F_n = \Delta f^{ls} n + \eta \gamma_{ls} n^{2/3}$$

   (*b*) Determine the number of atoms in a critical nucleus when the supercooling is 10°C.
7. (*a*) Using the result of Problem 6, determine the free energy required to form a critical nucleus when the supercooling is 10°K.
   (*b*) How many times larger than kT is ΔF_{nc}?
   (*c*) Is homogeneous nucleation of significance just below the melting point in a metal like tin? Explain.
8. Repeat Problem 7 assuming that the supercooling is 100° K. In this solution use the equation

$$\Delta F_{nc} = \frac{A}{\Delta T^2} \quad \text{where } A = \frac{4\eta^3 \gamma_{ls}{}^3 T_o{}^2}{27(\Delta h^{ls})^2}$$

   First evaluate A and then solve for ΔF_{nc}.
9. Now consider the equation giving the nucleation current in freezing on p. 492. It is normally assumed that the vibration rate of the atoms in solids is about 10^{13} vibrations per sec. The energy barrier Δf_a that an atom has to pass over in going from a liquid to the solid probably is around 0.2 electron volts.
   (*a*) compute the nucleation rate per mole of tin when the supercooling is 100°C.
   (*b*) If the supercooling was 150°C, what would be the nucleation rate per mole?
   (*c*) What is the significance of the answers to the above parts (*a*) and (*b*)?

10. (*a*) The maximum observed supercooling in silver is 230°C. Using the data in the accompanying table, compute the freezing nucleation rate of this metal when it has been supercooled by this amount.

a = 2.89 Å	$\nu = 10^{13}$ vib/sec
T_m = 960.5 °C	$N_o = 6 \times 10^{23}$ atoms/mol
γ_{ls} = 126 ergs/cm^2	Δf_a = 0.4 eV
ΔH_{ls} = 2700 cal/mol	

(*b*) With a supercooling of 250° C, what would be the nucleation rate?

11. This diagram represents a hypothetical solid embryo of silver growing against a mold wall.

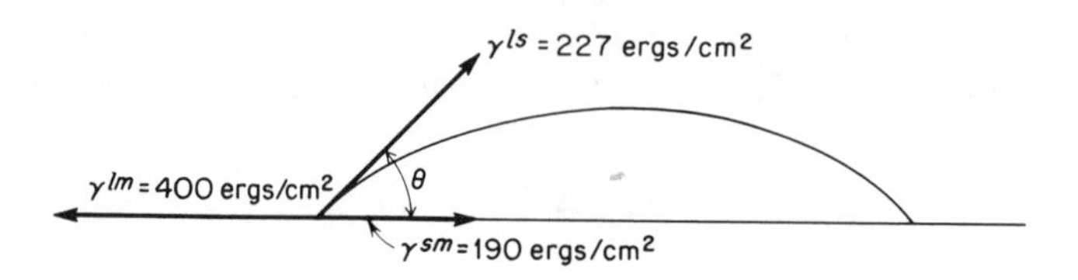

(*a*) Compute the angle of contact θ.

(*b*) Determine the magnitude of the factor used to convert the homogeneous nucleation free energy into a heterogeneous nucleation free energy.

12. (*a*) Under the conditions given in Problem 11, what would be the heterogeneous nucleation rate for silver when the metal is supercooled by 20°C? Use data for silver given in Problem 10 and determine I in nuclei per cm^2 per sec. In other words, express N^m as atoms per cm^2.

(*b*) Does this result imply that, under these conditions, it might be possible to supercool by about 20°C? If not, estimate the maximum possible attainable degree of supercooling, assuming it is not possible to go below the supercooling temperature where the nucleation rate becomes greater than 1 per second.

13. If the contact angle θ between the mold wall and a silver embryo is 2°, at what approximate supercooling will the heterogeneous nucleation rate equal 1 nucleus per cm^2 per sec?

14. With the knowledge that the growth rate equations given in Section 13.6 ignore a number of important factors, such as the strain energy and the surface energy, it is still interesting to apply this simple theory to an analysis of the growth of the alpha phase from the gamma phase in pure iron. For this purpose, take the data given in the accompanying table and make the following computations. *Note:* at 300°K, kT may be taken as 1/40 eV.

Let $\lambda = a$ and assume Δf_a to be independent of temperature.

$a = 2.48$ Å
$\nu = 10^{13}$ vib/sec
Heat of Transformation = 215 cal/mole
Transformation Temperature = 910 °C
$\Delta f_a = 1.3$ eV

(*a*) Determine the value of $\Delta f^{\gamma\alpha}$ at 20° intervals within the temperature range 710°C to 910°C.
(*b*) Compute the growth velocity at each of these temperatures.
(*c*) Plot your results as a curve of growth rate vs. temperature.

15. Assume that three pairs of iron specimens containing 0.09 atomic percent carbon are heated to 720°C and allowed to come to equilibrium. Following this, each pair is quenched to a pair of the temperatures as listed below. In each case, with the aid of Fig. 13.17, determine the ratio of the product of the two factors $\alpha_1{}^*$ and $\sqrt{D}$ in the Zener one-dimensional growth rate equation, that is, determine $(\alpha_1{}^*\sqrt{D})_{T_1}/(\alpha_1{}^*\sqrt{D})_{T_2}$.
(*a*) For temperature of quenching bath700°C and 660°C.
(*b*) 540°C and 500°C.
(*c*) 60°C and 100°C.
16. Compare the answers to the three parts of the preceding problem. What do they imply about the overall effect of temperature on the growth velocity?
17. (*a*) Consider an iron containing 0.09 atomic percent carbon that has been equilibriated at 720°C and then quenched to 300°C. How long would it take, according to Zener's equation, for a plate-shaped precipitate to increase its thickness by 1μ?
(*b*) How thick an expanse of the matrix would have to have its concentration depleted from 0.09 to that of $n_\alpha{}^e$ in order to form a plate of cemenite 1μ thick? Does this answer help explain the result in part (*a*) of this problem?
(*c*) How long would it take for an increase in thickness of $10^{-2}\mu$?
18. For an interface reaction to control the growth rate, the rate at which atoms diffuse up to the interface has to be much faster than the jump rate across the interface. Normally this means that the energy barrier Δf_a has to be large. As an exercise, compute a hypothetical Δf_a that would make the one-dimensional growth rate of cementite in iron 10^{-2} times slower than the diffusion-controlled growth rate given by Zener's relation $v = \alpha_1{}^*/2\sqrt{D/t}$. Take the temperature as 400°C, $D = 0.008e^{-19{,}800/kT}$, $t = 100$ sec, and assume that λ in the interface-controlled growth velocity equation can be represented by

$$\gamma = \{(n_B{}^\alpha)_t - (n_B{}^\alpha)_e\}\, a/2\,\{(n_B{}^\beta) - (n_B{}^\alpha)_t\}$$

where a is the atom diameter of iron, 2.48 Å.

14 Binary Phase Diagrams

Phase diagrams, also called equilibrium diagrams or constitution diagrams, are a very important tool in the study of alloys. They define the regions of stability of the phases that can occur in an alloy system under the condition of constant pressure (atmospheric). The coordinates of these diagrams are temperature (ordinate) and composition (abscissa). Notice that the expression "alloy system" is used to mean all the possible alloys that can be formed from a given set of components. This use of the word system differs from the thermodynamic definition of a system which refers to a single isolated body of matter. An alloy of one composition is representative of a thermodynamic system, while an alloy system signifies all compositions considered together.

The interrelationships between the phases, the temperature, and the composition in an alloy system are shown by phase diagrams only under equilibrium conditions. These diagrams do not apply directly to metals not at equilibrium. A metal quenched (cooled rapidly) from a higher temperature to a lower one (for example, room temperature) may possess phases that are more characteristic of the higher temperature than they are of the lower temperature. In time, as a result of thermally activated atomic motion, the quenched specimen may approach its equilibrium low-temperature state. If and when this occurs, the phase relationships in the specimen will conform to the equilibrium diagram. In other words, the phase diagram at any given temperature gives us the proper picture only if sufficient time is allowed for the metal to come to equilibrium.

In the sections that follow, unless specifically stated otherwise, it will always be implicitly implied that equilibrium conditions hold. Further, because phase diagrams are of great importance in solving problems in practical metallurgy, we shall conform to common usage and express all compositions in weight percent instead of atomic percent, as has been the case heretofore.

14.1 Isomorphous Alloy Systems In the discussion that follows, only two-component, or binary, alloy systems will be considered. The simplest of these systems is the *isomorphous,* in which only a single type of crystal structure is observed for all ratios of the components. A typical isomorphous

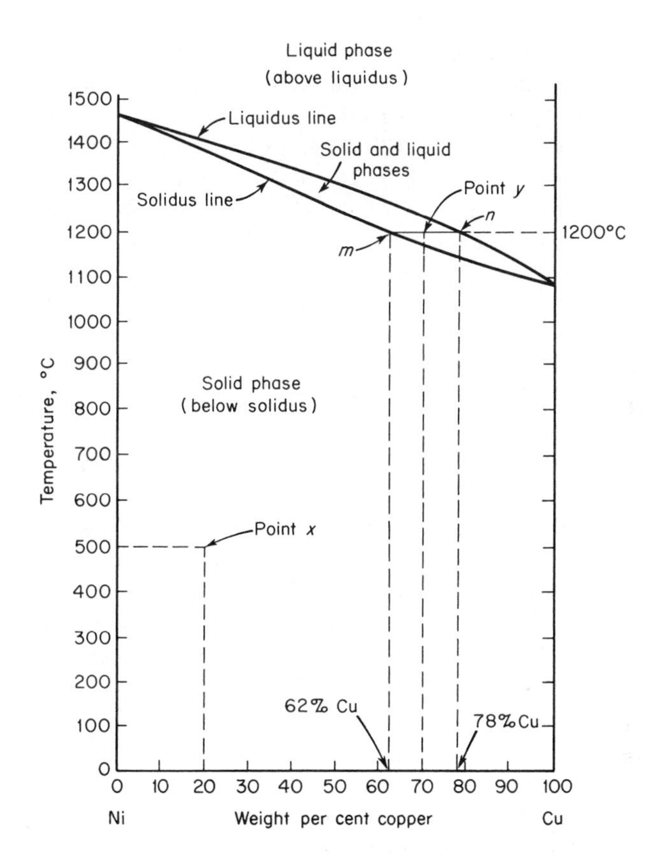

Fig. 14.1 The copper-nickel phase diagram.

system is represented in the phase diagram of Fig. 14.1. The binary alloy series in question is copper-nickel. These elements combine, as in all alloy systems of this type, to form only a single liquid phase and a single solid phase. Since the gaseous phase is generally not considered, there are only two phases involved in the entire diagram. Thus, the area in the figure above the line marked "liquidus" corresponds to the region of stability of the liquid phase, and the area below the solidus line represents the stable region for the solid phase. Between the liquidus and solidus lines is a two-phase area where both phases can coexist.

The significance of several arbitrary points located on Fig. 14.1 will now be considered. A set of coordinates – a temperature and a composition – is associated with each point. By dropping a vertical line from point x until it touches the axis of abscissae, we find the composition 20 percent Cu. Similarly, a horizontal line through the same point meets the ordinate axis at 500°C. Point x signifies an alloy of 20 percent Cu and 80 percent Ni at a temperature of

500°C. The fact that the given point lies in the region below the solidus line tells us that the equilibrium state of this alloy is the solid phase. The structure implied is, therefore, one of solid-solution crystals, and each crystal will have the same homogeneous composition (20 percent Cu and 80 percent Ni). Under the microscope, such a structure will be identical to a pure metal in appearance. In other respects, however, a metal of the given composition will have different properties. It should be stronger and have a higher resistivity than either pure metal, and it will also have a different surface sheen or color.

Let us turn our attention to the point, *y*, which falls inside the two-phase region, the area bounded by the liquidus and solidus lines. The indicated temperature in this case is 1200°C, while the composition is 70 percent Cu and 30 percent Ni. This 70 to 30 ratio of the components represents the average composition of the alloy as a whole. It should be remembered that we are now dealing with a mixture of phases (liquid plus solid) and that neither possesses the average composition.

To determine the compositions of the liquid and the solid in the given phase mixture, it is only necessary to extend a horizontal line through point *y* until it intersects the liquidus and solidus lines. These intersections give the desired compositions. In Fig. 14.1, the intersections are at points *m* and *n* respectively. A vertical line dropped from point *m* to the abscissa axis gives 62 percent Cu, which is the composition of the solid phase. In the same manner, a vertical line dropped from point *n* shows that the composition of the liquid is 78 percent Cu.

14.2 The Lever Rule. A very important relationship, one that applies to any two-phase region of a two-component, or binary, phase diagram, is the so-called *lever rule*. With regard to the lever rule, consider Fig. 14.2, which is an enlarged portion of the copper-nickel diagram of Fig. 14.1. Line *mn* is the

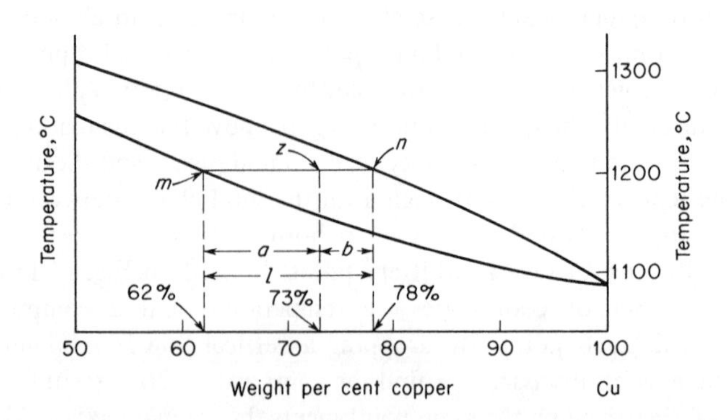

Fig. 14.2 The lever rule.

same in the new figure as in the old and lies on the 1200°C isothermal. One change has been made, however, and that is in the average alloy composition that is now assumed to be at point z, or at 73 percent Cu. This new composition, since it still lies on the line mn and therefore in a two-phase region must also be a mixture of liquid and solid at the given temperature. Further, since points m and n are identical in both Fig. 14.1 and 14.2, it can be concluded that this new alloy must be composed of phases whose compositions are the same as those in the previous example. The quantity that does change, when the composition is shifted along line mn from point y to z, is the relative amounts of liquid and solid. The compositions of the solid and the liquid are fixed as long as the temperature is constant. This conclusion is in perfect agreement with our thermodynamic deductions in Chapter 12, where it was shown that at constant temperature and pressure the compositions of the phases in a two-phase mixture are fixed.

While a composition shift of the alloy (taken as a whole) cannot alter the compositions of the phases in a two-phase mixture at constant temperature, it does change the relative amounts of the phases. In order to understand how this occurs, we shall determine the relative amount of liquid and of solid in an alloy of a given composition. For this purpose, take the alloy corresponding to point z, which has the average composition 73 percent Cu. In 100 grams of this alloy, 73 of them must be copper and the remaining 27 nickel, so that we have

Total weight of the alloy = 100 grams
Total weight of copper = 73 grams
Total weight of nickel = 27 grams

Let w represent the weight of the solid phase of the alloy expressed in grams, and $(100 - w)$ the weight of the liquid phase. The amount of copper in the solid form of the alloy equals the weight of this phase times the percentage of copper (62 percent) which it contains. Similarly, the weight of copper in the liquid phase is equal to the weight of liquid times the percentage of copper in the liquid (78 percent). Thus

Weight of copper in the solid phase $= 0.62w$
Weight of copper in the liquid phase $= 0.78(100 - w)$

The total weight of copper in the alloy must equal the sum of its weight in the liquid and in the solid phases, or

$$73 = 0.62w + 0.78(100 - w)$$

Collecting similar terms

$$73 - 78 = (0.62 - 0.78)w$$

and solving for w, the weight of the solid phase,

$$w = \frac{5}{0.16} = 31.25 \text{ grams}$$

Since the total weight of the alloy is 100 grams, the weight percent of the solid phase is 31.25 percent, and the corresponding weight percent of the liquid phase is 68.75 percent.

Now reexamine the equation given above

$$73 - 78 = (0.62 - 0.78)w$$

and divide each side by 100 grams, the total weight of the alloy. If this is done one obtains

$$0.78 - 0.73 = (0.78 - 0.62)\frac{w}{100}$$

or

$$\frac{w}{100} = \frac{0.78 - 0.73}{0.78 - 0.62}$$

where $w/100$ is the weight percent of the solid phase.

The above equation is worthy of careful consideration. In this expression the denominator is the difference in composition of the solid and the liquid phase (expressed in weight percent copper), that is, it is exactly the composition difference between points m and n. On the other hand, the numerator is the difference between the composition of the liquid phase and the average composition (composition difference of points n and z). The amount of the solid phase (expressed in weight percent of the total alloy) is therefore given by

$$\text{percent solid} = \frac{\text{composition of liquid} - \text{average composition}}{\text{composition of liquid} - \text{composition of solid}}$$

Similarly, it may be shown that

$$\text{percent liquid} = \frac{\text{average composition} - \text{composition of solid}}{\text{composition of liquid} - \text{composition of solid}}$$

The above equations represent the lever rule as applied to one specific problem. The same relationships can be expressed in a somewhat simpler form, for in Fig. 14.2, a, b, and l represent composition differences between the points zm, nz, and nm respectively, so that

$$\text{percent solid} = \frac{b}{l}$$

$$\text{percent liquid} = \frac{a}{l}$$

The above information will now be summarized.

Given a point such as z in Fig. 14.2, which lies inside a two-phase region of a binary-phase diagram,

(a) Find the composition of the two phases.

Draw an isothermal line (a *tie line*), through the given point. The intersections of the tie line with the boundaries of the two-phase region determine the composition of the phases. (In the present example, points m and n determine the compositions of the phases, solid and liquid respectively.)

(b) Find the relative amounts of the two phases.

Determine the three distances, a, b, and l (in units of percent composition), as indicated in Fig. 14.2. The amount of the phase corresponding to point m is given by the ratio b/l, while that corresponding to point n is given by a/l. Notice carefully that the amount of the phase on the left (m) is proportional to the length of the line segment (b) lying to the right of the average composition (point z), while the phase that lies at the right (n) is proportional to the line segment (a) lying to the left of point z.

14.3 Equilibrium Heating or Cooling of an Isomorphous Alloy.

Equilibrium heating or cooling means a very slow rate of temperature change so that at all times equilibrium conditions are maintained in the system under study. Just how slowly one has to heat or cool an alloy in order to keep it effectively in a state of equilibrium depends on the metal under consideration and the nature of the phase changes that occur as the temperature of the alloy is varied. For present purposes, it will arbitrarily be assumed that all temperature changes are made at a slow enough rate that equilibrium is constantly maintained. In Chapter 15 some aspects of nonequilibrium temperature changes, when the more intimate details of freezing are investigated will be considered. For the time being, however, our attention will be concentrated on the phase changes that occur as a result of temperature variations.

First, the phase changes that occur when a specific alloy of an isomorphous system has its temperature varied through the freezing range will be considered. For this purpose, the hypothetical phase diagram of Fig. 14.3 with assumed components A and B will be used. As an arbitrary alloy, consider the

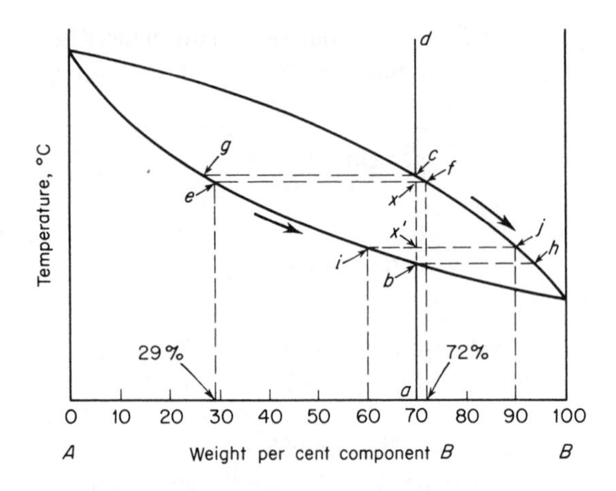

Fig. 14.3 Equilibrium cooling of an isomorphous alloy.

composition 70 percent B and 30 percent A. A vertical line is drawn through this composition in Fig. 14.3. Two segments of this line are drawn as solid lines (*ab* and *cd*), while the third segment between points b and c is shown as a dashed line. The solid portions fall in single-phase regions: *ab* in the solid-solution area, *cd* in the liquid-solution area. The dashed segment *bc* lies in a two-phase region of liquid-plus-solid phases. Any point on one of the solid sections corresponds to a single homogeneous substance. Points in the length *bc*, on the other hand, do not correspond to a single homogeneous form of matter, but to two – a liquid solution and a solid solution – each with a different composition. Furthermore, the compositions of these phases change as the temperature inside the two-phase region is varied. This can perhaps be best understood by considering the complete cycle of the phase change that occurs as the alloy is cooled from point d down to room temperature at point a.

At point d, the alloy is a homogeneous liquid and, until point c is reached, it remains as a simple homogeneous liquid. Point c, however, is the boundary of the two-phase region, which signifies that solid begins to form from the liquid at this temperature. The freezing of the alloy, accordingly, commences at the temperature corresponding to point c. Conversely, when the alloy has been cooled to point b, it must be completely frozen, for at all temperatures below b the alloy exists as a single solid-solution phase. Point b, therefore, corresponds to the end of the freezing process. An interesting feature of the freezing process is at once apparent: the alloy does not freeze at a constant temperature, but over a temperature range.

The analysis of phase-change phenomena that occur in the freezing range between points c and b can now be made rather easily with the aid of the two rules set down at the end of the last section. Suppose that the temperature of the alloy has been lowered to a position just below point c. This position is designated by the symbol x in Fig. 14.3, and at this position

(*a*) An isothermal line drawn through point x determines the composition of the phases by its intersections with the liquidus and solidus lines respectively.

Liquid phase (point *f*) has the composition 72 percent *B*.
Solid phase (point *e*) has the composition 29 percent *B*.

(*b*) By the lever rule the amount of the solid phase is:

$$\frac{x-f}{e-f}=\frac{70-72}{29-72}=\frac{2}{43}=4.7 \text{ percent}$$

and the amount of the liquid phase is

$$\frac{e-x}{e-f}=\frac{29-70}{29-72}=\frac{41}{43}=95.3 \text{ percent}$$

The above data show that the given alloy, at a temperature slightly below the temperature at which freezing starts, is still largely a liquid (95.3 percent) and the composition of the liquid (72 percent *B*) is close to the average composition. On the other hand, the solid phase, which is present in only a small quantity (4.7 percent), possesses a composition differing considerably from the average (29 percent *B*).

Now let us assume that the alloy is very slowly cooled from point x to point x'. If this operation is done slowly enough, equilibrium conditions will be maintained at the end of the drop in temperature and we can apply once more the two rules for determining the compositions and amounts of the phases. When this is done, it is found that the composition of the solid phase is now 60 percent *B* and that of the liquid phase is 90 percent *B*, while the amount of the liquid and the solid phase is 33 percent and 67 percent respectively. Comparing these figures with those for the higher temperature (corresponding to point x) shows two important facts. First, that as the temperature is lowered in the freezing range, the amount of the solid increases, while the amount of the liquid decreases. This result might well be expected. Second, as the temperature falls, the compostions of both phases change. This fact is not self-evident. It is also interesting to note that the composition shift of both phases occurs in the same direction. Both the liquid and the solid become richer in component *B* as the temperature drops lower and lower. This apparently anomalous result can only be explained by the fact that as the temperature falls there is also a corresponding change in the amounts of the phases and the fact that at all times the average composition of the liquid and the solid is constant.

Another important fact relative to the freezing process in a solid-solution type of alloy is that as the process continues and more and more solid is formed, there must be a continuous composition change taking place in the solid that has already frozen. Thus, at the temperature of point x, the solid has the composition 29 percent B, but at the temperature of x' it has the composition 60 percent B. The only way that this can occur is through the agency of diffusion. Since the composition change is in the direction of increasing concentrations of B atoms, this implies a steady diffusion of B atoms from the liquid toward the center of the solid, and a corresponding diffusion of A atoms in the reverse direction.

The equilibrium freezing of an isomorphous alloy can now be analyzed. On cooling the liquid solution, freezing starts when the liquidus line is reached (point c, Fig. 14.3). The first solid to form has the composition determined by the intersection of an isothermal line drawn through point c with the solidus line (point g). As the alloy is slowly cooled, the composition of the solid moves along the solidus line toward point c, while, simultaneously, the composition of the liquid moves along the liquidus line toward point h. However, at any given instant in the cooling process, such as typified by point x or x', the solid and the liquid must lie at the ends of an isothermal line. In addition, as the freezing process proceeds, the relative amounts of liquid and solid change from an infinitesimal amount of solid in a very large amount of liquid at the start of the process to the final condition of complete solidification.

The above discussion has been concerned with the freezing process. The reverse process, melting, in which the alloy is heated from a low temperature, where it is a solid, into the liquid phase, is equally simple to analyze. In this case, point b represents the temperature at which melting commences, point h the composition of the first liquid to form, and point g the composition of the last solid to dissolve.

It is well to note at this point that for all compositions of an isomorphous system, freezing or melting occurs over a range of temperatures. Thus, the melting and freezing points do not coincide as they do in pure metals.

14.4 The Isomorphous Alloy System from the Point of View of Free Energy. In an alloy system such as copper-nickel, the atoms of the components are so similar in nature that it can be assumed, at least as a first approximation, that they form an ideal solution in both liquid and solid phases. The free-energy-composition curves for both phases must, therefore, be similar in nature to that of Fig. 12.5. Consider now the situation that obtains at the melting point of pure nickel. At this temperature (1455°C), the free energies of the liquid and of the solid phases are equal at the composition of pure nickel. Figure 14.1 shows us, however, that for any other alloy composition of copper and nickel at this temperature the liquid phase is the stable phase and, therefore,

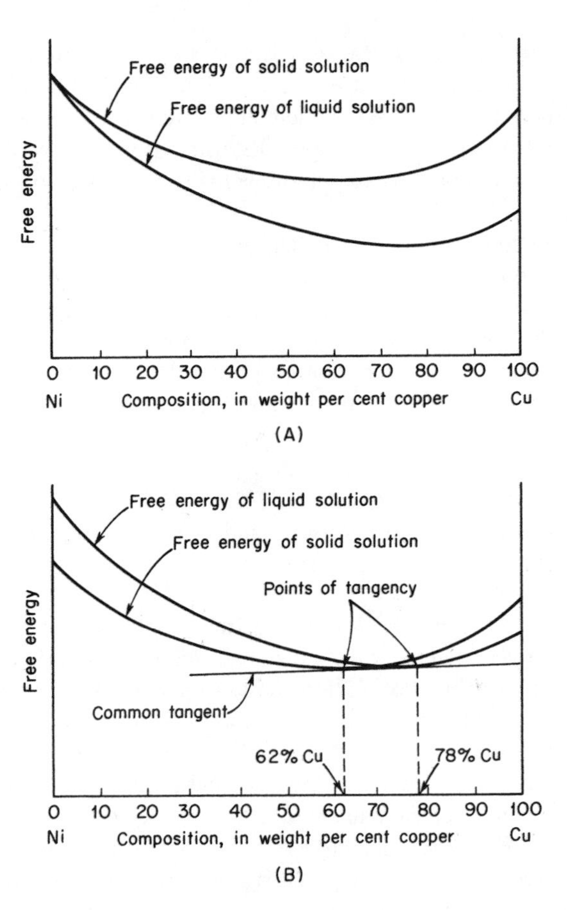

Fig. 14.4 Free-energy-composition curves for the copper-nickel alloy system. (A) Freezing point of pure nickel; 1455°C (B) 1200°C.

its free energy must lie below that of the solid phase. This relationship between the two free-energy curves is shown in Fig. 14.4A. Notice that the two curves intersect only at the composition of pure nickel. At any temperature higher than 1455°C, the two curves will separate so that the liquid-phase-free-energy-composition curve lies entirely below that of the solid phase. On the other hand, decreasing the temperature to a value that is still above the freezing point of pure copper (1083°C) has the effect of shifting the curve of the liquid phase upward with respect to that of the solid phase. The two curves now intersect at some intermediate composition, as shown in Fig. 14.4B. This figure represents the free energy of the two solutions at the same temperature (1200°C) as points y and z in Fig. 14.2 and 14.3 respectively. Notice that in this figure the common tangent drawn to the two free-energy curves has points of tangency at the compositions 62 percent Cu and 78 percent Cu. These values are the same as indicated in Fig. 14.2 and 14.3 for the compositions of the phases at 1200°C, which agrees with the results of Chapter 12, where it was shown that the points of contact of the common tangent determine the compositions of the phases in a two-phase mixture.

The complete copper-nickel-phase diagram can be mapped out by considering the relative motion of the two free-energy-composition curves as the temperature is dropped. With decreasing temperature, the effect, as indicated above is to raise the liquid-phase curve relative to that of the solid phase. As this happens, the point of intersection shifts continuously from the composition of pure nickel at the temperature 1455°C to the composition of pure copper at 1083°C. Concurrent with this movement of the point of intersection toward larger copper concentrations is a similar motion of both points of contact of the common tangent to the two free-energy curves. Below 1083°C, the two free-energy-composition curves no longer intersect, and the curve representing the solid phase lies entirely below the liquid curve.

14.5 Maxima and Minima. The copper-nickel phase diagram is typical of an alloy system in which the free-energy-composition curves, for a given temperature, of the two solution phases (liquid and solid) intersect at only one composition. Other alloy systems are known which have equivalent curves intersecting at two compositions. When this happens, it is generally observed that in the corresponding phase diagrams liquidus and solidus curves are so shaped as to form either a minimum or a maximum. This can be demonstrated graphically with the aid of Fig. 14.5 and 14.6. Figure 14.5 shows schematically the relationships between the free-energy curves that lead to a minimum configuration. In this case, the curve for the solid has less curvature than that for the liquid. The figure on the left of Fig. 14.5 shows that, with decreasing temperature, intersections of the two free-energy curves occur first at the compositions of pure components (A and B). These intersections then move inward toward the center of the figure and eventually meet at a single point. At the temperature where this occurs (T_c), the two free-energy curves are tangent. Continued lowering of the temperature (T_d) causes the free-energy curves to separate and makes the solid phase the only stable phase at all compositions. In an example such as this, it is possible to draw two common tangents to the free-energy curves. The points at which the tangents contact the free-energy curves determine the limits of the two-phase regions of the equilibrium diagram at any given temperature – for example, T_b. The motion of these points, with varying temperature, maps out the phase diagram shown on the right of Fig. 14.5.

A series of figures, similar to that of Fig. 14.5, is shown in Fig. 14.6. In this case, however, the solid solution is assumed to have the free-energy curve with

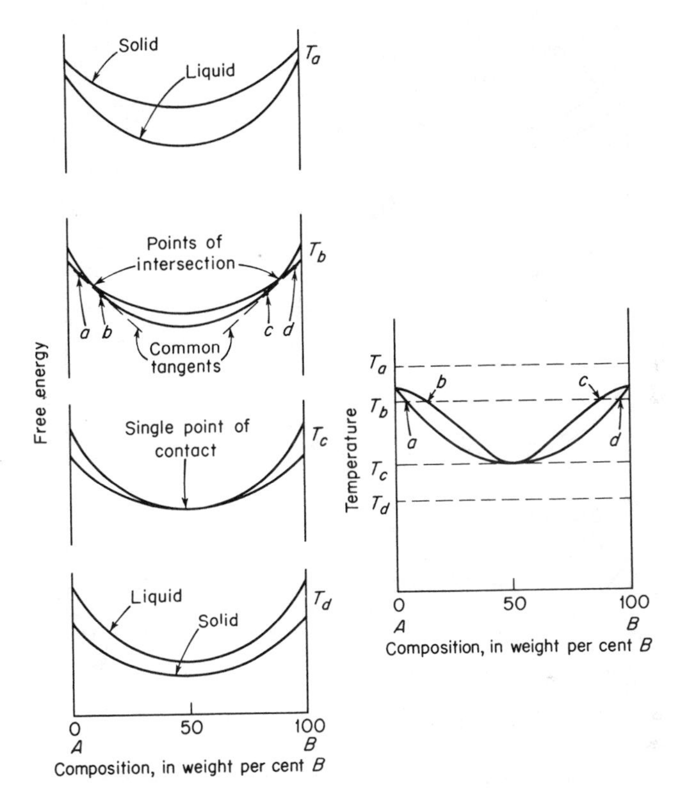

Fig. 14.5 Relationship of the free-energy curves that lead to a minimum.

the greater curvature, so that the free-energy curves, on falling temperature, meet first at a single central point at temperature T_b. This single intersection then splits into two, with the result that the phase diagram has the shape shown on the right of Fig. 14.6.

An important aspect of phase diagrams is shown in Fig. 14.5 and 14.6. When the boundaries of a two-phase region intersect, they meet at a maximum or minimum, and both curves (liquidus and solidus) are tangent to each other and to an isothermal line at the point of intersection. Such points are called *congruent points*. It is characteristic of congruent points that freezing can occur with no change in composition or temperature at these points. Thus, the freezing of an alloy at a congruent point is similar to the freezing of a pure metal. The resulting solid, however, is a solid solution and not a pure component.

It should be further emphasized that the boundaries that define the limits of a two-phase region in an equilibrium diagram can only meet at either congruent points, or at the compositions of pure components. These points, where the

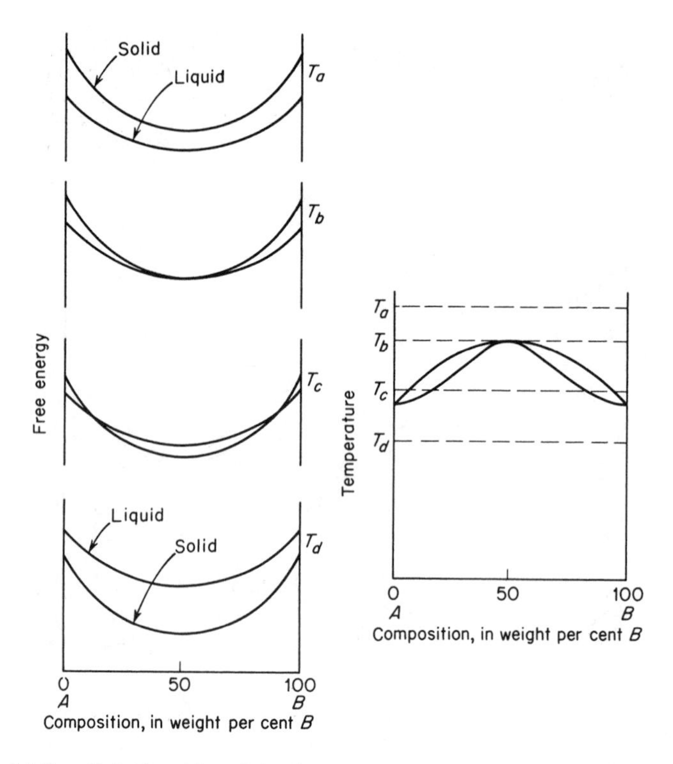

Fig. 14.6 Relationship of the free energy curves that lead to a maximum.

phase boundaries meet, are known as *singular points* and, accordingly, in an equilibrium diagram, the single-phase regions, or fields, are always separated by two-phase regions except at singular points.

A number of isomorphous phase diagrams show congruent points that correspond to minima of the liquidus and solidus curves. A typical example is shown in Fig. 14.7, which is the equilibrium diagram for the gold-nickel-alloy system. The significance of the solid line lying below the liquidus and solidus lines will be considered in a later section entitled "Miscibility Gaps."

Congruent points, at which the liquidus and solidus meet at a maxima, are not ordinarily found in simple isomorphous equilibrium diagrams. They are observed, however, in several more complicated alloy systems that possess more than a single solid phase. One of the simpler of the latter is the lithium-magnesium system whose phase diagram is shown in Fig. 14.8. The maximum appears at 601°C and 13 percent Li.

14.6 Superlattices. Copper and gold also freeze to form a continuous series of solid solutions, as shown in Fig. 14.9. The chemical

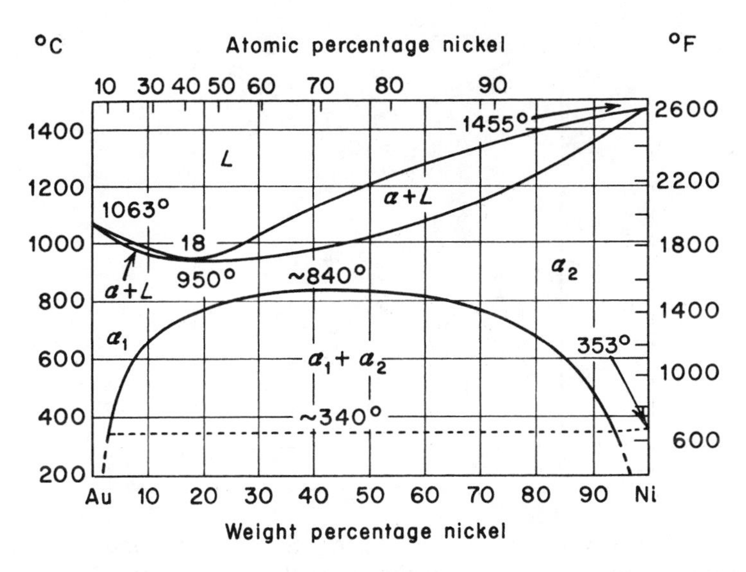

Fig. 14.7 Gold-nickel phase diagram. (From Carapella, Louis A., *ASM Metals Handbook*, 1948 ed.).

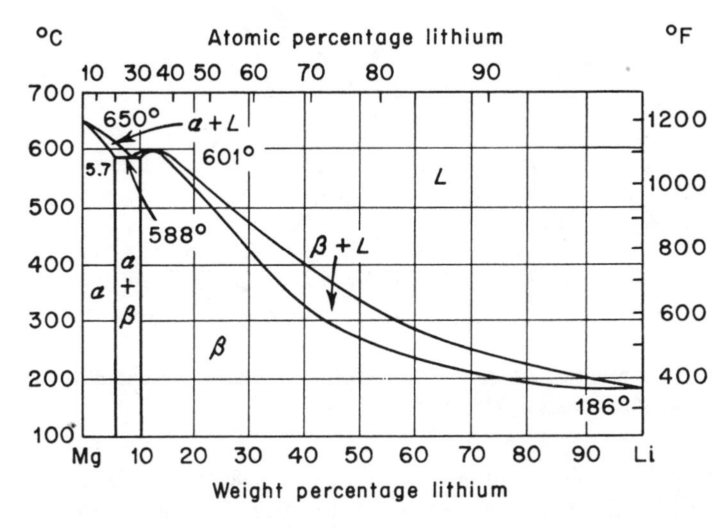

Fig. 14.8 Magnesium-lithium phase diagram. (From Sager, G. F., and Nelson, B. J., *ASM Metals Handbook*, 1948 ed.).

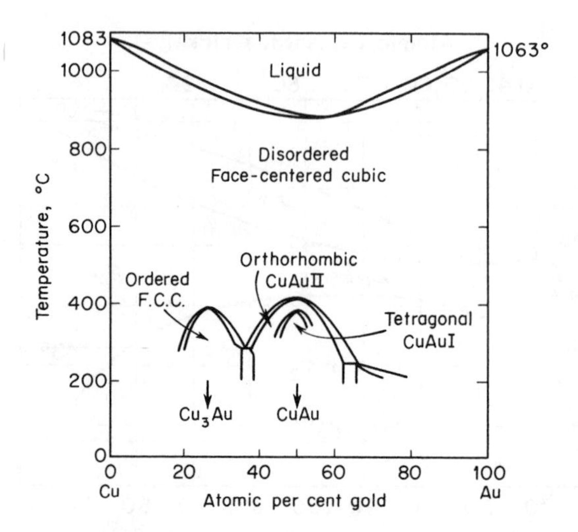

Fig. 14.9 The copper-gold phase diagram. (From *Constitution of Binary Alloys*, by Hansen, Max, and Anderko, Kurt. Copyright, 1958. McGraw-Hill Book Co., Inc., New York, p. 198. Used by permission.)

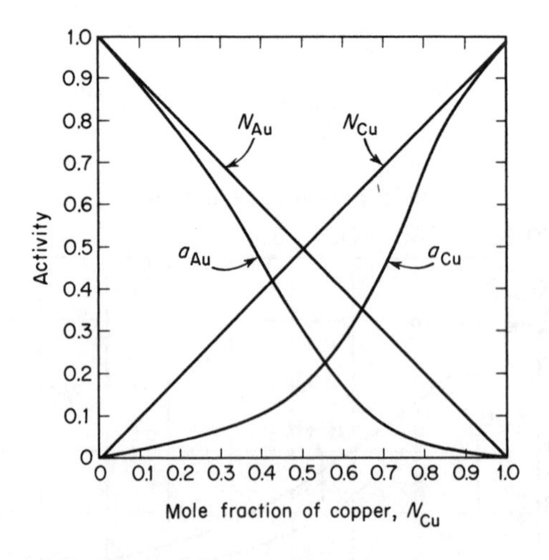

Fig. 14.10 Activities of copper and gold in solid alloys at 500°C. (After Oriani, R. A., *Acta Met.*, **2** 608 [1954].)

activities of these solid solutions exhibit a negative deviation. This fact is shown in Fig. 14.10 where it can be observed that at 500°C the activities of both gold and copper are smaller than the corresponding mole fractions. Negative deviations of the activities are generally considered as evidence that the components of a binary system possess a definite attraction for each other or at least a preference for opposite atomic forms as neighbors. An interesting result of this effect is the formation, in this system, of ordered structures in which gold and copper atoms alternate in lattice positions in such a way as to form the maximum number of gold-copper atomic bonds and the minimum number of copper-copper and gold-gold bonds. At higher temperatures, thermally induced atomic movements are too rapid to permit the grouping together of a large number of atoms in stable ordered structures. Two opposing factors, namely, the attraction of unlike forms for each other and the disrupting influence of thermal motion, therefore, lead to a condition known as *short-range order*. In this situation, gold atoms have a statistically greater number of copper atoms as neighbors than would be expected if the two atomic forms were arranged on the lattice sites of a crystal in entirely random fashion. The effectiveness of thermal motion in destroying an extensive periodic arrangement of the gold and copper atoms, of course, decreases with falling temperature, so that at low temperatures and at the proper compositions gold and copper atoms can arrange themselves in stable configurations that extend through large regions of a crystal. When this occurs, a state of long-range order is said to exist and the resulting structure is called a *superlattice* or a *superstructure*.

An ordered region of a crystal is known as a *domain*. The maximum theoretical size of a domain is determined by the size of the crystal in which the domain lies. Usually, however, a given metal crystal will contain a number of domains, and the relationship between crystals and domains is indicated in Fig. 14.11. A portion of two schematic ordered crystals is shown based on the assumption of equal numbers of black (A) and white (B) atoms. The grain at the upper left of the figure contains three domains, while that at the lower right has two. At the domain boundaries, which are outlined by dashed lines, A atoms face A atoms, and B atoms face B atoms. Inside of each domain, each A atom and each B atom is surrounded by atoms of the opposite kind. At the juncture

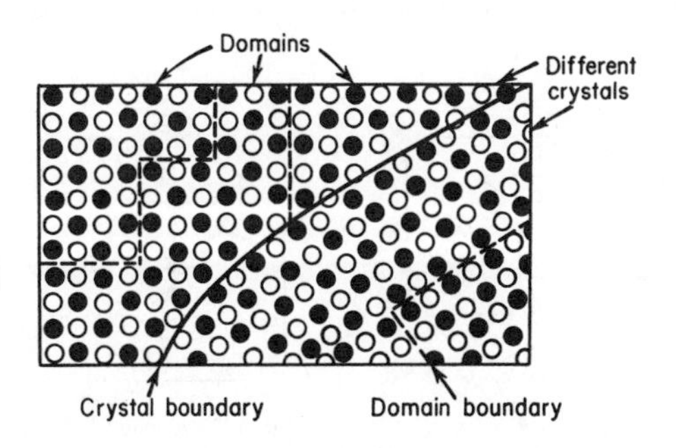

Fig. 14.11 Ordered domains in two different crystals. Ordering based on equal numbers of (A) black atoms and (B) white atoms.

between two domains, the sequence of the *A* and *B* atoms is reversed and it is common practice to call the domains *antiphase domains* and the boundaries *antiphase boundaries*.

In the copper-gold system, the transformation from short-range to long-range order produces superlattices within two basic composition ranges. One of these regions surrounds the composition corresponding to equal numbers of gold and copper atoms, and the other that of a ratio of three copper atoms for each gold atom. Each superlattice is a phase in the usual sense and is stable inside a definite range of temperature and composition. Three different superstructures have been identified in the copper-gold system: two corresponding to the composition CuAu and one to the composition Cu_3Au. The boundaries delineating the two-phase region surrounding each superlattice phase meet at congruent maxima. The congruent maximum of the Cu_3Au phase appears at 390°C, while the maxima of CuAu lie at 410°C and 385°C. Notice that, with respect to the CuAu phases, the lower temperature phase is formed on cooling from the higher temperature phase. Thus, when an alloy with a composition approximately equal to CuAu cools down from a temperature near the solidus, it first transforms from the high-temperature short-range ordered phase (disordered face-centered phase) to the phase designated in Fig. 14.9 as CuAuII (orthorhombic). On further cooling, this latter phase transforms into CuAuI (tetragonal).

The unit cells of two of the three superlattices of the copper-gold system are shown in Fig. 14.12. The left drawing shows the structure of Cu_3Au. It is merely a face-centered cubic unit cell with copper atoms at the face centers and gold atoms at the corners. That this configuration corresponds to the stoichiometric ratio Cu_3Au is easily proved. There are six face-centered copper atoms, each belonging to two unit cells, or a total of three copper atoms per unit cell. On the other hand, there are eight corner atoms, each belonging to eight unit cells, or one gold atom per unit cell. A perfect lattice, composed of unit cells of this nature, has a ratio of three copper atoms to each gold atom. Actually, as the

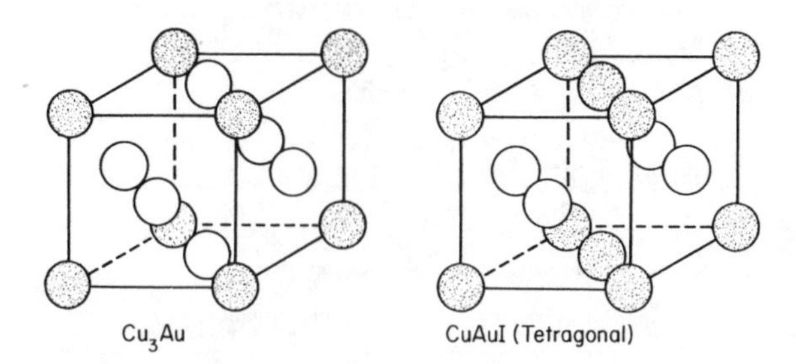

Fig. 14.12 Unit cells of two of the three known ordered phases in copper-gold alloys.

phase diagram shows, the Cu_3Au structure, as well as the CuAu structures, is capable of existing over a range of compositions. This range is somewhat limited because, as one deviates from the strict stoichiometric ratios, the perfection of the order decreases and with it the stability of the superlattices.

The right diagram in Fig. 14.12 represents the unit cell of the lower-temperature CuAu phase (CuAuI, or tetragonal phase). This structure is also a modification of the face-centered cubic lattice with alternating (001) planes completely filled with gold and copper atoms. The tetragonality of this phase is directly related to the alternating stacking of planes of gold atoms and planes of copper atoms. Such a structure has axes of equal length in a plane containing atoms of the same type, but an axis of different length in the direction perpendicular to this plane. The unit cell is thus distorted from a cube into a tetragon. The other phase, CuAuII, has a somewhat more complicated structure that will not be described, but it can be considered an intermediate step between the short-range-ordered, high-temperature face-centered cubic structure and the low-temperature-ordered tetragonal structure.

Ordered phases are found in a number of alloy systems. One of these that has been studied extensively occurs in the copper-zinc system and corresponds to an equal number of copper and zinc atoms. This phase will be considered briefly later in this chapter.

14.7 Miscibility Gaps. Gold and nickel, like copper and nickel and copper and gold, also form an alloy system that freezes into solid solutions in all proportions. See the phase diagram in Fig. 14.7. This system serves as an example of a completely soluble system in which the constituents tend to segregate as the temperature is lowered. Consider now the curved line that occupies the lower central region of the phase diagram. At all temperatures below 812°C and inside the indicated line, two phases are stable, α_1 and α_2. The first, α_1, is a phase based on the gold lattice with nickel in solution, and the other, α_2, nickel with gold in solution. Both phases are face-centered cubic, but differ in lattice parameters, densities, color, and other physical properties.

The two-phase field, where α_1 and α_2 are both stable, constitutes an example of what is commonly called a *miscibility gap*. A necessary condition for the formation of a miscibility gap in the solid state is that both components should crystallize in the same lattice form.

In Chapter 9, it was pointed out that many alloy systems have solvus lines that show increasing solubility of the solute with rising temperature. It is quite possible to consider the boundary of the miscibility gap as two solvus lines that meet at high temperatures to form a single boundary separating the two-phase field from the surrounding single-phase fields.

The gold-nickel system is particularly significant because it demonstrates that there is still much to learn about solid-state reactions. When a binary alloy is formed between A and B atoms, there are two possible types of atomic bonds between nearest neighbors; bonds between atoms of the same kind (A-A or B-B bonds), and bonds between unlike atoms (A-B bonds). Associated with each bond between a pair of atoms is a chemical bonding energy which may be written as ϵ_{AA} or ϵ_{BB} for pairs of like atoms, and ϵ_{AB} for a pair of unlike atoms. The total energy of the alloy may, of course, be written as the sum of the energies of all the bonds between neighboring atoms; the lower this energy the more stable the metal. If now the bonding energy between unlike atoms is the same as the average bonding energy between like atoms $\frac{1}{2}(\epsilon_{AA} + \epsilon_{BB})$, then there is no essential difference between the bonds, and the solution should be a random solid solution. When ϵ_{AB} is lower than the average bonding energy of like atoms, short-range order at higher temperatures and long-range order at lower temperatures are to be expected. On the other hand, segregation and precipitation are usually associated with the condition where ϵ_{AB} is greater than $\frac{1}{2}(\epsilon_{AA} + \epsilon_{BB})$.

The fact that gold-nickel alloys exhibit a miscibility gap would generally be construed as evidence that the bonding energy for a gold-nickel pair is larger than the average of the bond energies for gold-gold and nickel-nickel pairs because this effect is expected in segregation. Thermodynamic measurements of solid solutions of these alloys at temperatures above the miscibility gap show that the activities of both gold and nickel exhibit positive deviations. This fact also normally indicates that unlike pairs have a higher bond energy and that gold and nickel atoms prefer to segregate. However, X-ray diffraction measurements show a small but definite short-range order to exist in the solid solutions above the miscibility gap. This apparent contradiction is strong evidence that there is more than one factor involved in determining the type of structure that results. The simple so-called quasi-chemical theory, that pictures the development of segregation, or ordering, as the result of only the magnitudes of the interatomic bonding energies, is not sufficient to explain the observed results in the gold-nickel system. Other factors certainly must be involved in determining the nature of the solid-state reactions that occur in solid-solution alloys. In the gold-nickel system, the ambiguous results have been explained as being primarily due to the large difference in size of the gold and nickel atoms.[1] This difference amounts to about 15 percent and is at the Hume-Rothery limiting value for extensive solubility. When a random solid solution is formed of gold and nickel atoms, the fit between atoms is rather poor and the lattice is badly strained. One way that the strain energy associated with the misfit of gold and nickel atoms can be relieved is by causing the atoms to assume an ordered arrangement. Less strain is produced when gold and nickel atoms alternate in a crystal than when gold or nickel atoms cluster together. By this means, it is possible to explain the

1. Averbach, B. L., Flinn, P. A., and Cohen, Morris, *Acta Met.*, 2 92 (1954).

short-range order that exists at high temperatures. A still greater decrease in the strain energy of the lattice is possible if it breaks down to form crystals of the gold-rich and nickel-rich phases. Notice that, in this event, it is postulated that separate crystals of the two phases are formed with conventional grain boundaries between them, and that we are not talking about coherent clusters of gold atoms or nickel atoms existing in the original solid-solution crystals. Clustering in the coherent sense would raise, not lower, the strain energy. The nucleation of the segregated phases in a solid solution in which clustering does not occur is, of course, a difficult process, and the precipitation of the phases inside the miscibility gap of the gold-nickel system is a very slow and time-consuming process that is apparently only nucleated at the grain boundaries of the matrix phase.

Miscibility gaps are not only found in solid solutions, but also often in the liquid regions of phase diagrams. A particular liquid miscibility gap will be discussed in a later section.

14.8 Eutectic Systems. The copper-silver phase diagram, Fig. 14.13, can be taken as representative of eutectic systems. In systems of this

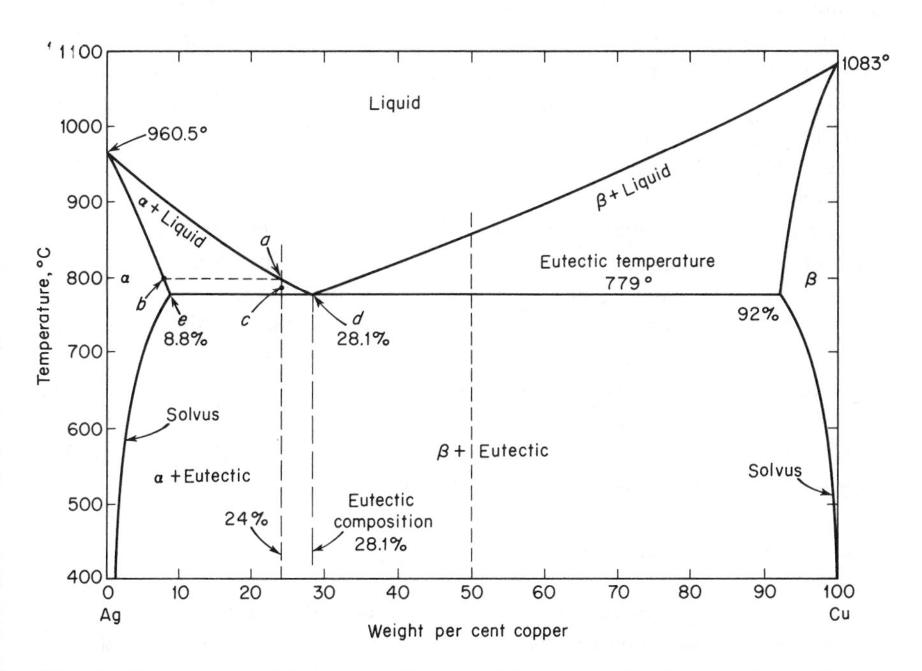

Fig. 14.13 Copper-silver phase diagram. (From *Constitution of Binary Alloys*, by Hansen, Max, and Anderko, Kurt. Copyright, 1958. McGraw-Hill Book Co., Inc., New York, p. 18. Used by permission.)

type, there is always a specific alloy, known as the *eutectic composition*, that freezes at a lower temperature than all other compositions. Under conditions approaching equilibrium (slow-cooling), it freezes at a single temperature like a pure metal. In other respects, the solidification reaction of this composition is quite different from that of a pure metal since it freezes to form a mixture of two different solid phases. Hence, at the eutectic temperature, two solids form simultaneously from a single liquid phase. A transformation, where one phase is converted into two other phases, requires that three phases be in equilibrium. In Chapter 12, it was shown that, assuming constant pressure, three phases can only be in equilibrium at an invariant point, that is, at a constant composition (in this case the eutectic composition) and at a constant temperature (eutectic temperature). The eutectic temperature and composition determine a point on the phase diagram called the *eutectic point*, which occurs in the copper-silver system at 28.1 percent Cu and 779.4°C.

At this time, it is perhaps well to compare Fig. 14.7, showing the gold-nickel equilibrium diagram and its miscibility gap, with the copper-silver diagram of Fig. 14.13. The close relationship between the two systems is evident. The copper-silver system corresponds to one in which the miscibility gap and the solidus lines intersect. The eutectic point, therefore, is equivalent to the minimum point in the gold-nickel system. In the gold-nickel system, however, an alloy with the composition of the minimum first solidifies as a single homogeneous solid solution and then, on further cooling, breaks down into two solid phases as it passes into the miscibility gap. An alloy possessing the eutectic composition of the copper-silver system, on the other hand, freezes directly into a two-phase mixture.

The fundamental requirement for a miscibility gap is a tendency for atoms of the same kind to segregate in the solid state. The same is also true of an eutectic system. A true miscibility gap, like that in the gold-nickel system, can only occur if the component metals are very similar chemically and crystallize in the same lattice form, since the components must be capable of dissolving in each other at high temperatures. In an eutectic system, the components do not have to crystallize in the same structure nor do they necessarily have to be chemically similar. If the two atomic forms are quite different chemically, however, then intermediate crystal structures are liable to form in the alloy system. Eutectics can still be observed in such systems, but they will not form between terminal phases, as shown in Fig. 14.13.

14.9 The Equilibrium Microstructures of Eutectic Systems. Any composition of an isomorphous system at equilibrium and in the solid state consists of a single homogeneous group of solid-solution crystals. When viewed under the microscope, such a structure does not differ essentially from that of a pure metal. It is usually difficult, therefore, to tell much about the composition of these single-phased alloys from a study of their microstructure alone. On the other hand, the appearance under the microscope of an alloy from the two-phase field (solid) of an eutectic system is very characteristic of its composition.

In discussing the microstructure and other aspects of the alloys of an eutectic system, it is customary to classify them with respect to the side of the eutectic composition on which they fall. Compositions lying to the left of the eutectic point are designated as *hypoeutectic*, while those to the right are called *hypereutectic*. These designations are easily remembered if one recalls the fact that *hypo-* and *hyper-* are Greek prefixes signifying below and above. Thus, reading the copper-silver equilibrium diagram in the usual way from left to right (increasing copper content), it is found that alloys with less than 28.1 percent Cu (eutectic composition) fall in the hypoeutectic class, whereas those containing more than 28.1 percent Cu belong to the hypereutectic group.

Figure 14.14 shows the microstructure of an alloy of 24 percent Cu and 76 percent Ag. Since this composition contains less copper than 28.1 percent (eutectic composition), the alloy is hypoeutectic. Two distinct structures are visible in the photograph: a more or less continuous gray area, inside of which is

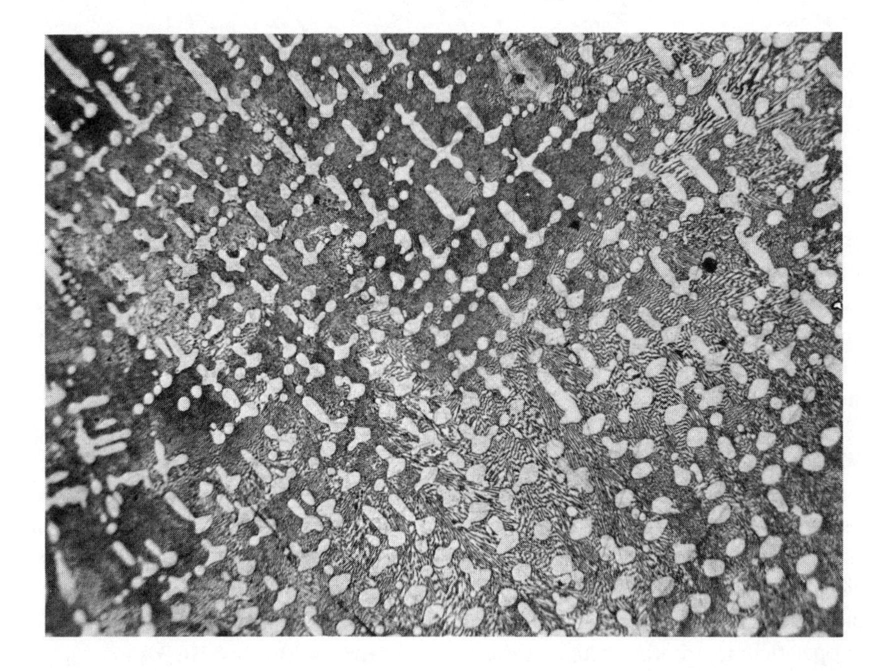

Fig. 14.14 A hypoeutectic structure from the copper-silver phase diagram containing approximately 24 percent copper. Lighter oval regions are proeutectic alpha, while the gray background is the eutectic structure.

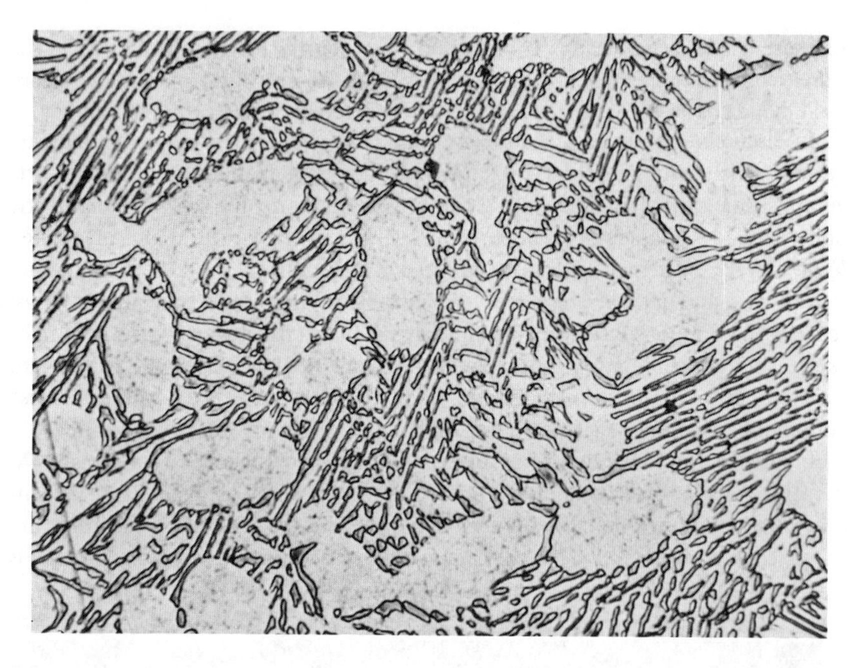

Fig. 14.15 The microstructure of Fig. 14.14 shown at a greater magnification. (White matrix is the alpha or silver-rich phase. Dark small particles are the beta or copper-rich phase. The eutectic structure is thus composed of beta particles in an alpha matrix.)

found a number of oval-shaped white areas. Notice that both the white and gray regions have their own characteristic appearance. Copper-silver alloys that contain between 8.8 percent and 28.1 percent Cu and are slowly cooled from the liquid phase, will show varying amounts of these two structures. The amount of the gray structure increases from zero to 100 percent as the composition of the alloy is changed from 8.8 percent to 28.1 percent. Figure 14.15 is another photomicrograph of the same alloy, but in this case the magnification is higher. What appears as a rough gray structure at lower magnification is shown by Fig. 14.15 to be an aggregate of small particles of one phase in a matrix of another phase. This is the eutectic structure of the copper-silver system. The small particles are composed of a copper-rich phase, while the continuous matrix is a silver-rich phase. The two phases forming the eutectic have colors that are characteristic of the element which is present in each in the greater amount. The copper-rich area is tinted red, while the silver-rich is white, so that both phases are clearly visible in a polished specimen that has not been etched. (In the present photographs, the specimens were etched to increase the contrast.) Careful study of Fig. 14.15 shows that the large white oval areas, that have no copper particles, are continuous with the white areas of the eutectic regions. These are, accordingly, extended regions of the silver-rich phase.

From the above, it can be concluded that hypoeutectic alloys in this alloy system possess a microstructure consisting of a mixture of the eutectic structure and regions containing only the silver-rich phase. The important thing to notice is that in a photograph of an eutectic alloy, such as Fig. 14.14, the features of the structure that stand out are not the phases themselves, but the contrast between the eutectic structure and the regions in which only a single phase exists. It is customary to call these parts of the microstructure that have a clearly identifiable appearance under the microscope, the *constituents* of the structure. Unfortunately, the term constituent is frequently confused with the term *component*. The words actually have quite different meanings. The components of an alloy system are the pure elements (or compounds) from which the alloys are formed. In the present case, they are pure copper and pure silver. The constituents, on the other hand, are the things that we see as clearly definable features of the microstructure. They may be either phases, such as the white regions in Fig. 14.14, or mixtures of phases, the gray eutectic regions of Fig. 14.14.

The microstructure in Fig. 14.14 is characteristic of an eutectic alloy that has been cooled (slowly) from the liquid phase. Cold-working and annealing this alloy will not change the amounts of the phases that appear in the microstructure, but such a treatment can well change the shape and distribution of the phases. In other words, Fig. 14.14 shows the structure of a hypoeutectic copper-silver alloy in the cast state, and it follows that what we see in this photograph is a function of the freezing process of the eutectic alloy.

Some of the details of the freezing process in alloys of an eutectic system will now be considered. For this purpose, we shall assume that the alloys are cooled from the liquid state to a temperature slightly below the eutectic temperature. At this temperature all compositions will be completely solidified, but considerations of the effect on the microstructures of the solvus lines, which show a decrease in solubility of the phases with decreasing temperatures, are eliminated. Actually, the change in the microstructure caused by the decreasing solubilities of the phases is, in general, slight and will be discussed briefly after we have explained the nature of the equilibrium freezing process.

A vertical line is shown in Fig. 14.13 that passes through the composition 24 percent Cu. This line represents the average composition of the alloy whose microstructure is shown in Fig. 14.14. At temperatures above point a, the alloy is in the liquid phase. Cooling to a temperature below point a causes it to enter a two-phase region where the stable phases are liquid and the alpha solid. The latter is a solid solution of copper in silver. Freezing of the given composition, accordingly, starts with the formation of crystals that are almost completely silver in composition (point b). Until the alloy reaches the eutectic temperature, the freezing process is similar to that of an isomorphous alloy with the liquid and solid moving along the liquidus and solidus lines respectively. The silver-rich crystals which form in this manner grow as many-branched skeletons called *dendrites* whose nature will be explained in Chapter 15. It suffices for the present to point out that when the alloy has cooled to just above the eutectic temperature (point *c*), it contains a number of skeleton crystals immersed in a liquid phase. Using the previously stated rules for analyzing a two-phase alloy at a given temperature, it is evident that the composition of the liquid at this temperature must correspond to the eutectic composition (point *d*). The solid phase, on the other hand, has a composition 8.8 percent Cu (point *e*). Since the average composition falls at a point about four-fifths of the way (24 percent–8.8 percent) from the composition of the solid phase to that of the liquid phase (28.1 percent–8.8 percent), it is to be expected, by the lever rule, that the ratio of liquid to solid will be approximately 4 to 1.

Immediately above the eutectic temperature the alloy is a mixture of liquid and solid. The solid is in the form of a number of many-branched skeleton crystals surrounded by the liquid phase of the eutectic composition. Further cooling of the alloy freezes this liquid at the eutectic temperature. Such a freezing process results in the formation of the eutectic structure: a mixture of two solid phases (copper particles in a matrix of silver) that freezes between the branches of the silver-rich crystals. A cross-section of a structure formed in this way has the appearance shown in Fig. 14.14. Since just above the eutectic temperature four-fifths of the alloy is liquid eutectic, it is to be expected that four-fifths of the completely solidified alloy will consist of eutectic solid. The remaining fifth of the structure corresponds to the silver-rich skeleton crystals that form during the freezing process at temperatures above the eutectic temperature. In Fig. 14.14 this part of the structure is the clear oval-shaped areas. The explanation of their shape is readily apparent if one considers that the plane of the photograph represents a cross-section of the arms of the skeleton crystal.

In a hypoeutectic alloy of the type shown in Fig. 14.14 or 14.15, the silver phase appears in two locations: in the eutectic, where it is present with the copper particles, and in the dendrite arms, where it is the only phase. The copper-rich phase, on the other hand, only appears in the eutectic structure. It is common practice to differentiate between the two forms of the silver-rich phase and to designate the dendritic silver-rich regions as "primary." The remaining silver-rich regions are designated *eutectic.* The primary regions are also commonly called the *proeutectic constituent* since they form at temperatures above the eutectic temperature.

All hypoeutectic compositions between 8.8 percent Cu and 28.1 percent Cu solidify in a manner similar to the 24 percent composition that has just been discussed. The ratio of the eutectic to the primary alpha phase in the alloys after they have passed through the eutectic temperature, of course, varies with the composition of the metal. The amount of eutectic in the microstructure increases directly with increasing copper content as we move from the composition 8.8 percent Cu to 28.1 percent Cu. At this latter value the structure will be entirely eutectic; at the former value it will consist of only primary alpha crystals.

At compositions below 8.8 percent Cu all alloys freeze in an isomorphous manner and are single-phased (as long as they are not lowered below the solvus line). When cooled below the solvus line, these compositions (0 to 8.8 percent Cu) become supersaturated and precipitate beta-phase (copper-rich) particles. This means that they are theoretically capable of age-hardening and the precipitation processes described in Chapter 9 apply. Notice that this phenomenon does not develop the eutectic structure that can only be formed when a liquid of the eutectic composition is solidified.

Let us now focus our attention on hypereutectic copper-silver alloys. Figure 14.15 shows a typical microstructure of one of these compositions (50 percent Cu–50 percent Ag). A vertical line is drawn through the appropriate composition in Fig. 14.13 to show where this alloy falls on the phase diagram. When hypereutectic alloys (28.1 percent Cu to 92 percent Cu) are cooled from the liquid state, the copper-rich beta phase forms from the liquid until the eutectic temperature is reached. This depletion of the copper content of the liquid shifts the liquid composition (along the liquidus line) toward the eutectic composition. After the eutectic temperature is passed, the resulting microstructure is a mixture of primary beta plus eutectic. The primary beta appears in Fig. 14.16 as oval-shaped dark areas. The structure of the eutectic in this alloy is the same as in the hypoeutectic alloys – copper particles in a matrix of silver. One very interesting feature can be seen by comparing this microstructure with the comparable hypoeutectic structure of Fig. 14.15. In both cases, the silver phase is the continuous phase, while the copper phase is discontinuous. Thus, the alpha, or silver-rich, phase of the eutectic is continuous with the primary silver dendrite arms in the hypoeutectic structure, but in the hypereutectic structure the primary copper-rich phase is not continuous with the copper phase in the eutectic. Similar results, where one phase tends to surround the other phase, are found in other eutectic systems besides the copper-silver system.

The amount of the eutectic structure in hypereutectic alloys varies directly with the change in copper concentration as one moves from 28.1 percent Cu to 92 percent Cu; decreasing from 100 percent at the former composition to 0 percent at the latter. Above 92 percent Cu to 100 percent Cu, all compositions freeze in an isomorphous manner to form single-phased (beta or copper-rich) structures. The latter, on cooling to room temperature, become supersaturated

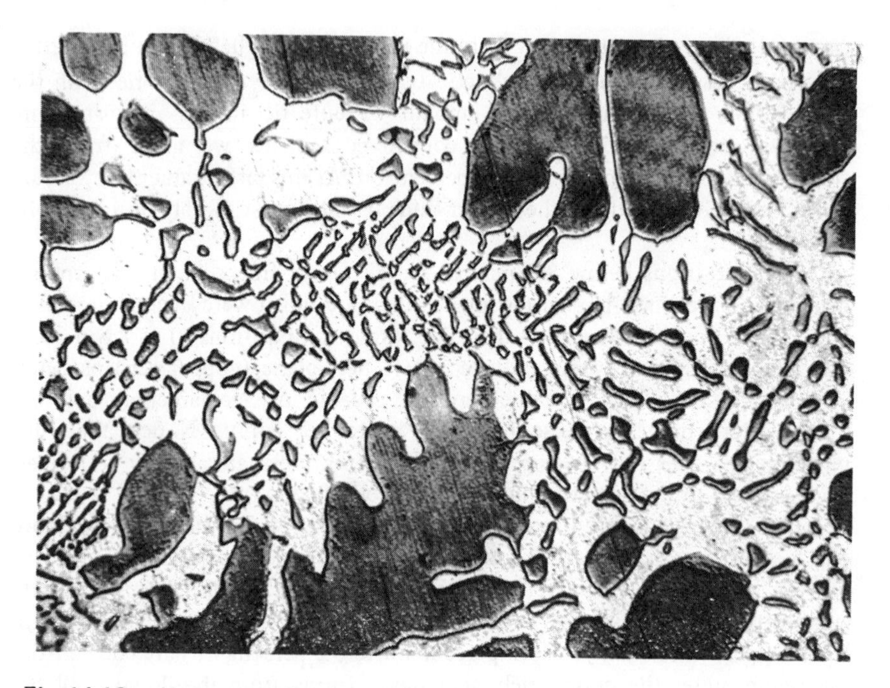

Fig. 14.16 Hypereutectic copper-silver structure consisting of proeutectic beta (large dark areas) and eutectic.

in silver and are subject to precipitation of the alpha phase. This effect is exactly analogous to that observed in the silver-rich alloys containing less than 8.8 percent Cu.

The effect of cooling the various compositions lying between 8.8 percent Cu and 92 percent Cu from the eutectic temperature to room temperature still remains to be described. In general, all of these alloys will contain part of the eutectic structure and are, therefore, two-phase structures. According to the phase diagram, both the alpha phase and the beta phase have decreasing solubilities with temperature and, on slow-cooling, both phases tend to approach the pure state, which means that copper will diffuse out of the silver-rich phase and silver will diffuse out of the copper-rich phase. With sufficiently slow-cooling, no new particles of either phase should form since it will be easier, for example, for the silver atoms leaving the beta phase to enter the alpha phase which is already present, than it will be to nucleate additional alpha particles. The net effect of the decreasing solubility of the phases (with falling temperature) upon the appearance of the microstructure is therefore small.

A final word should be said about the significance of the eutectic point. Hypoeutectic compositions, when cooled from the liquid phase, start to solidify as silver-rich crystals. The liquidus line to the left of the eutectic point can, therefore, be considered as the locus of temperatures at which various liquid compositions will start to freeze out the alpha phase. Similarly, the liquidus line to the right of the eutectic point represents the locus of temperatures at which the beta phase will form from the liquid phase. The eutectic point, which occurs at the intersection of the two liquidus lines, is, therefore, the point (with respect to temperature and composition) at which the liquid phase can change simultaneously into both alpha and beta phases.

14.10 The Peritectic Transformation. The eutectic reaction, in which a liquid transforms into two solid phases, is just one of the possible three-phase reactions that can occur in binary systems. Another that appears frequently involves a reaction between a liquid and a solid that forms a new and different solid phase. This three-phase transformation occurs at a peritectic point.

Figure 14.17 shows the constitution diagram of the iron-nickel system. A peritectic point appears at the upper left-hand corner. An enlarged view of this region is given in Fig. 14.18.

Before studying the peritectic reaction, some of the basic features of the alloy system in question should be considered. Iron and nickel have apparent atomic diameters that are almost identical (Fe, 2.476 Å and Ni, 2.486 Å). Since both iron and nickel belong to group VIII of the periodic table, these two elements are chemically similar. Both crystallize in the face-centered cubic system; nickel is face-centered cubic at all temperatures but iron only in the range 910°C to 1390°C. Conditions are thus ideal for the formation of a simple isomorphous system except for the fact that the stable crystalline form of iron is body-centered cubic at temperatures above 1390°C and below 910°C. It is not surprising, therefore, that iron-nickel alloys are face-centered cubic except for two small body-centered cubic fields at the upper and lower left-hand corners of the phase diagram.

Figure 14.17 also shows that a superlattice transformation, based on the composition $FeNi_3$, occurs in this system. The boundaries delineating this face-centered cubic order-disorder transformation appear at the lower right-hand side of the phase diagram.

The addition of nickel to iron increases the stability of the face-centered cubic phase. As a consequence, the temperature range in which this crystalline phase is preferred expands with increasing nickel content, and the boundaries separating the body-centered cubic fields from the face-centered field slope upward and downward respectively (with increasing nickel content).

With reference to Fig. 14.18, let us consider only that part of the diagram that lies above 1390°C. According to normal terminology, the high-temperature form of the body-centered cubic phase that appears in this temperature range is

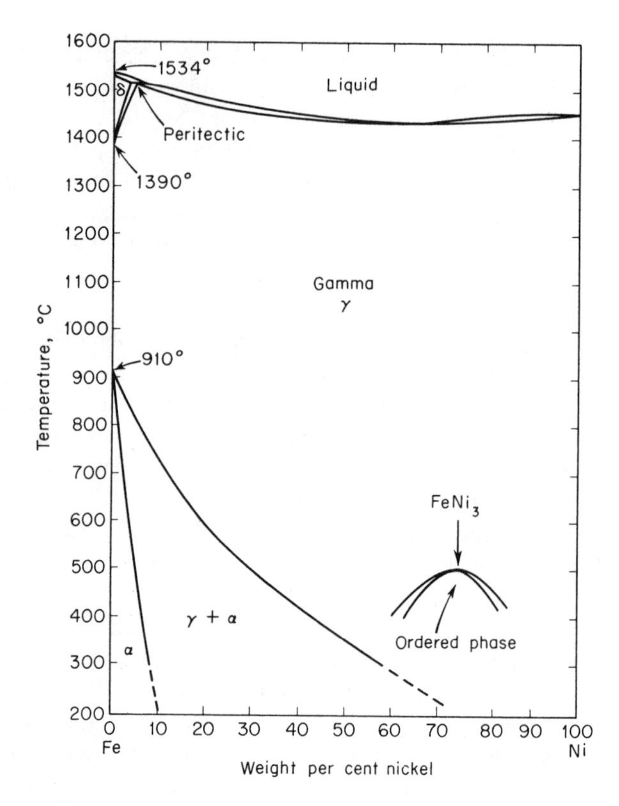

Fig. 14.17 The iron-nickel phase diagram. (From *Constitution of Binary Alloys*, by Hansen, Max, and Anderko, Kurt. Copyright, 1958. McGraw-Hill Book Co., Inc., New York, p. 677. Used by permission.)

called the delta (δ) phase, whereas the face-centered cubic phase is designated the gamma (γ) phase.

In the indicated temperature interval, when the composition of the alloy is very close to pure iron, delta is the stable phase. With increased nickel content, the stable structure becomes the gamma phase. Liquid alloys of very low-nickel concentrations (<3.4 percent Ni) freeze directly to the body-centered cubic phase, whereas those containing more than 6.2 percent Ni freeze to form the face-centered cubic phase.

The composition interval 3.4 percent Ni to 6.2 percent Ni represents a transition interval in which the product of the freezing reaction shifts from the delta to the gamma phase. The focal point of this part of the phase diagram is the peritectic point, which occurs at 4.5 percent Ni (the peritectic composition) and 1512°C (the peritectic temperature).

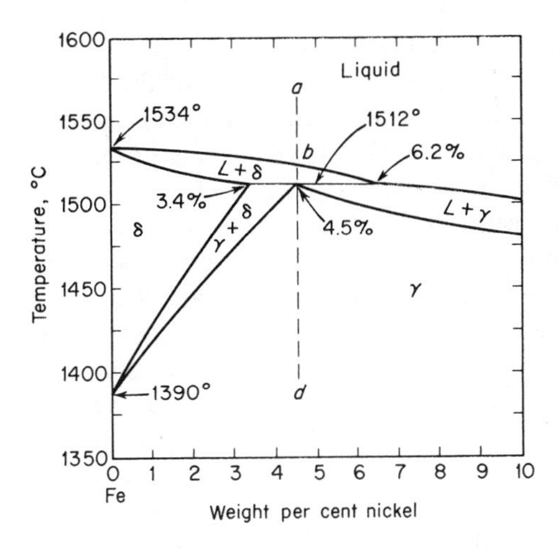

Fig. 14.18 The peritectic region of the iron-nickel phase diagram.

With the aid of the dashed line *ad* in Fig. 14.18, one can follow the freezing reaction of an alloy of peritectic composition. Solidification starts when the temperature of the liquid phase reaches point *b*. Between this point and the peritectic temperature, the alloy moves through a two-phase field of liquid and delta phases. Solidification, therefore, commences with the formation of body-centered cubic dendrites that are low in nickel. The liquid, accordingly, is enriched in nickel. At a temperature immediately above the peritectic temperature (1512°C), the rules for analyzing a two-phase mixture give us:

The composition of the phases:

delta phase	3.4 percent Ni
liquid phase	6.2 percent Ni

The amount of each phase (by lever rule):

delta phase	61 percent
liquid phase	39 percent

Directly above the peritectic temperature, an alloy of the peritectic composition has a structure composed of solid delta-phase crystals in a matrix of liquid phase. On the other hand, the phase diagram shows that just below the peritectic temperature the given alloy lies in a single-phase field (gamma), signifying a simple homogeneous solid-solution phase. Cooling through the peritectic temperature combines the delta and liquid phases to form the gamma phase. This is the iron-nickel peritectic transformation. In this particular system, the peritectic point is the direct consequence of the fact that the liquid phase freezes to form two different crystalline forms in adjacent composition ranges, and of the fact that one of the phases (gamma) is the more stable at lower temperatures and, therefore, displaces the other.

Notice that the peritectic reaction, like the eutectic, involves a fixed ratio of the reacting phases: 61 percent delta (3.4 percent Ni) combines with 39 percent liquid (6.2 percent Ni) to form the gamma phase (4.5 percent Ni). If an alloy at the peritectic temperature does not contain this exact ratio of liquid phase to delta phase, the reaction cannot be complete and some of the phase that is in excess will remain after the peritectic temperature is passed. Compositions in the interval 3.4 percent to 4.5 percent Ni, which lie to the left of the peritectic composition, contain an excess of the delta phase (more than 61 percent) immediately above 1512°C. On passing through the peritectic temperature, they enter a two-phase field: delta phase and gamma phase. Similarly, alloys to the right of the peritectic point, lying between 4.5 percent Ni and 6.2 percent Ni, after passing through the peritectic temperature, enter a two-phase field: gamma phase and liquid phase.

14.11 Monotectics. Monotectics represent another form of three-phase transformation in which a liquid phase transforms into a solid phase and a liquid phase of different composition. Monotectic transformations are associated with miscibility gaps in the liquid state. A reaction of this type occurs in the copper-lead system at 954°C and 36 percent Pb, as can be seen in Fig. 14.19. Notice the similarity between this monotectic and the eutectic of the copper-silver system (Fig. 14.13). The liquid miscibility gap lies just to the right of the monotectic point.

Note that the copper-lead phase diagram also possesses an eutectic point at 326°C and 99.94 percent Pb (0.06 percent Cu). Because it lies so close to the composition of pure lead, it is not possible to show it on the scale of the present diagram. Finally, this system is representative of one composed of elements that do not mix in the solid state; at room temperature the solubility of copper in lead is less than 0.007 percent, while the solubility of lead in copper is of the order of 0.002 to 0.005 percent Pb.[2] The terminal solid solutions are, accordingly, pure elements to a high degree of purity.

14.12 Other Three-Phase Reactions. The three basic types of three-phase reactions (eutectics, peritectics, and monotectics) that we have studied so far involve transformations between the liquid and solid phases. They are, therefore, associated with the freezing or melting processes in alloys. Several

2. Beck, Paul A., *ASM Metals Handbook* (1948), p. 1200.

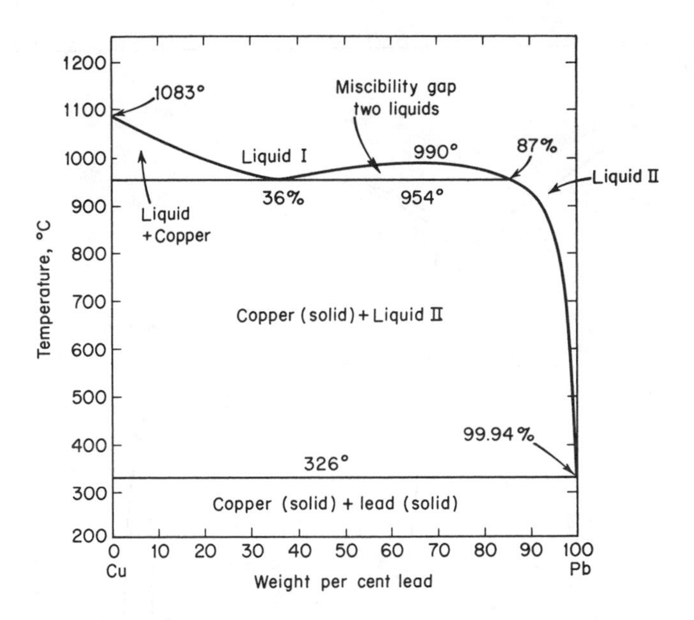

Fig. 14.19 The copper-lead phase diagram. (From *Constitution of Binary Alloys*, by Hansen, Max, and Anderko, Kurt. Copyright, 1958. McGraw-Hill Book Co., Inc., New York, p. 609. Used by permission.)

other important three-phase reactions involve only changes between solid phases. The most important of these occurs at eutectoid and peritectoid points. At an eutectoid point, a solid phase decomposes into two other solid phases when it is cooled. The reverse is true at a peritectoid point where, on cooling, two solid phases combine to form a single solid phase. The similarity between the eutectoid and peritectoid transformations on the one hand and the eutectic and peritectic transformations on the other hand is quite obvious.

A very important eutectoid reaction occurs in the iron-carbon system and will be considered in considerable detail at a later point.

14.13 Intermediate Phases. Intermediate phases are the rule rather than the exception in phase diagrams. Equilibrium diagrams given to this point were selected for their simplicity in order to demonstrate certain basic principles. Except for ordered or superlattice phases, no intermediate phases have been shown. Several important features of alloy systems that contain intermediate phases will now be treated briefly.

The silver-magnesium phase diagram (Fig. 14.20) is of interest in the study of intermediate phases since it shows two basic ways in which the single-phase fields associated with intermediate phases are formed. This alloy system possesses a total of four solid phases, of which two are terminal phases

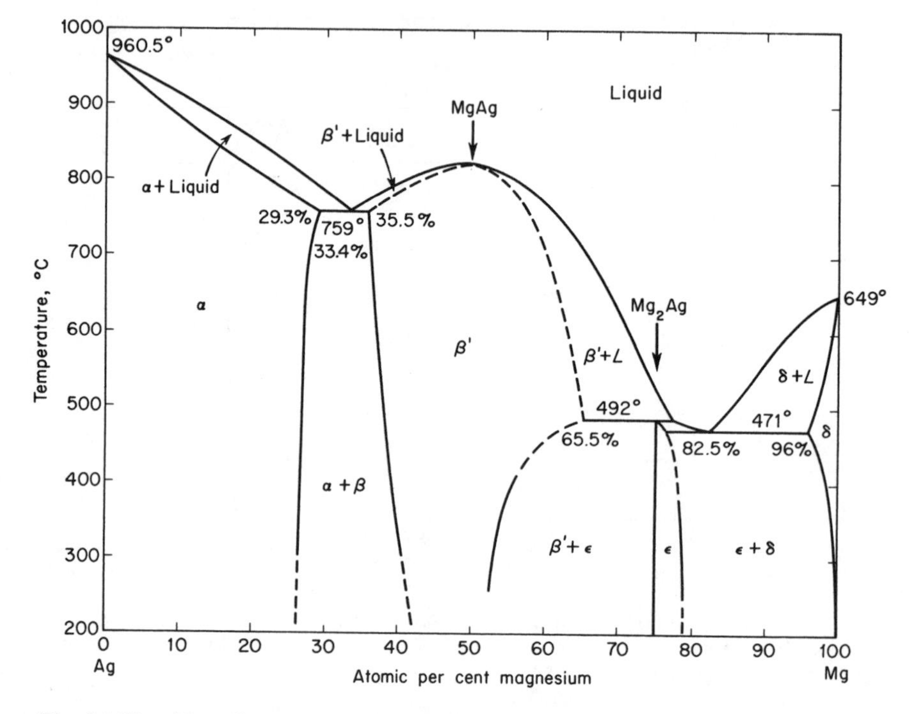

Fig. 14.20 The silver-magnesium phase diagram. (From *Constitution of Binary Alloys*, by Hansen, Max, and Anderko, Kurt. Copyright, 1958. McGraw-Hill Book Co., Inc., New York, p. 30. Used by permission.)

(alpha – f.c.c. based on the silver lattice, and delta – c.p.h. based on the magnesium lattice). Attention is called to the fact that both the beta prime and gamma phases (the intermediate phases) are stable over a range of composition and are, therefore, examples of true solid solutions.

Figure 14.20 shows that the β' phase is centered about a composition having equal numbers of magnesium and silver atoms. In order that this fact might be clearly visible, the diagram has been plotted in atomic percent. This phase is a superlattice based on the body-centered cubic structure. With exactly equal numbers of silver and magnesium atoms, a space lattice results, in which the corner atoms of the unit cell are of one form, while the atom at the center of the cell is of the other type. An example of this type of structure in a different system is shown in Fig. 14.24. The β' phase in the present alloy system is interesting in that the ordered structure is stable to the melting point. In order to indicate that the given structure is not a simple random solid solution, the symbol β' rather than β has been used.

The β' phase field is bounded at its upper end by a maximum having coordinates of 820°C and 50 atomic percent Mg. This body-centered cubic phase, therefore, forms on freezing by passing through a typical maximum configuration of liquidus and solidus lines. On the other hand, the upper limits of the single-phase field of the epsilon phase terminate in a typical peritectic point at 492°C and 75 atomic percent Mg. Formation of this latter phase (which has a complex, not completely resolved, crystalline structure), accordingly, results from a peritectic reaction.

The above paragraph has defined two basic ways that intermediate phases are formed during freezing – either by a transformation at a congruent maximum or through a peritectic reaction. Both types of reaction are common in phase diagrams. Intermediate phases may also form as a result of a transformation that takes place in the solid state. The formation of superlattices from solid solutions possessing a state of short-range order at congruent maxima has already been discussed. New solid phases may also form at peritectoid points that are the solid-state equivalent of peritectic points.

Returning now to the silver-magnesium system, Fig. 14.20 also shows that this system, in addition to the peritectic point and the congruent point, also possesses two eutectic points. The first lies at 759° C and 33.4 atomic percent Mg and corresponds to a reaction in which a liquid transforms into an eutectic mixture of the α and β' phases. The other eutectic point, at 471°C and 82.5 atomic percent Mg, yields an eutectic structure which is a mixture of the ϵ and δ phases. Notice that, in each case, the eutectic structures are composed of different sets of phases.

The intermediate phases in the silver-magnesium system are solid solutions stable over relatively wide composition ranges. Thus, the single-phase field of the β' phase extends from approximately 26 atomic percent Mg to 42 atomic percent Mg, while that of the ϵ phase covers the composition range from 75 atomic percent Mg to 79 atomic percent Mg. Many intermediate phases, on the other hand, have single-phase fields that are vertical lines. Such phases are commonly classed as *compounds.*

Intermediate phases of the compound type are formed during freezing in the same manner as the solid-solution phases; they may form either at congruent maxima or at peritectic points. The equilibrium diagram of the magnesium-nickel system, Fig. 14.21 which is also plotted in atomic percent, shows both a congruent point (1145°C and 33 percent Mg) and a peritectic point (760°C and 66.7 percent Mg). In this respect, it is analogous to the silver-magnesium system and, like the former, it also possesses two eutectic points. The only basic difference between the two systems lies in the absence of solubility in the phases of this latter system.

In the magnesium-nickel system, all phases, including the terminal phases, have very limited single-phase fields. The beta phase corresponds to the compound $MgNi_2$ and crystallizes in a hexagonal lattice. The gamma phase

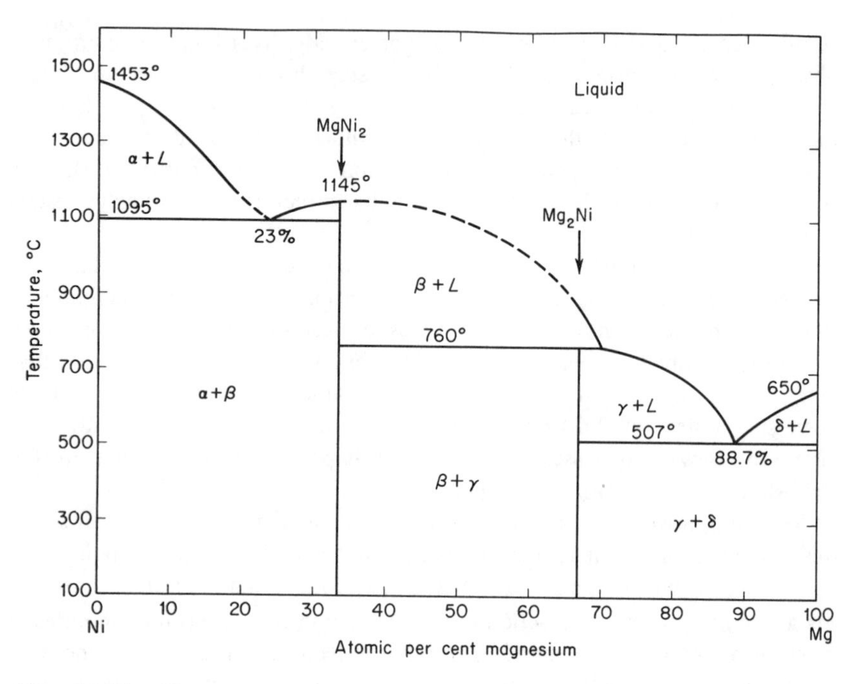

Fig. 14.21 The nickel-magnesium phase diagram. (From *Constitution of Binary Alloys,* by Hansen, Max, and Anderko, Kurt. Copyright, McGraw-Hill Book Co., Inc., New York, p. 909. Used by permission.)

appears at a ratio of the components equivalent to 1 nickel atom to 2 magnesium atoms or Mg_2Ni.

The first compound ($MgNi_2$) freezes or melts at a congruent maximum point. The vertical line representing this compound can be considered to divide the phase diagram into two independent parts. Each of these parts is a phase diagram in itself, as is shown in Fig. 14.22, where only the left section of the complete nickel-magnesium diagram is located. This partial diagram can be considered the nickel-$MgNi_2$ phase diagram: a system with one component an element and the other a compound.

14.14 The Copper-Zinc Phase Diagram. The equilibrium diagram of the copper-zinc system is shown in Fig. 14.23. Since copper-zinc alloys comprise the commercially important group of alloys known as brasses, the diagram is important for this reason alone. It is also significant because it is representative of a group of binary-equilibrium diagrams formed when one of the noble metals (Au, Ag, Cu) is alloyed with such elements as zinc and silicon. Attention is called to the fact that Fig. 14.23 is plotted in atomic percent. A

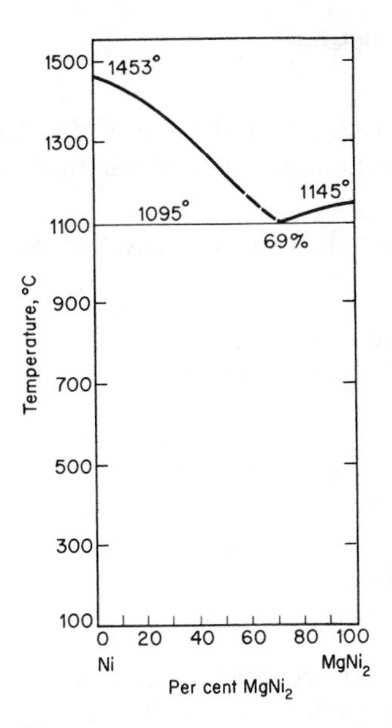

Fig. 14.22 The nickel-magnesium diagram can be divided at the composition of the compound $MgNi_2$ into two simpler diagrams. The above figure is the Ni–$MgNi_2$ diagram and corresponds to the left-hand portion of the Ni–Mg diagram.

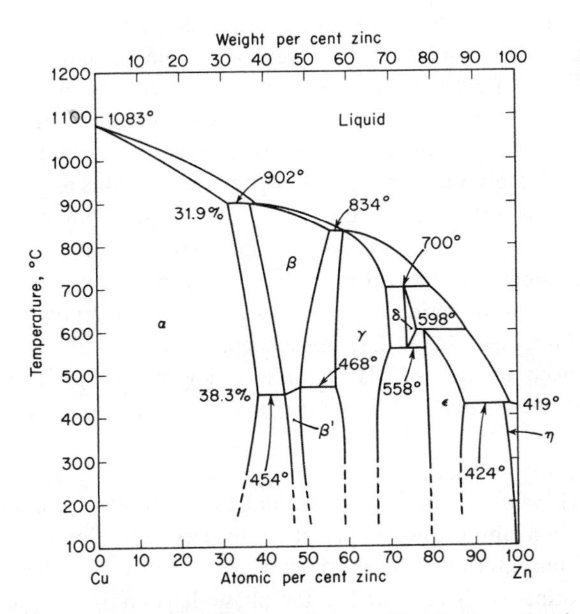

Fig. 14.23 The copper-zinc phase diagram. (From *Constitution of Binary Alloys,* by Hansen, Max, and Anderko, Kurt. Copyright, 1958. McGraw-Hill Book Co., Inc., New York, p. 649. Used by permission.)

weight-percent scale is given across the top of this diagram and it can be seen that copper-zinc compositions are almost identical when expressed in either weight or atomic percentages.

The seven solid phases in the copper-zinc system are classified as follows:

Terminal phases:

alpha (α) f.c.c. based on the copper lattice
eta (η) c.p.h. based on the zinc lattice

Intermediate phases:

beta (β) disordered body-centered cubic
beta prime (β') ordered body-centered cubic
gamma (γ) cubic, low symmetry
delta (δ) body-centered cubic
epsilon (ϵ) close-packed hexagonal

Except for the alpha and beta-prime phases, all single-phase fields terminate at their high-temperature extremities in peritectic points. There are, accordingly, five peritectic points in Fig. 14.23. The delta phase differs from the others in that it is stable over a rather limited temperature range (700°C to 558°C). Notice that the delta-phase field terminates at its lower end in an eutectoid point.

Another significant feature of this phase diagram is the order-disorder transformation that occurs in the body-centered cubic phase ($\beta - \beta'$). Near room temperature, the β' field extends from about 48 percent to 50 percent Zn. The stoichiometric composition CuZn falls at the edge of the β' field (50 percent Zn). The ordered body-centered cubic phase is, therefore, one that is based on a ratio of approximately one zinc to one copper atom. In this respect, it is similar to the copper-gold ordered phase CuAu, but here the structure is body-centered, whereas the CuAu phase is based on the face-centered cubic lattice. Ordering with equal numbers of two atomic forms in a body-centered cubic lattice produces a structure in which each atom is completely surrounded by atoms of the opposite kind (see Fig. 14.24). An example in the silver-magnesium system has already been mentioned. The atomic arrangement can be visualized by imagining that in each unit cell of the crystal the corner atoms are copper atoms, while those at the cell centers are zinc atoms. That such an arrangement corresponds to the formula CuZn is easily proved: since one-eighth of each corner atom (copper) belongs to a given cell, and there are eight corner atoms, the corner atoms contribute one copper atom to each cell. Similarly, there is one zinc atom at the center of each cell which belongs to this cell alone.

The transformation is indicated in the phase diagram by a single line running from 454°C to 468°C. Recent work[3] has indicated that the β and β' fields are

3. Rhines, F. N., and Newkirk, J. B., *Trans. ASM,* 45 1029 (1953).

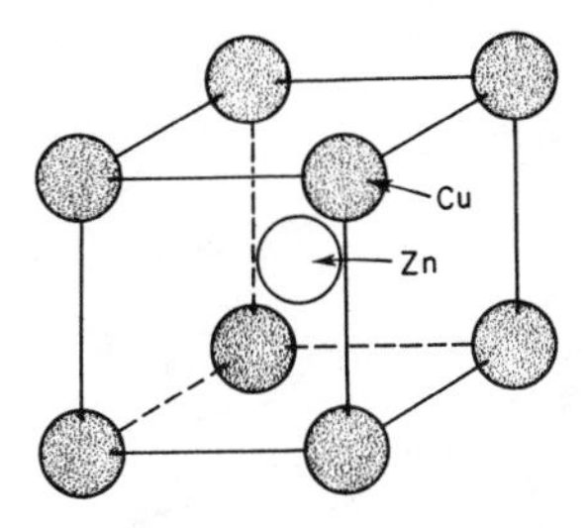

Fig. 14.24 The unit cell of the β' phase of the copper-zinc system. This type of structure is also found in the compound cesium-chloride and is usually called the CsCl structure.

separated by a normal two-phase field ($\beta + \beta'$), but the details of this region are still not completely worked out, which is due primarily to experimental difficulties connected with the rapidity of the transformation. The details of the transformation cannot be studied by quenching specimens from the two-phase region ($\beta + \beta'$), since any normal quench will not suppress the complete transformation to β'.

14.15 Diffusion in Non-Isomorphic Alloy Systems. When a thin sheet of copper is welded to a thin sheet of nickel and the diffusion couple thus formed is annealed at an elevated temperature, the composition should vary across the couple in a manner like that shown in Fig. 10.15. In other words, the composition should change smoothly and continuously from pure nickel on the one side of the couple to pure copper on the other side. A penetration curve of this type is obtained when an alloy system contains only a single solid phase. The corresponding penetration curve in an alloy system containing a number of different solid phases, such as the copper-zinc system, is basically different. For example, when a copper-zinc diffusion couple is annealed at a temperature of the order of 400°C, a layered structure is formed. Each layer in this couple corresponds to one of the five solid phases that can exist at this temperature. In order to explain the nature of such layered structures, a hypothetical alloy system will be considered whose phase diagram is shown in Fig. 14.25. This system has a single intermediate phase, the β phase, in addition to the terminal α and γ phases. A diffusion couple formed by welding a layer of pure metal A to a layer of pure metal B, after a diffusion anneal at a temperature T_1, can show three distinct layers corresponding respectively to the α, β, and γ phases, as shown in Fig. 14.26. If the composition-distance curve is determined for this couple, it also should have the form shown in Fig. 14.26. Note that where the penetration curve crosses either of the boundaries separating a pair of phases there is a sharp discontinuity in the composition. These sudden changes in composition are equal to the composition differences across the two-phase $\alpha + \beta$

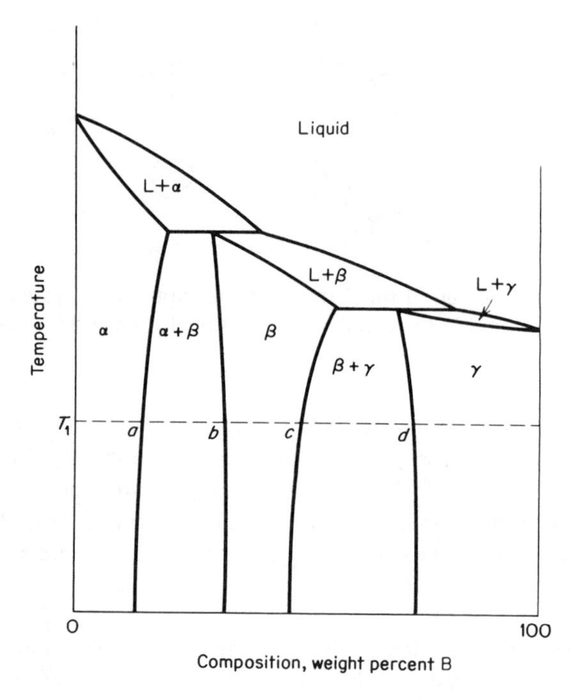

Fig. 14.25 The equilibrium diagram of a hypothetical alloy system with three solid phases.

and $\beta + \gamma$ fields. Thus, consider the boundary between the alpha and beta phases. Note that as the interface is approached from the alpha-phase side, the composition curve, corresponding to the B component, rises to the point marked a. This point represents the same composition as point a in the phase diagram of Fig. 14.25. On the other side of the boundary, the composition in the beta phase is given by point b, corresponding to point b on the phase diagram. Similarly, at the beta-gamma boundary, the composition changes from c to d, corresponding to points c and d on the phase diagram. In brief, the diffusion couple in an alloy system of this type is like a constant temperature section cut across the phase diagram in which each of the single-phase fields appears with a finite width, but in which the two-phase fields are represented by surfaces.

The reason why the two-phase fields appear as surfaces in a diffusion couple is not hard to understand. Two important factors need to be considered. First, for diffusion to occur, a concentration gradient must exist across the couple. More accurately, an activity gradient has to exist. If this gradient is missing at any position across the couple, the flux of A or B atoms past this position will

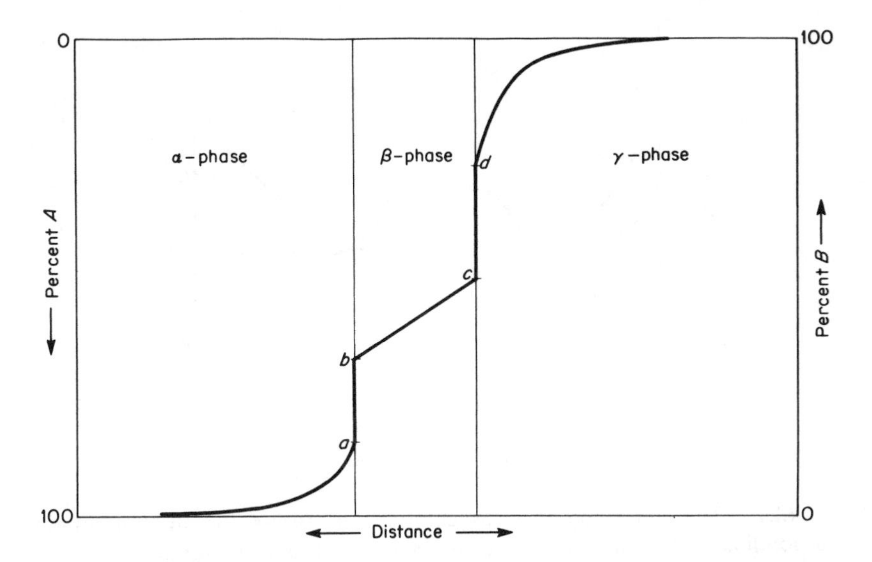

Fig. 14.26 A diffusion couple formed by welding a layer of pure metal A to one of pure metal B after an anneal at T_1 (see Fig. 14.25) will show a layered structure. Each layer in this structure corresponds to one of the phases in the equilibrium diagram. A curve showing the variation of the composition (B component) across the phase diagram is also shown in this figure.

stop. According to Section 10.2, we may write for the B component

$$\Delta \bar{F}_B = \bar{F}_B - F^{\circ}_B = RT \ln a_B$$

where $\bar{F}_B$ is the partial molal free energy of the B component, F°_B is the free energy per mole of pure B at the temperature in question, and a_B is the activity of the B component. This equation shows that at constant temperature a variation in a_B with distance across the couple corresponds to a variation in the partial molal free energy with distance. Thus, continuous diffusion across the couple can only occur if the partial molal free energy decreases continuously with distance. Now consider Fig. 14.27. This shows the hypothetical free energy composition curves for the three phases, α, β, and γ, at the temperature T_1. As discussed earlier in Section 12.8, a line drawn so that it is tangent to a pair of these free-energy curves defines the limits of a two-phase field. This is indicated in Fig. 14.27. Also by Section 12.8, the intersection with the sides of the figure of a tangent drawn to any point on a free-energy composition curve gives the partial molal free energy of the phase in question. By drawing a series of tangents to this set of three free-energy curves and selecting the minimum value

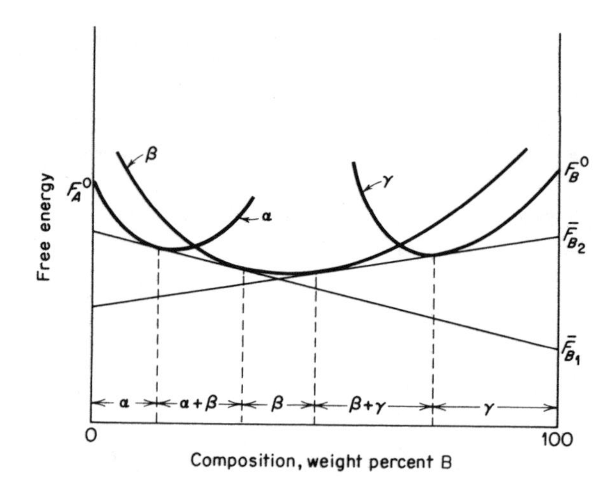

Fig. 14.27 The free energy versus composition curves for all three phases corresponding to the temperature T_1 are shown in the above diagram.

of the partial molal free energy corresponding to a given composition, it is possible to determine a curve corresponding to the variation of the partial molal free energy of the B component with composition across the whole phase diagram when the temperature is T_1. Such a curve is shown in Fig. 14.28. Note that starting with pure B, the partial molal free energy of the B component falls continuously with decreasing concentration of B until the two-phase $\gamma + \beta$ field is reached. In this region the partial molal free energy, defined by the common tangent to the gamma and beta energy curves in Fig. 14.28, is constant and equal to $\bar{F}_{B_2}$. A similar constant partial molal free-energy interval is also obtained in the alpha-beta two-phase region. Here the partial molal free energy is $\bar{F}_{B_1}$.

The significance of the above can be easily stated. The only way that a gradient in the partial molal free energy can be obtained in an interval where the partial molal free energy wants to remain constant is for the thickness of the interval to vanish. In other words, in the diffusion couple the two-phase regions have to appear as surfaces; that is, regions of zero thickness. If this did not occur, there could be no net diffusion flux across the couple. Finally, note that by Fig. 14.28 the partial molal free energy of the β and γ phases are equal at the interface. While a composition difference exists at the boundary, this is not true of the free energy. At the interface there is a change in slope of the partial molal free energy versus distance curve, but there is no discontinuity in $\bar{F}_B$.

The width of the single-phase layers in a diffusion couple will now be considered briefly. It is quite possible for a phase or phases to appear to be missing in a diffusion couple. Thus, as shown by Bückle,[4] when a diffusion

4. Bückle, H., *Symposium on Solid State Diffusion*, p. 170. North Holland Publishing Co. Copyright, Presses Universitaires de France, 1959.

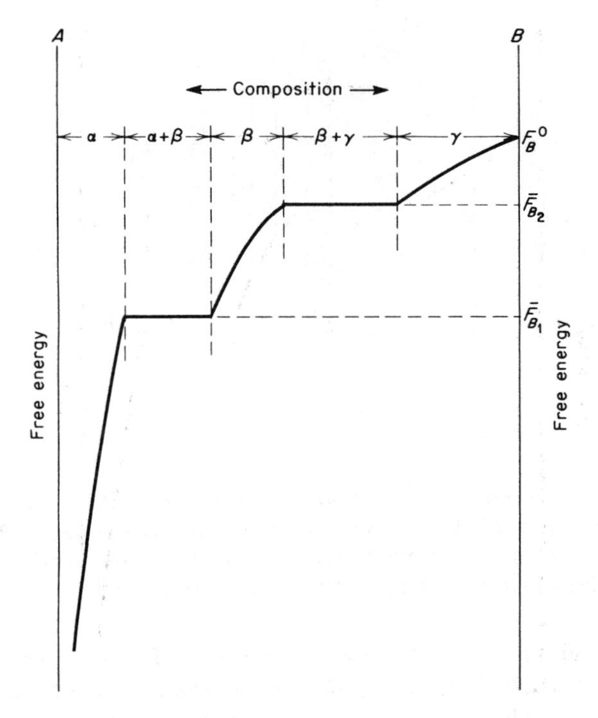

Fig. 14.28 The variation of $\bar{F}_B$ with composition across the phase diagram at T_1.

couple is formed from copper and zinc and annealed for about a half hour at 380°C, a layered structure, analogous to that shown schematically in Fig. 14.29, is obtained. Note that the β' phase layer is so thin that it is not visible at a magnification of about 150x. The thickness of a diffusion layer is determined by the relative velocities with which its two boundaries move. These boundary movements are controlled by diffusion and are therefore analogous to the diffusion-controlled planar interface growth phenomenon considered in Section 13.7. As an example, we will now consider the growth of the β-phase layer in Fig. 14.26. At the start of the anneal this layer has a zero thickness. With time it should grow in size. However, its net growth depends on the relative velocities with which its two boundaries move. The two boundaries will not normally move with the same velocity because the growth-controlling variables are not the same at the two boundaries. By analogy with the planar growth equation developed in Section 13.7, we may write, with respect to the boundary between the beta and alpha phases,

$$(n_B{}^b - n_B{}^a)\, A dx_{\alpha\beta} = AD_\alpha\left(\frac{dn_B{}^a}{dx}\right)dt - AD_\beta\left(\frac{dn_B{}^b}{dx}\right)dt$$

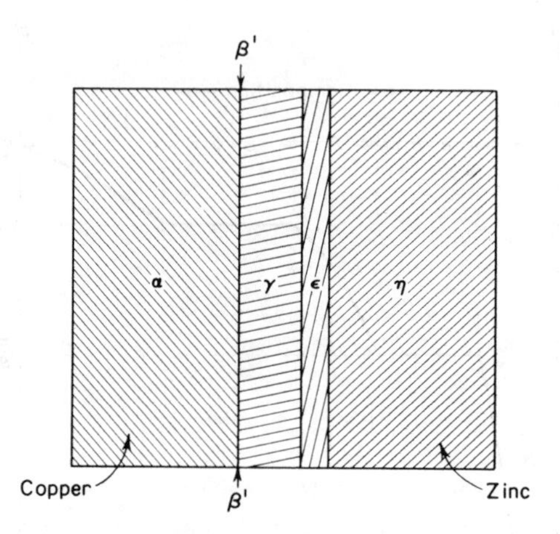

Fig. 14.29 A copper-zinc diffusion couple annealed for a short time at a temperature of about 380°C does not show a visible layer of the β' phase. (After Buckle, H., *Symposium on Solid State Diffusion*, p. 170. North Holland Publishing Co. Copyright, Presses Universitaires de France, 1959.)

where $n_B{}^b$ and $n_B{}^a$ are the compositions (number of B atoms per unit volume) of the two phases at the boundary, $dn_B{}^a/dx$ and $dn_B{}^b/dx$ are the concentration gradients in the alpha and beta phases at the boundary, and D_α and D_β are the corresponding diffusion coefficients. The left-hand side of this equation corresponds to the number of B atoms required to move the boundary through a distance $dx_{\alpha\beta}$, while the right-hand side represents the net flux that will give this number. In Section 13.7, the boundary was assumed to move only as the result of carbon atoms diffusing into cementite. In this case one has to consider both the flux of B atoms up to the interface in the β phase, as well as the flux of B atoms away from the interface in the α phase. The above expression may be simplified to

$$\frac{dx_{\alpha\beta}}{dt} = \frac{1}{(n_B{}^b - n_B{}^a)} \left[D_\alpha \left(\frac{dn_B{}^a}{dx} \right) - D_\beta \left(\frac{dn_B{}^b}{dx} \right) \right]$$

and a corresponding growth velocity expression written for the β to γ interface,

$$\frac{dx_{\gamma\beta}}{dt} = \frac{1}{(n_B{}^c - n_B{}^d)} \left[D_\gamma \left(\frac{dn_B{}^c}{dx} \right) - D_\beta \left(\frac{dn_B{}^d}{dx} \right) \right]$$

Note that the growth velocities of the two interfaces of the β phase depend on a number of parameters. These include the concentrations of the phases at the boundaries, the diffusion coefficients, and the concentration gradients. Since the diffusion coefficients are normally functions of the composition, the solution of layer-growth problems is usually very difficult. Also, it is quite possible to conceive of growth conditions that will not allow a phase to develop a layer of visible thickness. In fact, by suitable choices of diffusion coefficients, it is possible to make either boundary move in either direction.

The above analysis has assumed a condition of dynamic equilibrium and implies a diffusion couple of such a size that its outer layers do not change their compositions. In a diffusion couple of a finite size, the situation can be quite different. Thus, if in the above example the copper and zinc layers were very thin and present in a ratio of 48 percent zinc to 52 percent copper, a sufficiently long anneal should result in a specimen containing only a single phase. This would be the β' phase, or the phase corresponding to the average composition.

A word should be said about some practical examples of layered structures. A typical example is furnished by galvanized iron. When steel is dipped in molten zinc, the zinc diffuses into the iron and a layered structure is formed that contains four phases in addition to the base metal (steel). The outermost of these phases is a liquid. On cooling, this liquid passes through an eutectic point so that the outermost layer is basically an eutectic. Hot-dipped tin plate also has a layered alloy structure. In fact, in most cases where one metal is plated on another under conditions where diffusion can occur, it will be found that layered structures tend to develop.

Problems

1. A series of points, a, b, c, d, and e, have been placed on this hypothetical isomorphous phase diagram. All of these points lie on the 1300°C isotherm

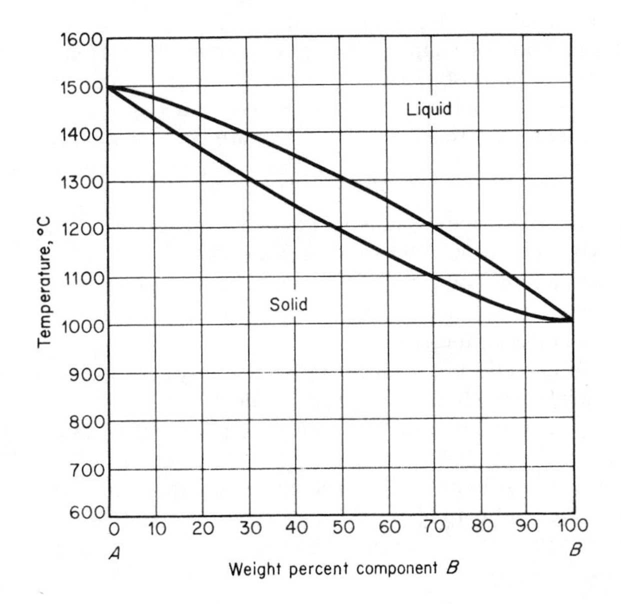

and fall at the compositions 20, 30, 40, 50, and 60 percent *B* respectively. In each case completely identify:

(*a*) The constituents.

(*b*) The components.

(*c*) The phases.

(*d*) The amount of each phase and the amount of each constituent.

2. A gold-nickel alloy containing 60 percent nickel is heated to 1200°C. Determine the composition and the amount of each phase in this metal when it reaches equilibrium at this temperature.
3. (*a*) A copper–15 percent silver alloy is heated to 1100°C and then very slowly cooled to 781°C. Describe the structure that should exist in the metal at this temperature. Give the composition and amount of each phase.
   (*b*) Make a sketch of the structure.
4. The alloy of Problem 3 is further cooled to 777°C and allowed to come to equilibrium. Make a sketch of the resulting microstructure.
5. Sterling silver contains 7.5 percent copper. Make a sketch of what you would expect to find in a metallographic specimen of this material if it were first annealed at 781°C for a long time and then slowly cooled to 400°C. Identify and give the amounts of the phases.
6. The United States silver coins used to contain 10 percent silver. If one of these coins was given the same heat treatment as described in the previous problem, what would the microstructure be like? Explain in detail.
7. With the aid of sketches representing the microstructure, describe the equilibrium freezing process in an alloy of iron containing 3 percent nickel. Carry the metal down into the gamma phase.
8. In the normal (nonequilibrium) freezing of an alloy containing 4.5 percent nickel, the final structure is often badly cored, or nonhomogeneous. Explain in detail why this should be so.
9. Consider an alloy of copper with 65 wt. percent lead. Describe the freezing process of this alloy starting at a temperature of 1100°C. Assume that the metal is cooled slowly in a crucible without stirring and brought down to room temperature. Use sketches to illustrate your answers corresponding to 956°C, 952°C, 328°C, and 324°C. In all cases give the amounts of the phases.
10. (*a*) Describe the reactions that should occur on cooling a crucible of $MgNi_2$ very slowly from 1300°C to room temperature.
    (*b*) Completely describe the constituents of the microstructure that should be obtained in an alloy of nickel with 30 percent magnesium on slow cooling from the liquid state.
11. (*a*) At the eutectic temperature in an alloy system like the copper-silver system, how many phases are in equilibrium?
    (*b*) What relationship exists between the partial molal free energies of these phases?
    (*c*) Draw a set of hypothetical free energy-composition curves corresponding to the eutectic temperature in the Cu–Ag system. Show how these curves can be used to define the phase relationships at this temperature.

12. (*a*) Assume a temperature about 50°C above the eutectic temperature and redraw the free energy curves for the Cu–Ag system.
    (*b*) Do the same for a temperature about 50°C below the eutectic temperature.
13. Imagine that a small piece of gold weighing one gram is placed on a piece of nickel weighing 99 grams and that the compact is heated to 1000°C. Describe the reactions that should occur as time goes by.
14. Assume that a diffusion couple is formed by placing a sheet of copper against a sheet of silver and that the couple is given a diffusion anneal at 800°C for a period of several hours. After this anneal, the specimen is then cooled to room temperature and examined metallographically. Describe the microstructure.

15 Solidification of Metals

Almost universally, commercial metal objects are frozen from a liquid phase into either their final shapes, called *castings*, or into intermediate forms, called *ingots*, which are then worked into the final product. Since the properties of the end-result are determined, in large measure, by the nature of the solidification process, the factors involved in the transformation from liquid to solid are of the utmost practical importance. Before the subject of solidification is considered, however, the nature of the liquid state will be treated briefly.

15.1 The Liquid Phase. In all cases of practical importance, liquid metal alloys occur as a single homogeneous liquid phase. To simplify the present discussion, we shall treat a special case – a pure metallic element with only a single solid phase. Figure 15.1 is a schematic phase diagram of a single-component system.

From what we have previously learned, we know that the solid is a crystalline phase in which the atoms are aligned in space in definite patterns over long distances. The regularity of crystal lattices makes it easy to study their structures with the aid of X-ray diffraction and electron microscopy; therefore, a great deal is known about the internal arrangements of atoms in metal crystals. At the

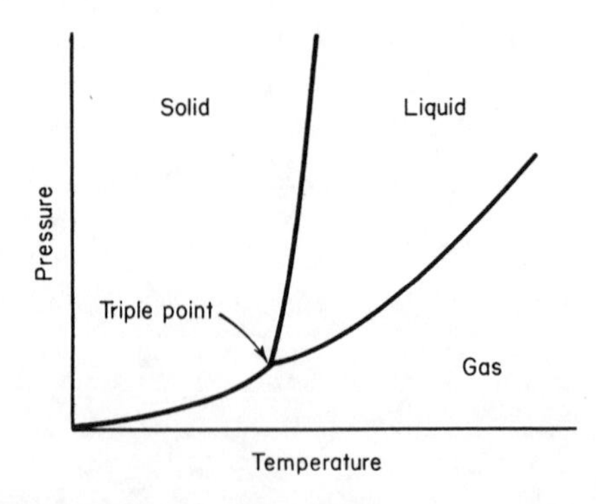

Fig. 15.1 Pressure-temperature phase diagram for a single-component system.

same time the uniformity of the structure of crystals makes it possible to employ mathematics in the study of their properties. On the other hand, the gas phase represents the other extreme from the solid phase where the structure is one of almost complete randomness, or disorder, instead of almost complete order. Here, in most cases, the atoms can be assumed to be removed far enough from each other that metallic gases can be treated as ideal gases. The physical properties of metallic gases, like those of metallic solids, are therefore capable of mathematical analysis.

While the solid crystalline phase is pictured as a completely ordered arrangement of atoms (neglecting defects such as dislocations and vacancies) and the gas phase as a state of random disorder (ideal gas), no simple picture has as yet been devised to represent the structure of the liquid phase. The principal trouble is the difficulty of the problem. The liquid phase possesses neither the long-range order of the solid nor the lack of interaction between atoms characteristic of the gas phase. It is, therefore, essentially an indeterminant structure. Actually, in a liquid, the average separation between atoms is very close to that in the solid. This fact is shown by the small change in density on melting, which for close-packed metals amounts to 2 to 6 percent only,[1] where part of this density change is probably associated with the formation of additional structural defects in the liquid phase (possibly additional vacancies, interstitials, or dislocations). Further, the latent heat of fusion released when a metal melts is relatively small, being only $\frac{1}{25}$ to $\frac{1}{40}$ of the latent heat of vaporization. Since the latter is a measure of the energy required to place the atoms in the gas phase, where it can be assumed that they are so far apart that they do not interact, it serves as a good measure of the energy binding atoms together. Both the small size of the energy released when a metal melts and the fact that the atoms in the liquid and solid phases have almost identical separations lead to the conclusion that the bonding of atoms in the solid and liquid phases are probably similar. X-ray diffraction studies of liquid metals tend to confirm this assumption. Time does not permit to delve into X-ray-diffraction techniques as applied to liquid metals, but the results of these investigations are interpreted as showing the following. Liquid metals possess a structure in which the atoms, over short distances, are arranged in an ordered fashion and have a coordination number approximately the same as the solid. The X-ray results also show that liquid metals do not possess long-range order. A plausible picture of the liquid structure might be the following: the atoms over short distances approach arrangements close to those exhibited in crystals, but due to the presence of many structural defects the exact nature of which is unknown, the long-range order typical of a crystal cannot be achieved. In a number of cases, it has been suggested that the liquid phase is essentially the same as that of the

1. Frost, B. R. T., *Prog. in Metal Phys.*, **5** 96 (1954). Pergamon Press, Inc., New York.

solid, with the incorporation into the structure of a large number of a specific form of structural defect, such as vacancies, interstitial atoms, or dislocations. However, these models have not been completely satisfactory and it must be admitted that the exact nature of the defects that exist in the liquid phase are not known. The fact that the atoms in the liquid have an extremely high mobility modifies this essentially static picture of the liquid phase. Diffusion measurements of liquids, made at temperatures just above the melting point, show that atomic movements in liquids are several orders of magnitude[2] (several powers of 10) more rapid than in solids just below the melting point. The rapid diffusion rates in liquids are undoubtedly a result of the structural defects characteristic of the liquid phase. Since the nature of the liquid state is unresolved, a simple picture of the way atoms move in a liquid cannot be presented. This is in contrast to solids, where the evidence strongly favors the vacancy-motion concept. One thing is clear, however: the energy barriers for atom movements in liquid diffusion are very small. Since diffusion in the liquid state must be associated with a corresponding motion of structural defects and the motion is very rapid we are led to the conclusion that the liquid phase is one of everchanging structure. The local order existing at any one position in space changes continually with time. This represents a basic difference between liquid and solid phases. In a solid, diffusion by vacancy movement does not, in general, alter the structure of the crystal, whereas in the liquid the structure is one of constant change resulting from atomic motion. Another important consequence of the ease of atom movements in the liquid phase is the development of the property that is most characteristic of the liquid phase: fluidity, or the inability of a liquid to support shear stresses of even very low magnitudes.

A surprising fact is that the liquid phases of most metals are very similar in their properties, even though their properties may be quite different in the solid state.[3] Thus, on melting, the close-packed metals are inclined to lower their coordination slightly, but metals of loose packing, such as gallium and bismuth, usually increase their coordination. The ultimate result is that liquid metals tend to have the same number of nearest neighbors. Similarly, the electrical and thermal conductivities of the liquid phases of metals also tend toward common values.

The above facts may be interpreted as demonstrating that the properties of the liquid phases of metals are more dependent on the structural-defects characteristic of liquids than they are upon the nature of the bonding forces between the atoms. On the other hand, in solid crystalline phases, the physical properties are very dependent on the nature of the bonding forces between atoms, since these interatomic forces determine the types of crystals that form and, as a consequence, the properties of the solids.

2. Nachtrieb, N. H., ASM Seminar (1958), *Liquid Metals and Solidification*, p. 49.
3. Frost, B. R. T., *Op. Cit.*, p. 96.

From the standpoint of the free energy of the phases and the reasons for the existences of transformations between solid, liquid, and gases, consider the following. Of the three phases, the solid crystalline phase possesses the lowest internal energy and the highest degree of order or lowest entropy. The liquid represents a phase with a slightly higher internal energy as measured by the heat of fusion, and a slightly larger entropy corresponding to its more random structure. Finally, the gas phase has the highest internal energy and the greatest entropy or disorder. The free-energy curves for these three phases may be plotted in a manner similar to those for the two solid phases of tin and, in each case, the slope will be steeper, in a more negative sense, the greater the inherent entropy of the phase. Thus, the free energy of the gas phase falls most rapidly with increasing temperature, the liquid phase next, and the solid phase least rapidly. The free energies of the three phases are also functions of the pressure, so that their positions relative to each other are not the same at different pressures. This fact is illustrated in Fig. 15.3 where the free-energy curves are drawn schematically for three isobaric lines (constant pressure) *aa, bb*, and *cc* of Fig. 15.2, which is based on Fig. 15.1. In the left-hand figure, corresponding to the isobaric *aa*, it can be seen that at temperatures below the melting point, T_m, the solid phase has the lowest free energy, but at T_m the free-energy curve of the liquid phase crosses that of the solid, and the liquid phase becomes the more stable until T_b, the boiling temperature, is reached. At this point, the free-energy curve of the gas phase crosses that of the liquid phase and for all higher temperatures the gas phase is the most stable. The middle figure corresponds to a constant-pressure line that passes through the triple point. Along this isobaric line (*bb*) the three free-energy curves cross at a single point (the triple point) where all three phases may coexist. At temperatures below the triple point, the solid phase has the lowest free energy, and at all temperatures above it the gas phase has the least free energy. At this particular pressure, the liquid phase is

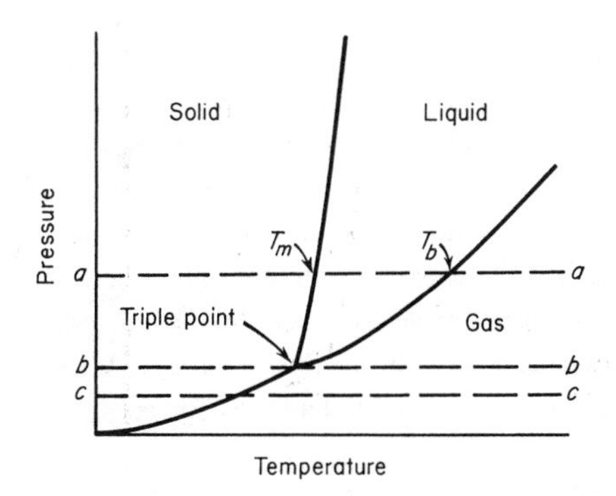

Fig. 15.2 Same as Fig. 15.1, but showing the isobars considered in Fig. 15.3.

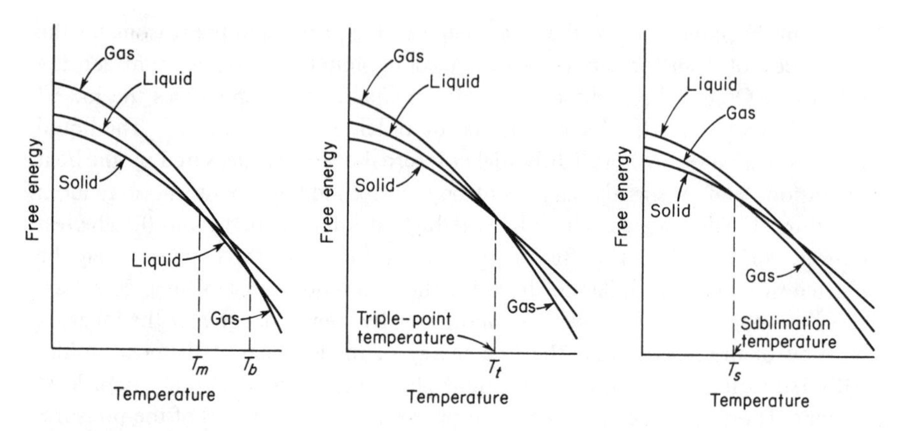

Fig. 15.3 Free-energy curves for the phases in a one-component system at three different pressures corresponding to isobars *aa, bb,* and *cc* respectively of Fig. 15.2.

only stable at one temperature – the triple-point temperature. Finally, on the right in Fig. 15.3, is shown the relative positions of the three free-energy curves when the pressure is very low (that is, at *cc*). In this case, the free-energy curve of the liquid phase lies above that of the gas phase, so that the liquid phase is not stable at any temperature.

15.2 Nucleation. As discussed in Section 13.3, the solidification of metals occurs by nucleation and growth. The same is also true of melting, but there is one important difference. Nucleation of the solid phase during freezing is a much more difficult process than the formation of nuclei in the liquid phase during melting. As a result, metals do not superheat to any appreciable extent when they are liquefied, whereas some supercooling occurs almost every time a metal is frozen. Further, if conditions during solidification are such that nucleation is homogeneous, liquid metals can be cooled to temperatures far below their equilibrium freezing points before solidification begins. This fact is illustrated in Table 15.1 which gives the maximum supercooling observed for a number of metals, as well as for the nonmetals water and benzene. It is not known whether all of the values listed in Table 15.1 correspond to freezing by homogeneous nucleation. However, heterogeneous nucleation, when crystals are formed on impurity particles, should reduce the magnitude of the observed supercooling, and we can conclude that the supercooling corresponding to homogeneous nucleation is at least as large, if not larger, than the value given in Table 15.1

The difficulty associated with the formation of homogeneous nuclei of the solid crystalline phase is typical not only of pure metals, but is also observed in

Table 15.1 Maximum Supercooling of Some Simple Liquid Substances.*

Substance	*Maximum Supercooling* °C
Mercury	77
Tin	76
Bismuth	90
Lead	80
Aluminum	130
Silver	227
Gold	230
Copper	236
Manganese	308
Nickel	319
Cobalt	330
Iron	295
Water	39
Benzene	48

* After Hollomon, J. H., and Turnbull, D., *Prog. in Metal Phys.*, 4, 356 (1953). Pergamon Press, Inc., New York.

alloys, as can be seen in Fig. 15.4 which shows that it is possible to supercool copper-nickel alloys approximately 300° C at all compositions. The very large magnitudes of the supercooling shown in Table 15.1 and Fig. 15.4 are only observed under rigidly controlled experimental conditions designed to prevent the occurrence of heterogeneous nucleation. In the usual metals of commerce, nucleation during freezing is almost always heterogeneous. Here the formation of nuclei is aided by the presence in the liquid melt of accidental impurity particles, or by the mold surfaces. When nucleation occurs in this manner, supercooling is greatly reduced and can only amount to a few degrees. The

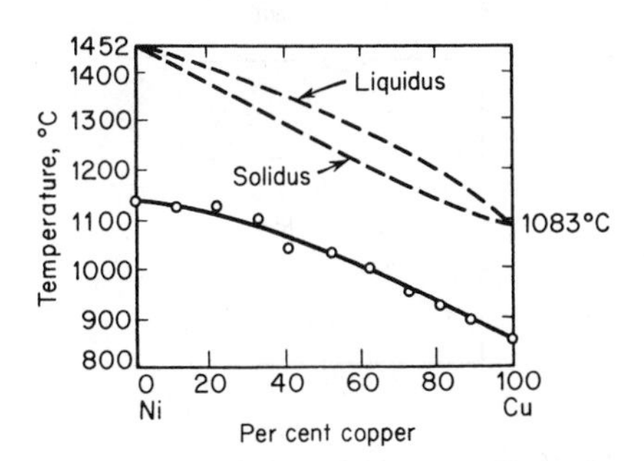

Fig. 15.4 Solidification temperatures of Cu-Ni alloys as a function of composition. (Cech, R. E., and Turnbull, D., *Trans. AIME,* **191** 242 [1951].)

subject of nucleation in commercial metals will be discussed in more detail when the freezing of ingots is discussed. Let us first, however, briefly treat the subject of nucleation during melting – the reverse of freezing.

In melting, as previously indicated, metals do not normally superheat. An explanation for this fact has been offered by Hollomon and Turnbull [4] based on the assumption that nucleation in melting occurs on the surfaces of solids. In this case, the nucleus, consisting of a region of the liquid phase, will be enveloped by two basically different surfaces. On one side is a liquid-solid interface and on the other, a gas-liquid interface. The total surface energy of this pair of surfaces is usually smaller than that of a single solid-gas interface. This conclusion is obtained from the experimental fact that when a solid and a liquid are in contact at the melting point, the liquid normally wets the surface of the solid so that, as proved in college physics textbooks,

$$\gamma_{gs} > \gamma_{sl} + \gamma_{gl}$$

where γ_{gs}, γ_{sl} and γ_{gl} are the surface tensions or surface energies of the gas-solid, solid-liquid, and gas-liquid interfaces respectively. What the above equation really signifies is that the surface energy associated with liquid nuclei can aid rather than hinder the formation of the nucleus. Thus, let Fig. 15.5A represent a liquid nucleus at an early stage of development, and Fig. 15.5B the same nucleus at a slightly later time. Spreading of the liquid nucleus along the surface, as shown in these figures, decreases the area on the gas-solid interface, while increasing those of the gas-liquid and liquid-solid interfaces. As a result, there is a decrease in surface energy, and since any rise of the temperature above the equilibrium melting temperature causes a decrease in the volume free energy favoring melting, it can be seen that both surface and volume free energies favor melting for even the slightest amount of superheating. There is also another

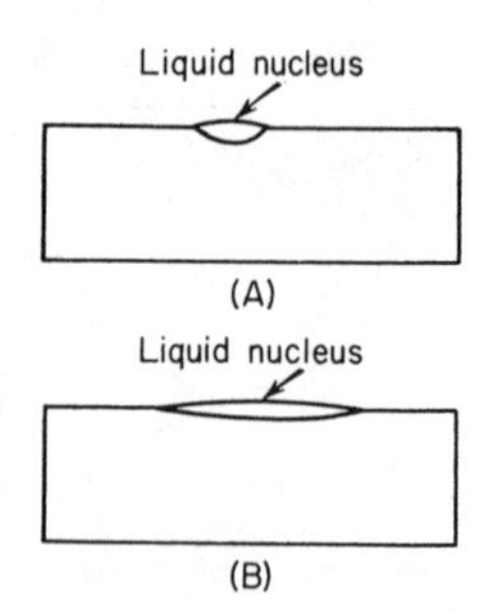

Fig. 15.5 The formation of a liquid nucleus during melting.

4. Hollomon, J. H., and Turnbull, D., *Prog. in Metal Phys.*, 4 333 (1953). Pergamon Press, Inc., New York.

factor, mentioned in Section 13.10; the diffusion rate also increases with rising temperature. The rate of a heating reaction should therefore tend to increase with superheating because of this fact.

15.3 Crystal Growth from the Liquid Phase. It has been shown[5] that the movement of the boundary separating a liquid from a solid crystalline phase, under a temperature gradient normal to the boundary, may be considered as the resultant of two different atomic movements. Thus at the interface, those atoms that leave the liquid and join the solid determine a rate of freezing, while those that travel in the opposite direction determine a rate of melting. Whether the boundary moves so as to increase or decrease the amount of solid depends on whether the rate of freezing or the rate of melting is the larger. This point of view is equivalent to viewing the movement of the interface as a two-directional diffusion problem in which

$$R_f = R_{f_0} e^{-Q_f/RT}$$

$$R_m = R_{m_0} e^{-Q_m/RT}$$

where R_f and R_m are the rates of freezing and melting respectively. R_{f_0} and R_{m_0} are constants, Q_f and Q_m are activation energies, and R and T have their usual significance.

The activation energy Q_f represents the energy required to take an atom that lies on the liquid side of the boundary to the saddle point, as shown in Fig. 15.6. Similarly, Q_m represents the energy required to bring an atom on the solid side

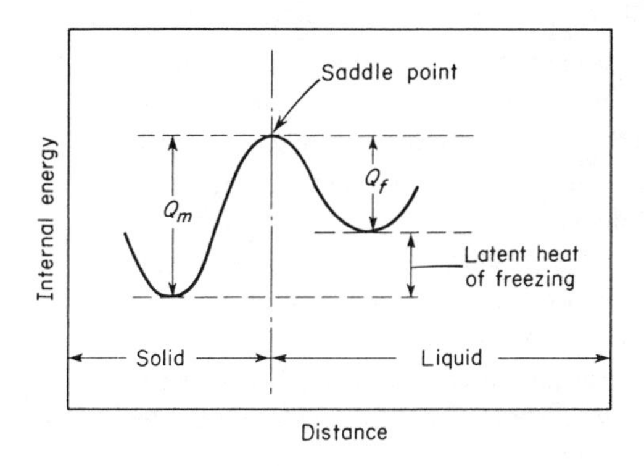

Fig. 15.6 The relationship between the activation energies for freezing and melting.

5. Chalmers, Bruce, *Trans. AIME*, 200 519 (1954).

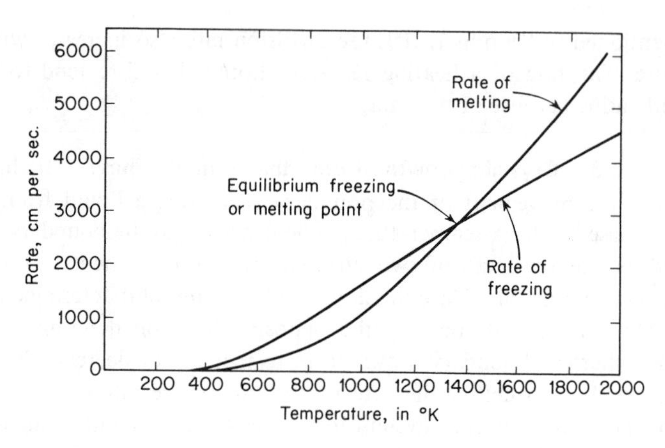

Fig. 15.7 Rates of freezing and melting for copper. (Chalmers, Bruce, *Trans. AIME*, **200** 519 [1954].)

of the interface up to the saddle point. In Fig. 15.6, the potential energy wells shown on each side of the saddle point differ by the latent heat of fusion. An atom in the solid possesses a lower energy than one in the liquid, but it should be noted that it also possesses a lower entropy.

The rates of freezing, R_f, and melting, R_m, may be expressed either as atoms per second crossing the boundary or as velocities of the boundary in centimeters per second. The constant R_{f_0} and R_{m_0} depend on a number of factors [6] that can be estimated. This has been done for copper by Chalmers, as may be seen in Fig. 15.7, which shows both the curves for the freezing rate and the melting rate as functions of temperature. Notice that they intersect at the equilibrium melting point of copper. Another interesting feature of Fig. 15.7 is that on either side of the true melting point the curves diverge sharply. The difference in the individual rates determines the actual rate with which the interface moves, which signifies that the observed growth rate, or melting rate, increases as the temperature deviates from the equilibrium freezing point. Because of the nucleation considerations that we have discussed previously, metals can be supercooled below the equilibrium freezing point, but cannot be superheated above it. The curves of Fig. 15.7 show that when supercooled metals freeze, the rate of solidification is very rapid.

We shall now return to a consideration of the constants R_{f_0} and R_{m_0} which appear in the rate equations. One of the factors that determine the magnitude of these constants is known as the *accommodation factor*, the chance that an atom in a given phase, either the liquid or the solid, can find a position on the other side of the boundary where it can attach itself. For the movement of atoms from

6. *Ibid.*

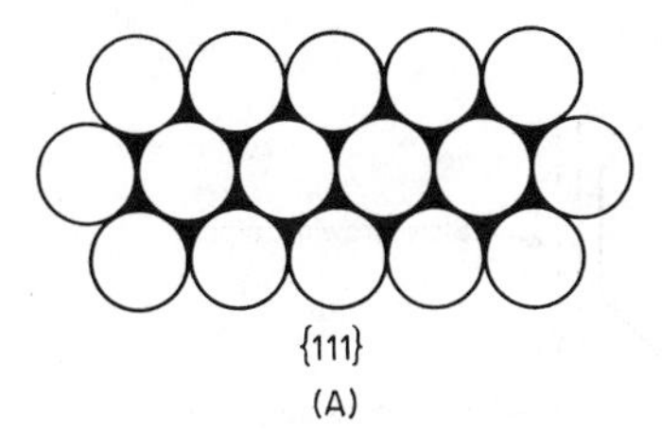

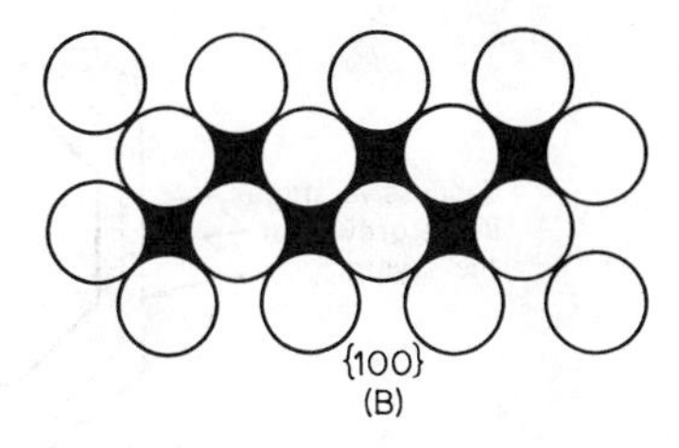

Fig. 15.8 Planes of looser packing, such as {100}, are better able to accommodate an atom that leaves the liquid to join the solid than a closer packed plane, such as {111}. Illustrated planes correspond to a face-centered cubic lattice. (After Chalmers, Bruce, *Trans. AIME,* **200** 519 [1954].)

the solid to the liquid, the accommodation factor should be more or less independent of the chemical nature of the atoms composing the liquid. This follows from the fact that the liquid phases of metals have very similar structures. On the other hand, different crystalline structures present entirely different types of surfaces toward the liquid phase, so that the accommodation factor for the motion of atoms from the liquid toward the solid varies with the nature of the solid. It is also true that the motion of atoms from the liquid to the solid depends on the indices of the particular crystalline plane which faces the liquid. The less close-packed the plane, the easier it is for liquid atoms to attach themselves to the crystal. This fact may be explained with the aid of Fig. 15.8 representing a face-centered cubic structure, which shows the atomic arrangements on {100} and {111} surfaces. The holes, or pockets, in the surface available for the accommodation of a liquid atom as it joins the crystal are larger for the less close-packed {100} plane than for the more closely packed {111} plane. As a result of this difference, for a given amount of supercooling there is a difference in the growth velocity of the two crystallographic planes: the least closely packed plane grows with the greater velocity.

The fact that low-atomic-density planes grow faster does not mean that a growing crystal assumes faces which are planes of this type. On the contrary, the tendency is for the crystal to assume faces that are close-packed, or slow-growing. The reason for this is easy to understand, for the low-density planes of fast growth tend to grow themselves out of existence, leaving behind only close-packed surfaces. This effect is shown schematically in Fig. 15.9 where several stages in the growth of the faces of a crystal are shown.

It has already been indicated that the rate with which the interface between liquid and solid moves during solidification is a function of the amount of supercooling. In addition, the interface motion is also affected by the sign of the temperature gradient in front of the interface. Thus, the crystal-growth process, when the temperature gradient in advance of the interface is a rising one, is

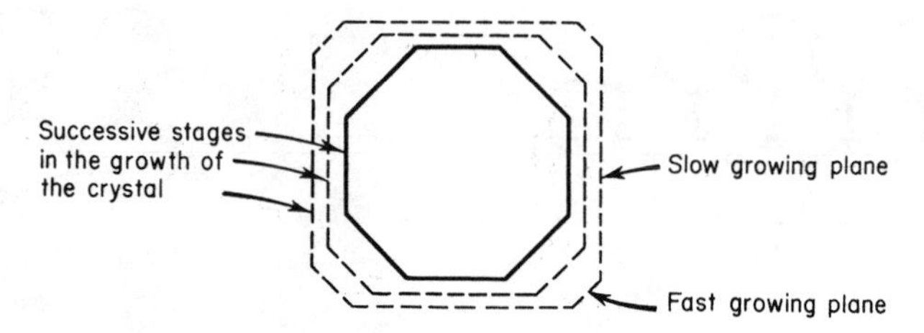

Fig. 15.9 A crystal growing in a liquid tends to develop faces that are slow-growing (close-packed).

different from that when the gradient is a falling one. These two cases will now be considered.

15.4 Stable Interface Freezing. First it will be assumed that the temperature rises as we move from the interface into the liquid, and that the temperature gradient is linear and perpendicular to the interface. Under these conditions, the interface is believed to maintain a stable shape and move forward as a unit. If by chance a close-packed plane (one of slow growth) lies perpendicular to the heat-flow direction, a strictly planar interface should, in theory, develop. However, the chance of finding a close-packed plane in exactly the right position is very small, and one has to consider those cases where a crystallographic plane of high-atomic density is almost parallel to the interface, as well as those situations where the interface does not approximate any high-density plane.

Consider the first of these possibilities. Here the interface tends to develop a series of close-packed planar steps. A schematic example of such a surface is shown in Fig. 15.10. Since each of the facets has an inclination in the heat-flow

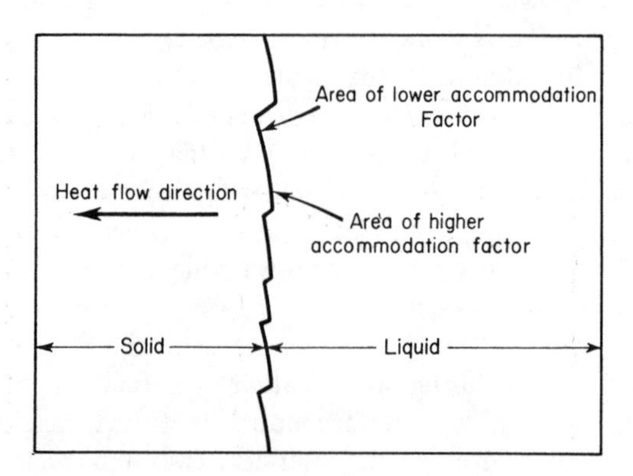

Fig. 15.10 A stable interface when a close-packed lattice plane is almost parallel to the interface.

direction, the temperature cannot be uniform over its entire area. Because of the rising temperature in advance of the interface, those portions of each facet which are most advanced will be in contact with hotter liquid than those portions to the rear. Since, for a given crystallographic plane, the rate of growth is a function of the degree of supercooling, it is not possible for the facets to maintain a strictly crystallographic surface and grow with a constant velocity. As a consequence, the steps assume a curved shape. The most advanced, or hotter, portion of each facet corresponds to a lower indice, or higher accommodation factor surface, while the most retarded, or cooler portion corresponds to a slow-growing, or lower accommodation-factor surface. In this manner, it should be possible to obtain an interface that grows at a constant speed.

The structure described in the above paragraph is that which attains when close-packed planes are nearly parallel to the interface. When there are no close-packed planes approximately parallel to the interface, a simpler interface, consisting of a random (crystallographic) planar surface, may be more stable. Such an interface is similar to one obtained when a close-packed crystallographic plane is strictly perpendicular to the heat flow direction, but differs in that it is not a close-packed plane, but a plane with high, or irrational indices.

15.5 Dendritic Growth. A very important type of crystalline growth occurs when the liquid-solid interface moves into a liquid, whose temperature falls, or decreases, in advance of the interface. One of the most important ways that a temperature gradient of this nature can form is as follows.

Suppose that Fig. 15.11 represents a region containing a liquid-solid interface and that the heat is flowing from right to left; the heat is being removed through the solid. At the same time, assume that a considerable degree of supercooling has been attained, so that the temperature of the liquid is well below the equilibrium freezing point. Because of the heat of fusion that is released at the interface, the temperature of the interface usually rises above that of both the liquid and the solid. Under these conditions, the temperature drops as one moves from the interface into the solid, because this is the heat-flow direction. It also

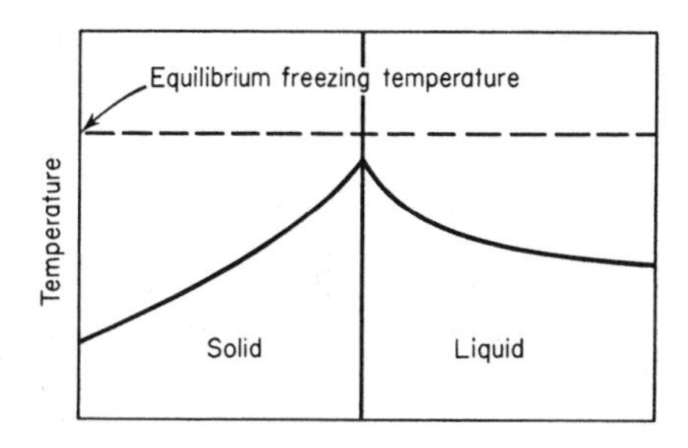

Fig. 15.11 Temperature inversion during freezing. (After Chalmers, Bruce, *Trans. AIME,* **200** 519 [1954].)

falls off into the liquid because there is a natural flow of heat from the interface into the supercooled liquid. The resulting temperature contour is shown in Fig. 15.11 and is known in general as a *temperature inversion.*[7]

When the temperature falls in the liquid in advance of the interface, the latter becomes unstable and crystalline spikes may shoot out from the general interface into the liquid. The resulting structure may also become quite complicated, with secondary branches forming on the primary spikes, and possibly with tertiary branches forming on the secondary ones. The resulting branched crystal often has the appearance of a miniature pine tree and is, accordingly, called a *dendrite* after the Greek word *dendrites* meaning "of a tree."

The reasons for the branched growth of a crystal into a liquid whose temperature falls in advance of the solid are not hard to understand. Whenever a small section of the interface finds itself ahead of the surrounding surface, it will be in contact with liquid metal at a lower temperature. Its growth velocity will be increased relative to the surrounding surface which is in contact with liquid at a higher temperature, and the formation of a spike is only to be expected. Associated with the development of each spike is the release of a quantity of heat (latent heat of fusion). This heat raises the temperature of the liquid adjacent to any given spike and retards the formation of other similar projections on the general interface in the immediate vicinity of a given projection. The net result is that a number of spikes of almost equal spacing are formed which grow parallel to each other in the fashion shown in Fig. 15.12. The direction in which these spikes grow is crystallographic and is known as the

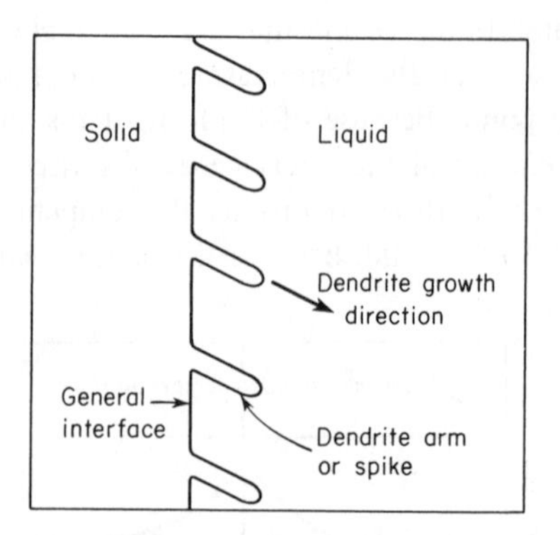

Fig. 15.12 Schematic representation of the first stage of dendritic growth. A temperature inversion is assumed to exist at the interface; that is, the temperature in the liquid drops in advance of the interface.

7. Weinberg, F., and Chalmers, B., *Canadian Jour. of Phys.*, **29** 382 (1951); **30** 488 (1952).

Table 15.2 Dendritic Growth Directions in Various Crystal Structures.*

Crystal Structure	*Dendritic Growth Direction*
Face-centered cubic	⟨100⟩
Body-centered cubic	⟨100⟩
Hexagonal close-packed	$\langle 10\bar{1}0 \rangle$
Body-centered tetragonal (tin)	⟨110⟩

* Chalmers, B., *Trans. AIME*, 200, 519 (1954).

dendritic growth direction. The direction of dendritic growth depends on the crystal structure of a metal, as may be seen in Table 15.2.

The branches, or spikes, shown in Fig. 15.12 are first order, or primary, in nature. How secondary branches may form from primary ones will now be considered. For this purpose consider Fig. 15.13 where section *aa* represents the general interface. Notice that in this figure the direction of dendritic growth is assumed to be normal to the general interface. This is done to simplify the presentation. Once the spikes have formed, growth at the general interface will be slow because here the supercooling is small and the latent heat of fusion associated with the formation of the spikes tends to further decrease its magnitude. At section *bb*, on the other hand, the average temperature of the liquid is, by definition, lower than at *aa*. However, even on this section at points in the liquid close to the spikes the temperature will be higher than midway between the spikes ($T_A > T_B$) because of the latent heat released at the spikes.

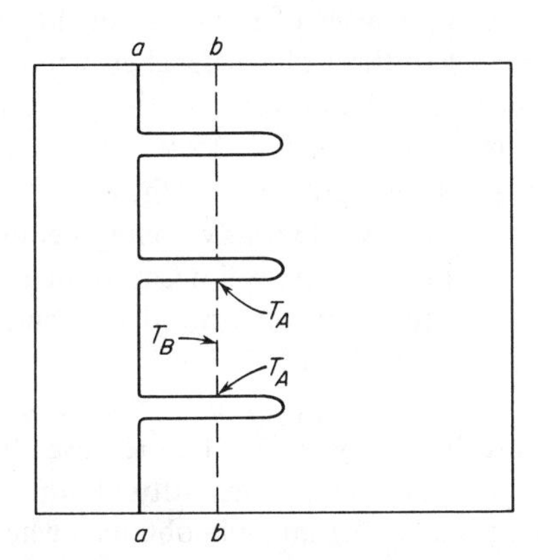

Fig. 15.13 Secondary dendrite arms form because there is a falling temperature gradient starting at a point close to a primary arm and moving to a point midway between the primary arms. Thus, $T_B < T_A$.

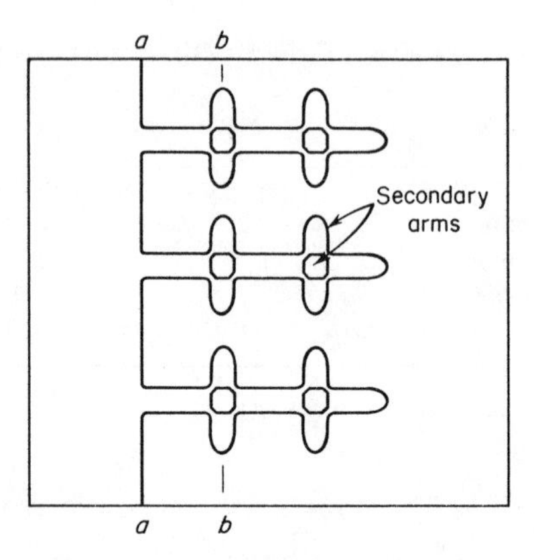

Fig. 15.14 In a cubic crystal, the dendrite arms form along ⟨100⟩ directions. Primary and secondary arms are thus normal to each other.

There is, therefore, a decreasing temperature gradient not only in front of the primary spikes, but also in directions perpendicular to the primary branches. This temperature gradient is responsible for the formation of secondary branches which form at more or less regular intervals along the primary branches, as shown in Fig. 15.14. Since the secondary branches form for the same basic reasons as the primary branches, their directions of rapid growth are along directions equivalent crystallographically to those taken by primary arms. In the case of cubic metals, dendrite arms may form along all of the ⟨100⟩ directions and the arms are perpendicular to each other. Further, in the case illustrated in Fig. 15.14 one would probably not find tertiary branches forming on the secondary branches. This is a matter of geometry of the growth pattern under consideration. In other cases of three-dimensional growth there is no reason why branches of higher order should not form if the space is available for their growth. Although it is not shown in Fig. 15.12 to Fig. 15.14, dendrite arms usually grow in thickness as they grow in length, with the eventual result that the arms grow together to form a single nearly homogeneous crystal.

Dendritic growth, as outlined above, will occur in the freezing of pure metals when the interface is allowed to move forward into sufficiently supercooled liquid. In metals of relatively high purity, however, it is almost impossible to obtain enough thermal supercooling, so that the entire freezing process is dendritic. This is because it is very difficult to remove all catalysts, or loci of heterogeneous nucleation, from a sizeable quantity of liquid metal. The very large supercoolings reported in Table 15.1 are only obtained when very small volumes of liquid metal (average particle size 10 to 50 microns [8]) are frozen. The smaller

8. Turnbull, D., and Cech, R. E., *Jour. Appl. Phys.*, 21 804 (1950).

the volume of liquid, the greater the chance that it will not contain a nucleation catalyst. On the other hand, a very large supercooling (of the order of 100°C) is required in pure metals for complete dendritic freezing. This large supercooling is necessary to overcome the effect of the latent heat of fusion released as the dendrites form, which naturally tends to remove the supercooling and the driving force for dendritic growth. In summary, it may be stated that it is usually not possible to cause more than 10 percent of a pure metal, in bulk, to freeze dendritically.[9]

15.6 Dendritic Freezing in Alloys. Dendritic freezing is a common phenomenon in many alloy systems. Here the supercooling which furnishes the driving force for dendritic growth is normally of a different type. The form discussed above is called *thermal supercooling* and can also be a factor in the freezing of alloys, but constitutional supercooling which we shall now consider, following in general the proposals of Rutter and Chalmers,[10] is of much greater importance. Constitutional supercooling results when a solid freezes with a composition different from that of the liquid from which it forms. Thus, consider an isomorphous-alloy system with the phase diagram shown in Fig. 15.15. In this system at the temperature T_1, solid of composition (a) is in equilibrium with liquid of composition (b). Let it now be assumed that Fig. 15.16 represents a volume of liquid of composition (b) placed in a long tubular mold and that heat is only removed from the left end of the mold so that the heat flow is linear and from right to left. Under these conditions, freezing will start at the left end of the liquid and the small volume element (dx) may be taken to

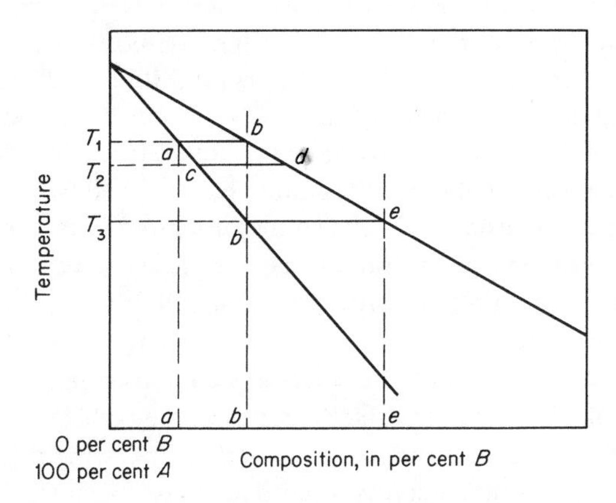

Fig. 15.15 Part of an isomorphous binary phase diagram.

9. Chalmers, Bruce, *Trans. AIME*, **200** 519 (1954).
10. Rutter, J. W., and Chalmers, B., *Canadian Jour. of Phys.*, **31** 15 (1953).

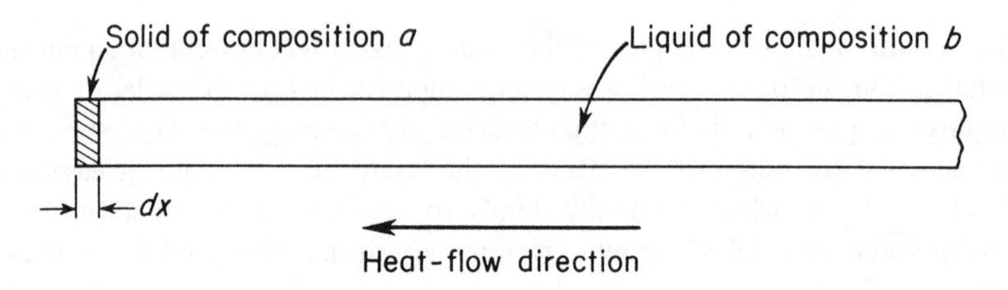

Fig. 15.16 A simple case of one-dimensional freezing.

represent the first solid to form. The freezing of this volume element will take place at temperature T_1 and the composition of the solid will correspond to point (a) in Fig. 15.15. If this layer of solid is frozen in a relatively short period of time, we can assume that it is formed from the liquid layer adjacent to the interface and not from the entire volume of the liquid. Since the solid contains a higher ratio of A to B atoms than does the liquid layer from which it forms, the latter is depleted in A atoms and enriched in B atoms. This composition change in the liquid just ahead of the interface increases as more solid is formed, but the increase eventually stops when a steady-state condition is attained. Concurrent with the change in composition of the liquid next to the interface is a similar change in the composition of the solid which can form from this layer of liquid. Thus, if the composition of the liquid corresponds to point d in Fig. 15.15, then the solid which freezes must have the composition of point c. It is also important to recognize that the indicated freezing process can only occur if the temperature at the interface is lowered from T_1 to T_2. The concentration of an excess of B atoms in the liquid adjacent to the interface reaches a maximum when the liquid next to the solid attains the composition of point e. At this instant the liquid is able to freeze solid of composition b. When this occurs, the steady state has been attained and the solid, which is formed from the liquid layer enriched in B atoms, is the same as the liquid drawn into this layer. At this time, the temperature at the interface must be T_3, as shown in Fig. 15.15. The composition as a function of distance along the mold is shown in Fig. 15.17A. This figure corresponds to the time when the steady-state freezing process has just been attained. A similar illustration is found in Fig. 15.17B, but pertains to a later time when the interface has moved further to the right. Notice that, in both cases, the composition of the solid rises from its original value a to that of the original liquid b. At the interface there is also a sudden rise in composition to the value e as one passes from solid into liquid. Following this sudden rise, the composition in the liquid decreases exponentially back to the value of the original liquid, which is b. The shape of the composition-distance curve in this exponential region is a function of the rate of freezing and the atomic diffusion rates in the liquid. It is important to note that in the above discussion it is

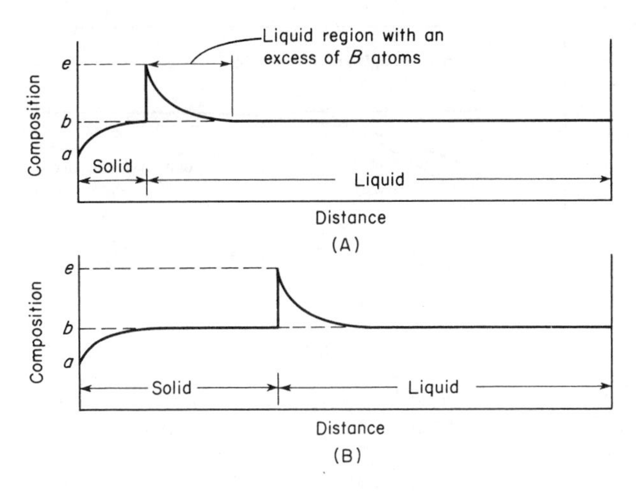

Fig. 15.17 Composition-distance curves corresponding to two different stages in the one-dimensional freezing problem of Fig. 15.16.

implied that there are no liquid convection currents. In the presence of such currents it is not possible to attain as large a concentration of B atoms in advance of the interface. There are other factors which may also limit the magnitude of the composition change in the liquid at the interface. However, for our purposes we can assume that the distance-composition curves shown in Fig. 15.17 are representative of what can actually happen under similar conditions during the freezing of an alloy.

For the sake of the present argument, let us assume that we can neglect considerations of convection currents and other complicating factors and that the steady-state composition variation shown in Fig. 15.17 holds. In most practical examples of freezing, liquid metal is poured into a mold cavity and freezes as a consequence of heat losses through the mold shell. As a result, the temperature is always lowest at the mold walls and rises toward the center of the mold. Solidification, accordingly, starts at the walls and proceeds inward. Applying these condisiderations to the present problem signifies that we should concentrate our attention on a freezing process in which the temperature rises in advance of the liquid-solid interface. This situation is illustrated in Fig. 15.18 where the temperature of the liquid is assumed to rise linearly with distance from the interface. A second curve in the figure shows the freezing point of the liquid alloy as a function of distance from the interface. This varies with distance from the interface due to the change in composition of the liquid as one moves away from the interface. At the interface, the freezing temperature, as shown previously, is T_3, but away from the interface it at first rises rapidly and then levels off to the temperature T_1, the temperature at which the bulk of the liquid

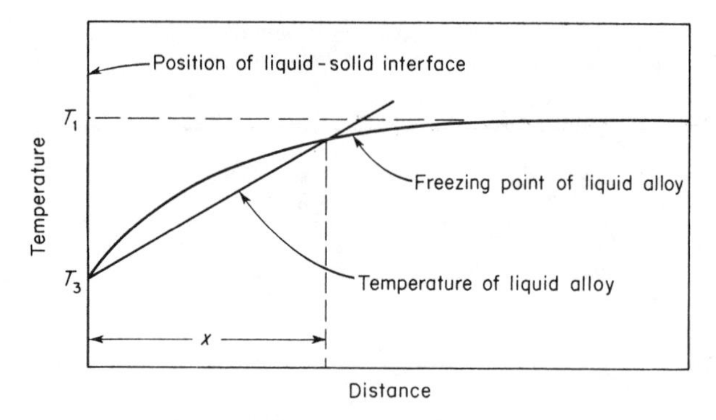

Fig. 15.18 Inside the distance (x), the temperature of the alloy is lower than its freezing point. This constitutes what is known as constitutional supercooling.

will begin to freeze. As shown in Fig. 15.18, the temperature of the liquid and the freezing point of the liquid intersect at two points: at the interface, and at the distance x from the interface. The pertinent point, however, is that within the distance x the liquid lies at a temperature below its freezing point. Inside this range it is effectively supercooled in spite of the fact that the temperature gradient is positive. This is the direct result of the concentration gradient in the liquid alloy in front of the interface.

Whether or not true dendritic freezing, with its associated shooting out of branches of primary and higher order, occurs when the liquid in advance of the interface is constitutionally supercooled depends on the amount of the supercooling. In large commercial castings, the supercooled layer (distance x in Fig. 15.18) is usually large, and dendritic freezing is an important factor. On the other hand, if the supercooled layer is thin, the growth of true spikes is not possible because of the limited depth of the supercooled layer into which they can grow. In this case, the instability of the interface may result in the formation of a surface composed of more or less oval projections of the type shown in Fig. 15.19. Because the movement of a surface of this type is coupled with the forward motion of the narrow supercooled region, its shape is stable. This leads to a very interesting result. In order for the surface to maintain its shape, freezing must occur uniformly over the entire surface. However, the solid lying on the surface at the centers of the projections, which are furthest to the right, lies at a temperature (T_1) higher than that at the cusps (T_2), which are furthest to the left. Coupled with this temperature difference is a difference in the composition of the liquid that freezes at the two positions. That which solidifies at the cusps has a higher concentration of solute than that at the center of the projections. The result of this freezing process is the formation of a cellular structure in which the cell walls (horizontal lines in Fig. 15.19) are defined as

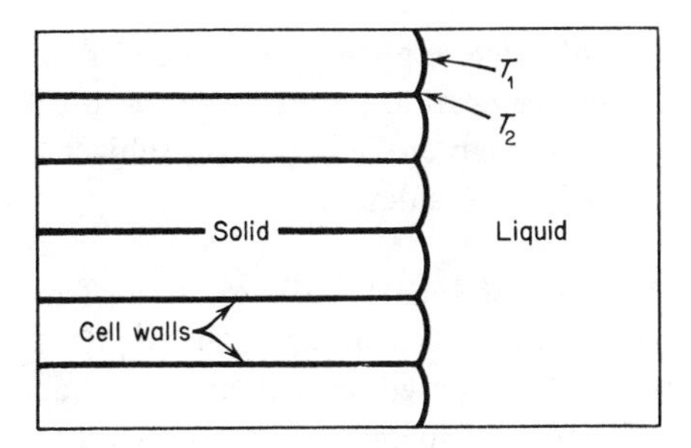

Fig. 15.19 When the region of constitutional supercooling is narrow, a cellular structure may form as the result of the movement of a stable interface of the type shown above.

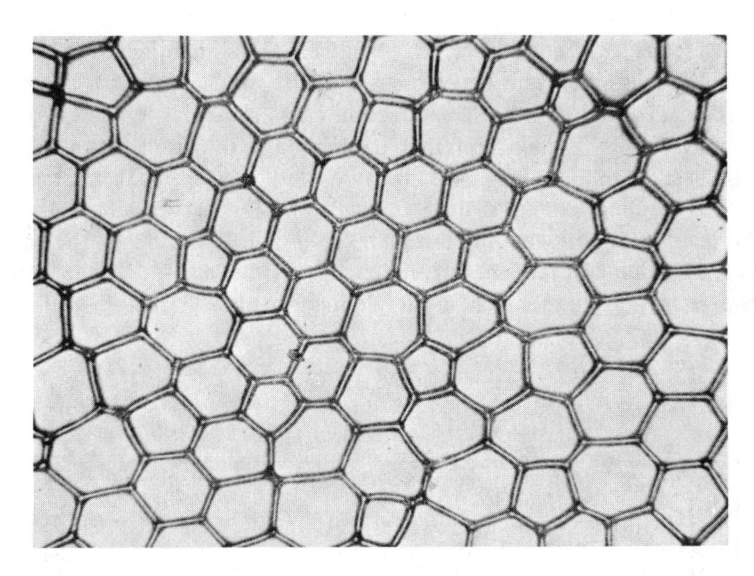

Fig. 15.20 Cellular structure in tin as viewed normal to the interface. 100 x . (Rutter, J. W., ASM Seminar, *Liquid Metals and Solidification*, 1958, p. 243.)

regions of high-solute concentration. Figure 15.20 shows an actual photograph taken normal to the interface of one of these cellular structures. Notice that what one sees in this figure is the surface of a single crystal which is not uniform in composition; the dark lines representing areas of increased solute concentration.

The cellular structure discussed above is noteworthy, for it shows how a nonuniform solute distribution can be obtained on a scale smaller than the crystal size. In general, this type of phenomenon is called *micro-segregation*. This is only one aspect of the segregation problem that occurs during the freezing of alloys. In a later part of this chapter the important subject of segregation in alloy castings will be considered in more detail.

15.7 Freezing of Ingots. The casting of ingots is a very important step in the manufacture of wrought products; items such as plates and beams which are plastically worked into their final shape. The size of these castings depends on the type of metal and its eventual use. In the manufacture of steel, large ingots, weighing 6 to 8 tons, are common.

When an ingot is frozen, three separate phases of the freezing process may occur, with each phase developing a characteristic arrangement of crystal sizes and shapes. The basic structures are illustrated in Fig. 15.21. In a narrow band following the contour of the mold lies the "chill zone," consisting of small equiaxed (equal-sized) crystals which usually have random orientations. Inside of this outer zone the crystals become larger in size, elongated in shape, with their lengths parallel to the heat-flow direction (normal to the mold walls). These grains have a very strong preferred orientation with a direction of dendritic growth parallel to their long axis. Because of the shape of the crystals in this zone, it is customary to call it the *columnar zone*. The last zone lies at the center of the ingot and represents the last metal to freeze. In this region the grains are again equiaxed and of random orientation.

The factors [11] instrumental in developing the above structures will now be

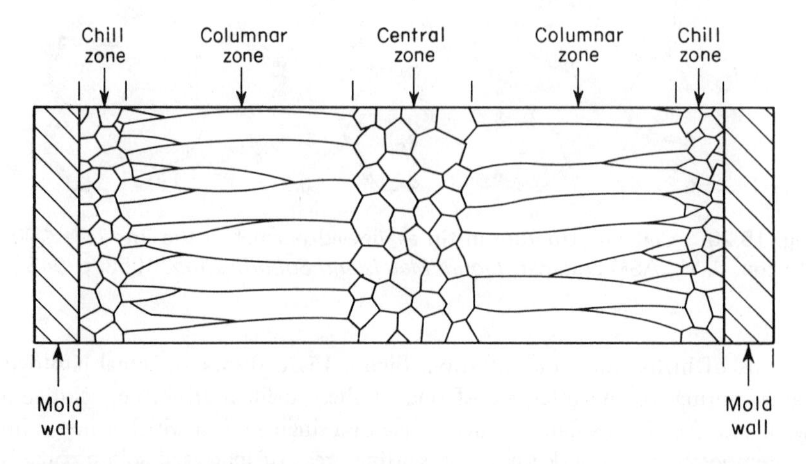

Fig. 15.21 A section through a large ingot, and the three basic zones of freezing that may be found in a casting.

11. Walton, D., and Chalmers, B., *Trans. AIME,* **215** 447 (1959).

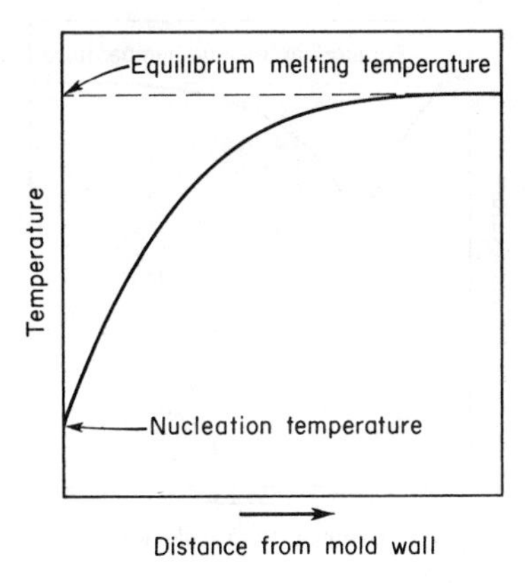

Fig. 15.22 Schematic temperature distribution in the liquid next to the mold wall just before nucleation. (After Walton, D., and Chalmers, Bruce, *Trans. AIME*, **215** 447 [1959].)

considered. First take the case of a pure metal. When liquid metal is poured into a mold, the mold walls, which are at a much lower temperature (usually room temperature) than the liquid, rapidly cool the layer of liquid with which they are in contact. As a result, the temperature of the liquid metal for a short distance away from the mold walls drops below the equilibrium freezing temperature in the manner shown in Fig. 15.22. Because of the rapidity with which the liquid temperature falls, a considerable magnitude of supercooling usually results. When nucleation (usually heterogeneous) occurs, its rate will be relatively rapid, with the result that the average size of the grain of this solid will be small. Because the crystals form independently, their orientation will be random. Finally, because their growth is limited by similar neighboring crystals nucleated at approximately identical times, their sizes will be nearly uniform and the structure is said to be *equiaxed*.

The crystals in the chill zone develop through both nucleation and growth. Crystal nuclei form in the liquid and grow in size until they make contact with neighboring crystals. The columnar zone forms in a different manner. Here crystal growth predominates and very little nucleation is observed. As soon as nucleation starts in the chill zone, the temperature in this region begins to rise again toward the equilibrium freezing temperature. This is the natural result of the release of the latent heat of fusion that changes the temperature contour shown in Fig. 15.22 to one more like that of Fig. 15.23. When this latter situation prevails, we see that an inversion in the temperature develops ahead of the crystals formed in the chill zone. Instead of the temperature rising ahead of the interface it now falls. This decreasing temperature of the liquid in advance of the interface is just the condition that promotes dendritic growth, and crystals

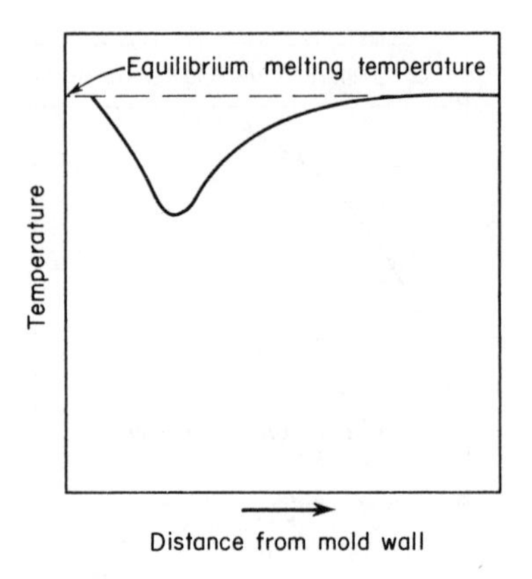

Fig. 15.23 Schematic temperature distribution in the liquid just after nucleation in the chill zone. (After Walton, D., and Chalmers, B., *Trans. AIME,* **215** 447 [1959].)

of the chill zone lying on the interface shoot out dendrite arms into the supercooled liquid, and the columnar zone begins to form. In a pure metal, the amount of this type of growth is usually strictly limited; the heat of fusion released by continued freezing soon removes the temperature inversion ahead of the interface. When this occurs, the temperature contour ahead of the liquid-solid interface changes to one with a rising temperature gradient and growth proceeds through the advance of a stable interface.

The columnar zone is thus composed of crystals that start at the chill zone and grow side by side in one direction; the heat-flow direction. In a pure metal, these crystals may continue to grow in this manner to the center of the ingot. Their growth stops when they meet the grains growing out from the opposite wall. The central equiaxed zone shown in Fig. 15.21 is not found in pure metal ingots. It is, however, a phenomenon commonly observed in the freezing of alloys. While it would appear that dendritic growth occurs only briefly during the freezing of pure metals, it has been proposed by Walton and Chalmers [12] that the preferred orientation found in the columnar zone results from the dendritic freezing phase. At the start of dendritic freezing, those crystals on the interface (chill zone) which possess a direction of rapid dendritic growth nearly normal to the interface will, in general, shoot out their spikes more rapidly than their less favorably oriented neighbors. The growth of these latter is also adversely affected by the heat of fusion released by the more rapidly growing crystals. In this manner, certain crystals are suppressed while others continue to grow in size, with the end result that only those crystals survive which have

12. *Ibid.*

dendritic-growth directions nearly parallel to the heat-flow direction. The general nature of the development of the preferred orientation is shown schematically in Fig. 15.21. Notice that as the columnar grains grow in length there is also an increase in their diameter. This is, above all, the result of the elimination of less favorably oriented crystals. Still, there is other evidence that grain coarsening in ingots can arise from other causes such as the grain boundary movements which would be due to grain-boundary surface-energy factors.

When a pure metal freezes in a mold, the supercooling promoting dendritic growth can only be of the thermal form. In alloys, on the other hand, in addition to the thermal supercooling, one has the possibility of producing constitutional supercooling. When this occurs, dendritic freezing is also observed and with it a corresponding development of a preferred orientation. The resulting preferred orientation is of the same type as that discussed above with a dendritic-growth direction in each columnar grain parallel to the heat-flow direction. The mechanism by which dendritic growth in alloy castings eliminates the less favorably oriented crystals differs[13] somewhat from that in pure metals, but need not concern us here.

One important result of constitutional supercooling in alloys is that it favors the development of the central equiaxed zone. As the solid-liquid interfaces that start from opposite sides of a mold approach each other at the center of a large ingot, their respective zones of constitutional supercooling eventually overlap. This is shown schematically in Fig. 15.24. The left-hand figure represents an early stage in the freezing process, while the right-hand figure represents a later stage when the zones have come together. At this point, a very large

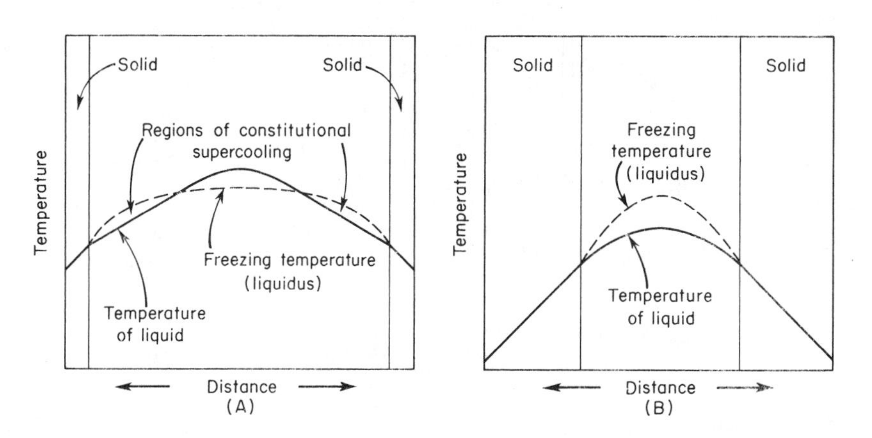

Fig. 15.24 The development of the constitutional supercooled region at the center of an alloy ingot that produces the central equiaxed zone of the ingot.

13. *Ibid.*

supercooling can occur in the liquid near the center of the ingot. There are two basic causes for this. First, the concentration of solute in the liquid just ahead of the interface tends to increase with increasing growth of the columnar zone, thus requiring lower and lower temperatures at the interface for continued solidification. Second, the temperature at the center of the ingot tends to approach that of the interfaces as the latter get closer together. This effectively flattens out the contour of the temperature-distance curve in the liquid. When a central equiaxed zone appears in an ingot, it is indicative of the fact that constitutional supercooling may have occured in this region and that it was able to develop to the point where nucleation could occur in the liquid at the center of the casting. Crystallization in this central zone may occur, therefore, through the appearance and growth of new crystals, and not through the continued growth of the elongated crystals of the columnar zone.

One other point is worth noting, relative to the formation of the central zone of an ingot. When nuclei form in this region, the heat of fusion which they release raises the temperature in the immediate surroundings of each nucleus. Each new crystal is thus surrounded by a temperature inversion and, as a result,

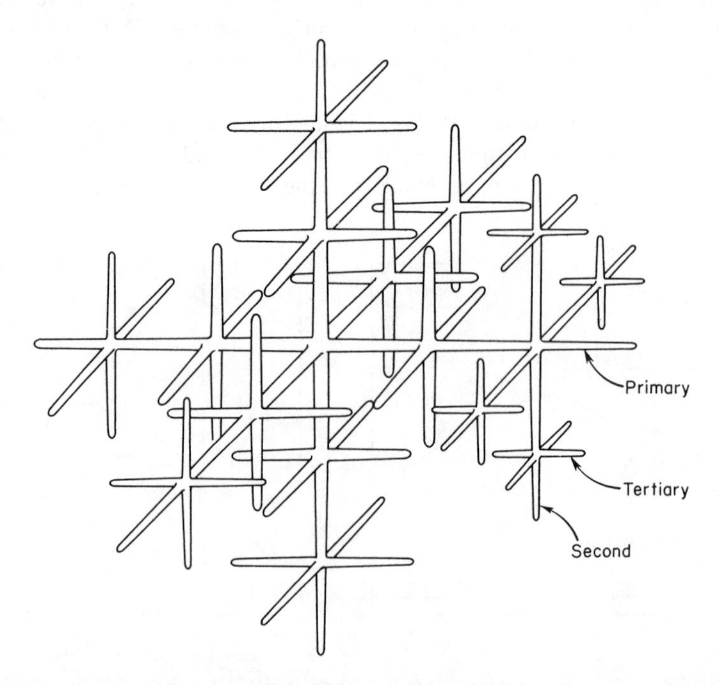

Fig. 15.25 A dendritic crystal formed by three-dimensional dendritic growth. Tertiary arms are shown on one set of secondary arms only (see right side of figure). (After Grignon {1775}and Tschernoff {1878}.)

begins its growth dendritically. However, since they form freely in the liquid and are completely surrounded by a temperature inversion, dendrite arms will shoot out along all the directions of dendrite growth. In a cubic crystal, arms form along all of the six ⟨100⟩ directions, as shown schematically in Fig. 15.25. Notice that geometrical considerations are such, in this case, that tertiary branches can form on the second-order branches.

Dendritic growth in the central zone only continues until the heat of fusion removes the constitutional supercooling. When this happens, freezing is completed by the filling in of the spaces between the dendrite arms and between neighboring crystals.

It should be mentioned that grain nucleation, as a result of constitutional supercooling, is just one way that the equiaxed grains can form at the center of an ingot. Another way they can form is by grain multiplication, which will be described in the next section.

15.8 The Grain Size of Castings. A detailed description of the factors in freezing that determine the final grain size of a casting is beyond our scope. Cole and Bolling[14] have summarized many of them in a paper that has particular reference to techniques used to obtain small grain sizes, primarily in the equiaxed central zone of a casting. Under the right conditions, this zone can become quite large. There are a number of significant reasons why it is desirable to produce castings with a very fine grain size. For example, it has been found that superplasticity (*see* Chapter 20) occurs primarily as a result of a very finely divided structure.

There are two basic ways of increasing the number of grains in a casting. One involves increasing the basic nucleation rate, while the other depends on breaking up the dendrites as they grow so as to form additional crystal seeds. This latter is called *grain multiplication*. Rapid cooling, as in die casting and casting into small cold molds, tends to increase the nucleation rate. It can also increase grain multiplication because of the enhanced convection currents associated with this type of casting. The nucleation rate can also be increased by adding catalysts to the melt. These inoculants are normally small, second-phase particles that increase the rate of heterogeneous nucleation. For example, boron and titanium are often added to aluminum alloys to increase nucleation. These elements can be considered to react with the aluminum in the alloy to form particles which act as catalysts.

Of the two basic methods of decreasing the grain size, the one of most general application is grain multiplication. This process probably occurs in most casting operations. As the dendrite arms grow and branch out, there is a continual

14. Cole, G. S., and Bolling, G. F., *Ultrafine Grain Metals*. Ed. Burke, J. J., and Weiss, V., p. 31, Syracuse University Press, Syracuse, N.Y., 1970.

release of the heat of fusion. This heat release, under the right conditions, can produce local melting and cause some dendrite arms to neck and pinch off. When this occurs, a small crystal fragment is formed. Agitation of the liquid metal due to convection currents tends to remove these crystallites from the liquid-solid interface and distribute them throughout the liquid. Many of them may eventually melt, but others grow into new, randomly-oriented crystals.

Artificial means can be employed to increase grain multiplication. In general, these involve fragmenting the growing dendrites at the growing liquid-solid interface and causing them to be redistributed throughout the melt by a forced fluid flow. Forced vibrations, alternating magnetic fields, ultrasonic vibrations, and a number of other methods may be used for this purpose. An effective method[15] involves the oscillation of a mold through several revolutions, followed by a change in the direction of rotation. This not only promotes the detachment of dendrite arms, but also causes them to be rather uniformly distributed throughout the liquid as a result of the agitation of the liquid.

15.9 Segregation. The liquids which are frozen to form industrial alloys usually contain, in addition to solute elements added intentionally for their beneficial effects, many impurity elements which find their way into the liquid metal by a number of different routes. Thus, impurity elements, present in the ores from which the basic metals were obtained, are frequently only partly eliminated during the smelting and refining operations. The refractory brick linings of furnaces used in melting or refining (purifying) and the gases in the furnace atmospheres may be other sources. In the latter case, elements enter the liquid metal in the form of dissolved gases. The various elements dissolved in commercial liquid metals can and often do react with each other to form compounds (oxides, silicates, sulfides, etc.). In many cases, the latter may be less dense than the liquid, and rise to the surface and join the slag which floats on top of the liquid metal. On the other hand, it is quite possible for small impurity particles (compounds) to exist in the liquid. Certain of the latter undoubtedly form nucleation centers for heterogeneous nucleation. This fact has been used to some extent to control the grain size of castings by artificially inoculating liquid metals with elements that combine to form nucleation catalysts. Increasing the number of nucleation centers will, of course, produce a finer grain size in the solidified casting.

When an alloy is frozen, a more or less general rule which applies is that solute elements, whether present as alloying elements or as impurities, are more soluble in the liquid state than in the solid state. This fact usually leads to a segregation of the solute elements in the finished casting. There are two basic ways of looking at the resulting nonuniformity of the solute. First, because the

15. *Ibid.*

liquid becomes progressively richer in the solute as freezing progresses, the solute concentrations in a casting tend to rise in those regions which solidify last (center of the ingot). This and similar long-range composition fluctuations fall in the classification of macrosegregation. In general, macrosegregation refers to the change in the average composition of the metal as one moves from place to place in an ingot. Segregation of this form is not always caused by the selective freezing out of high-melting-point constituents. Gravitational effects are often a factor in producing macrosegregation, especially during the formation of the central equiaxed zone. The crystals which form freely in the liquid often have a different density from that of the liquid. As a result, they may either rise toward the surface of the casting, or settle toward the bottom. An extreme example[16] of gravity induced segregation occurs in an alloy system somewhat different from the solid-solution alloys presently being considered. Thus, in the lead-antimony system, an eutectic forms at 11.1 percent Sb and 252° C. When alloys containing more than 11.1 percent Sb (say 20 percent) are frozen, crystals of almost pure antimony form from the liquid until the composition of the latter reaches the eutectic composition, at which time the eutectic mixture begins to freeze. Since antimony crystals have a lower density than the liquid from which they originate, they tend to rise toward the surface. Slow cooling of this alloy, therefore, results in a structure whose lower fraction is composed almost entirely of eutectic solid, with an upper portion containing the primary-alpha antimony crystals with some eutectic solid filling in the gaps between the primary crystals.

In castings not only do we find composition variations (macrosegregation) over large distances, but it is also possible to have localized composition variations on a scale smaller than the crystal size. This is called *microsegregation* and one form has already been described; the composition segregation associated with the cellular structure resulting from the combined movement of the liquid-solid interface and a very narrow zone of constitutional supercooling. A much more frequent form of microsegregation, commonly known as *coring*, is caused by dendritic freezing in alloys. The original dendrite arms, which shoot out into the supercooled metal, freeze as relatively pure metal. The liquid surrounding these arms is, accordingly, enriched in solutes and normally, when this liquid solidifies, the spaces between the arms become regions high in solute concentration.

Dendritic segregation, or coring, is very common in alloy castings. It can occur with a solute concentration as low as 0.01 percent under the proper conditions, and it cannot be suppressed by rapid cooling.[17] When a casting is sectioned and the surface prepared for metallographic examination, the exposed surface will usually be a planar section cutting through the forest of dendrite

16. Brick, R. M., and Phillips, Arthur, *Structure and Properties of Alloys*. McGraw-Hill Book Company, Inc., New York, 1949.
17. Tiller, W., and Rutter, J., *Can. J. Phys.*, 34 96 (1956).

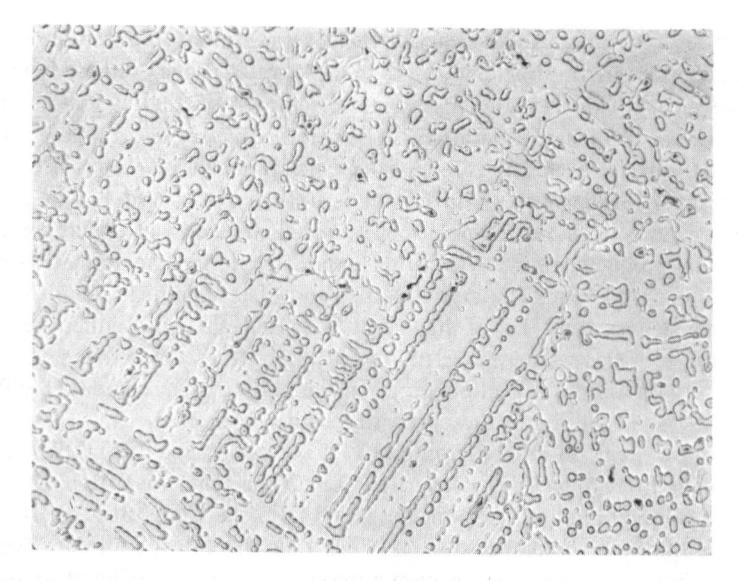

Fig. 15.26 Coring, or dendritic segregation, in a copper-tin alloy. Several different crystals are shown. Note that the dendrite arms have different orientations in each crystal. **200 x** .

arms. Since the composition at the center of an arm differs from that at points midway between arms, the dendrite arms can be revealed by etching with a suitable metallographic etch. Dendritic segregation in a copper-tin alloy is shown in Fig. 15.26.

15.10 Homogenization. The segregation of the solute due to coring can usually be removed if the equilibrium structure of the alloy is a single phase. This is accomplished by a heat treatment known as a *homogenizing anneal*. In this process, the metal is heated to a high temperature for a sufficient time to allow diffusion to homogenize the structure.

Before considering the kinetics of homogenization, the freezing process that leads to coring will be re-examined. Fig. 15.27 shows one corner of an isomorphous phase diagram. If a metal of composition x is frozen under equilibrium conditions, the solid will follow the solidus curve from point b to point c. This curve represents both the composition of the solid that is forming at any instant, and the average composition of all the solid that has previously formed. In normal freezing there is insufficient time for diffusion to homogenize the structure, and coring results. In this case, the solidus line from point b to point f represents only the composition of the solid that can form at any given temperature. Since the metal that has solidified earlier is always richer in

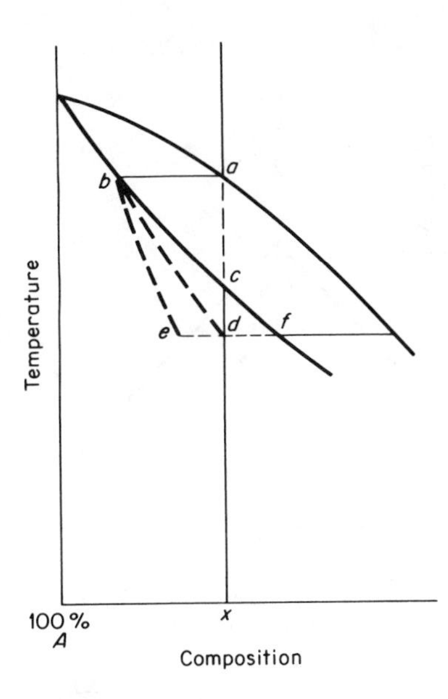

Fig. 15.27 During nonequilibrium freezing, the average composition of the solid follows a path such as *bd*.

component A than the solid that is just freezing, the average composition of the solid has to lie to the left of the solidus line. The average composition of the solid during normal freezing, therefore, follows a line such as bd in Fig. 15.27. At the same time, because there is always some diffusion, the composition of the first solid to form (at the center of the dendrite arms) also changes and follows a path such as be.

Note that one of the significant features of normal freezing is that it tends to occur over a greater temperature range than is required for equilibrium freezing. Thus, point d, which represents the end of the normal freezing process in Fig. 15.27, lies well below point c, corresponding to the completion of equilibrium freezing.

At the end of normal freezing the composition will vary over the range from point c to f in the phase diagram of Fig. 15.27. For the purpose of deriving an equation, it will now be assumed that the dendrites can be considered as a set of parallel plates whose composition varies linearly and periodically through this composition range, as indicated in Fig. 15.28. The horizontal dashed line in this figure represents the average composition of the alloy $n_{B_{av}}$. At the center of each dendrite arm, such as at point m, the composition is less than the average by the amount Δn_B, and midway between the dendrite arms, as at point n, it is greater than the average by Δn_B. The distance from the center of a dendrite arm to a

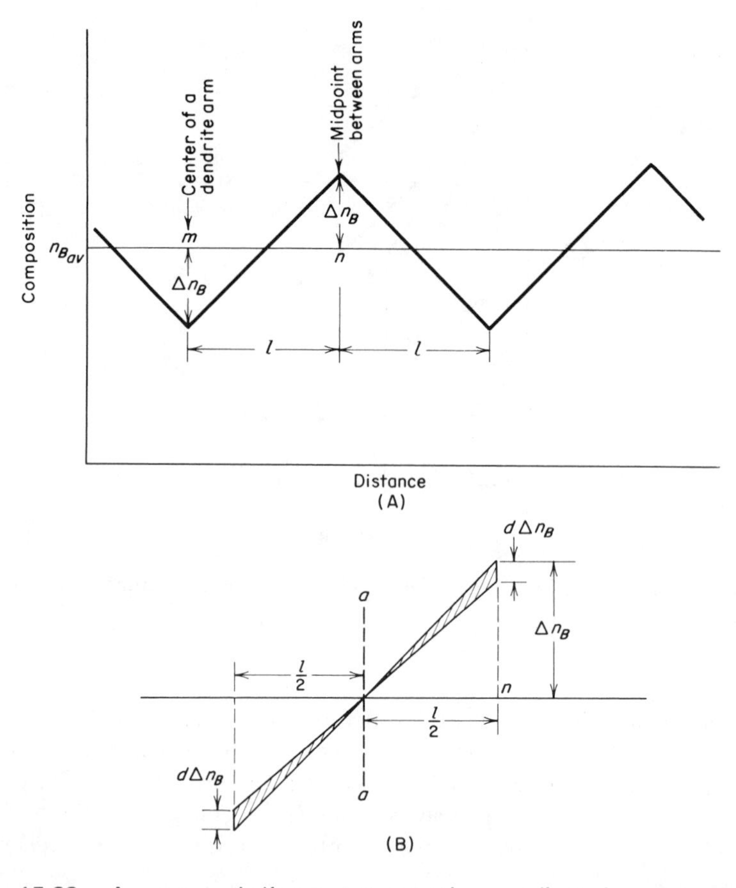

Fig. 15.28 An assumed linear concentration gradient in a cored metal specimen.

point midway between a pair of arms is assumed to equal l. This quantity is equal to one-half the dendrite arm spacing; the distance between the centers of the two arms.

In a time interval dt, the concentration at point m may be assumed to increase by Δn_B, while at point n it should decrease by the same amount. In order for this to happen, a number of B atoms equal to

$$d(\Delta n_B)\frac{l}{2}A$$

has to pass through the cross-section aa, where $d(\Delta n_B)l/2$ is the area of one of the small cross-hatched triangles in Figure 15.28, and A is the cross-section area of the specimen. This number can be equated to Jdt, where J is the flux of B atoms per second past section aa. Since the concentration gradient is assumed to be linear and therefore equal to Δn_B divided by $l/2$, we may write

$$d\,\Delta n_B\,\frac{l}{2}A = Jdt = -DA\left(\frac{\Delta n_B}{\frac{l}{2}}\right)$$

where D is the diffusion coefficient. This equation reduces to

$$\frac{d(\Delta n_B)}{dt} = -\frac{4D(\Delta n_B)}{l^2}$$

Note that the rate of change of Δn_B is proportional to Δn_B. This means that Δn_B must vary exponentially with the time, or

$$\Delta n_B = \Delta n_{B_0} e^{-(4D/l^2)t}$$

where Δn_{B_0} is the concentration differential when the time is zero. The quantity $l^2/4D$ is equivalent to a relaxation time τ, so that we may also write

$$\Delta n_B = \Delta n_{B_o} e^{-t/\tau}$$

A somewhat better approximation is to assume that the concentration varies sinusoidally with the distance rather than linearly. Figure 15.29 shows this type of variation which corresponds[18] to a relaxation time $\tau = l^2/\pi^2 D$ instead of $l^2/4D$.

As demonstrated in Section 11.2, the relaxation time is equal to the time required to make a function, that depends exponentially on the time, decrease its value by a factor $1/e$. In a typical cored specimen, the dendrite arm spacing may be of the order of 100μ or 10^{-2} cm. The diffusion coefficient (see Section 10.12) at elevated temperatures is, in many cases, equal to about 10^{-9}. Using these values gives a relaxation time

$$\tau = \frac{l^2}{\pi^2 D} = \frac{10^{-4}}{\pi^2\,10^{-9}} \approx 10^4 \text{ sec} \approx 3 \text{ hours}$$

This result clearly shows that for a relatively coarse dendrite arm spacing, a very long annealing time may be required in order to approach complete homogenization.

18. Shewmon, P. G., *Transformations in Metals*, p. 41, McGraw-Hill Book Co., New York, 1969.

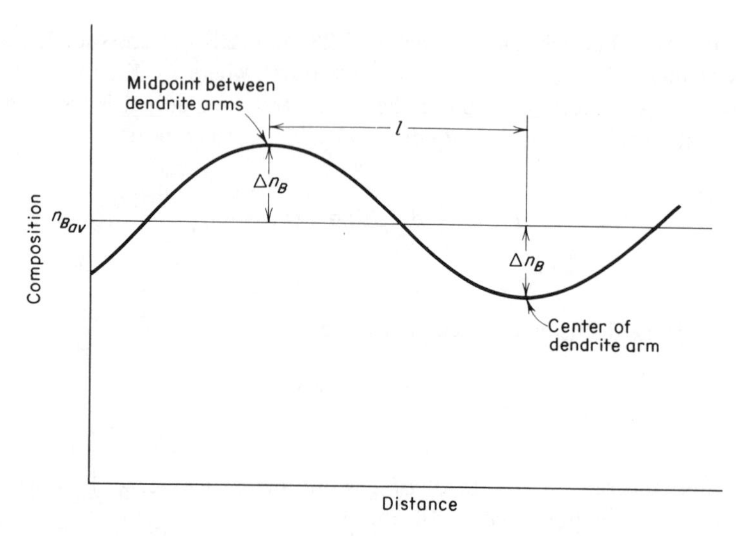

Fig. 15.29 A hypothetical sinusoidal concentration profile in a cored metal specimen.

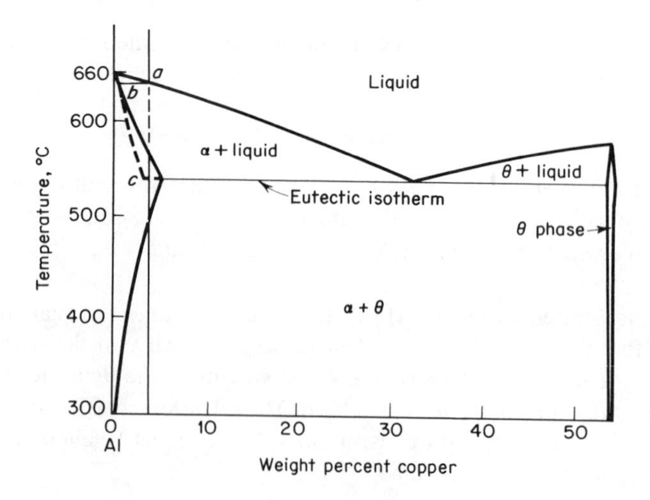

Fig. 15.30 Coring in an eutectic alloy system.

When only a single phase is present, whether or not complete homogenization is attained may not be of major importance. However, if coring yields a second nonequilibrium phase this may not be true. This is particularly the case in a number of precipitation hardening alloys. As an illustrative example, consider the alloy of aluminum with 4 percent copper. If equilibrium is maintained during freezing, this alloy will freeze as a homogeneous solid solution. In practice, equilibrium is never attained and the metal becomes badly cored. The aluminum side of the aluminum-copper phase diagram is shown in Fig. 15.30. During freezing, the average composition of the solid of a 4 percent copper alloy should move along a line such as *bc*. Point *c* lies on the eutectic isotherm. When the alloy is cooled to this temperature it is still a mixture of solid and a small quantity of liquid lying between the dendrite arms. The liquid has the eutectic composition and will freeze to form an eutectic composed of the alpha and theta phases. Sometimes when an eutectic forms and the primary phase (in this example the alpha phase) occurs as a widely distributed set of closely spaced dendrite arms, the corresponding phase (α) in the eutectic may form preferentially on the existing primary dendrites. This leaves the second phase (θ) as the only remaining visible phase in the eutectic. When this occurs, the eutectic is said to be a *divorced eutectic*.

In any case precipitation hardening alloys, such as the aluminum-4 percent copper alloy, can develop a two-phase structure as a result of coring. This is undesirable for several reasons. First, it ties up a large fraction of the hardening agent (the copper) in the precipitate where it is not available for strengthening the metal during a precipitation hardening treatment. Second, as will be shown in Chapter 19 on Fracture, precipitates can have a harmful effect on the ductility of a metal. Homogenizing a cored precipitation hardening alloy of this type will increase not only its strength, but also its ductility. This is clearly demonstrated in Fig. 15.31 where the tensile and yield strengths, as well as the

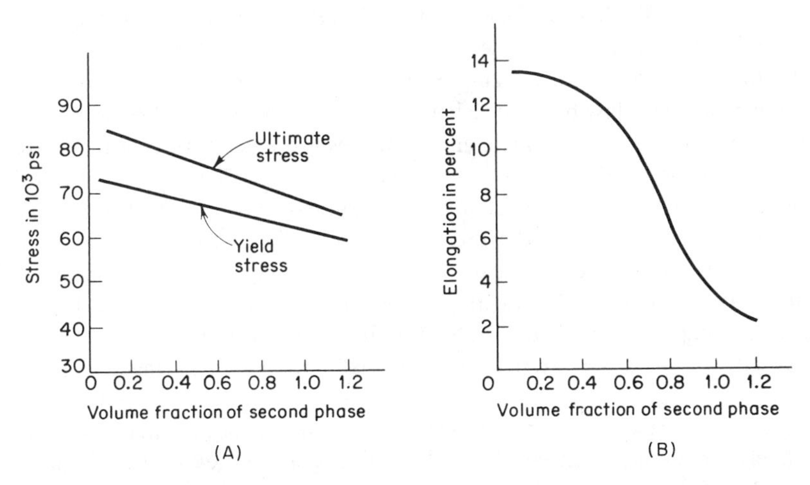

Fig. 15.31 (A) The ultimate stress and the yield stress of age-hardened 7075 aluminum alloy as a function of the volume fraction of second phase. (B) The elongation of age-hardened 7075 aluminum alloy as a function of the volume fraction of second phase. (From Singh, S. N. and Flemings, M. C., *TMS-AIME*, **245** 1811 [1969].)

elongation, of a commercial aluminum alloy are plotted as a function of the volume fraction of the second phase in the alloy.

Considerable attention is currently being devoted to attempts to develop casting techniques that will reduce the dendrite arm spacing. This is because the relaxation time for homogenization varies as the square of the dendrite arm spacing. The closer the dendrite arms, the easier it is to remove the second phase by homogenization. The gains to be achieved in precipitation hardening alloys by removing the second phase are clearly shown in Fig. 15.31. The most important factor in determining the dendrite arm spacing is the rate of cooling. Very rapid cooling can produce a dendrite arm spacing of the order of 5 to 10μ.[19] Other factors involved in considerations that lead to a reduction in the time required for homogenization cannot be covered here. Information about these is given in several papers.[20,21,22]

Finally, on the subject of homogenization, it should be mentioned that care should be exercised in selecting the temperature for the homogenization anneal. Partial melting or liquation of the metal may occur if it is heated to just under the solidus curve of an equilibrium diagram. Coring normally allows part of the structure to freeze at a temperature below that corresponding to the completion of freezing under equilibrium conditions. This part of the structure will melt if the metal is reheated to just below the solidus. In some cases, serious damage to a metal object may result from this cause.

15.11 Inverse Segregation. Dendritic freezing in the columnar zone of ingots may sometimes lead to a phenomenon known as *inverse segregation*. In the normal freezing of ingots, the center and top portions of the ingot which freeze last become richer in solutes than the outer portions of the ingot which freeze first. In some alloys, however, the dendrite arms may extend for long distances before the spaces between them are filled in by the freezing operation. Under the proper conditions, channels between the dendrite arms offer a path by which liquid from the center of the casting can work its way back toward the surface. The latter phenomenon is promoted by the fact that as solidification progresses, the casting as a whole can shrink away from the mold walls, causing a suction that pulls the liquid toward the surface. Another important factor is the internal pressure that develops in an ingot due to gas evolution during freezing. "Tin sweat," which occurs in tin bronzes (copper-tin alloys), can be explained in this manner. When the liquid metal contains a relatively large concentration of dissolved hydrogen, the release of this gas near the end of freezing forces the liquid enriched in tin through interdendritic pores

19. Antes, H. W., Lipson, S., and Rosenthal, H., *TMS-AIME*, **239** 1634 (1967).
20. *Ibid.*
21. Singh, S. N., and Flemings, M. C., *TMS-AIME*, **245** 1803, 1811 (1969).
22. Bower, T. F., Singh, S. N., and Flemings, M. C., *Met. Trans.* **1** 191 (1970).

to the surface where it coats the normally yellow-bronze-colored surface with a fine layer of white alloy (containing approximately 25 percent Sn).

15.12 Porosity. Neglecting shrinkage cracks resulting from unequal cooling rates in different sections of castings, there are two basic causes for porosity in cast metals: gas evolution during freezing, and the shrinkage in volume which accompanies the solidification of most metals.

Gases differ from other solutes in that their solubility in metals depends markedly on the applied pressure. Many gases with which metals come in contact are diatomic: O_2, N_2, H_2, etc. Providing that the solubility is small, it is often possible to express the relationship between the pressure and the solubility of diatomic gases in a simple form, known as Sievert's Law.

$$c_g = k\sqrt{p}$$

In the above equation, c_g is the solubility of the dissolved gas, p is the gas pressure, and k is a constant. Some experimental data corresponding to hydrogen dissolved in pure magnesium is shown in Fig. 15.32. The melting point of pure magnesium is 650°C. Notice that the data plots well as straight lines (on c_g and $\sqrt{p}$ coordinates) both above and below the melting point. This signifies that Sievert's law is probably valid for the solution of hydrogen in both liquid and solid magnesium.

The solubility of gases in metals is also a function of temperature and, in most cases, the solubility increases rapidly with temperature. If the maximum solubility is small, it is often possible to express the equilibrium concentration of

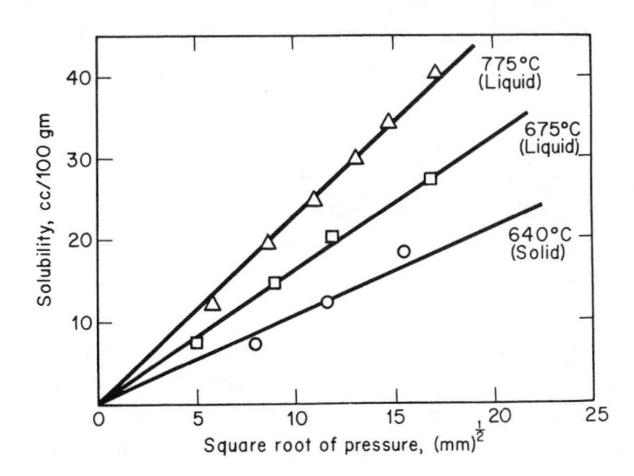

Fig. 15.32 Solubility of hydrogen in both liquid and solid magnesium as a function of the partial pressure of hydrogen. (Koene, J., and Metcalfe, A. G., *Trans. ASM,* **51** 1072 [1959].)

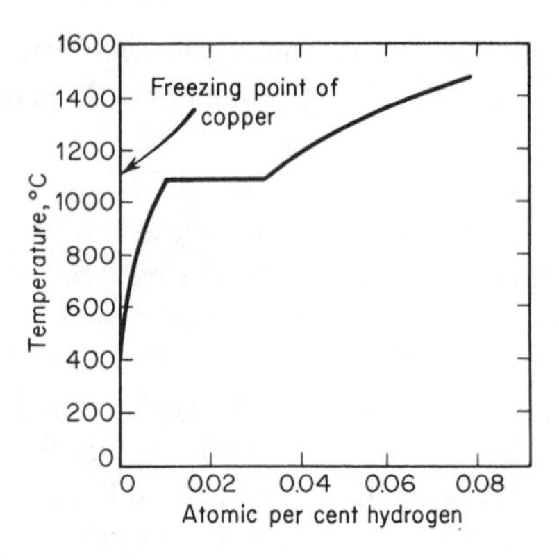

Fig. 15.33 Solubility of hydrogen in copper at one atmosphere pressure. (From *Constitution of Binary Alloys*, by Hansen, Max, and Anderko, Kurt. Copyright, 1958. McGraw-Hill Book Co., Inc., New York, p. 587. Used by permission.)

a gas in a metal, at a constant gas pressure, as an exponential function of the type previously derived for the solubility of carbon in iron; namely

$$c_g = Be^{-Q/RT}$$

where c_g is the concentration of the gas in the metal, B is a constant, Q the work to introduce a mole of gas atoms into the metal, while R and T have their usual significance.

The solubility of hydrogen as a function of temperature in copper at a constant pressure of one atmosphere is given in Fig. 15.33. This curve shows not only the rapid increase in solubility with increasing temperature, but also a large decrease in solubility when the metal transforms from a liquid to a solid at 1083°C. Similar changes in the solubility of gases in metals occur at the freezing points of most metals. It can be concluded that, in general, when a metal freezes its ability to hold gases is strongly decreased.

Let us consider these factors in relation to the freezing process. In almost all cases, the solubility of dissolved gases undergoes a large drop when metals freeze. For this reason, gaseous solutes segregate during freezing in a similar fashion to other solutes. Segregation signifies, therefore, localized increases in the gas concentration of a liquid. Gas bubbles may form in the liquid in those regions where the concentration of gas rises above the equilibrium concentration (saturation value). Homogeneous nucleation of bubbles, however, requires a high degree of supersaturation for the same reason as the nucleation of solid crystals in a liquid. In both cases, a surface with its inherent surface energy is formed between the old and new phases. As a result, gas bubbles usually form heterogeneously and, in most cases, the nucleation centers lie on the liquid-solid interface. Nucleation at the interface is also furthered by the build-up in concentration of the gaseous solutes at both the general interface and in between the dendrite arms as a result of segregation effects.

A very important factor in gas-bubble formation is the local pressure in the liquid. As previously mentioned, the equilibrium concentration of a gas in a liquid depends on the pressure. In metals with a given quantity of absorbed gas, gas-bubble nucleation is therefore promoted by a lowering of the pressure, and hindered by pressure increases. Thus, if liquid metals are frozen under sufficiently high pressures, gas evolution may be prevented. In this respect, die castings are a form of casting in which small metal parts of close dimensional tolerances are manufactured by forcing liquid metals (under very high pressures) into steel dies. Gas porosity is effectively eliminated in this casting method. On the other hand, it frequently happens that the normal shrinkage that occurs when a metal freezes causes pressure drops in the liquid. This situation may occur when the entire outer surface of a casting freezes over, leaving the liquid in the center surrounded by a shell of solid metal. A similar effect may also happen in localized areas of complicated castings wherein a small volume of liquid finds itself enclosed by solid. In either case, a vacuum may develop and, if it does, gas-bubble nucleation will be promoted.

Gas-bubble formation is a nucleation and growth phenomenon similar in many respects to other nucleation and growth phenomena previously considered. In particular, as mentioned above, the formation of the interface makes it hard to nucleate bubbles. However, once a bubble has formed and grown to a size greater than its critical radius, its growth becomes progressively easier. This fact is easily shown in the following manner. A gas bubble in a liquid is analogous to a soap bubble, except that it consists of a single liquid-gas interface separating a gas phase from a liquid phase. By analogy with the soap bubble, we may write

$$p_g - p_l = \frac{2\gamma}{r}$$

where γ is the surface energy in dyne/cm, p_g is the internal pressure in the gas bubble in dyne/cm^2, p_l is the pressure in the liquid in dyne/cm^2, and r is the radius of curvature of the bubble in centimeters. As the bubble grows in size, its radius of curvature increases. As a result, the pressure differential between the gas in the bubble and the pressure in the liquid falls. This implies (by Sievert's law) that the bubble, as it grows in size, is able to be in equilibrium with a decreasing concentration of gas atoms in the surrounding liquid. Continued growth of gas bubbles is also furthered by the solute concentration gradient which develops in the surrounding liquid as the gas bubble absorbs more and more gas atoms. The concentration falls toward the bubble, causing a diffusion of gas atoms from the surrounding liquid toward the bubble.

The effect of gas-bubble formation during freezing on the final solid metal structure depends upon whether or not the bubbles are trapped into the solid structure, and upon the shape of the entrapped cavities. Rapidly growing bubbles have the tendency to free themselves from the interface and rise toward the upper surface of the casting. If this surface is frozen over, a large cavity may form at the top of the ingot. If the upper surface is unfrozen, the gas is eliminated from the casting. In general, since gas bubbles that form near the top of an ingot grow under a lower hydrostatic head, they will be larger in size and more inclined to escape. This action is further increased by the sweeping action of bubbles which rise from points at greater depths in the mold.

When the growth rate is very slow, bubbles may be trapped by the solid growing around them. In this case, the cavity formed in the solid is approximately spherical in shape and is commonly called a *blowhole*. See Fig. 15.34A. In some cases, blowholes may not be particularly harmful to the properties of the metal. This is especially true in those metals that receive a high degree of hot-working during their manufacture, providing that the blowholes are formed at some distance below the surface of the ingot. This is required in order that the internal surfaces of the blowholes do not become oxidized through contact with oxygen from the surrounding air. If the surface of the blowhole is not oxidized, it is often possible, under the extreme pressures of hot-rolling, to collapse the blowholes and to weld their sides together. If the welds are successful, the blowholes are eliminated from the metal and any gas

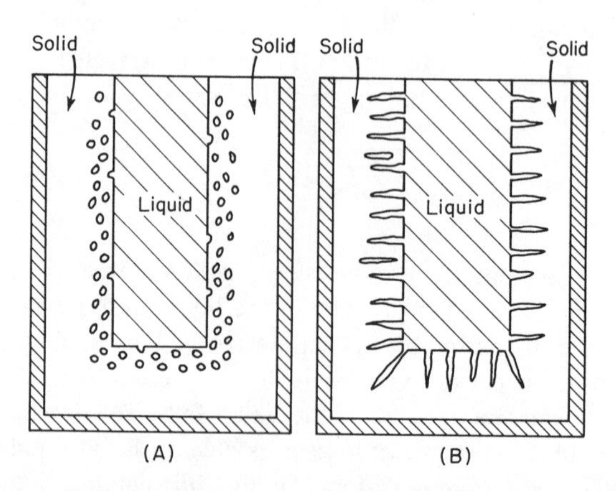

Fig. 15.34 Two forms of gas porosity: (A) Blowholes and (B) Wormhole porosity, both of which are macroscopic in size as contrasted to dendritic porosity which is microscopic.

remaining in the cavities is reabsorbed into the metal (now a solid). Conversely, if the surfaces of the collapsed blowhole do not weld, a long slitlike defect is formed in the metal which is called a *seam*.

Wormhole porosity is another form of cavity defect which may be found in castings. These are long tubular cavities produced when gas is evolved at an intermediate rate so that gas bubbles grow in length at the same rate as the liquid-solid interface moves. The nature of wormhole porosity is shown in Fig. 15.34B where it may be observed that the holes are elongated in the heat-flow direction. Wormhole porosity is easily observed in a nonmetallic ingot familiar to almost everyone: a large cake of ice frozen in a commercial icehouse from water containing dissolved air. The central whitish areas are usually regions of wormhole porosity.

Porosity of a much finer and more irregular shape than that described above occurs when shrinkage effects and gas evolution combine. Thus, in regions completely enclosed by solid metal which still contain some interdendritic liquid, pores can develop, which more or less parallel the dendrite arms. If the pores form primarily as a result of shrinkage, they will be very fine in cross-section and very numerous. On the other hand, if gas evolution occurs while the pores are forming, the cavities will be larger and less numerous.

Another form of interdendritic porosity may occur as a result of shrinkage effects alone. When very hot metal (metal superheated above its freezing point) is poured into a mold, an equiaxed layer of crystals is first formed next to the mold walls. The superheat in the metal then tends to melt this surface layer, with the final result that columnar crystals may develop at, or very close to, the surface. Under the right conditions, it is possible for liquid to remain in the spaces between the dendrite arms of these columnar grains long enough so that the shrinkage associated with continued freezing sucks this liquid into the interior of the casting. The result is interdendritic porosity in the surface layers of the casting. In a sense, this phenomenon is the reverse of the inverse segregation mentioned earlier. In the latter case, liquid from the interior, usually as the result of internal gas pressures, flows through the interdendritic passages toward the surface. In the present phenomenon the flow is in the opposite direction.

While considering the subject of gas evolution, it is worth noticing that controlled-gas evolution may sometimes be used to advantage. Thus, a very large percentage of the steel manufactured in this country is poured into ingots while it still contains a relatively high but controlled amount of gas. This gas content is adjusted so that when the ingot freezes enough, blowholes are trapped in the metal to compensate for the normal shrinkage of the metal as it freezes. In subsequent manufacturing stages (hot-rolling), the blowholes are eliminated as discussed earlier. This form of ingot casting greatly simplifies the problems associated with the casting of large steel ingots and is called *rimming*. Rimming is associated with a relatively high concentration of oxygen in the liquid steel that combines with some of the carbon dissolved in the steel to form carbon monoxide on freezing. This is the gas primarily associated with the formation of the blowholes. A high-oxygen content in the steel unfortunately leads to a relatively high concentration of oxides, silicates, and similar particles in the finished product. In many instances, nonmetallic inclusions may not be particularly detrimental in a steel. However, the higher the demands on the steel, the more detrimental are nonmetallic inclusions to the steel properties. Thus, high concentrations cannot be tolerated in highly stressed machinery steels. Furthermore, for steels containing more than about 0.3 percent carbon, it becomes very difficult to weld blowholes during the rolling processes, and high-quality steels and steels containing more than 0.3 percent carbon are usually deoxidized before casting. These degassed steels freeze without gas evolution and are said to be *killed*. The casting of ingots by this method is more expensive because the molds are more complicated and expensive. The principal factor to be avoided in a killed-steel ingot is the formation of large shrinkage cavities along the central axis of the casting, known as a *pipe*. A properly designed killed-steel ingot must, accordingly, feed liquid metal into the central regions of the casting as the freezing progresses.

Problems

1. (*a*) The specific heat of copper between 200°C and its melting point can be represented by the equation

$$C_p = 0.092 + 2.2 \times 10^{-5}\, T \text{ cal/gm-}^\circ\text{C}$$

where T is the temperature in degrees centigrade. The heat of fusion of copper is 50.0 cal/gm. Determine how large an amount of supercooling would be required (assuming that no heat is lost to the surroundings) in order that the freezing of copper will not be able to raise its temperature back to the equilibrium freezing temperature of 1083°C.
(*b*) Do you think that it would actually be possible to supercool copper to the point where it could not regain its melting point on freezing? Explain.
2. Consider the body-centered cubic crystal structure and the planes 100, 110, and 111. On the basis of the degree of close-packing associated with each plane, rate these planes in order of their growth velocity during freezing.
3. (*a*) Kattamis, Coughlin, and Flemings[23] make the statement that the dendrite arm spacing in aluminum alloys is only influenced by the cooling rate and alloy content. Rationalize the dependence of the dislocation arm spacing on the cooling rate.
(*b*) The above authors have plotted the dendrite arm spacing in an alloy of

23. Kattamis, T. Z., Coughlin, J. C., and Flemings, M. C., *TMS-AIME,* **239** 1505 (1967).

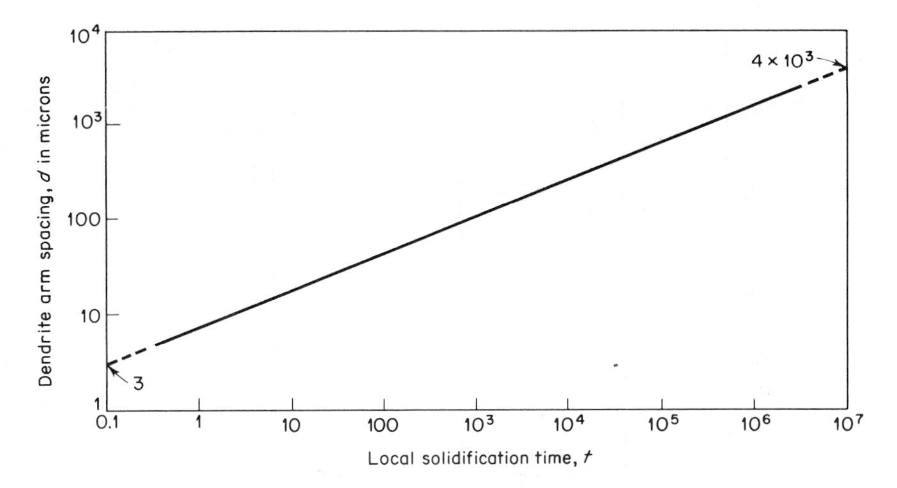

aluminum with 4.5 percent copper as a function of the local solidification time, which they define as the time at a given location in a casting or ingot between initiation and completion of solidification. This quantity may be assumed to be inversely proportional to the cooling rate at a given location. Their graph is shown in this figure. This data corresponds to an equation of the form

$$d = (t)^p$$

where d is the dendrite arm spacing, t is the local solidification time, and p is a constant. Determine p.

4. Doubling the cooling rate should have what effect on the dendrite arm spacing in an aluminum-4.5 percent copper alloy?
5. Assume that an alloy of nickel with 2 wt. percent aluminum is cast and, by metallographic examination, a dendrite arm spacing of 50 μ is observed. It is also determined that the composition difference between the center of an arm and the midpoint between two arms is one percent. Estimate the time required for homogenization if the annealing temperature is to be 1400°C and the composition difference is to be reduced to one tenth of its original value. *Note*: see Table 10.5.
6. In a precipitation hardening alloy, coring usually results in the precipitation of an eutectic between the dendrite arms. With regard to aluminum-copper alloys, Singh and Flemings[24] have assumed that this eutectic is divorced (consists only of the θ phase) and have derived an equation to describe the kinetics of homogenization that is similar to that discussed in the text for homogenization of a cored solid solution alloy. This equation which applies

24. Singh, S. N., and Flemings, M. C., *TMS-AIME,* **245** 1803 (1969).

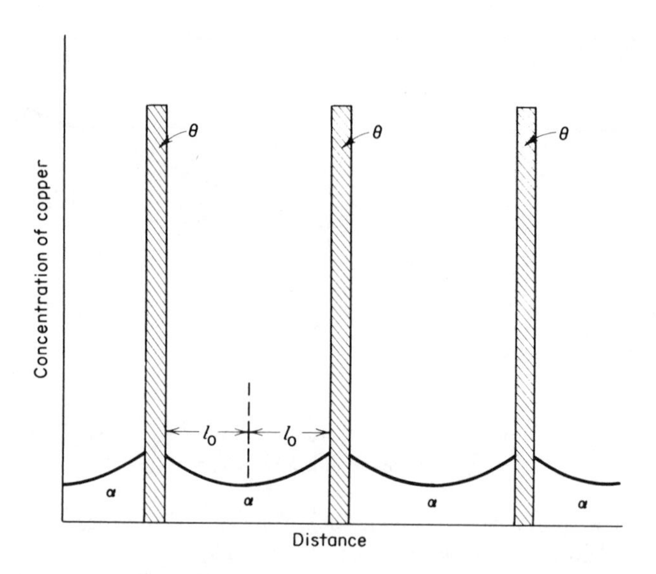

to a solution treatment close to the solvus is

$$g = g_0 e^{-\pi^2 Dt/4\, l_0^2}$$

where g is the volume fraction of the θ phase at time t, g_0 is the volume fraction of θ when t is zero, D is the diffusion coefficient, and l_0 is equal to half the dendrite arm spacing, as shown in this schematic figure. (Note that the concentration inside the precipitate is constant since it equals that of the θ phase. In the alpha phase, the concentration of copper is assumed to vary sinusoidally with distance and to be a minimum at the midpoint corresponding to the centers of the dendrites.)

(*a*) Assuming that at the annealing temperature D is 10^{-10} and that the original volume fraction of the θ phase is 1.0 percent, what solution time would be required to obtain a volume fraction of 0.01 percent if the dendrite arm spacing is 100μ?

(*b*) What would be the corresponding time if the dendrite arm spacing was 10μ?

7. Care must be taken in melting certain copper alloys to prevent absorption of hydrogen from the furnace gases or as a result of reaction with water vapor. The release of this gas on freezing can result in an open, porous structure of little value. As an exercise, compute the volume of hydrogen that should be released in 10 cc of copper melted in a hydrogen atmosphere at a pressure of one atmosphere when the metal is frozen. The atomic volume of copper is 7.09 cm^3 per gm atom. Assume the ideal gas law and refer to Fig. 15.33.

16 Deformation Twinning and Martensite Reactions

We shall now consider two apparently unrelated types of phenomena actually having much in common. One, deformation or mechanical twinning, is a mode of plastic deformation, while the other, comprising martensite reactions, is a basic type of phase transformation. Twinning, like slip, occurs as the result of applied stresses. Sometimes stress may be influential in partly triggering a martensitic transformation, but this is an effect of secondary importance. Martensite reactions occur in metals that undergo phase transformations, and the driving force for a martensite reaction is the chemical free-energy difference between two phases.

The similarity between martensite reactions and twinning lies in the analogous way twins and martensite crystals form, for, in both cases, the atoms inside finite crystalline volumes of the parent phase are realigned as new crystal lattices. In twins this realignment reproduces the original crystal structure, but with a new orientation. In a martensite plate, not only is a new orientation produced, but also a basically different crystalline structure. Thus, when martensite forms in steel quickly cooled to room temperature, the face-centered cubic phase, which is stable at elevated temperatures, is converted into small crystalline units of a body-centered tetragonal phase. On the other hand, when twinning occurs in a metal such as zinc, both the parent crystal and the twinned volumes still have the close-packed hexagonal zinc structure. In both twinning and martensite transformations, each realigned volume of material suffers a change in shape which distorts the surrounding matrix. The changes in shape are quite similar, so that martensitic plates and deformation twins look alike, taking the form of small lenses, or plates. Examples of deformation twins and martensite plates are shown in Fig. 16.1. Actually, as shall presently be seen, it is quite possible to convert large volumes of the parent structure into elements of one of the new structures, but the plate-like shapes shown in Fig. 16.1 are much more common.

16.1 Deformation Twinning. The deformation accompanying mechanical twinning is simpler than that associated with martensitic reactions. This is because there is no change in crystal structure, merely a reorientation of

Fig. 16.1 (A) Deformation twins in a polycrystalline zirconium specimen. Photographed with polarized light. (E. R. Buchanan.) 1500 x . (B) Martensite plates in a 1.5 per cent carbon-5.10 per cent nickel steel. (Courtesy of E. C. Bain Laboratory for Fundamental Research, United States Steel Corporation.) 2500 x .

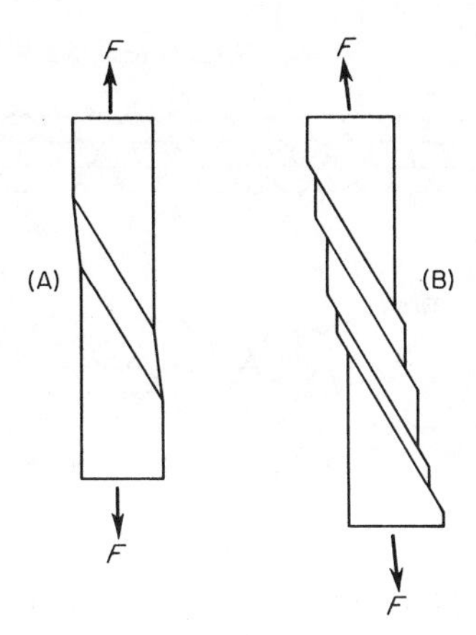

Fig. 16.2 The difference between the shears associated with twinning (A), and slip (B).

the lattice. The change of shape associated with deformation twinning is a simple shear, as shown in Fig. 16.2A where, for simplicity, it is assumed that the twin traverses the entire crystal. The difference between twinning and slip should be carefully recognized since, in both cases, the lattice is sheared. In slip, the deformation occurs on individual lattice planes, as indicated in Fig. 16.2B. When measured on a single-slip plane, this shear may be many times larger than the lattice spacing and depends on the number of dislocations emitted by the dislocation source. The shear associated with deformation twinning, on the other hand, is uniformly distributed over a volume rather than localized on a discrete number of slip planes. Here, in contrast to slip, the atoms move only a fraction of an interatomic spacing relative to each other. The total shear deformation due to twinning is also small, so that slip is a much more important primary mode of plastic deformation. It is also true that mechanical twinning is not readily obtained in metals of high symmetry (face-centered cubic). Nevertheless, the importance of mechanical twinning is becoming increasingly more apparent in explaining certain elusive mechanical properties of many metals. For example, when a metal twins, the lattice inside the twin is frequently realigned into an orientation where the slip planes are more favorably aligned with respect to the applied stress. Under certain conditions, a heavily twinned metal can be more easily deformed than one free of twins. On the other hand, lattice realignment, if confined to a limited number of twins, can induce fracture by permitting very large deformations to occur inside the confined limits of twins. Twins are also of importance in recrystallization phenomena, for the intersections of twins are preferred positions for the nucleation of new grains during annealing.

A good insight into the mechanics of twinning can be gained by studying the

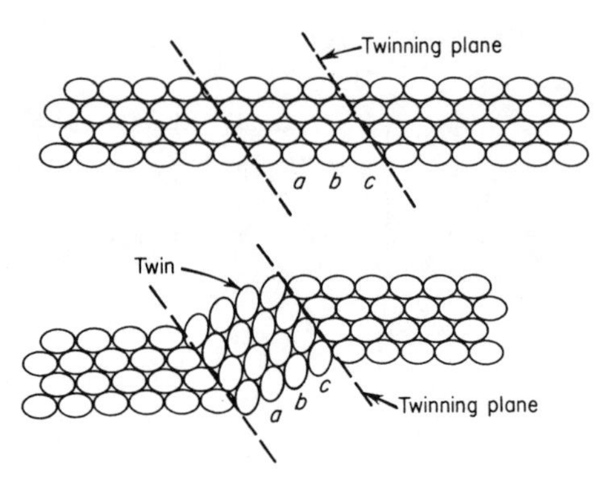

Fig. 16.3 Schematic representation showing how a twin may be produced by a simple movement of atoms.

simple diagrams of Fig. 16.3. The twinning represented in these sketches is only schematic and does not refer to twinning in any real crystal. The upper diagram represents a crystal structure composed of atoms assumed to have the shape of oblate spheroids. The lower diagram represents the same crystal after it has undergone a shearing action which produced a twin. The twin is formed by the rotation of each atom in the deformed area about an axis through its center and perpendicular to the plane of the paper. Three atoms are marked with the symbols *a, b,* and *c* in both diagrams to show their relative positions before and after the shear. Notice that individual atoms are shifted very little with respect to their neighbors. While it is in no way implied that the atom movements in a real crystal are the same as those shown in Fig. 16.3, it is true that in all cases the movement of an atom relative to its neighbors is very small. The two parts of Fig. 16.3 show another important characteristic of twinning: the lattice of the twin is a mirror image of the parent lattice. The lattices of twin and parent are symmetrically oriented across a symmetry plane called the *twinning plane*. The several ways that this symmetry can be attained will be discussed in the next section.

16.2 Formal Crystallographic Theory of Twinning. The formal crystallographic theory of twinning in metals has been summarized by Cahn[1] and the following treatment is largely based on his presentation.

Let us assume that shearing forces are applied to a single crystal specimen, as indicated on the left in Fig. 16.4, and that, as a result of this applied force, the crystal is sheared. The resulting shape of the crystal is shown in the second drawing. Furthermore, assume that after deformation, the sheared section still

1. Cahn, R. W., *Acta Met.*, 1 49 (1953).

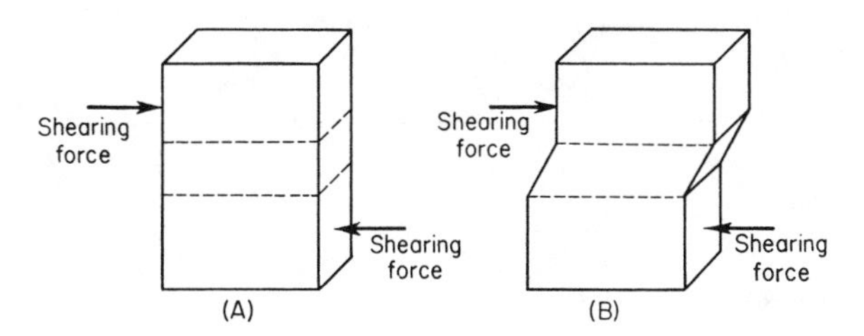

Fig. 16.4 (A) shows a single crystal in a position where it is subject to shearing forces. (B) shows a twinning shear that has occurred as a result of the force.

has the structure and symmetry of the original crystal. In other words, this region is to retain, after shearing, all the crystallographic properties of the metal of which it is composed, so that the size and shape of the unit cell must be unchanged. According to crystallographic theory, the size and shape of the unit cell will be unchanged only if it is possible to find three noncoplanar, rational lattice vectors in the original crystal that have the same lengths and mutual angles after shearing.

Figure 16.5 shows an edge view of a volume which has been sheared. The top surface is displaced a distance e to the right with respect to the bottom surface, and crystallographic planes with orientations perpendicular to the plane of the paper, such as B, C, and D, are rotated to new positions. Original positions of the planes, in each case, are indicated by dotted lines, and the final positions by solid lines. The only plane whose dimensions have not changed as a consequence of the shear is place C. Plane D has been shortened and plane B has been lengthened. The reason that C has not suffered a change in shape is because it makes the same angle θ with the base both before and after the shear. This plane is unique and, except for the plane defining the top and bottom surfaces of the sheared section, it is the only plane whose shape is not changed by the shear.

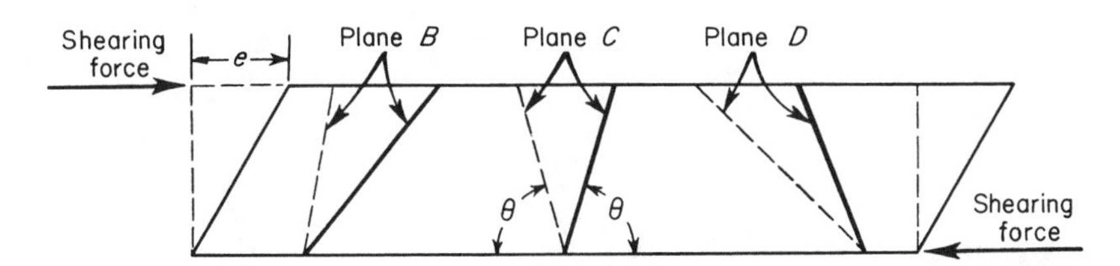

Fig. 16.5 The edge view of a small rectangular volume that has suffered a shear deformation as a result of the applied shearing forces. The dashed lines indicate the original shape of the volume; the solid lines its final shape. The figure is given to illustrate the change in shape of several planes as a result of the shear. The original positions of the planes are given with dashed lines. All three planes have orientations perpendicular to the paper.

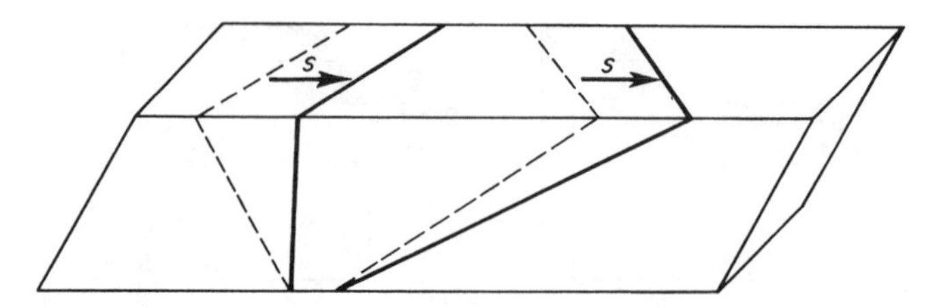

Fig. 16.6 A three-dimensional figure of a small sheared rectangular volume showing the distortion of two arbitrarily chosen planes caused by the shear. Dashed lines are the original positions of the planes, whereas the solid lines are the sheared positions. (The direction of the shear is indicated in each case by the arrow labeled *S*.)

The next three-dimensional figure, Fig. 16.6, shows several other arbitrarily chosen planes with more general orientations than those shown in Fig. 16.5. Again the positions of the planes before and after the shear are given (a change in shape is quite evident in each case).

From the above, it is evident that there are only two crystallographic planes in a shearing action that do not change their shape and size as a consequence of the shear. The first is the plane defining the upper and lower surfaces of the sheared volume. This plane contains the shear direction. In the case of shear by twinning, it is known either as the twinning plane or as the first undistorted plane, and is given the symbol K_1. The other plane, designated as B above, intersects K_1 in a line which is perpendicular to the direction of shear and makes equal angles with K_1 before and after shear occurs. This is called the second undistorted plane and its crystallographic designation is K_2. In Fig. 16.7, the

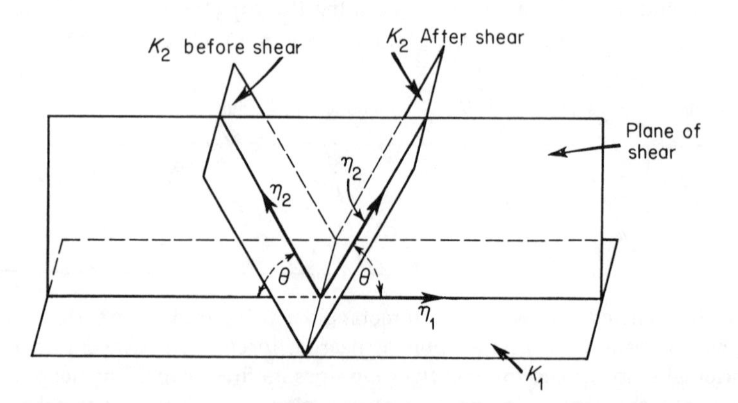

Fig. 16.7 The spatial relationships are given between K_1, K_2, η_1, η_2, and the plane of shear.

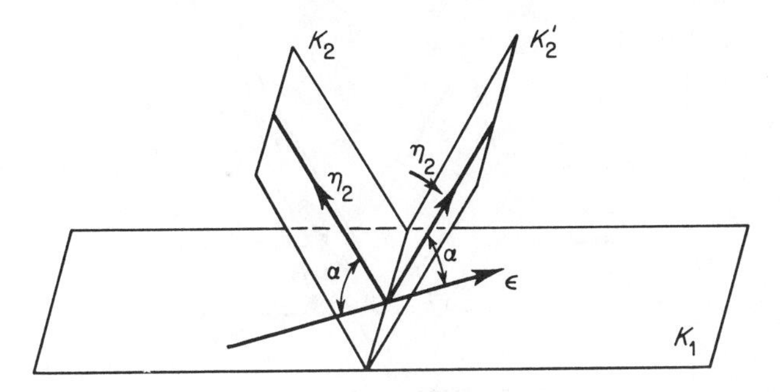

Fig. 16.8 In this figure, which corresponds to a shear of the first kind, notice that η_2 makes equal angles with an arbitrary vector ϵ, which lies in K_1, both before and after the shear. The sheared position of the second undistorted plane is designated K'_2.

relationship between K_1 and K_2 is graphically illustrated. Several other quantities can also be defined in terms of this figure. The shear direction is shown with an arrow and labeled with its customary designation η_1. The plane that lies perpendicular to K_1 and contains the shear direction is known as the *plane of shear.* The intersection of the plane of shear with the second undistorted plane K_2 is also an important direction. It is designated with an arrow and is given the usual symbol η_2. Notice that there are two positions for η_2 corresponding to the positions of K_2 before and after shearing.

Because all other planes suffer a change in size or shape during a shear, it is only in planes K_1 and K_2 that vectors can be found which will not be distorted by a twinning deformation. The basic problem is to find the possible combinations of three noncoplanar lattice vectors, lying in K_1 and K_2, which retain their mutual angles and magnitudes after the shear.

Let ϵ in Fig. 16.8 be any vector in plane K_1, then there is only one vector in K_2 which makes the same angle with ϵ before and after the shear. This is η_2 which is perpendicular to the intersection of K_1 and K_2. Since ϵ is any vector in K_1, η_2 must therefore make the same angle before and after shear with all vectors lying in K_1. Finally, if the assumption is made that K_1 is a rational plane (so that it contains rational directions) and that η_2 is a rational direction, then the conditions for the unit cell to have the same size and shape before and after shearing will have been realized.

In the same manner, it is easily seen that η_1 is the only direction in plane K_1 that makes the same angle with arbitrary vectors in K_2 before and after shearing. This follows from the fact that η_1 is also perpendicular to the intersection of K_1 and K_2. The relationship between δ (an arbitrary vector in K_2) with η_1 is shown in Fig. 16.9. It can, therefore, be concluded that another condition for

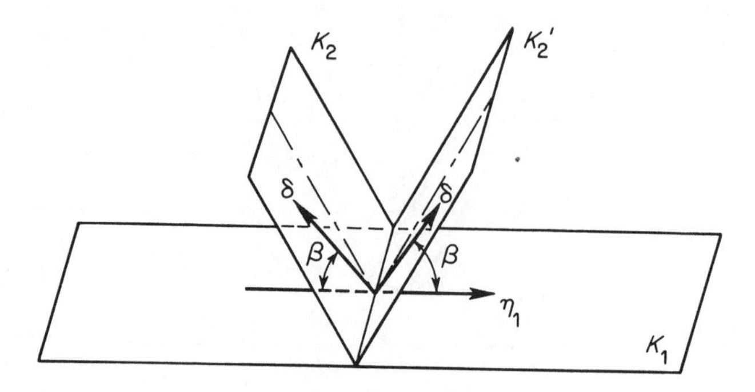

Fig. 16.9 This diagram represents a shear of the second kind. In this case δ represents an arbitrary vector in K_2 which makes the same angle with η_1 before and after shearing. The sheared position of the second undistorted plane is designated K'_2.

preservation of the lattice structure during twinning is that η_1 and K_2 be rational.

It follows from the above that there are three ways that a crystal lattice can be sheared while still retaining its crystal structure and symmetry:

(*a*) When K_1 is a rational plane and η_2 a rational direction. (A twin of the first kind)

(*b*) When K_2 is a rational plane and η_1 a rational direction. (A twin of the second kind)

(*c*) When all four elements K_1, K_2, η_1, and η_2 are rational. (A compound twin)

The lattice rotations that occur as a result of twinning are different, depending upon whether a twin of the first or the second kind is formed.

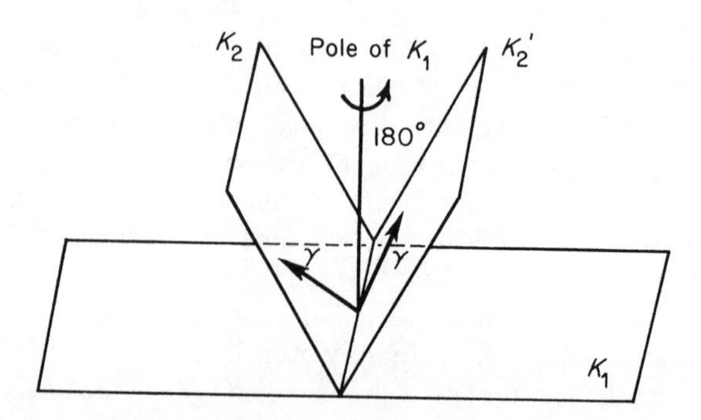

Fig. 16.10 The lattice rotation in a twin of the first kind is 180° about the normal to the twinning plane (K_1). Notice the rotation of the vector γ.

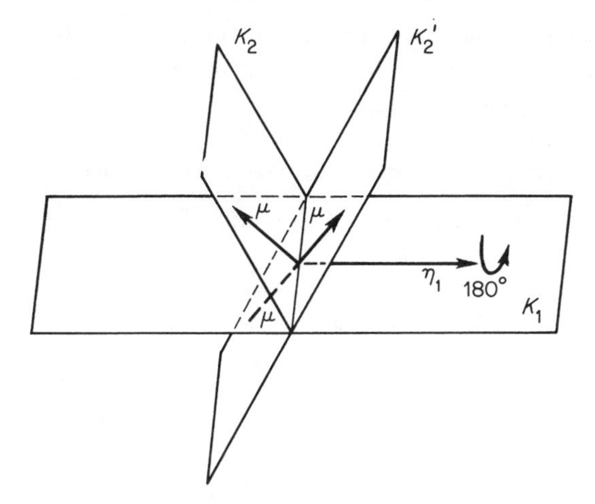

Fig. 16.11 In a twin of the second kind, the rotation is 180° about the shear direction η_1. In the drawing, the nature of this rotation is illustrated by projecting the sheared position of K_2 below the twinning plane (K_1). The 180° rotation of the vector μ is thus emphasized.

In a twin of the first kind, the lattice in the sheared section is related to that in the unsheared section by a rotation of 180° about the normal to K_1. This effect is shown in Fig. 16.10. The corresponding 180° rotation in a twin of the second kind is about η_1 as an axis. This is illustrated in Fig. 16.11. On a stereographic projection, the two rotations would appear as shown in Fig. 16.12. In a compound twin, due to symmetry considerations, both types of rotation lead to the same final orientation.

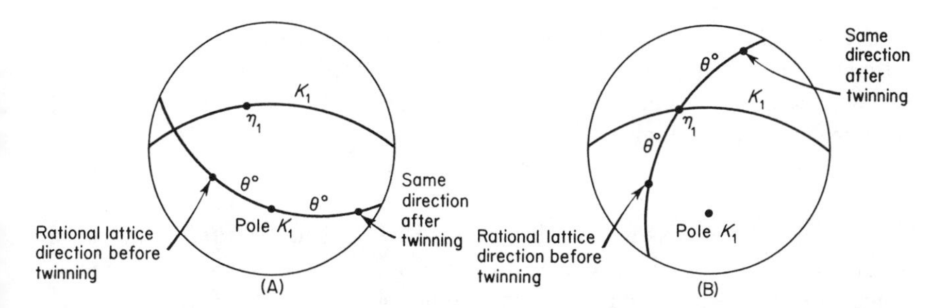

Fig. 16.12 Comparison of the lattice rotations in twins of the first and second kinds. In (A), which corresponds to a twin of the first kind, a rational lattice direction is rotated 180° about the pole of K_1. In (B), the rotation is 180° about η_1 for a twin of the second kind.

The following is a summary of the basic nomenclature relating to the theory:

K_1 = the twinning plane, or the first undistorted plane
K_2 = the second undistorted plane
η_1 = the shear direction
η_2 = the direction defined by the intersection of the plane of shear with K_2

The plane of shear is the plane which is mutually perpendicular to K_1 and K_2 and contains η_1 and η_2.

The composition plane is the plane separating the sheared and unsheared regions. In actual twins, it is usually close to K_1.

γ is the shear and equals the shear strain.

A twin that occurs inside another twin is called a *second-order twin.*

Twins of the second kind are relatively rare in metals and occur only in crystals of low symmetry. Cahn[2] has identified one such twin in uranium. They are also found in salts and minerals of orthohombic or lower symmetry. In metals of high symmetry, with cubic or close-packed hexagonal lattices, the twins generally belong to the compound classification.

From the above theory, it becomes apparent that the orientation of the lattice after twinning is given by a rotation of 180° about either η_1, or the normal to K_1, depending on the kind of twinning involved. On the other hand, the size and sense of shear are determined by the second undistorted plane K_2. It should also be noted that the shear in twinning, as contrasted to slip, is polar, that is, it can occur in only one direction.

The relationship between the shear and the second undistorted plane can be illustrated by a simple example. Consider the case of hexagonal metals. The common form of twinning in these metals is a compound twin in which we have

$$K_1 = (10\bar{1}2) \qquad K_2 = (\bar{1}012)$$

$$\eta_1 = [\bar{1}011] \qquad \eta_2 = [10\bar{1}1]$$

In zinc the c/a ratio is 1.856, while in magnesium it is 1.624. Because of this difference there is a 46.98° angle between the basal plane and the $\{10\bar{1}2\}$ planes in zinc, and a 43.15° angle in magnesium. Figure 16.13 shows how this affects twinning in these two metals.

The parts of Figure 16.13 are drawn so that the plane of the paper is the plane of shear. Symmetry conditions require that the second undistorted plane, K_2, be rotated clockwise in the case of zinc, and counterclockwise in magnesium. This has the effect, in the former case, of lengthening the crystal inside the twinned volume in a direction parallel to the basal plane. In magnesium on the other hand, the effect is reversed and the crystal is shortened in this direction. Figure 16.14 illustrates the fact that a tensile stress parallel to

2. *Ibid.*

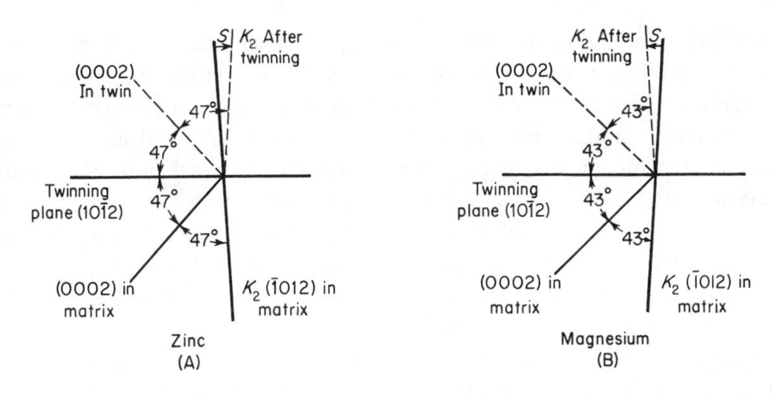

Fig. 16.13 The cause for the difference in shear sense for $\{10\bar{1}2\}$ twinning in zinc and magnesium. Notice that K_2, before twinning, lies to the left of the vertical in zinc, but in magnesium it is on the right. Symmetry conditions require that K_2 rotate in different directions for these two metals. The angles given in the above figures have been expressed in whole numbers for the sake of simplicity.

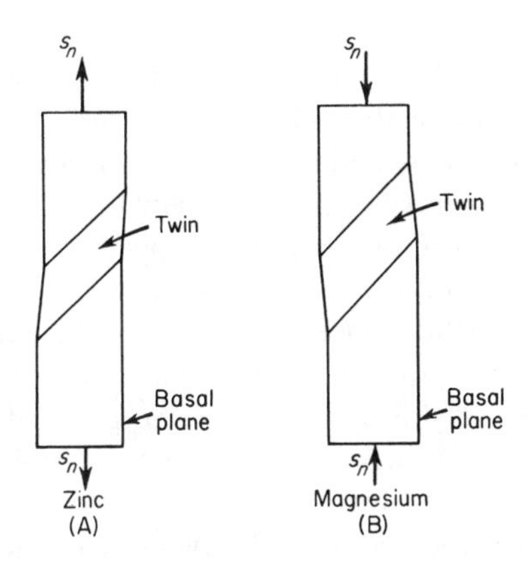

Fig. 16.14 The formation of a $\{10\bar{1}2\}$ twin in zinc increases the length of the crystal in a direction parallel to the basal plane, but decreases it in magnesium. A tensile stress applied parallel to the basal plane will twin a zinc crystal, but a compressive stress is needed in magnesium. In the above figures, the basal plane of the parent crystal lies parallel to the stress axes and perpendicular to the plane of the paper.

the basal plane in zinc favors $\{10\bar{1}2\}$ twinning. In magnesium, the sense of shear is such that twinning is favored by compression and not tension in this direction.

Mechanical twins have been observed in all the basic metal crystal structures. This includes both face-centered and body-centered cubic crystals. In some face-centered cubic metals, deformation twins are observed only at very low temperatures and at very large strains. Before Blewitt, Coltman, and Redman[3] observed twins in coppper deformed at 4.2°K in 1957, it was generally believed that f.c.c. metals probably did not twin. However since that time, mechanical twinning has been observed in many f.c.c. metals. Twins in face-centered cubic metals form on $\{111\}$ planes and the shear direction is $\langle 112 \rangle$.

In body-centered cubic metals, the twinning plane, K_1, is $\{112\}$, while η_1 is $\langle 111 \rangle$. Twins in body-centered cubic metals are also primarily observed at low temperatures where they may be a significant factor in the deformation and fracture mechanics. At ambient temperatures and above, mechanical twinning is normally considered to be of little importance in the plastic deformation of cubic metals, particularly if the deformation rate is low. Twinning is more important in hexagonal metals and in metals of low symmetry, such as tetragonal β-tin, orthorhombic uranium, and rhombohedral arsenic, antimony, and bismuth.

16.3 Identification of Deformation Twins. A mode of twinning is completely defined if K_1, K_2, η_1, and η_2 are all known. The identification of a new mode is not complete until all four quantities have been determined.

The composition plane of a twin has been defined as the plane which separates the sheared from the unsheared region. It is thus the visible boundary between a twin and the matrix. In very narrow twins (Fig. 16.1), the composition plane and the twinning plane K_1 usually have almost identical orientations because, as pointed out by Clark and Craig,[4] the energy of the twin boundary is a minimum when the composition plane coincides with the plane of twinning. Determinations of K_1, therefore, can be based on measurements of the orientation of the composition planes of thin twins. One method of determining the orientation of the composition, or habit plane of a thin twin requires that traces of the twin must be visible on at least two polished surfaces of a crystal. The orientation of the crystallographic axes of the crystal and their relations to these planes must also be known. If these conditions are satisfied, then the angular directions which the traces of the twins make on the reference planes can be measured and the data thus obtained plotted as points on a stereographic projection representing the orientation of the lattice. These points determine the orientation of the composition plane, and this, in turn, gives the orientation of K_1.

3. Blewitt, T. H., Coltman, J. K., and Redman, J. K., *J. Appl. Phys.*, 28 651 (1957).
4. Clark, R., and Craig, G. B., *Prog. in Metal Phys.*, 3 115 (1952).

The above method for determining K_1 is known as the *two-surface method.* Another method for determining K_1 requires that a Laue back-reflection photograph be made of an area in which the X-ray beam falls partly on the matrix and partly on a twin lamella. Thus, in one photograph the data for orienting both the parent crystal and the twin can be obtained. This method is the most conclusive for determining K_1. In addition, it will also yield η_1, the direction of shear. However, the method presents experimental difficulties which are hard to surmount when the size of the twins is small and the twinning shear is large, so that extensive plastic deformation has occurred inside and around the twinned area. This basic method is often employed in transmission electron microscope studies of twinning. Here the twin orientation relative to the matrix is determined by selected area electron diffraction (see Chapter 2).

Since η_1 is the shear direction of a twin, η_1 can often be deduced empirically by simple measurement of the direction along which a specimen is sheared when a twin forms. This will be discussed further in the next section.

The determination of K_2, the second undistorted plane, will also yield η_2 since this is the intersection of the second undistorted plane with the plane of shear. The rotation of the second undistorted plane from its position in the undistorted lattice to that which it assumes after the twinning process determines the magnitude of the shear in a twin. This fact is illustrated in Fig. 16.15. According to this diagram, we have

$$\frac{S/2}{h} = \tan(90° - \theta)$$

or

$$\frac{S}{h} = 2\tan(90° - \theta)$$

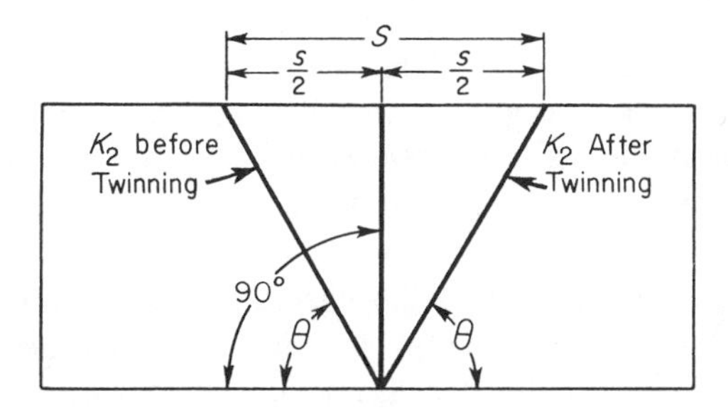

Fig. 16.15 The relationship between the twinning shear and the second undistorted plane. The plane of the paper corresponds to the plane of shear. The twinning-shear displacement is S. The angle between the first and second undistorted planes is θ.

where θ = the angle between K_1 and K_2
S = the shear displacement of the upper surface relative to the bottom
h = the width of the twin

Now the shear strain, or shear, is defined by

$$\gamma = \frac{S}{h}$$

and therefore

$$\gamma = 2 \tan (90^\circ - \theta) = 2 \text{ ctn } \theta$$

Where a twin lamella intersects the surface of a crystal there is always a distortion or tilt of the surface which is produced by the shearing action associated with the formation of the twin (Fig. 16.21). Where measurements of the surface tilt may be made accurately, it is frequently possible, with the aid of trigonometry, to determine the direction, sense, and magnitude of the shear. When this is possible, K_2 and η_2 can be determined.

The accuracy of the determination of K_2 and η_2 depends on the precision of the shear determination. The accuracy with which the latter may be determined depends, in large measure, on whether the surface tilt, at the junction of a twin with the surface, truly represents the twinning shear. If plastic flow has occurred inside the twin after it formed, then the surface tilt reflects both effects. The direction of the tilt of the surface determines the sense of shear in a twin. (See Fig. 16.14 for the case of $\{10\bar{1}2\}$ twins in zinc and magnesium.)

16.4 Nucleation of Twins. Twins form as the result of the shear-stress component of an applied stress which is parallel to the twinning plane and lies in the twinning direction η_1. The normal stress component (normal to the twinning plane) is unimportant in twin formation, a conclusion based on recent work[5] in which zinc crystals were deformed in tension while subjected to hydrostatic pressure. In these experiments, no measurable difference was found in the tensile stress required to form twins under hydrostatic pressures varying from 1 to 5000 atmospheres. Since the hydrostatic pressure has no shear component on the twinning plane, but does have a normal component, we must conclude that twins form only as a result of shear stress.

It has been estimated[6] that the theoretical shear stress necessary to *homogeneously* nucleate deformation twins in zinc crystals lies between 40 and 120 kg per mm^2 (56,000 to 168,000 psi). Experimentally measured values of the shear stress to form twins lie much lower than the theoretical values, ranging

5. Haasen, P., and Lawson. A. W., Jr., *Zeits. für Metallkunde,* 49 280 (1958).
6. Bell, R. L., and Cahn, R. W., *Proc. Roy. Soc.* (London), 239 494 (1957).

from 0.5 to 3.5 kg per mm^2. This is strong evidence for the belief that twins are nucleated heterogeneously.

A great deal of evidence suggests that nucleation centers for twinning are positions of highly localized strain in the lattice. Confirmation for this assumption is given by the fact that twins appear to form primarily in metals that have suffered previous deformation by slip.[6] It would further appear that the slip process must become impeded so that barriers are formed which prevent the motion of dislocations in certain restricted areas. Suitable barriers can be formed in a number of ways. Thus, under the proper geometrical conditions, the intersection of dislocations with each other can form barriers to the motion of other dislocations. Subgrain boundaries, grain boundaries, and deformation twins already present in the lattice can also act as impediments to the normal motion of dislocations, so that dislocations may pile up against them. Since a stress field exists around each dislocation, the concentration of large numbers of dislocations into relatively small volumes of a crystal should lead to an intensification of the stress in the neighborhood of dislocation pile-ups. In order to explain the dependence of the twinning stress on prior plastic deformation, it has been proposed that stresses associated with local concentrations of dislocations can lower the magnitude of the external stress needed to nucleate deformation twins.

Because the localized stress fields (of twin nucleation centers) can be formed in a number of different ways, depending on the geometry and orientation of the specimen, as well as the nature of the applied stress, it is probable that there can be no universal critical resolved shear stress for twinning as there is for slip. This explains the experimentally observed wide range of shear stresses required for twinning (that is, 0.5 to 3.5 kg per mm^2 in zinc).

There is an important factor relative to whether or not there is a critical resolved shear stress for twinning that apparently has been ignored. Normally, when one measures the stress required to nucleate a twin, an individual event is observed, whereas the critical resolved shear stress for slip represents the statistical average of many events. That is, when a crystal begins to deform macroscopically by slip, a very large number of dislocations are nucleated and begin to move. If one were to measure the average stress required to nucleate many thousands of dislocations, it is possible that for a given set of deformation conditions a relatively fixed value of the stress would be obtained as it is in slip.

16.5 Twin Boundaries. Let us now consider the interface between a twin and the parent crystal. The atomic arrangement at a twin boundary in a face-centered cubic metal is shown in Fig. 16.16. This diagram assumes that the twin interface is exactly parallel to the twinning plane K_1. In this structure, the two lattices (twin and apparent) match perfectly at the interface. Atoms on either side of the boundary have the normal interatomic

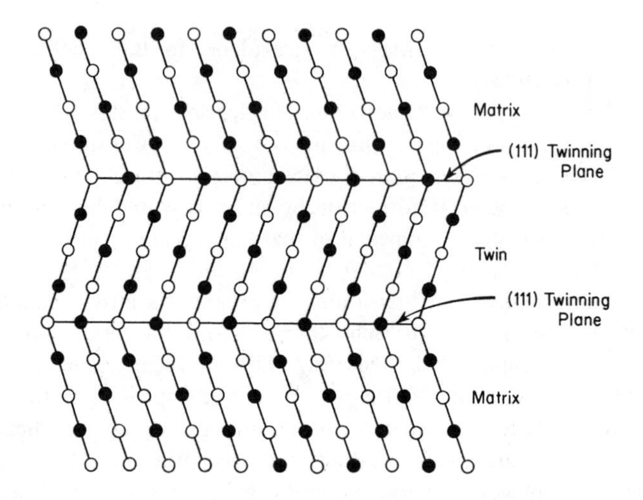

Fig. 16.16 Atomic arrangement at the twinning plane in a face-centered cubic metal. Black and white circles represent atoms on different levels (planes). (After Barrett, C. S., ASM Seminar, *Cold Working of Metals,* 1949, Cleveland, Ohio, p. 65.)

separation expected in a face-centered cubic lattice. The stacking sequence of the close-packed planes is

$$\overset{\downarrow}{\underset{\uparrow}{ABCACBACB}}\ldots\ldots\ldots\ldots$$

The interfacial energy of the boundary is very small. In the case of copper, it has been determined[7] to be 44 erg per cm^2 which is very much smaller than the surface energy of a copper-copper grain boundary,[8] which is 1725 $ergs/cm^2$. Barrett[9] has drawn diagrams similar to that of Fig. 16.16 for $\{10\bar{1}2\}$ twins in hexagonal metals and $\{112\}$ twins in body-centered cubic metals. He shows that in both cases a reasonable match between twin and parent lattices can be made across K_1. However, atoms on both sides of the interface are displaced small distances from their normal lattice positions. Because atomic bonds are strained in these twin interfaces, they must possess higher interfacial energies than the $\{111\}$ twin boundary of face-centered cubic metals. However, these energies are still much smaller than those of normal grain boundaries. The fact that

7. Valenzuela, C. G., *TMS-AIME,* 233 1911 (1965).
8. McLean, D., *Grain Boundaries in Metals*, p. 76, Oxford University Press, London, 1957.
9. Barrett, C. S., ASM Seminar (1949), *Cold Working of Metals.*

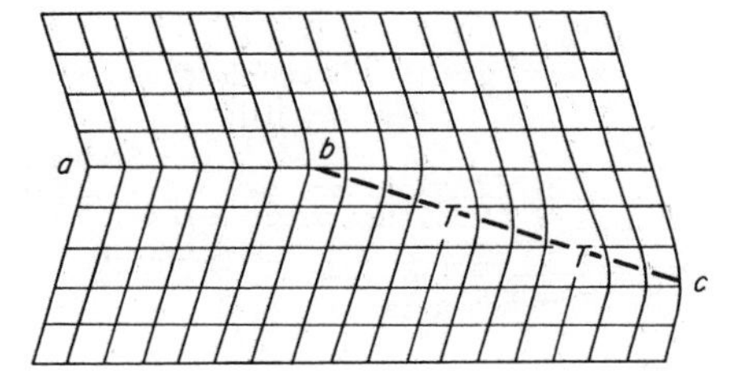

Fig. 16.17 The difference between a coherent twin boundary *ab* and an incoherent twin boundary *bc*. Notice the dislocations in the incoherent boundary. (After Siems, R., and Haasen, P., *Zeits, für Metallkunde,* **49** 213 [1958].)

deformation twins invariably form on planes of low indices may be explained in terms of the surface energy associated with the interface between twins and parent crystals. In general, the higher the indices, the poorer the fit at the interface, the higher the surface energy, and the lower the probability of twin formation.

A twin boundary that parallels the twinning plane is said to be a coherent boundary. Most twins start as thin narrow plates that become more and more lens-shaped as they grow. The average twin boundary is, accordingly, incoherent.

In a coherent boundary it is usually quite possible to match the two lattices without assuming the presence of dislocations in the boundaries. (See Fig. 16.16.) In an actual boundary where it is not coherent, it is generally accepted that a dislocation array is necessary in order to adjust the mismatch between lattices of parent and twin. This is demonstrated in Fig. 16.17, which shows schematically segments of both a coherent boundary and an incoherent boundary.

16.6 Twin Growth. The formation of a twin involves not only the creation of a pair of surfaces, but also the accommodation of the twinning shear by the lattice surrounding the twin. As the twin grows, it normally becomes more lenticular and its surface acquires an array of twinning dislocations. As may be seen by considering Fig. 16.17, the movement of the dislocations in an incoherent twin boundary can result in growth of the twin. Thus, a horizontal movement of the dislocations in Fig. 16.17 to the right will move the incoherent boundary to the right, increasing the size of the twin, which is assumed to be at the bottom of the figure.

It is theoretically possible to conceive of the array of twinning dislocations lying in the boundary of a lenticular twin as a single dislocation in the form of a

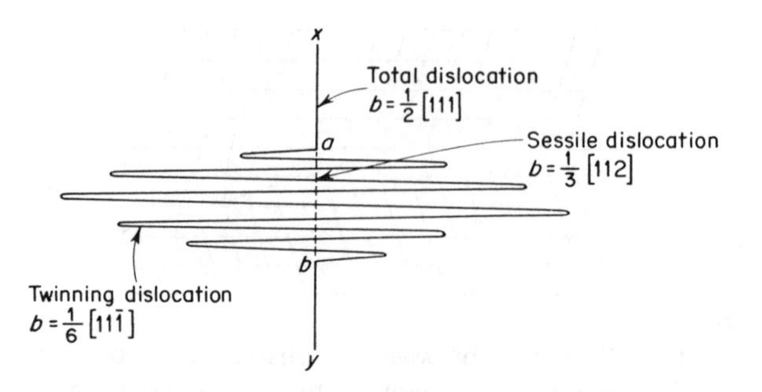

Fig. 16.18 The pole mechanism of Cottrell and Bilby for twinning in a body-centered cubic crystal. The helical twinning dislocation lies on the [$\bar{1}$21] plane which spirals around the dislocation *xy*. (After Cottrell, A. H. and Bilby, B. A., *Phil. Mag.*, **42** 573 [1951].)

helix. This is shown schematically in Fig. 16.18. Such a dislocation can result from the dissociation of a regular lattice dislocation.

In the case of body-centered cubic crystal, Cottrell and Bilby have proposed[10] that a ½[111] total slip dislocation should be able to dissociate into a glissile twinning partial dislocation of Burgers vector ⅓[11$\bar{1}$] and a sessile partial dislocation ⅓[112]. This reaction may be expressed by the equation

$$\tfrac{1}{2}[111] \rightarrow \tfrac{1}{3}[112] + \tfrac{1}{6}[11\bar{1}]$$

In the figure, line *xy* represents the total ½[111] dislocation which is assumed to be dissociated between points *a* and *b*. The straight dashed line between these two points represents the sessile ⅓[112] partial, while the helical dislocation, with b = ⅙[11$\bar{1}$], is the twinning dislocation. This latter is assumed to lie on the ($\bar{1}$21) plane. The key to this entire picture is the fact that the ($\bar{1}$21) plane spirals about the line *xy*. This occurs because all dislocations comprising the line (that is, line *xa* and *by* whose Burgers vector is ½[111], and line *ab* with Burgers vector ⅓[112]) have a component of their Burgers vector, normal to the ($\bar{1}$21) plane, equal to the interplanar spacing of this plane. In other words, as far as the ($\bar{1}$21) plane is concerned, line *xy* is a screw dislocation about which this plane spirals.

The schematic twin in Fig. 16.18 can grow in two ways. The twinning dislocation can increase its number of spirals. This would correspond to a growth in thickness. The other possibility is that the spirals themselves could expand, thereby increasing the length of the twin lamellae. Figure 16.18 only represents a part of the Cottrell-Bilby twinning mechanism. In their original paper, they

10. Cottrell, A. H., and Bilby, B. A., *Phil. Mag.*, **42** 573 (1951).

also considered the nature of the original dislocation dissociation that leads to this result. For the details, see the original paper.[11]

While there are some objections to this type of mechanism,[12] called a *pole mechanism*, it does indicate how a dislocation can be nucleated and grow. Pole mechanisms have also been proposed for other crystal structures, of which a notable example is that of Venables[13] for face-centered cubic crystals.

It is highly probable that if the pole mechanisms do work in metals, any given twin probably grows as a result of a number of pole mechanisms acting together. There is no reason why this should not be possible, and there is a good reason for assuming this; namely the macroscopic size of the average twin is such that it must intersect a large number of slip dislocations, each of which should constitute a pole.

Twins grow in size by increasing both their length and thickness. Large volumes of twinned material often form by the coalescence of separately nucleated twinned areas. The speed of twin growth is primarily a function of two variables that are not entirely independent. The first is the speed of loading that directly influences the rate of growth. The other is the stress required to nucleate twins. If twins nucleate at very low stresses, the stress required for their growth will be of the same order of magnitude as the nucleating stress. In this case, a small submicroscopic twin may form and grow more or less uniformly with an increasing stress until its growth is impeded by some means or other. On the other hand, if twins form under conditions that result in very high stress levels before nucleation, the stress for growth may be much smaller than that for nucleation. When this happens, twins grow at very rapid rates as soon as they are nucleated. Several interesting phenomena are associated with this rapid growth. First, the rapid deformations that ensue set up shock waves in the metal that can be heard as audible clicks. The crackling sound audible when a bar of tin is bent is due to deformation twinning. The other effect is visible in tensile tests of crystal specimens which undergo twinning during loading. The rapid formation of twins results in sudden increments in tensile strain. In a rigid testing machine, this causes the load to suddenly drop, giving the stress-strain curve a saw-tooth appearance in the region of twinning. A stress-strain curve of this nature is shown in Fig. 16.19.

16.7 Accommodation of the Twinning Shear. While the shear on a lattice plane is smaller in twinning than in slip, this shear becomes macroscopic when the twin attains an appreciable width. The shear of a very thin twin might possibly be accommodated elastically in the matrix surrounding a

11. *Ibid.*
12. Christian, J. W., *The Theory of Transformations in Metals and Alloys*, p. 792, Pergamon Press, Oxford, 1965.
13. Venables, J. A., *Deformation Twinning*, AIME Conf. Series, vol. 25, p. 77, Gordon and Breach Science Publishers, New York, 1964.

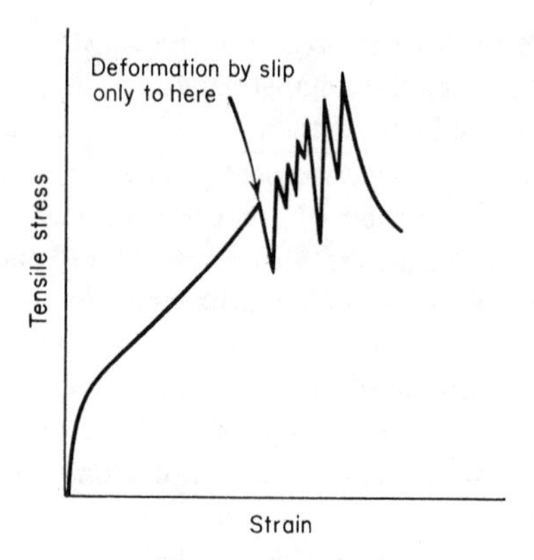

Fig. 16.19 Single-crystal tensile stress-strain curve showing discontinuous strain increments due to twinning. (After Schmid and Boas, *Kristallplastizität*, Julius Springer, Berlin, 1935.)

twin. However, as a twin grows in size, the accommodation normally requires plastic deformation in the surrounding metal. If the matrix is not able to accommodate the twinning shear of a twin of finite size, then a hole may develop at the point where the accommodation was not obtained. A well-known example occurs at twin intersections in iron and is known as a Rose's channel. This phenomenon was originally reported by Rose[14] in 1868. Figure 16.20 shows one way of forming such a channel.

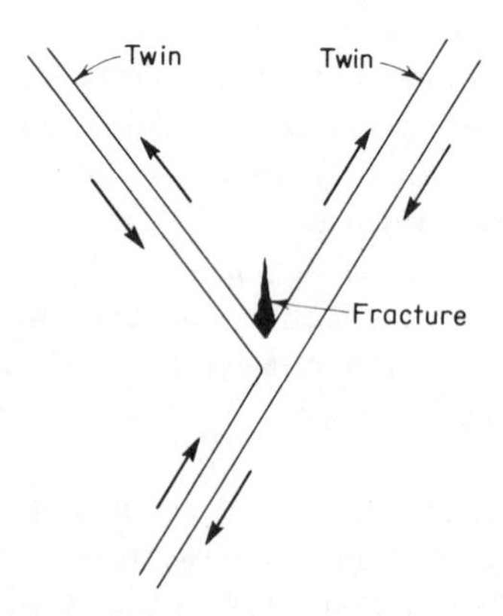

Fig. 16.20 A crack induced in an iron crystal at a twin intersection. Similar cracks were first reported by Rose in 1868. (After Priestner, R., *Deformation Twinning*, AIME Conf. Series, vol. 25, p. 321, Gordon and Breach Science Publishers, New York, 1964.)

14. Rose, G., *Berl. Akad.-Ber.*, 57 (1868).

An interesting example, where the shear is accommodated, occurs when a twin intersects a grain boundary. In a metal such as zirconium, where there is often a strong texture that results in neighboring grains having similar orientations, connected multiple twinning may occur, as shown in Fig. 16.21A. This photograph shows a number of twins in the lower central grain that are aligned with twins in the grain just above it. It is quite possible to follow twins connected in this manner through a large number of grains, and as many as 16 have been counted that were aligned in this way. This effect is heightened in zirconium by the large number of possible twinning systems. Another possibility is that the shear of the twin may be accommodated by slip in the next grain. In this case, a slip plane must exist in the second grain whose orientation is nearly the same as that of the twin in the first grain. A third possibility, if neither of the above two occurs, is kinking. In this case, if there is a slip plane in the next grain that is approximately perpendicular to the twin in the first grain, the lattice in the second twin might kink or bend to accommodate the twinning shear. An example of this is shown in Fig. 16.21B. Note the rather large, dark, horizontal twin in the grain at the right. Facing the twin is a horizontal white band in the grain on the left. This is a kink band inside which the crystal lattice has been rotated. This rotation of the crystal structure has changed the nature of the light reflected from the surface, thereby making the kink visible under the polarized light microscope. A close examination of the grain boundary shows that it has followed this rotation. Note the large displacement of the boundary where it is intersected by the twin.

If a twin lies entirely inside a grain, it will usually be lenticular and thus taper to a sharp edge at its periphery. This growth form may be assumed to be related to the shear accommodation process, since it should be easier to accommodate the twinning shear by slip in the matrix if the twin grows as a wedge. However, the twin shape is also undoubtedly influenced by the nature of the dislocation array that forms on the twin boundary. The more tapered the twin profile, the greater will be the separation between the dislocations lying in the boundary.

The accommodation of the twinning shear by slip is particularly easy in a body-centered cubic iron crystal, since the twinning plane $\{11\bar{2}\}$ is a possible slip plane and the twinning shear is parallel to the slip direction [111]. In this case it is quite possible for a twin to have a blunted end. Sleeswyk[15] has pointed out that the array of twinning dislocations on the surface of a lenticular twin should constitute a configuration of high energy. This array, as shown in Fig. 16.22A, is analogous to a pile-up of edge dislocations of the same sign along a slip plane. This arrangement could lower its energy if one-third of its dislocations should undergo the dissociation

$$\tfrac{1}{6}[111] \rightarrow \tfrac{1}{2}[111] - \tfrac{1}{3}[111]$$

15. Sleeswyk, A. W., *Acta Met.*, **10** 803 (1962).

(A)

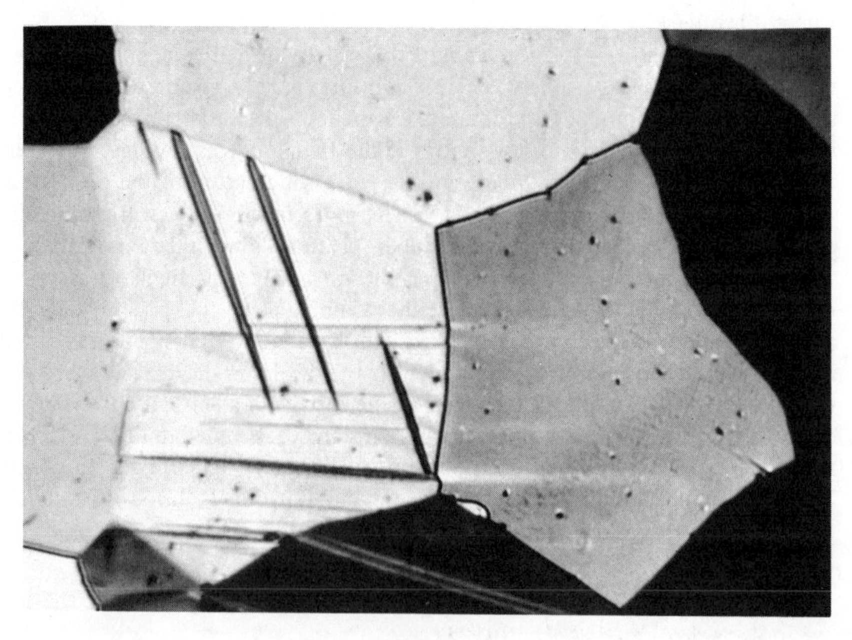

(B)

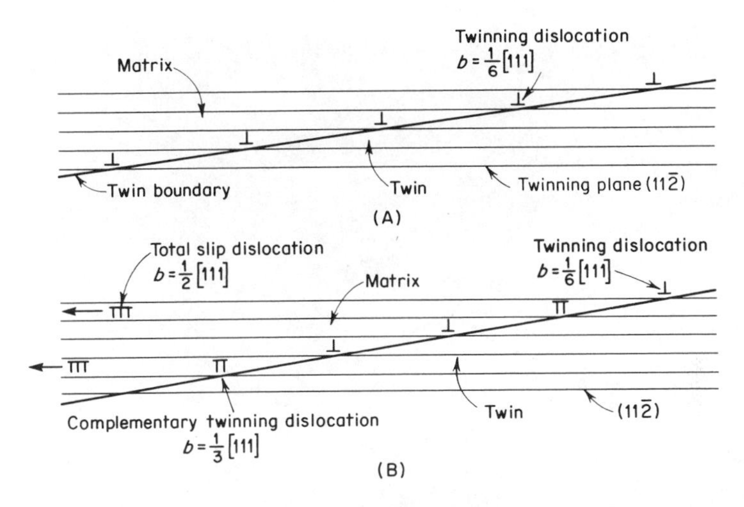

Fig. 16.22 Sleeswyk's emissary dislocation mechanism. (A) A lenticular twin boundary is equivalent to a high energy array of dislocations, all of the same sign. (B) If in an iron crystal every third dislocation dissociates and sends forth an (emissary) total slip dislocation, it will leave a boundary composed of twinning and complementary twinning dislocations. This boundary has a much lower energy. The emissary dislocation accommodates the twinning shear in the matrix.

In this reaction, as shown in Fig. 16.22B, a twinning dislocation (⅙ [111]) breaks down to form a total slip dislocation which glides off ahead of the twin, leaving a complementary twinning dislocation (⅓ [111]) behind in the boundary. This latter dislocation also produces the correct configuration of atoms for a twin, so that the reaction does not disturb the lattice arrangement in the twin. On the other hand, its presence in the twin boundary does have one important result: it drastically reduces the energy of the configuration. This can be understood by considering an analogous arrangement of simple positive and negative edge dislocations where pairs of positive edge dislocations of strength b are interspersed between negative edge dislocations of strength $2b$. Such an arrangement will not have a long-range strain field. In most cases of twinning in other metals, the slip planes that accommodate the twinning shear are not

---

Fig. 16.21 (A) Accommodation of the twinning shear at a grain boundary by twinning. $\{11\bar{2}1\}$ twins in zirconium. Specimen strained 0.65 percent by rolling at 77°K. Magnification 760 x. (B) In this case accommodation of the $\{11\bar{2}1\}$ twinning shear in zirconium has occurred across a grain boundary by kinking. The lattice rotation in the kink is revealed by polarized light. Specimen strained 3 percent by rolling at 77°K. Magnification 1650 x. (Hartt, W. H., and Reed-Hill, R. E., *TMS-AIME,* **239** 1511 [1967].)

Fig. 16.23 Deformation twins in an annealed magnesium crystal. Notice that the sharp edges of the lenses have been blunted.

aligned parallel to the twinning plane. However, it is quite possible that the accommodation of the twinning shear by slip results in a dislocation configuration in the boundary that lowers the boundary energy.

With regard to the sharp leading edge of a twin, it is interesting to note that when a twinned structure is annealed, the twins can tend to disappear by shrinking inward from their sharp edges, so that the latter become blunted or squared off (see Fig. 16.23). The sharp leading edge of a twin is also lost when a growing twin intersects a free surface. If a twin meets a single free surface, it assumes the shape of a half-lens (Fig. 16.23), and if it crosses a complete crystal, the twin acquires flat sides parallel to he twinning plane K_1 (Fig. 16.2A). In this last case, the twin can be accommodated in the lattice without appreciable distortion of the latter. This is not true for half-lens-shaped twins where the lattice around the twins is forced to accommodate the shearing strain. In the case of a half-lens, one commonly finds accommodation kinks adjacent to the twins. The diagrams of Fig. 16.24 show the nature of simple accommodation kinks in the hexagonal metals zinc and magnesium, as viewed on a surface normal to the basal plane and to the twins. In each case, the basal planes in the crystal adjacent to the twin are bent or kinked so as to allow the lattice of the parent to follow the shear strain of the twin.

When twins are observed on a free surface that has not been polished or etched

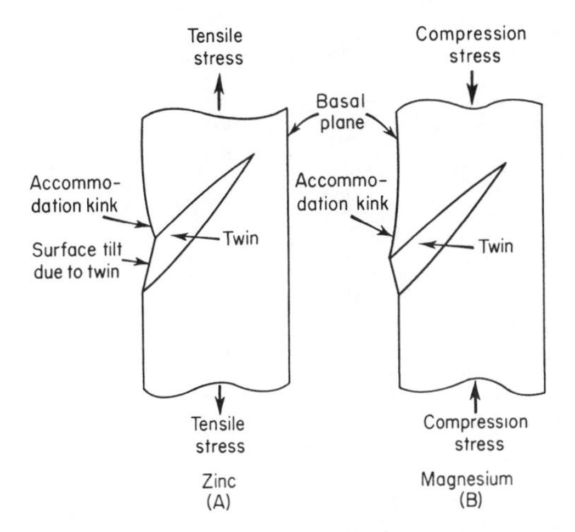

Fig. 16.24 Surface tilts and accommodation kinks resulting from intersections of half lens twins with the surface in two hexagonal metals.

subsequent to deformation, both the tilt of the surface due to the twin itself and the tilt of the surface due to the accommodation kink are usually quite visible. This fact is illustrated in Fig. 16.25. A number of interesting features of twinning are visible in this photograph. The specimen shown was first deformed to form small twins. It was then polished and etched, the etching delineating the original boundaries of the twins. It was then deformed again and rephotographed. Consider now the large horizontal twin. The original twin appears as a narrow dark band lying in the center of a lighter and wider band. This latter represents the twin after the second deformation and shows that the twin has grown as a result of the second loading. The fact that this area is lighter than the background is due to the surface tilt of twinning. The dark band below the twin corresponds to the accommodation kink. Finally, the difference in orientation in the twin and in the parent is clearly shown by the different inclination of the basal slip line traces in this twin and in the parent crystal.

In summary, because the twinning shear has to be accommodated, slip normally has to play a very active role in twinning. From a broad viewpoint, we should therefore consider twinning deformation in terms of the coupling between slip and twinning modes. It is highly probable that twinning would not be a significant practical means of plastic deformation in metals such as titanium and zirconium if there was a lack of this coupling. Generally speaking, this implies that twinning also involves slip. The reverse, of course, is not true.

16.8 Martensite. When the temperature of a metal capable of undergoing a martensite reaction is lowered, it eventually passes through an equilibrium temperature separating the stability ranges of two different phases.

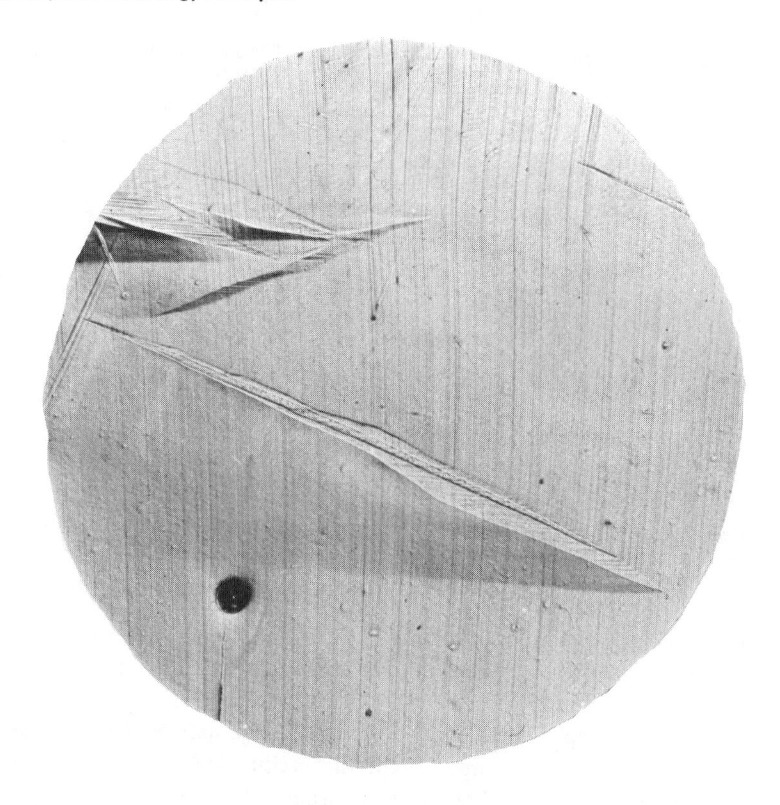

Fig. 16.25 Accommodation kinks at deformation twins in magnesium.

Below this temperature, the free energy of the metal is lowered if the metal changes its phase from that stable at high temperatures to that stable at low temperatures. This free-energy difference is the primary driving force for a martensite reaction.

The phase change that occurs in a martensite transformation is brought about by the movement of the interface separating parent phase from the product. As the interface moves, atoms in the lattice structure of the parent phase are realigned into the lattice of the martensite phase. The nature of the individual atomic movements in the region constituting the interface are not known, just as they are not known in deformation twinning. Nevertheless, it is undoubtedly true that the displacement of atoms, relative to their neighbors, is small in magnitude and probably more complicated than those in deformation twinning. Because of the manner in which martensite forms, no composition change occurs as the parent lattice is converted into the product phase and diffusion in either the parent phase or product phase is not required for the reaction to continue.

Martensite reactions are, accordingly, commonly referred to as *diffusionless phase transformations.*

The atomic realignments associated with martensite reactions produce shape deformations just as they do in mechanical twinning. Because the new lattice which is formed has a symmetry different from that of the parent phase, the deformation is, of necessity, more complicated. In mechanical twinning we have seen that the distortion is a simple shear parallel to the twinning plane or the symmetry plane between twin and parent crystal. As previously pointed out, the twinning plane is an undistorted plane; all directions in this plane are unchanged by twinning with respect to both their magnitudes and their angular separations. The *habit plane* or plane on which martensite plates form is also usually assumed to be an undistorted plane. The macroscopic shape deformation in the formation of a martensite plate is believed to be a shear parallel to the habit plane plus a simple (uniaxial) tensile or compressive strain perpendicular to the habit plane. A strain of this nature, known as an *invariant plane strain*, is the most general that can occur while still maintaining the invariance of the habit plane. Neither a shear parallel to the habit plane, nor an extension or contraction perpendicular to it will change the positions or magnitudes of vectors lying in the habit plane.

Data showing magnitudes of the shear and normal components of the invariant plane strains associated with martensite reactions are difficult to measure and relatively scarce. Most of the available information is given in Table 16.1. The normal component of the strain in most martensite transformations is also smaller and harder to measure than the shear component, which probably explains the lack of data for this component in Table 16.1. Because the shape deformation is primarily a shear, martensite plates deform the lattice of the matrix like deformation twins. Individual martensite plates, formed in the interior of a crystal, are therefore lens-shaped, and if a martensite plate crosses a crystal, its boundaries are flat and parallel to the habit plane.

While the atomic movements involved in martensite reactions are small compared to those in slip, there is considerable variation in their magnitudes from one alloy to the next. Both the crystallographic features of a martensite transformation and the kinetics of the reaction are more easily studied in an alloy involving the smallest possible atomic displacement. Two well-known martensite transformations fit this category: those in gold-cadmium alloys and that in an indium-thallium alloy. The very important martensite reactions that occur in the hardening of carbon and alloy steels involve relatively large atomic displacements. In the following sections we shall consider an example of both an alloy in which the atomic movements are small (indium-thallium) and another (iron-nickel, 70 percent – 30 percent) in which they are large. In this way, a better understanding of the various phenomena involved in martensitic transformations can be obtained.

Table 16.1 Habit Planes and Macroscopic Distortions in Martensite Transformations.*

System	*Phase Changes†*	*Habit Plane*	*Shear Direction*	*Shear Component of Strain*	*Normal Component of Strain*
Fe–Ni (30% Ni)	f.c.c. to b.c.c.	(9, 23, 33)	[156] ± 2°	0.20	0.05
Fe–C (1.35% C)	f.c.c. to b.c.t.	(225)	$[\bar{1}\bar{1}2]$	0.19	0.09
Fe–Ni–C (22% Ni, 0.8% C)	f.c.c. to b.c.t.	(3, 10, 15)	$[\bar{1}\bar{3}2]$ (approx.)	0.19	
Pure Ti	b.c.c. to h.c.p.	(8, 9, 12)	$[11\bar{1}]$ (very approx.)	0.22	
Ti–Mo (11% Mo)	b.c.c. to h.c.p.	4° from $[\bar{3}44]$ 4° from (8, 9, 12)	$[14\bar{7}]$ (approx).	0.28 ± 0.05	
Au–Cd (47.5% Cd)	b.c.c. to ortho-rhombic	[0.696, –0.686, 0.213]	[0.660, 0.729, 0.183]	0.05	
In–Tl (18–20% Tl)	f.c.c. to f.c.t	[0.013, 0.993, 1]⁹	$[01\bar{1}]$ (approx.)	0.024	

* Most of the data are abstracted from a comprehensive summary by Bilby, B. A., and Christian, J. W., *The Mechanism of Phase Transformations in Metals.* The Institute of Metals, London, 1956. For detailed references and explanation of the data in the table, the reader is referred to this summary

† The abbreviations signify: b.c.t. = body centered tetragonal
f.c.t. = face centered tetragonal

16.9 The Bain Distortion. A face-centered cubic lattice can also be considered as a body-centered tetragonal lattice. See Fig. 16.26 in which (in the upper diagram) a tetragonal unit cell is delineated in the face-centered cubic structure. In a tetragonal structure, the crystallographic axes lie at right angles to each other, as in a cubic structure, but one lattice constant c differs in magnitude from the other two. These latter dimensions of the unit cell are usually designated by the symbol a. When considered as a body-centered tetragonal structure, the normal face-centered structure has a c/a ratio of $\simeq 1.4$. Similarly, a body-centered cubic structure can be considered a body-centered tetragonal structure with a c/a ratio of unity. In 1924, Bain[16] suggested that a body-centered cubic lattice could be obtained from a face-centered cubic

16. Bain, E. C., *Trans. AIME,* 70 25 (1924).

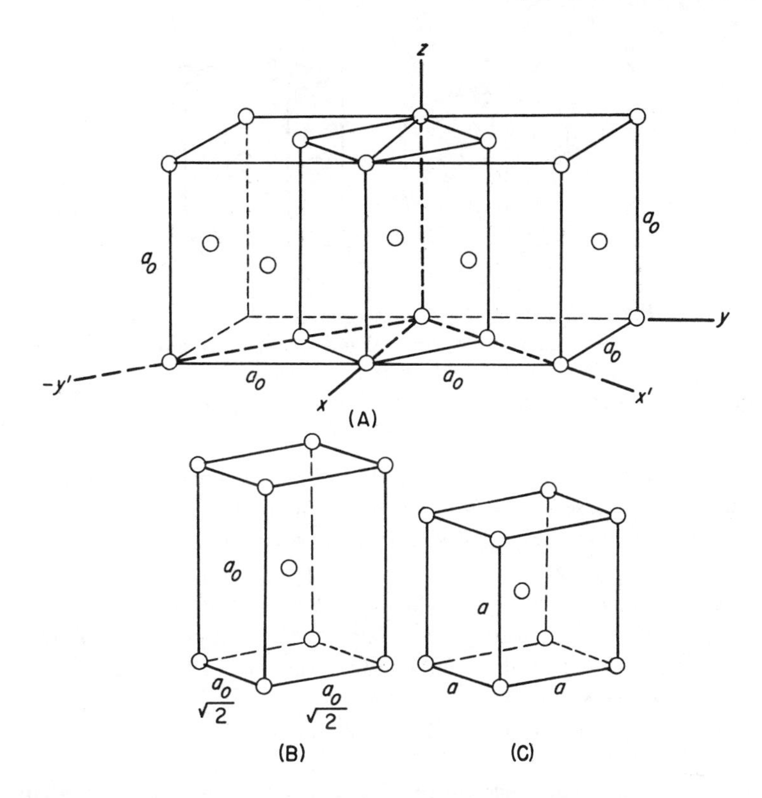

Fig. 16.26 Bain distortion for a face-centered cubic lattice transforming to a body-centered cubic lattice. The body-centered tetragonal cell is outlined in the face-centered cubic structure in (A), and shown alone in (B). The Bain distortion converts (B) to (C). (After Wechsler, M. S., Lieberman, D. S., and Read, T. A., *Trans. AIME,* **197** 1503 [1953].)

structure by a compression parallel to the c axis and an expansion along the two a axes. Any simple homogeneous pure distortion of this nature, which converts one lattice into another by an expansion or contraction along the crystallographic axes, belongs to a class known as *Bain distortions*.

The Bain distortion indicated above converts a face-centered cubic lattice into a body-centered lattice with a minimum of atomic movements. However, there is no undistorted plane associated with this Bain distortion (as we shall presently see), so that the invariant plane strain associated with martensitic transformations cannot be explained by a Bain distortion. The Bain distortion does, however, give an approximate measure of the magnitude of atomic movements involved in a transformation. Thus, in the iron-nickel alloy (30 percent Ni) of Table 16.1, which undergoes a face-centered cubic to body-centered cubic

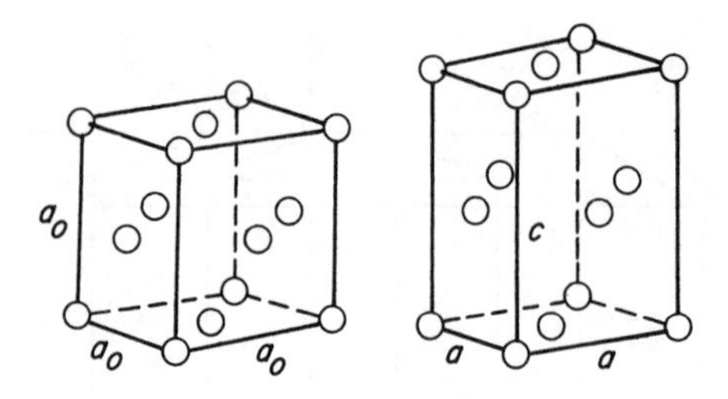

Fig. 16.27 Bain distortion in indium-thallium (18–20 percent Tl) alloy. Face-centered cubic transforms to face-centered tetragonal. In tetragonal structure, $c = 1.0238a_0$ and $a = 0.9881a_0$, where a_0 is cubic lattice constant. (Distortion of lattice is greatly exaggerated in above figures.)

transformation, the c axis is shortened by a factor of approximately 0.8, while the a axes are increased in length by a factor of about 1.14. On the other hand, in the indium-thallium alloy, where a face-centered cubic alloy transforms to a face-centered tegragonal structure, as shown in Fig. 16.27, the c (vertical axis) becomes 1.0238 a_0, where a_0 is the cubic lattice constant, while the a axes decrease to 0.9881 a_0. Clearly, much larger deformations are involved in the iron-nickel transformation than in the indium-thallium (18–20 percent Tl). This fact is reflected in the magnitudes of the macroscopic shears of the martensite transformations given in Table 16.1; 0.20 and 0.024 respectively.

16.10 The Martensite Transformation in an Indium-Thallium Alloy. We shall now consider in some detail the martensite transformation in the indium-thallium alloy, where the small size of the observed shear makes for easier understanding of the observed phenomena. Studies made of single crystals of this alloy are of special interest. Provided that the crystals are carefully annealed and not bent or damaged, they undergo martensitic transformations involving the motion of a single interface between the cubic (parent phase) and the tetragonal (product phase).[17,18] This transformation does not occur by the formation of lens-shaped plates, or even parallel-sided plates, but by the motion of a single planar boundary that crosses from one side of the crystal to the other. On cooling, an interface first appears at one end of a specimen and, with continued cooling, moves down through the entire length of the crystal. Because of dimensional changes accompanying the reaction, its progress can be followed[17] with the aid of a simple dilatometer that permits measurements of the specimen length to be made as a function of temperature. A typical set of data is shown in Fig. 16.28. Notice that, on cooling, the length of the specimen begins to change at approximately 72°C, signifying that the martensite

17. Burkart, M. W., and Read, T. A., *Trans. AIME*, 197 1516 (1953).
18. Basinski, Z. S., and Christian, J. W., *Acta Met.*, 2 148 (1954).

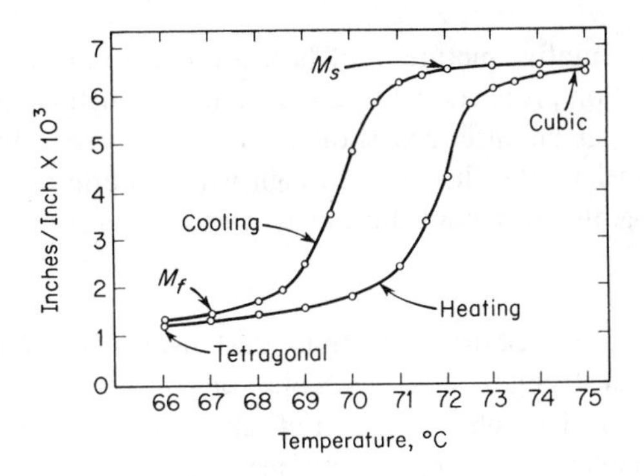

Fig. 16.28 The temperature dependence of the martensite transformation in the indium-thallium (18 percent Tl) alloy. Transformation followed by measurement of change in length of specimen. (From data of Burkart, M. W., and Read, T. A., *Trans. AIME,* **197** 1516 [1953].)

transformation started at this temperature. It is customary in all martensite transformations to designate, by the symbol M_s, the temperature corresponding to the start of the transformation. The curve also shows that the specimen did not transform completely until the temperature was lowered to about 67°C. This latter temperature is designated as the M_f, or martensite finish temperature. In order to move the interface from one end of the specimen to the other, it was necessary to drop the temperature 5° below the M_s temperature. During this temperature interval between M_s and M_f, the interface, or habit plane does not move steadily and smoothly, but rather in a jerky fashion. It moves very rapidly for a short distance in a direction normal to itself and then stops until a further decrease in temperature gives it enough driving force to move forward again. The irregular motion of the interface is not apparent in the dilatometer measurements of Fig. 16.28, but may be readily observed by watching the movement of the interface under a microscope. The necessity for an ever-increasing driving force to continue the reaction is an unusual phenomenon because it implies the existence of a volume relaxation,[19] or an energy term opposing the transformation which is proportional to the volume of metal transformed. An explanation for the effect can be given in terms of the intersection of the interface with obstacles. It is now believed that martensitic and deformation twin interfaces can be composed of dislocation arrays. It is also recognized that a moving screw dislocation, which cuts through another screw dislocation, acquires a jog or discontinuity which produces a row of vacancies or interstitials as the moving dislocation continues its advance. While the geometry of the dislocation arrays in the crystals and in the interfaces is not known, the screw-dislocation-intersection

19. Chang, L. C., *Jour. Appl. Phys.*, **23** 725 (1952).

picture gives us a tentative picture[20] of how a moving interface can develop a resistance to its motion proportional to the distance through which it moves. The screw components in an interface should cut other screw dislocations in a number proportional to the distance through which the interface moves: each intersection adding its own contribution to the total force holding back the boundary.

16.11 Reversibility of the Martensite Transformation. The indium-thallium transformation is reversible for, on reheating, the specimen reverts not only to the cubic phase, but also to its original single-crystal orientation. If recooled, the original interface reappears and the cycle can be completely reproduced (provided the specimen is not heated to too high a temperature or held for too long a time between cycles).

The reverse transformation is shown in Fig. 16.28 by the return of the specimen to its original dimensions. It does not, however, retransform in the same temperature interval as that in which the forward transformation occurred, but in one averaging about 2° higher. A hysteresis in the temperature dependence of the transformation is characteristic of most martensite transformations.

16.12 Athermal Transformation. In the single-crystal experiments discussed in the above paragraph, we have seen that the martensite transformation in indium-thallium occurs as a result of temperature changes that increase the driving force for the reaction (free energy). The progress of transformation is, accordingly, not time dependent. In theory, the faster one cools the specimen, the faster the interface moves, with equal amounts of transformation resulting at equal temperatures. Time does, however, have a secondary, though negative, effect on this transformation, for isothermal holding of the specimen at any temperature between the start and end of the transformation tends to stabilize the interface against further movement. This effect is demonstrated in Fig. 16.29 where the indicated transformation curve corresponds to a specimen whose heating cycle was interrupted at 71.6°C and held at this temperature 6 hr. Not only did the holding of the specimen at this temperature not induce additional transformation, but it stabilized the interface. In order to make the interface move again, it was necessary to increase the driving force by approximately 1°C. An equivalent phenomenon occurs on cooling, so that it can be concluded that the formation of martensite in the present alloy depends primarily on temperature. A transformation of this type is said to be athermal, in contrast to one that will occur at constant temperature (isothermal transformation). Although isothermal formation of martensite is

20. Basinski, Z. S., and Christian, J. W., *Acta Met.*, **2** 148 (1954).

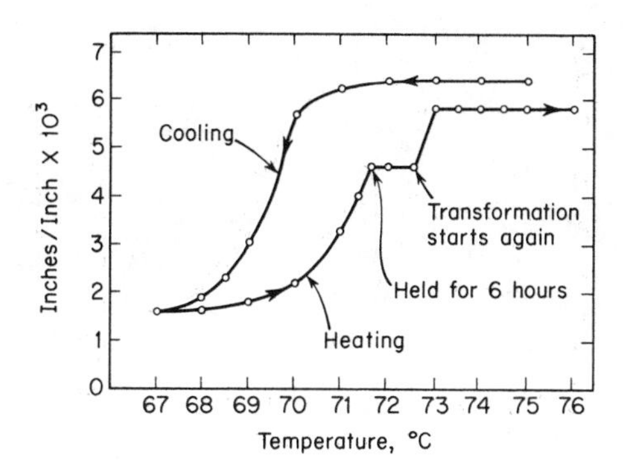

Fig. 16.29 Stabilization in the martensite transformation in an indium-thallium (18 percent Tl) alloy. (From data of Burkart, M. W., and Read, T. A., *Trans AIME,* **197** 1516 [1953].)

observed in some alloys, martensitic transformations tend to be predominantly athermal.

16.13 Wechsler, Lieberman, and Read Theory. The nature of the atomic mechanisms that convert one crystal structure into the other in the narrow confines which we define as the interface are not known. Wechsler, Lieberman, and Read[21] have shown, nevertheless, that the crystallographic features of martensite transformations can be completely explained in terms of three basic deformations enumerated below:

1. A *Bain distortion*, which forms the product lattice from the parent lattice, but which, in general, does not yield an undistorted plane that can be associated with the habit plane of the deformation.
2. A *shear deformation*, which maintains the lattice symmetry (does not change the crystal structure) and, in combination with the Bain distortion, produces an undistorted plane. In most cases, this undistorted plane possesses a different orientation in space in the parent and product lattices.
3. A *rotation of the transformed lattice*, so that the undistorted plane has the same orientation in space in both the parent and product crystals.

No attempt is made in this theory to give physical significance to the order of the steps listed above, and the entire theory is best viewed as an analytical explanation of how one lattice can be formed from the other.

The indium-thallium transformation will now be considered in terms of the Wechsler, Lieberman, and Read theory. The Bain distortion for the indium-

21. Wechsler, M. S., Lieberman, D. S., and Read, T. A., *Trans. AIME,* 197 1503 (1953).

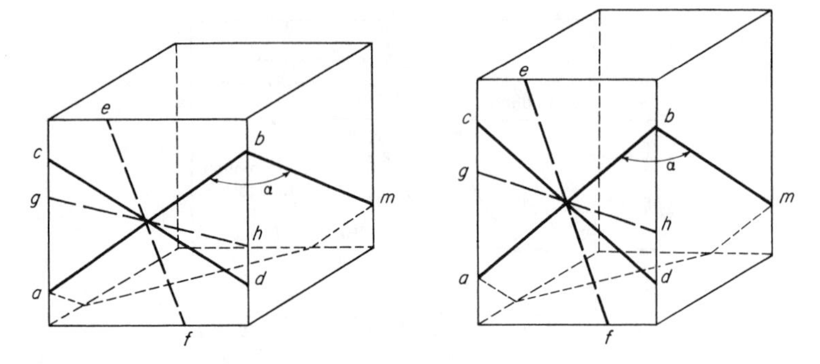

Fig. 16.30 In the cubic to tetragonal transformation, the Bain distortion does not possess an undistorted plane.

thallium transformation is shown in Fig. 16.27 where the very small actual deformation is magnified in order to show the effect more clearly. The original structure is cubic, while the final is tetragonal, so that the Bain distortion consists of an extension (2ϵ) along one axis (the *c* axis) and a contraction ($-\epsilon$) along the other two axes. Simple geometrical reasoning will show that the given distortion does not possess an invariant plane. Consider first the set of lines drawn on the front face of the cube in Fig. 16.30. Lines *ab* and *cd* represent those directions on this surface, the lengths of which do not change in the transformation. In the cube shown on the left and in the prism shown on the right, these lengths are the same. All other directions on this surface, such as *ef* and *gh*, have their lengths increased and shortened respectively by the distortion. Now assume that the distortion actually possesses an invariant plane that passes through the front surfaces of each of the two geometrical figures. To be an invariant plane, it must contain one of the two lines, whose lengths do not change. For our purpose, assume that this line is *ab* and that the trace of the plane on the right side of each geometrical figure is *bm*, where *bm* is a direction on these latter faces whose length is also not changed by the deformation. The plane defined by *abm* possesses the characteristic that before and after the distortion the two distances *ab* and *bm* are unchanged. However, the angle α between these two directions is obviously changed by the distortion. Plane *abm* cannot be an undistorted plane, for not only must the magnitudes of vectors be maintained, but also the angles between the vectors must be unchanged. If we now take into account the symmetry of the Bain distortion, it becomes quite evident that there is no undistorted plane associated with it.

The directions that are undistorted in the above Bain distortion form a cone whose positions in the cubic and tetragonal structures are indicated in Fig. 16.31. Notice that the vertex angle of the cone decreases during the Bain distortion as the cube is drawn into the right prism. In each of the two sketches

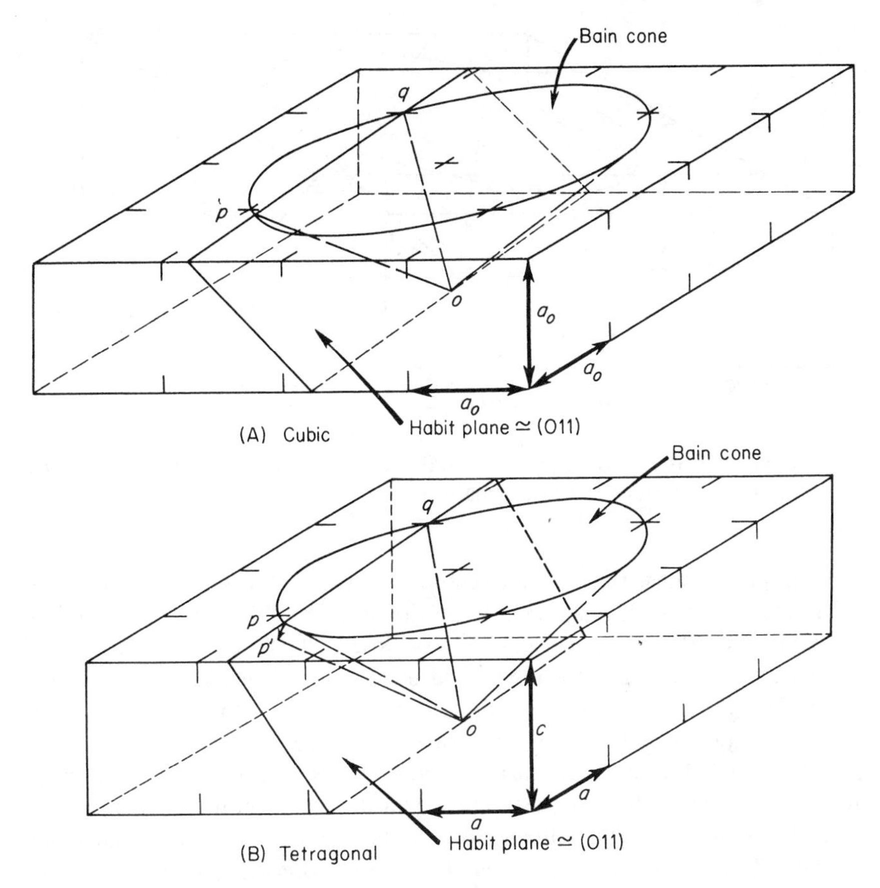

Fig. 16.31 The Bain cone in the cubic and tetragonal structures corresponding to the indium-thallium transformation. The distortion shown is exaggerated in order to emphasize its nature.

of Fig. 16.31, a plane is drawn intersecting the Bain cone in two lines designated as *op* and *oq*. The lengths of these lines are unchanged by the distortion since they lie on the Bain cone. Their angular separation, however, changes. The given plane is drawn to represent the habit, or interface, plane of the transformation in its positions before and after the Bain distortion, and lies very close to the (011) plane of both the cubic and tetragonal structures. Because the angle *poq* changes during the distortion, this plane is not an undistorted plane at the end of the Bain distortion.

We shall now turn our attention specifically to the tetragonal structure which is indicated again in Fig.16.32, where the position of the (101) plane which also passes through line *qo* is shown. Let us also consider that this is the first undistorted

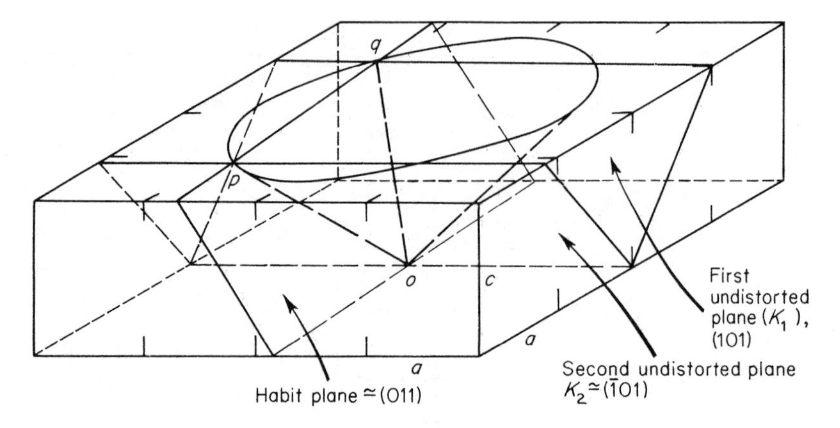

Fig. 16.32 Positions of the two undistorted planes, K_1 and K_2, in the indium-thallium martensite transformation.

plane (K_1) of a shear (in the same sense as in twinning). A second plane, which lies at almost 90° to K_1, is also given in the diagram. This plane belongs to the same zone as (101) and ($\bar{1}01$) and lies very close to the latter plane. This second plane contains line *po* and is the second undistorted plane (K_2) of the desired shear. The shearing action is indicated in Fig. 16.33. As a result of this operation, point *p* is translated to *p*′, with no change in the length of *op* as it rotates to *op*′. However, the shear is chosen so as to make angle *p*′*oq* equal angle *poq* in the original cubic structure, and the plane defined by *p*′*oq* is now an undistorted plane. This follows from the fact that the vectors *po* and *qo* are unchanged in length and angular relation to each other as a result of the two

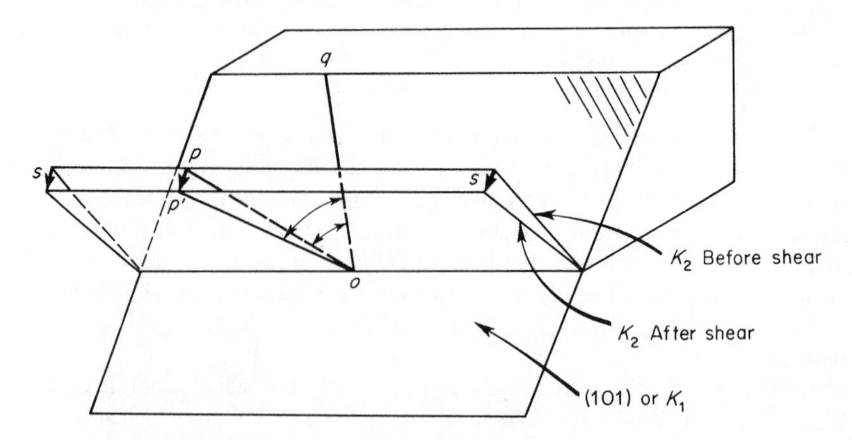

Fig. 16.33 Nature of the shear in the indium-thallium martensitic transformation.

deformations, which satisfies the theorem[22] "If two (noncollinear) vectors in a plane are undistorted (unchanged in length) and the angle between them is unchanged by a transformation, all vectors in that plane are undistorted (and all angles are unchanged), that is, the plane is one of zero distortion." Actually, line $p'o$ does not lie strictly in plane poq, so that this line and line oq define a new plane which is the true habit plane. In other words, plane poq is not the habit plane. However, the angle between plane $p'oq$ and poq is so small that we shall consider $p'o$ to lie in poq and that this latter is the true habit plane. The position of line $p'o$ is, accordingly, plotted in the lower diagram of Fig. 16.31. The plane containing both $p'oq$ and poq of this figure can now be taken as representing the approximate position of the undistorted plane after both the Bain distortion and the shear have occurred. A careful comparison of this sketch with the one above it shows that the undistorted plane in the cubic structure and in the tetragonal structure have different orientations in space. For this reason, $p'oq$ does not yet satisfy the requirements of a habit plane. It must not only be undistorted by the transformation, but it must also be unrotated. A rotation of the tetragonal structure must now be added, which brings plane $p'oq$ into coincidence with plane poq in such a manner that lines op' and oq of the tetragonal phase fall on lines op and oq of the cubic phase respectively. No effort will be made to demonstrate this rotation with diagrams, for the primary purpose of this discussion has only been to demonstrate the need for considering the three basic steps (Bain distortion, shear, and rotation).

There remains the problem of explaining the nature of the arbitrary shearing process which is needed during the second step of the deformation. In martensitic deformation, the required shear can occur in one of two basic ways: either as a result of slip or as a result of mechanical twinning. Assuming that the deformation is confined to the product lattice, Fig. 16.34 shows how this may be accomplished by slip. Glide occurs more or less homogeneously distributed

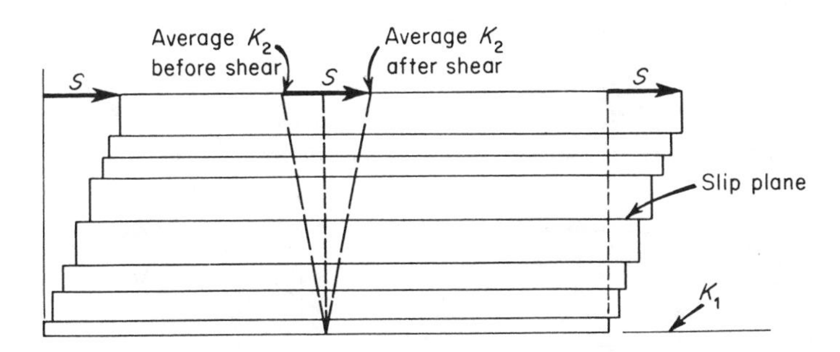

Fig. 16.34 Martensitic shear assuming that deformation occurs by slip.

22. Lieberman, D. S., *Acta Met.*, 6 680 (1958).

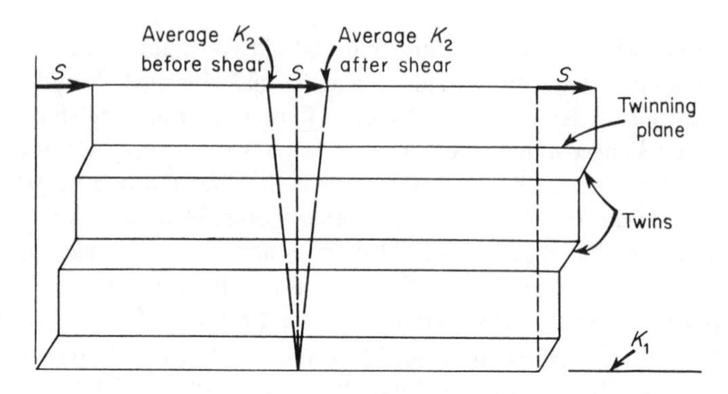

Fig. 16.35 Martensitic shear assuming that deformation occurs by deformation twinning. The required shear is met when only a fraction of the structure twins in a series of bands parallel to the twinning plane.

over a group of slip planes parallel to K_1. In the case of mechanical twinning, what usually happens is that the shear (strain) required to produce the habit plane of the deformation does not equal the shear (strain) which occurs when the product lattice undergoes mechanical twinning. In the present example, mechanical twinning, not slip, furnishes the shear. The twinning occurs in the tetragonal lattice of indium-thallium on {110} planes and with a shear of about 0.06. Notice that the indicated K_1 (101) is a twinning plane. The shear needed on this same plane in order to make angle $p'oq$ equal to poq is one-third of the twinning shear. Clearly, if the entire lattice is twinned after suffering a Bain distortion, it will not be possible to produce the desired undistorted habit plane. This difficulty is resolved in the tetragonal lattice for only part of the structure; twins taking the form of bands parallel to the twinning plane. Thus, the desired macroscopic shear will be obtained if the net width of the twinned areas is one-third that of the untwinned. This is shown in Fig. 16.35, and the resulting

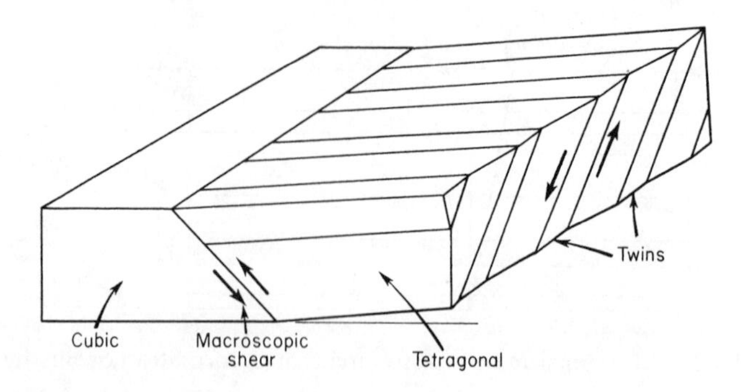

Fig. 16.36 The nature of the twinned tetragonal structure in the indium-thallium martensite transformation.

transformed structure is shown in Fig. 16.36. It is very important to note the difference between the two shears involved in the above discussion. In one case, we have considered the shear parallel to the tetragonal twinning planes. The sense of this shear is shown on the right face of the specimen shown in Fig. 16.36. The other shear is the macroscopic shear of the total transformation. This is the net deformation that we observe when we take the resultant of the three distortions (Bain, shear, and rotation) and represents motion parallel to the habit or interface plane. The latter (plus a distortion normal to the habit plane) is what one observes when one considers the specimen as a whole. As a result of the macroscopic shear, the specimen as a whole is tilted at the habit plane. This is demonstrated in the sketch of Fig. 16.36, which also shows the sense of this shear with the aid of vectors located on the front face of the specimen. The shear associated with twinning is on a very much finer scale than the macroscopic shear and is visible on the surface in the tetragonal structure only where it appears as a series of fine parallel surface tilts. Figure 16.37 shows an actual

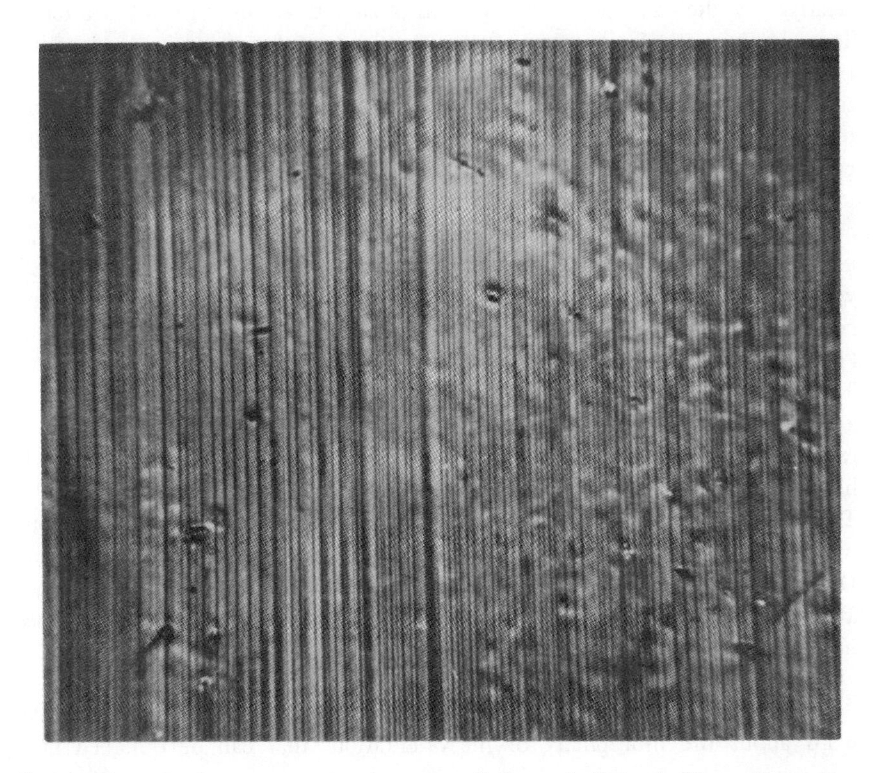

Fig. 16.37 Twinned structure (tetragonal) in an indium-thallium alloy after transformation. (Burkart, M. W., and Read, T. A., *Trans. AIME,* **197** 1516 1953.)

photograph of an indium-thallium alloy illustrating the fine twinned structure in the tetragonal phase. Finally, it should be noted that the twinning which occurs in the tetragonal region is on an extremely fine scale. This is necessary in order to minimize the distortion at the habit plane. A little thought shows that the habit plane on a microscopic scale is not a plane of zero distortion, for where each one of the tetragonal bands intersects the interface (Fig. 16.36), the lattice is distorted as the tetragonal structure tries to match the cubic lattice on the other side of the boundary. The habit plane is actually a plane of zero distortion only on a macroscopic scale. The distortion at the interface corresponding to any one of the tetragonal bands is compensated by that of the neighboring bands which are in a twinned relationship to the first band and distort the interface in the opposite sense. We may thus conclude that the finer the tetragonal twinned bands, the less intense the distortion in the interface.

16.14 Irrational Nature of the Habit Plane. One of the nice features of the Wechsler, Lieberman, and Read theory is that it shows quite clearly why the habit plane in martensitic transformations is usually irrational. This characteristic is in sharp contrast to the twinning plane in deformation twinning which is almost invariably a plane of low rational indices. Let us now reconsider line *oq* in Fig. 16.32. This line is actually defined as the intersection of the first undistorted (K_1) plane of the shear (twinning plane) and the Bain cone. In the figure it is a rational $[\bar{1}\bar{1}1]$ direction. However, line *op* is not a rational direction, but diverges slightly from the $[1\bar{1}1]$ direction. It would coincide with $[1\bar{1}1]$ if the shear were exactly equal to the simple twinning shear on the {101} plane, because then the second undistorted plane K_2 (Fig. 16.32) would be the $(\bar{1}01)$ plane. The required shear is only one-third the twinning shear, as we have seen, and the K_2 plane, therefore, makes a small angle with $(\bar{1}01)$. Since *op* is not a rational direction, neither is *op′* its position after shear. The final result is that since the habit plane is determined by *op′* and *oq* it is not a rational plane [that is, (011)]. The divergence of the habit plane from a rational plane in the indium-thallium alloy is, however, almost negligible, for it lies only 26′ away from the (011) plane[23] and has the indices (0.013, 0.993, 1). The figures illustrating the theory have, of course, been made considerably out of scale for the specific purpose of making the distortions more readily apparent. Distortions of the order used in the figure are, however, observed when larger atom displacements occur in the transformations. The indices of the habit planes in these alloys, accordingly, diverge widely from simple low-induce planes.

16.15 Multiplicity of Habit Planes. It is now necessary to say a word about the multiplicity of the orientations that can be obtained in a

23. Burkart, M. W., and Read. T. A., *Trans. AIME*, 197 1516 (1953).

martensitic transformation. The number of possible habit planes can be quite large in most reactions. In the indium-thallium alloy there are 24 possible habit planes, all of which lie very close to {110} planes (within 26′), which signifies that there are four habit plane orientations closely clustered about each of the six {110} planes.

16.16 The Iron-Nickel Martensitic Transformation. While the above discussion of the martensitic transformation in indium-thallium alloys has been somewhat lengthy, it has touched on many of the important aspects of martensitic transformations. The reversible characteristic of the transformation is duplicated in most other alloys that undergo martensitic transformations. The athermal nature of the transformation is also a typical martensitic property, and the same holds true for the stabilization phenomenon, the irrational nature of the habit plane, and the complexity of the deformation. There remains now, among other things, a consideration of the effect on the transformation of an increase in the deformation associated with the transformation. For this purpose, let us return to the previously mentioned iron-nickel alloy (70 percent Fe–30 percent Ni).[24] Figure 16.38 shows the temperature dependence of the transformation of this alloy (athermal characteristics). In this particular case, the transformation was followed by resistivity measurements. This is possible because the resistance of the body-centered cubic product is lower than that of the face-centered cubic parent phase and the resistance of a mixture of the phases is proportional to the amount of martensite formed. The diagram shows a much larger hysteresis effect than is found in the indium-thallium alloy. In

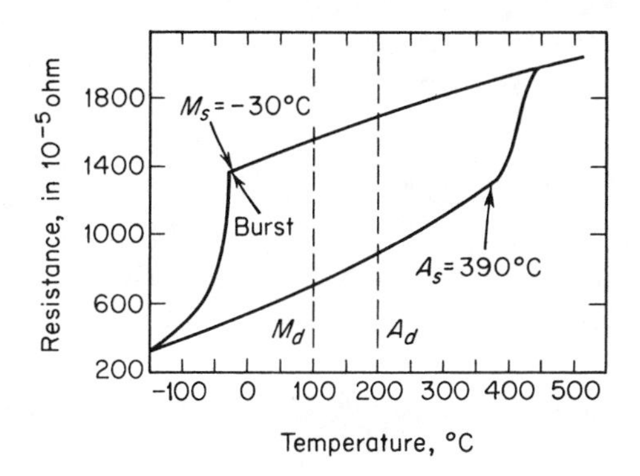

Fig. 16.38 The martensitic transformation in an iron-nickel (29.5 percent Ni) alloy. (From Kaufman, Larry, and Cohen, Morris, *Trans. AIME,* **206** 1393 [1956].)

24. Kaufman, L., and Cohen, M., *Trans. AIME,* **206** 1393 (1956).

indium-thallium the reverse transformation occurs only a few degrees above the forward transformation. In the iron-nickel alloy, martensite begins to form at $M_s = -30°C$. The reverse transformation commences at a temperature designated A_s (austenite start temperature) which is 420°C higher than M_s. The 420°C temperature difference is indicative of the large driving force needed in this alloy to nucleate the transformations.

The difficulty associated with the formation of martensite plates in the iron-nickel alloy appears quite plainly in another respect; the size of the plates which form. These are extremely small and typically ellipsoidal, or lens-shaped. At temperatures slightly below M_s, an interesting phenomenon occurs in these alloys: a large fraction (about 25 percent) of the austenite transforms to martensite in a burst.[25] It occurs so rapidly that a shock wave results which Machlin and Cohen report as capable of occasionally shattering the Dewar flask which contained the transforming specimen suspended by a thread in a refrigerating liquid. This is excellent evidence of the large driving force for the reaction. The plates that form in the burst (Fig. 16.39) are still small in size, but have a somewhat greater thickness than those formed prior to the burst.

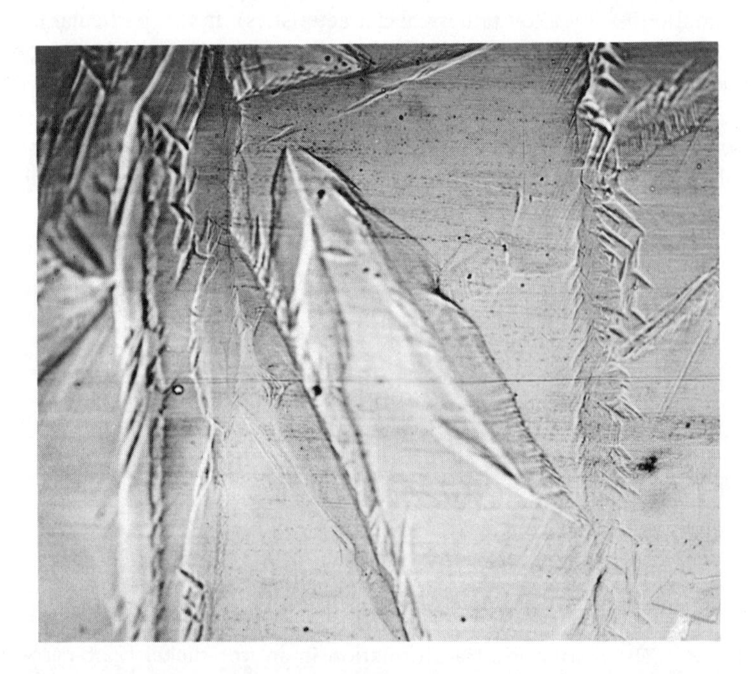

Fig. 16.39 Martensite plates that formed in a burst in an iron-nickel specimen. 500 x . (Courtesy of John F. Breedis.)

25. Machlin, E. S., and Cohen, Morris, *Trans. AIME*, **191** 746 (1951).

The burst phenomenon can be looked upon as an autocatalytic effect. At an early state in the transformation, a point is reached where the formation of a few more plates triggers the formation of a large number of additional plates.

The progress of the athermal transformation in this alloy is quite different from that in the indium-thallium alloy. In the latter, the transformation was accomplished by the movement of a single interface through the length of the entire specimen and a decreasing temperature was required in order to drive the interface. In the iron-nickel specimens, martensite plates, even in a single crystal of the parent phase, form and grow very rapidly to their final size. The transformation can only proceed with the nucleation of additional plates, which is accomplished by lowering the temperature and increasing the driving force.

The limited growth of the martensite plates in this alloy is undoubtedly associated with the magnitudes of the strains connected with their growth. These latter are large enough to cause plastic flow to occur in the matrix surrounding the martensite plate, and it has been proposed[26] that this deformation causes the plates to lose their coherency with the parent lattice. On this basis, it is proposed that growth stops with the loss of coherency.

16.17 Isothermal Formation of Martensite. Martensite has been observed to form isothermally,[27] as well as athermally, in 30 percent iron-nickel specimens. In either case, reaction proceeds as the result of the nucleation of additional plates, and not through the growth of existing plates. Curves showing the amount of isothermal martensite formed as a function of time are given in Fig. 16.40. These curves also serve another useful purpose because they demonstrate the interrelation between the athermal and isothermal transformations. The intersection of each isothermal curve with the ordinate axis gives the amount of martensite formed on the quench to the holding temperature.

16.18 Stabilization. The phenomenon of stabilization is also observed in iron-nickel alloys. The mechanism differs from that observed in the indium-thallium alloy, but the effect is the same; isothermal holding of the specimens at a temperature between M_s and M_f stabilizes the transformation so that additional supercooling is required in order to start the transformation again. In the indium-thallium specimens, stabilization occurs as a result of a retardation of the movement of the interface, whereas in the iron-nickel alloy, stabilization is manifested by an increased difficulty in the nucleation of additional plates. In order for the reaction to continue, an additional increment of driving force is needed to nucleate more plates. Stabilization is also observed during the reverse transformation when martensite is reacted to form austenite.

26. *Ibid.*
27. Machlin, E. S., and Cohen, Morris, *Trans. AIME*, 194 489 (1952).

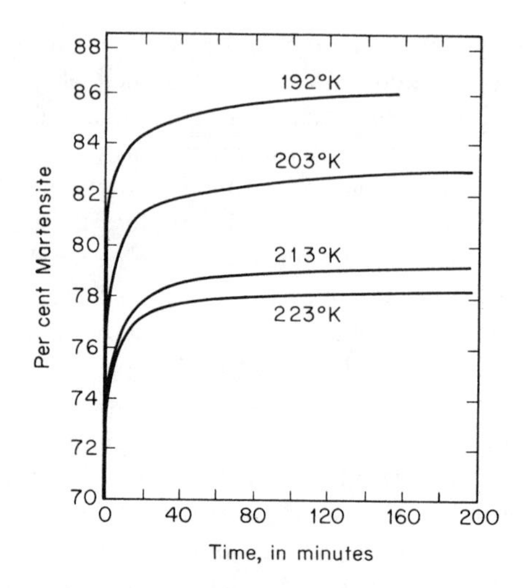

Fig. 16.40 Isothermal transformation subsequent to a direct quench to temperature shown. Notice that the amount of athermal martensite that forms prior to the isothermal transformation increases with decreasing holding temperature. (Fe-29 percent Ni). (From data of Machlin, E. S., and Cohen, Morris, *Trans. AIME*, **194** 489 [1952].)

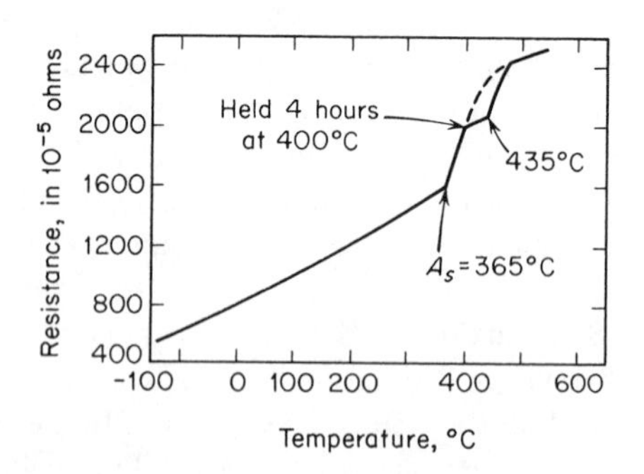

Fig. 16.41 Stabilization during reversed transformation (martensite to austenite) in Fe-29.7 percent Ni. Specimen held at 400°C for 4 hr. Transformation following holding period did not start again until temperature reached 435°C. (Kaufman, Larry, and Cohen, Morris, *Trans. AIME*, **206** 1393 [1956].)

In this case, no observable isothermal reaction occurs, so that the stabilization effect appears clearly in a resistivity-temperature curve. Such a curve is shown in Fig. 16.41.

16.19 Nucleation of Martensite Plates. The nucleation of martensite plates is a subject of great interest and also of considerable controversy. The available experimental evidence favors the belief that martensite nuclei form heterogeneously, as is the case with nucleation in most other transformations, or changes of phases.

The evidence is also strong for the assumption that the locations at which nuclei form are internal positions of high strain (dislocation configurations), commonly called strain embryos.

No simple theory of martensite nucleation is yet available. This is probably due to the fact that martensite can form under two extremely different sets of conditions. First, it can form athermally, which means that it can form in a very small fraction of a second provided the temperature is lowered sufficiently to activate those nuclei which will respond in this manner. This type of nucleation apparently does not need thermal activation. The ability of martensite to form without the aid of thermal energy is also demonstrated by the fact that martensite plates can be nucleated in some alloys at temperatures as low as 4°K, where the energy of thermal vibrations is vanishingly small. On the other hand, martensite is also capable of forming at constant temperature. The fact that, in this case, it has a nucleation rate which is time dependent shows that thermal energy can also be a factor in the nucleation of martensite.

16.20 Growth of Martensite Plates. The growth of martensite plates does not have the twofold nature of nucleation, for here it appears that thermal activation is not a factor. The evidence leading to this conclusion includes the fact that plates grow to their final sizes with great rapidity. Measurements show that the growth velocity is of the order of one-third the velocity of sound in the matrix.[28] To this must be added the fact that the growth velocity is independent of temperature. If the growth of martensite plates occurred as the result of atoms jumping over an energy barrier from the parent phase to the product phase, then the jump rate should be a decreasing function of the temperature, and at some finite temperature a noticeable dropping off in the martensitic growth rate should become apparent. This does not occur. Martensite has been observed to form with great rapidity even at 4°K.[29]

16.21 The Effect of Stress. Because the formation of martensite plates involves a change of shape in a finite volume of matter, applied stress can influence the reaction. This is entirely analogous to the formation of deformation twins by stress. However, because the formation of a martensite plate involves both shear and normal components of strain, the dependence on stress is more complicated than in the case of twinning. Nevertheless, theoretical predictions of the effect of various stress patterns on the formation of

28. Bunshah, R. F., and Mehl, R. F., *Trans. AIME,* 197 1251 (1953).
29. Kulin, S. A., and Cohen, M., *Trans. AIME,* 188 1139 (1950).

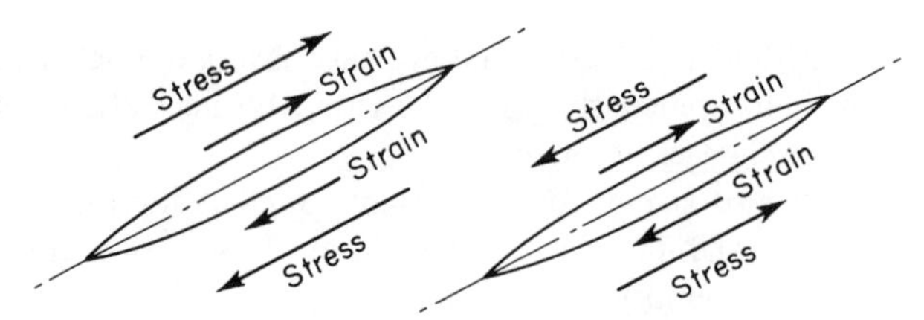

Fig. 16.42 The effect of external stress on martensite plate formation.

martensite have been found to agree quite well with experiment.[30] We shall not discuss this theory in detail, but it is important to note that the M_s temperature can be either raised or lowered as a result of applied stresses. This may be understood in terms of the following, perhaps oversimplified picture. Suppose, as shown in Fig. 16.42, the macroscopic strain associated with the formation of a martensite plate is a pure shear (normal component zero). Then if the sense of an applied shear stress is the same as the strain, the stress should aid the formation of the plate. A lower driving force for the reaction is to be expected and M_s should be raised. Similarly, if the shear-stress vector is reversed, the formation of the plate becomes more difficult and the temperature at which the plate forms should be lowered. It is important to note that, in respect to this latter, a simple applied shear stress may not necessarily lower M_s because of the multiplicity of the habit planes on which martensite can form. While the indicated plate might not be favorably oriented to the stress, it is quite probable that there are other plate orientations in the crystal that are.

16.22 The Effect of Plastic Deformation. Plastic deformation of the matrix also has an effect on the formation of martensite which is primarily to increase the sizes of internal strains and make the nucleation of martensite easier. As a result, martensite can form when the metal is plastically deformed at temperatures well above the M_s temperature. The amount of martensite thus formed decreases, however, as the temperature is raised, and it is common practice to designate the highest temperature at which martensite may be formed by deformation as the M_d temperature. In reversible martensitic transformations, plastic deformation usually has a similar effect on the reverse transformation. The temperature at which the reverse transformation starts is lowered by plastic deformation. The M_d and A_d (austenite start temperature for plastic deformation) are shown plotted as vertical dashed lines in Fig. 16.38. Notice that plastic deformation brings the start of the forward and reverse transformations much closer together – within approximately 100°C. The corresponding difference between M_s and A_s is 420°C.

30. Patel, J. R., and Cohen, Morris, *Acta Met.*, **1** 531 (1953).

Problems

1. With the aid of the data given near the end of Section 16.2, compute the twinning shear of a $\{10\bar{1}2\}$ twin in a
   (*a*) magnesium single crystal
   (*b*) zinc single crystal
   (*c*) beryllium single crystal (see Appendix B).
2. Would you expect a beryllium crystal to twin on a $\{10\bar{1}2\}$ plane if the stress axis is parallel to $<10\bar{1}0>$? Explain.
3. $\{10\bar{1}1\}$ and $\{10\bar{1}3\}$ have been observed as twinning planes in magnesium crystals. These twins are reciprocal twins, which means that K_1 and K_2 can be interchanged; that is, for $\{10\bar{1}1\}$ twinning, if K_1 is $(10\bar{1}1)$, then K_2 is $(10\bar{1}3)$. Similarly, if K_1 is $(10\bar{1}3)$, K_2 is $(10\bar{1}\bar{1})$.
   (*a*) Make a sketch of a unit cell showing K_1 and K_2 for twinning on the $(10\bar{1}\bar{1})$ plane.
   (*b*) Compute the twinning shear.
   (*c*) Compute the twinning shear for a $(10\bar{1}3)$ twin.
4. (*a*) For a stress axis lying in the basal plane, would you expect that $\{10\bar{1}\bar{1}\}$ twins should form under a tensile stress or a compressive stress? Explain.
   (*b*) Now consider $\{10\bar{1}3\}$. Will these twins form under a tensile or a compressive stress parallel to the basal plane? Explain.
5. Face-centered cubic metals twin on $\{111\}$ planes. Both K_1 and K_2 are of the same form; that is, $\{111\}$.
   (*a*) How many different ways can a twin form on a specific octahedral plane, say the (111) plane?
   (*b*) Give the Miller indices for η_1 for each of the answers to part (*a*) of this question. In working out your answer, the Thompson tetrahedron, shown in the figure for Problem 5.8, may be of help.
   (*c*) How many possible twinning $\{111\}$ modes are there in a face-centered cubic crystal?
6. (*a*) Draw a standard (111) stereographic projection of a face-centered cubic crystal showing the poles of the four $\{111\}$ planes. Assume that twinning occurs with (111) as K_1, and $(11\bar{1})$ as K_2. Draw in the great circle corresponding to the K_2 plane and identify the η_1 and η_2 directions; that is, label with the proper Miller indices.
   (*b*) Assume that the indicated twin forms as a twin of the first kind and rotate the plotted data in the stereographic projection into their proper orientations in the twin.
7. Repeat the above problem assuming that the twin forms as a twin of the second kind.
8. (*a*) Compute the magnitude of the twinning shear of a face-centered cubic twin.
   (*b*) How does the magnitude of this shear compare with that for the twins in the h.c.p. metals computed above? Do you think that it will be easier to accommodate in the matrix the strain of a f.c.c. twin or one of the hexagonal metal twins?

9. In body-centered cubic metals, twins form on {112} planes and <111> is the shear direction.
   (*a*) Make a standard (100) stereographic projection of a cubic crystal showing all of the <111> directions and all of the {112} plane poles.
   (*b*) How many {112} planes should pass through each <111> direction?
   (*c*) On the stereographic projection draw in the great circles corresponding to the {112} planes that pass through the [111] direction.
   (*d*) In a badly deformed iron crystal, what would be the maximum number of twin traces that you could see?
   (*e*) In a badly deformed f.c.c. crystal, what would be the maximum number of visible twin traces? Explain.
10. This diagram represents a face-centered cubic crystal with a rectangular cross-section. The crystal has twinned on three planes and the twin traces have been measured with respect to a vertical edge, or the stress axis of the

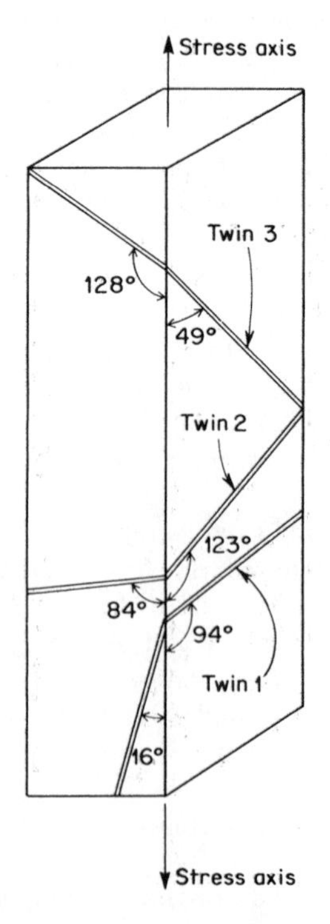

crystal. The angles thus obtained are shown in the figure. Orient this crystal by the two-surface technique following the steps listed below.
   (*a*) Lay out a stereographic projection on a sheet of tracing paper with the front face of the crystal as the basic circle and the top of this circle the stress axis.

(*b*) Around the basic circle, plot the twin trace orientations corresponding to the front face.
(*c*) Draw in the great circle corresponding to the side of the crystal and plot on the circle the corresponding twin trace orientations.
(*d*) Draw in the three great circles representing the three twinning planes. Plot the poles of these three planes.
(*e*) From the geometry of the f.c.c. crystal structure, determine the orientation of a cube pole; that is, {100}. Plot this on the figure.
(*f*) Rotate the stereographic projection thus obtained into a standard {100} projection, making sure that the stress axis is also rotated. In order to simplify the result, this last step is best performed on a second sheet of tracing paper.
(*g*) Draw in the boundaries of the standard stereographic triangle that surrounds the stress axis, thus defining the stress axis orientation.

11. This schematic diagram represents a cross-section cut through an iron crystal that was deformed at room temperature just before sectioning. The

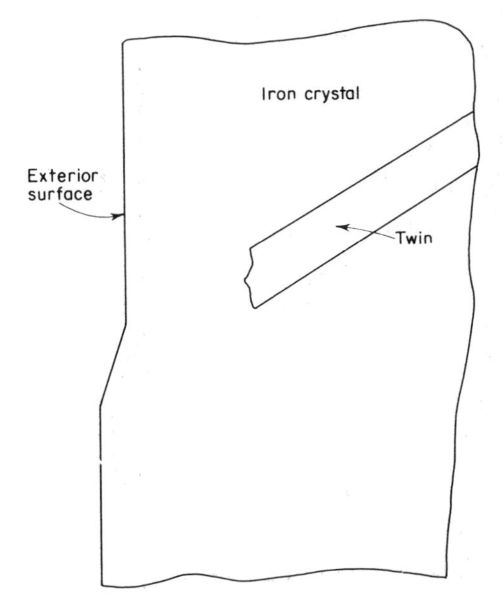

twin that is revealed has its shear direction parallel to the plane of the paper. Note that ahead of the twin, the surface has undergone a displacement.
(*a*) Explain how this displacement probably occurred.
(*b*) How is it possible for the twin boundary at the end of the twin to be blunted instead of tapering to a sharp edge?

12. (*a*) Compute the twinning shear of a $\{11\bar{2}1\}$ twin in zirconium. The angle between $\{11\bar{2}1\}$ and (0001) is 72.5°.

(*b*) Compute the tilt of the specimen surface where it is intersected by a $\{11\bar{2}1\}$ twin if the surface is the basal plane.

13. In a tetragonal crystal, twinning normally occurs with K_1 and K_2 both planes of the form $\{110\}$. Twins cannot form on all of the $\{110\}$ type planes. Identify the planes on which the twins can form by giving their Miller indices. (In describing the planes of this crystal, it is normal practice to let the third digit of the Miller indices represent the c, or longer axis).
14. In considering the phase-change mechanism known by his name, Bain showed that the face-centered cubic unit cell could be considered as a body-centered tetragonal unit cell.
    (*a*) What is the c/a ratio of this unit cell?
    (*b*) How many twinning modes should this tetragonal cell have?
    (*c*) Compute the twinning shear for this tetragonal cell.
    (*d*) When this same crystal structure is considered as face-centered cubic, what is the twinning shear?
    (*e*) How many twinning possibilities are there in the face-centered cubic structure?
    (*f*) Are the two types of twinning related?
    (*g*) Rationalize answers *b* and *e* with respect to each other.

17 The Iron-Carbon System

We shall now consider alloys of iron and carbon in some detail. There are several reasons for this: first, carbon steels constitute by far the greatest tonnage of metal used by man; second, no other alloy system has been studied in such detail; and third, the solid-state phase changes that occur in steel are varied and interesting. Further, it is becoming increasingly evident that the solid-state reactions of the iron-carbon system are similar in many respects to those that occur in other alloy systems. A study of the iron-carbon system is valuable, not only because it helps explain the properties of steels, but also as a means of understanding solid-state reactions in general.

The iron-carbon phase diagram, shown in Fig. 17.1, is not a complete diagram in that it is plotted only for concentrations (in weight percent) less than 6.67 percent carbon, the composition of Fe_3C, or cementite. The latter is an intermetallic compound with negligible solubility limits, and the iron-carbon diagram can be divided at its composition into two independent parts. That part of the diagram containing carbon concentrations higher than 6.67 percent has little commercial significance and is usually ignored.

Figure 17.1 is not a true equilibrium diagram because cementite is not an equilibrium phase. Graphite is more stable than cementite and, under the proper conditions, cementite decomposes to form graphite. In ordinary steels this decomposition is almost never observed because the nucleation of cementite in iron, supersaturated with carbon, occurs much more readily than the nucleation of graphite. Thus, when carbon is precipitated from solid solutions of alpha (body-centered cubic) or gamma (face-centered cubic) iron, the resulting precipitate is almost always cementite, or some other carbide, and not graphite. Once cementite has formed it is very stable and may be treated for practical purposes as an equilibrium phase. For these reasons, we can use Fig. 17.1 to predict phase changes in iron-carbon alloys undergoing slowly varying temperature cycles.

The iron-carbon diagram is characterized by three invariant points: a peritectic point at 0.16 percent C and 1493°C, an eutectic point at 4.3 percent C and 1147°C, and an eutectoid point at 0.80 percent C and 723°C. The peritectic transformation occurs at very high temperatures and in steels of very low-carbon concentration (the upper left-hand corner of the phase diagram).

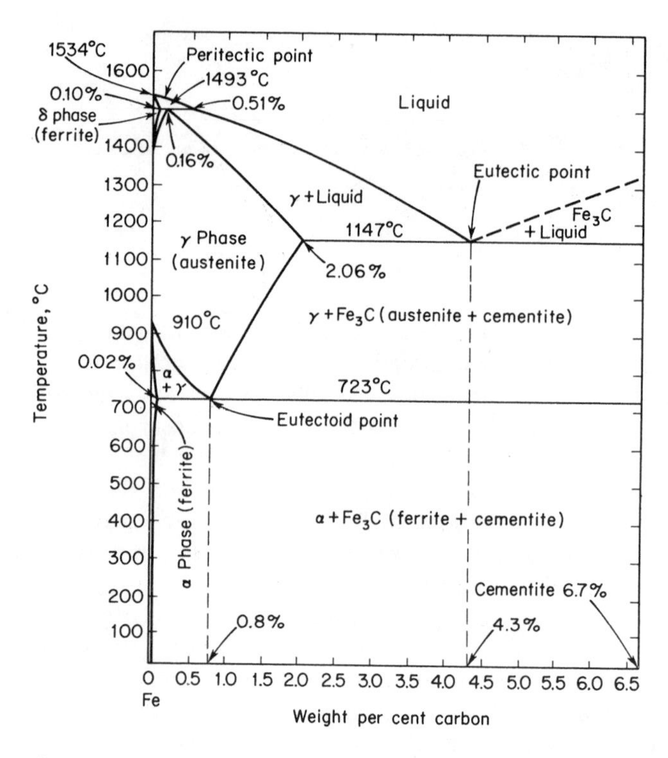

Fig. 17.1 The metastable system Fe-Fe_3C. (From *Constitution of Binary Alloys*, by Hansen, Max, and Anderko, Kurt. Copyright, 1958. McGraw-Hill Book Co., Inc., New York, p. 353. Used by permission.)

The peritectic transformation has only secondary effects on the structures of steels at room temperature. All compositions which, as they freeze, pass through the peritectic transformation region, enter the single-phase face-centered cubic field. Except for coring effects (segregation) caused by the complexities of the peritectic transformation, these alloys are equivalent to higher carbon compositions (above 0.51 percent C) that freeze directly to the gamma or face-centered cubic phase. With the assumption that sufficient time is allowed, as the alloys cool, to permit diffusion to bring about a homogeneous solid solution, we may disregard the peritectic transformation in considerations of phase transformations which occur at lower temperatures.

The face-centered cubic solid solution, or gamma phase, is given the identifying name *austenite*. A study of the phase diagram shows that all compositions containing less than 2.06 percent C pass through the austenite region on cooling from the liquid state to room temperature. Alloys in this interval are arbitrarily classed as steels. Actually, most carbon steels contain less than 1 percent carbon, with the greatest tonnages produced in the range 0.2 to 0.3 percent C (structural steels used in buildings, bridges, ships, etc.). Only in very rare instances is steel used with more than 1 percent C (razor blades, cutlery, etc.) and then the composition never rises more than a few tenths of one percent above one percent. Compositions above 2 percent are classed as cast irons. However, it is necessary to notice that cast irons of commerce are not simple alloys of iron and carbon, for they contain relatively large quantities of other elements, the most prevalent of which is silicon. In general, cast irons are better considered as ternary alloys of iron, carbon, and silicon. They also differ in another important respect from steels. The presence of silicon promotes the formation of graphite in them. As a result, cast irons may contain carbon in the form of both graphite and cementite. This fact signifies an important difference between cast irons and steels, for the latter contain only combined carbon in the form Fe_3C.

While the phase transformations associated with the eutectic point at 4.3 percent C are interesting and useful in a detailed study of the structures in cast irons, time does not permit us to consider this section of the phase diagram.

A detailed analysis of the phase transformations at the peritectic and eutectic points of the iron-carbon system are not essential in a study of steels. We shall focus our attention, therefore, on the phase transformations that occur in the eutectoid region of the phase diagram. This study is not simple, because we have to consider not only equilibrium phase transformations, but also those that occur under nonequilibrium conditions. With the assumption of slow-cooling, the phase changes in steels can be predicted with considerable accuracy using the equilibrium diagram. On the other hand, when the transformations do not occur under equilibrium conditions, such as when the metal is cooled rapidly, new and different metastable phases and constituents form. Since these structures are important to the theory of the hardening of steels, the kinetics and principles of their formation are of considerable importance.

Because the microstructures obtained when austenite is cooled to room temperature depend not only on the nature of the cooling cycle, but also on the original carbon concentration of the austenite, we shall first consider transformations in a single composition. For this we shall take the simplest to study, namely the eutectoid composition 0.8 percent C which, on slow cooling, completely transforms to the eutectoid structure. Afterwards, we shall treat those compositions lying to either side of the eutectoid point.

17.1 The Transformation of Austenite to Pearlite. In practice, steels are almost always cooled from the austenite region to room temperature in such a manner that the temperature falls continuously. Thus, annealed steels are slowly cooled by leaving them in a furnace after the power to the furnace has been shut off, or a normalized steel is cooled by removing the red-hot metal from the furnace and allowing it to cool in air. In continuous cooling processes, the nature of the reaction changes with the decreasing temperature. This change is particularly marked when the rate of cooling is appreciable. It goes almost without saying that the resulting microstructures are very difficult to analyze. Much more readily interpretable specimens are obtained when austenite is allowed to transform at constant temperature. True isothermal transformation is possible because of the relatively small heat of transformation involved; normally of the order of 1 kcal per mol, which is about one-fourth of the latent heat of fusion (L_F for pure iron = 3.7 kcal per mol). In addition, the use of small specimens and the relatively slow reaction rates observed when austenite decomposes, allow transformation heat to be removed fast enough to prevent appreciable temperature rises.

Figure 17.2 illustrates a simple experimental method of studying isothermal austenitic transformations. As may be seen by examining Fig. 17.3 (an enlarged section of the iron-carbon diagram containing the eutectoid point), a steel of eutectoid composition is austenitic at temperatures above 723°C. Thus, if a small furnace containing a crucible filled with a molten salt mixture is held at 730°C, eutectoid-composition steel specimens placed in this bath can be maintained in the austenite phase and at a temperature just above that at which they undergo the eutectoid reaction. This furnace is shown on the left in Fig. 17.2. Just to the right of this unit is a similar furnace which also contains a salt bath but maintains it at a temperature below 723°C. In an experiment of this type, it is convenient to use specimens in the shape of flat discs about the size of a dime. A short length of temperature-resistant wire forms a convenient handle for use in moving the specimens from furnace to furnace. Metal pieces of this size, when placed in liquid salt baths, very quickly attain the temperature of the bath. Consequently, if a specimen originally in the left-hand furnace is

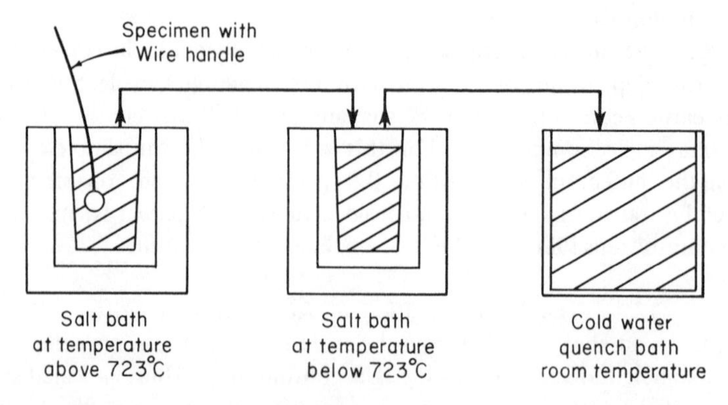

Fig. 17.2 Simple experimental arrangement for determining the kinetics of isothermal austenitic transformations.

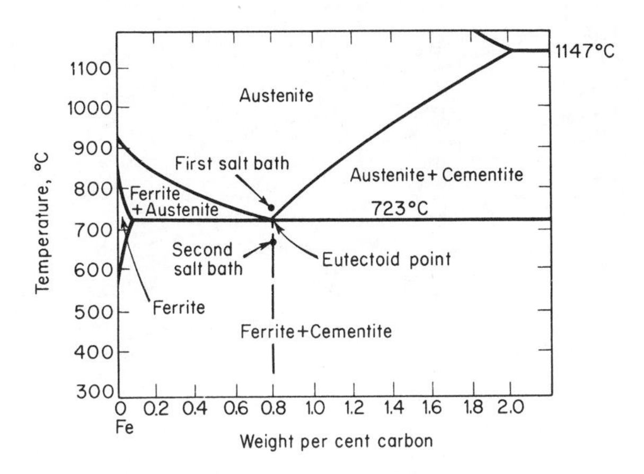

Fig. 17.3 The eutectoid section of the iron-carbon diagram.

quickly removed and inserted into the furnace to its right, we can assume that its temperature changes instantly from just above the eutectoid temperature to just below it. Since austenite is no longer stable below 723°C, it is now able to decompose into other phases. If this solid-state reaction is carried out at temperatures not too far below the eutectoid temperature, the reaction occurs by nucleation and growth and is, therefore, time dependent.

The isothermal decomposition of the austenite is usually followed with the aid of a number of specimens (generally about ten), all of which are quenched simultaneously from the upper temperature to the lower temperature. Individual specimens are then removed from the second bath at increasing time intervals (usually measured on a logarithmic scale) and quickly cooled to room temperature by quenching them in a bath of cold water. This latter fast-cooling effectively stops the isothermal reaction and any austenite still untransformed, at the instant of the quench, undergoes a martensitic (athermal) transformation as the specimen approaches room temperature. The nature of this martensitic transformation will be covered in detail later. It suffices to state here that it is of the same basic type as the iron-nickel transformation considered in Chapter 16. Fortunately, steel martensite has a different appearance under the microscope than the high-temperature reaction products of austenite. After a suitable metallographic polish and etch, the specimens, prepared as outlined above, may be examined metallographically and the relative amount of the isothermal-transformation product determined in each case.

17.2 Pearlite. If austenite is allowed to transform isothermally at temperatures just below 723°C, the reaction product is the same as that predicted by the iron-carbon diagram for a very slow, continuous cooling

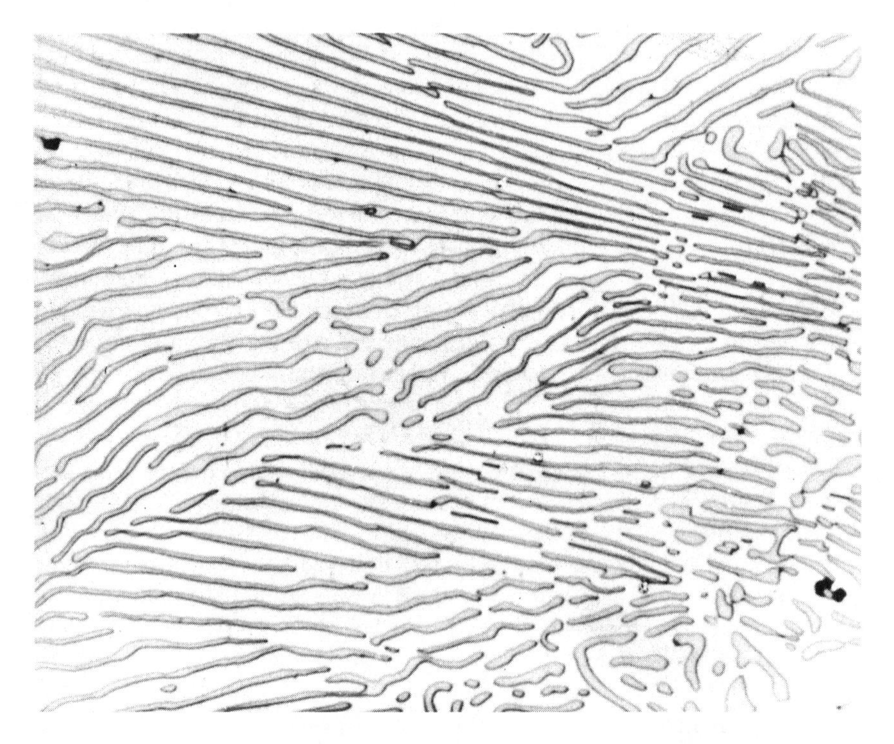

Fig. 17.4 Pearlite consists of plates of Fe_3C in a matrix of ferrite. (Vilella, J. R., *Metallographic Technique for Steel,* ASM Cleveland, 1938.) 2500x.

process. As may be seen by examining Fig. 17.3, the stable phases below the eutectoid temperature are ferrite and cementite, and the eutectoid structure is a mixture of these phases. This constituent, called pearlite, consists of alternate plates of Fe_3C and ferrite, with ferrite the continuous phase. An example of a pearlite structure is shown in Fig. 17.4. Pearlite is not a phase, but a mixture of two phases – cementite and ferrite. It is, nevertheless, a constituent because it has a definite appearance under the microscope and can be clearly identified in a structure composed of several constituents. (See Fig. 17.28.) It should be further recognized that when austenite of eutectoid composition is reacted to form pearlite just below the eutectoid temperature, the two phases appear in a definite ratio. This ratio is easily computed by using the lever rule and assuming that ferrite contains zero percent carbon.

$$\text{Percent ferrite} = \frac{6.67 - 0.8}{6.67} \simeq 87.5\%$$

$$\text{Percent cementite} = \frac{0.8}{6.67} \simeq 12.5\%$$

Since the densities of ferrite and cementite are approximately the same (7.86 and 7.4 respectively), the lamellae of iron and Fe_3C have respective widths of about 7 to 1.

The decomposition of austenite to form pearlite occurs by nucleation and growth. As in almost all other cases, nucleation occurs heterogeneously and not homogeneously. If the austenite is homogeneous (uniform in composition), nucleation occurs almost exclusively at grain boundaries. When it is not homogeneous, but has concentration gradients and contains residual iron carbide particles, nucleation of pearlite can occur both at the grain boundaries and in the centers of austenite grains.[1] This would be the case if austenite, once reacted to pearlite, were re-austenitized for too short a time and then decomposed to pearlite a second time.

No one apparently has observed the actual nucleation of a pearlite colony, so that the mechanism by which pearlite forms can only be surmised. It is also not generally agreed as to whether the first step in the formation of a colony of pearlite is the appearance of a small plate of cementite or of a small plate of ferrite.

Let us assume, after the early proposals of Mehl,[2] that the active nucleus is a small lamella of cementite which forms at an austenite grain boundary and grows into one of the austenite grains. As this small plate, which is shown as part 1 in Fig. 17.5, grows in length and thickness, it removes carbon atoms from the

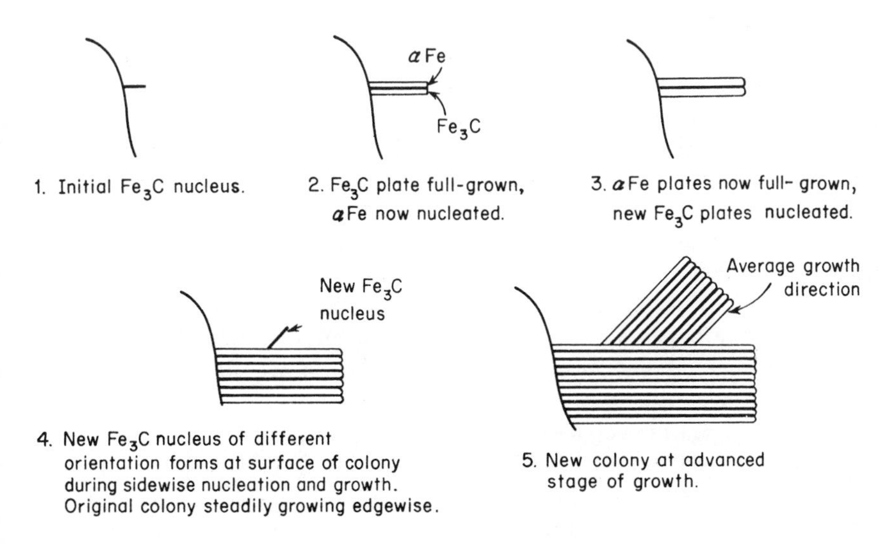

Fig. 17.5 The nucleation and growth of pearlite. (Mehl, R. F., *Trans. ASM*, **29**, 813 [1941].)

1. Mehl, R. F., and Hagel, W. C., *Prog. in Metal Phys.*, 6 74 (1956).
2. Mehl, R. F., *Trans. ASM,* 29 813 (1941).

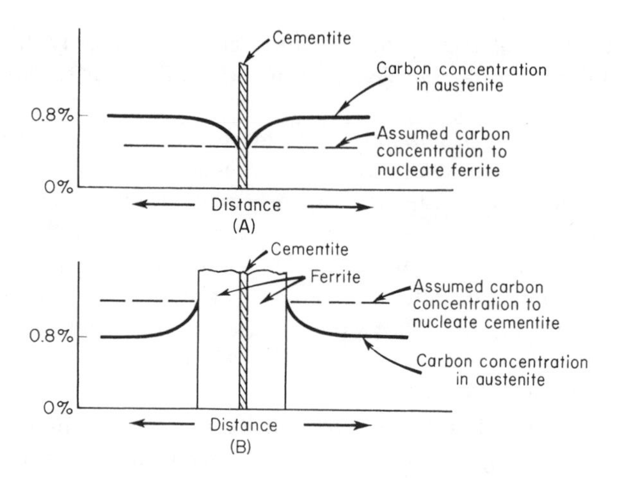

Fig. 17.6 Growing cementite and ferrite lamellae may nucleate each other.

austenite on either side of it. The carbon concentration of the austenite in contact with the cementite falls as a result, and eventually a carbon-concentration contour is achieved in the austenite that has the general shape of the curve shown in Fig. 17.6A. When the composition of the austenite next to the cementite reaches some more or less fixed value, ferrite nucleates and grows along the surface of the cementite plate (Fig. 17.5 at 2). Since these ferrite lamella can contain almost no carbon, their continued growth is associated with a build-up of carbon at the ferrite-austenite interface, as is indicated in Fig. 17.6B. This build-up continues until a new layer of cementite nucleates. Growth of this new cementite layer will, in turn, induce the formation of a new layer of ferrite, and as this process continues, the alternating lamellae of a pearlite colony are formed.

When austenite is reached to form pearlite at a constant temperature, the spacings between adjacent lamellae of cementite are very nearly constant. This is true not only for plates in a given colony, but for the plates in different colonies. An interesting experimental fact[3] is that in a given colony of pearlite, the cementite plates have a common orientation in space, and the same is also true of the ferrite plates.

Growth of pearlite colonies occurs not only by the nucleation of additional lamellae, but also through an advance at the ends of the lamellae. This fact is implied in the parts of Fig. 17.5, where, in each case, at a more advanced state in the development of the pearlite colony, both the number and the length of the lamellae are increased. In general, it has been observed that the two forms of

3. Mehl, R. F., and Hagel, W. C., *Prog. in Metal Phys.*, 6 74 (1956).

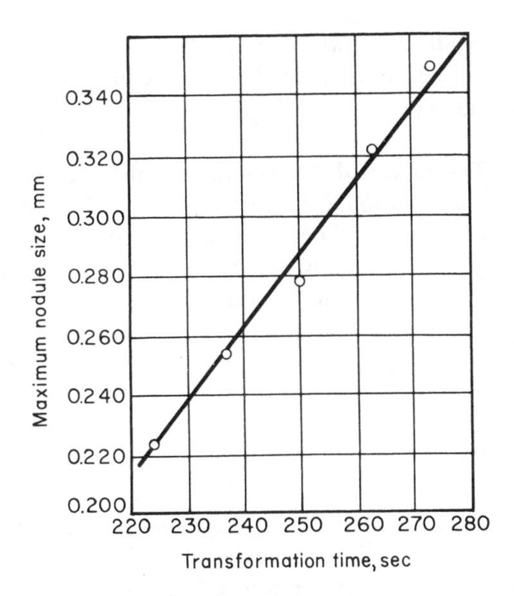

Fig. 17.7 Plot of data showing the linearity of the growth rate of pearlite. Eutectoid high-purity steel transformed at 708°C. (From Frye, J. H., Jr., Stansbury, E. E., and McElroy, D. L., *Trans. AIME,* **197** 219 [1953].)

growth are approximately equal. Finally, pearlite growth also involves the nucleation of new colonies at the interfaces between established colonies and the parent austenite. This step is indicated in part 4 of Fig. 17.5, where a single cementite lamella is shown growing outward from a ferrite lamella of the original colony. The new colony, once it has nucleated, grows by the same procedure as the original colony. (See part 5 of Fig. 17.5.) In this manner, through the addition of new colonies in combination with the growth of older colonies, a single pearlite nucleus located at an austenite grain boundary may grow into a large group of contiguous colonies. The resulting structure is called a *pearlite nodule*. Because pearlite colonies have almost equal rates of growth in directions parallel and perpendicular to the lamellae, a pearlite nodule is usually spherical in shape. On a plane surface, such as that seen under a microscope, a fully developed pearlite colony usually has a circular shape.

Pearlite colonies grow unimpeded until they impinge on adjacent colonies. During this period, the rate of growth, as determined experimentally, is constant. This fact is illustrated in Fig. 17.7. Data for a curve of this type are obtained by reacting, at a given temperature, a number of specimens for different lengths of time. The specimens are prepared for metallographic examination and the diameter of the largest pearlite nodule in each specimen measured. The slope of the curve, obtained when the diameter is plotted against the reaction time, equals the growth rate G. After impingement, pearlite nodules can only grow into austenite remaining between nodules, which constitutes the last stage of pearlite growth.

17.3 The Effect of Temperature on the Pearlite Transformation.

Interlamellar Spacing and the Rate of Growth. The interlamellar spacing S_p is a constant in pearlite formed from austenite at a fixed temperature and in a given specimen, varying only slightly about a mean. This mean value is generally referred to in speaking of the interlamellar spacing of pearlite. The actual variation about the average is considerably less than one would assume from an examination of a metallographic surface, where the spacing between lamellae may appear to vary widely. This is caused by the fact that the plane of the surface does not intersect all pearlite colonies at the same angle. The true spacing is only observed when the lamellae are perpendicular to the surface. When the lamellae make small acute angles with the surface, the separation between adjacent plates of cementite appears to be increased and the slight curvature that often occurs in pearlite lamellae is also amplified. Finally, the interlamellar spacing of pearlite is not structure sensitive.[4] In simple eutectoid iron-carbon alloys reacted to pearlite at a fixed temperature, the distance between the lamellae will be independent of the original size of the austenite grains and of any reasonable lack of homogeneity or uniformity in the composition of austenite.

The temperature at which austenite is transformed has a strong effect on the interlamellar spacing of pearlite. The lower the reaction temperature, the smaller S_p. The spacing of the pearlite lamellae has a practical significance because the hardness of the resulting structure depends upon it; the smaller the spacing, the harder the metal. Pearlite formed from austenite at temperatures just below the eutectoid temperature (700°C) has a pearlite spacing of the order of 10^{-3} mm. The hardness of this structure is about Rockwell C-15, while pearlite formed at 600°C has a spacing of the order of 10^{-4} mm and a correspondingly greater hardness, Rockwell C-40.

The rate of growth of pearlite (G) is also a strong function of temperature, as can be seen by examining Fig. 17.8. At temperatures just below the eutectoid, the growth rate increases rapidly with decreasing temperature, reaching a maximum at 600°C, and then decreases again at lower temperatures.

The functional relationships between rate of growth and temperature, as well as that between interlamellar spacing and temperature, have been of considerable interest to theoreticians. A number of theories have been developed,[5] but there is no general agreement with regard to the relationships governing the pearlite spacing and the rate of growth. In nearly all cases, the problem has been

4. *Ibid.*
5. *Ibid.*

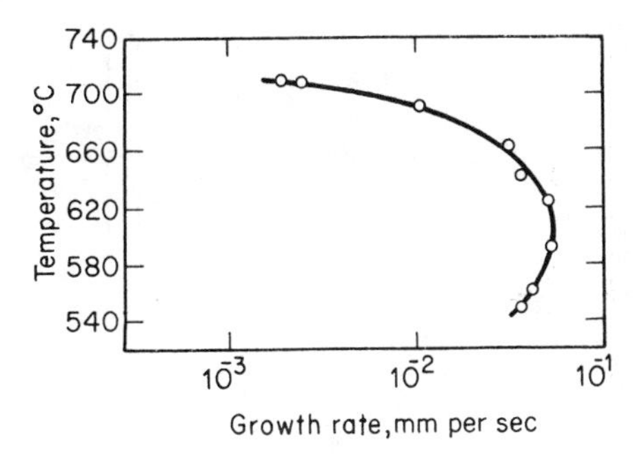

Fig. 17.8 Rate of growth (G) of pearlite as a function of reaction temperature in a high-purity iron-carbon alloy of eutectoid composition. (From Frye, J. H., Jr., Stansbury, E. E., and McElroy, D. L., *Trans. AIME,* **197** 219 [1953].)

approached from a consideration of the factors involved in the edgewise growth of pearlite colonies (growth occurring at the ends of the lamellae). At this interface between the pearlite colony and the parent austenite, growth is primarily a steady-state phenomenon. The interface moves into austenite of constant composition with a constant velocity and leaves behind it parallel bands of ferrite and cementite of constant width. In spite of this characteristic, the boundary conditions are not simple or even completely defined, as may be seen by the following argument. At a relatively small distance into the austenite, the carbon concentration is constant or uniform. Yet, on passing through the interface into the pearlite colony, the carbon atoms become segregated into cementite lamellae. Thus, in a very small volume near the interface, a large movement of carbon atoms must occur in order to form the cementite layers. It has been variously suggested that the necessary flow occurs either in the austenite, through the ferrite, or possibly along the austenite-pearlite interface. All three of these possibilities are indicated in Fig. 17.9. Most theories have been based on the first possibility, where diffusion is considered to be almost entirely through the austenite. A good argument can also be made for the ferrite paths. While the carbon content of ferrite is extremely low, diffusion rates in ferrite are of the order of 100 times more rapid than in austenite.

For the sake of the present discussion, let us consider that the necessary flux of carbon atoms occurs in the austenite ahead of the moving interface. This flow of carbon atoms is caused by concentration gradients that develop on the austenite side of the interface. As iron atoms cross the interface to form ferrite, carbon atoms are rejected back into the austenite. This leads to an increase in carbon content in the austenite facing the ferrite. At the same time, at those positions of the interface where cementite forms, large numbers of carbon atoms are absorbed into this phase, causing the austenite facing the cementite to

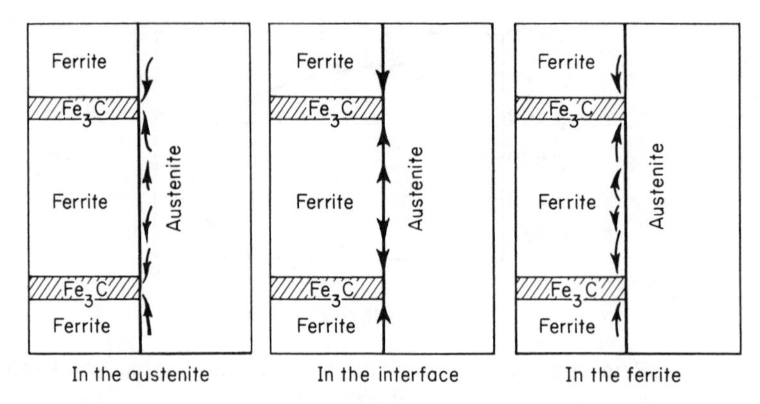

Fig. 17.9 Possible diffusion paths for carbon atoms during pearlite growth.

become depleted in carbon. A falling concentration gradient thus develops between a point opposite the center of a ferrite lamella to a point opposite a neighboring cementite lamella. The magnitude of this concentration gradient can be estimated with the aid of Hultgren's extrapolation. Thus, in Fig. 17.10, which is a reproduction of the eutectoid region of the iron-carbon diagram, the lines corresponding to equilibrium between cementite and austenite, and ferrite and austenite, are extrapolated below the eutectoid temperature. The horizontal dashed line in this figure corresponds to an isothermal-reaction temperature. The intersection of this isothermal with the extrapolated equilibrium lines gives two compositions. That on the right, marked C_a, represents the composition of the austenite that should be in equilibrium with ferrite, while that on the left,

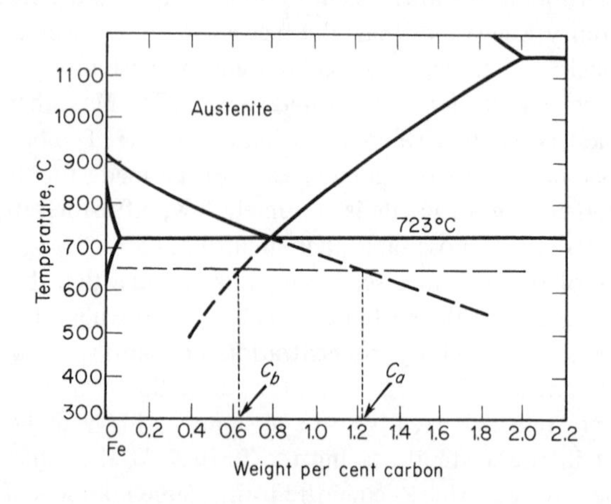

Fig. 17.10 Hultgren's extrapolation.

marked C_b, represents the composition of austenite that should be in equilibrium with cementite. Compositions C_a and C_b can be assumed to represent rough approximations of the compositions of the austenite opposite the centers of ferrite and cementite lamellae respectively.

In Chapter 10 it was shown that the basic equation of diffusion is Fick's first law:

$$J = -DA \frac{dn}{dx}$$

where J is the flux or amount of matter crossing an area A per second under the action of a concentration gradient dn/dx, and D is the diffusion coefficient. We shall now apply this expression to derive Mehl and Hagel's[6] simplified equation for the growth rate of pearlite. An advancing pearlite colony is shown schematically in Fig. 17.11. Two dashed lines, each marked by the symbols xy, are shown in the figure extending from the interface into the austenite. These lines intersect the midpoints of two ferrite lamellae surrounding a single cementite lamella. To a good approximation, it can be assumed that during the growth of pearlite all of the carbon atoms inside of the volume defined by the lines xy diffuse at a steady rate into the cementite lamella. Because the carbon concentration of the cementite is fixed, its growth rate (and that of pearlite) is directly proportional to the number of carbon atoms per second leaving the austenite and joining a cementite lamella. The growth rate is also inversely proportional to the cross-section area of the cementite lamella, so that

$$G \simeq \frac{J}{A}$$

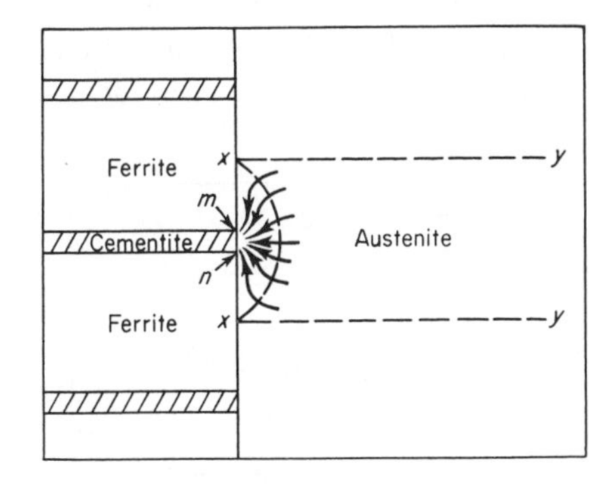

Fig. 17.11 Flow of carbon atoms into a cementite plate during growth of pearlite.

6. *Ibid.*

where G is the growth rate, J is the flux of carbon atoms into the lamella, and A is the cross-section area of the lamella. The growth rate, accordingly, may be assumed to be proportional to the flux of carbon atoms per unit area at the austenite-cementite interface. Now consider a surface, back in the austenite and away from the interface, such as that indicated by the curved dashed line **xx**. The growth rate of the pearlite is also proportional to the flux per unit area at this surface, so that we can also write with respect to this surface

$$G \simeq \frac{J}{A}$$

But by Fick's Law

$$G \simeq \frac{J}{A} = -\frac{AD}{A}\frac{dn}{dx} = -D\frac{dn}{dx}$$

While, strictly speaking, we do not know how the concentration gradient varies across the area **xx**, it is reasonable to assume that at any given temperature the average concentration gradient is proportional to $(C_a - C_b)$ and inversely proportional to the interlamellar spacing S_p, so that

$$G \simeq -\frac{D_c^{\gamma}(C_a - C_b)}{S_p}$$

where G is the pearlite growth rate, S_p the interlamellar spacing of pearlite, $(C_a - C_b)$ the concentration difference obtained by the Hultgren extrapolation, and D_c^{γ} the diffusivity of carbon in austenite. The above very approximate equation does not agree well with experimental data, as can be seen in Fig. 17.12. However, the shapes of the theoretical and experimental curves are similar and this equation makes possible certain deductions about the de-

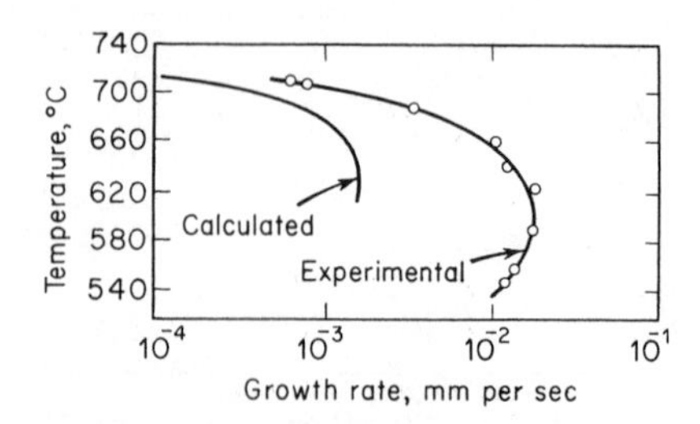

Fig. 17.12 Comparison of the experimental and calculated curves for the rate of growth of pearlite as a function of temperature. (After Mehl, R. F., and Hagel, W. C., *Prog. in Metal Phys.*, **6** 113 [1956].)

pendence of the growth rate on the temperature. First, as the temperature falls, $(C_a - C_b)$ increases and S_p decreases. Both effects act to increase the growth rate with falling temperatures. At high temperatures, however, these factors slow down the growth rate; large spacings and small concentration differences result in small diffusion rates. Conversely, the diffusivity is strongly temperature-dependent (varying as $e^{-Q/RT}$) and this term becomes very small at low temperatures, so that the growth rate decreases again at very low temperatures.

A more exact treatment of the growth rate of pearlite would have to treat the following factors. First, the driving force for the transformation is not just the free-energy difference between austenite and its reaction products, cementite and ferrite. From this value it is necessary to subtract the energy stored in the interfaces between the cementite and ferrite lamellae. The remaining free energy is then used up in three basic irreversible processes, the first of which is the diffusion of carbon through the austenite into the cementite. The others are the reaction of austenite to ferrite and to cementite respectively. In other words, at temperatures below the eutectoid, the transfer of iron atoms across the ferrite-austenite interface is an irreversible chemical reaction that undoubtedly plays a part in determining the overall growth rate. The same must also be true at the austenite-cementite boundary. An analysis of this nature might be carried out by using the theory of the thermodynamics of irreversible processes. For further information on this subject one is referred to a paper by Machlin.[7]

In general, the interlamellar spacing S_p is involved in any theory of growth. It may appear either in the growth equation, or in the assumptions leading to the final equation. The dependence of S_p on the temperature is, therefore, important, not only because it measures an important characteristic of the pearlite transformation, but also because it helps to determine the growth rate. Unfortunately, the experimental data available to the present date are not of sufficient accuracy to determine unambiguously the functional relationship between S_p and the temperature. Figure 17.13 shows $\log S_p$ plotted against $\log (T_E - T_R)$ where T_E is the eutectoid temperature and T_R is the reaction temperature. Both are expressed in degrees centigrade. The data conform reasonably well to a straight line that has a slope of minus one. On a log-log plot, a straight line with a slope of unity is equivalent to a functional relationship of the type

$$S_p \approx \frac{1}{T_E - T_R}$$

This inverse proportion between the interlamellar spacing and the temperature difference between the reaction temperature and the eutectoid temperature has

7. Machlin, E. S., *Trans. AIME,* 197 437 (1953).

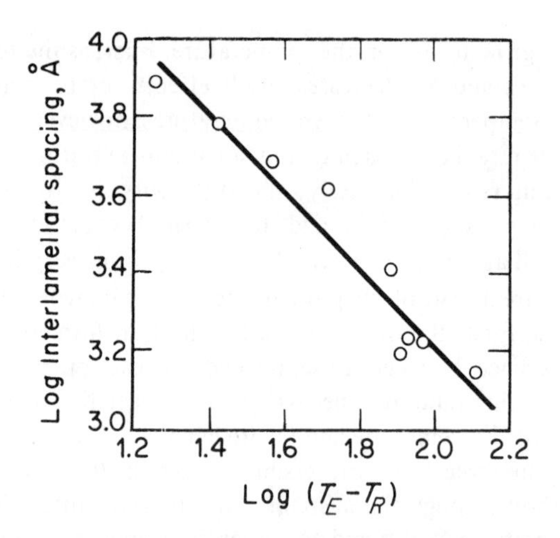

Fig. 17.13 Plot of logarithm of the pearlite interlamellar spacing as a function of the logarithm of the difference in temperature between the reaction temperature (T_R) and the eutectoid temperature (T_E). (Fisher, J. C., ASM Seminar, *Thermodynamics in Physical Metallurgy*, 1950, p. 201.)

been predicted by Zener.[8] However, the same data shown in Fig. 17.13 can be plotted in a number of different ways with equally good agreement, so that the experimental data cannot be said to prove unambiguously Zener's relationship.

In summary, the available experimental data are not of sufficient accuracy to yield empirical relations giving the dependence on time of the growth rate, or the interlamellar spacing. Further, the theoretical determinations of these same quantities involves a complex diffusion problem coupled with two irreversible reactions – austenite to cementite and austenite to ferrite. In the diffusion problem, the concentration gradients are not exactly defined and, further, the diffusivity of carbon is a function of these unknown gradients.

The Rate of Nucleation. The rate of nucleation N is the number of nuclei that form in a unit volume (usually a cubic millimeter) per second. As previously mentioned, in austenite of homogeneous composition, nuclei form heterogeneously at grain boundaries. Unlike the isothermal growth rate, the isothermal rate of nucleation is a function of time. This is shown in Fig. 17.14. In order to compare nucleation rates at different temperatures, it is necessary to consider the average nucleation rate for each temperature. The variation of the average N with temperature is shown in Fig. 17.15. Plotted on this same figure is the rate of growth as a function of temperature. Much about the changes in pearlitic structure that occur as a function of temperature can be deduced from a study of these curves.

8. Zener, C., *Kinetics of the Decomposition of Austenite, Trans. AIME,* **167** 550 (1946).

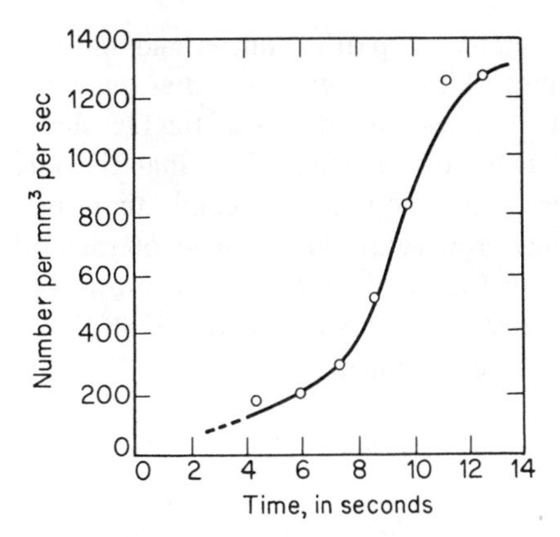

Fig. 17.14 Rate of nucleation (*N*) of pearlite as a function of time. Eutectoid steel transformed at 680°C. (Mehl, Robert F., and Dube, Arthur, *Phase Transformations in Solids*, John Wiley and Sons, Inc., New York, 1951, p. 545.)

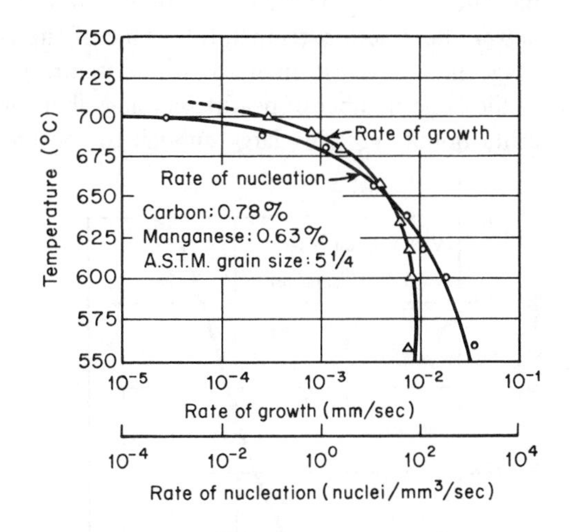

Fig. 17.15 Variation of *N* and *G* with temperature in an eutectoid steel. (Mehl, R. F., and Dube, A., *Phase Transformations in Solids*, John Wiley and Sons. Inc., New York, 1951, p. 545.)

Let us first consider a temperature only slightly below the eutectoid temperature, 700°C. At this temperature, the rate of nucleation is very small, approaching zero. The growth rate, on the other hand, has a finite value between 10^{-3} and 10^{-4} mm per sec. Only a few pearlite nuclei form and, because of the relatively high growth rate, the nuclei grow into large pearlite nodules. The nodules, in fact, grow to sizes much larger than the original austenite grains; nodule growth proceeds across austenite grain boundaries.

Because of the scarcity of pearlite nuclei and the long distances between nuclei forming at these high temperatures, they may be considered to form randomly throughout the austenite in spite of the fact that they actually form at grain boundaries. If it is now assumed, first, that the rate of nucleation N is constant (the average nucleation rate); second, that the nodules maintain a spherical shape as they grow (until they impinge on each other); and, third, that the growth rate G is constant, then it has been shown by Johnson and Mehl[9] that the fraction of austenite transformed to pearlite is given as a function of time by the following reaction equation

$$f(t) = 1 - e^{(-\pi/3)NG^3t^4}$$

where $f(t)$ is the fraction of austenite transformed to pearlite, N is the nucleation rate, G is the growth rate, and t is the time. The typical sigmoidal curve which results when the reaction curve is plotted is shown in Fig. 17.16.

As the reaction temperature is lowered, the rate of nucleation increases at a much faster rate than the rate of growth, and more and more pearlite colonies are nucleated the lower the reaction temperature. One of the consequences is that, at an early stage in the transformation, the austenite grain boundaries become outlined by the large number of pearlite colonies that form along them. Instead of one pearlite nodule growing large enough to consume a number of

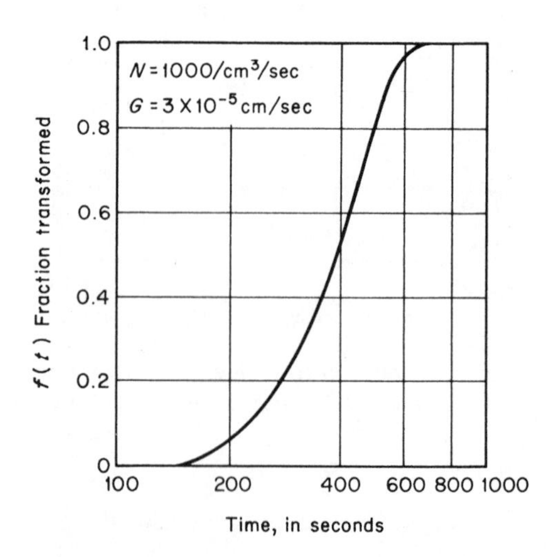

Fig. 17.16 Theoretical reaction curve obtained from Johnson and Mehl equation. (Mehl, R. F., and Dube, A., *Phase Transformations in Solids*, John Wiley and Sons, Inc., New York, 1951, p. 545.)

9. Johnson, W. A., and Mehl, R. F., *Trans. AIME*, 135 416 (1939).

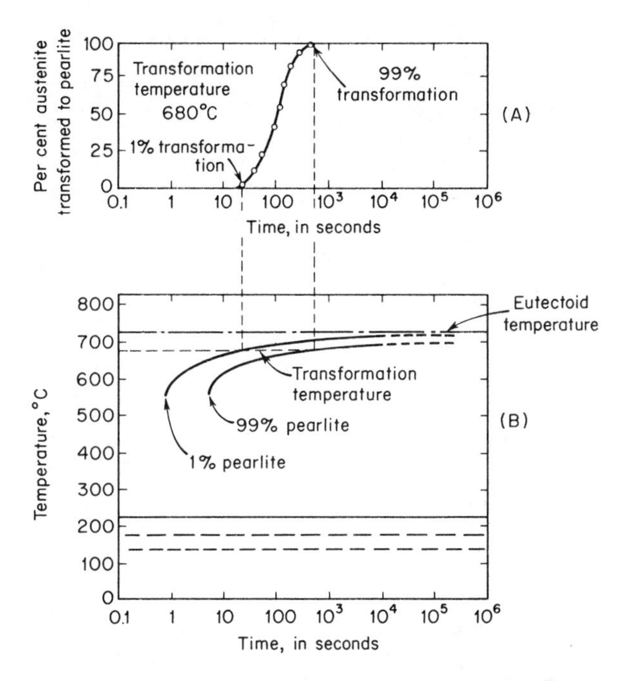

Fig. 17.17 (A) Reaction curve (schematic) for isothermal formation of pearlite. (B) Time-temperature-transformation diagram obtained from reaction curves. (Adapted from *Atlas of Isothermal Transformation Diagrams*, United States Steel Corporation, Pittsburgh, 1951.)

austenite grains, there is now a number of pearlite nodules growing into a single austenite grain. Under these conditions, it is no longer possible to think in terms of random nucleation. We should think, rather, in terms of grain-boundary nucleation. The analysis of the transformation is now mathematically more difficult, but it has been carried out.[10]

17.4 Time-Temperature-Transformation Curves. Information of a very important and practical nature can be obtained from a series of isothermal reaction curves determined at a number of temperatures. First, consider the theoretical curve of Fig. 17.16. Experimental counterparts of this curve are obtained by isothermally reacting a number of specimens for different lengths of time and determining the fraction of the transformation product in each specimen. A plot of these data as a function of the reaction time gives the desired curve, one of which is shown in the upper diagram of Fig. 17.17.

From this reaction curve the time required to start the transformation and

10. *Ibid.*

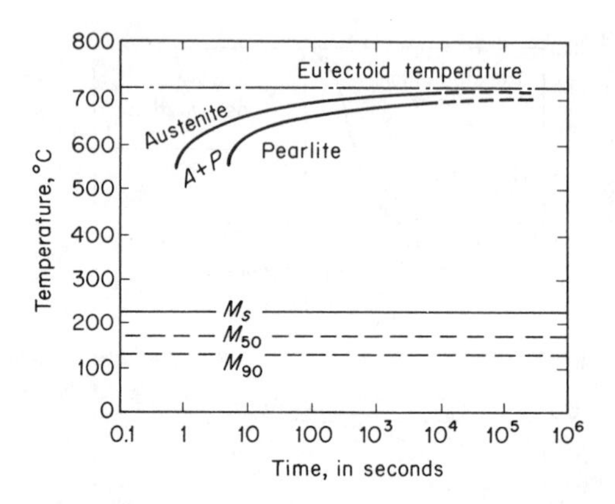

Fig. 17.18 The partial isothermal transformation diagram for an eutectoid steel: 0.79 percent carbon, 0.76 percent manganese. (Adapted from *Atlas of Isothermal Transformation Diagrams,* United States Steel Corporation, Pittsburgh, 1951.)

the time required to complete the transformation may be obtained. This is done in practice by observing the time to get a finite amount of transformation product, usually 1 percent, corresponding to the start of the transformation. The end of the transformation is then arbitrarily taken as the time to transform 99 percent of the austenite to pearlite. A plot of these data for a series of temperatures gives Fig. 17.18, which is clearly not an equilibrium diagram, but still a form of phase diagram. It shows the time relationships for the phases during isothermal transformation. The area to the left of the first C-shaped curve corresponds to structures that are austenitic. Any point to the right of the second of these two curves represents a pearlitic structure and is, consequently, a mixture of two phases (cementite and ferrite). Between the two curves is a region of pearlite and austenite with the relative ratios of these two constituents varying from all austenite to all pearlite as one moves from left to right.

One of the significant factors about the pearlite transformation is the very short time required to form pearlite at temperatures around 600°C. This, of course, could have been deduced from the information of Fig. 17.15 where it can be seen that at this temperature both the rate of nucleation and the rate of growth have a maximum.

The time-temperature-transformation (*T-T-T*) diagram of Fig. 17.18 corresponds only to the reaction of austenite to pearlite. It is not complete in the sense that the transformations of austenite, which occur at temperatures below about 550°C, are not shown. To complete this study, it is necessary to consider two other types of austenitic reactions: austenite to bainite and austenite to martensite.

The bainite reaction will be covered in the next section and the martensite reaction in steels in Chapter 18. With regard to the martensite reaction, it can be stated that in the present eutectoid steel, the reaction is primarily athermal. The horizontal lines drawn in Fig. 17.18 show the M_s (martensite start) temperature and the temperatures at which certain indicated fractions of martensite are obtained. It goes without saying that martensite only forms if the steel, on cooling, does not transform to pearlite at higher temperatures.

17.5 The Bainite Reaction. The bainite reaction is perhaps the least understood of all the austenite reactions. It is also particularly difficult to investigate in simple iron-carbon alloys because it overlaps the region of the pearlite transformation (at temperatures of the order of 500°C). Steels transformed in this temperature range have structures containing both pearlite and bainite. In certain alloy steels, the presence of certain elements in substitutional solid solution (in the austenite) has the effect of separating the temperature ranges in which the respective pearlite and bainite reactions occur. The bainite reaction is much more readily studied in these alloys. Some of the conclusions obtained from these investigations are characteristic of bainite reactions in general and can be applied to simple iron-carbon alloys.

The most puzzling feature of the bainite reaction is its dual nature. In a number of respects, it reveals properties that are typical of a nucleation and growth type of transformation such as occurs in the formation of pearlite, but, at the same time, it shows an equal number that would classify it as a martensitic type of reaction. Like pearlite, the reaction product of the bainite transformation, called *bainite*, is not a phase but a mixture of ferrite and carbide. In the bainite transformation, the carbon which is uniformly distributed in the austenite is concentrated into localized regions of high-carbon content (the carbide particles) leaving an effectively carbon-free matrix (the ferrite). The bainite reaction involves composition changes and requires diffusion of carbon. In this it differs markedly from a typical martensitic transformation. Conversely, the compositional changes that occur during a bainite transformation do not involve the substitutional alloying elements which may be present in the metal. These elements are not redistributed during the basic transformation, so that the composition of both the ferrite and carbide with respect to these elements remains the same as that of the original austenite. Another property of the bainite reaction that differentiates it from martensitic reactions is that it is not athermal. The formation of bainite requires time, and when austenite is reacted to bainite isothermally, a typical S-shaped reaction curve is obtained, as can be seen in Fig. 17.19. The similarity of this curve to those obtained in simple nucleation and growth transformations can be seen by comparing it with that previously presented in the discussion dealing with the pearlite reactions.

While bainite and pearlite are both mixtures of ferrite and carbide, the

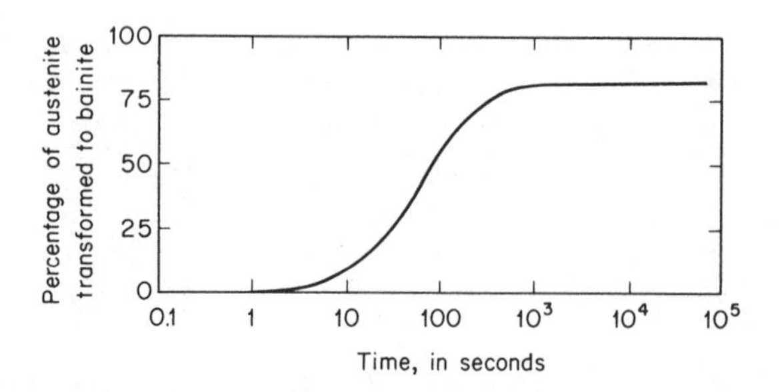

Fig. 17.19 A characteristic of the isothermal bainite transformation is that it may not go to completion. (After Hehemann, R. F., and Troiano, A. R., *Metal Progress*, **70**, No. 2, 97 [1956].)

mechanism of bainite formation differs from that of pearlite. The resulting product also does not have the alternating arrangement of parallel lamellae of ferrite and cementite found in pearlite. As we have seen, pearlite, due to almost equal growth rates in all directions, tends to develop in the form of spheres. This is not true of bainite, which grows as plates – a typical martensitic characteristic. When observed on a metallographic section, bainite has a characteristic aciculer (needlelike) appearance, in many respects similar to that of deformation twins and martensite plates. The formation of bainite plates is also accompanied by surface distortions[11] (surface tilts and accommodation kinks), so that it is quite probable that lattice shear is involved in the formation of the plates. A basic difference, however, between bainite and martensite plates appears in the speed with which they form. Martensite plates are, in most cases, formed under conditions of high-driving force and grow to their final sizes in a small fraction of a second, whereas bainite plates grow slowly and continuously. The growth of the latter is apparently retarded by the time required for the diffusion processes that accompany the reaction.

Each bainite plate is composed of a volume of ferrite in which carbide particles are embedded. Because of the extremely fine structure in the bainite plates, it is necessary to use an electron microscope to resolve its components. Figure 17.20 shows several bainite plates photographed with an electron microscope at a magnification of 15,000 times. The irregular dark areas are regions of ferrite, while the light areas inside these darker patches are carbide particles. The bainite shown in this photograph was formed at a relatively high transformation temperature (approximately 460°C). This structure is commonly called *upper bainite*. When bainite is formed at lower temperatures (approximately 250°C), the ferrite plates assume a more regular needlelike shape than shown in Fig. 17.20 and the product is called *lower bainite*. At the same time, the carbide particles, which appear in the present example as rather coarse

11. Ko, T., and Cottrell, J. A., *Jour. of the Iron and Steel Inst.*, **172** 307 (1952).

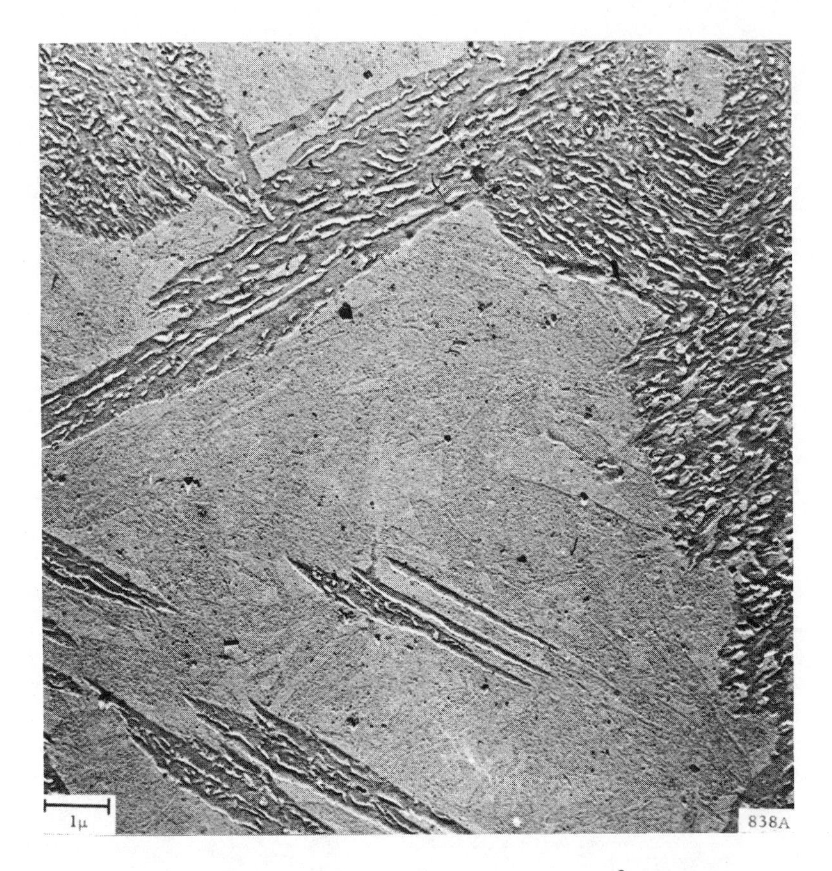

Fig. 17.20 The structure of bainite transformed at 460°C revealed by the electron microscope. Original magnification 15,000x reduced 30 percent in above photograph. (*Trans. ASTM,* **52**, 540, Fig. 20 [1952]. [Second Progress Report of Subcommittee XI of Committee E–4.]])

structures paralleling the length of the ferrite plate, assume a different aspect. They become smaller in size and appear as cross-striations[12] making an angle of about 55° to the axis of the plate. (See Fig. 17.21.) The net result of this change in the structure of bainite with transformation temperature is a variation in its appearance under the light microscope (that is, at lower magnifications). To illustrate this point, Fig. 17.22 shows the appearance of bainite formed at two different isothermal temperatures.

Something should be said about the nature of the carbides which appear in bainite. It would appear that in carbon steels transformed to bainite at

12. *Trans. ASTM,* 52 543 (1952).

Fig. 17.21 The structure of bainite transformed at 250°C as revealed by the electron microscope. Original magnification 15,000x reduced 30 percent in above photograph. (*Trans. ASTM,* **50**, 444, Fig. 23 [1950]. [Second Progress Report of Subcommittee XI of Committee E–4.])

temperatures above 300°C, the carbides are simply cementite, Fe_3C.[13] It has been recognized for some time that in bainite formed at lower temperatures (below 300°C), another carbide other than cementite may form. Recent investigations have shown that this carbide is often epsilon (ϵ) carbide,[13]

13. Hehemann, R. F., ASM Seminar, *Phase Transformations*, Amer. Soc. for Metals, Metals Park, Ohio, 1970.

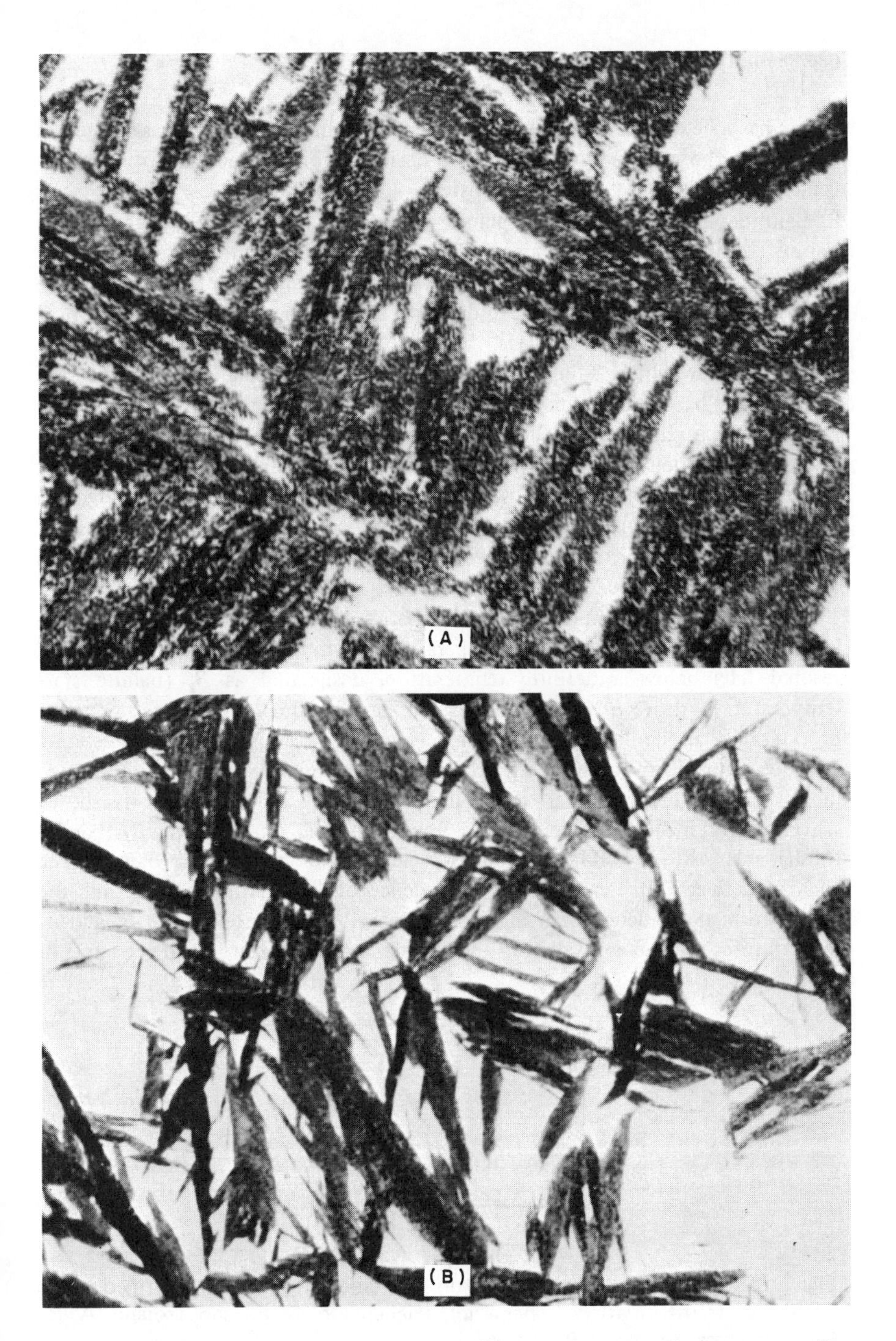

Fig. 17.22 (A) Bainite formed at 348°C. (B) Bainite formed at 278°C. (Photographs courtesy of Edgar C. Bain Laboratory for Fundamental Research, United States Steel Corporation.) 2500x.

which has a hexagonal crystal structure instead of the orthorhombic structure of cementite. The carbon concentration in epsilon carbide also differs from that of cementite and is about 8.4 percent instead of 6.7 percent.

Bainite plates possess habit planes that are characteristically irrational in nature. In this regard they are similar to the plates formed in a martensitic reaction. Bainite plates thus tend to form along crystallographic planes of the parent austenitic phase, but the indices of these planes are not simple whole numbers. It is further observed[14] that the habit-plane indices vary with the temperature at which bainite is formed. Another important fact is that the ferrite in the bainite plates possesses a basically different orientation relationship relative to the parent austenite than does the ferrite that appears in pearlite. Because this orientation is similar to that found in simple ferrite nucleating directly from austenite (that is, in a low-carbon steel above the eutectoid temperature), it is generally believed that bainite is nucleated by the formation of ferrite. This is in contrast to pearlite which many authorities believe to be nucleated by cementite.

One of the most important characteristics of the bainite reaction is that bainite does not form until the temperature at which austenite is isothermally reacted falls below a definite temperature, designated as B_s (bainite start temperature). Above B_s austenite does not form bainite except in the presence of externally applied stresses. Further, at temperatures just below B_s, austenite does not transform completely to bainite. The amount of bainite formed increases as the isothermal reaction temperature is lowered, as is shown schematically in Fig. 17.23. Below a lower limiting temperature, B_f (bainite finish), it is thus possible to transform austenite completely to bainite. The analogy between the temperature dependence of the bainite reaction and the temperature dependence of a martensitic reaction is apparent. The B_s and B_f temperatures are the equivalent of the M_s and M_f temperatures, and the curve of

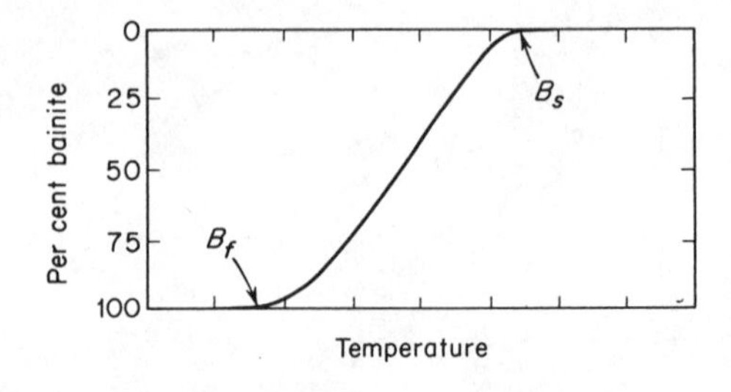

Fig. 17.23 Effect of temperature on the amount of bainite formed in an isothermal transformation (schematic). (Hehemann, R. F., and Troiano, A. R., *Metal Progress*, **70**, No. 2, 97 [1956].)

14. Greninger, A. B., and Troiano, A. R., *Trans. AIME*, **140** 307 (1940).

Fig. 17.23 is very similar to those shown in Chapter 16 giving the amount of athermal martensite formed as a function of temperature.

In those alloy steels where the pearlite and bainite reactions do not overlap, the fraction of austenite which is untransformed when the steel is reacted between B_s and B_f is capable of remaining as austenite for indefinitely long periods of time. This is true provided the metal is held at the transformation temperature. If cooled to room temperature, however, the remaining austenite, or some fraction of it, undergoes the martensite transformation. In simple iron-carbon alloys which undergo isothermal transformation in the region where the pearlite-bainite overlap occurs, it can be assumed that the austenite that is not transformed to bainite goes to pearlite.

The fact that the bainite transformation does not go to completion in the temperature interval B_s to B_f implies[15] that in this range of temperatures both nucleation and growth stop before all of the austenite is consumed. Thus, a finite number of nuclei must form, which grow to typical bainite plates. The limitation on the growth of these plates is not difficult to understand. A bainite plate can grow until it intersects another plate, or an austenite grain boundary. Alternatively, the growth of bainite plates can be limited by a loss of coherency between the ferrite in the plate and the parent austenite. On the other hand, the reasons why nucleation stops are not easy to determine.

In view of the complexities of the bainite transformation, they will not be discussed here. Current views on this subject can be found in a review paper by Hehemann.[16]

17.6 The Complete *T-T-T* Diagram of an Eutectoid Steel. The time-temperature-transformation (*T-T-T*) diagram of Fig. 17.18 is shown again in Fig. 17.24. In the latter figure, however, the curves corresponding to the start and finish of transformation are extended into the range of temperatures where austenite transforms to bainite. Because the pearlite and bainite transformations overlap in a simple eutectoid iron-carbon steel, the transition from the pearlite reaction to the bainite reaction is smooth and continuous. Above approximately 550°C to 600°C, austenite transforms completely to pearlite. Below this temperature to approximately 450°C, both pearlite and bainite are formed. Finally, between 450°C and 210°C, the reaction product is bainite only.

An interesting feature of the bainite reaction is that as the reaction temperature is lowered, the rate at which bainite forms decreases. Accordingly, very long times are required to form bainite just above the M_s temperature.

The significance of the dotted line, which runs between the two curves, marking the beginning and end of isothermal transformations, should be

15. Hehemann, R. F., and Troiano, A. R., Metal Prog., 70, No. 2, 97 (1956).
16. Hehemann, R. F., ASM Seminar, *Phase Transformations,* Amer. Soc. for Metals, Metals Park, Ohio, 1970.

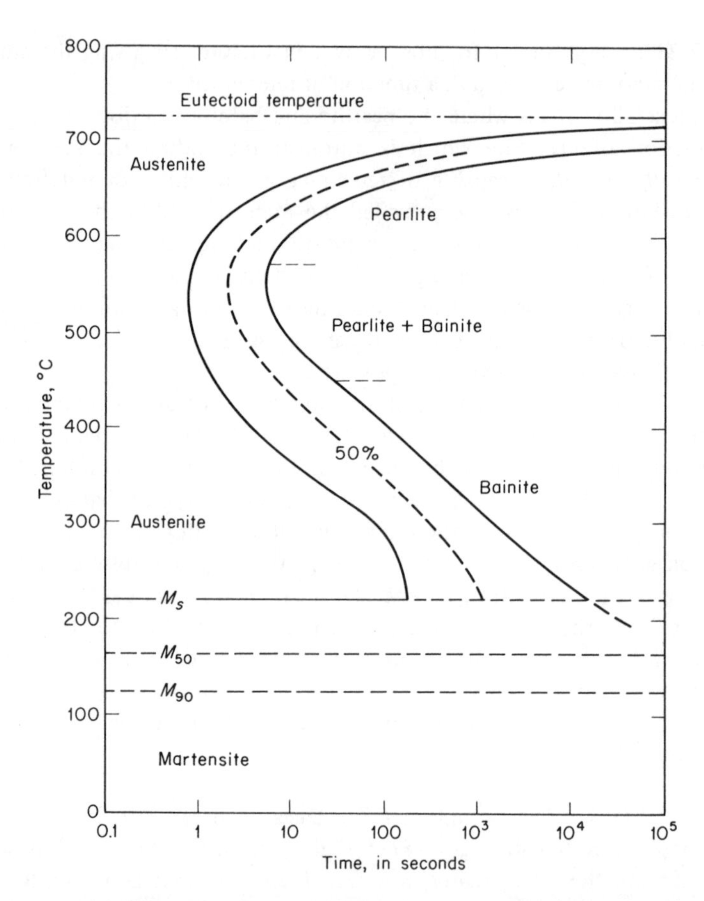

Fig. 17.24 The complete isothermal transformation diagram for an eutectoid steel. Notice that this steel is not a high-purity iron-carbon alloy, but a commercial steel (AISI 1080) containing 0.79 percent carbon and 0.76 percent manganese. The effect of the manganese will be discussed in Chapter 18. (Adapted from *Atlas of Isothermal Transformation Diagrams*, United States Steel Corporation, Pittsburgh, 1951.)

mentioned. This represents, at any given temperature, the time to transform half the austenite to bainite or austenite to pearlite, as the case may be.

Let us consider some arbitrary time-temperature paths along which it is assumed austenitized specimens are carried to room temperature. These are shown in Fig. 17.25 and represent exercises for showing the principles of the use of time-temperature-transformation (*T-T-T*) diagrams.

Path 1. The specimen is cooled rapidly to 160°C and left there 20 min. The rate of cooling is too rapid for pearlite to form at higher temperatures, therefore

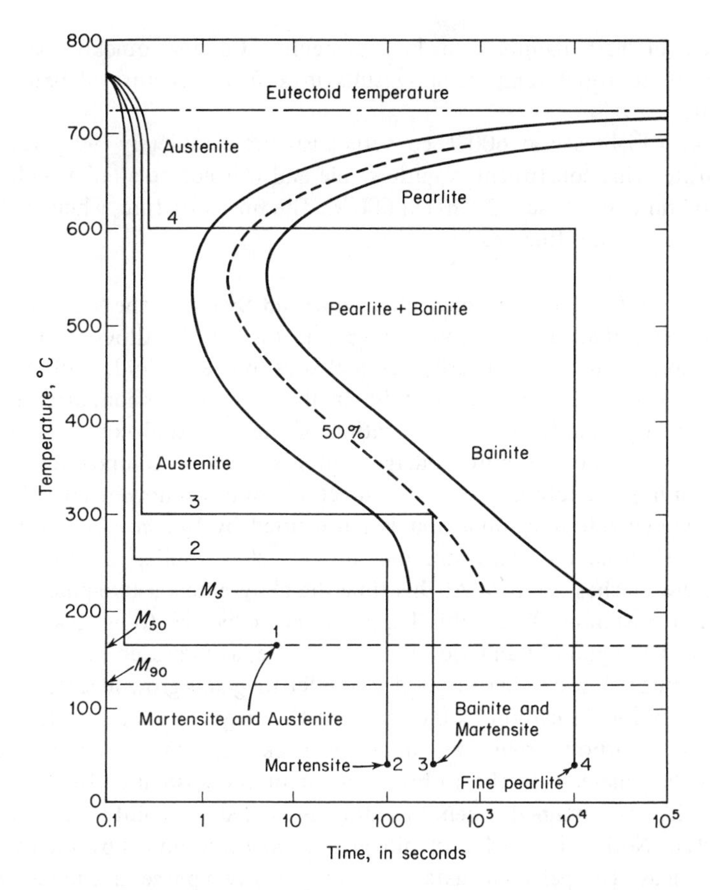

Fig. 17.25 Arbitrary time-temperature paths on the isothermal transformation diagram.

the steel remains in the austenitic phase until the M_s temperature is passed, where martensite begins to form athermally. Since 160°C is the temperature at which half of the austenite transforms to martensite, the direct quench converts 50 percent of the structure to martensite. Holding at 160°C forms only a very small quantity of additional martensite because in simple carbon steels isothermal transformation to martensite occurs only to a very limited extent. At point 1, accordingly, the structure can be assumed to be half martensite and half retained austenite.

Path 2. In this case, the specimen is held at 250°C for 100 sec. This is not sufficiently long to form bainite, so that the second quench from 250°C to room temperature develops a martensitic structure.

Path 3. An isothermal hold at 300°C about 500 sec produces a structure composed of half bainite and half austenite. Cooling quickly from this temperature to room temperature results in a final structure of bainite and martensite.

Path 4. Eight sec at 600°C converts austenite completely (99 percent) to fine pearlite. This constituent is quite stable and will not be altered on holding for a total time of 10^4 sec (2.8 hr) at 600°C. The final structure, when cooled to room temperature, is fine pearlite.

17.7 Slowly Cooled Hypoeutectoid Steels. The steel region of the iron-carbon diagram is reproduced again in Fig. 17.26. Alloys to the left of the eutectoid point are arbitrarily designated as hypoeutectoid, while those to the right are known as hypereutectoid. In the above discussion, attention has been directed primarily to a consideration of the eutectoid composition. The transformations that austenite undergoes in steels, whose compositions fall on the low, or hypoeutectoid, side of the eutectoid point will now be considered. A typical hypoeutectoid composition is represented by line *ac*. At point *a* this alloy is austenitic. Its transformation, on very slow cooling, starts when the temperature reaches point *b*. At this time the alloy enters a two-phase field of ferrite and austenite. When this happens, ferrite begins to nucleate heterogeneously at the grain boundaries of the austenite, as indicated in Fig. 17.27A. With continued slow-cooling to point *c*, the ferrite grains grow in size. Since the ferrite is very low in carbon (<0.02 percent C), its growth is associated with a rejection of carbon from the interface back into the austenite, and a corresponding increase in the carbon content of the austenite. The two-phase mixture that is obtained when the alloy is cooled to point *c* is shown in Fig. 17.27B. Notice that each austenite grain is now surrounded by a network of ferrite crystals. The pertinent data concerning this two-phase mixture are easily

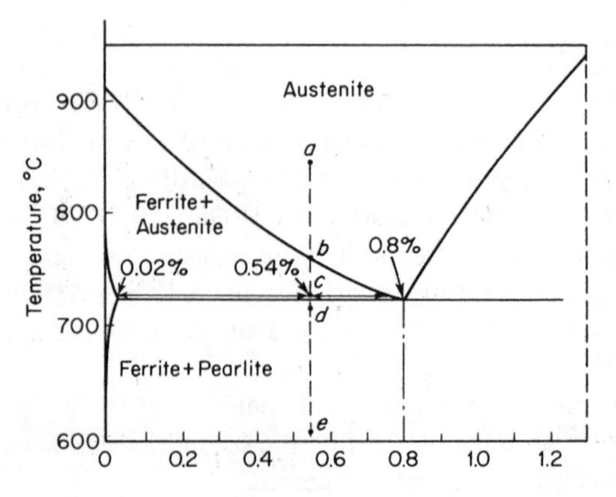

Fig. 17.26 Transformation of a hypoeutectoid steel on slow cooling.

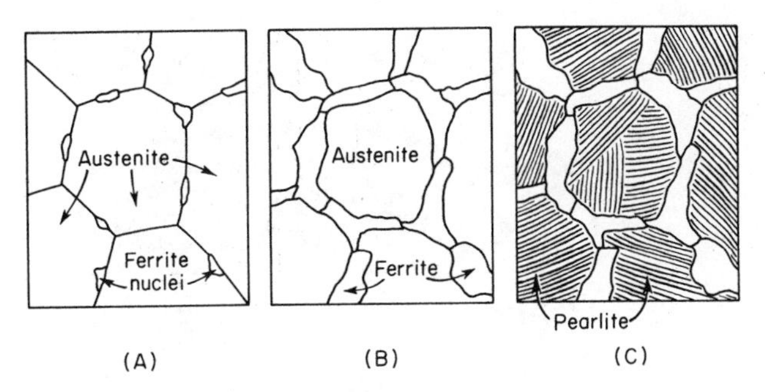

Fig. 17.27 Three stages in the formation of a slowly cooled hypoeutectoid structure corresponding to points *b, c,* and *d* respectively in Fig. 17.26.

deduced from Fig. 17.26. By the lever rule, we have:

$$\text{The amount of ferrite} = \frac{0.8 - 0.54}{0.8 - 0.02} = \frac{0.26}{0.78} = \frac{1}{3}$$

$$\text{The amount of austenite} = \frac{0.54 - 0.02}{0.80 - 0.02} = \frac{0.52}{0.78} = \frac{2}{3}$$

and the intersections of the tie lines with the single-phase boundaries show us that the ferrite must contain 0.02 percent C and the austenite 0.8 percent C. The structure thus consists of two-thirds austenite and one-third ferrite in which the austenite has the eutectoid composition and is just above the eutectoid temperature. Slow cooling of this metal through the eutectoid temperature transforms the remaining austenite to pearlite, so that the structure at point *d* consists of a mixture of ferrite and pearlite. (See Fig. 17.27C.) The ratio of ferrite to pearlite is, of course, the same as the ratio of ferrite to austenite which obtains at point *c*, namely, 1 to 3. Continued cooling of the alloy to room temperature causes no visible change in the microstructure. Theoretically, a change should occur since the solubility of carbon in ferrite decreases with temperature, but the quantity of carbon involved is very small, since at the eutectoid temperature ferrite only dissolves 0.02 percent C.

For most practical purposes, in making lever-rule computations one can assume that the carbon content of ferrite is zero. If this assumption is made, then the part of the structure of a hypoeutectoid steel that is pearlite varies directly as the ratio of the carbon content of the steel divided by the eutectoid composition 0.8 percent. Thus, a 0.2 percent C steel should have a structure with approximately one-fourth pearlite, and a steel with a structure composed of half pearlite should have a carbon content of 0.4 percent C. The microstructures of several slowly cooled hypoeutectoid steels are shown in Fig. 17.28.

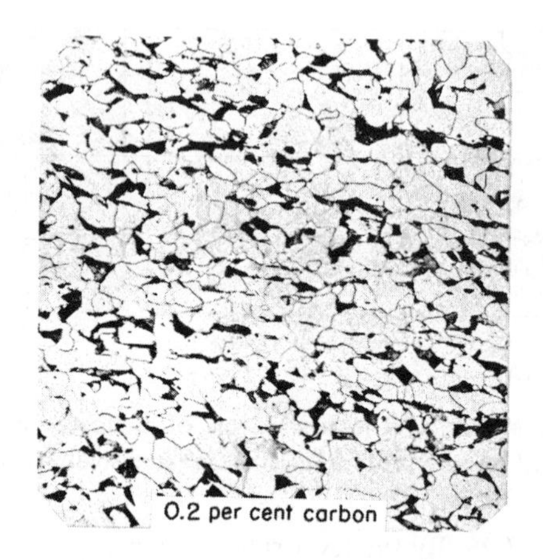

Fig. 17.28 Hypoeutectoid-steel microstructures. Black areas are pearlite, white areas are ferrite. Approximately 300x.

The structure of a hypoeutectoid steel serves to show the difference between the phase and constituent concepts. In any one of the photographs of Fig. 17.28, two basic types of structures are clearly evident: the white ferrite areas and the dark pearlitic areas. The constituents of these specimens are, accordingly, pearlite and ferrite. The phases in these structures are, however, ferrite and cementite. In each specimen, cementite is localized in the pearlitic areas, while the ferrite occurs both in the pearlite and in the simple ferrite grains. While there is no basic difference between the two forms of ferrite, it is sometimes convenient to differentiate between them. Ferrite that appears in the pearlite is called the *eutectoid ferrite*, while the other is the *proeutectoid ferrite*.

The Greek prefix "pro" is used to designate the latter because, on cooling, it forms before the eutectoid-structure (pearlite) forms.

The relative amounts of the two phases in a steel are also found by the lever rule. In this case, the levers extend to the compositions of the phases (ferrite 0 percent and cementite 6.7 percent), instead of to the compositions of the constituents (proeutectoid ferrite 0 percent C and pearlite 0.8 percent C).

17.8 Slowly Cooled Hypereutectoid Steels. Steels having a carbon content that falls above the eutectoid point transform in a fashion similar to those lying below the eutectoid composition. The proeutectoid constituent in this case is cementite instead of ferrite which, for a typical composition (1.2 percent C) (Fig. 17.29), forms between points *h* and *i* as the temperature is lowered along line *gk*. At point *i*, the structure is a mixture of cementite (6.7 percent C) and austenite (0.8 percent C). A lever-rule computation shows that 93.2 percent of the structure is austenite and 6.8 percent is cementite. The precipitation of the proeutectoid cementite reduces the carbon content of the austenite to that of the eutectoid point. As a result, slowly lowering the temperature of the specimen through the eutectoid temperature converts the remaining austenite to pearlite. Continued cooling of the specimen to room temperature in this case also causes no visible change in the microstructure, so that the structure obtained at point *j* may be taken as representative of that visible at room temperature, which is 6.7 percent proeutectoid cementite and 93.3 percent pearlite, which are the constituents of the structure. The phases are cementite and ferrite. The latter can be determined with the aid of the lever rule

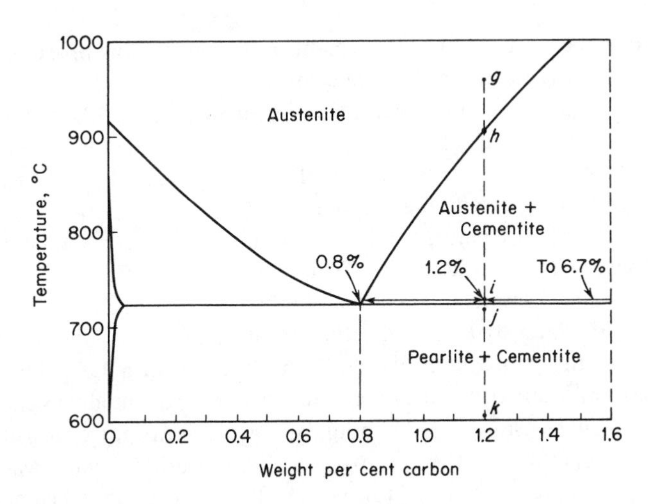

Fig. 17.29 Transformation of a hypereutectoid steel on slow cooling.

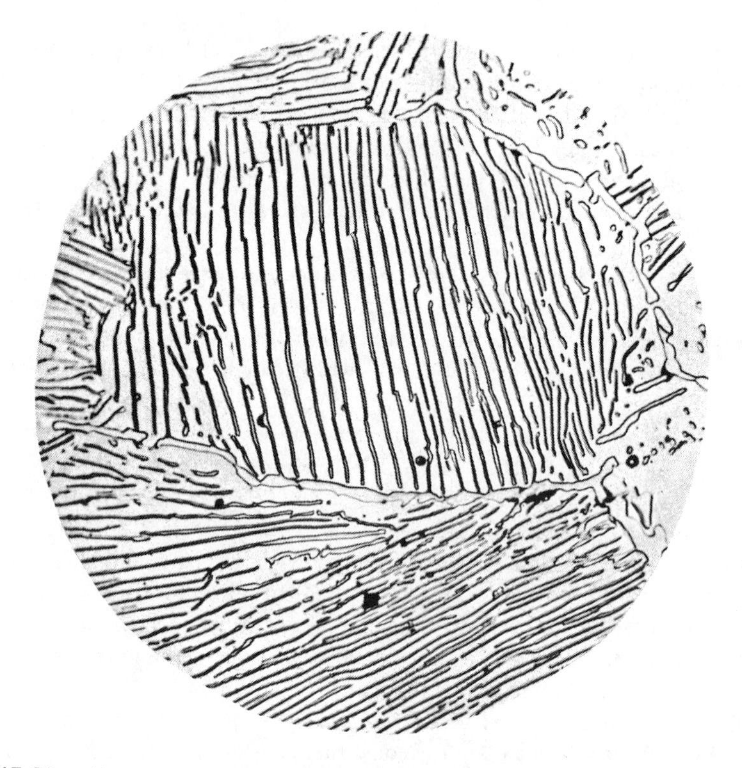

Fig. 17.30 Hypereutectoid-steel microstructure. Notice the band of cementite plates outlining the pearlite colony in the center of the photograph. 1000x.

and shown to be 18 percent total cementite (eutectoid plus proeutectoid) and 82 percent ferrite (all contained in the pearlite).

A typical hypereutectoid structure is shown in Fig. 17.30. Hypereutectoid microstructures usually differ in appearance from hypoeutectoid ones. The most important difference is in the amount of the proeutectoid constituent. In the above example (1.2 percent C steel that lies 0.4 percent above the eutectoid point) there is only 6.7 percent proeutectoid cementite. At an equal distance in weight percent on the other side of the eutectoid point, 0.4 percent C, the structure contains 50 percent proeutectoid ferrite. A particular characteristic of hypereutectoid microstructures is the small amount of the proeutectoid constituent. In the photograph of Fig. 17.30, which is of a 1.1 percent C steel, the proeutectoid cementite appears as a fine network surrounding pearlite areas, as may be seen by studying the boundary around the large pearlite colony occupying the center of the photograph. Each pearlite region was a single austenite grain before the transformation. The proeutectoid cementite thus forms heterogeneously at the austenite grain boundaries.

Another important difference between the structures of low- and high-carbon steels is that in steels ferrite is the continuous phase and cementite the surrounded phase. Thus, the proeutectoid cementite, while outlining the austenite grain boundaries, still consists of a number of disconnected plates. Conversely, proeutectoid ferrite often completely surrounds pearlite areas in the form of a number of contiguous grains.

17.9 Isothermal Transformation Diagrams for Noneutectoid Steels. Isothermal transformation diagrams (*T-T-T*) for both a hypoeutectoid steel (0.35 percent C) and a hypereutectoid steel (1.13 percent C) are shown in Figs. 17.31 and 17.32 respectively. The similarity between these diagrams and that of the eutectoid steel are at once apparent, but there are, nevertheless, important differences. Among the latter is the presence in both figures of the line *mn* corresponding to the start of the isothermal transformation of the proeutectoid constituent. In each case, the designated line lies above and to the left of those defining the isothermal transformation of austenite to pearlite. Notice that these lines are asymptotic (at large times) to constant temperature

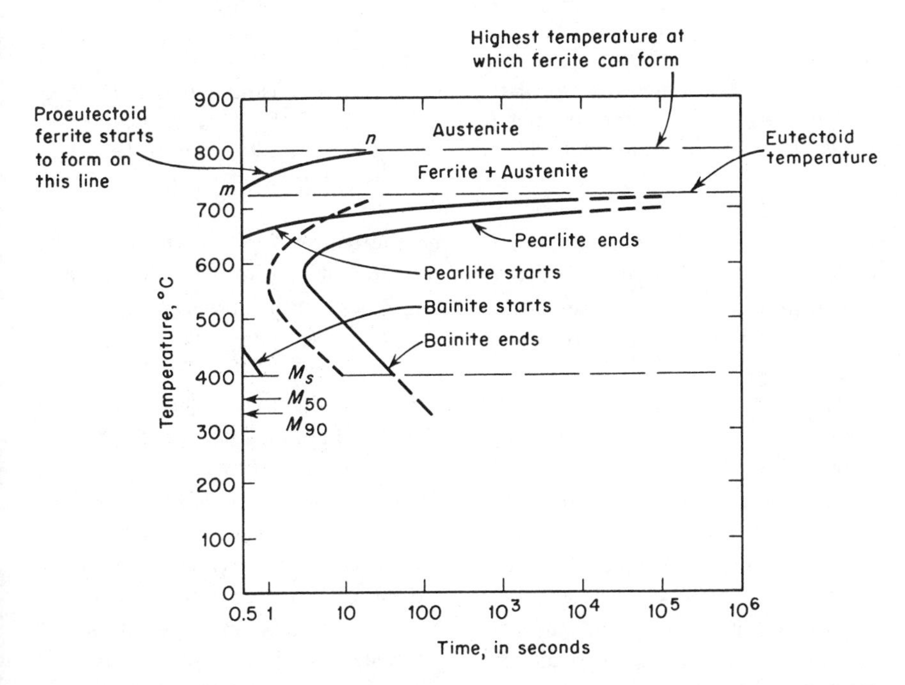

Fig. 17.31 Isothermal transformation diagram for a hypoeutectoid steel: 0.35 percent carbon, 0.37 percent manganese. (Adapted from *Atlas of Isothermal Transformation Diagrams,* United States Steel Corporation, Pittsburgh, 1951.)

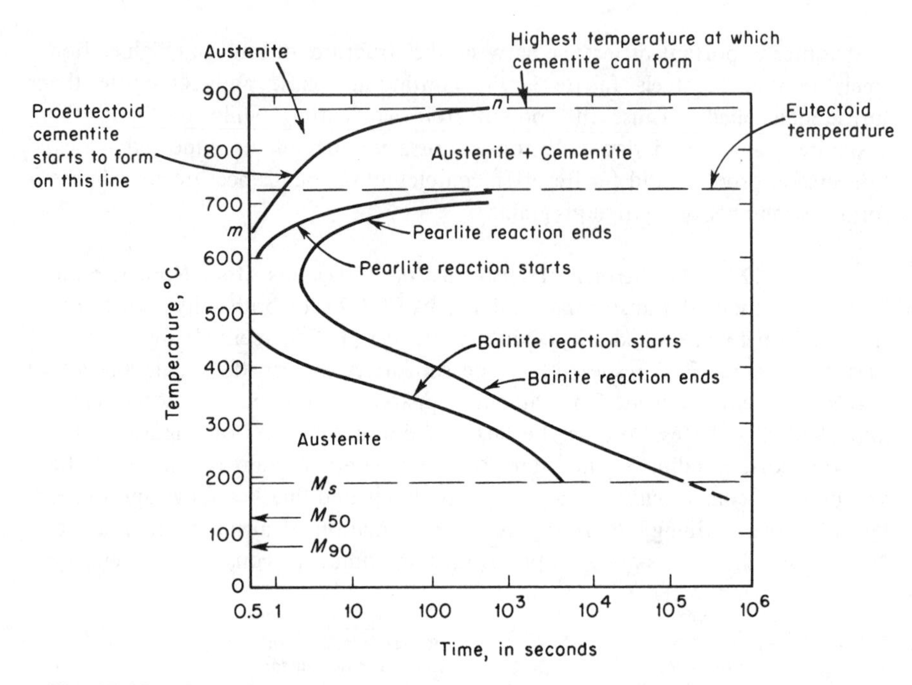

Fig. 17.32 Isothermal transformation diagram for a hypereutectoid steel: 1.13 percent carbon, 0.30 percent manganese. (Adapted from *Atlas of Isothermal Transformation Diagrams*, United States Steel Corporation, Pittsburgh, 1951.)

lines made on each figure to pass through the temperatures at which the two given alloys are first able to form the proeutectoid constituents (ferrite and cementite respectively) on very slow cooling. These temperatures are equivalent to those designated by point *b* in Fig. 17.26 and point *h* in Fig. 17.29 respectively.

The isothermal-transformation diagram for hypoeutectoid steel is reproduced again in Fig. 17.33. Several arbitrary cooling paths are made in this diagram in order to illustrate the complete significance of all the lines shown.

In each case, it is assumed that the specimens are austenitized at 840°C, which is some 40° above the temperature at which ferrite is able to first form in this composition. Along Path 1 it is assumed that the specimen is instantly quenched to 750°C and held at this temperature for 1 hr. During the first second of this isothermal treatment, the structure remains entirely austenitic, but at the end of this second the curve designating the start of the ferrite nucleation is crossed and ferrite begins to form. From this point on to the end of 10,000 sec (2.8 hr) the structure lies in the two-phase austenite-ferrite region. Because of the large length of time at this temperature, the amount of ferrite formed should

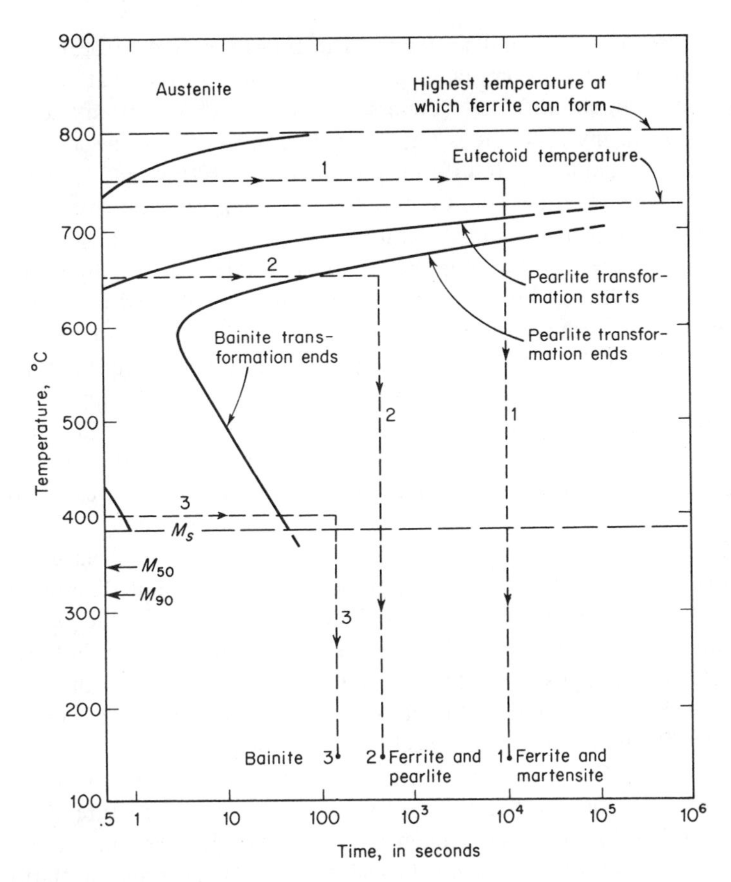

Fig. 17.33 Arbitrary time-temperature paths on the isothermal diagram of a hypoeutectoid steel.

be close to that predicted by the equilibrium diagram for this temperature. No pearlite should form because we are still above the eutectoid temperature (723°C). In the diagram, Path 1 is completed by a quench to room temperature, which should transform any austenite left at 750°C almost completely to martensite, so that the final structure can be assumed to consist of ferrite and martensite.

Path 2 represents one in which the specimen is assumed to be isothermally transformed at a temperature below the eutectoid temperature, and for this purpose 650°C has been selected. Because of the extreme rapidity with which ferrite forms from austenite in this temperature range, even a very rapid quench (cooling time less than 0.5 sec) cannot suppress the formation of some ferrite during the quench. As a result, the specimen starts its isothermal transformation as a mixture of ferrite and austenite. The transformation to pearlite in this temperature range is also very rapid and the latter begins to form at once. During this period, from about 0.5 sec (the assumed start of the transformation) to the end of 100 sec, the austenite is transformed to pearlite. The specimen may be assumed to be completely transformed at the end of 100 sec and to consist of a mixture of ferrite and pearlite. Cooling to room temperature at any normal rate of cooling does not change this structure.

Transformations of the type indicated above, in which a hypoeutectoid specimen is quickly cooled to a fixed temperature and allowed to transform at this temperature, represent a nonequilibrium irreversible transformation. One of the results is that the ratio of ferrite to pearlite that is obtained is not the same as that obtained on a very slow continuous cooling cycle approaching an equilibrium transformation. The amount of ferrite in the irreversible process is usually smaller, which means that the amount of pearlite in the final microstructure is greater. This increase cannot occur without a corresponding change in the composition of pearlite: normally, on slow cooling, $\frac{1}{8}$ cementite and $\frac{7}{8}$ ferrite. Transformations of hypoeutectoid steels at temperatures below the eutectoid point, therefore, tend to suppress the amount of proeutectoid ferrite and lower the carbon content of the eutectoid structure (the pearlite). At this point it is well to point out that there is a corresponding effect in the case of hypereutectoid compositions. Transformations that occur below the eutectoid temperature tend to suppress the amount of the proeutectoid cementite and, accordingly, to raise the carbon concentration of the pearlite.

The great rapidity with which ferrite and pearlite form in the particular alloy under consideration precludes the formation of a microstructure that is all bainite. As Fig. 17.33 shows, a quench which takes even as short a period as 0.5 sec to reach 400°C still passes through the lines designating the start of the ferrite and pearlite transformations. A specimen quickly cooled and held at 400°C for about 100 sec (Path 3) accordingly contains bainite mixed with a small amount of ferrite and pearlite. Finally, a direct quench to room temperature should furnish a hardened specimen containing a high percentage of martensite, but with also unavoidable small percentages of ferrite and pearlite.

An analysis similar to the above can be carried out for the hypereutectoid steel shown in Fig. 17.32. The principal difference will be in the nature of the proeutectoid constituent – cementite instead of ferrite.

Problems

1. (*a*) With the aid of the equation

$$f(t) = 1 - e^{(-\pi/3)NG^3t^4}$$

where $f(t)$ is the fraction of austenite transformed to pearlite, N is the nucleation rate, G is the growth rate, and t is the time, compute the length of time needed to obtain 50 percent transformation under the following conditions

$$N = 1000/\text{cm}^3/\text{sec}$$

$$G = 3 \times 10^{-5} \text{ cm/sec}$$

(*b*) Check your answer with Fig. 17.16.

2. A slowly cooled steel shows a microstructure containing 50 percent pearlite and 50 percent ferrite.
   (*a*) Estimate the carbon concentration in this alloy.
   (*b*) Describe the microstructure that would be obtained (at equilibrium) if this metal is heated to 730°C.
   (*c*) What structure should exist in the metal if it is heated to and held at 800°C?
   (*d*) Make sketches illustrating the above microstructure.
3. In hypoeutectoid steels, the ferrite grains normally surround the pearlite colonies. Rationalize this fact giving a sketch showing how the microstructure should develop during cooling.
4. A steel of 0.5 percent carbon, when examined metallographically, shows a structure that is 80 percent pearlite and 20 percent ferrite.
   (*a*) What should be the equilibrium amount of pearlite in this alloy?
   (*b*) How might it be possible to attain the above structure?
   (*c*) In the equilibrium structure, the widths of the ferrite and cementite lamellae are in a ratio of 7 to 1. What should they be in the material described above?
5. Consider an alloy of iron with 1.1 percent carbon. Describe its microstructure, giving the percentages of the phases and of the constituents:
   (*a*) when the metal is slowly cooled to room temperature from the austenite phase
   (*b*) if the metal is cooled at a moderately rapid rate so that the proeutectoid constituent is present only to an extent of about 1.5 percent.
6. A number of small steel specimens are austenitized at 760°C and then brought down to room temperature along one of the following time-temperature paths. Describe the microstructure that should be obtained in each specimen. In particular, explain how the microstructures for paths *c* and *d* would differ.
   (*a*) Cooled to room temperature in a period less than 1 second.
   (*b*) Cooled to 160°C in less than 1 second and maintained at this temperature indefinitely.
   (*c*) Quenched to 650°C and held at this temperature for 1 day, then quenched to room temperature.
   (*d*) Quenched to 550°C and held at this temperature for 1 day, then quenched to room temperature.

7. A specimen of 1.1 percent carbon steel with a structure like that shown in Fig. 17.30 is heated to 730°C and then quenched back to room temperature. Make a sketch of the resulting microstructure, identifying the constituents.
8. Assume that the 0.54 percent carbon steel of Fig. 17.26 is slowly cooled to point C and then quenched to 450°C and held at this temperature for one day, whereupon it is cooled to room temperature. What microstructure would you expect to find? Estimate the percentages of the constituents.
9. Transpose paths 1, 2, and 3 in Fig. 17.33 to Fig. 17.32 and determine the resulting microstructures.
10. Describe a cooling path by which one could obtain a structure consisting of fine pearlite, lower bainite, and martensite in an eutectoid steel specimen.
11. (*a*) Demonstrate that the pearlite growth rate equation

$$G \simeq -\frac{D_c^{\gamma}(C_a - C_b)}{S_p}$$

given in Section 17.3 may also be written

$$G \simeq - D_c^{\gamma}(\Delta T)^2$$

where ΔT is the temperature difference between the eutectoid temperature and the temperature at which the austenite is reacted to form pearlite.
(*b*) With the aid of this relationship, explain qualitatively the shape of the growth temperature curve shown in Fig. 17.8.

18 The Hardening of Steel

18.1 Continuous Cooling Transformations. The isothermal transformation diagram is a valuable tool for studying the temperature dependence of austenitic transformations. In even a single reaction, such as the austenite to pearlite transformation, the product varies in appearance with the transformation temperature. A specimen that is allowed to transform over a range of temperatures therefore has a mixed microstructure very difficult to analyze without prior information. Before the original work on isothermal transformations by Davenport and Bain,[1] the basic austenitic reactions were not clearly defined and our knowledge of them was, to say the least, confused. The time-temperature relationships which are mapped out on an isothermal transformation diagram, however, are strictly applicable only to transformations carried out at constant temperatures. Unfortunately, very few commercial heat treatments occur in this manner. In almost all cases, the metal is heated into the austenite range and then continuously cooled to room temperature, with the cooling rate varying with the type of treatment and the size and shape of the specimen. The difference between isothermal transformation diagrams and continuous cooling transformation diagrams are perhaps most easily understood by comparing these two forms for a steel of eutectoid composition. This particular composition is chosen because of its simplicity, and the pertinent diagrams are given in Fig. 18.1. Two cooling curves, corresponding to different rates of continuous cooling, are also shown in Fig. 18.1. In each case, the cooling curves start above the eutectoid temperature and fall in temperature with increasing times. The inverted shape of these curves is due to plotting the time coordinate (abscissa) according to a logarithmic scale. On a linear time scale the curves would be concave toward the right, signifying a decreasing rate of cooling with increasing time.

Now consider the curve marked 1. At the end of approximately 6 sec this curve crosses the line representing the beginning of the pearlite transformation. The intersection is marked on the diagram as point *a*. The significance of point *a* is that it represents the time required to nucleate pearlite isothermally at 650° C (the temperature of point *a*). A specimen carried along line 1, however only

1. Davenport, E. S., and Bain, E.C., *Trans. AIME*, 90 117 (1930).

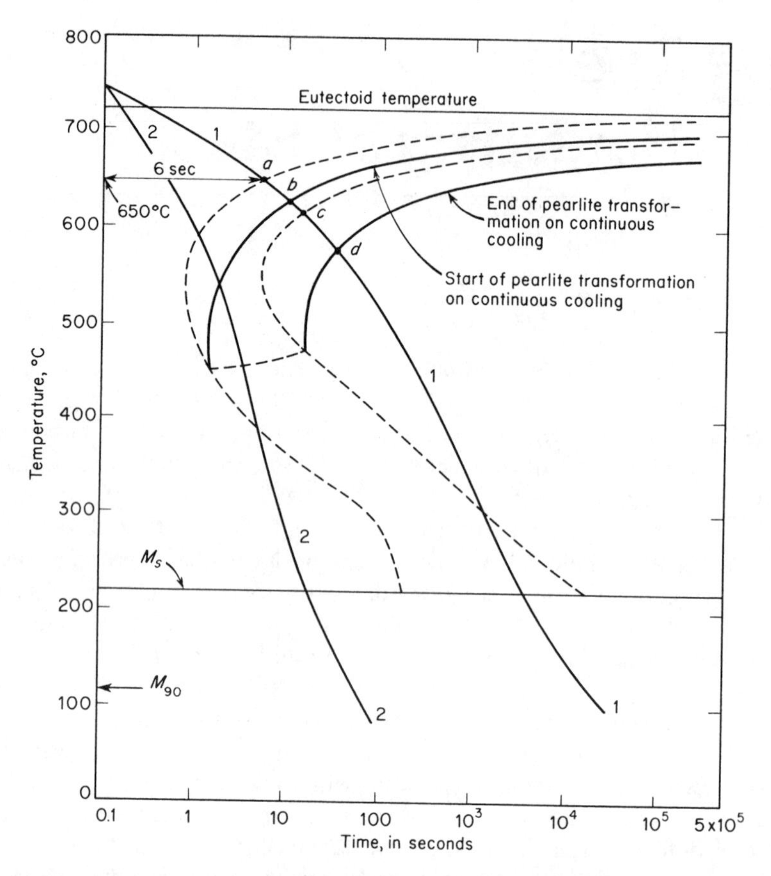

Fig. 18.1 The relationship of the continuous cooling diagram to the isothermal diagram for an eutectoid steel (schematic). (After *Atlas of Isothermal Transformation Diagrams,* United States Steel Corporation, Pittsburgh, 1951.)

reached the 650° C isothermal at the end of 6 sec and may be considered to have been at temperatures above 650° C for the entire 6-sec interval. Because the time required to start the pearlite transformation is longer at temperatures above 650° C than it is at 650° C, the continuously cooled specimen is not ready to form pearlite at the end of 6 sec. Approximately, it may be assumed that cooling along path 1 to 650° C has only a slightly greater effect on the pearlite reaction than does an instantaneous quench to this temperature. In other words, more time is needed before transformation can begin. Since in continuous cooling an increase in time is associated with a drop in temperature, the point at which transformation actually starts lies to the right and below point *a*. (The location of this point may be estimated with the aid of several appropriate assumptions.[2]) This position is designated by the symbol b. In the same fashion, it can be shown that the finish of the pearlite transformation, point d, is depressed downward and to the right of point c, the point where the continuous cooling curve crosses the line representing the finish of isothermal transformation.

The above reasoning explains qualitatively why the continuous-cooling-transformation lines representing the start and finish of the pearlite transformations are shifted with respect to the corresponding isothermal-transformation lines. Why the bainite reaction in this metal does not appear on continuous cooling also needs to be explained. This is not too hard to understand and is related to the fact that the pearlite reaction lines extend over and beyond the bainite-transformation lines. Thus, on slow or moderate rates of cooling (curve 1), austenite in the specimens is converted completely to pearlite before the cooling curve reaches the bainite-transformation range. Since the austenite has already been completely transformed, no bainite can form. Alternatively, as shown by curve 2, the specimen is in the bainite-transformation region for too short a period of time to allow any appreciable amount of bainite to form. An element in this latter conclusion is the fact that the rate at which bainite forms rapidly decreases with falling temperatures. It is generally assumed (as a first approximation) in drawing the continuous-cooling-transformation diagram for this particular alloy, that the transformation, along a path such as curve 2, stops in the region where the bainite and pearlite transformations overlap on the isothermal diagram. As a result, the microstructures corresponding to path 2 should consist of a mixture of pearlite and martensite, with perhaps a small amount of bainite which we shall ignore. The martensite, of course, forms from the austenite which did not transform to pearlite at higher temperatures.

The continuous-cooling-transformation diagram of an eutectoid steel is shown again in Fig. 18.2. A number of cooling curves are also shown on the diagram. The given curves are not quantitative, but rather qualitative, representations of how various cooling rates can produce different microstructures. The curve marked "full anneal" represents very slow-cooling and is usually obtained by cooling specimens (suitably austenitized) in a furnace which has its power supply shut off. This rate of cooling normally brings the steel to room temperature in about a day. Here the transformation of the austenite takes place at temperatures close to the eutectoid temperature and the final structure is, accordingly, a coarse pearlite and close to that predicted for an equilibrium transformation. The second curve, marked "normalizing," represents a heat treatment in which specimens are cooled at an intermediate rate by pulling them out of the austenitizing furnace and allowing them to cool in air. In this case, cooling is accomplished in a matter of minutes and the specimen transforms in the range of temperatures between 550°C and 600°C. The structure obtained in this manner

2. Grange, R. A., and Kiefer, J. M., *Trans. ASM,* 29 85 (1941).

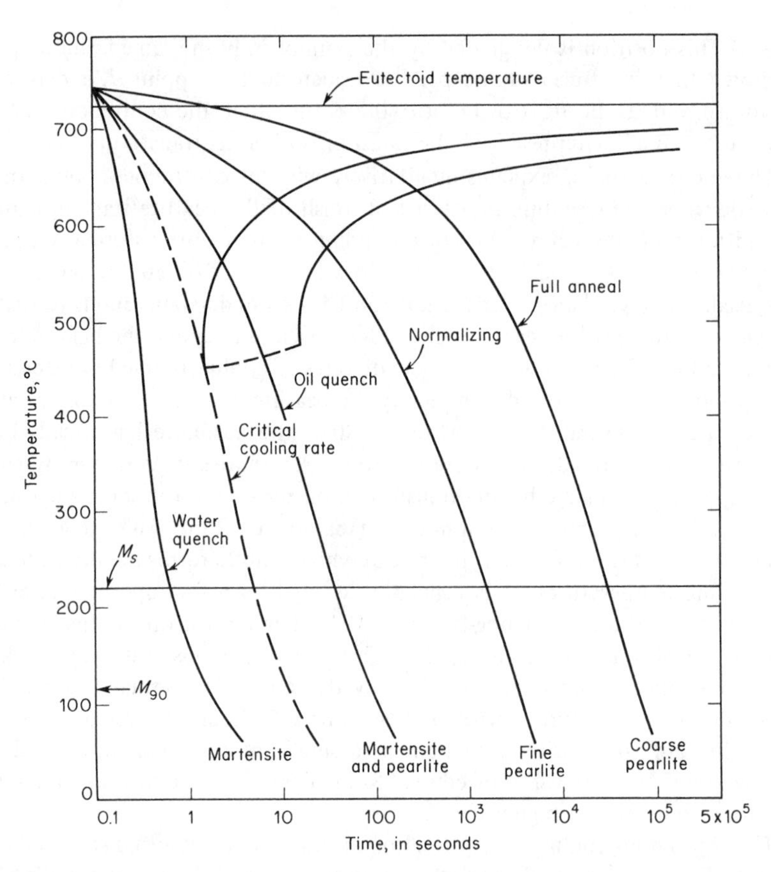

Fig. 18.2 The variation of microstructure as a function of cooling rate for an eutectoid steel.

is again pearlite, but much finer in texture than that obtained in the full annealing treatment. The next cooling curve represents a still faster rate of cooling, such as might be obtained when a red-hot specimen is quenched directly into a bath of oil. Cooling at this rate produces a microstructure which is a mixture of pearlite and martensite. Finally, the curve farthest to the left and marked "water quench" represents a rate of cooling so rapid that no pearlite is able to form and the structure is entirely martensitic.

18.2 Hardenability. Another curve is also shown in Fig. 18.2 as a dashed line – the critical-cooling-rate curve. Any rate of cooling faster than this produces a martensitic structure, while any slower rate produces a structure containing some pearlite.

In a steel specimen of any appreciable size, the cooling rates at the surface and at the center are not the same. The difference in these rates naturally increases with the severity of the quench, or the speed of the cooling process. Thus, there will be little difference, at any instant, between the temperature at the surface and at the center of a bar of some size when it is furnace-cooled (fully annealed). On the other hand, the same bar, if quenched in a rapid coolant such as iced brine, has markedly different cooling rates at the surface and at the center, which are capable of producing entirely different microstructures at the surface and center of the bar. This effect is demonstrated in Fig. 18.3, where cooling curves are plotted to represent paths followed by the surface and the center of a sizable bar (perhaps 2 in. in diameter) assumed to be quenched in a very rapid cooling medium. Two other curves are also plotted on the diagram representing the critical cooling rate, and the rate that gives a structure 50

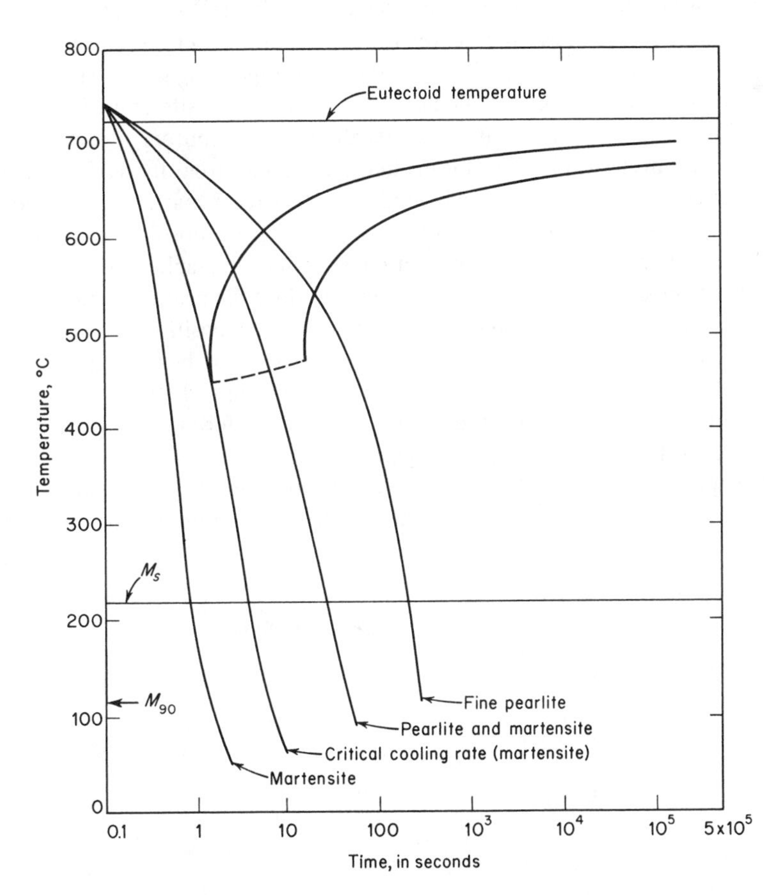

Fig. 18.3 The effect of the difference in the cooling rate at the surface and at the center of a cylindrical bar on the resulting microstructure (schematic).

percent martensite and 50 percent pearlite. It is significant that in the present example these two rates fall within the extremes denoted by the surface and the center of the bar. We may conclude that this specimen will have a structure at the surface which is martensitic, while the center will have a pearlitic structure. The change in the microstructure with distance along a diameter is accompanied by a corresponding variation in the hardness of the metal. This fact can be easily demonstrated by cutting a quenched bar in two on an abrasive cutoff wheel, while taking care to see that the specimen is cooled properly during the sawing so that the metal does not become overheated and the microstructure altered. The circular cross-section thus exposed is subjected to a number of hardness tests made at equal intervals along a diameter, and the results plotted to give a hardness contour of the type shown in Fig. 18.4. The hardness traverse shows that the martensitic structure near the surface is very hard (Rockwell C-65), while the pearlitic structure near the center is considerably softer (Rockwell C-40). Also plotted on the diagram is a horizontal line corresponding to the hardness of an eutectoid-steel structure containing 50 percent martensite and 50 percent pearlite (C-54). Notice that this intersects the hardness contour in the regions where it rises most steeply. This means that the distance from the surface where the specimen cooled at a rate which produced 50 percent martensite is capable of rather precise experimental measurement. The same position can also be determined by preparing the cross-section as a metallographic specimen and observing it under the microscope. Alternatively, since the pearlite etches darker than martensite, a macroscopic measurement of the position at which the structure effectively changes from martensite to pearlite can be made in terms of a color change. At any rate, the position corresponding to half-martensite and half-pearlite is easy to measure and is used as a criterion for measuring the depth to which a steel hardens with a given type of quench.

The depth at which the 50 percent martensite structure is obtained in a bar of steel is a function of a number of variables that include the composition and

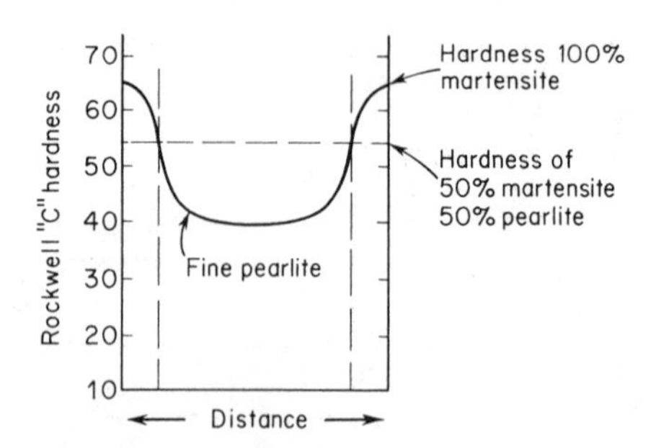

Fig. 18.4 Typical hardness test survey made along a diameter of a quenched cylinder (after sectioning the cylinder).

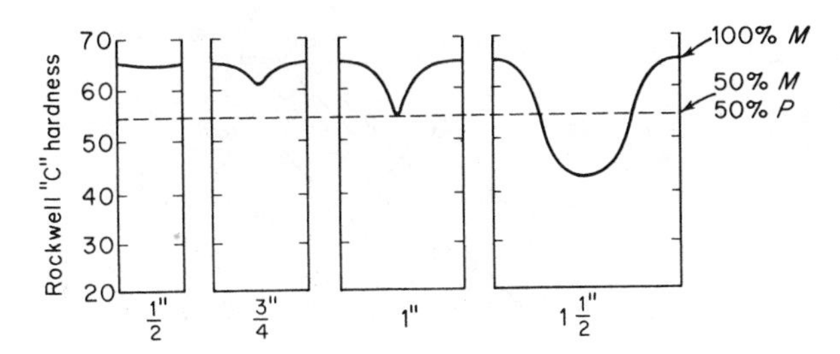

Fig. 18.5 Hardness test traverses similar to that of Fig. 18.4 made on a series of steel bars of the same composition, but with different diameters (schematic).

grain size of the metal (austenitic), the severity of the quench, and the size of the bar. Let us consider first the effect of changing the diameter of the bar of steel. Suppose that a number of bars of the same steel are given an identical quench in a brine solution and then sectioned so as to obtain hardness contours. The results are shown schematically in Fig. 18.5. An investigation of these curves shows that there is one unique diameter, the value of which is 1 in. The bar with this diameter hardens so that it has the 50 percent pearlite-50 percent martensite structure just at its center. All bars with smaller diameters are effectively hardened throughout, while any bar with a larger diameter has a soft core containing pearlite. This particular diameter is called the *critical diameter*. Its value depends on the steel in question and the means of quenching, and its importance lies in the fact that it gives a measure of the ability of the steel to respond to a hardening heat treatment. The particular steel being discussed has a moderate ability to harden, or, as it is more properly stated, it has a moderate hardenability. According to Fig. 18.5, its critical diameter D is 1.0 in. The addition of suitable alloying elements to steels can greatly increase their hardenability and this is shown by corresponding increases in their critical diameters. The critical diameter D of a steel is, consequently, a measure of its hardenability (ability to harden), but it also depends on the rate of cooling (the type of quench). In order to eliminate this latter variable, it is general practice to refer all hardenability measurements to a standard cooling medium. This standard is the so-called *ideal quench*, which uses a hypothetical cooling medium assumed to bring the surface of a piece of steel instantly to the temperature of the quenching bath, and maintain it at this temperature. The critical diameter corresponding to an ideal quench is called the *ideal critical diameter* and is designated D_I.

The ideal quenching medium is assumed to remove the heat from the surface as fast as it can flow out from inside of the bar. While such a quenching medium does not exist, its cooling action on steels is capable of computation and comparison with those of ordinary commercial quenching media, such as water, oil, and brine. Information of this type is frequently presented in the form of

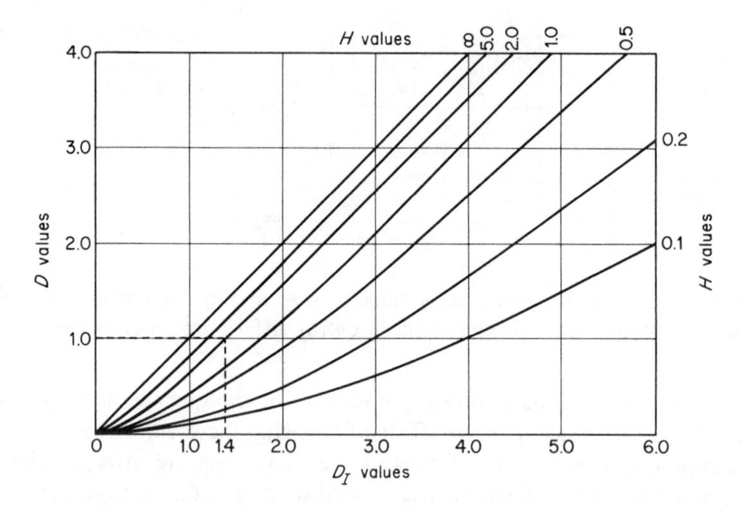

Fig. 18.6 Relationship of the critical diameter D to the ideal critical diameter D_I for several rates of cooling (H values). (After Grossman, M. A., *Elements of Hardenability*, ASM, Cleveland, 1952.)

curves such as those shown in Fig. 18.6 where the ideal critical diameter D_I is plotted as the abscissa, and the critical diameter D is plotted as the ordinate. A number of different curves are plotted on this chart, each corresponding to a different rate of cooling. In each case, the rate of cooling is measured by a number known as its H value, or the severity of the quench. Some of the values of this number for commercial quenches are given in Table 18.1. The use of the chart is easily illustrated by considering the example given above (Fig. 18.5) where it was determined that the critical diameter D, obtained in a brine quench ($H = 2$), was 1.0 in. The H value for a brine quench with no agitation is 2.0. The intersection of the curve for this severity of quench with the 1.0-in. ordinate occurs at a D_I value (abscissa) 1.4. The ideal critical diameter, or hardenability of the steel in question, is $D_I = 1.4$ in.

Table 18.1 Severity of Quench Values for Some Typical Quenching Conditions.

H Value	*Quenching Condition*
0.20	Poor oil quench – no agitation
0.35	Good oil quench – moderate agitation
0.50	Very good oil quench – good agitation
0.70	Strong oil quench – violent agitation
1.00	Poor water quench – no agitation
1.50	Very good water quench – strong agitation
2.00	Brine quench – no agitation
5.00	Brine quench – violent agitation
∞	Ideal quench

Table 18.1 shows the value of agitation during a quench. When hot metal is placed in a liquid cooling medium, gas vapors are formed at the surface of contact between the hot metal and the liquid, which retard the heat flow from metal to liquid. Agitation, or movement of the specimen relative to the liquid, is instrumental in removing the bubbles from the surface and increasing the rate of cooling. The fact that brine, water, and oil are better cooling agents in descending order is closely related to the removal of vapor bubbles from the surface. The lower inherent viscosity of water allows the bubbles to be moved faster in a water quench than in an oil quench. In a brine quench, the presence of salt in the water causes a series of small explosions to occur near the hot surface and therefore violently agitates the cooling solution in the vicinity of the quenched material.

The Grossman [3] method of determining the ideal critical diameter D_I, outlined above, is too time consuming to be of wide practical application. It has been described as a means of introducing the concept of hardenability and the quantity which we use to measure it, namely, D_I. A much more convenient and widely used method of determining hardenability employs the Jominy End Quench Test.

In the Jominy test, a single specimen takes the place of the series of specimens required in the Grossman method. The standard Jominy specimen consists of a cylindrical rod 4 in long and 1 in. in diameter. In making a test, the specimen is first heated to a suitable austenitizing temperature and held there long enough to obtain a uniform austenitic structure. It is then placed in a jig and a stream of water allowed to strike one end of the specimen. The experimental arrangement is shown in Fig. 18.7. The advantage of the Jominy

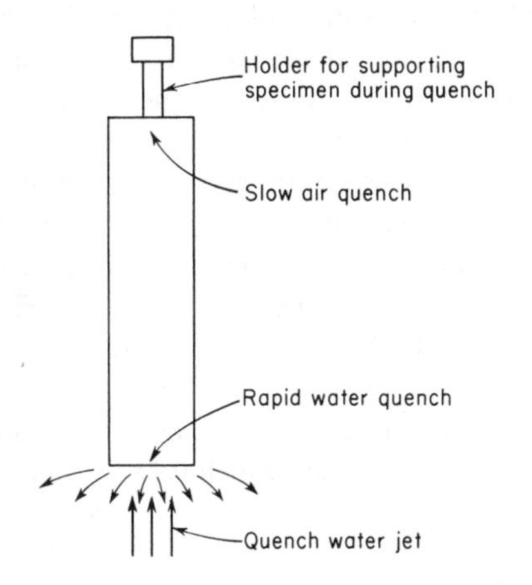

Fig. 18.7 Jominy hardenability test.

3. Grossman, M. A., *Elements of Hardenability,* ASM, Cleveland, 1952.

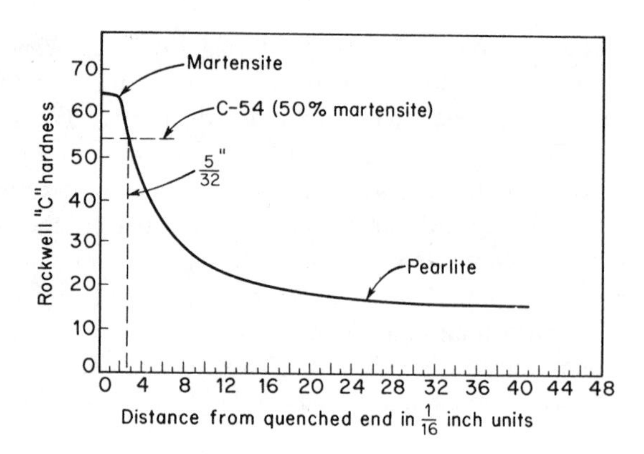

Fig. 18.8 Variation of hardness along a Jominy bar. (Schematic for steel with 0.8 percent carbon.)

test is that in a single specimen one is able to obtain a range of cooling rates varying from a very rapid water quench at one end to a slow air quench at the other end. Following the complete transformation of the austenite in the bar, two shallow flat surfaces are ground on opposite sides of the bar and a hardness-test traverse is made running from one end of the bar to the other along one of these prepared surfaces. The data thus obtained are plotted to give the Jominy hardenability curve. A typical example is shown in Fig. 18.8 where it can be seen that the hardness is greatest where the cooling is most rapid – near the quenched end. A great deal of effort has gone into the determination of the cooling rates expected at various distances from the quenched end of a Jominy bar and to correlating these data with cooling rates inside circular bars and other shapes. Of particular importance to our present discussion is the relationship between the size or diameter, of a steel bar quenched in an ideal quenching medium which has the same cooling rate at its center as a given position along the surface of a Jominy bar. This information is furnished in the curve of Fig. 18.9. Its importance is associated with the fact that if the position on the Jominy bar where the structure is half martensite is known, then this curve makes possible the determination of the ideal critical diameter. For example, consider the Jominy curve shown in Fig. 18.8. In a 0.80 percent C steel, the hardness of 50 percent martensite (Fig. 18.23) should be about Rockwell C-54. This value occurs in the Jominy curve at $\frac{5}{32}$ in. from the quenched end. According to Fig. 18.9 the D_I of the given steel is 1.4 in., which is in agreement with our previous value, 1.4 in.

18.3 The Variables that Determine the Hardenability of a Steel. The hardenability of a steel, which is expressed by its D_I, is a function of its chemical composition and the size of the austenite grains that it contains at the

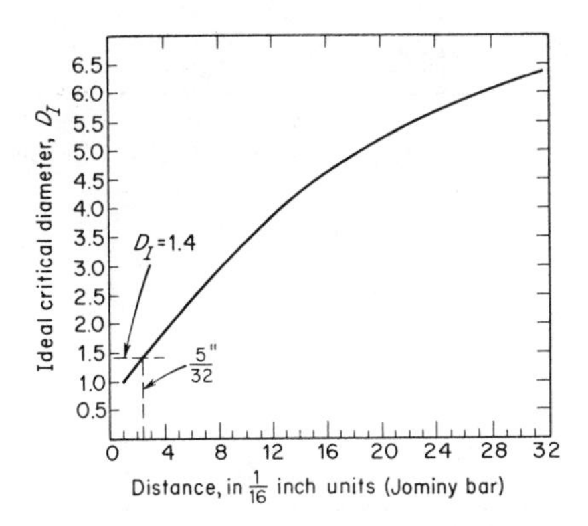

Fig. 18.9 Location on Jominy bar at which the cooling rate is equivalent to the center of a circular bar quenched in an ideal quenching medium. (After Lamont, J. L., *Iron Age*, **152**, Part 2, 1943, Copyright 1943, Chilton Co.)

instant of quenching. We shall now discuss the effect of these factors on the hardenability, but first a word should be said about the principles of changing hardenability. A metal with a high hardenability is one in which austenite is able to transform to martensite without forming pearlite, even when the rate of cooling is rather slow. Conversely, high rates of cooling are required to form martensite in steels of low hardenability. In either case, the limiting factor is the rate at which pearlite forms at elevated temperatures. Any variable that moves the pearlite transformation lines to the right in a continuous-cooling-transformation diagram, such as Fig. 18.3, makes it possible to obtain a martensite structure at a slower rate of cooling. A movement of the pearlite transformation nose to the right is thus associated with an increase in hardenability. From another viewpoint we can say that anything that slows down the nucleation and growth of pearlite increases the hardenability of a steel.

18.4 Austenitic Grain Size. When steel is heated into the austenite region in order to austenitize the metal, the low-temperature structure which is transformed to the gamma phase is, in general, an aggregate of cementite and ferrite (that is, pearlite, or decomposed martensite). In this reversed transformation, the austenite grains form by nucleation and growth; the nuclei form heterogeneously at cementite-ferrite interfaces. Because of the large interfacial area available for nucleation, the number of austenite grains that appear is usually large. The transformation of steel upon heating is, therefore, characterized initially by a small austenitic grain size. However, in the austenite range, thermal movements of the atoms are rapid enough to cause grain growth, so that extended times and high temperatures in the austenite range are capable of greatly increasing the size of the initial austenite grains.

Table 18.2 ASTM Grain-Size Numbers.

ASTM Grain-Size Number	*Average Number of Grains per Square Inch as Viewed at 100 Diameters*
1	1
2	2
3	4
4	8
5	16
6	32
7	64
8	128

The size of the austenite grain that is attained before a metal is cooled back to room temperature is important in determining a number of the physical properties of the final structure, including the hardening response of the steel. Before describing this latter effect, let us explain the most commonly accepted method of designating the austenitic grain size. The ASTM grain-size number is defined by the relationship

$$n = 2^{N-1}$$

where n is the number of grains per square inch as seen in a specimen viewed at a magnification of 100 times, and N is the ASTM grain-size number. The usual range of austenitic grain sizes in steels lies between 1 and 9. The number of grains per square inch in this interval is given in Table 18.2. Notice that as the grains get smaller (more numerous), the grain-size number increases.

18.5 The Effect of Austenitic Grain Size on Hardenability. The effect of grain size on hardenability has been explained[4] on the basis of the heterogeneous manner in which pearlite nucleates at austenitic grain boundaries. While the rate of growth G of pearlite is independent of the austenite grain size, the total number of nuclei that form per second varies directly with the surface available for their formation. Thus, in a fine-grain steel, ASTM No. 7, there is four times as much grain-boundary area as in coarse-grain steel of grain size No. 3. The formation of pearlite in the fine-grain steel is, therefore, more rapid than it is in the coarse-grain steel and, accordingly, the fine-grain steel has a lower hardenability.

The use of a coarse austenitic grain size in order to increase the hardenability of steel is not generally practiced. The desired increase in hardenability is accompanied by undesirable changes in other properties, such as an increase in

4. *Ibid.*

brittleness and a loss of ductility. Quench cracks, or cracking of the steel due to thermal shock and the stresses incident to the quenching operation, are also more common in large-grain specimens.

18.6 The Influence of Carbon Content on Hardenability. The hardenability of a steel is strongly influenced by its carbon content. This fact is shown in Fig. 18.10 where the variation of the ideal critical diameter D_I with carbon content is plotted for three different grain sizes. In addition to showing that the hardenability increases with increasing carbon content, these curves demonstrate the very low hardenability of simple iron-carbon alloys. For example, an eutectoid steel with about 0.8 percent C and a small grain size (No. 8) possesses an ideal diameter 0.28 in. This means that the maximum theoretical diameter of a steel bar of this relatively high-carbon content that can be hardened to its center (in an ideal quench) is about ¼ in. Any ordinary quench will, accordingly, not harden even this size bar to its center. Fortunately, so-called commercial carbon steels always contain some manganese and sometimes small amounts of other elements that increase their hardenability. The manganese in these steels is required in order that they can be manufactured economically. In this respect, the isothermal diagrams previously described were for steels containing manganese. The hardenabilities of the latter are considerably higher than if simple iron-carbon alloys had been considered.

Because increasing carbon content is associated with an increase in hardenability, it is evident that the formation of pearlite and proeutectoid constituents

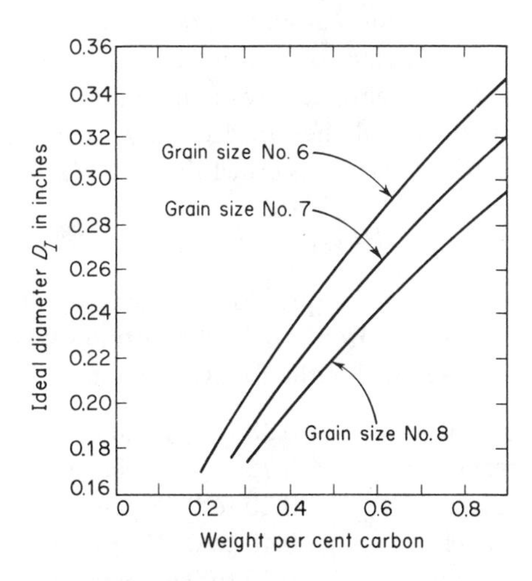

Fig. 18.10 Ideal critical diameter as a function of carbon content and austenite grain size for iron-carbon alloys. (After Grossman, M. A., *Elements of Hardenability*, ASM, Cleveland, 1952.)

becomes more difficult the higher the carbon content of the steel. This statement is true not only for hypoeutectoid steels, but also for those with carbon contents greater than the eutectoid composition (hypereutectoid steels), provided that each steel is completely transformed to austenite before its hardenability is measured. It frequently happens in practice that hypereutectoid steels are austenitized in the two-phase cementite plus austenite region. When this happens, nearly all of the structure becomes austenitic, but a small amount of cementite is stable and does not dissolve. On cooling, the residual carbide particles encourage pearlite nucleation, resulting in lowered hardenability.

18.7 The Influence of Alloying Elements on Hardenability. Each and every one of the chemical elements in a steel has an influence on its hardenability. The degree, of course, varies with the element in question. Of the common alloying elements added to steel, cobalt is the only one known to decrease hardenability. The presence of cobalt in steel increases both the rate of nucleation and the rate of growth of pearlite,[5] and steels containing this element are more difficult to harden than those that do not contain it.

Other common alloying elements, to the extent that they are soluble in iron, increase the hardenability of steels. There are a number of ways that this effect can be demonstrated. One of the simplest is by empirical hardenability multiplying factors.[6] These factors make possible a first-order approximation of the hardenability of a steel when only its chemical composition and its austenitic grain size are known. A short list of the needed factors is given in Table 18.3. In this table, columns 2, 3, and 4 give the same information as that shown graphically in Fig. 18.10 for the austenitic grain sizes 6, 7, and 8. Thus, for example, the ideal critical diameter of a steel with a carbon content of 0.40 and an austenitic grain size of 8 is obtained by running down the first column (the percentage column) to 0.40 and then horizontally to the third column. The hardenability obtained in this manner is called the *base diameter* and is

$$D_{I_{\text{carbon}}} = 0.1976$$

Suppose now that the steel in question conforms to that classified as an AISI 8640 steel in the American Iron and Steel Institute's system. In this case its composition limits should fall within the following bounds:

carbon	0.38 to 0.43 percent
manganese	0.75 to 1.00 percent
silicon	0.20 to 0.35 percent
nickel	0.40 to 0.70 percent
chromium	0.40 to 0.60 percent
molybdenum	0.15 to 0.25 percent

5. Mehl, R. F., and Hagel, W. C., *Prog. in Metal Phys.*, 6 74 (1956).
6. Grossmann, M. A., *Trans. AIME*, 150 242 (1942).

Let us also assume, for the purpose of a simple illustrative calculation, that the steel under consideration contains 0.40 percent carbon and the maximum concentration of all the other indicated elements. The next step is to find the value of the multiplying factor for each element. This is done in the same manner as is used for finding the base diameter. Thus, opposite 1.00 percent as read in the percent column, there appears in the manganese column the figure 4.333. The base diameter is multiplied by this factor to give the hardenability of the steel as determined by its austenitic grain size, its carbon content and its manganese content. In the same manner we find that the multiplying factors for the other elements are

Percentage of Element	*Multiplying Factor*
1.00% Mn	4.333
0.35% Si	1.245
0.70% Ni	1.255
0.60% Cr	2.296
0.25% Mo	1.75

The total hardenability of the steel in question is obtained by multiplying the base diameter by every one of these factors, or

$$D_I = 0.1976 \times 4.333 \times 1.245 \times 1.255 \times 2.296 \times 1.75$$

$$D_I = 5.40$$

The significance of the above numerical value is worth considering. The addition of a total amount of alloying elements equal to less than 3 percent produces a steel with an ideal critical diameter of 5.40 in. Even in a poor water quench ($H = 1.0$), a steel of this hardenability should have a critical diameter of nearly 5 in. (4.7). An ordinary carbon steel of the same carbon content (C AISI 1040) contains manganese in the limits 0.60 to 0.90 percent Mn as the only basic alloying element, Assuming the maximum manganese content, D_I is 0.8 in. and the D (critical diameter) for a poor water quench ($H = 1.0$) is less than ½ in. The importance of the alloying elements in low-alloy steels in developing the hardenability is quite apparent.

Refer again to the allowable compositional limits in the low-alloy steel discussed above. It is interesting to compute its hardenability in terms of the minimum compositional limits rather than the maximum values. When this is done, one obtains a D_I of 2.44 in. It is clear that the hardenability of commercial steels vary within rather wide limits, corresponding to the compositional variations dictated by the problems of manufacture.

The fact that different alloying elements have greatly different effects on hardenability is clearly indicated in Table 18.3. An element which has no effect would have a multiplying factor of unity. For the elements listed in the table, nickel shows the least effect and manganese the greatest. Phosphorus and sulfur,

Table 18.3 Hardenability Multiplying Factors.*

Percent	*Carbon-Grain Size*			Mn	Si	Ni	Cr	Mo
	#6	#7	#8					
0.05	0.0814	0.0750	0.0697	1.167	1.035	1.018	1.1080	1.15
0.10	0.1153	0.1065	0.0995	1.333	1.070	1.036	1.2160	1.30
0.15	0.1413	0.1315	0.1212	1.500	1.105	1.055	1.3240	1.45
0.20	0.1623	0.1509	0.1400	1.667	1.140	1.073	1.4320	1.60
0.25	0.1820	0.1678	0.1560	1.833	1.175	1.091	1.54	1.75
0.30	0.1991	0.1849	0.1700	2.000	1.210	1.109	1.6480	1.90
0.35	0.2154	0.2000	0.1842	2.167	1.245	1.128	1.7560	2.05
0.40	0.2300	0.2130	0.1976	2.333	1.280	1.146	1.8640	2.20
0.45	0.2440	0.2259	0.2090	2.500	1.315	1.164	1.9720	2.35
0.50	0.2580	0.2380	0.2200	2.667	1.350	1.182	2.0800	2.50
0.55	0.273	0.251	0.231	2.833	1.385	1.201	2.1880	2.65
0.60	0.284	0.262	0.241	3.000	1.420	1.219	2.2960	2.80
0.65	0.295	0.273	0.251	3.167	1.455	1.237	2.4040	2.95
0.70	0.306	0.283	0.260	3.333	1.490	1.255	2.5120	3.10
0.75	0.316	0.293	0.270	3.500	1.525	1.273	2.62	3.25
0.80	0.326	0.303	0.278	3.667	1.560	1.291	2.7280	3.40
0.85	0.336	0.312	0.287	3.833	1.595	1.309	2.8360	3.55
0.90	0.346	0.321	0.296	4.000	1.630	1.321	2.9440	3.70
0.95				4.167	1.665	1.345	3.0520	
1.00				4.333	1.700	1.364	3.1600	

* Abstracted from *U.S.S. Carilloy Steels*, U.S. Steel Corp., Pittsburgh, 1948.

which occur in steels as impurities, are generally considered to have factors of unity.

The hardenability of alloy steels is also visible in isothermal-transformation diagrams. As an example, consider the steel designated AISI 4340. The isothermal transformation diagram for this steel is shown in Fig. 18.11. The hardenability of this steel, for the composition indicated in the figure, is 6.55 in. A significant characteristic of the transformation diagram of this steel is that the pearlite and bainite transformations both exhibit noses. At the upper nose, the diagram shows that the minimum time required to form a visible amount of proeutectoid ferrite is about 200 sec (650° C), and just below this temperature the minimum time to form pearlite is somewhat more than 1800 sec (30 min). In the same manner, the minimum time for the formation of a visible amount of bainite is slightly over 10 sec at 450° C. This transformation diagram should be compared with those for plain carbon steels given earlier (Figs. 17.24, 17.31, and 17.32 in Chapter 17). The difference in the rapidity with which the transformations occur is quite evident.

The continuous cooling transformation for the AISI 4340 steel is shown in

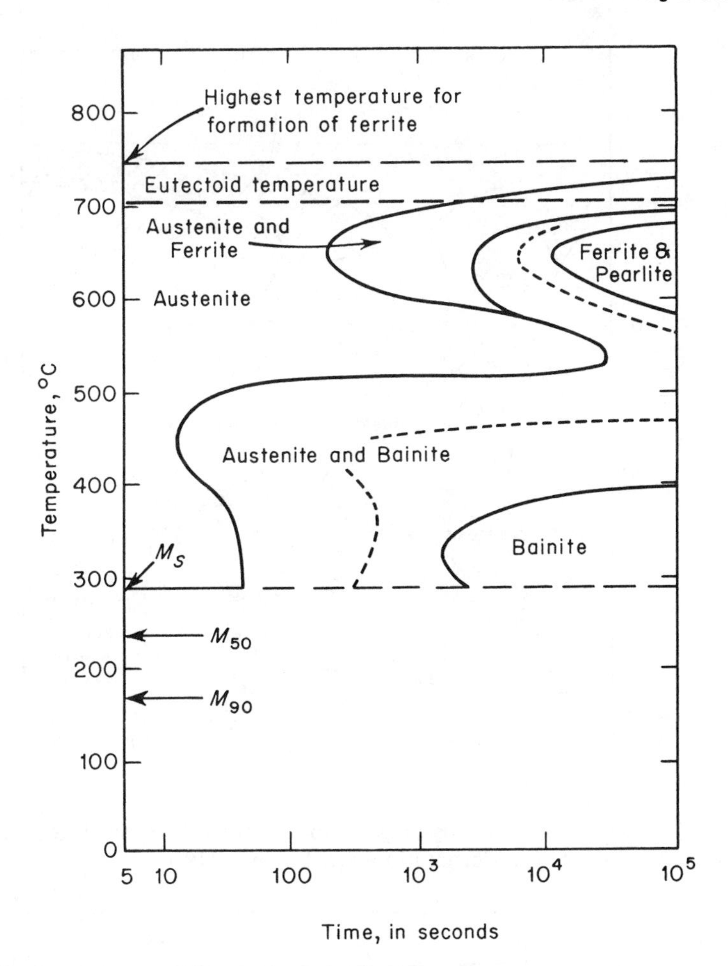

Fig. 18.11 Isothermal transformation diagram of a low alloy steel (4340): 0.42 percent carbon, 0.78 percent manganese, 1.79 percent nickel, 0.80 percent chromium, 0.33 percent molybdenum. Grain size 7–8. Austenitized at 1550°F. (From *U.S.S. Carilloy Steels*, United States Steel Corporation, Pittsburgh, 1948.)

Fig. 18.12. It is apparent in the figure that any cooling rate which brings the steel to room temperature in less than 90 sec produces a martensitic structure. The effect of the high hardenability is also clearly shown in the corresponding Jominy curve for this steel, which is shown in Fig. 18.13. At a distance greater than 2 in. from the quenched end of the bar the structure is still 95 percent martensite.

The isothermal diagram of Fig. 18.11 is characteristic of those steels in which bainite can be obtained during continuous cooling. In the diagram for a plain

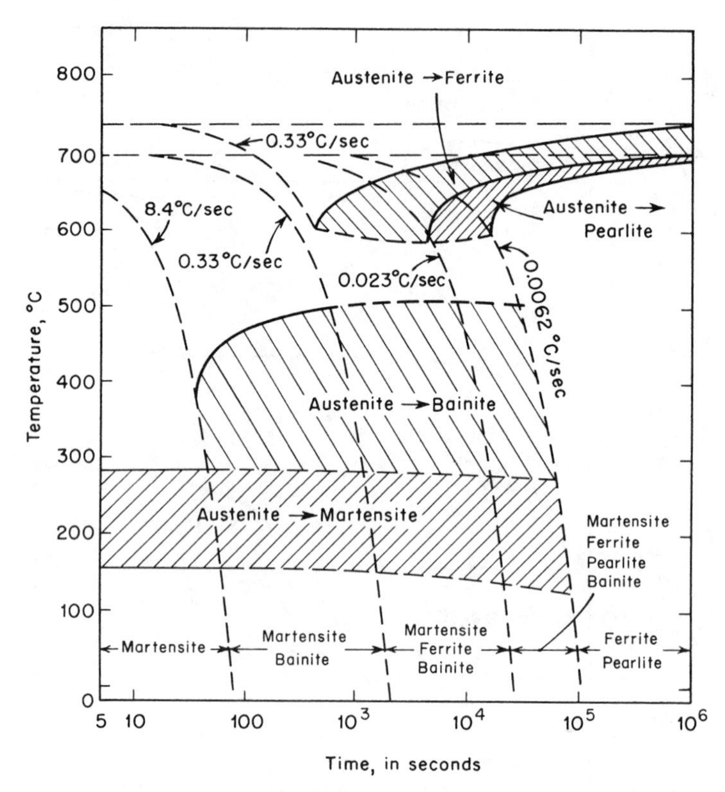

Fig. 18.12 Continuous cooling diagram for 4340 steel. (From *U.S.S. Carilloy Steels*, United States Steel Corporation, Pittsburgh, 1948.)

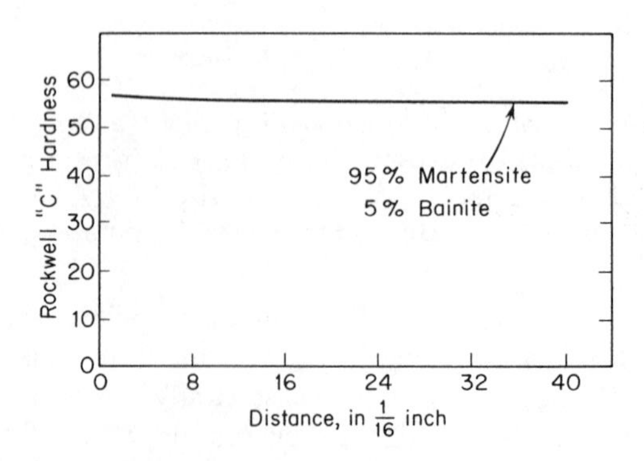

Fig. 18.13 Jominy hardenability curve for 4340 steel. (From *Atlas of Isothermal Transformation Diagrams,* United States Steel Corporation, Pittsburgh, 1951.)

carbon steel considered earlier, measurable amounts of bainite do not form on continuous cooling because the pearlite transformation region extends over the corresponding bainite region. In the present alloy steel, the bainite nose extends beyond the pearlite nose, thus making possible the formation of bainite on continuous cooling. The possible structures that can be obtained in this alloy with different cooling rates are shown at the bottom of the diagram.

18.8 The Significance of Hardenability. Is high hardenability desirable in a steel? The answer is that it is not always desirable. This is especially true when steels are to be welded. Steels containing any appreciable amount of added elements are notoriously difficult to weld successfully. In making a weld, two pieces of steel are joined by casting a section of molten metal between them. This operation naturally heats the metal adjacent to the weld and, for some distance to each side of the weld center, steel is raised into the austenitic region. If the hardenability of the metal is high, hard brittle martensite may form on cooling to room temperature. This is accentuated by the quenching effect of the cold steel surrounding the heat-affected zone, for heat flows rapidly from the heated region into the surrounding metal. Because of the above considerations, structural steels, intended for construction of bridges, buildings, and ships, are usually designed to have moderate hardenabilities.

High hardenabilities are desirable in steels that are to be hardened as one of the final steps in their manufacture. It is generally believed that the best combination of physical properties (strength plus ductility) are obtained if the metal is transformed completely to martensite during the hardening treatment. In some cases, if the M_f temperature is below room temperature, this may require that the metal be cooled to subzero temperature in order to remove most of the residual austenite. However, the primary requirement for obtaining a martensitic structure in a given steel object is that the steel have a sufficiently high hardenability to harden under the required quenching conditions. Unfortunately, hardenability is associated with alloying elements in the steel, the presence of which raises its cost. As a result, the basic problem is economic and requires the avoidance of the use of steel with too high a hardenability (and too high a cost) for the job at hand. An important factor in determining the required hardenability is the speed of quench that must be used. Of course, the more rapid the quench, the lower the required hardenability, but high cooling rates are also associated with severe thermal shocks. These can cause quench cracks and warping of the finished object, leading to its rejection. Thus, in commercial practice, oil and even air quenches are commonly used in order to reduce damage to steel parts caused by the more rapid quenching processes. The added cost of the high-hardenability steel in these cases is justified by the savings in the number of parts rejected.

18.9 The Martensite Transformation in Steel. The martensitic structure in steels is a simple phase which marks it from the aggregates of ferrite and carbides, which we call pearlite and bainite. The martensitic crystal structure is body-centered tetragonal and can be assumed to be an intermediate structure between the normal phases of iron – face-centered cubic and body-centered cubic. The relationship between the three structures has already been discussed in Chapter 16. The Bain distortion in steel is shown in Fig. 18.14. In these drawings, the positions that carbon atoms occupy are shown by black dots. It must be recognized, however, that actually in any given steel specimen only a very small percentage of the possible positions are ever filled. In the face-centered cubic structure there are as many possible positions for carbon atoms as there are iron atoms. This means that if all positions were filled, the alloy would have a composition containing 50 atomic percent carbon. The maximum actually observed is 8.9 atomic percent (2.06 weight percent). Figure 18.14A represents face-centered cubic austenite. In this structure, the carbon atoms occupy the midpoints of cube edges and the cube centers. These are equivalent positions for in each case a carbon atom finds itself located between two iron atoms along a ⟨001⟩ direction. The equivalent positions in the austenite, when it is considered a body-centered tetragonal structure, are shown in Fig. 18.14B. Notice that in this cell the carbon positions occur between iron atoms along c-axis edges and in the centers of the square faces at each end of the prismatic cell. Finally, the martensitic structure is shown in Fig 18.14C. In this last case, the tetragonality of the cell is greatly reduced, but the carbon atoms are still in the

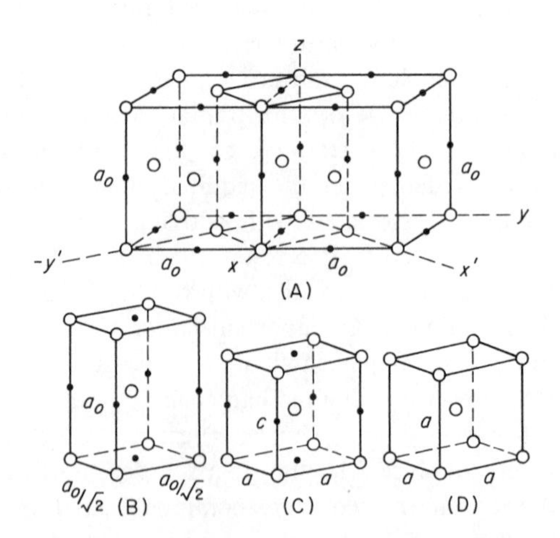

Fig. 18.14 The Bain distortion in the martensite transformation of steels. Black dots represent positions that carbon atoms can occupy. Only a small fraction are ever filled. (A) Face-centered cubic. (B) Tetragonal representation of austenite. (C) Tetragonal martensite. (D) Body-centered cubic.

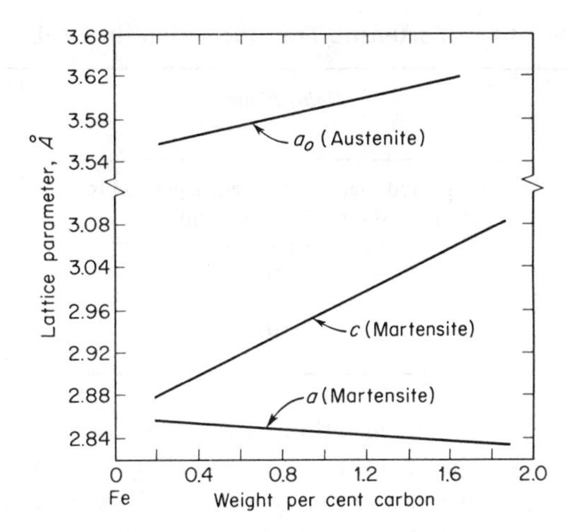

Fig. 18.15 Variation of the lattice parameters of austenite and martensite as a function of carbon content. (Roberts, C. S., *Trans. AIME,* **197**, p. 203 [1953].)

same relative positions with respect to their iron-atom neighbors as in the austenite unit cell. The resulting structure is tetragonal only because carbon atoms are inherited from the austenite, and the transformation which would normally proceed to body-centered cubic is not able to go to completion. The carbon atoms can be considered to strain the lattice into the tetragonal configuration and the extent of the tetragonality which occurs may be deduced from Fig. 18.15. Notice that the lattice parameters are plotted as a function of carbon content in both austenite and martensite, and, in each case, the parameters vary linearly with carbon content. In martensite with increasing carbon content the c-axis parameter increases, while the parameter associated with the two a axes decreases. At the same time, the cubic parameter of austenite (a_0) increases with increasing carbon content. These relationships may be expressed in terms of simple equations [7] where x is the carbon concentration.

Martensite parameters (Ångstroms)

$$c = 2.861 + 0.116x$$

$$a = 2.861 - 0.013x$$

Austenite parameter (Ångstroms)

$$a_0 = 3.548 + 0.044x$$

7. Roberts, C. S., *Trans. AIME,* 191 203 (1953).

Table 18.4 Martensite Transformations in Steel.

Carbon Content	*Habit Plane*	*Orientation Relationship*
0–0.4%	Elongated martensite needles parallel to ⟨110⟩ austenite directions and lying on {111} austenite planes	?
0.5–1.4%	$(225)_A$	$(111)_A$ $(101)_M$ $[1\bar{1}0]_A$ $[11\bar{1}]_M$
1.5–1.8%	(259)	?

A simple computation of the c/a ratio at 1.0 percent carbon yields 1.045. This number should be compared to the corresponding ratio when face-centered cubic austenite is considered as a tetragonal lattice. As mentioned previously, this ratio is 1.414, so that in most steels (less than 1.00 percent C), the martensitic lattice is certainly much closer to body-centered cubic than to face-centered cubic.

The lattice parameter curves of Fig. 18.15 show that the tetragonality of martensite varies with the carbon content. This should have a bearing on the crystallographic characteristics of the transformations. With reference to the Wechsler, Lieberman, and Read theory of martensite formation, a change in tetragonality of the product phase means a difference in magnitude of the Bain distortion. This, in turn, implies a difference in the size of the required shear and rotation. As a result, it might be expected that both the habit plane and the orientation relationship between the parent and product should vary with carbon content. This is actually observed, as may be seen in Table 18.4.

The best data in Table 18.4 relates to the concentration range between 0.5 and 1.4 percent C, which fortunately includes almost all of the commercially important carbon steels that are hardened to martensite. In this range of carbon contents, the habit plane of the martensite plates is very close to a {225} plane in the austenite. Since there are 12, {225} planes and two possible twin-related orientations of the martensite for each habit plane, there are 24 possible ways that an austenite crystal can form a martensite plate. The corresponding orientation relationship between the lattices in the martensite and in the austenite of these steels is known as the *Kurdyumov-Sachs relationship.* It states that the (101) plane of the martensite is parallel to the (111) plane of the parent austenite, and, at the same time, the $[11\bar{1}]$ direction in the martensite is parallel to the $[1\bar{1}0]$ direction of the austenite.

The martensite plates which form in the middle carbon region of steels are quite similar in appearance to those that occur in iron-nickel alloys. They are, accordingly, small in size and lenticular in shape. The martensitic transformation in iron-carbon alloys is primarily athermal. However, small amounts of austenite

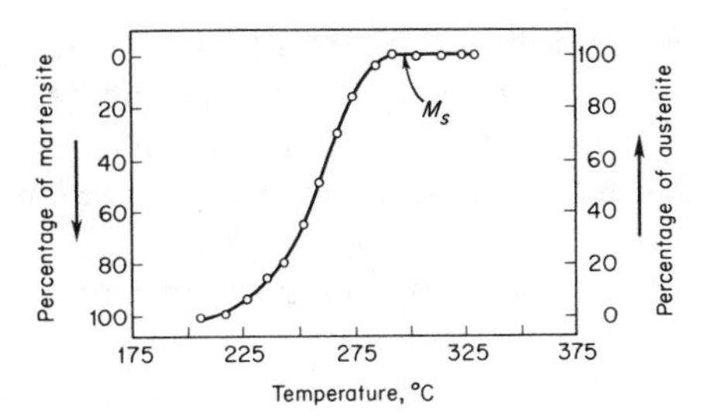

Fig. 18.16 The formation of martensite in a 0.40 percent carbon low-alloy steel (2340) as a function of temperature. (After Grange, R. A., and Stewart, H. M., *Trans. AIME*, **167**, 467 [1946].)

may be transformed to martensite isothermally.[8] A typical curve showing the formation of athermal martensite as a function of temperature is shown in Fig. 18.16. The similarity of this curve to that for the formation of martensite on cooling given for other alloys in Chapter 16 is apparent. However, one important difference should be observed. There is no curve given for the reverse transformation. The reason for this is quite simple: the martensitic reaction in steels is not reversible. Iron-carbon martensite represents an extremely unstable structure with a high free energy relative to the more stable phases – cementite and ferrite. The entrapment of carbon atoms in what should be a body-centered cubic structure, capable of holding at equilibrium an infinitesimal amount of carbon, produces a lattice with a high internal strain. Even moderate reheating promotes its decomposition. The study of the phenomena of martensite decomposition will be considered in the section entitled "Tempering."

Both the M_s (martensite start) and M_f (martensite finish) temperatures in steel are functions of the carbon content, as shown in Fig. 18.17. The M_f

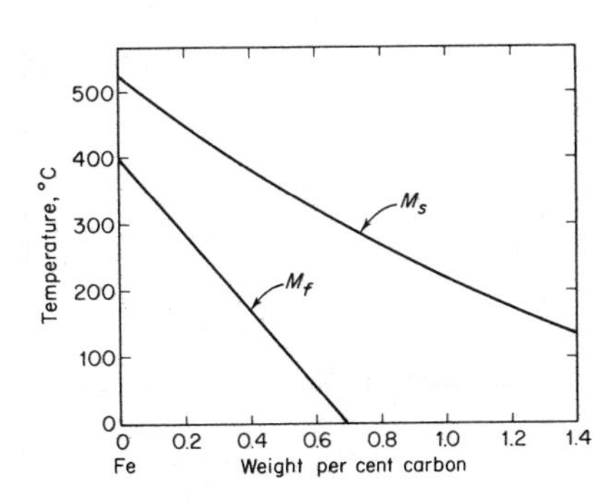

Fig. 18.17 Variation of M_s and M_f with carbon concentration in steel. (After Troiano, A. R., and Greninger, A. B., *Metal Progress*, **50**, 303 [1946].)

8. Averbach, B. L., and Cohen, Morris, *Trans. ASM*, **41** 1024 (1949).

temperature, however, is usually not clearly defined. By this it is meant that the martensite reaction can never be theoretically complete even at absolute zero temperature. The transformation of the last traces of austenite becomes more and more difficult the smaller the total amount of austenite remaining. Curves such as the M_f line shown in Fig. 18.17 are based on visual estimations of the structures of metallographic specimens and small amounts of retained austenite are very difficult to measure in a structure composed of many small overlapping martensite plates. The M_f temperature shown in Fig. 18.17 should thus be interpreted as the temperature at which the reaction is completed as far as one can determine by visual means. One technique for the measurement of retained austenite in quenched steels involves quantitative X-ray diffraction measurements and is capable of measuring retained austenite in amounts as small as 0.3 percent. This method[9] has been used to determine the amount of retained austenite in carbon steels quenched to room temperature and to liquid-air temperatures (–196°C). The results are shown in Fig. 18.18. Notice that the M_f temperature in Fig. 18.17 occurs at room temperature (20°C) at about 0.6 percent C. The actual amount of retained austenite under these conditions is over 3 percent and, after cooling to –196°C, there is still almost 1 percent of austenite remaining.

Substitutional alloying elements in steels also affect the martensitic transformation. This may be reflected in the habit plane indices and the orientation relationships between the martensite and the parent austenite. Unfortunately, there is very little data available on these effects. The influence of alloying

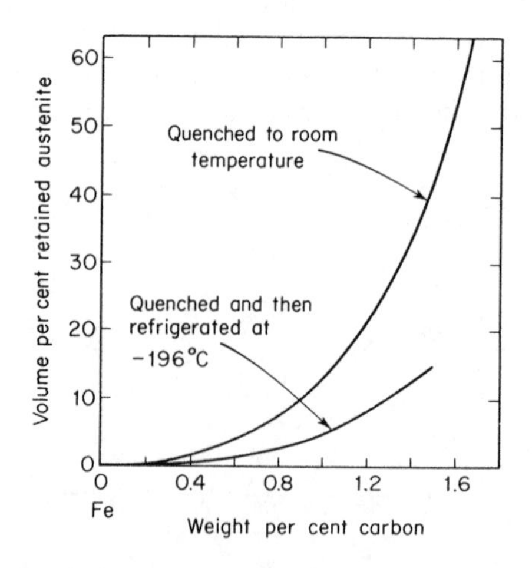

Fig. 18.18 Retained austenite in steel as a function of carbon content. (After Roberts, C. S., *Trans. AIME*, **197**, 203 [1953].)

9. Roberts, C. S., *Trans. AIME*, 197 203 (1953).

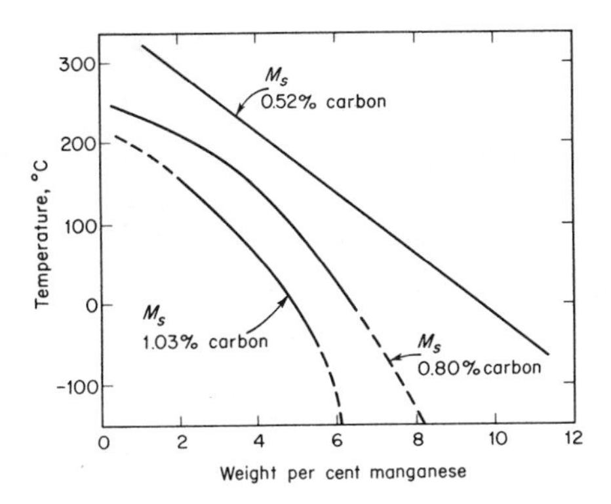

Fig. 18.19 Variation of M_s with manganese content in three series of steels with different carbon concentrations. (After Russell, J. V., and McGuire, F. T., *Trans. ASM*, **33**, 103 [1944].)

elements on the M_s temperature is much more easily recognized and measured. As a typical example, consider the addition of manganese to steels containing carbon where a very decided lowering of the M_s temperature occurs. This is shown in Fig. 18.19. One of the interesting features of this set of curves is the very rapid decline of the M_s temperature with increasing manganese content in steels that also contain 1.0 percent carbon. For those compositions containing more than 6 percent manganese, the M_s is so far below room temperature that we can assume that a quench to room temperature produces a structure that remains austenitic indefinitely at room temperature. The famous Hadfield manganese steel (10–14 percent Mn and 1–1.4 percent C) takes advantage of this fact to produce a steel with an austenitic structure which work hardens very rapidly and has an initial very high strength due to the presence of the manganese and carbon in solid solution. This combination of properties makes for a very tough, hard, abrasion-resistant metal. A typical application is in the buckets and teeth of power shovels.

The effect of other solid-solution elements on the M_s temperature varies with the element concerned. Manganese has the strongest effect followed by chromium. All common alloying elements (substitutional) lower the M_s temperature, except for cobalt and aluminum which raise it.

18.10 The Hardness of Iron-Carbon Martensite. The hardness of iron-carbon martensites as a function of carbon content is shown in Figs. 18.20 and 18.21. In the first case, the hardness is plotted on the Vickers diamond-pyramid hardness scale, and in the other the Rockwell-C scale. It is frequently felt that, because the Rockwell curve levels off at about 0.6 percent C, there is no increase in the hardness of martensite once the carbon content exceeds 0.6

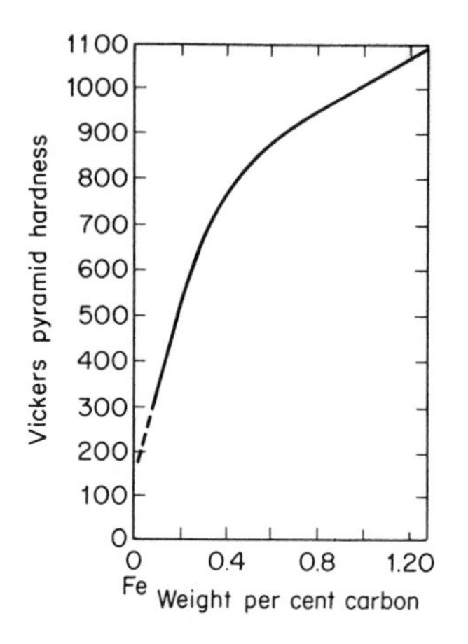

Fig. 18.20 Variation of hardness of martensite as a function of carbon content as measured on the Vickers scale. (After Bain, E. C., *Functions of the Alloying Elements in Steels,* ASM, Cleveland, 1939.)

percent C. That this is not true is shown by the Vickers relationship. Actually, the apparent leveling off of the hardness on the Rockwell instrument is due largely to the fact that this machine is somewhat insensitive in the range of hardnesses encountered in hardened high-carbon steels. The Rockwell-C scale uses a cone-shaped indentor which has a rounded point, and hardness is measured by the depth to which this penetrator moves under a fixed load. The elastic component of the deformation is subtracted from the total movement. In very hard materials of the type presently considered, the penetrator moves into

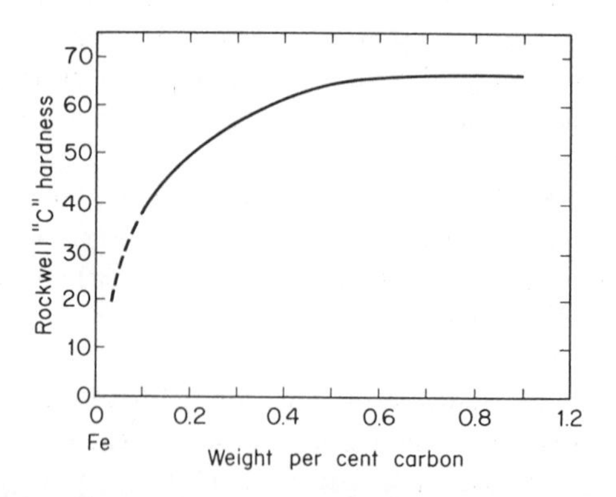

Fig. 18.21 Variation of the hardness of martensite as a function of carbon content as measured on the Rockwell-C scale. (After Burns, J. L., Moore, T. L., and Archer, R. S., *Trans. ASM,* **26**, 1 [1938].)

the metal only a very small distance and the blunted tip requires relatively large hardness differences in order to make an appreciable difference in the depth of penetration and in the hardness number. In the Vickers test, on the other hand, the hardness is measured not by the depth of penetration, but by dividing the load by the area of an indentation formed by pressing into the metal a diamond penetrator with the shape of an inverted four-sided pyramid. Since this penetrator is not blunted at its tip and the hardness is measured in terms of the area of indentation, Vickers or diamond-pyramid hardness values are more characteristic of the actual hardness of the metal. One might reasonably ask, therefore, why the Rockwell data are given at all. This is easily answered, for the Rockwell machine is very widely used and most of the data which appear in the literature is based on Rockwell tests. The principal advantage of the Rockwell test is that it is very rapid and inexpensive and, at the same time, relatively accurate and reproducible.

The hardness of martensite results from the presence of carbon in the steel. In this respect, it is significant that the martensite product in alloys other than steels are not necessarily hard. It is also evident from Fig. 18.20 that an appreciable amount of carbon (about 0.4 percent) is needed in steel to cause a marked degree of hardening. In order to obtain a truly hardened steel, the two required factors are: first, an adequate carbon concentration in the metal, and second, rapid cooling to produce a martensitic structure.

Recent experimental work[10] based on transmission electron microscopy has clearly shown that the martensite in carbon steels can form by two reactions. One gives a structure known as *lath martensite*; the other is a *lenticular martensite* that is internally twinned. The primary factor controlling the relative volume fractions of these two forms is apparently the transformation temperature. A lower transformation temperature favors a higher concentration of the twinned lenticular martensite. Since increasing the carbon concentration. generally lowers the M_s temperature, higher carbon steels tend to have large volume fractions of the twinned component. This is demonstrated in Fig. 18.22. On the other hand, in low-carbon steels the martensite is primarily of the lath type.

The lath martensite is characterized by a high internal dislocation density of the order of 10^{11} to 10^{12} cm^{-2}. These dislocations are arranged in plate-shaped cells. On the other hand, the twinned martensite normally does not contain a high density of dislocations. The observed two basic forms of martensite are consistent with an assumption that in the lath martensite the macroscopic shear is accomplished by slip, while in the lenticular martensite it occurs by twinning (see Figs. 16.34 and 16.35). The difference in the defect concentration between the two martensite forms has an important bearing on the respective distribution

10. Speich, G. R., *TMS-AIME,* 245 2553 (1969).

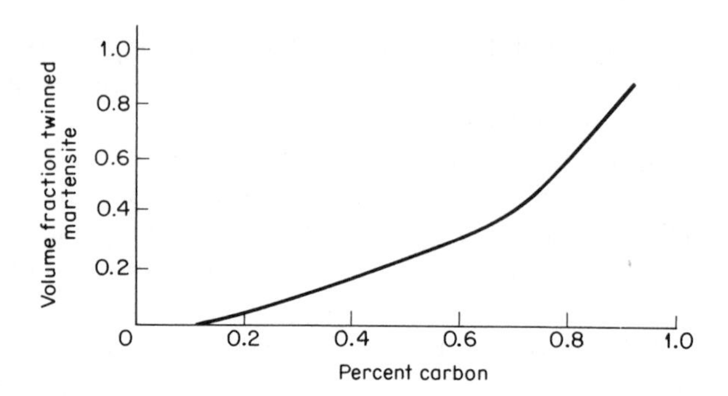

Fig. 18.22 Curve showing the volume fraction of twinned martensite as a function of the carbon concentration. (From Speich, G. R., *TMS-AIME*, **245**, 2553 [1969].)

of the carbon atoms inside them. In lath martensite, the carbon atoms tend to diffuse and segregate around the dislocations. Even in a very rapid quench there is ample time for this diffusion to occur. On the other hand, in the twinned structure where there is a much lower density of dislocations, the carbon atoms are forced to occupy normal interstitial sites.

As may be seen in Fig. 18.20, the hardness of martensite in pure iron is about 200 DPH. An annealed pure iron has a hardness of the order of 100 DPH. Speich[11] indicates that the high hardness of martensitic pure iron is caused by the substructural strengthening effect of the finely spaced cell walls and lath boundaries in the lath martensite. In effect, this is attributing the hardness to the high dislocation density in this structure.

In iron containing carbon there is additional hardening due to the carbon. In this case the carbon is believed to increase the hardness through its interaction with the dislocations. This can occur by segregation of the dislocations in the cell walls and by solid-solution strengthening. With regard to this latter effect, the martensitic transformation can be viewed[12] as trapping an abnormally high concentration of carbon atoms in solid solution, this very high concentration thereby producing a large hardening component.

A very important practical fact about the hardness of martensite in steels is that in all of the so-called low-alloy steels (less than about 5 percent total alloying elements) the hardness of martensite can be assumed to depend only on the carbon concentration of the metal. Consequently, if one knows the carbon content of any low-alloy steel, he is able to determine (with the aid of one of the curves such as Fig. 18.20 or 18.21) the approximate hardness of the steel when it has a martensitic structure.

11. *Ibid.*
12. Hirth, J. P., and Cohen, M., *Met. Trans.*, 1 3 (1970).

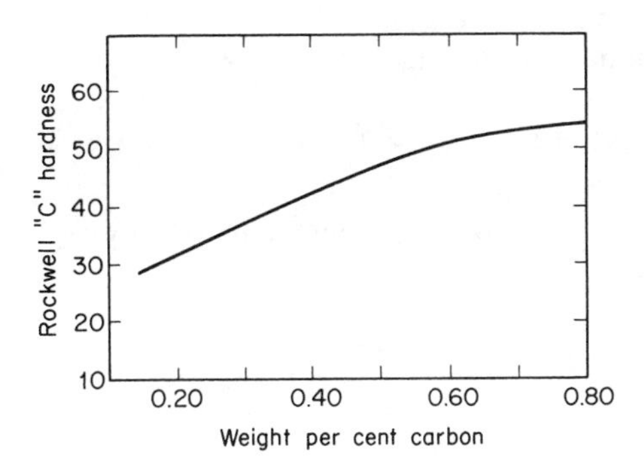

Fig. 18.23 Average hardness of 50 percent martensite structures in low-alloy steels. (After Hodge, J. M., and Orehoski, M. A., AIME, TP 1800, 1945.)

The hardness of a steel containing some fixed fraction of martensite, say 50 percent, is also approximately a function of only its carbon content. In this respect, Fig. 18.23 shows the variation of the Rockwell-C hardness with carbon content in steels containing 50 percent martensite structures. In practice, the hardness of commercial steels usually lie within ± 4 Rockwell-C hardness units of this curve.

18.11 Dimensional Changes Associated with the Formation of Martensite. When austenite transforms to martensite, there is a change in volume that can be computed by considering the Bain distortion and the lattice parameters of austenite and martensite. With reference to Fig. 18.15 and considering a 1 percent carbon steel, in the austenite the lattice parameter is

$$a_0 = 3.548 + 0.044(1.0) = 3.592 \text{ Å}$$

and the volume of the austenite unit cell (tetragonal form) is

$$V_A = a_0 \cdot \frac{a_0}{\sqrt{2}} \cdot \frac{a_0}{\sqrt{2}} = \frac{(3.592)^3}{\sqrt{4}} = 23.15 \text{ Å}^3$$

In martensite, the lattice parameters are:

$$a = 2.861 - 0.013(1.0) = 2.848 \text{ Å}$$

$$c = 2.861 + 0.116(1.0) = 2.977 \text{ Å}$$

The volume of the martensite unit cell is

$$V_M = c \times a \times a = 2.977(2.848)^2 = 24.14 \text{ Å}^3$$

The change in volume is, accordingly,

$$\Delta V = V_M - V_A = 24.14 - 23.15 = 0.99 \text{ Å}^3$$

and the relative change in volume, assuming martensite to form from austenite at room temperature, is

$$\frac{\Delta V}{V_A} = \frac{0.99}{24.14} = 4.3 \text{ percent}$$

When a 1 percent carbon steel transforms to martensite, there is a volume increase of about 4.3 percent that can be considered to be an average value, representative of steels in general, that does not vary widely with the carbon content. This is because we are transforming from austenite with a c/a ratio of 1.414 to martensite with c/a ratios lying between 1.0 and 1.090, corresponding to the maximum range of carbon contents (0 to 2 percent C).

Because of the many orientations which martensite plates can take in a single austenite crystal, it can be assumed that the volume expansion is isotropic in a specimen of sufficient size. Length changes may, accordingly, be used to measure the deformation associated with the martensitic reaction. In this respect, as is shown in calculus, a small isotropic length change is approximately equal to one-third the corresponding volume change. Therefore,

$$\frac{\Delta l}{l} = \frac{\Delta V}{3V} = \frac{4.3 \text{ percent}}{3} = 1.4 \text{ percent}$$

18.12 Quench Cracks. When a piece of steel is cooled in such a way as to form martensite, two basic dimensional changes occur. First, there is the normal thermal contraction due to cooling, but superimposed on this is the expansion of the metal as it transforms from austenite to martensite. Under the right conditions, these volumetric changes can produce very high internal stresses. If these stresses become large enough, they can produce plastic deformation and the steel will be deformed or warped. While plastic deformation tends to reduce the severity of the quenching stresses, the degree to which this is accomplished depends on a number of factors, and it is quite possible to have large enough residual stresses remaining in the metal to actually cause rupture. These localized fractures are called *quench cracks.*

The approximate nature of the residual stress pattern that develops in an actual steel shape on quenching to martensite will now be considered. For this purpose, steel specimens in the form of round cylinders will be considered. When quenched, the surface always cools faster than the center and undergoes the martensitic transformation first, which hardens the surface relative to the center.

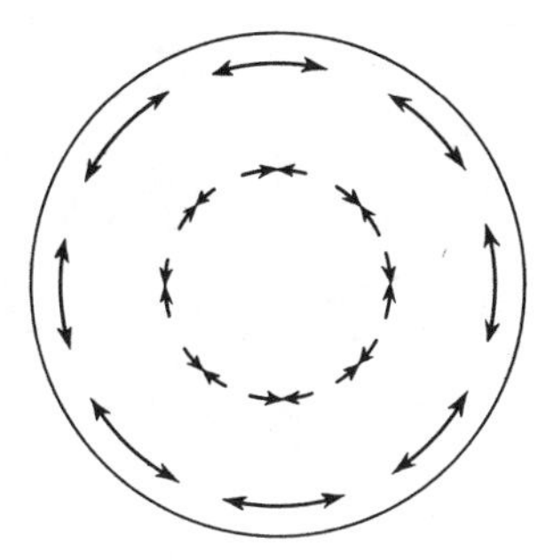

Fig. 18.24 Schematic tangential stress pattern in a quenched cylinder when the surface is left in a state of tension and the center in compression.

Supplementing this hardening is the fact that the yield strength, or plastic flow stress of the metal, increases with decreasing temperature. Now, whether or not the surface will be set in tension relative to the center of the bar depends on the sign of the net volumetric change that occurs in the interior of the bar after the surface has hardened. If the expansion in this region, due to the martensitic transformation, is larger than the remaining thermal contraction, the surface will attain a residual state of tension. The center will, of course, remain in a state of compression and the nature of this stress pattern is implied in Fig. 18.24. The reversed stress pattern, in which the surface ends in a state of residual compressive stress and the center in a state of tensile stress, occurs when the thermal contraction in the center of the bar subsequent to the hardening of the surface exceeds the martensitic expansion. Which of these two basic stress distributions eventuates depends on the relative cooling rates at the surface and at the center of the bar. This, of course, is a function of both the size of the bar and the speed of the quench. When the product of these two variables is large (large diameter and fast cooling), the surface hardens although the center is still at a very high temperature. The magnitude of the thermal contraction in this case is large and determines the sign of the volume change in the central regions of the bar. In other words, thermal contraction usually exceeds the expansion of the martensite transformation. When the difference between the cooling rates at the surface and center are only moderate, the center is at a temperature only slightly above the surface when the latter hardens. The thermal contraction in the central areas subsequent to the hardening of the surface is then smaller than the expansion due to the formation of martensite. Figure 18.25 shows schematically this variation of the stress at the surface and center as a function of the difference in the rates of cooling of the two regions. Notice that the curves in Fig. 18.25 are actually for axial stresses. However, the surface tangential stress approximately equals the axial stress.[13]

13. Holloman, J. H., and Jaffe, L. D., *Ferrous Metallurgical Design*. John Wiley and Sons, Inc., New York, 1947.

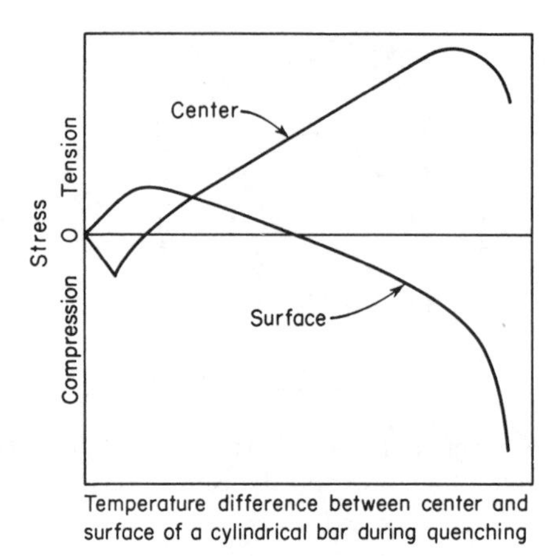

Fig. 18.25 Residual axial stresses in a hypothetical steel cylinder quenched to form martensite as computed by Scott. (Scott, H., *Origin of Quenching Cracks*, Sci. Papers Bur. Standards, **20**, 399 [1925].)

Quench cracks occur as a result of tensile stress. Figure 18.25 shows that, starting with equal rates of cooling at center and surface and increasing this difference, a residual tensile stress results at the surface. In practice, many steel specimens fall in this range where the surface attains a tensile stress. It is significant that over a large range of specimen diameters, the magnitude of the surface tensile stress increases with increasing difference in cooling rate. For a fixed rate of cooling, this is equivalent to saying that increasing the diameter increases the magnitude of the residual tensile stress. With very large diameters, however, the surface obtains a residual compressive stress.

Several factors help to determine the magnitudes of the residual stresses. It was shown in the previous section that the expansion that occurs when steel transforms to martensite at room temperature involves a dilation of about 1.4 percent. Actually, during a quench, austenite does not transform completely to martensite at room temperature, but over a range of temperatures starting at the M_s temperature. It is also true that the higher the temperature at which the transformation occurs, the less the expansion due to the formation of martensite. This follows from a corresponding change in the austenite and martensite lattice parameters. In steels in which the M_s temperature is high, the specific-volume changes are smaller and, as a consequence, there is a reduced tendency to form quench cracks. High-carbon steels and those containing alloying elements that lower M_s are, conversely, more subject to quench cracking. In addition to lowering M_s, high-carbon contents also increase the danger of cracking because of the increased hardness or brittleness that high-carbon concentrations impart to the steel.

The length of time that a steel is maintained at temperatures of the order of room temperature to 100°C after quenching also has an effect on the formation of quench cracks: the longer the time, the more liable the formation of cracks. Several different explanations of this phenomenon have been offered. One of the most attractive is related to the isothermal transformation of retained austenite to martensite.[14] The formation of isothermal martensite adds an additional volumetric strain to an already badly strained metal, thereby increasing the probability of crack formation.

18.13 Tempering. Steels that have undergone a simple hardening quench are usually mixtures of austenite and martensite, with the latter constituent predominating. Both of these structures are unstable and slowly decompose, at least in part, if left at room temperature; the retained austenite transforms to martensite and the martensite undergoes a reaction which will be described shortly. Since specific-volume changes are connected with both reactions, hardened steel objects undergo dimensional changes as a function of time when left at room temperature. Of more general importance is the fact that a structure which is almost completely martensite is extremely brittle and is also very liable to develop quench cracks if aged at room temperature. These factors lead to the conclusion that steels with a simple martensitic structure are of little useful value, and a simple heat treatment called *tempering* is almost always used to improve the physical properties of quenched steels. In this treatment, the temperature of the steel is raised to a value below the eutectoid temperature, held there for a fixed length of time, after which the steel is cooled again to room temperature. The obvious intent of tempering is to allow diffusion processes time to produce both a dimensionally more stable structure and one that is inherently less brittle.

It is common practice to subdivide the reactions that occur during tempering into five categories, called the five stages of tempering.[15] In the first stage, a carbide which is not cementite precipitates in the martensite. In consequence, the carbon content of the martensite is reduced and a two-phase structure of carbide and low-carbon martensite results. This reaction occurs at rates that can be measured in the temperature interval between room temperature and approximately 200°C. The second stage corresponds to the decomposition of retained austenite to bainite. The bainite reaction becomes measurable at about 100°C and extends to about 300°C, while for temperatures below 100°C austenite may be assumed to react to martensite and not bainite. The formation of ferrite and cementite from the reaction products of stages one and two constitute what is known as the third stage of tempering. This reaction occurs in

14., Averbach, B. L., and Cohen, Morris, *Trans. ASM,* 41 1024 (1949).
15. *Ibid.*

a reasonable length of time at temperatures above 200°C. The fourth stage involves the growth of the cementite particles, and the fifth stage (see Section 18.17) applies primarily to alloy steels. In this last stage, intermetallic compounds and complex carbides are found.

The First Stage of Tempering. In low-carbon steels (that is, steels containing less than 0.2 percent carbon) the reactions during tempering differ somewhat from those above this concentration of the solute. The amount of carbon that can segregate at the dislocations in the lath martensite component of the structure is about 0.2 percent.[16] On quenching and during the initial stages of tempering, the carbon diffuses to and collects around the dislocations. For steels containing less than 0.2 percent, the carbon can be assumed to be nearly completely associated with the dislocations and the lath boundaries. For compositions above 0.2 percent carbon there is an increasing fraction of the solute that is not able to segregate in this manner. These carbon atoms remain in normal interstitial sites. This fraction grows with increasing carbon concentration not only as a result of the greater abundance of solute atoms, but also because, with increasing carbon concentration, the fraction of twinned martensite grows. Since this component of the structure has a much lower dislocation density, there is a corresponding decrease in the number of dislocations at which the carbon may segregate as the solute concentration is increased.

During the first stage of tempering of carbon steels containing less than 0.2 percent carbon, the principal effect is merely the segregation of additional carbon atoms to the dislocations and lath boundaries. However, for concentrations greater than 0.2 percent carbon, a carbide is observed to precipitate. This is epsilon (ϵ) carbide, which forms from that fraction of the carbide not segregated at the dislocations and lath boundaries. This constituent should not be considered a preliminary step in the formation of cementite (Fe_3C), but rather as another phase which, under conditions existing in the first stage of tempering, nucleates and grows more rapidly than cementite.[17] The indicated conditions are low temperatures and the highly strained condition of the martensite lattice in a high-or medium-carbon steel. The crystal structure of epsilon carbide is hexagonal close-packed with lattice parameters.[18]

$$c = 4.33 \text{ Å} \quad a = 2.73 \text{ Å} \quad c/a = 1.58$$

The carbon content of epsilon carbide also differs from that of cementite and has not been determined with precision. The chemical composition of this phase is believed[19] to lie close to $Fe_{2.4}C$ and its carbon content is, accordingly, about one-fifth greater than that of cementite.

16. Speich, G. R., *TMS-AIME,* **245** 2553 (1969).
17. Roberts, C. S., Averbach, B. L., and Cohen, Morris, *Trans. ASM,* **45** 576 (1953).
18. Jack, K. H., *Jour. of the Iron and Steel Inst.,* **169** 26 (1951).
19. Roberts, C. S., Averbach, B. L., and Cohen, Morris, *Trans. ASM,* **45** 576 (1953).

It would thus appear that high-carbon martensite decomposes, by nucleation and growth, into a two-phase mixture of carbide and low-carbon martensite. The carbon content of the martensite in the product is independent of the carbon content of the original martensite and has a value equal to approximately 0.20 percent carbon. Therefore the first stage of tempering produces a mixture of martensite and a second phase, epsilon carbide, that increases linearly in amount with increasing carbon content (above 0.2 percent C) of the steel.

The epsilon-carbide particles are extremely small in size, as is to be expected considering the low temperature range at which they form. They have, however, been identified with the aid of the electron microscope[20] and it appears that the preferred position for their nucleation is at subgrain boundaries inside martensite plates. The average diameter of these subgrains is about 10^{-4} to 10^{-5} cm. The thickness of the epsilon-carbide subgrain-boundary network is less than 2×10^{-6} cm (200 Å).

The first stage of tempering is accompanied by a change in specific volume of the metal. It is thus possible to follow the kinetics of the reaction with the aid of precision measurements of the change in length that occurs during the process. In making these measurements, the tempering cycle is interrupted at certain time intervals and the specimen cooled to room temperature for the length determination. Following each reading, the specimen is returned to the tempering temperature. The curve obtained when a 0.68 percent carbon steel is tempered at 200°F (93°C) is shown in Fig. 18.26. Also shown in the same figure is the corresponding curve obtained when the steel is aged at room temperature. Even in the latter case there is an appreciable dilation of the specimen corresponding to the first-stage reaction. The length change in the specimen tempered at 93°C

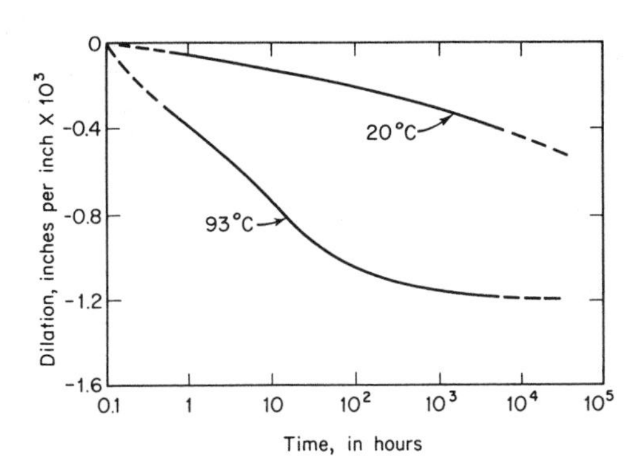

Fig. 18.26 Length changes during the first stage of tempering a martensite structure in a 0.68 percent steel. (Roberts, C. S., Averbach, B. L., and Cohen, Morris, *Trans. ASM*, **45**, 576 [1953].)

20. Lement, B. S., Averbach, B. L., and Cohen, Morris, *Trans. ASM*, **46** 851 (1954).

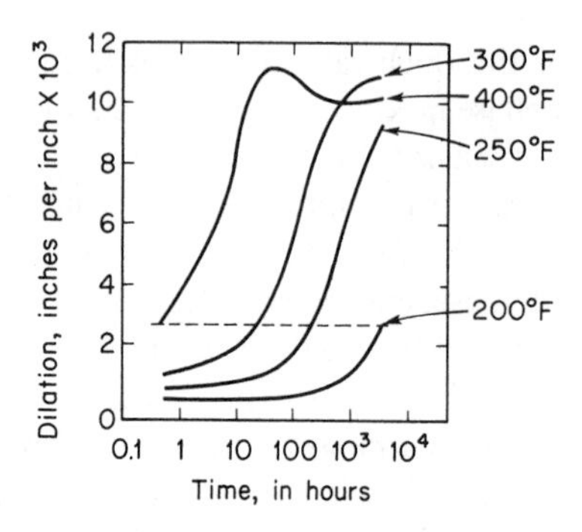

Fig. 18.27 Length changes corresponding to second stage of tempering. Curves correspond to decomposition of 100 percent austenite to bainite. 1.43 percent carbon steel. (Roberts, C. S., Averbach, B. L., and Cohen, Morris, *Trans. ASM*, **45**, 576 [1953].)

is considerably larger, and the fact that it approaches asymptotically the value 1.2×10^{-3} in. per in. signifies that this value probably corresponds to the end of the first stage of tempering in the given alloy. An important fact should be recognized relative to these length changes; they are in the opposite direction to those that occur when austenite transforms to martensite. The specimen shrinks in volume (or length) instead of expanding.

The Second Stage of Tempering. During the second stage, retained austenite transforms to bainite. At these low temperatures (100°C to 300°C), the microstructure of the bainite that forms consists of ferrite and epsilon carbide.[21] Large, positive dimensional changes accompany this transformation. A set of curves corresponding to the decomposition of the austenite component in a high-carbon steel are given in Fig. 18.27. The amount of the reaction can be assumed to vary linearly with the magnitude of the length change. The horizontal dashed line shown in the figure thus stands for a fixed amount of transformation. The intersection of this line with the dilation curves for each temperature gives the length of time to obtain a fixed amount of transformation. A plot of the logarithm of these time increments against the reciprocal of the absolute temperatures gives the straight-line relationship shown in Fig. 18.28. It can be concluded that the transformation appears to follow a simple exponential rate law of the form

$$\text{rate} = \frac{1}{t} = Ae^{-Q/RT}$$

where t is the time required to form a definite amount of bainite, A is a constant, Q is the empirical activation energy for the reaction, and R and T have

21. Mentser, Morris, *Trans. ASM*, **51** 517 (1959).

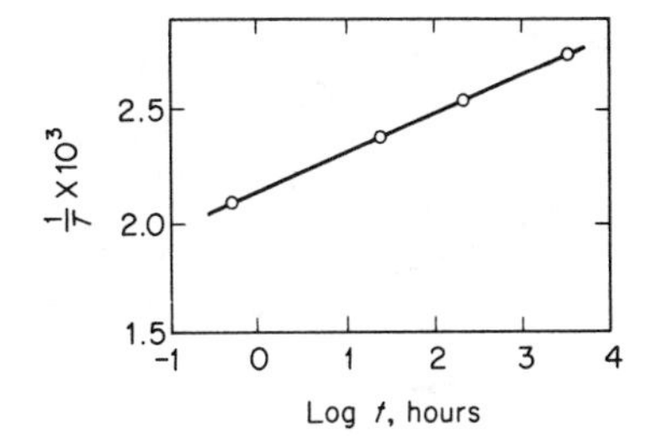

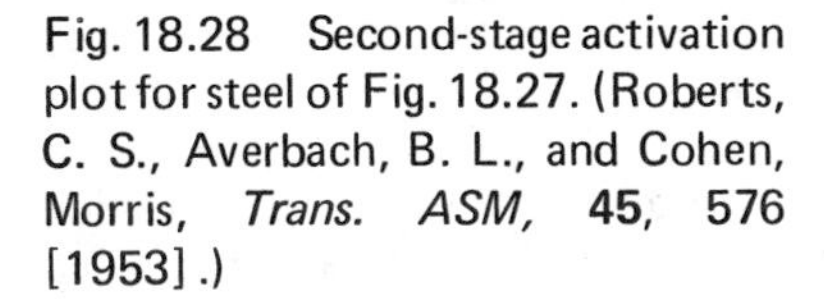
Fig. 18.28 Second-stage activation plot for steel of Fig. 18.27. (Roberts, C. S., Averbach, B. L., and Cohen, Morris, *Trans. ASM,* **45**, 576 [1953].)

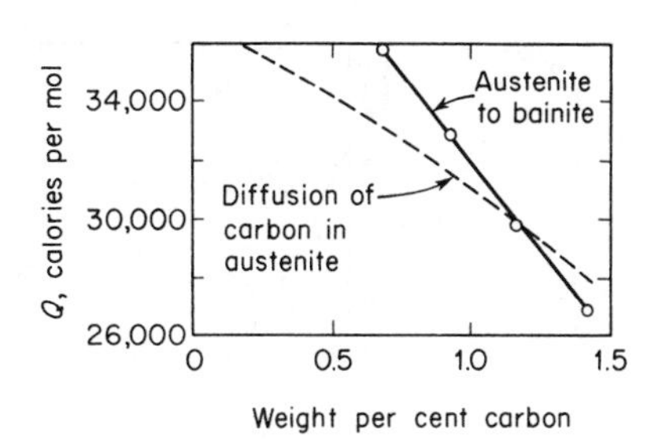

Fig. 18.29 Variation of second-stage activation energy with carbon content. (Roberts, C. S., Averbach, B. L., and Cohen, Morris, *Trans. ASM,* **45**, 576 [1953].)

their usual significance. The value of the activation energy depends on the carbon content of the steel. (See Fig. 18.29). The dashed line in Fig. 18.29 shows the variation of the activation energy for diffusion of carbon in austenite. The similarlity between these two curves has been interpreted[22] as indicating that the decomposition of bainite is controlled by diffusion of carbon in the austenite.

A word should be said about the apparent similarity between bainite and the decomposed martensite obtained in the first stage of tempering. Both constituents contain epsilon carbide, but the matrix differs in the two cases. In bainite the matrix is ferrite that is cubic in structure (c/a is unity), while in decomposed martensite the matrix is a low-carbon tetragonal martensite with a c/a ratio of approximately 1.014.

The Third Stage of Tempering. Speich has shown that in carbon steels very rapidly quenched so that their structure is essentially all martensite, a rod-shaped carbide particle can develop between 200 to 250°C. This particle has not been completely identified as to its structure.[23] It has been observed to form in all steel compositions (below 0.57 percent C). Whereas carbon atoms do not leave their sites at dislocations to form ϵ carbide, this is not true of the rod-shaped carbides. These latter can be formed at the expense of the carbon segregated to the dislocations and lath boundaries. It would therefore appear that, qualitatively, the energy of a carbon atom relative to its position decreases in the following order: (1) an interstitial lattice site; (2) a site in an epsilon carbide particle: (3) a site at the dislocation; (4) a position in a rod-shaped carbide. Temperature may be a factor in determining this order.

22. Roberts, C. S., Averbach, B. L., and Cohen, Morris, *Trans. ASM,* **45** 576 (1953).
23. Speich, G. R., *TMS-AIME,* **245** 2553 (1969).

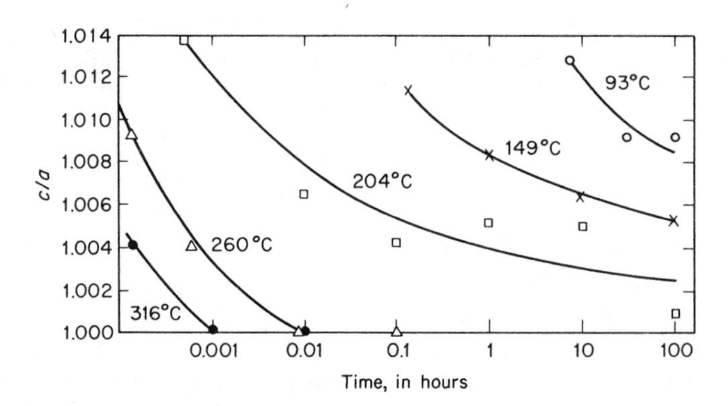

Fig. 18.30 Low-temperature kinetics of the third stage of tempering: variation of the martensite matrix *c/a* ratio with time. (Werner, F. E., Averbach, B. L., and Cohen, Morris, *Trans. ASM*, **49**, 823 [1957].)

In the third stage, the epsilon carbides also dissolve and the low-carbon martensite loses both its carbon and tetragonality and becomes ferrite, or cubic iron. The general kinetics of the third stage transformation can be followed by the changes in the *c/a* axial ratio of the low-carbon martensite as it decomposes. This type of measurement is based on X-ray diffraction measurements of the positions of martensite diffraction lines. Figure 18.30 shows the decrease in the martensite *c/a* ratio as a function of time for a number of different temperatures. The extreme rapidity with which the low-carbon martensite becomes cubic at the higher temperatures, 500°F (262°C) and 600°F (316°C), is readily apparent.

A dimensional change also accompanies the third stage of tempering which is in the same sense as that of the first stage of tempering: the specimen undergoes an additional shrinkage in volume, or length. These changes are, of course, in the opposite direction to those that occur during the transformation of austenite to martensite or bainite where expansions occur. The total shrinkage in length which occurs in a 1.0 percent carbon steel during tempering, considering first- and third-stage effects, is of the order of ¼ percent.

At a relatively low temperature, of the order of 250°C, the third-stage processes develop a structure, composed of ferrite and cementite, in a relatively short time.

At 400°C, according to Speich, the rod-shaped carbides dissolve and are replaced by a spheroidal Fe_3C precipitate. The particles of this latter tend to nucleate preferentially at lath boundaries and former austenite grain boundaries, but there may also be general nucleation.

In the third stage, reactions also occur in the ferrite matrix. The first of these takes place between 500 and 600°C and involves the recovery of the dislocations in the lath boundaries. This produces a low-dislocation-density acicular ferrite structure. On further heating to 600 to 700°C, the acicular ferrite grains then recrystallize to form an equiaxed ferrite structure. This recrystallization occurs with greater difficulty at higher carbon concentrations because of the pinning action of the carbide particles on the ferrite boundaries.

The end result of the third stage of tempering in a plain carbon steel is an aggregate of equiaxed ferrite grains containing a large number of spheroidal iron carbide particles. In the fourth stage the spheroidal carbide particles grow. This is accomplished by diffusion processes and results from the fact that larger particles have lower free energies than do smaller particles. A recent study[24] of the growth kinetics of the carbide particles in iron-carbon alloys indicates that the growth rate is diffusion-controlled. However, the problem is complex, as indicated by the fact that the effective diffusion constant for growth lies between the diffusion coefficients for the diffusion of carbon and for the diffusion of iron in the ferrite. Figure 18.31 indicates some of the problems associated with studying the growth of a precipitate. Note that the precipitate particles have a range of particle sizes and that with increasing annealing time, while the average particle grows in size so that the number of particles decreases (Fig. 18.32), the range of particle sizes increases. An excellent series of pictures is shown in Fig. 18.33. These are electron micrographs of the structures of steel specimens tempered in the range where the carbide particles are coalescing.

18.14 Spheroidized Cementite. Spheroidized cementite is the name given the structure consisting of cementite spheroids embedded in a matrix of ferrite when the particles become large enough to be easily visible under the light microscope. Such a structure is readily attained in a moderate length of time, if the third stage of tempering is carried out at a temperature just below the eutectoid temperature (723°C). A typical photograph (made with a light microscope) of a spheroidized cementite structure is shown in Fig. 18.34. This structure is perhaps the most stable of all the ferrite and cementite aggregates. Martensite, bainite, and even pearlite can be transformed into this microstructure by holding the metal for a sufficiently long time just below the eutectoid temperature. The formation of the spheroidized cementite is, of course, slowest when the starting structure is pearlite, and it is also true that the coarser the pearlite structure the more difficult it is to spheroidize it.

The spheroidized cementite structure is especially desirable in softened high-carbon steels, for steels containing this microstructure are more readily machined and give better results when heat-treated. Consequently, high-carbon steel sold under the label "annealed" will almost always possess a spheroidized structure.

24. Vedula, K. M., and Heckel, R. W., *Met. Trans.*, 1 9 (1970).

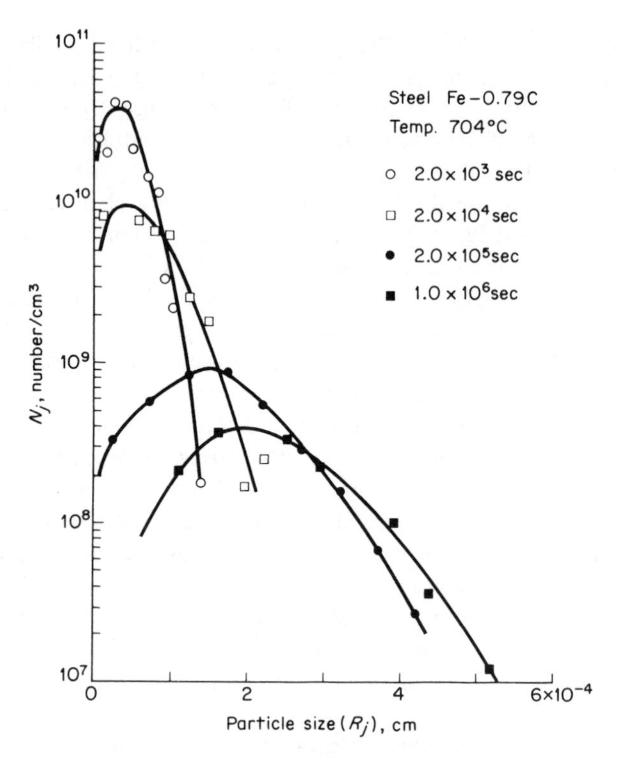

Fig. 18.31 Number of particles per unit volume, N_j, in a given size class as a function of the mean particle size of the class, R_j, for an eutectoid steel spheroidized at 704°C. (From, Vedula, K. M., and Heckel. R. W., *Met. Trans.*, **1**, 9 [1970].)

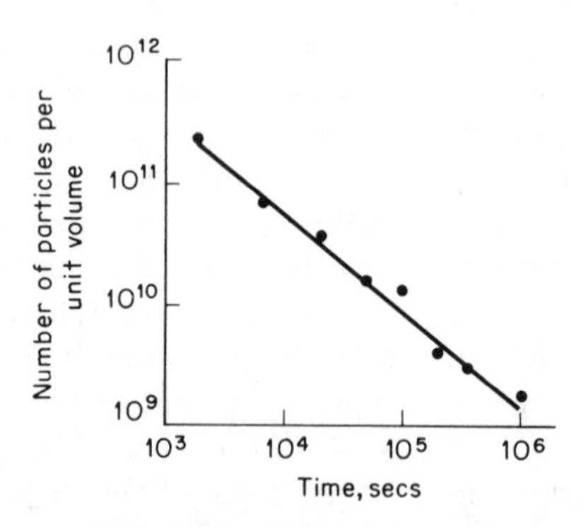

Fig. 18.32 The total number of particles per unit volume as a function of the time of spheroidization for the steel of Fig. 18.31. (From Vedula, K. M. and Heckel, R. W., *Met. Trans.*, **1**, 9 [1970].)

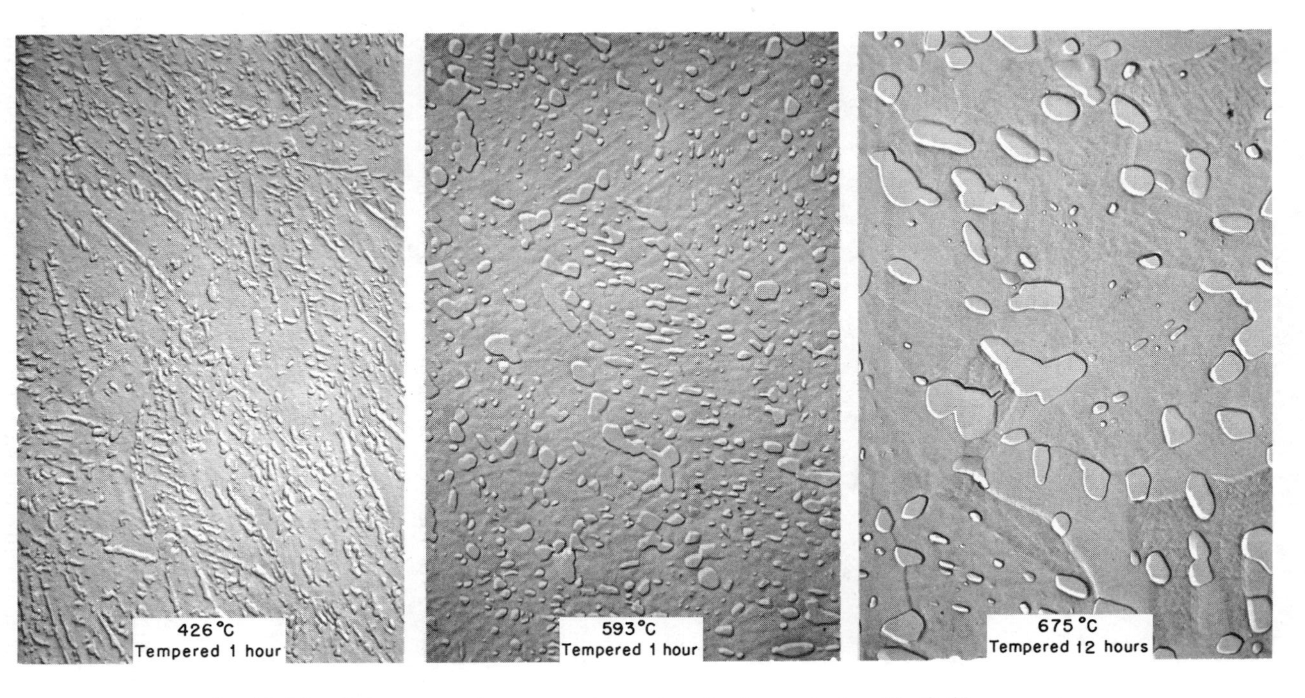

Fig. 18.33 Structure of tempered martensite in a steel with 0.75 percent carbon. Electron microscope. 15,000x. (From Turkalo, A. M., and Low. J. R., Jr., *Trans. AIME*, **212**, 750 [1958].)

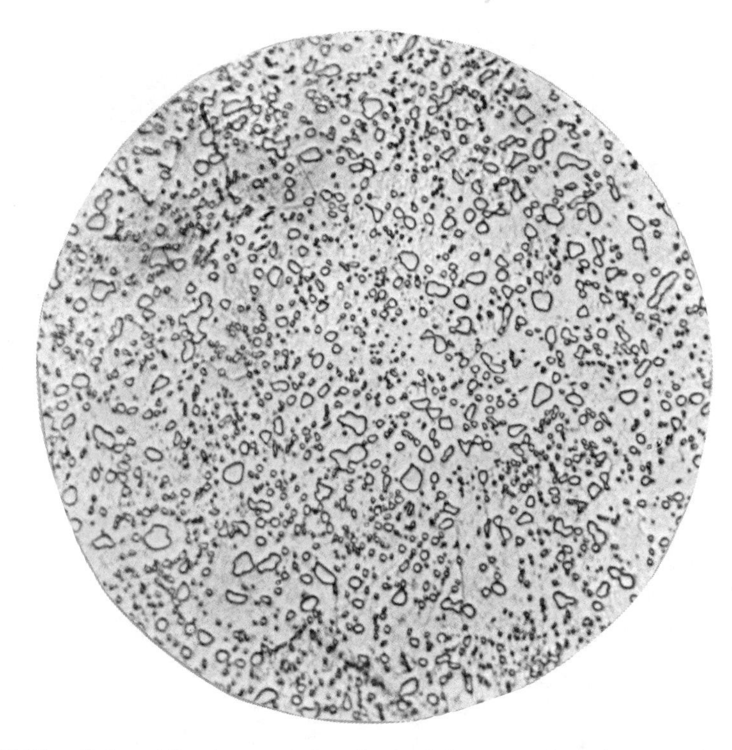

Fig. 18.34 Spheroidized cementite in a 1.1 percent carbon steel. Light microscope. 1000x.

18.15 The Effect of Tempering on Physical Properties. The microstructure changes that occur during tempering greatly change the physical properties of steel. The hardness of tempered steels will now be considered. The changes in this property are functions of both tempering time and tempering temperature. Curves such as those of Figs 18.35 and 18.36 where the hardness (as measured at room temperature after the tempering heat treatment) is plotted as a function of tempering temperature, specifically refer to constant tempering times (1 hr) at each tempering temperature. Curves for both medium- and high-carbon steels are shown in Fig. 18.35 and in each case the specimens employed were refrigerated at -196°C before tempering. The purpose of this treatment was to reduce the retained austenite in the quenched metal to a negligible quantity, so that the results plotted in Fig. 18.35 are truly representative of the effects of tempering martensite. If the specimens had contained a finite amount of retained austenite, an additional hardening component would have been introduced due to the transformation of austenite to martensite or bainite. This is frequently observed in hardness versus tempering temperature curves and appears as a rise in hardness just above room temperature.

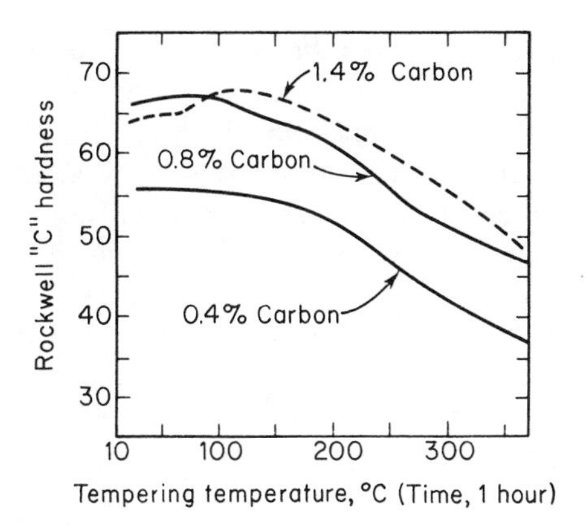

Fig. 18.35 Effect of tempering temperature on the hardness of three steels of different carbon contents. (Lement, B. S., Averbach, B. L., and Cohen, Morris, *Trans. ASM*, **46**, 851 [1954].)

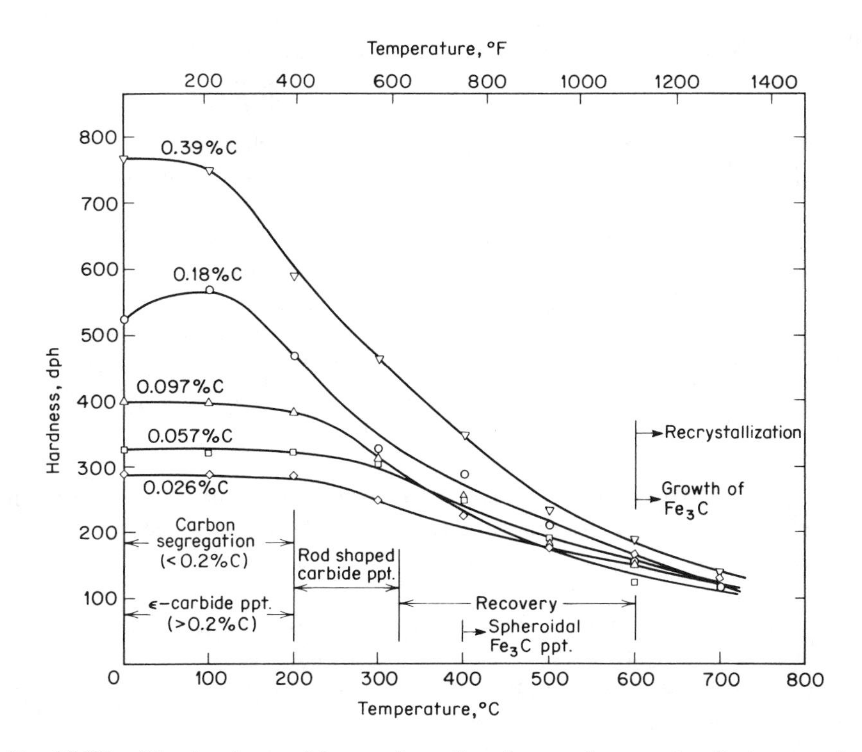

Fig. 18.36 The hardness of low and medium iron-carbon martensite tempered for 1 hour between 100°C and 700°C. (From Speich, G. R., *TMS-AIME*, **245**, 2553 [1969].)

A slight increase in hardness (not due to retained austenite) may be observed in the higher carbon steel (1.4 percent) as the tempering temperature is carried to about 200°F (93°C). This is undoubtedly associated with the precipitation of epsilon carbide during the first stage of tempering. A similar rise is not observed in the lower carbon steel (0.4 percent) because of the much smaller quantity of epsilon carbide that can precipitate in this composition. It should also be mentioned that while the precipitation of epsilon carbide undoubtedly contributes a hardening component to the steel, the depletion of carbon from the martensite matrix can be expected to contribute a softening component. The observed hardness, therefore, reflects the result of these two effects. A decided softening of the specimen, however, occurs when reactions associated with the third stage of tempering become appreciable. This is shown by the marked drop in hardness that starts at about 200°C. In the early parts of the third stage, both the solution of epsilon carbides and the removal of the carbon from the martensite (low-carbon form) should soften the metal. However, at the same time, cementite precipitates contribute a hardening effect.

When the steel has attained a simple ferrite and cementite structure, further softening results from the growth or the coalescence of cementite particles. This softening, due to growth in size and decrease in number of the cementite particles, continues and becomes more rapid the closer one approaches the eutectoid temperature (723°C). In effect, this means that for fixed tempering times the hardness of tempered martensite will be lower the closer one comes to the eutectoid temperature. The curves of Fig. 18.35 are only drawn for tempering temperatures below 375°C. Above this temperature to 723°C the three given curves can be expected to continue to fall in hardness with approximately the same slope as they exhibit in the temperature range from 200°C to 375°C. Figure 18.36 is from the more recent work of Speich.[25] It shows the effect of tempering on the hardness (DPH) of steels of low to medium carbon concentration, and complements the data given in Fig. 18.35. In addition, this diagram also outlines the reactions that occur during tempering.

18.16 The Interrelation Between Time and Temperature in Tempering. The fact that time and temperature have corresponding effects in tempering of steels, especially in the third stage where aggregates of cementite and ferrite are involved, has been known for a long time. This is demonstrated in a simple fashion in Fig. 18.37, where a number of curves corresponding to different tempering temperatures are given. In each case, the curve shows the dependence of hardness on the time at the indicated tempering temperature. The same data can also be related by a simple rate equation

$$\frac{1}{t} = Ae^{-Q/RT}$$

25. Speich, G. R., *TMS-AIME*, **245** 2553 (1969).

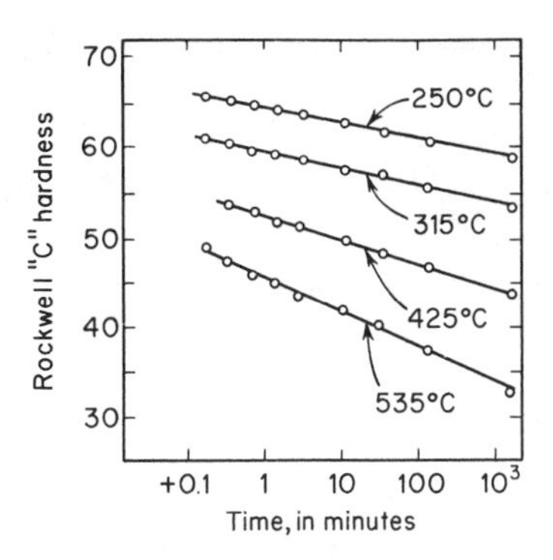

Fig. 18.37 The effect of time and temperature on the hardness of a tempered steel: 0.82 percent carbon, 0.75 percent managanese. (After Bain, E. C., *Functions of the Alloying Elements in Steel*, ASM, Cleveland, 1939.)

where t is the time to attain a given hardness, Q is an empirical activation energy for the process, A is a constant, and R and T have their usual significance.

18.17 Secondary Hardening and the Fifth Stage of Tempering. Most steels are used with either a pearlitic, or a quenched and tempered structure consisting of an aggregate of carbide particles embedded in a matrix of ferrite. In either case, a two-phase ferrite and carbide structure is involved. When alloying elements are added to steels, they may enter the ferrite or the carbides in varying amounts depending on the alloying element concerned. Some elements, however, are not found in the carbides. These include: aluminum, copper, silicon, phosphorus, nickel, and zirconium. Other elements are found in both the ferrite and the carbides. A number of these elements in the order of their tendency to form carbides (manganese having the least and titanium the greatest) are: manganese, chromium, tungsten, molybdenum, vanadium, and titanium.

Most alloying elements in steels tend to increase the resistance of the steel to softening when it is heated, which means that for a given time and temperature of tempering, an alloy steel will possess a greater hardness after tempering than a plain carbon steel of the same carbon content. This effect is especially significant in a steel which contains appreciable amounts of carbide-forming elements. When these latter are tempered at temperatures below 1000°F, the tempering reactions tend to form cementite particles based on Fe_3C, or, more accurately, $(Fe, M)_3C$ where M represents any of the substitutional atoms in the steel. In general, the alloying elements are present in the cementite particles only in about the same ratio as they are present in the steel as a whole. However, when the tempering temperature exceeds 1000°F (540°C), appreciable amounts of alloy carbides are precipitated. This precipitation of alloy carbides is now

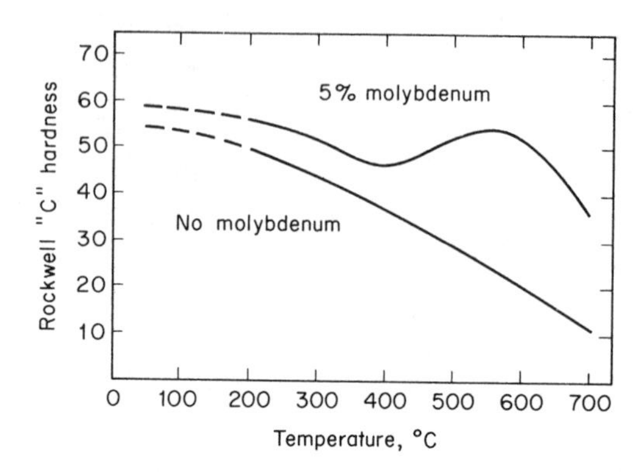

Fig. 18.38 Secondary hardening in a steel containing 0.35 percent carbon. (After Bain, E. C., *Functions of the Alloying Elements in Steels*, ASM, Cleveland, 1939.)

designated the fifth stage of tempering. The precipitation of these new carbides which, in general, do not conform to the formula $(Fe, M)_3C$, induces a new form of hardening which is believed due to coherency.[26,27] A schematic comparison of the tempering curves of a plain carbon steel and a steel with large amounts of carbide-forming elements is shown in Fig. 18.38.

That the alloy carbides do not form readily at lower tempering temperatures is undoubtedly related to the fact that at these tempering temperatures the rate of diffusion of substitutional elements is too slow to permit their formation. That cementite can form is because the diffusion rate of carbon is still very large at temperatures below 1000°F. The formation of cementite depends only on the diffusion of carbon.

The use of carbide-forming elements in order to produce a steel with a high resistance to tempering is best illustrated by the so-called high-speed steels. These are tool steels, whose original purpose was for use as tool bits in lathes and other machines where the cutting edges often get very hot during machining operations. Steels of this type retain their hardness for very long periods of time even at a red heat. A typical high-speed composition is represented by the composition designated 18-4-1 which contains approximately 18 percent tungsten, 4 percent chromium, 1 percent vanadium in addition to approximately 0.65 percent carbon.

26. Payson, P., *Trans. ASM,* **51** 60 (1959).
27. Seal, A. K., and Honeycombe, R. W. K., *Jour. of the Iron and Steel Inst.* (London), 188 9 (1958).

Problems

1. Figure 18.3 is a schematic diagram representing a set of cooling paths that might be obtained at various distances below the surface when a 2-inch diameter steel bar is quenched in iced brine. Make a similar sketch showing the corresponding distribution of cooling curves that might be obtained in:
   (*a*) a water quench;
   (*b*) an oil quench.
2. With regard to the three possible quenching media of iced brine, water, and oil (see Problem 1), answer the following questions:
   (*a*) Which cooling medium will give the most nearly uniform final structure? Explain.
   (*b*) Which will produce the smallest temperature gradient between the surface and the center of the bar during the time the metal is undergoing austenite decomposition?
   (*c*) Make sketches showing the expected hardness traverses for the specimens quenched in the three media.
3. Given a steel of carbon concentration 0.4 percent,
   (*a*) What is the hardness of a 50 percent martensite structure in this material?
   (*b*) If the ideal critical diameter of this steel is 1.0, at what distance from the quenched end of a Jominy hardenability test bar should this hardness be observed?
   (*c*) What is the maximum diameter of a round steel bar in which one could obtain a 50 percent martensite structure at its center using a water quench with a strong agitation?
4. AISI CI050 has the following composition in percent: Carbon, 0.48–0.55; Manganese, 0.60–0.90; Phosphorus, 0.040 max; Sulfur, 0.050 max.
   (*a*) Compute the ideal critical diameter of this steel based on its minimum composition and an ASTM grain size of 8.
   (*b*) Compute D_I assuming the composition conforms to the maximum limits and the grain size is ASTM 6.
5. A typical high-strength structural steel has a composition in percent: C, 0.15; Mn, 0.75; P, 0.08; Si, 0.25; Cu, 0.35; Ni, 0.75; Cr, 0.25; Mo, 0.30.
   (*a*) Compute the D_I for this steel assuming that the grain size is ASTM 7 and that the multiplying factor for Cu is 1.00.
   (*b*) A typical carbon structural steel ASTM A–36 has an approximate composition: C, 0.25; Mn, 1.00; Cu, 0.20; Si, 0.20; P, 0.04; S, 0.05. Compute its D_I assuming ASTM grain size of 7.
   (*c*) Is the difference in hardenability significant in these two steels as far as the problems of welding them are concerned? Explain.
   (*d*) Structural steel is normally hot-rolled and air-cooled. Measurements[28] have shown that an air-cooled bar of 1-inch diameter takes about 5 minutes for its surface to drop from 1550°F to 900° F. What microstructure would

28. *ASM Metals Handbook*, 8th Ed., vol. 2, p. 10. American Society for Metals, Metals Park, Ohio, 1964.

you expect to find in a 1-inch bar of each of these steels? In what way will the microstructures differ? Should this have any effect on the relative strengths of the two materials?

6. The grain size of a metal is often expressed in microns. Some superplastic metals have an average grain size of about one micron. Approximately what is the ASTM grain size in these materials?
7. (*a*) The composition of a AISI 4340 steel is given in the caption to Fig. 18.11. Compute its D_I. Assume the grain size is ASTM 7 and the multiplying factor for the nickel is 1.70.
   (*b*) Discuss your answer in relation to Figs. 18.9 and 18.13.
8. Would it be possible to obtain martensite in a 1-inch diameter bar of 4340 that was air-cooled from 1600°F? Explain (see Problem 5).
9. (*a*) Estimate the relative change in volume when an eutectoid steel is reacted to form martensite.
   (*b*) What is the corresponding relative change in a linear dimension?
   (*c*) Assuming $E = 30 \times 10^7$, how large a tensile stress would be needed to give a piece of steel a strain of the magnitude of your answer to part (*b*) of this problem?
   (*d*) The compressibility of iron at room temperature is

$$\frac{\Delta V}{V} = 5.66 \times 10^{-7} \text{ per kilograms per cm}^2$$

   How large a pressure would be required in order to decrease the volume of an iron specimen by the same amount that an eutectoid steel specimen expands when it transforms to martensite?
10. You are given a steel of eutectoid composition with a very coarse pearlite structure and are instructed to heat-treat it so as to obtain a relatively fine spheroidized cementite structure. The steel is in the form of a bar 2 inches in diameter, and its ideal critical diameter is 1 inch. Could this be done? If so, describe a suitable heat treatment.
11. (*a*) What diameter AISI 8640 steel round bar can be hardened to a 50 percent martensite structure at its center using a very good oil quench ($H = 0.50$)?
    (*b*) If the bar whose diameter was computed in part (*a*) were to be quenched in a violently agitated brine quench, do you think there might be any problems? Explain.

19 Fracture

Fracture is considered the end result of plastic-deformation processes, and there are many different ways that plastic deformation leads to failure, as can be seen by considering just a few ways that single crystals fracture.

19.1 Failure by Easy Glide. Let us consider the hexagonal metals that deform by easy glide on the basal plane (Zn, Cd, and Mg). In these metals, the single-crystal stress-strain curves depend upon the temperature of the test, as shown in Fig. 19.1A for magnesium. At low temperatures (18°C) the slope of the stress-strain curve rises very rapidly, while at higher temperatures (250°C) the slope of the stress-strain curves is small. The rate at which the metal hardens with strain at lower temperatures is much greater than that at high temperatures. This difference is undoubtedly due to dynamic recovery, with

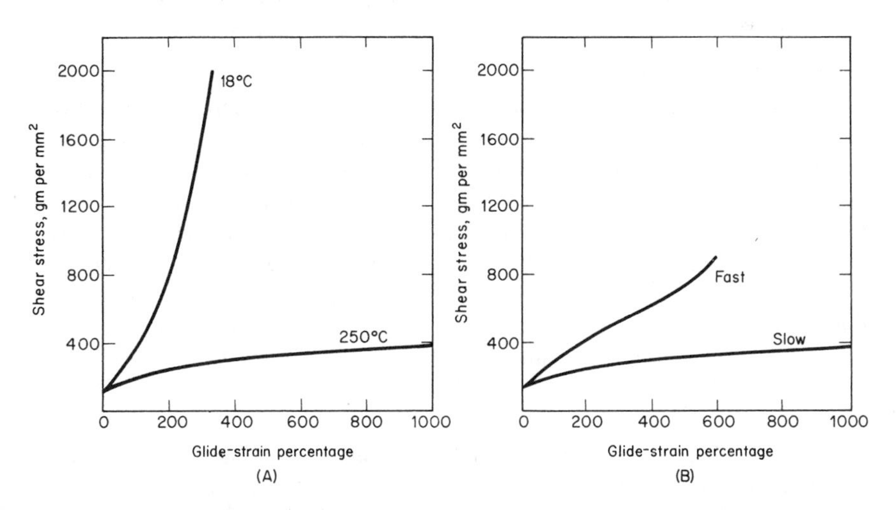

Fig. 19.1 Stress-strain curve of magnesium crystals. (A) The effect of temperature on the stress-strain curves. (B) The effect of strain rate. The curve marked "fast" represents a rate of straining 100 times faster than that for the curve marked "slow." (From Schmid, E., and Boas, W., *Kristallplastizität*, Julius Springer, Berlin, 1935.)

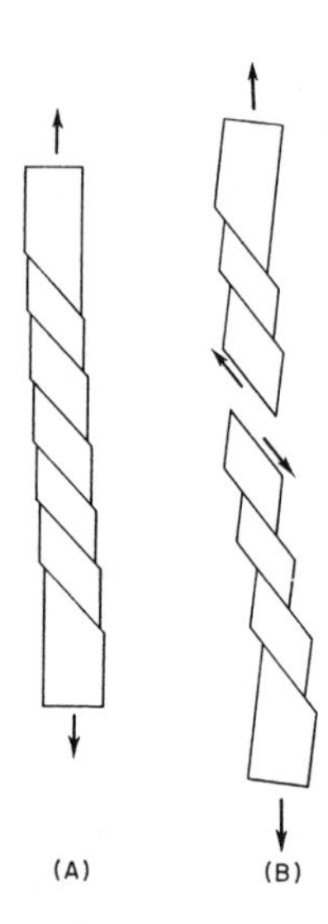

Fig. 19.2 (A) High temperatures and low strain rates promote extensive slip on a few slip bands. (B) Fracture can occur by shear along these bands.

softening occurring more rapidly during a high-temperature test than during a low-temperature test. This conclusion is further emphasized by the fact that increasing the rate of testing has much the same effect on the stress-strain curve as has lowering the temperature (Fig. 19.1B).

From the above it can be concluded that at higher temperatures and at very slow strain rates basal glide involves very little, if any, strain hardening. Once deformation starts along those slip bands in which it is easiest to activate dislocation sources, it can continue on them without appreciable increase in their resistance to slip. This produces coarse, widely separated slip bands, as is shown schematically in Fig. 19.2A. The notches formed where slip bands cut the surface are a factor in their development. The stress concentration at these notches aids the slip process in the bands. Finally, as more and more deformation occurs in the slip bands, the cross-section area of the bands becomes smaller and smaller. This raises the effective shear stress, thereby increasing the tendency for the slip to remain confined to the operating bands.

Provided that another plastic deformation process, such as mechanical twinning or slip on a plane other than the basal plane, does not occur at a late stage in the deformation process, it is quite possible for a crystal, such as that indicated in Fig. 19.2, to shear completely in two along one of the coarse slip bands. Shearing in two along a slip band is one possible mode of single-crystal failure.

19.2 Rupture by Necking (Multiple Glide). Outside of certain hexagonal metals, single glide is the exception rather than the rule in single crystals. In cubic metals, after only a relatively small deformation, easy glide breaks down into multiple glide on two or more systems. When deformation occurs by slip on several slip systems in a single-crystal tensile test, the rate at which the metal strain hardens can have an important effect on the fracture mechanism. What generally happens is that, early in the test, some part of the gage length deforms at a slightly more rapid rate than the rest. The cross-section area in this region becomes slightly smaller than in the rest of the specimen, with a corresponding increase in the shear stress. If the rate at which the slip planes harden with increasing strain is small, then continued slip in this reduced section will occur more easily than in the remainder of the specimen. In this manner a neck can form in the gage length. The various steps involved in the development of a neck are outlined schematically in Fig. 19.3 for an assumed case of double glide. The slip planes are indicated in Fig. 19.3A and are assumed perpendicular to the plane of the paper, with the slip directions in the plane of the paper. The neck is shown in Fig. 19.3B, and the ultimate shape of the specimen can be observed in Fig. 19.3C. Here it can be seen that continued growth of the neck results in a chisel-edge type of failure. Such a failure is frequently observed in metal crystals.

If more than two slip systems operate during the deformation of a metal

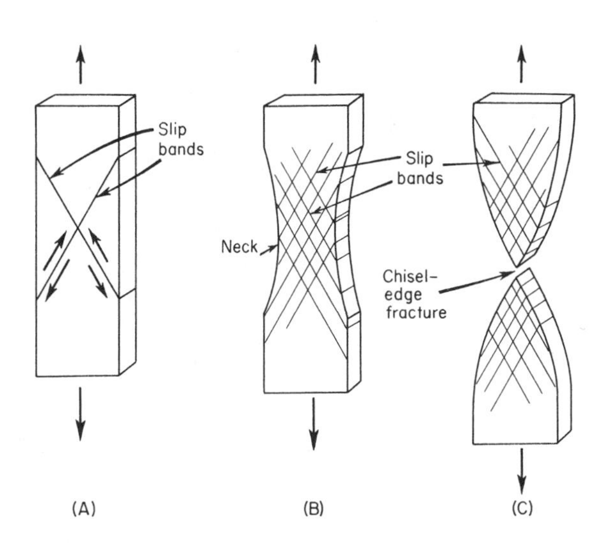

Fig. 19.3 (A) Crystal oriented for double slip. (B) Development of a neck. (C) Chisel edge fracture.

crystal and a neck is formed, the final fracture can occur when the cross-section at the neck pulls down to a point rather than a chisel edge.

It should not be inferred that the end result of multiple glide is always a simple chisel edge, or point type of fracture. The original orientation of the crystal is an important factor in determining the character of the fracture. It may well be that, as the metal is deformed and the lattice of the crystal is reoriented by the deformation, other slip systems will become active in addition to those that were activated at the start of the deformation. It is thus quite possible, for example, for necking, as a result of multiple glide, to give way to a final fracture by shear on a single plane.

19.3 The Effect of Twinning. While the shear associated with mechanical twinning is, in general, small, twinning always involves a lattice reorientation in the twins. This may place new slip systems in a favorable position relative to the stress axis, so that plastic deformation occurs more readily inside the twin than in the parent crystal outside the twin. Thus, when zinc or cadmium crystals are strained in tension, the basal plane of the crystal rotates toward the stress axis, as shown in Fig. 19.4. The effect of this rotation is to decrease the shear-stress component on the basal plane. In order to continue the deformation of the crystal, the tensile force must be increased. A point is eventually reached where the metal undergoes deformation twinning on a $\{10\bar{1}2\}$

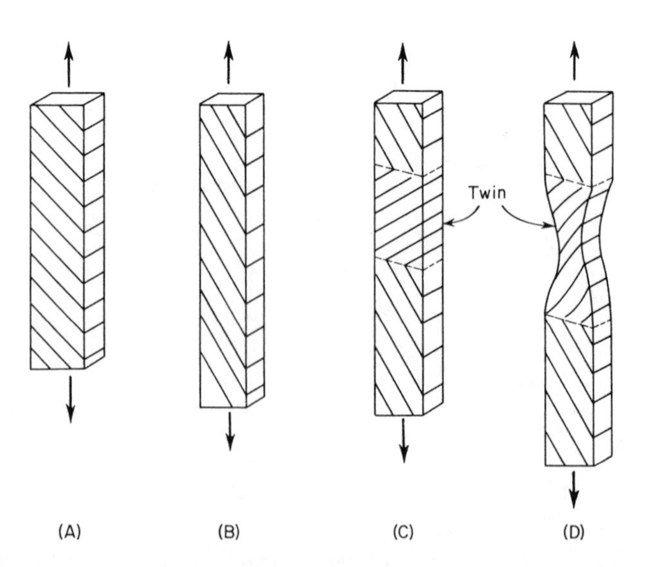

Fig. 19.4 The lattice reorientation due to twinning may induce failure. (A) Zinc or cadmium crystal. (B) Rotation of lattice due to slip. (C) Formation of twin. (D) Deformation leading to fracture in twin.

plane. When this occurs, the basal plane in the twin is found to be in a position favorable for slip. Fracture then occurs as a result of secondary basal slip inside the twin.

The fracture methods illustrated above, which involve large plastic deformation by slip, differ somewhat from the normal concept of fracture which usually involves the spreading of a crack. It is perhaps better, therefore, to classify the above forms of metal failure as ruptures rather than as fractures.

19.4 Cleavage. Under certain conditions, it is possible to split crystals in two pieces along planes of low indices. Let us assume that the block in Fig. 19.5A is a single crystal of zinc. Now suppose that a wedge or knife edge is laid along a basal-plane trace in the manner indicated in the figure and that the knife edge is given a sharp blow with a small hammer. If the temperature at which this operation is carried out is sufficiently low, the crystal will split, or be cleaved, into two parts, the separation following the basal plane. (See Fig. 19.5B.) This operation is called *cleavage* and the plane on which it occurs is known as the *cleavage plane* of the crystal. In the case of zinc, it is, of course, the basal plane (0001).

Zinc crystals are capable of being cleaved at room temperature, but only with some difficulty, and the cleaved surfaces are usually badly distorted. The splitting of zinc crystals becomes progressively easier the lower the temperature at which cleavage is attempted. Very nice cleavages may be readily obtained at the temperature of liquid air (–196°C). In fact, if a carefully grown, undistorted zinc crystal is rapidly cleaved at this temperature, the surface may be so perfect that it becomes impossible to focus on the surface with a light microscope.[1]

An interesting fact relative to the cleavage of zinc crystals is that the fracture follows the basal plane even in a bent or a distorted crystal. Thus, if a crystal is first bent and then cleaved, as shown in Fig. 19.6, the fracture surface will show the curvature of the basal plane. This fact has frequently been used in the study of plastic-deformation phenomena (using zinc crystals), to follow the distortion inside crystals.

Because basal cleavage is such a well-known phenomenon in zinc crystals, it is commonly believed that other hexagonal metals cleave on the basal plane as readily as zinc. However, this is not the case, except possibly for beryllium, which has been observed[2] to cleave on (0001) and $\{11\bar{2}0\}$ planes. Magnesium does not cleave easily on the basal plane[3] or on any other plane, nor is there any

1. Low, J. R., Jr., *Fracture*, p. 68. The Technology Press and John Wiley and Sons, Inc., New York, 1959.
2. Kaufman, A. R., *The Metal Beryllium*, p. 367. American Society for Metals, Cleveland, Ohio, 1955.
3. Reed-Hill, R. E., and Robertson, W. D., *Acta Met.*, **5** 728 (1957).

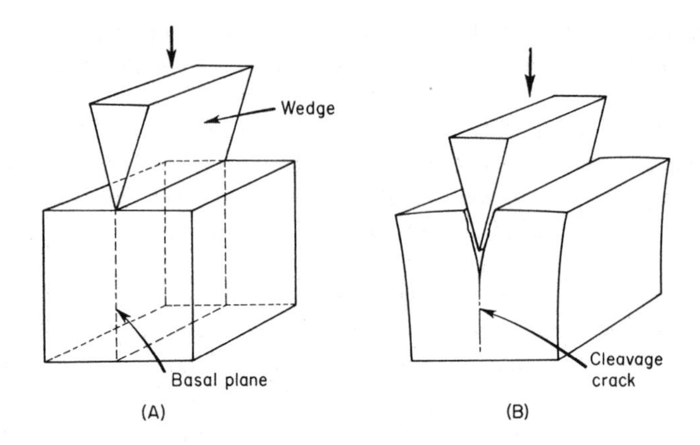

Fig. 19.5 Cleavage of a zinc crystal.

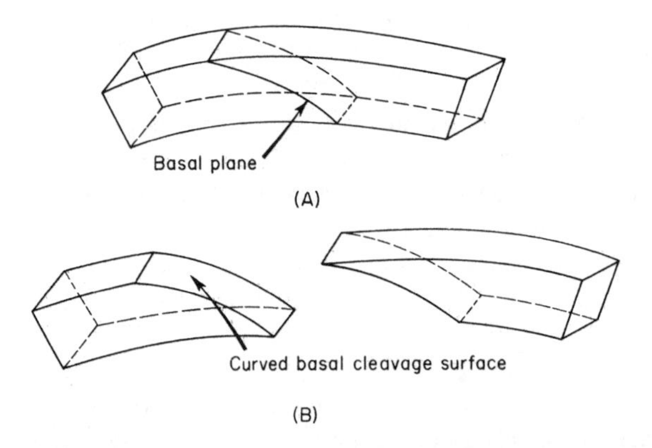

Fig. 19.6 A distorted zinc crystal cleaved. The cleavage follows the curved surface of the basal plane.

information in the literature relative to cleavage of cadmium crystals. There is also no evidence that anyone has observed true cleavage in a face-centered cubic metal. The one important class of metals in which cleavage is most frequently observed is the body-centered cubic metals, although the alkali metals (sodium, potassium, etc.) are body-centered cubic, and evidently do not cleave. The cleavage plane in the body-centered lattice form is usually {100}, although there are examples in which it has been indicated that cleavage along {110} is preferred.[4]

While most of the commercially important metals are not subject to cleavage, it is still a significant subject because of the fact that iron is a body-centered

4. Barret, C. S., and Bakish, R., *Trans. AIME*, **212** 122 (1958).

cubic metal that cleaves. The low-temperature brittleness of steels can be directly attributed to this fact. When fractures occur in polycrystalline iron or steel by transcrystalline cleavage, very little energy is expended in propagating the fractures, so that they closely resemble those that occur in glass or other brittle isotropic elastic solids.

Very little fundamental research has been done on the mechanics of cleavage in single crystals. The most comprehensive study of crystalline cleavage has not been done on metal crystals, but on crystals of an ionic salt, LiF, by Gilman and his associates. Some of the findings of the lithium fluoride work will be discussed presently.

19.5 Fracture in Glass. Much about the mechanics of fractures involving cleavage can be obtained from studying fracture in elastic materials, such as glass. Cleavage fractures in crystals are often complicated by the fact that plastic deformation occurs around and ahead of the crack. Under the conditions of high-crack velocity and low temperatures, plastic deformation tends to be suppressed and crystalline cleavage fractures are then equivalent to fractures in brittle materials, for example, cold glass. There is, however, one important difference. Cleavages tend to follow specific crystallographic planes of low indices, whereas in glass, which is amorphous, this condition has no meaning.

Although there is evidence that glass can deform plastically at room temperature,[5] the amount of this deformation during the period of a brittle fracture is undoubtedly very small. We may assume, therefore, that this type of material deforms elastically to the instant of fracture.

Let us assume that Fig. 19.7 represents a truly elastic solid (one which fractures without plastic flow of any kind), placed in a state of tensile stress by the forces marked f, and that the vertical plane mn is the fracture plane. Now, if there are no flaws in the solid, fracture must occur by breaking bonds between atoms which face each other across the plane of fracture. The two vertical rows of circles represent atoms, and the bonds between pairs of atoms are indicated by short horizontal lines. As a rough approximation, these lines can be thought of as springs that permit atom pairs to be displaced toward or away from each other, depending on whether the applied external force is compressive or tensile. If the force is small, atomic displacements will vary linearly with the force (stress).

Let us use the symbol x to represent the change in mean interatomic distance that occurs as a result of the applied stress. The strain can now be written

$$\epsilon = \frac{x}{a}$$

5. Anderson, O. L., *Fracture*, p. 331. The Technology Press and John Wiley and Sons, Inc., New York, 1959.

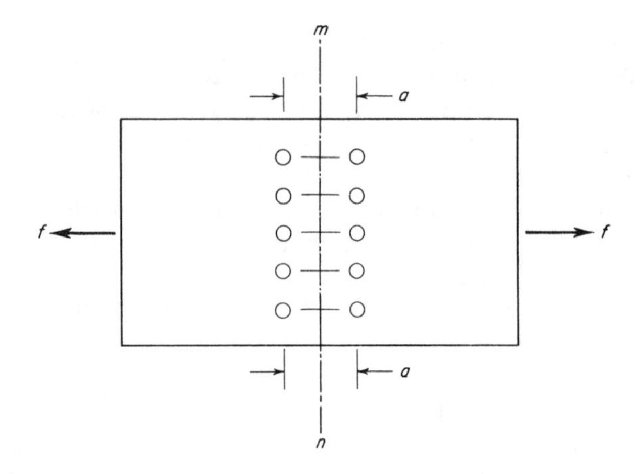

Fig. 19.7 The two vertical rows of circles represent a pair of lattice planes of a crystal. The fracture plane is indicated by line *mn*.

and for small strains

$$\sigma = E\epsilon = \frac{Ex}{a}$$

where σ is the stress taken normal to the fracture plane, ϵ is the strain, E is Young's modulus, and a is the interatomic distance (Fig. 19.7) in the absence of stress.

For large displacements, however, the relationship between the displacement and the applied stress is not linear. When the atoms are forced close together, very large repulsive forces are brought into play, causing the magnitude of the compressive stress to rise more and more rapidly with increasing deformation. Large positive displacements, due to tensile stresses, also cause a deviation from linearity, but in this case the restoring force decreases rather than grows in effectiveness as the displacement increases. These effects are shown schematically by the curve relating the stress and displacement. (See Fig. 19.8.) Notice that when atomic bonds are lengthened, the restoring stress rises to a maximum and then falls. The value of this maximum stress, designated by σ_{th}, may be taken as the theoretical fracture stress. At this stress a point of instability is reached where additional strain occurs under decreasing stress and with the normal methods of loading tensile specimens this condition leads to fracture.

In the study of fracture, the interesting part of the curve of Fig. 19.8 lies above the zero stress axis. Inside this region the curve may be represented approximately by a simple sine function of the form

$$\sigma = \sigma_{th} \sin \frac{2\pi x}{\lambda}$$

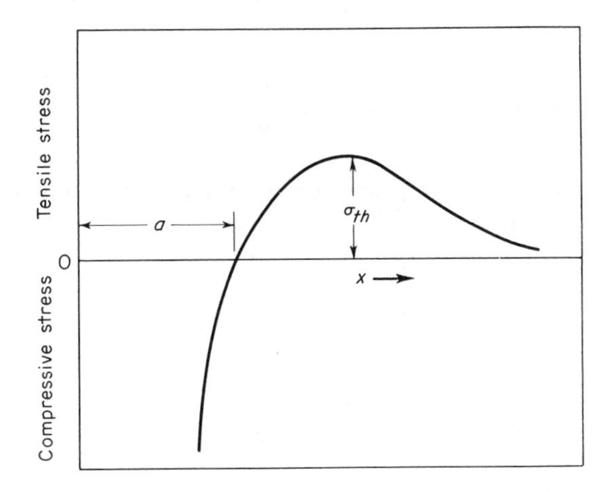

Fig. 19.8 Cohesive forces in a solid as a function of interatomic distance. (After Orowan, E., *Reports on Prog. in Physics,* **12** 185 [1948–1949].)

where σ is the applied stress, σ_{th} is the stress at the instant of fracture, x is the displacement or change in the mean interatomic distance, and $\lambda/4$ is the value of x when the stress equals σ_{th}. The work per unit area of the fracture surface (W) which is expended in bringing the specimen to the point of fracture is, accordingly,

$$W = \int_o^{\lambda/2} \sigma_{th} \sin \frac{2\pi x}{\lambda} \cdot dx = \frac{\sigma_{th}\,\lambda}{\pi}$$

From the equation relating the stress to the displacement x, we have

$$\frac{d\sigma}{dx} = \frac{d}{dx}\,\sigma_{th} \sin \frac{2\pi x}{\lambda} = \frac{2\pi}{\lambda}\,\sigma_{th} \cos \frac{2\pi x}{\lambda}$$

and for small values of x, $\cos 2\pi x/\lambda = 1$, and therefore

$$\frac{d\sigma}{dx} = \frac{2\pi}{\lambda}\,\sigma_{th}$$

but at small displacements

$$\sigma = E\,\frac{x}{a}$$

or

$$\frac{d\sigma}{dx} = \frac{E}{a}$$

and thus

$$\frac{2\pi}{\lambda}\sigma_{th} = \frac{E}{a}$$

Solving for λ/π in this equation and substituting it in the equation for W, the work per unit area, gives

$$W = \frac{2\,\sigma_{th}^{2}\,a}{E}\ \text{ergs/cm}^2$$

When fracture occurs, this element of strain energy can be assumed to be transformed into the energy of the surfaces that are created by the fracture. In other words,

$$\frac{2\sigma_{th}\,a}{E} = 2\gamma$$

or

$$\sigma_{th} = \sqrt{\frac{\gamma E}{a}}$$

where γ is the surface energy (surface tension), E is Young's modulus, and a is the mean interatomic distance across the fracture plane at zero stress.

The above equation predicts extremely high theoretical fracture stresses for isotropic elastic solids. Thus, in many solids, a is about 3 Å, Young's modulus about 10^{11} dynes per cm^2, and γ is approximately 10^3 dynes per cm. Substituting these values in the relation yields

$$\sigma_{th} = \sqrt{\frac{10^3 \cdot 10^{11}}{3 \cdot 10^{-8}}} \simeq 10^{11}\ \text{dynes/cm}^2$$

or

$$\sigma_{th} \simeq 10^6\ \text{lb/in.}^2$$

Tensile strengths of this order of magnitude are found only in very rare cases. Two notable examples are freshly drawn small glass fibers[6] and mica sheets loaded so as not to stress their edges.[7]

6. Griffith, A. A., *Phil. Trans. Roy. Soc.*, **A221** 163 (1924).
7. Orowan, E., *Zeits. für Phys.*, 82 235 (1933).

19.6 The Griffith Theory. The observed strength of ordinary window glass is less than one-hundredth of its theoretical strength. This discrepancy between the strength ordinarily observed and the theoretical strength led Griffith[8,9] to postulate that the low observed strengths were due to the presence of small cracks or flaws in low-strength glass. In considering this problem, we shall first use the approach of Orowan. That of Griffith will be given in Section 19.10. Because the extremities of cracks have the ability to act as stress raisers, Griffith assumed that the theoretical stress was obtained at the ends of a crack, even though the average stress was still far below the theoretical strength. Fracture, according to this concept, occurs when the stress at the ends of the cracks exceeds the theoretical stress. When this occurs, the crack is able to expand catastrophically.

In calculating the stress at the ends of a crack we shall not use the approach of Griffith, but that of Orowan.[10] In both cases, however, a flat plate containing a crack of elliptical cross-section was considered. Such a flaw is illustrated in Fig. 19.9, where it can be observed that the length of the crack is $2c$ and that the stress axis is perpendicular to the major axis of the ellipse. The stress and strain around an elliptical hole of this type, with the indicated orientation of the tensile stress, have been computed by Inglis.[11] According to these calculations, the stress σ_e at the end of the crack is

$$\sigma_e = 2\sigma\left(\frac{c}{\rho}\right)^{1/2}$$

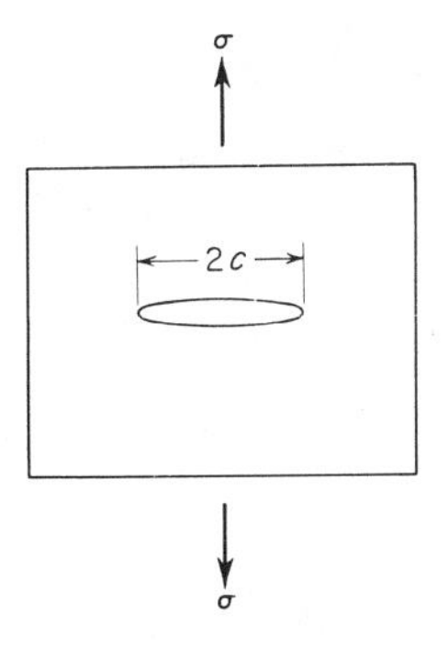

Fig. 19.9 A Griffith crack.

8. Griffith, A. A., *Op. Cit.*, p. 163.
9. Griffith, A. A., First International Conf. Appl. Mech., Delft, **55A** (1924).
10. Orowan, E., *Reports on Prog. in Phys.*, **12** 185 (1948–49).
11. Inglis, C. E., *Trans. Inst. Naval Arch.*, **55** 219 (1913).

where $2c$ is the length of the major axis of the elliptical hole, σ is the average applied stress, and ρ is the radius of curvature at the ends of the ellipse.

We have already seen that

$$\sigma_{th} = \left(\gamma \frac{E}{a}\right)^{1/2}$$

If the fracture is to spread, σ_e must equal σ_{th}, and

$$2\sigma\left(\frac{c}{\rho}\right)^{1/2} = \left(\frac{\gamma E}{a}\right)^{1/2}$$

This relationship can now be solved for σ, yielding

$$\sigma_f = \left[\frac{\gamma E}{4c}\left(\frac{\rho}{a}\right)\right]^{1/2}$$

where σ_f is the average applied stress at which the crack will spread, γ is the specific surface energy, $2c$ is the crack length, ρ is the radius of curvature at the end of the crack, and E is Young's modulus. This relationship is Orowan's version of the Griffith criterion for brittle fracture. It differs only slightly (in a small numerical factor) from Griffith's original relation. Notice that as the crack length increases, the stress to keep it moving decreases. This signifies that once the crack starts moving it is able to accelerate to high velocities.

19.7 Griffith Cracks in Glass. A great deal of effort has gone into attempts to prove the existence of Griffith cracks. Unfortunately, a detailed knowledge of the nature of the flaws, or cracks, and their relation to the strength of glass has been only partly attained. The evidence for Griffith cracks is usually indirect. This does show, however, that cracks that lower the strength of glass are probably on the surface, where a surface crack of depth c is equivalent to an interior crack of length $2c$.

Griffith's original work on glass fibers is strong evidence for his theory. When freshly drawn soda-glass fibers were tested in bending, tensile strengths as high as 900,000 psi were obtained, which is very close to the theoretical strength of glass ($\simeq 10^6$ psi).

Another classic experiment in fracture is the work of Orowan[12] who showed that the tensile strength of mica plates could be increased by a factor of 10. Mica ordinarily has a tensile strength of 30,000 to 40,000 psi. Orowan assumed that this low strength was due to cracks located at the edges of the sheets, and that the flat surfaces of the sheets should be free of flaws since they are cleavage

12. Orowan, E., *Zeits. für Phys.*, 82 235 (1933).

surfaces, and should normally be without defects. In testing this hypothesis, he made tensile-test grips the widths of which were smaller than the sheet width. In this manner he loaded the mica specimens so that their edges were not stressed. The result of this experiment was an observed tensile strength of more than 400,000 psi.

19.8 Crack Velocities. When an elastic material is stressed, potential energy is stored in the material. The magnitude of this strain energy, by simple elasticity theory, is

$$\text{Strain energy} = \frac{\sigma^2}{2E} \text{ ergs/cm}^3$$

where E is Young's modulus and σ is the applied tensile stress. According to the Inglis[13] solution for the stress and strain around an elliptical crack in a flat plate of unit thickness, the presence of a crack reduces the total strain energy by the amount,

$$\text{Decrease in strain energy} = -\frac{\pi c^2 \sigma^2}{E}$$

In other words, as the crack length ($2c$) increases, more and more energy is made available for the propagation of the crack. Part of this energy is expended in forming the surfaces of the crack. This amounts to $4\gamma c$ in a plate of glass of unit thickness where γ is the specific surface energy and $2c$ is the crack length. The remaining energy is transformed into kinetic energy. As the ends of the crack move forward, the material at the sides of the crack moves apart with a finite velocity and a kinetic energy can be associated with this movement of material near the end of the crack. At any instant, therefore, we have

Gain in kinetic energy = loss in strain energy − gain in surface energy

or

$$\frac{d}{dt}(\text{kinetic energy}) = \frac{d}{dt}\left(\frac{\pi c^2 \sigma^2}{E} - 4\gamma \dot{c}\right)$$

This relationship has been solved[14,15] and the solution has the form

$$v_c = k v_l \left(1 - \frac{c_o}{c}\right)^{1/2}$$

13. Inglis, C. E., *Trans. Inst. Naval Arch.*, **55** 219 (1913).
14. Mott, N. F., *Engineering*, **165** 16 (1948).
15. Anderson, O. L., *Fracture*, p. 331. The Technology Press and John Wiley and Sons, Inc., New York, 1959.

where v_c is the velocity of the crack, v_l is the longitudinal velocity of sound in the solid, c_o is the Griffith critical half crack length, c is the half crack length at any instant, and k is a dimensionless constant.

The above equation shows that the crack gains speed as it expands from the critical crack length. It also shows that the velocity approaches a maximum value (kv_l) when the crack length grows very large. The constant in the above equation has been evaluated[16] on the basis of an assumed crack shape, stress distribution, and value of Poisson's ratio. The calculated value is 0.38, which means that the computed limiting speed with which the crack could move in an elastic solid is 0.38 the velocity of sound in the same material. The fact that the crack velocity is limited by the speed of sound is not surprising. The displacements that occur in the material adjacent to the crack can only be transmitted from one atom to the next at the speed of elastic sound waves.

19.9 The Griffith Equation. We shall now consider Griffith's derivation of the fracture stress which was derived on a thermodynamical basis. Griffith noted that when a crack is able to propagate catastrophically, the gain in surface energy must be equal to the loss in strain energy. Thus, again considering an elliptical slot in a flat plate and using the Inglis relationship for the strain energy given in the preceding section, we have

$$\frac{\partial}{\partial c}\left(\frac{\pi \sigma_f^2 c^2}{E}\right) = \frac{\partial}{\partial c}(4c\gamma)$$

or

$$\frac{2\pi \sigma_f^2 c}{E} = 4\gamma$$

and

$$\sigma_f = \left[\frac{2\gamma E}{\pi c}\right]^{1/2}$$

This relationship should now be compared with the following equation that was derived earlier from considerations of the stress concentration at the ends of the crack

$$\sigma_f = \left[\frac{\gamma E}{4c}\left(\frac{\rho}{a}\right)\right]^{1/2}$$

It is now normally considered that these two relationships represent two different criteria for the unstable expansion of a crack, both of which must be

16. Roberts, D. K., and Wells, A. A., *Engineering,* 178 820 (1955).

satisfied. The two equations yield the same fracture stress when the radius of curvature at the crack root $\rho = (8/\pi)\, a \simeq 3a$. For very sharp cracks ($\rho < 3a$), the strain energy criterion (Griffith's relationship) should control the crack advance, whereas for cracks with $\rho > 3a$, the stress concentration at the end of the crack (Orowan's relation) should determine the point at which the crack is able to move unstably.

The Griffith relation given above applies specifically to a flat elliptical crack laying in a flat plate. Calculations for cracks of other shapes in solids with different material geometries, as well as calculations based on more accurate atomic considerations, have been made. The results of all these calculations quite generally confirm the functional relation:

$$\sigma_f \approx \left(\frac{\gamma E}{c}\right)^{1/2}$$

and can, therefore, be considered as one of general significance with regard to brittle fracture.

19.10 The Nucleation of Cleavage Cracks. Glass at room temperature can be considered to be a material incapable of plastic deformation. The same is not true of metals that are capable of deforming by slip and twinning even at temperatures approaching absolute zero. It has also been observed that when a metal fails by brittle cleavage, a certain amount of plastic deformation almost always occurs prior to the fracture. This has been interpreted by many as evidence that metals do not fracture as the result of pre-existing Griffith cracks, but that cleavage cracks are probably nucleated by plastic-deformation processes. In confirmation of this point of view is the fact that cleavage can sometimes be made to occur in the interior of annealed polycrystalline specimens; specimens in which there is little likelihood of the existence of small cracks before the application of stress. In a well-annealed metal, cracks should be self-healing and disappear.

While it is quite possible to nucleate cleavage fractures in metal crystals with a chisel and hammer, this approach tells us nothing about the formation of fracture nuclei in metal specimens under normal conditions of loading. Of more interest is how a crack starts in a single-crystal tensile specimen. A number of these studies have been performed in which zinc or iron crystals were used.

The fracture surfaces of iron single crystals are never perfect cleavages, even though the metal can fail in a completely brittle manner. Thus, while a surface can be a macroscopically flat plane, observation under a microscope at even relatively low magnification shows a considerable amount of surface detail. A typical cleavage fracture surface is shown in Fig. 19.10. In this picture, the rectangular block pattern is caused by the intersection of mechanical twins with the surface of cleavage. In cases such as this, it would appear that the fracture

Fig. 19.10 Cleavage-fracture surface of an iron crystal. (Biggs, W. D., and Pratt, P. L., *Acta Met.,* **6** 694 [1958].)

front moves across the specimen discontinuously, being impeded by the twins that form in front of it, with the crack having to be continuously renucleated on the far side of the twins in order to keep on moving.

When tested in tension at temperatures above –100°C, annealed iron single crystals of all orientations are completely ductile. Failure occurs by slip mechanisms of the type outlined at the start of the present chapter. As the temperature is lowered below –100°C, the nature of the fracture becomes orientation-dependent, so that at –183°C a relatively large fraction of the crystals fail by brittle cleavage. The remainder still fail in a ductile fashion by slip. The orientations that cleave are those in which the stress axis of the specimen lies close to a ⟨100⟩ direction.

In analyzing brittle fracture in iron, it is important to consider the effect of temperature on the critical resolved shear stress (Fig. 19.11). In body-centered cubic metals, such as iron, the temperature-dependence of the critical resolved shear stress for slip is very large, varying by a factor of almost eight in the temperature interval from room temperature to –196°C. Both cleavage and twinning require nucleation, and high stress levels needed to deform iron plastically at low temperatures favor these nucleation processes.

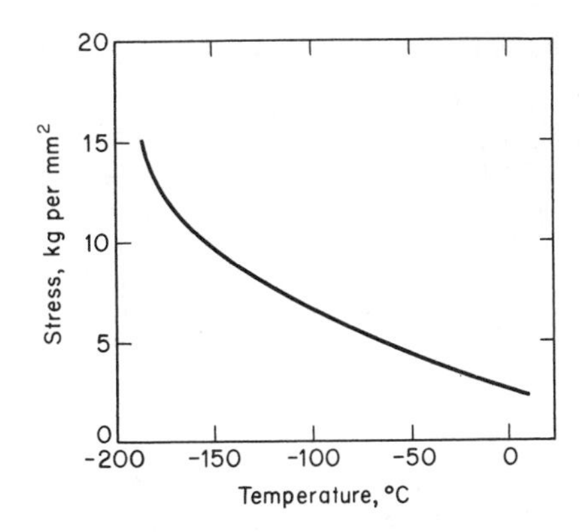

Fig. 19.11 Temperature dependence of the critical resolved shear stress in iron single crystals. Data corresponds to the stress at the upper yield point. (Biggs, W. D., and Pratt, P. L., *Acta Met.*, **6** 694 [1958].)

19.11 Crack Nucleation in Iron. Zener[17] was the first to propose that an obstacle to slip might nucleate a crack. According to his model, the coalescence of a number of edge dislocations lying on the same slip plane should open up a crack on a plane normal to the slip plane. The Zener model is outlined in Figs. 19.12 and 19.13, where the first figure shows the effect of bringing three dislocations together, and the second indicates the orientation relationship of the crack and the slip plane. In the given figures it is assumed that a twin boundary acts as the barrier to the movement of the dislocations.[18] In a polycrystalline material, grain boundaries might also be considered to act in a similar manner.

A large number of other mechanisms for the nucleation of fracture based on dislocation interactions have been proposed. Cottrell[19] has proposed one based on the gliding together of dislocations lying on intersecting slip planes. This mechanism is capable of accounting for cleavage on {100} cube planes as a result of slip on {110} ⟨111⟩ slip systems. Figure 19.14A shows the basic concept: two slip dislocations combine to form an edge dislocation. A three-dimensional view of the same dislocation interaction is given in Fig. 19.14B in which the summation of the slip Burgers vectors to form the Burgers vector of the crack is shown. This vector sum may be written

$$\tfrac{1}{2}[1\bar{1}\bar{1}] + \tfrac{1}{2}[\bar{1}1\bar{1}] = [00\bar{1}]$$

where $\frac{1}{2}[1\bar{1}\bar{1}]$ is the Burgers vector of the slip dislocation in the $(0\bar{1}1)$ plane,

17. Zener, C., ASM Seminar (1948), *Fracturing of Metals*, p. 3.
18. Biggs, W. D., and Pratt, P. L., *Acta Met.* **6** 694 (1958).
19. Cottrell, A. H., *Trans. AIME,* **212** 192 (1958).

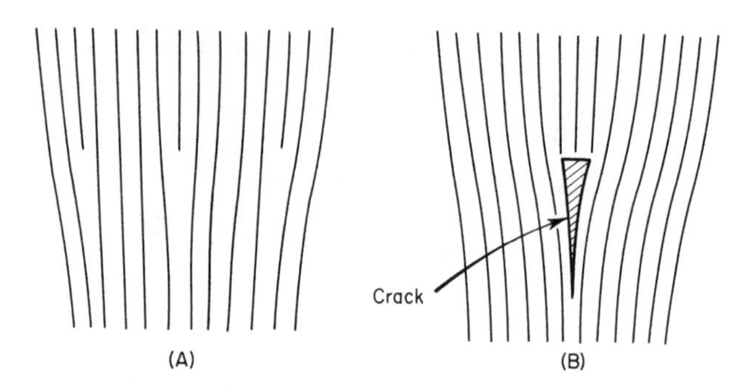

Fig. 19.12 If a number of edge dislocations of the same sign are forced together, a small cracklike defect results.

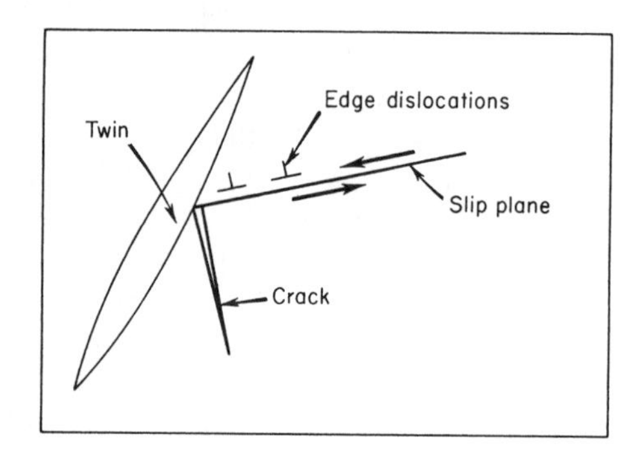

Fig. 19.13 Zener's mechanism for crack formation applied where a twin boundary acts as the barrier to slip.

and ½ $[\bar{1}1\bar{1}]$ is the Burgers vector of the slip dislocation in the (011) plane, and $[00\bar{1}]$ is the Burgers vector formed by the dislocation reaction. The opening of the crack in the (001) plane can be considered to result from the coalescence of a number of these $[00\bar{1}]$ dislocations. This is indicated in Fig. 19.14C. Cottrell's mechanism has not been experimentally verified in iron, but crack nucleation at the intersection of slip bands in a manner analogous to Cottrell's mechanism has been observed in a ceramic material, magnesium oxide.[20] This ionic material is cubic and crystallizes in the rocksalt structure. It deforms by slip on {101} planes in ⟨101⟩ directions. Dislocations are readily revealed in MgO by etching and the amount of slip that occurs before cracking takes place is, in general, small.

Figure 19.15 shows a small crack that has formed at the intersection of two slip bands in a MgO crystal. The {110} slip bands have been made visible by

20. Parker, Earl R., *Fracture,* p. 181. The Technology Press and John Wiley and Sons, Inc., New York, 1959.

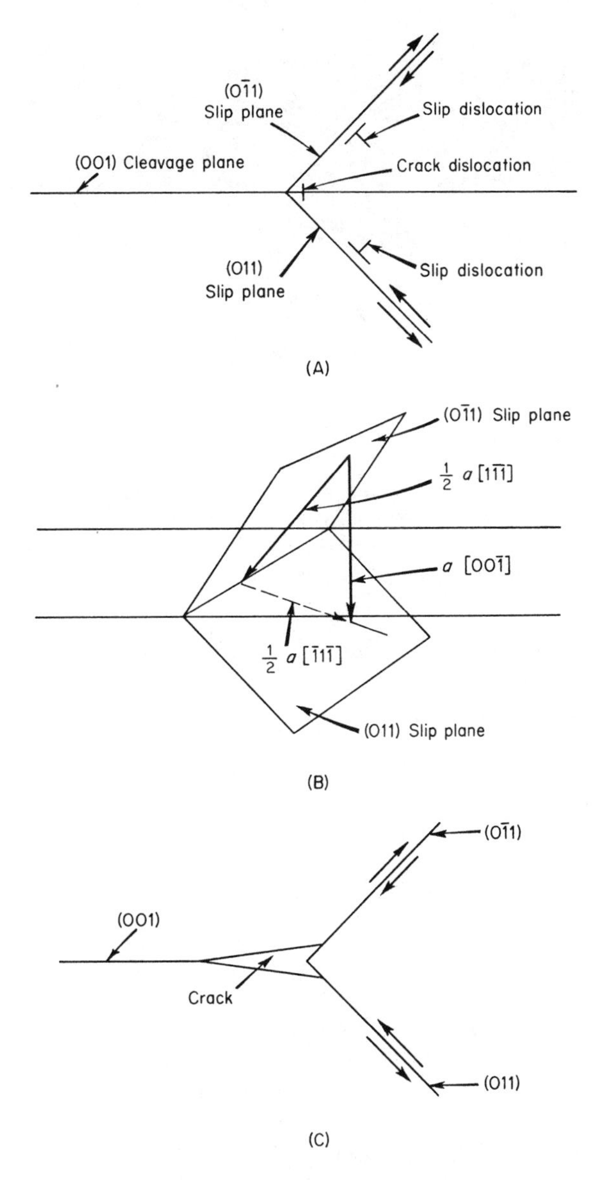

Fig. 19.14 Cottrell's dislocation intersection mechanism to explain {100} crack nuclei in iron crystals. (A) Two slip dislocations combine to form crack dislocation. (B) The addition of the Burgers vectors. (C) Orientation of the crack.

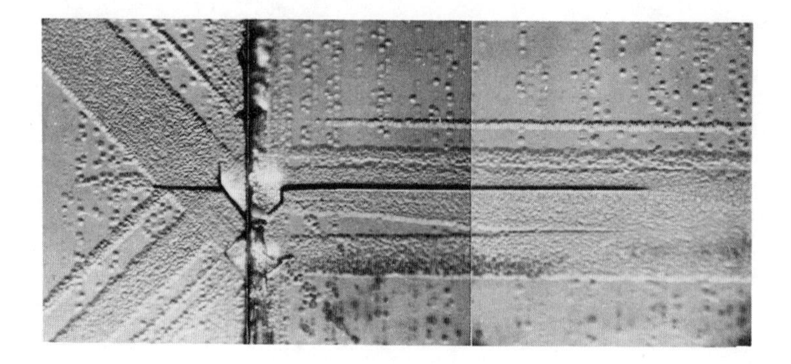

Fig. 19.15 Crack in MgO that formed at intersection of slip bands. This compositive photograph shows two adjacent sides of a rectangular crystal which meet at a common edge (vertical dark line). (Parker, Earl R., "Fracture of Ceramic Materials," *Fracture*, The Technology Press and John Wiley and Sons, Inc., New York, 1959, p. 181.)

etching the surface. Each slip band makes an angle of 45° with the horizontal and consists of a number of slip planes, which explains their finite width. The bands are filled with small etch pits, each of which corresponds to the intersection of a dislocation line with the surface. The crack is the horizontal dark line that has formed at the intersection of the large upper band with one of the lower bands. It runs out to the surface of the specimen on the right and lies on a {100} plane.

19.12 Crack Nucleation in Zinc. Cleavage cracks in zinc differ from those in iron in that the basal plane, which is the slip plane, is also the cleavage plane. As indicated above, the cleavage plane in iron is a cube plane {100}, while the preferred slip plane is {110}. It is at once clear that the dislocation pile-up mechanism shown in Fig. 19.13 cannot apply to crack nucleation in zinc. The indicated mechanism requires the cleavage plane to be perpendicular, or at least nearly perpendicular, to the slip plane. It has been shown experimentally, however, that obstacles to slip can nucleate cracks in zinc. An excellent example is shown in Fig. 19.16 where the specimen is a bicrystal tested in tension in which cracks have formed at the boundary between crystals. This boundary can be considered an obstacle to slip against which dislocations should pile up as the metal is deformed. In this type of fracture, the formation of the crack in the basal plane evidently relieves the strain energies associated with the dislocation array next to the boundary and the applied external stress.

Fig. 19.16 Crack in zinc bicrystal nucleated at the grain boundary which runs horizontally. (Gilman, J. J., *Trans. AIME*, **212** 783 [1958].)

The foregoing discussion of the various ways that cleavage nuclei can form is far from complete. Several general conclusions can be made. As far as it is presently known, cleavage fracture is probably nucleated by some form of plastic deformation. Present theories favor the concept of dislocation interactions as inducing cleavage nuclei. Dislocations on different slip planes can combine to form new dislocations on the fracture plane, thereby opening up a crack. Alternatively, slip on a given plane can be impeded by some sort of a barrier leading to a pile-up of dislocations which, in turn, nucleate the fracture. It is more or less generally agreed that the obstacle to slip must be very strong so that it can stand the high stress at the head of the dislocation pile-up. Opinion favors deformation twins and grain boundaries as obstacles with sufficient strength to satisfy this condition.

19.13 Propagation of Cleavage Cracks. In elastic solids (cold glass), the strain energy that is released as a crack propagates is converted into the surface energy of the crack surfaces and the kinetic energy of the moving material at the sides of the crack.

An additional energy term must be considered for the propagation of crystalline cleavage cracks. This term is that associated with the plastic deformation which usually accompanies the motion of cracks through crystals. Low temperature and high strain rates tend to raise the yield points of metals and to suppress plastic deformation. The energy term associated with plastic

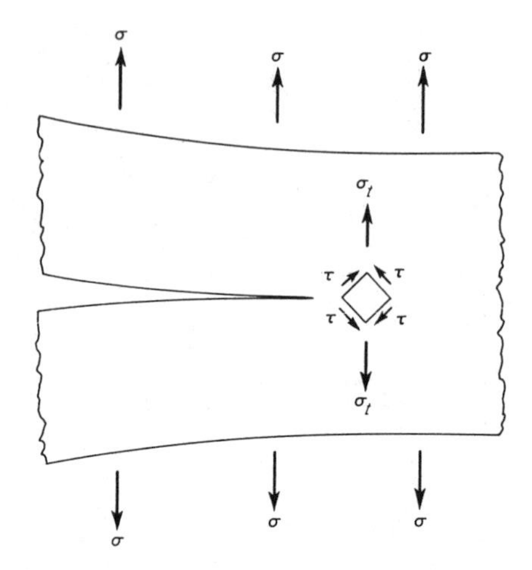

Fig. 19.17 When an external tensile stress (σ) is applied to an elastic body containing a crack, the material just ahead of the crack is subjected to a very large tensile stress (σ_ϵ). This in turn, is equivalent to shear stresses (τ) on planes at 45° to the plane of the crack.

deformation, therefore, becomes less important the lower the temperature of testing and the higher the velocity with which the crack moves. In the following sections certain of the experimental aspects of these effects will be discussed.

Plastic deformation associated with a moving crack is most liable to occur just ahead of the crack. The metal in this region is effectively in a state of very high uniaxial tensile stress, with the stress axis normal to the plane of the crack. A simple tensile stress of this type is equivalent to a set of shear stresses on planes at 45° to the tensile-stress axis, as indicated in Fig. 19.17. Because these shear stresses are large, it is quite possible to nucleate dislocations ahead of the crack on slip planes that are favorably oriented with respect to the shear stress.

Gilman and his associates[21] have reported some interesting work dealing with plastic deformation during cleavage fracture, using single crystals of lithium fluoride, which is an ionic cubic solid similar to MgO in which dislocations can be readily revealed by a reliable etching technique. It has been observed that in this material, when a crack moves slowly or stops, dislocations are nucleated in advance of the crack. Actually, there is a limiting velocity below which dislocations are nucleated and above which no evidence of plastic deformation is to be found. The surface of a cleaved LiF crystal is shown in Fig. 19.18. In this particular case, the crack slowed down and then accelerated. After the cleavage the fracture surface was etched. This formed an etch pit at each intersection of a dislocation with the surface. In Fig. 19.18, the long array of closely spaced pits

21. Gilman, J. J., Knudsen, C., and Walsh, W. P., *Jour. Appl. Phys.*, 29 601 (1958).

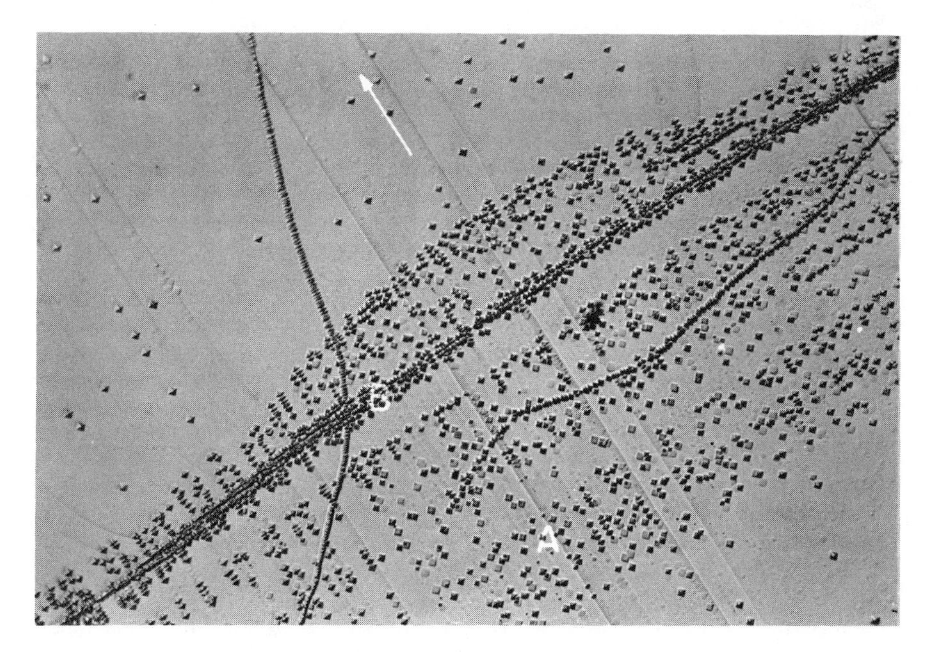

Fig. 19.18 Dislocations that formed in LiF as a result of a temporary reduction in the cleavage-fracture velocity. (Gilman, J. J., Knudsen, C., and Walsh, W. P., *Jour. Appl. Phys.,* **29** 601 [1958].)

represent the dislocations that formed in front of the crack when its velocity fell below the critical velocity for forming dislocations.

An interesting feature of the work on LiF is that the maximum velocity with which a crack can move in this material has actually been measured and is 0.31 the velocity of sound.[22] This compares quite favorably with the theoretical value for fracture in an elastic material which is 0.38, as was previously pointed out.

Now consider the effect of plastic deformation on cleavage crack propagation. When slip takes place during the movement of a crack, energy is absorbed in nucleating and moving dislocations. This energy comes at the expense of the elastic strain energy that drives the crack. If the work to overcome plastic deformation is too large, the crack may decelerate and stop, which implies that in those crystalline materials which cleave, a minimum velocity must be achieved before the crack can move freely. If the velocity is too slow then too much energy will be absorbed in the form of slip. The plastic deformation term may well be one of the most important factors which make some metals incapable of cleavage. In this regard, it is interesting to remember that face-centered cubic

22. *Ibid.*

metals, with their many equivalent slip systems, have not been observed to cleave.

In addition to the energy associated with the formation and growth of dislocations, there is another way that dislocations absorb energy from a moving crack. The cutting of dislocation lines by cracks involves energy losses. This is true particularly when the intersected dislocations are in a screw orientation. When a cleavage fracture passes a screw dislocation, the fracture surface receives a step whose height equals the Burgers vector of the dislocation. The nature of the development of this step is shown in Fig. 19.19. As a crack progresses, the steps that it obtains, due to its intersection with dislocations, tend to run together. If the steps are from dislocations of the same sign they combine to form larger steps, whereas, if they are from dislocations of opposite sign, they cancel.

The step produced on the cleavage plane by a single screw dislocation is, of course, too small to be seen. However, if the crack intersects a large number of screw dislocations of the same sign, then, by combining, steps can be developed large enough to be easily visible. As grown, LiF crystals often contain low-angle boundaries which have a twist component. Such boundaries contain a cross-grid of closely spaced screw dislocations, all of the same sign, and when a cleavage crack crosses one of these boundaries, large-sized steps develop in the cleavage plane. An excellent example of this is shown in Fig. 19.20. Notice how the steps progressively run together to form still steeper steps. The pattern that the steps

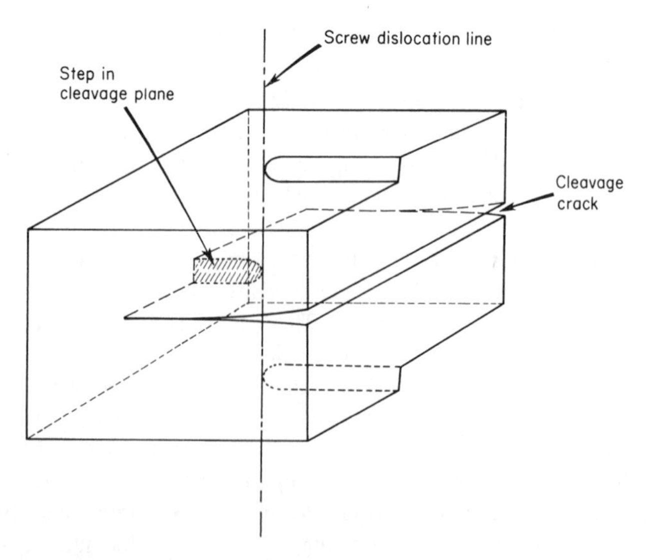

Fig. 19.19 Production of a step in the cleavage plane when a cleavage fracture intersects a screw dislocation.

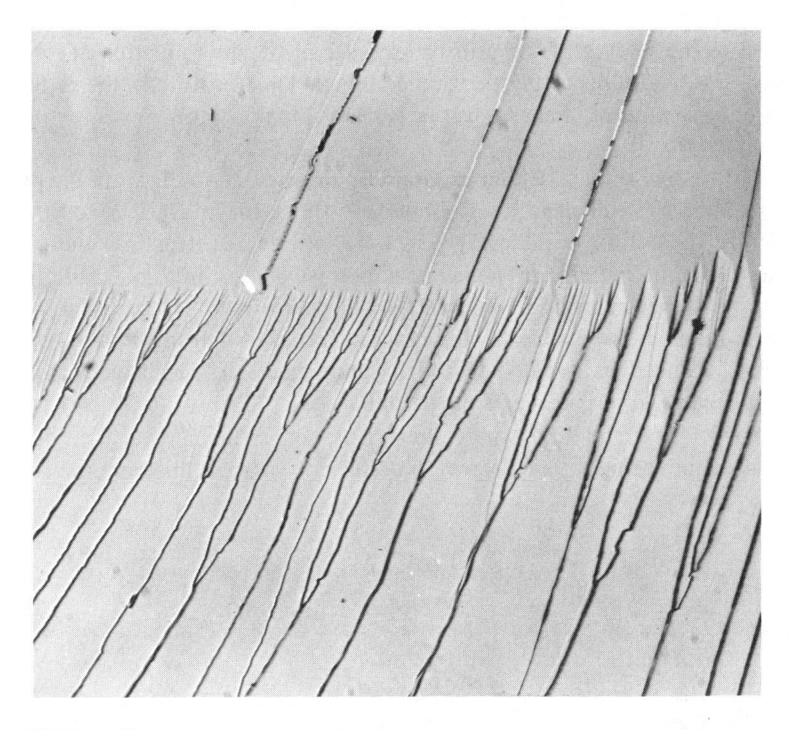

Fig. 19.20 Cleavage steps resulting from the intersection of a cleavage crack with a low-angle boundary containing a grid of screw dislocations of the same sign. (Gilman, J. J., *Trans. AIME,* **212** 310 [1958].) 250x.

form on the cleavage surface is known as a *river pattern.* In general, the latter run normal to the crack front, and it is often possible to locate the point of origin of a cleavage crack by following the river pattern back to its source.

The river pattern on brittle fracture surfaces can arise from a number of causes other than the cutting of screw dislocations by a crack. Thus, if a cleavage crack is started, say by a hammer and chisel, in such a manner that the crack is not strictly parallel to the cleavage plane, the fracture must contain steps in order to accommodate the misalignment. River patterns are also observed on the surfaces of glass fractures. Since glass is normally amorphous, the crack does not follow a crystallographic plane and dislocations in the normal sense cannot exist.

There are several reasons why it is more difficult for a fracture to move along a stepped surface. First, a cleavage plane that contains a large number of steps has a larger surface area and therefore a larger surface-energy term. Second, the advance of the crack not only involves separation of the crystal along cleavage plane segments, but also entails the continued growth of the surfaces of the steps, or small *cliffs*. Unless a secondary cleavage plane, or slip plane, is nearly normal to the surface of the primary cleavage plane, the formation of steps will involve what amounts to plastic tearing of metal in order to form the surfaces of the step. A great deal of energy can be expended in doing this.

19.14 The Effect of Grain Boundaries. As shown in the preceding section, a small-angle twist boundary (in a single crystal) adds to the difficulty of motion of a cleavage crack by introducing steps into the fracture plane. Grain boundaries in polycrystalline metals also impede the motion of cracks, and there is sound evidence that the magnitude of the effect is much larger because it is possible to find cleavage cracks in deformed polycrystalline tensile specimens that are no larger than a grain diameter. Such a microcrack in a polycrystalline iron specimen is shown in Fig. 19.21.

Consider first the case where the orientation difference between adjacent crystals is more than a few degrees, but still not large. In this case, the cleavage

Fig. 19.21 A cleavage crack that stops at the boundaries of a single grain in a polycrystalline iron specimen. (Hahn, G. T., Averbach, B. L., Owen, W. S., and Cohen, Morris, *Fracture*, The Technology Press and John Wiley and Sons, Inc., New York, 1959, p. 91.) 200x.

Fig. 19.22 Large-cleavage steps that develop when a cleavage crack passes from one crystal to another. Specimen 3 percent silicon-iron alloy, cleaved at 78° K, direction of crack propagation top to bottom. (Low, J. R., Jr., *Fracture*, The Technology Press and John Wiley and Sons, Inc., New York, 1959, p. 68.) 250x.

planes in the two crystals, while approximately aligned, still make a finite angle with each other. It is not possible, under these conditions, for the fracture surface to pass smoothly through the boundary, and what probably happens is that a series of parallel cleavage surfaces are nucleated on different levels. The ultimate result is that the fracture surface develops a series of steps originating at the grain boundary. A typical example can be seen in Fig. 19.22.

In the average polycrystalline specimen, the misalignment between grains is larger than that between the crystals in Fig. 19.22 and, in general, the fracture surface is much more irregular. A study[23] of the river patterns on the fracture surfaces of polycrystalline specimens shows that the crack also propagates in an erratic fashion, sometimes moving in a direction opposite to the mean direction of movement. Evidence is also found for discontinuous crack propagation, meaning that failure is not merely by the movement of a single-crack front, but that a number of crack segments form and then join together. Since the

23. Low, John R., Jr., *Fracture*, p. 68. The Technology Press and John Wiley and Sons, Inc., New York, 1959.

individual segments may not be on the same level, this usually involves plastic tearing between fracture surface segments.

The above information demonstrates that cleavage cracks are more difficult to propagate through polycrystalline material than through a single crystal. However, this should not be taken as evidence that brittle cleavage fractures cannot move easily through polycrystalline material. Rather, it may limit the initial size to which a cleavage crack can grow, so that cracks can form which grow in diameter only to the size of a single grain, or to several grains in diameter. Further growth of these microcracks is prevented by the difficulty of passing the crack through a large-angle grain boundary. At a sufficiently high stress, however, it is conceivable that the crack will spread catastrophically. The conditions under which this occurs are of interest because expansion of microcracks leads to brittle fracture. As an approximation, we may consider that the average microcrack has a length equal to a grain diameter. The Griffith criterion for the expansion of a crack

$$\sigma_f = \left[\frac{2\gamma E}{\pi c}\right]^{1/2}$$

may now be applied if we make the assumption that γ, the specific surface energy, can be replaced by a corresponding effective surface energy γ_p. This latter quantity takes into account not only the true surface energy, but also the energy of plastic deformation expended in forcing the crack through a polycrystalline aggregate. For the given conditions, the Griffith criterion becomes

$$\sigma_f = \left[\frac{4\gamma_p E}{\pi d}\right]^{1/2}$$

where γ_p is the effective specific surface energy, d the average grain diameter, σ_f the applied stress, and E is Young's modulus. This relation predicts that the strength of polycrystalline metals which fail by brittle cleavage should vary as the reciprocal of the square root of the average grain diameter. Empirical relations of this nature have been reported for both zinc and iron.[24,25]

As the diameter becomes larger and larger, a critical diameter is reached above which the stress needed to expand a microcrack becomes smaller than the stress to form or nucleate a crack inside a crystal. When this occurs, it is conceivable that the first crack that forms causes the failure of the specimen. By the time the crack has grown to a size equal to the grain diameter, it will be large enough to pass through the boundary and continue to grow. Above the critical diameter the fracture stress is controlled by the stress required to nucleate cracks in

24. Low, J. R., Jr., *Trans. ASM*, **46A** 163 (1954).
25. Greenwood, G. W., and Quarrell, A. G., *Jour. Inst. of Metals*, 82 551 (1954).

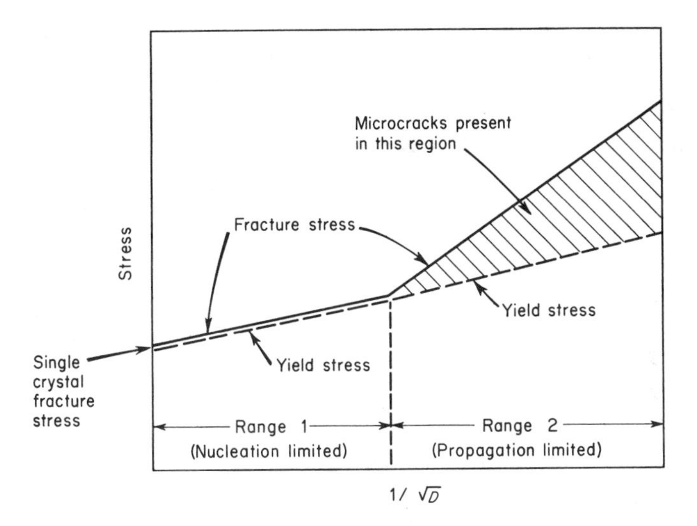

Fig. 19.23 Summary of experimental results showing the effect of grain size D on the yield and fracture stresses of polycrystalline metals. (After Gilman, J. J., *Trans. AIME,* **212** 783 [1958].)

crystals, and below it by the stress required to propagate them through a polycrystalline aggregate. Since cleavage cracks are not nucleated until the yield point of the metal has been reached, fracture should occur almost simultaneously with yielding at diameters larger than the critical diameter. Gilman has summarized the above conclusions in the form of a simple schematic diagram. (See Fig. 19.23.) Notice that this figure is plotted in terms of the reciprocal of the square root of the grain diameter.

Schematic stress-strain curves are shown in Fig. 19.24 corresponding to the two regions of Fig. 19.23. Figure 19.24A gives the shape of the curve when fracture is nucleation-limited, and Fig. 19.24B, when fracture is propagation-limited. Notice that there is a small but definite plastic strain in the second case. Figure 19.23 corresponds to a relatively low temperature where brittle fracture is possible. At higher temperatures, because slip occurs more easily, all fractures tend to become ductile. In this case the stress-strain curves should have the appearance of Fig. 19.24C.

19.15 The Effect of the State of Stress. Both cleavage-crack nucleation and propagation are favored by high tensile stresses. On the other hand, slip requires shear stress. When deformation occurs by slip, however, the applied stresses tend to be relieved. In other words, it is difficult to achieve large stresses when a metal deforms easily by slip. From these considerations we can conclude that any stress system capable of producing a combination of large

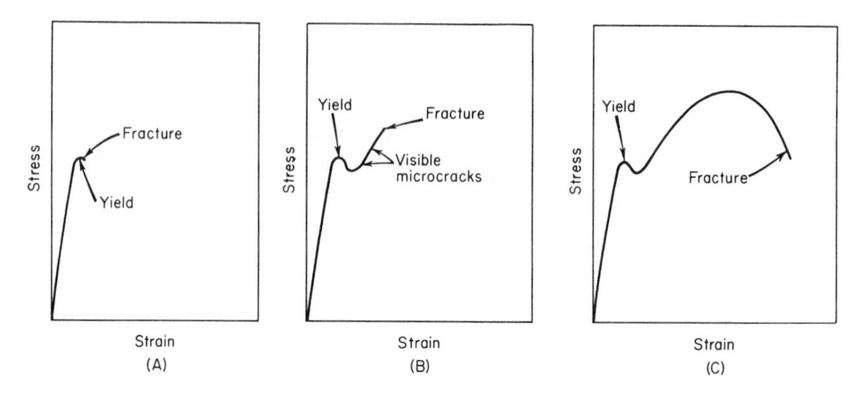

Fig. 19.24 Tensile stress-strain curves. (A) Brittle fracture, nucleation limited. (B) Brittle fracture, propagation limited. (C) Ductile fracture.

tensile stresses and small shear stresses favors cleavage. The nature of the state of stress in a metal specimen is clearly an important consideration in the fracture process.

In simple uniaxial tension, the stress can be viewed as equivalent to a set of shear stresses oriented at 45° to the tensile stress axis. This relationship is shown in Figs. 19.25A and 19.25B, where the tensile stress in one case is assumed horizontal and in the other vertical. If the two tensile stresses (at 90° to each other, as shown in these drawings) are applied simultaneously to the same specimen, the shear-stress components will oppose each other. In the two-dimensional case illustrated, it is clear that under a state of biaxial tension the shear stress in the material is reduced. Also, if a third tensile stress is applied

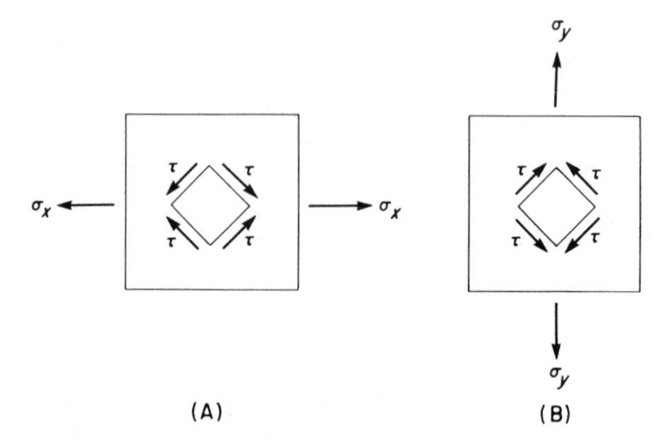

Fig. 19.25 Tensile stresses at right angles produce shear stress components which oppose each other.

normal to the plane of the above two stresses, and all tensile stresses are assumed equal, a state of hydrostatic tension will occur in which the material experiences no shear stress whatsoever. The equivalent state of hydrostatic compression is well known – for example, a liquid under pressure.

From the above it can be concluded that whenever a metal specimen is tested under conditions of biaxial or triaxial tension, slip that requires shear stress will tend to be suppressed. Because of this, a high level of tensile stress can be attained and brittle fracture by cleavage can be promoted. Conversely, if a metal specimen is pulled in tension while immersed in a fluid under pressure, the above conditions will be radically altered. In this case, the applied tensile stress will be complemented by two compressive stresses, having shear-stress components in the same direction as those of the applied tensile stress. Deformation by slip in this case is highly favored, and very high ductilities in polycrystalline specimens have actually been attained in this manner.

An easy way of approximating the state of triaxial tension involves the placing of a simple V-notch around the girth of a cylindrical tensile-test specimen. Figure 19.26 represents such a specimen. When it is loaded in tension, the reduced section at the notch will be the first position to yield. As this region elongates (in the direction of the applied stress), its natural tendency is to shrink in

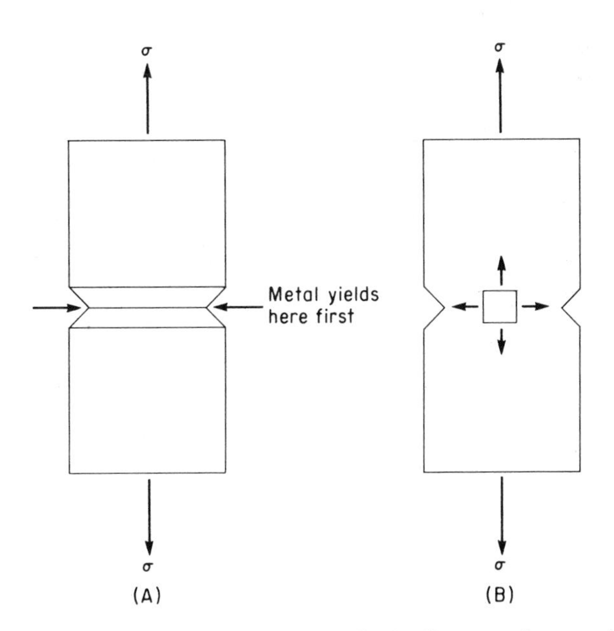

Fig. 19.26 A notched tensile specimen yields first at the notch, which develops a state of triaxial tension in the cross-section at the notch. (A) Three-dimensional view. (B) Cross-section showing stress distribution.

the horizontal plane. This, however, is resisted by the metal lying above and below the notch that has not yet yielded. The metal in the cross-section at the notch is thus placed under three tensile stresses; the vertical applied stress and two induced horizontal stresses at 90° to each other.

19.16 The Impact Test. The most noticeable effect of a notch in a tensile test, or other specimen, is that it raises the temperature at which fracture can occur by cleavage. The number of engineering tests that have been devised to evaluate the transition from brittle to ductile fracture in steels is almost legion. One of these, however, stands out because of its simplicity and almost universal acceptance – the Charpy impact test with a V-notch specimen. The shape and dimensions of a typical specimen are shown in Fig. 19.27. It can be seen that it consists of a steel bar with a square cross-section 1 centimeter on a side. A V-notch is cut across one of its faces. In making a test (Fig. 19.28), the specimen is supported at its two ends in the fashion of a simple beam. It is then struck a blow with a blunt hardened knife edge on the side opposite, and directly behind, the notch. The effect of this mode of stressing is to place the metal at the base of the notch is a state of triaxial tension. The rate of loading in the impact test is much faster than in a normal tensile test, being of the order of 10 million times faster. This rate of straining results from the fact that the knife edge is mounted at the center of percussion of a heavy pendulum hammer which

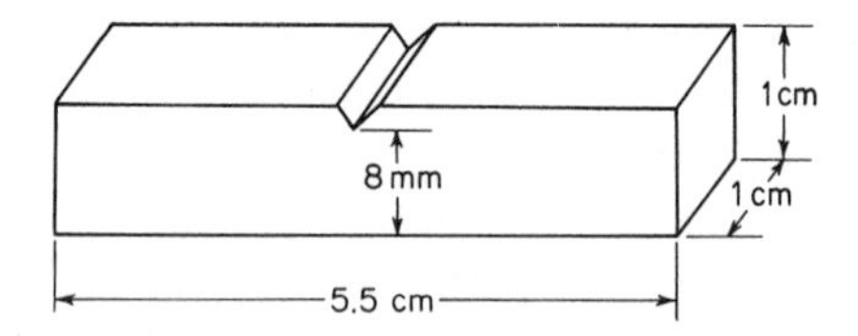

Fig. 19.27 V-notch Charpy impact test specimen.

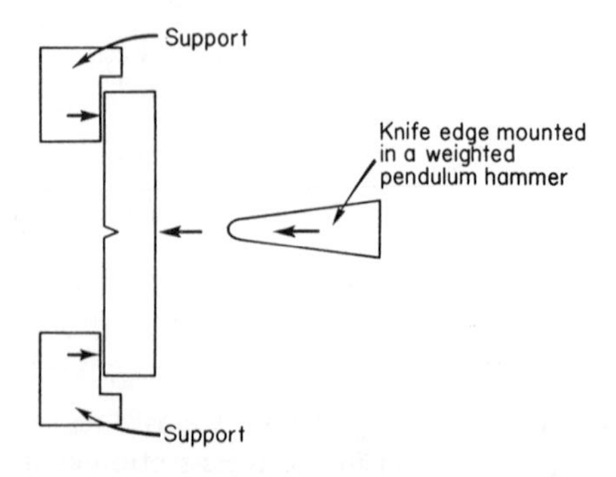

Fig. 19.28 Method of applying the impact load to a Charpy specimen.

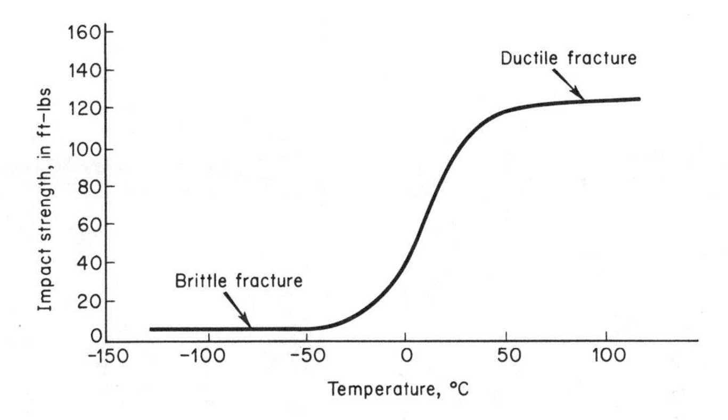

Fig. 19.29 Representative Charpy impact ductile to brittle fracture transition curve.

is dropped from a fixed height. Because the hammer always falls the same distance, it contains a fixed amount of energy when it strikes the specimen, usually of the order of 200 ft-lb. Fracturing the specimen removes energy from the hammer, which is measured on the testing machine by the height to which the hammer rises after it has broken the specimen. The energy expended in fracturing the Charpy impact specimen is the quantity measured in the test. If the fracture is completely ductile, the energy expended will be high; when it is completely brittle, the energy expended will be low.

The impact test furnishes us with a simple method of following the change in the fracture mode of a steel as a function of temperature. A representative curve, showing the transition from ductile to brittle behavior, is given in Fig. 19.29. One of its important features is that the transition is not sharp but occurs over a range of temperatures. If the fractured surfaces of the impact specimen are examined after fracture, it is generally found that there is a reasonable correlation between the amount of the cross-section that has broken in a ductile fashion and the energy expended in breaking the specimen. Therefore the shift from ductile to brittle behavior can also be followed by examining fractured surfaces of impact specimens. Completely ductile specimens exhibit surfaces that are rough or fibrous, while those of brittle specimens contain an irregular array of small bright facets, each corresponding to the surface of a cleaved crystal. In those specimens where the fracture is part ductile and part brittle, the brittle, or bright, area is found at the center of the cross-section. Figure 19.30 shows some typical cross-sections of impact-test specimens. Notice that the photograph of the completely ductile specimen shows that a very large distortion of this specimen occurred during fracture. In the completely brittle specimen, the fracture cross-section is almost a perfect square, showing that fracture occurred, in this case, with negligible plastic deformation.

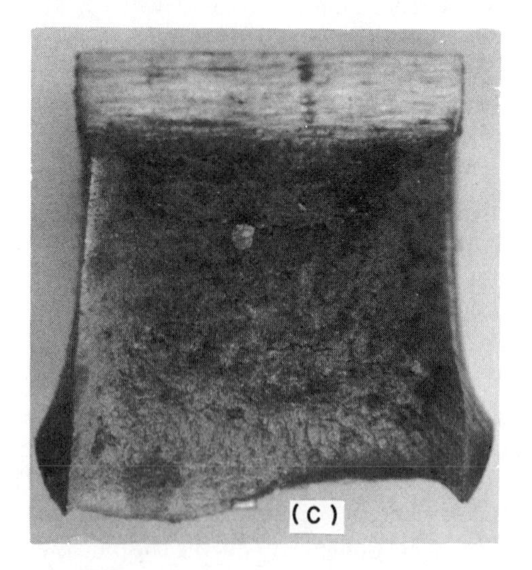

Fig. 19.30 Typical Charpy v-notch fracture surfaces: (A) completely brittle fracture; (B) part ductile, part brittle; (C) completely ductile.

19.17 The Significance of the Impact Test. Brittle fractures in engineering structures have been a subject of considerable concern ever since it became the practice to weld ships and other large structures. The hull of a welded ship is really one continuous piece of steel. A crack that starts in such a structure can pass completely around the girth of the ship, causing it to break in two, and a number of failures of this nature have occurred. Similarly, a welded gas pipeline is also a large continuous piece of steel, and brittle fractures have been known to travel in them with high velocities for distances as long as half a mile.[26] Brittle fractures in ships have received the most extensive study. In general, these show that cracks start at some notch or stress raiser. These may be due to faulty design or to accidents of construction, such as arc strikes; points where a welder started his arc, leaving behind a notch in the steel. It has further been observed that brittle failures have almost universally occurred at low ambient temperatures – the middle of winter. Finally, the hull has to be in a state of stress, which may be caused by heavy seas. Ship failures, which occurred while the ship lay at a dock, however, have been recorded. In the latter case, thermal expansion due to the sun hitting the deck early in the morning can account for the stresses required to propagate fracture.

The importance of the Charpy impact test lies in the fact that it reproduces the ductile-brittle transformation of steel in about the same temperature range as it is actually observed in engineering structures. In an ordinary tension test, the transition for iron or steel occurs at a much lower temperature. This fact has been known for some time, as can be seen in Fig. 19.31, which corresponds to data reported in 1935. These curves, which apply to a particular composition, measure the transition in terms of the energy absorbed in the impact test, the elongation in the tension test, and the angle of twist (to failure) in the torsion test. In each case, the measured property decreases with increasing brittleness of fracture. The transition temperature, as measured in the impact test, is at least 100°C higher than in the tension test. It is also significant that in the torsion test

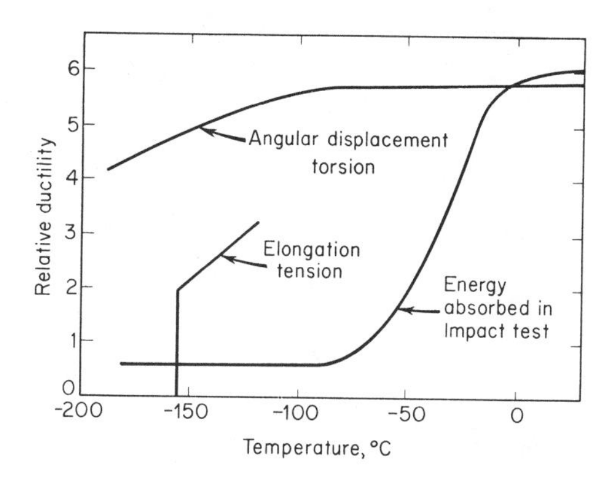

Fig. 19.31 The ductile-brittle transition of iron as measured in three types of tests: impact, tension, and torsion. (After Heindlhofer, K., *Trans. AIME,* **116** 232 [1935].)

26. Shank, M. E., *ASTM Special Technical Publication No. 158* (1953), p. 45.

the ductile-brittle transition was not attained because the measurements were not carried out at sufficiently low temperatures.

Figure 19.31 emphasizes clearly the importance of the state of stress in brittle fracture. In simple torsion, the maximum shear stress is equal to the maximum tensile stress. In simple tension, the maximum shear stress is one-half the maximum tensile stress, and, finally, in an impact test, because of the notch, the shear stress is a still smaller fraction of the maximum tensile stress. The higher the ratio of the tensile stress to the shear stress in a given type of loading, the higher the transition temperature.

The Charpy impact test has been widely used to measure the effect of a number of variables on the brittle to ductile transition. As may be seen in Fig. 19.29, there is no single temperature at which an average ferrous metal suddenly becomes brittle; the transition occurs more or less over a range of temperatures. Still, as a matter of convenience, it is common practice to speak of the transition temperature of a metal. This term, however, needs to be carefully defined, as there are a number of different ways of expressing it. One is to take the temperature at which an impact specimen fractures with a half-brittle and half-ductile fracture surface. A second way of defining the transition temperature uses the average energy criterion: the temperature at which the energy absorbed falls to one-half the difference between that needed to fracture a completely ductile specimen, and that needed to fracture a completely brittle specimen. The temperature at which a Charpy specimen breaks with a fixed amount of energy, usually 15 or 20 ft-lb, is also a widely employed basis for the transition temperature. The last two criteria are illustrated in Fig. 19.32.

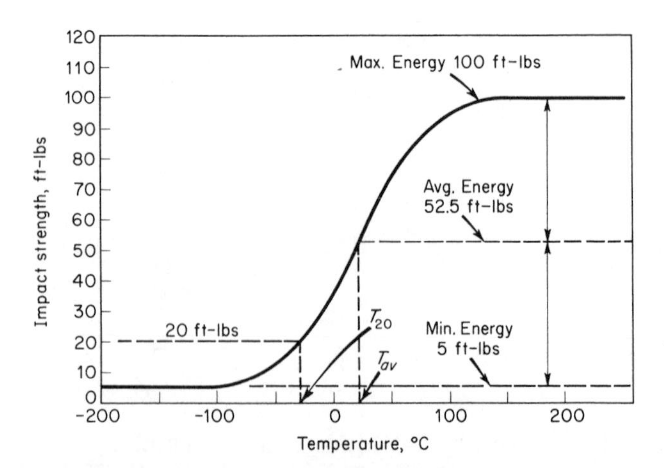

Fig. 19.32 The transition temperature can be defined in several ways, two of which are shown above. T_{20} is the transition temperature using the 20 ft-lb criterion, while T_{av} is that for the average energy criterion.

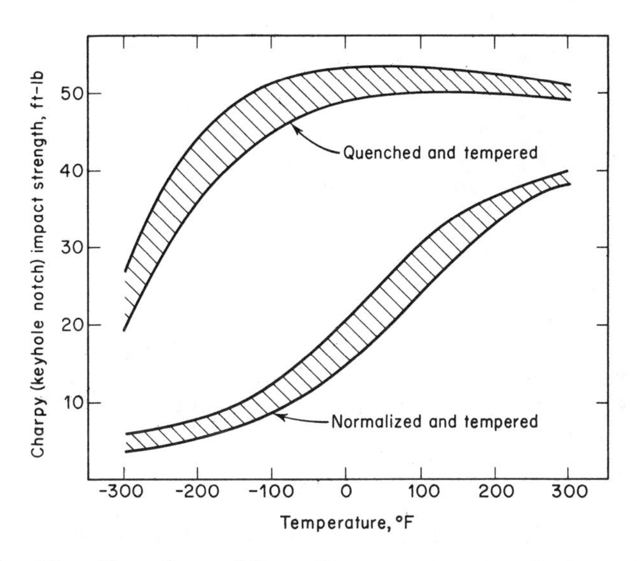

Fig. 19.33 The effect of two different heat treatments on the low-temperature impact strength of a medium carbon alloy steel (4340). Both heat treatments produce the same hardness in the steel, but widely different transition temperatures. Data are plotted as bands to indicate scatter of experimental measurements. (After Society Automotive Engineers, Inc., *Low Temperature Properties of Ferrous Materials*, Special Publication [SP-65] .)

The number of variables that affect the transition temperature in steel are numerous. One of the most important is the microstructure. A typical example of the effect of microstructure is shown in Fig. 19.33, where a comparison is made between two impact-strength versus temperature curves for the same steel after two different heat treatments. Both heat treatments gave the metal the same hardness. In one, the steel was quenched and tempered to produce spheroidized cementite, and in the other, the steel was normalized (air-cooled) and then tempered to produce pearlite. The difference in the two structures is very apparent. The transition temperature is approximately 300°F lower for the spheroidized cementite structure, which is visible proof of why machinery parts, in which toughness is an important factor, are heat-treated so as to produce a spheroidized cementite structure.

The composition of a ferrous alloy has a very pronounced effect on the transition temperature. Figure 19.34 shows the effect of carbon concentration on the transition temperature of a typical steel containing 1.0 percent manganese and 0.30 percent silicon in addition to carbon. Notice that the transition temperature rises rapidly with increasing carbon content. In general, other variables being constant, the higher the carbon content of a commercial

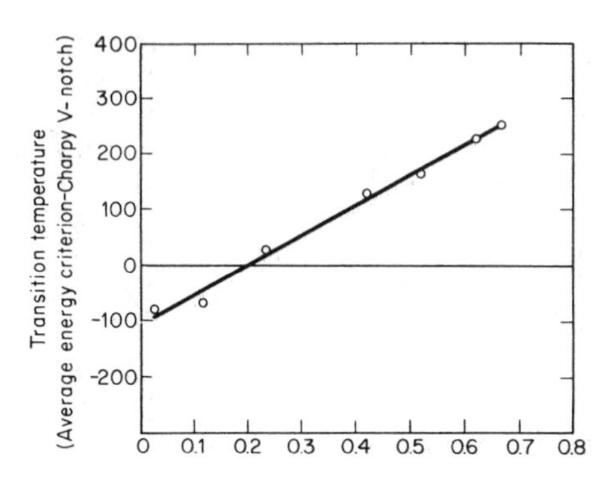

Fig. 19.34 The effect of carbon concentration on the transition temperature of a typical steel. (After Rinebolt, J. A., and Harris, W. J., Jr., *Trans. ASM,* **43** 1175 [1951].)

steel, the more liable it is to brittle fracture near room temperature. Phosphorus has an even stronger effect on the transition temperature of steel. This is one reason why phosphorus is not desirable in ordinary carbon steels. Some elements, for example, manganese, actually seem to have the reverse effect and lower instead of raise the transition temperature. Recent trends in the design of steel for use in the hulls of ships have therefore tended to the more liberal use of manganese in order to take advantage of this tendency.

19.18 Ductile Fractures. There is considerable confusion as to how to differentiate explicitly between brittle and ductile fractures. This is mainly because one tends to think in terms of the entire deformation process leading up to the final act of fracture. To this must be coupled the fact that the word "brittle" is associated with a minimum of plastic deformation, while the word "ductile" connotes large plastic deformation. However, metals can fail by cleavage (a basically brittle process) after a relatively large preceding macroscopic deformation. In the same manner, it is quite possible to have a negligible macroscopic strain in a metal that fails by a ductile mechanism. In the last case, the fracture usually occurs in some localized region in which the deformation is very high. In viewing the problem of fracture, it would seem best, therefore, to reserve the terms "ductile fracture" and "brittle fracture" to refer to the actual act of propagating a crack. Thus, a *brittle fracture* is one in which the movement of the crack involves very little plastic deformation of the metal adjacent to the crack. Conversely, a *ductile crack* is one that spreads as a result of intense localized plastic deformation of the metal at the tip of the crack. There can, of course, be no sharp division between ductile and brittle failures, but the extremes of these two methods of failure are quite distinguishable. A completely

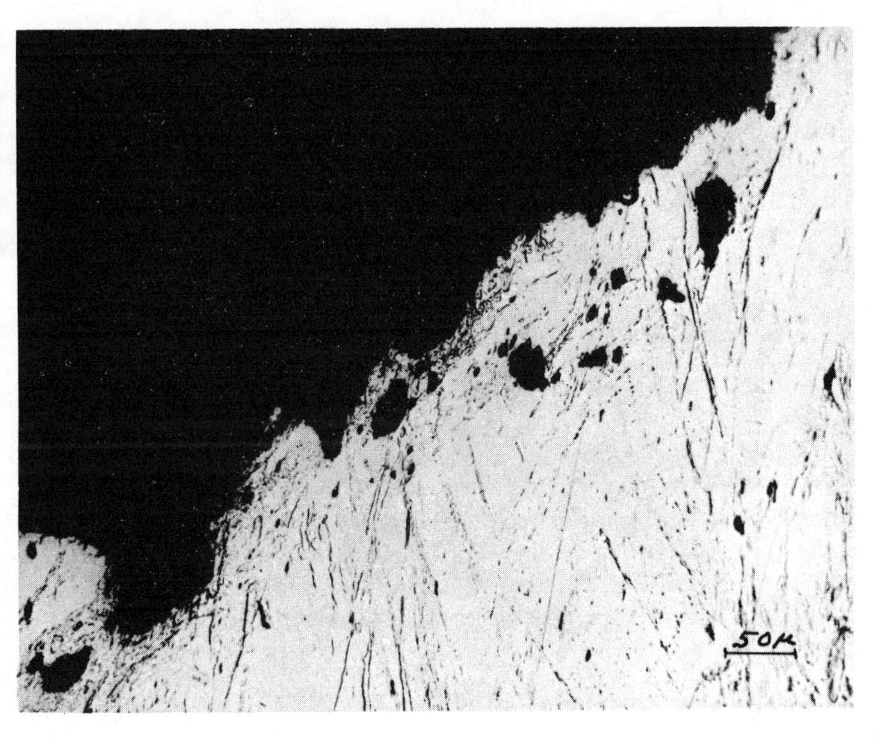

Fig. 19.35 Irregular surface contour of a ductile fracture. (Rogers, H. C., *TMS-AIME,* **218** 498 [1960].)

brittle cleavage fracture will show sharp planar facets which reflect light, while a completely ductile fracture presents a rough dirty-gray surface. The reason for the latter is that a ductile fracture surface has a rough, irregular contour where much of the surface is inclined sharply to the average plane of the fracture. Figure 19.35 shows a cross-section through a ductile fracture that illustrates this characteristic of ductile fracture.

The failure of most ductile materials in the polycrystalline form occurs with a *cup-and-cone* fracture. The appearance of a typical specimen that has failed in this fashion is illustrated in Fig. 19.36. This type of fracture is closely associated with the formation of a neck in a tensile specimen. As indicated in Fig. 19.37, fracture begins at the center of the necked region on a plane that is macroscopically normal to the applied tensile-stress axis. As deformation progresses, the crack spreads latterly toward the edges of the specimen. Completion of the fracture occurs very rapidly along a surface that makes an angle of approximately 45° with the tensile-stress axis. In a perfect example, this final stage leaves

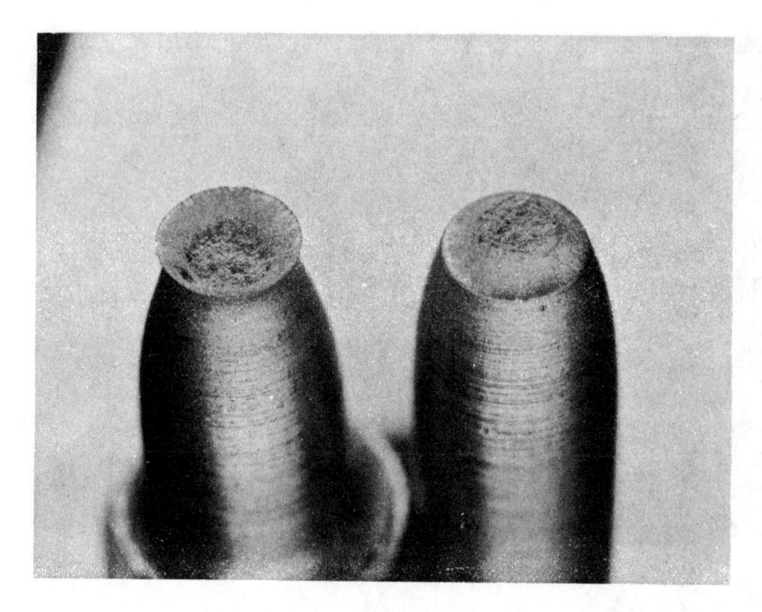

Fig. 19.36 Cup and cone fracture.

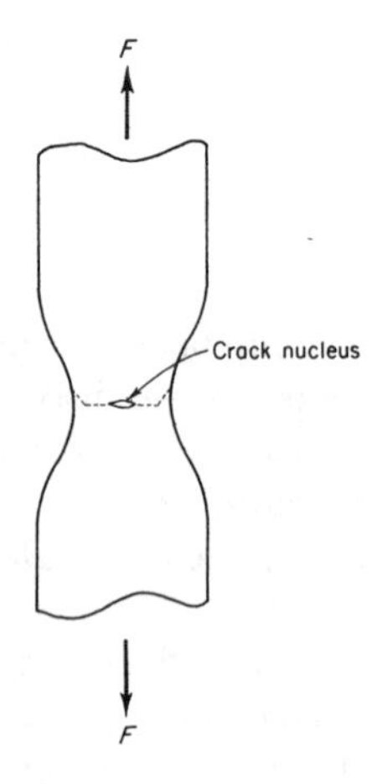

Fig. 19.37 Cup and cone fracture. Crack starts at the center of the specimen and spreads radially, with final completion of the fracture on a cone-shaped surface making an angle of approximately 45° with the tensile axis.

a circular lip on one-half of the specimen and a bevel on the surface of the other half. Thus, one half has the appearance of a shallow cup and the other half, a cone with a flattened top.

Recent work[27] has shown that in the necked region of a tensile-test specimen, small cavities may form in the metal near the center of the cross-section before a visible crack is found. The density of these pores depends strongly on the amount of deformation; increasing with increasing deformation. Thus, Rogers observed that the number of pores per cm^3 was of the order of 10^3 greater in the region of the neck in his copper specimens than in a position in the gage section well away from the neck. Coalescence of these cavities, as a result of their growth under the applied stress, can be assumed to lead to the formation of a ductile crack. Such a condition is actually shown in Fig. 19.38. There is good evidence that, in most commercial metals, these internal cavities probably form at non-metallic inclusions. The manner in which inclusions nucleate pores may only be surmised, but certainly hard brittle inclusions will impede the natural plastic flow of the matrix surrounding them. The belief that inclusions play a strong role in nucleating ductile fractures is supported by the fact that extremely pure metals are much more ductile than those of slightly lower purity. In a tensile test very pure metals often draw down to almost a point before they fracture.

Once a crack of the type shown in Fig. 19.38 has developed it can propagate by the *void-sheet mechanism.* This mechanism acts as follows. The stress

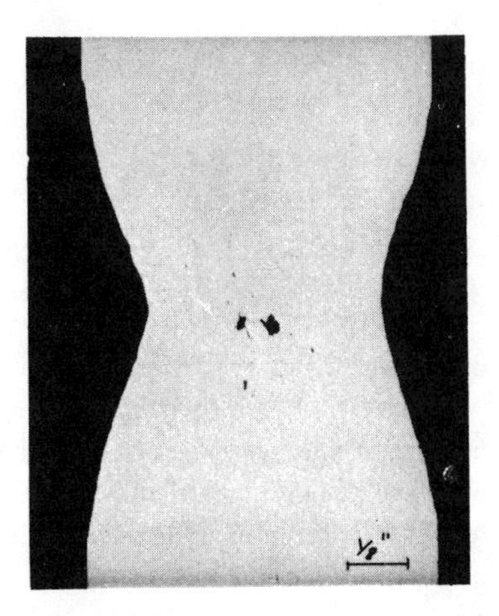

Fog; 19.38 Voids at the center of a copper tensile-test specimen. The two large voids at the center are joined by a crack. (Rogers, H. C., *TMS-AIME,* **218** 498 [1960].)

27. Rogers, H. C., *TMS-AIME,* **218** 498 (1960).

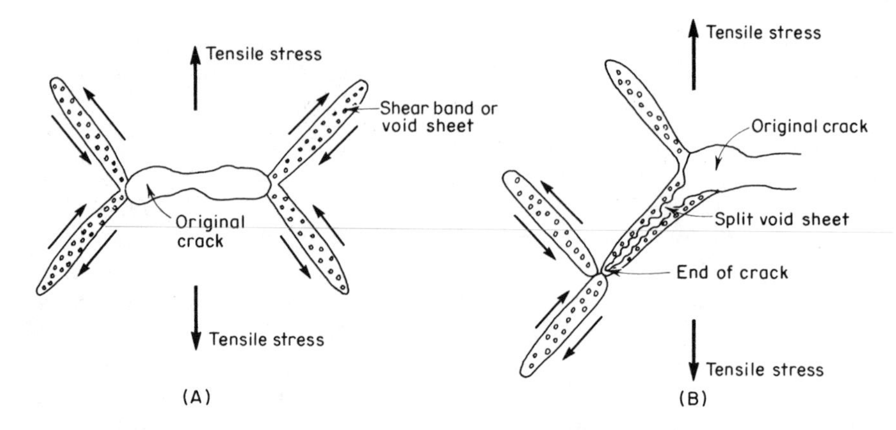

Fig. 19.39 The void sheet mechanism in ductile fracture. (A) The stress concentration at the ends of a small crack initiated by void coalescence induces bands of shear strain at its ends. The highly concentrated strain in these bands nucleates pores inside the bands. (B) The crack advances when a void sheet splits.

concentration at the ends of the crack localizes the plastic deformation in these regions into shear bands that make angles of 30° to 40° with the stress axis. The relationship of the shear bands to the crack is shown in Fig. 19.39A. Because the deformation inside the bands is very intense, the bands become filled with voids. Rogers has used the term *void sheets* to describe these bands at this point. As the voids in these bands grow, they eventually impinge on each other, with the result that a void sheet splits in two and the crack advances as indicated in Fig. 19.39B. The movement of the crack down the void sheet changes the location of the end of the crack so that new shear bands are formed, as shown in this diagram. At this point, the crack could theoretically move along either of these two new bands, one of which is an extension of the original band that split, while the other is inclined in the opposite direction. However, if the crack continues to move forward in its original direction, it will move away from the specimen cross-section at the center of the neck and into a region of decreasing stress. On the other hand, if it moves into the other shear-band orientation, it moves back into the region of maximum stress concentration. This is what normally occurs. Repetition of this process acts to spread the crack across the specimen cross-section. In a ductile metal, the final fracture can occur by several methods: one of these produces a cup and cone, and the other a double cup. When the final fracture occurs so as to form a typical cup-cone failure, as in iron, brass, or duraluminum, it is probable that the shear lip of the cup forms by the void-sheet mechanism. As the central fracture advances toward the surface, the specimen cross-section available to carry the load is progressively decreased. Eventually a condition is reached where the shear bands may extend to the surface. A

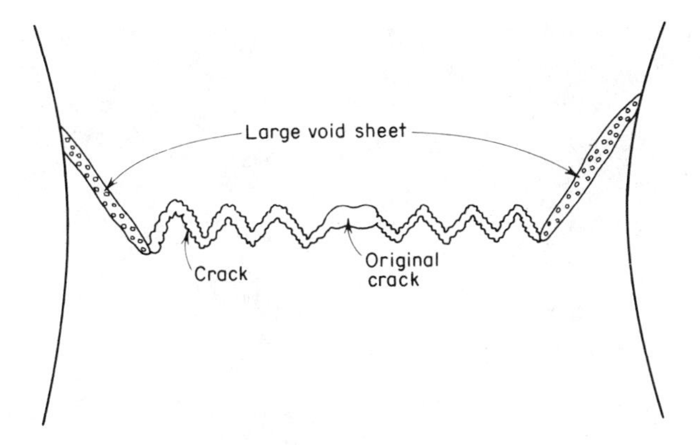

Fig. 19.40 The development of large void sheets that extend to the surface can explain the lip on a cup and cone fracture.

catastrophic splitting of these bands can then result in the shear lip. This case is illustrated in Fig. 19.40.

An interesting aspect of the fracture that occurs by the void-sheet mechanism is that it produces a fracture surface with a characteristic appearance. At low magnifications the surface has a rough, spongy texture. However, at higher magnifications, when observed by the scanning electron microscope, the true nature of the fracture surface can be seen. Such a photograph is shown in Fig. 19.41. The significant feature of this photograph is the rather uniformly ordered array of cuplets that are found on the surface. Each of these cuplets corresponds to a pore that appeared in the void sheet. The fact that the cuplets are oriented in a single direction is a direct result of the deformation that occurs in the shear band. As indicated in Fig. 19.42A, this tends to elongate the pores in a direction approximately parallel to the tensile stress axis. When the void sheet splits, both of the resulting fracture surfaces will shown cuplets, but they will point in opposite directions on the two sides, as shown in Fig. 19.42B. A close examination of Fig. 19.41 will also reveal that small holes have opened up at the bottom of some cups. These represent a connection with other pores below the surface. These subsurface pores are clearly visible in Fig. 19.35, which is a cross-section cut through a specimen that failed by the void sheet mechanism. The stress axis was in the vertical direction. Notice the inclination of the fracture surface characteristic of a void sheet.

Rogers[28] has proposed a mechanism for ductile-crack propagation which is quite interesting and is based on the following concepts. It has been demonstrated that the rate of plastic strain at the surface of a notch is larger than that

28. Rogers, H. C., *Acta Met.*, 7 750 (1959).

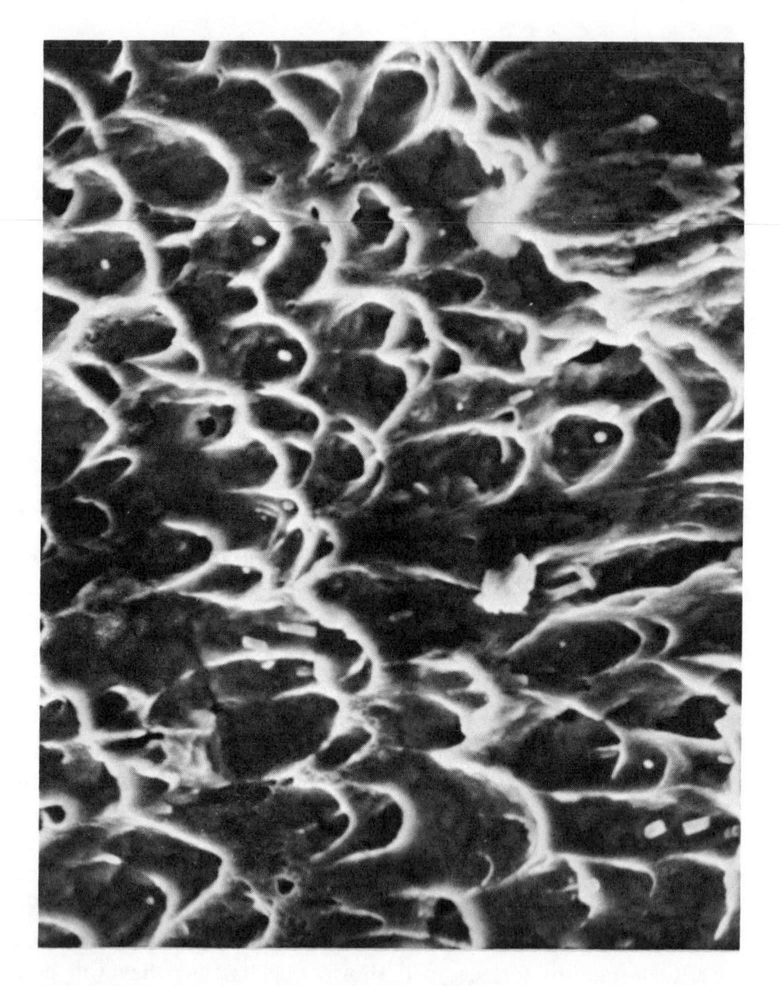

Fig. 19.41 A scanning electron microscope photograph of the fracture surface of an austenitic stainless steel (304) specimen that failed by the void-sheet mechanism. Notice the oriented arrangement of cuplets. Magnification 28,000x. Photograph courtesy of Ellis Verink.

at a distance inside the metal away from the notch. For example, in Fig. 19.43 the metal should flow more rapidly at points a than at point b. This difference in strain rate may be quite large and depends on the sharpness of the notch. We can conclude that, during a tensile test, crystals lying at the surface of the notch are deformed more rapidly than those in the interior of the specimen. Rogers suggests that these crystals should deform like a single crystal that thins down to a chisel edge and ruptures. Ductile-crack movement is thus pictured as the

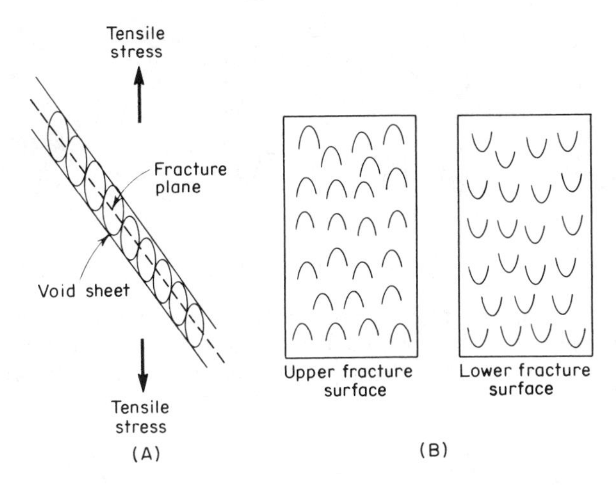

Fig. 19.42 (A) The pores in a void sheet are elongated in a direction roughly parallel to the tensile-stress axis. (B) When the void sheet splits, this produces oppositely pointing cuplets on the two fracture surfaces. (After Rogers, H. C., *TMS-AIME,* **218** 498 [1960].)

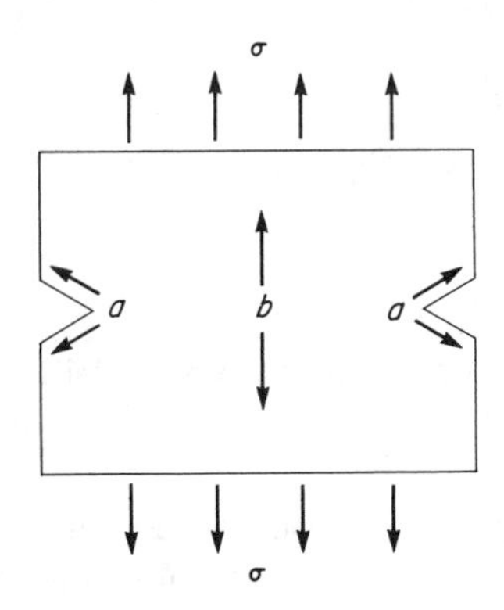

Fig. 19.43 The strain rate at the root of the notch (points *a*) is larger than at a point such as *b* in the interior of a tensile specimen.

successive pulling apart of crystals that find themselves in succession at a notch surface. This propagation mechanism for a ductile crack is indicated schematically in Fig. 19.44.

The above method of crack propagation should hold for any ductile-crack extension whether it is the inward movement of an external notch (or neck), or the outward movement of an internal crack that starts at a void. It can also explain the final stages of a tensile fracture in which a double cup-cone fracture is obtained. This type of fracture is shown schematically in Fig. 19.45 and is

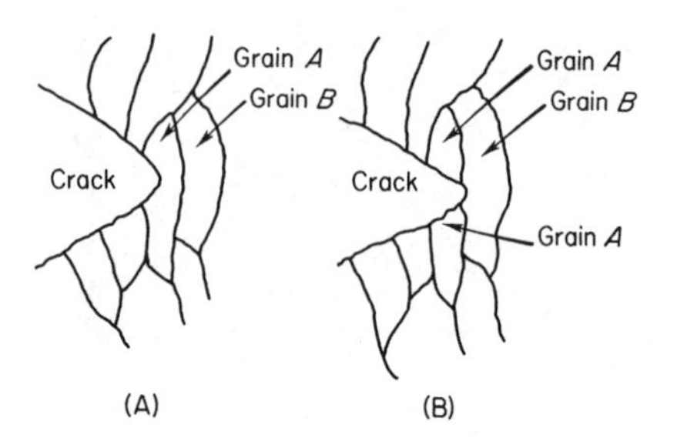

Fig. 19.44 Propagation of a ductile fracture according to Rogers. (B) represents a later time than (A). Notice that crystal *A* has been pulled apart in the indicated time interval. (After Rogers, H. C., *Acta Met.*, **7** 750 [1959].)

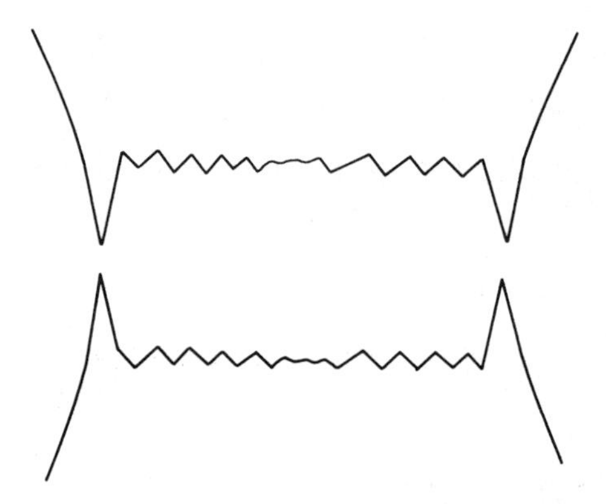

Fig. 19.45 Double cup fracture. This type of failure is observed in many pure face-centered cubic metals. (After Rogers, H. C., *TMS-AIME*, **218** 498 [1960].)

observed[29] in some ductile metals such as copper, aluminum, silver, gold, and nickel. In this type of failure the crack also originates in the center of the specimen and propagates by the void-sheet mechanism until a thin ring of metal remains. This then fails by the progressive slipping apart of the grains, rather than by the void-sheet mechanism. The result is two cups instead of a cup and cone.

19.19 Intercrystalline Brittle Fracture. Fracture by cleavage is not the only form of brittle fracture that occurs in metals. Brittle fracture can also occur in which the fracture passes along grain boundaries. In some instances this type of fracture can be caused by grain-boundary films of a hard-brittle second phase, like that formed by bismuth in copper. In other cases it may not

29. Rogers, H. C., *TMS-AIME*, 218 498 (1960).

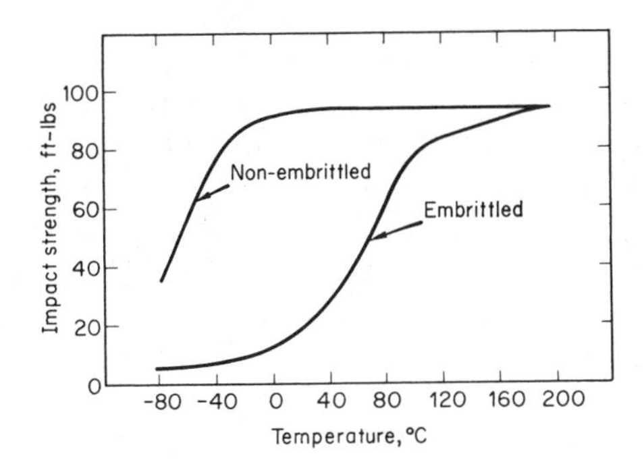

Fig. 19.46 The effect of temper brittleness on the impact properties of a low-alloy steel (3135). (After Hollomon, J. H., Jaffe, L. D., McCarthy, D. E., and Norton, M. R., *Trans. ASM,* **38** 807 [1947].)

be due to an actual precipitate, but to a concentration of solute in the metal close to crystal boundaries. Why this concentration of solute should lower the cohesion across the boundaries is not known.

19.20 Temper Brittleness. When certain low-alloy steels are tempered after a quench (which produces a martensitic structure) in the temperature range 450°C to 500°C and then slowly cooled, they become susceptible to a form of intercrystalline brittle fracture. The cause of this phenomenon is not known, but it is thought to be due to a segregation of some element, possibly in the form of a carbide at the grain boundaries. However, carbide precipitation has not been proved and the embrittlement may be due to a grain-boundary segregation. The effect of temper brittleness becomes quite apparent when one makes a Charpy-impact transition-temperature determination for embrittled steel and compares it with one for steel in the nonembrittled condition. Metal in the latter state can be obtained by quenching specimens from the tempering temperature. Curves showing the impact strength of embrittled and nonembrittled steel are shown in Fig. 19.46. It is known that the tendency to temper brittleness may be reduced in those alloy steels in which it appears by the addition of small percentages of molybdenum. This constitutes the primary reason for the inclusion of this alloying element in the chemical composition of many low-alloy steels.

The curves of Fig. 19.46 resemble those of Fig. 19.33, which shows the difference in the transition-temperature characteristics of tempered martensitic and pearlitic structures. A pearlitic structure, however, has both a low-impact toughness and a low ductility in the tensile test. Temper brittleness lowers only the impact strength.[30]

30. Lorig, C. H., *ASTM Special Technical Publication No. 158* 147 (1953).

19.21 Blue Brittleness. This is a phenomenon that has been recognized in steels for many years. Its name is a result of the fact that it occurs in the temperature range (several hundred degrees centigrade above room temperature) where a blue oxide film is formed on the surface of a steel specimen. The term "blue brittleness" is somewhat of a misnomer, as the metal does not become brittle in the normal sense. What actually happens is that the elongation in a tensile test undergoes a minimum at the blue brittle temperature. This is shown in Fig. 19.47 for the case of commercial purity titanium. However, the reduction in area of a tensile specimen does not normally show a pronounced minimum at the blue brittle temperature. This signifies that the fracture is not brittle.

The blue brittle phenomenon is associated with dynamic strain-aging (see Section 8.14). The interaction between dislocations and impurity atoms acts to bring about the point at which necking starts in a tensile specimen at a relatively small strain and, once necking begins, the strain becomes highly concentrated in the neck. Both effects act to reduce the elongation that is obtained. That this phenomenon is closely associated with the impurity atoms in the lattice is

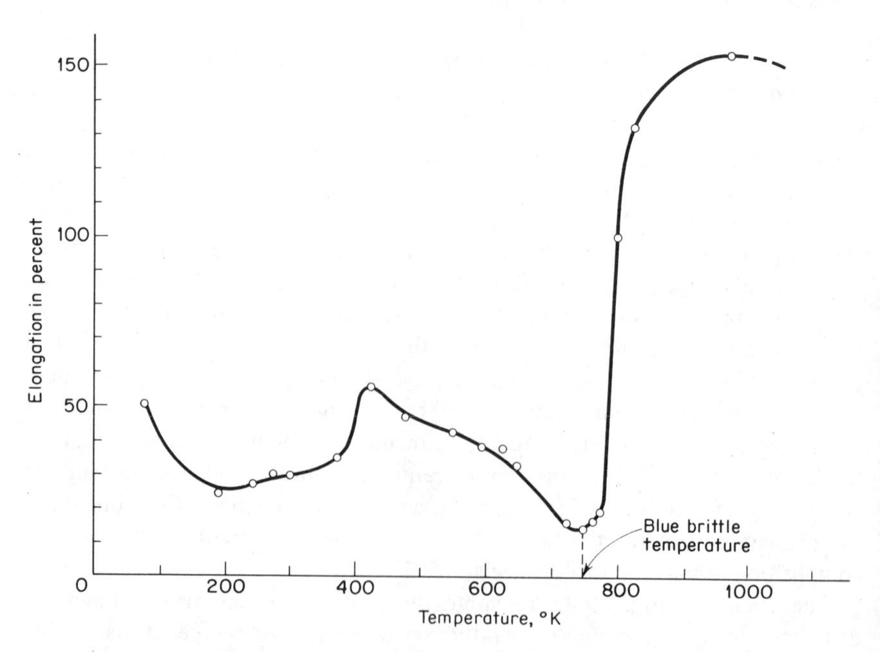

Fig. 19.47 The "blue brittle" phenomenon represents a loss of tensile elongation at the blue brittle temperature. The phenomenon is best known in steels, but occurs in other metals. This curve shows the effect in commercial purity titanium. Strain rate 3×10^{-4} sec^{-1}. (From the data of A. T. Santhanam.)

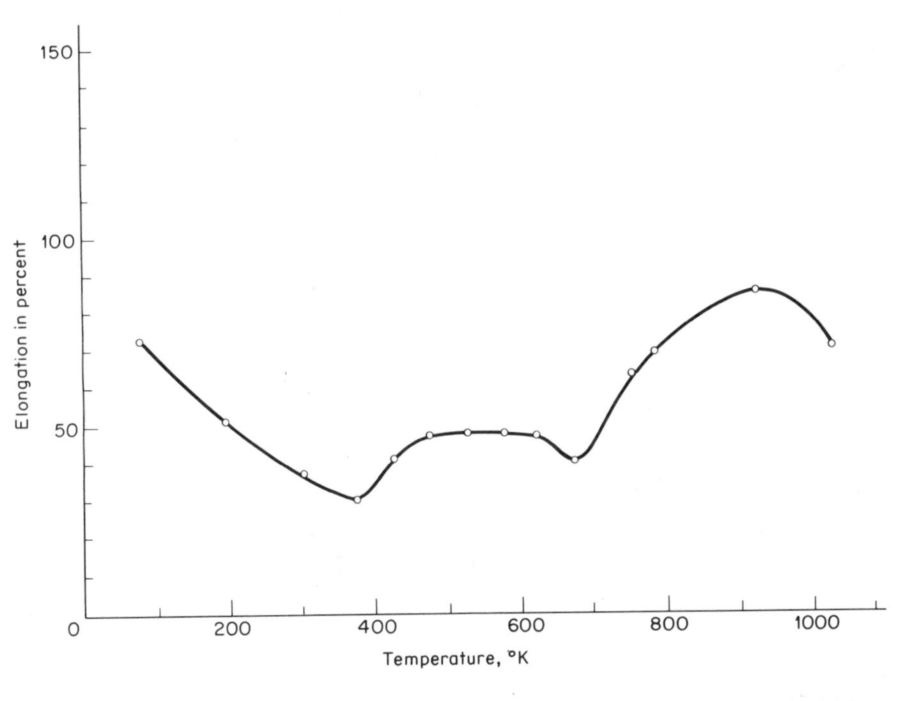

Fig. 19.48 Blue brittleness is associated with dynamic strain-aging. In high purity titanium, as shown by this curve, the elongation minimum almost disappears. Strain rate 3×10^{-4} $\sec^{-1}$. (From the data of Anand Garde.)

demonstrated in Fig. 19.48. This curve shows the elongation as a function of temperature in high purity titanium specimens. Notice that the minimum observable in the corresponding curve for commercial purity titanium has nearly disappeared.

19.22 Fracture Mechanics. A new field has recently developed known as *fracture mechanics*. In this area one attempts to obtain information about the relative toughness of engineering materials under conditions similar to those encountered in engineering practice. In considering this subject we shall use the approach of Irwin,[31] who considers the work done in moving a crack a small distance, Δx, in order to determine the effective force resisting its movement. In computing this work, he assumes, for ease of presentation, that the crack moves backward and closes rather than opens. Thus, let us assume that Fig. 19.49A represents one end of a crack that extends through the thickness of an infinitely wide plate of unit thickness, and that the plate is loaded by a uniform tensile stress σ. Now, as indicated in Fig. 19.49B, a hypothetical tensile

31. Irwin, G. R., *Encyclopedia of Physics*, Vol. VI, Springer, Heidelberg, 1958.

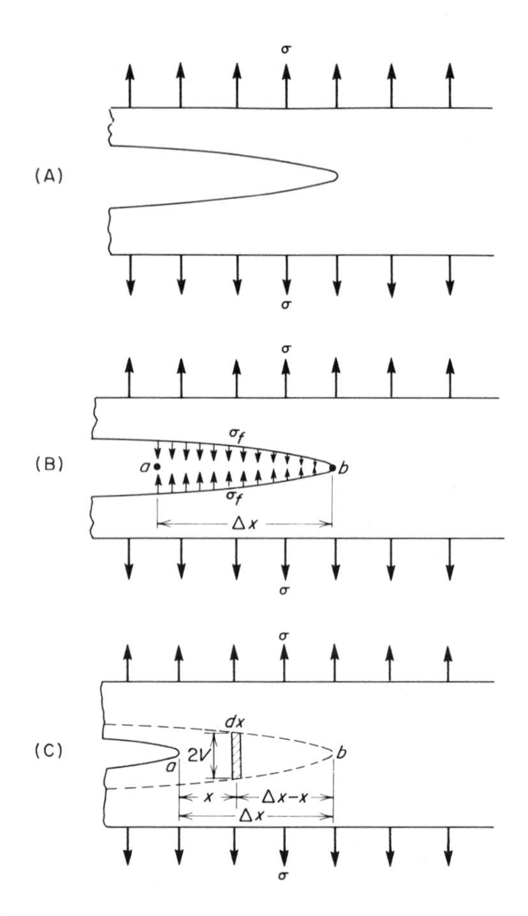

Fig. 19.49 These drawings illustrate the Irwin method of analyzing the conditions associated with fracture.

stress, σ_y, is applied over the upper and lower surfaces of the crack for a distance, measured backward from the edge of the crack, equal to Δx. This stress should close up the crack in the given interval if σ_y is increased at all points to the same value that it had when the crack was originally at point a. That is, before it advanced from a to b through the distance Δx. This situation is indicated in Fig. 19.49C. The required stress distribution is capable of being computed using elasticity theory.

Now consider a small element, dx, of the interval Δx. As shown in Fig. 19.49C, the opposite sides of this element move through a distance $2V$ as the crack closes. The theory assumes that the displacement of the crack surfaces varies linearly with the hypothetical stress σ_y applied to the crack surfaces. The work done on the small element dx as σ_y is raised to the value that closes the crack is therefore given by

$$dW = \frac{2V}{2} \sigma_y dx = V\sigma_y dx$$

while the work required to close the crack over the entire distance Δx is

$$\Delta W = \int_0^{\Delta x} V\sigma_y dx$$

It is now assumed that the material ahead of the crack is subjected to a condition of plane stress. This means that the stress component σ_z, parallel to the edge of the crack, is zero so that all the stress components lie in the plane of the plate. With this condition, elasticity theory predicts that

$$\sigma_y = \frac{\sigma\sqrt{c}}{\sqrt{2x}}$$

and

$$V = \frac{2\sigma\sqrt{c}}{E} \sqrt{2(\Delta x - x)}$$

where x is the distance measured from point a in Fig. 19.49C, c is the half-crack length, σ is the applied external stress, and E is the modulus of elasticity. Evaluating the integral for ΔW gives

$$\Delta W = \frac{\sigma^2 c\pi\Delta x}{E}$$

In general, the derivative of the work with respect to the distance corresponds to a force, so that Irwin defines the crack extension force as

$$G = \frac{\Delta W}{\Delta x} = \sigma_f^{\,2} \frac{c\pi}{E}$$

where σ_f is the applied stress at fracture. This force may be viewed as that necessary to move the edge of the crack through the metal. Note that if this expression is compared to the Griffith relationship given in Section 19.10, we see that

$$G = \frac{\sigma_f^{\,2} c\pi}{E} = 2\gamma$$

and it may be deduced that the force per unit length exerted on the crack front as it advances is equal to twice the specific surface energy γ. This is only to be expected since, according to the Griffith concept, the work expended in advancing the crack is converted into the surface energy of the two crack surfaces.

In the above example, the Irwin analysis has been applied to a crack in a brittle elastic solid where its movement does not involve plastic deformation ahead of the crack. It is significant that the Irwin approach can also be applied to problems where the movement of the crack involves plastic deformation. In this case we may write

$$G = 2\gamma_p$$

where γ_p is the plastic work done in forming a square centimeter of surface. Actually, this expression should more properly be written

$$G = 2(\gamma + \gamma_p)$$

where γ is the true surface energy and γ_p is the plastic work per cm^2 involved in the creation of the surface. However, in most practical examples, $\gamma_p >> \gamma$, and γ can be neglected in comparison to γ_p.

19.23 The Stress Intensity Factor. As indicated above for fracture in a brittle elastic plate, the y or vertical component of the stress ahead of the crack is given by the expression

$$\sigma_y = \frac{\sigma\sqrt{c}}{\sqrt{2x}}$$

where x is the distance ahead of the crack. Note that in the numerator of the right hand term we have the product of the applied stress, σ, and the square root of c, the half-crack length. These two factors determine the general level of the stress at points ahead of the crack. The larger the applied stress or the greater the crack length, the more severe will be the stress at any point in front of the crack. Because of this, it is common practice in fracture analysis to use a parameter known as the *stress intensity factor* which contains the product $\sigma\sqrt{c}$. In the case of a crack in a brittle elastic plate where the stress components ahead of the crack correspond to a condition of plane stress, the stress intensity factor K is defined as

$$K = \sigma\sqrt{c\pi}$$

When the applied stress is raised to the point where the crack is able to move rapidly, a critical value of the stress intensity factor is obtained. This condition is written

$$K_c = \sigma_f\sqrt{c\pi}$$

The quantity K_c is called the fracture toughness and is rather simply related to the crack extension force since

$$G_c = \frac{\sigma_f^2 c\pi}{E} = \frac{K_c^2}{E}$$

Both G_c and K_c may be considered to be parameters that measure the resistance of a substance to crack extension. It is important to note, however, that the above relation between G_c and K_c refers specifically to a particular set of fracture conditions: where the crack advances elastically in a plate under the action of a simple tensile stress, and the stress components in front of the crack correspond to a condition of plane stress ($\sigma_z = 0$). For other fracture conditions, the relationship between G_c and K_c will normally be different. In this regard, it is important to realize that there are three basic modes of fracture. These are shown in Fig. 19.50. Mode I corresponds to fracture where the crack surfaces are displaced normal to themselves. This is a typical tensile type of fracture. In the second mode, the crack surfaces are sheared relative to each other in a direction normal to the edge of the crack; while in Mode III, the shearing action is parallel to the edge of the crack.

In the example described above, the fracture is obviously of the first type, and to indicate this fact, it is normal practice to add the subscript I to the symbols K_c and G_c. The relationship given earlier is, accordingly, more properly written

$$G_{Ic} = \frac{\sigma_f^2 c\pi}{E} = \frac{K_{Ic}^2}{E}$$

19.24 Fracture Toughness Measurements. A number of different types of tests have been developed to measure G_c and K_c values under different fracture conditions. These are described in standard fracture mechanics texts and reference works. For our present purposes we shall only consider the failure of sheet or plate material subject to tensile loading. In this type of test, several factors have to be carefully considered. First, laboratory specimens are normally finite in size, so it is no longer possible to assume that the crack moves in a plate of infinite width. Second, the fracture of real metals always involves, to a varying degree, plastic deformation ahead of the crack. A significant factor

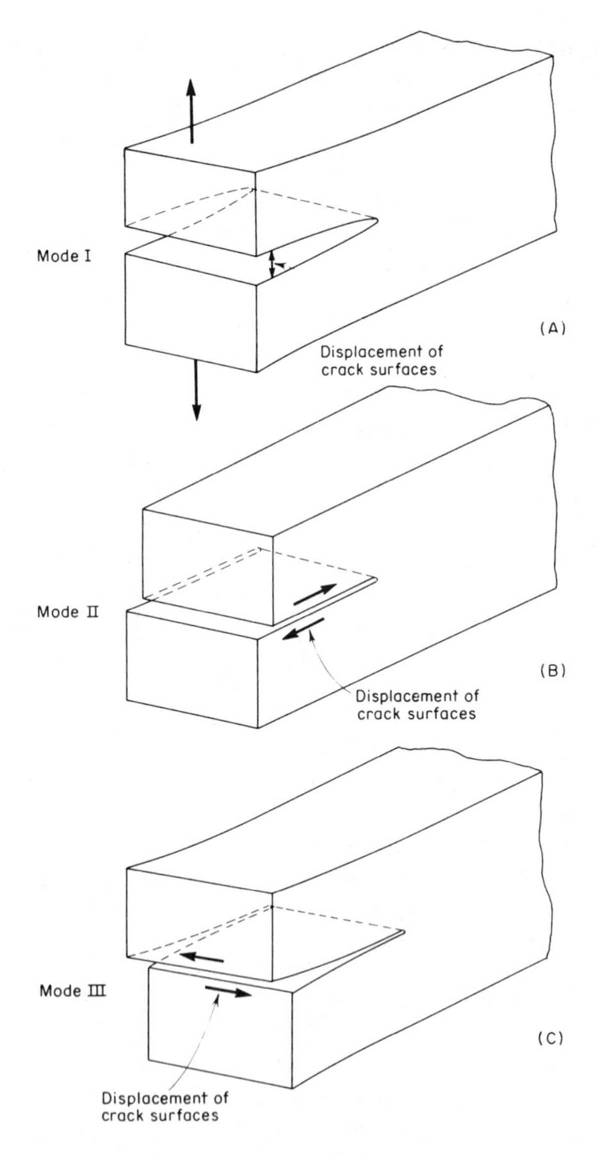

Fig. 19.50 The three basic fracture modes.

in these considerations is the thickness of the plate. In a thick plate, the large depth of metal parallel to the crack front tends to restrict plastic flow parallel to the crack. On the other hand, a crack in a thin plate does not feel this restriction. As a consequence, a crack that passes through a thin plate can draw in the plate

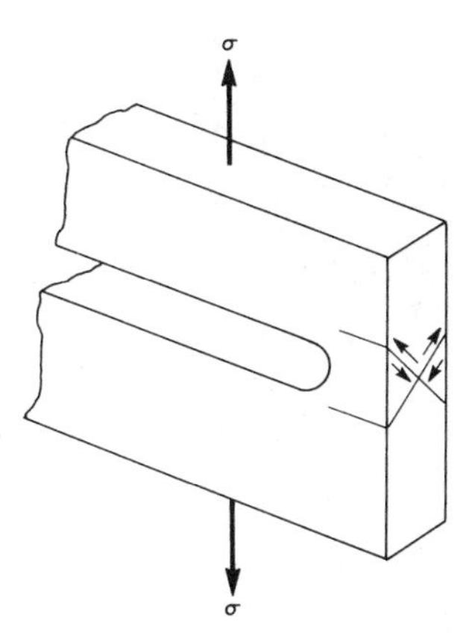

Fig. 19.51 The metal in advance of a crack in a thin plate tends to deform by plane stress and to shear on planes at 45° to the surface of the plate.

or cause it to neck as a result of shears occurring on planes inclined at 45° to the plane of the plate, as shown in Fig. 19.51. This deformation in the direction parallel to the crack front tends to relax the corresponding stress component, σ_z. As a result, the applied stress σ tends to develop a state of plane stress in front of the crack. On the other hand, in a thick plate the inability of the metal to deform in the direction parallel to the crack results in the development of a third stress component, σ_z, parallel to the edge of the crack. In this case the material ahead of the crack is deformed under a condition of triaxial stress and the deformation, as indicated schematically in Fig. 19.52, is confined to shears lying in the plane of the plate. It is customary, therefore, to state that the crack in a thick plate moves under plane-strain conditions.

When the crack moves in a thick plate and the strains ahead of the crack tend to develop a state of plane strain, the fracture tends to occur by Mode I (see Fig. 19.50). The actual microscopic fracture mechanism producing this normal type of fracture can be of several types including cleavage, as in steel at low temperatures. Another possibility is normal rupture, which occurs as a result of the opening up of voids, their growth, and coalescence in the material ahead of the crack front. As discussed in Section 19.18, the central flat fracture surface in a cup and cone fracture of a tensile specimen is often of this latter type. On the other hand, when the plate is thin and fracture occurs under a condition of plane stress, the fracture tends to occur by a shear mode where the fracture surface lies at approximately 45° to the stress axis. This type of failure, which also results from void formation and coalescence, is indicated schematically in Fig. 19.53. The inclined surfaces of the cup and cone fracture of a tensile specimen are usually of this type.

For plates of intermediate thickness, the tendency is for the fracture to occur

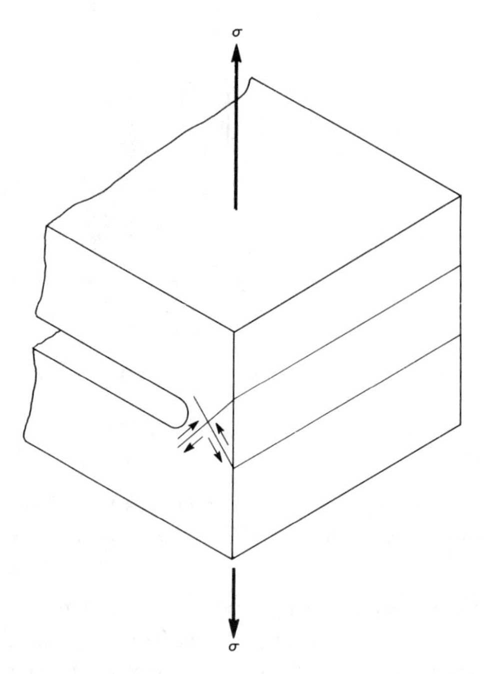

Fig. 19.52 In a thick plate, the metal in advance of the crack tends to deform by plane strain and to shear on planes at 45° to the fracture plane.

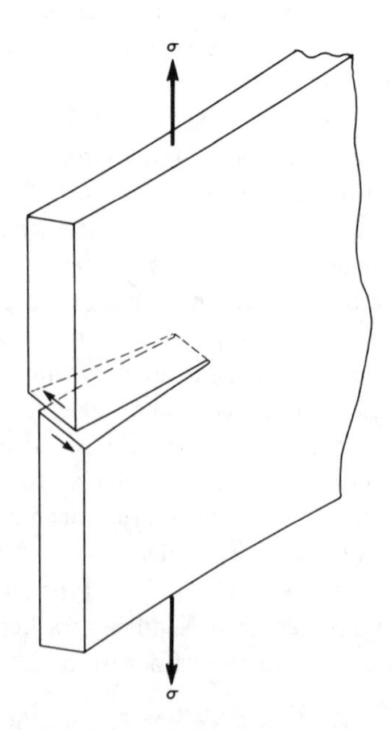

Fig. 19.53 Mode III fracture in a thin plate.

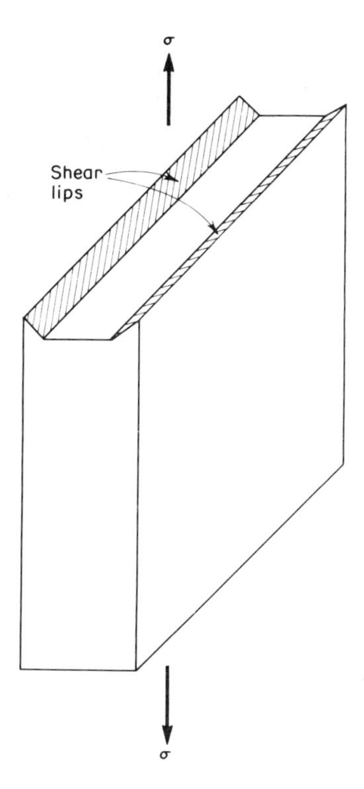

Fig. 19.54 Mixed fracture in a plate of intermediate thickness. Central flat surface corresponds to Mode I plane-strain fracture, shear lips to plane-stress Mode III fracture.

as a mixture of plane strain and plane stress types. The resulting shape of the fracture surface is shown in Fig. 19.54. The central flat fracture surface corresponds to failure under plane strain. On either side of this surface there will normally be two shear lips inclined at about 45° to the central flat area. As the crack advances, the central flat area tends to run ahead of the two inclined surfaces. This running ahead of the central normal fracture causes the load to be supported by two much thinner sections which deform under conditions of plane stress. These surfaces subsequently fracture by Mode III shearing.

A type of specimen frequently employed to measure fracture toughness properties is shown in Fig. 19.55. This consists of a plate of length four times its width (W), and whose thickness (t) may vary with the size of the plate stock to be investigated. The load is applied through pins inserted into holes bored at the top and bottom of the specimen as indicated. A horizontal slot is cut into the center of the specimen. Sharp cracks, equivalent to those existing at the front of a moving crack, are formed at the ends of the slot by repeatedly loading the

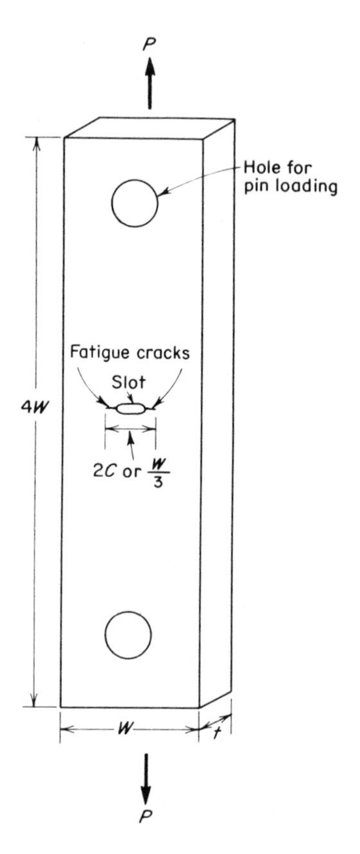

Fig. 19.55 One form of fracture toughness specimen. (After Brown, W. F., Jr., and Srawley, J. E., *Fracture Toughness Testing*, ASTM, Philadelphia, STP 381 133 [1965].)

specimen to a load not large enough to fracture the plate, but sufficient to open up small cracks. These are fatigue cracks, and their nature will be discussed in the final sections of this chapter. The total length of the slot and the fatigue cracks ($2c$) is made equal to one-third of the plate width, as indicated in Fig. 19.55.

If the thickness of the specimen is sufficiently large, it will fracture by plane strain and, under these conditions, the fracture toughness is given by the relation

$$K_{Ic} = \frac{P_f}{Wt}\sqrt{\frac{W}{(1-\nu^2)}\tan\frac{\pi c}{W}} = \sqrt{\frac{EG_{Ic}}{(1-\nu^2)}}$$

where W is the specimen width, t is its thickness, P_f is the load on the specimen at the point of catastrophic failure, ν is Poisson's ratio, and c is the half-crack length at the instant of fracture. Ordinarily one has to add two corrections to the initial value of c existing in the specimen before loading. One accounts for any growth of the crack that may occur before the crack accelerates and fractures the specimen. The other allows for the fact that the plastic deformation that develops ahead of the crack effectively increases the crack length. These corrections for simple plane-strain fracture may be small enough to be neglected. In any event, the application and nature of these corrections are described in standard references.[32]

If the plate is made thin enough, fracture will occur under plane stress conditions on a 45° plane (see Fig. 19.53). In this case, the fracture toughness is determined by the relation

$$K_c(45^\circ) = \sigma_f \sqrt{W \tan \frac{\pi c}{W}} = \sqrt{E\, G_c(45^\circ)}$$

For plates of intermediate thickness, the fracture is normally of the mixed shear and normal types, as shown in Fig. 19.54. In this case, the fracture toughness is computed with the equation given above, but the designation 45° is dropped, so that

$$K_c = \sigma_f \sqrt{W \tan \frac{\pi c}{W}} = \sqrt{E\, G_c}$$

In the above two equations, σ_f is the maximum load divided by the gross cross-section area, and c is the effective half-crack length at the point of maximum load: corrected for slow growth occurring before the maximum load is attained and the development of plastic flow ahead of the crack. Since in plane stress fracture there is usually an appreciable growth of the crack before the fracture becomes catastrophic, some means of determining the value of c at the maximum load must be obtained. There are a number of standard techniques for accomplishing this.[33] This fracture toughness, as determined using a specimen of the type shown in Fig. 19.55, is a function of the plate thickness. In general, below some finite plate thickness (t_0), fracture occurs only by the shear mode. Let us first consider specimens falling in this class. As a rule, the thinner the plate, the easier it is to propagate the fracture. This is because the vertical depth of the plastic zone ahead of the crack varies almost directly as the thickness of the plate. In a ductile fracture, the total work done per unit volume of the material that is plastically deformed is roughly a constant. This is

32. Tetelman, A. S., and McEvily, A. J., Jr., *Fracture of Structural Materials*, John Wiley and Sons, Inc., New York, 1967.
33. *Ibid.*

equivalent to saying that the area under the stress-strain curve is essentially independent of the specimen size if the specimens are geometrically similar. Accordingly, if the crack moves ahead a distance dx, the work done is equal to W_v, the constant work per unit volume times a volume element dx long, by t thick, by t in depth, or $W_v t^2 dx$. This work should also equal $\gamma_p 2\sqrt{2}tdx$, where γ_p is the work per unit of fracture surface created, and $2\sqrt{2}tdx$ is the fracture surface created. Therefore, since the crack extension force $G_c = G_c(45°)$ equals $2\gamma_p$, we have

$$G_c(45°) = 2\gamma_p = \frac{t^2\,dx\,W_v}{tdx\sqrt{2}}$$

or

$$G_c(45°) \approx t$$

This states that when the plate is thin, the force required to move the crack is proportional to the plate thickness. Thus extremely thin sheets of metal can fail by fractures involving very little energy. Although these fractures are basically ductile, the significant fact is that thin sheet material can fail brittlely.

Since the fracture toughness varies as the square root of the crack extension force $G_c(45°)$, $K_c(45°)$ should vary as the square root of the plate thickness for values of t below t_o. Above t_o the fracture is mixed, and since the central normal fracture requires less work than the shear fractures on either side of it, the fracture toughness tends to fall again above t_o as the relative amount of normal fracture grows and the shear lips become smaller. The thickness t_o thus corresponds to a maximum fracture toughness G_c. Above t_o, the fracture toughness falls until the plate becomes thick enough so that the entire fracture occurs by the normal mode. At this point, the fracture toughness becomes equal to K_{Ic} and remains equal to this value for all larger values of t. All of these features are indicated schematically in Fig. 19.56.

In the range of plate thickness where the fracture is mixed, it is normally possible to obtain values for both K_{Ic} and K_c. In this regard, consider Fig. 19.57, which shows a curve of the specimen elongation plotted as a function of the load P applied to the specimen. The features of this curve are as follows. As the load increases, the elongation increases slowly and almost in proportion to the load. The initial part of the curve is not necessarily strictly linear, since with the growth of the load there may be some slight increase in the effective crack length. Eventually there is a small but rapid increase in the deflection. This corresponds to a small advance of the crack at its center by the normal fracture mode. This is known as a *pop-in*. The load at this point may be used to compute the value of K_{Ic} by the equation indicated earlier. General fracture does not occur until the maximum load is reached. When this occurs, the combined flat

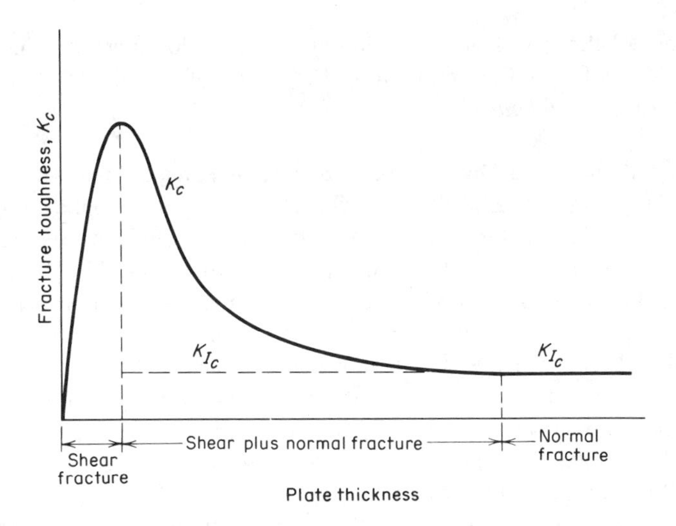

Fig. 19.56 Schematic variation of the fracture toughness as a function of the plate thickness.

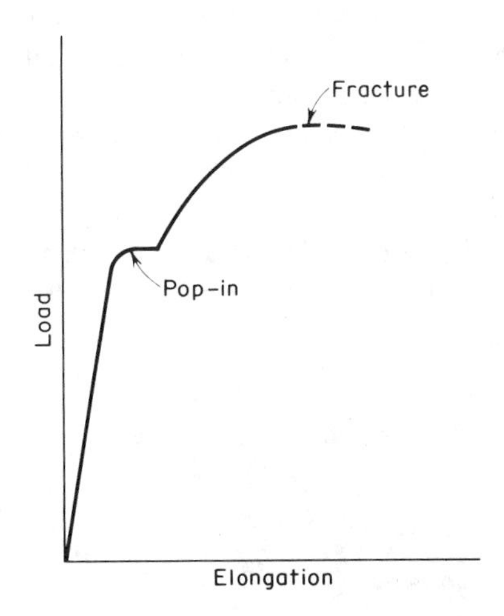

Fig. 19.57 Load-elongation curve for a plate specimen that fails by a mixed fracture.

central crack with its accompanying shear lips spreads rapidly across the specimen. The load at this point may be used to compute K_c.

19.25 Fatigue Failures. Failures that occur in machinery parts are almost always fatigue fractures. When a metal is subjected to many applications of the same load, fracture occurs at much lower stresses than would be required for failure in a tensile test. The failure of metals under alternating stresses is known as *fatigue.*

19.26 The Macroscopic Character of Fatigue Failure. A fatigue fracture always starts as a small crack which, under repeated applications of the stress, grows in size. As the crack expands, the load-carrying cross-section of the specimen is reduced, with the result that the stress on this section rises. Ultimately the point is reached where the remaining cross-section is no longer strong enough to carry the load, and the spread of fracture becomes catastrophic. Because of the manner in which the fracture develops, the surfaces of a fatigue fracture are divided into two areas with distinctly different appearances, as is shown schematically in Fig. 19.58. In most cases, the surface will have a polished or burnished appearance in the region where the crack grew slowly. This texture arises because the metal surfaces of the crack rub against each other as the specimen is deformed back and forth through each stress cycle. In the last stage, when the specimen finally fractures, there is no rubbing action and the surfaces developed at this time are rough and irregular. Because the latter area usually has a granular appearance, an erroneous conclusion is frequently made relative to fatigue fractures, namely, that the metal crystallized in service and thereby became embrittled.

In machinery components, the amplitudes of the stress cycles are not always of the same magnitude. For example consider an automobile drive shaft in which the stress cycles are much larger during periods of fast acceleration than when the car moves at a steady speed. Under variable stress amplitudes, the crack may stop spreading when the stress is low and continue to grow when it rises. This

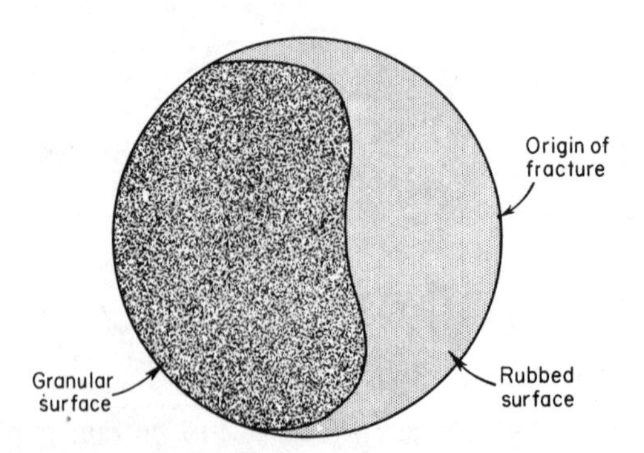

Fig. 19.58 The surface of a fatigue fracture usually shows two regions with characteristically different aspects. One is smooth, or burnished, corresponding to the slow spreading of the fatigue crack, and the other is rough, or granular, belonging to the metal that failed as a result of overload.

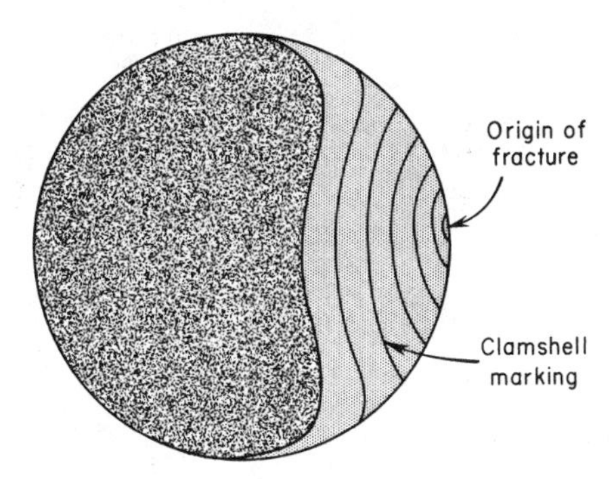

Fig. 19.59 Fatigue fractures in machinery parts subjected to stress cycles that vary in magnitude with time are liable to show a clamshell pattern in the burnished area.

alternation of periods of rapid growth with periods of slow or no growth changes the degree of rubbing which the surfaces of the crack undergo, with the result that the surface may attain a "clamshell" appearance. (See Fig. 19.59.) Usually these ring-shaped markings on the fracture surface are concentric with the origin of fracture and make possible its determination. Their existence on the fracture surface of a metal object is also good evidence that the failed part fractured by a fatigue mechanism.[34] Other evidence in this regard is certainly the presence of the burnished and smooth areas on the fracture surface. Finally, fatigue fractures are almost always characterized by a lack of macroscopic plastic deformation at small distances from the immediate fracture surface. In this respect, they resemble typical brittle fractures, and if a fractured machinery part indicates that a large degree of plastic flow occurred just prior to fracture, it is usually indicative of the fact that failure occurred as a result of a temporary overload rather than because of a uniform repeating load.

19.27 The Rotating-Beam Fatigue Test. There are as many ways of running a test to measure fatigue as there are ways of applying repeated stresses to a metal. A specimen may be first stretched in simple tension and then the stress direction reversed so that the specimen is placed in compression. Alternating the direction of twist in a torsion specimen is another type of repeated stress. A relatively simple alternating stress is obtained by reversed bending. In some instances fatigue has been studied by using combined stress loadings in which the obvious intent was to subject the metal to conditions approaching those found in machinery parts. Since, in many cases, machine

34. Grover, H. J., Gordon, S. A., and Jackson, L. R., *The Fatigue of Metals and Structures*. U.S. Government Printing Office, 1954.

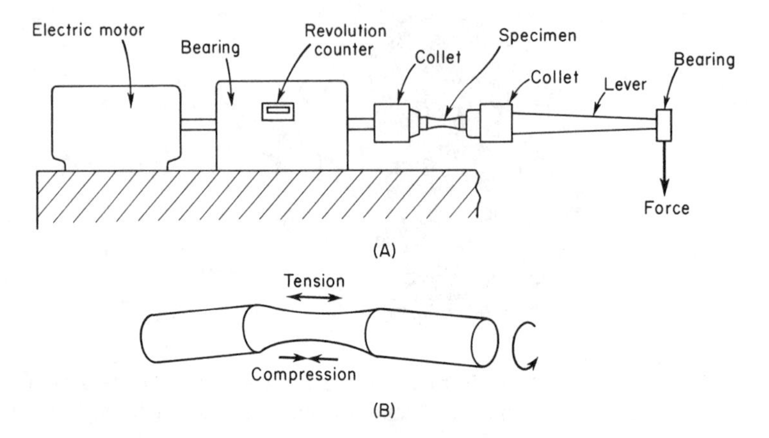

Fig. 19.60 (A) One form of rotating-beam fatigue-testing machine. (B) Fatigue-test specimen. Specimen is bent while it rotates. Any point in the reduced middle section alternates between states of tensile and compressive stress.

elements and structures are not subjected to a loading that completely reverses the stress magnitude, tests are often run in which the specimen is given a steady load, say in tension, and then an additional alternating load is superimposed on the steady load.

A rotating beam is the type of fatigue specimen most generally used. One of its greatest advantages is its relative simplicity.

Figure 19.60 shows schematically the basic components of one type of rotating-beam fatigue-testing machine. Its main element is a small high-speed electric motor capable of running at a speed of 10,000 rpm. Speeds of this order do not greatly influence the data, while they markedly reduce the time needed to obtain the necessary information. Next to the motor is a large bearing, whose purpose is to relieve the motor of the large bending moment which is applied to the specimen. The specimen proper is mounted in collets that serve as grips. One collet is attached to the shaft driven by the motor while the other is attached to a rotating lever arm. At the end of the latter is a small bearing, used to apply a downward force to the lever arm. The application of this force places the small circular cross-section specimen in a state of bending, so that its upper surface is in tension while its lower surface is in compression. As the specimen is rotated by the action of the motor, any given position on the surface of the specimen alternates between a state of maximum tensile stress, and a state of maximum compressive stress.

In making a test, one measures the number of cycles required to fracture the specimen at a given stress. The stress, of course, is the maximum fiber stress

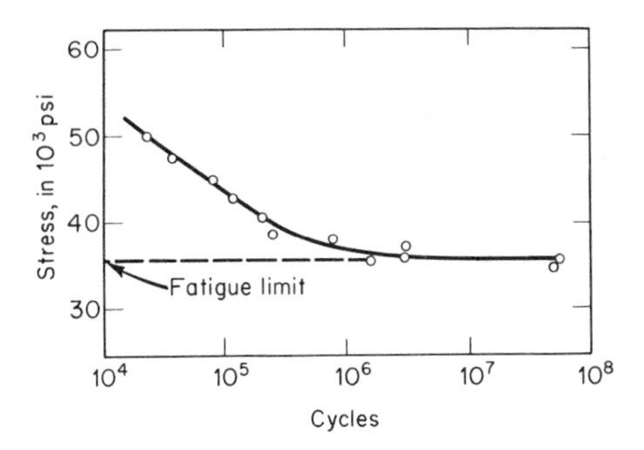

Fig. 19.61 A typical *S-N* curve for a steel. Individual data points, however, usually have a larger scatter than is observed in the above figure. (From *Prevention of Fatigue of Metals*, The Staff of the Battelle Memorial Institute, John Wiley and Sons, Inc., New York, 1941, p. 46.)

developed on the specimen surface by the bending moment which is created by hanging a weight on the end of the lever arm. This stress can be easily computed in terms of the magnitude of the applied weight, the length of the lever arm, and the diameter of the specimen at its minimum cross-section. If the maximum tensile stress applied by bending is only slightly less than that which will break the specimen in a simple tensile test, the fatigue tester will run only a few cycles before the specimen fractures. Continued reduction of the stress greatly increases the life of the specimen, so that in plotting fatigue-test results, it is common practice to plot the maximum bending stress against the number of cycles to fracture, using a logarithmic scale for the latter variable. Figure 19.61 shows the type of curve that is usually obtained when the fatigue specimens are steel. This form of curve is significant because there is a stress below which the specimens do not fracture. At this particular stress, called the *fatigue* (or endurance) *limit*, the *SN* (stress-number of cycles) curve turns and runs parallel to the N axis. This is an important effect, for it implies that if a steel is only loaded to a stress which is below its fatigue limit, no matter how many times the stress is applied, it will not fail.

There appears to be a reasonably good correlation between the fatigue limit of a steel and its ultimate tensile strength.[35] The ratio of these two quantities, known as the *endurance ratio*, is therefore of some value and usually falls in the range 0.4 to 0.5. Since the yield strength of some steels is often close to one-half the ultimate strength, the endurance limit and the yield stress are often approximately equal. It should not be inferred, however, that the two are equal, because there is no good overall correlation between the two quantities.

35. Crussard, C., Plateau, J., Tamhankar, R., Henry, G., and Lajeanesse, D., *Fracture*, p. 524. The Technology Press and John Wiley and Sons, Inc., New York, 1959.

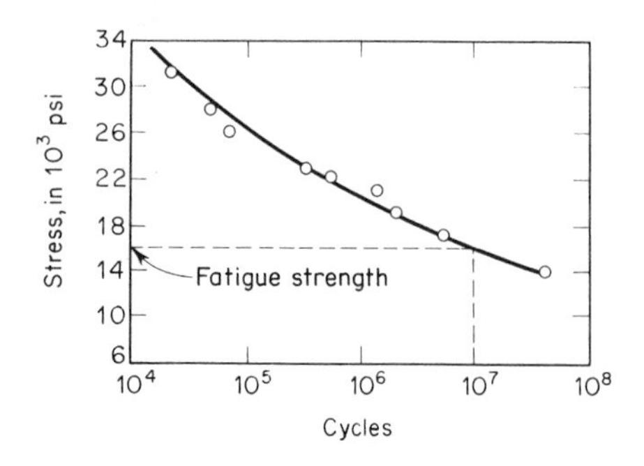

Fig. 19.62 A typical *S-N* curve for a nonferrous metal (aluminum alloy). (From *Prevention of Fatigue of Metals*, The Staff of the Battelle Memorial Institute, John Wiley and Sons, Inc., New York, 1941, p. 48.)

Unlike steels, most nonferrous metals do not appear to have an endurance limit. The *SN* curves for these metals generally have the appearance shown in Fig. 19.62 where, with decreasing stress, the curve continues to fall steadily, although at a decreasing rate. In speaking of the ability of nonferrous alloys to resist fatigue failure, one generally has to specify how many stress cycles the metal should withstand. The stress that will cause fracture at the end of the specified number of stress alterations, normally 10^7, is known as the *fatigue strength* of the metal. This is illustrated in Fig. 19.62.

19.28 The Microscopic Aspects of Fatigue Failure. It is customary to subdivide the fatigue process into three phases: crack initiation, crack growth, and the ultimate catastrophic failure. We shall now consider crack initiation.

It is almost universally agreed that fatigue failures start at the surface of a fatigue specimen. This is true whether the test is made in a rotating-beam machine where the maximum stress is always at the surface, or in a push-pull machine that gives a simple tensile-compressive stress cycle. Furthermore, fatigue fractures start as small microscopic cracks and, accordingly, are very sensitive to even minute stress raisers. It is quite apparent, in the light of these considerations, that a fatigue specimen will give results that are representative of the metal tested only if its surface is free of defects. Tool or grinding marks left on the surface make the formation of fatigue cracks easier and may result in low apparent values of the fatigue limit or fatigue strength. A fundamental condition for making a fatigue test is that the surface of the specimen be carefully polished.

In considering the problem of fatigue, the fact that fatigue failures have been observed in specimens tested at 4°K is significant. At this extremely low temperature, thermal energy cannot make any appreciable contribution to the mechanism of fatigue fractures. It can therefore be concluded that it is possible to have fatigue failure without thermal activation. This means that while diffusion processes may be involved in some cases of fatigue, they are not necessary to the formation of fatigue cracks.

Slip and twinning are plastic deformation processes that are believed to be able to occur without thermal activation, and it is felt that they are strongly involved in fatigue-failure mechanisms. Because fatigue failures are promoted by the presence of minute stress raisers, the study of the mechanisms of fatigue is best done on metals or alloys that have a minimum of nonmetallic inclusions. (The role of nonmetallic inclusions will be briefly discussed presently.) In most metals at temperatures close to room temperature, slip seems to be the predominant factor in fatigue. At very low temperatures, and with steel or iron specimens, mechanical twinning is probably of importance.

A number of dislocation mechanisms have been proposed[36,37] to explain the experimentally observed phenomena of fatigue. None of these mechanisms, however, is completely satisfactory, therefore they will not be considered. Rather, a brief discussion will be given of some of the experimental information that is now available.

In polycrystalline metals, slip bands (groups of closely spaced and overlapping slip lines) are observed to form on specimens prior to fracture. The nature of these bands differs somewhat with the metal under consideration and with whether slip takes place by single or multiple glide. Figure 19.63 shows that in

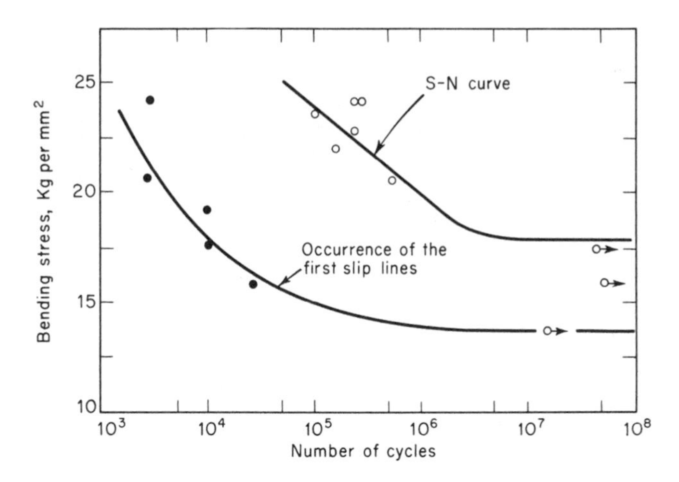

Fig. 19.63 Slip lines may become visible well before a fatigue specimen fractures. Above data for flat specimens of low-carbon steel (0.09 percent C) stressed in bending. (After Hempel, M. R., *Fracture*, The Technology Press and John Wiley and Sons, Inc., New York, 1959, p. 376.)

36. Cottrell, A. H., and Hull, D., *Proc. Roy. Soc.* (London), **A242** 211 (1957).
37. Mott, N. F., *Acta Met.*, **6** 195 (1958).

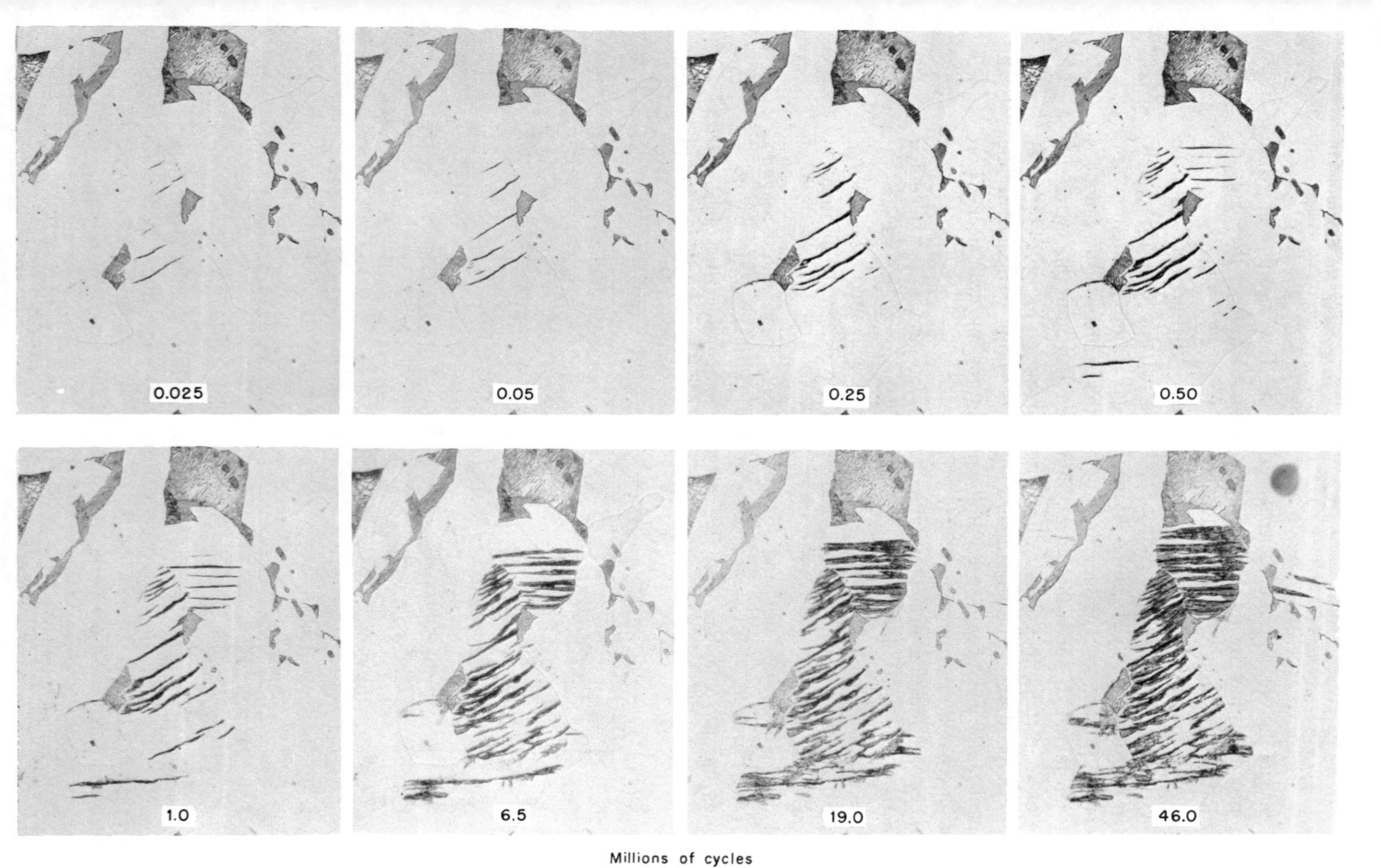

Fig. 19.64 Formation of slip bands in a low-carbon steel (0.09 percent C) as a function of the number of cycles. Plane-bending fatigue-test specimen stressed below fatigue limit. (Hempel, M. R., *Fracture*, The Technology Press and John

the case of a particular low-carbon steel (0.09 percent C), slip lines first become visible at about 1/100 the number of cycles required to fracture the specimens. It can also be seen that at large values of N, the curve representing the first appearance of slip lies below that for the fracture of the specimens. Slip can thus occur in this metal at stresses well below the fatigue limit. This result has also been observed in other metals, but it is not a universal effect, for there are also metals in which slip lines only appear when the stress is equal to or above the fatigue limit.[38]

During a fatigue test slip lines appear first in those crystals of a specimen whose slip planes have the highest resolved shear stress. As time goes on and the number of stress cycles increases, the size and number of slip bands rises. Figure 19.64 shows the variation in the slip bands as a function of the number of cycles of testing in this low-carbon steel. The extent and number of the slip bands are also a function of the amplitude of the applied stress; higher stresses give larger values.

In fatigue, the direction of the strain is reversed over and over again. The slip lines that appear on the surface reflect this fact. When the strain is in a single basic direction, the slip steps that appear on a crystal surface have a relatively simple topography. This is particularly true if only a single slip plane is active. The nature of these slip markings is indicated in Fig. 19.65A. On the other hand, under cyclic loading the slip bands tend to group into packets or striations.[39] The surface topography of these striations is more complex and is indicated schematically in Fig. 19.65B. Note that both ridges and crevices tend to be formed. There is good evidence[40] that the crevices are closely associated with the initiation of cracks. Whether or not crevices are formed in a particular specimen or a specific grain of a given specimen is largely a function of the crystal orientation. If the shear direction in the striation is nearly normal to the surface, crevice formation may be favored. However, if the slip direction is parallel to the surface, crevices are not normally developed. It is still possible, even in this case, for damage to occur inside a striation since pores or holes can also open up inside the slip-band packet.

One consequence of dislocation movements during fatigue is that small localized deformations which are called *extrusions* may occur in the slip bands. As shown in Fig. 19.66, an extrusion is a small ribbon of metal which is apparently extruded from the surface of a slip band. Because extrusions are normally accompanied by cracks in the slip packet, they may be of significance[41] in crack initiation. The inverse of extrusions, which are narrow

38. Hempel, M. R., *Fracture*, p. 376. The Technology Press and John Wiley and Sons, Inc., New York, 1959.

39. Forsyth, P. J. E., *The Physical Basis of Metal Fatigue*, p. 36. American Elsevier Pub. Co., Inc., New York, 1969.

40. *Ibid.*

41. *Ibid.*

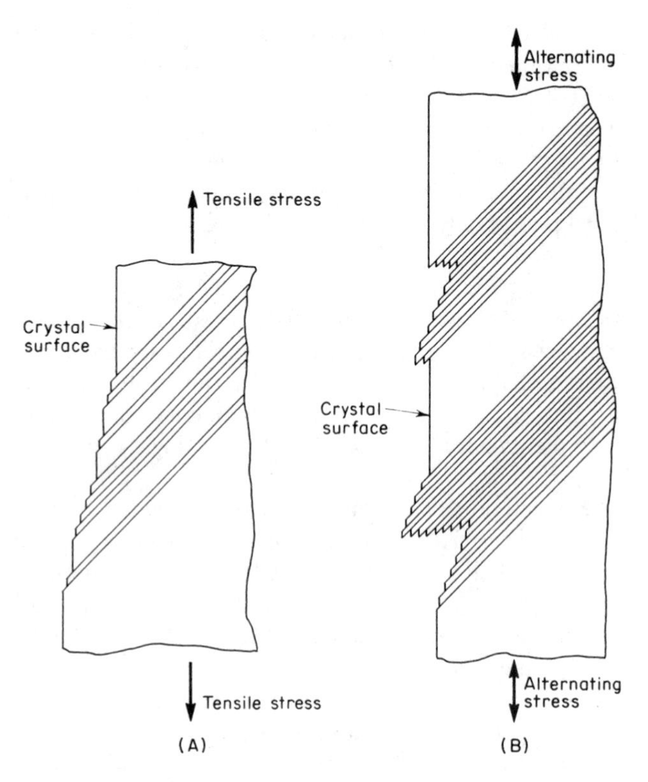

Fig. 19.65 The difference in the surface contours where slip bands intersect the surface. (A) One-directional deformation. (B) Alternating deformation.

crevices called *intrusions*, have also been observed. These surface disturbances (intrusions and extrusions) are approximately 10^{-4} to 10^{-5} cm in height and appear as early as $\frac{1}{10}$ of the total life of a specimen.

19.29 Fatigue Crack Growth. With increasing numbers of cycles, the surface grooves deepen and the crevices and intrusions take on the nature of a crack. When this happens, Stage I of the crack-growth process has begun. Fracture initiation can start at a relatively early point in the fatigue life of a specimen and, under favorable circumstances, Stage I crack growth can continue for a large fraction of the fatigue life. If the specimen has a preferred orientation in which neighboring grains have slip planes with nearly equivalent orientations, it is possible for slip-band cracks to extend across many grains.[42] Low applied stresses and deformation by slip on a single slip plane favor Stage I

42. *Ibid.*

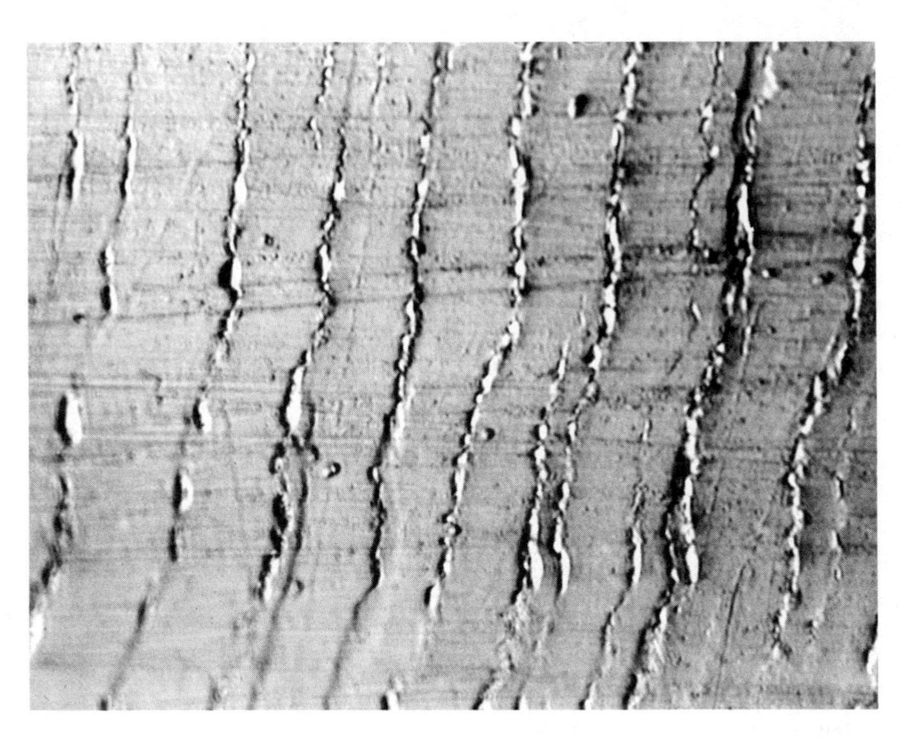

Fig. 19.66 Extrusions on the surface of a copper crystal. (Backofen, W. A., *Fracture*, The Technology Press and John Wiley and Sons, Inc., New York, 1959, p. 435.)

growth. On the other hand, multiple-slip conditions favor Stage II growth. From a practical viewpoint, Stage I is normally of secondary importance to Stage II. It should be noted that in addition to fracture initiation at slip planes, it is also possible for cracks to form at grain boundaries or subgrain boundaries. Slip-band initiation, however, appears to be of the greatest overall importance.

Stage I growth follows a slip plane, whereas Stage II growth does not have this crystallographic character. In this case, the fracture conforms rather closely to fracture mechanics conditions. Thus, if the applied stress favors plane-strain deformation, the fracture surface in Stage II will follow a plane normal to the principal applied tensile stress. On the other hand, in a thin plate with increasing size of the crack, the fracture surface tends to shift into a plane at 45° to the specimen surface, as indicated in Fig. 19.53. When this happens, the deformation conforms to plane stress.[43] The transition from Stage I to Stage II is often induced when a slip-plane crack meets an obstacle such as a grain boundary.

43. Tetelman, A. S., and McEvily, A. J., Jr., *Fracture of Structural Materials*, p. 367. John Wiley and Sons. Inc., New York, 1967.

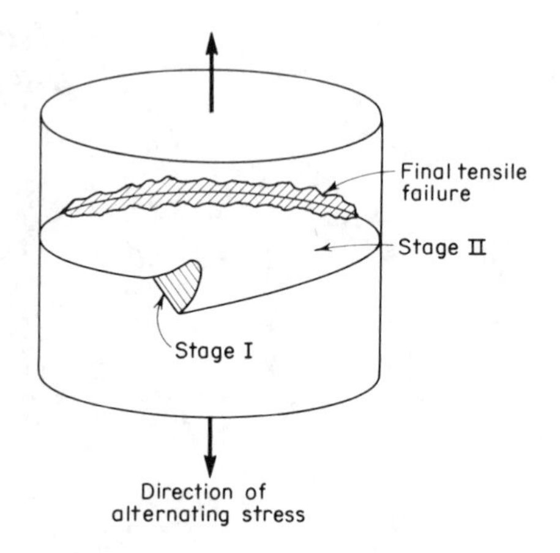

Fig. 19.67 This drawing illustrates the relationship of the Stage I and Stage II crack growth processes in a fatigue fracture. (From Forsyth, P. J. E., *The Physical Basis of Metal Fatigue*, p. 90, American Elsevier Pub. Company, Inc., New York, 1969.)

Figure 19.67 is a schematic drawing illustrating the fracture process resulting from Stage I and Stage II crack growth, followed by the final catastrophic failure.

The growth of a fatigue crack is relatively easy to follow in a sheet specimen, and there is considerable data in the literature[44,45] showing a dependence of growth rate on the square root of the crack length and the applied stress. Normally this takes the form of a power law in which the exponent is close to four. Stated in the form of an empirical equation, this relationship is

$$\frac{dc}{dN} = A(\sigma\sqrt{c})^4$$

where N is the number of cycles, c is the half-crack length, σ is the peak gross stress, and A is a constant. It is interesting to note[46] that the stress intensity factor for a crack traveling in a thin plate under Mode I tensile deformation, is

$$K = \sigma\sqrt{\pi c}$$

where σ is the tensile stress, while for Modes II and III crack propagation, the

44. *Ibid.*
45. Pelloux, R. M., *Ultra Fine Grain Metals*, p. 235. Ed. by Burke, J. J., and Weiss, V. Syracuse University Press, Syracuse, N.Y., 1970.
46. Tetelman, A. S., and McEvily, A. J., Jr., *Op. Cit.*, p. 367.

stress intensity factor is

$$K = \tau\sqrt{\pi c}$$

where τ is the shear stress across the fracture plane of the shear failures. It would therefore appear that, in general, the growth velocity of a fatigue crack in Stage II is proportional to approximately the fourth power of the stress intensity factor, or

$$\frac{dc}{dN} \approx K^4$$

This result clearly implies the role of fracture mechanics in fatigue, or the fact that the stress ahead of the crack, as measured by the stress intensity factor, largely controls the crack growth rate.

19.30 The Effect of Nonmetallic Inclusions. Nonmetallic inclusions reduce the fatigue strength of metals. This fact is illustrated in Fig. 19.68, which gives three *SN* curves for a high-strength medium-carbon-alloy steel, AISI

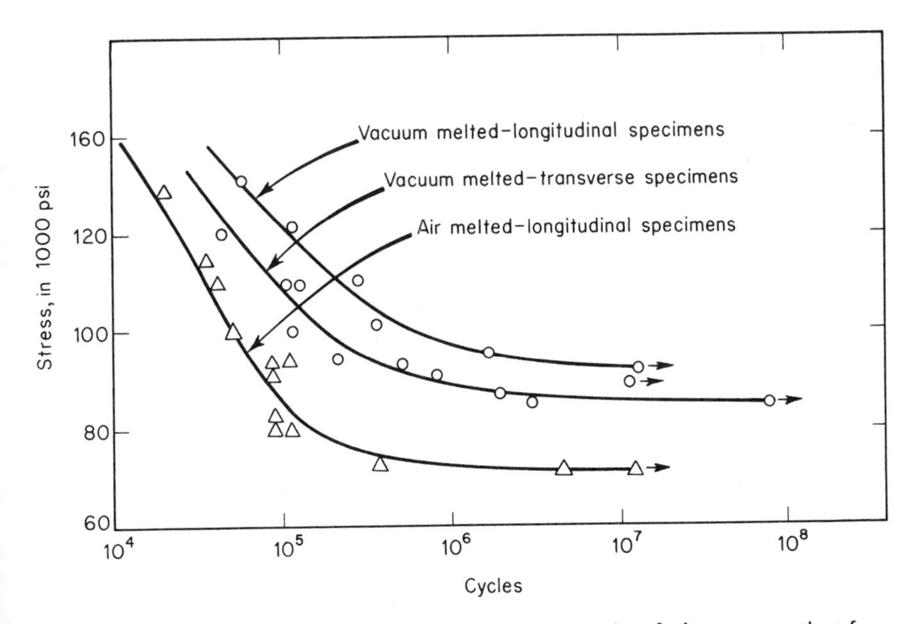

Fig. 19.68 Effect of nonmetallic inclusions on the fatigue strength of a low-alloy steel (4340). All specimens heat-treated to same ultimate strength (230,000 psi). (Aksoy, A. M., *Trans. ASM,* **49** 514 [1957].)

(4340). Steel specimens heat-treated to the same strength (ultimate strength 230,000 psi) were used to obtain each of the three curves. The upper and middle curves correspond to vacuum-melted steel, while the lower curve corresponds to steel that was air-melted. The inclusions in the vacuum-melted metal are smaller in size and much less numerous than in the air-melted steel. This is because gaseous elements are absorbed into the air-melted steel, increasing the number of inclusions. Notice that the air-melted steel has the lower fatigue limit.

The middle curve of the three represents data for vacuum-melted specimens machined from fabricated-steel stock in a direction transverse to the hot-rolling direction. The other two curves are for specimens cut with their long axes parallel to the hot-rolling direction. The middle curve should be compared to that above it, which shows that the fatigue strength or fatigue limit is less when specimens are cut transverse to the rolling direction. The reason for this is probably as follows.

When a metal is fabricated by rolling at a red heat, the inclusions tend to be deformed and elongated in the rolling direction. In a specimen cut transverse to the rolling direction, the long axis of the inclusions lies in a plane perpendicular to the bending stress. On the other hand, in the longitudinal specimens, the long axis of the inclusions is parallel to the bending stress. In the former case, the cross-section of the inclusions normal to the stress is larger than it is in the latter case. This difference in the area of the inclusions normal to the stress is believed to be the biggest factor in the lower fatigue strength of transverse specimens.

Microscopic examinations of metals with high concentrations of nonmetallic inclusions, or second-phase particles, show that small cracks form readily at inclusions.[47] These cracks may appear almost at the start of the test and well before visible slip lines are observed. Not only do inclusions nucleate cracks, but they aid in their propagation, for the cracks readily jump from one inclusion to the next.

19.31 The Effect of Steel Microstructure on Fatigue. In an earlier section it was shown that a steel with a quenched and tempered (spheroidized cementite) structure has a lower transition temperature for brittle fracture than a steel with the same strength and a pearlitic microstructure. The fatigue properties of a quenched and tempered structure are also superior to those of a pearlitic structure. Almost twenty years ago it was shown[48] that the ratio of the fatigue limit to the ultimate strength (endurance ratio) was about 0.60 for steels with tempered martensitic structures, and about 0.40 with austenitic or pearlitic structures.

47. Hempel, M. R., *Fracture*, p. 376. The Technology Press and John Wiley and Sons, Inc., New York, 1959.

48. Caz, F., and Persoz, L., *La Fatigue Des Metaux*, 2nd ed., Chap. V, (1943), Dunod, Paris.

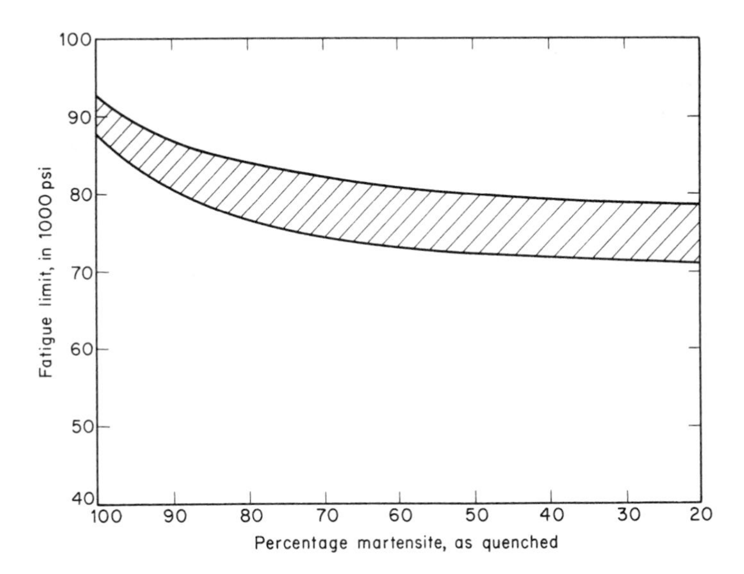

Fig. 19.69 Fatigue limit versus percentage of martensite in "as quenched" specimens tempered to same hardness (Rockwell C-36). Notice that the curve is plotted as a band in order to indicate the scatter of experimental data. (Borik, F., Chapman, R. D., and Jominy, W. E., *Trans. ASM,* **50** 242 [1958].)

More recent work has shown that the amount of martensite obtained during a quench is very important in determining the fatigue properties of steel. Figure 19.69 shows the result obtained by quenching a number of different alloy steels. All specimens had the same carbon content and were quenched at several different rates of cooling in order to obtain various percentages of martensite in the quenched metals. Each specimen was then tempered to the same hardness (Rockwell C-36) and, following this, its fatigue limit was determined. The relationship between the fatigue limit and the amount of martensite obtained on quenching is shown in Fig. 19.69. The drop in fatigue limit for even as little as 10 percent of structures other than martensite is quite noticeable. This drop constitutes another important reason why steels in machinery parts are almost universally heat-treated before they are tempered, in order to obtain as nearly as possible a 100-percent martensitic structure.

19.32 Low-Cycle Fatigue. Low-cycle fatigue is a term that defines fatigue studies at high stress where the fatigue life is normally short (order of 10^4 cycles). Many objects that are subjected to alternating or cyclic loads are only loaded through a relatively small number of cycles in their useful lifetime. A typical example might be the spring on the self-starter of an automobile engine. In a ten-year period, the average car will probably be started about 10,000 times. Savings in cost and weight can be made by designing such a spring in terms of its useful life rather than in terms of its endurance limit (10^7 cycles). Low-cycle fatigue is of primary interest in the aircraft and other industries where weight is a major concern.

19.33 Certain Practical Aspects of Fatigue. Most fatigue failures in actual engineering structures are the result of macroscopic stress raisers which were unintentionally incorporated in the object that failed. Any sharp corner or angle on the surface of an object which undergoes repeated stresses is a potential danger point for a fatigue failure. Especially is this true if the sharp corner lies in a region in which the stress cycle places the metal in a state of tension during part of the cycle. A well-known example of an inadvertent fatigue failure is that of metal airplane propellers which were originally stamped with the manufacturer's name in the middle of the blades. The grooves caused by the stamping served as stress raisers, causing early fatigue failures. Keyways cut into shafts are frequently a source of fatigue damage. Even holes drilled into a metal part can provide stress raisers at the sharp lip of the holes. The V-notches at the bottom of threads cut on bolts are still another source of fatigue cracks.

Since fatigue failures originate at the surface of metallic specimens, strengthening the surface generally improves the fatigue life. Surface hardening for this purpose can be accomplished by a variety of methods. One method often used is shot-peening, where the surface of a metal shaft or other object is bombarded with steel shot. This cold-works the surface layers and leaves them in a state of residual compressive stress with a considerable improvement in fatigue properties. Surface hardening by carburizing or nitriding may also be used to harden the surface of metal subjected to repeated stresses.

Problems

1. Polycrystalline zinc has very limited ductility below room temperature. Why should this be so? Explain in terms of the mechanical properties of zinc crystals.
2. Consider the cleavage cracks in the zinc bicrystal shown in Fig. 19.16. For simplicity, assume that one of these cracks is the result of a pile-up of edge dislocations at the grain boundary and that the dislocations lie on the same slip plane. Would you expect the cleavage to occur on the slip plane containing the dislocations, on a slip plane above this plane, or on one below it?
3. Brenner[49] reports that single-crystal iron whiskers of diameter 1.6μ have a tensile strength of 1340 kg/mm^2. Assuming that these crystals have a [111] direction parallel to their axis

49. Brenner, S. S., *J. Appl. Phys.*, 27 1484 (1956).

(*a*) Compute the resolved shear stress on one of the most highly stressed {110} ⟨111⟩ slip systems when fracture occurred.
(*b*) The shear modulus μ measured on a {110} plane in a ⟨111⟩ direction is 6100 kg/mm^2. Compute the ratio of τ_{max}/μ.
(*c*) Theoretical calculations generally give $\mu/30$ as a theoretical shear strength. How does the answer obtained in part (*b*) of this question compare to $\mu/30$?
(*d*) What can you deduce about the perfection of the whisker crystals?

4. (*a*) Compute the theoretical tensile strength of iron using the equation in Section 19.5. Assume $E = 30 \times 10^6$ psi, $\gamma = 2000$ ergs/cm^2, and $a = 2.5$Å.
(*b*) How does this value correspond with the observed tensile strength of an iron whisker?
5. (*a*) Compute a theoretical tensile strength for a perfect lead crystal. $E = 2 \times 10^6$, $a = 3.5$Å, and assume $\gamma = 1000$ ergs/cm^2.
(*b*) Make a similar calculation for tungsten with $E = 50{,}000{,}000$ psi, $a = 2.7$Å, and $\gamma = 2900$ ergs/cm^2.
6. (*a*) Some high-strength steels have an ultimate stress of the order of 350,000 psi. What is the maximum sized internal flaw in the form of a sharp-edged crack that an otherwise perfect iron crystal would possess at this stress level? Use the Griffith fracture criterion and assume $\gamma = 2000$ ergs/cm^2 and $E = 30 \times 10^6$ psi.
(*b*) What would be the corresponding flaw size at a stress level of 3.5×10^6 psi.
7. Indicate two different dislocation reactions that could produce a fracture in an MgO crystal on the (100) plane.
8. In an AISI 4340 steel, γ_p, the effective surface energy that takes into account the work done in expanding a crack, is about 2×10^7 ergs/cm^2. Assume a grain size of ASTM 8 and compute a stress to propagate a crack equal in length to an average grain diameter.
9. For the crack in the preceding problem, how long would it have to become before it was traveling with a velocity equal to 20 percent of the longitudinal velocity of sound?
10. A fracture toughness measurement conducted at -40°C gave a crack extension force G_{Ic} value of 0.8 lbs/inch for a low-carbon steel of grain size 0.025 mm and $E = 30 \times 10^6$ psi.
(*a*) Compute the effective surface energy σ_p in ergs/cm^2 associated with the crack propagation.
(*b*) Assuming that a Griffith crack criterion equation is valid and that fracture is propagation limited, at what tensile stress should this steel fracture at -190°C?
11. The low-alloy steel shaft of a new propeller-driven aircraft failed the first time it was used in subzero arctic weather. The fracture surface showed a completely brittle fracture, while a microstructural investigation showed a spheroidized cementite structure of the type expected in this part. In considering this failure, it was found that similar parts on other aircraft had never failed in this manner even when operated under identical conditions.

As an engineer given the responsibility of pin-pointing the cause of this failure, what aspects or aspect of the manufacturing processes used in making this part would you check on? Explain.

12. (*a*) Steigerwald and Hanna,[50] using a specimen of the type shown in Fig. 19.55, give K_{Ic}, for a high-strength maraging steel of 275,000 psi yield-stress, as 68,000 psi. Compute an effective surface energy for crack propagation in this material, assuming Poisson's ratio is 0.33. Express your answer in ergs/cm^2.
(*b*) How does this compare with the value of the true surface energy of iron (see Problem 4)? What is the principal cause for this difference?
13. (*a*) For AISI 4340, Steigerwald and Hanna[51] give $K_c45°$ as 160,000 psi/in$^{1/2}$ for a plate thickness of 0.05 inches. Assuming a shear fracture, compute W_v, the work done per unit volume when the crack forms.
(*b*) The density of iron is 7.87 gm/cm^3 and its specific heat at room temperature is 0.11 cal/g/°C. Compute the indicated temperature rise due to this fracture.
14. A low-alloy steel truck driveshaft failed after the vehicle had been driven some 40,000 miles. The result was a very serious accident. At the time of the accident, the truck was being driven at a moderate speed.
(*a*) What would be the most probable fracture mechanism in a case such as this?
(*b*) What steps would you take in order to verify the fracture mechanism?
(*c*) Do you think that there might be cause for legal action against the manufacturer of the driveshaft if a metallographic examination shows a structure of mixed pearlite and ferrite? Explain.
(*d*) What other factors should be looked for in trying to determine the cause of the failure of the driveshaft?

50. Steigerwald, E. A., and Hanna, G. L., *TMS-AIME*, **242**, 320 (1968).
51. *Ibid.*

20 Creep

20.1 Work Hardening. The drawing in Fig. 20.1 is assumed to represent a typical engineering stress-strain curve of a polycrystalline pure metal. Note that after the metal begins to deform plastically above the proportional limit at point a, the stress still continues to rise. This rise in the flow-stress reflects an increase in the strength of the metal caused by the deformation. That this strength increase is real can be shown by unloading the specimen from a point such as b on the curve and then by reloading the specimen. Provided that this experiment does not occur at a temperature where recovery rates are rapid, the stress on the specimen will have to be returned to the value σ_b that it had prior to the unloading before macroscopic plastic flow will again occur. Deforming the specimen to a plastic strain ϵ_b has, accordingly, raised the stress at which the metal will flow from σ_a to σ_b. As will be shown shortly, the stress

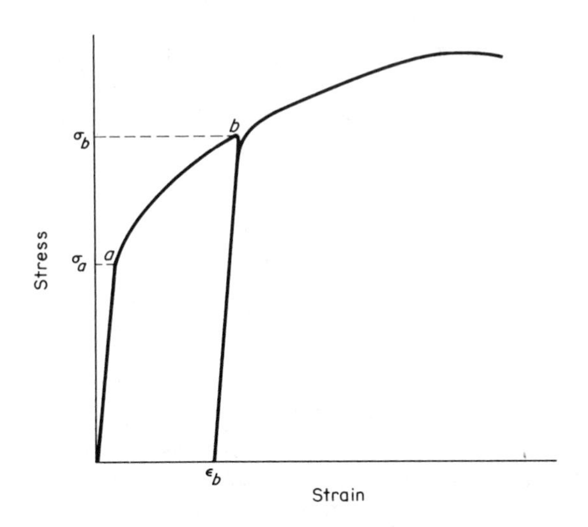

Fig. 20.1 Normally when a metal is deformed to a strain such as ϵ_b and then it is unloaded, it will not begin to deform until the stress is raised back to σ_b. The strain ϵ_b raises the flow-stresses from σ_a to σ_b.

at which a metal will flow, the *flow stress*, is intimately connected with changes in the dislocation structure in the metal resulting from deformation. However, it is necessary that we first consider a different way of representing the stress-strain data. The engineering stress-strain curve expresses both stress and strain in terms of the original specimen dimensions, a very useful procedure when one is interested in determining strength and ductility data for purposes of engineering design. On the other hand, this type of representation is not as convenient for showing the nature of the work hardening process in metals. A better set of parameters are true stress and true strain. True stress (σ_t) is merely the load divided by the instantaneous cross-section area, or

$$\sigma_t = \frac{P}{A}$$

where P is the load and A is the cross-section area. If one assumes that during plastic flow the volume remains effectively constant, then we have $A_o l_o = Al$, where A_o and l_o are the original cross-section area and gage length, respectively, and A and l are the corresponding values of these quantities at any later time. With this assumption, it follows that

$$\sigma_t = \frac{P}{A} = \frac{Pl}{A_0 l_0} = \frac{P}{A_0} \frac{(l_0 + \Delta l)}{l_0} = \sigma(1 + \epsilon)$$

where Δl is the increase in length of the specimen, σ is the engineering stress, and ϵ is the engineering strain. This equation simply states that the true stress is equal to the engineering stress, times one, plus the engineering strain. True strain is defined by the relationship

$$\epsilon_t = \int_{l_0}^{l} \frac{dl}{l} = \ln \frac{l}{l_0} = \ln(1 + \epsilon)$$

which states that the true strain is equal to the natural logarithm of one plus the engineering strain.

The above equations for the true stress and true strain are valid as long as the deformation of the gage section is essentially uniform. In this regard, it is generally assumed that the gage section deforms uniformly to approximately the point of maximum stress (point a in Fig. 20.2) on an engineering stress-strain diagram. Beyond this point the specimen begins to neck and the strain is restricted to the necked region. It is also possible to follow the relation between the true stress and true strain beyond the point of maximum load by considering the stress-strain behavior of the metal only in the necked region. One has,

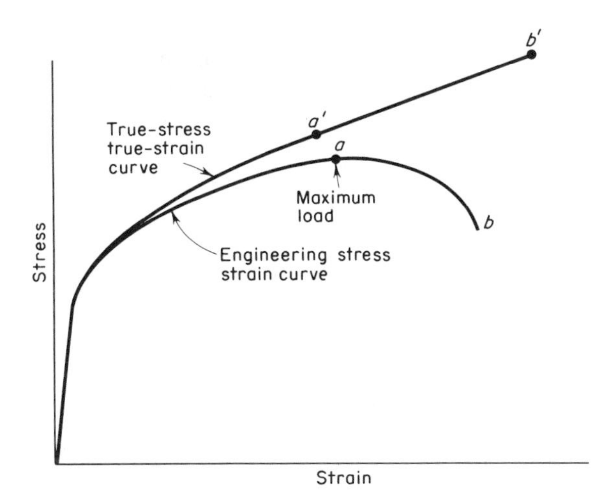

Fig. 20.2 A comparison between an engineering stress-strain curve and the corresponding true stress and true strain curve.

however, to correct for the triaxiality of the stress due to the effective notch produced by the neck; and the strain should be measured in terms of the specimen diameter at the neck, rather than in terms of the specimen gage length. If these factors are considered, one obtains a stress-strain curve of the form of the upper curve in Fig. 20.2. Note that the true stress continues to rise with increasing stress until the point of fracture. There seems to be good evidence for assuming that between the point of maximum load and the point where fracture occurs, the true-stress true-strain curve is approximately a straight line.[1,2] The curve of Fig. 20.2 shows that in a simple tensile test, the true strength of the metal normally increases with increasing strain until it fractures.

20.2 Considère's Criterion. A condition for the onset of necking was originally proposed by Considère[3] based on the assumption that necking begins at the point of maximum load. At the maximum load

$$dP = A d\sigma_t + \sigma_t dA = 0$$

where P is the load, A is the cross-section area, and σ_t is the true stress. In effect, this relationship states that for a given strain increment, the resulting decrease in specimen area reduces the load-carrying ability of the cross-section by the same

1. Glen, J., *J. of the Iron and Steel Institute,* **186** 21 (1957).
2. MacGregor, C. W., *J. Franklin Inst.*, **238** 111 (1944).
3. Considère, *Ann. ponts et chaussees,* **9**, ser. 6 p. 574 (1885).

amount that the load-carrying ability of the specimen is increased by the rise in its strength due to strain hardening. The above relationship may be rearranged to read

$$d\sigma_t = -\sigma_t \frac{dA}{A}$$

In plastic deformation, a reasonable assumption is that the volume remains constant, so that

$$dV = d(Al) = l dA + A dl = 0$$

or

$$\frac{dA}{A} = -\frac{dl}{l}$$

where l is the specimen gage length and dl/l is ϵ_t, the true strain. Therefore

$$\frac{d\sigma_t}{d\epsilon_t} = \sigma_t$$

This is Considère's criterion for necking. When the slope of the true-stress true-strain curve, $d\sigma_t/d\epsilon_t$, is equal to the true stress σ_t, necking should begin. As will be shown presently, this relationship can help to rationalize a basic difference in the observed stress-strain behavior of face-centered cubic and body-centered cubic metals.

20.3 The Relation between Dislocation Density and the Stress. With the development of the transmission electron microscope technique, it has been possible to make direct studies of the dislocation structure in deformed metals. These investigations have indicated that for a very wide range of metals there exists a rather simple relationship between the dislocation density and the flow-stress of a metal. Thus, let us assume that Fig. 20.3 represents the general shape of the stress-strain curve of a metal and that a series of specimens are deformed to different strains, as indicated by the marked points along the curve. Furthermore, let us assume that on reaching the specified strains, they are unloaded, sectioned for observation in the electron microscope, and that dislocation density measurements are made on the foils. Figure 20.4 shows the actual experimental results obtained using a set of titanium specimens. This data corresponds to specimens of three different grain sizes. Note that all of the data plots on the same straight line. Data such as this supports the assumption that

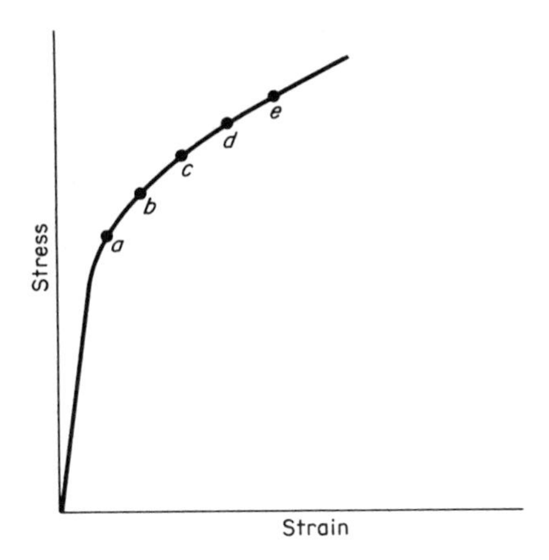

Fig. 20.3 To determine the variation of the dislocation density with strain during a tensile test, a set of tensile specimens are strained to a number of different positions along the stress-strain curve, such as points *a* to *f* in the above diagram. These specimens are then sectioned to obtain transmission electron microscope foils.

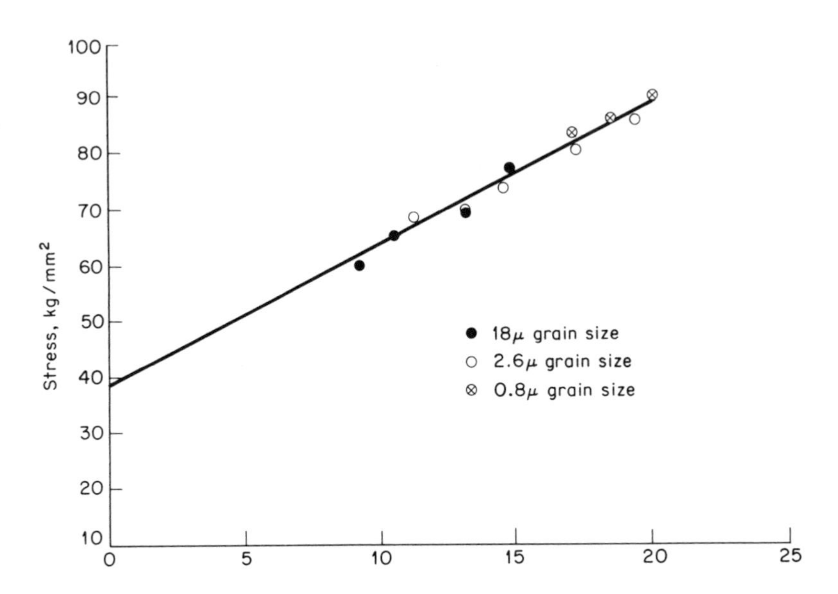

Fig. 20.4 The variation of the flow-stress σ with the square root of the dislocation density $\rho^{1/2}$ for titanium specimens deformed at room temperature, and at strain rate of 10^{-4} $\sec^{-1}$. (After Jones, R. L., and Conrad, H., *TMS-AIME*, **245** 779 [1969].)

the stress varies directly as the square root of the dislocation density, or

$$\sigma = \sigma_0 + k\rho^{1/2}$$

where ρ is the measured dislocation density in centimeters of dislocation per unit volume, k is a constant, and σ_o is the stress obtained when $\rho^{1/2}$ is extrapolated to zero. This result is good evidence that the work hardening in metals is directly associated with the build-up of the dislocation density in the metal. While the above relationship corresponds to data from polycrystalline specimens, the relationship has also been observed in single-crystal specimens. In this case, it is more proper to express the relationship in terms of the resolved stress on the active slip plane τ. This gives us

$$\tau = \tau_0 + k\rho^{1/2}$$

where τ_o is the extrapolated shear stress corresponding to a zero dislocation density. Actually, if the dislocation density were zero, then the metal could not be deformed. As a consequence, σ_o or τ_o are best considered as convenient constants rather than as simple physical properties.

20.4 Taylor's Relation. In 1934, Taylor[4] proposed a theoretical relationship that is basically equivalent to the experimentally observed functional relationship between the flow-stress and the dislocation density. In the model that he used, it was assumed that all the dislocations moved on parallel slip planes and the dislocations were parallel to each other. This model has since been elaborated by Seeger[5] and his collaborators. In brief, this approach assumes that if the dislocation density is expressed in numbers of dislocations intersecting a unit area, then the average distance between dislocations is proportional to $\rho^{-1/2}$. As shown earlier on p. 333, the stress field of a dislocation varies as $1/r$, or in general we may write

$$\tau \approx \frac{\mu b}{r}$$

where μ is the shear modulus, b is the Burgers vector, and r is the distance from the dislocation. Now consider two edge dislocations on parallel slip planes. If they are of the same sign, they will exert a repulsive force on each other. If they

4. Taylor, G. I., *Proc. Roy. Soc.*, **A145** 362, 388 (1934).
5. Seeger, A., *Dislocations and Mechanical Properties of Crystals*, p. 243. John Wiley and Sons, New York, 1957.

are of opposite sign, the force will be attractive. In either case, this interaction must be overcome in order to allow the dislocations to continue to glide on their respective slip planes. Since, as shown above, the average distance between dislocations is proportional to $\rho^{-1/2}$, we have

$$\tau = \alpha\mu b\rho^{1/2}$$

or

$$\tau = k\rho^{1/2}$$

where k is a constant of proportionality equal to $\alpha\mu b$.

As will be demonstrated presently, the dislocation density developed at a given strain is frequently a function of the test temperature. This means that when a metal is strained a fixed amount, the increase in its strength may depend on the deformation temperature. In most cases, the amount of work hardening obtained when the specimen is strained a fixed amount decreases with increasing temperature. However, if the applied strain is small, as at the critical resolved shear stress of single crystals, the work hardening should be small and the temperature variation of the contribution of work hardening may be small enough to be neglected, particularly if the stress is expressed in terms of the parameter τ/μ, where μ is the shear modulus. The shear modulus normally increases with decreasing temperature, and this has a corresponding effect on the flow-stress, since the stress associated with a dislocation is always proportional to the shear modulus. The use of the parameter τ/μ for single-crystal data (or σ/E in the case of polycrystalline specimens, where E is Young's modulus) is recommended in order to remove from flow-stress considerations the effect of the temperature-dependence of the elastic constants. If one now plots the tensile yield stress of a metal divided by E for a given type of metal, one often obtains a curve of the type shown in Fig. 20.5. This curve shows a rise in the yield stress as the temperature decreases. Since, under the stated conditions, the contribution to the flow-stress of the dislocation density is effectively constant, the curve of Fig. 20.5 clearly indicates that there must be a second basic component of the flow-stress that is temperature-dependent. As a result of this line of reasoning, it is common practice to consider that the flow-stress of a pure metal consists of two basic components.

$$\tau = \tau^* + \tau_\mu$$

or

$$\sigma = \sigma^* + \sigma_E$$

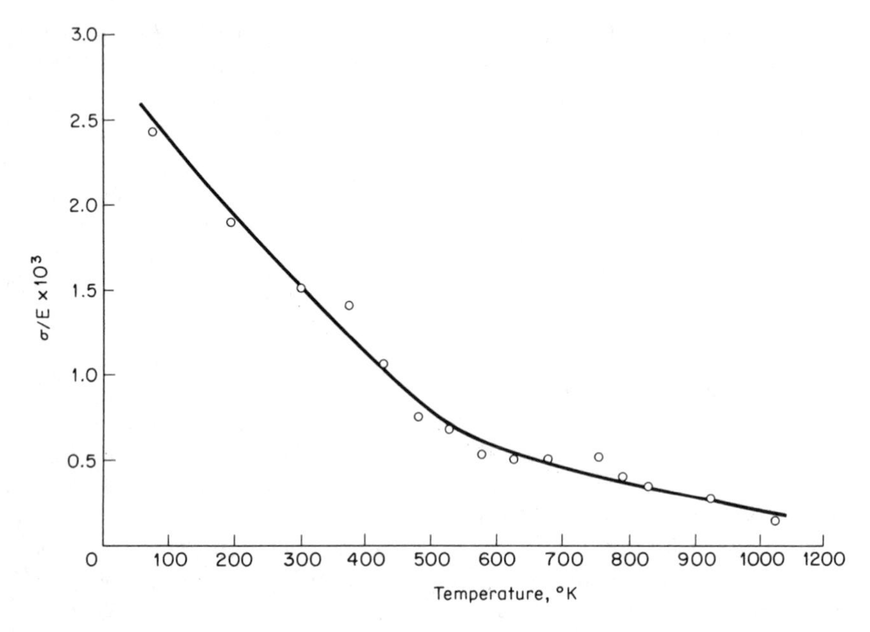

Fig. 20.5 Variation of σ/E with temperature for high purity titanium. Strain rate 3×10^{-4} sec^{-1}. Data of Anand Garde.

where τ^* and σ^* are the components that are temperature-dependent, and τ_μ and σ_E are the components that reflect the effect of the dislocation structure existing in the metal. The subscripts μ or E are used with these latter components to indicate that their temperature-dependence is primarily due to the temperature-dependence of the moduli. This part of the stress is often called the *athermal flow-stress component*, implying that, except for its dependence on the modulus, it is completely temperature-independent. This view is, however, only approximately correct, since any factor, such as recovery, that modifies the dislocation structure can alter τ_μ or σ_E. As an example, increasing the temperature tends to cause recovery effects that can alter the dislocation structure.

20.5 The Nature of the Flow-Stress Components. Let us now consider the difference between the two flow-stress components of a pure metal in more detail. The work hardening component is associated with the interaction of the stress fields of the dislocations in the metal and should occur primarily between dislocations that are nearly parallel. As indicated above, the average separation between dislocations is proportional to $\rho^{-1/2}$ and in a deformed metal, the measured dislocation density is normally of the order of 10^{10}. This means that the average separation of dislocations is about 10^{-5} cm, or about 300 atomic distances. Thus, the work hardening component involves the movement of dislocations through stress fields whose origins are rather distant on an atomic scale. This effect should be little influenced by thermal vibrations of the lattice. It is therefore more proper to call τ_μ a long-range component of the flow-stress. On the other hand, τ^*, the thermal component of the flow-stress, is that which involves thermal activation. In general, this component is associated with the movement of dislocations past obstacles whose strain fields are more localized. A typical example might be the cutting of one dislocation by another. In this case, thermal vibrations are believed to be effective in helping the applied stress to overcome the barrier. Since the thermal component of the flow-stress involves short-range interactions, it is often designated as the *short-range component*, or it may be called the *thermal component*.

20.6 The Effect of Alloying on the Flow-Stress Components. Alloying of a pure metal may affect both the long-range and the short-range components of the flow-stress. A given solute element may alter both components. However, the trend of the present data indicates that in h.c.p. metals such as zirconium and titanium, interstitial elements tend to have a stronger effect on the short-range component than on the long-range component. Evidence for this is shown in Fig. 20.6, which shows the critical resolved shear stress as a function of temperature of several grades of zirconium, differing in the amount of interstitials (oxygen) that they contain. Note that as the

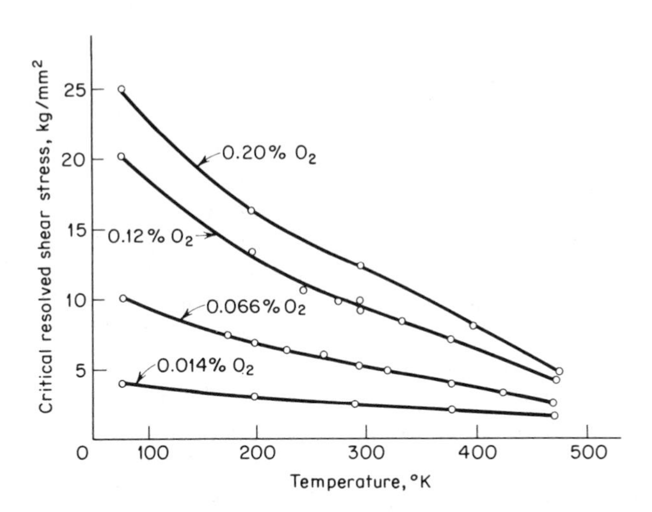

Fig. 20.6 The effect of oxygen on the temperature-dependence of the critical resolved shear stress of zirconium crystals. Oxygen concentrations are in weight percent. (From Soo, P., and Higgins, G. T., *Acta Met.*, **16** 177 [1968].)

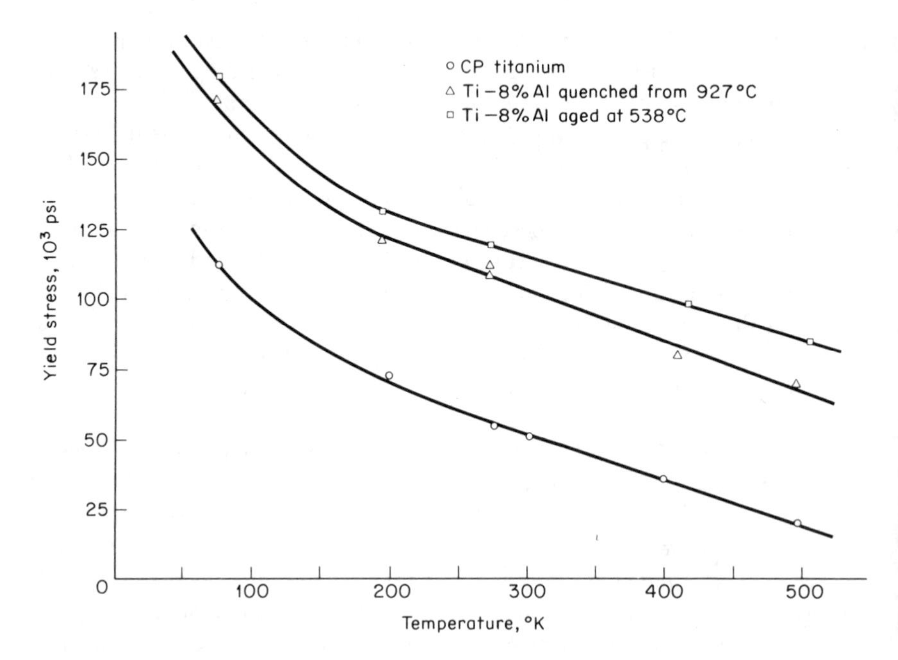

Fig. 20.7 When titanium is alloyed with 8 percent aluminum, the principal effect on the flow-stress is to increase the athermal component. (From Evans, K. R., *TMS-AIME*, **242** 648 [1968].)

concentration of interstitial elements increases, the temperature-dependence of the yield stress increases. In body-centered cubic metals, the role of interstitial elements is still a subject of controversy.[6] A basic problem with these metals is usually their low solubility for interstitial atoms. For example, it is very difficult to purify iron sufficiently to make conclusive tests as to the effect of interstitials on the flow-stress components. On the other hand, substitutional solid solutes may have a greater influence on the long-range component. An example of the effect of substitutional solute (8 percent Al) on the flow-stress of titanium is shown in Fig. 20.7. In this case, the effect of the solute is to increase the level of the flow-stress by roughly the same amount at all temperatures. This diagram also shows a third curve corresponding to the alloy in a precipitation-hardened state. Note that precipitation hardening also raises the flow-stress by roughly the same amount at all temperatures. This is generally true of precipitation hardening. That is, it tends to add a temperature-independent component to the flow-stress. The increase in the flow-stress due to precipitation hardening will, of course, largely vanish if the temperature is raised to the point where the

6. Christian, J. W., *Second International Conf. on the Strength of Metals and Alloys*, p. 31. Am. Soc. for Metals, Metals Park, Ohio, 1970.

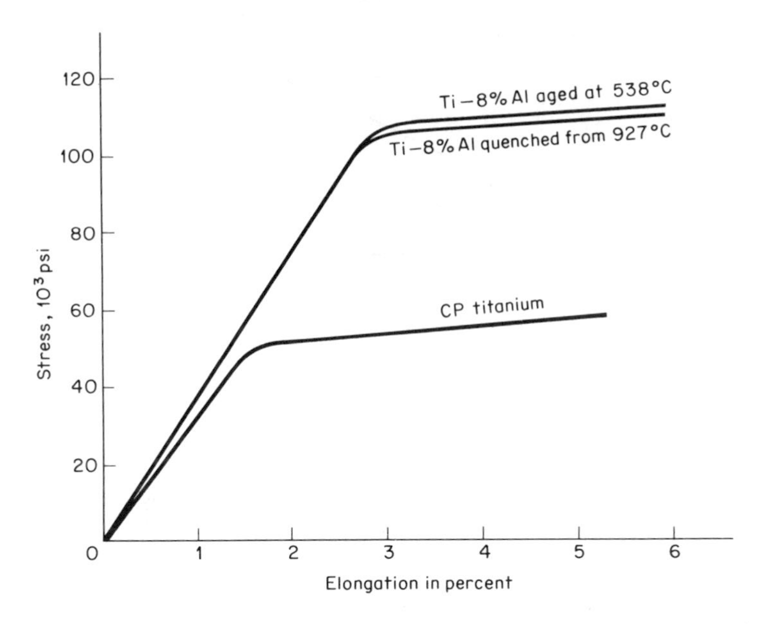

Fig. 20.8 The basic work-hardening rate is nearly unchanged when titanium is alloyed with 8 percent aluminum. (From Evans, K. R., *TMS-AIME,* **242** 648 [1968].)

precipitate becomes unstable and undergoes coalescence. It is important to note that the effect of alloying on the long-range component is basically over and above the effects of work hardening. This is clearly shown in Fig. 20.8. Note that in both the annealed and aged-hardened condition, the 8 percent alloy shows almost the same work hardening response.

20.7 The Relative Roles of the Flow-Stress Components in Face-Centered Cubic and Body-Centered Cubic Metals. Figures 20.9 and 20.10 show two sets of schematic engineering stress-strain curves characteristic of a pure face-centered cubic metal of moderate- to high-stacking fault energy (Al, Cu, Ag) on the one hand, and of a pure body-centered cubic metal on the other. Both sets represent tests performed over a range of temperatures. The significant point of difference between the two sets of curves is that as the temperature is lowered in the case of the face-centered cubic metal, the elongation or tensile ductility increases; whereas it decreases in the case of the body-centered cubic metal. This difference in behavior can be largely explained in terms of Considère's criterion. In the face-centered cubic metals, the thermally activated component of the flow-stress (σ^*) is small and only slightly temperature dependent. Notice that the yield stress, as shown in Fig. 20.9, increases only

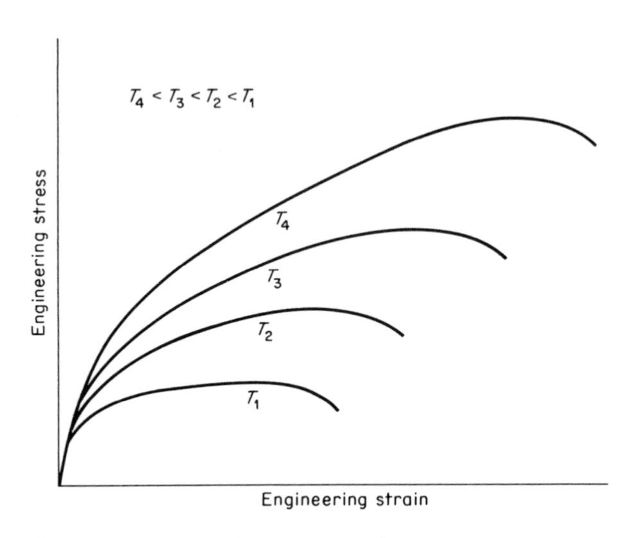

Fig. 20.9 Schematic engineering stress-strain curves for a face-centered cubic metal of high stacking fault energy. Note that the elongation increases with decreasing temperature.

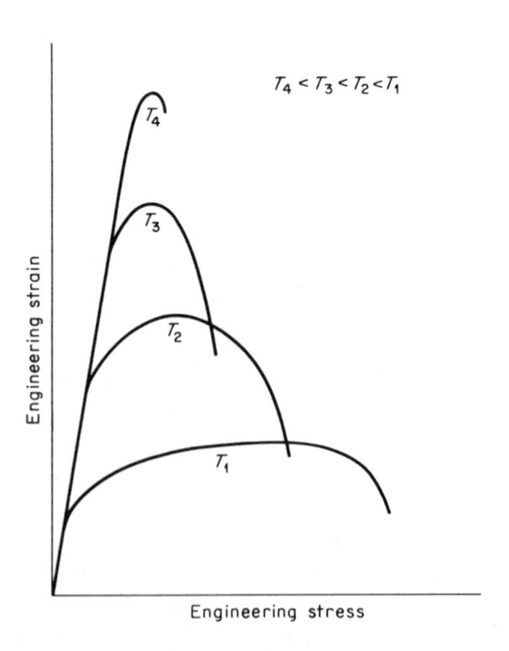

Fig. 20.10 Schematic engineering stress-strain curves for a body-centered cubic metal. In this case, the elongation decreases with temperature. Normally, curves of this type will also show a pronounced yield point which is eliminated to simplify the presentation.

slightly with decreasing temperature. In addition, high stacking-fault energy f.c.c. metals are subject to a high degree of dynamic recovery that increases in intensity with increasing test temperature. This acts to give these metals a work hardening rate that is strongly temperature-dependent; decreasing with increasing temperature. In other words, at a given strain the slope of the true stress–true strain curve $d\sigma_t/d\epsilon_t$ decreases with increasing temperature. Since all of the stress-strain curves start at approximately the same stress level (the yield stress), a progressively larger strain is therefore needed with decreasing temperature in order to bring about the condition where the flow-stress σ_t equals the slope of the stress-strain curve $d\sigma_t/d\epsilon_t$. This means that the start of necking is displaced to larger and larger strains with decreasing temperature. This makes both the uniform strain (that achieved before necking) and the total strain to fracture, larger the lower the temperature. Figure 20.11 shows a set of true stress–true strain curves for silver specimens, which are in good agreement with the above considerations. These curves are plotted only to the strains at which necking commenced. Note the good correspondence between the magnitudes of the slopes at the ends of the curves and the level of the corresponding flow-stress when necking began.

In contrast to face-centered cubic metals, body-centered cubic metals of commercial purity are characterized by a very strong temperature-dependent

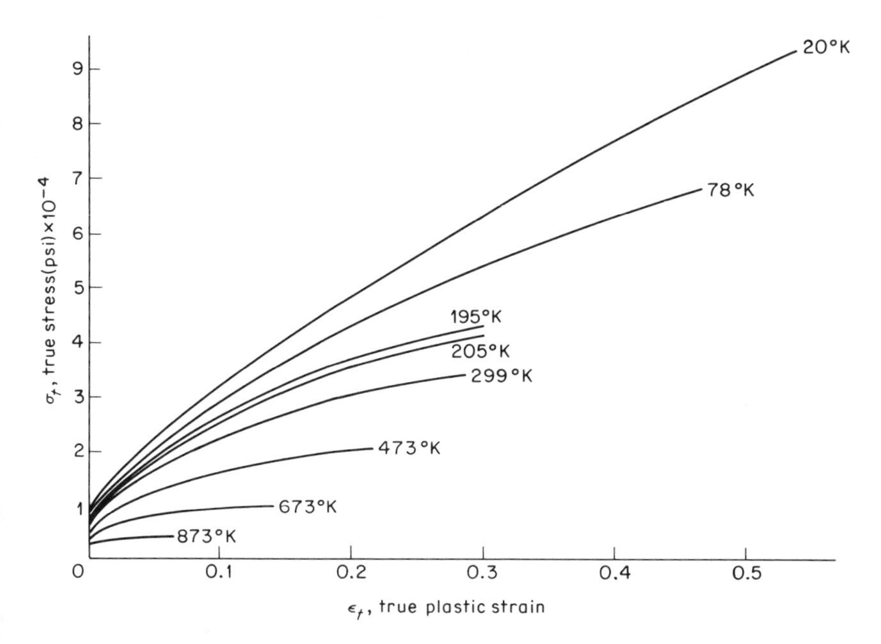

Fig. 20.11 True stress–true-strain curves for silver specimens of 17μ grain size. (From Carreker, R. P., Jr., *Trans. AIME*, **209** 112 [1957].)

component of the flow-stress. Note that in Fig. 20.10 the yield stress increases rapidly with decreasing temperature. This means that at low temperatures the general level of the flow-stress is normally much higher in a nearly pure b.c.c. metal than in a f.c.c. metal of the type discussed above. A basic result of this is that Considère's criterion is met at smaller and smaller strains as the deformation temperature is lowered in the b.c.c. metals. An interesting feature of the curves in Fig. 20.10 is that many b.c.c. metals actually show a reasonable amount of ductility after necking commences. This is represented by the part of the stress-strain curves where the engineering flow-stress decreases with increasing strain. All of this strain occurs in the neck. Such specimens will normally show a high reduction in area and are not basically brittle. The smaller observed elongation at low temperatures is therefore primarily a matter of premature necking induced by the high level of the flow-stress, and a work hardening rate that is not high enough to compensate for the high flow-stress.

In what has been stated above, the tendency for b.c.c. metals to cleave and fail in a brittle fashion has not been considered. This can be an important factor (under certain conditions) in determining the extent of the elongation.

20.8 Superplasticity. Under certain conditions a metal can exhibit an effect called *superplasticity*. When this occurs, the elongation in a simple tensile test becomes extraordinarily large and may be of the order of 1000 percent. Almost all of this strain occurs during the necking stage, so that superplasticity is primarily a necking phenomenon. However, there is one significant point that should be noted. In a normal tensile test the neck tends to be highly localized, as indicated in Fig. 20.12A; but when the metal deforms superplastically, the neck is spread out over the entire gage section, as shown schematically in Fig. 20.12B. The primary condition for an extended or *diffuse neck* to form is that the flow-stress become strongly strain-rate dependent. The reason for this is not hard to understand. When a sharply defined neck forms, all of the strain may be assumed to be concentrated in the neck. All of the

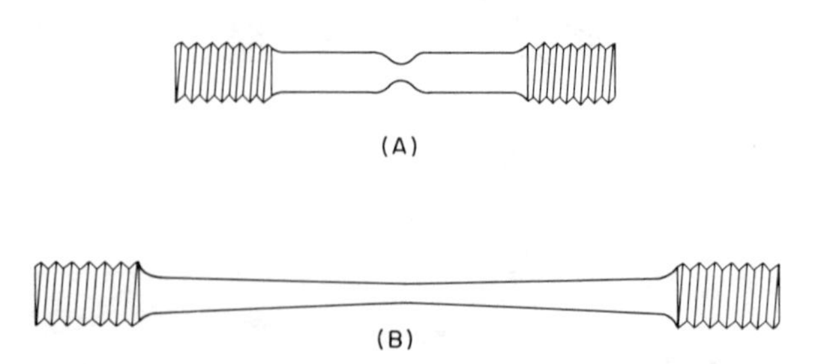

Fig. 20.12 (A) A well-defined sharp neck. (B) An extended or diffuse neck such as is observed in superplasticity.

specimen cross-sections above and below the neck effectively cease to deform. In an extended neck, on the other hand, all cross-sections within the gage section deform during necking. The strain rate, however, varies along the gage section and is an inverse function of the cross-section area. The minimum cross-section at the center of the gage length deforms faster than a cross-section near the shoulders. However, since at both locations the load P on the cross-section is the same, the applied stress at the center cross-section with the smaller area is larger than that near the shoulders. Dynamic equilibrium is nevertheless maintained by the difference in strain rates at the two locations. The faster strain rate at the center gage section raises the flow-stress in the metal at this point to compensate the larger applied stress at this location.

There are a number of experimental conditions that favor superplasticity. Normally is is observed at elevated temperatures. It is also a function of the strain rate, being observed within a certain range of strain rate at a given temperature. Finally, it is primarily associated with metals that have a very fine structure. This means either a very fine grain size ($< 10\ \mu$) or, in a two-phase structure such as an eutectic, a very finely divided eutectic structure.

Under the stated conditions, the deformation becomes relatively simple. At elevated temperatures, dynamic recovery becomes strong enough so that the work hardening is very small and may be neglected. At the same time, the thermally activated component of the flow-stress can be approximated by the power law stated earlier in Section 8.14

$$\sigma = B(\dot{\epsilon})^n$$

where σ is the flow-stress, $\dot{\epsilon}$ is the strain rate, B is a constant, and the exponent n is called the strain rate sensitivity. The larger the value of n, the more sensitive the flow-stress is to a change in strain rate. In most metals of average grain size, n increases with temperature from zero at absolute zero to about 0.2 near their melting points. In materials showing a superplastic response, n is normally between 0.3 and 0.8. Experimental measurements have shown a close correspondence between n and the elongation in a tensile test, with the elongation increasing with increasing values of n. This is clearly shown in Fig. 20.13.

20.9 The Nature of the Time-Dependent Strain. The fact that the flow-stress contains a component that responds to thermal activation implies that plastic deformation can occur while both the temperature and the stress are maintained constant. This is actually true, and the deformation that occurs under these conditions is known as *creep*. Creep deformation (constant stress) is possible at all temperatures above absolute zero. However, since it depends on thermal activation, the strain rate at a given stress level is extremely temperature sensitive. As a result, the higher the temperature, the more significant becomes

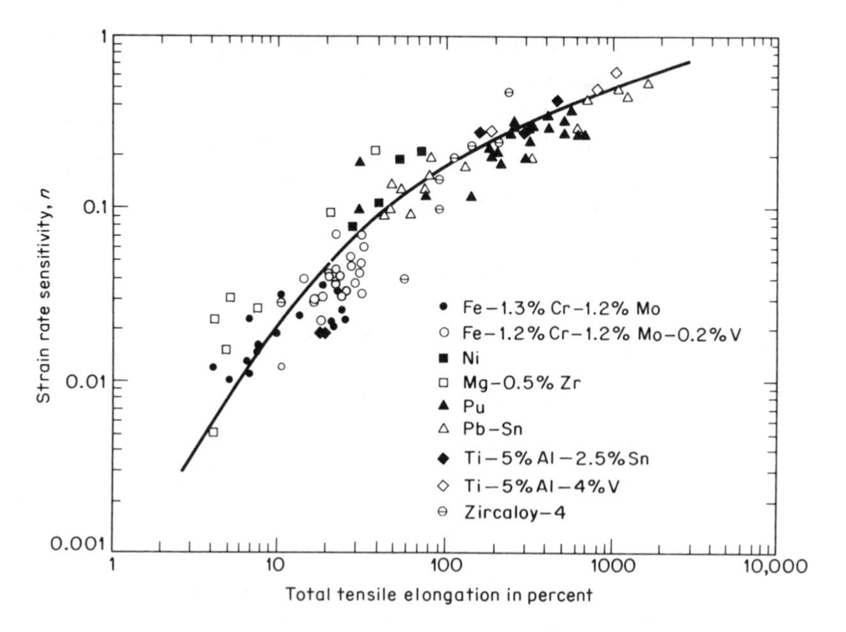

Fig. 20.13 Correlation between the strain rate sensitivity and the elongation in a specimen deformed by tension. (From Woodford, D. A., *Trans. ASM,* **62** 291 [1969].)

the creep phenomena. Thus creep is usually not considered to be important in the application of steel to bridges, ships, or other large structures, and the design of these items (for use at ambient temperatures) is based primarily on elasticity theory. However, steel objects for use at elevated temperatures (above about 900°F) are designed with due consideration of the fact that they will deform slowly under load. An excellent example is furnished by the steel tubes in a petroleum cracking still, used for making gasoline, which are subject to both high temperatures and high stresses (pressures). It is common practice to design these tubes for an expected life of several years. By the end of this period the expanded walls of the tube will no longer be able to carry the load, and the tube is removed and replaced by a new one.

Two empirically observed facts concerning creep phenomena are significant. One is simply that plastic flow can occur while stress and temperature are maintained constant; the other is that the flow or strain rate ($\dot{\epsilon}$) is extremely temperature sensitive. It is frequently represented by an equation of the form

$$\dot{\epsilon} = Ae^{-q/kT}$$

where A and q are constants, and k and T have their usual significance.

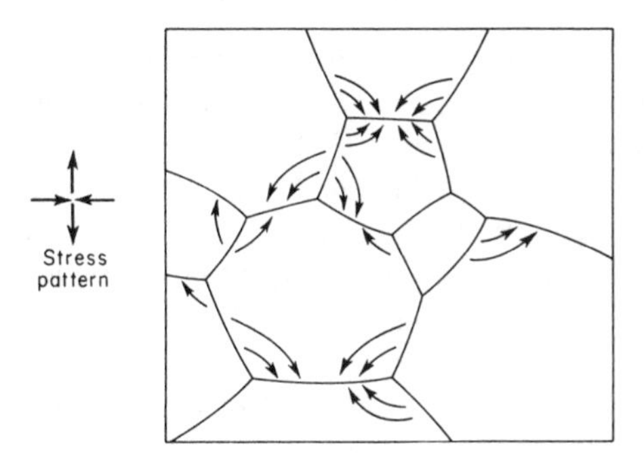

Fig. 20.14 Schematic representation of diffusion creep. Self-diffusion should result in plastic flow if matter is carried from boundaries subject to compressive stress (vertical boundaries) over to boundaries under a tensile stress (horizontal boundaries). (After Herring, C., *Jour. Appl. Phys.*, **21** 437 [1950].)

It is a well-documented fact that crystalline metals deform by the three following types of mechanisms: (*a*) slip, (*b*) climb, and (*c*) shear on, or adjacent to, grain boundaries (polycrystalline). A fourth method of deformation, based on vacancy diffusion inside the grains of a polycrystalline material, has been proposed by Nabarro[7] and Herring.[8] Self-diffusion inside grains can produce plastic flow if matter is carried from grain boundaries, which are subject to a net compressive stress, over to grain boundaries where there is a net tensile stress, which is indicated schematically in Fig. 20.14. Notice that, in this figure, the assumed macroscopic stress places horizontal boundaries under tensile stress and vertical boundaries under compressive stress. In the given mechanism, grain boundaries are considered to be both sources and sinks for vacancies, and the rate of creep deformation should be dependent on the grain size of the metal. Herring[9] has calculated that the strain rate for this mechanism varies inversely as the square of the grain diameter. He also suggests that this form of plastic flow, if it exists, should only be significant at very high temperatures and under very small stresses. These are more or less the conditions that occur during the sintering of powdered metal compacts, where small stresses due to surface tension cause metal particles to flow and consolidate themselves. In most practical engineering applications of metals at elevated temperature, the Nabarro-Herring mechanism should be of secondary importance to other creep mechanisms because the normal operating temperatures are too low and the usual operating stresses are too high.

7. Nabarro, F. R. N., *Proceedings of Conference on Strength of Solids* (1948), Physical Society, London, p. 75.
8. Herring, C., *Jour. Appl. Phys.*, 21 437 (1950).
9. *Ibid.*

If we may neglect Nabarro-Herring diffusion creep, then, with the possible exception of grain-boundary shear, creep deformation in metals that do not undergo deformation twinning should depend primarily on dislocation movements. Further, while the mechanism by which grains slide relative to one another is still not completely defined, there is evidence, as we shall presently see, that grain-boundary shearing is controlled by plastic deformation inside the grains. From this it is apparent that the problem of understanding creep deformation resolves itself, in most cases, into the identification of the obstacles that dislocations meet and into analyzing the ways that thermal energy can help to overcome these obstacles.

The problem of defining creep mechanisms, even in single crystals, is still far from accomplished. For the present, it would seem that the best that can be done is to attempt to gain an understanding of the nature and difficulty of the problem. In this regard, let us briefly review, first from a general point of view, and later in terms of specific mechanisms, some of the possible ways that it is believed thermal energy can activate or aid in the motion of dislocations.

The interaction between thermal energy and stress, in certain simpler cases, may be viewed in the following manner. In the absence of a stress, it is assumed that there is a deformation mechanism available capable of acting locally under the influence of temperature. In considerations of this nature, it is well to remember the random nature of thermal vibrations. Because of this irregularity, only a very limited number of adjoining atoms in a crystal can ever be considered as vibrating, even approximately in the same manner. It thus becomes a basic precept of creep theories that thermal energy can activate deformation mechanisms involving only a relatively small number of atoms. Thermal energy is therefore not capable of causing movement of long dislocation segments, but rather is only able to activate dislocation reactions on a very limited scale. A simple illustrative example is furnished by dislocation climb, where the addition or removal of a single atom at a jog on an edge dislocation results in a small contraction or expansion of the lattice in the direction normal to the extra plane of the dislocation.

Let us suppose that, for an assumed creep mechanism to operate, an energy barrier of the type shown in Fig. 20.15A must be overcome. Points A and C in this diagram represent free-energy minima corresponding to a single operation of the deformation mechanism and, in a shift from A to C, it may be assumed that the crystal suffers a small unit strain in a particular sense. A reversed movement from C to A would give a unit strain in the opposite sense.

Now assume that, under a given stress, the mechanism acts from A to C and that the strain is in the direction of the stress, so that positive work is performed. In an elementary case, this may be considered as effectively lowering the height of the energy barrier in the direction from A to C, while raising it for a change from C to A. (See Fig. 20.15B.) In the equations which are to be

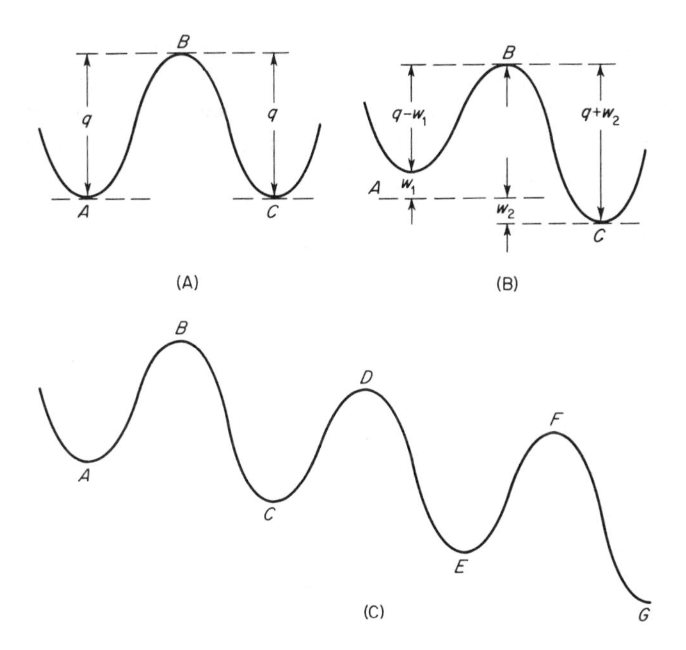

Fig. 20.15 Effect of stress on the activation energy of a hypothetical creep mechanism.

developed, the energy barrier and the effect of stress on the barrier should be expressed in terms of free energy. While this is the theoretically more correct approach, creep theory is not at the state where it is convenient to do this. In general, the entropy functions of activation processes are unknown and are often considered to be of relatively minor importance.[10] For these reasons, the entropy will be neglected and it will be assumed that energy barriers can be expressed in terms of the work to overcome the barriers (activation energies).

Let the change in energy, due to the stress, be w_1 when the mechanism operates from state A to the activated state B, and w_2 when it operates from the activated state B to state C. Then, the energy barrier for a change of state from A to C is $q - w_1$, and is $q + w_2$ for the change from C to A, where q is the height of the barrier in the absence of a stress. The respective frequencies with which the mechanism should operate from A to C and from C to A are, accordingly,

$$(A \text{ to } C) \qquad \nu_1 = \nu_0 e^{-(q - w_1)/kT}$$

$$(C \text{ to } A) \qquad \nu_2 = \nu_0 e^{-(q + w_2)/kT}$$

10. Schoeck, G., ASM Seminar (1957), *Creep and Recovery*, p. 199.

where ν_1 and ν_2 are the respective frequencies of the mechanism, ν_0 is a constant which is assumed to have the same value for both directions of the mechanism, k is Boltzmann's constant, and T the absolute temperature.

Let us assume that, once the deformation mechanism has passed the energy barrier from A to C, it is possible for it to operate again in the same sense and pass over a similar barrier from state D to state E, and then continue on in this same manner over a series of barriers, as indicated in Fig. 20.15C. In a sequence of barriers, the deformation mechanism always has the alternative of going over a barrier in the direction of the stress, or of going in the direction opposite the stress. The relative frequencies with which the mechanism should operate in the two directions are given by the two frequency equations stated above. In general, the frequency corresponding to strain in the direction of the stress will be larger than that opposed to the stress. If a large number of identical mechanisms are capable of operating in this same fashion, then the net rate at which creep strain, $\dot{\epsilon}$, occurs should be proportional to the difference between the forward and reverse frequencies, thus

$$\dot{\epsilon} \approx \nu_1 - \nu_2 \approx \nu_0 (e^{-(q-w_1)/kT} - e^{-(q+w_2)/kT})$$

or

$$\dot{\epsilon} = Ae^{-q/kT}(e^{+w_1/kT}\, e^{-w_2/kT})$$

where A is a constant containing ν_0. Let us further assume that $w_1 = w_2 = w$, which with the definition of the hyperbolic sine function

$$\sinh x = \frac{e^x - e^{-x}}{2}$$

gives us

$$\dot{\epsilon} = Ae^{-q/kT} \cdot 2 \sinh \frac{w}{kT}$$

This is an interesting equation because it expresses the creep rate in a form comparable to that often obtained empirically in creep tests. Here the temperature dependence is represented by an exponential factor

$$e^{-q/kT}$$

and the stress dependence by a hyperbolic sine function whose argument is a function of the stress; the value of w depending on the magnitude of the stress.

The simplest assumption might be that w varies directly as the stress τ^*

$$w = \upsilon\tau^*$$

where υ is a constant depending on the nature of the mechanism and is called the *activation volume*. The strain rate becomes, in this case,

$$\dot{\epsilon} = Ae^{-q/kT} \cdot 2 \sinh \frac{\upsilon\tau^*}{kT}$$

For a very small applied stress, the hyperbolic sine function may be replaced by its argument, with the result that the strain rate becomes directly proportional to the stress

$$\dot{\epsilon} = Ae^{-q/kT} \frac{2\upsilon\tau^*}{kT}$$

Alternatively, if q is large compared to kT, and the stress is also large, then the backward frequency of the deformation mechanism ν_2 will be negligible compared with the forward frequency ν_1, and the mechanism may be considered to act only in the forward direction. In this case, the strain rate becomes directly proportional to ν_1 and we may write

$$\dot{\epsilon} = Ae^{-(q - w)/kT} = Ae^{-q/kT}e^{w/kT}$$

If we again assume a direct dependency between w and the stress, this leads to an exponential relationship between the strain rate and the stress

$$\dot{\epsilon} = Ae^{-q/kT}e^{\upsilon\tau^*/kT}$$

The assumptions on which this last equation are based are particularly significant because the energy barrier in a typical dislocation reaction is usually of the order of one eV or larger and thus appreciably larger than the thermal energy, which is approximately $\frac{1}{40}$ eV at room temperature, or $\frac{1}{10}$ eV at 900°C. In addition, most fundamental creep investigations have been concerned only with tests carried out under relatively high stresses.

20.10 Creep Mechanisms. Let us now turn to a consideration of some mechanisms that have been proposed as having significance during the creep of metals.

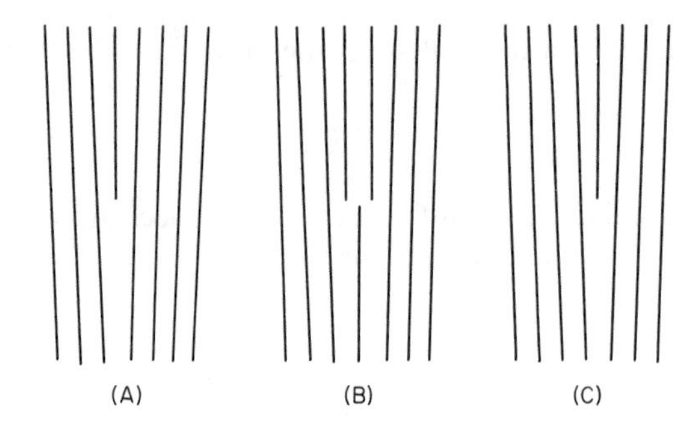

Fig. 20.16 Three steps in the movement of an edge dislocation. (B) represents a high energy state.

Activation of Dislocation Sources. It is now well recognized that dislocations are created in a metal during the process of plastic deformation. Since it takes work to form a dislocation loop at any type of source, it is possible that thermal energy can aid an applied stress in overcoming this energy barrier. Whether or not this is an important effect is still to be determined. It would now appear that most dislocations are nucleated heterogeneously at impurity particles. In this case the role of thermal energy would be difficult to evaluate. On the other hand, it has been suggested that the formation of a dislocation loop at a Frank-Read source involves a cooperative movement of too many atoms to be successfully aided by random thermal vibrations.[11]

Overcoming the Peierls Stress. An example of how thermal energy can aid a dislocation to move through a crystal will now be considered. As explained in Chapter 4, a dislocation moves in a stepwise fashion. Thus, consider a dislocation in an edge orientation, as shown in Fig. 20.16 where three intermediate stages in the forward movement of the extra plane through an atomic distance are illustrated. As the dislocation passes from stage a to c it travels through the high energy stage b and the dislocation, therefore, passes over a typical energy barrier of the type shown in Fig. 20.15A. Continued motion of this dislocation involves the passing of a series of similar energy barriers. For the reason already mentioned, thermal energy and the stress are incapable of working together in such a way as to activate the movement of a long section of an edge dislocation over such barriers. Thermal vibrations may, however, cause short segments of this dislocation to move forward one unit, placing a double kink in it, as is shown in Fig. 20.17A. In this figure, the edge dislocation is seen as if one were looking at Fig. 20.16 from below. The ends of this kinked dislocation can move easily to right and left under an applied stress, as indicated in Fig. 20.17B, with the result that the entire dislocation moves forward one atomic distance. The

11. *Ibid.*

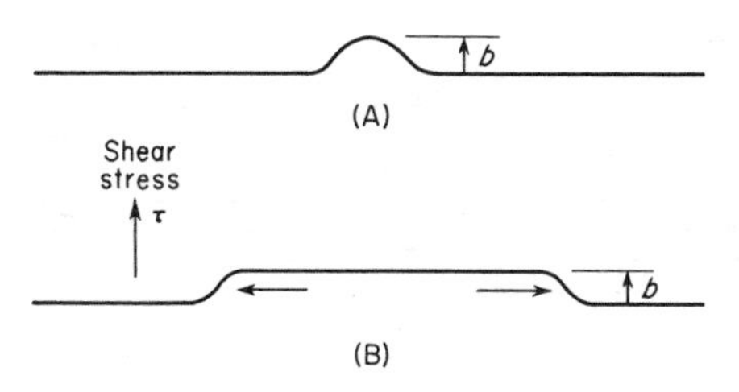

Fig. 20.17 Thermal activation of the movement of an edge dislocation. Thermal energy aids the stress in the forming of a kink, which then spreads under the action of the stress.

effect of thermal energy is to help in the formation of the original kink, which then expands under the applied stress, thereby advancing the dislocation and shearing the crystal.

The force holding a dislocation in its low-energy position in a lattice is known as the *Peierls force*, and the above mechanism is one of those which have been advanced for overcoming this force. The question of whether or not the Peierls force is significant in the thermally activated plastic deformation of real metal crystals is still subject to debate. It was originally proposed[12] on theoretical grounds that the Peierls stress was too small to be significant. It may also be argued that the Peierls stress should only be important when dislocations lie parallel to close-packed crystallographic directions. Thus when a dislocation makes a small angle with a close-packed direction, it should consist of a series of connected kinks,[13] as is shown in Fig. 20.18A. The dislocation will assume this shape as a compromise between two opposing factors. First, the energy is lower when the dislocation lies along a close-packed direction; second, every dislocation possesses an effective line tension tending to make its length a minimum. The latter tries to straighten out the dislocation into the shape of Fig. 20.18B, while the former tends to make it take the form of Fig. 20.18C. The resulting shape is, of course, that of Fig. 20.18A which, because of its kinks, should move readily under an applied shear stress in the manner of Fig. 20.17, while thermal vibrations can be expected to have little effect on its motion. Also, if the dislocation makes a large angle with the close-packed direction, the kinks should overlap, with the result that the dislocation will effectively straighten out. In this case, because it does not lie along an energy minimum at any point, the dislocation should also be little influenced in its movement by thermal energy.

12. Cottrell, A. H., *Dislocations and Plastic Flow in Crystals*. Oxford University Press, London, 1953.
13. Read, W. T., Jr., *Dislocations in Crystals*. McGraw-Hill Book Company, Inc., New York, 1953.

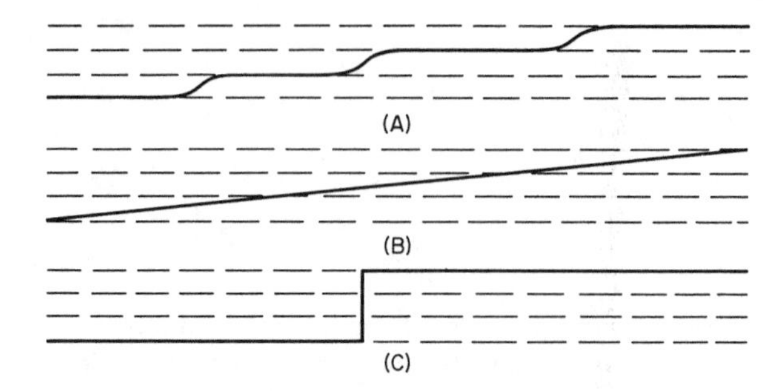

Fig. 20.18 The shape of a dislocation when it is nearly parallel to a close-packed lattice direction. Dashed horizontal lines represent close-packed lattice planes. (From *Dislocations in Crystals*, by Read, W. T., Jr. Copyright, 1953. McGraw-Hill Book Co., Inc., New York. Used by permission.)

In spite of the above objections, it appears that in certain metals at very low temperatures there is experimental evidence favoring the view that a Peierls type of stress may be a controlling factor in dislocation movement. Internal-friction measurements of face-centered cubic metals at low temperatures exhibit an energy-absorption peak – the *Bordoni peak,*[14] which has been explained[15,16] on the basis of thermal activation of dislocation kinks of the type described in the preceding paragraphs. These computations predict a Peierls stress of the order of 10^{-4} G where G is the shear modulus of the crystal. This would make the Peierls stress equal to about several hundred psi, a value large enough to affect the movement of dislocations since it is of the order of the critical resolved shear stress in face-centered cubic metals at low temperatures. Conrad,[17] however, has discussed the question as to whether or not the Peierls stress mechanism can explain the observed experimental data for copper, and indicates that the operation of this mechanism is not completely confirmed.

Dislocation Intersection. A more probable factor in the control of the motion of dislocations through a crystal under constant stress and temperature is the intersection of dislocations with each other. All real crystals contain a built-in network of dislocations, that grows in complexity with increasing plastic deformation. This network has been conveniently referred to as the *forest of dislocations*. Because of the forest, any slip dislocation does not travel far before it intersects other dislocations passing through its slip plane at various angles. The intersection process is important for two reasons. The forcing of a dislocation through the stress field of another dislocation involves an element of

14. Bordoni, P. G., *Jour. Acoust. Soc. Amer.*, 26 495 (1954).
15. Seeger, A., *Phil. Mag.*, 1 651 (1956).
16. Seeger, A., Donth, H., and Pfaff, F., *Trans. Faraday Society*. Submitted.
17. Conrad, H., *Acta Met.*, 6 339 (1958).

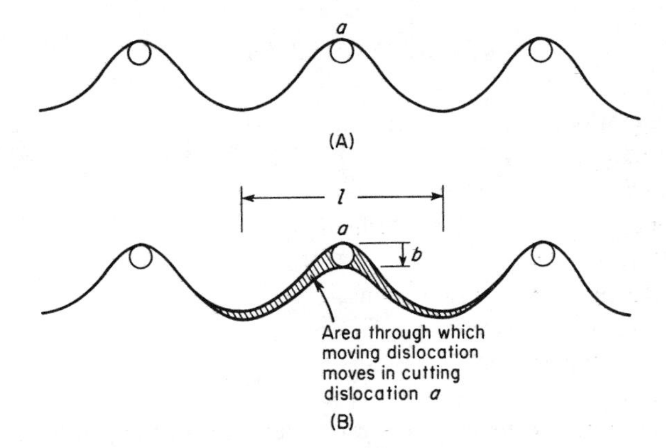

Fig. 20.19 A simplified representation of the geometrical factors involved in a dislocation intersection. (After Cottrell, A. H., *Dislocations and Plastic Flow in Crystals,* Oxford University Press, London, 1953.)

work and, second, because of the intersection, the dislocations in question can receive jogs, the movement of which through the lattice can be thermally aided.

The work to drive one dislocation through another has been estimated[18] to vary in different metals from several electron volts to a fraction of an electron volt (one electron volt equals approximately 23,000 cal per mole). Thermal energy ($1/40$ eV at 300°K or $1/10$ eV at 1200°K) should be energetic enough to aid an applied stress in driving one dislocation through another. According to Cottrell's[19] original proposal, a dislocation meeting another dislocation might appear as shown in Fig. 20.19A.

Now let us suppose that the moving dislocation moves through the dislocation marked a under an applied shear stress τ^*. The effective distance through which the dislocation moves during the intersection can be taken as equivalent to the Burger's vector b. The force on the dislocation is equal to $\tau^* b$ per unit length of the dislocation. Further, the length of the dislocation that moves during the intersection can be taken as the spacing between dislocations (l). The force on this section is $\tau^* bl$ and, since it moves a distance equal to b, the work done by the force during the intersection is $\tau^* b^2 l$. If, in the absence of the stress, the energy barrier corresponding to the driving of one dislocation through another is q, then an expression can be written for the strain rate in analogy with our previous derivations, assuming $q >> kT$ and that τ^* is large.

18. Seeger, A., *Report of a Conf. on Defects in Crystalline Solids*, p. 391, The Physical Society, London (1954).
19. Cottrell, A. H., *Dislocations and Plastic Flow in Crystals*. Oxford University Press, London, 1953.

$$\dot{\epsilon} = Ae^{-(q - \tau^* b^2 l)/kT}$$

In the above equation,

$\dot{\epsilon}$ = the strain rate
A = a constant
q = the energy barrier for the intersection
b = the Burgers vector
l = the spacing between dislocations
k = Boltzmann's constant
T = absolute temperature
τ^* = the stress

The latter is to be interpreted as the effective shear stress acting on the dislocation where

$$\tau^* = \tau - \tau_\mu$$

and τ is the applied stress and τ_μ is an average internal stress due to the long-range stress fields of all the other dislocations.

The above strain-rate equation is an interesting example of a theoretical equation corresponding to a specific creep mechanism. It has been used[20] very recently, with certain minor modifications, to explain the creep data of magnesium single crystals deforming by basal slip.

Movement of Dislocations with Jogs. The second effect of a dislocation intersection is the formation of jogs on the dislocations. This formation is particularly significant when the dislocations are in the screw orientation because the jogs have an edge orientation, as is shown in Fig. 20.20. This type of jog can move by slip along the dislocation, but must move by climb in the glide direction of the screw dislocation. Depending on the direction of motion and the sign of the jog, a movement of the jog in the direction of motion of the screw dislocation should result in the creation behind the jog of either a row of interstitial atoms or a row of vacancies. Because the energy to form a vacancy is much smaller than that to form an interstitial atom, it is generally conceded that only vacancies should be created by the movement of screw dislocations with jogs.

The movement of dislocations with vacancy producing jogs can be aided by thermal energy in two ways. First,[21] assume that a jog has no vacancies behind it. Then for it to move forward one unit, a vacancy must be created. The work corresponding to this process can be represented by an activation energy q_j.

20. Conrad, H., Hays, L., Schoeck, G., and Wiedersich, H., *Acta Met.*, 9 367 (1961).
21. Mott, N. F., *Creep and Fracture of Metals at High Temperatures*. Philosophical Library, New York, 1957.

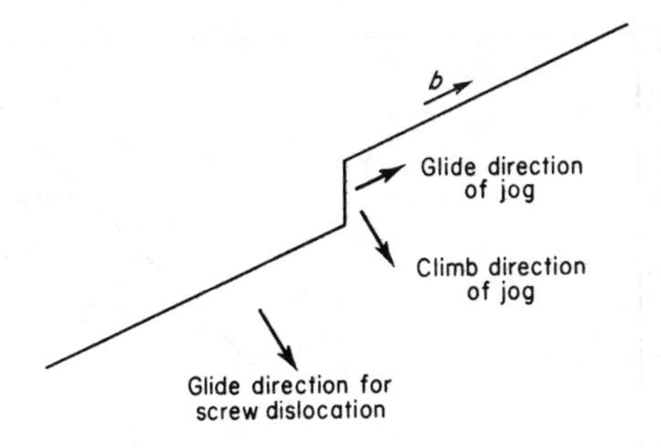

Fig. 20.20 Movement of a screw dislocation with a jog.

When a vacancy is formed behind a jog, the former creates a back stress on the latter which tends to restrain it from further movement. If the vacancy should diffuse off into the lattice, however, the jog is free to move another step. For this reason, the motion of a jog should involve both the energy to form a vacancy and the energy to move it away ($q_f + q_m$). The sum of these two values is, of course, the activation energy for self-diffusion, q_d.

A strain-rate equation can be derived for the movement of a dislocation with vacancy producing jogs in a similar manner to that used for the intersection of dislocations. If a single jog moves forward one unit, a dislocation segment (x), approximately equal to the distance between jogs, advances a distance equal to a Burgers vector b. The work done by the stress is thus $(\tau^* bx)b$, where τ^* is the effective shear stress, x the distance between jogs, and b is the Burgers vector. The creep rate of a metal due to the movement of jog-producing vacancies are thus expressed as

$$\dot{\epsilon} = Ae^{-(q_d - \tau^* b^2 x)/kT}$$

The above Mott equation has been used by him to explain certain aspects of creep in face-centered cubic metals.

Dislocation Climb. We shall next consider briefly the role of dislocation climb in the plastic flow of metals at constant stress and temperature. At high temperatures, experimentally determined activation energies for creep frequently equal the activation energy for self-diffusion, as is shown in Fig. 20.21.

In the preceding section, a mechanism for creep was discussed that gave an activation energy for creep equal to that for self-diffusion. Since simple climb of edge dislocations depends on the diffusion of vacancies either toward edge dislocations (positive climb), or away from edge dislocations (negative climb), creep deformation resulting from climb should also involve the activation energy

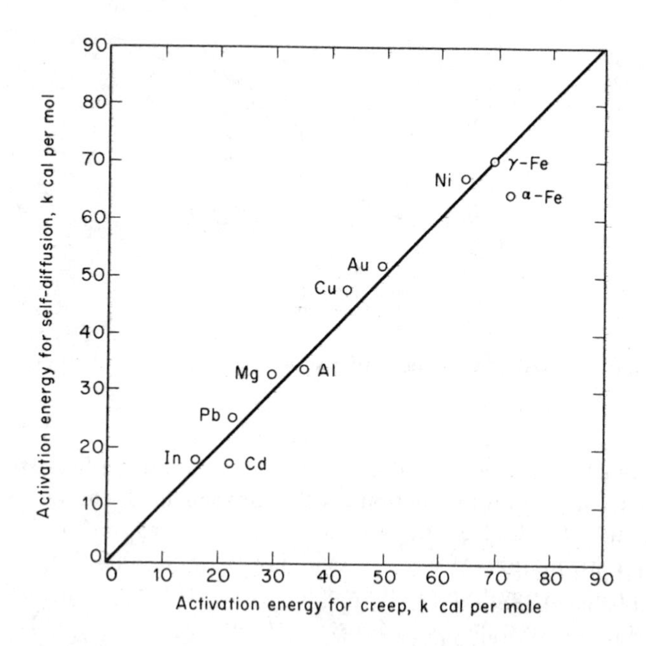

Fig. 20.21 Comparison of the activation energies for creep with the activation energies for self-diffusion in a number of metals. (After Dorn, J. E., ASM Seminar, *Creep and Recovery*, p. 255 [1957].)

for self-diffusion. There is, however, one additional factor that should be taken into account. According to recent theories, vacancies do not normally attach themselves to straight sections of the extra plane of an edge dislocation, but rather to jogs along the edge. Jogs can form in the extra plane by thermal activation or as a result of dislocation intersections. In the former case, the observed creep rate should show an activation energy which includes both the activation energy for self-diffusion and activation energy for the formation of jogs on the dislocation, or

$$q = q_d + q_j$$

where q is the activation energy for creep, q_d the activation energy for self-diffusion, and q_j the activation energy for the formation of the jogs. However, if the energy to form a jog is relatively high the number of jogs formed by thermal activation should be quite small. Then after any appreciable amount of plastic deformation, the number of jogs due to dislocation intersections should be many times more numerous than those formed thermally. If this number is large enough, it can be assumed that the time for a vacancy to find a jog will be small compared to the time for it to diffuse to a dislocation. Under

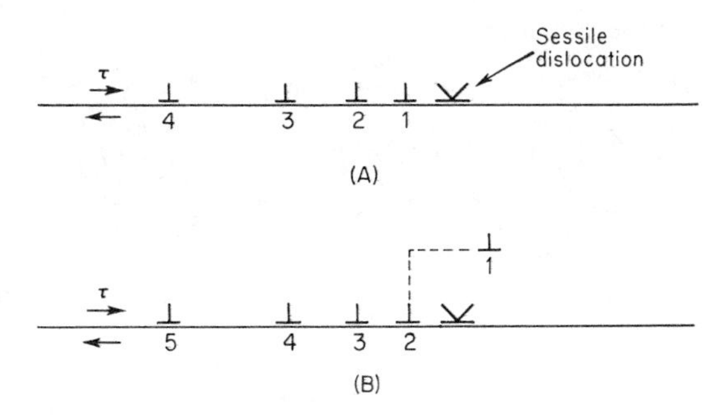

Fig. 20.22 (A) A dislocation pile-up at a sessile dislocation. (B) Climb of the leading slip dislocation over the sessile dislocation permits all slip dislocations to advance.

these conditions, the thermal formation of jogs can be neglected and the activation energy for creep (due to the climb of dislocations) assumed equal to that for self-diffusion.[22]

It is possible that the most important function of climb during creep is to help dislocations to overcome obstacles to slip. In other words, the deformation is primarily accomplished by slip, but the factor controlling the amount of slip is the climb of dislocations over obstacles. Figure 20.22 shows one possible mechanism. In the first sketch, a sessile dislocation is shown blocking the motion of other dislocations in its slip plane, with the result that a pile-up has formed in front of the sessile dislocation. Climb of the leading blocked dislocation over the sessile dislocation, as shown in Fig. 20.22B, permits deformation by slip to continue. In the second case (Fig. 20.23), dislocations of opposite sign on parallel slip planes are shown climbing toward each other, thereby lowering the dislocation density of the metal and reducing the back stress on the dislocation sources. This should permit additional dislocation loops to form and grow. Finally, in the third illustrated case in Fig. 20.24A, dislocation loops are shown

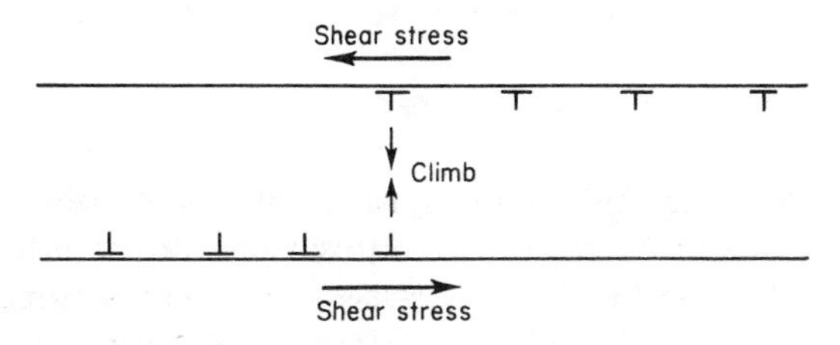

Fig. 20.23 Annihilation of dislocations of opposite sign will permit glide to continue.

22. Schoeck, G., ASM Seminar (1957), *Creep and Recovery,* p. 199.

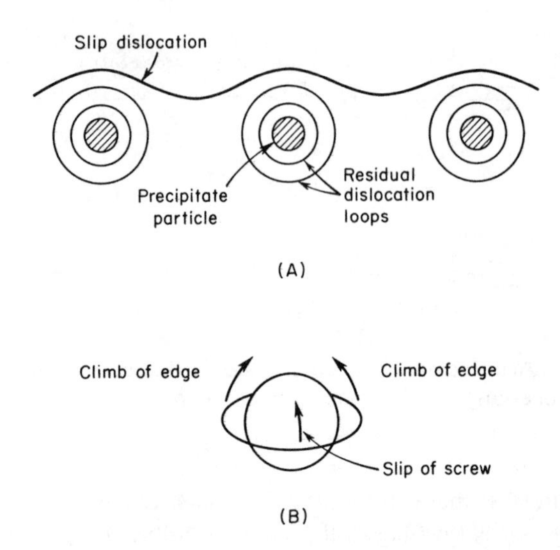

Fig. 20.24 Climb over precipitate particles. (A) Residual loops left around precipitate particles block movement of additional slip dislocations. (B) Collapse of loop by climb of edge components of loop and slip of screw components.

surrounding precipitate particles, or inclusions. As discussed earlier, according to Orowan (Chapter 9), loops should form around incoherent particles as the dislocations move among them. However, the progressive development of more and more loops around precipitate particles blocks the motion of succeeding dislocations through the lattice. By the aid of climb, Fig. 20.24B, the edge components of the loops can move up and over a particle to annihilate each other, while the remaining parts of the loops, which are basically in screw orientations, may aid the collapse by slip.

Weertman[23] has used the concept of dislocations climbing over obstacles to slip to derive a theoretical equation for the creep rate. Assuming the obstacles are sessile dislocations and that the stress is small, this equation is reduced to

$$\dot{\epsilon} = A(\sigma)^{\alpha} e^{-q_d/kT}$$

where A and α are constants, with α equal to about 4, σ is the applied normal stress, q_d the activation energy for self-diffusion, and k and T have their usual significance. It is interesting that this stress dependence is different from that given in previous equations. Such a stress dependence has actually been

23. Weertman, J., *Jour. Appl. Phys.*, **26** 1213 (1955).

observed[24] under low stresses and at high temperatures in the creep of aluminum.[24,25] However, at higher stresses, the empirical strain-rate equation is[25]

$$\dot{\epsilon} = Ae^{\beta\sigma}e^{-q_d/kT}$$

where β is a constant. No theoretical explanation has yet been obtained[26] for this exponential relationship which applies specifically to aluminum at high stresses and high temperatures.

Movement of Dislocation Atmospheres. Except in metals of very high purity, solute atmospheres probably form around dislocations. The effect of these atmospheres on the motion of dislocations under an applied stress has been considered in detail in Section 8.10. In the presence of an atmosphere, the rate of dislocation movement and the corresponding observed strain rate of a creep specimen should be proportional to the diffusion rate of the solute atoms, and for a fixed stress[27] the following equation may be written:

$$\dot{\epsilon} \approx D \approx D_o e^{-q_s/kT}$$

where $\dot{\epsilon}$ is the strain rate, D is the diffusion coefficient for the diffusion of solute atoms, and D_o and q_s are the corresponding frequency factor and activation energy. Notice that the above activation energy is for the diffusion of solute atoms and is not the self-diffusion activation energy of the matrix atoms. It has also been proposed[28] that a dislocation impeded by an atmosphere should behave viscously. This implies that the rate of dislocation movement and the creep rate, at a given temperature, should vary directly with the stress, or that

$$\dot{\epsilon} \approx \sigma D_o e^{-q_s/kT}$$

or

$$\dot{\epsilon} = A\sigma e^{-q_s/kT}$$

where A is a constant which includes the frequency factor D_o and σ is the applied normal stress. The significance of this mechanism has been discussed in Section 8.10.

24. Weertman, J., *Jour. Appl. Phys.*, **26** 1213 (1955).
25. Dorn, J. E., *Jour. of the Mechanics and Physics of Solids*, **3** 85 (1954).
26. Schoeck, G., ASM Seminar (1957), *Creep and Recovery*, p. 199.
27. Cottrell, A. H., *Dislocations and Plastic Flow in Crystals*. Oxford University Press, London, 1953.
28. *Ibid.*

Activated Cross-Slip. A mechanism that has special application to creep in face-centered cubic metals, and perhaps to creep in metals of other crystal structures, involves thermal activation of cross-slip.

A general characteristic of a dislocation in a screw orientation is its ability to move on any plane of a zone that has the line of the screw dislocation as its zone axis. The interchange of slip planes by screw dislocations leads to the phenomenon known as *cross-slip*. In face-centered cubic metals, cross-slip normally involves the movement of screw components between pairs of {111} planes. Cross-slip cannot normally occur, however, when a screw dislocation is dissociated into a pair of partial dislocations with a connecting layer of stacking fault. The movement of partial dislocations from one {111} slip plane to another places atoms in high-energy positions and is not considered to be energetically feasible. However, a method by which cross-slip in face-centered cubic metal can occur has been theoretically treated by Schoeck and Seeger.[29] A qualitative description of the mechanism follows.

First, the extended dislocation has to undergo a constriction; the partials unite to form a total dislocation of length l, as is shown in Fig. 20.25A. This is assumed to occur by thermal activation. The constricted dislocation then expands or breaks down into partial dislocations on the cross-slip plane (Fig. 20.25B). Finally, in the presence of a sufficiently large resolved shear stress on the cross-slip plane, the extended dislocation moves by glide in the cross-slip plane (Fig. 20.25C).

The length of dislocation which can move on the cross-slip plane has a critical value. Below this value the dislocation is unstable and should collapse and return to its original slip plane. This critical length depends on the stress in the cross-slip plane. The higher the resolved shear stress on the cross-slip plane, the more difficult it will be for the dislocation to collapse. The critical length, accordingly, is smaller the higher the stress. In other words, the size of the thermally activated constricted segment required for cross-slip to occur becomes smaller the larger the stress in the cross-slip plane. Since the work to form the constriction decreases as the constriction becomes smaller, the greater the stress the smaller the activation energy for cross-slip. Calculated values for the activation energy for cross-slip, when the shear stress on the cross-slip plane equals the critical resolved shear stress of the metal concerned (100 gm/mm^2 for Al, and 140 gm/mm^2 for Cu), give 1.05 eV for aluminum and 10 eV for copper. The much higher activation energy for copper means that cross-slip should be much less prevalent in copper than in aluminum, or that higher temperatures and higher stresses are required for extensive cross-slip in copper. The reason for the higher activation energy in copper is related to the lower stacking-fault energy of

29. Schoeck, G., and Seeger, A., *Defects in Crystalline Solids*, p. 340. The Physical Society, London, 1955.

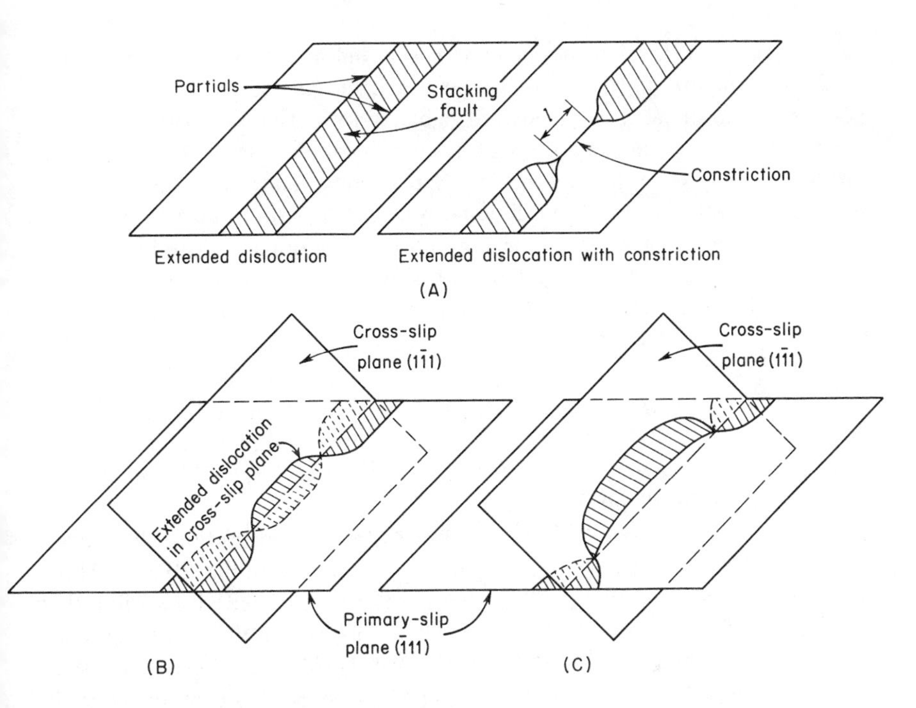

Fig. 20.25 The stages in the cross-slip of an extended screw dislocation. (A) An extended screw dislocation in the primary slip plane undergoes a thermally activated constriction. (B) The constricted dislocation splits into a pair of partials on the cross-slip plane. (C) The dislocation moves under an applied stress in the cross-slip plane. (After Schoeck, G., and Seeger, A., *Defects in Crystalline Solids*, p. 340. The Physical Society, London, [1955].)

this metal and the correspondingly greater stacking-fault width. More work is required to constrict a wider stacking fault than a narrow one, which accounts for the difference in activation energy between copper and aluminum.

The cross-slip mechanism that has just been discussed is believed to be the principal dislocation mechanism involved in the dynamic recovery of face-centered cubic metals (see Section 7.9). It may also be of significance in other crystal forms.

20.11 Creep When More than One Mechanism is Operating. It is usually difficult to conceive of a metal deforming under creep conditions in such a manner that only one creep mechanism is operative. Thus, if dislocations intersect, it is probable that both the mechanism of intersection as well as the movement of jogs caused by the intersection will have to be considered. At the same time, the Peierls stress can play a role and, if the metal has elements in solid solution, the movement of dislocation atmospheres must be considered. Besides, if the metal is face-centered cubic, activated cross-slip may enter the picture. Under certain conditions of testing, however, one or the other of the various mechanisms can control the creep rate, and the major effort in fundamental research has been directed toward finding the specific mechanism that controls the creep rate under a given set of deformation conditions. This, however, is not usually an easy task, for different creep mechanisms ordinarily exhibit different functional relationships between the stress and the creep rate, as well as between the temperature and the creep rate.

Let us assume, for simplicity and the purpose of illustration, that creep in a given metal is dependent on only two mechanisms and that we would like to determine how the resulting creep rate varies with temperature under constant stress. Also, since the creep rate depends on the structure of the metal, let us assume that all creep specimens are to have identical structures at the instant their creep rates are measured. This condition could be obtained, in principle at least, by predeforming a number of identical specimens at the same temperature and stress to the same total strain. At the end of this primary deformation, the temperature of each specimen could then be suddenly changed to some other predetermined value and the specimen allowed to undergo further deformation at the new temperature. During this change of temperature, the stress is assumed to be maintained at its original value. In each case, the change in temperature will result in a change in creep rate, the magnitude of which should be a function of only the temperature.

Let us assume that we are concerned with only two simple mechanisms and that their creep rates can be represented by the following equations:

$$\dot{\epsilon}_1 = A_1 e^{-q_1/kT}$$

$$\dot{\epsilon}_2 = A_2 e^{-q_2/kT}$$

where $\dot{\epsilon}_1$ and $\dot{\epsilon}_2$ are the creep rates, q_1 and q_2 are the activation energies, and A_1 and A_2 are constants (at constant stress and structure).

The two simplest ways that we can portray a pair of creep mechanisms working together are: (1) that they act in parallel, that is, independently of each other or (2) that they act in series, so that the first mechanism must operate before it is possible for the second one to function. In the first, the net strain rate equals the sum of the individual rates, while in the second it equals the total strain, obtained when the mechanisms operate one after the other, divided by the total time needed for both mechanisms to function, or

$$\dot{\epsilon}_t = \frac{\epsilon_1 + \epsilon_2}{\tau_1 + \tau_2}$$

where $\dot{\epsilon}_t$ is the net strain rate, ϵ_1 and ϵ_2 are the unit strains obtained when the respective mechanisms operate individually, and τ_1 and τ_2 are the average time intervals between operations of the respective mechanisms. The above equation shows clearly that when creep mechanisms are in series, the one with the slower rate, or longer delay period, controls the creep rate. On the other hand, the faster rate should control when mechanisms work in parallel, for then

$$\dot{\epsilon}_t = \dot{\epsilon}_1 + \dot{\epsilon}_2$$

where $\dot{\epsilon}_1$ and $\dot{\epsilon}_2$ are the respective creep rates for the individual mechanisms.

Statements are often seen in the literature to the effect that, when mechanisms are in series, the one with the slower rate controls and that when they are in parallel, the one with the faster rate controls. However, the fact that the relative magnitude of the strain rates of any two mechanisms is a function of a specific set of experimental conditions is not always specified. A change in temperature, stress, or structure may very well change the relative importance of two mechanisms. An illustrative example follows, based on a consideration of mechanisms in parallel (independent).

According to the assumption that the mechanisms are independent of each other, we have the observed creep rate $\dot{\epsilon}_t$ equal to the sum of the individual rates, or

$$\dot{\epsilon}_t = \dot{\epsilon}_1 + \dot{\epsilon}_2$$

Now let us rewrite the exponentials in terms of powers of 10, and the activation energies as Q_1 and Q_2 expressed in calories per mole:

$$\dot{\epsilon}_t = A_1 10^{-Q_1/2.3RT} + A_2 10^{-Q_2/2.3RT}$$

or

$$\dot{\epsilon}_t = A_1 10^{-Q_1/4.6T} + A_2 10^{-Q_2/4.6T}$$

Also, for the sake of convenience, let $Q_1 = 2Q_2 = 46{,}000$ cal per mole. These are reasonable assumptions in terms of observed activation energies in real crystals. On this basis we have

$$\dot{\epsilon}_t = A_1 10^{-10{,}000/T} + A_2 10^{-5000/T}$$

If A_1 is equal to or smaller than A_2, then the term on the right, corresponding to the second mechanism, will be larger at all temperatures than that corresponding to the first mechanism. In this case, the measured strain rate would appear to be that of the second mechanism and it would be said that the second mechanism controlled the creep rate.

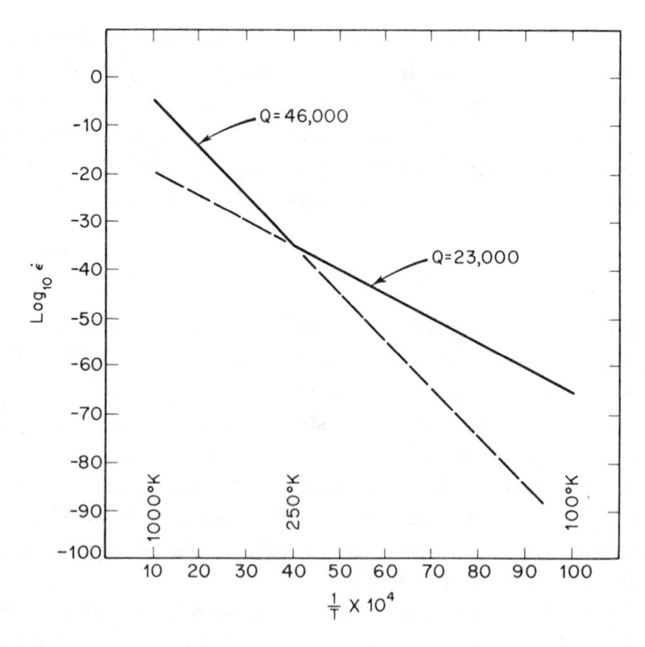

Fig. 20.26 Creep rate as a function of temperature for two different hypothetical mechanisms.

Next consider the alternative case where A_1 is much larger than A_2. The term with the larger activation energy now has the larger coefficient. Under this condition, it is possible for the controlling mechanism to change with temperature. Thus, for example, let us choose A_1 equal to 10^{+5}, and A_2 equal to 10^{-15}. The temperature variation of the two strain rates using these assumptions is shown in Fig. 20.26, where the logarithm of the strain rate is plotted against the reciprocal of the absolute temperature. This figure shows a change in mechanism at 250°K. The mechanism with the lower activation energy has the higher rate below 250°K, while that with the higher activation energy has the higher rate above 250°K. In regard to Fig. 20.26, it is important to notice that if we were to add the two straight lines so as to obtain the total strain rate, $\dot{\epsilon}_t$, the resultant curve would differ very little from the curve defined by the two solid lines that intersect at 250°K. This is because we are dealing with exponential functions which vary rapidly with $1/T$. As an example, at 250°K the individual strain rates $\dot{\epsilon}_1$ and $\dot{\epsilon}_2$ both equal 10^{-35}. Their sum is 2×10^{-35}, or $10^{-34.7}$, which plots on the logarithmic scale of the figure as a point at −34.7. To the scale of the figure, this point is not significantly different from −35, the plotted value of $\log_{10} \dot{\epsilon}_1$ or $\log_{10} \dot{\epsilon}_2$. At any other temperature, it will be found that the total strain is even more closely represented by the solid lines in Fig. 20.26.

A shift in the controlling mechanism similar to that discussed in the above paragraph is frequently observed in polycrystalline diffusion studies. At high temperatures, volume diffusion with a higher activation energy controls the diffusion rate and, at lower temperatures, surface diffusion with the lower activation energy controls. In this regard, it is interesting that grain-boundary diffusion yields to volume diffusion at high temperatures because there are many more atoms capable of taking part in bulk diffusion than there are capable of contributing to the grain-boundary component. As in the above-assumed creep case, both types of diffusion add their effects, with the result that a plot of diffusion data usually exhibits a break like that in Fig. 20.26. Finally, with respect to a change in the controlling mechanism for creep, it is necessary to assume that the dislocations of the crystal are capable of undergoing the reaction with the higher activation energy in many more ways than they are capable of undergoing the reaction with the lower activation energy. More accurately, it should perhaps be said that if all possible dislocation mechanisms of the two types should act simultaneously, the net strain due to all mechanisms with the higher activation energy would have to be much larger than that due to the mechanisms with the lower activation energy.

Let us consider the curves of Fig. 20.26 as though they represented a plot of experimentally determined creep data. These data, taken on their face value and knowing nothing of the mechanisms, would indicate that the activation energy for creep undergoes a sudden change at 250°K, somewhat in the fashion shown in Fig. 20.27. Experimentally determined activation energies for creep in real metals sometimes show such sudden changes. One metal in which systematic and thorough studies of the activation energies for creep have been made is aluminum. The results of some of these investigations, made by Dorn and his

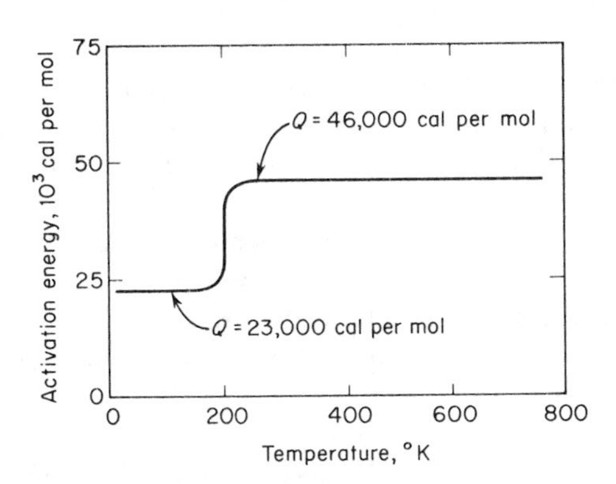

Fig. 20.27 Plot of activation energy versus temperature for hypothetical data of 20.25.

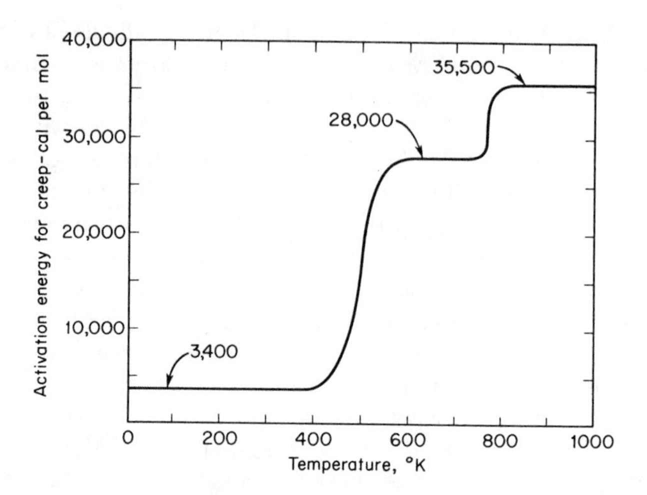

Fig. 20.28 Activation energy for creep in single crystals of pure aluminum deforming by simple glide on octahedral planes, {111} ⟨110⟩. (After Lytton, J. L., Shepard, L. A., and Dorn, J. E., *Trans. AIME,* **212** 220 [1958].)

co-workers,[30] are shown in Fig. 20.28. These data represent creep studies under carefully controlled conditions using single-crystal specimens so oriented that they could slip only on octahedral planes, {111}⟨110⟩. It is interesting that three different discrete activation energies were observed, with values 3400, 28,000, and 35,500 cal per mole, corresponding to the temperature intervals 0 to 400°K, 600 to 750°K, and 800°K to the melting point. The largest activation energy (35,500 cal per mole), corresponding to the highest temperature range, is very close to the activation energy for self-diffusion in aluminum. For this reason, the rate controlling creep mechanism, at high temperatures in these single-crystal specimens, has been ascribed to dislocation climb. The 28,000-cal-per-mole value in the intermediate temperature range is close to the computed activation energy for activated cross-slip in aluminum. Creep specimens in this temperature range also exhibit slip lines showing a high degree of cross-slip activity, giving further credence to the belief that the controlling mechanism at intermediate temperatures is cross-slip. A specific mechanism has not yet been clearly identified for the 3400-cal-per-mole activation energy observed at very low temperatures. Both the Peierls stress and dislocation intersections have been suggested[31] as possible controlling mechanisms in this temperature range, with the most recent point of view favoring the former.[32]

30. Lytton, J. L., Shepard, L. A., and Dorn, J. E., *Trans. AIME,* 212 220 (1958).
31. *Ibid.*
32. Walton, D., Shepard, L. A., and Dorn, J. E., *Trans. AIME,* 221 458 (1961).

20.12 Grain-Boundary Shear. It is well known that metal grains in polycrystalline material are able to move relative to one another. Under ideal conditions, this form of deformation can be confined to a very narrow region adjacent to the grain boundaries, so that it would appear that flow actually occurred along the boundary surface. The shear direction is the direction lying in the boundary with the maximum resolved shear stress. Furthermore, grain-boundary shearing in polycrystalline metals has been found to be discontinuous; that is, the flow is not smooth and continuous under load, but spasmodic and irregular; and it occurs to different degrees at different points along a boundary, and in varying amounts at a given point with respect to time. The boundary thus yields, then stops moving, and then, at a later time, may move again. Frequently, however, the deformation occurs in a region extending a finite distance into one of the grains adjoining the boundary.[33] In general, this effect becomes greater the higher the temperature of testing.

It is relatively easy to demonstrate grain-boundary shear by engraving a grid, or network of lines, on the surface of a creep specimen. It is necessary, however, to prepare the surface of the specimen beforehand by a suitable metallographic technique in order to make the grain structure of the metal visible. Upon testing the specimen under conditions favorable for grain-boundary shear, the lines of the grid are sheared where they cross grain boundaries, as is indicated schematically in Fig. 20.29 for a bicrystal. The magnitude of the grain-boundary shear in

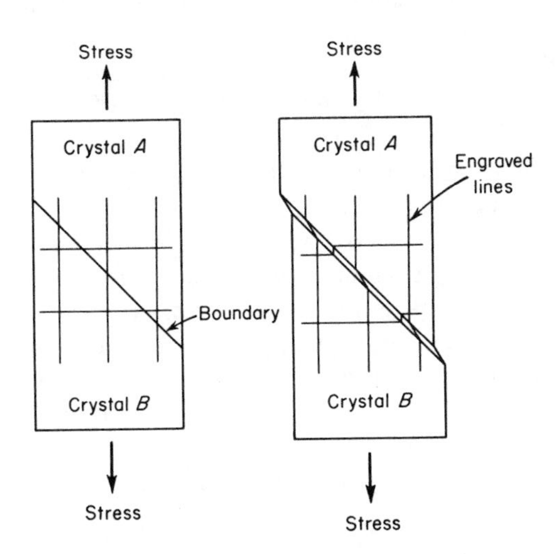

Fig. 20.29 Grain-boundary shear can be revealed by the shear displacements of grid lines engraved on a specimen surface.

33. Rhines, F. N., Bond, W. E., and Kissel, M. A., *Trans. ASM,* 48 169 (1956).

polycrystalline specimens can also be estimated, from measurements at the surface, of the relative displacements of grains in a direction normal to the surface. Both methods of measurement suffer from the fault that they are based on surface measurements. While they may give a reasonably accurate picture of the shear at the surface, they really tell us very little about what happens to grains in the interior of the specimen. An attempt to estimate the magnitude of grain-boundary sliding in the center of a specimen has been made by Rachinger,[34] using as an index the change in shape of the grains during deformation. If there is no grain-boundary sliding and all the deformation occurs by slip inside the grains, then the average grain should suffer the same relative elongation and contraction as the specimen itself. On the other hand, if all the deformation occurs by shear and rotation of the grains relative to one another, the average grain should maintain its original shape. Working with aluminum and specimens slowly strained (0.1 percent per hr) at 300°C, he observed no appreciable change in the shape of the grains after approximately 50 percent strain, and concluded that the contribution of grain-boundary sliding (under the test conditions) was of the order of 90 to 95 percent of the total strain. This is an interesting observation, but should not be taken as generally illustrative of the order of magnitude of the internal grain-boundary-shearing component, since Rachinger's results may be capable of alternative interpretations[35,36] that predict a lower magnitude for the grain-boundary component.

Except for Rachinger's results, most measurements of the grain-boundary component of the total strain come from surface measurements. These usually report[37] the grain-boundary component as varying from a small percentage to as high as 30 percent of the total strain. Occasionally surface measurements have been made which have indicated a grain-boundary component approaching Rachinger's measurements. In brief, the present picture seems to be that grain-boundary shearing is a significant deformation mechanism at high temperatures, the importance of which has not been completely evaluated.

The magnitude of the grain-boundary shear in a particular metal is a function of several variables – temperature, strain, stress, etc. – with the functional relationships to these variables only poorly established. One relation about which there seems to be reasonable agreement is that between the grain-boundary component and the total strain. It has been observed[37] in a number of cases that the ratio of grain-boundary shear to the total shear (ϵ_{gb}/ϵ_t) is approximately a constant over the period of a creep test.

34. Rachinger, W. A., *Jour. Inst. of Metals,* **81** 33 (1952).
35. McLean, D., *Grain Boundaries in Metals*. Oxford University Press, London, 1957.
36. Conrad, H., *Mechanical Behavior of Materials at Elevated Temperatures,* p. 218. McGraw-Hill Book Company, Inc., New York, 1961.
37. Gifkins, R. C., *Fracture*, p. 579. Technology Press and John Wiley and Sons, Inc., New York, 1959.

Some experimental results bearing directly on grain-boundary shearing have been reported by Rhines, Bond, and Kissel.[38] According to them, grain-boundary shear should not be considered as simple sliding of one grain over another, but rather as plastic deformation which occurs in the material alongside the boundaries proper. In this regard, they point out that grain-boundary shearing does not become a significant deformation mechanism until the metal is heated into the temperature range where recovery can occur. Also, it is well recognized that in the deformation of polycrystalline materials, the greatest amount of plastic distortion usually occurs in the region adjoining the grain boundaries. Recovery should normally occur first in the areas where the strain energy is greatest. Thus, during deformation at elevated temperatures, it is to be expected that the regions adjacent to grain boundaries will undergo softening due to recovery effects long before the material in the centers of the grains. This softening of the metal next to boundaries permits additional deformation to occur in this region, which appears as what is commonly called grain-boundary shear.

20.13 Intercrystalline Fracture. At low temperatures (below one-half the absolute melting point) metals customarily fail by fractures passing through the interiors of grains, that is, by transcrystalline fractures. Intercrystalline fractures at these temperatures, therefore, are the exception and are usually associated with some structural irregularity, for example, the presence of a brittle intercrystalline film or a form of corrosion that weakens grain boundaries. At high temperatures, intercrystalline fractures or fractures that run along grain boundaries, are the rule rather than the exception, and a metal, which at low temperatures fails with a normal transcrystalline failure, is apt to fail by a fracture that passes along the grain boundaries at elevated temperatures. These intercrystalline fractures are closely related to grain-boundary shearing.

High-temperature intercrystalline fractures, at their first inception, can have quite different aspects, depending on the metal and conditions of the test. There are two well-defined forms, however, which are commonly encountered. These are shown schematically in Fig. 20.30 and indicate, as is shown in the left-hand figure, that cavities can form at grain corners (edges where three grains meet), or, as in the right-hand figure, as small oval pores lying along the grain boundaries. These two types of incipient fractures do not necessarily form exclusive of each other, but appear simultaneously in the same specimen under the proper conditions. Complete rupture of a specimen is usually considered to be caused by the growth and coalescence of the small openings. This may result from continued operation of the shearing mechanisms which started the cavities, or, as it has also been proposed, by the condensation of vacancies onto them. In

38. Rhines, F. N., Bond, W. E., and Kissel, M. A., *Trans. ASM,* 48 169 (1956).

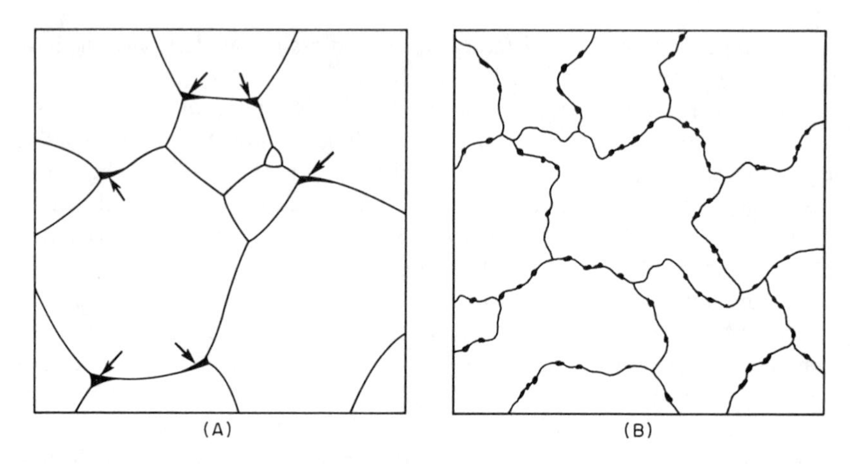

Fig. 20.30 Two ways that intercrystalline fractures start in metals. (A) Wedge-shaped cracks at grain corners (indicated by arrows). (B) Oval cavities along the boundaries.

general, it can be stated that the best present evidence now points[39] to both wedge- and oval-shaped cavities as resulting from grain-boundary sliding.

A number of years ago Zener[40] proposed a mechanism for the formation of wedge-shaped cavities at grain edges. He suggested that if a grain boundary subject to shear stress were to shear (Fig. 20.31), then a sufficiently high-stress concentration might result at the end of the relaxed boundary (at the grain corner) to cause the formation of a microcrack. A cavity nucleation cannot occur in this manner unless the stress concentration exceeds the cohesive strength of the grain boundary. It also cannot occur if the stress concentration at the end of the boundary is relieved by plastic flow in the grain ahead of the boundary. One way that the stress concentration can be relieved is shown in Fig. 20.32, where slip on properly oriented planes in the grain ahead of the boundary resulted in lattice bending which accommodates the shear strain along the boundary. This type of phenomenon has often been observed on the surface of a creep specimen where the resulting upheaval of the surface in the grain ahead of the slipped boundary is called a *fold.*[41] An example of a double fold is shown in Fig. 20.33. An alternate mechanism which can prevent the opening of a crack at a grain-boundary corner is for the boundary to migrate away from the stressed region before the stress concentration has risen to the point where a pore can form.

39. Conrad, H., *Mechanical Behavior of Materials at Elevated Temperatures*, p. 149. McGraw-Hill Book Company, Inc., New York, 1961.

40. Zener, C., *Elasticity and Anelasticity of Metals*. The University of Chicago Press, Chicago, Ill., 1948.

41. Chang, R. C., and Grant, N. J., *Trans. AIME*, 194 619 (1952).

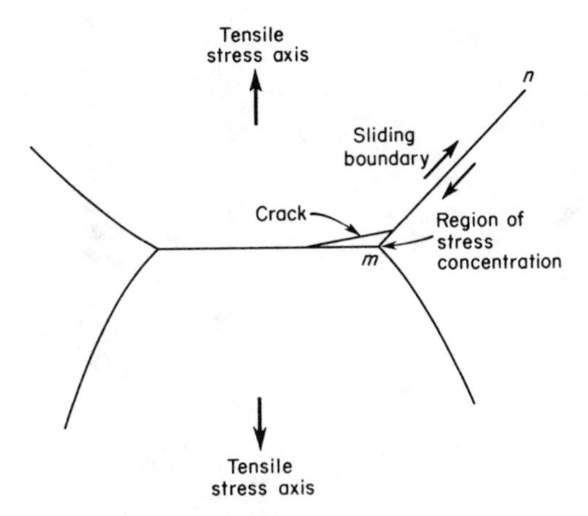

Fig. 20.31 Zener's method for the formation of wedge-shaped cavities. Sliding along the boundary *mn* relaxes the shear stress along the boundary and concentrates the stress at the grain corner.

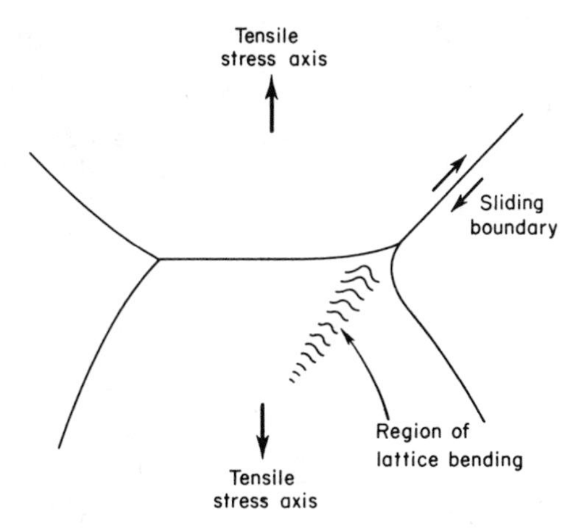

Fig. 20.32 The stress concentration at a grain corner may be relieved by plastic deformation in the grain ahead of the sliding boundary.

Now consider the small oval-shaped fractures that form along grain boundaries. These are believed to be due to discontinuities lying on or in the grain boundary. Figure 20.34 shows one mechanism[42] that has been proposed for the formation of the type of pore under discussion. Here the boundary is assumed to have a preexisting jog that interferes with the normal shear deformation along the boundary. Displacement along the boundary causes stress concentrations at

42. Chen, C. W., and Machlin, E. S., *Acta Met.*, **14** 655 (1956); *Trans. AIME*, **209** 829 (1957).

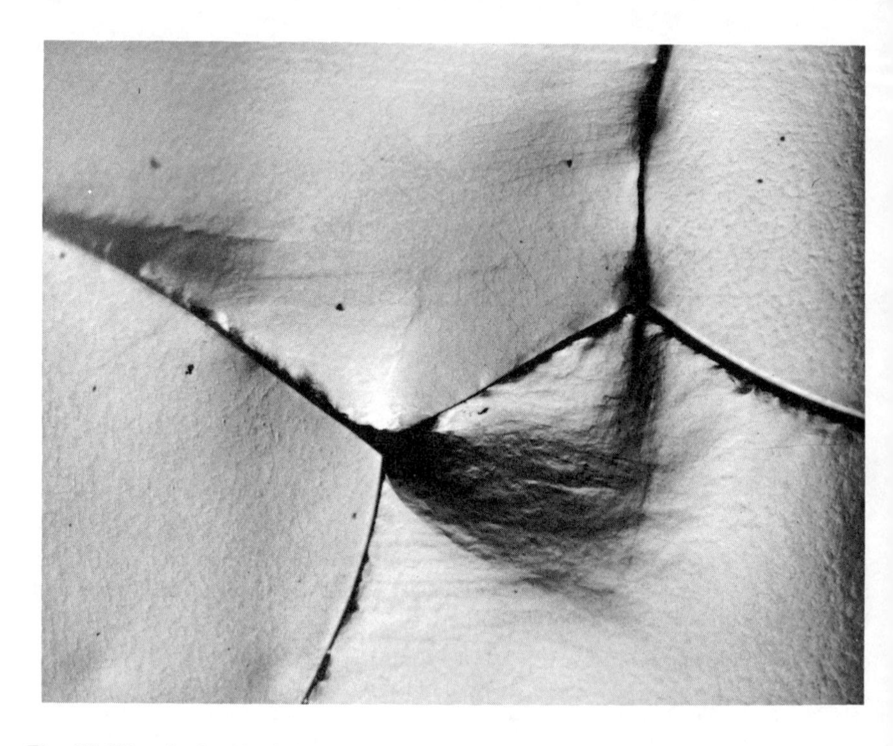

Fig. 20.33 A double fold on the surface of a 20 percent Zn-Al creep specimen tested at 500° F and 2300 psi. 75x. (Chang, H. C., and Grant, N. J., *Trans. AIME,* **206** 544 [1956].)

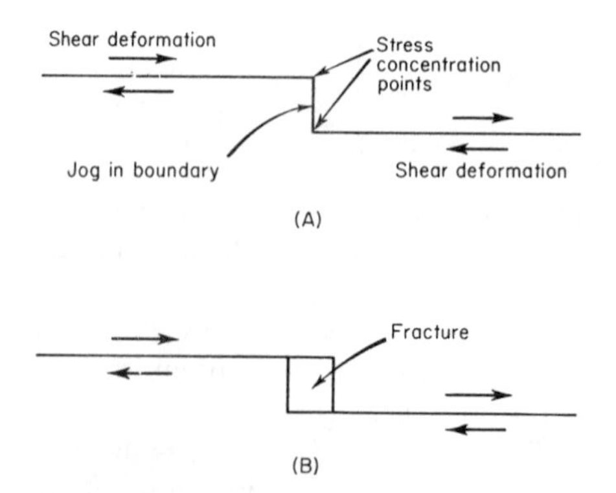

Fig. 20.34 One method for the formation of cavities along grain boundaries.

the jog, as indicated in Fig. 20.34A, with the result that a crack develops, as shown in Fig. 20.34B. As in the case of wedge-shaped openings at grain corners, the formation of these pores can be prevented either by plastic flow in the grains to relieve the stress concentration, or by migration of the boundary away from the point of stress concentration. It should be noticed that boundary migration is also a form of stress relief.

Any factor that tends to increase the resistance to shear inside the grains relative to that at the grain boundaries, and makes boundary migration more difficult, tends to promote grain boundary fracture. In general, slip inside grains can be made more difficult by work hardening, solid-solution hardening, precipitation hardening, etc. Precipitation and solid-solution hardening can also greatly restrict boundary migration. The magnitude of these effects depends on the alloy in question. It is also quite possible to have a combination of factors that can cause grain-boundary fracture to be prevalent in some intermediate elevated-temperature range and then, on further increase in the temperature, for a change to occur in the balance of strength between the grains and the boundaries, that causes a reversion to the transcrystalline fracture mechanism. An increase in ductility will normally accompany the reappearance of transcrystalline fracture.

The tendency for transcrystalline fracture to occur in a limited temperature interval has been shown by Rhines and Wray[43] to be able to produce a ductility minimum in a plot of tensile elongation against the temperature. They have designated this as the *intermediate temperature ductility minimum*. Figure 20.35 shows their data for monel metal (nominal composition 65 percent Ni–35 percent Cu). Notice the similarity of this curve to that associated with "blue brittleness" shown in Fig. 19.47. The fracture mechanisms, however, differ. At the intermediate temperature ductility minimum, the intergranular fracture also greatly reduces the reduction in area, so that the effect is truly an embrittlement. The blue brittle phenomenon is associated with necking, so that there is always a rather large reduction in area at the blue brittle temperature.

20.14 The Creep Curve. The ordinary creep test is a constant-temperature, constant-load tensile experiment during which strain is measured as a function of time. When a creep test is run for the purpose of determining creep mechanisms, it is usually necessary to vary or to decrease the load during the test in order to compensate for the reduction in the cross-section area of the specimen as it deforms, and thus maintain a constant stress on the metal. Most creep tests are not run at constant stress, but rather under a constant, or fixed, load. These tests are those whose primary purpose is to obtain engineering design data. A dead-load test is easier to perform since a rather simple loading arrangement can be used, whereas a constant-stress apparatus may take consider-

43. Rhines, F. N., and Wray, P. J., *Trans. ASM,* 54 117 (1961).

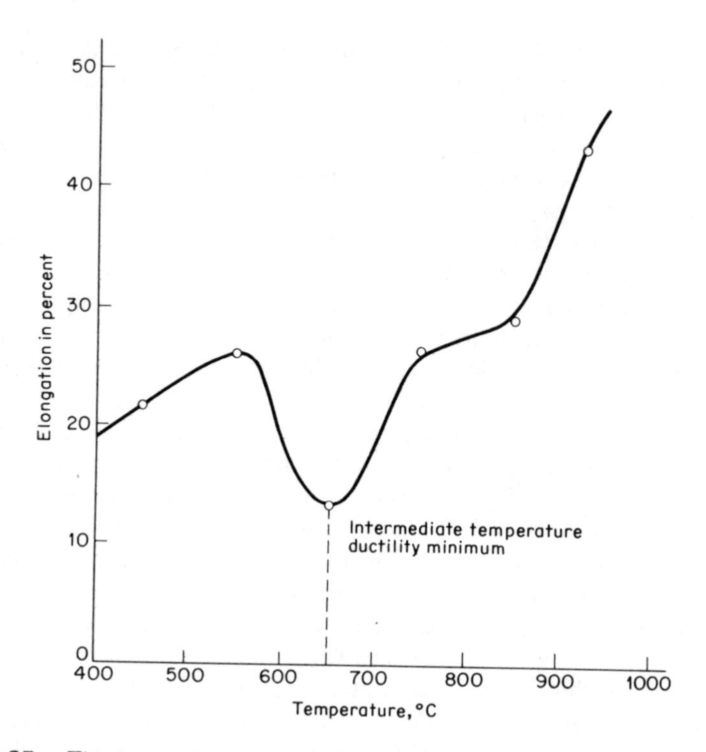

Fig. 20.35 The intermediate temperature ductility minimum in monel metal. This elongation minimum is associated with intergranular fracture and is accompanied by a corresponding minimum in the reduction in area. Specimens deformed rapidly at an average loading rate of 600 lbs. per sec. (From data of Rhines, F. N., and Wray, P. J., *Trans. ASM*, **54** 117 [1961].)

ably more instrumentation. Constant-load tests are usually acceptable for engineering purposes for two reasons. First, most engineering data involve very slow creep rates, so that during the period of an ordinary creep test only a nominal reduction in the cross-section of the specimen occurs and the stress is nearly constant over the period of the test. Second, a constant-load test agrees with the usual engineering practive of plotting short-time tensile-test data using a stress equal to the instantaneous load divided by the original cross-section area of the specimen.

Constant-load creep tests usually give curves which contain, in varying degree, the three basic stages designated in Fig. 20.36. These stages occur during the progress of the creep test and do not include the instantaneous strain that appears while the load is being applied to the specimen. The latter strain is given by the line *oa*. Stage I, at the beginning of a test, is a region of decreasing slope,

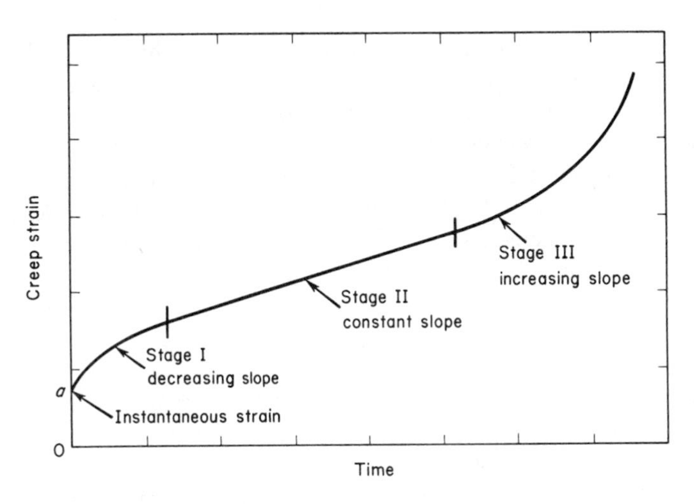

Fig. 20.36 The three stages of the creep curve.

while stage II is that part of the creep curve with an approximately constant slope. Stage III follows stage II and represents the final section of the curve in which the slope rises rapidly until the specimen fractures.

At the start of the test (stage I), the strain rate decreases rapidly from a very large initial value, which is indicative of the fact that structural alterations, occurring in the metal as it deforms, act to retard normal-flow processes. These changes occur primarily in the number, type, and arrangement of dislocations. In brief, at the start of a creep test, strain hardening decreases the flow rate. In order to develop a basic understanding of the phenomena involved in creep tests, it is necessary to point out at this time that, simultaneously with the the natural tendency of the metal to harden during deformation, softening reactions may occur which will tend to oppose or nullify strain hardening. Examples of typical softening processes are recovery and recrystallization.

Let us now assume that we are specifically concerned with a constant-load creep test in which softening reactions play only a minor role, so that they may be neglected, and also that when failure occurs, it is transcrystalline and not intercrystalline. Such a creep test might be performed on many metals at relatively low temperatures (below the normal recovery and recrystallization range). Throughout a test of this type a metal should continually increase in hardness, but it should be noticed that at the same time the cross-section area of the specimen is being constantly reduced as a result of the strain which the specimen is undergoing. Because the load on the specimen is fixed, the stress in the gage section must rise. Since the creep rate of a metal is a sensitive function of the stress, a rise in stress will normally cause a corresponding increase in the creep rate. Two opposing factors thus operate: (1) strain hardening which tends to decrease the creep rate and (2) a rise in stress which tends to increase it. During that part of the test where the slope becomes effectively constant (second-stage creep), an approximate balance may be assumed to have been achieved between these factors. This balance cannot go on indefinitely, however, especially if the specimen begins to undergo necking. A point will eventually be reached where the increase in creep rate caused by the rising stress overcomes the strain hardening. At this point, stage three begins and the flow process becomes catastrophic, with deformation accelerating until the specimen fractures.

The above analysis is predicated on the assumption that the material comprising the specimen is metallurgically stable, meaning that it does not undergo softening reactions during the course of the test. As emphasized by Lubahn and Felgar,[44] the second-stage creep rate in a constant-load test of a structurally stable material usually has no real significance. The apparent straight-line part of the creep curve may disappear if one runs a corresponding test under constant-stress conditions. That this is true can be seen in Fig. 20.37. Notice that the constant-stress creep curve in this figure shows an ever-decreasing slope, signifying a constantly decreasing strain rate, or a continuously increasing amount of strain hardening. In effect, the constant-stress creep curve of a structurally stable material should never reach the second stage of creep.

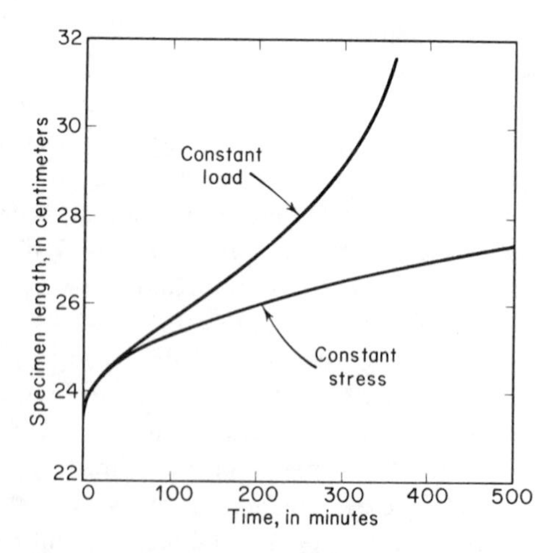

Fig. 20.37 Comparison of a constant-load creep test with a constant-stress creep test. Specimens were lead wires with the same initial lengths and loads. (After Andrade, E. N. daC., *Proc. Roy. Soc.,* **84** 1 [1910].)

44. Lubahn, J. D., and Felgar, R. P., *Plasticity and Creep of Metals.* John Wiley and Sons, Inc., New York, 1961.

At this point it is well to consider the significance of the concept of a structurally stable material. In such a metal, since recovery processes are precluded, the density of dislocations and other lattice defects should normally increase with increasing strain. As a result, dislocation motion may be expected to become progressively more difficult and, while thermal activation may still permit flow to occur, the rate of flow should decrease.

In theory, a constant-stress creep test should be much more meaningful than a constant-load test. However, there is a serious factor which more or less limits its applicability – this is in regard to the problem of necking. As long as a specimen undergoes uniform elongation, the stress on the gage section is the same at all cross-sections. When necking occurs, the stress can no longer be the same at all cross-sections. If the load is reduced so as to keep the stress at the neck at the original value, then the stress is lowered at all other points along the gage length. Under these conditions, the strain measured over the length of the gage section is no longer a representative strain. Further, since a neck can act as a stress concentrator, the strain at the point of the neck is also not simply related to the applied stress. In general, it can be concluded that constant-stress tests are basically limited in their applicability to that part of a test in which the elongation is uniform.

Let us now consider the effects which metallurgical reactions that occur during deformation can have on a creep test. Most of these, for example, recovery, recrystallization, and overaging in a precipitation-hardened alloy, tend to increase the creep rate. However, it is possible, as in the case of an alloy that precipitates a second phase during deformation, to have a reaction that lowers the creep rate.

For convenience, let us assume that a given specimen undergoes only one reaction during deformation and that this is recovery. Then, in a typical constant-load test, strain hardening will be opposed not only by the increased flow rate caused by the increasing stress associated with the reduction in the specimen cross-section, but also by the softening due to recovery. The second-stage creep rate now represents an effective balance among these three factors. Also, it is important to recognize that in the range of temperature where recovery occurs, grain-boundary shearing and intercrystalline fracture are liable to occur. When fracture by this mode begins, it has an effect equivalent to necking in that the cross-section area capable of carrying the load is reduced. This will, in general, accelerate the creep rate and promote the third stage. In this regard, the sudden occurrence of recrystallization, because it may be assumed to remove strain hardening rapidly, can also induce the third stage of creep.

In any given constant-load creep experiment, the three stages of creep vary in importance or magnitude according to the factors enumerated above. The variations in the shape of the creep curve may alternatively be considered in

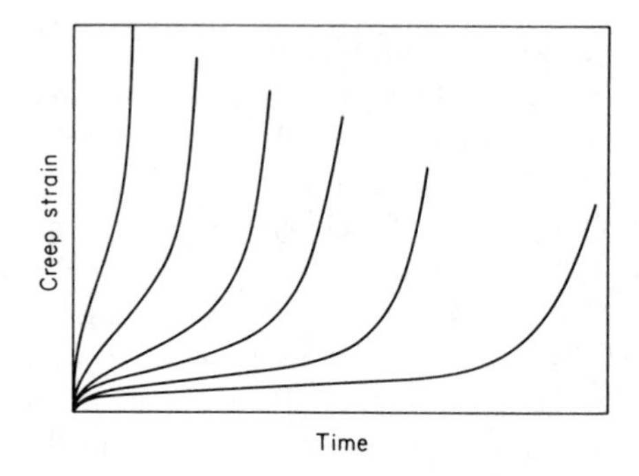

Fig. 20.38 Creep curves for one material under various stresses at one temperature. (Griffiths, W. T., *Jour. Roy, Aero. Soc.*, **52** 1 [1948].)

terms of the temperature and the initial applied stress as variables. Both variables tend to alter the shape of the creep curve in a similar manner. Thus, Fig. 20.38 shows a series of creep curves corresponding to different stresses, all measured at the same temperature, while Fig. 20.39 shows a set corresponding to creep at different temperatures in which the specimens were subject to the same stress. These figures clearly show that high stresses and high temperatures reduce the extent of the primary stage and practically eliminate the secondary stage, with the result that the creep rate accelerates almost from the beginning of the test. At intermediate stresses (Fig. 20.38), or intermediate temperatures (Fig. 20.39), the primary and secondary stages become more clearly defined. Next, the curves for the lowest temperature and stress in Figs. 20.39 and 20.38 respectively show long, well-defined second-stage portions. The second stage of creep, therefore, becomes most pronounced under conditions that favor a very slow creep rate. Finally, the curves of Figs. 20.38 and 20.39 show another important charac-

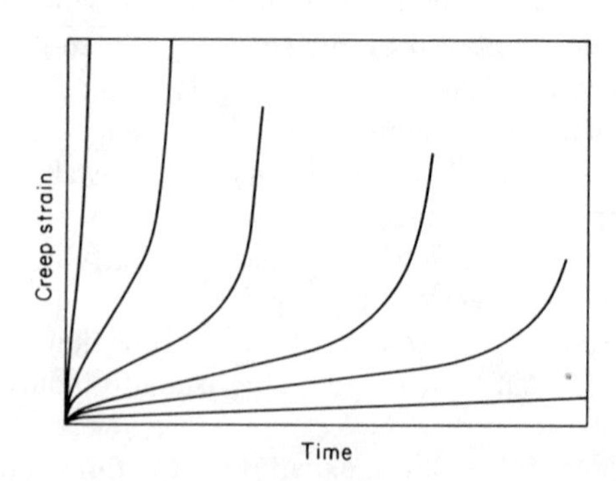

Fig. 20.39 Creep curves for one material at various temperatures under the same stress. (Griffiths, W. T., *Jour. Roy. Aero. Soc.*, **52** 1 [1948].)

teristic of most creep phenomena, namely, the longer the life of a creep test, the smaller the total extension that the metal undergoes.

20.15 Practical Applications of Creep Data. The engineering uses for metals at high temperatures frequently demand materials that can be used at moderate stresses for long periods of time, while the parts made from these materials maintain rather close dimensional tolerances. As an example, the blades in a steam turbine designed for very high-temperature application might be expected to last about eleven years, with an allowable change in length not to exceed 0.1 percent. Since eleven years is approximately 100,000 hr, the material cannot creep at a rate in excess of 10^{-8} in. per in. per hr. In an application of this nature, we are, of course, dealing with a creep curve of the type shown near the bottom of either Fig. 20.38 or 20.39. Here the second stage of creep extends for long periods of time and the slope of the second stage has practical significance. The slope in this region is commonly expressed in percentage of creep per 1000 hr, or some similar unit. Because it is not generally feasible to extend creep tests for periods as long as eleven years, it is common practice to extrapolate the information obtained from shorter tests in order to estimate the expected amount of deformation (at a given stress and temperature) in the longer time interval. Figure 20.40 illustrates the basic method used in obtaining the desired information. In the figure, it is assumed that a creep test has been carried out to 10,000 hr and that the plotted data show that a constant slope has been attained. At any point within this region of constant slope, the total creep strain is given by the equation

$$\epsilon = \epsilon_0 + \epsilon' t$$

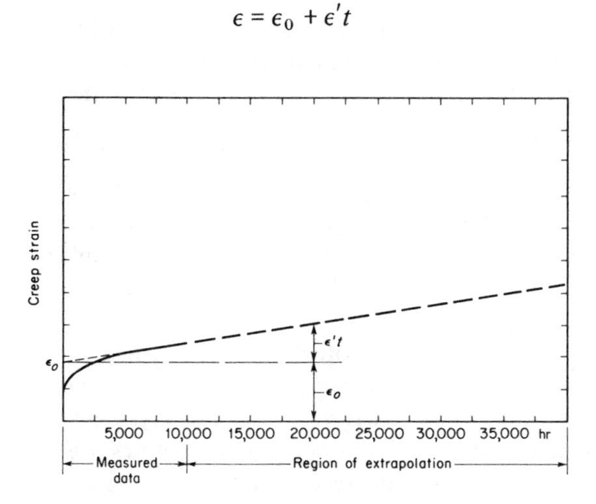

Fig. 20.40 Creep data are often extrapolated to far beyond the limits of actual measurement.

where ϵ_0 is defined in the figure as the ordinate intercept of the extended second-stage creep line, ϵ' is the slope of this line, and t is the time. If it is assumed that the second stage of creep lasts as long as the expected life of the material in its practical application (eleven years in our illustrative problem), then the creep strain for the given time interval at the specified temperature and stress can be obtained directly from the above equation. Extrapolations of this nature are not without danger, as it is possible for a metal to undergo an unpredicted structural modification well before the end of the period (as long as eleven years). This, of course, might induce the third stage of creep and lead to premature fracture. This hazard is greater the greater the magnitude of the extrapolation, or the shorter the period of testing used in obtaining the original creep data. For this reason, it is common practice to test specimens at least 1000 hr and, in some cases, as long as 10,000 hr. The compilation of data for design purposes is costly, since for a single material it may be necessary to make five to eight tests at one temperature. This will usually define the effect of stress on the creep rate at this temperature. Similar series are then carried out at a number of different testing temperatures to define the role of temperature. To investigate a single alloy may thus involve from as many as 30 to 40 specimens, each of which ties up a unit of creep apparatus for a period of approximately 6 wk (1000 hr).

Creep data of the type discussed above involve experiments in which specimens are tested for a small fraction of their expected lives. The principal item of information obtained from the tests is the second-stage creep rate. Another creep test often used for obtaining data of engineering significance is the stress-rupture test. In this case, the specimen is deformed to complete fracture at constant load and temperature, and the principal item of information obtained is the time to rupture. In obtaining stress-rupture data, several series of specimens are tested at different temperatures. In these series, each specimen is subjected to a different load. Hypothetical data for a set of specimens ruptured at the same temperature are shown in Fig. 20.41. Notice that it gives two intersecting straight lines on a log-log plot. The line to the right corresponds to specimens with lower loads and longer times in which to rupture, while that to the left corresponds to higher loads and shorter times in which to rupture. The most common cause for a break in the slope of the stress-rupture curve is a shift in the fracture mechanism as conditions of the test change. This is indicated in the diagram, but it should be noted that the change in mechanism with decreasing stress does not occur sharply. Specimens corresponding to points near the break often show mixed fractures; partly intercrystalline and partly transcrystalline. Finally, notice that transcrystalline fractures are favored by short (high strain-rate) tests, while intercrystalline fractures are promoted by long (slow strain-rate) tests. This is in agreement with what has been said earlier about the effect of stress on intercrystalline fracture: low stresses and slow strain rates favor intercrystalline fractures.

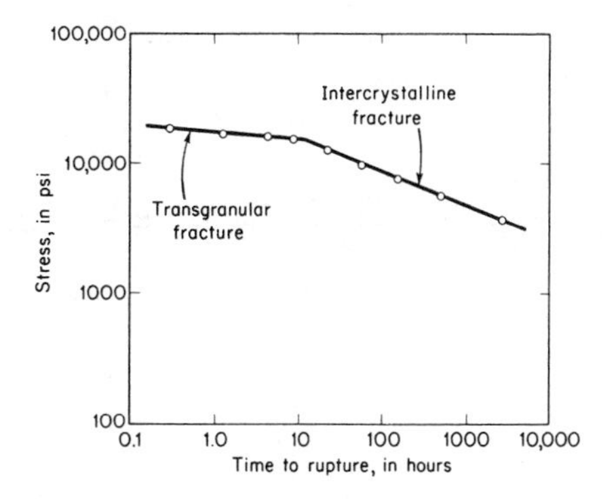

Fig. 20.41 Typical stress-rupture diagram showing the data from a series of specimens tested at the same temperature under different loads.

Coupled with the shift in fracture mechanism indicated in Fig. 20.41 is a corresponding change in the ductility of the specimens. Transcrystalline fractures are usually characterized by necking of the specimen and high ductility. Intercrystalline fractures show little necking and often fail with little total elongation. The difference in ductility and nature of the fractures is schematically indicated in Fig. 20.42. It should be pointed out, however, that if extensive cracking occurs in the specimen away from the point of actual rupture, a specimen that fractures intercrystallinely may still show an apparently large elongation.

Another important characteristic of intercrystalline fracture may be deduced from Fig. 20.41, namely, that this type of fracture greatly reduces the effective life of the metal. Thus, if the line in Fig. 20.41, corresponding to specimens rupturing with transcrystalline fractures, is extrapolated to the right of the break, one obtains a relationship between stress and rupture time which should

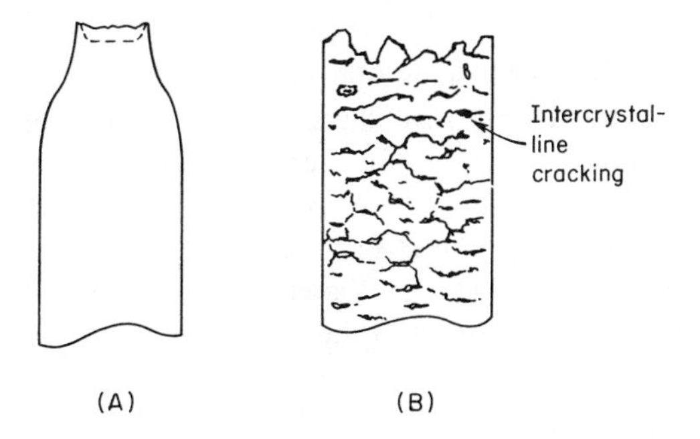

Fig. 20.42 The difference between (A) transcrystalline and (B) intercrystalline fracture. Transcrystalline, or ductile fracture is usually accompanied by necking and a cup and cone fracture, whereas intercrystalline fracture is characterized by a general disintegration of the structure as it parts along grain boundaries.

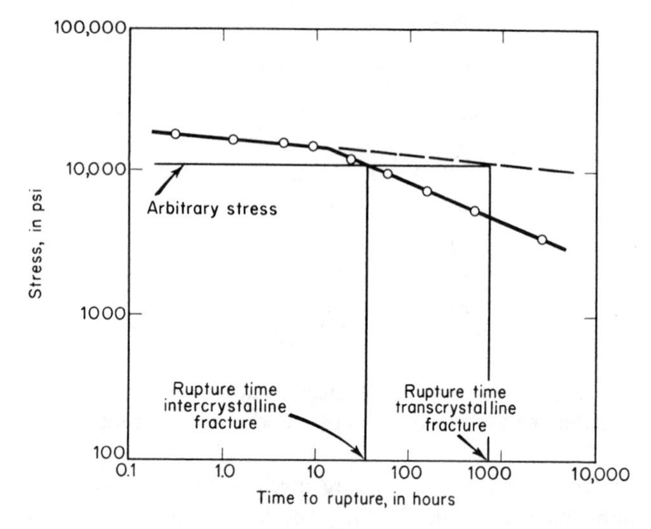

Fig. 20.43 Intercrystalline fracturing effectively reduces the rupture life of a creep specimen.

be valid if intercrystalline fracture did not occur. This extrapolated line (Fig. 20.43) shows that at a given stress the rupture time for the extrapolated transcrystalline fracture is much longer than that for the actually observed intercrystalline fracture.

Stress-rupture data are often extrapolated to determine the total life of a given material (when the desired life is very long) in a manner similar to that discussed above for creep data. There is also danger in this type of extrapolation, for an unexpected break in the curve of the type shown in Fig. 20.41 may well occur somewhere between the end of the measured data and the expected life of the material.

20.16 Creep-Resistant Alloys. Most successful commercial high-temperature metals are complicated alloys, for it has been found that, as a rule, alloys have better over-all properties for use at elevated temperatures than do pure metals. It is beyond the scope of this text to discuss these materials in detail. Typical compositions can be found in handbooks.[45] An understanding of the principles on which these alloys are based has only recently been possible. Several recent papers have summarized the principles.[46,47]

45. See *ASM Metals Handbook*, 1961.
46. Freeman, J. W., and Corey, C. L., *Creep and Fracture of Metals at High Temperatures*, p. 157. Philosophical Library, New York, 1957.
47. Constant, A., and Delbart, G., *Creep and Fracture of Metals at High Temperatures*, p. 191. Philosophical Library, New York, 1957.

The problem of developing creep-resistant alloys is basically twofold: the resistance of both the grains and the grain boundaries to flow should be increased, while recovery or softening effects should be minimized. In the temperature ranges where most commercial alloys are used, plastic deformation may be considered to be controlled by the motion of dislocations inside grains. Most of the well-known methods of hardening crystals are, therefore, applicable to creep-resistant alloys. These include solid-solution hardening, precipitation hardening, and hardening by cold work.

In general, commercial heat-resistant alloys are based on a matrix which is a solid solution. This matrix is normally more creep resistant than a pure metal because the elements in solid solution make the movement of dislocations through the lattice more difficult. The basic strengthening mechanism is probably that associated with the formation of solute atom atmospheres around dislocations. In most cases, the matrix grains are further strengthened by precipitation hardening. As at low temperature, the precipitate is most effective in interfering with the motion of dislocations when the particle size is very small and the particles are widely distributed in large numbers throughout the lattice. The precipitates in commercial alloys are frequently carbides, nitrides, oxides, or intermetallic compounds. Finally, prior plastic deformation of the matrix is another important method of increasing the high-temperature strength of the matrix. This cold-working for added creep resistance does not have to be carried out at room temperature, but must occur at some temperature below that at which the metal recrystallizes during deformation. Work hardening, at an intermediate elevated temperature, is often called *warm-working*. There is evidence[48] that the amount of work hardening for best results is limited, at least in certain alloys, to about 15 to 20 percent deformation. Overworking should lead to a lower recrystallization temperature and a greater tendency to soften during a creep test.

All three of the above-mentioned hardening mechanisms (solution, precipitation, and work hardening) are unstable relative to a rise in temperature. Their use for the purpose of increasing the creep resistance of metals is therefore limited to temperature ranges within which the strengthening mechanisms are stable.

In solid-solution hardening, an increase in temperature increases the diffusion rates of the solute atoms in the dislocation atmospheres and at the same time tends to disperse the atoms of the atmosphere. Both effects make the movement of dislocations easier. Recrystallization, of course, can completely remove the effects of cold-working (warm-working) and thereby increase the creep rate at a given stress and temperature. The occurrence of recrystallization during the progress of a creep test can be considered harmful to creep strength even in

48. Freeman, J. W., and Corey, C. L., *Creep and Fracture of Metals at Elevated Temperatures*, p. 157. Philosophical Library, New York, 1957.

metal that was not work hardened prior to the start of the test. Recrystallization is a softening phenomenon and will remove work hardening which has occurred during the progress of a test. In general, we can conclude that alloying elements that raise the recrystallization temperature of the matrix should be beneficial to high-temperature strength.

The strengthening of a creep-resistant alloy by a finely dispersed precipitate is subject to the same softening processes as occur in any normal precipitation-hardening alloy. Heating of the alloy to too high a temperature can cause re-solution of the precipitates. Alternatively, at somewhat lower temperatures, overaging of the precipitate can occur. In either case, the creep resistance of the metal will be sharply decreased. An interesting aspect of certain creep-resistant alloys is that the original precipitate may go back into solution while a second precipitate forms. If this second precipitate forms with the proper particle, size, and coherency, the alloy may be strengthened rather than weakened. This dissolving of one precipitate while another forms is the same type of phenomenon as that which occurs in the secondary hardening in high-speed tool steels. In this latter case, it was shown earlier that cementite particles, formed on tempering the steel at low temperatures, dissolved at higher temperatures while, simultaneously, a precipitation of complex alloy carbides occurred. These latter precipitates give the metal its excellent high-temperature hardness.

In increasing the creep resistance of an alloy, one must not lose sight of the grain boundaries. If the metal near the boundaries is not strengthened at the same time as that in the interior of the grains, then intercrystalline fracture is promoted, with accompanying losses in ductility and strength. High-strength creep-resistant alloys, as a class, tend to have limited ductility because of a greater improvement in the strength of the grains than in the strength of the boundaries. In some cases it has been possible, however, to control the compositions of an alloy in such a manner that the resistance to grain-boundary shear is increased as well as the resistance to slip in the grains. Certain alloys with high-creep resistance accordingly exhibit good ductility.

20.17 Alloy Systems. As the temperature of a metal is raised, thermal vibrations become more and more intense and atoms have an increasing tendency to shift lattice positions either by vacancy diffusion, or by jumping across grain boundaries from one crystal to another. The effectiveness of thermal vibrations in bringing about atomic movements is inversely related to the strength of the bonds holding an atom in its lattice site. The stronger the bonds, the higher the temperature needed to overcome the bonds. In a pure metal a rough measure of the strength of its atomic bonds is its melting point. In the same manner, the recrystallization temperature of a pure metal serves as an approximate measure of the limiting temperature, below which atomic movements are normally too slow to allow appreciable deformation to occur by

Table 20.1 Highest Temperature at which Certain Heat-Resistant Alloys Can Be Used.*

Base Metal	*Melting Point*	*Temperature for Useful Strength of Best Alloys*	*Percentage of Absolute Melting Point*
		Light Alloys	
	°F	°F	%
Mg	1200	650	67
Al	1220	550	60
Ti	3100	1200	46
		Superalloys	
Fe (martensitic)	2800	1350	56
Fe (austenitic)	2800	1600	63
Ni	2650	1960	78
Co	2720	1900	74
		Refractory Alloys	
Cb	4470	2200	54
Mo	4760	2650	59
W	6170	2550	45

* From Jahncke, L. P., and Frank, R. G., *Met. Prog.*, 74 77 (Nov. 1958).

creep. It is not surprising that in a pure metal the recrystallization temperature and the melting point are nearly proportional to each other, with the recrystallization temperature usually falling between 0.35 to 0.45 of the absolute melting point.[49] Thus, the more or less upper limit of high-temperature usefulness of a pure metal may be defined by its recrystallization temperature, and lies somewhere between 0.35 and 0.45 of its absolute melting point.

According to the principles outlined in the preceding section, the alloying of pure metals permits us to raise their effective temperature of usefulness. A measure of the maximum useful temperature in an alloy can be arbitrarily defined as the temperature at which it just has the ability to sustain a stress of 10,000 psi for 100 hr without fracturing. Table 20.1 lists certain alloys according to this criterion. While this measure of usefulness has no absolute significance, it does give an approximate means of evaluating the effectiveness of alloying on certain pure metals. Notice that in Table 20.1, the best results, as measured in percentage of the absolute melting point, have been obtained in alloys of nickel and cobalt with values of 74 and 76 percent. Alloys of these

49. Jahncke, L. P., *Metal Prog.*, 72 113 (1957).

metals are much used in the temperature range between 1000°F and 2000°F, but above this range one is more or less forced to consider alloys of the refractory metals (metals with very high melting points), namely, columbium, molybdenum, and tungsten.

The use of refractory metals at high temperatures poses serious difficulties. First, they are very difficult to fabricate into engineering components. This is due, in part at least, to their very high melting points, which is the primary factor making them interesting in the first place. High melting points imply high recrystallization temperatures and, therefore, high hot-working temperatures. Hot-working of these metals is, therefore, very difficult. Since they are also body-centered cubic, they possess, in general with iron, a tendency toward brittleness at low temperatures. This limits their ability to be easily cold-worked. Finally, however, the biggest present unsolved problem is their very high oxidation rates at elevated temperatures. In particular, molybdenum has a very high oxide vapor pressure, so that as the oxide forms at elevated temperatures, it vaporizes off the surface, causing a rapid wasting away of the metal. A number of approaches have been tried in an attempt to control the oxidation problem of refractory metals. One of these is by alloying, another by the use of oxidation-resistant coatings to protect the surface of the metal. One form of coating that has received serious attention is a ceramic layer bonded to the surface of the metal.

Problems

1. (*a*) At an engineering strain of 0.05, the engineering stress is 50,000 psi. Compute the corresponding values of the true stress and true strain.
   (*b*) Do the same for an engineering strain of 0.30 and an engineering stress of 70,000 psi.
2. (*a*) Derive an expression for the true strain in terms of measurements of the specimen diameter.
   (*b*) A specimen whose original diameter was 0.50 in. necked down to a diameter of 0.20 in. before fracture. The fracture stress (engineering) was 50,000 psi. Compute the true fracture stress and true fracture strain (at the neck).
   (*c*) What was the percentage reduction in area of the specimen?
3. A well-known empirical equation for describing a tensile stress-strain curve is

$$\sigma_t = k\epsilon_t^{\ m}$$

where m is called the work hardening exponent. In some face-centered cubic metals this relationship gives a reasonable approximation of the stress-strain curve. In the case of silver, m has been shown[50] to vary linearly

50. Carreker, R. P., Jr., *Trans. AIME*, 209 112 (1957).

with decreasing temperature; increasing from 0.10 at 900°K to about 0.60 at 78°K. The true stress-true strain curves in Fig. 20.11 correspond to this data.

(*a*) Determine values of k for deformation at 78°K and at 673°K. Does k vary with deformation temperature?

(*b*) Determine a simple physical interpretation of the parameter k.

4. Demonstrate that if $\sigma = k\epsilon_t^m$ describes the stress-strain curve, then the uniform elongation, or the strain obtained prior to necking, is equal to m. That is,

$$\epsilon_{t_{\text{necking}}} = m$$

5. The normal experimental procedure for obtaining the work hardening exponent m involves plotting the logarithm of the true stress against the logarithm of the true strain. If $\sigma = k\epsilon_t^m$ holds, then a straight line will be obtained whose slope is m. In most cases, the plotted data give curves that are only approximately linear. There is also another important problem concerned with the use of m as an index of work hardening. This can be seen with the aid of the following problem.

(*a*) The relationship $\sigma_t = k\epsilon_t^m$ assumes σ_t, is zero when ϵ is zero. Suppose that a given metal conforms to this relationship and its flow-stress at a strain of 0.01 is 5000 psi, and at a strain of 0.10 it is 25,000 psi. Compute its indicated work hardening parameter using the relation

$$m = \frac{\ln \dfrac{\sigma_{t_2}}{\sigma_{t_1}}}{\ln \dfrac{\epsilon_2}{\epsilon_1}}$$

(*b*) Now assume that the same metal has an initial yield stress which raises the level of the flow-stress during the entire test so that at $\epsilon_t = 0.01$, $\sigma_t = 50{,}000$ psi, and at $\epsilon_t = 0.10$, $\sigma_t = 70{,}000$ psi. Compute m.

(*c*) Compare the two values of m and the actual amounts of work hardening. What can you conclude about m as a reliable index of work hardening?

6. (*a*) Write an equation for the data in Fig. 20.4. Extrapolate this equation to a stress of 100,000 psi and determine the predicted value of the dislocation density.

(*b*) The relationship between flow-stress and dislocation density is often written in the form

$$\sigma = \sigma_o + \alpha E b \rho^{1/2}$$

where E is the modulus of elasticity. If $E = 16.8 \times 10^6$ and $b = 2.95$ Å, compute α.

7. At a constant temperature the strain rate relationship given in Section 20.9 reduces to $\dot{\epsilon} = A' \sinh v\tau^*/KT$, where $A' = 2Ae^{-q/kT}$. A given metal crystal deforms at a strain rate of 10^{-10}sec^{-1} when the shear stress τ^* is 100 gm/mm^2 and the temperature is 900°C. The activation volume for the mechanism controlling the creep rate is $20\ b^3$ where b is the Burgers vector and equal to 3 Å. Compute the creep rate when τ^* is (1) doubled, (2) increased by a factor of 10, and (3) increased by a factor of 50. (*Note:* assume v remains constant).
8. If the forest dislocation density is 10^{10} cm/cm^3, compute the activation volume for dislocation intersections. Assume $b = 3$ Å and give your answer in units of b^3.
9. Assuming that the strain-rate relationship

$$\dot{\epsilon} = Ae^{-(q - v\tau^*)/KT}$$

is valid and that A is a constant, show that (at a given value of the stress)

$$v = kT \frac{\ln \frac{\dot{\epsilon}_2}{\dot{\epsilon}_1}}{\tau^*_2 - \tau^*_1}$$

where $\dot{\epsilon}_1$ and τ^*_1 are the strain rate and the effective stress at one strain rate, and $\dot{\epsilon}_2$ and τ^*_2 are the corresponding values at another strain rate.
10. The above expression for the activation volume is often employed to measure v, using a change in strain rate experiment. Normally it is assumed that the change in flow-stress $\Delta\tau$ (or $\Delta\sigma$), caused by a change in strain rate, is the same as $\Delta\tau^*$; that is, $(\tau^*_2 - \tau^*_1)$. What is the justification for this assumption?
11. In a face-centered cubic metal that is subject to a number of cycles wherein the strain rate in a tensile test is varied back and forth by a factor of 10, the

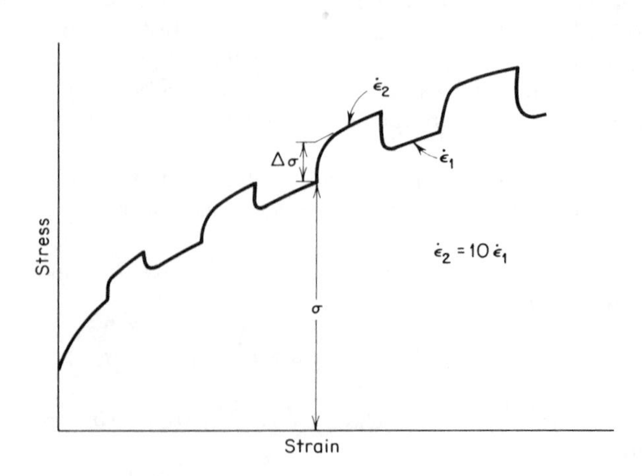

stress-strain curve will appear like that in the accompanying figure. Note that, with increasing deformation, the change in flow-stress $\Delta\sigma$ increases. In some cases it has been found that $\Delta\sigma/\sigma$ is a constant, where σ is the flow-stress at the time the strain rate is changed. This is known as the Cottrell-Stokes law.[51]

Furthermore, in f.c.c. metals, $\Delta\sigma$ is normally smaller than, for example, in b.c.c. metals where, as a rule, $\Delta\sigma$ is not a function of σ.

Discuss these factors in relation to the size of the activation volume in b.c.c. metals and in f.c.c. metals. In other words, in which metals should v be larger?

Also, in f.c.c. metals, how should the activation volume vary with strain (or stress)? Is this consistent with the assumption that in these metals the controlling mechanism is dislocation intersections?

51. Cottrell, A. H., and Stokes, R. J., *Proc. Roy. Soc.*, **A233** 17 (1955).

Appendices

APPENDIX A

Angles Between Crystallographic Planes in the Cubic System*
(in degrees).

HKL	*hkl*					
100	100	0.00	90.00			
	110	45.00	90.00			
	111	54.74				
	210	26.56	63.43	90.00		
	211	35.26	65.90			
	221	48.19	70.53			
	310	18.43	71.56	90.00		
	311	25.24	72.45			
	320	33.69	56.31	90.00		
	321	36.70	57.69	74.50		
110	110	0.00	60.00	90.00		
	111	35.26	90.00			
	210	18.43	50.77	71.56		
	211	30.00	54.74	73.22	90.00	
	221	19.47	45.00	76.37	90.00	
	310	26.56	47.87	63.43	77.08	
	311	31.48	64.76	90.00		
	320	11.31	53.96	66.91	78.69	
	321	19.11	40.89	55.46	67.79	79.11
111	111	0.00	70.53			
	210	39.23	75.04			
	211	19.47	61.87	90.00		
	221	15.79	54.74	78.90		
	310	43.09	68.58			
	311	29.50	58.52	79.98		
	320	36.81	80.78			
	321	22.21	51.89	72.02	90.00	

*Abstracted from IMD Special Report Series, No. 8, "Angles Between Planes in Cubic Crystals," R. J. Peavler and J. L. Lenusky, The Metallurgical Society, *AIME*, 29 W. 39 St., New York, N.Y.

APPENDIX A *(continued)*

HKL	*hkl*								
210	210	0.00	36.87	53.13	66.42	78.46	90.00		
	211	24.09	43.09	56.79	79.48	90.00			
	221	26.56	41.81	53.40	63.43	72.65	90.00		
	310	8.13	31.95	45.00	64.90	73.57	81.87		
	311	19.29	47.61	66.14	82.25				
	320	7.12	29.74	41.91	60.25	68.15	75.64	82.87	
	321	17.02	33.21	53.30	61.44	68.99	83.14	90.00	
211	211	0.00	33.56	48.19	60.00	70.53	80.40		
	221	17.72	35.26	47.12	65.90	74.21	82.18		
	310	25.35	40.21	58.91	75.04	82.58			
	311	10.02	42.39	60.50	75.75	90.00			
	320	25.06	37.57	55.52	63.07	83.50			
	321	10.89	29.20	40.20	49.11	56.94	70.89	77.40	83.74
		90.00							
221	221	0.00	27.27	38.94	63.51	83.62	90.00		
	310	32.51	42.45	58.19	65.06	83.95			
	311	25.24	45.29	59.83	72.45	84.23			
	320	22.41	42.30	49.67	68.30	79.34	84.70		
	321	11.49	27.02	36.70	57.69	63.55	74.50	79.74	84.89
310	310	0.00	25.84	36.87	53.13	72.54	84.26		
	311	17.55	40.29	55.10	67.58	79.01	90.00		
	320	15.26	37.87	52.12	58.25	74.74	79.90		
	321	21.62	32.31	40.48	47.46	53.73	59.53	65.00	75.31
		85.15	90.00						
311	311	0.00	35.10	50.48	62.96	84.78			
	320	23.09	41.18	54.17	65.28	75.47	85.20		
	321	14.76	36.31	49.86	61.09	71.20	80.72		
320	320	0.00	22.62	46.19	62.51	67.38	72.08		
	321	15.50	27.19	35.38	48.15	53.63	58.74	68.24	72.75
		77.15	85.75	90.00					
321	321	0.00	21.79	31.00	38.21	44.41	49.99	64.62	69.07
		73.40	85.90						

APPENDIX B

Angles Between Crystallographic Planes for Hexagonal Elements.*

		Be	Ti	Zr	Mg	Zn	Cd
HKiL	*hkil*	*c/a* = 1.5847	1.5873	1.5893	1.6235	1.8563	1.8859
0001	$10\bar{1}8$	12.88	12.90	12.92	13.19	15.00	15.23
	$10\bar{1}7$	14.65	14.67	14.69	14.99	17.03	17.28
	$10\bar{1}6$	16.96	16.99	17.01	17.35	19.66	19.95
	$10\bar{1}5$	20.10	20.13	20.15	20.55	23.21	23.53
	$10\bar{1}4$	24.58	24.62	24.65	25.11	28.19	28.56
	$20\bar{2}7$	27.60	27.64	27.67	28.17	31.48	31.89
	$10\bar{1}3$	31.38	31.42	31.45	32.00	35.55	35.98
	$20\bar{2}5$	36.20	36.25	36.29	36.87	40.61	41.06
	$10\bar{1}2$	42.46	42.50	42.54	43.15	46.98	47.43
	$20\bar{2}3$	50.66	50.70	50.74	51.31	55.02	55.44
	$10\bar{1}1$	61.34	61.38	61.41	61.92	64.99	65.33
	$20\bar{2}1$	74.72	74.74	74.76	75.07	76.87	77.07
	$10\bar{1}0$	90.00	90.00	90.00	90.00	90.00	90.00
	$21\bar{3}2$	67.55	67.59	67.61	68.04	70.57	70.86
	$21\bar{3}1$	78.33	78.35	78.36	78.60	80.00	80.15
	$21\bar{3}0$	90.00	90.00	90.00	90.00	90.00	90.00
	$11\bar{2}8$	21.61	21.64	21.71	22.09	24.89	25.24
	$11\bar{2}6$	27.85	27.88	27.91	28.42	31.75	32.16
	$11\bar{2}4$	38.39	38.44	38.47	39.07	42.87	43.32
	$11\bar{2}2$	57.75	57.79	57.82	58.37	61.69	62.07
	$11\bar{2}1$	72.50	72.52	72.54	72.93	72.92	75.18
	$11\bar{2}0$	90.00	90.00	90.00	90.00	90.00	90.00
$10\bar{1}0$	$21\bar{3}0$	19.11	19.11	19.11	19.11	19.11	19.11
	$11\bar{2}0$	30.00	30.00	30.00	30.00	30.00	30.00
	$01\bar{1}0$	60.00	60.00	60.00	60.00	60.00	60.00

*A. Taylor, and Leber, Sam, *Trans., AIME*, **200**, 190, (1954).

APPENDIX C

Indices of the Reflecting Planes for Cubic Structures.

Simple Cubic	*Body-centered Cubic*	*Face-Centered Cubic*
{100}	—	—
{110}	{110}	—
{111}	—	{111}
{200}	{200}	{200}
{210}	—	—
{211}	{211}	—
{220}	{220}	{220}
{221}	—	—
{300}	—	—
{310}	{310}	—
{311}	—	{311}
{222}	{222}	{222}
{320}	—	—
{321}	{321}	—
{400}	{400}	{400}
{322}	—	—
{410}	—	—
{330}	{330}	—
{411}	{411}	—
{331}	—	{331}
{420}	{420}	{420}
{421}	—	—
{332}	{332}	—

APPENDIX D

Conversion Factors and Constants

Conversion Factors

Electron volts to calories	1 eV = 23,000 cal/mol
Electron volts to ergs	1 eV = 1.6×10^{-12} erg
Calories to joules	1 cal = 4.185 joule
Joules to ergs	1 joule = 10^7 erg
Psi to gm/mm^2	1 psi = 0.704 gm/mm^2

Constants

Constant	*Symbol*	*Value*
Avogadro's number	N	6.02×10^{23} molecules/mole
Boltzmann's constant	k	1.38×10^{-16} erg/°K
Charge on the electron	e	4.80×10^{-10} statcoulombs
Electron volt	eV	1.6×10^{-12} erg
Mass of the electron	m	9.11×10^{-28} gm
Planck's constant	h	6.63×10^{-27} erg sec
	$h/2\pi$	1.06×10^{-27} erg sec
Universal gas constant	$R = kN$	2 cal/mol/°K

APPENDIX E

Twinning Elements of Several of the More Important Twinning Modes.

Type of Metal	K_1	η_1	K_2	η_2	*Observed in*
Body-centered cubic	{112}	$\langle 11\bar{1} \rangle$	$\{11\bar{2}\}$	⟨111⟩	
Face-centered cubic	{111}	$\langle 11\bar{2} \rangle$	$\{11\bar{1}\}$	⟨112⟩	
Hexagonal close-packed	$\{10\bar{1}1\}$	$\langle 10\bar{1}\bar{2} \rangle$	$\{10\bar{1}3\}$	$\langle 30\bar{3}2 \rangle$	Mg, Ti
	$\{10\bar{1}2\}$	$\langle 10\bar{1}\bar{1} \rangle$	$\{10\bar{1}\bar{2}\}$	$\langle 10\bar{1}1 \rangle$	Be, Cd, Hf, Mg, Ti, Zn, Zr
	$\{10\bar{1}3\}$	$\langle 30\bar{3}\bar{2} \rangle$	$\{10\bar{1}\bar{1}\}$	$\langle 10\bar{1}2 \rangle$	Mg
	$\{11\bar{2}1\}$	$\langle 11\bar{2}\bar{6} \rangle$	(0002)	$\langle 11\bar{2}0 \rangle$	Hf, Ti, Zr
	$\{11\bar{2}2\}$	$\langle 11\bar{2}\bar{3} \rangle$	$\{11\bar{2}\bar{4}\}$	$\langle 22\bar{4}3 \rangle$	Ti, Zr

APPENDIX F

*Selected values of intrinsic stacking-fault energy γ_I, twin-boundary energy γ_T, grain-boundary energy γ_G, and crystal-vapor surface energy γ for various materials in ergs/cm^2.**

Metal	γ_I	γ_T	γ_G	γ
Ag	17[1,*]		790[8]	1,140[4]
Al	~200[2]	120[2]	625[8]	
Au	55[1,*]	~10[10]	364[8]	1,485[4]
Cu	73[1,*]	44[9]	646[5]	1,725[4]
Fe		190[4]	780[8]	1,950[8]
Ni	~400[1,3]		690[8]	1,725[8]
Pd	180[3]			
Pt	~ 95[3]	196[6]	1,000[6]	3,000[6]
Rh	~750[3]			
Th	115[3]			
W				2,900[7]

1. T. Jøssang and J. P. Hirth, *Phil. Mag.*, 13 657 (1966).
2. R. L. Fullman, *J. Appl. Phys.*, 22 448 (1951).
3. I. L. Dillamore and R. E. Smallman, *Phil. Mag.*, 12 191 (1965).
4. D. McLean, "Grain Boundaries in Metals," Oxford University Press, Fair Lawn, N.J., 1957, p. 76.
5. N. A. Gjostein and F. N. Rhines, *Acta Met.*, 7 319 (1959).
6. M. McLean and H. Mykura, *Surface Science*, 5 466 (1966).
7. J. P. Barbour et al., *Phys. Rev.*, 117 1452 (1960).
8. M. C. Inman and H. R. Tipler, *Met. Reviews*, 8 105 (1963).
9. C. G. Valenzuela, *Trans. Met. Soc. AIME*, 233 1911 (1965).
10. T. E. Mitchell, *Prog. Appl. Mat. Res.*, 6 117 (1964).

* From Hirth, J. P. and Lothe, J., *Theory of Dislocations*, p. 764, McGraw-Hill Book Company, New York, 1968. Used by permission.

APPENDIX G

The International System of Units

In January 1972 the Metallurgical Transactions announced that the use of the International System of Units would be required for papers to be published in this journal. It is possible that other journals will follow suit. The essentials relating to these units, as published* in *Metallurgical Transactions*,** 1972, vol. 3, pp. 356-358, follow.

* Based on NBS Special Publication 330 (now out of print) which contains additional information on the background and use of the SI system. Another useful reference is the ASTM Metric Practice Guide E380-70.

** Metallurgical Transactions is a joint publication of the Metallurgical Society of the American Institute of Mining, Metallurgical and Petroleum Engineers and the American Society for Metals. Permission granted for the reproduction of this material by AIME.

I. Base Units and Symbols The base units of the International System are collected in Table 1 with their names and their symbols.

The general principle governing the writing of unit symbols:

Roman (upright) type, in general lower case, is used for symbols of units; if however the symbols are derived from proper names, capital roman type is used [for the first letter]. *These symbols are not followed by a full stop (period).*

Unit symbols do not change in the plural.

TABLE 1 SI base units

Quantity	*Name*	*Symbol*
length	metre	m
mass	kilogram	kg
time	second	s
electric current	ampere	A
thermodynamic temperature	kelvin	K
luminous intensity	candela	cd
amount of substance	mole	mol

II. Derived Units *Expressions* Derived units are expressed algebraically in terms of base units by means of the mathematical symbols of multiplication and division. Several derived units have been given special names and symbols which may themselves be used to express other derived units in a simpler way than in terms of the base units.

Derived units may therefore be classified under three headings. Some of them are given in Tables 2, 3, and 4.

TABLE 2 Examples of SI derived units expressed in terms of base units

Quantity	SI unit: Name	SI unit: Symbol
area	square metre	m^2
volume	cubic metre	m^3
speed, velocity	metre per second	m/s
acceleration	metre per second squared	m/s^2
wave number	1 per metre	m^{-1}
density, mass density	kilogram per cubic metre	kg/m^3
concentration (of amount of substance)	mole per cubic metre	mol/m^3
activity (radioactive)	1 per second	s^{-1}
specific volume	cubic metre per kilogram	m^3/kg
luminance	candela per square metre	cd/m^2

TABLE 3 SI derived units with special names

Quantity	SI unit: Name	Symbol	Expression in terms of other units	Expression in terms of SI base units
frequency	hertz	Hz		s^{-1}
force	newton	N		$m \cdot kg \cdot s^{-2}$
pressure	pascal	Pa	N/m^2	$m^{-1} \cdot kg \cdot s^{-2}$
energy, work, quantity of heat	joule	J	$N \cdot m$	$m^2 \cdot kg \cdot s^{-2}$
power, radiant flux	watt	W	J/s	$m^2 \cdot kg \cdot s^{-3}$
quantity of electricity, electric charge	coulomb	C	$A \cdot s$	$s \cdot A$
potential difference, electromotive force	volt	V	W/A	$m^2 \cdot kg \cdot s^{-3} \cdot A^{-1}$
capacitance	farad	F	C/V	$m^{-2} \cdot kg^{-1} s^4 \cdot A^2$
electric resistance	ohm	Ω	V/A	$m^2 \cdot kg \cdot s^{-3} \cdot A^{-2}$
conductance	siemens	S	A/V	$m^{-2} \cdot kg^{-1} \cdot s^3 \cdot A^2$
magnetic flux	weber	Wb	$V \cdot s$	$m^2 \cdot kg \cdot s^{-2} \cdot A^{-1}$
magnetic flux density	tesla	T	Wb/m^2	$kg \cdot s^{-2} \cdot A^{-1}$
inductance	henry	H	Wb/A	$m^2 \cdot kg \cdot s^{-2} \cdot A^{-2}$
luminous flux	lumen	lm		$cd \cdot sr$
illuminance	lux	lx		$m^{-2} \cdot cd \cdot sr$

TABLE 4 Examples of SI derived units expressed by means of special names

	SI unit		
Quantity	*Name*	*Symbol*	*Expression in terms of SI base units*
dynamic viscosity	pascal second	Pa·s	$m^{-1} \cdot kg \cdot s^{-1}$
moment of force	metre newton	N·m	$m^{2} \cdot kg \cdot s^{-2}$
surface tension	newton per metre	N/m	$kg \cdot s^{-2}$
heat flux density, irradiance	watt per square metre	W/m^2	$kg \cdot s^{-3}$
heat capacity, entropy	joule per kelvin	J/K	$m^{2} \cdot kg \cdot s^{-2} \cdot K^{-1}$
specific heat capacity, specific entropy	joule per kilogram kelvin	J/(kg·K)	$m^{2} \cdot s^{-2} \cdot K^{-1}$
specific energy	joule per kilogram	J/kg	$m^{2} \cdot s^{-2}$
thermal conductivity	watt per metre kelvin	W/(m·K)	$m \cdot kg \cdot s^{-3} \cdot K^{-1}$
energy density	joule per cubic metre	J/m^3	$m^{-1} \cdot kg \cdot s^{-2}$
electric field strength	volt per metre	V/m	$m \cdot kg \cdot s^{-3} \cdot A^{-1}$
electric charge density	coulomb per cubic metre	C/m^3	$m^{-3} \cdot s \cdot A$
electric flux density	coulomb per square metre	C/m^2	$m^{-2} \cdot s \cdot A$
permittivity	farad per metre	F/m	$m^{-3} \cdot kg^{-1} \cdot s^{4} \cdot A^{2}$
current density	ampere per square metre	A/m^2	
magnetic field strength	ampere per metre	A/m	
permeability	henry per metre	H/m	$m \cdot kg \cdot s^{-2} \cdot A^{-2}$
molar energy	joule per mole	J/mol	$m^{2} \cdot kg \cdot s^{-2} \cdot mol^{-1}$
molar entropy, molar heat capacity	joule per mole kelvin	J/(mol·K)	$m^{2} \cdot kg \cdot s^{-2} \cdot K^{-1} \cdot mol^{-1}$

Note.—The values of certain so-called dimensionless quantities, as for example refractive index, relative permeability or relative permittivity, are expressed by pure numbers. In this case the corresponding SI unit is the ratio of the same two SI units and may be expressed by the number 1.

Recommendations

Additional recommendations:

(*a*) The product of two or more units is preferably indicated by a dot. The dot may be dispensed with when there is no risk of confusion with another unit symbol

for example: N·m or N m *but not:* mN

(*b*) A solidus (oblique stroke, /), a horizontal line, or negative powers may be used to express a derived unit formed from two others by division *for example:* m/s, $\frac{m}{s}$ or $m \cdot s^{-1}$

(*c*) The solidus must not be repeated on the same line unless ambiguity is avoided by parentheses. In complicated cases negative powers or parentheses should be used

for example: m/s^2 or $m \cdot s^{-2}$ — *but not:* m/s/s

$m \cdot kg/(s^3 \cdot A)$ or $m \cdot kg \cdot s^{-3} \cdot A^{-1}$ — *but not:* $m \cdot kg/s^3/A$

Supplementary Units

These may be regarded either as base units or as derived units.

TABLE 5 SI supplementary units

Quantity	*SI unit*	
	Name	*Symbol*
plane angle	radian	rad
solid angle	steradian	sr

The radian is the plane angle between two radii of a circle which cut off on the circumference an arc equal in length to the radius.

The steradian is the solid angle which, having its vertex in the center of a sphere, cuts off an area of the surface of the sphere equal to that of a square with sides of length equal to the radius of the sphere.

Supplementary units may be used to form derived units. Examples are given in Table 6.

TABLE 6 Examples of SI derived units formed by using supplementary units

Quantity	*SI unit*	
	Name	*Symbol*
angular velocity	radian per second	rad/s
angular acceleration	radian per second squared	rad/s^2
radiant intensity	watt per steradian	W/sr
radiance	watt per square metre steradian	$W \cdot m^{-2} \cdot sr^{-1}$

III. Decimal Multiples and Sub-Multiples of SI Units

TABLE 7 SI prefixes

Factor	*Prefix*	*Symbol*	*Factor*	*Prefix*	*Symbol*
10^{12}	tera	T	10^{-1}	deci	d
10^{9}	giga	G	10^{-2}	centi	c
10^{6}	mega	M	10^{-3}	milli	m
10^{3}	kilo	k	10^{-6}	micro	μ
10^{2}	hecto	h	10^{-9}	nano	n
10^{1}	deka	da	10^{-12}	pico	p
			10^{-15}	femto	f
			10^{-18}	atto	a

Recommendations

(*a*) Prefix symbols are printed in roman (upright) type without spacing between the prefix symbol and the unit symbol.

(*b*) An exponent affixed to a symbol containing a prefix indicates that the multiple or sub-multiple of the unit is raised to the power expressed by the exponent,

for example: $1\ \text{cm}^3 = 10^{-6}\ \text{m}^3$
$1\ \text{cm}^{-1} = 10^{2}\ \text{m}^{-1}$

(*c*) Compound prefixes are to be avoided,

for example: 1 nm *but not:* 1 mμm

The Kilogram

Among the base units of the International System, the unit of mass is the only one whose name, for historical reasons, contains a prefix. Names of decimal multiples and sub-multiples of the unit of mass are formed by attaching prefixes to the word "gram."

IV. Outside Units Used with the International System Users of SI will also wish to employ certain units which, although not part of it, are in widespread use. These units play such an important part that they must be retained for general use with the International System of Units. They are given in Table 8.

TABLE 8 Units in use with the International System

Name	*Symbol*	*Value in SI unit*
minute	min	1 min = 60 s
hour	h	1 h = 60 min = 3 600 s
day	d	1 d = 24 h = 86 400 s
degree	°	$1° = (\pi/180)$ rad
minute	′	$1' = (1/60)° = (\pi/10\ 800)$ rad
second	″	$1'' = (1/60)' = (\pi/648\ 000)$ rad
litre	l	1 l = 1 $dm^3 = 10^{-3}\ m^3$
tonne	t	1 t = 10^3 kg

It is likewise necessary to recognize, outside the International System, some other units which are useful in specialized fields of scientific research, because their values expressed in SI units must be obtained by experiment, and are therefore not known exactly (Table 9).

TABLE 9 Units used with the International System in specialized fields

Name	*Symbol*
electronvolt	eV
unified atomic mass unit	u
astronomical unit	AU
parsec	pc

TABLE 10 Units to be used with the International System for a limited time

Name	*Symbol*	*Value in SI units*
nautical mile		1 nautical mile = 1 852 m
knot		1 nautical mile per hour = (1852/3600) m/s
ångström	Å	1 Å = 0.1 nm = 10^{-10} m
are	a	1 a = 1 $dam^2 = 10^2\ m^2$
hectare	ha	1 ha = 1 $hm^2 = 10^4\ m^2$
barn	b	1 b = 100 $fm^2 = 10^{-28}\ m^2$
bar	bar	1 bar = 0.1 MPa = 10^5 Pa
standard atmosphere	atm	1 atm = 101 325 Pa
gal	Gal	1 Gal = 1 $cm/s^2 = 10^{-2}\ m/s^2$
curie	Ci	1 Ci = $3.7 \times 10^{10}\ s^{-1}$
röntgen	R	1 R = 2.58×10^{-4} C/kg
rad	rad	1 rad = 10^{-2} J/kg

V. Units Not Recommended *CGS Units to Avoid* It is in general preferable not to use, with the units of the International System, CGS units which have special names. Such units are listed in Table 11.

TABLE 11 CGS units with special names

Name	*Symbol*	*Value in SI units*
erg	erg	1 erg = 10^{-7} J
dyne	dyn	1 dyn = 10^{-5} N
poise	P	1 P = 1 dyn·s/cm^2 = 0.1 Pa·s
stokes	St	1 St = 1 cm^2/s = 10^{-4} m^2/s
gauss	Gs, G	1 Gs corresponds to 10^{-4} T
oersted	Oe	1 Oe corresponds to $\frac{1000}{4\pi}$ A/m
maxwell	Mx	1 Mx corresponds to 10^{-8} Wb
stilb	sb	1 stilb = 1 cd/cm^2 = 10^4 cd/m^2
phot	ph	1 ph = 10^4 lx

Other Units to Avoid

As regards other units outside the International System, it is in general preferable to avoid them, and to use instead units of the International System. Some of those units are listed in Table 12.

TABLE 12 Other units generally deprecated

Name	*Value in SI units*
fermi	1 fermi = 1 fm = 10^{-15} m
metric carat	1 metric carat = 200 mg = 2 x 10^{-4} kg
torr	1 torr = $\frac{101\ 325}{760}$ Pa
kilogram-force (kgf)	1 kgf = 9.806 65 N
calorie (cal)	1 cal = 4.186 8 J
micron (μ)	1 μ = 1 μm = 10^{-6} m
X unit	
stere (st)	1 st = 1 m^3
gamma (γ)	1 γ = 1 nT = 10^{-9} T
γ	1 γ = 1 μg = 10^{-9} kg
λ	1 λ = 1 μl = 10^{-6} l

List of Important Symbols

a	lattice constant of a crystal
a_A, a_B	activities
b	Burgers vector
c	c-axis constant in hexagonal and tetragonal crystals
c	half-crack length
$\bar{d}$	crystalline interplanar spacing
d	diameter, grain diameter
e	charge on the electron
f	force
f	fraction transformed
f	free energy per atom
h	Planck's constant
$\hbar$	Planck's constant divided by 2π
	Miller indices
	wave number $2\pi/\lambda$
k	Boltzmann's constant
l	distance
m	mass
n	grain-growth exponent
n	number
n	strain-rate sensitivity
n_x, n_y, n_z	quantum numbers
p	pressure
p	momentum
q	activation energy
r	radius or distance
r	rate
r	amount of recovery
s	entropy
t	time
v	velocity
w	weight
x	distance
z	atomic number

A	area
A	Madelung number
A	vibration amplitude
Å	Ångstrom units
B	magnetic flux density
B	mobility
B_s	temperature at which Bainite transformation starts
B_f	temperature at which Bainite transformation ends
C	components
C	concentration
C_p	specific heat at constant pressure
D	grain diameter

D	diffusion coefficient (also $\tilde{D}$, D^*)
E	electric field intensity
E	energy, internal energy
E_F	Fermi energy
F	Gibbs' free energy
F	degrees of freedom
F	force
G	rate of growth
G	crack extension force
G_c	critical crack extension force
G_{Ic}	mode I critical crack extension force
H	magnetic field intensity
H	enthalpy
I	moment of inertia
I	rate of nucleation
J	flux
K	stress intensity factor
K_c	fracture toughness
K_{Ic}	mode I fracture toughness
K_1	first undistorted plane of a twin
K_2	second undistorted plane of a twin
L	length
M	bending moment
M	magnetic moment per unit volume
M_f	martensite finish temperature
M_s	martensite start temperature
N	number
N	Avogadro's number
N	rate of nucleation
N_1	number of grain-boundary intercepts per centimeter
N_A, N_B, etc.	mole or atom fractions
P	pressure
P	probability
P	number of phases
Q	activation energy (per mole)
Q	quantity of heat (per mole)
R	universal gas constant
R	rate
R	radius
S	entropy
S	distance
T	temperature (usually absolute)
T_F	Fermi temperature
U	lattice or crystalline energy
V	potential energy
V	volume
W	total energy
W	work
Y	distance
Z	coordination number

List of Greek Letter Symbols

Greek letters, starting at the beginning of the alphabet, are used to represent the various solid phases in an alloy system. Angles are also represented by Greek letters.

In addition to the above, Greek letters are used for:

Symbol	Meaning	Symbol	Meaning
α (alpha)	polarizability	μ	micron (10^{-6} meter)
γ (gamma)	activity coefficient	μ_B	Bohr magneton
γ	surface energy	ν (nu)	frequency
γ	shear strain	ν	Poisson's ratio
ϵ (epsilon)	strain	ρ (rho)	dislocation density
ϵ_t	true strain	ρ	electrical resistivity
$\epsilon_{AA}, \epsilon_{AB}$, etc.	bonding energy between atoms	ρ	radius of curvature
		σ (sigma)	stress
ξ (zeta)	volume fraction	σ_t	true stress
η (eta)	shape factor	τ (tau)	period, relaxation time
η_1	shear direction in a twin	τ	shear stress
η_2	fourth twinning element in a twin	$\tau_\sigma, \tau_\epsilon$	relaxation times at constant stress and constant strain
θ (theta)	angle of incidence		
λ (lamda)	distance	ϕ (phi)	potential energy
λ	wave length	χ (chi)	magnetic susceptibility
μ (mu)	dipole moment	ψ(psi)	wave function
μ	shear modulus	ω(omega)	angular frequency

Index

Acceptor level, semiconductor, 123
Accommodation factor in freezing, 576
Accommodation kinks, twinning, 634
Accommodation of the twinning shear, 629
Activated cross slip, 858
Activation of
  dislocation intersections, 850
  dislocation kinks or jogs, 849, 852
  dislocation movement, 844
  dislocation sources, 848
  motion of screw dislocation with jogs, 852
Activation energy
  creep, dislocation climb controlled, 853
  creep in aluminum, 864
  creep vs. self-diffusion, 854
  decomposition of bainite, 736
  diffusion at low solute concentrations, 413
  diffusion of carbon in alpha iron, 449
  free surface diffusion, 419
  freezing, 575
  grain boundary diffusion, 419
  grain growth, 308
  interstitial diffusion, 432
  melting, 575
  movement of vacancies, 258
  recovery, 274
  recrystallization, 287
  self-diffusion, 409
  tempering of steel, 745
  to form vacancies, 254
  transfer of carbon atoms from cementite to alpha iron, 331
  transfer of carbon atoms from graphite to alpha iron, 331
Activation volume, 847
Activity coefficients, 381
  curves, positive and negative deviations of, 380
Age hardening (*see also* Precipitation hardening), 358
Aging treatment, 361
Alkali halides, 76
Alkali metals, 81
Allotropic (*see also* Polymorphic), 454
Allotropy, certain elements, 465
Alloy systems, 480
Alloying elements
  carbide forming in steel, 745
  effect on M_s and M_f, 724
  non-carbide forming in steel, 745
Alloys, creep resistant, 880
Aluminum, creep mechanisms, 864
Anelastic strain, 434
  variation with time, 435
Anelasticity, 434
Anisothermal annealing, 269
Anisotropy, 12
Annealed high carbon steels, 739
Antiferromagnetism, 133
Antiphase boundaries, 538
Antiphase domains, 538
Apparent diffusion coefficient, 421
Arhennius law, 418
Asterism, 45
  effect of recovery on, 275
Athermal flow stress component, 834
Atmospheres
  dislocation, 336
  movement of, 857
Atomic jump frequency, 382
Austenite, 662
  isothermal transformation, 664
  lattice parameter, 721
  retained, 724
Austenite grain size, 711
  effect on hardenability, 712
  relation to quench cracks, 713
Austenite to bainite transformation, 681
Austenite to martensite transformation, 681, 720
Austenitic steel, Hadfield's manganese, 725
Avogadro's number, 254
Bain cone, 644
Bain distortion, 638

indium-thallium, 640
iron-nickel, 639
lack of invarient plane, 644
steel, 720
Bainite, 681
acicular structure, 681
activation energy for decomposition, 736
carbides in, 683
finish temperature (B_f), 686
formation on continuous cooling, 717
growth of, 687
habit planes, 686
lower, 682
nucleation, 686
start temperature (B_s), 686
upper, 682
Bainite and decomposed martensite, difference, 737
Bainite transformations in steel, 681
deformation associated with, 682, 736
temperature dependence of, 686
Baryons, 85
Base diameter, hardenability, 714
Becker-Doring theory (nucleation), 488
Bend gliding, 163
Bending of crystals, 163, 275
Beta to beta′ transformation in brass, 558
Binary alloys, 455
Binary systems 455
three-phase transformations, 474
Bismuth embrittlement of copper, 223
Bloch, zone theory, 101
Blowholes, 606
Blue brittleness, 351, 796
Body-centered cubic metals, cleavage of, 754
Body-centered cubic structure, 5, 6
as tetragonal, 638
Bohr magneton, 128
Bohr theory of the atom, 81
Boltzmann, L., 401
Boltzmann's constant, 892
Boltzmann's equation, 246
Boltzmann's solution of Fick's second law, 401
Bonds, 62
covalent, 77
homopolar, 77
ionic, 62
metallic, 77
Bordoni peak, 850
Born, M., 62
Born exponent, 67
Born theory of ionic crystals, 62
Bragg angle, 48
Bragg law, 39
electrons, 101
X-rays, 39
Brass, 327, 386, 556, 563
beta, 456
Brillouin zones, 105, 117, 118
Brittle fracture,
appearance of surface, 786
compared to ductile fracture, 786
relation to yield stress, 777
welded steel structure, 782
Burgers vector, 156
energy considerations, 172
f.c.c. partial dislocations, 203
vector notation, 200

Carbon concentration
effect on hardenability, 713
effect on M_s and M_f, 723
effect on transition temperature, 786
Carbon steels, 661
low hardenability of, 713
Cast iron, 663
Castings, 568
grain size in, 593
preferred orientation in, 591
Cellular structure, formation of during freezing, 587
Cementite, 331, 359, 661, 739
stability of, 661
Central zone of an ingot, freezing of, 592
Ceramic coating, 884
Cesium-chloride lattice, 62
Charge on the electron, 83
Charpy impact test, 780
Chemical diffusion, 412
Chill zone of an ingot, freezing of, 588
Chisel edge rupture, 751
Cleavage, 753
bent crystals, 754
hexagonal metals, 753
iron crystals, 763
lithium fluoride, 755
plastic deformation during, 769
relationship to critical resolved shear stress in iron, 764
significance of, 754
temperature dependence in iron crystals, 764
Cleavage cracks
interaction with screw dislocations, 772
nucleation, 763
propagation, 769
Cleavage plane
body-centered cubic metals, 753
zinc, 753

Cleavage surface, distortion of, 763
Climb of edge dislocations, 169
  in polygonization, 277
  positive and negative, 170
Cobalt, effect on hardenability, 714
Coherency, effect on hardness, 371
Coherent particle, 371
Cohesion of solids, 60
Cohesive energy, 60, 71
Cold working, 267
Columnar zone of an ingot, freezing of, 588
Complementarity, principle of, 83
Components, 454, 455, 545
Compound twin, 618
Compounds, 555
Compressibility, 66
Conduction bands of semiconductors, 121
Conductors, electrical, 116
Congruent points, 533
Conjugate slip system, 198
Constituent, 545
Constitution diagrams (*see* Phase diagrams)
Constitutional supercooling, 583
Continuous cooling transformation of steel, 701
Control of grain size during freezing, 593
Coordination number, 7
  body-centered cubic, 7
  close-packed hexagonal, 11
  close-packed lattices, 11
Copper embrittlement by bismuth, 224
Copper-lead phase diagram, 553
Copper nickel
  phase diagram, 523
  porosity due to diffusion, 390
Copper-silver phase diagram, 541
Copper-tin alloy, dendritic segregation in, 596
Copper-zinc
  alloys, 327
  phase diagram, 563
  system, 456, 556
Coring, 595
Cottrell
  crack nucleation mechanism, 765
  dislocation intersection model, 851
  theory of the sharp yield point, 344
Cottrell-Bilby pole mechanism of twin growth, 628
Coulomb
  energy, 62, 64
  force, 62
Covalent
  bonds, 77
  molecules, 75
Crack nucleation
  Cottrell mechanism, 765
  dislocation mechanisms of, 765
  in fatigue, 818
  in iron, 765
  in magnesium oxide, 766
  in zinc, 768
  Zener mechanism, 765
Crack velocity
  equation, 761
  in lithium fluoride, 771
  limiting, 762
Creep, 841
  activation energies for aluminum, 864
  by diffusion, 843
  change in activation energy, 859
  climb of dislocations over obstacles, 855
  definition of, 841
  dislocation atmosphere controlled, 857
  dislocation climb, 853
  dislocation intersection controlled, 850
  effect of grain boundaries, 882
  energy barriers in, 845
  extrapolation of data, 877
  mechanisms, 847
  more than one mechanism, 859
  motion of dislocation jogs controlled, 852
  Mott equation, 853
  parallel mechanisms, 860
  practical applications, 877
  precipitation during, 882
  recrystallization during, 882
  series mechanisms, 860
  shift in controlling mechanism, 863
  strain rate equations (idealized), 846
  tests, 841
  thermal activation, 842
  use of refractory metals, 884
Creep curve, 872
  effect of stress, 876
  effect of temperature, 876
Creep resistant alloys, 880
Creep test
  constant load vs. constant stress, 874
  effect of metallurgical reactions on, 875
  structurally stable material, 874
  three stages of, 872
Critical cooling rate of steel, 705
Critical deformation in recrystallization, 294
Critical diameter, 707
  ideal, 707
Critical nucleus size
  in freezing, 482
  in precipitation, 369
Critical plane, 199

Critical radius, 369
Critical resolved shear stress, 173
  composition dependence of, 177
  face-centered cubic metals, 178
  hexagonal metals, 180
  relationship to cleavage in iron, 764
  temperature dependence of, 177
Cross slip, 190
  activated, 858
  double, 193
  effect of extended dislocations on, 207
  in aluminum and copper, 858
Crystal(s), 5
  binding, 60
  Born theory of ionic, 62
  classifications, 60
  internal energy of, 60, 76
  ionic, 60
  molecular, 75
  splitting of, 753
Crystal growth from liquid phase, 575
Crystal structure, 1, 5
  body-centered cubic, 6
  close-packed hexagonal, 11
  face-centered cubic, 7
  of metallic elements, 21, 22
  rotation during tensile and compressive deformation, 195
Crystallographic features of martensite transformation, 643
Cup and cone fracture, 787

Darken's equations, 392
Debye frequency, 71
Debye theory, 71
Deformation bands, 166
Deformation twinning, 611
  accommodation kinks, 635
  appearance of twins, 611
  comparison with slip, 613
  composition plane, 620
  critical resolved shear stress for, 625
  first undistorted plane, 620
  formal crystallographic theory of, 614
  identification of twins, 622
  in indium-thallium, 649
  in magnesium, 621
  in martensite transformation, 647
  lattice reorientation, 611
  nature of deformation, 611
  nucleation of twins, 624
  plane of shear, 620
  relationship between shear and second undistorted plane, 621
  relationship to cleavage, 763
  relationship to rupture, 752
  second undistorted plane, 616
  shear direction, 617
  simple model, 614
  twinning elements (K_1, η_1, K_2, η_2), 620
    determination of, 622
  twinning plane, 620
Degenerate states, 91
Dendrites, 546, 579
  growth directions, 581
  growth of, 579
  in alloys, 583, 595
  in pure metals, 582
Dendritic segregation, 595
Density change on melting, 569
Density of dislocations, 345, 830
Density of states, 98
  curve, zone theory, 111
  effect of temperature on, 99
Diamagnetic, 125
Diamond, energy gap, 122
Diamond lattice
  apparent size of carbon, 328
  structure, 78
Die castings, 605
Diffraction
  electrons, 51, 83
  light, 83
  X-rays, 38
Diffraction patterns, selected area, 58
Diffuse neck, 840
Diffusion, 378, 431
  as function of composition, 385, 433
  copper-zinc, 386
  couple, 386
  diffusivity constant, 398
  dilute nickel base solid solutions, 414
  flux of atoms in, 383
  frequency factor, 410
  gold-nickel system, 417
  ideal solutions, 381
  interrelation between time and distance, 400
  interstitial, 431
  liquids, 570
  low solute concentrations, 412
  mechanisms, 388
  net flow of atoms, 391
  net flow of vacancies, 391
  non-isotropic alloy system, 559
  penetration curve, 399
  porosity in copper-nickel, 390
  radioactive isotopes, 406, 414
  self-diffusion, 406
  silver, 421
  solid state, 250
  state of strain produced by, 391

substitutional solid solutions, 378
thermodynamic factor, 416
Diffusion coefficient, 384
carbon in face-centered cubic iron, 433
computation of from relaxation time, 440
experimental data for interstitial diffusion, 449
in binary gas solutions, 384
in gold alloys with nickel, palladium, and platinum, 385
interstitial equation, 431
interstitial equation, high precision of, 432
intrinsic, 392, 407
temperature dependence, 409
tracer, 415
Diffusion controlled growth, 504
Diffusion creep, Nabarro-Herring, 843
Diffusionless phase transformations (*see also* Martensite reactions), 637
Diffusivity (*see also* Diffusion coefficient), 384
Digonal axes, hexagonal crystals, 20
Dihedral angle, 221
Dimensional changes in steel during tempering, 735, 736, 738
Dipole
electrical, 67
field of, 68
induced, 69
moment, 69
Dipole-quadrupole interaction, 75
Dislocation climb, 169, 277, 392
in creep, 853
overcoming obstacles to slip, 855
Dislocation density, 177, 268, 345, 830
relation to stress, 830
Dislocation intersections, 209
Cottrell's model, 851
creep rate equation, 852
thermal activation, 851
work of, 851
Dislocation jogs, association with vacancies, 854
Dislocation kinks, thermal activation, 848
Dislocation mechanism of crack nucleation, 765
Dislocation movement
creation of interstitials, 852
creation of vacancies, 852
thermal activation of, 844
Dislocation sources, thermal activation of, 160, 848
Dislocation(s), 146
atmospheres, 337, 339
movement, 857
Burgers vector, 156
climb of edge, 169
curved, 155
edge, 149, 157, 334
extended, 207
forest, 850
in face-centered cubic metals, 201
in subboundaries, 279
interactions with solute atoms, 333
intersections, 209
left-hand and right-hand screws, 152
line tension, 849
loops, 156
model of a small angle grain boundary, 214
multiplication, 158, 193
negative and positive edges, 152
nucleation in lithium fluoride, 159
Orowan equation, 338
partial, 201
Peierle's stress, 848, 850
screw, 152, 157, 333
stair-rod, 209
vector notation, 200
Divacancies, 260
Divorced eutectic, 601
Domains, 537
magnetic, 132
Donor level, semiconductor, 123
Double cross slip, 193
Driving force
bubble growth in a soap froth, 305
diffusion, 470
grain growth, 298
martensite reaction, 611, 636
mechanical twinning, 611
recrystallization, 292
Dual nature of matter, 83
Ductile fracture, 786
appearance of surface, 787
compared to brittle fracture, 786
in a tensile specimen, 786
nucleation of, 789
void sheet mechanism, 789
Ductile to brittle fracture transition, 781
relation to stress state, 783
Dynamic recovery, 282
Dynamic strain-aging, 347
blue-brittleness, 351, 796
role of drag stress in, 351

Easy glide
face-centered cubic metals, 179
hexagonal metals, 182
recovery after, 272
rupture by, 749

Edge dislocations, 149
  interaction with solute atoms, 336
  stress field of, 334
(8 – N) rule, 80
Elastic after effect, 441
Elastic solids, fracture in, 755
Elastic strain, compared to anelastic, 434
Electro-etching, 5
Electro-polishing, 5
Electron(s)
  charge on, 83
  diffraction of, 51, 83
  energy of, 87
  mean kinetic energy, 97
  rest mass, 83
  sharing of, 77
  shells, 62
  spins, 78
  wave length, 83, 101
Electron compounds, 113
Electron interchange in hydrogen molecule, 77
Electron Theory of Metals, 81
Electron volt, 851
Electronic energy as a function of wave number, 108
Electronic specific heat, 100
Electrotransport, 425
Embrittlement
  copper by bismuth, 223
  iron by sulfur, 224
Embryo, 369
Endurance limit, 813
Endurance ratio, 813
Energy
  cohesive, 60
  coulomb, 62
  gaps, 109
  levels, 91
  of a grain boundary, 219
  zero point, 60, 74, 76
Energy barrier
  for motion of atoms into vacancies, 257
  in creep, 845
Entropy, 240
  associated with interstitial diffusion, 431
  associated with solute atoms in interstitial solid solution, 329
  intrinsic, 329
  statistical mechanics, definition of, 244
Entropy of mixing, 245, 378
  associated with vacancies, 252, 330
  contribution to free energy of solution, 467
Epsilon carbide, 684, 734, 744
Equilibrium between two phases, 460
Equilibrium concentration
  of carbon atoms in alpha iron, 331
  of vacancies, 254
Equilibrium configuration of grain boundaries at a three-grain juncture, 221
Equilibrium diagrams (*see* Phase diagrams)
Equilibrium heating and cooling of an isomorphous alloy, 527
Error function, 398
Etch pits
  lithium fluoride, 161
  magnesium oxide, 771
Etching, 4
Eutectic
  composition, 475
  copper-silver system, 541
  divorced, 601
  freezing of, 549
  reaction, 475
  temperature, 475
  ternary, 478
Eutectic point
  iron-carbon system, 661
  significance of, 542
Eutectic system, 541
  equilibrium microstructure of, 542
  freezing in, 542
Eutectoid point, 552
  iron-carbon system, 662
Eutectoid reaction, 553, 663
Eutectoid steel, time-temperature-transformation diagram, 687
Eutectoid transformation, heat of reaction, 664
Exaggerated grain growth, 318
Extended dislocations, 207
  activated constriction of, 858
  face-centered cubic, 204
  hexagonal metals, 207
  movement of, 205
  relation to cross slip, 207
Extrinsic stacking fault, 206
Extrusions, 817

Face-centered cubic cell, 7
Face-centered cubic structure, 7, 9
  considered as tetragonal, 638
Face-centered cubic metals
  extrinsic stacking fault in, 206
  intrinsic stacking fault in, 206
Failure
  by easy glide, 749
  by multiple glide, 751
  by necking, 751
Fatigue
  effect of steel microstructure, 822

effect of stress raisers, 824
effect of surface strengthening, 824
practical aspects, 824
properties of quenched and tempered steel, 823
stress vs. number of cycles to failure curve, 813
Fatigue failure, 809
at 4°K, 814
definition, 809
erroneous explanation of, 811
microscopic aspects of, 814
proof of, 811
significance of diffusion process, 814
surface appearance, 810
clamshell structure of surface, 811
Fatigue limit, 813
correlation with tensile strength, 813
relationship to percent martensite on quenching, 823
Fatigue strength, 814
effect of non-metallic inclusions, 821
effect of shot-peening, 824
Fatigue test
development of slip bands during, 815
rotating beam, 811
various methods, 811
Fermi-Dirac statistics, 100
Fermi energy, 96, 99
silver, 97
Fermi level, 97
Ferrimagnetism, 133
Ferrite, 331, 359
Ferromagnetic, 125
Ferromagnetism, 130
Fick, Adolf, 384
Fick's first law, 384
Fick's second law, 397
Boltzmann's solution, 400
50% martensite, hardness, 729
50% martensite – 50% pearlite criterion, 706
Final polishing, 4
Fine grinding, 3
First law of thermodynamics, 245
Five degrees of freedom of a grain boundary, 216
Flow-stress, 827
long-range component, 834
short-range component, 835
thermal component, 834
Fluidity of liquids, 570
Flux of atoms in diffusion, 383
direction, 384
unequal in binary system, 390
Folds, 868
Force, coulomb, 62
Forest of dislocations, 850
Fracture
critical grain diameter, 776
cup and cone, 787
effect of small angle twist boundary, 772
effect of surface energy, 760, 776
in elastic solids (glass), 755
nucleation and propagation limited, 776
spread of in polycrystalline metals, 774
three modes, 802
Fracture mechanics, 797
fracture toughness, 801
pop-in, 808
stress intensity factor, 800
Fracture mode
effect of notches, 779
effect of stress state, 777
triaxial tension effect, 779
Fracture propagation, effect of grain boundaries, 774
Fracture surface of impact specimens, 782
Fracture toughness, 801
Frank-Read source, 158
Free electron theory, 88
Free energy
curves of liquid, solid and gas phases, 572
of crystal due to vacancies, 251
of deformed metal, 268
precipitate particle, 368, 373
variation during precipitation of second phase, 359
variation of with temperature, 463
Free energy diagram
binary isomorphous system, 531
copper-nickel system, 531
maxima, 534
minima, 533
single phase, 468
two phases in equilibrium, 474
Free energy of solution, 458, 467
dependence on composition, 468
Free surface diffusion, 418
Freezing, 490, 568
isomorphous alloys, 527
solid solutions, 527
Frequency of atomic jumps, 383
Frequency factor (diffusion), 410
interstitial diffusion, 432
Full anneal, 703

Gas bubbles
effect on metal structures, 606
growth during freezing, 604
Gas phase, 568

Geiger counter, 50
Germanium, 123
Geometrical coalescence, 302
Gibbs free energy, 242
Gibbs phase rule, 477
Glass, fracture, 755
Gold-copper
  activities, 534
  phase diagram, 536
  superlattices, 538
Gold-nickel
  diffusion, 417
  phase diagram, 535, 539
Grain boundary, 213
  atmospheres, 310
  effect of inclusions on motion, 311
  effect of pores on motion, 314
  effect on creep resistance, 882
  effect on fracture propagation, 774
  effect on mechanical properties, 228
  energy, 219
  equilibrium configuration at three-grain juncture, 220
  five degrees of freedom, 216
  shear, 865
  small angle, dislocation model, 214
  surface energy, 220
  surface tension, 219
Grain boundary diffusion, 418
  activation energy, 419
  rate of, 421
  relation to volume diffusion, 421
Grain boundary movement, 221
  in grain growth, 304
Grain boundary shear, 865
  intercrystalline fracture, 867
  measurement, 865
  nature, 867
  relation to total strain, 866
Grain geometry, three-dimensional changes in, 303
Grain growth, 221, 271, 298
  boundary movement, 304
  effect of free surfaces, 315
  effect of impurity atoms in solid solution, 310
  effect of inclusions, 311
  effect of pores, 314
  effect of preferred orientation, 318
  effect of specimen dimensions, 315
  effect of thermal grooves, 315
  exaggerated or abnormal, 318
  geometrical coalescence, 302
  law, 304
  soap froth analogy, 298
Grain growth exponent, 309
Grain growth law, 304
Grain multiplication, 593
Grain size, 305
  austenite, 711
  of castings, 593
Grain size measure, ASTM, 712
Graphite, stability, 661
Gray tin, 122, 461, 503
Griffith cracks, 759
  evidence for, 760
Griffith criterion for fracture, 760, 762
  in polycrystalline metals, 776
Griffith experiments on glass, 759
Griffith theory of fracture, 762
  Orowan's approach, 759
Grinding, 3
Grossman hardenability, 709
Growth
  diffusion controlled, 504
  interface controlled, 511
  of bainite, 687
  of martensite plates (steel), 655
  of pearlite, 667, 670
Growth kinetics, 501
Growth velocity in freezing, 575
Grube, G., 397
  analysis of diffusion data, 397, 407, 415, 432
Guinier-Preston zones, 366
Gyromagnetic ratio, 130

Habit plane
  bainite, 686
  indium-thallium, 650
  martensite, 638, 643, 650, 722
  steel, martensite, 722
Hard-ball model, 6
Hardenability, 704
  base diameter, 714
  carbon concentration effect, 713
  computation, 714
  definition, 707
  dependence on carbon content, 713
  effect of alloying elements, 714
  effect of austenite grain size, 712
  Grossman method, 709
  in terms of T-T-T diagram, 716
  multiplying factors, 716
  of hypereutectoid steels, 714
  significance, 710, 719
Hardening of steel, secondary, 745
Hardness
  of 50 percent martensite, 729
  of martensite in low alloy steel, 728
  of martensite in steel, 725
  of tempered steels, 742

Hardness contour of quenched steel bar, 706
Hardness tests, Rockwell and Vickers, 726
Heat
  of formation of a compound, 60
  of sublimation, 60
Heisenberg uncertainty principle, 81
Heterogeneous nucleation, 368, 496
  during freezing, 572, 573
  in precipitation, 374
  of pearlite, 667
Hexagonal metals
  c/a ratio, 181
  cleavage, 753
  pyramidal slip, 182
  slip in polycrystalline, 183
High carbon steel, annealed, 742
Hole conduction, 123
Homogeneous nucleation during freezing, 572
Homogenization, 597
Homopolar bonds, 77
Hot short, 225
Hultgren's extrapolation, 672
Hume-Rothery rules for substitutional solid solutions, 332
Hydrogen molecule, 78
Hypereutectic composition, 543
  freezing, 547
Hypereutectic structure, 547
Hypereutectoid steel
  microstructures of, 694
  slowly cooled, 693
  time-temperature-transformation diagram, 696
Hypoeutectic composition, 543
  freezing, 545
Hypoeutectic structure, 543
Hypoeutectoid steel
  lever rule applied to, 690
  microstructures, 691
  time-temperature-transformation diagram, 695

Ideal critical diameter (D_I), 707
  as a function of critical diameter, 708
  computation from Jominy data, 710
Ideal quench, 707
Ideal solutions, 378, 407
Impact test, 780
  Charpy, 780
Inclusions
  effect on fatigue strength, 821
  effect on grain growth, 311
  effect on nucleation during freezing, 593
Incubation period, precipitation hardening, 361
Indices (*see* Miller indices)
Indium-thallium martensite transformation, 640, 643
  Bain distortion, 644
  habit plane, 650
  motion of interface, 640
  twinning in, 649
Inert gas solids, lattice energy of, 71
Inert gases, 69
Inglis equation, 759
Ingots, 568
  freezing of, 585
Insulator, electrical, 116
Intercrystalline fracture, 214
  brittle, 223, 795
  creep, 867
  relation to grain boundary shear, 868
  stress-rupture test, 879
  Zener mechanism, 868
Interdendritic porosity, 607
Interdiffusion coefficient, 407
Interface controlled growth, 511
Interlamellar spacing of pearlite, 670
  effect of hardness, 670
  relation to pearlite growth, 671
  temperature dependence, 671
  Zener's equation for, 676
Intermediate phases, 327, 553
Intermediate solid solutions, 327, 553
Intermediate temperature ductility minimum, 871
Intermetallic compound, 456
Internal energy, 240
  of crystals, 60, 76
Internal friction (*see also* Anelasticity)
  Bordoni peak, 850
  iron-carbon alloys, 366
  Q^{-1}, 445
  use in interstitial diffusion, 432
Interplanar spacing, 47
Interstitial atoms, 326, 327
  carbon in iron, maximum number, 720
  effect of stress on distribution, 433
  lattice strain due to, 433
  relation to stored energy of cold work, 268
Interstitial diffusion, 431
  contrasted to vacancy diffusion, 431
Interstitial solid solutions, 326, 327
  carbon in body-centered cubic iron, 328
  conditions controlling formation, 327
Interstitial solute atoms, 260
Interstitials
  creation by dislocation jogs, 213
  creation by dislocation movement, 852

Intrinsic diffusion coefficients, 392, 407
determination, 404
relation to tracer coefficients, 415
temperature dependence, 412
Intrinsic stacking fault, 207
Intrusions, 818
Invarient plane strain, 637
Invarient points in iron-carbon phase diagram, 661
Inverse segregation, 602
Ionic crystals, 61
Iron
crack nucleation, 765
ferro-magnetic, 466
magnetic modifications, 466
magnetic transformations, 455
para-magnetic, 466
phases, 454
relationship of critical resolved shear stress to cleavage, 764
second order transformation, 466
Zener's interpretation of allotropy, 466
Iron carbide (cementite), 661
Iron-carbon alloys, 328
intermetallic phase, 330, 331
sharp yield point, 342
Iron-carbon phase diagram, 661
Iron-carbon system, 661
Iron crystals, cleavage, 765
Iron-nickel alloy
Bain distortion, 639
isothermal martensite transformation, 653
martensite transformation, 651
size of martensite plates, 652
stabilization of martensite, 653
Iron-nickel system, 549
Irreversible reaction, 242, 244
Isomorphous alloy, freezing and melting points, 530
Isomorphous binary alloy systems, 522
Isothermal annealing, 270
Isothermal transformation
austenite, experimental method, 664
of martensite, 653

Jogs in dislocations, 221
Johnson and Mehl equation, pearlite transformation, 678
Jominy end quench test, 709
computation of D_I, 710

Killed steel, 608
Kinetic theory, 238, 239
Kinks in dislocations, 211
Kirkendall, K. O., 386
Kirkendall diffusion couple, 386
Kirkendall effect, 386
significance (vacancy movements), 388
Kirkendall markers
movement, 386
velocity, 393
Kurdyumov-Sachs relation, 722

Latent heat of fusion, 569
Latent heat of vaporization, 569
Lattice
energy, 60
vibrations, 73
Lattice constants (*see* Lattice parameters)
Lattice parameters
austenite, 721
epsilon carbide, 734
iron, 329
steel, martensite, 721
Lattice sites occupied by carbon atoms in iron, 434
Lattice structures
body-centered cubic, 6, 7
cesium chloride, 62
close-packed hexagonal, 11
diamond, 78
face-centered cubic, 7, 8
sodium chloride, 61
superlattices (*see* Superlattice)
zinc blende, 62
Laue photographs, asterism, 275
Laue X-ray diffraction techniques, 43
use of to identify twins, 623
Lead-antimony system, gravity segregation, 595
Leptons, 85
Lever rule
applied to hypoeutectoid steel, 691
binary systems, 524
Light, diffraction, 84
Limiting grain size, 316
Liquid phase, 568
bonding of atoms, 569
diffusion, 569
fluidity of, 570
nucleation, 572
Liquidus line, 523
Lithium fluoride
cleavage, 755, 770
double cross-slip mechanism, 193
nucleation and observation of dislocations, 160
Logarithmic decrement, 445
Long-range order, 537
Low angle grain boundary, 214
Lüders bands, 343

Macrostructure, 1
Madelung, 62
Madelung number, 66
Magnesium, solubility of hydrogen, 603
Magnesium-lithium phase diagram, 535
Magnesium-nickel phase diagram, 556
Magnesium oxide, crack nucleation, 768
Magnesium-silver phase diagram, 554
Magnetic moment, 125
Magnetism, 124
  antiferromagnetism, 133
  Bohr magneton, 128
  diamagnetism, 125
  domains, 132
  ferrimagnetism, 133
  ferromagnetism, 130
  gyromagnetic ratio, 130
  magnetic moment, 125
  paramagnetism, 125, 127
  saturation magnetization, 130
Manganese
  effect on martensite transformation in steel, 725
  effect on transition temperature, 786
Markers (*see* Kirkendall markers)
Martensite
  athermal, 642, 653, 723
  burst formation, 652
  finish temperature, M_f, 641, 723
  habit planes, 638, 643, 650, 722
  hardness in steel, 725
  irrational habit planes, 650
  isothermal transformation, 653, 723
  lattice parameters in steel, 721
  multiplicity of habit planes, 650
  plates, appearance, 611, 652, 722
  reaction (*see* Martensite transformation)
  shears, table, 638
  stabilization, 653
  start temperature, M_s, 641, 723
  tetragonality in steels, 721
Martensite transformation(s), 635
  at 4°K, 655
  crystallographic features, 643
  dimensional changes in steel, 729
  effect of manganese in steel, 725
  effect of plastic deformation, 656
  effect of stress, 655
  hysteresis, 643, 651
  indium-thallium, 638, 640
  iron-nickel, 638, 651
  irreversible in steel, 723
  motion of interface, indium-thallium, 640
  nature of deformation, 636
  reversibility, 642
  shear, 643, 647, 648
  slip or twinning in, 647
  steels, 720
  Wechsler, Lieberman and Read theory, 643
Mass of the electron, 83
Matano, C., 401
Matano analysis of diffusion data, 400
Matano interface, 401
Matthiessen's rule, 135
Maxima
  in phase diagram, 532
  free energy curves in phase diagram, 534
Mean time of stay
  atom in lattice site, 382
  interstitial atom, 436, 449
  relation to relaxation time in an interstitial atom, 436
Mechanical twinning (*see* Deformation twinning)
Meissner state, 136
Melting
  isomorphous alloys, 530
  solid solutions, 530
Metallic bonding, 77
Metallographic specimen preparation, 2
Metals
  electron theory of, 81
  polymorphic (allotropic), 21
Microsegregation, 588, 595
Microstructure, 1, 2
Miller indices, 15
  cubic directions, 15
  cubic planes, 17
  hexagonal directions, 19
  hexagonal planes, 19
Minima
  free energy curves, 533
  in phase diagrams, 532
Miscibility gaps, 539
  liquid state, 552
Mold walls, effect on freezing, 585
Mole fraction, 457
Molecular crystals, 75
Molybdenum
  effect on reducing temper brittleness, 795
  high temperature use, 883
Monotectic, 552
Mott, creep rate equation, 853
Mullens, W. W., 315
Multiple glide, rupture by, 751

Nabarro-Herring diffusion creep, 843
Necking
  diffusion neck, 840

rupture by, 751, 789
Nonideal solutions, 379
Non-isomorphic alloy systems, diffusion in, 559
Nonmetallic inclusions
ductile fracture nucleation, 789
effect on fatigue strength, 821
fatigue crack nucleation, 822
Nonpolar molecules, 76
Normalizing, 703
Notches, effect on fracture mode, 779
Nucleation
bainite, 687
Becker-Döring theory, 488
cleavage cracks, 763
cracks in zinc, 768
dislocations, 159
ductile fracture, 789
freezing of liquids, 572
gas bubbles during freezing, 605
heterogeneous, 368, 496
homogeneous, 368
martensite plates, 654
melting of liquids, 574
new grains at twin intersections (annealing), 613
pearlite, 667
precipitate particles, 365, 374
recrystallization, 290, 291
Volmer-Weber theory, 480, 488
Nucleation and growth
liquids, 572
pearlite, 667
precipitation, 361
recrystallization, 290
Nucleus growth (recrystallization), 291

Number
Avogadro's, 254
coordination, 7

Oil quench, 704
One-component system, 454, 461, 483
Orange peel effect, 295
Order of X-ray reflection, 40, 47
Ordered structures (*see also* Superlattice), 537
Orowan
approach to the Griffith theory, 759
mica fracture experiment, 760
theory of hardening by precipitates, 371, 856
Orowan equation, 338
Overaging, 363

Paramagnetic, 125, 127
Partial dislocations
face-centered cubic metals, 202
hexagonal metals, 207
Partial molal free energy, 379
graphical determination of, 471
Particle in a box problem, 88
Pauli exclusion principle, 77
Pearlite, 665
composition change with cooling rate, 698
growth mechanisms, 667
growth rate, 669, 712
Mehl and Hagel equation, 673
temperature dependence, 670, 674
heterogeneous nucleation, 667
interlamellar spacing (*see also* Interlamellar spacing of pearlite), 670, 676
nodule, 669
nucleation, 667
nucleation rate, 676
temperature dependence, 677
time dependence, 677
ratio of cementite to ferrite, 666
widths of cementite and ferrite lamellae, 666
Peierl's stress, 848
Penetration curve, 398
Penetration depth, 137
Peritectic composition and temperature, 551
Peritectic transformation, 549
iron-carbon system, 662
iron-nickel system, 550
Peritectoid point, 553
Peritectoid reaction, 553
Phase angle, stress and strain, 445
Phase diagrams
binary, 522
copper-gold, 536
copper-lead, 553
copper-nickel, 523
copper-silver, 541
copper-zinc, 557
gold-nickel, 535
iron-carbon, 662
iron-nickel, 550
isomorphous, 523
magnesium-lithium, 535
nickel-magnesium, 556
one-component system, 483, 568
silver-magnesium, 554
Phase rule, 477
Phases, 222, 454
continuous and discontinuous, 456
copper, 454

difference between phases and constituents, 692
factors controlling number, 461
factors involved in phase changes, 464
gaseous, 455
in tin, 461
iron, 454
liquid, 455
solid, 455
Phase transformations on heating, 514
Phonons, 134
Phosphorus, effect on transition temperature, 786
Pipe (ingots), 608
Planck's constant, 82
Plane
basal, 11
octahedral, 8, 10
of shear (twinning), 620
prism, 20
Plastic deformation
accompanying cleavage, 763
effect on crack propagation, 769
Point defects, 268, 282
Polar molecules, 76
Pole mechanism, twin growth, 628
Polishing, 3, 4
Polycrystalline metals
Griffith fracture criterion, 776
spread of fracture, 774
Polygon walls, 280
Polygonization, 274, 277
Polymorphic 454
metals, 22
Pop-in, 808
Porosity
due to diffusion, 390
resulting from freezing, 603
stress dependence, 392
Powder metallurgy, 314
Precipitate
dissolution of, 515
nucleation of, 365
Precipitate particles
growth, 373
interference of growing, 510
nucleation, 368
shapes, 375
Precipitation during creep, 882
Precipitation from solid solution, 361
effect of composition, 364
effect of temperature, 363
effect on hardness, 363
incubation period, 361
rate, 362
recrystallization during, 375
Precipitation hardening, 358
Preferred orientation, 13
in castings, 591
Preformed nuclei, 291
Primary slip system, 197
Prismatic slip in hexagonal metals, 181, 190
Proeutectic, 546

Quadrupole, electrical, 67
Quantum number space, 94
Quantum numbers, 90
Quench
ideal, 707
oil, water, 704
severity (H), 708
Quench cracks, 730
relation to austenite grain size, 713
Quenching, residual stresses, 730
Quenching action of various media, 708

Radioactive isotopes in diffusion studies, 406, 415
Rate of
freezing, 575
growth, pearlite, 669, 670, 713
melting, 575
nucleation, pearlite (*see also* Pearlite nucleation rate), 676
nucleation in recrystallization, 290
nucleus growth in recrystallization, 290
recovery, 273
recrystallization, 290, 297
twin growth, 629
Ratio of nucleation rate to growth rate in recrystallization, 295
Recovery, 271
after easy glide, 272
dynamic, 282
polygonization, 274
Recrystallization, 271, 284
activation energy, 287
after easy glide, 272
driving force for, 294
during creep, 875, 882
effect of strain, 288
effect of time and temperature, 286
in single crystals, 272, 294
preformed nuclei, 291
secondary, 318
Recrystallization temperature, 288
effect of metal purity, 297
Recrystallized grain size, 294
Refractory metals, 884
Relaxation time, 435
experimental determination, 441
Relaxed modulus, 443

Residual stresses due to quenching, 732
Resistivity ratio, 135
Resonance effect, 78
Reversible reaction, 241
Rimming of steel, 607
River pattern on fracture surfaces, 773
Rockwell hardness test, 725
Rogers, mechanism for ductile fracture propagation, 789, 791
Rotational slip, 167
Rough polishing, 3
Rupture
  by easy glide, 749
  by multiple glide, 751
  by necking, 751
  chisel edge, 751
  relationship of twinning to, 752

Saturation magnetization, 130
Schoeck and Seeger theory of activated cross-slip, 858
Schottky defect, 251
Schrödinger equation, 87
Screw dislocation
  interaction with cleavage cracks, 772
  interaction with solute atoms, 336
  stress fields, 333
  with jogs, 211
Seams, 607
Second order transformation in iron, 466
Second undistorted plane
  martensite transformation, 646
  twinning, 616
Secondary recrystallization, 318
Segregation, 540, 594
  coring, 595
  dendritic, 595
  inverse, 602
  macrosegregation, 595
  microsegregation, 595
  tin sweat, 602
Self-diffusion, 406
  activation energy, 410
  coefficient, 409
  temperature dependence, 409
Semiconductors, 120
  impurity, 121
  intrinsic, 121
  n-type, 124
  p-type, 124
Sharp yield point, 341
Shear
  martensite transformation, 643, 647
  twinning, 620
Shear failure by slip, 749
Shear strength of crystals, 144
Ship failures, 782
Shockley-Read equation, low angle boundary, 219
Short-range flow stress component, 835
Short-range order, 537
Sievert's law, 603
Silicon, 123
Silver, self-diffusion, 420
Silver-magnesium phase diagram, 554
Single crystals, growth, 319
Singular points, 534
Sinks of vacancies, 392
Slip, 144
  in martensite transformation, 647
  in polygonization, 277
  on intersecting slip planes, 179
  on widely spaced bands, 750
  shear failure by, 749
Slip bands, 192
Slip direction, 171
  not along closest packed direction, 182
Slip lines, 145
  wavy, 183, 192
Slip plane, 158, 171
Slip systems, 173
  body-centered cubic metals, 183
  crystallographically equivalent, 177
  face-centered cubic metals, 178
  hexagonal metals, 180
Small angle twist boundary, effect on fracture, 772
Smigelskas, A. D., 386
Smith, C. S., 300
S-N (fatigue) curve, 813
Snoek effect, 433
Soap bubble froth, 299
Sodium chloride lattice, 62
Solid solutions, 222, 326
  β-brass, 456
  intermediate, 456
  interstitial, 327
  substitutional, 326
  supersaturated, 361
  terminal, 456
Solid state reactions in condensed systems, 493
Solidification, 568
Solidus line, 523
Solubility
  gases in metals, 603
  hydrogen in copper, 604
  hydrogen in magnesium 603
  interstitial solid solutions, 358
Solution treatment, 360
Solvus (line), 359
Sommerfeld theory of free electron, 88

Sources of vacancies, 392
Specific damping capacity, 445
Specific heat, solid forms of tin, 462
Specimen size, effect on grain growth, 316
Spheroidized cementite, 739
  fatigue properties of, 822
  low transition temperature of, 785
Spontaneous reactions, 241, 242
Stability of graphite and cementite, 661
Stable interface freezing, 578, 586
Stacking fault
  extrinsic or double, 206
  face-centered cubic, 204, 206
  hexagonal metals, 207
  intrinsic, 206
Stacking fault energy, 205
  copper vs. aluminum, 858
  relation to cross-slip, 209
Stacking sequence
  close-packed planes, 9
  face-centered cubic metals, 204, 206
  hexagonal metals, 207
Stair-rod dislocation, 209
Standard projections, 31
State functions, 241
Statistical mechanics, 239
  definition of entropy, 244
Steady state freezing, 586
Steel
  annealed high carbon, 739
  appearance of martensite plates, 722
  athermal martensite, 723
  Bain distortion, 720
  bainite (*see* Bainite)
  carbide forming alloying elements, 745
  carbon content range, 662
  continuous cooling transformations, 701
  critical cooling rate, 705
  dimensional changes during tempering, 735, 736, 738
  effect of manganese on martensite transformation, 725
  Hadfield manganese, 725
  hardness of martensite, 725
  high speed, 746
  isothermal martensite, 723
  isothermal transformation, 664
  low alloy, hardness of martensite, 726
  martensite finish temperature (M_f), 723
  martensite habit planes, 722
  martensite lattice parameters, 721
  martensite start temperature (M_s), 723
  martensite transformation, 720
  non-carbide forming alloying elements, 745
  secondary hardening, 745
  sharp yield point, 341
  S-N (fatigue) curves, 813
  strain aging, 346
  tempered, hardness of, 742
  tempering (*see also* Tempering), 733
  tetragonality of martensite, 721
  thermal contraction, 731
  variables determining hardenability, 710
  warping, 730
Stereographic projection, 22
  standard projection, 22
  standard triangle, 33
Stereographic triangle, for cubic crystals, 33
Stored energy of cold work, 267
Strain, time dependent, 841
Strain aging, 346
Strain induced boundary migration, 319
Strains due to diffusion, 391
Strength of glass, discrepancy between actual and theoretical, 759
Stress
  effect on martensite transformation (M_s), 655
  interaction with thermal energy, 845
Stress intensity factor, 800
Stress-rupture test, 878
  extrapolation of data, 880
  intercrystalline fracture, 878
Stress-strain curves
  face-centered cubic single crystal, 179
  magnesium single crystal, 144
  strain aging of steel, 346
  with sharp yield point, 342
Stresses due to diffusion, 391
Stretcher strains, 343
Subboundaries, 277, 278
Subgrains, 277
Substitutional solid solutions, 326
  Hume-Rothery rules, 332
Substructure, 277
Superconductivity, 135
  Meissner state, 136
  penetration depth, 137
  Type I superconductor, 136
  Type II superconductor, 139
Superconductor
  Type I, 136
  Type II (hard), 139
Supercooling of liquids during freezing, 572
  maximum supercooling of certain metals, 573
Supercooling of nickel-copper alloys, 573
Superheat of solids during melting, 572, 574
Superlattice(s), 534
  copper-zinc system, 558

gold-copper system, 538
iron-nickel system, 549
Superstructures (*see* Superlattice)
Surface tension
of grain boundaries, 220
soap films, 299
System
alloy, 455
thermodynamic, 455

Temper brittleness, 795
lowering of transition temperature, 795
molybdenum effect on reducing, 795
Temper roll, 344
Temperature gradients in freezing
falling, 579
rising, 578
Temperature inversion, 580
Tempering (*see also* Steels), 733
activation energy of steels, 736
definition, 733
effect on physical properties, 742
fifth stage, 745
first stage, 734
fourth stage, 739
interrelation between time and temperature, 744
need for, 733
resistance to in steel, 745
second stage, 736
softening of steel due to, 744
third stage, 737
Tensile strength, correlation with fatigue limit, 813
Ternary alloys, 455
Ternary systems, 478
Textures, 13
Theoretical fracture stress, 758
Thermal activation (*see* Activation)
Thermal component, 835
Thermal contraction, steel, 731
Thermal energy, interaction with stress, 845
Thermal grooves, effect on grain growth, 315
Thermodynamic factor (diffusion), 416
Thermodynamic properties, 457
Thermodynamic state, 457
Thermodynamics, 238
of solutions, 457
Thermomigration, 425
Tie line, 527
Tilt boundary, 216
Time-temperature-transformation (T-T-T) curves, 679
Time-temperature-transformation (T-T-T) diagram
eutectoid steel, 687
hypereutectoid steel, 696
hypoeutectoid steel, 695
relation to hardenability, 716
Tin
allotropy, 461
alpha (gray), 461, 503
beta (white), 461, 503
free energy of solid forms, 464
specific heats, 316
Tin sweat, 602
Torsion pendulum, 441
Tracer diffusion coefficients, 415
Transcrystalline fracture, 214
Transformation
austenite to bainite, 681
austenite to martensite, 720
austenite to pearlite, 663
eutectoid steel, 665
Transition metals, 327
Transition temperature
ductile to brittle fracture, 784
lowering of by temper brittleness, 795
variables affecting, 785
Triple point, 571
Twin
compound, 618
of the first kind, 618
of the second kind, 618
Twin boundaries, 625
Twin growth, 627
rate of, 629
Twinning (*see* Deformation twinning)
Twinning shear, accommodation of, 629
Twist boundary, 216
Two-component system, 455, 466
single phase equilibrium, 466
three-phase equilibrium, 474
two-phase equilibrium, 470
Two-surface method, 622

Uncertainty principle, 81
Unit cells, 5
body-centered cubic, 6
close-packed hexagonal, 11
face-centered cubic, 7
Universal gas constant, 258
Unrelaxed modulus, 443

Vacancies, 169, 238
association with jogs, 853
creation by dislocation jogs, 212
creation by dislocation movement, 852
entropy associated with, 252
equilibrium concentration, 254
free energy, 251

internal energy, 251
jumping of atoms into, 257
mean distance between, 260
motion, 257, 381
sources and sinks, 392
Valence band of semiconductors, 123
van der Waal's binding, 67, 71
van der Waal's crystals, 67, 71
Vickers hardness test, 725
Void sheet mechanism, 789
Voids due to diffusion, 390
Volmer-Weber theory (nucleation of liquid from a vapor), 480, 488

Water quench, 704
Wave equation, vibrating string, 85
Wave function ψ, 86
Wave length of an electron, 52, 83, 102
Wave mechanics, 84
Wave number, 87, 102
critical, 108
dependence on energy of electrons, 107
Wave properties of electrons, 83
Wechsler, Lieberman and Read theory of martensite transformations, 643
Weertman creep equation, 856
Wert, C., 362, 363, 368, 448, 449, 450, 504
Widmanstätten structure, 375
Work function, 98
Work to form a vacancy, 251
Work to form one mole of vacancies, 254
Work to move interstitial solute atoms, 432
Wormhole porosity, 607
in ice, 607
Wulff net, 27
rotations on, 28

X-ray
scattering of, 38
spectrometer, 50
white, 41
X-ray diffraction, 38
data index, 50
Debye-Scherrer, 46
Laue techniques, 43
powder method, 46
rotating crystal technique, 45
X-ray reflection, order of, 40, 47

Yield point, 342
lower, 342
relation to strain aging, 346
upper, 342
Yield stress
of crystals, theoretical, 144
relation of brittle fracture stress, 777

Zener
crack nucleation mechanism, 765
diffusion controlled growth theory, 504
interpretation of the allotropy of iron, 466
mechanism for intercrystalline fracture, 868
pearlite lamellae spacing equation, 676
ring mechanism in diffusion, 389
theory of interaction between grain boundaries and inclusions, 311, 318
Zero point energy, 60, 74, 76
Zinc
cleavage plane, 753
crack nucleation, 768
recovery, 272
Zinc blende lattice, 62
Zone axis, 24
Zone theory, 101
Zones of planes, 24